2008 European Microwave Integrated Circuits Conference

(EuMIC)

Amsterdam, Netherlands
27 – 28 October 2008

IEEE Catalog Number: CFP08GAS-PRT
ISBN: 978-2-8748-7007-1

Copyright © 2008 European Microwave Association
All Rights Reserved

This publication is a representation of what appears in the IEEE Digital Libraries. Some format issues inherent in the e-media version may also appear in this print version.

IEEE Catalog Number: CFP08GAS-PRT
ISBN 13:   978-2-8748-7007-1

Additional Copies of This Publication Are Available From:

Curran Associates, Inc
57 Morehouse Lane
Red Hook, NY 12571 USA
Phone: (845) 758-0400
Fax: (845) 758-2633
E-mail: curran@proceedings.com

TABLE OF CONTENTS

A Survey on 60 GHz Broadband Communication: Capability, Applications and System Design 1
    M. Peter, W. Keusgen, J. Luo

Co-Design of Fully Integrated 60GHz CMOS Digital Radio in QFN Package 5
    J. Laskar, S. Pinel, D. Dawn, S. Sarkar, P. Sen, B. Perunama, D. A. Yeh, F. Barale

Integrated 60 GHz Circuits and Systems for High-Speed Communications 9
    M. Abbasi, S. E. Gunnarsson, R. Kozhuharov, C. Karnfelt, C. Fager, D. Kuylenstierna, C. Stoij, H. Zirath

A 60 GHz SiGe HBT Chip Set 13
    G. Boeck, V. Subramanian, W. Keusgen, V. Do

EuMIC: A 40-to-76 GHz Balanced Distributed Doubler in 0.13-(mu)m CMOS Technology 17
    Bo-Jiun Huang, Bo-Jr Huang, Chung-Chun Chen, Kun-You Lin, Huei Wang

Monolithic Integrated Oscillator with Silicon IMPATT Diode for Automotive Radar Applications 20
    H. Xu, M. Morschbach, J. Werner, E. Kasper

High Voltage RF LDMOS Technology for Broadcast Applications 24
    S. J. C. H. Theeuwen, W. J. A. M. Sneijers, J. G. E. Klappe, J. A. M. de Boet

Analysis of Bias Effects on VSWR Ruggedness in RF LDMOS for Avionics Applications 28
    G. Formicone, F. Boueri, J. Burger, W. Cheng, Y. Kim, J. Titizian

On-Chip S-Shaped Rat-Race Balun for Millimeter-Wave Band Using Wafer-Level Chip-Size Package Process 32
    C. Inui, Y. Manzawa, M. Fujishima

Impact of Si Substrate Resistivity on the Non-Linear Behaviour of RF CPW Transmission Lines 36
    C. R. Neve, D. Lederer, G. Pailloncy, D. C. Kerr, J. M. Gering, T. G. McKay, M. S. Carroll, J. P. Raskin

Innovative and Complete Dummy Filling Strategy for RF Inductors Integrated in an Advanced Copper BEOL 40
    C. Pastore, F. Gianesello, D. Gloria, E. Serret, P. Benech

Miniaturized Multilayer CPW pHEMT Amplifiers 44
    Q. Sun, V. T. Vo, J. Tan, R. A. Davies, A. A. Rezazadeh

A Miniaturized Wafer-Scale Package Demonstrated with Three Enhancement Mode Amplifiers 48
    K. Phan, J. Kessler, H. Morkner, M. Vice, L. Nguyen, J. Roland

Millimetre-Wave Hot-Via Interconnect-Based GaAs Chip-Set for Automotive RADAR and Security Sensors 52
    P. F. Alleaume, C. Toussain, C. Auvinet, D. Domnesque, P. Quentin, M. Camiade

Process Stabilization and Sensitivity Analyses of a Single Recess GaAs pHEMT Process Using Device Simulations 56
    P. Abele, M. Schafer, J. Splettstosser, M. Thinnes, H. Stieglauer, D. Behammer

60 GHz Frequency Conversion 90 nm CMOS Circuits 60
    M. Kantanen, J. Holmberg, T. Karttaavi, J. Volotinen

Analysis and Design of a 50-GHz 2:1 CMOS CML Static Frequency Divider Based on LC-Tank 64
Y. Mo, E. Skafidas, R. Evans, I. Mareels

A Sub-1V 22-GHz CMOS Injection-Locked Frequency Divider.................................68
C. Chen, C. C. Tzuang

A 24 GHz Indirect VCO in 0.18 (mu)m CMOS Technology 71
G. Bu, A. R. Tavakoli, K. Entesari

A Compact 23 GHz Hartley VCO in 0.13(mu)m CMOS Technology 75
M. Bao, Y. Li

S-Band AlGaN/GaN Power Amplifier MMIC with over 20 Watt Output Power 79
M. Van Heijningen, G. C. Visser, J. Wurfl, F. E. van Vliet

2--6 GHz GaN MMIC Power Amplifiers for Electronic Warfare Applications...........83
M. A. Gonzalez-Garrido, J. Grajal, P. Cubilla, A. Cetronio, C. Lanzieri, M. Uren

Efficient AlGaN/GaN HEMT Power Amplifiers 87
R. Quay, F. van Raay, J. Kuhn, R. Kiefer, P. Waltereit, M. Zorcic, M. Musser, W. Bronner, M. Dammann, M. Seelmann-Eggebert, M. Schlechtweg, M. Mikulla, O. Ambacher, J. Thorpe, K. Riepe, F. van Rijs, M. Saad, L. Harm, T. Rodle

AlGaN/GaN HEMT with over 110 W Output Power for X-Band 91
S. Zhong, T. Chen, C. Ren, G. Jiao, C. Chen, K. Shao, N. Yang

Balanced Microstrip AlGaN/GaN HEMT Power Amplifier MMIC for X-Band Applications 95
J. Kuhn, F. van Raay, R. Quay, R. Kiefer, W. Bronner, M. Seelmann-Eggebert, M. Schlechtweg, M. Mikulla, O. Ambacher, M. Thumm

Opportunities at mm-Wave Frequencies: SiGe or CMOS? 99
H. Veenstra, M. Notten

Perspectives of (Sub-) 32nm CMOS for Analog/RF and mm-Wave Applications ...103
M. Dehan, B. Parvais, A. Mercha, V. Subramanian, G. Groeseneken, W. Sansen, S. Decoutere

Development of Ultrahigh-Speed InP/GaAsSb/InP DHBTs: Are Terahertz Bandwidth Transistors Realistic?..107
C. R. Bolognesi, H. Liu, O. Ostinelli, Y. Zeng

Technology Features and Design Tools for mm-Wave Applications in CMOS...........111
J. J. Pekarik

Design Aspects of 65-nm CMOS MMICs...115
M. Karkkainen, M. Varonen, D. Sandstrom, T. Tikka, S. Lindfors, K. A. I. Halonen

Characterization and Modeling of Impact Ionization Effects on Small and Large Signal Characteristics of AlGaAs/GaInAs/GaAs PHEMTs...119
C. Teyssandier, F. De Groote, R. Sommet, J. Teyssier, C. Chang, E. Leclerc, B. Carnez, R. Quere

A Methodology to Characterize the Low-Frequency Noise of InP Based Transistors.........123
A. A. Lisboa de Souza, J. C. Nallatamby, M. Prigent

Revised RF Extraction Methods for Deep Submicron MOSFETs..............................127
J. C. Tinoco, J. P. Raskin

Robust Packaged Diode Modelling with a Table-Based Approach 131
A. Rodriguez-Testera, O. Mojon, M. Fernandez-Barciela, E. Sanchez

X-Parameter Measurement and Simulation of a GSM Handset Amplifier 135
J. M. Horn, J. Verspecht, D. Gunyan, L. Betts, D. E. Root, J. Eriksson

An Integrated IQ Demodulator with Integrated Low-Power Multi-Gigabit BPSK / ASK Analog Signal Processor in 90nm CMOS 139
D. A. Yeh, S. Sarkar, S. Pinel, P. Sen, J. Laskar

Performance Analysis of Balanced and Unbalanced Feed-Forward Equalizer Structures for Multi-Gigabit Applications in 0.18(mu)m CMOS Process 143
H. S. Kim, D. Bhatta, K. H. Lee, C. Scholz, E. Gebara, J. Laskar

A 20GS/s 8-Bit Current Steering DAC in 0.25(mu)m SiGe BiCMOS Technology 147
S. Halder, H. Gustat, C. Scheytt, A. Thiede

A 1.2V Programmable ADC for a Multi-Mode Transceiver in 0.13(mu)m CMOS 151
D. Chang, S. Javvadi, C. Munoz, J. Abele, A. Hadiashar, M. Lugin, R. Quintal, G. Dawe

A 52GHz Millimeter-Wave PLL Synthesizer for 60GHz WPAN Radio 155
J. Lee, H. Kim, H. Yu

Digitally-Assisted Linearization of Wideband Direct Conversion Receivers..........................159
E. A. Keehr, A. Hajimiri

Wideband CMOS Receivers Exploiting Simultaneous Output Balancing and Noise/Distortion Canceling 163
S. C. Blaakmeer, E. A. M. Klumperink, D. M. W. Leenaerts, B. Nauta

SiGe BiCMOS High Linearity Mixers for Base-Station Applications.....................................167
T. Tikka, V. Saari, J. Jussila, K. Halonen, J. Ryynanen

5.5 GHz Low Voltage and High Linearity RF CMOS Mixer Design.......................................171
S. Lu, J. Guo

High Linearity Down-Conversion CMOS Mixers 175
D. Manstretta

Advanced Modeling of MISHFET Devices and their Performance in Current-Mode Class-D Power Amplifiers 179
J. Cumana, C. Lautensack, M. Eickelkamp, J. Goliasch, A. Noculak, A. Vescan, Rolf H. Jansen

Decrease in Slow Current Transients and Current Collapse in GaN-Based FETs with a Filed Plate 183
A. Nakajima, K. Itagaki, K. Horio

Increased Reliability of AlGaN/GaN HEMTs Versus Temperature Using Deuterium 187
G. Astre, J. G. Tartarin, J. Chevallier, S. Delage

X-Band GaN SPDT MMIC with over 25 Watt Linear Power Handling..................................190
J. Janssen, M. van Heijningen, K. P. Hilton, J. O. Maclean, D. J. Wallis, J. Powell, M. Uren, T. Martin, F. E. van Vliet

High Power Microstrip GaN-HEMT Switches for Microwave Applications...........................194
V. Alleva, A. Bettidi, A. Cetronio, W. Ciccognani, M. De Dominicis, M. Ferrari, E. Giovine, C. Lanzieri, E. Limiti, A. Megna, M. Peroni, P. Romanini

InAs/In_1-xGa_xAs Composite Channel High Electron Mobility Transistors for High Speed Applications 198
E. Y. Chang, C. Kuo, H. Hsu, C. Chang

Integrated Schottky Structures for Applications Above 100 GHz...202
B. Alderman, H. Sanghera, B. Thomas, D. Matheson, A. Maestrini, H. Wang, J. Treuttel, J. V. Siles, S. Davies, T. Narhi

Heterostructure Barrier Varactor Quintuplers for Terahertz Applications 206
J. Stake, T. Bryllert, A. O. Olsen, J. Vukusic

Metamorphic MMICs for Operation Beyond 200 GHz..210
A. Tessmann, I. Kallfass, A. Leuther, H. Massler, M. Schlechtweg, O. Ambacher

Industrial MHEMT Technologies for 80 - 220 GHz Applications 214
D. Smith, G. Dambrine, J. Orlhac

Characterization and Numerical Simulations of High Power Field-Plated pHEMTs 218
A. Chini, M. Esposto, G. Verzellesi, S. Lavanga, C. Lanzieri, A. Cetronio

A Computational Load-Pull Method for TCAD Optimization of RF-Power Transistors in Bias-Modulation Applications 222
O. Bengtsson, L. Vestling, J. Olsson

Thermal Model Extraction of GaN HEMTs for Large-Signal Modeling 226
S. Dahmani, E. S. Mengistu, G. Kompa

Modeling of Radiation, Conductor, and Dielectric Losses in SIW Components by the BI-RME Method 230
M. Bozzi, L. Perregrini, K. Wu

Tapped Integrated Inductors: Modelling and Application in Multi-Band RF Circuits 234
M. Dehan, J. Borremans, P. Wambacq, S. Decoutere

6-24 GHz Mixer Using 0.25(mu)m Enhancement Mode PHEMT Technology in a Low Cost Chip Scale Package 238
S. Kumar, J. Kessler, H. Morkner

35-65GHz MMIC QPSK Modulator 242
V. F. Fusco, C. Wang, T. Pochiraju

Wideband CMOS LC VCOs for IEEE 802.15.4a Applications 246
C. Stagni, A. Italia, G. Palmisano

A 60-GHz LNA and a 92-GHz Low-Power Distributed Amplifier in CMOS with Above-IC 250
C. Pavageau, O. Dupuis, M. Dehan, B. Parvais, G. Carchon, W. De Raedt

20 GHz Power Amplifier Design in 130 nm CMOS 254
M. Ferndahl, T. Johansson, H. Zirath

Advanced Components for Applications in S-Band and X-Band Radars 258
T. Boles

Low-Cost S-Band Multi-Function MMIC 262
A. P. de Hek, M. Rodenburg, F. E. van Vliet

Silicon-Germanium for Phased Array Radars 266
H. Berg, H. Thiesies, E. Hemmendorff, G. Sidiropoulos, J. Hedman

Thales Components and Technologies for T/R Modules 270
Y. Mancuso

GaN MMIC Based T/R-Module Front-End for X-Band Applications 274
P. Schuh, H. Sledzik, R. Reber, A. Fleckenstein, R. Leberer, M. Oppermann, R. Quay, F. van Raay,
M. Seelmann-Eggebert, R. Kiefer, M. Mikulla

Current Trends and Challenges in III-V HBT Compact Modeling 278
M. Rudolph

A Nonlinear Drain Resistance pHEMT Model for Millimeter-Wave High Power Amplifiers 282
A. Inoue, H. Amasuga, S. Goto, M. Miyazaki

Widebandgap Semiconductor HFET Models for Microwave CAD 286
R. J. Trew, W. Kuang, Y. Liu, G. L. Bilbro

A Non-Quasi-Static SOI MOSFET Model 290
D. R. Burke, T. J. Brazil

Accurate Nonlinear Electron Device Modelling for Cold FET Mixer Design 294
V. Di Giacomo, A. Santarelli, A. Raffo, P. A. Traverso, D. Schreurs, J. Lonac, D. Resca, G. Vannini,
F. Filicori, M. Pagani

Gate Width Dependence of Noise Parameters and Scalable Noise Model for HEMTs...........................298
Y. Zhu, C. Wei, O. Klimashov, B. Li, C. Zhang, Y. Tkachenko

A 70GHz VCO with 8GHz Tuning Range in 0.13(mu)m CMOS Technology 302
Z. Liu, E. Skafidas, R. Evans

A Low-Voltage, Fully-Integrated (1.5--6)GHz Low-Noise Amplifier in E-Mode pHEMT
Technology for Multiband, Multimode Applications...........................306
Z. Hasan-Abrar, Y. H. Chow, Y. W. Eng

Space Mapping Algorithm with Improved Convergence Properties for Microwave
Optimization 310
S. Koziel, J. W. Bandler

Full W-Band High-Gain LNA in mHEMT MMIC Technology...........................314
W. Ciccognani, F. Giannini, E. Limiti, P. E. Longhi

An Electrothermal Model of High Power HBTs for High Efficiency L/S Band Amplifiers 318
A. Xiong, R. Sommet, O. Jardel, T. Gasseling, A. A. Lisboa de Souza, R. Quere, S. Rochette

Regenerative Frequency Divider SiGe-RFIC with Octave Bandwidth and Low Phase Noise 322
T. Wallin, J. Hellen, H. Berg, S. Elfgren

New Macromodeling Approach to Phase Noise Analysis of Locked Oscillators 326
M. M. Gourary, S. G. Rusakov, S. L. Ulyanov, M. M. Zharov, B. J. Mulvaney, K. K. Gullapalli

Performance of Unstuck (Gamma) Gate AlGaN/GaN HEMTs on (001) Silicon Substrate at
10 GHz...........................330
J. C. Gerbedoen, A. Soltani, N. Defrance, M. Rousseau, C. Gaquiere, J. C. De Jaeger, S. Joblot, Y.
Cordier

A Reflection-Type Biphase Modulator with Balanced Loads 334
W. Ciccognani, F. Di Paolo, M. Ferrari, F. Giannini, E. Limiti

Design Manufacturing and Packaging of a 5-Bit K-Band MEMS Phase Shifter 338
S. Bastioli, F. Di Maggio, P. Farinelli, F. Giacomozzi, B. Margesin, A. Ocera, I. Pomona, M. Russo,
R. Sorrentino

Microwave Compact Passive Circuit Model of Isolated Interconnect over a Silicon
Substrate with a Through-Silicon Via (TSV) Ground Supply Network...........................342
W. Woods, G. Wang, H. Ding

Spectral Response Modelling of Hetcrojunction Phototransistors for Short Wavelength
Transmission...........................346
H. A. Khan, A. A. Rezazadeh, S. C. Subramaniam

A Low-Power Ultra-Compact CMOS LNA with Shunt-Resonating Current-Reused
Topology...........................350
M. Wei, S. Chang, Y. Liu

Efficient Frequency Domain Plus Spatial Expansion Method for Semiconductor Devices
Modeling 354
G. Leuzzi, V. Stornelli

Ka-Band Wide-Bandwidth Voltage-Controlled Oscillators in InGaP-GaAs HBT
Technology...........................358
C. Chiong, H. Chang, M. Chen

Circuital Modelling of Shunt Capacitive RF MEMS Switches 362
G. Bartolucci, R. Marcelli, S. Catoni, B. Margesin, F. Giacomozzi, A. Lucibello, V. Mulloni, P. Farinelli

Alumina and LTCC Technology for RF MEMS Switches and True Time Delay Lines..........366
R. Buttiglione, S. Catoni, G. De Angelis, M. Dispenza, A. Fiorello, K. Kautio, M. Ladhes, R. Marcelli, K. Ronka

The Fundamental Design Approach of the RF-DC Conversion Circuit for Optimizing its Characteristics...................370
T. Yamamoto, K. Fujimori, M. Sanagi, S. Nogi

The Impact of Technology Node Scaling on nMOS SPDT RF Switches 374
T. K. Thrivikraman, W. L. Kuo, J. P. Comeau, J. D. Cressler

Design and Temperature Dependent Analysis of GaAs Multilayer Transmission Lines..........378
J. Yuan, A. A. Rezazadeh, J. Lu, Q. Sun, V. T. Vo

DC-Contact RF MEMS Switches Using Thin-Film Cantilevers 382
H. Shen, S. Gong, N. S. Barker

Impact of Diode Geometry on Local Oscillator Breakthrough in Sub-Harmonic Mixers 386
V. Gutta, A. Fattorini, A. E. Parker, J. T. Harvey

RF Noise Shielding Method and Modelling for Nanoscale MOSFET 390
J. Guo, Y. Lin, Y. Tsai

An Optimum Cascode Topology for High Gain Micro/Millimeter Wave CMOS Amplifier Design 394
M. Nezhad-Ahmadi, B. Biglarbegian, H. Mirzaei, S. Safavi-Naieini

Efficient Design Methodology of RF-MEMS Based Tuner 398
D. Dubuc, C. Bordas, K. Grenier

Experimental Study of Ground Plane Width Effect in Multilayer MCM CPW Lines 402
K. K. Samanta, I. D. Robertson

A Low-Power 3-5-GHz UWB Down-Converter with Resistive-Feedback LNA in a 90-nm CMOS Process...................406
G. Sapone, G. Palmisano

A Fully Integrated 60 GHz SiGe BiCMOS Mixer 410
S. Lee, J. Lee, H. Kim

Millimeter-Wave Low Spurious Quadruple Harmonic Image Rejection Mixer with 90-Degree LO Power Divider 414
K. Kawakami, K. Nishida, M. Hieda, M. Miyazaki

A Highly Linear (40.5--43.5) GHz MMIC Single Balanced pHEMT Resistive Up-Converter Mixer for LMDS Applications 418
A. Khy, B. Huyart, H. Teillet

A 94-GHz Monolithic Front-End for Imaging Arrays in SiGe:C Technology...................422
E. Ojefors, U. Pfeiffer

Multiple-Throw Millimeter-Wave FET Switches for Frequencies from 60 up to 120 GHz 426
I. Kallfass, S. Diebold, H. Massler, S. Koch, M. Seelmann-Eggebert, A. Leuther

120-GHz-Band Low-Noise Amplifier with 14-ps Group-Delay Variation for 10-Gbit/s Data Transmission...................430
H. Takahashi, T. Kosugi, A. Hirata, K. Murata, N. Kukutsu

A Rigorous Assessment of Electro-Thermal Device Instabilities via Harmonic Balance Modeling 434
F. Cappelluti, F. L. Traversa, F. Bonani

Efficient Circuit-Level Nonlinear Analysis of Interference in UWB Receivers 438
V. Rizzoli, F. Mastri, A. Costanzo, D. Masotti, F. Donzelli

Large-Signal Performance of Resonant Tunnelling Diodes in K-Band Oscillators 442
B. Munstermann, A. Matiss, W. Brockerhoff, F. J. Tegude

Analysis and Synthesis of the Microstrip Lines Based on Support Vector Regression 446
N. T. Tokan, F. Gunes

Global Digital-Analog Co-Simulation Methodology for Power and Signal Integrity Aware Design and Analysis 450
S. Wane, G. Boguszewski

24 GHz LTCC I/Q Mixer Using Packaged HEMTs 454
V. Napijalo, V. Cojocaru

A Triple Tuned Ultra-Wideband VCO in X-K Band 458
M. Tsuru, K. Kawakami, K. Tajima, K. Miyamoto, M. Nakane, M. Hieda, M. Miyazaki

24 GHz LTCC Amplifier Using Packaged HEMTs 462
V. Napijalo, V. Cojocaru, T. Yokoyama

K-Band Frequency Synthesizer with Subharmonic Signal Generation and LTCC Frequency Tripler 466
T. Baras, A. F. Jacob

Demonstration of Heterogeneous Integration of Technologies for a Ku-Band SiP Doppler Radar 470
X. Sun, S. Brebels, S. Stoukatch, R. Jansen, L. Dussopt, M. A. Dubois, C. O'Mahony, S. Berberich, R. Houlihan, W. De Raedt

A Wide Tuning Range MEMS Varactor Based on a Toggle Push-Pull Mechanism 474
P. Farinelli, F. Solazzi, C. Calaza, B. Margesin, R. Sorrentino

Phase Shifter Design Based on Fast RF MEMS Switched Capacitors 478
B. Lacroix, A. Pothier, A. Crunteanu, P. Blondy

Microwave Tunable Bandpass Filter with MEMS Thermal Actuators 482
S. Fouladi, W. D. Yan, R. R. Mansour

Monolithic MEMS T-Type Switch for Redundancy Switch Matrix Applications 486
K. Y. Chan, M. Daneshmand, A. A. Fomani, R. R. Mansour, R. Ramer

Reliability of Dielectric Less Electrostatic Actuators in RF-MEMS Ohmic Switches 490
D. Mardivirin, A. Pothier, M. El Khatib, A. Crunteanu, O. Vendier, P. Blondy

EuMIC: A 400 MHz -- 1600 MHz SiGe MMIC Beam-Former for the Square Kilometre Array 494
K. Visser, E. van der Wal, M. Ruiter, D. Kant

X- and K-Band Tunable Phase Generation Circuits for Monolithic mm-Wave Phased Arrays 498
C. Carta, M. Seo, U. Madhow, M. Rodwell

A Four-Antenna Transceiver MIMIC for 60 GHz Wireless Multimedia Applications 502
S. Koch, I. Kallfass, A. Leuther, M. Schlechtweg, S. Saito, M. Uno

A Multi-Channel S-Band FMCW Radar Front-End 506
A. P. M. Maas, F. E. van Vliet

A 0.13-(mu)m SiGe BiCMOS LNA for 24-GHz Automotive Short-Range Radar 510
E. Ragonese, A. Scuderi, G. Palmisano

Optimization of Class E Power Amplifier Design Above Theoretical Maximum Frequency 514
E. Cipriani, P. Colantonio, F. Giannini, R. Giofre

Compact Concurrent Dual-Band Power Amplifier for 1.9GHz WCDMA and 3.5GHz OFDM Wireless Systems 518
  A. Cidronali, N. Giovannelli, I. Magrini, G. Manes

Doherty Amplifier Design for 3.5 GHz WiMAX Considering Load Line and Loop Stability 522
  M. A. Y. Medina, D. Schreurs, I. Angelov, B. Nauwelaers

GaN Doherty Amplifier with Compact Harmonic Traps 526
  P. Colantonio, F. Giannini, R. Giofre, L. Piazzon

An Innovative Time-Domain Simulation Technique for Strongly Nonlinear Heterogeneous RF Circuits Operating in Diverse Time-Scales 530
  J. F. Oliveira, J. C. Pedro

Vertical RF Transition with Mechanical Fit for Three-Dimensional Heterogeneous Integration 534
  L. Chen, J. Wood, S. Raman, N. S. Barker

The Effect of Dielectric Height and Ground Plane Width on Multilayer MCM FGCPW Lumped Elements 538
  K. K. Samanta, I. D. Robertson

EuMIC: 3D Packaging Technology for Integrated Antenna Front-Ends 542
  B. Bonnet, P. Monfraix, R. Chiniard, J. Chaplain, C. Drevon, H. Legay, P. Couderc, J. Cazaux

Design of 77 GHz Interconnects for Buried SiGe MMICs Using Novel System-in-Package Technology 546
  M. D. Richter, K. F. Becker, L. Bottcher, M. Schneider

Packaging Aspects of Photodetector Modules for 100 Gbit/s Ethernet Applications 550
  C. Jiang, G. G. Mekonnen, V. Krozer, T. K. Johansen, H. G. Bach

A 75--95 GHz Wideband CMOS Power Amplifier 554
  B. Wicks, E. Skafidas, R. Evans

A 19.1-dBm Fully-Integrated 24 GHz Power Amplifier Using 0.18-(mu)m CMOS Technology 558
  J. Kuo, Z. Tsai, H. Wang

10 Watt High Efficiency GaAs MMIC Power Amplifier for Space Applications 562
  F. Scappaviva, R. Cignani, C. Florian, G. Vannini, F. Filicori, M. Feudale

A 20 Watt Micro-Strip X-Band AlGaN/GaN HPA MMIC for Advanced Radar Applications 566
  C. Costrini, M. Calori, A. Cetronio, C. Lanzieri, S. Lavanga, M. Peroni, E. Limiti, A. Serino, G. Ghione, G. Melone

A Novel Silicon High Voltage Vertical MOSFET Technology for a 100W L-Band Radar Application 570
  B. Battaglia, D. Rice, P. Le, B. Gogoi, G. Hoshizaki, M. Purchine, R. Davies, W. Wright, D. Lutz, M. Gao, D. Moline, A. Elliot, S. Tran, R. Neeley

Author Index

WELCOME MESSAGES

Welcome from the President of the *European Microwave Association*

European Microwave Association

I am privileged to welcome you to the 11th European Microwave Week (EuMW), an initiative of the *European Microwave Association* (EuMA), founded in 1998 as a non-profit organisation based in Belgium. Amsterdam is a special place for EuMA, since it is the venue of the first EuMW, initiated in 1998 by complementing the European Microwave Conference (EuMC), the most prestigious microwave event in Europe since 1969, with two other major conferences in related areas, workshops, short courses and a large technical exhibition. Amsterdam is also the venue of the first EuRAD, which was added to the EuMW in 2004. I am pleased to acknowledge the fruitful collaboration with the IEEE MTT Society (technical co-sponsor of EuWiT) and the GAAS® Association (co-sponsor of EuMIC, formerly GAAS) in creating such event, which has now become the Europe's premier microwave, RF, wireless, and Radar event, attracting on average 1,500 delegates, with a record of about 1100 papers submitted and a 7500 m² exhibition.

Originally created in order to run on a firm legal basis the EuMW, EuMA has more recently undertaken initiatives for better serving the microwave community worldwide. The association has been opened to general membership, so that all microwave engineers from all over the world can join the EuMA as Members or Student Members. Many EuMA members now come from US and Asian countries. I am glad to announce that the General Assembly, the highest governing body of the association gathering representatives from European countries, has recently approved a motion to include also representatives from North America and Asia Pacific. This is a first step towards the creation of a truly trans-national association. I expect that this move will trigger similar and coordinated actions by other technical communities in Europe with whom EuMA has various collaborations.

If you are not a EuMA member yet I encourage you to join. You will be entitled to discounted fees for attending EuMA and EuMA-sponsored conferences and workshops, participate in EuMA activities, receive the EuMA Newsletter, and pay a discounted

Roberto Sorrentino
President
European Microwave Association

price both for the 2 EuMC DVD Archives which include all papers from the EuMC (1969 - 2003 and 2004 - 2008 respectively) and subscription to the EuMA Proceedings, a quarterly peer-reviewed journal. The journal is being totally reshaped to become a truly authoritative European Journal, under a new name and a new publisher. To join EuMA or to renew your membership simply go to EuMA Desk located in the registration area or to the web page www.eumwa.org, follow the instruction on the screen and pay a modest fee of € 20 (€ 10 for students) after choosing your preferred payment method. This will be more than compensated for by the discount you will get in registration fees (non member registration fee is 30% more expensive), DVD and journal subscription prices. For new members and EuMW delegates it will be our pleasure to give you an EuMA welcome gift, that you can receive at the EuMA Desk.

EuMA is proud to acknowledge those individuals who have given exceptional contributions to the microwave community. In 2004 the Distinguished Service Award, was established to "recognize an individual who has given outstanding service for the benefit of the European microwave community and, in particular, for the advancement of the European Microwave Association". In addition, a new award, the Outstanding Career Award will be presented for the first time this year to recognize an individual "whose career has exemplified outstanding achievements in the field of Microwaves". Both awards for 2008 will be presented during the opening ceremony of the EuMW.

This is the third EuMW held in Amsterdam after the successful events in 1998 and 2004, within the third 5-year cycle through Europe, continuing on to Rome 2009, Paris 2010 and Manchester 2011. Horizon House is arranging the EuMW, together with the EuMA for the sixth year now and we are proud to offer the scientific community such an outstanding conference week.

A special word of welcome to the organizers and delegates of the IQPC Military Radar Conference, which is collocated with the EuMW in the RAI this year. EuMA looks forward to a successful cooperation and hopes for fruitful networking and welcomes the IQPC delegates who participate in the EuRAD conference.

I would like to conclude my welcome message by extending my sincere thanks to TNO for making available people and resources to perform the local organization of the EuMW. Furthermore I would like to thank Frank van den Bogaart, 11th EuMW General Chairman, Peter Hoogeboom, Chairman of the 38th EuMC, Herbert Zirath, Chairman of EuMIC 2008, Homayoun Nikookar, Chairman of EuWiT 2008, and Alexander Yarovoy, Chairman of EuRAD 2008, for setting up four outstanding technical and scientific programs for this memorable event in Amsterdam.

Roberto Sorrentino
President
European Microwave Association   October 2008

WELCOME MESSAGES

Welcome to EuMIC 2008

We are very pleased to welcome you to the 2008 European Microwave Integrated Circuits (EuMIC) Conference in Amsterdam. The conference is chaired this year by Herbert Zirath. John Long and Frank van Vliet are Co-Chairs. This conference is the successor of the well-known GAAS symposium, which was first held in Rome in 1990. Since 2007 the conference is organized under the umbrella of both the European Microwave Association and GAAS® Association. We have an outstanding technical programme this year, with 4 plenary presentations and more than 100 technical papers. The papers are distributed across more than 20 sessions, scheduled for Monday and Tuesday of the European Microwave Week. During the rest of the week there will be 7 workshops and one short course. A significant number of sessions and workshops are joint sessions with the European Microwave Conference (EuMC) and the European Wireless Technology (EuWiT) conference, indicating the strong interaction between these conferences.

Be sure not to miss our eminent plenary speakers, Dr. Mehmet Souyer from IBM, Dr. Thijs de Graauw of SRON, Professor Gabriel Rebeiz from UCSD, and Dr. Bill Deal from Northrop Grumman. The plenary invited talks reflect the state-of-the-art in different areas; Dr Souyer will share the latest SiGe-development for integrated microwave circuits, Dr. Thijs de Graauw will present the status of the European Space instrument, HIFI, for space research, Professor Gabriel Rebeiz will share the recent development of SiGe and CMOS based phased arrays up to 60 GHz at UCSD, and Dr Bill Deal with present the latest results from the new 1 THz f_{max} InP HEMT device, and circuits.

This year we have emphasized a programme that should interest both industry and academia, with a large number of focussed sessions ranging from Phased Array Technology, Millimeter-Wave IC-Design for Commercial Applications, Submm-Wave Circuits, Linear Receiver Design, Advanced Power Amplifiers, to Active Device Modelling.

On Monday evening, we will continue the well known Foundry Round Table Discussions, organized by Massimo Comparini from the GAAS® Association. Selected European and non-European foundries will discuss trends and emerging technologies in the area of Microwave and Mm-Wave Integrated Circuits.

To acknowledge the high quality of papers presented at EuMIC2008, the technical program committee of EuMIC and the general assembly of the European Microwave Association will be awarding a Best Paper Prize and a Best Student Paper Prize. These winners receive a plaque commemorating their achievements. In additon the GAAS® Association will provide 3 student fellowships.

Together with the other conferences, we have planned several social events and some interesting tours to venues in and around Amsterdam. See page 24 - 25 for more information.

Finally, we would like to thank all of the people involved in the organization of this event, in particular the EuMW General Chair Frank van den Bogaart and his team, the organizers of the other conferences, the EuMIC TPC members, all reviewers, presenters of the invited talks, the workshop organizers and finally all people who submitted papers to this conference. Their combined efforts result in an outstanding event with an interesting, varied and exciting programme, which can be found on the following pages.

We are looking forward to seeing you in Amsterdam, and wish you an enjoyable and productive stay.

Herbert Zirath
Chalmers University, Sweden
Chair EuMIC

John Long
TU Delft
Co-Chair EuMIC

Frank van Vliet
TNO Defence and University of Twente
Co-Chair EuMIC and TPC

WELCOME MESSAGES

Welcome to the 11th European Microwave Week 2008

The European Microwave Week 2008 continues the series of successful microwave events, since its start in 1998 with its 11th Conference and Exhibition Week. From October 27th through 31st, 2008 the event will be held again in Amsterdam, The Netherlands. This conference series has continued to grow and is now the premier event in this field in Europe. The European Microwave Week consists of four conferences:

- The 38th European Microwave Conference (EuMC)
- The 3rd European Microwave Integrated Circuits Conference (EuMIC)
- The 1st European Wireless Technology Conference (EuWiT)
- The 5th European Radar Conference (EuRAD)

The Week provides the opportunity to attend these four conferences, various workshops and short courses given by leading experts in the field. In addition to the conferences the European Microwave Exhibition provides a unique market place for the delegates to engage with the leading companies on the largest trade show on RF and Microwaves in Europe.

For the first time EuMW2008 also offers a number of very attractive side events. IQPC's Military Radar Conference will be collocated and accessible to EuMW delegates. Forum and panel discussions will be organized on topics of special interest. We will host a Women in Engineering event and also a special competition, the EuMW2008 Student Challenge.

The ambition of the 11th European Microwave Week 2008 is to introduce a lead theme to bind the 4 conferences, the trade show and the application areas. For EuMW2008 the central theme will be "Bridging Gaps" to encourage interaction between research and application development.

The EuMW2008 event is the first to focus on the needs of engineers and researchers for the creation of innovative applications, products and services by providing an inspiring environment for discussion between academia and industry. We hope to enable the two communities to share the latest trends and developments that are widening the field of application of microwaves. Microwave devices, systems for telecommunications (both terrestrial and space borne), transportation, medical, radar as well as new fields of application will be present in the programme. Special attention has been put on the coordination of areas of common interest between the different conferences, workshops and short courses.

The four conferences specifically target ground breaking innovation in microwave research through a call for papers explicitly inviting to submit presentations on the latest trends in the field driven by the roadmaps of the industry.

The tradeshow area will accommodate floor space for "Bridging Gaps" exhibition pavilions where EuMW Science and Technology "meets" applications.

The location of EuMW2008 will be the Amsterdam RAI. The venue is easy to access with very good transportation from Schiphol Amsterdam Airport and railway stations. Hotel accommodation is available throughout Amsterdam with good connection to the Amsterdam RAI convention centre by public transportation.

We are proud to present to you the programme of EuMW2008 which could not have been achieved without the valuable contribution of many enthusiastic colleagues in industry, research centres and universities all over the world who have helped to organize this year's event in Amsterdam.

On behalf of the Local Organizing Committee we would like to express our gratitude to the four international Technical Programme Committees and the hundreds of reviewers who worked hard to shape the record number of individual contributions into the final programmes.

Also we like to thank those who organized the workshops, the focussed sessions, the short courses and the special events which we believe are an essential element to fulfil the needs of our delegates.

And last but not least it has been a great pleasure to cooperate with Horizon House. Their experience and support again proved invaluable for guiding the chair and his team through all the steps of organizing a challenging event.

We are convinced that you will find the European Microwave Week 2008 to be unique, refreshing and inspiring offering many opportunities to learn about new ideas and to share some of your own with other delegates and colleagues. We look forward to welcome you in Amsterdam at the 11th European Microwave Week.

Frank van den Bogaart
General Chair

Frank van den Bogaart
TNO Defence, The Hague
General Chair

Leon Kaufmann
TU Eindhoven
EuMW Co-Chair

Marc van Heijningen
TNO Defence, The Hague
EuMW Secretary

Huib Pasman
TNO Defence, The Hague
Commercial Affairs

CONFERENCE COMMITTEES

**EuMIC
Technical Programme
Committee**

Shmuel Auster,
Elta Systems Ltd, Israel

Klaus Beilenhoff,
UMS, Germany

Carlos Camacho Peñalosa,
University de Malaga, Spain

Massimo Claudio Comparini,
Thales Alenia Space (TAS), Italy

Gilles Dambrine,
IEMN, France

Fabio Filicori,
University of Bologna, Italy

Giovanni Ghione,
Politecnico di Torino, Italy

Franco Giannini,
Università degli Studi di Roma
'Tor Vergata', Italy

Hans Hartnagel,
Technical University
Darmstadt, Germany

Tom Johansen,
Technical University of
Denmark, Denmark

Hiroshi Kondoh,
Hitachi, Central Research lab,
Japan

John Long,
Delft University of Technology,
The Netherlands

Gottfried Magerl,
Technical University Vienna,
Austria

Steve Marsh,
Midas Consulting, UK

Angel Mediavilla,
University de Cantabria, Spain

Tapani Närhi,
ESA-ESTEC, The Netherlands

Ali Rezazadeh,
University of Manchester, UK

Michael Schlechtweg,
FhG-IAF, Germany

Derek Smith,
OMMIC, France

Andreas Thiede,
University of Paderborn,
Germany

Frank van Vliet,
TNO and University of Twente,
The Netherlands

Huei Wang,
NTU, Taiwan

Herbert Zirath,
Chalmers University of
Technology, Sweden

The 2008 EuMW TPC team

WELCOME MESSAGES

The European Microwave Week 2008 Local Organizing Committee

Appearing from left to right: André Verweij, Leon Kaufmann, Dominique Schreurs, Homayoun Nikookar, George Heiter, John Long, Jan Geralt bij de Vaate, Frank van den Bogaart, Frank van Vliet, Erik Fledderus, Peter Hoogeboom, Herbert Zirath, Alexander Yarovoy, Emiel Stolp, Marc van Heijningen.
(Photograph taken at the TPC-Meeting on April 12, 2008 in The Hague, courtesy Shmuel Auster.)

The EuMW Core Team

General Chair	**Frank van den Bogaart,** TNO Defence, Security and Safety, The Hague
Co-Chair	**Leon Kaufmann,** Technical University Eindhoven
General TPC Coordinator	**Jan Geralt bij de Vaate,** ASTRON, Dwingeloo
Workshops Coordinator	**Dominique Schreurs,** Katholieke Universiteit Leuven (Belgium)
EuMW Secretary	**Marc van Heijningen,** TNO Defence, The Hague
Treasurer	**Paul Belt,** TNO Defence, The Hague
Publication Chair	**Peter Hoogeboom,** TNO Defence, The Hague and Delft University of Technology
Electronic Submissions Chair	**George Heiter,** Heiter Microwave Consulting, USA
Local Arrangements	**André Verweij,** TNO Defence, The Hague
Visa Affairs	**Rowena Sardjoe,** Delft University of Technology
Commercial Affairs	**Huib Pasman,** TNO Defence, The Hague

The Conference Teams

EuMC 2008

Chair	**Peter Hoogeboom,** TNO Defence and Delft University of Technology
Co-Chair	**Dominique Schreurs,** Katholieke Universiteit Leuven (Belgium)
Co-Chair and TPC	**Jan Geralt bij de Vaate,** ASTRON, Dwingeloo

EuMIC 2008

Chair	**Herbert Zirath,** Chalmers University of Technology (Sweden)
Co-Chair	**John Long,** Delft University of Technology
Co-Chair and TPC	**Frank van Vliet,** TNO Defence and University of Twente

EuWiT 2008

Chair	**Homayoun Nikookar,** Delft University of Technology
Co-Chair and TPC	**Erik Fledderus,** TNO ICT and Eindhoven University of Technology

EuRAD 2008

Chair and TPC	**Alexander Yarovoy,** Delft University of Technology
Co-Chair	**Emiel Stolp,** Thales Nederland, Hengelo

EuMA GENERAL ASSEMBLY AND STEERING COMMITTEE 2008

EuMA
General Assembly 2008

Board of Directors

Roberto Sorrentino,
President, GA Chairman

Heinrich Dämbkes

Richard Ranson

Antti Räisänen

Paul-Alain Rolland

Peter Russer

André Vander Vorst

Alexander Yarovoy

EuMW Chairs

Frank van den Bogaart
(2008)

Wolfgang Heinrich
(2007), 1st Vice-Chair

Paolo Lampariello
(2009), 2nd Vice-Chair

Ordinary Members

Tom Brazil,
Group 4

Spartak Gevorgian,
Group 6

Levent Gürel,
Group 11

Victor Fouad Hanna,
IEEE Region 8

Tatsuo Itoh,
Group 15

Bumman Kim,
Group 16

Viktor Krozer,
Group 7

Andrzej Kucharski,
Group 9

Yuri Kuznetsov,
Group 12

Gottfried Magerl,
Group 13

Andrea Massa,
Group 3

Francisco Medina Mena,
Group 14

Ferenc Mernyei,
Group 8

Robert Plana,
Group 1

Lorenz-Peter Schmidt,
Group 2

Guy Vandenbosch,
Group 5

Dmytro Vavriv,
Group 10

Founder Members

Leo Ligthart

Asher Madjar

Holger Meinel

Steve Nightingale

Roberto Sorrentino

André Vander Vorst

EuMA Proceedings Editor

Robert Weigel

Honorary Secretary

Andrew F Wilson

Countries represented

Group 1 – France, Monaco

Group 2 – Germany

Group 3 – Italy, San Marino, Vatican City

Group 4 – United Kingdom, Ireland, Gibraltar, Malta

Group 5 – Belgium, The Netherlands, Luxemburg

Group 6 – Iceland, Norway, Sweden

Group 7 – Finland, Denmark

Group 8 – Bulgaria, Czech Republic, Hungary, Romania, Slovakia

Group 9 – Poland, Estonia, Latvia, Lithuania

Group 10 – Ukraine, Moldova, Georgia

Group 11 – Albania, Bosnia-Herzegovina, Croatia, Cyprus, FYR Macedonia, Greece, Israel, Slovenia, Turkey, Yugoslavia

Group 12 – Russia, Belarus

Group 13 – Austria, Liechtenstein, Switzerland

Group 14 – Spain, Portugal, Andorra

Group 15 – North America

Group 16 – Asia Pacific

EuMW
Steering Committee 2008

Official Members

EuMA President: Roberto Sorrentino

for EuMC: Asher Madjar, Danielle Vanhoenacker, Steve James Nightingale

for EuMIC: Paul-Alain Rolland (EuMA), Franco Giannini (GAAS)

for EuWiT: Petteri Alinikula (EuMA), Fred Schindler (IEEE MTT-S)

for EuRAD: Alexander Yarovoy

Invited Members

EuMA Treasurer – André Vander Vorst

GAAS Treasurer – Massimo Claudio Comparini

IEEE MTT-S – Shmuel Auster

Current year Chairs:
EuMW – Frank van den Bogaart
EuMC – Peter Hoogeboom
EuMIC – Herbert Zirath
EuWiT – Homayoun Nikookar
EuRAD – Alexander Yarovoy

EuMW Treasurer – Paul Belt

EuMW General Chairs of past and next two years:
Wolfgang Heinrich, Paolo Lampariello, Gilles Dambrine

REVIEWERS 2008

The conference organizers would like to thank the persons listed below and the TPC members. All of them have been indispensable as EuMW2008 Reviewers and in drafting an outstanding Technical Programme:

Bernd Adelseck
Mahmoud Al Ahmad
Federico Alimenti
José I. Alonso
Ann-Marie V. Andersson
Mykhaylo I. Andriychuk
Paolo Arcioni
Myriam Ariaudo
Holger Arthaber
Abdullah Atalar
Herve Aubert
Ozlem Aydin Civi
Dominique Baillargeat
Peter G. Baltus
John W. Bandler
Afonso M. Barbosa
Geneviève B. Baudoin
Tibor Berceli
Corinne Berland
Pierre O. Bertram
Alexandre Bessemoulin
Stephan Biber
Stéphane Bila
Filiberto Bilotti
Hans L. Bloecher
Markus Boeck
Sylvain Bollaert
Fabrizio Bonani
Vicente E. Boria-Esbert
Hermann F. Boss
Maurizio Bozzi
Mike Brookbanks
Ibrahim Budiarjo
Vittorio Camarchia
Marc M.C. Camiade
Nuno B. Carvalho
Johan A. Catrysse
Heinz J. Chaloupka
Young-Kai Chen
Victor Chernyak
Siou Teck Chew

Alessandro Chini
Alessandro Cidronali
Paolo Colantonio
Juan-Mari Collantes
Iñigo Cuiñas
David Daniels
Plamen I. Dankov
François Danneville
René de Jongh
Pedro De Paco
Andrew W. Dearn
Pierre Degauque
Anders Derneryd
Yann Deval
Liam M. Devlin
Hervé Dillenbourg
Simona Donati Guerrieri
M'Hamed Drissi
David Dubuc
Philippe Duême
Christopher Duff
Christoph Ernst
Helmut Essen
Christian Fager
Novak E.S. Farrington
Anna Maria Fiorello
Gerhard Fischerauer
Marc J. Franco
Wolfgang Freude
Piotr Z. Gajewski
Heyno Garbe
José A. Garcia
Peter Gardner
Giampiero Gerini
Raphaël Gillard
Roberto Gomez-Garcia
Jean-Marie Gorce
Katia Grenier
Wilhelm Grüner
Yasar Gurbuz
Mehrdad Hajivandi
Svein-Erik Hamran
Victor F. Hanna
Stephen A. Harman
Alan Harrison
Dirk Heberling
Marc van Heijningen
Peter de Hek
Matti H.A.J. Herben

Ralf Hocke
Jia-Sheng Hong
Frederick Huang
Gernot Hueber
Peter G. Huggard
Isabelle M. Huynen
Igor Immoreev
Per Ingvarson
Dieter Jäger
Goutoule Jean-Marc
Andrzej Karwowski
Danielle Kettle
Marek J. Kitliński
Jens Klare
Peter Knott
Boris Kutuza
Nathalie Labat
Madan Kumar Lakshmanan
Michèle Lalande
Torben Larsen
Tuami Lasri
David D.L. Lautru
Gabriel Lellouch
Friedrich Lenk
Yoke Choy Leong
Pascal Leuchtmann
Giorgio Leuzzi
Ernesto Limiti
Fujiang Lin
Olivier Llopis
Pierfrancesco Lombardo
Cyril Luxey
Jan Machac
John C. Mahon
Robert J. Malmqvist
Gianfranco Manes
Cyril Mangenot
Javier Marti Canales
Villegas Martine
Teresa M. Martin-Guerrero
Alex R. Masidlover
Markus L. Mayer
Milos Mazanek
Alexander Megej
Chafik Meliani
Chinchun Meng
Adriano Meta
Brian Minnis
Mauro Mongiardo

REVIEWERS 2008

Gilles Montoriol
Hiroshi Murata
Bart Nauwelaers
Jean Michel Nebus
Ioan Nicolaescu
Kenjiro Nishikawa
Máirtín O'Droma
Morten Olavsbråten
Ban Leong Ooi
Giancarlo Orengo
Damien Pacaud
Maurizio Pagani
Jean-Paul Parneix
Daniel Pasquet
Matteo Pastorino
Regina Paszkiewicz
Parbhu Patel
Graham A. Pearson
Jose C. Pedro
Christian Person
Ullrich R. Pfeiffer
Odile Picon
Massimiliano Pieraccini
Jerzy K. Piotrowski
Robert Plana
Dirk Plettemeier
Ignazio Pomona
Lluis Pradell
Nigel E. Priestley
Michel Prigent
Jérôme Puech
Rüdiger Quay
Raja Abdullah Raja Syamsul Azmir
Richard G. Ranson
Marten G. Risling
Eric Rius
Vittorio Rizzoli
Ian D. Robertson
Duncan A. Robertson
Neil Roddis
Vauzelle Rodolphe
António J. Rodrigues
Hendrik Rogier
Yves Rolain
Ilona Rolfes
Anders Rydberg
Christian Sabatier
Krzysztof Z. Sachse
Pierre Saguet

Paulius Sakalas
Atsushi Sanada
David M. Sanchez-Hernandez
Timofey Savelyev
Vladimir V. Sazonov
Christian G. Schaeffer
Wolfgang Schiller
Martin Schneider
Bernhard Schönlinner
Martin Schüßler
Jonathan B. Scott
Daniel Segovia-Vargas
Stefano Selleri
Uwe Siart
Robin Sloan
Jonathan L. Sly
Bart Smolders
Paul Snoeij
Christopher M. Snowden
Jacques Sombrin
Raphael R.S. Sommet
Andreas Springer
Jan Stake
Benoît Stockbroeckx
Andrew G. Stove
Malcolm G. Stubbs
Daniel G. Swanson
Jean-Guy Tartarin
Paul J. Tasker
John E. Tattersall
Joseph L. Tauritz
Franz J. Tegude
Tsuneo Tokumitsu
Serge Toutain
Sergei Tretyakov
Robert J. Trew
Wendy Van Moer
Gerard C. van Rhoon
Dirk van Troyen
Andre Vander Vorst
Giorgio Vannini
Serge Verdeyme
Valerie Vigneras
Roberto Vincenti Gatti
Hubregt J. Visser
Stefan Vossen
Mark R. Walbridge
Matthias Weiß
Wolfgang Winkler

Marian Wnuk
King Wah Wong
Karl Woodbridge
Paul van Zeijl
Haibin Zhang

A Survey on 60 GHz Broadband Communication: Capability, Applications and System Design

Michael Peter [#1], Wilhelm Keusgen [#2], Jian Luo [#3]

[#] *Fraunhofer-Institut für Nachrichtentechnik, Heinrich-Hertz-Institut*
Einsteinufer 37, 10587 Berlin, Germany
[1] michael.peter@hhi.fraunhofer.de
[2] wilhelm.keusgen@hhi.fraunhofer.de
[3] jian.luo@hhi.fraunhofer.de

Abstract— **This paper gives a review of important aspects and current developments in 60 GHz broadband communication. We present channel measurement results for a conference room and analyze the channel capacity. The results clearly demonstrate the capability of the 60 GHz approach. Finally, we outline a concept for a wireless 60 GHz in-flight entertainment system.**

I. INTRODUCTION

Indoor broadband radio communication is playing an increasing role in our today's life. However, the growing number of electronic devices and broadband multimedia applications give rise to high requirements on wireless systems for short-range data transmission [1]. Fig. 1 illustrates that the rate requirements reflected in the specifications of transmission standards have increased tenfold every five years and will shortly break the 1 Gbit/s limit in the area of Wireless Local Area Networks (WLANs). Several technical approaches are currently discussed to address this challenging task: Ultra-Wideband (UWB) transmission, 60 GHz transmission and optical free space transmission.

UWB uses a large bandwidth exceeding 500 MHz or 20% of the arithmetic center frequency. Within the frequency band of 3.1–10.6 GHz the FCC power spectral density emission limit is −41.3 dBm/MHz in order to avoid interferences with narrowband systems operating in the same band. However, in Europe the same limit is only allowed in the range between 6.0 and 8.5 GHz in the long term, which poses a severe limitation on wireless networks basing on UWB.

During the past years infrared (IR) and LED-based visible-light transmission have attracted increasing attention for short-range wireless communication [2], [3], whereas visible light communication is only possible in the downlink. Virtually there is an unlimited amount of unregulated available bandwidth, and optical systems do not interfere with existing radio systems. However, high-rate optical transmission is reliant on a Line-of-Sight (LOS) between optical transmitter and receiver and is only feasible over very short distances. Besides the fact that daylight and other lighting sources have a great impact on visible light communications, Intersymbol Interference (ISI) seems to be a strictly limiting factor and the practically achievable data rate is still unknown [3].

Finally, 60 GHz communication is a very promising approach to open up the bottleneck of wireless short range transmission.

Future systems are supposed to operate in the unlicensed band from 57 to 64 GHz and could support data rates above 1 Gbit/s. The main advantage of this frequency band is the large available bandwidth without comparably restrictive power limits as for UWB transmission. In addition, small antenna dimensions – also in the case of multi-antenna systems – result from the short wavelength. Recent developments in silicon semiconductor technology give rise to the expectation that powerful and marketable integrated transceiver chips are within the grasp [4].

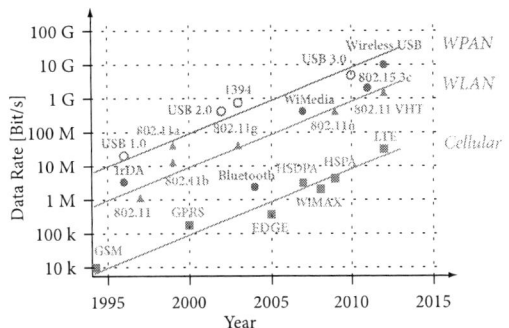

Fig. 1. Trend of data rates in wireless communications. Some wired WPAN-standards are noted for comparison

II. APPLICATIONS AND 60 GHz STANDARDIZATION

Several commercial applications of 60 GHz transmission are considered as particularly promising. In the context of WLAN standardization an extension of the existing and widely used IEEE 802.11a, IEEE 802.11b and IEEE 802.11g standards is discussed. The 60 GHz VHT (Very High Throughput) variant is supposed to support rates beyond the 600 Mbit/s as planned for the IEEE 802.11n extension. Adding the VHT variant to the 802.11 family would ensure a high compatibility with current WLAN systems and significantly facilitate the market introduction of 60 GHz products since only minor modifications on the network and the link layer are necessary.

Currently the IEEE 802.15.3 Task Group 3c is developing a millimeter-wave-based alternative Physical Layer (PHY)

for the existing 802.15.3 Wireless Personal Area Network (WPAN) Standard 802.15.3. These millimeter-wave WPANs shall provide a data rate of at least 2 Gbit/s for applications as high speed internet access and HDTV video streaming.

Another interesting application is to provide access to multimedia contents at a data kiosk. Users are enabled to connect their PDAs or cell phones to a public access point which serves as a central control entity and can download city information or videos within a few seconds.

Closely related to the data kiosk application is radio transmission inside vehicle cabins. The main purpose is to establish wireless links between displays and a central media server. Principally, one must distinguish between applications where the number of terminals is comparably small (e.g. in cars) and systems that must be capable to serve a huge number of users like in aircraft passenger cabins. In the first case one central unit is controlling the communication, whereas in the latter case multiple access points are needed resulting in a cellular network.

Finally, numerous special applications in the field of medical and safety technology are conceivable with the main objective to get rid of impractical cable connections.

III. 60 GHz Wideband Channel Characterization

The characterization of the 60 GHz multipath radio channel is an important task because the knowledge of the channel impacts on the transmitted signal are crucial for wireless system design. Two approaches can be pursued in order to get insight into site-specific wave propagation: channel measurements and deterministic propagation prediction by Ray-Tracing (RT) simulations [5], where RT is a promising technique to investigate the wideband radio.

A. Channel Description

The time-invariant multipath channel can be described by its impulse response (CIR). It is commonly modeled as a sum of time-shifted complex-weighted Diracs $\delta(\tau)$:

$$h(\tau) = \sum_{l=0}^{L-1} h_l \cdot \delta(\tau - \tau_l), \qquad (1)$$

where τ_l, h_l denote the delay and the complex weight of the lth channel tap, respectively. Since practical systems have a limited bandwidth, each channel tap may result from a superposition of multiple Multipath Components (MPCs), and the channel is subjected to small-scale fading.

B. Wideband Channel Measurements

With regard to 60 GHz channel characterization we performed measurements in various environments like an office, a conference room and an aircraft passenger cabin [6], [7] using a modular measurement setup [8].

Fig. 2 shows an exemplary small-scale set resulting from a measurement in a conference room with omnidirectional antennas. During the measurement procedure the antenna was moved on a linear track over a distance of 10 cm by the help of a controllable positioning platform. We can see that a multitude

of significant multipath components arrive at the receiver with delays of up to 75 ns. Furthermore, the effect of small-scale fading is clearly visible.

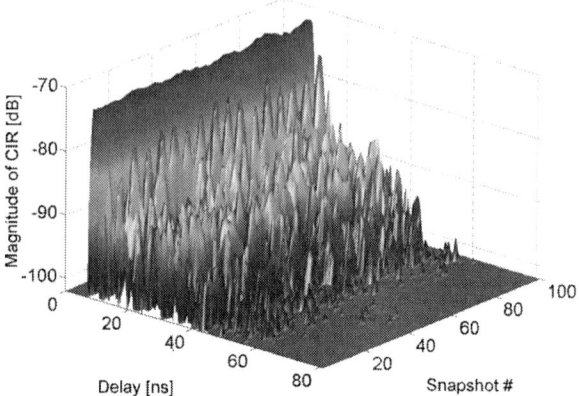

Fig. 2. Exemplary small-scale sets of CIRs under LOS conditions.

In general, millimeter wave systems are proposed for LOS transmission, but in practical scenarios intervisibility between transmitter (Tx) and receiver (Rx) might not be guaranteed. Therefore, it is important to analyze the impact of an Obstructed Line-of-Sight (OLOS). In the measurements OLOS was achieved by blocking the direct path with an absorber mat. Our investigations show that the original LOS-MPC drops below the noise threshold. However, the impact on other components is negligible so that a considerable multipath power still arrives at the receiver. The drop in total average power is in the range of 4 to 8 dB for omnidirectional antennas. If the communication system is highly capable to adapt the transmission to the channel state, it is possible to sustain a link even under OLOS conditions.

C. Path Loss and Time Dispersion

The path loss (PL) is defined as the ratio of the effective transmitted power to the received power:

$$\mathrm{PL}\,[\mathrm{dB}] = (P_T - P_R)\,[\mathrm{dBm}] + (G_T + G_R)\,[\mathrm{dBi}], \qquad (2)$$

where P_T, P_R, G_T and G_R denote the transmitted power, the received power, the transmitter antenna gain and the receiver antenna gain, respectively.

The path loss exponent model (also referred to as log-distance law) describes the average path loss as a function of the distance d between the Tx and the Rx [9]:

$$\overline{\mathrm{PL}}(d)\,[\mathrm{dB}] = \overline{\mathrm{PL}}(d_0)\,[\mathrm{dB}] + 10\,n\log\left(\frac{d}{d_0}\right), \qquad (3)$$

where $\overline{\mathrm{PL}}(d_0)$ and n denote the intercept point and the path loss exponent, respectively. Typical values for n and $\overline{\mathrm{PL}}(d_0)$ resulting from indoor measurements in a conference room are given in Table I together with the average RMS delay spread $\overline{\tau}_{\mathrm{rms}}$ characterizing the time dispersion [6].

978-2-8748-7007-1/08 $25.00 © 2008 EuMA

TABLE I
CHANNEL PARAMETERS FOR THE CONFERENCE ROOM SCENARIO.

Conditions	$\overline{\mathrm{PL}}(d_0 = 1\,\mathrm{m})$ [dB]	n	τ_{rms} [ns]
LOS	68.5	1.34	6.29
OLOS	78.7	0.61	8.75

D. Channel Capacity

In multiple antenna (MIMO) systems [10], the ergodic capacity C gives an upper bound of the achievable data rate with subject to the sum transmit power constraint P_T

$$C = \max_{\boldsymbol{R}_x:\mathrm{tr}\{\boldsymbol{R}_x\}=P_T} \mathrm{E}\left\{\log_2 \det\left(\boldsymbol{I} + \frac{1}{\sigma_n^2}\boldsymbol{H}\boldsymbol{R}_x\boldsymbol{H}^{\mathrm{H}}\right)\right\}. \quad (4)$$

Herein $\boldsymbol{R}_x$ is the covariance matrix of the transmit vector $\boldsymbol{x}$ with $P_T = \mathrm{tr}\{\boldsymbol{R}_x\}$ ($\mathrm{tr}\{\boldsymbol{A}\}$: trace of $\boldsymbol{A}$) and σ_n^2 is the mean noise power per receive antenna. The matrix $\boldsymbol{H}$ stands for the MIMO-channel. $\mathrm{E}\{X\}$ denotes the expectation of X, $\boldsymbol{H}^{\mathrm{H}}$ the conjugate transpose of $\boldsymbol{H}$ and $\boldsymbol{I}$ the identity matrix. Without channel state information at the transmitter, it is optimal to distribute P_T equally to the N_T transmit antennas, which leads to

$$C = \mathrm{E}\left\{\log_2 \det\left(\boldsymbol{I} + \frac{\Phi_{xx}}{N_0 N_T}\boldsymbol{H}\boldsymbol{H}^{\mathrm{H}}\right)\right\}. \quad (5)$$

There Φ_{xx} is the constant spectral transmit power density with $P_T = \Phi_{xx}\,\mathrm{BW}$ (BW: bandwidth) and N_0 is the white noise power density per receive antenna.

When using measured channel parameters, (5) gives realistic estimates of the achievable data rates in 60 GHz communication sytems: Fig. 3 shows the capacity C as a function of the distance d between Tx and Rx for single antenna (SISO) and 2×2 MIMO configurations, both for LOS and OLOS channels. The variances (powers) of the entries in $\boldsymbol{H}$ have been determined according to the log-distance law with the values given in Table I. For the LOS case the distribution of the elements of $\boldsymbol{H}$ was Ricean with a Rice-factor $K = 2$, where equal deterministic parts and statistically independent fading parts were used. For the OLOS channels $\boldsymbol{H}$ was modeled with statistically independent complex normally distributed (Rayleigh) entries. The density Φ_{xx} was chosen to $-77\,\frac{\mathrm{dBm}}{\mathrm{Hz}}$, which corresponds to a transmit power $P_T - 10\,\mathrm{dBm}$ for BW $= 500\,\mathrm{MHz}$. The noise power density was $-174\,\frac{\mathrm{dBm}}{\mathrm{Hz}}$. A noise figure was not considered in (5). Therefore in a real system the transmit power would have to be increased by the amount of the noise figure to compensate for the performance loss.

Generally it can be noticed from Fig. 3, that very high capacity values can be achieved in 60 GHz communication systems even for OLOS channels with non-directive transmission and that the utilization of MIMO configurations gives a significant improvement. It can be recognized that a capacity of 4 bit/s/Hz i.e. $2\,\frac{\mathrm{Gbit}}{\mathrm{s}}$ for BW $= 500\,\mathrm{MHz}$ can be ensured for instance by transmitting one data stream and applying 16 QAM-Modulation in a 2×2 MIMO configuration.

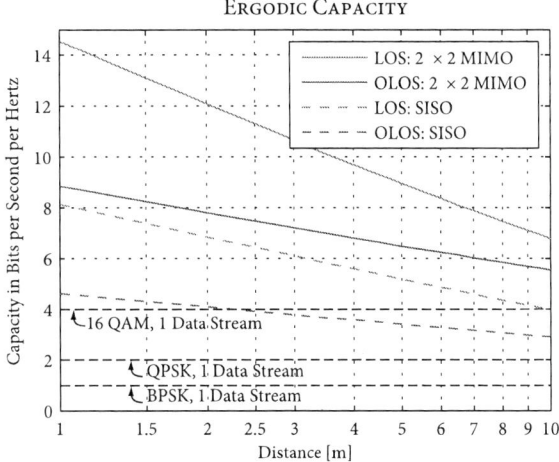

Fig. 3. Ergodic capacity for SISO and 2×2 MIMO systems, utilizing 60 GHz LOS and OLOS channels.

IV. IFE SYSTEM CONCEPT

A wireless In-Flight Entertainment (IFE) network is to provide access to multimedia content like video on demand in aircrafts and has to be able to support up to 1000 users in full-duplex mode. Despite the utilization of sophisticated techniques for data compression, the aggregate data rate is several Gbit/s or even tens of Gbit/s.

A. Architecture

In view of the high aggregate data rate we propose a cellular system architecture [11]. Fig. 4 illustrates the cell arrangement in a wide-bodied aircraft for a frequency reuse factor of 8, where up to 32 seat-terminals (TM), i.e. passengers, are served by one Access Point (AP), utilizing a 60 GHz air interface. The user links are allocated to the APs dynamically, depending on the current channel state. If the channel quality decreases sharply, e.g. if the line-of-sight is blocked, or in the case of an AP breakdown, a handover of the affected TMs to other APs can be accomplished. Multi-antenna techniques are deployed primarily to achieve diversity gains.

B. Duplex Mode and Multiple Access

Up- (UL) and Downlink (DL) operate in frequency division duplex (FDD) mode. In order to enable TMs at reasonable costs OFDMA/TDMA is used as broadcast channel scheme in the DL, whereas FDMA/CDMA is deployed in the UL. OFDM allows for low complexity receivers, while the crest-factor of CDMA signals is quite low, which limits the requirements on the power amplifiers at the TMs. In addition, the TDMA component lowers the requirements on the signal processing hardware, since the TMs have multiple time slots to process a data frame. Power control can be applied across multiple time slots to reduce co-channel interference. The additional use of FDMA in the UL limits the length of the CDMA code and

978-2-8748-7007-1/08 $25.00 © 2008 EuMA

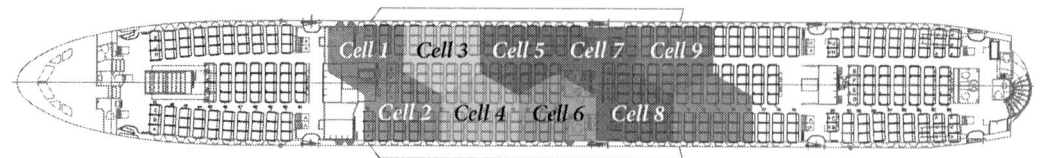

Fig. 4. Cellular system architecture.

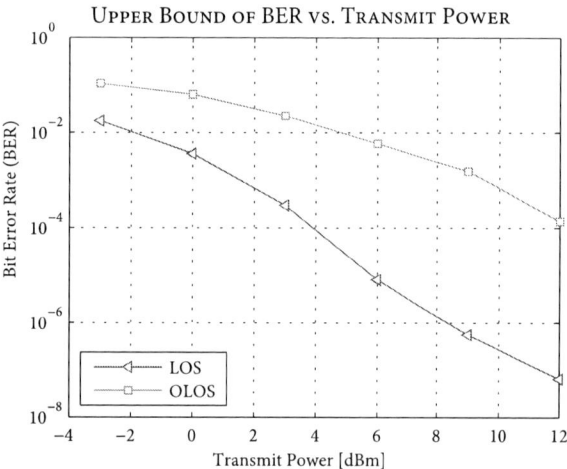

Fig. 5. Upper bound of BER-performance vs. sum Tx-power .

hence reduces the complexity of the multi-user detection at the AP significantly. Furthermore it allows for adaptive subband allocation in the UL.

C. Simulation Results

We simulated the performance of the IFE system with measured 60 GHz channels assuming a 2×2 MIMO configuration. Space-Time Code (STC) [12] and Maximum Ratio Combining (MRC) have been applied to achieve MIMO diversity gain, where an additional antenna gain of 10 dBi was assumed. Fig. 5 shows that a transmit power of 1.5 dBm and 9.5 dBm is required under LOS and OLOS conditions, respectively, in order to achieve an (uncoded) BER of 10^{-3}.

V. CONCLUSIONS

In this paper, we have shown that millimeter wave transmission is a promising candidate for future systems that are supposed to achieve data rates beyond 1 Gbit/s. On the basis of channel measurement results we have determined a channel capacity between 5.6 bit/s/Hz and 14.5 bit/s/Hz for applicable 2×2 MIMO systems in realistic indoor scenarios. Since wireless IFE is a promising application for 60 GHz transmission, we

have developed a system concept that is capable to supply several hundreds of passengers inside an aircraft cabin with data rates of approximately 10 Mbit/s in downlink. Simulation results show that for a 2×2 MIMO configuration a transmit power around 10 dBm is sufficient to ensure a reliable downlink transmission, even under OLOS conditions.

REFERENCES

[1] P. Smulders, "Exploiting the 60 GHz band for local wireless multimedia access: prospects and future directions," *Communications Magazine, IEEE,* vol. 40, no. 1, pp. 140–147, 2002.

[2] A. J. Moreira, A. M. Tavares, R. J. Valadas, and A. M. de Oliveira Duarte, "Modulation methods for wireless infrared transmission systems: performance under ambient light noise and interference," in *Proc. SPIE Vol. 2601, Wireless Data Transmission,* Dec. 1995, pp. 226–237.

[3] T. Komine and M. Nakagawa, "Fundamental analysis for visible-light communication system using LED lights," *Consumer Electronics, IEEE Transactions on,* vol. 50, no. 1, pp. 100–107, 2004.

[4] S. K. Reynolds, B. A. Floyd, U. R. Pfeiffer, T. Beukema, J. Grzyb, C. Haymes, B. Gaucher, and M. Soyuer, "A silicon 60-GHz receiver and transmitter chipset for broadband communications," *Solid-State Circuits, IEEE Journal of,* vol. 41, no. 12, pp. 2820–2831, 2006.

[5] M. Peter, W. Keusgen, and R. Felbecker, "Measurement and ray-tracing simulation of the 60 GHz indoor broadband channel: Model accuracy and parameterization," in *Antennas and Propagation, EuCAP 2007. The second European Conference on,* Nov. 2007.

[6] M. Peter and W. Keusgen, "Impact of antenna configuration and shadowing on the characteristics of the 60 GHz indoor wideband radio channel," in *XXIXth General Assembly of the International Union of Radio Science, 2008. URSI GA 2008,* Aug. 2008, to be published.

[7] M. Peter, W. Keusgen, and M. Schirrmacher, "Measurement and analysis of the 60 GHz in-vehicular broadband radio channel," in *Vehicular Technology Conference, 2007. VTC 2007-Fall. 2007 IEEE 66th,* Sept.–Oct. 2007.

[8] M. Peter and W. Keusgen, "A component-based time domain wideband channel sounder and measurement results for the 60 GHz in-cabin radio channel," in *Antennas and Propagation, EuCAP 2007. The second European Conference on,* Nov. 2007.

[9] M. Feuerstein, K. Blackard, T. Rappaport, S. Seidel, and H. Xia, "Path loss, delay spread, and outage models as functions of antenna height for microcellular system design," *Vehicular Technology, IEEE Transactions on,* vol. 43, no. 3, pp. 487–498, 1994.

[10] A. Paulraj, D. Gore, R. Nabar, and H. Bolcskei, "An overview of MIMO communications - a key to gigabit wireless," *Proceedings of the IEEE,* vol. 92, no. 2, pp. 198–218, 2004.

[11] J. Luo, W. Keusgen, and A. Kortke, "Design concepts of a 60 GHz broadband wireless communication system," in *Vehicular Technology Conference, 2008. VTC 2008-Fall. 2008 IEEE 68th,* Sept. 2008, to be published.

[12] S. M. Alamouti, "A simple transmit diversity technique for wireless communications," *Selected Areas in Communications, IEEE Journal on,* vol. 16, no. 8, pp. 1451–1458, 1998.

Co-design of fully integrated 60GHz CMOS digital radio in QFN package

J. Laskar, S. Pinel, D. Dawn, S. Sarkar, P. Sen, B. Perunama, D. Yeh, and F. Barale

Georgia Electronic Design Center, School of Electrical and Computer Engineering,
Georgia Institute of Technology, Atlanta, GA 30332-0269 USA

joy.laskar@ece.gatech.edu

Abstract — **The past few years has witnessed the emergence of CMOS based circuits operating at millimeter wave-frequencies. Co-design of fully integrated 60Ghz CMOS single chip digital radio with low cost QFN package is the promise for high volume low cost fabrication, opening huge commercial markets. As standardization efforts catalyzed the interest and investment of industry and agencies, one can be assured of ubiquitous millimeter-wave technology in the consumer electronic market place in the fairly near future.**

I. INTRODUCTION

In the past few years, the interest in the millimeter wave spectrum at 30–300 GHz has drastically increased. The emergence of low cost high performance CMOS technology and low loss, low cost organic packaging material has opened a new perspective for system designers and service providers because it enables the development of millimeter wave radioes at the same cost structure of radios operating in the gigahertz range or less. In combination with available ultra wide bandwidths, this makes the millimeter wave spectrum more attractive than ever before for supporting a new class of systems and applications ranging from ultra-high speed data transmission, video distribution, portable radar, sensing, detection and imaging of all kinds.

Recently, the availability of standard CMOS technology enabling the design of MMIC circuits operating efficiently up to 100 GHz has revived the interest and investment in the use of the 60GHz spectrum, targeting indoor ultra-high speed short-range wireless communications for multimedia applications.

Similarly, numerous opportunities exist for low cost commercial millimeter-waves systems at higher frequency such as 77GHz for automotive radar, 71-76 and 81-86 GHz outdoor 10Gbps networks, 94GHz for medical and security imaging providing a prelude to the possibility of terabit (in aggregate throughput) systems operating well into the sub-THz regime (several hundred GHz).

The emergence of a multitude of "bandwidth hungry" multimedia applications has exacerbated the need for multi-gigabit wireless solutions, out of reach of conventional WLAN technology (802.11a,b and g), or even more recent emerging UWB and MIMO systems. Uncompressed high-definition video distribution and massive data synchronization are driving data-throughput requirements well beyond gigabit/s, and already demanding up to 10Gbps with the

introduction of, for example, HDMI 1.3 standard. The availability of the 7 GHz unlicensed bandwidth in the 60GHz spectrum represents a unique opportunity to address such data-throughput requirements. As standardization efforts catalyzed the interest and investment of the industry, one can count on the spreading of the 60GHz technology in the consumer electronic marketplace in a fairly near future [1].

In this paper, we discuss the integration of a 60 GHz digitally controlled single-chip 90nm CMOS fully integrated radio [2-3] scalable to phased array system and co-designed with integrated antenna [4-5] in standard QFN package. This co-design has been optimized for robustness against process variation and temperature, and verified by measurement results.

A wideband super-heterodyne architecture combined with a high-speed digital signal processor has been designed to support the whole 57 to 66 GHz bandwidth available and enable data throughput exceeding 7Gbps QPSK and 15Gbps 16QAM for a total DC power budget below 200mW

In addition, an on-chip signal processor provides optional high speed PHY processing (ADC/DAC, pulse shaping, equalization, demodulation and bit synchronization, etc...). The 60 GHz CMOS single-chip radio is fully digitally controlled. This is the highest overall performance (power consumption, data rate, and bandwidth) and the highest level of integration for a 60GHz single-chip radio reported till date providing the lowest energy per bit transmitted wirelessly at multi-gigabit rate, to meet the very stringent low-power specifications for battery operated consumer portable electronic devices, and enable scalability towards low-power portable radar.

II. 60GHz SINGLE-CHIP CMOS 90NM RADIO ARCHITECTURE AND IMPLEMENTATION

With the proper transistor layout employed for 60GHz circuits, 90nm CMOS technology has been established as a technology of choice to enable robust and low-power implementation of CMOS-based multi-gigabit transceiver. Numerous recent publications [6-8] reported examples of efficient implementations of 60GHz CMOS building blocks, thus demonstrating the potential of CMOS technology at 60GHz. When integrated on a low-cost organic packaging, this technology promises efficient high-volume fabrication, lowering the cost (same cost structure as radios operating in the gigahertz range or less), and opening huge commercial impact and opportunities for consumer electronics.

The block level architecture of the single-chip 60GHz radio is described in Fig. 1 and the fabricated die photo is shown in Fig. 2. The wideband super-heterodyne architecture has been designed to support the whole 57 to 66 GHz band. The band is divided into 4 channels of 2160 MHz. However, there are possibilities of two or three channel bonding to support ultra-high throughput. Thus, in order to support the ultra-high throughput, the IF frequency needs to be carefully chosen, and needs to be high enough to simultaneously prevent baseband spectrum aliasing. In this particular work, the chosen 7GHz-13GHz IF incorporates ultra-wide-band design techniques.

The 60 GHz CMOS single-chip radio is externally controlled by a standard Serial-to-Parallel interface. Digital controls mainly include: Transmit/Receive/Sleep operation modes, channel/frequency selection, Transmitter/Receiver chain gain and phase (providing scalability to phased array), on-chip signal processor operations modes/bypass.

The performances of the fully integrated single-chip 60GHz radio are summarized in table I.

TABLE I
MEASURED PERFORMANCE SUMMARY OF 60GHZ SINGLOE CHIP RADIO

Transceiver Blocks	PA	IQ Modulator QVCO PLL	VCO PLL	LNA Mixer IF amplifier	IQ Demod QVCO PLL
Frequency (GHz)	57-65	7-13	49-55	57-65	7-13
Gain (dB)	17	-	-	32	17
P_{Sat} (dBm)	8.4	-	-	-	-
P1-dB (dBm)	5.1 (output)	3 (output)	-	-30 (input)	-
Matching (dB)	>12 (output)	>25 (input)	-	>15 (input)	>25 (output)
Phase noise @1MHz offset	-	-	-95 dBc /Hz	-	-
DC power consumption (mW)	54	17+22+20	20+40	70	19+20+ 20
Total DC power consumption	Tx Mode: 173 mW Rx Mode: 189mW				

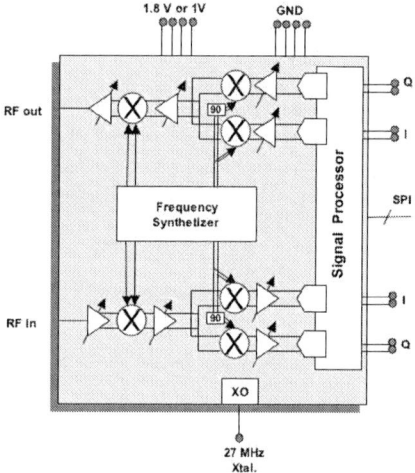

Fig. 1: 60GHz single-chip 90nm CMOS radio architecture.

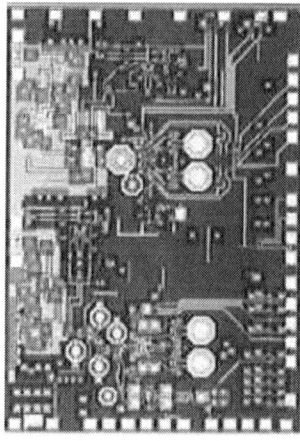

Fig. 2: 60GHz single-chip 90nm CMOS radio die photograph.

In addition, an on-chip signal processor provides optional high speed PHY processing (ADC/DAC, pulse shaping, equalization, demodulation, and bit synchronization).

III. FR4-HIGH FREQUENCY LAMINATES BASED MODULE AND ANTENNA TECHNOLOGY

The combination of FR4 and high frequency low cost laminates appears as a platform of choice for the packaging of the future 60 GHz gigabit radio. Multi-layers substrates (2ft. x 1.5ft. panels) are fabricated using high volume standard PWB production lines. An example of a large panel area multi-layer substrate including compact embedded filter, wideband millimeter-wave feed-through transition and antenna array is shown in figure 3.

Fig. 3: Large panel area multi-layer substrate.

978-2-8748-7007-1/08 $25.00 © 2008 EuMA

Compact filter designs using planar and integrated waveguide (IWG) techniques have been validated and measured (figure 4), exhibiting less than 2 dB minimum insertion for a relative bandwidth of 8 % at 61.5 GHz and a rejection greater than 20dB at 6 GHz offset. Wideband millimeter-wave feed-through transitions exhibiting less than 0.2 dB insertion loss have been also implemented[4].

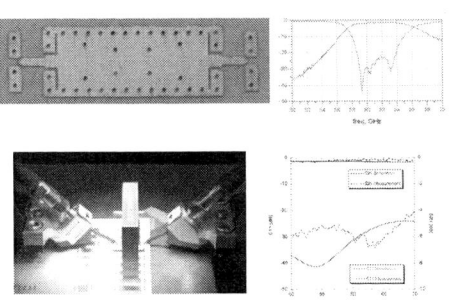

Fig. 4: Multi-layer compact embedded filter and wideband millimeter-wave feed-through transition.

Numerous antenna arrays solutions have been developed to address various applications scenarios ranging from VSR (very short reach) omni-directional to point-to-point links [5]. Such a generic packaging platform provides a path of choice toward the low cost integration of scalable SISO-MIMO radio systems (SM radio) using compact multi-sector phased array architecture that overcomes simultaneously the fundamental limitations of millimeter wave signal propagation and CMOS technology.

The multi-sector architecture can either be integrated on single large panels or in a compact 3D integrated millimeter-wave module [8], including embedded filter and antenna phased arrays as shown in figure 5.

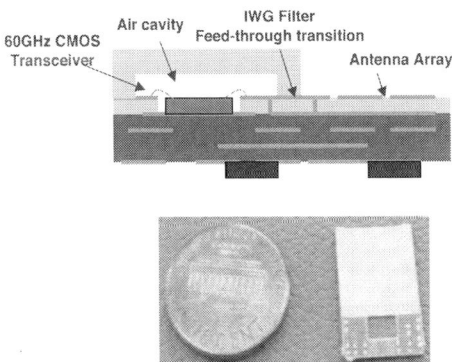

Fig. 5: Compact 3D integrated millimeter-wave modules, including embedded filter and antenna phased array.

Extended azimuth and elevation coverage, provided by conformal multi-sector configurations (shown in figure 6), and extended range (including non-LOS scenario) provided by high gain adaptive phased array technology, are the breakthrough attributes of the future commercial millimeter-wave systems.

Fig. 6: Conformal multi-sectors phased array architecture for extended azimuth and elevation coverage.

IV. 15 GBPS AND HD-VIDEO MILLIMETER-WAVE TESTBED.

The GEDC has established an experimental millimeter-wave wireless test-bed, using 60GHz as a demonstrator vehicle to study the channel characteristic of a real indoor environment. Researchers recently established a new world record for the highest data rate transmitted wirelessly at 60GHz, achieving a peak data transfer rate of 15 Gigabit/s at a distance of 1 meter, 10 Gigabit/s at a distance of 2 meters and 5 Gigabit/s at a distance of 5 meters. In addition, high definition video streaming running at 1.485 Gb/s has been demonstrated through a 1 inch thick wood table. Special efforts have been dedicated to the complete transceiver module implementation operating at a power budget well below the one hundred pico-joules/bit range. In figure 7, we show the experimental set-up of the video transmission through a one inch thick wood table.

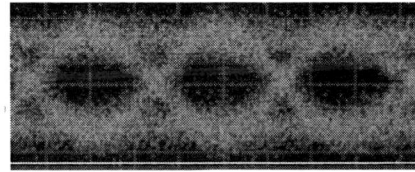

Fig. 7: Demodulated transmission of a multi-gigabit signal and experimental set-up of the video transmission through a one inches thick wood table.

V. CONCLUSION

The development of millimeter waves radios at the same cost structure of consumer electronic radios operating in C-Band opens a new field of innovation for systems designers. The convergence of FR4 based Modules, CMOS MMIC, Signal Processing and high efficiency PHY-MAC technologies will enable a new generation of low cost high performances millimeter-wave systems. The feasibility of ultra high-speed wireless transmission beyond 10Gbps has been demonstrated on a low power low cost platform. Power budgets well below the one hundred pico-joules/bit range have been achieved. One may expect that 100Gbps serial transmission with a femto-joule/bit power budget can be developed in the near future. These advances will enable a variety of volume millimeter-wave CMOS applications including: peer-to-peer ultra fast synchronization and adaptive WPAN, automotive radar, out-door point-to-point/point-to-multi-point links, portable radar, security, sensing and imaging and medical applications.

ACKNOWLEDGEMENT

This material is based upon work supported by SPAWAR under Award No. N66001-06-1-2033. Any opinions, findings, and conclusions or recommendations expressed in this publication are those of the author (s) and do not necessarily reflect the views of the SPAWAR. Authors would like to acknowledge DARPA for their support in this work.

REFERENCES

[1] J. Laskar, S. Pinel, D. Dawn, S. sarkar, B. Perumana and P. Sen, "The Next Wireless Wave is a Millimeter Wave", *Microwave Journal*, pp22-36, August 2007.

[2] S. Pinel, S. sarkar, P. Sen, B. Perumana, D.Yeh, D. Dawn, J. Laskar, "60GHz Single chip 90 CMOS Radio", *ISSCC Dig. Tech Papers 2008*, Feb 2008.

[3] S. sarkar, P. Sen, B. Perumana, D.Yeh, D. Dawn, S. Pinel, J. Laskar, "60GHz Single chip 90 CMOS Radio with integrated signal processor", to be presented at IMS 2008.

[4] KiSeok Yang, Stephane Pinel, Il Kwon Kim, and Joy Laskar. "Low-Loss Integrated Waveguide Passive Circuits Using Liquid Crystal Polymer System-on-Package (SOP) Technology for Millimeter-Wave Applications," *IEEE Transactions on MTT*, vol. 54, Issue 12, Part 2, Page(s):4572 - 4579 , Dec. 2006.

[5] Il Kwon Kim, Stephane Pinel, Joy Laskar, and Jong-Gwan Yook, "Circularly & Linearly Polarized Fan Beam Patch Antenna Arrays on Liquid Crystal Polymer Substrate for V-band Applications", *APMC Dig. Tech Papers 2005*, vol. 4, Dec. 2005.

[6] Babak Heydari, et al., "Low-Power mm-wave Components up to 104GHz in 90nm CMOS", *ISSCC Dig. Tech Papers 2007*, pp 200-201, Feb. 2007.

[7] Sohrab Emami et al., "A Highly Integrated 60GHz CMOS Front-End Receiver", ISSCC Dig. Tech Papers, pp 190-191, Feb. 2007.

[8] Terry Yao, et al., "Algorithmic Design of CMOS LNAs and PAs for 60-GHz Radio", *IEEE J. Solid-State Circuits*, vol. 42, No. 5, pp. 1044-1057, May 2007.

Integrated 60 GHz Circuits and Systems for High-Speed Communications

Morteza Abbasi [1], Sten E. Gunnarsson [1], Rumen Kozhuharov [1],
Camilla Kärnfelt [1], Christian Fager [1], Dan Kuylenstierna [1], Christer Stoij [2], Herbert Zirath [1,3]

[1]*Microwave Electronics Laboratory, Chalmers University of Technology, Gothenburg, Sweden*
[2]*Sivers IMA AB, KISTA, Sweden*
[3]*Ericsson AB, Microwave and High Speed Electronics Research, Mölndal, Sweden*
abbasi@chalmers.se

Abstract— Integrated 60 GHz transmitter and receiver GaAs front-end MMICs as well as a packaged transceiver module used for wireless communication link are presented. The data transmission capacity of the chips is examined using binary ASK modulation. A broadband 60 GHz direct carrier quadrature modulator is proposed to be integrated with the transmitter for realizing complex modulation schemes to increase the data rate.

I. INTRODUCTION

High data rate transmission is the key element of the future communication systems. The possibility of transmission of data, voice and high definition videos require broadband communication systems together with advanced modulation and error correction techniques. This broad bandwidth is freely available at mm-wave bands where the dimensions are also smaller and therefore the integration level can be increased to a high extent. In particular, due to the high oxygen absorption, the frequency band around 60 GHz is even more attractive for wireless personal and local area networks, WLAN and WPAN.

The performance of 60 GHz receiver and transmitter RF front end units has been demonstrated with different integration levels and in different technologies [1]–[5]. However, the complexity and therefore the cost of these systems can be further reduced by implementing more effective architectures. One way of doing this would be to combine the modulation and the up-conversion in the transmitter, commonly referred to as *Direct Carrier Modulation* technique. Therefore at the transmitter module, instead of analog up-conversion of a digitally modulated IF signal, the information directly modulates the RF carrier. Most of the sophisticated modulation schemes where phase and amplitude of the carrier are changing, such as M-ary Phase Shift Keying (PSK) and Quadrature Amplitude Modulation (QAM) can be implemented by direct quadrature modulators. In section II the performance of the 60 GHz integrated RF front end circuit previously developed at our laboratory is reviewed. The packaged chips in form of a transceiver module are presented in sections III. In section IV a direct carrier quadrature modulator is proposed for improving the performance of the next generation transceiver chips.

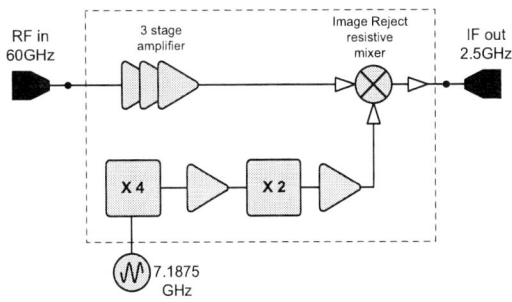

(a)

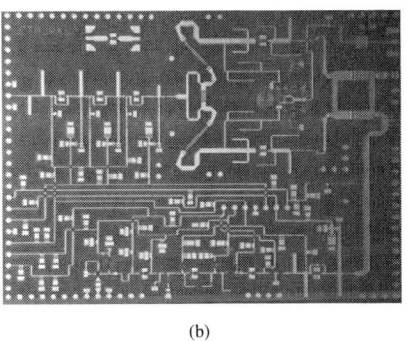

(b)

Fig. 1. The block diagram (a) and photo (b) of the 60 GHz receiver chip. The chip measures $5.5 \times 4.0 mm^2$.

II. 60 GHz TRANSMIT AND RECEIVE CHIPSETS

The 60 GHz transmitter and receiver chips made in GaAs based High Electron Mobility Transistor(HEMT) have been reported both in $0.15\mu m$ pHEMT and mHEMT versions [1], [2], [6]. In this section, the performance of the chips using the mHEMT process is briefly reviewed. The performance of these chips in high data rate transmission is presented and related to the circuit parameters.

A. Receiver

The schematic block diagram and photograph of the receiver chip are shown in figure 1. In this case the external local oscillator is designed with a balanced colpitts topology and is

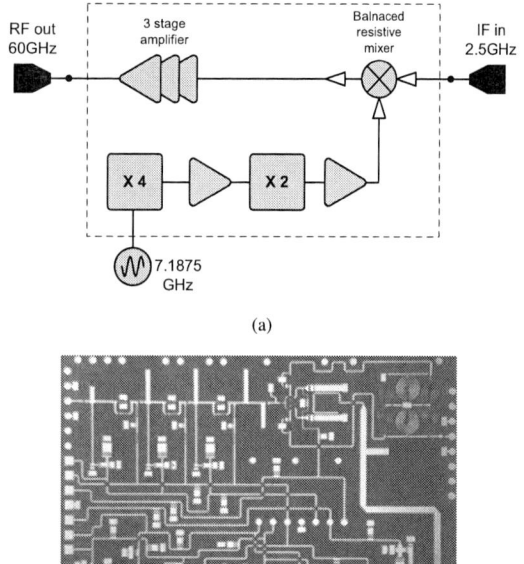

(a)

(b)

Fig. 2. The block diagram (a) and photo (b) of the 60 GHz transmitter chip. The chip measures $4.0 \times 3.0mm^2$

reported in [7]. As can be seen from the block diagram the chip provides an IF signal centered at 2.5 GHz. This signal would then be down-converted and demodulated.

B. Transmitter

The schematic block diagram and photograph of the transmitter chip are shown in figure 2. Here, again the IF signal is generated by up-converting the output of the modulator to 2.5 GHz. In order to deliver enough power to the transmitting antenna, the RF output needs to be amplified with an additional power amplifier. Due to the on-chip balun at the input, the 3dB IF bandwidth is from 1.25 GHz to 2.75 GHz (1.5 GHz of bandwidth).

Assuming a NRZ rectangular pulse for each bit, the bandwidth of the main lobe in the spectrum of an Amplitude Shift Keying (ASK) modulated signal is equal to $2 \times r_b$ where

r_b represents the bit rate [8]. Therefore using simple binary ASK, transmission of 1Gbps without any major intersymbol interference (ISI) should be possible. Multi-level modulations can be used to increase the bit rate.

C. Data Transmission Demonstration

The transmitter and receiver are connected back-to-back with the insertion of an attenuator in the middle to model the channel loss. The IF signal at the transmitter side is generated by up-converting the output of a pseudo random binary sequence (PRBS) generator to 2.5 GHz. The modulation scheme is therefore limited to binary Amplitude Shift Keying (ASK). However, this gives a representation of the maximum transfer rate and bit error rate (BER). On the receiver side the IF signal is down-converted and amplified and measured on an oscillator. The bit error rate is measured with an error detector synchronized with the input PRBS. The measurement setup is shown in figure 3.

Figure 4 shows the measured eye diagram of the received signal at four different data rates. From the figure, it can be seen that the eye is getting more closed in vertical direction for rates higher than 1 Gbps as was expected from the IF bandwidth. However, transmission up to 1.5 Gbps is still possible with acceptable BER. It should be possible to transmit higher bit rates using more bandwidth efficient multi-level modulation schemes.

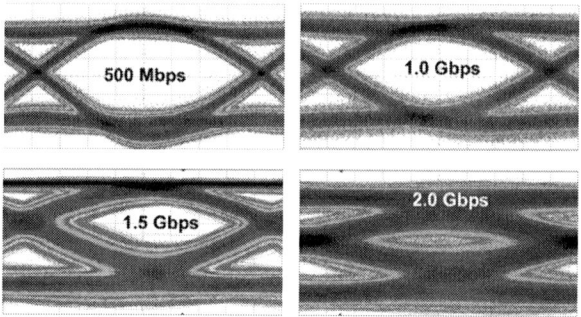

Fig. 4. Measured eye diagrams of the received signal at four different data rates.

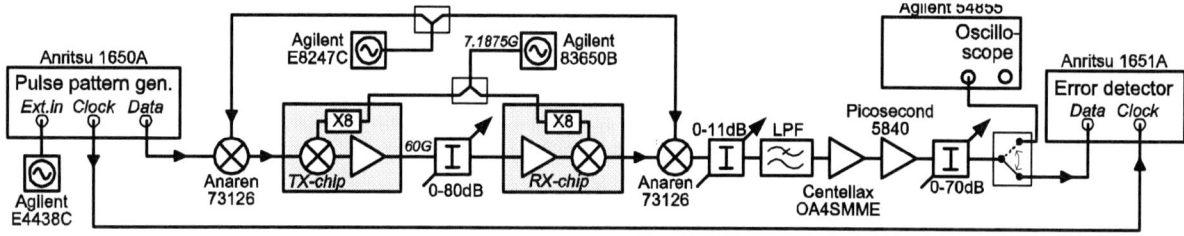

Fig. 3. The setup used to measure the data rate capability of the transceiver chips.

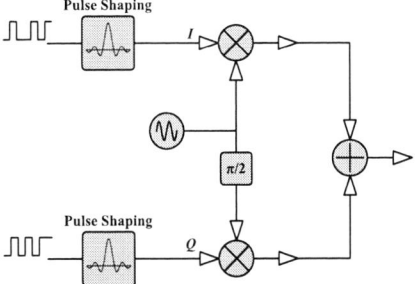

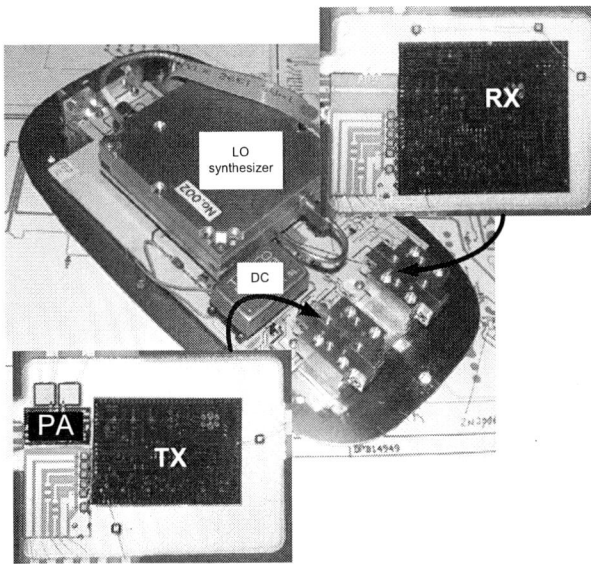

Fig. 5. Packaged 60GHz pHEMT transmitter and receiver chips.

Fig. 6. Schematic diagram of the quadrature modulator.

III. PACKAGING AND INTERFACING

Recently the pHEMT versions of the RX and TX chips [1] have been packaged and used in a 60GHz transceiver module. The module is to be used for secure wireless communications. The transceiver unit includes a computer controlled synthesizer for generation of the LO signals for the chips as well as DC voltage regulators, converters and distribution circuitry. The RF interface is in waveguide with appropriate transitions. An additional power amplifier chip is included after the TX MMIC to increase the power level. The transmitter can deliver > 10 dBm power at the waveguide interface. Figure 5 shows the packaged transceiver with the designation of the TX and RX MMICs.

IV. 60 GHz DIRECT CARRIER QUADRATURE MODULATOR

As mentioned before, in order to reduce the number of hardware components and therefore reduce the cost one solution would be to use direct carrier modulation. In this scheme, instead of up-converting the IF signal which contains the digitally modulated signal, the baseband data directly modulates the RF carrier. The block diagram of the modulator is shown schematically in figure 6. The information is divided into two sets of the so-called in-phase (I) and quadrature-phase (Q) elements. The carrier is also divided into two quadrature parts each being modulated by one of the I and Q channels. The outputs will be added in phase.

Quadrature modulators using reflection type phase shifters have been reported previously [9]. However, in order to achieve acceptable output constellations the control voltages should be carefully tuned for each data point and the calibration data is saved in a look-up table.

The modulator presented here uses resistive mixers for mixing the data with the carrier. There are some advantages with this approach compared to the reflection type phase shifter. Resistive mixers are more linear and therefore less intermodulation products are generated. Furthermore, provided that the mixers are not in saturation, there is a linear relation between the input IF and the output RF signals. Therefore, there would be no need to do much correction on the data points. Figure 7 shows the chip photograph of the 60 GHz quadrature modulator.

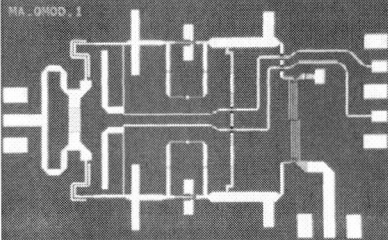

Fig. 7. Chip photograph of the 60GHz direct carrier quadrature modulator.

A major practical problem with direct modulators is the leakage of the carrier signal, which is at the same frequency as the LO of the mixers, to the output. This signal can drive the following power amplifier, if there is any, into saturation. The generated intermodulation products can make problems with LO pulling effect and distortion of the signal. As shown in figure 8(a), one solution to avoid LO leakage is to use a balanced structure instead of single ended. However, in order to keep the circuit compact and to reduce the total loss, the LO leakage is canceled by the use of an inductive feedback in the mixer cells. The inductive path resonates with the intrinsic capacitor from the gate to drain of the transistor which is the path for the leakage of the signal. This is schematically shown in figure 8(b).

A. Performance

Different modulation schemes are implemented and measured using the quadrature modulator. The static output constellations for PSK and QAM schemes with different levels are examined. The measured constellations for 16 PSK and 256 QAM are shown in figure 9.

978-2-8748-7007-1/08 $25.00 © 2008 EuMA

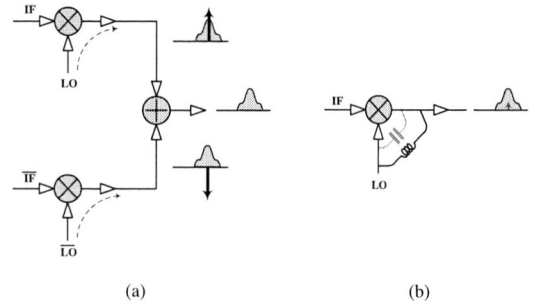

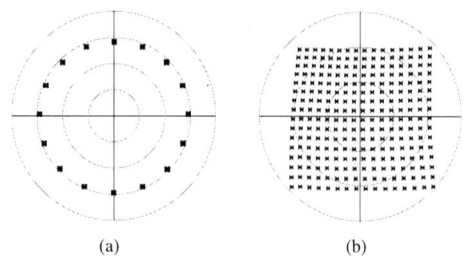

Fig. 8. LO leakage reduction (a) Using balanced structure (b) Using inductive resonating loop.

Fig. 9. Difference Modulation Schemes Implemented and Measured with the Quadrature Modulator.

TABLE I

PERFORMANCE SUMMARY FOR THE 60GHz MODULATOR

LO power	5dBm
Input-Output return loss	$\geq$10dB
IF 3dB bandwidth	0-5GHz
RF 1dB bandwidth	52-68GHz
Uncorrected amplitude - phase error	5% - 1°

The constellations are measured with the receive port of the network analyzer while the LO signal is taken from port 1. The baseband control voltages are changed and applied one at a time. Since the power from the network analyzer was not sufficient to pump the mixers, an external amplifier has been added to provide high enough LO signal to the mixers. It is assumed that the amplifier has similar amplitude and phase properties for different states. The input voltages as well as the data from the network analyzer are controlled with a computer.

Apart from a fixed correction to all the control voltages, no individual adjustment is made for the points in the constellations. The fixed error stems from the phase imbalance of the lange coupler on the LO path. This results in the constellation being tilted.

Table I summarizes the performance of the 60GHz direct carrier quadrature modulator.

V. CONCLUSIONS

60 GHz transceiver front end MMICs realized in GaAs mHEMT and pHEMT technologies have been realized. The fully packaged 60 GHz transceiver module was presented for secure wireless communication link. The capability of transmitting data with the chips has been demonstrated using simple binary ASK modulation. A 60 GHz direct carrier quadrature modulator was presented as an alternative to two step up-conversion. Having a broad IF bandwidth, with this modulator more bandwidth efficient schemes can be realized and therefore the data transmission rate can considerably be increased.

ACKNOWLEDGMENT

WIN semiconductors is acknowledged for fabrication of the chips presented in this paper.

REFERENCES

[1] S. E. Gunnarsson, C. Kärnfelt, H. Zirath, R. Kozhuharov, D. Kuylen-stierna, A. Alping, and C. Fager, "Highly integrated 60 GHz transmitter and receiver MMICs in a GaAs pHEMT technology," *IEEE Journal of Solid-State Circuits*, vol. 40, no. 11, pp. 2174 – 2185, 2005.

[2] S. E. Gunnarsson, C. Kärnfelt, H. Zirath, R. Kozhuharov, D. Kuylen-stierna, C. Fager, M. Ferndahl, B. Hansson, A. Alping, and P. Hallbjorner, "60 GHz single-chip front-end MMICs and systems for multi-Gb/s wireless communication," *IEEE Journal of Solid-State Circuits*, vol. 42, no. 5, pp. 1143 – 1156, 2007.

[3] S. Reynolds, B. Floyd, U. Pfeiffer, T. Beukema, J. Grzyb, C. Haymes, B. Gaucher, and M. Soyuer, "A silicon 60-GHz receiver and transmitter chipset for broadband communications," *IEEE Journal of Solid-State Circuits*, vol. 41, no. 12, pp. 2820 – 31, 2006/12/.

[4] B. Razavi, "A 60-GHz CMOS receiver front-end," *IEEE Journal of Solid-State Circuits*, vol. 41, no. 1, pp. 17 – 22, 2006.

[5] A. Yamada, E. Suematsu, K. Sato, M. Yamamoto, and H. Sato, "60GHz ultra compact transmitter/receiver with a low phase noise PLL-oscillator," vol. vol.3, Philadelphia, PA, USA, 2003, pp. 2035 – 8.

[6] "Mp15-01 - //http://www.winfoundry.com."

[7] H. Zirath, R. Kozhuharov, and M. Ferndahl, "Balanced Colpitt oscillator MMICs designed for ultra-low phase noise," *IEEE Journal of Solid-State Circuits*, vol. 40, no. 10, pp. 2077 – 2085, 2005.

[8] J. G. Proakis and M. Salehi, *Communication systems engineering*. Harlow: Prentice Hall, cop., 2002.

[9] A. E. Ashtiani, S.-i. Nam, A. d'Espona, S. Lucyszyn, and I. D. Robertson, "Direct multilevel carrier modulation using millimeter-wave balanced vector modulators," *IEEE Transactions on Microwave Theory and Techniques*, vol. 46, no. 12 pt 2, pp. 2611 – 2619, 1998.

978-2-8748-7007-1/08 $25.00 © 2008 EuMA

A 60 GHz SiGe HBT Chip Set

Georg Boeck[1], Viswanathan Subramanian[1], Wilhelm Keusgen[2] and Van-Hoang Do[3]

[1] *Microwave Engineering Lab, Berlin University of Technology, Einsteinufer 25, Berlin, Germany*
[2] *Fraunhofer Institut für Nachrichtentechnik, Heinrich-Hertz-Institut, Einsteinufer 37 Berlin, Germany*
[3] *TES Electronic Solutions GmbH, Berlin, Carnotstraße 7, Berlin, Germany*
[1] boeck@tu-berlin.de

Abstract—**A 60 GHz SiGe HBT chipset for high speed wireless communication systems has been developed. The functionalities of LNA, up-converter, down-converter and PA have been realized with good performance. Design strategy, achieved results and comparison with state-of-the-art work will be presented. The work proves that single chip integration of the whole 60 GHz RF-frond-end will be possible using silicon based technologies.**

I. INTRODUCTION

The requirements for future broadband multimedia applications exceed the capacity of the currently used standards like 802.11a and 802.11b/g. Due to special attenuation properties of 60 GHz signals and availability of a large license-free bandwidth, the frequency band 57 GHz to 64 GHz is becoming a favorable region of choice for short range high data rate implementation [1]. This along with the continuous advancements of SiGe BiCMOS technology [2]-[3] resulted in increased realization of millimeter wave (mmW) transceiver circuits [4] – [5].

Successful and efficient Si-RFIC design in the 60 GHz regime and beyond requires accurate models of all active and passive components. Especially all transmission lines, interconnects and parasitic structures have to be modeled very precisely because of its serious influence on the frequency response.

In [6]-[8], certain methods and techniques are applied for the lumped element modeling and characterization of the passive on-chip elements up to 90 GHz. Based on these simulation methods the passive elements in the circuits are designed and included in the circuit design environment.

The subsequent paper is divided into three sections. Section II briefs the circuit design and simulation strategy. Section III presents the realized 60 GHz SiGe RICs with the achieved performance. Section IV concludes this article.

II. CIRCUIT DESIGN STRATEGY

The design methodology is a typical bottom-up approach where the layout of the circuit is developed from firstly simulated schematic. Special care is taken in the layout design and all interconnects are implemented as microstrip transmission lines by providing ground metals of sufficient width. These microstrip lines formed by interconnects at various metal levels of the technology stack are used very effectively as stand alone, matching, filtering or impedance transforming elements in the circuit design.

III. REALIZED FRONT END CIRCUITS

This part of the article addresses the realized LNA, down-converter, up-converter and PA circuits. Design aspects, circuit details, and simulation and measurement results in comparison with state-of-the-art results are presented briefly. For detailed information on each circuit, readers are requested to refer to the corresponding publication.

A. LNA

Fig. 1 shows the simplified schematic of the LNA [9]. It consists of two cascaded cascode stages in order to achieve a sufficient gain and a high reverse isolation. The first stage is designed with the goal of the best noise performance and fulfilling the input impedance matching requirements at the same time. The second stage of the LNA is designed for maximum power gain. Current mirrors are used for biasing the two amplifier stages. The whole biasing circuit is on chip but not shown in Fig. 1. The active chip area is 0.5×0.9 mm^2.

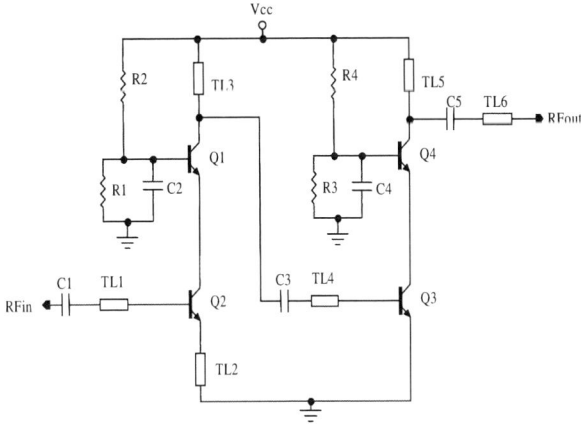

Fig. 1. Simplified schematic of the LNA [9].

Table I compares the important results of this work with previously published work in the same frequency range. The comparison shows that the achieved performance is at the same level as the best results reported so far in this field.

TABLE I
COMPARISON OF OUR WORK
WITH PREVIOUSLY PUBLISHED 60 GHZ SIGE LNAS [9]

S_{21} [dB]	NF_{50} [dB]	P_{DC} [mW]	IP_{1dB} [dBm]	References
18	6.8*	66	n.a	[10]
15	4.5	10.8	-20	[11]
18 dB	**7.3**	**20.25**	**-18**	**This work**

* Simulated

B. Down-converter

Fig. 2 shows the simplified schematic of the designed down-converter [12] with mixer core and on-chip active push pull balun at the output [13]. The trans-conductance part Q1 of the single balanced mixer is a resistively degenerated common emitter transistor improving the linearity. The final value of this resistance is chosen through a proper trade-off between linearity improvements on the one hand and worsening gain and noise figure on the other hand.

The LO switching transistors Q2 and Q3 are loaded with a serial combination of resistance and inductance. The gain drop vs. frequency due to a resistive load is compensated by the inverse response of the inductance. This idea is used instead of tuned loads [14] to avoid a shift of the IF frequency band due to parasitic effects at the IF-output. An on-chip active push pull balun formed by a serially connected emitter follower

with a common emitter transistor with degeneration is used as an output buffer. The balun just provides the differential to single ended conversion without any additional gain. The achieved performance of the down-converter is summarized in Table II.

TABLE II
SUMMARY OF DOWN-CONVERTER PERFORMANCES [12]

Ouput freq. [GHz]	RF-BW [GHz]	C.G. [dB]	P_{dc} [mW]	IP_{1dB} [dBm]	Chip size [mm²]
5	20	2.5	15.5	-5	1x0.7

C. Up-converter

Fig. 3 shows the simplified schematic of the designed upconverter [15] with all the major sub-circuit blocks. The core of the up-conversion mixer is based on the Gilbert micromixer [16]. All baluns for IF, RF and LO are integrated on chip.

The LO-RF suppression depends on the common-mode gain of the LO balun. The higher the output impedance of the current source, the better is the LO suppression. Therefore, a λ/4 shorted stub is inserted (see Fig. 3) between the current source and ground, which improves the LO suppression at 60 GHz by 15 dB.A driver amplifier is added at the mixer output. It is a single-stage amplifier in cascode configuration operating in class AB. This buffer amplifier was designed for high linearity performance according to the guidelines of power amplifiers. Thus, its output is not matched to 50 Ω. Table III summarizes the measured performances of the up-converter.

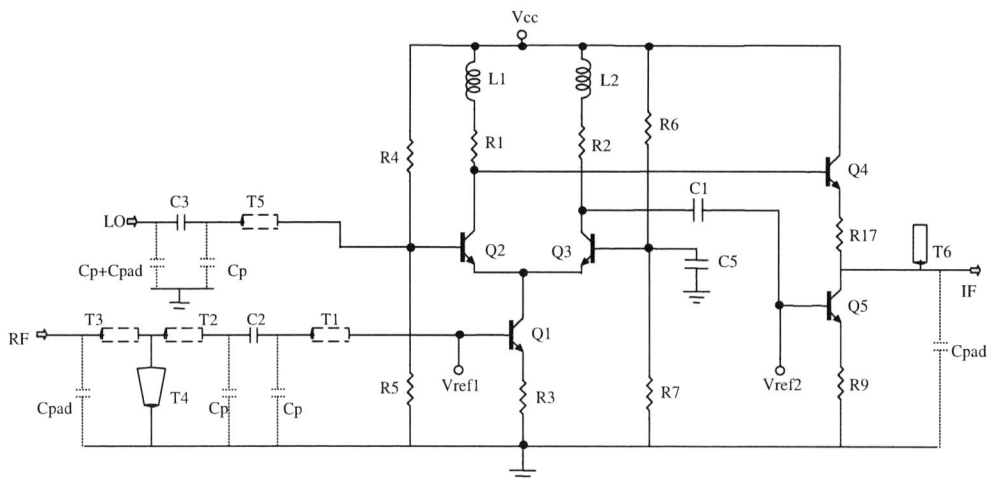

Fig. 2.   Simplified schematic of the down-converter with dominating parasitic elements [12].

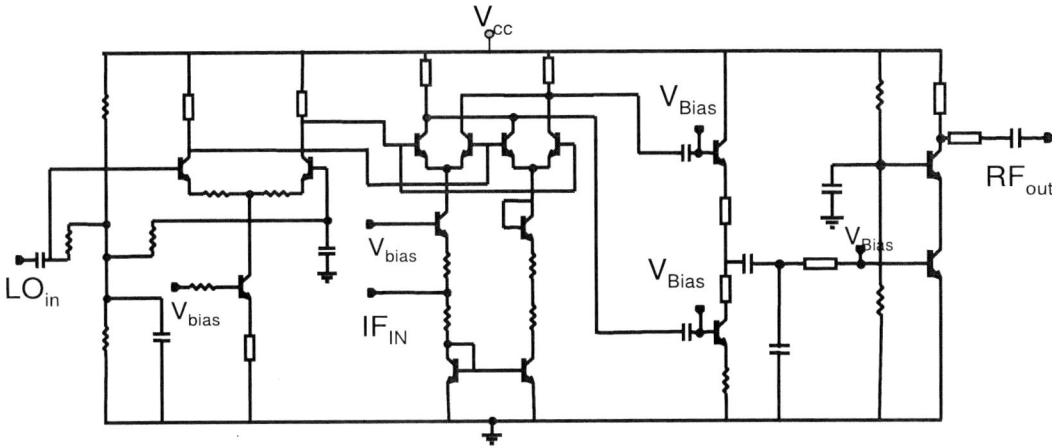

Fig. 3. Simplified schematic of the complete up-converter including all sub-circuits [15]

TABLE III
SUMMARY OF UP-CONVERTER PERFORMANCES [15]

Output Freq [GHz]	IF BW [GHz]	C.G [dB]	P_{dc} [mW]	IP_{1dB} [dBm]	Chip size [mm^2]
60	10	-3	80	0	1.1x1.1

D. Power Amplifier

Several works on power amplifiers in SiGe technologies have been published in the past. In 2004 first SiGe-HBT PAs at V-band frequencies (50 – 75 GHz) were reported [19] - [20]. Some works at 60 GHz [21] - [24] and at 77 GHz [25] with saturated output power (P_{sat}) up to 17.5 dBm and maximum power added efficiency (PAE) of 12 % followed. In [24] a power amplifier in push-pull topology having P_{sat} of 20 dBm and a peak PAE of 12.7 % was presented. Unfortunately its output power under 1 dB gain compression (P_{1dB}) is just 13 dBm.

Fig. 4 shows the simplified schematic of the PA published in [18]. It consists of two cascaded cascode stages operating in class AB mode in order to provide high gain, isolation and efficiency. On-chip microstrip transmission lines are extensively used for matching and filtering. All the implemented transmission structures are electromagnetically modeled using a 2.5 dimensional EM solver. The first stage (driver stage) consists of two transistors Q1 and Q2 in cascode configuration. A current mirror, composed of three transistors and two resistances is used for biasing the driver stage [17]. Compared to the traditional way of mirroring the collector current, in this design the base current has been mirrored in the RF transistor and the bias point has been set. The reason for such implementation is the better performance of the amplifier with respect to PAE and P_{1dB}. The core of the second stage (power stage) is formed by transistors Q3, Q4. The emitter sizes of these transistors are chosen 4 times larger than the emitter sizes of the ones in the driver stage. There a biasing circuit similar to the driver stage has been utilized.

The input transmission lines with the shunt metal-insulator-metal (MIM) capacitor are used to provide a broad band input matching of the circuit. The inter-stage matching is formed by a transmission line and the shunt MIM capacitor. The resonant

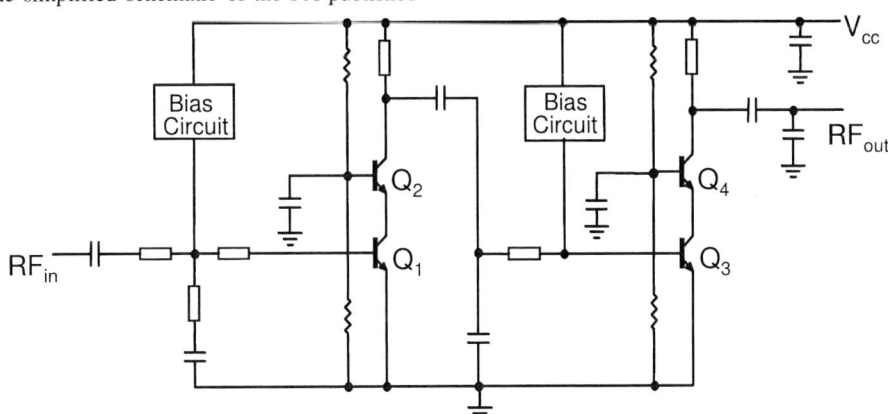

Fig. 4. Simplified schematic of the PA [18]

978-2-8748-7007-1/08 $25.00 © 2008 EuMA

loads at the collectors of the two stages are realized using the lines together with the parasitic capacitance at these nodes. In this design, no separate output matching is required. Therefore output capacitor is just a DC blocking capacitor. A chip micrograph of the PA is shown in Fig. 8. The total chip size is 0.75 x 1.06 mm^2, the active chip area is much lower. The input and output signals are conducted at the bottom and top side of the chip in ground-signal-ground configuration. The supply and reference voltages are applied to the pads at the left and right side.

Table IV provides a comparison of our work [18] with previously published power amplifiers in SiGe-HBT technology at millimeter-wave frequencies. With respect to PAE, P$_{1dB}$ and gain the achieved performance belongs to the best ever reported in this field.

TABLE IV.
COMPARISON WITH PREVIOUSLY PUBLISHED V-BAND PA IN SIGE

Freq. [GHz]	Power gain [dB]	P$_{1dB}$ [dBm]	Psat [dBm]	peak PAE [%]	Ref.
60.0	10.8	11.2	16.2	3.0	[19]
77.0	-	-	18.6	5.4	[20]
61.5	12.0	8.5	14.0	4.2	[21], [22]
60.0	11.5	11.2	15.8	16.8	[23]
77.0	17.0	14.5	17.5	12.8	[25]
60.0	18.0	13.1	20.0	12.7	[24]
61.0	**18.8**	**14.5**	**15.5**	**19.7**	**This work**

IV. CONCLUSION

A 60 GHz SiGe HBT based chipset for future high speed wireless communication systems is presented. All important circuit blocks have been demonstrated with good functionality, LNA, down-converter, up-converter and PA. Especially for the most critical circuit, the PA excellent performance has been achieved. This proves that single chip integration of the whole 60 GHz system (SoC) will be possible in the near future.

ACKNOWLEDGMENT

The authors would like to thank Dr. Dietmar Warning and colleagues from TES Electronic Solutions GmbH in Berlin for supporting this 60 GHz chip development project. The project was funded by Investitionsbank Berlin (IBB) under contract No. 10022937 in using means of EFRE.

REFERENCES

[1] Peter Smulders, "Exploiting the 60 GHz Band for Local Wireless Multimedia Access: Prospects and Future Directions," *IEEE Commun. Mag.*, vol. 2, no. 1, pp. 140-147, Jan. 2002.

[2] B. Jagannathan, et al., "Self-Aligned SiGe NPN transistor with 285 GHz fMAX and 207 GHz fT in a manufacturable technology," *IEEE Electronic Device Letters*, vol. 23, no. 5, 2002.

[3] M. Khater, et at., "SiGe HBT technology with f$_{max}$/f$_t$ = 350/300 GHz and gate delay below 3.3 ps," *IEEE International Electron Devices Meeting*, pp. 247-250, Dec. 2004.

[4] B. Floyd, et al., "A Silicon 60 GHz Receiver and Transmitter Chipset for Broadband Communications," IEEE ISSCC, pp. 220-221, Feb. 2006.

[5] Wolfgang Winkler, Johannes Bomgraber, Hans Gustat, Falk Komdorfer, "60 GHz Transceiver Circuits in SiGe:C BiCMOS Technology," *ESSCIR*, pp. 83-86, Sept. 2004.

[6] V. Subramanian, Zihui Zhang, D. Gruner, F. Korndoerfer, Georg Boeck, "Analysis and Characterization of Microstrip Structures up tp 90 GHz in SiGe BiCMOS," *37th European Microwave Conf.*, October 2007.

[7] D. Gruner, Z. Zhang, V. Subramanian, F. Korndoerfer, G. Boeck, "Lumped Element MIM Capacitor Model for Si-RFICs," *International Microwave and Optoelectronics Conference*, Oct. 2007.

[8] D. Gruner, Z. Zhang, F. Korndoerfer, G. Boeck, "Lumped Element Coplanar-Microstrip-Transition Model for Si-RFICs up to 90 GHz," *IEEE International Workshop on Radio-Frequency Integration Technology (RFIT2007)*, Dec. 2007.

[9] Van-Hoang Do, Viswanathan Subramanian, Wilhelm Keusgen, Georg Boeck, "60 GHz SiGe LNA," *IEEE ICECS*, Dec. 2007.

[10] Y. Sun, J. Borngräber, F. Herzel, W. Winkler, "A fully integrated 60 GHz LNA in SiGe:C BiCMOS technology," *IEEE BCTM*, pp. 14-17, Oct. 2004.

[11] B. Floyd: "V-Band and W-Band SiGe Low-Noise Amplifiers and Voltage-Controlled Oscillators," *IEEE RFIC Symposium*, pp. 295-298, Feb. 2005.

[12] V. Subramanian, Van-Hoang Do, Wilhelm Keusgen, Georg Boeck, "60 GHz SiGe HBT Downconversion Mixer," *37th European Microwave Conf.*, October 2007.

[13] D. Kim, S. Lee, J. Lee, "Up-conversion mixer for PCS application using Si BJT," *Proc. 2nd ICMMT*, pp. 424-427, Sep. 2000.

[14] S. K. Reynolds, "A 60-GHz super heterodyne downconversion mixer in silicon-germanium bipolar technology," *IEEE journal of solid-state circuits*, vol. 39, pp. 2065-2068, Nov. 2004.

[15] Van-Hoang Do, Viswanathan Subramanian, Wilhelm Keusgen, Georg Boeck, "A 60 GHz Monolithic Upconversion Mixer in SiGe HBT Technology," *IEEE RFIT*, Dec. 2007.

[16] B. Gilbert, "The MICROMIXER: A Highly Linear Variant of the Gilbert Mixer Using a Bisymmetric Class-AB Input Stage," *IEEE Journal of Solid-State Circuits*, vol. 32, pp. 1312-1423, Sept. 1997.

[17] Van-Hoang Do, Viswanathan Subramanian, Wilhelm Keusgen, Georg Boeck, "Design and Optimization of a High Efficiency 60 GHz SiGe-HBT Power Amplifier," *IEEE RFIT*, Dec. 2007.

[18] Van-Hoang Do, Viswanathan Subramanian, Wilhelm Keusgen, Georg Boeck, "A 60 GHz SiGe-HBT Power Amplifier with 20 % PAE at 15 dBm Output Power," *IEEE MWCL, To be published on* Mar. 2008.

[19] U. R. Pfeiffer, S. K. Reynold, and B. A. Floyd, "A 77 GHz SiGe Power Amplifier for Potential Applications in Automotive Systems," *IEEE RFIC Symposium*, pp. 91-94, June 2004.

[20] H. Li, and H. M. Rein, "Fully Integrated VCOs with Powerful Output Buffer for 77 GHz Automotive Radar Systems and Applications Around 100 GHz," *IEEE J. Solid-state Circuits*, vol. 39, pp. 1650-1658, Oct. 2004.

[21] U. R. Pfeiffer, D. Goren, etc. "SiGe Transformer Matched Power Amplifier for Operation at Millimeter-Wave Frequencies," *in Proc. 31st ESSCIRC*, pp. 141-144, Sept. 2005.

[22] B. Floyd, S. K. Reynolds and U. R. Pfeiffer, "A Silicon 60 GHz Receiver and Transmitter Chipset for Broadband Communications," *in IEEE ISSCC Dig.*, pp. 649-648, Feb. 2006.

[23] C. Wang, Y. Cho, et al., "A 60 GHz Transmitter with Antenna in a 0.18 μm SiGe BiCMOS Technology", *in IEEE ISSCC Dig.*, pp. 186-187, Feb. 2006.

[24] U. R. Pfeiffer, "A 20 dBm Fully-Integrated 60 GHz SiGe Power Amplifier with Automatic Level Control," *in Proc. ESSCIRC*, pp. 356-359 Sept. 2006

[25] A. Komijani and A. Hajimiri, "A Wideband 77 GHz 17.5 dBm Power Amplifier in Silicon," *Proc. IEEE CICC*, pp. 571-574 Sept. 2005.

Proceedings of the 3rd European Microwave Integrated Circuits Conference

A 40-to-76 GHz Balanced Distributed Doubler in 0.13-μm CMOS Technology

Bo-Jiun Huang, Bo-Jr Huang, Chung-Chun Chen, Kun-You Lin, and Huei Wang

Dept. of Electrical Engineering and Graduate Institute of Communication Engineering
National Taiwan University, 1Roosevelt Road, Sec. 4, 10617 Taipei, Taiwan, R.O.C.
hueiwang@ew.ee.ntu.edu.tw

Abstract—**A miniature 40- to 76-GHz monolithic balanced distributed frequency doubler is developed in a commercial 0.13-μm CMOS process. This balanced doubler consists of a reduced-size broadside-coupled Marchand balun and two distributed doublers, and suppresses fundamental signals better than 25 dB. The measured conversion losses are 8-11 dB for the output frequencies from 40 to 76-GHz under 6-dBm input drive, with a low dc power consumption of 12 mW. The chip size is 0.64 × 0.65 mm². To the best of our knowledge, this doubler achieves the widest bandwidth among all the CMOS doublers reported to date.**

I. INTRODUCTION

In wireless systems, the frequency doubler cascaded with a low frequency oscillator is usually utilized to implement the millimetre-wave (MMW) source. This approach has the advantages of higher output power and low phase noise compared with a higher frequency oscillator. Most of frequency doublers are implemented using GaAs- or InP-based high mobility transistors (HEMTs) or field effect transistors (FETs) based on GaAs/InP process [1]-[7], with high dc power consumptions and larger chip sizes. On the other hand, CMOS doublers demonstrated the features of low dc power and compact size [8].

In this paper, a broadband frequency doubler is developed to cover multi-band applications. This doubler adopts the distributed architecture to achieve a broadband conversion gain, with a broadband 180° balun to form a balanced doubler. In addition, reflectors are added at the input terminations to enhance the output signals. This distributed doubler demonstrates a conversion loss of 8-11 dB in a broadband output frequency of 40 to 76 GHz under 6-dBm input drive. With a reduced-size broadside coupled Marchand balun, it achieves a fundamental rejection of better than 25 dB with a compact size of 0.416 mm².

II. CIRCUIT DESIGN

The CMOS distributed doubler designed in a commercial 0.13-μm 1P8M CMOS process with an f_T of 98-GHz [9]. The schematic diagram is shown in Fig. 1. The balanced doubler is composed of four identical common-source transistors, each with a 24-finger NMOS and total gate width of 72-μm. The transistors are all biased at the near pinch-off region, where the nonlinearity of transconductance versus gate-to-source voltage is used to generate harmonic signals in the drain line.

Each gate is biased through a 5-kΩ resistor, and all the drain supply voltages are given via the bypass circuits.

The input fundamental signal is fed into the two distributed blocks by the Marchand balun with equal magnitude and a phase difference of 180°. As the harmonic signals are generated via the device in nonlinear region, the forward second harmonic signals are combined in-phase at the output port. Based on the balance structure, the fundamental signal and odd harmonics would also be cancelled out-of-phase [4]. Then, a good harmonic suppression can be achieved. Thin film microstrip lines (TFMS) [8] are used to implement all the matching networks, including the artificial gate and drain transmission lines.

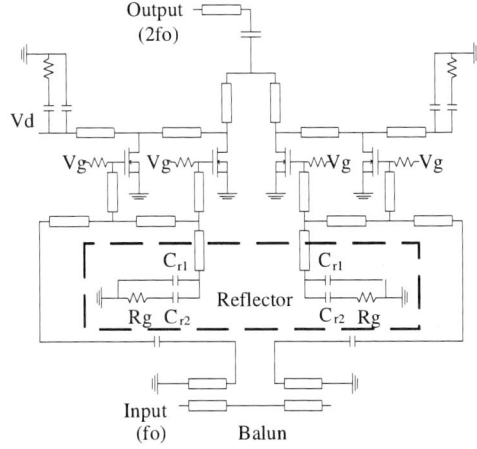

Fig. 1 Schematic diagram of the balanced distributed doubler.

In CMOS process, the broadside coupled balun suffers a high insertion loss of 6-7 dB compared with the ideal case of 3 dB. In order to reduce the conversion loss, the backward waves of the second harmonic signals can be reflected to the circuit output port by using the reflectors. The reflectors are placed in the input termination ports of the individual distributed doublers; each consists of two bypass capacitors, C_{r1} and C_{r2}, and a gate termination resistor R_g. As the nonlinear harmonic signals are generated, the fundamental signal will pass through C_{r1} and then are absorbed by R_g. The second harmonic signals are directly reflected to the output by

978-2-8748-7007-1/08 $25.00 © 2008 EuMA   17

C_{r2}. With the reflectors, the excess losses due to the balun are compensated and the doubler conversion loss is reduced from 11.5 dB to 8 dB. Figure 2 presents the chip photograph of the distributed doubler with a die size of 0.64 mm $\times$ 0.65 mm. As can be observed, all the input and output TFMS are meandered for a compact layout.

III. MEASUREMENT

The circuit was measured via on-wafer probing. The transistors were all biased near the pinch-off region with V_{gs} = 0.4 V and V_{ds} =1 V. The conversion loss versus input power from $-$5 to 15-dBm at input RF frequency of 30-GHz is plotted in Fig. 3. It is observed that the conversion loss saturates at input power of 6-dBm. The input and output 1-dB suppression point of the second harmonic signal is 9-dBm and 0-dBm, respectively. The simulated and measured conversion losses versus ouput frequency with an input power of 6-dBm are shown in Fig. 4. It is observed that the simulated performance agrees with the measured results except the frequency over 76 GHz. The measured conversion loss is between 8 and 11 dB from 40 to 76 GHz. The measured and simulated conversion losses exhibit discrepancies above 76-GHz. This might be due to the uncertainty in the device model at such a high frequency. The simulated and measured fundamental rejections versus output frequency at 6-dBm input drive are illustrated in Fig. 5. The simulation result presents a fundamental rejection of 23 to 50 dB, and the measured fundamental suppression is from 25 to 47 dB. The simulation and measurement results are also quite consistent from 40 to 76 GHz. The maximum dc power consumption is about 12 mW. Table I summarizes the performance of previously published doubler in millimetre-wave regime and this work [1]-[8]. It is observed that the proposed balanced distributed doubler achieves an impressed performance among CMOS doublers, even comparable to those in GaAs-based HEMT processes. Moreover, it demonstrates the widest bandwidth and the smallest chip size among all the CMOS doublers reported to date.

IV. CONCLUSION

A balanced distributed frequency doubler using standard-bulk 0.13-μm CMOS technology has been presented in this paper. The doubler suppressed fundamental signals more than 25 dB based on the balanced circuit architecture. At 6 dBm input power, the doubler features a minimum conversion loss of 8 dB over 40 to 76 GHz output frequencies by using the distributed topology with reflectors. The input and output 1-dB suppression points are 9-dBm and 0-dBm, respectively. The dc power dissipation is only about 12 mW under 1-V supply voltage. It also demonstrated the widest bandwidth, and the smallest size doubler reported to date.

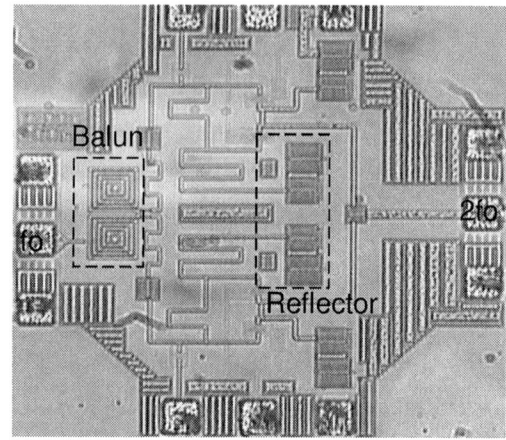

Fig. 2. Chip photo of the distributed doubler with a die size of 0.64 x 0.65 mm²

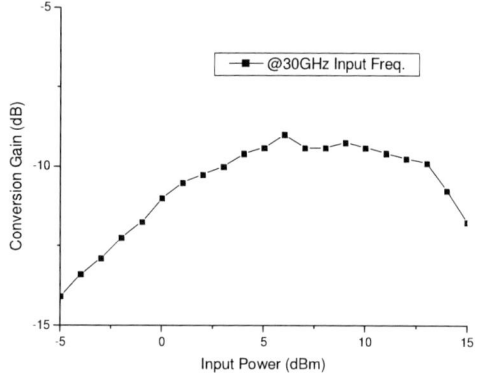

Fig. 3. The measured conversion loss versus input power at 30-GHz input frequency.

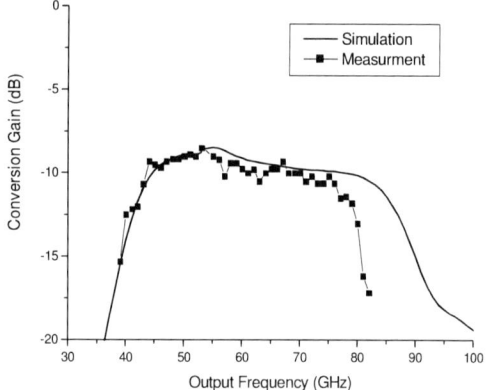

Fig. 4. Simulated and measured conversion losses.

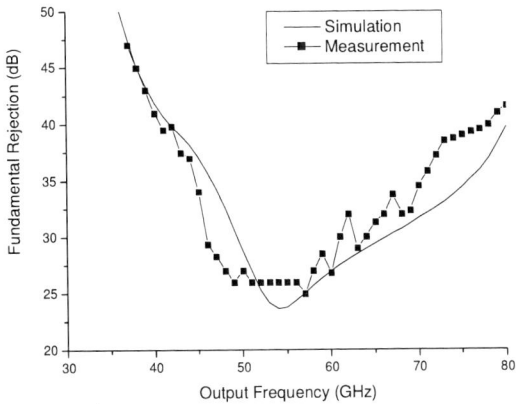

Fig. 5 Simulated and measured fundamental rejections versus output frequency.

ACKNOWLEDGMENT

The chip was fabricated by TSMC through the Chip Implementation Center (CIC) in Taiwan, ROC. This work was supported by the National Science Council and Ministry of Education of Taiwan, R.O.C., under Projects NSC 96-2752-E-002-003-PAE, NSC 96-2219-E-002-015, NSC 96-2219-E-002-020, and 95R0062-AE00-01.

REFERENCES

[1] J. B. Beyer and S.N. Prasad, "MESFET distributed amplifier design guidelines," *IEEE Trans. Microwave Theory Tech.*, vol. 32, pp. 268-275, Mar. 1984.

[2] C. Fager, L. Landen, and H. Zirath, "High output power, broadband 28-56 GHz frequency doubler," in *IEEE MTT-S IMS Symp. Dig.*, vol. 3, June 2000, pp. 1589-1591.

[3] F.van Raay and G. Kompa, "Design and stability test of a 2-40 GHz frequency doubler with active balun," in *IEEE MTT-S IMS Symp. Dig.*, vol. 3, June 2000, pp.1573-1576.

[4] K.-L. Deng, and H. Wang, "A miniature broadband pHEMT MMIC balanced distributed doubler," *IEEE Trans. Microwave Theory Tech.*, vol. 51, pp. 1257-1261, Apr. 2003.

[5] B. Piernas, H. Hayashi, K. Nishikawa, K. Kamogawa, and T. Nakagawa, "A broadband and miniaturized V-band PHEMT frequency doubler," *IEEE Microwave Guided Wave Lett.*, vol. 10, pp. 276-278, July 2000.

[6] Y.-L. Tung, P.-Y. Chen, and H. Wang, "A broadband pHEMT MMIC distributed doubler using high-pass drain line topology," *IEEE Microwave Wireless Components Lett.*, vol. 14, pp. 201-203, May 2004.

[7] K. Nishikawa, T. Enoki, S. Sugitani, I. Toyoda, and K. Tsunekawa, "Low-voltage and broadband V-band InP HEMT frequency doubler MMIC," in *IEEE MTT-S IMS Symp. Dig.*, June 2005, pp. 45-48.

[8] M. Ferndahl, B. Motlagh, and H. Zirath, "40 and 60 GHz frequency doublers in 90-nm CMOS," in *IEEE MTT-S IMS Symp. Dig.*, vol. 1, June 2004, pp. 179-182.

[9] Chieh-Min Lo, Chin-Shen Lin, and Huei Wang, "A miniature V-band cascode LNA in 0.13-μm CMOS," *ISSCC Dig. Tech. Papers*, pp. 402-403, Feb. 2006.

TABLE I
PERFORMANCE SUMMARY OF PREVIOUS REPORTED MMW DOUBLERS AND THIS WORK

Process	Output frequency (GHz)	Power Consumption (mW)	Fundamental Suppression (dB)	Chip Area (mm²)	ConversionGain (dB)	Reference
0.15-μm GaAs pHEMT	48 ~ 56	275	21 ~32	3	0 ~ 4	[2]
0.15-μm GaAs pHEMT	2 ~ 40	N/A	27 ~ 37	2.2	-13 ~ -10	[3]
0.15-μm GaAs pHEMT	30 ~ 50	132	9 ~ 25	1.5	-7 ~ -5	[4]
0.15-μm GaAs pHEMT	63 ~ 75	50	33 ~ 48	1	-20 ~ -8.5	[5]
0.15-μm GaAs pHEMT	22 ~ 42	N/A	13.4 ~ 26	2	-10 ~ -2	[6]
0.1-μm InP HEMT	54 ~ 70	8	22 ~ 30.5	0.89	-5 ~ -2	[7]
90-nm CMOS	36 ~ 43 / 50 ~ 69	4 / 4	18 / 23	0.65 / 0.5	-19 ~ -16 / -18 ~ -15	[8]
0.13-μm CMOS	**40 ~ 76**	**12**	**25 ~ 47**	**0.416**	**-8 ~ -11**	**This Work**

978-2-8748-7007-1/08 $25.00 © 2008 EuMA

Monolithic Integrated Oscillator with Silicon IMPATT Diode for Automotive Radar Applications

H. Xu[#1], M. Morschbach[#2], J. Werner[#3] and E. Kasper[#4],

[#] *Institute for Semiconductor Engineering, Universität Stuttgart*

Pfaffenwaldring 47, 70569 Stuttgart, Germany

[1]xu@iht.uni-stuttgart.de
[2]morschbach@iht.uni-stuttgart.de
[3]werner@iht.uni-stuttgart.de
[4]kasper@iht.uni-stuttgart.de

Abstract - **This paper describes properties of monolithic W-band IMPATT oscillator, their integration into SIMMWICs and proves their basics functionality as mW power sources for simple cost effective and flexible mm-wave systems. In this oscillator system monolithic integrated IMPATT diodes are combined with coplanar waveguide resonator on unthinned silicon wafers for mm-wave operation (around 90GHz).**

I. INTRODUCTION

Different frequencies are devoted to different applications, e.g. broadband communication (60GHz), automotive radar (77-81GHz), military (94GHz), and industrial applications (122GHz). In Europe, the operating frequency for automotive radar will be increased to 79 GHz in the near future – latest 2013. The need for fast, small and flexible solutions will increase with growing demand on Si mm-wave circuits. This demand could be satisfied by silicon monolithic mm-wave integrated circuits (SIMMWIC) based on two terminal actives and passive waveguide structures. In this paper we report on a SIMMWIC W-Band oscillators with coplanar waveguide resonators and driven by integrated IMPATT diodes. Monolithic integration of IMPATT diodes and usage of unthinned silicon substrates are key elements.

II. INTEGRATION TECHNOLOGY

For SIMMWIC processingen in house research fabrication process was used with molecular beam epitaxy (MBE) of the active layers and a low temperature (<600°C) device formation with relaxed optical lithography rules (1μm). As substrate high resistivity silicon (>1000Ωcm) was used in order to minimize waveguide losses from a conductive substrate. The active layers of IMPATT diode, in this case a n^+-n-p^+-stack for a single-drift (SD) IMPATT diode with a n-doped avalanche/drift region, are grown with molecular beam epitaxy (MBE) [1] [2]. This method facilitates the growth of high-quality layers in corporation with the abrupt change of the doping level, more than four orders of magnitude within a few nanometers, which is needed at np^+ junction of the device.

The device formation (Fig. 1) is based on a double mesa structure without planarization because of the relaxed lithography rules, low temperature processing is provided by dry etching of Si (nominal room temperature, actual <

200°C), passivation by plasma enhanced chemical vapor deposition (PECVD, 350°C) of oxid, sputter deposition of contact metals (Al, Ni) and annealing (450°C for NiSi contacts [3]).

The process consists of two etch steps, one to define the mesas of the active devices, the second to remove the high doped buried layer beneath the passive parts (CPWs and resonators). Fig. 2 shows an integrated IMPATT diode with the double-mesa and the metallization.

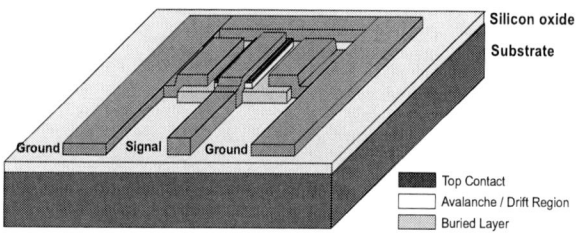

Fig 1 Schematic of an IMPATT diode integrated in a coplanar waveguide.

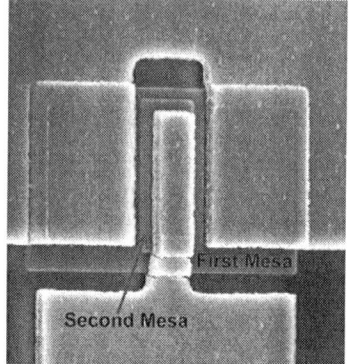

Fig. 2 Scanning electron micrograph (SEM) of the monolithic integrated IMPATT diode. (Double mesa process)

III. IMPATT DIODE CHARACTERIZATION

IMPATT diodes are known as powerful (Watt regime) discrete microwave sources. Their application is often limited to military systems because of complicated impedance matching (low resistance level) and sophisticated thermal management (upside down mounting on diamond heat sinkers). Integrated IMPATTs for lower power levels (mW regime) should be able to provide higher negative differential resistance levels, to improve matching of the passive network and to relax thermal management problems.

The IMPATT diode is operated in the avalanche breakdown. Therefore, the breakdown characteristic is the most important property. Fig. 3 is the DC- characteristics of the IMPATT diode ($8x8\mu m^2$). The ideal breakdown has a nearly infinite steep slope, in reality the slope is finite (effective series resistance) because of heating (negative thermal coefficient of avalanche breakdown), of compensation of space charge (space charge effect) and because of the finite series resistance.

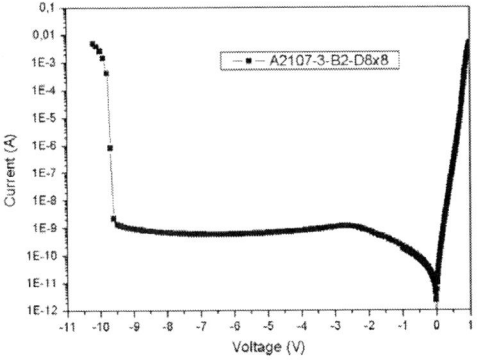

Fig. 3 DC- characteristics of the IMPATT diode ($8x8\mu m^2$)

The negative differential resistance is the most important microwave property. S-parameter measurements were performed to characterize IMPATT diodes integrated in coplanar waveguides, up to a maximum frequency of 110GHz. With a de-embedding procedure the real- and imaginary part of the impedance of the inner diodes were calculated. Above the avalanche frequency the diodes showed the expected negative real- and imaginary parts of the impedance. Fig. 4a and Fig. 4b show the real- and imaginary part of the diode for three different currents (20mA, 25mA and 30mA) which be measured with a 110GHz on-Wafer S-parameter measurement system [4].

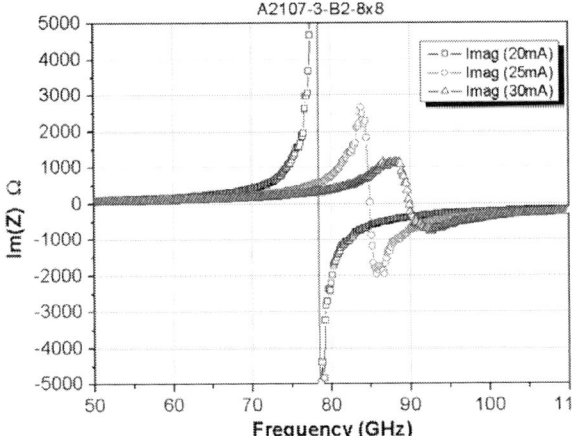

Fig.4a Imaginary part of the diode for three different currents, assessed with S-parameter measurements.

Maximum negative resistances larger than 5000Ω are obtained at resonance frequencies from 79GHz to 90GHz (diode dimensions: $8x8\mu m^2$).

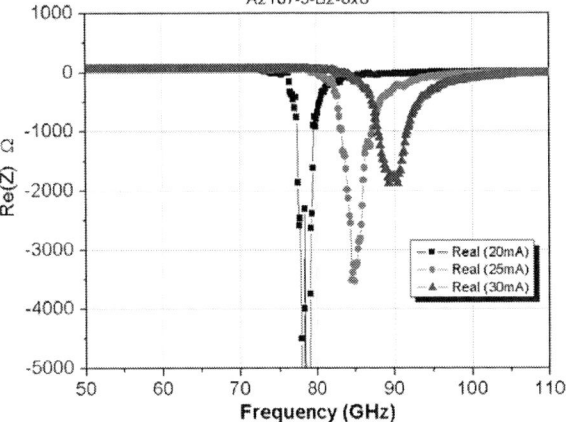

Fig.4b Real and Imaginary part of the diode for three different currents, assessed with S-parameter measurements.

IV. COPLANAR W-BAND OSCILLATOR DESIGN

As a coplanar W-band oscillator circuit (Fig. 5a), IMPATT diode is embedded in a coplanar wave guide (CPW) resonator. This is a simple fabrication process of the circuit, as no discontinuities or air bridges are necessary. For oscillation the impedance of the resonator has to be adapted to the impedance of the IMPATT diode. The oscillation condition for an operating oscillator is given by:

$$\Gamma_R = 1 / \Gamma_D \qquad (1)$$

Where Γ_R is the reflection coefficient of the coplanar resonator circuit and Γ_D is the reflection coefficient of the IMPATT diode at operational current.

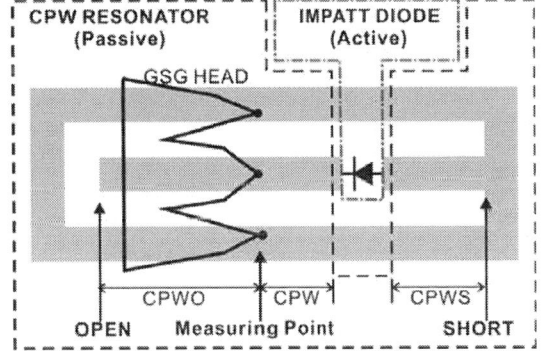

Fig. 5a Schematic diagram of a coplanar W-band oscillator

Fig. 5a shows the schematic diagram of a coplanar W-band oscillator which is designed for spectrum analyzer measurements with GSG (ground signal ground) on-wafer prober head (50Ω). The IMPATT diode is contacted between an open end coplanar wave guide with an on-wafer prober GSG head and a short end coplanar wave guide.

978-2-8748-7007-1/08 $25.00 © 2008 EuMA

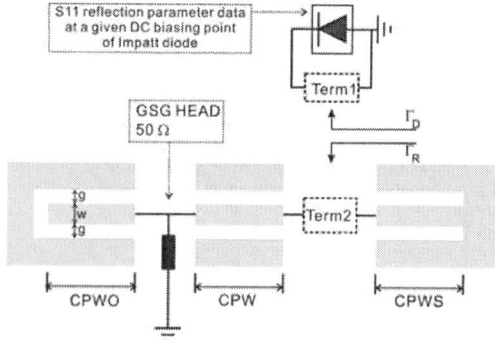

Fig. 5b The schematic diagram of the W band Oscillator design system.

Fig. 5b shows the schematic diagram of a coplanar W band oscillator design and simulation system model. The impedance of the resonator is determined by the lengths at the three coplanar waveguides (CPWO, CPW and CPWS). The GSG prober head is placed between the waveguides CPWO and CPW. The prober head is used to supply a drive current and measure the output signal but also to adjust the impedance of the resonator in order to exact match impedance of IMPATT diode. By moving Measuring Point of prober head the impedance can be adjusted, matching between resonator circuit and IMPATT diode. When at the same frequency f_o the Γ_R and $1/\Gamma_D$ curves cross each other in the Smith-diagram (see Fig 6), the oscillating condition is filled.

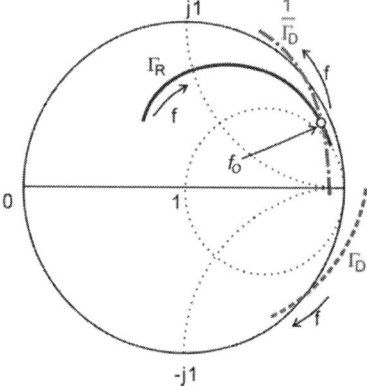

Fig. 6 The principal representation of W band Oscillator design system. Reflection coefficient Γ_R of the resonator and diode $1/\Gamma_D$ intersect at the oscillation frequency f_o.

V. COPLANAR W-BAND OSCILLATOR MEASUREMENT

To measure the W-band oscillator in the millimeter wave range, an on wafer prober station and a spectrum analyzer (R&S FSEK30) with a mixer have been used. An additional 6dB attenuator is placed between the prober head and the mixer, in order to match between the W-band oscillator and the mixer. [5][6]

A photo of the Coplanar W-band Oscillator with the GSG prober head can be seen in figure 7. The IMPATT diode is visible as a black square in the middle.

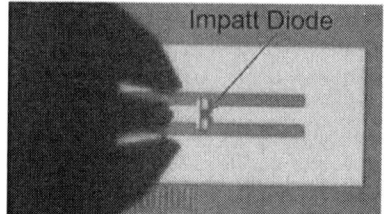

Fig. 7 Optical image of a Coplanar W-band Oscillator with the GSG prober head.

The measured output spectrum for a DC biasing at 60mA and measuring point at CPWO=280µm, CPW=50µm is shown in Fig. 8. An oscillation frequency of 88.74 GHz with output power of 7.2 dBm was obtained.

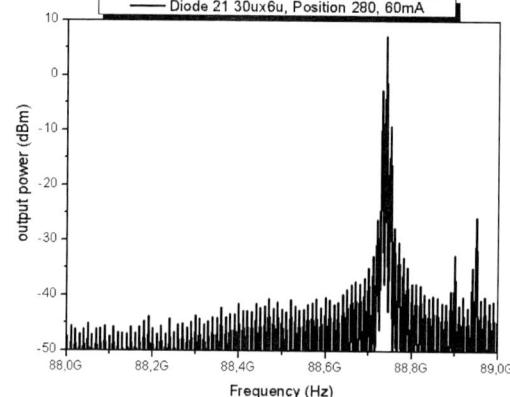

Fig.8 The measured output spectrum for the oscillator. (Measuring point at CPWO=280µm, and a DC biasing of 60mA)

With different DC bias currents the frequency may be shifted as shown in Fig. 9. The oscillation frequency is decreased with increasing bias currents. The ratio of frequency change over current is about:

$$\frac{\Delta f}{\Delta I} \approx \frac{-0.07 GHz}{mA}$$

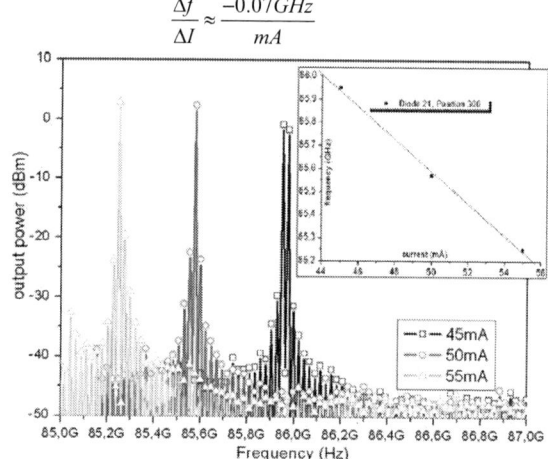

Fig. 9 Frequency shift with different DC bias currents 45mA to 55mA (measuring Point at CPWO=300µm).

With different measuring points the frequency may also be adjusted as shown in Fig. 10 by CPWO=280µm and CPWO=250µm at DC bias current I=50mA.

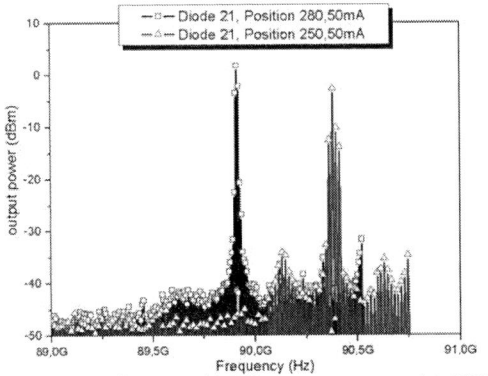

Fig. 10 Frequency adjustment with different measurement points (CPWO= 280μm, 250μm).

VI. OUTLOOK

Monolithic integrated IMPATT diodes in combination with coplanar waveguides have demonstrated generation of mm-wave power (W-band, mW range) in simple SIMMIC oscillators. A basic module of RF systems is the transmitter with the microwave oscillator and the connected antenna [7]. A complete integrated IMPATT Transmitter will be designed and processed with a W-band VCO IMPATT Oscillator and a CPW-Slot antenna. Fig. 11 shows the scheme of a monolithic integrated W-band Transmitter which is integrated with a VCO IMPATT oscillator and a CPW-Slot antenna. The same technology as for the IMPATT diode will be used for the varactor diode.

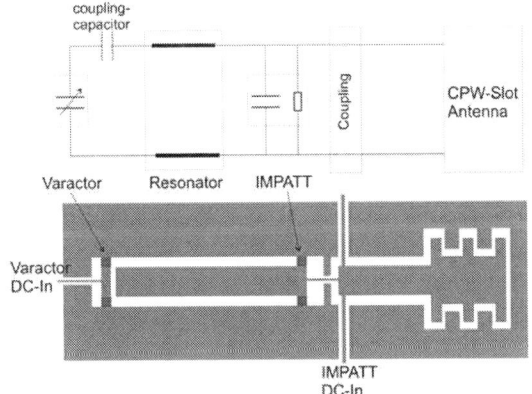

Fig. 11 Schematic diagram of a monolithic integrated W-band Transmitter.

The high electric permittivity of the silicon substrate, ε_r= 11.8, together with the thickness of the substrate lead to a limited quality of the planar antenna. Directivity, gain and efficiency can be increased by thinning the substrate to a thickness of 200μm. Fig. 12 shows the layout of a W band CPW-Slot antenna. Fig. 13 shows the measurement of the return loss of a CPW-Slot antenna on a thinned silicon substrate. The minimum return loss is achieved -33dB at 84.4GHz.

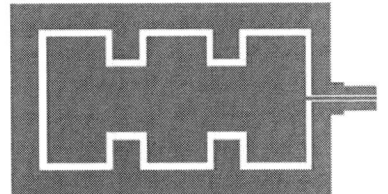

Fig. 12 Layout of 80 GHz CPW-SLOT antenna.

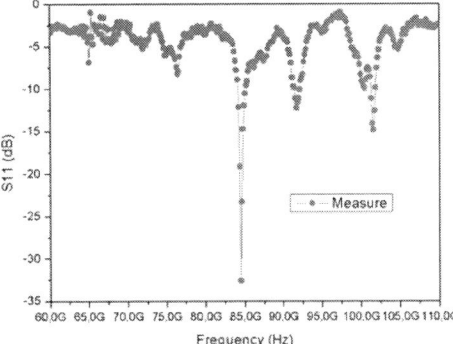

Fig. 13 Results of measured input reflection coefficient (S_{11}) of the CPW-Slot antenna.

ACKNOWLEDGEMENT

The technical support by Mr. Matthies and Mrs. Rohmer is acknowledged. Thanks are due to Dr. M. Oehme, Dr. J. Hasch, C. Schöllhorn for help and discussion about MBE, technology and measurement procedures.

REFERENCES

[1] M. Oehme and E. Kasper, "MBE Growth Techniques", in: *Silicon Heterostructure Handbook: Materials, Fabrication, Devices, Circuits, and Applications of SiGe and Si Strained-Layer Epitaxy*; edited by John D. Cressler, CRC PRESS, 2000 NW Corporate Blvd., Boca Raton, FL 33431-9868, USA (2006) p. 85-94

[2] E. Kasper and M. Oehme, "IMPATT diodes", in: *Silicon Heterostructure Handbook: Materials, Fabrication, Devices, Circuits, and Applications of SiGe and Si Strained-Layer Epitaxy*; edited by John D. Cressler, CRC PRESS, 2000 NW Corporate Blvd., Boca Raton, FL 33431-9868, USA (2006) p. 661-678

[3] J. Eberhardt, F. Kasper, "Ni Ag metallization for SiGe HBTs using a Ni silicide contact", *Semicond. Sci. Technol.* 16, L47-L49 (2001)

[4] C. J. Schöllhorn, M. Morschbach, H. Xu, W. Zhao and E. Kasper, "S-Parameter Measurements of the Impedance of mm-Wave IMPATT Diodes in Dependency on the Current Density", *Journal of Microwave and Optoelectronics*, Vol.3, No.5, Special Issue, pp.81-96 (2004)

[5] J. Hasch, E. Kasper. "S-parameter characterization of mm-wave IMPATT oscillators". *Topical Meeting on Silicon Monolithic Integrated Circuits in RF Systems 2006*, p.213-217

[6] E. Kasper, H. Xu, E. Dörner, J. Werner, "Monolithic integrated coplanar W-band IMPATT oscillator", *Topical Meeting on Silicon Monolithic Integrated Circuits in RF Systems, 2008* Page(s):222 - 225

[7] C. Schöllhorn, H. Xu, M. Morschbach, E. Kasper, "Monolithically integrated IMPATT diodes for K-band transmitters", *Topical Meeting on Silicon Monolithic Integrated Circuits in RF Systems 2004*, p.207-210

High Voltage RF LDMOS Technology for Broadcast Applications

S.J.C.H. Theeuwen, W.J.A.M. Sneijers, J.G.E. Klappe, J.A.M. de Boet

NXP Semiconductors, Gerstweg 2, 6534 AE, Nijmegen, The Netherlands

steven.theeuwen@nxp.com, walter.sneyers@nxp.com, jos.klappe@nxp.com, jan.de.boet@nxp.com

Abstract— **We present high voltage (40-50V) RF LDMOS technologies to realize 300-500W power levels for frequencies up to 1.0 GHz. This technology has an extremely good ruggedness, one octave wide band operation, and reliable circuit matching.**

I. INTRODUCTION

RF power amplifiers are key components in base stations for personal communication systems (GSM, EDGE, W-CDMA, Wimax). For these power amplifiers, RF Laterally Diffused MOS (LDMOS) transistors are the standard choice of technology because of their excellent power capabilities, gain, efficiency, linearity, reliability and low cost. LDMOS transistors are also the device choice in broadcast applications, where additional requirements like bandwidth, ruggedness, and thermal resistance are important.

In broadcast applications the bandwidth (BW) requirement is almost one octave for UHF: 470-860 MHz, more than a factor of 10 larger than for WCDMA signals in base stations. Furthermore the LDMOS transistor should be designed to have, besides a VSWR of 10:1, the ability to withstand an abrupt mismatch in the transmitter at high power. The thermal resistance of the transistor and package is also very critical due to the high power levels requiring values below 0.4 K/W.

There is a continuous demand for higher power levels to reduce the transistor board space to a minimum. In order to develop a cost effective broadcast transmitter, twice the state of the art power (150W) is required to make a direct 2 to 1 replacement of an analogue amplifier possible, while even higher power levels are desirable for digital broadcast signals.

The maximum power is limited by the low impedance output-matching stage, giving problems with reproducibility and reliability. The way around is to develop a device, which can be operated at a higher supply voltage. This increases the load impedance making a more reliable circuit design possible, which ensures long lifetime of the matching components.

In this paper we present a 40V LDMOS technology and the evolution to a 50V technology. Both technologies have very wide band operation, extremely good ruggedness and very high output power for UHF broadcast applications. The power of these high voltage devices (300-500W) has more than doubled compared to the state of the art 32V LDMOS device.

II. DEVICE DESCRIPTION

The high voltage RF LDMOS technology is based on the base station RF LDMOS technology [1]. This technology, as shown in figure 1, is processed in an 8-inch CMOS-fab capable of lithography down to 0.14 um, where the LDMOST process is derived from C075 CMOS (0.35um gate) process

with LOCOS isolation. Additions to this C075 process are the source sinker to the substrate, CoSi2 gate silicidation, tungsten shield, mushroom-type drain structure with thick 2.8 μm fourth AlCu metallization layer. The layout of the device is given in figure 2. It consists of many (typically 50) parallel fingers connected by a drain and gate bond bar. Four of these dies are combined in a ceramic push-pull package to realize a power product. A specially developed ESD device [2] protects the gate.

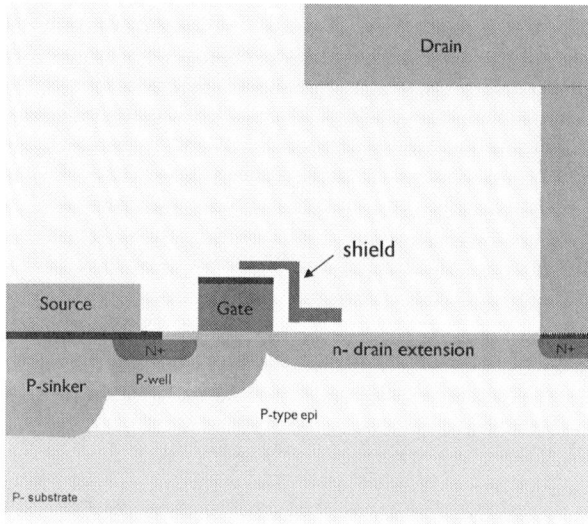

Figure 1: Schematic cross-section of state of art RF LDMOST fabricated in an 8 inch CMOS fab.

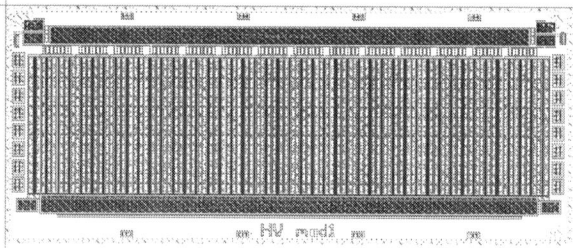

Figure 2: Top view of the layout of a single high voltage RF LDMOS device. Many finger are put in parallel to realize a 500W power device.

The breakdown voltage has further been increased to 90 and 110V for the 40V and 50V, respectively, by engineering the drain extension, epitaxial layer thickness, and shield construction to open the way to high voltage operation. The transistor has been designed to optimize the on-resistance and breakdown voltage. The success of this optimization can be recognized from the extremely flat electric field distribution across the drain extension region with only very weak peaks at the gate and drain edges. This distribution is shown in figure 3. Also the ruggedness and reliability properties of the transistor radically improve due to this engineered field distribution.

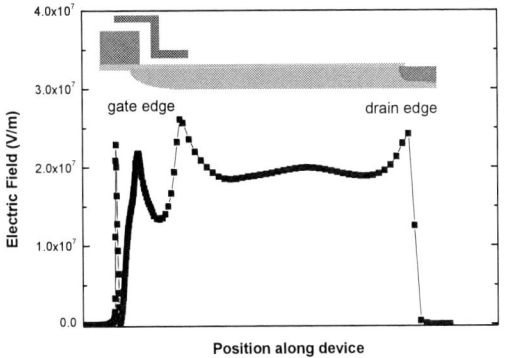

Figure 3: Simulated electric field distribution for the high voltage RF LDMOS technology at off-state breakdown.

III. RF PERFORMANCE

Figure 4 shows the narrow band CW class AB RF performance of the high voltage LDMOS at 0.86 GHz for a supply voltage of 42V. The gain of the device is typically 21dB, being limited by a series gate resistor to achieve good stability for low frequencies and easy tuning. The peak-efficiency is 65% at a power level of 400W, while 500W is achieved for the 50V supply voltage technology (not shown).

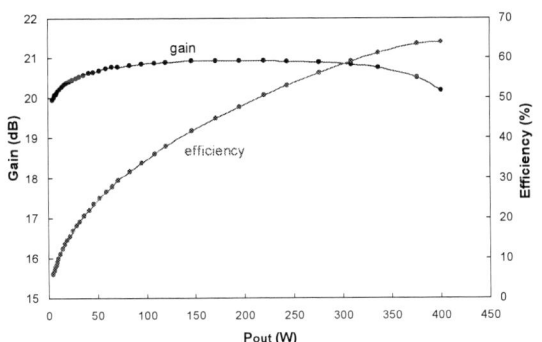

Figure 4: CW performance of the high voltage LDMOS transistor for a supply voltage of 42V at a frequency of 0.86 GHz.

In figure 5 the 2-tone performance is plotted for a tone spacing of 100 kHz in a narrow band test fixture. The third order inter modulation product IMD3 crosses the –30dBc close to 200W average output power. The corresponding drain efficiency is 52%.

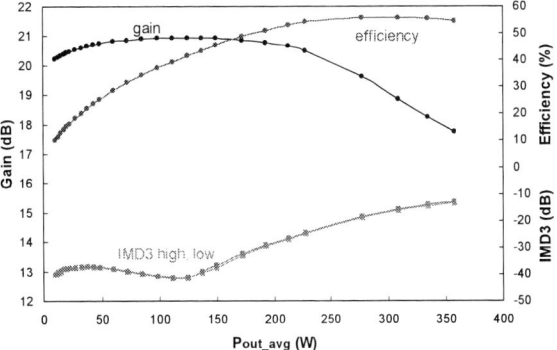

Figure 5: 2-tone performance at 0.86 GHz with a tone spacing of 100 kHz for a supply voltage of 42V.

The broadcast industry is introducing digital broadcasting massively now, making broadcasters and service providers keen to push forward the move to digital transmission. This digital DVB-T signal is an OFDM modulated signal with a CCDF at 0.01% probability of 9.6 dB (envelope approach). At the output of the transistor approximately 8 dB CCDF-0.01% is required for linearization. The transistor should be able to handle both analog and digital signals. The device performance for DVB is shown in figure 6. The shoulder distance for this signal (measured at 4.3MHz from center frequency) stays below –32dBc up to an average output power of 85W with a CCDF > 8dB. The gain is about 20dB and the efficiency is 32%. For the 50V technology 120W output power is achieved and same efficiency.

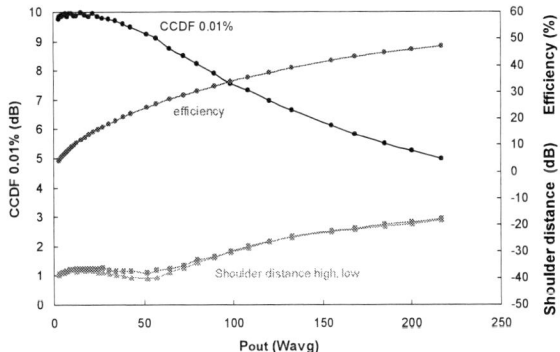

Figure 6: Performance at 42V for an 8k DVB-T signal.

IV. BROADBAND OPERATION

The broadband operation is measured for the push-pull transistor in a broadband water-cooled circuit. In order to achieve ultra wide broadband power, the device can internally be matched, as has been done for the 42V technology. The broadband operation is shown in figures 7, 8 and 9. For CW operation the power levels for ½dB compression are above 300W. A broadband efficiency of 50-62% is achieved.

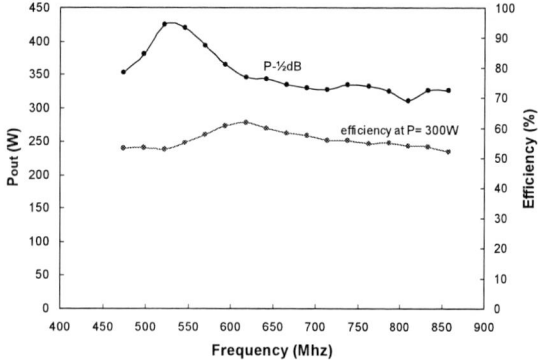

Figure 7: Broadband CW operation of the high voltage LDMOS showing the power at 0.5dB gain compression and the efficiency at 300W output power over the whole frequency band. Supply voltage is 42V.

For 2-tone, the average power level at –30dBc is far above 150W in combination with an efficiency of 40-50%. This results in a broadband linearity far below –30dBc ensuring easy digital pre-correction.

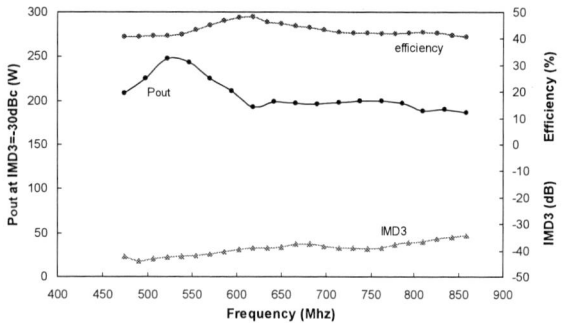

Figure 8: 2-tone power at an IMD3 of –30dBc, Gain, efficiency, and IMD3 versus frequency. Supply voltage is 42V.

The broadband device performance for DVB is shown in figure 9. The DVB average power is more than doubled compared to the previous broadcast technology showing 75W at 42V and 110W at 50V with a CCDF of typically 8dB. The efficiency is 30-32% and the gain is 19dB.

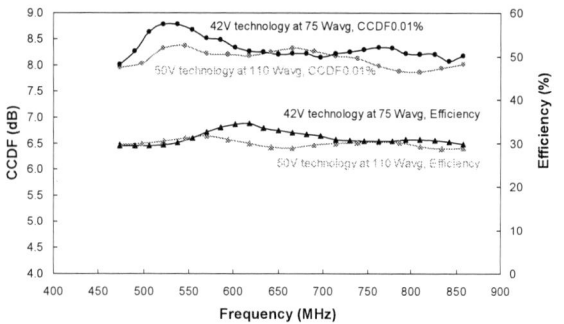

Figure 9: Performance of 42V and 50V technologies for an 8k DVB-T signal.

V. RF PERFORMANCE ANALYSIS

From the series and parallel loss mechanisms [1] in the LDMOS we can analyze the realized measured DVB efficiency and compare this to the theoretical maximum efficiency in class B. The maximum efficiency of the high voltage technologies is around 75% at 0.86 GHz as deduced from measurement on small devices. The high voltage efficiency curve versus back-off power is plotted in figure 10 and we see that this curve is very close to the theoretical maximum class B efficiency. The measured 30-32% DVB efficiency is achieved at 8 dB back-off. At a frequency of 0.86 GHz and supply voltages of 40-50V we have a combination of both parallel and series losses due to the on resistance and output capacitance. Based on their technology values we have estimated the losses in class AB, which are indicated by the blue curve in figure 10. The realized efficiency is larger than is estimated from this loss mechanism theory. The difference can be attributed to higher harmonic contributions, which become important at low frequencies compared to the cut-off frequency.

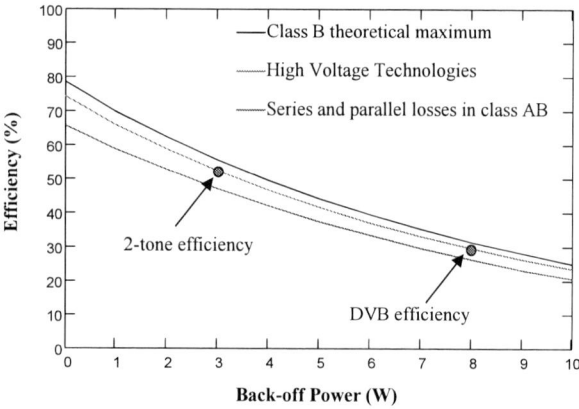

Figure 10: Efficiency versus power back-off for the high voltage LDMOS compared to the class B theoretical maximum and a device modelled with series and parallel losses in class AB.

VI. RUGGEDNESS

A crucial design requirement of the high voltage transistor for broadcast is the ability to withstand an abrupt mismatch at the output at full power. In this situation the transistor must be able to sink extremely large drain currents without fusing. In figure 11 a fast pulse I-V measurement for a small on-wafer test transistor (gate width Wg=0.6 mm) is shown. It shows that the 42V device can tolerate drain voltages up to 130V, sinking a current of more than 0.4 A/mmWg before the parasitic bipolar transistor is triggered and the device breaks down. The 50V device can withstand 150V supply voltage peaks and sinks more than 0.8 A/mmWg. This extremely good ruggedness is confirmed by the ability to bias the transistor with supplies up to 60V, far above the 40-50 V operation range. This superior ruggedness has been realized by tailoring the electric field of the high voltage technologies.

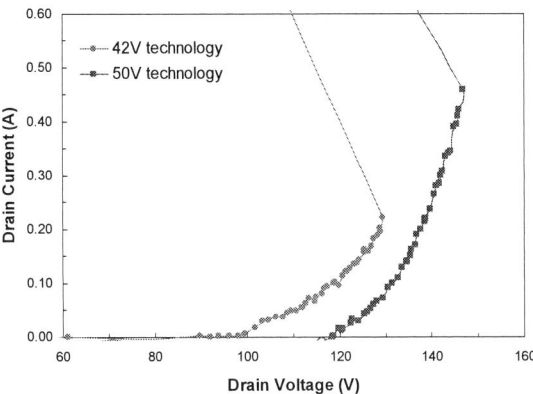

Figure 11: Pulsed drain current versus drain voltage measurements of 0.6 mm on-wafer 42V and 50V high voltage devices.

VII. RELIABILITY

The high voltage LDMOS process is qualified based on the standard procedures as used by NXP Semiconductors [3], which comply with the standards of industry. Special attention is paid to the hot carrier degradation. Hot carrier degradation is caused by high electric fields in combination with high current densities. The degradation is measured while the transistor is biased to quiescent conditions, which is for the high voltage technology typically at a current of 5mA per mm gate width and a drain-source voltage of 40-50V. A degradation of this quiescent current could lead to a change in device performance. The high voltage device shows improved behaviour, which is shown in figure 12. In the same plot there is degradation data of a broadcast transistor from a 32V generation. Despite the higher bias conditions of the new high voltage transistors, they show hardly any degradation. The degradation is only 2% after an extrapolation to 20 years.

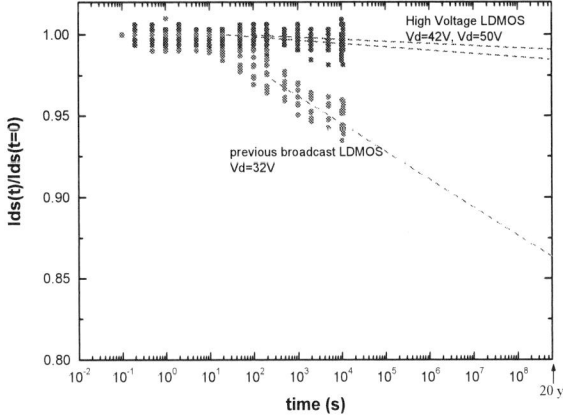

Figure 12: Degradation of the bias current as a function of time at room temperature. The high voltage LDMOS is biased at 42 V and 50V, respectively. The previous broadcast LDMOS technology is biased at 32 V. The current in the bias points is 5 mA/mmWg.

The thermal resistance becomes very important for a 300-500W device. So therefore the thermal resistance has been simulated using ANSYS [4]. An example of the simulation is shown in figure 13. Using this ANSYS tool the thermal resistance of die and package was optimized. This has resulted in a typical measured value of 0.35 K/W, giving a junction temperature below 160°C allowing reliable operation.

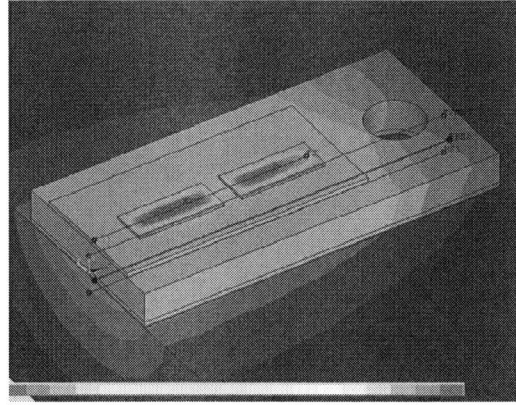

Figure 13: Thermal simulations with ANSYS have been done to optimise the thermal resistance of the high voltage LDMOS device. This figure shows one half of the power device.

VIII. CONCLUSIONS

To conclude, we have shown new high voltage (40-50V) RF LDMOS technologies, which can deliver 300-500W of CW power and 75-110 W of linear average DVB power for frequencies up to 1.0 GHz. These technologies have very good ruggedness, one octave wide band operation, and are extremely reliable.

ACKNOWLEDGEMENTS

The authors wish to acknowledge R. van den Heuvel, T. Manders, R. Dumoulin, and M. Bigcas for their support during the development.

REFERENCES

[1] F. van Rijs, S.J.C.H. Theeuwen, "Efficiency improvement of LDMOS transistors for base stations: towards the theoretical limit", *International Electron Devices Meeting, IEDM2006*, pp. 205-208 (2006).

[2] T. Smedes, J. de Boet, T. Rödle, "Selecting an Appropriate ESD Protection for Discrete RF Power LDMOSTs", *2005 EOS/ESD Symposium*, 1A1 (2005).

[3] P.J. van der Wel, S.J.C.H. Theeuwen, J.A. Bielen, Y. Li, R.A. van den Heuvel, J.G. Gommans, F. van Rijs, P. Bron, H.J.F. Peuscher, "Wear out failure mechanisms in aluminium and gold based LDMOS RF power applications", *Microelectronics Reliability*, 46, pp. 1279-1284, (2006).

[4] Z. Radivojevic, K. Andersson, J. A. Bielen, P. J. van der Wel and J. Rantala, "Operating limits for RF power amplifiers at high junction temperatures", *Microelectronics Reliability*, Vol. 44, 6, pp. 963-972 (2004).

978-2-8748-7007-1/08 $25.00 © 2008 EuMA

Analysis of Bias Effects on VSWR Ruggedness in RF LDMOS for Avionics Applications

G. Formicone[#1], F. Boueri[#2], J. Burger[#3], W. Cheng[#4], Y. Kim[#5], J. Titizian[#6]

Integra Technologies, Inc. 321 Coral Circle, El Segundo, CA 90245, USA

[1]gformicone@integratech.com
[2]fboueri@integratech.com
[3]jburger@integratech.com
[4]wcheng@integratech.com
[5]young_kim@integratech.com
[6]john_titizian@integratech.com

Abstract — **A 210W RF LDMOS power transistor optimized for pulsed applications has been used to characterize VSWR ruggedness as a function of bias and gain compression. The ruggedness test used is the 10:1 VSWR load mismatch. The transistor, operated in class AB power amplifier, can deliver 210W of output power when biased at 32V (P3dB) or 36V (P1dB), having a minimum breakdown voltage of 85V. In both conditions the transistor passes 4:1 VSWR mismatch without degradation. We also found that when operated at 32V and 210W (3dB compression) the transistor passes 10:1 VSWR load mismatch without any degradation. On the contrary, when operated at 36V (1dB compression), the transistor either goes into catastrophic failure or it survives the mismatch test with a severe power rating degradation in excess of 5%. Measured electrical data and simulated junction temperature data help explaining the different results on the VSWR ruggedness.**

I. INTRODUCTION

In the present paper we characterize the effects of gate and drain bias on the transistor ruggedness as measured by the VSWR load mismatch test in high power RF LDMOS. VSWR is the ratio of the optimum load impedance for maximum power and the actual load impedance in a mismatch condition for a given frequency. In 10:1 VSWR output mismatch 67% of the output power is reflected from the load back to the transistor and only 33% is transmitted to the load [1]. A rugged transistor must be able to absorb and dissipate into heat this reflected power without suffering any damage or degradation. This is usually accomplished during the design stage of the transistor making sure that there is enough drain resistance to absorb the reflected power. In the case of RF LDMOS, this desirable feature is part of the optimization process in the drain engineering design. Usually, the drain engineering design aims at minimizing the on-resistance for a given breakdown voltage, for CW operation. If ruggedness is indeed another major constrain, as in the case of avionics or radar pulsed applications, more trade-off conditions must be taken into consideration.

Integra Technologies, Inc has designed its proprietary RF LDMOS transistor with a grounded Faraday shield to isolate the gate from the drain and therefore reduce the feedback capacitance Cdg, as shown in the drawing of Fig. 1. Integra's LDMOS technology also adopts an all gold metallization stack for maximum ruggedness and reliability.

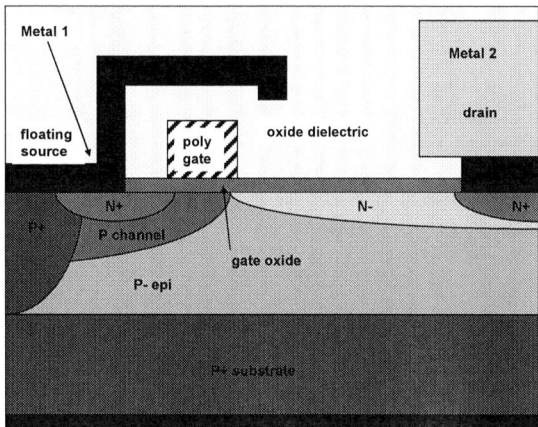

Fig. 1 Cross-section of an LDMOS design with Faraday Shield.

II. RESULTS AND DISCUSSIONS

A 210W single ended RF LDMOS transistor, designed and optimized for pulse signal operation, has been used to study the effects of gate and drain bias on RF ruggedness as measured by the die ability to pass a 10:1 VSWR load mismatch test. The transistor, suitable for avionics applications in the 1.0-1.1GHz band, has a minimum breakdown voltage of 85V, and is typically biased at 10mA or 50mA of quiescent current. The threshold voltage is about 3.5V. The on-resistance was measured on this large die at Vds=0.1V and Vgs=10V, yielding 0.25 ohms. The input and output capacitances were measured on this large periphery LDMOS chip using a capacitance meter by grounding the source (substrate of the die) and applying a 1MHz probe signal at the gate (with floating drain) and at the drain (with floating gate) respectively. Fig. 2 shows the input capacitance versus gate-to-source voltage, including the 200pF input match capacitor, and Fig. 3 shows output capacitance versus drain-to-source voltage.

978-2-8748-7007-1/08 $25.00 © 2008 EuMA

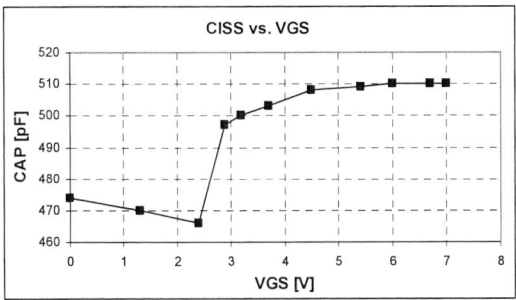

Fig. 2 Input capacitance profile.

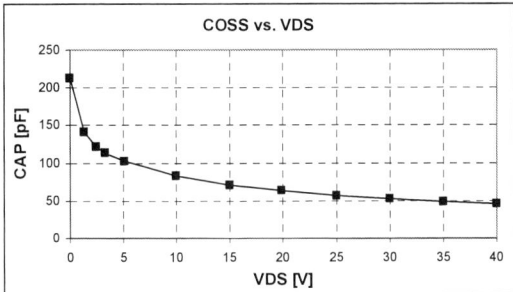

Fig. 3 Output capacitance profile.

A single die in a standard LDMOS package with input match, delivers in excess of 200W output power when biased at either 32V or 36V. Notice that in this voltage range the output capacitance is almost constant at 50pF. Biased at a drain voltage of 32V, the output power of 210W is achieved with 3dB gain compression; with the drain bias at 36V, the same power level is now achieved at 1dB gain compression. Fig. 4 shows a picture of the transistor single die packaged with input match.

Fig. 5 shows a picture of a larger package with two LDMOS dice. In both cases the transistor passes a 4:1 VSWR load mismatch test, where 36% of the output power is reflected from the load back to the transistor [1]. This result confirms that the die has been designed to sustain a minimum amount of load mismatch in excess of 3:1 VSWR load mismatch.

Next the transistor ruggedness has been characterized for a higher level of load mismatch, such as 10:1 VSWR. This is considered a good measure of ruggedness test for most practical applications encountered in avionics and radar power amplifiers. Operated in class AB with a supply voltage of 32V and 10mA quiescent drain current, the RF LDMOS transistor has a 3dB output power of 210W, 60% drain efficiency, 13dB gain and it passes 10:1 VSWR load mismatch at all phase angles without any power degradation. Fig. 6 shows the gain and drain efficiency versus output power for this bias condition.

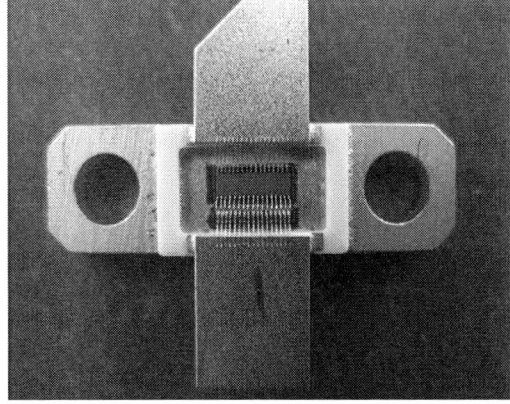

Fig. 4 Picture of one LDMOS die packaged with input match.

Fig. 5 Picture of two LDMOS dice in a single package with input match.

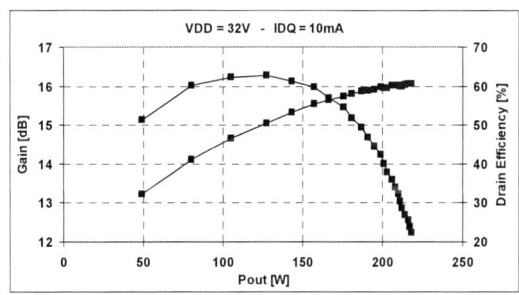

Fig. 6 Pulsed CW data at 1030MHz for gain and drain efficiency vs. output power for one LDMOS transistor biased at 10mA and 32V.

Similarly, two chips in a larger package biased at 32V and 20mA yield over 350W of output power 3dB into compression with 50% drain efficiency and 12dB gain. The RF data are shown in Fig. 7. The device still passes 10:1 VSWR load mismatch at all phase angles without degradation. The signal pulse has a duty cycle of 2% and is 50us long.

978-2-8748-7007-1/08 $25.00 © 2008 EuMA

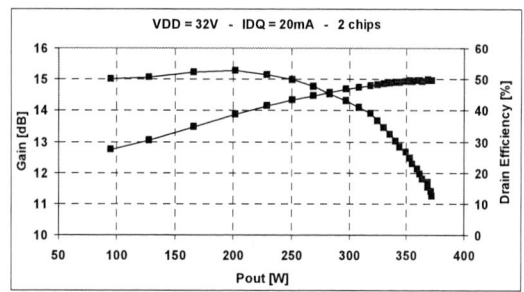

Fig. 7 Pulsed CW data at 1030MHz for gain and drain efficiency vs. output power for two LDMOS transistors in the same package biased at 20mA and 32V.

In Fig. 8 we report the effect of gate bias by increasing the quiescent current from 10mA to 50mA on the RF data, since both options are used in actual avionics applications. The supply voltage is still 32V. Notice that the output power compression feature is the same, but higher quiescent current yields a higher gain and lower efficiency, as expected.

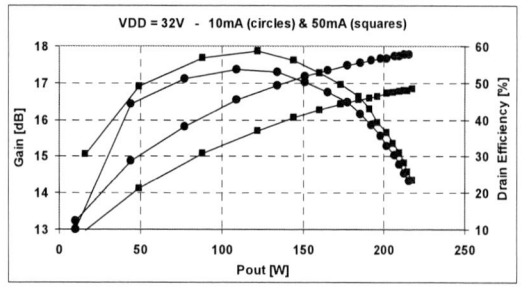

Fig. 8 Pulsed CW data at 1030MHz for gain and drain efficiency vs. output power for one LDMOS transistor biased at 32V for 10mA and 50mA quiescent current.

Next, the single chip in the smaller package has been tested at 36V with 50mA quiescent current and a 2% 50us pulse signal. Here 210W is the output power at 1dB compression and although the transistor can absorb a medium level of mismatch (it passes 4:1 VSWR at all phase angles without any appreciable degradation) it fails to pass a 10:1 VSWR load mismatch. Some transistor units pass the test at 36V, but with 5% to 10% degradation in the RF output power. In Fig. 9 it is shown the RF data at 32V and 36V for a direct comparison between the two cases. Although the power level of the transistor used in the 10:1 VSWR test is the same in the two conditions, the ruggedness results are very different and must be attributed to the different bias conditions, either the supply voltage or the quiescent current.

Since we use very short pulse signals, the thermal effects have been minimized [2]. In fact, the thermal time constant τ is found to be about 25us, based on the following information [3]: $\tau=(2h/\pi)^2/\alpha$, where h=75um (~3mils) after silicon wafer thinning, and $\alpha=0.92$ cm^2/s is the silicon thermal diffusivity. Considering that the duty cycle of the pulse signal used in our measurements is 2%, the time between each pulses is 2.5ms with 50us pulse width. With such short pulse width and long inter-pulse intervals, the junction temperature has the time to relax to ambient temperature before a new pulse starts [4, p.243, case 2].

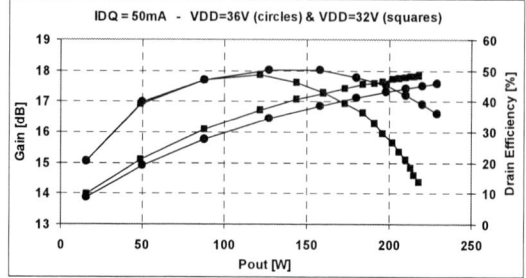

Fig. 9 Pulsed CW data at 1030MHz for gain and drain efficiency vs. output power for one LDMOS transistor biased at 50mA quiescent current at 32V and 36V.

With the pulse width (50us) longer than the thermal time constant (25us), when the pulse signal is on, the junction temperature increases to its peak value, estimated to be about 82 °C if the transistor is biased at 10mA 32V, whereas it is about 126 °C when biased at 50mA 36V. See Table I for the measured electrical data and the extracted dissipated power and junction temperature increase in a matched load condition. We have used a die thermal resistance R_{TH} of 0.126 °C/W [calculated based on ref. 4, p. 234 and die layout data] and a package thermal resistance of 0.25 °C/W. Therefore from the dissipated power the junction temperature increase is $\Delta T = P_{diss} R_{TH}$.

TABLE I
MEASURED ELECTRICAL DATA WITH MATCHED LOAD AND ESTIMATED
JUNCTION TEMPERATURE INCREASE.

IDQ [mA]	Pout [W]	Vdd [V]	Pin [W]	Id [A]	Pdiss [W]
10	210	32	10	11	152
50	210	36	4	13.2	269.2

IDQ [mA]	Rjc [C/W]	ΔT [C]	Rpkg [C/W]	Tc [C]	Tj [C]
10	0.126	19.2	0.25	63	82.2
50	0.126	33.9	0.25	92.3	126.2

From the measured electrical data is it worth noticing that the higher quiescent current leads to a larger drain current drawn from the supply voltage source and to about 120W difference in dissipated power between the two cases. The junction temperature difference in the two cases is about 45 °C. Although the estimated peak junction temperature seems pretty high in both cases, the junctions operates at these temperatures for a very short time period, since the device is operated in pulse conditions with 2% duty cycle. Because of the pulsed conditions, the average junction temperature is only a few degrees above the room temperature of 25 °C [4] and it cannot have a significant contribute to the different ruggedness behaviour reported in our measurements.

Based on our data and test conditions we attribute the different ruggedness reported in the different operation of the transistor to the different voltage swing and its relationship with the onset of avalanche breakdown at high VDS. Our measurements show that, although higher voltage operation leads to higher saturated power, the transistor's ruggedness at high levels of load mismatch is compromised faster than the improvement in output power. A lower supply voltage yields the same power level achieved with a higher voltage (although at different gain compression levels), but with a better ruggedness measure. The compromise is that the same power rating is achieved at a higher level of gain compression and therefore of nonlinearity. It is well established that RF power transistors operated at higher supply voltage lead to higher output power [2]. However, unless the transistor is redesigned for a higher voltage operation, i.e. with a higher breakdown voltage, over-biasing leads to die failure or accelerated performance degradation.

We want to emphasize that for Integra's LDMOS, the optimization of the die design process, through drain and channel engineering, and layout techniques, is driven by achieving highest pulse peak power operation with safest ruggedness margins, and not best linearity in CW operation as is done by other manufactures whose expertise remains focused on base stations applications.

Although the paper highlights a measurable ruggedness difference in high VSWR load mismatch for Integra's LDMOS when operated at 32V versus 36V, we want to remind the reader that medium levels of VSWR load mismatch (4:1) are indeed absorbed without any measurable degradation in both supply voltage operations and at power levels well above the 100W mark. For comparison, it is also noteworthy that the BJT (Bipolar Junction Transistor) technology, traditionally used for avionics and radar applications, is typically tested to pass a 5:1 VSWR load mismatch at power levels below 100W [5], with 3:1 VSWR being more typical at higher power levels [6]. RF LDMOS, on the other hand, has shown much higher ruggedness measures at higher power levels, and it can also be easily designed or optimized to meet more stringent ruggedness conditions.

III. CONCLUSIONS

The ruggedness of high power (>200W) RF LDMOS transistors to VSWR load mismatch has been characterized. We used a 210W RF LDMOS die manufactured by Integra Technologies, Inc. It is an all gold LDMOS technology specifically designed to address requirements for avionics and radar pulsed applications. The high power transistor (>200W) can be biased at either 32V or 36V, and in both cases it is able to survive a medium amount of load mismatch, such as 4:1 VSWR. However, we found that at higher levels of load mismatch, such as 10:1 VSWR, the transistor passes the test when operated at 32V but at 36V it either fails catastrophically or it survives with severe degradation, most likely due to excessive drift of the on-resistance. Finally, when biased at 36V and used at a power level corresponding to peak gain, typically around 150W, the transistor does pass the 10:1 VSWR load mismatch at all phase angles without any degradation.

ACKNOWLEDGMENT

The authors would like to thank Michael Yoo for his support in the fab for the development of Integra's LDMOS.

REFERENCES

[1] P. Malloy and J. Smith, "Importance of Mismatch Tolerance for Amplifiers used in Susceptibility Testing," *Application Note #27A, AR.*

[2] N. Dye and H. Granberg, "Radio Frequency Transistors: Principles and Practical Applications," Newnes, second ed., *2001*, p.81.

[3] R. Anholt, "Electrical and Thermal Characteri-zation of MESFETs, HEMTs, and HBTs," Arthech House, *1995*, p.69.

[4] John L. B. Walker, "High-Power GaAs FET Amplifiers," Arthech House., *1993*, p.234..

[5] K. Shenai and E. McShane, "Current Status and Emerging Trends in RF Power FET Technologies," *2001 IEEE MTT-S.*

[6] See product literature at www.integratech.com

On-Chip S-Shaped Rat-Race Balun for Millimeter-Wave Band Using Wafer-Level Chip-Size Package Process

Chiaki Inui, Yasuo Manzawa, and Minoru Fujishima

School of Frontier Sciences, The University of Tokyo
7-3-1 Hongo, Tokyo, 113-8656, Japan
{chiaki,yasuo,fuji}@rfic.t.u-tokyo.ac.jp

Abstract— In millimeter-wave CMOS circuits, a balun is useful for connecting off-chip single-end devices and on-chip differential circuits to improve noise immunity. However, an on-chip balun occupies large chip area. To reduce the chip area required for the on-chip balun, a new rat-race balun using rewiring technology with a wafer-level chip-size package (W-CSP) is proposed. The W-CSP balun occupies no area in a die because it is placed over integrated circuits. In the proposed balun, an S-shaped structure is adopted in order to directly connect the balun to differential GSGSG pads on a chip with small area. The S-shaped W-CSP balun was fabricated on a silicon-on-insulator (SOI) substrate. The core area of the S-shaped rat-race balun is 480μm × 735μm, which is 22.4% that of a square rat-race balun. As a result of measurement, we found that the minimum insertion loss is 1.7dB and the operating frequency range is 40 to 61GHz.

I. INTRODUCTION

Recently, millimeter-wave applications have received much attention in fields such as high-speed wireless personal area networks (WPANs) using the 60GHz band aimed at a wireless high-definition multimedia interface (HDMI) and automotive radar using the 76GHz band. In particular, a millimeter-wave transceiver with CMOS circuits is expected to realize low-power operation that will expand battery duration and mobility in consumer-electronic appliances. Here, in CMOS circuits, differential topology is often adopted in analog and radio-frequency circuits. For example, power-supply noise is cancelled by a differential low-noise amplifier (LNA) and local-oscillator (LO) leakage is also cancelled by a double balanced mixer. Additionally, in large signal operation, linearity is improved with differential topology because the second-harmonic signals are balanced out [1]. On the other hand, off-chip devices in the millimeter-wave band generally have single-end ports. Moreover, tests for single-end devices are much simpler than those for differential devices. To realize single-end devices using differential CMOS circuits while retaining the features in Table 1, on-chip baluns are required.

Two types of millimeter-wave on-chip baluns have been reported: rat-race baluns [2]-[4] and Marchand baluns [5]-[9]. Rat-race baluns realize low insertion loss even with single-layer signal paths, since electromagnetic coupling of two transmission lines, as required for the Marchand baluns, is unnecessary, as shown in Fig. 1. However, rat-race baluns occupy larger area than Marchand baluns because the total length of the transmission line of the rat race is 1.5 times the wavelength. To overcome this issue, a rewiring technology to fabricate a lead frame utilizing the W-CSP is adopted to form the rat-race balun in this study. The new balun requires no additional die area because devices made by the W-CSP technology are placed over a chip. However, since the W-CSP balun is connected to differential GSGSG pads on the chip, differential ports in the W-CSP balun should be aligned with the pads. Additionally, the area occupied by the balun should be small so as to be unobtrusive in the package lead frame. From these viewpoints, the form of the conventional rat-race baluns is unsuitable for the W-CSP balun, because the differential ports are not aligned with the GSGSG pads even though the meandering structure may reduce the area. In this work, the newly proposed S-shaped structure having differential ports aligned with the GSGSG pads is used, and the meandering structure reduces the area. In the following sections, after the principle of the W-CSP S-shaped balun is explained, measurement results are discussed.

II. S-SHAPED RAT-RACE BALUN USING W-CSP

A basic rat-race balun has four ports with a circular transmission line, as shown in Fig. 2. When the rat race is used for a balun, only three ports are used: single-end port 1 and two differential ports 2 and 4. The distances between ports are proportional to wavelength λ, and the total length of the circular transmission line is 3λ/2.

Because the rat-race balun requires only a single layer, other layers are freely utilized for the circuit, and the full chip area can be utilized effectively for the circuits by using the rewiring layer in the W-CSP. A cross-sectional view of the W-CSP balun with CMOS circuits is shown in Fig. 3. Since the rewiring layer is made of a thick copper film on a thick polyimide insulator film, the loss of the transmission line and the influence by the wiring of other layers are small because the thick insulator layer increases the distance between lossy silicon substrate and the rewiring layer.

Here, since the W-CSP balun is connected to differential GSGSG pads, the differential ports in the rat-race balun should be aligned so as to connect directly to the GSGSG pads on the chip, as shown in Fig. 4(a). In the conventional on-chip rat-race baluns, although a meandering structure has often been adopted for area reduction, the differential ports are not aligned with the GSGSG pads, as shown in Fig. 4(b). Since the terminal impedance of port 3 does not affect the performance of the balun [10], port 3 can be removed, and the rat race is deformed into an S shape to be aligned with the differential GSGSG pads, as shown in Fig. 5. Figure 5(a) shows the basic shape of the rat-race balun, where the space in the center of the circular transmission line occupies a large

area. To remove the center space, the circular transmission lines can be modified into parallel straight transmission lines, as shown in Fig. 5(b). The removed of port 3 renders the shape of the rat race flexible. With ports 2 and 4 placed close to each other, when the parallel straight line is folded, the rat race becomes an S shape, as shown in Fig. 5(c). As a result, the core size of the S-shaped rat-race balun becomes 22.4% that of the square rat-race balun.

TABLE I

FEATURE SUMMARY OF SINGLE- AND DIFFERENTIAL-END STRUCTURES. A BALUN IS REQUIRED TO IMPROVE TESTABILITY WHILE MAINTAINING CIRCUIT ROBUSTNESS.

	Testability	Noise immunity
Single-end	◯	△
Differential	△	◯

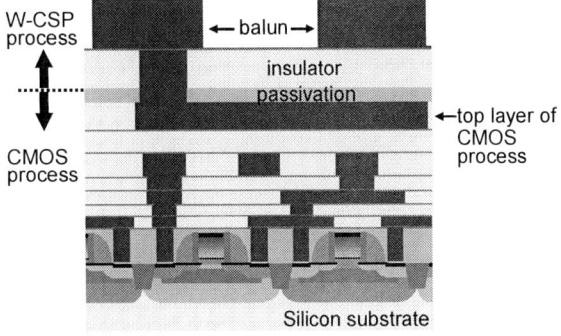

Fig. 3. Simplified crosssection of W-CSP and CMOS process.

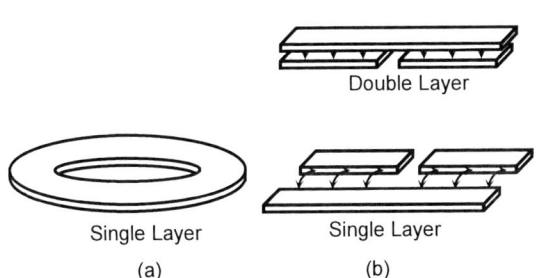

Fig. 1. (a) Rat-race balun realizes low-insertion loss even with single-layer single path. (b) Electromagnetic coupling of two transmission lines is required for the planar and stacked Marchand baluns.

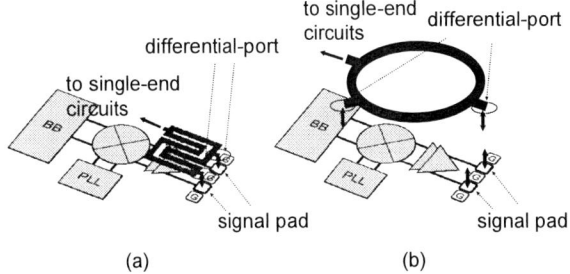

Fig. 4. (a) Balun with differential ports aligned with the GSGSG pads can be connect to the chip directly, but (b) conventional balun cannot be connected directly to the GSGSG pads.

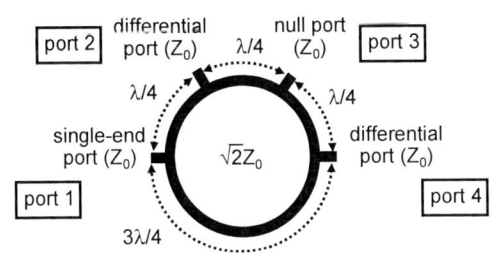

Fig. 2. Basic rat-race balun.

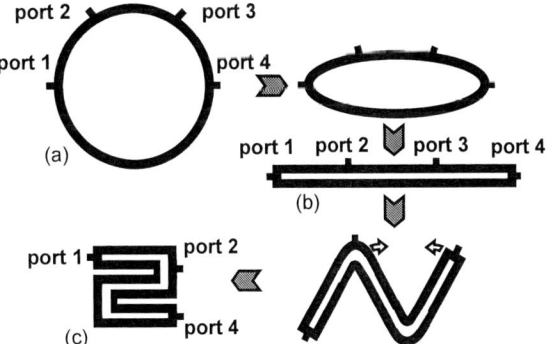

Fig. 5. (a) The basic structure of the rat race is circular. (b) The rat race is flattened to remove the center space. (c) The flattened rat race is folded to align the differential ports to the GSGSG pads.

III. MEASUREMENT RESULTS

A millimeter-wave S-shaped rat-race balun is fabricated using the rewiring technology of the W-CSP on a 0.15μm SOI-CMOS substrate. A slow-wave transmission line (SWTL) [11], as shown in Fig. 6, is adopted for the S-shaped W-CSP balun. In the SWTL, signal lines and coplanar grounds are prepared using a rewiring layer in the W-CSP, and the ground shield under the signal line is formed using the top layer of the CMOS region. The width of the SWTL is 15μm and the space between the signal line and the coplanar ground is 20μm. The characteristic impedance and transmission loss are 52.5Ω and 0.5dB/mm at 50GHz, respectively, as shown in Fig. 7. It is noted that the measurement results are coincident in the cases of a high-resistivity substrate and a low-resistivity substrate, owing to the presence of the ground shield. Figure 8 shows a micrograph of the fabricated chip. The core size without pads is 480μm × 735μm. Figure 9 shows the measured amplitude imbalance and phase imbalance between the differential ports to be ±1.5dB and ±15° from 40 to 61GHz, respectively.

Here, differential-mode gain G_{diff} and common-mode gain G_{com} are obtained as

$$G_{\text{diff}} = \left| S_{12} - S_{13} + S_{21} - S_{31} \right| / 2\sqrt{2} \qquad (1)$$

and

$$G_{\text{com}} = \left| S_{12} + S_{13} + S_{21} + S_{31} \right| / 2\sqrt{2} . \qquad (2)$$

G_{diff} represents the signal penetration with conversion from single-end signals to differential signals. In the case of a passive balun, since G_{diff} theoretically becomes 0dB, the insertion loss of the balun is estimated using the G_{diff}. G_{com}, on the other hand, indicates the signal leakage without conversion from single-end signals to differential signals. Figure 10 shows G_{diff} and G_{com} of the fabricated balun. The measured insertion loss is from 1.7dB to 3.7dB at operating frequencies from 40 to 61GHz. Since G_{com} is below −19dB at the operating frequency, it is concluded that no side effects due to the removal of port 3 are observed. Figure 11 shows minimum insertion losses as a function of operating frequencies for various on-chip baluns. The minimum insertion loss of the proposed balun is the smallest compared with those of other millimeter-wave baluns.

IV. CONCLUSIONS

The rewiring technology of the W-CSP was applied to the rat-race balun to suppress the area penalty. To overcome the alignment issue of the differential GSGSG pads on a chip, an S-shaped rat-race balun is proposed. The proposed S-shaped rat-race balun not only reduces the chip area but also is connected directly to the differential GSGSG pads. The core size of the S-shaped rat-race balun becomes 22.4% that of a square rat-race balun. The S-shaped rat-race balun was fabricated using the W-CSP technology on a 0.15μm SOI-

CMOS substrate. From the measurement results, it was found that the minimum insertion loss is 1.7dB. The operating frequency is from 40 to 61 GHz when the amplitude and phase imbalances between the differential ports are ±1.5dB and ±15° and the insertion loss is from 1.7dB to 3.7dB. The adoption of the proposed S-shaped rat-race balun using the W-CSP can eliminate the need for the additional area between on-chip differential circuits and off-chip single-end circuits. When the proposed balun is adopted, device integration will be improved in millimeter-wave CMOS circuits, which is expected to contribute to low-cost 60GHz short-range high-speed wireless transmission used for wireless HDMI and 76GHz automotive radar.

ACKNOWLEDGMENT

This study is supported by the joint research project with Semiconductor Technology Academic Research Center (STARC). The authors would like to thank Mr. Tani and Mr. Terui of OKI Electric Industry Co., Ltd. for the chip fabrication.

REFERENCES

[1] B. Razavi, "Design of analog CMOS integrated circuits," McGraw-Hill Higher Education, 2001.

[2] M. K. Chirala and B. A. Floyd, "Millimeter-wave lange and ring-hybrid couplers in a silicon technology for E-band applications," IEEE MTT-S International Microwave Symposium Digest, pp. 1547-1550, June. 2006.

[3] S. Wang and C.-K. C. Tzuang, "Compacted Ka-band CMOS rat-race hybrid using synthesized transmission lines," MTT-S International Microwave Symposium Digest, pp. 1023-1026, 2007.

[4] C. Y. Ng, M. Chongcheawchamnan, and I. D. Robertson, "Miniature 38GHz couplers and baluns using multilayer GaAs MMIC technology," 33rd European Microwave Conference Proceedings, vol. 3, pp. 1435-1438, 2003.

[5] I. C. H. Lai, C. Inui and M. Fujishima, "CMOS on-chip stacked Marchand balun for millimeter-wave applications," IEICE Electronics Express, vol. 4, no. 2, pp. 48-53, 2007.

[6] J. –X. Liu, C. -Y. Hsu, H. –R. Chuang, and C. –Y. Chen, "A 60-GHz millimeter wave CMOS Marchan balun," IEEE RFIC Symposium, pp. 445-448, 2007.

[7] K. S. Ang, I. D. Robertson, K. Elgaid, and I. G. Thayne, "40 to 90 GHz impedance-transforming CPW Marchand balun," IEEE MTT-S International Microwave Symposium Digest, vol. 2, pp. 1141-1144, 2000.

[8] K. W. Hamed, A. P. Freundorfer, and Y. M. M. Antar, "A monolithic double-balanced direct conversion mixer with an integrated wideband passive balun," IEEE Journal of Solid-State Circuits, vol. 40, no. 3, pp. 622-629, Mar. 2005.

[9] P.-S. Wu, C.-S. Lin, T.-W. Huang, H. Wang, Y.-C. Wang, and C.-S. Wu, "A millimeter-wave ultra-compact broadband diode mixer using modified Marchand balun," 13th European Gallium Arsenide and Other Compound Semiconductors Application Symposium, pp. 349-352, 2006.

[10] H. Bex, "New broadband balun," Electronic Letters, vol. 11, no. 2, pp. 47-48, Jan 23, 1975.

[11] I. C. H. Lai, H. Tanimoto, and M. Fujishima, "Characterization of high Q transmission line structure for advanced CMOS processes," IEICE Transactions on Electronics, vol. E89-C, no. 12, pp. 1872-1879, Dec. 2006.

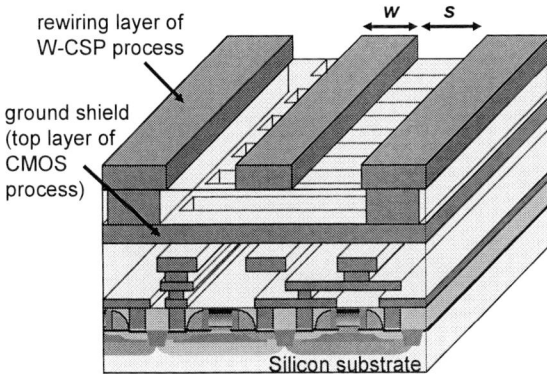

Fig. 6. Structure of slow wave transmission line (SWTL).

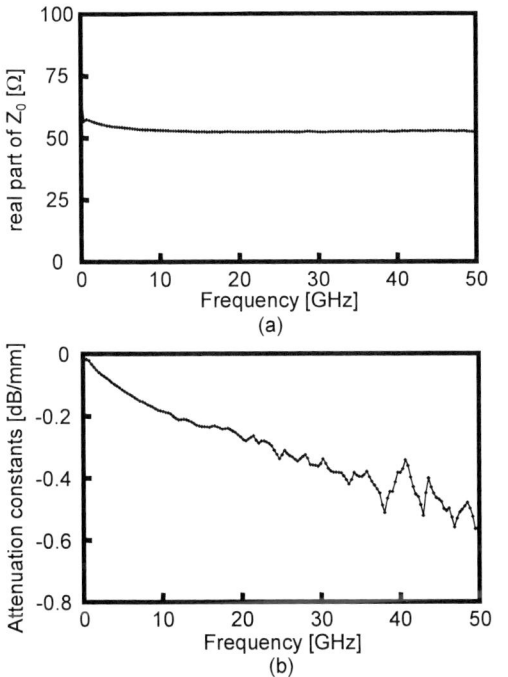

Fig. 7. Measured results of (a) real part of the characteristic impedance Z_0 and (b) attenuation constants of the transmission line.

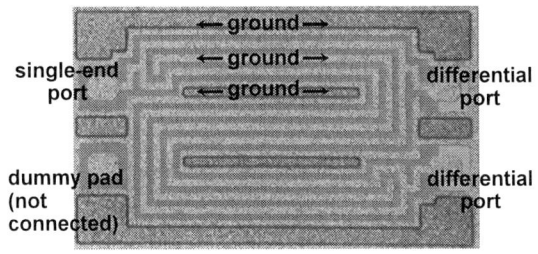

Fig. 8. Micrograph of the proposed balun.

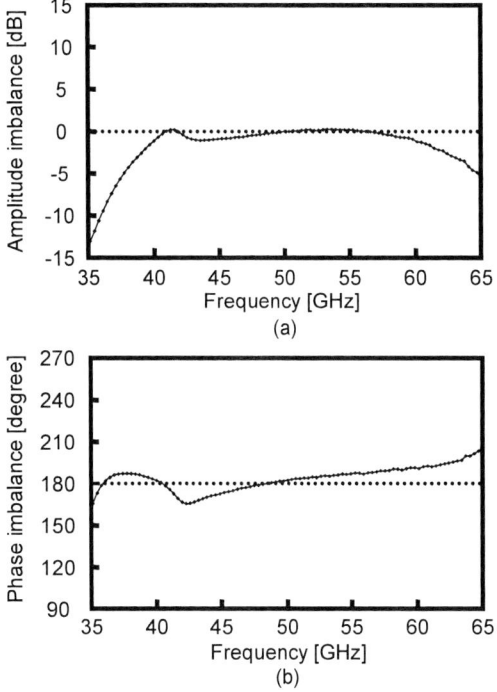

Fig. 9. Measured (a) amplitude and (b) phase imbalances.

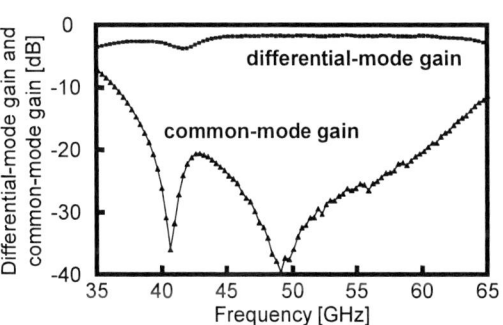

Fig. 10. Measured differential-mode gain and common-mode gain.

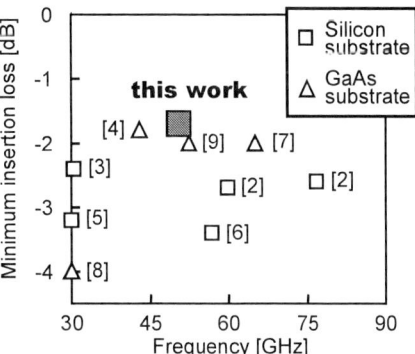

Fig. 11. Minimum insertion losses of proposed and other millimeter-wave on-chip baluns.

978-2-8748-7007-1/08 $25.00 © 2008 EuMA

Impact of Si substrate resistivity on the non-linear behaviour of RF CPW transmission lines

C. Roda Neve[1], D. Lederer[2], G. Pailloncy[3], D.C. Kerr[4], J.M. Gering[4], T.G. McKay[5], M.S. Carroll[5] and J.-P. Raskin[1]

[1]Microwave Laboratory, Université catholique de Louvain, Place du Levant, 3, B-1348 Louvain-la-Neuve, Belgium
cesar.rodaneve@uclouvain.be, Tel: +32 (0)10 47 80 96
[2]Farran Technology Ltd, Ballingcollig Cork, Ireland
[3]NMDG Engineering, C. Van Kerckhovenstraat 110, B-2880 Bornem, Belgium
[4]RFMD, Greensboro, NC, USA, [5]RFMD Scotts Valley, CA, USA

Abstract— Non-linear behaviour of RF coplanar transmission lines is analyzed for various values of Si substrate resistivitiy. Based on small-signal measurements performed under different DC bias conditions, voltage dependent capacitance and conductance per unit length of the transmission line are extracted and compared for several silicon substrates. Harmonic distortion of large RF signal at 900 MHz along CPW lines is measured using a spectrum analyzer based setup as well as with a LSNA which gives us access to the phase of the harmonic components. For an input power of +25dBm, the highest harmonic component (2nd) is as high as -15, -57, -37 and -63 dBm for resistivity substrates of 20, 500, 5k and 2kΩ-cm, respectively. A reduction of 45 and 15 dB for all harmonic components was obtained for the 5 and 2kΩ-cm HR-Si substrates, respectively, when a trap-rich passivation layer was used at the Si/SiO₂ interface, and for both characterization setups. The impact of the resistivity value on signal distortion with its relation to the bias-dependence substrate characteristic and the efficiency of the trap mechanism of the passivation layer are for the first time introduced from experimental result considerations.

I. INTRODUCTION

Although high resistivity silicon (HR-Si) has slightly higher losses than other classical RF substrates, such as III-V compounds components, its introduction has converted silicon into a suitable technology for high frequency applications [1]. It indeed offers hybrid integration capability, improved RF performance with CMOS device scaling, it is perfectly compatible with highly performant Silicon-on-Insulator (SOI) technologies as well as low cost. All these features make HR-Si a very attractive handle wafer for mobile integrated systems.

For small RF signals, it is known that in order to maintain low RF attenuation levels, the effective resistivity (ρ_{eff}) of the HR-Si wafer must be higher than 3 kΩ-cm [2]. However, oxidized HR-Si suffers from parasitic surface conduction (PSC) at the Si/SiO₂ interface, thereby reducing ρ_{eff} of the wafer by more than one order of magnitude [2]. In addition, surface charge concentration also depends on the applied bias voltage, leading to bias-dependent losses. Both issues can be overcome by introducing a trap-rich passivation layer, such as polycrystalline Silicon (PSi) or crystallized amorphous silicon (α-Si) [3, 4] between the oxide and the HR-Si substrate. The traps capture the free carriers at the Si-SiO₂ interface, thereby enabling the substrate to recover its nominal resistivity value [2].

The RF harmonic distortion (HD) behaviour of coplanar (CPW) lines on Si substrates has also been recently unveiled [5]. Moreover, it has been experimentally demonstrated that HD can be strongly reduced for 5 kΩ-cm HR-Si substrate if a trap-rich passivation layer is introduced at the Si/SiO₂ interface. Knowing the growing interest of Si technologies for large RF signal devices [6], a deep understanding of the harmonic behaviour of the Si substrate is of first importance.

In this paper, we study the generation and propagation of harmonic signals into various values of Si substrate resistivity. Thanks to their high sensitivity to PSC and to its bias dependence behaviour [7], CPW lines provide a perfect tool to investigate the non linear behaviour of Si substrates.

To this effect, we first introduce the non-linear mechanisms that take place along CPW lines on an oxidized Si substrate. The bias-dependent characteristics of CPW lines lying on various Si substrates are then presented and the efficiency of a passivation layer to greatly reduce such dependence is demonstrated. Finally, HD is characterized for RF CPW transmission lines on different silicon substrates. Measurements were made using two different setups, one using a spectrum analyzer based setup [5] at a fixed frequency (900 MHz), and a second using a Large-Signal Network Analyser (LSNA) which allows the de-embedding of the non-linear environment to extract the harmonic components only related to the CPW line.

II. NON-LINEAR MECHANISM IN OXIDIZED HR-SI

Typically, non-linearity effects are characterized by means of HD analysis and/or by intermodulation distortion (IMD) of two different RF tones. In large RF signal applications low PIM substrates are typically used. In that case, the main sources of non-linearity in RF transmission lines are the metal-to-metal joints and the metal performance under large RF signals. In the case of HR-Si, the source of non-linearity has been proved not to come from the metallic conductors but from the semiconductor substrate itself [5].

In Fig. 1, the steady state of the carrier distribution in the cross section of a CPW line on top of an oxidized p-type HR-Si substrate is shown under various bias conditions. In Fig. 1(a), for zero voltage ($V_S = 0$ V), fixed charges inside the oxide induce a shallow inversion layer at the Si-SiO₂ interface where the carriers can laterally move, hence creating a conductive layer and increasing RF losses (i.e. reducing the effective substrate resistivity). In Figs. 1(b) and 1(c) we can see that when applying a positive or negative voltage, a larger inversion or an accumulation layer is created, varying the conductive characteristic at the interface and having a bias dependence conductive layer. For highly positive or negative bias the conduction characteristic remains stable and the equivalent capacitance is that of the insulator.

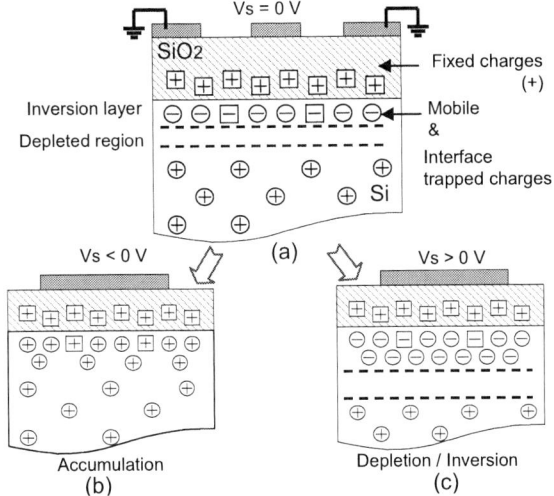

Fig. 1. Cross section of a coplanar MOS p-type structure under zero, (a) negative (accumulation) (b) and positive bias voltage (depletion/inversion).

For small signals this bias dependence translates only into a different attenuation level at different bias points and in a reduction of the effective resistivity of the HR-Si [2]. However, in the case of large signals the amplitude of the signal may become large enough to vary the conductive characteristic of the substrate depending on its voltage. The PSC carriers at the interface have a response time fast enough to follow signals at high frequencies (> 100 MHz). However, interface traps and fixed charges at the SiO2/Si interface as well as substrate resistivity and DC bias also significantly influence the spatial charge distribution inside the Si substrate and are therefore expected to have a substantial impact on the response time of carriers.

The equivalent circuit per unit length of a transmission line is presented in Fig. 2, where the conductance (G) and capacitance (C) are the parameters related to the substrate, and the resistance (R) and inductance (L) to the metallic line. The substrate characteristics are bias dependent, hence the values of C and G will be function of the voltage of the input signal. For a large RF input signal (at a frequency such that the electrical length of the line is smaller than half of the signal wavelength) non-linearities lead to a distorted output signal with several harmonic components. The degree of non-linearity depends on the RF signal amplitude and, as is demonstrated in [5], on the DC bias point.

Based on this simplified equivalent circuit, and on some hypotheses, we investigate here the influence of the Si substrate resistivity on the HD of CPW lines. The bias depend-

ence of the CPW line parameters on HR Si substrates is first presented in the coming section.

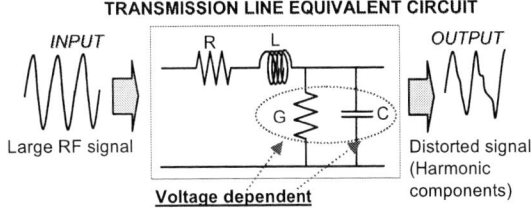

Fig. 2. Equivalent circuit of a transmission line on HR-Si for large RF signals.

III. TRANSMISSION LINE BIAS DEPENDENCE ON HR-SI

In order to determine the bias dependence of a transmission line on HR-Si, we extract the capacitance and conductance per unit length at different bias points of a 50 Ω CPW line. The CPW line dimensions are 58 μm for the width (W) of the central conductor, 36 μm for the slot width (S), and 208 μm for the width of the planar ground conductors (Wg). The cross-section of the CPW line is presented in Fig. 3, with and without passivation layer. Two different 300 nm-thick passivation layers and two HR-Si resistivity substrates (2 and 5 kΩ-cm) were considered. In all cases front and back metal are a 1 μm-thick aluminum layer and the thermal SiO2 layer is of 50 nm.

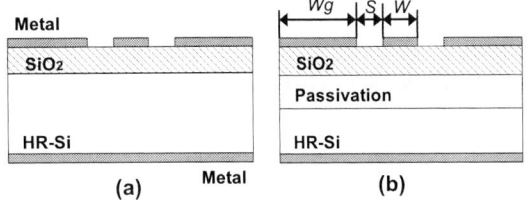

Fig. 3. Cross section of the CPW line (a) without and (b) with a trap-rich passivation layer.

On-wafer small-signal S-parameter measurements from 40 MHz to 40 GHz were made for thru-open-line CPW calkit standards. A TRL de-embedding technique was used to de-embed the effect of the pad contacts and to obtain the propagation constant of the line. The characteristic impedance of the de-embedded line was calculated using the method described in [8]. Figs. 3 and 4 show the extracted capacitance and conductance per unit length, respectively, versus DC bias at 900 MHz. Both HR-Si substrates (2 and 5 kΩ-cm) present a strong bias dependence for C and G. C and G variations larger than 35% and 150%, respectively, were measured for the 5 kΩ-cm wafer taking the zero bias point as the reference. Bias dependence is less important, not higher than 5%, in the case of the 2 kΩ-cm substrate. These differences can be related to the different Si/SiO2 interface and oxide quality (concentration of fixed charges and interface traps) between both wafers which were not processed at the same time. However, it is quite interesting to note that when introducing any of the two passivation materials, the bias dependence nearly vanishes for both wafers except for the 2 kΩ-cm substrate when a bias higher than +12 V or lower than -12 V is applied.

978-2-8748-7007-1/08 $25.00 © 2008 EuMA

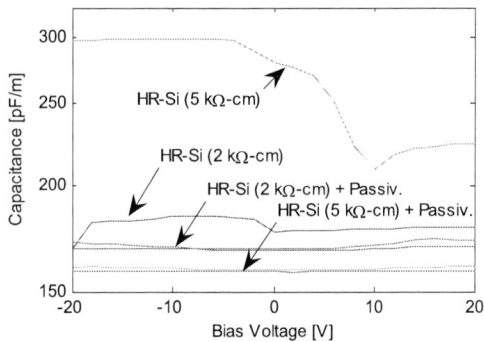

Fig. 4. Extracted capacitance per unit length of a 8000 μm-length CPW line versus DC bias from small-signal S-parameter measurement at 900 MHz.

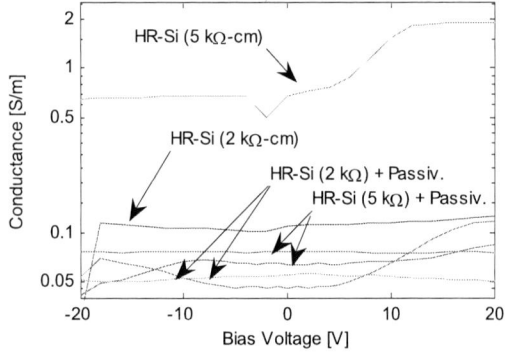

Fig. 5. Extracted conductance per unit length of a 8000 μm-length CPW line versus DC bias from small-signal S-parameter measurement at 900 MHz.

IV. HARMONIC DISTORTION CHARACTERIZATION FOR HR-SI SUBSTRATES WITH AND WITHOUT PASSIVATION LAYER

Two different setups have been used to characterize the HD of the studied CPW lines. The first setup, described in [5], offers a high dynamic range for power sweep measurements at a fixed frequency, 900 MHz. Several filters and attenuators are used to minimize system harmonics, and only power amplitude information of the harmonic components is obtained. The second setup is based on a Large-Signal Network Analyser (LSNA). Extending the capabilities of a VNA, a LSNA accurately measures the amplitude and phase of the incident and reflected waves at the ports of the device under test (DUT), both at the fundamental and harmonics frequencies. Moreover, it not only provides information about the performance of the DUT but also about the surrounding environment. In such way, we are able to de-embed non-linearities due to other components, e.g. from the source. It simplifies the setup, and allows a large frequency band.

Figs. 6 and 7 show the 2nd and 3rd output harmonic power measurements, respectively, when using the first setup for different values of Si substrate resistivity, with and without passivation layer, for 2.2 mm-long CPW lines. Characterization was done for a power sweep from +10 to +35 dBm and for a fixed CW signal at 900 MHz. As expected, in all cases the 2nd harmonic has higher values than the 3rd one. The highest 2nd and 3rd harmonic components appear for the 20Ω-cm

resistivity Si, -15 and -37dBm for +25 dBm input power, and a reduction of 22, 42 and 48 dB is measured for the 2nd harmonic for the 5kΩ, 500Ω and 2 kΩ-cm HR-Si, respectively. It might be expected that the performance of a HR-Si substrate should increase with resistivity, however from the bias dependence characterization we have seen than the 2 kΩ-cm HR-Si behaves better than the 5 kΩ-cm, probably due to higher interface trap density and the thicker oxide layer (2.5 μm) used for the 2 kΩ-cm HR-Si substrate. In a similar way the 500 Ω-cm Si substrate behaves better than both HR-Si substrates as it is less sensible to PSC, remaining the effective substrate resistivity closer to its nominal value.

The figures clearly show that the use of a passivation layer (PSi) reduces the harmonic power components at the output. All harmonics seem to be reduced in a similar level, around 15 dB for the 2 kΩ-cm HR-Si and more than 40 dB for the 5 kΩ-cm HR-Si. This reduction, due to the bias independence when using a trap-rich layer at the Si/SiO₂ interface, seems to be also related to the efficiency of the trapping with resistivity values lower than 3 kΩ-cm: the highest reduction in harmonics is seen on the highest resistivity substrates. Further investigation must be done in the future to clarify this point.

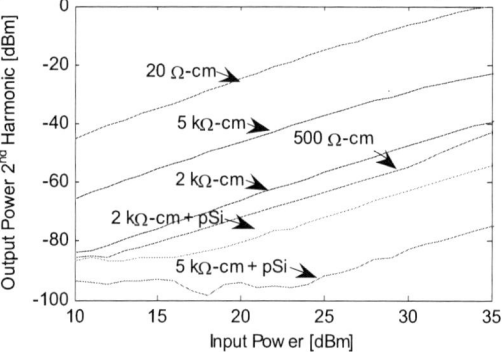

Fig. 6. Measured 2nd harmonic component using setup in [5], for different Si substrate resistivities, with and without passivation layer, at 900 MHz. For HR-Si 2 kΩ-cm metal is Au and SiO₂ thickness is 2.5 μm.

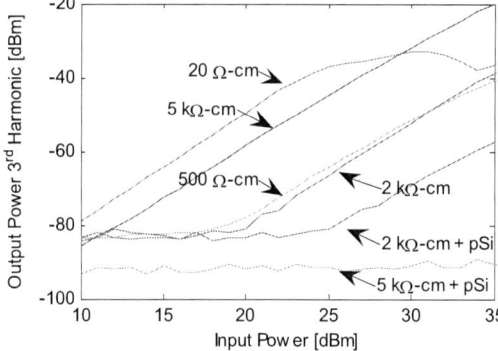

Fig. 7. Measured 3rd harmonic component using setup in [5], for different Si substrate resistivities, with and without passivation layer, at 900 MHz. For HR-Si 2 kΩ-cm metal is Au and SiO₂ thickness is 2.5 μm.

Fig. 8 and 9 shows the HD measurements using a LSNA for a 2 kΩ-cm HR-Si substrate, with and without passivation layer (α-Si), at 900 MHz and 4 GHz, respectively. First, an on-wafer LRRM calibration using a calkit substrate was made. Then, the small-signal scattering parameters were measured under linear operation of our CPW line. Finally, the HD of our CPW line was measured for a power sweep from -10 to +25 dBm of a CW signal from 900 MHz to 4 GHz.

Using a Volterra series decomposition [9], and assuming that the harmonics generated by the CW source are low enough to not contribute to the non-linearities of the CPW line, the extracted harmonic components due to the RF transmission line only were calculated using the following formula:

$$b2extr.(\varpi_i) = b2(\varpi_i) - S21(\varpi_i) \times a1(\varpi_i) - S22(\varpi_i) \times a2(\varpi_i)$$

With S11, S12, S21 and S22 the measured small-signal scattering parameters of the line, and $a1(\omega_i)$, $a2(\omega_i)$ and $b2(\omega_i)$ the incident and reflected waves for each harmonic component i at each power input level and each fundamental frequency.

At 900 MHz, only the 2nd harmonic was detected above the LSNA measurement noise limit, the 3rd harmonic falls below or too close to this limit to be accurately resolved. Measured harmonic power levels are 10 dB higher than those measured with the first setup, probably due to the difference attenuation due to connectors and cables for both setups, from the different oxide thickness used (50 nm in this case versus 2.5 µm for the first setup) and from extraction limitations that should be studied with more detail. However the same harmonic reduction, 12-15 dB, is measured when using a passivation layer, validating our results. It is worth noticing that the harmonics power of the CW source are of the same order of magnitude than the HD introduced by the HR-Si non-linearity, and that the noise floor measurement is higher than the one when using the first setup.

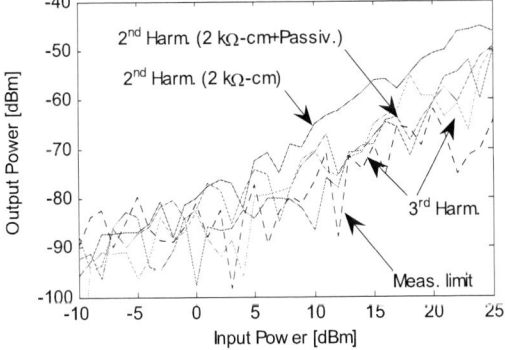

Fig. 8. Extracted HD for a 2 kΩ-cm HR-Si substrate with and without passivation, at 900 MHz. Metal is Al on a 50 nm-thick SiO₂.

Moreover, the LSNA setup is not limited to one fixed frequency, allowing HD characterization for a wide frequency range. At 4 GHz similar power levels were measured for the 2nd harmonic. However when using the passivation layer harmonic reduction is not present at input powers higher than +20 dBm, or only 3 dB for input powers lower than +20 dBm. The origin of these different behaviours at higher frequencies must be identified with future investigations and characterizations.

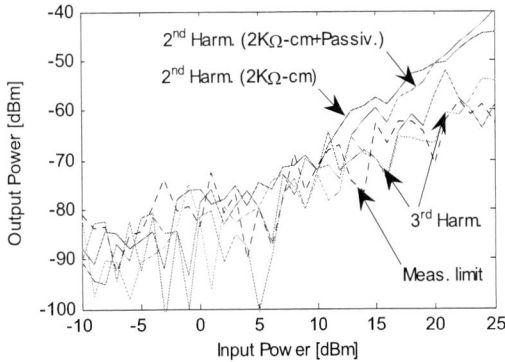

Fig. 9. Extracted HD for a 2 kΩ-cm HR-Si substrate with and without passivation, at 4GHz. Metal is Al on a 50 nm-thick SiO₂.

V. CONCLUSIONS

Non-linear behaviour of oxidized HR-Si substrates is presented and explained in terms of a conductive bias-dependent carrier layer at the Si/SiO₂ interface. Its relation with the equivalent circuit of a transmission line is also introduced. Bias dependence of a CPW line lying on different HR-Si substrates causes variations in the capacitance and conductance larger than 35% and 150%, respectively, taking the zero bias point as the reference.

Harmonic distortion of CPW structures on different resistivity Si substrates at 900 MHz is measured using two different setups: the one proposed in [5] and a LSNA. In both cases the results show harmonic power levels higher than -50 dBm for +20 dBm input power. A reduction of 45 dB and 15 dB of the harmonic components was achieved when using a passivation layer for 5 and 2 kΩ-cm HR-Si substrates, respectively. Such reduction is due to the suppression of the bias dependent effects at the Si/SiO2 interface when using a trap-rich layer. It is worth noticing that a relation between the efficiency on the trapping mechanism and the resistivity value related to the bias dependence of the substrate seems to exist for HR-Si substrates with not high enough resistivity value.

REFERENCES

[1] H.S. Gamble, et al., "Low-loss CPW lines on surface stabilized high-resistivity silicon", *IEEE MGWL*, vol. 9, no. 10, pp. 395-397, 1999.

[2] D. Lederer and J.-P. Raskin, "Effective resistivity of fully-processed SOI substrates", *Solid-State Electr.*, vol.49, no.3, pp.491-496, 2005.

[3] D. Lederer and J.-P. Raskin, "New substrate passivation method dedicated to high resistivity SOI wafer fabrication with increase substrate resistivity", *IEEE EDL*, vol. 26, no. 11, pp. 805-807, November 2005.

[4] M. Norling, et al., "Comparison of High-Resistivity Silicon Surface Passivation Methods", *Proc. 2nd European Microwave Integrated Circuits Conference*, pp. 215-218, Oct. 2007.

[5] Daniel C. Kerr, et al., "Identification of RF Harmonic Distortion on Si Substrates and its Reduction Using a Trap-Rich Layer", *IEEE SiRF 2008*, pp.151-154, 2008.

[6] T. G. McKay, et al., "Linear cellular antenna switch for highly integrated SOI front-end," *IEEE Intl. SOI Conf.*, 2007.

[7] D. Lederer, "Fabrication and characterization of advanced SOI material and devices", Ph.D. Thesis, UCL, Louvain-la-Neuve, Belgium, 2006.

[8] M. Dehan, "Characterization and Modeling of SOI RF integrated components", Ph.D. Thesis, UCL, Louvain-la-Neuve, Belgium, 2003.

[9] F. Verbeyst and M, Vanden Bossche, "VIOMAP, the S-parameter Equivalent for Weakly Nonlinear RF and Microwave Devices", *IEEE T-MTT*, Vol. 42, No. 12, December 1994

Innovative and Complete Dummy Filling Strategy for RF Inductors Integrated in an Advanced Copper BEOL

Carine Pastore [#*1], Frédéric Gianesello [#2], Daniel Gloria [#3], Emmanuelle Serret[#4], and Philippe Bench[*5].

#STMicroelectronics, 850 rue Jean-Monnet, 38 926 Crolles Cedex, France.
[1]carine.pastore@st.com
[2]frederic.gianesello@st.com
[3]daniel.gloria@st.com
[4]emmanuelle.serret@st.com

*IMEP, 3, rue Parvis Louis Néel, BP 257, 38 016 Grenoble Cedex 1, France.
[5]Philippe.Bench@minatec.inpg.fr

Abstract — A complete strategy to manage dummy fills inside and underneath a large spectrum of integrated RF inductors realized in a 0.13 µm CMOS technology using a Damascene Copper Back End of Line (BEOL) is presented here. The main motivation of this paper is first to evaluate through a Design Of Experiment (DOE) modeling, the impact on RF inductor performances of dummy fills inserted inside or underneath the coils, and then determine the right metal fill density to insert to be compliant with Digital metal density rules without degrading their electrical performances.

I. INTRODUCTION

In today's advanced IC processes based on Damascene Copper technology, fill metals are required to meet metal density rules and guarantee component and circuit integrity. This is a direct consequence of the Chemical Mechanical Polishing (CMP) process use which is actually sensitive to the layout density [1] and can create topographical and electrical defects called erosion in high density areas, and dishing in wide lines [2]. In order to avoid these kinds of over-polishing and provide a dielectric layer planarity as uniform as possible, more or less severe design rules are applied to all BEOL structures [3], in particular for integrated inductors. Actually, dedicated metal density rules are set for integrated inductors in order not to degrade the electrical performances of these key RF components. In most BEOL, the inductor centre is wide and kept metallization free. As coils are made of top metal levels in most inductor layouts, the space under the coils is lower metal level free too, see in Fig. 1. However, these requirements are area consuming and difficult to apply in Advanced processes.

Thus, the main motivation of this paper is to use dedicated test structures, RF characterizations and DOE modeling [4] to provide a relevant and efficient dummy metal filling strategy inside, and underneath various inductors integrated in an Advanced Copper BEOL, in respect with the digital density rules and without degrading the RF device performances.

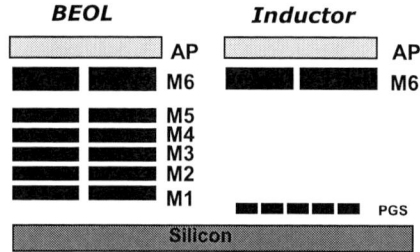

Fig. 1. BEOL and inductor architecture descriptions

II. INDUCTOR DESCRIPTION

A. Manufacturing Process

Octagonal symmetric inductors with a Patterned Ground Shield (PGS) were used in this study. This set of inductors was fabricated in an industrial 0.13 µm node CMOS technology with a Double Damascene Copper Back End on a conventional 10 Ω. cm silicon wafer. The technology consists in six metallization levels, an aluminum capping layer AP, dielectric layers between all conductive layers with a dielectric constant k equal to 4, and a top passivation layer, Fig. 1.

B. Reference Inductors

In order to provide the most accurate dummy metal filling strategy for most RF applications, in comparison with earlier work [5]-[7], a large spectrum of integrated inductors in terms of geometrical and electrical parameters was studied, see in Table 1.

Inductor1 is made of a unique coil built at the upper two metal levels (M5-M6) stacked with the aluminum capping layer, see in Fig. 1. It is characterized by a high Q-factor and a low inductance L_s value, as shown in Table 1.

Inductor2 is made of two turns and wide conductors. Its coils are built using the last metal level (M6) stacked with the aluminum capping layer, see in Fig. 1, and present a localized underpass with the stack M4-M5. This device is characterized by a high Q-factor and a medium inductance value L_s, as shown in Table 1.

Inductor3 and Inductor4 are multi-turn inductors with narrow conductors. Their coils are built using the upper three metal levels (M4-M5-M6) without the aluminum capping layer, and present a continuous underpass with the stack M4-M5, and an upperpass M6, see in Fig. 1. This device targets medium inductance L_s value and compact size, as shown in Table 1.

TABLE I

ELECTRICAL AND GEOMETRICAL PARAMETERS OF THE REFERENCE TEST STRUCTURES

	Inductor1	Inductor2	Inductor3	Inductor4
Q_{max}	14.0 @ 13.25GHz	13.5 @ 4.5 GHz	5.0 @ 2.3 GHz	5.7 @ 1.1 GHz
L_s [nH] @ 100 MHz	0.7	1.1	9.3	18.3
Cut-off F_c [GHz]	30.50	9.75	4.10	1.90
Serial R_s [Ω] @ 100 MHz	1.9	1.0	17.5	13.5
Radius [μm]	110	62.68	15	35
Coil width/spacing [μm]	5	34/20	2/0.6	5/0.6
Number of turns	1	2	12	11

C. Inductors with Dummy Fill Cells

The objective of this work is to define a way to manage fill metals inside and under this set of inductors without degrading RF performances, and being compliant with digital design rules. In order to evaluate the effect of dummy fills on the device electrical characteristics, squared fill metals were additionally inserted separately in the centre, and underneath the inductor coils, as explained in Fig. 2.

The variable parameters are the density D, the distances d1 (inside inductor), d2 (outside inductor), and the dummy configuration: aligned A or crossed C dummy fills, as shown in Fig. 2 and Table 2. A larger dummy fill density D range from 25% up to 80% and a larger metal fill-coil distance d1 (inside inductor) and d2 (outside inductor) range were considered in two separated studies, compared to [5]-[7]. The used dummy fill widths are 0.46 μm, 1.21 μm and 3.9 μm for a fixed 0.46 μm spacing value. The dummy configurations: aligned A or crossed C dummy fills, never studied in literature, was investigated too.

Only 120 samples (all inductors included) were chosen to perform this novel evaluation, using the DOE methodology with a D-Optimal matrix and employing D, d1, d2 and the type of dummy fill stacks (A or C) as input factors in Statgraphics software, see in Table 2.

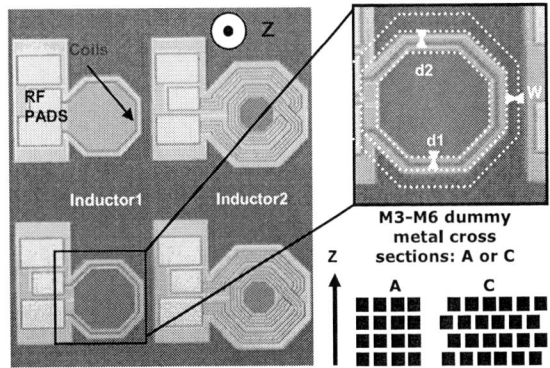

Fig. 2. Used RF pads, Inductor1 (on the top: D=25% inside, d1=1 μm, A; at the bottom: D=80% inside, d1=1 μm, A) and Inductor2 (on the top: D=52.5%, inside, d1=1 μm, A; at the bottom: D=52.5%, d1=5.5 μm, A) images showing the considered DOE input factors

TABLE. II

DOE DESCRIPTION WITH INPUT FACTOR RANGES.

	DOE 1 : IN THE CENTER OF THE COILS	DOE 2 : UNDER THE COILS
Inductor types	All	All
Density range	D=[25%-80%]	D=[25%-80%]
Metals	[M3-M6]	Inductor1 : [M3-M4] Inductor2: [M3-M5] Inductor3 and Inductor4: [M3]
Distances	d1=[1 μm-10 μm]	Inductor1 :d1=d2=[2 μm-12 μm] Others: d1=d2=[0 μm-10 μm]
Orientation	Aligned A or Crossed C	Aligned A or Crossed C (except for Inductor3 and Inductor4)

III. RF MEASUREMENTS AND INDUCTOR PARAMETER EXTRACTION

A. Measurements

S-parameter measurements up to 50 GHz were performed on the 120 structures using an Agilent HP8510C VNA and Cascade Microtech Infinity GSG RF probes on the achieved inductors including the reference structures and those where fill metals were inserted. De-embedding was performed using a dedicated open structure to remove pad contribution.

In order to determine the influence of dummy fills on the inductor RF performances, three main parameters have been extracted: the quality factor Q, the coil inductance L_s and the coil resistance R_s values as given in [6]. Cut off frequency F_c has been extracted too when Q=0.

No significant variation has been observed on DC resistance R_s value, as illustrated in Fig. 3 and inductance L_s value, see in Fig.4. Consequently, only results on Q-factor are presented hereafter.

B. Dummy Metal fills in the Center of the Coils

The main objective of this subsection is to evaluate the impact of inserted dummy fill cells in the center of the coils. Main results are given in [8]. Concerning high Q inductors (Inductor1 and Inductor2), aggressive design rules (D=80%, d1=1 μm) are very critical and can impact Q_{max} value up to 30% because of an increase of parasitic capacitance [6]. Regarding Inductor 3 and 4, whatever the used design rules (aggressive or relaxed ones), the impact on Q factor is low with a reduction on Q_{max} value less than 5%.

C. Dummy Metal fills Underneath the Coils

In this subsection, the impact of inserted dummy fill cells underneath the coils on RF inductor performances is evaluated. Concerning Inductor1, if aggressive design rules are used (D=80%, d1=d2=12 μm, C), the impact on Q factor is high, see in Fig. 3, with a reduction on Q_{max} value of 26% and a decrease on the Q peak frequency of 18% because of a parasitic capacitance increase [6]-[7]. In the opposite, if relaxed design rules are applied (D=25%, d1=d2=7 μm, A), the impact is less than 5% on Q_{max} value.

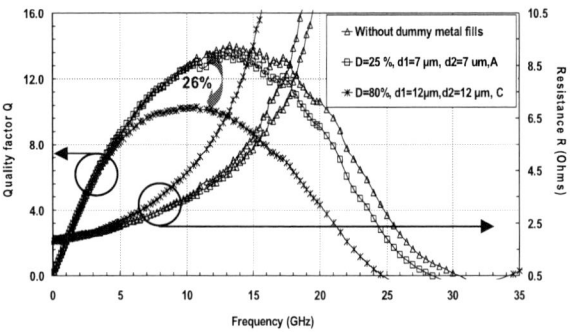

Fig. 3. Measured quality Q-factor of Inductor1 with and without dummy cells underneath the coils.

Concerning Inductor2, if aggressive design rules are used (D=80%, d1=d2=10 μm, C), the impact on Q factor is high, see in Fig. 4, with a reduction on Q_{max} value of 12% and a decrease on the Q peak frequency of 15% because of a parasitic capacitance increase [6]-[7]. In the opposite, if relaxed design rules are applied (D=25%, d1=d2=10 μm, A), the impact is less than 5% on Q_{max} value.

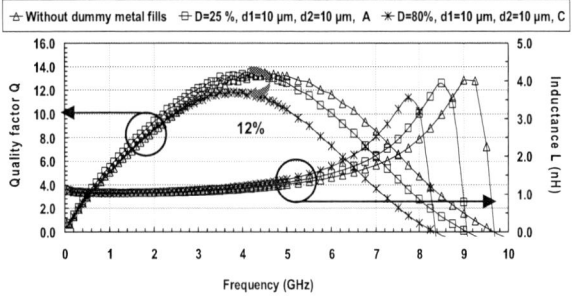

Fig. 4. Measured quality Q-factor of Inductor2 with and without dummy.

Regarding Inductor3, if aggressive design rules are used (D=80%, d1=d2=10 μm), the impact on Q factor is moderated, in Fig. 5, with a reduction on Q_{max} value of 7.3 % and a decrease on the Q peak frequency of 8.6% because of a parasitic capacitance increase [6]-[7]. However, if relaxed design rules are applied (D=25%, d1=d2=10 μm), the impact is less than 2% on Q_{max} value.

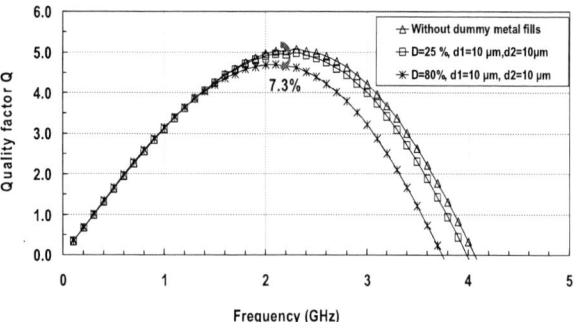

Fig. 5. Measured quality Q-factor of Inductor3 with and without dummy cells underneath the coils.

Concerning Inductor4, if aggressive design rules are used (D=80%, d1=d2=10 μm), the impact on Q factor is moderated, see in Fig. 6, with a reduction on Q_{max} value of 6.3 % and a decrease on the Q peak frequency of 18.1% because of a parasitic capacitance increase [6]-[7]. However, if relaxed design rules are applied (D=25%, d1=d2=10 μm), the impact is less than 2% on Q_{max} value.

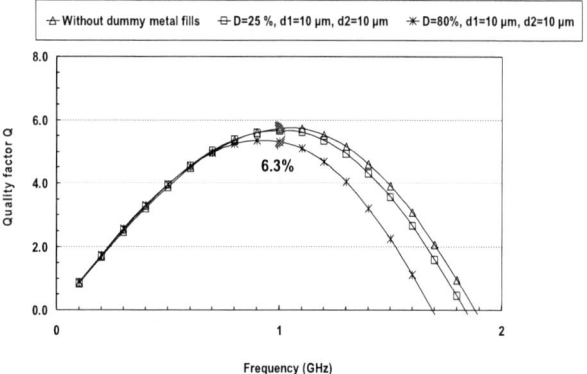

Fig. 6. Measured quality Q-factor of Inductor4 with and without dummy cells underneath the coils.

IV. DOE MODELING

As it has been demonstrated in the previous subsections, the sensitivity of inductor to dummy metal filling is structure dependant. Actually, high Q inductors are more impacted by metal fills whereas moderate Q inductors seem to be less sensitive because of the high turn-to-turn parasitic capacitance. That is why, only the analysis results obtained for Inductor1 using Statgraphics software are presented hereafter.

978-2-8748-7007-1/08 $25.00 © 2008 EuMA

A. Dummy Metal fills in the Center of the Coils

The standardized pareto chart for Q_{max} with an adjusted R-squared equal to 99%, indicates that the first significant parameter among the considered input factors is the dummy fill density D [8]. However, the distance d1 and its combination with the density also have a strong influence on the Q-factor. The bigger the distance is, the lower the impact on Q_{max} value is. Moreover, the DOE analysis reveals that the stack configuration (aligned A or crossed C fill metals) is not a critical parameter [8].

B. Dummy Metal fills Underneath the Coils

The standardized pareto chart for Q_{max} represented in Fig. 7, with an adjusted R-squared equal to 98%, indicates that the first significant parameter among the considered input factors is the dummy fill density D. However, the distances d1 and d2 and their combination with the density D also have a strong influence on the Q-factor. Moreover, the DOE analysis reveals that the stack configuration (aligned A or crossed C fill metals) is not a critical parameter.

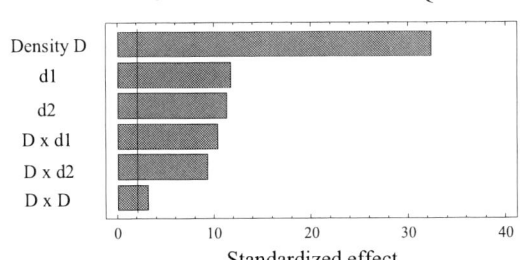

Fig. 7. Pareto diagram of Inductor1 with adjusted R^2=98%.

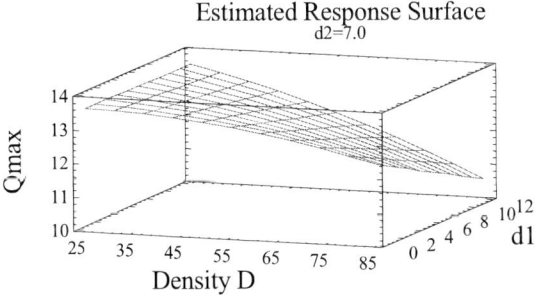

Fig. 8. Estimated response surface graph for Inductor1.

For a low density (D=25%) the distances d1 and d2 have no effect on the Q factor, Fig. 8. For a high density (D=80%), the best performances are observed for the lowest distances (d1=2 μm for d2 fixed at 7μm), as it is shown in the estimated response surface graph, see in Fig. 8. This illustrates the importance of fringe capacitance for high fill metal density.

V. DISCUSSION

Thanks to the DOE modelling, an upper limit for density filling has been first determined (~25%) in a minimal number of experiments and fulfil metal density requirements without decreasing the electrical performances. The density D and the (d1, d2) distances are firstly recognized as the most significant parameters in terms of RF performance inductor degradation whereas the stack configuration (A or C) is not a critical parameter.

VI. ACKNOWLEDGEMENT

The authors would like to thank the TILT team of STMicroelectronics for its valuable help and support during the realization of test structures.

VII. CONCLUSION

For the first time, a complete dummy filling strategy based on dedicated test structures, DOE analysis and RF measurements using STMicroelectronics 0.13 μm bulk CMOS technology has been proposed to evaluate the impact on RF performances of a large spectrum of integrated inductors. The DOE modeling confirms that it is possible to introduce dummy cells (up to 25%) in the center, and underneath the coils controlling the distance between the inserted fill metals and the coils, without degrading the electrical RF performances. Thanks to this new approach, it seems now possible to reduce the number of design rules dedicated to inductors, by introducing dummy fills inside and under the inductors which would enable inductor layout to fulfill digital density rules without any impact on the electrical inductor performances.

REFERENCES

[1] S Kordic, "Fixing Today's Issues with Copper Electroplating and Copper CMP", SEMICON Europa, pp. 50-59, 2000.

[2] T. Park. And al., "Pattern and Process Dependencies in Copper Damascene Chemical Mechanical polishing Processes, VMIC, pp.437-442, 1998.

[3] A. B. Kahng, G. Robins, A. Singh, and A. Zeliikovsky, "Filling Algorithms and Analyses for Layout Density Control", IEEE Trans. On Computer-Aided Design of Integrated Circuits and Systems, vol. 18, no. 4, pp. 445-462, 1999.

[4] A. Dean and D. Voss, "Design and Analysis of Experiments", Springer-Verlag New York, Inc., 1999.

[5] L. F. Tiemeijer, R. J. Havens, H.J. Pranger, "Physics-Based Wideband Predictive Compact Model for Inductors with High Amounts of Dummy metal fills", IEEE Transaction on Microwave Theory and Technique, Vol. 54, No. 8, pp. 3378-3386, 2006.

[6] Lan Nan, K. Mouthaan, Y. Z. Xiong, J. Shi, S. C. Rustagi, B. L. Ooi, "Experimental Characterization of the Effect of Metal Dummy Fills on Spiral Inductors", IEEE RFIC Symposium, pp. 307-310, 2007.

[7] X. Sun, G. Carchon, Y. Kita, K. Chiba., T. Tani, and W. De Raedt "Experimental Analysis of Above IC Inductor Performances with Different Patterned Ground Shield Configuration and Dummy Metals", 36th European Microwave Conference Proceeding, pp. 40-43, 2006.

[8] C. Pastore, F. Gianesello, D. Gloria, E. Serret, Ph. Benech, "Test Structure Definition for Dummy Metal Filling Strategy Dedicated to Advanced Integrated RF Inductors", IEEE ICMTS Proceeding, pp. 214-219, 2008.

Miniaturized Multilayer CPW pHEMT Amplifiers

Q. Sun[#1], V. T. Vo[*1], J. Tan[#2], R. A. Davies[*2], A. A. Rezazadeh[#3]

[#]*The Electromagnetics Research Centre, School of Electrical and Electronics, The University of Manchester,*
Sackville Street, P.O. Box 88, Manchester, M60 1QD, United Kingdom
q.sun@postgrad.manchester.ac.uk

[*]*Filtronic Compound Semiconductors Ltd,*
Heighington Lane Business Park, Newton Aycliffe, County Durham, DL5 6JW, United Kingdom

Abstract— **Compact pHEMT amplifiers, which are composed of newly developed miniaturized multilayer inductors and capacitors, have been designed, fabricated and characterised. Their measured performances are presented and compared with those of amplifiers composed of conventional planar components. The results show that multilayer technology reduces the size of the amplifiers by approximately 50% while maintaining the same performance. In this paper we demonstrate that the developed compact components are integrated with pre-fabricated GaAs pHEMTs to form 3D MMICs providing excellent performance and space saving resulting in low cost manufacturing of future MMICs.**

I. INTRODUCTION

Lumped inductors and capacitors are key fundamental components in MMIC applications. They are widely used in filters and matching networks for active components such as pHEMTs and HBTs. The main disadvantage of conventional planar microstrip lumped inductors and capacitors is their comparatively large circuit area especially for the case where high inductance or capacitance is required. For example, in conventional amplifier designs, inductors normally take 80% circuit area. Such disadvantage directly results in increased circuit fabrication cost. To reduce circuit area, stacked inductors [1-3] and MIM capacitors using very thin isolator [4, 5] were demonstrated. However the stacked inductors reported in [1] achieved just 50% area saved and the resonant frequency is not high enough due to the high parasitic capacitance brought out by microstrip substrate. Although MIM capacitors with high capacitance were reported in [4], the very thin dielectric film with the typical thickness from 20nm to 100nm introduces more complexities and cost during fabrication process, and capacitance is very sensitive to the dielectric thickness. Moreover since in microstrip designs grounding has to be realized by making via holes through thin substrate, unexpected parasitic parameters cannot be neglected, which also increase the difficulty during the design stage and the total circuit cost.

In order to utilize the chip area in more efficient manner, we presented 3-metal-2-dielectric sandwitched multilayer inductors and capacitors [6, 7]. The stacked inductors are composed of one planar spiral stacked on top of the other. The two spirals are separated by a polyimide layer and joined at the centre using a via-hole through the polyimide layer. MIMIM folded capacitors are developed from normal planar MIM capacitors. The performance of both planar and 3-D inductors and capacitors were analysed and compared. The results show that the overlayed stacked spiral inductors have been shrunken to 1/4 of the size of the conventional planar designs and the multilayer folded capacitors save as much as a half of the MIM capacitors while maintaining similar performance.

By applying the above compact components developed in The University of Manchester into the matching networks of amplifiers, we successfully designed and fabricated planar and multilayer CPW amplifiers using GaAs pHEMTs supplied by Filtronic. Both types of amplifiers were characterised by on-wafer RF measurements from 45MHz to 10GHz. The measured results show good agreements with the simulated ones using small-signal model while a half circuit area is saved using the multilayer CPW technology. The effect of DC probes in on-wafer RF measurement is also analyzed to provide practical suggestions for amplifier designers.

II. DESIGN AND OPTIMIZATION

A. Multilayer CPW overlayed inductors

Fig.1 shows the cross-sectional views and micrographs of both planar and multilayer stacked inductors. In directly overlayed inductors, two identical spirals are stacked and connected using a via-hole through the polyimide.

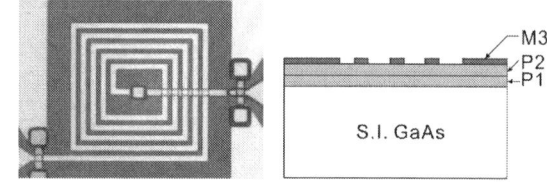

(a) Planar CPW inductor (280⎡m x 280⎡m)

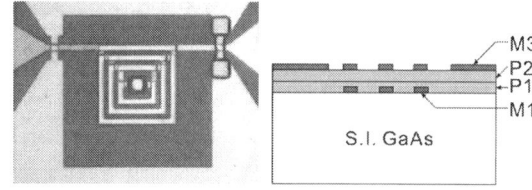

(b) Multilayer CPW inductor (170μm x 170μm)

Fig. 1 The micrographs and cross-sectional views of two types of inductors

978-2-8748-7007-1/08 $25.00 © 2008 EuMA

Fig.2 presents the area of the conventional planar and multilayer inductors as the function of inductance for comparison. It is apparent that for a given inductance value a significant area reduction can be achieved using multilayer technology. The directly overlayed stacked spiral inductors take up only a quarter of the space of the conventional planar inductors while maintaining similar inductor performance.

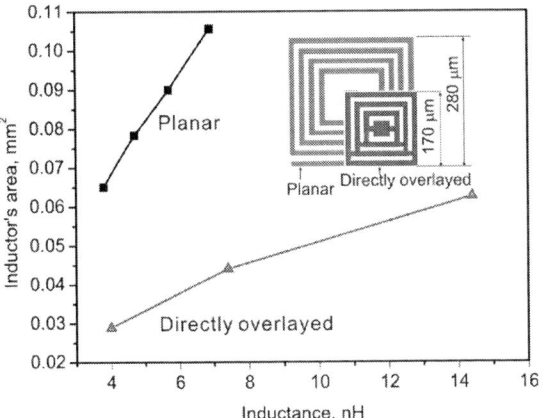

Fig. 2 The inductor area as the function of extracted inductances deduced from measured S parameters of directly overlayed and planar CPW inductors

B. Multilayer CPW folded capacitors

Fig.3. shows capacitance extracted from measured S parameters as the function of the area of capacitors. As illustrated in the cross-sectional view of folded capacitors, an electrode is composed by connecting M1 to M3 layer which results in double effective area. From Fig. 3 it is seen that for the same area almost 100% more capacitance was achieved by the folded configuration.

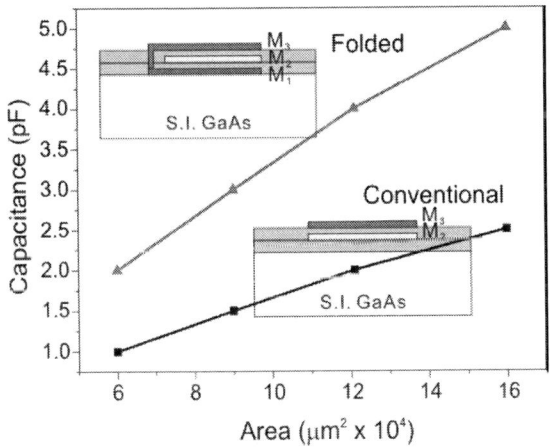

Fig.3 Extracted capacitance from measured S parameters of conventional MIM and multilayer folded CPW capacitors as the function of capacitor area

C. Compact layout in multilayer CPW design

When placing components together in circuit layout the separation between adjoining components should be as small as possible in order to minimize the MMIC chip area. This however could create parasitic coupling between the components and therefore degrade the performance of the whole circuit.

Fig. 4 shows layouts of two sets of adjoining 50Ω CPW transmission lines and 50Ω microstrip (MS) transmission lines. The separation (S) varies from 90μm to 150μm. Isolation (S21) between the two adjoining CPW or MS transmission lines is shown in the same figure. It can be clearly seen that CPW lines need only a 90μm separation for -30 dB isolation at 10 GHz, while the MS lines require at least 150 μm for the same isolation. This clearly demonstrates the great advantage of employing CPW layout design over the MS in terms of compactness and thus MMIC space saving.

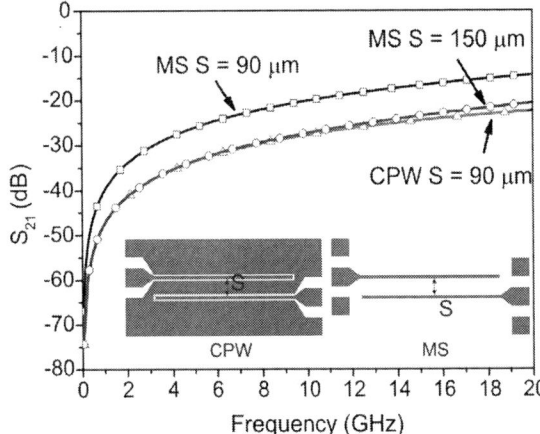

Fig. 4 Comparison of isolation characteristics between the CPW and MS designs

D. Design of compact 3D MMIC amplifiers

Fig. 5 shows a 2GHz amplifier circuit with a stabilizing shunt resistor and simultaneous conjugate matching networks, which are realized with miniaturised multilayer inductors and capacitors. The matching networks are designed to match impedance and determine frequency response simultaneously as introduced in [8, 9]. The parameters of inductor and capacitor models which are presented in the dashed frames in Fig.5 were extracted from measured S parameters.

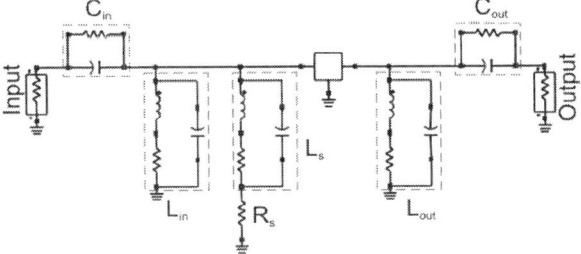

Fig. 5 Amplifier under simultaneous conjugately matched condition realised with miniaturised inductors and capacitors

Three 2GHz circuit layouts were designed using microstrip, planar CPW and 3D CPW technology respectively as shown in Fig. 6. Multilayer directly overlayed inductors and folded capacitors are employed in the 3D CPW layout. In planar CPW and microstrip designs, polyimide with thickness of 2.5 µm is applied as the insulator dielectric for MIM capacitors. In the microstrip design, grounding is realized by via holes through the thin substrate, which introduces extra fabrication cost. It is clearly seen that the 3D multilayer circuit area is almost half the size of the conventional planar CPW demonstrating the compactness of the 3D technology. The reasons for the advantages of 3D components over the planar and MS are as follows:

• 3D directly overlayed inductors (double spirals) consume a fourth size of planar ones.
• 3D capacitors (folded) consume a half size of planar ones.
• Separation of 90µm is needed for the -30dB isolation in CPW circuits, while 150µm for the same isolation in the MS circuits.

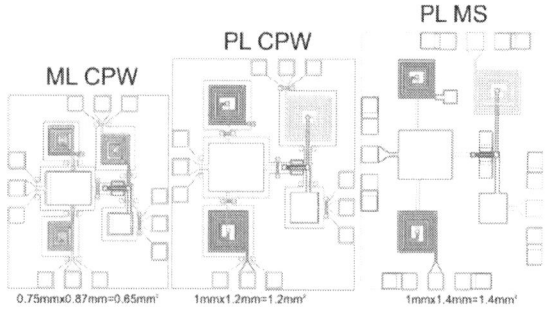

Fig. 6 Three layout designs from the same amplifier circuit shown in Fig. 5: Multilayer CPW (ML CPW), Planar CPW (PL CPW), and Planar Microstrip (PL MS) designs. Note that MS layout was design for comparison purpose only and was not fabricated.

III. RESULTS AND DISCUSSION

The components and circuits in this work have been fabricated using three layers of metals and two layers of sandwich dielectrics. In these multilayer structures, different metal layers need properly interconnecting through the etched windows of the polyimide insulating layers. The thickness of Au layers (M1, M2 and M3) is about 0.8 µm. The isolating polyimide layers between metal layers is 2.5 µm thick, and the semi-insulating GaAs substrate is about 600 µm.

The polyimide used in this work has a dielectric constant of about 3.7. The polyimide interconnection windows were formed by oxygen plasma reactive ion etching (RIE) through a photoresist protecting layer patterned using the photolithography process.

Fabricated 3D and conventional 2GHz amplifiers shown in Fig.7 have been characterized using on-wafer probing from 45 MHz to 10 GHz. It was found that carefully arranging the DC probes, which were used to feed Vg and Vd to the amplifiers, is necessary in order to accurately measure the performance of these amplifiers. As Vg and Vd are fed through shunt inductors, which are parts of matching networks, bypass capacitors are required to provide AC ground for the matching

networks and to avoid oscillation. Thus off-chip capacitors are required for measurements, which however are not available for on-wafer measurements. To solve this problem we successfully managed to make DC probes for Vg and Vd which have 1nF capacitors mounted to the probes. The capacitors are located about 6mm from the tip of the probes.

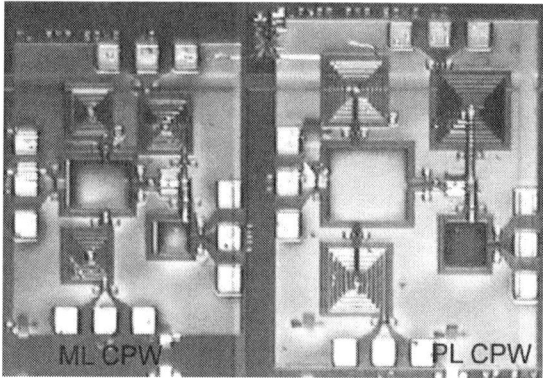

Fig. 7 Micrograph of fabricated 3D and planar CPW amplifiers

Fig. 8 shows the measured and simulated S parameters of the designed multilayer and planar CPW amplifiers. There are two distinctive features observed between the measured and the simulated data of Fig. 8. First the measured S21 at 2GHz are about 2-3dB smaller than the simulated result and the second observation is that the measured data shifted to lower frequency as compared to the simulated result.

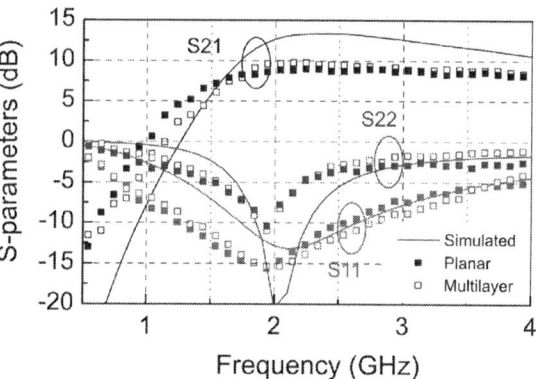

Fig. 8 Measured and simulated S parameters of designed multilayer and planar CPW amplifiers

The discrepancies came from the effect of extra inductance of the DC probe. A close examination of the DC probe suggested that the lead inductance of the DC probes have a significant effect on the performance of the measured circuits. In order to show the effect of the DC probes we added extra 6nH inductors to the gate and drain feeding shunt inductors in the amplifier circuit in order to model the inductance of the probe leads which are of 6mm long. The performances of the amplifiers then are simulated again. After taking into account the effect of the probe inductance we have re-plotted the S21 in Fig.9 for both the CPW planar and CPW 3D multilayer amplifiers together with the measured one. From this figure it is presented that both amplifiers have similar measured performances. It is also noted that the new simulation results are very close to the measured data especially considering the effect of the DC probe inductance.

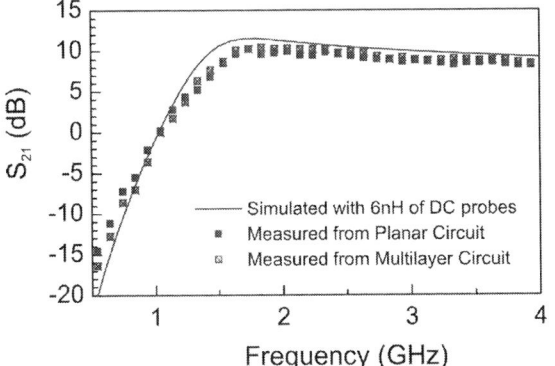

Fig.9 Measured and simulated S21 for the two amplifiers (CPW planar and CPW 3D multilayer). The effect of DC probe inductance have been taken in to account.

IV. CONCLUSIONS

This paper has demonstrated the integration of compact multilayer CPW inductors and capacitors developed by The University of Manchester with the prefabricated GaAs pHEMTs from Filtronic. The 3D CPW amplifier achieved miniaturized circuit size as small as half of conventional planar designs due to the advantages of 3D CPW technology while maintaining similar performance. Such experience on multilayer circuit designs is helpful for MMIC designers to achieve compact circuit size in order to reduce the fabrication cost. The measured results show good agreements with the simulated results by taking into account the effect of the extra inductance of DC probes.

ACKNOWLEDGMENT

This work was supported by the Electro-Magnetic Remote Sensing Defense Technology Centre, established by the UK Ministry of Defence and run by a consortium of SELEX Sensors and Airborne Systems Ltd, Thales Defence, Roke Manor Research and Filtronic.

The authors would like also to thank Mr. Keith Williams for the help and advices in circuit design, RF and MM-wave simulations and measurements.

REFERENCES

[1] M. W. Geen, G. J. Green, R. G. Arnold, J. A. Jenkins, and R. H. Jansen, "Miniature multilayer spiral inductors for GaAs MMICs," in 11th Gallium Arsenide Integrated Circuit (GaAs IC) Symposium Technical Digest, 1989, pp. 303-306.

[2] I. J. Bahl, "High-performance inductors," IEEE Transactions on Microwave Theory and Techniques, vol. 49, pp. 654-664, 2001.

[3] W. Y. Yin, S. J. Pan, L. W. Li, and Y. B. A.-G. Gan, Y.B., "Modelling on-chip circular double-spiral stacked inductors for RFICs," IEE Proceedings - Microwaves, Antennas and Propagation, vol. 150, pp. 463-469, 2003.

[4] J.-H. Lee, D.-H. Kim, Y.-S. Park, M.-K. Sohn, and K.-S. Seo, "DC and RF characteristics of advanced MIM capacitors for MMIC's using ultra-thin remote-PECVD Si3N4 dielectric layers," IEEE Microwave and Guided Wave Letters, vol. 9, pp. 345-347, 1999.

[5] X. Z. Xiong and V. F. Fusco, "A comparison study of EM and physical equivalent circuit modeling for MIM CMOS capacitors," Microwave and Optical Technology Letters, vol. 34, pp. 177-181, 2002.

[6] L. Krishnamurthy, V. T. Vo, and A. A. Rezazadeh, "Thermal Characterization of Compact Inductors and Capacitors for 3D MMICs," in 36th European Microwave Conference, Manchester, UK, 2006, pp. 52-55.

[7] V. T. Vo, L. Krishnamurthy, Q. Sun, and A. A. Rezazadeh, "Miniature CPW Inductors for 3-D MMICs," in IEEE MTT-S International Microwave Symposium Digest, 2006, pp. 1377-1380.

[8] D. M. Pozar, Microwave Engineering, 3rd ed: John Wiley & Sons Inc, 2004.

[9] E. d. Silva, High Frequency and Microwave Engineering: Elsevier Health Sciences, 2001.

A Miniaturized Wafer-Scale Package Demonstrated with Three Enhancement Mode Amplifiers

Khanhtran Phan [1], Julie Kessler [1], Henrik Morkner [1], Mike Vice [1], Lan Nguyen [1], Jim Roland [2]

[1] *Wireless Semiconductor Division, Avago Technologies*
350 West Trimble Road, San Jose, CA 95131, USA

[2] *Wireless Semiconductor Division, Avago Technologies*
4380 S. Ziegler Road, Fort Collins, CO 80525, USA

Abstract— This paper presents a 1mm x 0.5mm x 0.25mm miniaturized package fabricated using internal E-PHEMT 6" wafers and industry-first Wafer-Scale Packaging technology. To demonstrate this technology, a set of three distinct amplifiers were developed. The 1-12GHz 50-Ohm gain block operates with a 5V, 55mA supply and achieves 14 dB gain, 3.7 dB NF, 28dBm OIP3 and 17dBm OP1dB. The 0.5-6GHz Bypass LNA requires a 5V, 24mA supply and has 14.5dB gain, 2dB NF in the gain state. The low-voltage LNA operates with a 1.8V, 20mA and produces 14.3dB gain, 2.2dB NF and -3dBm IP-1dB. These three SMT compatible products are unique and industry-leading in all aspects of performance, packaging and cost.

I. INTRODUCTION

As the popularity of mobile devices grows stronger, the quest to miniaturize system components continues to intensify. To stay competitive and on the winning edge of today's mobile market, device manufacturers face the difficult task of bundling many applications into their products while having to reduce the physical size and cost. Now more than ever, today's mobile devices pack more features that require not only the sophistication of system software, but also the complexity of the hardware backbone. Although some applications can have their designs integrated in the same IC, others must be fully self-contained in their own packages. As a result, it is evident that the fascinating growth of mobile devices has to be enabled by the ongoing development of two parallel approaches: the high level, multi-function circuit integration on one side; and the breakthroughs in packaging technology on the other. When these two approaches are combined, they enable the manufacture of incredibly small form-factor products that offer amazing functional complexity.

This paper describes an innovative miniaturized Wafer Scale Package (WSP); shown in Figure 1, that provides a breakthrough in packaging technology and 3 amplifiers that are housed in this package. The package is extremely small, substantially low cost, yet delivers exceptional performance. It effectively reduces PCB area by almost 50% compared to industry leading plastic package alternatives such as CEL's 6-pin lead-less MiniMold (1511 PKG) or JRC's USB6-A8. The 3 amplifiers are unquestionably suitable for many commercial wireless applications in the 0.5-12GHz range but will also prove their worth in other markets such as wireline, fiber-optic, base station, CATV, instrumentation etc. As packaged, these ready-to-use, SMT compatible amplifiers measure a mere

Fig. 1. 0402 Wafer Scale Packages (shown top and bottom view)

1mm x 0.5mm x 0.25mm, the size of the industry standard 0402 components. The three amplifiers and the WSP are fabricated using internal PHEMT technology.

II. PROCESS DESCRIPTION

Avago Technologies E-PHEMT process uses a dual recessed 0.25µm optical gate. The device has a f_T of 45 GHz, and f_{max} of 65 GHz at Vds=2V, Vgs=0.7V. DC parameters for the device are shown in Table 1. The wafer material is selected for high power and breakdown with some compromises for noise. A Ti/Pt/Au T-gate is used for low input resistance and high-reliability. The process is designed to operate with a DC drain voltage up to 5.5V. All steps are defined using stepper lithography on 6-inch wafers.

The MMIC process is equipped with passive components which include a 213Ω/ bulk resistor, 0.39 fF/μm^2 Si_3N_4 MIM capacitor, 27pH backside via and two metal layers for transmission lines.

TABLE 1.

TYPICAL DC PARAMETERS FOR E-PHEMT

Parameter	Mean
Gm (mS/mm)	580
Vgs @ peak Gm (V)	0.7
Ids @ peak Gm (mA/mm)	171
Imax (mA/mm) @Vgs=Vto	331
BVgd @ 1mA/mm (V)	-17
Vto @1mA/mm (V)	0.97
Vth @1mA/mm (V)	0.25

III. WAFER SCALE PACKAGE

The Avago Technologies Wafer Scale Package is a new and exciting technology that can carry SMT packaging to potentially 100 GHz yet add minimal cost. A Wafer Scale Package is composed of a pair of bonded GaAs wafers, in which all I/Os are routed through via-holes to the backside of the device wafer. The combination of the gasket and the cap wafer provides an air cavity, structural support and protection for the device wafer. A cross section of the wafer scale package is shown in Figure 2. When the wafer is sawn between the gaskets, it becomes thousands of individually packaged parts. The base wafer is a standard processed GaAs wafer. The backside vias and I/O pads serve as the lead-frame of the wafer scale package. The backside metal is thick enough to ensure adequate coverage in the bottom of the vias without inhibiting the use of standard solder paste in assembly. The pad size and separation are adequate to ensure that no special package footprint concessions need to be made by customers and their board houses. Lead free solder paste is the recommended adhesive, as it will help the parts to self-align on the board.

The package is designed with a 0402 (1mmx0.5mm) compatible dimension and full surface mount capability. Because all I/Os are routed to the backside of the device wafer through via-holes, the RF transitions suffer almost no signal loss and minimal parasitics. As a result, the wafer scale package performance could potentially reach into the 100GHz and beyond. This is a significant improvement over conventional plastic packages where bond-wires exhibit substantial parasitics that limit the operating frequency. Since the designed wafer scale package employs a 0402 foot print, it is fully compatible with modern, low cost, high volume assembly and test. Figure 3 shows 6" capped GaAs wafer and its die sawing process as well as the bottom view of the 0402 package.

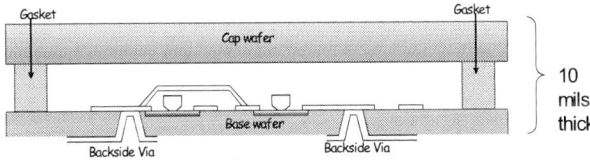

Fig. 2.Cross-section of the Wafer Scale Package

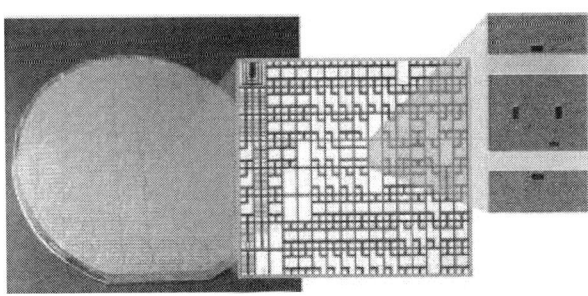

Fig. 3.Wafer Scale Packaging process illustration

IV. AMPLIFIERS' DESIGNS

A. 1-12GHz 50-Ohm Gain Block

The gain block design is based on the conventional Darlington configuration [3] with a few modifications, shown in Fig. 4. The most significant addition is the introduction of the series-tuned inductor in the form of a microstrip line at the source of the input FET [2], [4] to help improve input match and NF at high frequency. The biasing of the amplifier does not use the conventional method of feedback resistor and external voltage-drop resistor in the power supply. Instead, it is done through the use of a resistive voltage divider directly from the power supply. A bias and temperature stabilizing circuitry was not integrated, thus, the performance is only measured for Vdd=5V.

The FETs used in this design have a turn-on voltage of around 0.4V, which allows for the use of resistive bias for the second FET. Fig. 5 shows the microphotograph of the gain block without the cap wafer. The dimension of the die is 1mm x 0.5mm, but the die size could greatly be reduced by almost 50% if not for the pre-determined package size of 1mm x 0.5mm, as evidenced by the empty space around the core.

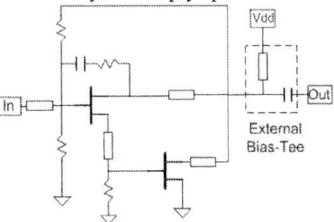

Fig. 4. Simplified schematic of the Gain Block

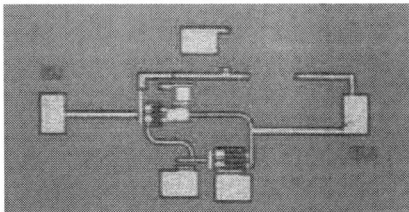

Fig. 5. Microphotograph of the uncapped Gain Block

B. 0.5-6GHz Bypass LNA

The core of the Bypass LNA is a cascode amplifier in Avago's low noise enhancement mode PHEMT technology. The bias circuit is fitted with a power down feature which is accessed from the input port by way of a large value external resistor. In addition, a TEE pad style bypass switch is connected between input and output ports. The switch represents a feedback path when the amplifier is on, hence the shunt FET in the TEE topology is used to enhance isolation and prevent unwanted feedback which could cause instability. Careful choice of the sizes of the switch FETs allows good port match to be achieved in the bypass state. The switch is controlled by an integrated driver which works in conjunction with the power-down circuit so that the switch closes in power-down mode and opens in the gain state. A simplified representation of the arrangement is given in Fig. 6.

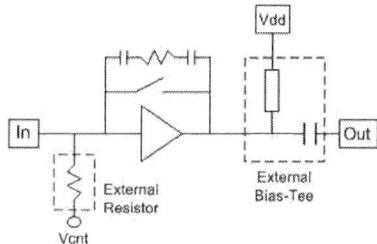

Fig. 6. Simplified schematic of the Bypass LNA

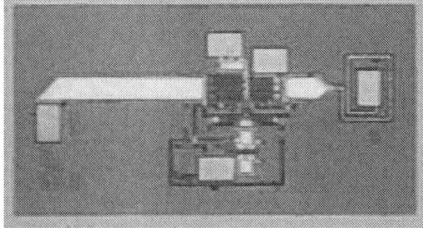

Fig. 7. Microphotograph of the uncapped Bypass LNA

C. 1-6GHz Low-voltage LNA

For low voltage operations that are CMOS compatible with 1.8V or 3V battery supply, Avago offers another 1-6 GHz LNA optimized for this voltage requirement and is shown in Fig. 8. The LNA is a 1-stage amplifier with on-chip matching networks at input and output. At 1.8V, the LNA typically draws 20mA current. The bias circuit integrates a shut-down feature which can be accessed from the input port through a large value external resistor. No matching coefficients are required for impedance matching to 50 Ω systems.

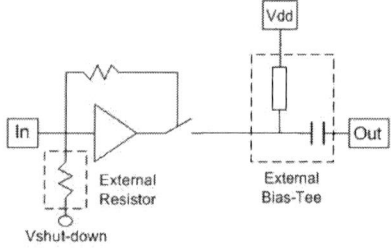

Fig. 8. Simplified schematic of the low-voltage LNA

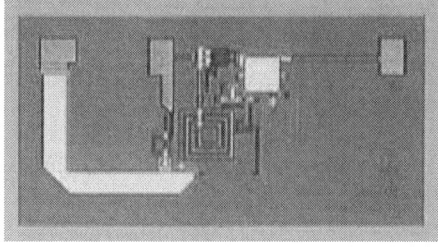

Fig. 9. Microphotograph of the uncapped low-voltage LNA

V. MEASUREMENT RESULTS

The packaged amplifiers measurements were made using an evaluation board as shown in Fig. 10. The demonstration board uses 50-ohm coplanar launches into both sides of the package, and industry standard SMA connectors are used. Measurements are made on Agilent 8510 Network Analyzer, 8975 Noise Figure Meter and 8565 Spectrum Analyzer.

Fig. 10. Evaluation board for the three amplifiers

The 1-12GHz Gian Block's bias conditions in these measurements are 5V and 60mA. S-parameter results are shown in Fig. 11. The gain block shows more than 12dB of gain across the specified band, where most of the lower half shows typical gain of 15dB. Input and output return losses are more than 10dB at most of the frequencies in the band. Reverse isolation is approximately -20dB.

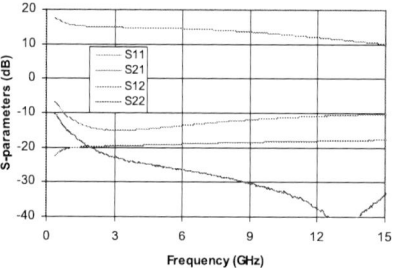

Figure 11. 1-12GHz Gain Block's S-parameters

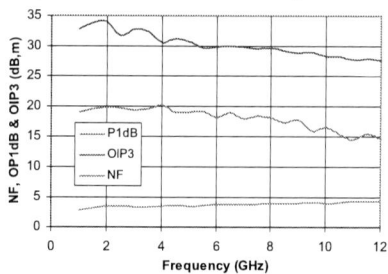

Figure 12. 1-12GHz Gain Block's NF, OP1dB & OIP3

Fig. 12 shows measured Noise Figure into 50-ohm load. The typical Noise Figure performance of about 2.4dB shows an improvement of about 1dB over typical wide-band resistive feedback Darlington amplifiers [1]. The output P1dB and output IP3 are also shown in Fig. 10. Below 6GHz, the gain block is capable of providing over 30dBm of OIP3. Above 6GHz, the OIP3 averages 28dBm. The output P1dB of more than 15dBm is enough to drive most passive mixers.

The 0.5 – 6GHz Bypass LNA is biased from a 5 V supply and typically draws a current of 24mA. In the gain state the typical gain is 14.5 dB, with 2 dB noise figure and 0 dBm input P1dB. The input and output return losses are more than 10dB across the band. Output IP3 is approximately 23.5dBm when in gain state. Fig. 13 shows the S-paramters and Fig. 14 shows NF and OIP3 in the gain state.

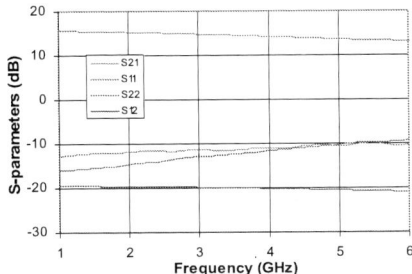

Figure 13. 0.5-6GHz Bypass LNA's S-parameters in gain state

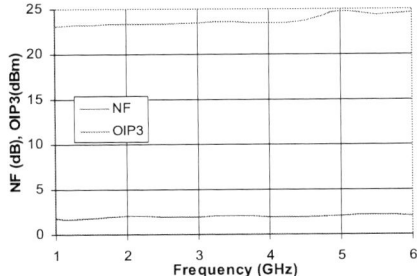

Fig. 14. 0.5-6GHz Bypass LNA's NF and OIP3

In the bypass state the loss is 5 dB and the compression point increases to +12 dBm. The S-parameters in the bypass state is shown in Fig. 15.

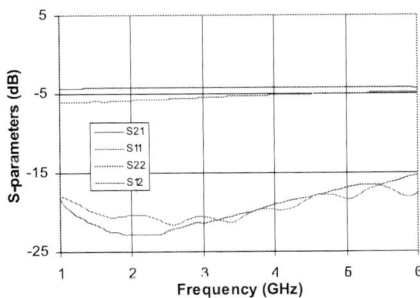

Fig. 15. 0.5-6GHz Bypass LNA's S-parameters in bypass state

The low-voltage LNA is measured with a Vdd=1.8V with a corresponding current of 20mA. At 2.4GHz, its typical gain is 14.3 dB with noise figure of 2.2dB, input P1dB of -3dBm and OIP3 of 22.5dBm. As Vdd increases towards 3V, the low frequency gain improves slightly but the high frequency gain degrades about the same amount. The OIP3, however, has peak performance around Vdd=2.5V, although the improvement is only 1-2dB compared to Vdd-1.8V. Fig. 16 shows the low-voltage LNA's gain, NF and OIP3 for Vdd=1.8V and Idd=20mA.

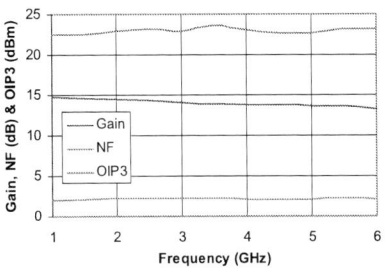

Fig. 16. 1-6GHz Low-voltage LNA's Gain, NF & OIP3

VI. CONCLUSION

This paper describes a new and exciting Wafer Scale Package and three enhancement mode amplifiers that use this industry-first packaging technology. The package uses state-of-the-art technology and innovative fabrication process to minimize signal loss, potentially pushing the limit to the 100GHz range. It is panel-processed through assembly and test to allow total automation.

The amplifiers provide high gain, high IP3, low NF and integrated 50-Ω input and output matching networks to simplify system design process. Having broadband performance, high power, linearity and low noise figure, these amplifiers can be used in many positions in any radio architecture. Additionally, these amplifiers are housed in a 3-I/O miniaturized package with the size of 1mm x 0.5mm, which is the industry standard for 0402 components. This SMT compatible package does not require any special tooling other than traditional mass production pick-and-place machines. These amplifiers have the unique combination of bandwidth, performance, miniaturized package and low cost. Research products such as these are true enablers of pushing the mobile device market to more impressive milestones in the future.

ACKNOWLEDGMENT

The authors wish to thank all the people with technical discussion and support that made these products possible. This includes, Hue Tran, Kohei Fujii.

REFERENCES

[1] K. Phan, H Morkner, "A High Performance Yet Easy to Use Low Noise Amplifier in SMT Package for 6 to 20GHz Low Cost Applications," *European Microwave Conference, 12th GaAs Symposium*, Amsterdam pp. 603-606, 2004.

[2] K. Phan, K. Fujii, H. Morker, "Two High Dynamic Range mmW Amplifiers in SMT Package with ESD Protection, " *37th European Microwave Conference*, Munich, pp. 1209-1212, 2007

[3] K. Kobayashi, "Improved Efficiency, IP3-Bandwidth and Robustness of a Microwave Darlington Amplifier using 0.5um ED PHEMT and a New Circuit Topology," IEEE CSIC Digest, pp. 93-96, 2005

[4] C. Armijo, R. Meyer, "A New Wideband Darlington Amplifier," IEEE Journal of Solid State Circuits, Vol. 24, No. 4, Aug. 1989, pp. 1105-1109

[5] H. Morkner, "Miniature 3V GaAs LNA, VGA, and PA for Low Cost .5-4 Ghz Wireless Applications", *1997 Wireless Symposium*, Feb. 1997,pp.190- 19

Millimetre-wave Hot-Via interconnect-based GaAs chip-set for automotive RADAR and security sensors

PF. Alléaume[#1], C. Toussain[#,*], C. Auvinet[#], D. Domnesque[#], P. Quentin[#], M. Camiade[#]

[#]*United Monolithic Semiconductors, route départementale 128 – BP46, 91401 Orsay Cedex, France*
[1]pierre-franck.alleaume@ums-gaas.com
[*]*XLIM - UMR CNRS n°6172 - 123, avenue Albert Thomas - 87060 Limoges CEDEX, France*

Abstract—New microwave front-ends are more and more integrated and operate commonly at very high frequencies above 40GHz. So the chip to sub-system interconnections at millimetre-wave frequencies play as key factor to address the main system requirements which are generally a low cost, a high reliability level with of course the best electrical performances. UMS has developed a 77GHz chip-set for automotive RADAR using a wire free connection called hot-via. This very broad band transition was optimised in W band in order to remain compatible with the existing chip designs and to be realistic in term of industrialisation, cost and assembly in heterogeneous microwave sub-systems.

I. INTRODUCTION

With the generalization of high data-rate wire-less telecom networks, and the rise of new microwave applications as automotive RADARs, the occupancy of the radio-frequency spectrum from 1GHz to 40GHz becomes critical. New frequency bands in a range from 50GHz to 120GHz were recently attributed in order to solve this situation. So, the challenge for the new generation of microwave front-ends is now to operate at very high frequencies and to be as cheap as possible to enter mass markets. Three main parameters must be optimized in parallel with approximately the same contribution on the final product cost. Firstly the semi-conductor process, secondly the chip assembly process, and finally the polyvalence of the chip-set to comply with several applications in order to increase the global production volume.

The development of a new generation of wire-free GaAs chip-set in W band for automotive RADAR application at 77GHz is reported in this paper. Wire-free connections also called hot-vias are used to interconnect the GaAs chips to the sub-system's mother board. The hot-via concept directly addresses the three cost reduction axis defined above. The cost of the hot-via process including the bumping stage is low and don't impact strongly the price of the MMIC. The transition is defined to simplify the assembly flow at the sub-system level, especially when a lot of connexions are considered, by removing the expensive wire bonding operation. Furthermore, the hot-via transition is a real answer to the device frequency band extension thanks to its very low serial parasitic inductance in comparison with a standard 1 mil wire bonding.

The concept of the hot-via transition proposed by UMS in the 90's [3] [7] has been optimized and modelled during the last two years. Thanks to the chip and board co-design a self matched hot-via transition has been developed and implemented on demonstrators.

A low cost assembly process based on a conductive adhesive epoxy solution has been also investigated [5]. Two demonstrators were developed in the frame of the MEDEA+ program called HiMission and orientated to address the automotive RADAR market in the frequency band 76GHz to 81GHz. This extended frequency band covers the long range (LLR) and short range (SRR) RADARs applications, and finally should lead on a very interesting cost decrease of the microwave front-end thanks to the reusability of the GaAs chip-set and bigger production volumes. A 77GHz hot-via transmitter chip designed with the UMS's 0.15µm PHEMT process (PH15) and a double channel sub-harmonic down-converter based on the UMS's Schotky diode technology (BES) have been developed and measured. The electrical performances of this chip-set assembled on heterogeneous modules are reported in this paper. More than +15dBm of output power at 77GHz have been measured for the transmitter module. About the receiver module, conversion losses better than 8dB have been also demonstrated.

II. THE HOT-VIA TRANSITION

The basic concept to design a broad band and high frequency chip-set is to create a very short electrical path from the chip's RF ports to the microwave board. Then, wire bonding must be replaced by metallic bumps.

Two options are conceivable: a standard flip-chip configuration where the chip is flipped down on the substrate or the hot-via solution, where the chip remains in the same position as for the standard microelectronic configuration using wire bonding (Figure 1). This second option has several advantages at the microwave frequencies in comparison with the flip-chip mode. Firstly, there is no risk of chip detuning since the active side of the chip where are located all the devices is not facing to the substrate. Then electrical models and the experience built with the previous generations of chips are directly re-usable. Secondly, a visual inspection of the MMIC active side after assembly is still possible. Thirdly, by design, the chip thermal management is simplified. Several bumps added on the chip back-side (chip ground pad) will act as efficient thermal drains. Then, the thermal resistance of the structure is very similar with the one of a standard solution where the chip is directly attached to the carrier with a conductive adhesive epoxy.

Only the electrical model of the in/out transition at the chip accesses is specific. Thanks to a co-design of the chip accesses and of the microwave mother-board, a self-matched (50Ω/50Ω) transition has been defined over a very broad frequency band from DC to 100GHz. This transition, thanks to its natural matching, can be directly implemented at the ports of an existing chip originally designed to be wire bonded. Only few minor modifications are necessary. The frequency band width of the new chip resulting from this evolution will be significantly enlarged.

Standard wire bonding configuration

Gold wire bonding (parasitic inductance)
Chip's bond pad
Active side of the GaAs chip (transistors, passive devices...)
GaAs chip
Transmission lines
Metallic chip's back-side (for micro-strip chips)
Long electrical path from the chip's bond pad to the carrier's transmission line
Microwave carrier

Flip-Chip configuration

Gold bump (low inductance)
Chip bond's pad
Metallic chip's back-side (for micro-strip chips)
GaAs chip
Microwave carrier
Transmission lines
Short electrical path from the chip's bond pad to the carrier's transmission line
Active side of the GaAs chip (transistors, passive devices...)

Hot-Via configuration with thermal drain

Active side of the GaAs chip (transistors, passive devices...)
GaAs chip
Gold signal bump (low inductance)
Hot-Via (connection through the chip)
Chip's bond pad
Transmission lines
Thermal and ground bumps
Microwave carrier
Metallic chip's back side (for micro-strip chips)
Short electrical path from the chip bond pad to the carrier's transmission line

Figure 1 : Micro-strip chips assembly options.

Electromagnetic 3D simulations have been done in order to define a self matched structure. Since the hot-via transition constituted by the via-hole and the bumps has a very low serial parasitic inductance in the RF path (10pH to 20pH), in comparison for example with a standard wire bonding of 1 mil, and 400µm long that is equivalent to ~0.4nH, it is obvious that it is necessary to consider also all parasitic capacitor which could affect the hot-via transition model at the first order. So, only a rigorous chip transition and substrate co-design can help to define the optimal transition.

Two options have been simulated and optimized. The first one is defined to be implemented onto a micro-strip carrier. In this case, micro-strip lines are used at the modules accesses. The same ground plane is used as reference for the micro-strip lines and for the hot-via transition. But a small ground discontinuity is managed at the interface of the hot-via transition and of the active core of the chip, because the chip is designed as a micro-strip circuit using its own back-side metal plane as ground reference (Figure 2 A/). The second option is defined to be used with a coplanar structure. Then a substrate including only one single metal layer is usable and the ground discontinuity is managed at the mother-board / MMIC back side interface through the ground bumps (Figure 2 B/). This second option is very attractive for many reasons. It is compatible with low cost single metal layer substrates, and it is less sensitive to the bumps height variations. Then several bumping techniques can be considered without effect on the electrical performances.

Several test cells have been fabricated in order to validate the 3D electromagnetic models and two types of technologies for the microwave substrate have been evaluated. First, thin-film NiCrAu coplanar alumina carriers have been fabricated. This material was selected for its very good mechanical compatibility with the GaAs chip. They are very similar in term of coefficient of thermal expansion (5.73ppm/°C for GaAs and 8.4ppm/°C for Al2O3), but, also for the very high electrical performances at millimetre-wave frequencies. In this study the ceramic solution was considered as reference. The second and much advanced option is based on a thin-film BCB process on silicon wafer developed by ACREO AB. It integrates three copper and two BCB layers (BCB thickness=14µm). This process is suitable for highly integrated microwave modules requiring also high performance passives off-chip

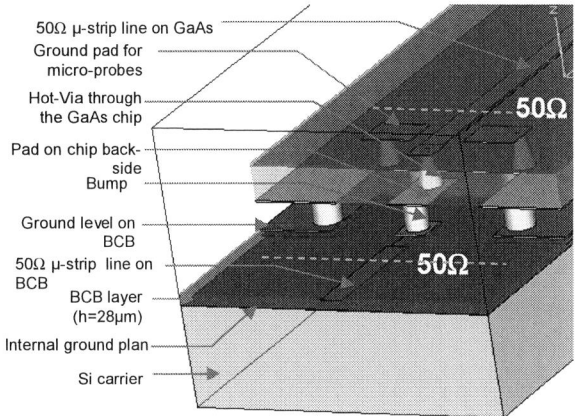

50Ω µ-strip line on GaAs
Ground pad for micro-probes
Hot-Via through the GaAs chip
Pad on chip back-side
Bump
Ground level on BCB
50Ω µ-strip line on BCB
BCB layer (h=28µm)
Internal ground plan
Si carrier

A/ 3D model of a hot-via chip co-designed with a micro-strip BCB/silicon carrier.

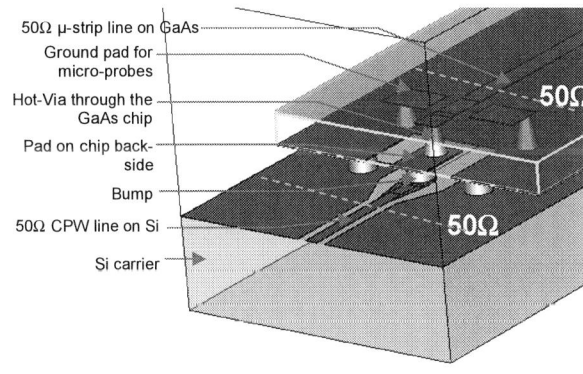

50Ω µ-strip line on GaAs
Ground pad for micro-probes
Hot-Via through the GaAs chip
Pad on chip back-side
Bump
50Ω CPW line on Si
Si carrier

B/ 3D model of a hot-via chip co-designed with a coplanar silicon carrier.

Figure 2 : 3D models of a hot-via GaAs chip attached onto a microwave carrier.

A 50Ω micro-strip line etched on GaAs chip has been selected as test cell to optimize in simulation the hot-via interconnections. The carrier foot-print, the patterning of the chip's back-side metal and the via-hole ending on the top side of the chip have been optimized to get a 50Ω/50Ω transition from DC to 100GHz. This test vehicle has been fabricated and assembled onto a coplanar ceramic carrier (Figure 3). The scattering parameters obtained in simulation and measure are compared on the Figure 4. A very good accordance is observed up to 100GHz.

978-2-8748-7007-1/08 $25.00 © 2008 EuMA

Figure 3 : Microphotograph of a CPW test module. GaAs chip bonded onto a ceramic carrier.

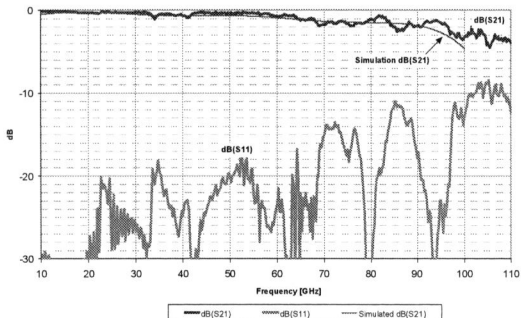

Figure 4 : EM simulations and measure scattering parameters comparison for the structure shown on the Figure 3

III. ASSEMBLY PROCESS

The hot-via chips are designed to comply with existing assembly processes. A stud-bumping process was selected to build the first stage of interconnection between the GaAs chip and the microwave carrier. This existing technology is compatible with high production volumes and assures a high repeatability at low cost. The bumping is performed at the wafer level before sawing in one step and does not require any modification of the standard front end process. Four inches GaAs wafers thinned to 100μm were stud-bumped with more than 50.000 bumps per wafer in less than 30 minutes. The tolerance on the bump's height is better than +/-3μm for a 45 μm bump. Stacked bumps are also available with a finish height of 80μm (Figure 5).

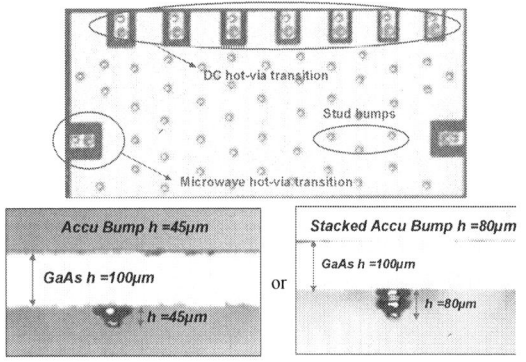

Figure 5 : Chip back-side views of a stud-bumped hot-via chip. Side views showing the gold stud-bumps shape. (AccuBumps™ developed by Kulicke&Soffa)

A large quantity of bumps are performed on the chip (about 50 bumps for a chip of 6mm²). The majority of the bumps are used as ground connections, thermal drain and mechanical absorbers to get a good compatibility between the GaAs chip and any substrate mismatched in term of thermal expansion coefficient as silicon. This configuration makes compatible

the die with several final assembly processes. Two processes which are thermo-compression and dipping have been applied for the demonstrators. For dipping the heads of the stud-bumps are dipped into a conductive adhesive (C/A) epoxy. After curing, the C/A ensures the mechanical contact between the head of the bumps and the pads on the carrier. This process is completed with an under-filling step. The complete process flow is presented on the Figure 6. It has been evaluated with the support of ACREO AB. A very good mechanical contact is obtained between the bumps and the metal pads thanks to the vertical force generated by the under-fill shrinkage after final curing. The contact resistance for a single bump has been measured from 200mΩ to 400mΩ.

In a near future, a new process compatible with the SMT techniques will also be investigated. The gold stud-bumps should be dipped into an appropriate solder past (Sn/Au) offering small metal particles (<35μm). After reflow, a Sn/Au alloy cap recovers the gold pillar bump, and is ready to be assembled on the sub-system mother-board like a BGA.

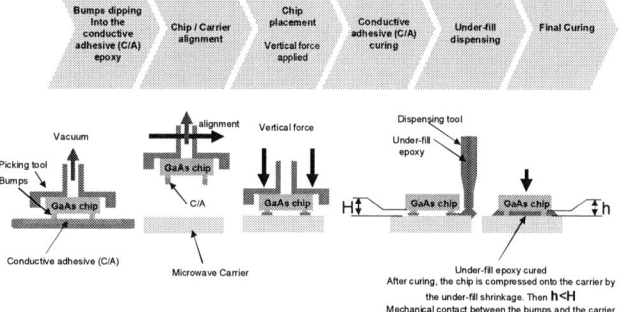

Figure 6 : Conductive adhesive epoxy assembly process flow.

IV. DEMONSTRATORS AND MEASUREMENT RESULTS.

In order to demonstrate the capability of the hot-via concept to achieve broad band performances at high frequency, a first tape-out has been launched. A hot-via transmitter (Figure 7) and receiver modules (Figure 9) designed for automotive RADAR applications, were assembled onto ceramic and BCB/silicon carriers.

A. Transmitter module.

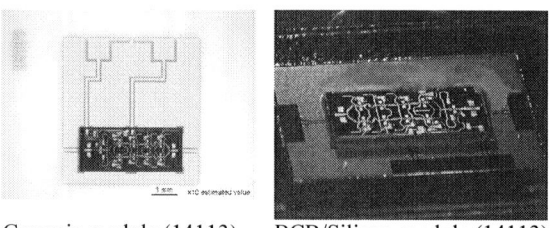

Ceramic module (14113)   BCB/Silicon module (14113)

Figure 7 : Microphotograph of a Hot-Via transmitter module.

The transmitter output power has been measured from 75GHz to 82GHz. More than 15dBm output power is obtained at 77GHz at the nominal bias conditions (V+=+4.5V, V-=-4.5V, I+=166mA) and +16dBm for V+=+5V, V-=-5V, I+=184mA.

Despite of a frequency shift of about 4% observed between the chip measurements and the design goals, this first demonstrator confirms the benefit of the hot-via transition. No difference is observed on the measured output power between a hot-via chip assembled onto the module and the same chip

978-2-8748-7007-1/08 $25.00 © 2008 EuMA

where the hot-via connections are removed and replaced by ground-signal-ground test pads for probe station measurements (Figure 8). The frequency shift is only due to the intrinsic chip design and doesn't come from the hot-via accesses. Furthermore, this result confirms that the hot-via transition complies to the frequency band extension requested by several applications since no low pass filtering effect is observed between the bare die and the hot-via modules as it would be observed with a standard wire bonding technique. A second tape-out will correct the frequency shift in order to cover the long and short range automotive RADAR frequency bands at the same time.

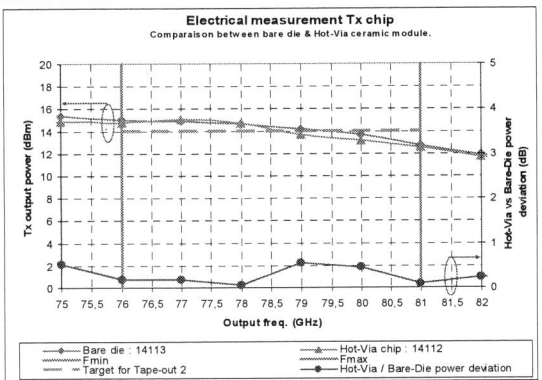

Figure 8 : Measurement results of the Hot-Via transmitter module.

B. Receiver module.

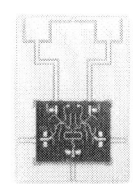 

Ceramic module: Top view   Ceramic module: Side view

Figure 9 : Microphotograph of a Hot-Via receiver module.

The characterisation of the receiver module concluded on 8dB of conversion losses at 77GHz after assembly on a ceramic carrier (Figure 9). As for the transmitter, the effect of the hot-via transition on the microwave performances is negligible. Less than 0.6dB of conversion losses deviation between the module and the equivalent bare die without hot-via accesses, were measured (Figure 10).

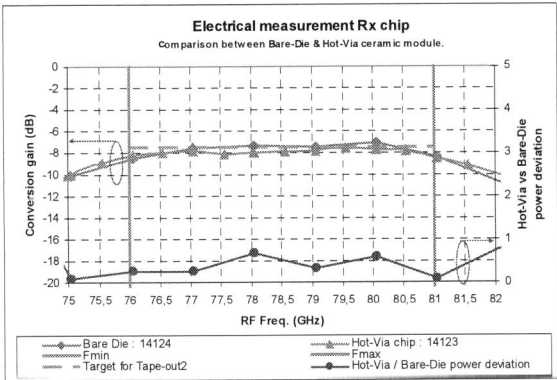

Figure 10 : Measurement results of the Hot-Via receiver module.

V. CONCLUSIONS

A broad band GaAs chip-set for automotive RADAR application at 77GHz has been developed to satisfy with a wire-free assembly process. The full automotive band 76-77GHz (LRR) and 77-81GHz (SRR) is now covered with one single chip-set. The benefit of the hot-via connection is to achieve exactly the same performances at the module level than obtained with a simple bare die before assembly. For the fist time, it was not observed any frequency band degradation coming from the assembly process. Furthermore the hot-via configuration, fully compatible with MMIC micro-strip approach, helps also to solve the frequency detuning phenomenon and improve the thermal management of the active device in comparison with a flip-chip solution.

A low cost assembly process using gold stud bumps and conductive adhesive epoxy was investigated. This process is orientated to the consumer markets requiring low cost assembly solutions and was defined to be applicable on a large family of microwave substrates. It was used with success on a thin-film BCB / silicon substrate developed by ACREO AB. The modules fabricated with this new technology platform are defined to answer to the growing needs of high integration level and high performance passives.

A second-tape out is under development in order to extend the frequency band of the first version to take advantage of the natural very broad-band capability of the hot-via transition. Finally a full SMT assembly process will also be investigated.

ACKNOWLEDGMENT

This work has been done in the frame of the HiMission MEDEA+ project. The assembly process based on a conductive adhesive epoxy has been developed conjointly with ACREO AB (Norrköping, Sweden) who is also developing the thin film BCB process on silicon. The electromagnetic simulations were supported by XLIM (Limoges, France).

I would like also to thank Dr. Hoppermann (Fraunhofer IZM, Berlin, Germany) for very fruitful discussions.

VII. REFERENCES

[1] "GHz Flip Chip – An Overview." Katarina Boustedt. Ericsson Microwave Systems AB.

[2] "40 GHz Hot-Via Flip-Chip Interconnects". F.J. Schmückle. FBH Berlin.

[3] Patent "Procede d'interconnexion entre un circuit integre et un support, et circuit integre adapte a ce procede", July 1990, Patent no. French patent number: 2 665 574, U.S. patent number US: 5 158 911, Pierre Quentin, Thomson-CSF

[4] P. Monfraix, T. Adam, J.L.Lacoste, C. Drevon, G. Naudy, B. Cogo, J.L. Cazaux, J.J.Roux, "Design to reliability shielded vertical interconnection applied to microwave chip scale packaging",30th *European Microwave Conference.*,vol. 1, pp.409-412, 2000.

[5] Anisotropic Conductive Adhesives for Millimeterwave Flipchip Interconnections. Johann Heyen*, and Arne F. Jacob. Institut für Hochfrequenztechnik, Technische Universität Braunschweig, Germany. 12th GAAS Symposium- Amsterdam, 2004

[6] 60GHz Broadband MS-to-CPW Hot-Via Flip Chip Interconnects. Wu, W.-C. Hsu, L.-H. Chang, E.Y. Kärnfelt, C. Zirath, H. Starski, J.P. Wu, Y.-C. Microwave and Wireless Components Letters, IEEE, Nov. 2007.

[7] Design data for hot-via interconnects in chip scale packaged MMICs up to 110 GHz.
A. Bessemoulin, Microwave Conference, 2004. 34th European Volume 1, Issue 11-15 Oct. 2004 Page(s): 97 – 100.

978-2-8748-7007-1/08 $25.00 © 2008 EuMA

Proceedings of the 3rd European Microwave Integrated Circuits Conference

Process Stabilization and Sensitivity Analyses of a Single Recess GaAs pHEMT Process using Device Simulations

Peter Abele[#1], Michael Schäfer[*2], Jörg Splettstößer[#3], Martin Thinnes[#4], Hermann Stieglauer[#5], Dag Behammer[#6]

[#] *United Monolithic Semiconductors GmbH,*
Wilhelm-Runge-Strasse 11, D-89081 Ulm, Germany
[1]abele@ums-ulm.de, [3]splettstoesser@ums-ulm.de, [4]thinnes@ums-ulm.de,
[5]stieglauer@ums-ulm.de, [6]behammer@ums-ulm.de

[*]*CADwalk, Engineering Office Heinrich Walk,*
Stegäckerstraße 7, D-89604 Allmendingen, Germany
[2]m.schaefer@cadwalk.de

Abstract— **In this work we investigate device simulations for a sensitivity analyses on the PH25 single recess pHEMT process. The relation of the most critical process and epitaxial parameters on the electrical DC parameters are presented and discussed. The control of the recess etching is an important process module in stabilizing the electrical parameters. Improving the recess etching resulted in a significant reduced spread of the electrical parameters.**

I. INTRODUCTION

Sensitivity analyses are used to estimate how much the output variability depends on each of the input variables. Device simulations represent a good mean in performing sensibility analyses for process control. These simulation tools describe the device based on physical equations solving Poison's equation, carrier continuity equation and transport equations. The geometry of the device is also considered. This results in an accurate link between process parameters and electrical parameters. This work shows the influence of some critical process parameters of a single recess pHEMT process (PH25) on the electrical device characteristics and the improvement of the spread of the electrical parameters by optimizing critical process steps.

II. SINGLE RECESS TECHNOLOGY

The epitaxial layers are grown by MBE with an InGaAs channel and delta doping in the barrier. Source and drain contacts are defined by a lift off process. After recess formation by wet chemical etching using AlAs as stop layer for a well defined distance between Schottky contact and channel a dielectric assisted Al-T-gate is defined by lift off. The gate foot is written by e-beam. The transistor is finally passivated using a nitride. The gate length for these transistors is Lg=220nm. Fig. 1 shows a schematic cross section of the device.

III. DEVICE SIMULATION

The simulations were performed with ATLAS from Silvaco. In a first step the simulation tool must be calibrated to the process. Therefore lots of measurements have to be performed

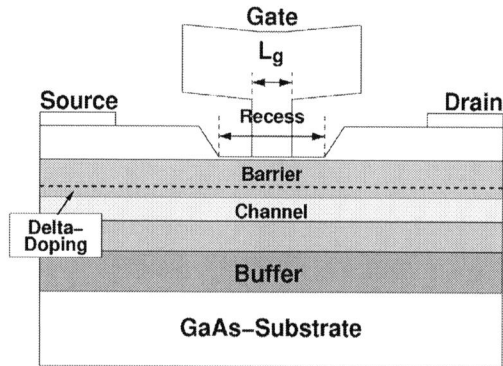

Fig. 1 Schematic cross section of single recess pHEMT transistor (PH25)

on the device and compared to device simulation results. Several parameters like permeability or band diagram of the different materials are described in literature [1]. Often these parameters have to be modified to get good agreement between simulation and measurement results. Others can be obtained from the manufacturer of the epitaxial layers like background doping and some, like surface states, are process dependent.

The energy balance model was used for the transport equations to get a good description of the Gm-curve. For the Schottky contact the UST-model was used [2] [3]. This model gives a good representation of the Schottky-contact for forward and reverse basing. This is important for the modeling of the leakage currents and the breakdown voltage. Fig. 2 shows a comparison between device simulation results and measurements of the leakage currents.

One critical simulation parameter with respect to the gate leakage current and the breakdown voltage is the resolution of the mesh especially at the drain side of the gate foot. This is important to consider when doing the variations with respect to the geometry parameters of the device.

To ease work a program was written that automatically generates the device structure for the simulation by entering the geometry parameters including the epitaxial layer stack.

978-2-8748-7007-1/08 $25.00 © 2008 EuMA

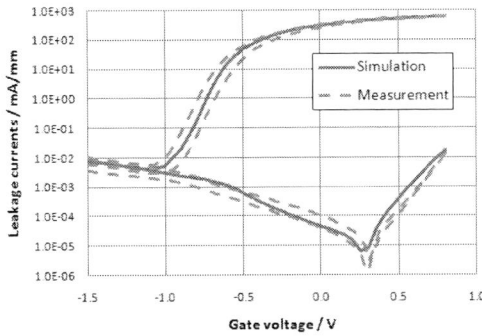

Fig. 2 Simulation and measurement of the gate and drain leakage current versus gate voltage

During the calibration of the device simulator it turned out that different parameter sets resulted in similar results. To get more confidence in the device simulation results for all simulated variations three different parameters set were used. For example the gate leakage current is mainly influenced by the parameters of the Schottky contact between gate and semiconductor and the surface states along the recess. Changing the density of the surface states can be compensated by the parameters of the Schottky contact.

IV. SENSITIVITY ANALYSES

After the simulation parameters were calibrated to the measurements of the pHEMT transistors a sensitivity analyses was performed. The variation of the following input variables for the sensitivity analyses are discussed in this work:

- gate length,
- recess size,
- barrier thickness and
- delta doping.

The gate length and the recess size are determined by the process parameters. The delta doping is fixed during the growth of the epitaxial stack. The barrier thickness is influenced by both, the epitaxy and the process if the etching of the recess does not stop at the stop layer.

Beside these parameters also the background doping and the doping of the cap–layer were varied, but not discussed here. The surface states are fixed for the variations.

The output variability of the four most important electrical parameters on the above mentioned input variables will be shown and discussed:

- Idss,
- Vg100,
- Vbds and
- Gm.

Idss is the drain current at Vgs=0V and Vds=2.5V. Vg100 is the gate voltage when the drain current is reduced to Idss/100. Vbds is the transistor breakdown voltage and Gm is the transconductance derived from DC measurements.

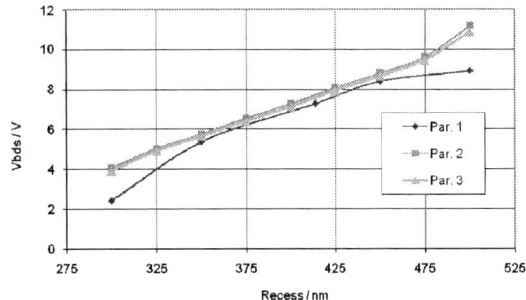

Fig. 3 Simulation of the breakdown voltage Vbds dependent on the size of the recess using three different parameters sets

Fig. 3 and Fig. 4 show device simulation results of the breakdown voltage Vbds and the drain current Idss dependent on the size of the recess. The size of the recess is controlled by process parameters.

The three different parameter sets result in a similar dependency of the breakdown voltage on the size of the recess. They quantitatively show the increase of the breakdown voltage on the size of the recess. In the range from 350nm to 450nm all three parameter sets are in good agreement.

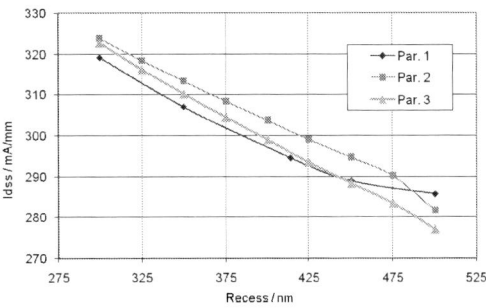

Fig. 4 Simulation of the drain current Idss dependent on the size of the recess using three different parameter sets

For the drain current Idss a decrease with increasing recess size can be seen. A critical parameter in the growth of the epitaxial layers is the delta doping. A variation in this parameter shows its influence on the breakdown voltage Vbds and strong influence on the drain current Idss.

Fig. 5 shows the dependence of the breakdown voltage on the variation of the delta doping. The slope of all three parameter sets show a comparable slope of the break down voltage with respect to the delta doping.

While the dependence of the breakdown voltage on the delta doping is moderate the drain current Idss shows a strong dependence on the doping.

In Fig. 6 can be seen that all simulation parameters result in nearly the same dependence of the drain current on the delta doping. Assuming a variation of the delta doping of just ±5% results in a variation in the drain current Idss of 74mA.

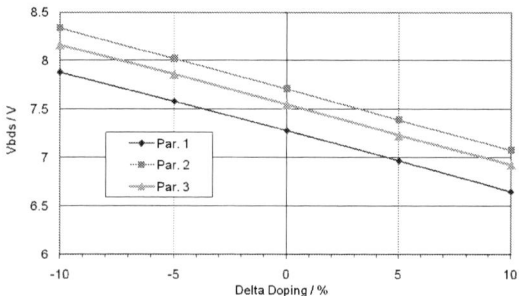

Fig. 5 Simulation of the breakdown voltage Vbds dependent on the delta doping for three different parameter sets

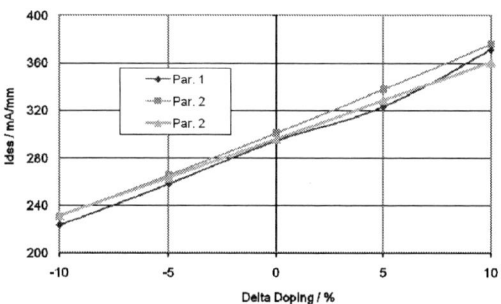

Fig. 6 Simulation of the drain current Idss dependent on the delta doping for three different parameter sets

In Table I the result of the sensitivity analyses is summarized. The number in parentheses below "Varied Parameters" name the assumed range of the different parameters.

TABLE I
SUMMARY OF THE RESULTS OF THE SENSITIVITY ANALYSES

Electrical Parameter	Varied Parameters			
	Gate length (+/-10%)	Recess length (+/-30nm)	Delta doping (+/-5%)	Barrier thickness (+/-5%)
Idss Slope	-0.55 mA/nm	-0.2 mA/nm	7.4 mA/%	4.96mA/%
Idss Variation	11mA	12mA	74mA	49mA
Vg100 Slope	1mV/nm	0.3mV/nm	-10 mV/%	-11mV/%
Vg100 Variation	0.02V	18mV	0.1V	0.11V
Vbds Slope	7.5mV/nm	0.0325 V/nm	63 mV/%	4.9mV/%
Vbds Variation	0.15V	1.95V	0.63V	49mV
Gm Slope	-0.325 mS/nm	0.17 mS/nm	3.28 mS/%	-1.7mS/%
Gm Variation	6.5mS	10.2mS	32.8mS	17mS

The most critical parameters, as a result from device simulation, are shaded. The breakdown voltage is mostly influenced by the recess length and the delta doping. The barrier thickness shows most influence on the drain current and Vg100. The delta doping has strong influence on Vbds, Idss, and Vg100.

V. OPTIMIZING THE RECESS MODULE

The variation of the delta doping is in the hand of the manufacturer of the epitaxy and can not be influenced by the process. The barrier thickness depends on both, the growth of the epitaxial layer and the process. If the etch stop layer does not withstand the recess etching, the Schottky contact of the gate will be closer to the channel. The recess length is only influenced by process parameters. As the etching of the recess has significant influence on the electrical characteristics of the device, this process module was subjected to several improvements. Also the removal of the nitride used for the dielectric assisted gate and the subsequent passivation were modified. As a result of these main modifications a remarkable stabilization of the process was achieved.

Fig. 7 shows the probability distribution of the drain current which is according to Table I mostly dependent on the delta doping level and the barrier thickness. The measurements were collected over one year.

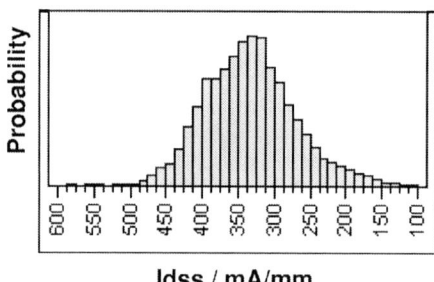

Fig. 7 Spread of Idss within one year before the process optimizations

The average value of the drain current is 345mA/mm with a standard deviation of 59mA/mm. After the process modifications were applied the spread of the drain current shows a drastic reduction. This can be seen in Fig. 8.

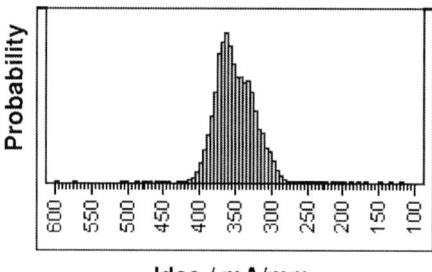

Fig. 8 Spread of Idss with improved process modules within one year after the process optimizations

The mean value of the drain current is 350mA/mm with a standard deviation of 29.5mA/mm, which corresponds to 8.4%.

The other parameter influenced by both, the barrier thickness and the delta doping, is Vg100. Fig. 9 shows the probability distribution before the process modifications.

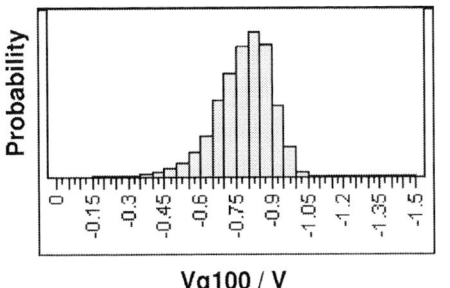

Fig. 9 Spread of Vg100 within one year before the process optimizations

The mean value is -0.78V with a standard deviation of 0.11V. Fig. 10 shows the improved probability distribution after the process optimization.

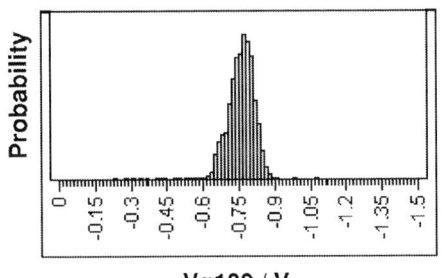

Fig. 10 Spread of Vg100 with improved process modules within one year after the process optimizations

The spread is reduced with an average value of -0.76V and a standard deviation of 0.052V. Besides Idss and Vg100 Table II also lists Vbds and Gm after process optimization.

From the device simulation results listed in Table I Vbds is strongly dependent on the size of the recess. The low standard deviation of this parameter indicates the good lateral control of the recess etching. The remaining standard deviation can mainly be attributed to the delta doping.

TABLE II
AVERAGE VALUE AND STANDARD DEVIATION OF DC PARAMETERS AFTER PROCESS IMPROVEMENTS

	Electrical Parameters			
	Idss	**Vg100**	**Vbds**	**Gm**
Unit	mA/nm	V	V	mS/mm
Average	350	-0.760	7.96	587
Std. Dev.	29.5 (8.4%)	0.052 (6.8%)	0.59 (7.4%)	18.3 (3.1%)

VI. SUMMARY

Device simulations were used for a sensitivity analyses on the PH25 single recess pHEMT process. To get more confidence in the simulation results three different sets of parameters were used showing comparable results.

The sensitivity analyses show that the recess module is from the process point of view a very critical step and the delta doping from the epitaxial.

After process optimization, especially the recess module, the remaining spread of the electrical parameters can be mainly attributed to the variation in the delta doping during the growth of the epitaxial layers.

ACKNOWLEDGMENT

The authors would like to thank Dr. Zandler from Silvaco (www.silvaco.com) for lots of discussions and the support during the device simulations.

REFERENCES

[1] V. Palankovski, "Analyses and Simulation of Heterostructure Devices", Springer-Verlag ISBN 3-211-40537-2

[2] Kazuya Matsuzawa, Ken Uchida, and Akira Nishiyama ,"A Unified Simulation of Schottky and Ohmic Contacts", IEEE Transactions on Electron Devices, Vol. 47, No. 1, January 2000.

[3] MeiKei Leong, Paul M. Solomon, S.E. Laux, Hon-Sum Philip Wing, and Dureseti Chidambarrao, "Comparison of Raised and Schottky Source/Drain MOSFETs Using a Novel Tunneling Contact Model" IEDM 1998

Proceedings of the 3rd European Microwave Integrated Circuits Conference

60 GHz Frequency Conversion 90 nm CMOS Circuits

Mikko Kantanen[#1], Jan Holmberg[#2], Timo Karttaavi[□3], Juha Volotinen[*4]

[#]*VTT Technical Research Centre of Finland*
Tietotie 3, FI-02150 Espoo, Finland
[1]mikko.kantanen@vtt.fi
[3]jan.holmberg@vtt.fi

[□]*Nokia Research Centre*
Itämerenkatu 11-13, FI-00180 Helsinki, Finland
[3]timo.karttaavi@nokia.com

[*]*AWR- Aplac*
Lars Sonckin kaari 16, FI-02600 Espoo, Finland
[4]juha.volotinen@aplac.com

Abstract— This paper presents design and a characterisation of an active single-stage single-ended 30 to 60 GHz frequency doubler and a resistive down conversion mixer with differential buffer stage. These MMICs are realised using 90-nm CMOS process. The doubler exhibit 7.1 dB conversion loss and 10.8 dB fundamental frequency suppression with 0 dBm input power and 13.7 mW power consumption. Maximum output power of -4.2 dBm is achieved with 5 dBm input power. The mixer has 9.8 dB conversion gain with +5 dBm local oscillator level. The compression point P_{1dB} is -2 dBm with 14 mW power consumption.

I. INTRODUCTION

During the past couple of years CMOS circuits operating in millimeter wave range have been under intense development. Main focus has been on circuits for short range communication at unlicensed band around 60 GHz.

Frequency conversion circuits are essential part of any radio transceiver. Although millimeter wave oscillator designs have been reported up to 410 GHz [1], in sophisticated transceiver it may be advisable to use lower frequency oscillator with frequency multipliers to obtain higher stability and lower noise. According to authors' knowledge, only one CMOS frequency doubler for 60 GHz has been presented so far [2].

II. TECHNOLOGY

The MMIC circuits were fabricated in STMicroelectronics 90-nm bulk CMOS technology. This process is aimed at digital applications. Seven copper metal layers were available with two topmost layers thicker than the others. Dedicated Metal-Insulator-Metal (MIM) capacitor processing was not used in this design since it is not available when striving for the best possible cost efficiency with a standard digital process option.

III. CIRCUIT DESIGNS

A. Doubler

The circuit schematics of an active single-stage single-ended doubler design is presented in Fig. 1. Reactive matching network consisting RF short circuited shunt stub and series transmission line at the input provides power match for the 30 GHz input signal. Output matching network has two purposes, to provide power match for 60 GHz output frequency while suppressing the 30 GHz fundamental frequency. Fundamental suppression is achieved with RF short circuited shunt stub which has very short length; only 0.02λ at 30 GHz. Series transmission line and open ended shunt stub provide the 60 GHz power match. RF shorts are realised using finger capacitors using stacked fingers on the five thin metal layers.

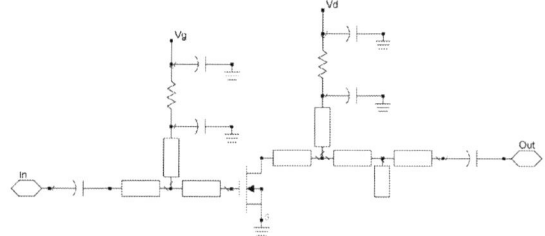

Fig. 1 Schematics of the doubler circuit.

NMOS device with 80 parallel 1 μm wide fingers was selected due to extensive millimetre-wave modelling work from previous processing runs [3]. Modelling measurements indicate a transition frequency f_t of 130 GHz and f_{max} of 180 GHz for the 80x1 μm device. For optimal second harmonic generation the transistor is biased close to pinch-off [4], [5]. For this transistor size this means $V_d = 1.0$ V and $V_g = 0.2$ V. BSIM3 model provided by the foundry with additional external parasitic components based on earlier

978-2-8748-7007-1/08 $25.00 © 2008 EuMA

measurements was used in non-linear simulations during the design phase.

Transmission lines are realised as a coplanar waveguides (CPWs). The signal and ground conductors consist of two topmost metals connected to each other using dense via arrays. The top metal is used to extend the ground plane over the entire circuit. The transmission line used in the design has 50.8 Ω characteristic impedance and a loss of 1.5 dB/mm at 60 GHz. Micrograph of the doubler circuit is shown in Fig. 2.

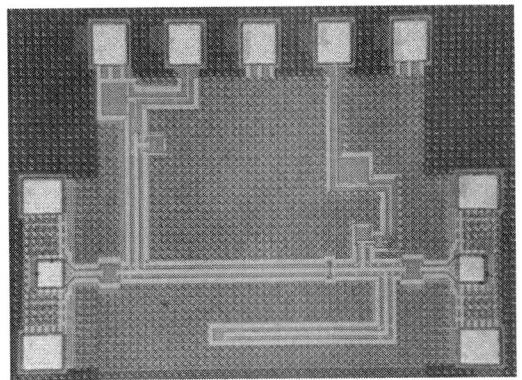

Fig. 2 Micrograph of the doubler circuit. Chip size is 1.0 mm x 0.7 mm.

B. Mixer

From the various topologies resistive mixers [6] [7] have recently become popular in microwave applications due to simpler structure and lower power consumption compared to e.g. Gilbert cell mixers [8]. A resistive downconversion mixer for the receiver was designed. The schematics of the implemented single balanced mixer is shown in Fig. 3. The circuit consists of a mixer core, a buffer amplifier and a Marchand balun in the local oscillator (LO) input port. The buffer amplifier provides differential baseband or low intermediate frequency (IF) signal. In order to exploit the earlier modelling work the NMOS devices with 80 parallel 1 μm wide fingers are used as the mixing transistors. Radio frequency (RF) signal is fed into the sources and balanced IF to the zero biased gates. Coplanar waveguides are used as transmission lines.

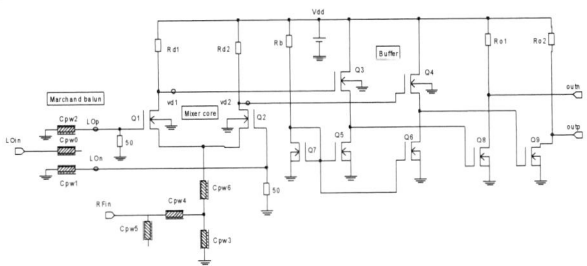

Fig. 3 Schematics of the mixer circuit.

A micrograph of the mixer is shown in Fig. 4. In this prototype circuit a passive balun was used in the local port. However, in the next run doubler and mixer blocks will be integrated together and the passive balun will be replaced by an active one resulting in significant space saving. The mixer with the buffer amplifier consumes roughly only 0.2 mm x 0.25 mm area before any layout optimization.

Fig. 4 Micrograph of the mixer circuit. Chip size is 1.1 mm x 0.7 mm.

IV. MEASUREMENT RESULTS

A. Doubler

Conversion loss and fundamental suppression of the doubler were measured on-wafer using Agilent E8257C analog signal generator and Agilent E4419B power meter with suitable power sensors. The input and output match were measured with Agilent N5250A PNA millimeter wave network analyzer. The measured conversion loss and fundamental suppression as a function of input power with comparison to simulated values are presented in Fig. 5.

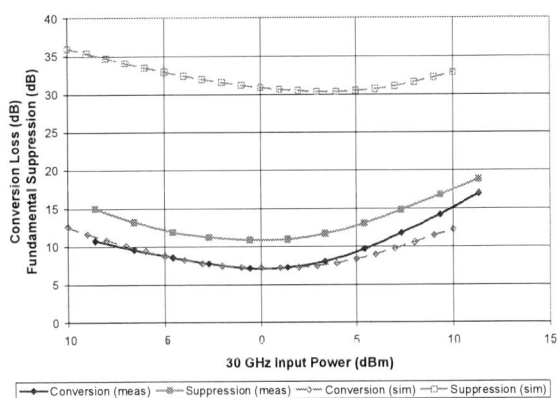

Fig. 5 Measured (solid lines) and simulated (dashed lines) conversion loss and fundamental suppression.

Measurements show the conversion loss of 7.1 dB and fundamental suppression of 10.8 dB with 0 dBm input power. The measured power consumption is 13.7 mW. The comparison to simulation show excellent agreement for conversion loss, but the measured fundamental suppression is almost 20 dBs lower than simulated. This is probably due to very short stub used to suppress the fundamental frequency. Very short stubs are prone to modelling errors.

The measured output power at 60 and 30 GHz as a function of input power are shown in Fig. 6. Maximum output power at 60 GHz is -4.2 dBm.

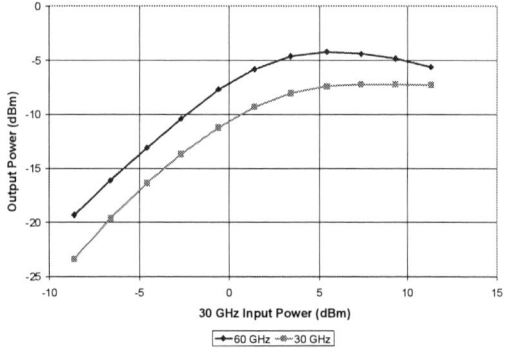

Fig. 6 Measured output power at 60 and 30 GHz as a function of input power.

Frequency response of the doubler is presented in Fig. 7. The conversion loss is 7-8 dB over 57-64 GHz range. The frequency response was not optimised during circuit design.

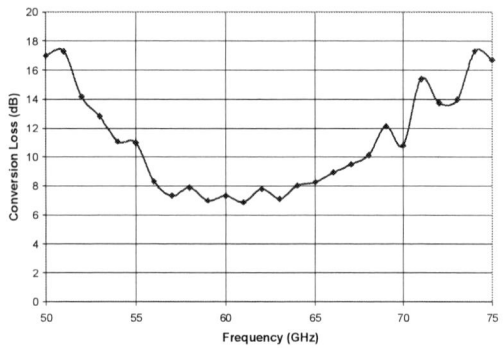

Fig. 7 Measured conversion loss over V-band. Input power 0 dBm.

Measured input and output match with comparison to simulated values are presented in Fig. 8. The best input match is achieved at 30 GHz as expected. The best output match is shifted about 5 GHz towards lower frequencies.

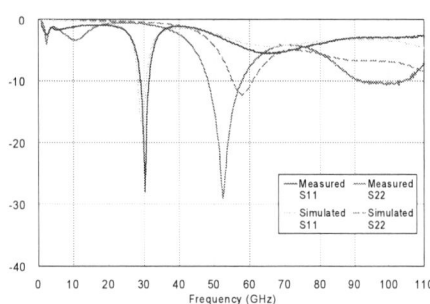

Fig. 8. Measured (solid lines) and simulated (dashed lines) input and output match for the doubler.

B. Mixer

The conversion gain, compression, and S-parameter measurements were carried out using the earlier described equipment. The differential conversion gain of the mixer depends on the LO-level and the optimum of 9.8 dB is reached around +5 dBm. The measured and simulated conversion gain versus LO-level are shown in Fig. 9. The measured results deviate less than one decibel from the simulated ones for the LO-levels less than 0 dBm. In higher levels the used simulation models seem to be pessimistic.

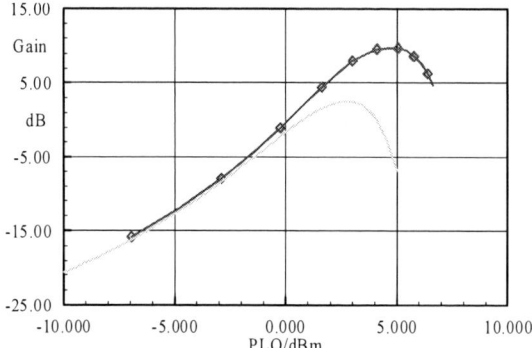

Fig. 9 Measured (with markers) and simulated conversion gain of the mixer.

The measured differential one decibel compression point P_{1dB} is -2 dBm, which is enough for the aimed applications. The measured RF- and LO-port matchings are depicted in Fig. 10. Optimum RF-port match fall on about 10 GHz lower frequency than anticipated, but it is still satisfactory around 60 GHz. The baseband bandwidth of the circuit is over 400 MHz. The measured LO to RF port isolation is about 34 dB.

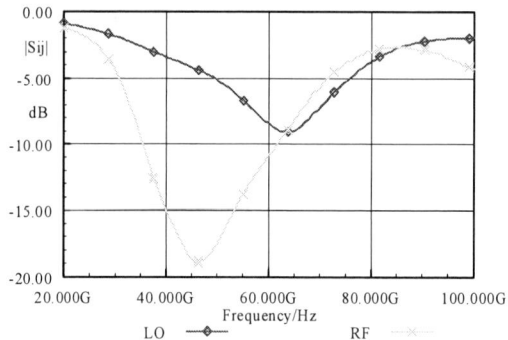

Fig. 10 Measured input matches of the mixer.

The mixer itself is zero biased and the buffer amplifier is the only power consuming block. The reported results are achieved with 0.9 V supply voltage and 15 mA current.

V. Conclusions

A 30 to 60 GHz CMOS frequency doubler and down conversion mixer using 90-nm bulk CMOS technology were designed and measured. The doubler circuit has a single-stage single-ended topology and exhibit 7.1 dB conversion loss and

13.3 dB fundamental frequency suppression with 0 dBm input signal and 13.7 mW power consumption. Fundamental suppression of the doubler is worse than simulated due to unsuccessful matching circuit. Fundamental suppression can be improved in future designs by replacing short shunt stub with a small capacitor or by redesign of the output matching network.

The down conversion circuit is based on a resistive mixer with differential buffer amplifier. It provides reasonable performance with low power consumption (14 mW). The mixer achieves the maximum conversion gain of 9.8 dB at +5 dBm LO-level. Over 400 MHz wide baseband enables the use in high data rate applications.

ACKNOWLEDGMENT

The authors wish to thank STMicroelectronics for chip fabrication and Berkeley Wireless Research Centre for their cooperation in characterising the passive and active circuit elements.

REFERENCES

[1] E. Seok, and K. O, "A 410 GHz CMOS push–push oscillator with an on-chip patch antenna," in *IEEE ISSCC 2008 Digest of Technical Papers*, San Francisco, CA, 2008.

[2] M. Ferndahl, B. m. Mothlag, and H. Zirath, "40 and 60 GHz frequency doublers in 90-nm CMOS," in *IEEE MTT-S IMS Digest*, Fort Worth, TX, 2004, p.179-182.

[3] T. Karttaavi and J. Holmberg, "100 GHz push-push oscillator in 90 nm CMOs technology," in *Proc. EuMIC2007*, München, Germany, 2007, pp. 112-114.

[4] S. A. Maas, *Nonlinear Microwave Circuits*, Norwood, MA: Artech House Inc., 1998.

[5] C. Fager, L. Landén, and H. Zirath, "High output power, broadband 28-56 GHz MMIC frequency doubler," in *IEEE MTT-S IMS Digest*, Boston, MA, USA, 2000, pp. 1589-1591.

[6] F. Ellinger, "26.5-30 GHz resistive mixer in 90 nm VLSI SOI CMOS technology with high linearity for WLAN," *IEEE Tr. on Microwave Theory and Techniques*, vol, 53, pp. 2559-2565, August 2005.

[7] F. Ellinger et al., "30-40 GHz drain-pumped passive mixer MMIC fabricated on VLSI SOI CMOS technology," *IEEE Tr. on Microwave Theory and Techniques*, vol. 52, pp. 1382-1391, May 2004.

[8] B. Floyd, S. Reynolds, U. Pfeiffer, T. Zwick, T. Beukema, and B. Gaucher, "SiGe bipolar transceiver circuits operating at 60 GHz," *IEEE J. of Solid-State Circuits, vol. 40*, pp. 156-167, January 2005.

Proceedings of the 3rd European Microwave Integrated Circuits Conference

Analysis and Design of a 50-GHz 2:1 CMOS CML Static Frequency Divider Based on *LC*-tank

Yuan Mo [1], Efstratios Skafidas [2], Rob Evans [3], Iven Mareels [4]

National ICT Australia, Department of Electrical and Electronic Engineering, University of Melbourne
Parkville, VIC 3010, Australia
[1]ymo@ee.unimelb.edu.au
[2]stan.skafidas@nicta.com.au
[3]rob.evans@nicta.com.au
[4]i.mareels@unimelb.edu.au

Abstract—In this paper, a 2:1 current model logic (CML) frequency divider operating at frequencies up to 50 GHz is reported. A novel circuit topology is employed, which consists of the conventional CML structure with *LC*-tank components as the output load of the divider. An analytical model of the proposed frequency divider is developed and a new method is presented to estimate the divider's performance. The proposed CML frequency divider contains four spiral inductors and is fabricated on standard 130-nm CMOS technology. The division range of the proposed divider was measured from 30 GHz to 50 GHz with 11.7 mW power dissipation at a 1.5-V supply voltage.

I. INTRODUCTION

High-speed and wideband frequency dividers are both critical and essential for building high-performance phase-locked loops (PLLs) and frequency synthesizers in millimeter-wave and gigabit wireless communications. These applications require the frequency divider to work at high frequencies with very large bandwidth while consuming as little power as possible.

CML frequency divider's architecture has the capability to fulfill most of these requirements. However, since the first millimeter-wave CMOS static frequency divider was published in [1], the research activities involved in increasing the divider's speed have mostly relied on reduced transistor's gate size and increased supply voltage and circuit's power consumption [1]–[3].

In this paper, a 2:1 static CML frequency divider is proposed operating up to 50 GHz. This is achieved with a new architecture that has an *LC*-tank as the divider's load instead of the traditional resistive load. A new analytical model is presented to calculate the division bandwidth. Scattering parameters (S-parameters) are used in design to estimate the operating frequency of the divider without the conventional self-oscillation condition. The conventional self-oscillation condition is conservative and results in a divider which has lower maximum operating frequency. The methods proposed in this paper are used to design a high-speed CML static frequency divider on 130-nm CMOS. The measurement shows that the maximum operating frequency of proposed frequency

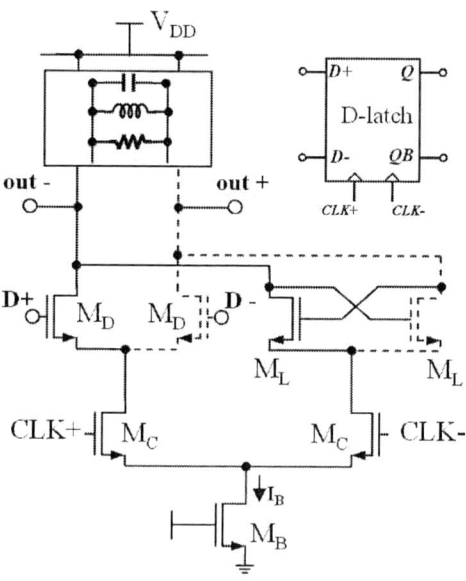

Fig. 1. Block diagram and schematic of the *LC*-tank D-latch current model logic.

divider is 50 GHz. The operating bandwidth is 20 GHz with 11.7 mW power dissipation from a 1.5-V supply voltage.

II. MODEL OF *LC*-TANK CML FREQUENCY DIVIDER

The typical schematic of D-latch current model logic (CML) circuit is shown in Fig. 1 where the conventional resistive load is replaced by the *LC*-tank in this design. For correct operation of the D-latch, the D signals (D+ and D-) need to be stable before the input signals (CLK+ and CLK-) begin to change. It is assumed that only one of the transistors M_D is turned on and the other is off. The same is assumed for the transistors M_L. In Fig. 1, the dash line indicates the off transistors.

A static frequency divider is formed with two CML D-latches in master-slave connection and the D inputs of the first latch are the outputs of the second latch, as shown in Fig. 2(a). Meanwhile, one of the differential transistors M_C is off and the other is completely switched under the control of

978-2-8748-7007-1/08 $25.00 © 2008 EuMA

input signals CLK+ and CLK-, which makes the bias current I_B go through only one transistor of M_C. Therefore, with the above assumptions, the signal D+ is equal to V_{DD} and D- = $V_{DD} - |I| |Z(\omega)|$, where $Z(\omega)$ is the impedance of the LC-tank and I denotes the current passing through the LC-tank in the latch. To ensure the transistor controlled by the D- signal is off, the gate-source voltage should be smaller than the threshold voltage of the transistor M_D, this is equivalent to

$$V_{DD} - |I| |Z(\omega)| - V_S \leq V_T, \tag{1}$$

where V_S and V_T are the source and threshold voltage of the M_D, respectively. V_S is dominated by the turned on transistor of M_D whose gate voltage is V_{DD}. Hence, as in [4], V_S can be approximated by

$$V_S = V_{DD} - \frac{2I_B}{g_{mD}} - V_T, \tag{2}$$

where g_{mD} is the transconductance of M_D. Combining (1) and (2), we obtain

$$\frac{2I_B}{g_{mD}} \leq |I| |Z(\omega)|. \tag{3}$$

Denoting Q as the finite quality factor of the LC-tank, the impedance $Z(\omega)$ can be expressed as [5]

$$Z(\omega) = \frac{Z_o}{1 + j2Q\frac{\Delta\omega}{\omega_o}}, \tag{4}$$

where ω_o is the resonant frequency of the LC-tank and $\Delta\omega$ is the frequency deviating from ω_o. The current I injected into LC-tank is a square wave at the half of input signals' frequency. The expression of $|I|$ can be written as

$$|I| \approx \frac{2}{\pi}I_B, \tag{5}$$

since the higher harmonics of $|I|$ are filter out by the LC-tank. Using (3), (4), (5) and transposing for $\Delta\omega$, we are able to obtain the division bandwidth ω_B is equal to

$$\omega_B = 2 \cdot \Delta\omega \leq \frac{\omega_o}{Q} \sqrt{\left(\frac{g_{mD}Z_o}{\pi}\right)^2 - 1}. \tag{6}$$

From (6), it is shown that the division bandwidth for the new architecture depends on LC-tank quality factor Q, the transconductance of transistor M_D and the impedance Z_o of LC-tank at resonant frequency. The division bandwidth can be increased by reducing Q, consuming more power or increasing the width of M_D to enhance g_{mD}.

III. CIRCUIT DESIGN

The proposed LC-tank based 2:1 CML static frequency divider for high-speed operation is shown in Fig. 2. The divider's structure is based on common topology of 2:1 CML static frequency divider, which is composed of two D-latches and the D inputs of the first latch are the feedback of the outputs of the second latch. The current source in the bottom (M_B in Fig. 1) is omitted in the proposed circuit to facilitate

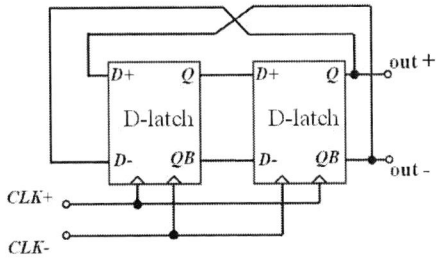

(a)

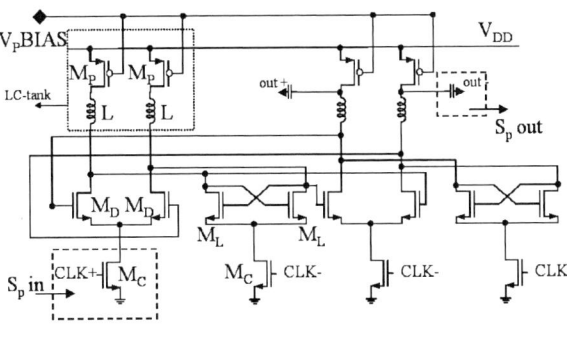

(b)

Fig. 2. (a) Block diagram of 2:1 CML static frequency divider. (b) Schematic of the proposed 2:1 CML static frequency divider.

low-voltage and low-power operation. The care needs to be exercised with this circuit to ensure CLK+ and CLK- are 180 degrees (minimal clock skew) out of phase to ensure minimal disruption of the current flowing through the latch.

The LC-tank components, as indicated in Fig. 2(b), consist of PMOS transistors M_P and spiral inductors L. The output ports (out+ and out-) are set between M_P and L in the second latch, which can protect the divider from the influence of outside parasitic capacitors. However, the output power will be degraded since only parts of the LC-tank are used as the output load.

In the conventional CML static frequency design procedure, the self-oscillation conditional is necessary, as it is used to estimate the divider's performance. Usually, the higher operating frequency is realized by achieving a higher self-oscillation frequency [6]. However, satisfying this condition, it is required the divider have sufficient gain for oscillation. Hence, the large values of the load and transistor size are needed for the oscillation gain, which will reduce the operating frequency of the divider. In our design, the self-oscillation is not used as it is a conservative sufficient but not necessary condition. The design procedure utilized to optimize the transistors size in Fig. 2(b) without self-oscillation condition is presented in [7]. For the purpose of estimating the divider's behavior, S-

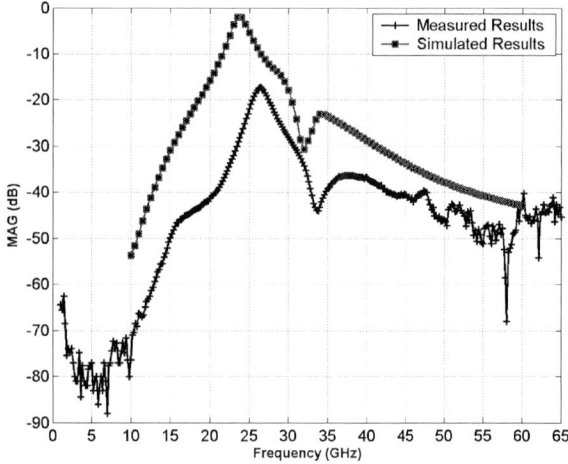

Fig. 3. Simulated and measured S_{21} parameter of the divider.

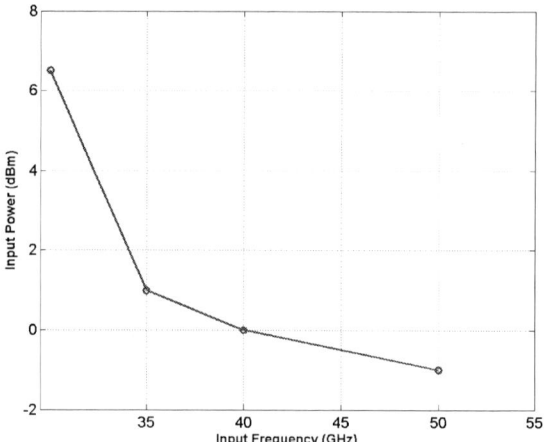

Fig. 4. Input sensitivity curve of the divider.

parameters are used to indicate the resonant frequency of the divider. The maximum operating frequency is approximately two times the resonant frequency. The S-parameters' input and output ports are shown in Fig. 2(b), denoted as "S_p in" and "S_p out" in estimating the divider's resonant frequency and only S_{21} is utilized.

In the proposed divider, the value of each spiral inductor L is 283 pH and the width of M_P is 120 μm. Fig. 3 shows the simulation result of S_{21}, where the resonant frequency appears near 25 GHZ and the divider's operating frequency is expected to be about 50 GHz. In the final design, the widths of M_L and M_D are 0.96 μm and 31.2 μm, respectively. At 1.5-V supply voltage, 7.51 mA flows through the whole divider.

IV. MEASUREMENT RESULTS

The proposed divider circuit is fabricated in 130-nm bulk CMOS with 3 thin and 2 thick copper layers and 3 thick RF layers (2 copper and 1 aluminum). The die micrograph is shown in Fig. 5. The chip size is 0.96×1.145 mm^2 including the pads frame. The divider core with spiral inductors only occupy 270×550 μm^2.

On-wafer probing is used to measure the frequency divider circuit. The divider is firstly measured by the Anritsu 110 GHz Vector Network Analyzer (VNA) for S_{21} parameter. The measured value of S_{21} is given in Fig. 3. Compared with the simulated result, the divider's resonant frequency is approximately 27 GHz while the magnitude of S_{21} is -18 dB. This low power is due to the direct connection to VNA without buffering. Two synchronized signal generators provide differential input signals for the divider. Fig. 4 shows the measured input sensitivity, where 50 GHz is the maximum operating frequency of the divider with about 20 GHz division bandwidth. Fig. 6 shows the output spectrum of the proposed frequency divider operating at 50 GHz. It is interesting to observe that this input sensitivity curve of the proposed divider does not exhibit the usual "V" curve as shown in [1], [6]. The power consumption is 11.7 mW at the 1.5-V voltage supply.

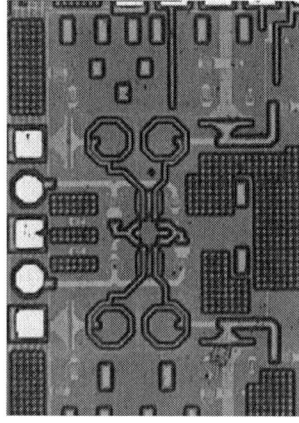

Fig. 5. Micrograph of the divider (die size: 0.96×1.145 mm^2).

Table I summaries the performances of recently published static frequency dividers. The 50 GHz operating frequency is much higher than those of all 120/130-nm CMOS static frequency dividers [1], [6], [8] and also higher than the 80-nm divider's maximum operating frequency reported in [9].

V. CONCLUSION

By using LC-tank as the load of a D-latch current model logic circuit, a high-speed 2:1 static frequency divider in the 130-nm bulk CMOS process is demonstrated. The proposed divider can operate up to 50 GHz with 11.7 mW power dissipation from 1.5-V voltage supply. Compared with recently published results, this divider with a new topology exploits the static frequency divider's high-speed capability without the constrain of self-oscillation condition.

978-2-8748-7007-1/08 $25.00 © 2008 EuMA

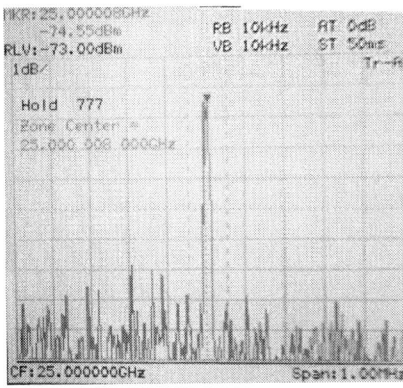

Fig. 6. Output spectrum of the divider operating at 50 GHz.

TABLE I

SUMMARY OF THE PROPOSED DIVIDER PERFORMANCE COMPARED WITH
OTHER RECENT STATIC FREQUENCY DIVIDERS

Ref.	Technology	Max. Freq.	Input Power*	Vdd	Power
[8]	120-nm CMOS	27 GHz	5 dBm	1.5 V	45 mW
[1]	120-nm SOI CMOS	33 GHz	-6 dBm	2.4 V	22.1 mW
[6]	130-nm CMOS	26 GHz	0 dBm	1.5 V	3.88 mW
[9]	80-nm CMOS	44 GHz	$\leq$ 0 dBm	1.1 V	3.2 mW
This	130-nm CMOS	50 GHz	-1 dBm	1.5 V	11.7 mW

*the input power needed in the maximum operating frequency

ACKNOWLEDGMENT

The authors acknowledge IBM, MOSIS and the support of National ICT Australia. The authors also wish to thank the whole 60-GHz RF design group in Electrical and Electronic Department, the University of Melbourne for useful discussion. National ICT Australia is funded by the Australian Government's Department of Communications, Information Technology, and the Arts and the Australian Research Council through Backing Australia's Ability and the ICT Research Centre of Excellence programs.

REFERENCES

[1] J.-O. Plouchart *et al.*, "A power efficient 33 GHz 2:1 static frequency divider in 0.12-μm SOI CMOS," in *IEEE Radio Frequency Integrated Circuit Conference*, June 2003, pp. 329 – 332.

[2] J.-O. Plouchart, J. Kim, V. Karam *et al.*, "Performance variation of a 66 GHz static CML divider in a 90nm SOI CMOS," in *ISSCC Dig. Tech. Papers*, Feb. 2006, pp. 526 – 527.

[3] L. Daihyun, J. Kim *et al.*, "Performance variation of a 90 GHz static CML divider in a 65nm SOI CMOS," in *ISSCC Dig. Tech. Papers*, Feb. 2007, pp. 542 – 543.

[4] B. Razavi, *Design of Analog CMOS Integrated Circuits*. McGraw-Hill Science/Engineering/Math, 2000, 1st Edition.

[5] A. Mazzanti, P. Uggetti, and F. Svelto, "Analysis and design of injection-locked LC dividers for quadrature generation," *IEEE J. Solid-State Circuits*, vol. 39, no. 9, pp. 1425 – 1433, Sept. 2004.

[6] C. Cao and K. O. Kenneth, "A power efficient 26-GHz 32:1 static frequency divider in 130-nm bulk CMOS," *IEEE Microw. Wireless Compon. Lett.*, vol. 15, no. 11, pp. 721 – 723, Nov. 2005.

[7] Y. Mo, E. Skafidas, R. Evans, and I. Mareels, "A 40 GHz power efficient static CML frequency divider in 0.13 μm CMOS technology for high speed millimeter wave wireless systems," in *IEEE International Conference on Circuits and Systems for Communications (ICCSC)*, May 2008.

[8] H. D. Wohlmuth and D. Kehree, "A high sensitivity static 2:1 frequency divider up to 27 GHz in 120 nm CMOS," in *Proc. Eur. Solid-State Circuits Conf.*, Sept. 2002, pp. 741 – 744.

[9] C. Kromer *et al*, "A 40-GHz static frequency divider with quadrature outputs in 80-nm CMOS," *IEEE Microw. Wireless Compon. Lett.*, vol. 16, no. 10, pp. 564 – 566, Oct. 2006.

Proceedings of the 3rd European Microwave Integrated Circuits Conference

A Sub-1V 22-GHz CMOS Injection-Locked Frequency Divider

Chih-Chia Chen and Ching-Kuang C. Tzuang

Graduate Institute of Communication Engineering, National Taiwan University
Taipei 106, Taiwan

r94942022@ntu.edu.tw
cktzuang@cc.ee.ntu.edu.tw

Abstract — In this paper, a sub-1V low power CMOS injection-locked frequency divider is presented. The divide-by-2 frequency divider is fabricated using standard 0.18-μm CMOS technology. By utilizing a NMOS cross-coupled pair with a capacitive feedback in the oscillator core, the ultra-low voltage operation and a reasonable locking range is achieved at K-band. The measurement results show that at the incident power of 3-dBm the locking range is about 1.8 GHz, from the incident frequency 20.8 to 22.6 GHz, with a power consumption of 0.39 mW (excluding output buffers) in the 0.65 V supply-voltage.

I. INTRODUCTION

With the increasing demands on high speed communication services, the development of low-cost and low-power CMOS RF integrated circuits for applications such as RFID and wireless sensor networks have attracted great attention in recent years. For the circuit implementations, low-voltage operations are desirable in order to reduce power consumption and extend the battery duration of portable wireless devices. In addition, as the CMOS technology continues to shrinks in size, the reduced supply voltage is also an inevitable trend for CMOS designs. Unfortunately, due to the required threshold voltage and the inherently low transconductance of the MOSFETs, it is extremely challenging to operate RF front-end circuits with reduced supply voltage and power dissipation, especially at frequencies greater than 20 GHz.

Frequency dividers (FDs) are one of the key components in high speed communication systems. There are several existing topologies for high speed FDs such as using current-mode logic (CML) [1], dynamic logic [2], Miller divider [3], and the injection-locked divider. Miller dividers and CML dividers can be realized up to very high frequencies [1], [3], unfortunately with a large power consumption. Furthermore, these types of FDs require staking of at least two transistors, thus poses a drawback for low voltage operation. Dynamic logic dividers feature a very small power consumption [2], but the maximum operation frequency is limited to a few gigahertz. For high speed and low power operation, the injection-locked FD [4]-[6] is the most suitable one among various types of FDs. Although the locking range of a

conventional injection-locked LC-resonator divider is generally narrow, it fortunately is not very relevant in LC-VCO-based PLLs, as LC-VCOs feature a limited tuning range.

In this paper, a 22GHz injection-locked frequency divider is realized using 0.18-μm CMOS technology. By incorporating the capacitive feedback in the cross-coupled CMOS VCO, the requirements of low power, low voltage, and high speed as well as sufficient locking range are achieved. Section II describes the circuit topology and design considerations of the proposed FD. Experimental results are shown in Section III. Finally, Section IV concludes this paper.

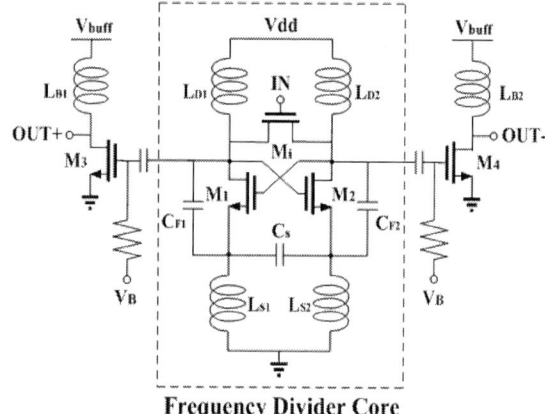

Fig. 1 Schematic of the proposed low voltage injection-locked frequency divider topology.

II. CIRCUIT DESIGN

A. The Proposed Circuit Topology

Fig. 1 shows the schematic of the proposed injection-locked frequency divider. In order to operate at low voltage, an ultra low voltage oscillator with capacitive feedback is used [7]. In order to reduce the required supply voltage and to eliminate

978-2-8748-7007-1/08 $25.00 © 2008 EuMA

additional noise contribution, the tail current transistor in a conventional cross-coupled VCO is replaced by on-chip inductors (L_S). The capacitive feedback loop (C_F and C_S) provides an in-phase relationship across the drain and source terminals. Noted that, the signals between the gate and source terminals are now anti-phase, thus the effective transconductances of M_{1-2} are enhanced. Due to the use of on-chip inductors (L_S) and capacitive feedback loop established by capacitor C_F and C_S, the drain and source voltage can swing above the supply voltage and below the ground potential. The designed oscillator topology can also be view as a G_m-boosted Colpitts oscillator [8] with inductor current source. In this work, the stacked structure [9] is adopted to implement on-chip inductors, thus a more compact chip size is achieved.

The conventional LC-resonator injection-locked FD employs a tail current source as a signal injector [5]. However, this injection path is inefficient especially at higher frequencies [6]. In the proposed injection-locked FD, an improved direct injection scheme based on a MOS transistor switch M_i over the tank is used, as shown in Fig. 1. The switch M_i directly modulates the oscillation state of the oscillator without additional voltage headroom. The dc gate bias voltage of the injector M_i is also carefully biased to get enough locking ranges.

B. Design Considerations

The locking range for an oscillator in first-order approximation is given by [10]

$$\Delta\omega = \frac{\omega_0}{2Q}\frac{I_{in}}{I_{osc}}$$

where Q is the quality factor of the LC-tank, and I_{in} is the injected current. It is found that for larger locking ranges I_{in} should be maximized. In this sense, wider input devices (M_i) improve the locking range, but cause higher input capacitances. Too wide devices deteriorate the parallel tank resistance, which ultimately causes the oscillation to die. On the contrary, if Q is lowered, to increase the divider locking range, the power dissipation rises. Therefore, power is traded off not only with speed, but also with the input locking range. At the same time, the oscillation conditions with respect to both gain and phase must be fulfilled since injection locking is based on an oscillator. Consequently, all device parameters are carefully sized in order to obtain reasonable locking ranges at ultra-low voltage and ultra-low power operations.

III. MEASUREMENT RESULTS

The frequency divider has been designed and fabricated in a 0.18-μm CMOS technology. Fig.2 shows the micrograph of the proposed FD with a chip area of 0.56×0.48 mm^2, excluding the pads. The circuit is measured via on-wafer probing. The injection signal is provided by an Agilent

E8257D signal generator. The output signal is measured single-ended, using an Agilent E4448A spectrum analyzer, and the other end terminated to a 50Ω on-chip resistor.

With a supply voltage of 0.65 V, the current and power consumption excluding out buffers are 0.6 mA and 0.39 mW, respectively. Fig. 3 shows the input sensitivity curve with a free-running frequency at 21.6 GHz. For an input injection with a power level of 3-dBm, a maximum locking range of 1.8 GHz is obtained from the measurement. Thus a reasonable locking range is achieved without the use of varactors to track the free-running frequency of the injection-locked oscillator. Fig. 4 shows the output spectrum when the injection signal is 22.2 GHz.

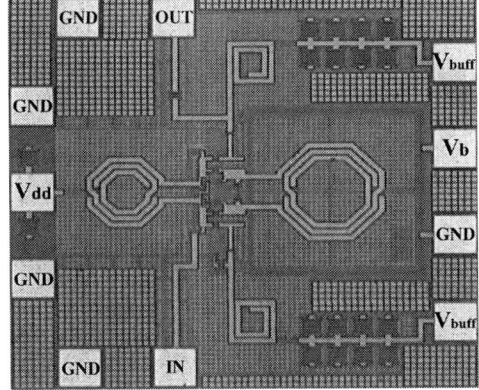

Fig. 2 Chip photo of the proposed 22-GHz sub-1V frequency divider.

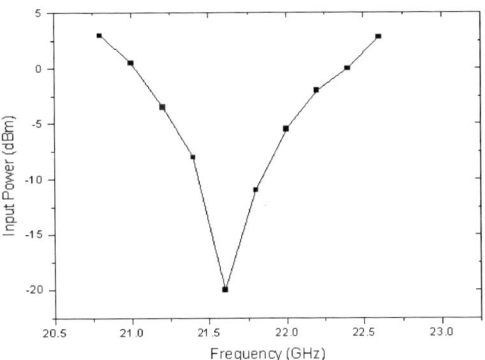

Fig. 3 Measured input sensitivity at the supply voltage of 0.65 V.

The phase noise measurements in Fig. 5 are another proof that this direct injection locking scheme really leads to injection locking. Fig. 5(a) is the phase noise of the divider when locked on to the injection signal and Fig. 5(b) is in free-running mode. The phase noise in locked condition at 1 MHz offset is 45.1 dB lower than that in free running condition as shown in Fig. 5 (a) and (b).

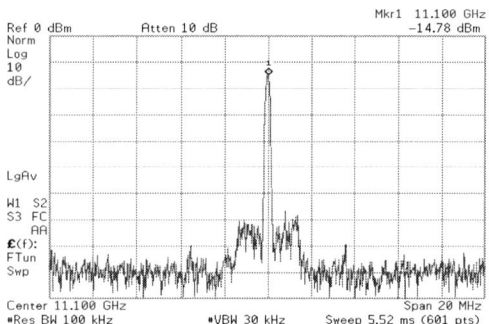

Fig. 4 Spectrum of the 11.1 GHz output.

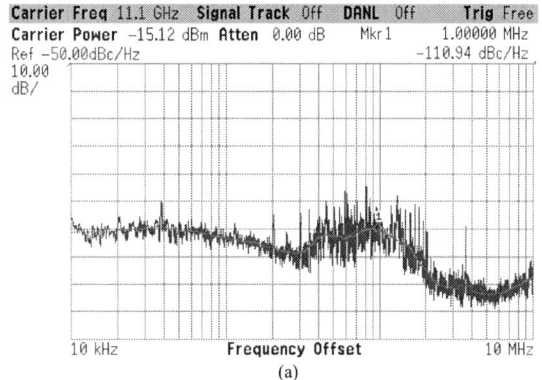

(a)

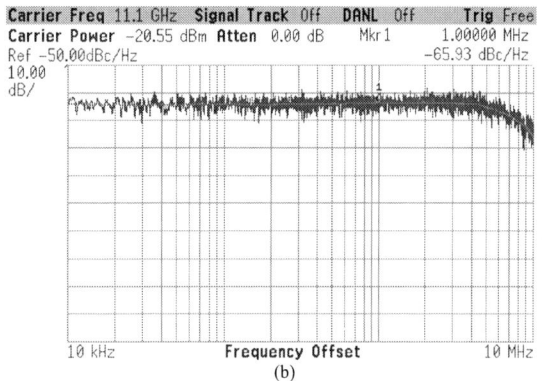

(b)

Fig. 5 The phase noise performance of the divider circuit in (a) locked mode, and (b) free running mode.

The performance of the fabricated injection-locked FD is summarized in Table I along with other fully integrated CMOS implementations. The proposed circuit has the advantage of ultra-low voltage and ultra-low power operation, while maintaining a reasonable locking range.

TABLE I
COMPARISON WITH OTHER HIGH SPEED CMOS FREQUENCY DIVIDERS

work	V_{dd} [V]	Power [mW]	Input Power [dBm]	Locking Range [GHz]	Process
[3]	2.5	16.8	4	38.3-40.6	0.18-μm CMOS
[4]	0.75	4.5	0	3.14-4.63	0.18-μm CMOS
[5]	1.2	1.2	5	18.0-19.3	0.35-μm CMOS
[6]	1.5	23	7	14.2-17.2	0.13-μm CMOS
[11]	1.8	12	12	17.4-27.0	0.13-μm CMOS
This work	0.65	0.39	3	20.8-22.6	0.18-μm CMOS

IV. CONCLUSIONS

A K-band injection-locked frequency divider is designed and fabricated in 0.18-μm CMOS technology. The proposed circuit employs a capacitive feedback cross-coupled pair oscillator and a MOS switch directly coupled to the tank. This circuit draws 0.6 mA at a supply voltage of 0.65 V with a locking range from 20.8 to 22.6 GHz when the input power is 3-dBm. The frequency divider demonstrates high performance in low voltage operation.

ACKNOWLEDGMENT

This work was supported by the National Science Council of Taiwan under Grant NSC 96-2752-E-002-009-PWE and NSC 95-2221-E-002-084-MY2.

REFERENCES

[1] H. Knapp, H.-D. Wohlmuth, M. Wurzer, and M. Rest, "25 GHz static frequency divider and 25 Gb/s multiplexer in 0.12-μmCMOS," in *ISSCC Dig. Tech. Papers*, Feb. 2002, pp. 302–303.

[2] Q. Huang and R. Rogenmoser, "Speed optimization of edge-triggered CMOS circuits for gigahertz single-phase clocks," *IEEE J. Solid-State Circuits*, vol. 31, pp. 456–463, Mar. 1996.

[3] J. Lee and B. Razavi, "A 40-GHz frequency divider in 0.18-μm CMOS technology," *IEEE J. Solid-State Circuits*, vol. 39, no. 4, pp. 594–601, Apr. 2004.

[4] Y.-H. Chuang, S.-H. Lee, R.-H. Yen, S.-L. Jang, J.-F. Lee, and M.-H. Juang, "A wide locking range and low voltage CMOS direct injection-locked frequency divider," *IEEE Microw. Wireless Compon. Lett.*, vol. 16, no. 5, pp. 299–301, May 2006.

[5] H. Wu and A. Hajimiri, "A 19GHz 0.5mW 0.35-μm CMOS frequency divider with shunt-peaking locking-range enhancement," in *ISSCC Tech. Dig.*, Feb. 2002, pp. 412–413.

[6] M. Tiebout, "A CMOS direct injection-locked oscillator topology as high-frequency low-power frequency divider," *IEEE J. Solid-State Circuits*, vol. 39, no. 7, pp. 1170–1174, Jul. 2004

[7] H.-H. Hsieh and L.-H. Lu, "A Low-Phase-Noise K-Band CMOS VCO," IEEE Microw. Wireless Compon. Lett., vol.16, no. 10, pp. 552–554, Oct. 2006.

[8] X. Li, S. Shekhar, and D. Allstot, "Gm-boosted common-gate LNA and differential Colpitts VCO/QVCO in 0.18-μm CMOS," *IEEE J. Solid-State Circuits*, vol. 40, no. 12, pp. 2609–2619, Dec. 2005.

[9] A. Zolfaghari, A. Chan, and B. Razavi, "Stacked Inductors and Tansformers in CMOS Technology", *IEEE Journal of Solid-State Circuits*, vol. 36, pp. 620-628, April 2001.

[10] R. Adler, "A study of locking phenomena in oscillators," *Proc. IEEE*, vol. 61, pp. 1380–1385, Oct. 1973.

[11] U. Singh and M. M. Green, "High-frequency CML clock dividers in 0.13-μm CMOS operating up to 38 GHz," *IEEE J. Solid-State Circuits*, vol. 40, no. 8, pp. 1658–1661, Aug. 2005.

978-2-8748-7007-1/08 $25.00 © 2008 EuMA

A 24 GHz Indirect VCO in 0.18 μm CMOS Technology

Gang Bu[1], Ahmad Reza Tavakoli[2], Kamran Entesari[3]

Electrical and Computer Engineering Department, Texas A&M University
College Station, TX 77843, USA

[1]gang.bu@gmail.com
[2]tavakoli@ece.tamu.edu
[3]kentesar@ece.tamu.edu

Abstract— **A 24 GHz LO signal is generated indirectly by proposing a 12 GHz VCO followed by an active mixer which acts as a frequency doubler. This technique improves the tunability and reduces the power consumption of the VCO/Active doubler compared to a stand-alone VCO at 24 GHz considerably while providing the same amount of phase noise. The circuit is implemented in 0.18 μm CMOS technology. The resulted indirect VCO has 20 % tuning range, 15 mW power consumption, -8.5 dBm output power and a phase noise around -100 dBc/Hz @ 1 MHz offset for a 23.5 GHz carrier. This VCO is suitable for use in low power 24 GHz PLLs and also direct conversion receiver applications.**

I. INTRODUCTION

With the increasing demand for high data rate communication systems and automotive radars, millimeter-wave transceivers are becoming more attractive for commercial applications. The 24 GHz ISM band [1] and 22 - 29 GHz band for short-range automotive radar applications [2] are among the future targets of commercial section. With recent advances in the short-channel CMOS technology, the cut-off frequency (f_t) of CMOS transistors is becoming comparable with traditional expensive technologies such as GaAs and SiGe. Therefore, in addition to provide relatively low cost of fabrication, CMOS technology allows the integration of the millimeter-wave transceiver and analog/digital baseband circuitry on a single chip [3].

The VCO, as the main part of the frequency synthesizer, requires passive elements with high quality factor to provide low phase noise at the frequency of operation. The overall quality factor of the integrated tank circuit available in CMOS process drops dramatically at higher frequencies due to MOS varactors quality factor degradation (f > 20 GHz). Therefore, more power is required to provide the same amount of phase noise compared to a low frequency VCO [4], [5], [6].

One solution to this problem is employing the indirect method. In this method, the VCO operates at a fraction of the desired frequency followed by an active/passive frequency multiplier which provides the required frequency of operation. As a result, the overall power consumption of the VCO is reduced considerably [6]. An operating VCO at lower frequencies also increases the capacitance ratio of a MOS varactor over the parasitics which results in higher tuning range for the overall circuit.

The other advantage of indirect method shows itself inside the frequency synthesizer architecture (Fig. 1). The divider is the most power hungry section of the frequency synthesizer and its power consumption increases dramatically with increase in the frequency of operation. By employing indirect method (Fig. 1(b)), the divider in the loop is running at a fraction of the output frequency which results in much lower power consumption compared to the traditional frequency synthesizer (Fig. 1(a)).

Different implementations of indirect VCO in silicon have been reported. A 24 GHz indirect VCO with an active tripler in SiGe HBT is reported in [7]. Another indirect implementation using passive doubler in 0.18um CMOS has achieved lower power consumption compared to a stand-alone VCO at 24 GHz with the price of considerably lower output power [6].

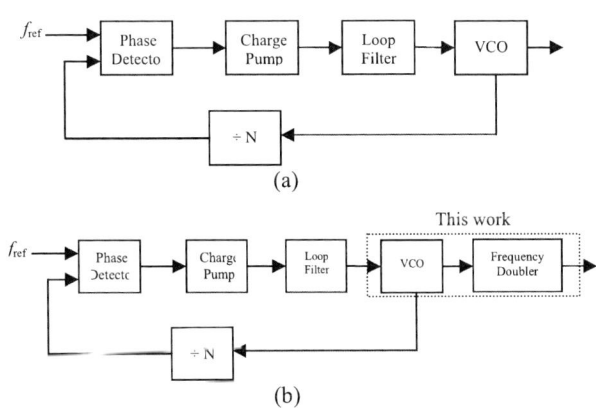

Fig. 1 (a) Traditional, (b) indirect frequency synthesizer architecture

To the best of out knowledge, this work is the first implementation of an indirect VCO at 24 GHz employing an active doubler in CMOS technology. Although the power consumption of this work (15 mW) is slightly higher than the one with passive doubler in [6] (11 mW), this circuit takes advantage of higher output power because of the gain provided by the active doubler with the same amount of phase noise at 1 MHz offset from the carrier. It should be mentioned that the additional flicker/thermal noise due to the active

978-2-8748-7007-1/08 $25.00 © 2008 EuMA

doubler is negligible compared to the phase noise of the on chip VCO [8].

One of the main problems in direct conversion receivers is the frequency pulling/pushing due to the leakage of RF to LO which modulates the VCO operating frequency [4]. The active doubler provides much higher output to input isolation compared to the passive implementation reported in [6] which makes the proposed VCO more suitable for direct conversion receivers.

Also higher tuning range is achieved compared to [6] because of less parasitic capacitance effect in the VCO implementation due to avoiding cross-coupled complementary PMOS transistors in VCO circuit. No need to mention that this approach improves the overall tuning range (%20) and power consumption (15 mW) compared to the reported stand-alone CMOS VCOs at 24 GHz, having the same amount of phase noise.

II. CIRCUIT DESIGN

The 24 GHz indirect VCO consists of a 12 GHz LC-tank oscillator cascaded with a Gilbert-cell mixer wherein both LO and RF inputs are connected to the differential output of the 12 GHz VCO, therefore, acting as a frequency doubler. Capacitive coupling has been used to connect the cascaded stages.

A. VCO design

Fig. 2 (a) shows the circuit for 12 GHz LC tank VCO. The transistor dimensions and bias currents of cross-coupled transistors (M_1) have been chosen to optimize the phase noise while providing the safe margin of negative trans-conductance to guarantee the oscillation start-up and operation of the VCO [4]. The bias current of the VCO is fed through the PMOS transistor on the top (M_{bias}). Since PMOS transistors have less overall flicker noise, they reduce the effect of flicker noise modulation at the output [9]. M_{bias} is designed large enough to reduce the overdrive voltage of M_{bias}, and providing more swing of the output which also provides more reduction in the flicker noise.

The value of inductors in the VCO (L_1=230 pH) are selected in a way to force the VCO operating at the edge of inductive limited/voltage limited region (with small margin of safety toward inductive limited region), therefore, optimizing phase noise and maximizing output signal of the VCO [9]. The physical shape and dimension of inductors have been modelled and optimized using Sonnet [10] to improve their quality factor ($Q_{Ind.}\sim16$) and reduce the overall phase noise while ensuring the accuracy of the inductor models in the VCO.

Available PMOS varactors in the 0.18 μm CMOS process have been employed to implement the varactor tank (Fig. 2(a)). C_2 and C_3 are tuned digitally (V_{bit0}, V_{bit1}: 0 or 1.8 v) for coarse tuning while C_1 is tuned with an analog signal (V_{tune}: 0 to 1.8 v) for fine tuning of the oscillation frequency. The size of C_3 is considered twice of C_2 to provide more resolution for digital tuning and the size of the analog varactor, C_1, is considered to be the same as C_3 to provide enough linear tuning range in

each digital state for the indirect VCO ($Q_{Var}\sim$ 15-20). The tuning property will be discussed in more detail in the measurement section.

B. Doubler design

Fig 2 (b) shows the schematic for the Gilbert-cell doubler. The dimension and bias current of trans-conductance transistors (M_2) have been optimized for high trans-conductance (G_m) and low flicker/thermal noise. The switching transistors (M_3) have been designed to reduce the gain drop associated with poor switching [11]. All the transistors specially the current source transistors (M_{tail}), have been designed to have the least possible overdrive voltage and provide maximum amount of voltage swing at the output with proper selection of voltages.

The inductor located at the drain of M_3 (L_2=290 pH) has been employed as a load for the mixer to resonate out the parasitic capacitors at the output of the doubler and provide the filtering of out of band signals while increasing the output swing compared to resistive load. Additional resistor (R_1=1 kΩ) has been used to reduce the quality factor of the output tank, therefore providing more flat gain in the wide tuning range.

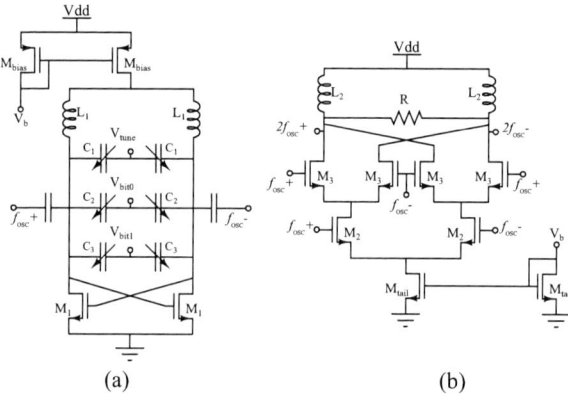

Fig. 2 (a) 12 GHz VCO, and (b) Gilbert-cell mixer schematics

An open drain differential buffer has been used in order to drive the 100 Ω differential input impedance of an external balun. The buffer is just for measurement purposes and providing a low-impedance output to make the probing feasible. There is no need for an open drain buffer in on-chip implementation of the indirect VCO as part of a receiver.

III. MEASUREMENT RESULT

All circuits are fabricated in a 6-metall TSMC 0.18μm CMOS process with f_t of NMOS devices around 50 GHz. Fig. 3 shows the photograph of fabricated chip. The VCO/Active doubler combination sinks 8.3 mA current from a 1.8 V supply voltage. On wafer probing has been employed to measure the output signal using a GSGSG differential probe.

The output of the open drain differential buffer is biased with two Bias-Ts. The differential output signal is then fed in to a balun (0-180⁰, hybrid coupler) and the single-ended output is measured by using the spectrum analyser (Agilent E4446A). The loss of the test setup is measured to be around 8.0 dB at the frequency of operation which is high due to the fact that the available cables provided good loss performance below 18 GHz. Also, the loss of the buffer is around 1 dB due to simulation results.

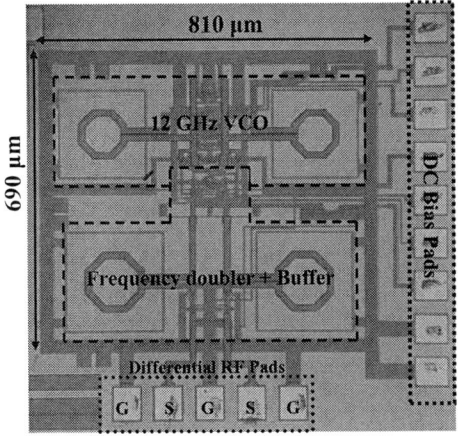

Fig. 3 Chip photograph of the combined VCO/active doubler

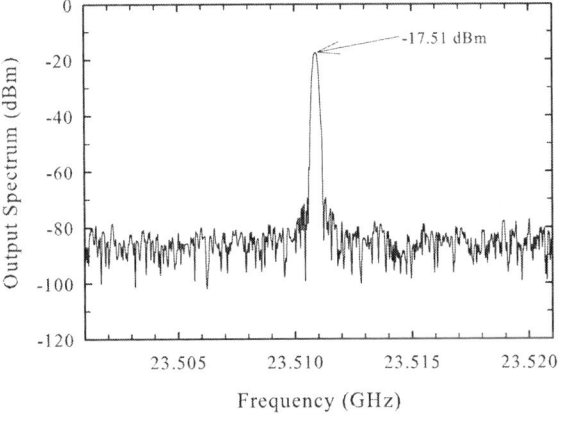

Fig. 4 The output spectrum of the VCO/Active doubler combination (indirect VCO) cantered at 23.5 GHz (Span: 20 MHz, VBW: 180 kHz, RBW: 180 kHz)

Fig. 4 shows the output power spectrum at 23.5 GHz. By de-embedding the loss effect of the buffer/measurement setup, the real output power of the VCO is around -8.5 dBm. This value is 2-3 dB lower than the simulation results. This is because of the poor modelling of the doubler transistors at 24 GHz which causes parasitic capacitors of fabricated transistors to be different from the models. Therefore, the inductor load of the mixer (L_2) does not resonate out the parasitic capacitors properly. More accurate model of the transistors at 24 GHz will result in closer simulated and measured VCO output power.

The measured phase noise of the indirect VCO is shown in Fig. 5. The phase noise of the output is around -77, -100, -109 dBc/Hz at 100 kHz, 1 MHz and 5 MHz offset from the carrier, respectively at 23.5 GHz carrier frequency.

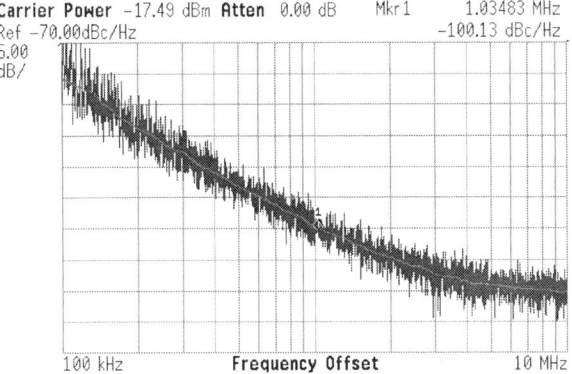

Fig. 5 Measured phase noise of the indirect VCO

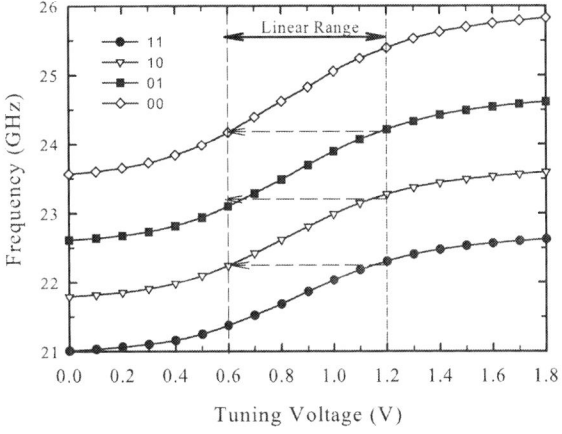

Fig. 6 Measured tuning range of the indirect VCO for different tuning bits and tuning voltages

The tuning range of the VCO is shown in Fig. 6. This high tuning range (%20) is a result of employing high capacitance ratio of varactors over parasitics, which is hard to achieve in a stand-alone VCO at 24 GHz in 0.18 μm CMOS technology. Besides, by keeping the analog tuning voltage between 0.6 - 1.2 V a linear analog tuning is achieved for each digital state. It is also possible to connect the control signal of one of the coarse tuning varactors (C_2 or C_3) or both of them to the same analog voltage used to perform fine tuning for C_1, therefore covering wider range of tuning with the analog adjustment. The frequency synthesizer can take advantage of this flexibility in order to select the required sensitivity for VCO in the PLL structure. The output power has around 3 dB variations in the tuning range mostly because of the fixed resonant tank in the frequency doubler.

978-2-8748-7007-1/08 $25.00 © 2008 EuMA

TABLE I

COMPARISON TO OTHER PUBLISHED VOLTAGE CONTROLED OSCILATORS AT THIS FREQUENCY RANGE

Reference	Approach	Frequency (GHz)	L {1 MHZ} (dBc/Hz)	Tuning Range	Technology	P$_{DC}$ (mW)	P$_{OUT}$ (dBm)
This work	VCO /Active doubler	23.5	-100.13	20 %	0.18 µm CMOS	15	-8.5
[6]	VCO /Passive doubler	25.1	-99.94	12 %	0.18 µm CMOS	11	-18.8
[6]	Direct VCO	21.6	-101.7	6.25 %	0.18 µm CMOS	45	-4.2
[7]	VCO /Active Tripler	23.5	-100	1.8 %	SiGe	180	-10

Table 1 provides the comparison between this design and other related works at this frequency range. Comparing to the direct VCO in [6], this design has the same phase noise at 1 MHz offset while consuming much less power and providing more than three times the tuning range with a price of 4 dB less output power. In comparison with the indirect VCO/passive doubler in [6], this design has slightly more power consumption but 10 dB higher output power and 1.7 times higher tuning range by employing a 0 to 1.8 V tuning voltage compared to the -2 to 2 V tuning voltage employed in [6].

IV. CONCLUSION

An indirect VCO/Active doubler at 23.5 GHz has been presented using 0.18 um CMOS technology. With an output power of -8.5 dBm, power consumption of 15 mW and a phase noise of around -100 dBc/Hz at 1 MHz offset while maintaining a high tuning range (20 %), the resulted VCO is suitable for use in low power 24 GHz PLLs and driving 24 GHz direct conversion mixers.

ACKNOWLEDGMENT

The authors would like to thank Dr. Kuan Zhou from University of New Hampshire for his support in this project.

REFERENCES

[1] International Telecommunication Union [Online]. Available: http://www.itu.int/.

[2] "Federal Communications Commission," FCC 02-04, Section XV.515.15.521.

[3] ITRS Roadmap [Online]. Available: http://public.itrs.net

[4] B. Razavi, RF Microelectronics. Prentice-Hall, Englewood Cliffs, NJ, 1997.

[5] D. B. Leeson, "A simple model of feedback oscillator noise spectrum," Proceedings of the IEEE, vol.54, no.2, pp.329-330, Feb. 1966.

[6] D. Ozis, N. M. Neihart, and D. J. Allstot, "Differential VCO and passive frequency doubler in 0.18 µm CMOS for 24 GHz applications," IEEE Radio Frequency Integrated Circuits (RFIC) Symposium, pp.33-36, 11-13 June 2006.

[7] M. Danesh, F. Gruson, P. Abele, and H. Schumacher, "Differential VCO and frequency tripler using SiGe HBTs for the 24 GHz ISM band," IEEE Radio Frequency Integrated Circuits (RFIC) Symposium, pp. 277-280, 8-10 June 2003.

[8] K. Pulgia, "Phase noise analysis of component cascades," IEEE Microwave Magazine, pp. 329—330, Feb. 1966.

[9] D. Ham and A. Hajimiri, "Concepts and methods in optimization of integrated LC VCOs," IEEE Journal of Solid-State Circuits, vol. 36, pp. 896–909, June 2001.

[10] Sonnet 10, Sonnet Software Syracuse, NY, 2005.

[11] H. Darabi and A. A. Abidi, "Noise in RF-CMOS mixers: A simple physical model," IEEE Journal of Solid-State Circuits, vol. 35, pp. 15–25, Jan. 2000.

Proceedings of the 3rd European Microwave Integrated Circuits Conference

A Compact 23 GHz Hartley VCO in 0.13μm CMOS Technology

Mingquan Bao and Yinggang Li

Microwave and High Speed Electronics Research Centre, Ericsson Research, Ericsson AB
Flöjelbergsgatan 2a, Mölndal, Sweden
mingquan.bao@ericsson.com

Abstract— **A 23GHz balanced Hartley VCO is designed and manufactured in 0.13μm COMS technology. In this VCO design, a pair of coupled inductors are employed to make the whole circuit compact. Its size is only 0.17 mm². Very low phase noise, -108 dBc/Hz at 1MHz offset frequency, is measured and a wide tuning range of 13% is achieved. In addition, the VCO consumes only 5.7 mW dc power at 0.79 V supply voltage[*].**

I. INTRODUCTION

In RF and microwave monolithic integrated circuits (MMICs), the size of passive components, e.g., inductors, capacitors and resistors, is much larger than that of a transistor. Therefore, technology scaling towards smaller and smaller transistors does not actually result in reduction of chip area with a same scaling factor. Thus, reducing the size of passive components is attractive for cost reduction. This becomes even more crucial as the technology scaling continues which results in considerable increase of manufacturing cost in terms of dollars per unit chip area.

If not always, inductors often consume most of the chip area. The area of a VCO (voltage controlled oscillator), for example, is certainly always dominated by inductors.

Hartley VCOs have potential to achieve very low phase noise at high frequencies (>15GHz) [1], [2], [3]. But unlike the Cross-coupled VCOs [4] or Colpitts VCOs [5], which are also popular topologies in use, the Hartley VCO uses two inductors in its resonator. When chip size is at a premium, an instinctive choice is using as less inductors as possible. Thus, the Hartley topology seems unsuitable for VCOs in RF or MMIC implementations.

In this paper, we propose a method to reduce the physical size of a Hartley VCO by making use of positive mutual inductance between two inductors.

II. SCHEMATIC OF THE COMPACT HARTLEY VCO

A. Mutual inductance

For two coupled inductors, the inductance L_i (i=1,2) of the primary and secondary inductors is given by

$$L_i = L_{s,i} \pm M \qquad (1)$$

where, $L_{s,i}$ is the self-inductance for a standing-alone inductor and M is the magnitude of the mutual inductance. The sign of the mutual inductance depends on the relative current directions of the two inductors: positive if in parallel and

negative if in opposite. Thus, for a given inductance L_i, the positive mutual inductance allows reduction of the physical size of the inductors.

B. Schematic

Based on the discussion above, controlling inductor's current direction becomes crucial in the design. This can be achieved by properly connecting transistors and inductors. As shown in Fig.1, the currents I_{d1} and I_{g1} flowing through the gate and the drain inductors, L_{d1} and L_{g1}, are always in opposite. So are the drain/gate currents of the two transistors in a balanced VCO:

$$I_{d2} = -I_{d1}; \qquad I_{g2} = -I_{g1} \qquad (2)$$

Thus, one transistor's gate inductor L_g should be coupled with the drain inductor L_d of the other transistor. Such an arrangement guarantees that the currents in the two coupled inductors flow in the same direction when the VCO operates in differential mode.

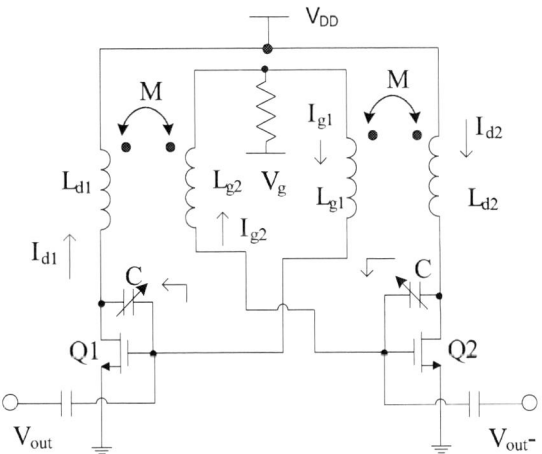

Fig. 1 Schematic of the Hartley VCO utilizing coupled inductors

C. Analyses of the impact Hartley VCO

The proposed VCO as shown in Fig.1 is formed by symmetrically synthesizing two single-ended Hartley VCOs. A better insight into the effect of mutual inductance on VCO performance can be obtained by analysing a single-ended Hartley VCO plus the coupled inductors, as shown in Fig. 2, where a resonator consists of two coupled inductor pairs, a

---

[*] Patent pending

978-2-8748-7007-1/08 $25.00 © 2008 EuMA

varactor C and a virtual resistor R representing various losses in the resonator.

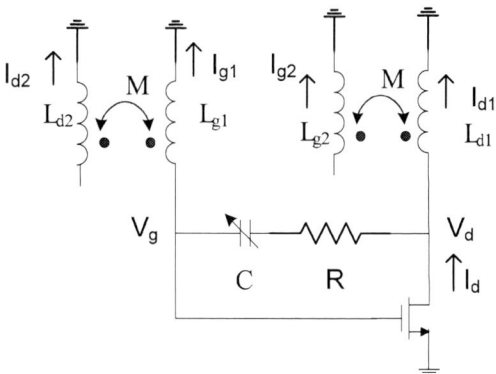

Fig. 2 Schematic of a single-ended Hartley VCO including mutual inductance

The trans-impedance defined by $Z_t = V_g / I_d$ represents the intensity and the relative phase of the feedback. Due to the coupled inductors, the gate and drain voltages are given by

$$V_d = j\omega L_{d1} I_{d1} + j\omega M I_{g2}$$
$$V_g = j\omega L_{g1} I_{g1} + j\omega M I_{d2} \qquad (3)$$

Ignore the current flowing into the gate of the transistor, we get

$$I_d = I_{g1} + I_{d1} \qquad (4)$$

Inserting (2) and (4) into (3) yields the trans-impedance

$$Z_t = -j\omega M + \frac{-\omega^2 (L_{d1} + M)(L_{g1} + M)}{R + \dfrac{1}{j\omega C} + j\omega \left[(L_{d1} + M) + (L_{g1} + M) \right]} \qquad (5)$$

From (5) it can be seen that the self-inductance L_{d1} and L_{g1} are replaced, respectively, by their individual sum with the mutual inductance M. In most cases $1/R \gg \omega M$ and the effect of the term $-j\omega M$ on phase can be ignored and the oscillation frequency is given by

$$\omega_o = \frac{1}{\sqrt{(L_{d1} + L_{g1} + 2M)C}} \qquad (6)$$

Under this condition the phase of the trans-impedance compensates exactly the phase difference between gate and drain.

In our design, two central tapped inductors L_d (outer one) and L_g (inner one) are used. The layout is shown in Fig.3. The two inductors are made mostly in a same metal layer except for the parts close to the 4 ports where the metal one-level lower is used for the 2 ports of L_g. The dc biases are applied at the centre tapped points.

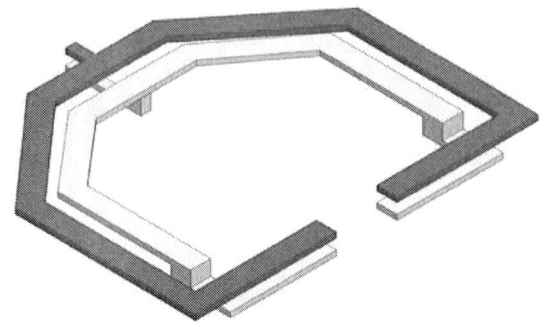

Fig. 3 Coupled inductors of L_d and L_g

The die photo of the Hartley VCO is shown in Fig.4. The circuit size is 0.25×0.14 mm^2 without pads and is 0.53×0.33 mm^2 if pads are included. This size is only 12.5% of that for an 18 GHz, 90nm CMOS VCO where the same balanced Hartley topology is used but the inductors are "isolated" from each other [2].

The VCO circuit is manufactured in Infineon 0.13µm CMOS process which offers 6 metal layers for interconnect. The top two metal layers are used for the coupled inductors as shown in Fig.3.

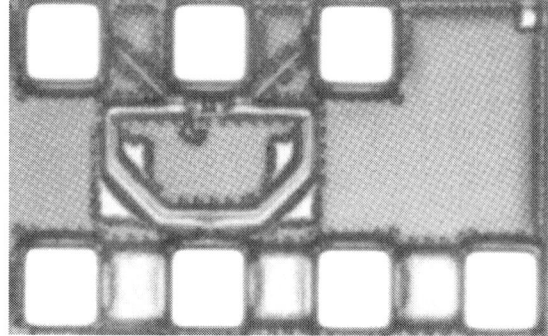

Fig. 4 Photo of Hartley VCO, size 0.53x0.33 mm^2 including pads

III. MEASUREMENT RESULTS

On-chip measurement is carried out. One of the differential outputs is connected to the instrument and the other is terminated with 50Ω resistive load. Spectral analyser is used to measure the oscillation frequency and output power. Fig.5 shows the measured results. It can be seen that the VCO oscillates from 20 GHz to 22.8 GHz, corresponding to 13% tunability. It operates with low supply voltage of 0.79V and consumes 5.7 mW dc power only.

The VCO's phase noise is measured in a setup based on a delay line discriminator technique. As an example, Fig. 6 shows the measured phase noise versus offset frequency for tuning voltage of -1.05V.

978-2-8748-7007-1/08 $25.00 © 2008 EuMA

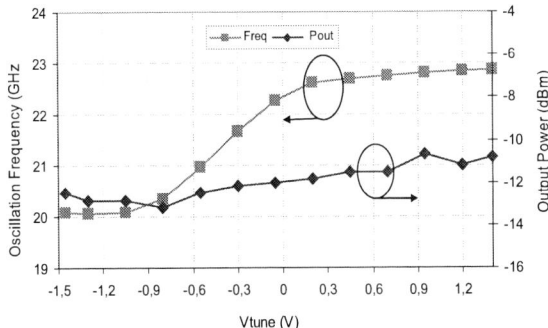

Fig. 5 Oscillation frequencies and output power versus tuning voltage

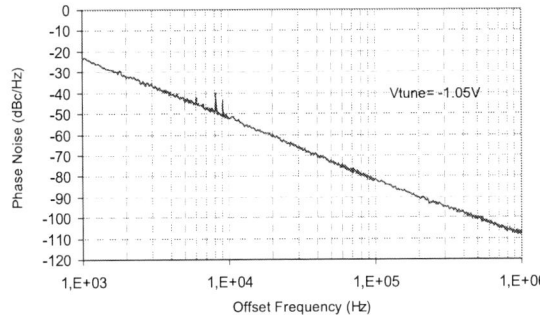

Fig. 6 Phase noise versus offset frequency

The phase noise as a function of tuning voltage is shown in Fig. 7. Within the whole tuning range, the phase noise is quite constant, varying from -108 dBc/Hz to -103 dBc/Hz at 1MHz offset

A. Comparison with state-of-the-art

A commonly used Figure-of-Merit (FOM) is defined as

$$FOM = -L(f_{off}) + 20\log\left(\frac{f_o}{f_{off}}\right) - 10\log\left(\frac{P_{DC}}{1mW}\right) \quad (7)$$

where $-L(f_{off})$ is the phase noise in dBc/Hz at f_{off} from the carrier, f_0, and P_{DC} is the VCO dc power dissipation in mW. A FOM of -186 dBc/Hz is calculated for the present VCO.

When tuning range is taken into account, the FOM is modified as:

$$FOM_T = -L(f_{off}) + 20\log\left(\frac{f_o \cdot tuning(\%)}{f_{off} \cdot 10}\right) - 10\log\left(\frac{P_{DC}}{1mW}\right) (8)$$

and -190 dBc/Hz is achieved for the VCO. These numbers are compared in Table I with some of the published data for fully integrated VCOs operating at above 11 GHz.

The results demonstrate that the VCO presented in this work achieves a good performance balance in terms of phase noise, tuning range, and dc power consumption. Especially, the circuit size is significantly smaller than the published VCOs listed in Table I. This is mainly attributed to the size reduction of the inductors based on a proper use of the mutual inductance.

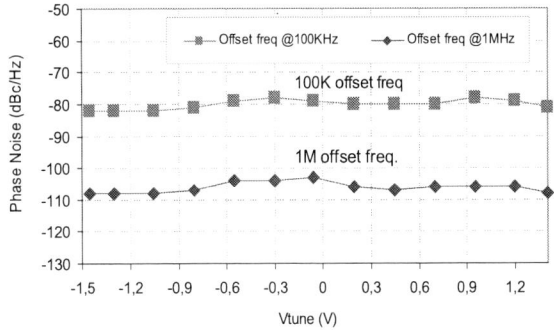

Fig. 7 Phase noise versus tuning voltage

IV. CONCLUSIONS

A compact Hartley VCO is presented. The compactness is realized by introducing proper couplings between inductors whose currents are carefully controlled in favor of positive mutual inductances. In addition to the advantages in chip size and hence in cost, the Hartley VCO demonstrates also good performance: low phase noise, wide tuning range, as well as low supply voltage and low power consumption. All these features are desirable in wireless transceiver applications.

TABLE I
SUMMER OF REPORTED VCOS OPERATING FREQEUNCY ABOVE 11 GHz

Ref.	f_c (GHz)	PN@1MHz (dBc/Hz)	Tuning range (%)	V_{DD} (V)	P_{DC} (mW)	FOM (dBc/Hz)	FOM_T (dBc/Hz)	Technology	Size (mm²)
This work	22	-108	13	0.79	5.7	-186	-190	CMOS 0.13μm	0.17 ; 0.036*
[1]	25	-130	1.6	9	90	-195	-183	GaAs HBT	0.66
[2]	18	-119	10	0.8	4.2	-199	-197	CMOS 90 nm	1.39
[3]	41.6	-105	2.1	2.5	20	-185	-171	GaAs HBT 2μm	1
[5]	21.5	-113	4.9	3.2	130	-178	-172	SiGe HBT 0.25μm	0.4 ; 0.1*
[6]	18	-117	5.6	1.2	14.4	-191	-186	CMOS 0.13μm	0.12*
[7]	11.5	-110.8	5.5	1.8	8.1	-183	-177	CMOS 0.18μm	0.45
[8]	16	-111	6.7	1.8	8.1	-187	-182	CMOS 0.18μm	0.36

*chip area without pads

ACKNOWLEDGMENT

The authors would like to acknowledge the support of the MEDEA+ project 2T401 "HiMission" and Infineon for manufacturing the CMOS circuits. The authors would also like to thank L. Aspemyr, H. Jacobsson and T. Johansson for their support at various stages of this work.

REFERENCES

[1] M. Bao, Y. Li, and H. Jacobsson, "A 25-GHz ultra-low phase noise InGap/GaAs HBT VCO", *IEEE Microwave and Wireless Components Letters*, vol. 15, pp. 751-753, Nov., 2005.

[2] H. Jacobsson, M. Bao, L. Aspemyr, A. Mercha and G. Carcho, "Low phase noise sub-1V supply 12 and 18 GHz VCO in 90 nm CMOS", *Proc. of IEEE MTT-S 2006*, pp. 573-576, 2006

[3] C. H. Lin, K. H. Liang, H. Y. Chang, Y. J. Chan, C. C. Chiong, and E. Bryerton, "A Q-band low phase noise voltage controlled oscillator using balanced π-feedback in 2-μm GaAs HBT process", *Proc. of*

Radio Frequency Integrated Circuits (RFIC) Symposium 2007, pp. 119-122, 2007.

[4] T. Lee, "Design of CMOS radio frequency integrated Circuits", *Cambridge University press*, pp. 504-514, 1998

[5] M. Bao, Y. Li, and H. Jacobsson, "a 21.5/43 GHz dual-frequency balanced Colpitts VCO in SiGe technology", *IEEE J. solid-state circuits*, vol. 39, pp.1352-1355, Aug. 2004

[6] G. Le Grand de Mercey, "A 18GHz rotary traveling wave VCO in CMOS with I/Q outputs", *Proc. European Solid-State Circuits, 2003*, pp: 489 – 492, 2003.

[7] B. Park, S. Lee, S. Choi, and S. Hong, "A 12 GHz fully integrated cascade CMOS LC VCO with Q-enhancement circuit", *IEEE Microwave and Wireless Components Letters*, vol. 18, pp. 133-135, Feb., 2008.

[8] C. L. Yang, and Y. C. Chiang, "Low phase noise and low power CMOS VCO constructed in current-reused configuration" *IEEE Microwave and Wireless Components Letters*, vol. 18, pp. 136-138, Feb., 2008

S-Band AlGaN/GaN Power Amplifier MMIC with over 20 Watt Output Power

M. van Heijningen[1], G.C. Visser[1], J. Würfl[2], F.E. van Vliet[1]

[1]TNO Defense, Security and Safety
Oude Waalsdorperweg 63, 2597 AK, The Hague, The Netherlands
Marc.vanHeijningen@tno.nl

[2]FBH Ferdinand-Braun-Institut für Höchstfrequenztechnik,
Gustav-Kirchhoff-Straße 4, 12 489 Berlin, Germany
joachim.wuerfl@fbh-berlin.de

Abstract— This paper presents the design of an S-band HPA MMIC in AlGaN/GaN CPW technology for radar TR-module application. The trade-offs of using an MMIC solution versus discrete power devices are discussed. The MMIC shows a maximum output power of 38 Watt at 37% Power Added Efficiency at 3.1 GHz. An output power of more than 20 Watt has been simulated from 2.5 to 3.7 GHz. The robustness against high output VSWR values up to 4:1 has been checked and simulations show a maximum drain-gate voltage of around 60 V.

I. INTRODUCTION

The wide-bandgap semiconductor technology Gallium-Nitride (GaN) is a relatively new and very promising technology for high frequency power applications. The advantages of GaN are already described by many authors, such as [1], and are mainly the high power density, high breakdown voltage and good thermal properties of the mostly used Silicon-Carbide (SiC) substrate. Although the processing technology is still under development and continuously improving, the first GaN commercial products are already available. These products are mainly individually packaged transistors or power-bars for telecommunication infrastructure applications, such as WiMAX base-station amplifiers. GaN MMIC technology is available but has not yet been used much for S-band power amplifiers. This paper presents an S-band GaN MMIC power amplifier, together with the trade-offs and comparison to power-bars, and the application of GaN MMIC in S-band radar Transmit-Receiver (TR) module front-ends.

II. GaN TECHNOLOGY IN TR-MODULES

Due to the excellent power performance and robustness of GaN technology it is expected that GaN MMICs will replace a number or all GaAs MMICs that are currently used in TR-modules for radar front-ends. Not only the power amplifier, but also the low-noise amplifier [2] can be realized in GaN technology. The high power handling of the GaN technology also makes it possible to replace the isolator/circulator with a GaN switch. This will however require a careful analysis and design of the output matching network of the HPA, which is no longer isolated from the antenna [3]. Because of the robustness of the technology it will be possible to remove, or shrink, the limiter, which is often needed in front of a GaAs

LNA. Overall, the use of GaN MMICs will lead to smaller and more light-weight TR-modules.

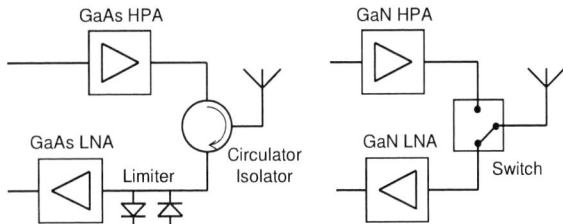

Fig. 1 Example of a classical GaAs frontend and robust GaN frontend.

For radar applications in L-band mostly discrete devices or power-bars are being used for the power amplification. For X-band mostly MMIC HPAs are being used. S-band is in between these two other bands and the choice for a discrete power device [4] or an MMIC is not easy. At these frequencies GaN also has to compete with LDMOS power devices, which are widely used today. The major advantage of GaN over LDMOS is the output impedance, which is for GaN much closer to 50 Ω and allows wide-band designs.

The choice between an MMIC solution and a discrete power device for S-band applications is mostly determined by the system design considerations such as the available area for the power amplification part and weight: Discrete devices often need external matching, which uses much more area than an MMIC solution. When designing for example a tile-based radar panel an MMIC HPA will be the preferred solution.

The choice whether or not to use GaN technology in radar systems is determined by the benefits at system level. The higher output power alone is not necessarily a reason to use a GaN HPA, since this will also lead to higher power dissipation and changes the thermal system design. Keeping the output power the same as used in a GaAs HPA but using a smaller MMIC size will be advantageous for many systems. Several radar performance improvements, when using higher output power per TR element, are presented in [5].

The current GaN development is mostly driven by commercial applications for broadband wireless access, using discrete devices, and only very few HPA MMICs around S-band are presented so far [6].

III. ALGaN/GaN PROCESSING TECHNOLOGY

The AlGaN/GaN structures are grown on a 2-inch semi-insulating SiC substrate by metal–organic vapour phase epitaxy. The principle process flow towards GaN devices is very comparable to standard III/V devices. Of course the type of ohmic contact and Schottky gate metallization and, very important, the surface passivation has to be matched to the GaN/AlGaN material system and to the epitaxial layer sequence. The technology of the surface passivation is decisive, since any surface and interface charges directly control the concentration of the 2DEG in the channel. Additionally any traps introduced by epitaxy and/or processing may lead to strong dispersion effects and compromise microwave performance. Fig. 2 shows a schematic cross section of active and passive elements that may be integrated in a GaN based MMIC process. The FBH GaN-HEMT process relies on mixed wafer stepper and electron beam lithography to define the structures [7]. The passivation and the MIM dielectrics consist of an optimized SiNx layers deposited by PECVD. Typically, gate field plates are applied to eliminate dispersion and to enhance breakdown voltage.

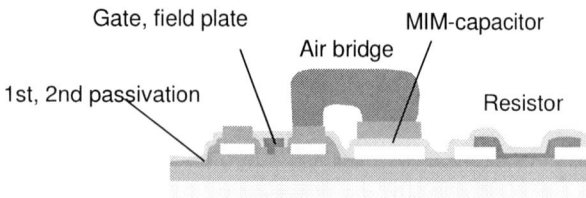

Fig. 2 Front end processing techniques at FBH.

IV. MMIC Design

Starting point for the MMIC design is a collection of models and measurement data for inductors, capacitors and resistors, processed on a previous run. 2.5D EM Momentum simulations have been used to simulate other passives and parts of the matching networks.

For the active devices, S-parameter measurement data and a modified Angelov large signal model for an 8x125 um device are available. Design target for the S-band HPA MMIC is at least 20 Watt output power around a center frequency of 3.1 GHz. This performance is close to the current state of the art that can be realized on a GaAs HPA MMIC in this band. However, the goal is to realize this on an MMIC area that is at least half of what would be needed to realize the same output power on GaAs, which is estimated to be around 30 mm^2.

The design has started by analysing the performance of the 8x125um basic FET size. Loadpull simulations have been performed to determine the load for maximum output power and maximum efficiency. A compromise load in between these two impedance values has been chosen and corresponds to 47.4 +32.2 j Ω at 3.1 GHz. The power sweep simulation performed at this load at 3.1 GHz is shown in Fig. 3.

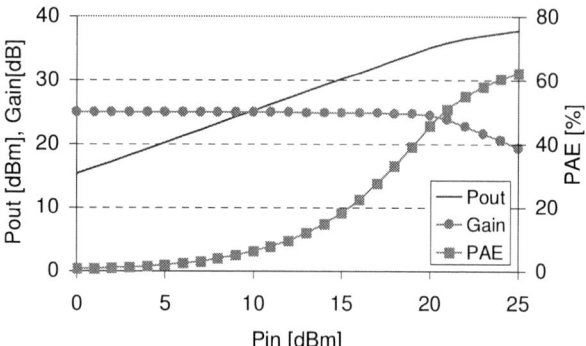

Fig. 3 Simulated output power, gain and power added efficiency of the 8x125um FET at the compromise load at 3.1 GHz and 26 V drain bias.

The small signal gain is around 25 dB and the 1 dB compression output power is 36 dBm at 51 % power added efficiency. The maximum PAE of over 60 % occurs at a gain compression of 5 dB. For S-band applications a unit gate width of 125um is not the optimum choice. Therefore two 8x125um FETs have been placed in parallel, creating an 8x250um FET. The output-stage of the HPA consists of four 8x250um FETs and has a total gate width of 8mm. This output-stage should deliver at least 20 Watts.

The next step has been to make the unit FET unconditionally stable, for frequencies above approximately 500 MHz. At lower frequencies the stability can be improved by off-chip measures, e.g. decoupling in the biasing network. The unit cell has been made unconditionally stable by adding a parallel R-C network in series with the gate and, to increase low-frequency stability, a shunt resistor to ground of 370 Ω at each gate-bias pad. Because the used processing technology is still under development a large safety margin has been used in the design of the stability network, resulting in a gain loss of about 6 dB per stage.

The resulting MMIC layout is shown in Fig. 4, and has a size of 5.0mm x 3.3mm. This is about half the size of a GaAs S-band HPA MMIC for the same output power. After the complete circuit and layout design extra stability analyses have been done to check for odd-mode oscillations and loop gains over the power supplies. These analyses have shown that extra external power supply decoupling and filtering of especially the gate bias lines are required to prevent possible oscillations.

In this first stage of the HPA two 8x250um FETs have been used. The output power of this first stage has to compensate the relatively large losses of the stability network, but still there is a large power margin between the first and second stage. In a future redesign the dimension of this first stage can be reduced.

The losses of the input, interstage and output matching network are respectively 1.0 dB, 1.5 dB and 1.5 dB, not counting the extra loss of the stability network. Monte Carlo simulations have been performed to guarantee the performance of these networks over a wide spreading range for the passive components.

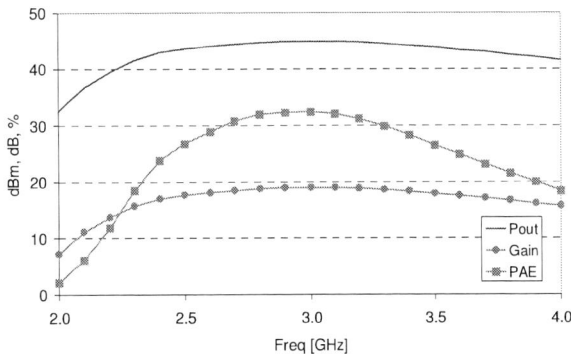

Fig. 6 Simulated output power, gain and power added efficiency at 27 dBm input power.

Fig. 7 and Fig. 8 show the small signal input and output matching and gain of the MMIC. The matching is better than 10 dB from 2.6 to 4.3 GHz. The extra gain peak around 1.5 GHz is not intentionally, but appears because the input is matched at this frequency and because of the large gain of the single devices. It has been verified that this peak does not cause any stability problems.

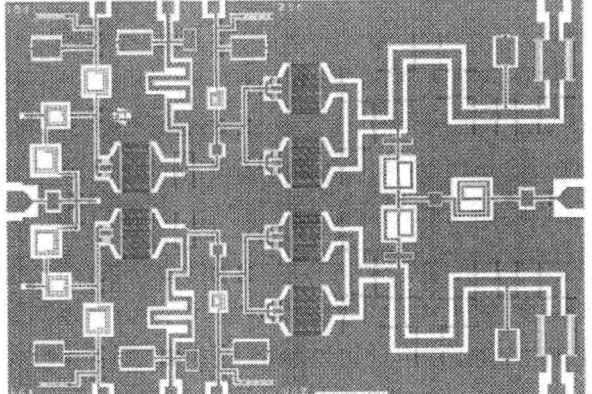

Fig. 4 Layout of the 2-stage S-band HPA MMIC (5.0 mm x 3.3 mm).

V. SIMULATION RESULTS

Currently the design is in processing and only simulation results can be shown. The large signal power sweep is shown in Fig. 5. The saturated output power is 45.8 dBm with 37 % PAE at 4 dB gain compression and 26 V drain bias. The overall gain is around 20 dB, which seems to be low compared to the gain of a single device. This is mainly caused by the fact that large safety margins have been used in the design of the stability network, to cope with large process spreading. Once the processing technology has been stabilized, a redesign can be made with more gain.

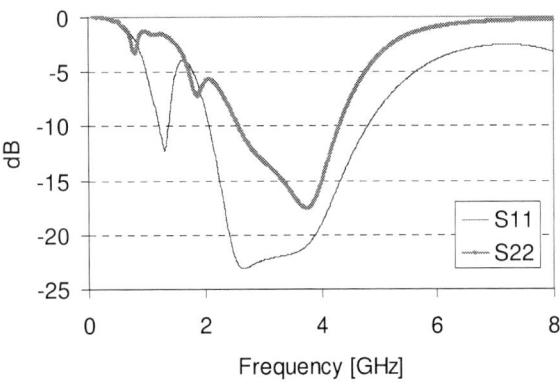

Fig. 7 Simulated small signal input and output matching.

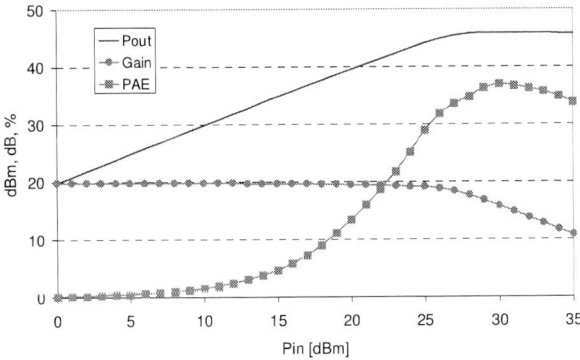

Fig. 5 Simulated output power, gain and power added efficiency at 3.1 GHz.

The large signal performance versus frequency is shown in Fig. 6, for a fixed input power level of 27 dBm. It can be seen that the HPA delivers at least 43 dBm output power over a wide frequency range from 2.5 to 3.7 GHz.

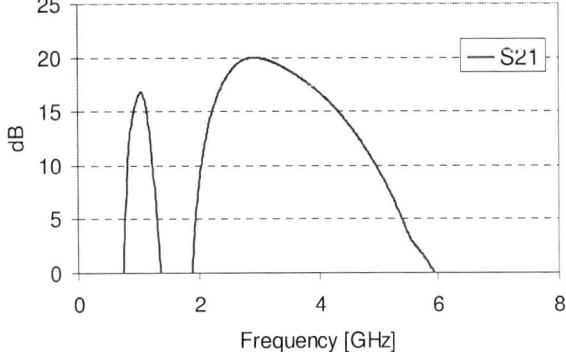

Fig. 8 Simulated small signal gain.

978-2-8748-7007-1/08 $25.00 © 2008 EuMA

Also the behaviour of the amplifier at high output VSWR ratios has been checked. This has been simulated in the middle of the frequency band at 3.1 GHz with a source power of 26 dBm. The output load has been swept from a VSWR value of 1.0 to 4.0 for all load angles. The result is shown in Fig. 9 and it can be seen that the output power drops to around 41 dBm for a VSWR of 4.0. Fig. 10 shows the maximum drain-gate voltage for this same simulation. Even at the very high VSWR value of 4.0, the maximum voltage is only around 60 V, which is still below the expected breakdown voltage of this process.

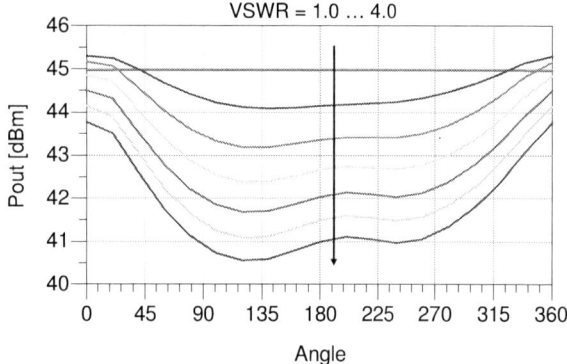

Fig. 9 Simulated output power at 3.1 GHz and 26 dBm source power for an output VSWR from 1.0 to 4.0 in steps of 0.5.

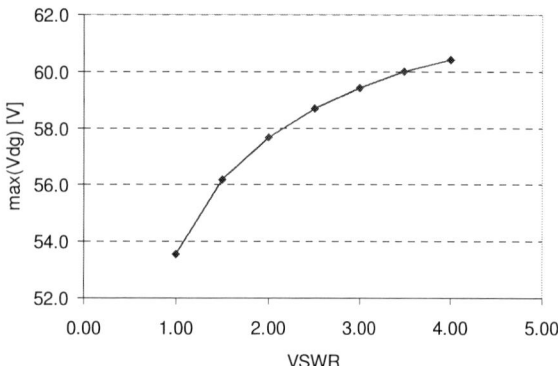

Fig. 10 Simulated maximum drain-gate voltage of the output FETs.

Finally, thermal simulations have been performed to check the maximum junction temperature. In the simulation the MMIC has been soldered on a CuMo carrier. A very pessimistic thermal conductivity of 330 W/m/K has been used for the SiC substrate. The result is a maximum temperature of 210°C, at a worst case base plate temperature of 80°C, assuming a constant DC dissipation of 8 Watt per FET. Using a more optimistic thermal conductivity of 490 W/m/K the maximum temperature is 180°C, also at a base plate temperature of 80°C. These simulations assume CW input power, while the amplifier is intended for pulsed operation only.

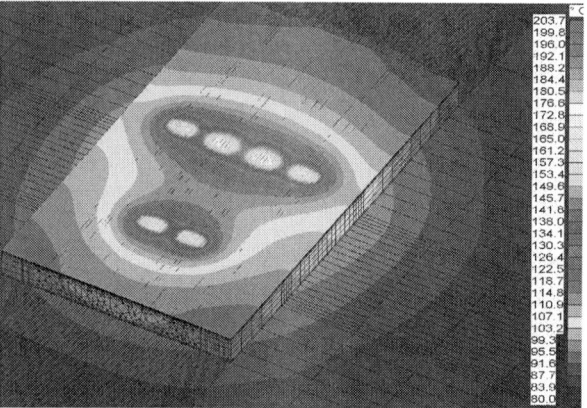

Fig. 11 Thermal simulation at 80°C base plate temperature.

VI. CONCLUSIONS

The design and simulation results of an S-band HPA MMIC in AlGaN/GaN technology have been presented. First the trade-offs between using an MMIC or discrete power device have been discussed. The final choice for a discrete power device or an HPA MMIC will be determined by system considerations. For applications where size and weight are important, such as airborne radars, GaN HPA MMICs will become the preferred solution.

The MMIC shows a maximum output power of 38 Watt at 37% Power Added Efficiency at 3.1 GHz. An output power of more than 20 Watt has been simulated from 2.5 to 3.7 GHz. No breakdown is expected for high output VSWR values up to 4:1.

ACKNOWLEDGMENT

This work has been support by Thales Nederland BV and the Dutch Ministry of Economic Affairs.

REFERENCES

[1] U.K. Mishra, L. Shen, T.E. Kazior, and Y.-F. Wu, "GaN-Based RF Power Devices and Amplifiers", Proc. of the IEEE, vol. 96, no. 2, pp. 287-305, Feb. 2008.

[2] M. Rudolph, R. Behtash, R. Doerner, K. Hirche, J. Würfl, W. Heinrich, and G. Tränkle, "Analysis of the Survivability of GaN Low-Noise Amplifiers", IEEE Trans. Microwave Theory Tech., vol. 55, no. 1, 37-43, Jan 2007.

[3] G. van der Bent, M. van Wanum, A.P. de Hek, M.W. van der Graaf and F.E. van Vliet, "Protection Circuit for High Power Amplifiers Operating Under Mismatch Conditions", Proc. 2nd European Microwave Integrated Circuits Conference, pp. 158-161, Munchen, Oct. 2007.

[4] E. Mitani, M. Aojima, and S. Sano, "A kW-class AlGaN/GaN HEMT Pallet Amplifier for S-band High Power Applications", Proc. 2007 European Microwave Int. Circuits. Conf, EuMIC2007, pp. 176-179.

[5] M. E. Russell, "Future of RF Technology and Radars", IEEE Radar Conference 2007, pp.11- 16.

[6] J.W. Milligan, S. Sheppard, W. Pribble, Y.-F. Wu, St.G. Muller, and J.W. Palmour, "SiC and GaN Wide Bandgap Device Technology Overview", IEEE Radar Conference 2007, pp.960-964.

[7] R. Lossy A. Liero, J. Würfl, G. Tränkle, "High power, high gain AlGaN/GaN HEMTs with novel power bar design", IEDM 2005, Technical digest, pp. 589-591.

978-2-8748-7007-1/08 $25.00 © 2008 EuMA

2-6 GHz GaN MMIC Power Amplifiers for Electronic Warfare Applications

M. Angeles Gonzalez-Garrido [1*], Jesus Grajal [1*], Pablo Cubilla [2], Antonio Cetronio [3], Claudio Lanzieri [3], Mike Uren [4]

[1] *ETSIT, Universidad Politecnica de Madrid. Ciudad Universitaria s/n, 28040, Madrid, Spain*
*angeles,jesus@gmr.ssr.upm.es

[2] *INDRA Sistemas S.A. Ctra. Loeches 9, 28850, Torrejon de Ardoz, Madrid, Spain*

[3] *SELEX-SI s.p.A. Engineering Division, Via Tiburtina Km 12.4, 00131 Rome, Italy*

[4] *QinetiQ Ltd. Malvern Technology Centre, St. Andrews Road, WR14 3PS, Malvern, United Kingdom*

Abstract—This paper presents two MMIC broadband high power amplifiers of 4 mm of periphery at the output stage in the frequency band 2-6 GHz. The amplifiers are based on Al-GaN/GaN high electron mobility transistor (HEMT) technology on SiC substrate. They have been fabricated in two different european foundries: SELEX Sistemi Integrati and QINETIQ. SELEX has a gate process technology of $0.5\mu m$, and devices of $10 \times 100\mu m$ periphery in microstrip technology and QINETIQ has a gate-length of $0.25\mu m$, and devices of $8 \times 125\mu m$ in coplanar technology. The coplanar amplifier from QINETIQ has demonstrated an output power of 8W in continuous wave at V_{ds}=20V which confirm model predictions. On the other hand, SELEX microstrip amplifier has a saturation power of 10W CW at V_{ds}=25V and 4 GHz. This amplifier measured on-wafer in pulsed conditions exhibits a maximum power of 17W at V_{ds}=30V.

I. INTRODUCTION

Monolithic microwave integrated circuits (MMIC) based on gallium nitride (GaN) high electron mobility transistors (HEMT) have the advantage of providing wideband power performance [1]. This capability has a great potential for communication and electronic warfare (EW) systems. The benefits of this technology are related to the material high breakdown voltage, high current density, and compatibility with semi-insulating silicon carbide (SiC) substrate, which has a high thermal conductivity [2]. The expected improvements in the performance and reliability will prevent from using vacuum tubes (TWTAs) in many applications [3]. GaN has important advantages in comparison with traditional solid state technologies like GaAs. GaN offers higher power densities as has been demonstrated in multiple works at different frequency bands [4]-[5]. Furthermore, GaN HEMTs can operate at much higher drain voltages than GaAs allowing the use of smaller devices for the same output power [1]. Therefore, device impedances are higher, and improvement on broadband matching can be performed [6].

This paper presents the design procedure and the characterisation of two broadband 2-6 GHz high power amplifiers (HPAs) in GaN monolithic technology. The same design methodology has been used in both HPAs, but one is microstrip (MS) and fabricated at Selex foundry and the other one is coplanar (CPW) and processed at Qinetiq. Four devices of 1mm periphery were combined to achieve high output power in continuous wave (CW). QINETIQ-HPA delivers a maximum output power of 8W at drain voltage V_{ds}=20V. SELEX-HPA was designed to deliver 10W CW at V_{ds}=25V, which was obtained on test-jig measurements with a base plate temperature of 0C. Also, SELEX-HPAs were tested on-wafer in pulsed condition, and a maximum power of 17W at $V_{ds} = 30$V was measured. These amplifiers are developed in the frame of the european KORRIGAN project [7]. The main objective is to develop a stand alone european supply chain supporting GaN technology, which will provide all major european defence industries with reliable state-of-the-art GaN foundries services.

II. DEVICE TECHNOLOGY

We have design our HPAs with two different GaN technologies from Selex si and Qinetiq. The most relevant properties of the selected devices are summarized in Table I. Large-signal models of the transistors have been extracted by the University of Rome Tor Vergata for Selex and by XLIM for Qinetiq.

TABLE I
GAN HEMT TECHNOLOGY

	SELEX-MS	QINETIQ-CPW
Substrate thickness (μm)	85	400
Gate process	Stepper	E-beam
Gate length (μm)	0.50	0.25
Device periphery (μm)	10x100	8x125
Saturation current (mA/mm)	950	900
Transconductance (mS/mm)	120	220

III. WIDEBAND HIGH POWER AMPLIFIER DESIGN

The HPAs operate in class AB. The common specifications for the designs are a small-signal gain higher than 15dB and an input reflection coefficient better than -10dB in the frequency band 2-6 GHz. SELEX-HPA will deliver 10W CW output power and, QINETIQ-HPA 8W CW. The transistor size is 1mm in both desings.

The targeted gain requires a two-stage design. Besides, it was necessary to combine several devices in order to achieve high output power level. In the following, the common design steps and considerations are described.

A. Amplifiers Topology

A corporate topology is selected for power combination. The first stage consists of two unit cells which drive the output stage composed of four equal cells. The amplifier has three matching networks: input-, inter-, and output-stage. The HPA schematic is displayed in Fig. 1, where Li represents the losses of each matching networks, and N_i, PAE_i, G_i, P_i represents respectively the number of combined cells, the power added efficiency, the gain, and the output power, at the i-stage. The difficulty in the design of broadband matching networks increases with the number of cells to be combined. Equation (1) describes the power added efficiency PAE of a corporative topologies with 2^n transistors at the output stage. An analysis of equation (1) shows that the output-stage losses (L_3) have a great impact on the total PAE, which is an important specification in an HPA to reduce thermal effects. L_3 is also critical to the output power, therefore output-stage has been designed to achieve minimum losses.

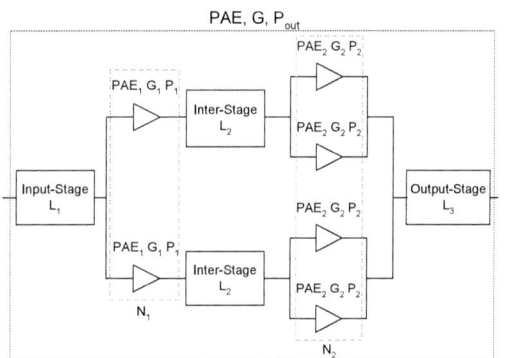

Fig. 1. Two-stage corporate topology amplifier.

$$PAE = \frac{PAE_1 PAE_2 (L_1 L_2 L_3 G_1 G_2 - 1)}{PAE_1 G_1 L_1 L_2 (G_2 - 1) + PAE_2 L_1 (G_1 - 1)} \quad (1)$$

B. Networks synthesis

The matching networks are synthesized from filter theory and implemented with lumped elements. The design process follows this steps: First, the output network synthesizes the optimum impedances provided by load-pull simulations to maximize second-stage transistors output power in a broad frequency band. Next, the inter-stage network synthesizes the optimum loads for the first-stage and provides the input matching for the second-stage. Finally, the input network matchs the amplifier input to 50 Ω.

Stabilization networks and DC bias networks are included and considered in the synthesis process. Drain bias networks have been designed with transmission lines due to technological limits in the DC current through inductors.

C. Stability

An in-depth analysis of the stability of the amplifiers was performed to guarantee that no oscillation phenomena arise. The transistors were analysed using the classical approach based on Rollet's K factor. RC cells were selected to stabilize both designs. Moreover, an odd stability analysis of the amplifiers is carried out, based on pole-zero identification of the frequency response obtained at several nodes of the circuit [8]-[9]. Parametric simulations performed for expected working conditions do not exhibit odd oscillation risk. However, resistors were added between transistors in order to increase the margin for oscillations due to the low yield of this technology.

D. Thermal analysis

Another important issue in an HPA design lies in the high power dissipation occurring at high output power levels. Thermal management is critical, because around 40W should be dissipated. Thermal resistances, R_{TH}, have been calculated with COMSOL Multiphysics (FEMLAB) for the basic transistor cells and a dissipated power of 8W. QINETIQ-HPA has a R_{TH} in the range of 8 C/W for the $400\mu m$ substrate thickness. On the other hand, SELEX-HPA has a R_{TH} around 14 C/W. From these calculations, the estimated temperature gradient between the channel temperature and the backside temperature ($\Delta T = Tchannel - Tbackside = Pdiss \cdot R_{TH}$) is around 64 C in the CPW QINETIQ design and 112 C in the MS SELEX. Therefore, a test-jig mounted on a cooling platform is necessary in order to provide the amplifier with a proper heat dissipation system to fix a low backside temperature.

IV. PERFORMANCE

The CPW QINETIQ-HPA has an area of 6.3×3.75 mm^2 and, MS SELEX-HPA size is 6.4×3.6 mm^2. A photo of each amplifier is shown in Fig. 2. and Fig. 3 respectively. Both amplifiers have been measured on the test-jig of Fig. 4 for CW characterization. The test-jig was mounted on a dissipation system to reduce thermal influence.

A. QINETIQ-HPA

QINETIQ CPW amplifier has demonstrated good thermal resistance. However, this CPW design exhibits problems of stability and RF response degradation due to inefficient ground planes connection. To solve this, it is very important to ensure equal potential in every ground plane of the circuit. This may be improved using enough air-bridges in layout, reducing circuit size and getting better contact between circuit ground and the test carrier.

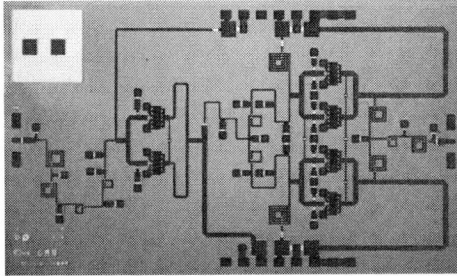

Fig. 2. Picture of microstrip broadband HPA from SELEX (6.4×3.6mm²).

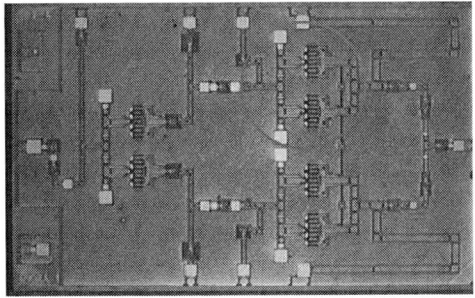

Fig. 3. Picture of CPW broadband HPA from QINETIQ (6.3×3.75mm²).

Fig. 4. Measurement test-jig.

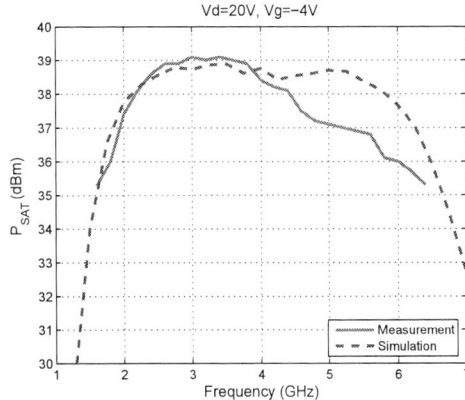

Fig. 5. QINETIQ-HPA saturation power.

at 3GHz, at a bias condition V_{ds}=20V, V_{gs}=-4.5V. Agrement between simulation and measurement is good, the small-signal gain is 18 dB, the saturation output power is 38.5 dBm and the PAE is higher than 20%.

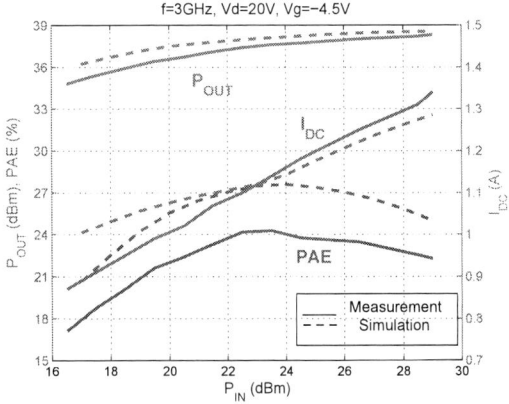

Fig. 6. QINETIQ-HPA output power, PAE and DC current at 3 GHz.

B. SELEX-HPA

SELEX-HPAs were assembled on carriers for RF characterization after on-wafer pulsed power screening. The chips were tested at V_{ds}=20 and 25V under CW conditions and the base plate temperature fixed at 0 C. Typical measured small signal gain and input reflexion coefficient are shown in Fig. 7. Over the 2-6 GHz frequency range the gain is greater than 15 dB, and the input reflexion coefficient is lower then -7 dB. Simulated data is also shown for comparison.

This amplifier was designed to operate at V_{ds}=25V and I_{ds}=1.3A to deliver an output power of 10W as is depicted in the simulation of Fig.8. The HPA was measured at V_{ds}=20V and I_{ds}=1.1A and V_{ds}=25V and I_{ds}=0.9A. The saturation power at V_{ds}=20V is better than 38.5 dBm in band and higher than 39 dBm at V_{ds}=25V. This values are between

QINETIQ has performed load-pull measurements of a power combiner composed o two unit cells (2x8x125), which is shown inside a circle in Fig.3. That is half the output stage with its stabilization networks at the input and its combination network at the output. This network delivers a saturated output power of 6.5W (3.25W/mm) to the optimum load at $V_{ds} = 25V$, $V_{gs} = -4V$ and 2.8GHz.

The saturation power of the QINETIQ-HPA at a bias condition $V_{ds} = 20V$, $V_{gs} = -4V$ is reported in Fig. 5. Good agrement between simulation and measurement is obtained in the 2-4GHz band, where the 8W output power target is fulfilled. However, power performance in the 4-6GHz band has a mismatch of 1-2dB.

Fig. 6 presents the power performance, PAE and DC current

978-2-8748-7007-1/08 $25.00 © 2008 EuMA   85

simulation predictions as it is reported in Fig.8. Thermal effect is observed at V_{ds}=25V because the drain current decreases. The maximum saturation power is 10W at 4 GHz.

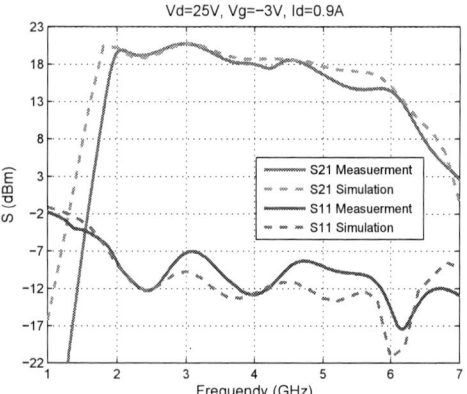

Fig. 7.   SELEX-HPA S-Parameters.

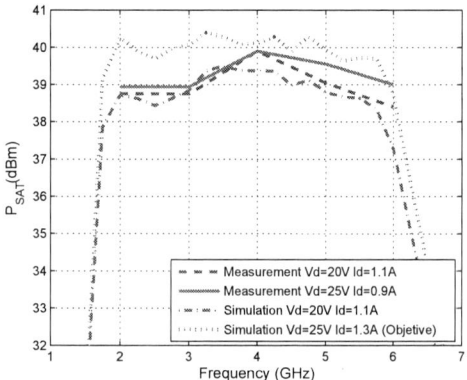

Fig. 8.   SELEX-HPA saturation power.

Measured P_{OUT} and PAE for the SELEX-HPA at $V_{ds} = 20V$ and 4 GHz are shown in Fig.9. The amplifier has 10W of saturation power and better than 25% of PAE. Also, pulsed on-wafer measurements are presented in Fig. 9. Thermal influence did not exist in this measurements because we used a short pulse of 1.1% duty cycle and 20μs of width. In this conditions, 10W were measured at $V_{ds} = 20V$ and 17W at $V_{ds} = 30V$.

V. CONCLUSION

This work presents two AlGaN/GaN broadband high-power MMIC Amplifiers operating from 2 to 6 GHz. A common design methodology has been followed for two different technologies: coplanar approach at Qinetiq's foundry; and microstip approach at Selex's foundry. In the coplanar design there is good agreement between simulations and measurements below 4 GHz, and 8 W saturation output power and 18 dB small-signal gain are achieved. The microstrip amplifier

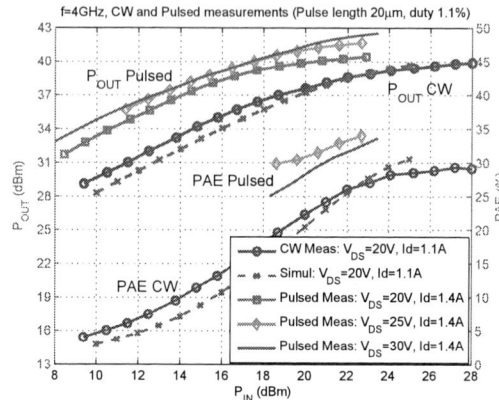

Fig. 9.   SELEX-HPA output power and PAE at 4 GHz measured in CW and pulsed conditions.

has demonstrated a CW saturated output power higher than 8 W and better than 15 dB gain in the 2-6 GHz frequency range. A maximum saturation power of 10 W in CW and 17 W in pulsed conditions have been measured. From preliminary measurements, it seems that the coplanar design has better thermal behaviour, however it exhibits problems of stability and RF response degradation due to unequal potential in circuit ground planes. Microstirp design is in good agrement with simulation predictions and seems more versatile in a broadband design.

ACKNOWLEDGMENT

The authors wish to acknowledge the financial assistance of the Spanish Ministry of Defence and INDRA Sistemas S.A, in the frame of the KORRIGAN project.

REFERENCES

[1] D. E. Meharry, R. J. Lender, K. Chu, L. L. Gunter, and K. E. Beech, "Multi-Watt Wideband MMICs in GaN and GaAs," *2007 IEEE MTT-S Int. Microwave Symp. Dig.*, pp. 1251-1254, June 2007.
[2] U. K. Mishra, P. Parikh, W. Yi-Feng, "AlGaN/GaN HEMTs-an overview of device operation and applications," *Proceedings of the IEEE*, vol. 90, No. 6, pp. 1022-1031, 2002.
[3] J. W. Milligan, S. Sheppard, W. Pribble, Y.-F. Wu, G. Muller, and J. W. Palmour, "SiC and GaN Wide Bandgap Device Technology Overview," *2007 IEEE Radar Conference*, pp. 960-964, April 2007.
[4] K. Takagi, Y. Kashiwabara, K. Masuda, K. Matsushita, H. Sakurai, K. Onodera, H. Kawasaki, Y. Takada, and K. Tsuda, "Ku-band AlGaN/GaN HEMT with Over 30W", *2007 EUMC*, pp. 169-172, October 2007.
[5] K. W. Kobayashi, Y. C. Chen, I. Smorchkova, R. Tsai, M. Wojtowicz, and A. Oki, "1-Watt Conventional and Cascoded GaN-SiC Darlington MMIC Amplifiers to 18 GHz", *IEEE RFIC Symp.*, vol. 3, pp. 1721-1724, 2007.
[6] K. Yamanaka, K. Mori, K. Iyomasa, H. Ohtsuka, H. Noto, M. Nakayama, Y. Kamo, Y. Isota, "C-band GaN HEMT Power Amplifier with 220W Output Power," *IEEE MTT-S Int. Microwave Symp.*, pp. 1251-1254, 2007.
[7] G. Gauthier, Y. Mancuso, and F. Murgadella, "KORRIGAN - a comprehensive initiative for GaN HEMT technology in Europe," *EGAA Symp. 2005*, pp. 361-363, October 2004.
[8] C. Barquinero, A. Suarez, A. Herrera, J. L. Garca, "Complete Stability Analysis of Multifunction MMIC Circuits," *IEEE Tran. on Microwave Theory and Thechniques*, vol. 55, No. 10 pp. 2024-2033, October 2007.
[9] J. Jugo, A. Anakabe and J. M. Collantes, "Control design in the harmonic domain for microwave and RF circuits," *IEE Proceedings on Control Theory and Applications*, vol. 150, No. 2 pp. 127-131, March 2003.

Efficient AlGaN/GaN HEMT Power Amplifiers

R. Quay [*1], F. van Raay[*], J. Kühn[*], R. Kiefer[*], P. Waltereit[*], M. Zorcic[*], M. Musser[*], W. Bronner[*],
M. Dammann[*], M. Seelmann-Eggebert[*], M. Schlechtweg[*], M. Mikulla[*], O. Ambacher[*], J. Thorpe[†],
K. Riepe[†], F. van Rijs[‡], M. Saad[‡], L. Harm [‡], and T. Rödle[‡]

Fraunhofer Institute for Applied Solid-State Physics,
Tullastr. 72, D-79108 Freiburg, Germany
phone: ++49-761-5159-843, fax:++49-761-5159-565
[1]ruediger.quay@iaf.fraunhofer.de

† *United Monolithic Semiconductors, Wilhelm-Runge-Strasse 11, 89081 Ulm, Germany*
‡ *NXP Semiconductors, Gerstweg 2, 6534 AE, Nijmegen, The Netherlands*

Abstract— **This paper describes efficient GaN/AlGaN HEMTs and MMICs for L/S-band (1–4 GHz) and X-band frequencies (8–12 GHz) on three-inch s.i. SiC substrates. Dual-stage MMICs in microstrip transmission-line technology yield a power-added efficiency of $\geq$40% at 8.56 GHz for a power level of $\geq$11 W. A single-stage MMIC yields a PAE of $\geq$55% with 6 W of output power at V_{DS}= 20 V. The related mobile communication power HEMT process yields an average power density of 10 W/mm at 2 GHz and V_{DS}= 50 V. The average PAE is 61.3% with an average linear gain 24.4 dB and low standard deviation of all parameters. The devices yield more than 25 W/mm of output power at 2 GHz when operated in cw at V_{DS}=100 V with an associated PAE of $\geq$60%. The GaN HEMT process with 0.5 μm gate-length yields an extrapolated lifetime of 10^5 h when operated at V_{DS}= 50 V at a channel temperature of 90 °C. When operated at 2 GHz devices with 480 μm gate-width yield a change of the RF power-gain of less than 0.2 dB under high gain-compression at V_{DS}= 50 V and a channel temperature of 250 °C.**

I. INTRODUCTION

GaN/AlGaN HEMTs find various applications for energy-efficient high-power microwave amplification. Typical examples are 3rd generation microwave communications base stations [1] and military radar applications [2]. Industrialization of the technology is continuing and reproducibility, cost, robustness, and reliability are key issues addressed worldwide. This paper gives examples of GaN/AlGaN X-band MMICs with PAE values $\geq$ 40% comparable to GaAs PHEMT MMICs while improving the power density of the MMICs. Further, a powerbar process is given which is suitable for operation at V_{DS}=50 V and yields promising reliability at this bias.

II. EPITAXY AND PROCESS TECHNOLOGY ON THREE-INCH SUBSTRATES

GaN/AlGaN single heterostructures are grown by MOCVD in a 12×3-inch multi-wafer reactor. Reliability and reproducibility of the epitaxy to device quantities are the key issues addressed [3].

A. GaN/AlGaN HEMT MMIC Technology

The three-inch process technology is based on a 0.25μm gate technology and includes optimized field plates. The two-terminal breakdown voltages of the power HEMTs are $\geq$100 V.

The MMIC technology yields a cut-off frequency f_T of 33 GHz at V_{DS}= 28 V. A loadpull power mapping performed at V_{DS}= 28 V in cw-mode for a device with a gate width of 0.48 mm is given in Fig. 1. The mapping covers 21 transistors from a three-inch wafer. The average maximum PAE of the cells is 54%, while the associated RF output power density at -3 dB compression is $\geq$4.5 W/mm at 10 GHz. An average high-gain performance is achieved with $\geq$15 dB linear gain at 10 GHz. This result is reproducible and obtained for five wafers within the process run, as illustrated in Table I showing mean values of 21 cells for each wafer. No additional harmonic matching is applied in this case. After the frontside processing the full three-inch s.i. SiC wafer is thinned to 100 μm thickness and a via-hole backside process is applied featuring source viaholes.

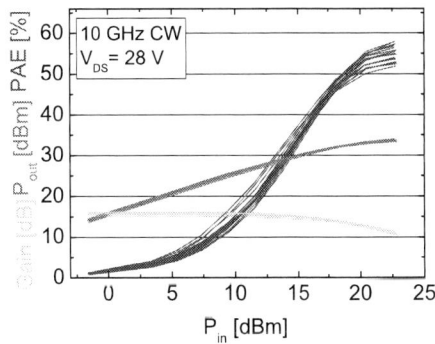

Fig. 1. Cw-loadpull mapping of an AlGaN/GaN HEMT power cell at 10 GHz with W_g= 0.48 mm, PAE (blue), gain (green), output power (red).

TABLE I

CW-AVERAGE LOADPULL PERFORMANCE FOR FIVE THREE-INCH WAFERS
AT 10 GHz AT V_{DS}= 28 V

Wafer	mean per wafer		
	power [W/mm]	lin. gain [dB]	PAE [%]
1	4.5	16	51.5
2	4.75	15.9	54.6
3	4.5	15.7	53.8
4	5	15.2	54
5	5	15	53.1

978-2-8748-7007-1/08 $25.00 © 2008 EuMA

B. Device and MMIC Modeling

The large-signal GaN HEMT modeling is based on an in-house two-dimensional voltage-lag model to accurately describe thermal effects and low-frequency dispersion, and their impact on gain as well as PAE. Following the general theory presented in [5] long term memory effects are described by internal states. This approach facilitates a correct description of retarded responses (such as the DC characteristics) as well as the instantaneous responses relevant for RF power performance. For the MMIC design a library of passive microstrip components is available, including all technology specific elements like MIM capacitors, resistors, and inductances.

III. EFFICIENT ALGAN/GAN MMICS IN MICROSTRIP TRANSMISSION LINE TECHNOLOGY

High-power results with output power levels beyond 20 W at X-band and 3 GHz bandwidth have been demonstrated with previous technology generations [6]. However, the demonstration of multistage MMICs with high-efficiency operation at PAE$\geq$ 40% comparable to GaAs MMICs under realistic operating conditions is so far rare [4]. Fig.2 shows the chip image of the single-stage HPA realized with the optimized AlGaN/GaN MMIC technology. The chip size is $1.75 \times 2.75 \, \mathrm{mm}^2$. The device is matched for bandwidth between 8.5-10 GHz. Fig.3 gives the chip image of the dual-

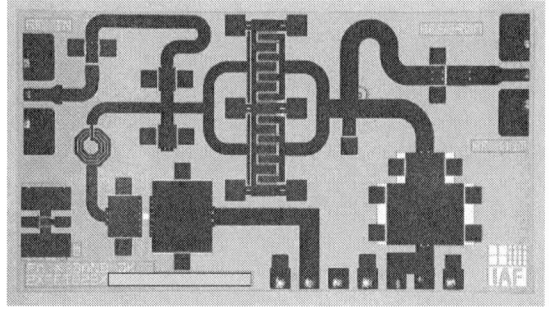

Fig. 2. Chip image of the single-stage X-band HPA, chip size $1.75 \times 2.75 \, \mathrm{mm}^2$.

stage HPA realized with the optimized AlGaN/GaN MMIC technology. The chip size is $4.5 \times 3 \, \mathrm{mm}^2$ in this case.

A. MMIC RF Small-Signal and Power Performance

Fig. 4 gives the measured S-parameters of the dual-stage HPA measured in cw-operation at V_{DS}=28 V. The device yields a S_{21} (red) of 18 dB between 8.5 and 11 GHz, an output match S_{22} (blue) of better 10 dB between 9.5 and 12 GHz, and an input match S_{11} (green) of better than 8 dB between 8 and 11 GHz.

In the following, all large-signal MMIC RF results are given for a duty-cycle of 10% with long DC-pulses of $100 \, \mu s$ pulse length. Fig. 5 gives the output power measured at 8.24 GHz for the single-stage amplifier. The linear gain is 12 dB and the compression level at the reported efficiency is -3 dB. Biased at V_{DS}= 20 V, the single-stage MMIC yields a maximum PAE of

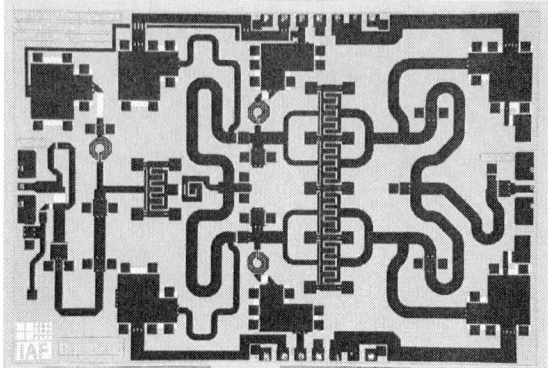

Fig. 3. Chip image of the dual-stage HPA, chip size $4.5 \times 3 \, \mathrm{mm}^2$.

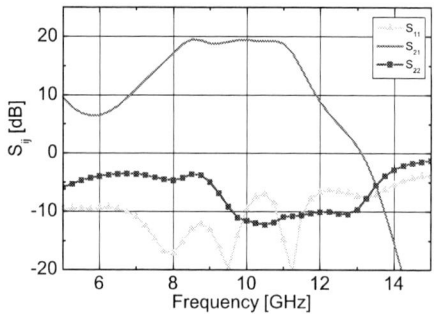

Fig. 4. Cw- S-parameters of the dual-stage HPA measured at V_{DS}= 28 V.

$\geq$55% with 6 W of output power. When the device is biased at V_{DS}= 28 V for maximum efficiency, the PAE is $\geq$45 % with an associated output power of 40 dBm and a power density of 5 W/mm for the same frequency, gain, and the same pulsed conditions.

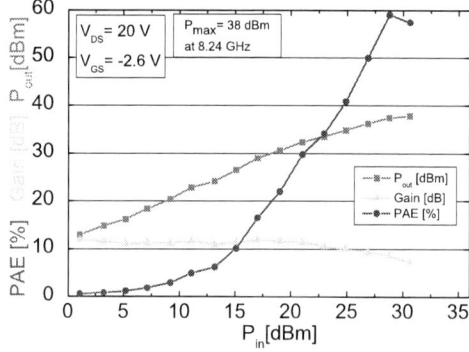

Fig. 5. Measured pulsed output power, gain and PAE vs. input power of the single-stage amplifier at $V_{DS} = 20 \, V$.

978-2-8748-7007-1/08 $25.00 © 2008 EuMA

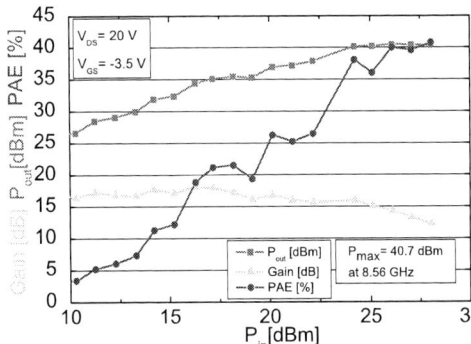

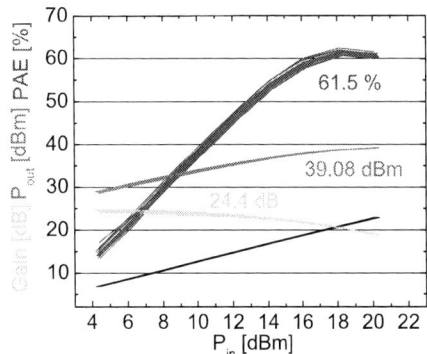

Fig. 6. Measured pulsed output power, gain and PAE vs. input power of the dual-stage amplifier at $V_{DS} = 20\,V$.

Fig. 7. Cw-loadpull mapping of an AlGaN/GaN HEMT power cell at 2 GHz with W_g= 0.8 mm with the mean values given for 21 cells biased at V_{DS}= 50 V.

The dual-stage MMIC biased at V_{DS}= 20 V for maximum efficiency in class-A/B yields a PAE of 40% with an associated output power of 40.7 dBm in pulsed operation, as shown in Fig. 6. The linear gain is 17 dB, the compression level of the reported efficiency is -3 dB.

IV. L-/S-BAND GAN POWER HEMTS

The increase in operation voltage, the switching speed, and the robustness are attractive for new architectures of RF amplifiers in base station applications for mobile communication.

A. GaN/AlGaN HEMT Powerbar Technology

The power bar HEMT technology is dedicated to operation between 0.5–4 GHz. It is based on a gate length of l_g= 0.5 μm. The operation bias is 50 V with an off-state breakdown voltage of 160 V. The current-gain cut-off frequency is 16 GHz measured at V_{DS}=50 V. Further details on epitaxy and device technology are given in [3].

Fig. 7 gives the on-wafer cw-mapping of an AlGaN/GaN HEMT power cell biased at V_{DS}= 50 V, again given for 21 cells per wafer. The average values of maximum PAE, linear gain, and output power are given. The devices yield an average PAE of $\geq$60% which has been reproduced for $\geq$20 three-inch wafers. Furthermore, the associated output power density is 10 W/mm. The average linear gain is 24.4 dB.

HEMTs with a periphery of 0.8 mm yield cw-output-power densities of beyond $\geq$26 W/mm at a bias of V_{DS}= 100 V and 2 GHz with an associated PAE of $\geq$ 60%, as shown in Fig. 8. The backside of the wafer is kept at a constant temperature of 25°C to assure constant conditions. This result has been observed repeatedly without device failure, although no longterm reliable is claimed under these extreme conditions. However, such measurements shows the electrical potential and the robustness of the technology.

On large periphery devices with 32 mm gate-width packaged in standard ceramic housings a cw-output power beyond 125 W is obtained with a PAE above 50%, a drain efficiency above 55%, and a linear gain of $\geq$15 dB at 2.14 GHz at V_{DS}=50 V.

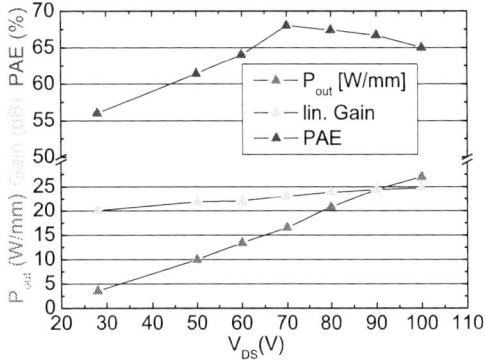

Fig. 8. Cw-loadpull results of an AlGaN/GaN HEMT power cell at 2 GHz with W_g= 0.8 mm for various V_{DS}.

B. DC-Device Reliability

Longterm device reliability is a key requirement any application. Fig. 9 gives extrapolated reduction of the drain current as a function of time for an AlGaN/GaN HEMT with a gate length of 0.5 μm operated at V_{DS}= 50 V and a low quiescent drain current of I_{Dq}=50 mA/mm under cw-DC-conditions. The test is performed for 1000 hours with four devices each from two different subsequent processing iterations. No burn-in procedure is used. The results of the two iterations do intermix well and are consistent. For a failure criterion of a drain current reduction of -20%, we observe an extrapolated lifetime of 2×10^5 h (20 years) for a channel temperature of T_{chan}=90°C. The development of the gate leakage currents over time (not shown) gives no significant change at a low leakage level of $I_g \leq$30 μA/mm.

C. RF-Device Reliability

Fig. 10 gives the change of the difference of output and input power for an aging test of a 0.48 mm device under

978-2-8748-7007-1/08 $25.00 © 2008 EuMA

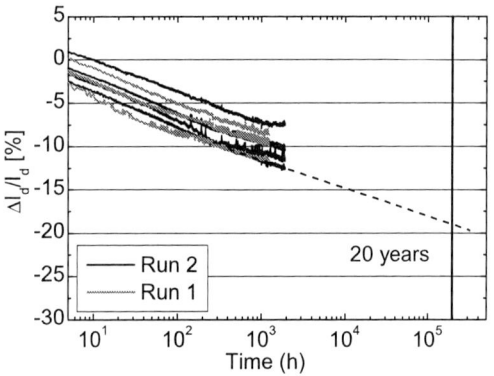

Fig. 9. Measured and extrapolated change of drain currents of 2×4 HEMTs from two processing iterations operated at $V_{DS} = 50\ V$ and $T_{chan} = 90°C$.

RF-compression at 2 GHz in continuous-wave operation. The channel temperature amounts to 250 °C at an operation bias $V_{DS} = 50\ V$. The overall change is in the range of -0.2 dB for the four devices tested. The associated change of the drain currents is less than 5%, the average leakage currents change less than 10%. These results are promising for the application of this technology.

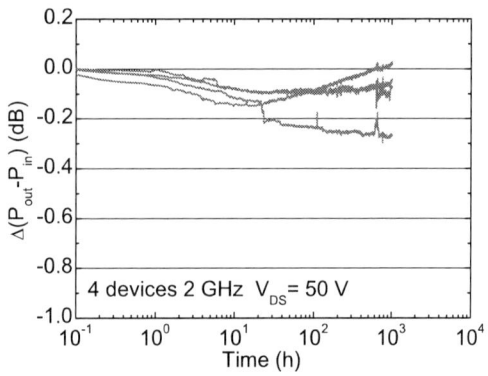

Fig. 10. Change of the difference output and input power at 2 GHz when operated at $V_{DS} = 50\ V$ and $T_{chan} = 250°C$.

V. CONCLUSIONS AND OUTLOOK

This paper describes efficient GaN/AlGaN HEMTs and MMICs for L/S- band (1-4 GHz) and X-band frequencies (8–12 GHz) on three-inch s.i.SiC substrates.

Dual-stage MMICs in microstrip transmission line technology yield a power-added efficiency of $\geq$40% at 8.56 GHz for power levels of $\geq$11 W. A single-stage MMIC yields a PAE of $\geq$55%, a linear gain of 12 dB %, and 7 W of output power at $V_{DS} = 20\ V$ and 8.24 GHz. The related mobile communication power HEMT process yields an average power density of 10 W/mm at 2 GHz and $V_{DS} = 50\ V$. The average PAE is 61.3% with a linear gain of 24.4 dB and low standard deviation. Very high operation bias of 100 V demonstrate the robustness of the devices with maximum output power densities of 25 W/mm and PAE $\geq$60% at 2 GHz. The process yields an extrapolated lifetime of 10^5 h when operated at $V_{DS} = 50\ V$ at a channel temperature of 90 °C. The results are promising with respect to the sustainable industrialization of efficient and reliable GaN/AlGaN HEMTs and MMICs.

ACKNOWLEDGMENTS

The authors acknowledge the continuous support of the Federal Ministry of Defense (BMVg), Bonn, the Bundeswehr Technical Center for Information Technology and Electronics (WTD 81), Greding, and the support of the Federal Ministry of Education and Research (BMBF), Bonn.

REFERENCES

[1] S. Singhal, A.W. Hanson, A. Chaudhari, P. Rajagopal, T. Li, J.W. Johnson, W. Nagy, R. Therrien, C. Park, A.P. Edwards, E.L. Piner, K.J. Linthicum, I.C. Kizilyalli, Qualification and Reliability of a GaN Process Platform, *Proc. CS Mantech, Austin*, 2007.

[2] M.J. Rosker, The Present State of the Art of Wide-Bandgap Semiconductors and Their Future,*RFIC Tech Dig.* 2007, Honolulu.

[3] P. Waltereit, W. Bronner, R. Quay, M. Dammann, S. Müller, H. Kiefer, H. Walcher, F. van Raay, O. Kappeler, M. Mikulla, F. van Rijs, T. Rödle, S. Murad, J. Klappe, P. van der Wel, P. Henriette, B. Aleiner, I. Blednov, J. Thorpe, R. Behtash, H. Blanck, K. Riepe, A Uniform, Reproducible and Reliable GaN HEMT Technology with Breakdown Voltages in Excess of 160 V Delivering more than 60% PAE at 80 V, *Proc. CS Mantech 2008*, Chicago.

[4] J.S. Moon, D. Wong, M. Antcliffe, P. Hashimoto, M. Hu, P. Willadsen, M. Micovic, H. P. Moyer, A. Kurdoghlian, P. MacDonald, M. Wetzel, and R. Bowen, High PAE 1 mm AlGaN/GaN HEMTs for 20 W and 43% PAE X-band MMIC Amplifiers *IEDM Technical Digest*, pp. 385–388, San Francisco, 2006.

[5] M. Seelmann-Eggebert, T. Merkle, F. van Raay, R. Quay, M. Schlechtweg, A Systematic State-space Approach to Large-signal Transistor Modelling, *IEEE Transactions on Microwave Theory and Techniques* 55 (2007), no. 2, pp. 195–205.

[6] F. van Raay, R. Quay, R. Kiefer, W. Bronner, M. Seelmann-Eggebert, M. Schlechtweg, M. Mikulla, and G. Weimann, X-Band High-Power Microstrip AlGaN/GaN HEMT Amplifier MMICs, *IEEE MTT-S International Microwave Symposium Digest*, pp. 1368–1371, San Francisco, 2006.

AlGaN/GaN HEMT with over 110 W Output Power for X-Band

ShiChang Zhong, Tangsheng Chen, Chunjiang Ren, Gang Jiao, Chen Chen, Kai Shao, Naibin Yang

National Key Laboratory of Monolithic Integrated Circuits and Modules, P.O.Box 1601,Nanjing 210016, P.R.China

Nanjing Electronic Devices Institute, P.O.Box 1601,Nanjing 210016, P.R.China

superwindcn@yahoo.com.cn

Abstract— **AlGaN/GaN HEMT using field plate and recessed gate for X-band application was developed on SiC substrate. Internal matching circuits were designed to achieve high gain at 8GHz for the developed device with single chip and four chips combining, respectively. The internally matched 5.52mm single chip AlGaN/GaN HEMT exhibited 36.5W CW output power with a power added efficiency (PAE) of 40.1% and power density of 6.6W/mm at 35V drain bias voltage (Vds). The device with four chips combining demonstrated a CW over 100W across the band of 7.7-8.2GHz, and an maximum CW output power of 119.1W with PAE of 38.2% at Vds =31.5V. This is the highest output power for AlGaN/GaN HEMT operated at X-band to the best of our knowledge.**

I. INTRODUCTION

There is a growing requirement for high solid-state power amplifiers in wireless communication and military applications. AlGaN/GaN HEMT has attracted much research interest due to the inherent advantages of high voltage, high frequency and high power density and is a candidate for next generation power transistor used in solid-state power amplifier. In recent years, many papers reported high performance AlGaN/GaN HEMTs with output power over 100W and even 1kW at L-/S-/C-band [1-4]. X-band AlGaN/GaN HEMTs with 87.1W output power have also been exhibited [5]. However no papers reported their output power with over 100W at X-band or above.

In this paper, we present our result of a high output power device combined with four 5.52mm single chip AlGaN/GaN HEMT for X-band application. The developed internally matched AlGaN/GaN HEMT has excellent performance. It exhibited an output power of over 100 W across the band of 7.7-8.2GHz at 31.5V drain bias voltage and a maximum output power of 119.1W with a PAE of 38.2% at 8GHz. The chip has no via holes structure.

To the best of authors' knowledge, this is the highest power AlGaN/GaN HEMT ever reported for X-band application.

II. DEVICE STRUCTURE

Fig.1 shows schematic cross-section of the developed AlGaN/GaN HEMT [6,7]. The epitaxial structure including a AlN nucleation layer, a GaN buffer, a AlN interlayer, a AlGaN barrier layer, and a thin GaN cap was grown on a semi-insulating SiC substrate by MOCVD. All semiconductor layers were not intentionally doped and the nominal Al composition of AlGaN was chosen to be 0.25. Filed plate and recessed gate were employed during the AlGaN/GaN HEMT processing to improve device performances. Gate length and field plate of the HEMT were chosen to 0.35μm and 0.5μm, respectively.

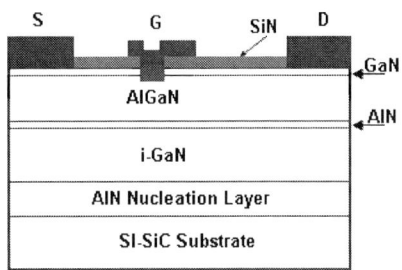

Fig.1 Schematic cross-section of the developed AlGaN/GaN HEMT

Fig.2 shows DC and pulse I-V characteristics of the manufactured 1mm gate-periphery device. The HMET has a saturation drain current of about 1.25A/mm with a pinch-off voltage of -2.5V. Pulse I-V measurements were carried out under a quiescent bias of (Vgs=-4V, Vds=30V), and the pulse width is 200ns. Excellent agreement of the DC and pulse I-V characteristics indicates that the device is almost current collapse free.

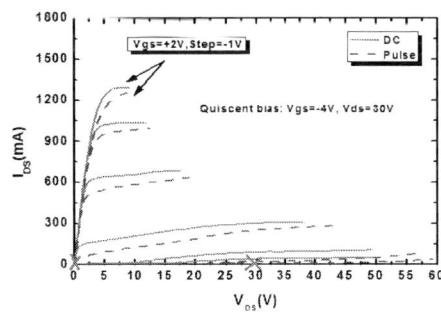

Fig.2 DC and pulse I-V characteristics of a 1mm AlGaN/GaN HEMT

978-2-8748-7007-1/08 $25.00 © 2008 EuMA

Top view of the finished device with a total gate-width of 5.52mm is shown in Fig.3. The device has a unit finger width of 115μm. To lower channel temperature and improve reliability, the gate-to-gate space of the chip was extended to 55μm and the SiC substrate was thinned to 100μm. The size of chip is 2.7mm×0.6mm×0.1mm.

Fig.3 Photography of the developed AlGaN/GaN HEMT with a gate width of 5.52mm. Chip size: 2.7mm×0.6mm×0.1mm

III. MATCHING CIRCUIT DESIGN

The internal matching circuits design of the developed device were based on an empirical methodology [8,9] and the circuits were optimized to achieve high gain characteristics at 8GHz. Fig.3a is the matching circuit network for a single 5.52mm AlGaN/GaN HEMT. Fig.3b is the matching circuit network for 4×5.52mm chips. The input and output matching circuits were designed using large-signal load impedance determined from load-pull measurements of a 1mm AlGaN/GaN HEMT. The results were then scaled to model a 5.52mm chip. The input and output impedances of the 5.52mm AlGaN/GaN HEMT under large signal operation are about 2Ω and 10Ω.

For the circuits for single 5.52mm single chip, a quarter wave transformer and a L-C-L section transformer were used to transform the input impedance from 2Ω to 50Ω. Another quarter wave transformer and a L-C section transformer were used to transform the output impedance from 10Ω to 50Ω.

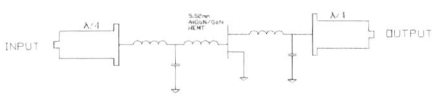

a. circuit network of a single 5.52mm chip

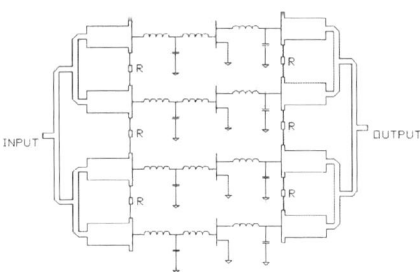

b. circuit network of 4×5.52mm chips

Fig.4 Matching circuit network of AlGaN/GaN HEMT

In the 4×5.52mm unit circuit network, two Wilkinson dividers with the same impedance levels at each branch port

as those in the single 5.52mm unit were employed to combine four 5.52mm chips. The quarter wave transformer and the Wilkinson combiners were realized with microstrip line on 380μm thick Al2O3 ceramic (εr=9.6) material. The lumped capacitors were realized on 150μm thick BaTiO3 (εr=78) substrate. The resistances of about 10Ω between the terminals of circuit closed to chips were employed to prevent the odd oscillation. Fig.4 shows the photograph of internally matched 4×5.52mm AlGaN/GaN HEMT. The circuits are housed in a conventional hermetic copper package. The package size excluding flange and leads is 17.4mm×24mm.

Fig.5 Photography of the internally matched AlGaN/GaN HEMT with 4 chips. Package size: 17.4mm×24mm

IV. RF PERFORMANCE

Fig.6 shows the operating drain voltage(Vds) dependence of saturated output power(Psat) , power-added efficiency(PAE), gain and power density of a single 5.52mm device at 8 GHz. Fig.7 shows the output power, PAE and linear gain versus input power at a drain voltage of 35V and frequency of 8 GHz. It is noticed that single device has exhibited 6.6W/mm power density with saturated output power of 36.5W. The power density is very high at this power level.

Fig.8 shows the frequency response of saturated output power, PAE, Gain and saturated power density of the 4×5.52mm internally matched AlGaN/GaN HEMT at a drain voltage of 31.5V and an input power level of 44dBm. It is noticed that the four-chip units have demonstrated over 100W and the flatness less than ±0.4dB in range of 7.7-8.2GHz. Fig.9 shows the output power, PAE and linear gain versus input power of the 4×5.52mm device at a drain voltage of 31.5V and frequency of 8 GHz. You may notice that the internally matched device combined four 5.52mm AlGaN/GaN HEMTs demonstrated an output power of 50.76dBm(119.1W) with 8.5dB linear gain and 38.2% PAE under CW operation and an input power level of 44dBm. To the best of authors' knowledge, the saturated output power of 119.1W under CW operation   is highest power AlGaN/GaN HEMT ever reported for X-band application.

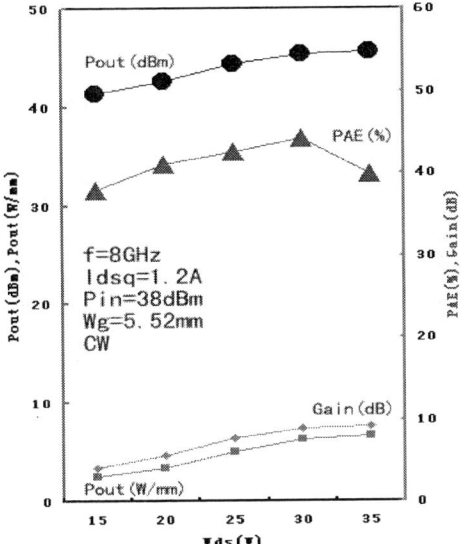

Fig.6 Operating voltage dependence of Psat, PAE, Gain, power density under CW condition at 8GHz. Wg=5.52mm

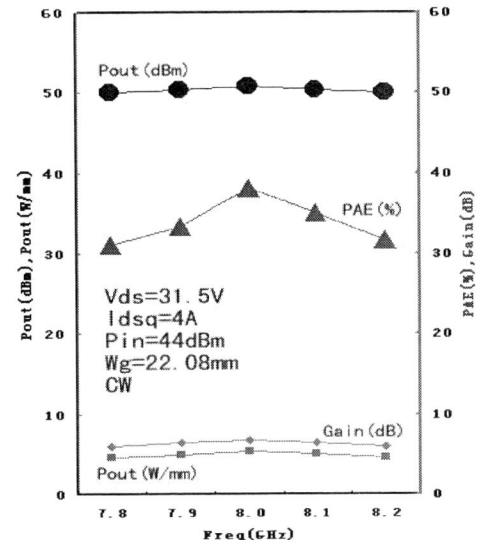

Fig.8 Psat, PAE, Gain and power density across the band of 7.8-8.2GHz at drain bias voltage of 31.5V, CW operation. Wg=4×5.52mm

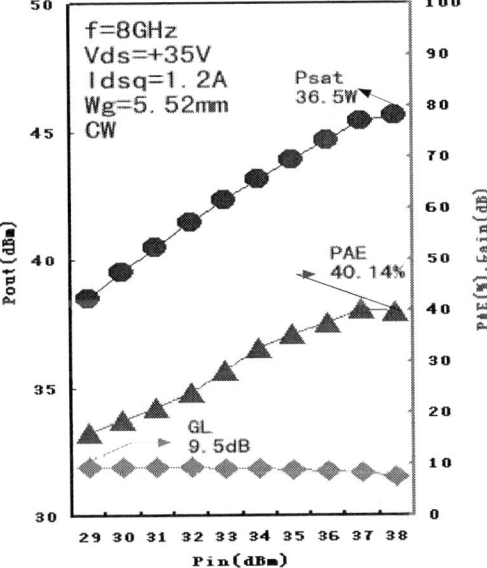

Fig.7 Input power dependence of Psat, PAE and Gain under CW operation at 8GHz. Wg=5.52mm

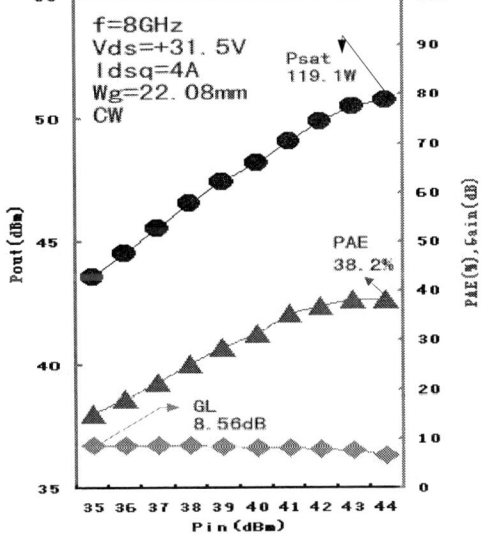

Fig.9 Input power dependence of Psat, PAE and Gain under CW operation at 8GHz. Wg=4×5.52mm

V. CONCLUSION

In this study, the developed 4×5.52mm AlGaN/GaN HEMT exhibited a 50.76dBm(119.1W) output power with 38.2% power-added efficiency and 8.5 dB linear gain at 8GHz. Especially the developed single 5.52mm AlGaN/GaN HEMT has demonstrated a power density of 6.6W/mm which is about ten times than the GaAs FET. It fully shows the advantage of AlGaN/GaN HEMT as a promising candidate for the next generation of microwave power transistor.

ACKNOWLEDGMENT

The authors would like to thank Prof. Wang Xiaoliang from the Institute of Semiconductors, Chinese Academy of Science for the support of GaN epi wafers. Additionally, Thanks are given to Yunsheng Luo, Bin Zhang, Liyun Wu and Shijun Tang for their supports and suggestions.

REFERENCES

[1] N. Ui, S. Sano, "A 100W class-E GaN HEMT with 75% Drain Efficiency at 2GHz," *Proc. 1nd European Microwave Integrated Circuits Conf.*, pp. 72-74, 2006

[2] E. Mitani, M. Aojima, S. Sano, "A kW-class AlGaN/GaN HEMT Pallet Amplifier for S-band High Power Application，" Proc. 2nd *European Microwave Integrated Circuits Conf.*, pp. 176-179, 2007.

[3] K. Yamanaka, K.mori, K.Iyomasa, H. Ohtsuka, H. Noto, M.Nakayama, Y. Kamo, "C-band GaN HEMT Power Amplifier with 220W output power," *IEEE MTT-S digest*, pp.1251-1254, 2007.

[4] K. Yamanaka, K. Iyomasa, H. Ohtsuka, M. Nakayama, Y. Tsuyama, T. Kunii, Y. Kamo, T. Takagi, "S and C band over 100W GaN HEMT 1chip high power amplifiers with cell division configuration," *Gallium Arsenide and Other Semiconductor Application Symposium, EGAAS 2005*. European, pp. 241-244, 2005.

[5] M. Nishihara, T. Yamamoto, K. Inoue, M. Nishi, S. Sano, "An 87W AlGaN/GaN HEMT for X-band Pulse Operation," *IEICE technical Report,* ED2007-211, pp. 29-31, 2008.

[6] Chen Tangsheng. Jiao Gang, Li Zhonghui, Li Fuxiao, Shao Kai, Yang Naibin, "AlGaN/GaN MIS HEMT with AlN Dielectric," *GaAs MANTECH Conf Proc.*, 2006.

[7] Chen Tangsheng, Wang Xiaoliang, Jiao Gang, Zhong Shichang, Ren Chunjiang, Chen Chen, Li Fuxiao. "Recessed-Gate AlgaN/GaN HEMTs with Field-modulating Plate, " *Chinese Journal of Semiconductors*, to be published.(in Chinese)

[8] S. T. Fu, J. J. Komiak, L.F. Lester, K. H. G. Duh, P. M. Smith, P.C. Chao, and T. H. Yu, "C-band 20W Internally Matched GaAs Based Pseudomorphic HEMT Power Amplifiers," *1993 GaAs IC Symposium,* pp. 355-358, 1993.

[9] Zhong Shichang, Chen Tangsheng, Lin Gang, Li Fuxiao, "8-Watt Internally Matched GaAs Power Amplifier for 16-16.5 GHz Band," *ICSICT*, 2006

978-2-8748-7007-1/08 $25.00 © 2008 EuMA

Proceedings of the 3rd European Microwave Integrated Circuits Conference

Balanced Microstrip AlGaN/GaN HEMT Power Amplifier MMIC for X-Band Applications

J. Kühn, F. van Raay, R. Quay, R. Kiefer, W. Bronner, M. Seelmann-Eggebert, M. Schlechtweg, M. Mikulla, O. Ambacher, and M. Thumm*

Fraunhofer Institute for Applied Solid-State Physics, Tullastr. 72, D-79108 Freiburg, Germany
phone: ++49-761-5159-842, fax:++49-761-5159-565
jutta.kuehn@iaf.fraunhofer.de

*Universität Karlsruhe (TH), Institut für Höchstfrequenztechnik und Elektronik (IHE),
Kaiserstr. 12, D-76131 Karlsruhe, Germany

Abstract—This paper describes a balanced AlGaN/GaN HEMT single-stage power amplifier demonstrator for X-band frequencies in microstrip line technology on thinned s.i. SiC substrates. The design features a modular circuit concept and microstrip MMIC directional couplers with low impedance levels. These 3 dB-couplers designed for a center frequency of 10 GHz show a coupling factor of 3.5 dB ± 0.4 dB and a low net insertion loss of 0.3 dB. The balanced amplifier reaches 11 W pulsed output power at 3 dB compression level and a maximum gain of 10 dB at 8.56 GHz with an input and output match of better than 14 dB from 8.3 to 13 GHz. This 0°/90° balanced microstrip AlGaN/GaN HEMT power amplifier MMIC demonstrator may be an interesting alternative to existing hybrid solutions.

I. INTRODUCTION

During the last decade, impressive performances have been achieved with balanced power amplifiers using GaAs-based HEMT MMIC technology for microwave applications, e.g. published in [1]. For the emerging AlGaN/GaN material system, a recent example of a low noise balanced MMIC amplifier using coplanar waveguide (CPW) Lange couplers is published in [2] and a high power amplifier in [3]. Balanced Amplifiers are advantageous for application fields where strong output (and/or input) mismatches occur, e.g. high power measurement setups or X-band near-field radars. Compared to CPW technology, the use of microstrip technology in power amplifiers yields smaller chip size, simple RF-grounding scheme, and direct MMIC compatibility with commercial hybrid microstrip radar modules [4], [5]. Balanced microstrip MMIC power amplifiers using AlGaN/GaN HEMTs may be an interesting alternative to existing hybrid solutions. In this work, a first demonstrator for balanced microstrip MMIC HPAs with saturated output power beyond 10 W was designed.

II. CIRCUIT DESIGN

The 0°/90° balanced approach allows the realization of X-band radar HPAs with superior output matching and bandwidth, because the output return loss is nearly independent of the output reflection coefficient occurring at the output terminals of the active devices in the 0°/90° balanced configuration. A key feature of the design is the use of low-impedance microstrip Lange couplers on input and output side. This ensures a sufficient power and current handling capability

and avoids critical microstrip line width and gap dimensions. Furthermore, the output coupler impedance of 25 Ω fits to the total optimum load impedance level of the output stage of a GaN HEMT HPA in the 10 to 20 W power region, in contrast to the much lower impedance levels in, e.g., GaAs pHEMTs. This leads to simple and broadband matching circuits when this concept is applied to GaN.

The block diagram of the balanced amplifier is shown in Fig. 1, illustrating the modular circuit concept using single-stage amplifiers, impedance level transformers, and low-impedance Lange couplers. The balanced amplifier consists of two identical single-stage amplifiers PA 1 and PA 2 connected by two 3 dB directional couplers. The Lange couplers are located at input and output port of the amplifiers to improve the return losses and to achieve simple and easy matching of the amplifier stages. In the following subsections, the single amplifier modules consisting of the amplifier PA 1-2 and the Lange couplers, as well as the resulting whole balanced amplifier are described in detail.

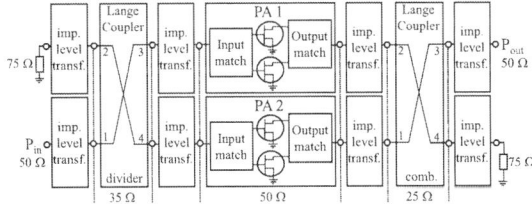

Fig. 1. Block diagram of the Balanced AlGaN/GaN HEMT power amplifier MMIC.

A. Single-Stage Amplifier Design

At first, a single-stage power amplifier PA 1-2 consisting of two matching networks at the input and output of the amplifier, and two parallel 8×125 µm GaN HEMT cells using a T-gate technology with a gate length of 250 nm was designed. Further descriptions of the AlGaN/GaN HEMT technology, and the MMIC and backside process can be found in [6]. The design is based on load-pull simulations of a 1 mm GaN FET base cell using an in-house GaN-specific large-signal model. The

978-2-8748-7007-1/08 $25.00 © 2008 EuMA   95

power amplifier is dimensioned for maximum output power operation in the frequency range from 8 to 10 GHz and has a total gate width W_g of 2 mm. Both input and output matching networks are designed for 50 Ω external load for easy on-wafer testability and comprise simple narrowband line-capacitor structures.

B. Microstrip Lange Coupler Design

The Lange coupler is used both as a power divider at the input and a power combiner at the output of the balanced amplifier, as described in [7]. To achieve a sufficient current handling capability on the coupling lines, the Lange coupler of the balanced amplifier's input operates at an impedance level of 35 Ω while the output impedance of the coupler is chosen to be 25 Ω. The interdigitated microstrip couplers consist of four parallel lines at the input and six parallel lines at the output of the balanced amplifier with alternate lines connected together. The mechanical lengths of input and output coupler are 2717 μm and 2671 μm, respectively. Both lengths realized on SiC substrate with an ϵ_r = 9.7 correspond approximately to a quarter wave length ($\lambda/4$) at the center frequency of 10 GHz. Other critical dimensions of Lange couplers are the coupled line width w which determines the current carrying capacity, and the gap s between the single lines which strongly influences the coupling factor and which therefore turns out to be sensitive to MMIC technology deviations, especially in galvanic metallization structures. For a metal thickness of t = 7 μm, both dimensions are calculated with Agilent ADS LineCalc to w = 29.3 μm and s = 13.3 μm for the 25 Ω output coupler. For a 50 Ω 6-line coupler on SiC centered to 10 GHz, the length, width and gap geometries are: l = 2980 μm, w = 10.5 μm and s = 10 μm. For the same coupler on GaAs (ϵ_r = 12.9), the values amount to l = 2588 μm, w = 6.7 μm and s = 9.6 μm. In spite of the shorter length of the coupler on GaAs, the w and s geometries turn out to be too tolerance sensitive to galvanic metallizations, especially for GaAs. In addition, the width w would be too small concerning the required current loading for RF output powers beyond 10 W.

C. Balanced Power Amplifier Design

The block configuration of the balanced amplifier displayed in Fig. 1 was chosen with regard to a good testability of the single amplifier modules. This modular concept leads to different impedance levels of the power amplifiers and Lange couplers. Therefore, impedance matching networks between the power amplifiers and couplers are necessary. These impedance level transformations are simple single-stage line-capacitor networks operating over 8 to 10 GHz bandwidth with good performance. Due to the approximately $\lambda/8$ line length of the transformers, these line-capacitor networks fit just between coupler and PA 1-2 and matching network on input and output side.

The isolated ports of the Lange couplers are terminated by Z_L = 75 Ω resistors. The higher impedance level leads to a better current drive capability due to an increase of Z_L by the factor 1.5 with a given maximum line width. On the other hand, the higher resistance leads to a slightly lower bandwidth, because an additional matching network from 25/35 Ω to 75 Ω is required. The required chip size of the whole balanced amplifier is 4×3 mm^2. The chip photograph is shown in Fig. 2.

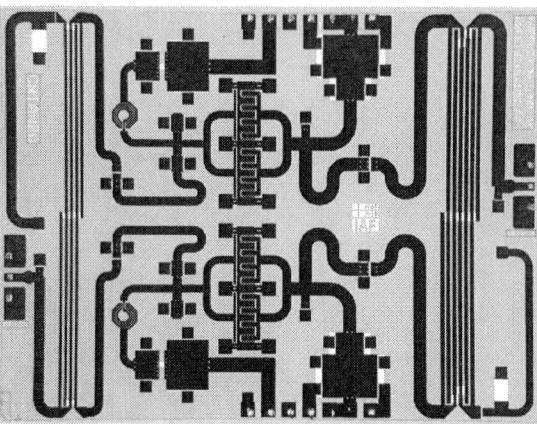

Fig. 2. Chip photograph of the balanced amplifier MMIC with a total gate width of W_g = 4 mm and a chip size of 4×3 mm^2.

III. CHARACTERIZATION

All measurements are performed in an on-wafer configuration with a thermo-chuck held at a temperature of 297 K. In the following subsections, small-signal and RF power simulation and measurement results will be discussed.

A. Small-Signal Performance

1) Single-Stage Power Amplifier: The simulated and measured small-signal results of the single-stage power amplifier PA 1-2 are shown in Fig. 3. The bias was V_{GS} = -2.6 V

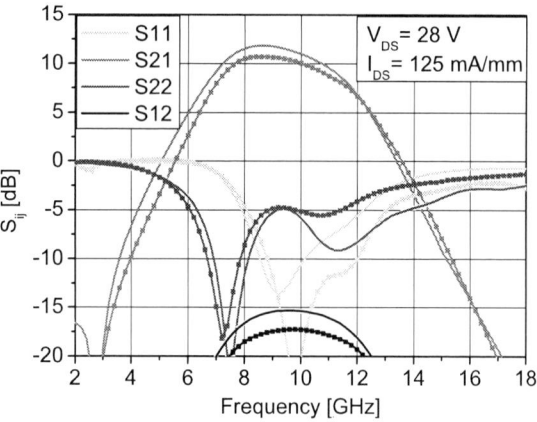

Fig. 3. Simulated and measured S-parameters of the single-stage power amplifier at V_{DS} = 28 V. Simulation: Lines, Measurement: Lines with squares.

and V_{DS} = 28 V and a drain current of I_D = 125 mA/mm.

Simulation and measurement are in good agreement over the whole frequency range. A measured maximum gain of 10.6 dB is achieved for a frequency of 8.8 GHz with an input match of better than 10 dB and an output match of better than 5 dB. The measured reverse isolation is better than 17 dB over the whole frequency range.

2) Microstrip Lange Coupler: The test structure of the output microstrip Lange coupler was measured on-wafer in a two-port 50 Ω environment with the two unused ports terminated by 50 Ω resistors. The 50 Ω two-port data was transformed into a 25 Ω four-port matrix. Within ADS, the coupler was simulated with a 25 Ω port impedance. In Fig. 4, the simulated and measured through connection (S_{31}) and coupling factor (S_{32}) of the output coupler are compared. Simulated and measured data show a good agreement for lower frequencies. However, for higher frequencies the measured S-parameters show a decreasing coupling effect which can be explained by the difference between the simulated and realized gap between the single coupler lines. The symmetry performance of the coupler is still acceptable. In a frequency range from 8 to 10 GHz, the coupling factor for both simulation and measurement is 3.5 dB $\pm$ 0.4 dB. The coupler achieves a simulated directivity of 22 dB and a measured value of 17 dB at 10 GHz. The simulated and measured net insertion loss of the output coupler amount to 0.23/0.29 dB and are in good agreement. Low net insertion loss of the output Lange coupler is extremely important for optimum power transfer from the amplifier stages to the output load. This was achieved in spite of compact realization on a 100 μm thin SiC MMIC substrate with rather small line cross-section.

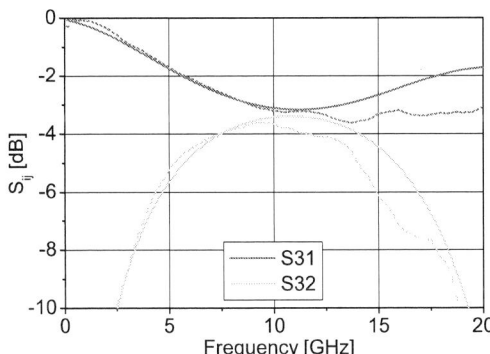

Fig. 4. Overlay of simulated and measured through connection (S_{31}) and coupling factor (S_{32}) of the output microstrip Lange coupler referenced to 25 Ω. Simulation: Line, Measurement: dashed Line.

3) Balanced Power Amplifier: Small-signal simulation and measurement of the balanced power amplifier were performed as well at V_{DS} = 28 V with V_{GS} = -2.8 V and I_D = 115 mA/mm. The simulated and measured S-parameters are plotted in Fig. 5. Simulation and measurement reveal

an excellent agreement over the whole frequency range. A maximum small-signal gain of 10.7 dB is measured at 8.8 GHz in CW mode. The amplifier modules PA 1-2 are connected to the couplers via rather simple impedance matching circuits optimized for 8 to 10 GHz (see above). Therefore, the gain of the balanced amplifier is limited to this frequency range. These additional impedance level transformers cause further losses. In spite of this, the simulated/measured S_{21} performance of the balanced amplifier is in good agreement with the S_{21} of the PA 1-2 shown above. The difference of 0.5 dB less gain in the balanced amplifier at the band edges 8 GHz and 10 GHz corresponds to the net insertion losses of the couplers and the impedance level transformer networks. The input and output matching of the balanced amplifier are better than 14 dB from 8.3 to 13 GHz which is a considerable advantage compared to the single-stage amplifiers PA 1-2, and reverse isolation S_{12} is better than 17 dB for all frequencies.

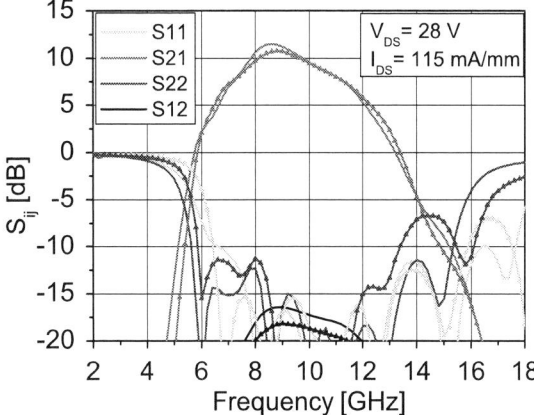

Fig. 5. Simulated and measured small-signal gain and input and output return losses versus frequency of the balanced amplifier at V_{DS} = 28 V. Simulation: Lines, Measurement: Lines with Triangles.

B. RF Power Performance

Fig. 6 presents a comparison between large-signal results of the simulations in CW mode and pulsed RF measurements with 10 % duty cycle and 100 μs pulse width. Simulated and measured data are displayed for a frequency of 8.56 GHz using V_{GS} = -2.9 V with V_{DS} = 20 V over an input power range from 15 dBm to 35 dBm. A maximum pulsed output power of 11 W (40.5 dBm) at 3 dB compression level is reached for 34 dBm input power. The measured power added efficiency (PAE) at maximum power is approx. 40 % and the associated gain amounts to 7 dB. The balanced amplifier features a maximum small signal gain of 10 dB. For comparison, the single-stage power amplifier PA 1-2 reaches a maximum pulsed output power of 4.5 W (36.7 dBm) at 3 dB gain compression level for V_{DS} = 20 V, a PAE of approx. 45 % at maximum power, and a small signal gain of 11.5 dB.

978-2-8748-7007-1/08 $25.00 © 2008 EuMA

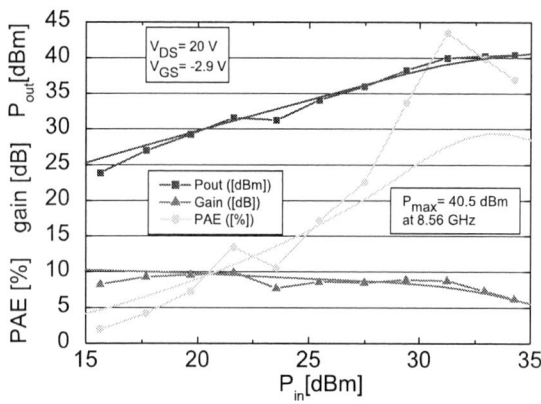

Fig. 6. Simulated and measured pulsed output power, gain and PAE vs. input power of the balanced amplifier at $V_{DS} = 20\ V$. Simulation: Line, Measurement: Line with Symbols.

IV. CONCLUSION AND OUTLOOK

A balanced AlGaN/GaN HEMT power amplifier MMIC for X-band frequencies was designed and fabricated in microstrip technology on thinned s.i. SiC substrate. This demonstrator serves as proof-of-concept for a first single-stage microstrip MMIC 0°/90° balanced HPA using low-impedance Lange couplers. The chip size of the MMIC is $4 \times 3\ mm^2$. The developed amplifier delivers a pulsed output power of 11 W with 3 dB compression level at 8.56 GHz at a low drain voltage of $V_{DS} = 20$ V and a duty cycle of 10 %. The associated power gain is 7 dB.

Further optimization of the MMIC circuit design will be performed to optimize the performance of the balanced power amplifier. A direct impedance matching between the power amplifiers PA 1-2 and the couplers (Fig. 1) as well as an optimization of the Lange couplers should deliver a noticeable improvement of the performance of the balanced amplifier.

The elimination of the additional matching networks between the amplifiers and couplers should result in a reduction of the required MMIC chip size. Consequently, the reduced circuit design equipped with the optimized Lange couplers should improve the output power, power gain and efficiency of the circuit.

ACKNOWLEDGMENT

The authors acknowledge the continuing support of the Federal Ministry of Defense (BMVg), the support of the Federal Ministry of Education and Research (BMBF), the Federal Office of Defense Technology and Procurement (BWB), and the Bundeswehr Technical Center for Information Technology and Electronics (WTD 81).

REFERENCES

[1] J. Komiak, W. Kong, and K. Nichols, "High Efficiency Wideband 6 to 18 GHz PHEMT Power Amplifier MMIC", *IEEE MTT-S International Microwave Symposium Digest*, pp. 905-907, June 2002.

[2] S. Seo, D. Pavlidis, and J.S. Moon, "Wideband balanced AlGaN/GaN HEMT MMIC low noise amplifier", *IEEE Electronic Letters*, vol. 41, issue 16, pp. 37-38, August 2005.

[3] D.E. Meharry, R.J. Lender, K. Chu, L.L. Gunter, and K.E. Beech, "Multi-Watt Wideband MMICs in GaN and GaAs", *IEEE MTT-S International Microwave Symposium Digest*, pp. 631-634, June 2007.

[4] A. Bessemoulin, R. Quay, S. Ramberger, H. Massler, and M. Schlechtweg, "A 4-Watt X-Band Compact Coplanar High-Power Amplifier MMIC With 18-dB Gain and 25 % PAE", *IEEE Journal of Solid-State Circuits*, vol. 38, no. 9, September 2003.

[5] J. W. Palmour, S. T. Sheppard, R. P. Smith, S. T. Allen, W. L. Pribble, T. J. Smith, Z. Ring, J. J. Sumakaris, A. W. Saxler, and J. W. Milligan, "Wide Bandgap Semiconductor Devices and MMICs for RF Power Applications", *IEDM Technical Digest*, pp. 385-388, Dec. 2001.

[6] F. van Raay, R. Quay, R. Kiefer, W. Bronner, M. Seelmann-Eggebert, M. Schlechtweg, M. Mikulla, and G. Weimann, "X-Band High-Power Microstrip AlGaN/GaN HEMT Amplifier MMICs", *IEEE MTT-S International Microwave Symposium Digest*, pp. 1368-1371, June 2006.

[7] R.C. Waterman, W. Fabian, R.A. Pucel, Y. Tajima, and J.L. Vorhaus, "GaAs Monolithic Lange and Wilkinson Couplers", *IEEE Transactions on Electron Devices*, vol. 28, no. 2, February 1981.

Opportunities at mm-Wave frequencies: SiGe or CMOS?

Hugo Veenstra, Marc Notten

Philips Research
High Tech Campus 37, 5656 AE, Eindhoven, The Netherlands
Hugo.Veenstra@philips.com

Abstract— In contrast to what is generally assumed, the evolution of f_T beyond the 45nm CMOS generation may not be able to follow the ITRS roadmap. Moreover, the gap between intrinsic device and circuit performance is expected to increase with new generations, due to an increase in interconnect parasitic capacitance in the transistor pcell area. Such problems are not expected for SiGe processes. The move to higher frequencies for new applications leads to a shift in system partitioning, since the receiver front-end must be located physically close to the antenna. Emerging mm-Wave applications such as radar and high data-rate wireless need to apply beam-forming which can be realized at low cost using phased arrays. These RF requirements justify a 2-chip solution: one analog phased array front-end plus one digital SoC. RF signal distribution on chip will be a determining factor in the choice of technology. For this reason, SiGe is and will remain leading over CMOS for mm-Wave.

I. INTRODUCTION

An important question for every application is, what IC technology to use. For applications below 10GHz, CMOS technology is becoming standard practice nowadays, including the RF front-end. Driven by the need for low-cost and thus single-chip implementations, the RF front-end is combined with the digital part to form a System-on-Chip (SoC) solution. However, at frequencies above 10GHz, signal distribution becomes more difficult. Today, SiGe technology still offers superior RF performance over CMOS in terms of generally applied metrics such as f_T, f_{max} and breakdown voltages. Besides, bipolar RF circuits are usually less sensitive to interconnect parasitics due to a higher transconductance and favourable transistor terminal impedances for a given power consumption.

For mm-Wave circuit design, the complete link budget must be analysed. It is not sufficient to only optimize receiver noise figure and transmitter output power. For example, the signal attenuation from the RF interconnect between IC and antenna must also be minimized, which implies minimum distance between IC and antenna and low-loss board material. Also, many applications at mm-Wave frequencies use beam-forming. If the antenna beam is not fixed but must be controlled, active beam-forming is an attractive alternative for mechanical beam-forming. The realization of active beam-forming leads to antenna arrays, often in combination with multiple RF front-ends. For all the above-mentioned reasons, we are still far away from SoC for mm-Wave applications, and it is questionable if we will get there in the coming decade. This paper addresses the suitability of mm-Wave ICs (e.g.,

with operating frequencies between 10-100GHz) in SiGe and CMOS technologies.

In Section II, emerging mass-market applications in the 10-100GHz range are briefly described, with focus on radar. Section III will discuss the authors' view on the suitability of SiGe BiCMOS and CMOS for mm-Wave applications, starting from the ITRS roadmap. In Section IV, the fundamentals for our view are explained on the basis of circuits from a recently realized 60GHz Doppler Radar IC in SiGe technology. The suitability of circuits with similar functionality in CMOS technology will be discussed. The paper ends with conclusions in Section V.

II. EMERGING CONSUMER APPLICATIONS

A. Automotive radar at 22-29 and 76-81GHz

Long-Range Radar (LRR) at 76-77GHz is already available on high-end cars for several years. These radar systems are usually built from discrete components. Recently, a SiGe alternative for the traditional Gunn diode oscillator has become available [1]. A wide range of SiGe ICs for a 76-81GHz automotive radar have been realized within the German KOKON project [2].

While LRR is mainly used for adaptive cruise control, short-range radar (SRR) has many applications related to comfort enhancement and safety. Since for SRR, range resolution requirements are more stringent than for LRR, more RF bandwidth is needed. Frequency bands for SRR are available at 22-29GHz (in Europe until 2013) and 77-81GHz. SiGe ICs for 24GHz SRR are available that offer a high degree of front-end integration [3], while SiGe ICs for 77-81GHz SRR can be expected within a few years. For automotive radar, no CMOS ICs are on the market yet. To the authors' knowledge, CMOS ICs for SRR and LRR front-ends are also not under development.

B. 60GHz: not only for high data-rate wireless

Since the automotive bands are exclusively reserved for automotive applications, other applications must operate at alternative frequencies. The 60GHz ISM band is often proposed for high data-rate short-range wireless links. This is a logical application, although it is facing competition from emerging Ultra-Wide Band (UWB) wireless communication at 3.4-10.3GHz. If all the available UWB bands are used, UWB can offer comparable data rates across longer distances at potentially lower costs. A 60GHz radio can be made physically smaller due to a smaller antenna size, but for applications such as wireless HDMI, the module size is not too critical.

The availability of several GHz of bandwidth near 60GHz makes this band also attractive for short-range radar applications. The range resolution ΔR of a radar system is inversely proportional to the bandwidth used via

$$\Delta R = \frac{c}{B} \qquad (1)$$

with c the speed of light and B the bandwidth of the transmitted signal. A bandwidth of B=3GHz enables a typical range resolution of ΔR=10cm. Such a resolution is sufficient for applications such as indoor presence and position detection. Information about presence and position of persons inside a room can be used for the control of lighting, heating and for security applications. Another promising short-range radar application is virtual clothing [4]. The image from a camera is fused with a radar image to construct a 3D-image. To construct a real 3D camera, three properties must be detected: distance, relative speed and azimuth. The distance is derived from the signal delay. Relative speed is found from the Doppler frequency shift of the received signal. In zero-IF receivers, a positive Doppler frequency shift (approaching object) must be distinguished from a negative Doppler frequency shift (object moving away). This is usually done based on quadrature down-conversion. The azimuth is derived using beam-forming and/or beam-steering. An example phased array receiver is shown in Fig. 1.

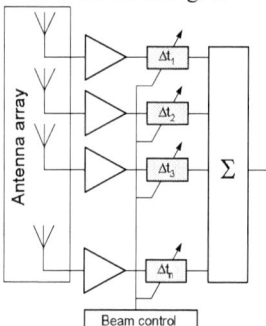

Fig. 1: Typical phased array receiver block diagram.

Phased arrays are becoming popular for active beam-forming. In phased arrays, antenna arrays with multiple receivers are needed with individual phase control, either in the VCO or in the RF paths.

In radar systems, the link budget is based on Friis equation:

$$P_r = P_t \frac{G_t G_r \sigma \lambda^2}{(4\pi)^3 d^4} \qquad (2)$$

with P_r the received signal power, P_t the transmit power, G_r and G_t the receive and transmit antenna gains, σ the radar cross-section of the object and d the distance to the object. On top of the free space loss, an additional loss up to 20dB/km can occur at 60GHz due to fog or rain. This makes radar operation at 60GHz not attractive for outdoor applications.

With only 1mW of transmit power and a receiver Noise Figure of around 20dB, a range of several meters is already feasible.

III. STATUS AND OUTLOOK FOR SiGe AND CMOS

Based on the metrics f_T and f_{max} for the naked devices, recent CMOS generations achieve competitive performance compared to SiGe BiCMOS. For example, the predicted evolution of f_T is shown in Fig. 2 [5]. The minimum feature sizes, L_g for CMOS and W_e for SiGe, are indicated in the figure.

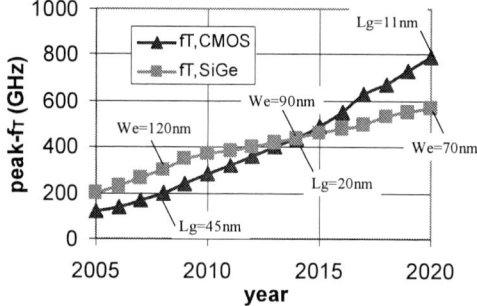

Fig. 2: ITRS 2006 roadmap for SiGe and CMOS.

In CMOS it is difficult to relate the f_T of the naked device to final circuit performance. In the first place, the peak-f_T is usually obtained at a high current density, requiring multiple contacts at the drain and source terminals. This leads to a substantial increase in parasitics from the metal connections, an effect that gets more pronounced in newer generations. In current 65nm generations, a typical reduction in f_T by 50% occurs due to metal in the pcell area. In the second place, the f_T in CMOS shows an important dependency to the bias conditions. The peak-f_T in CMOS is obtained by applying the full available supply voltage VCC between gate and source terminals. In most applications, the headroom will be limited. For example, if the gate-source voltage is limited to VCC/2, the available f_T drops by another 40%.

In contrast to what is predicted by the ITRS roadmap, the evolution of f_T beyond the 45nm generation is likely to show a reduction. To reduce leakage currents and boost mobility in digital CMOS platforms, new transistor topologies are being developed. The FinFET shows a substantial increase in gm/gds and thus allows high gain at low frequencies. However, the demonstrated f_T of a 45nm FinFET is currently a factor of 3 below the f_T of a 45nm planar bulk device, due to additional fringing capacitances of the device. Moreover, high series resistances and high fringing capacitance are inherent to nanowire devices. Thus, it is likely that RF performance will degrade with future CMOS generations, unless technological breakthroughs are found [6].

In SiGe technology, the relation between performance metrics f_T, f_{max}, f_A, f_{cross} and circuit performance is relatively well understood [7]. Moreover, the low impedance level of the circuits makes the impact of interconnect inside circuits less pronounced as in CMOS. The currently available performance of SiGe technologies is already sufficient to realize highly-integrated transceiver front-ends for operating frequencies at 60GHz, as will be demonstrated in the next section.

Passives

A wide range of passives is required for mm-Wave front-end circuits. Thin-film resistors are preferred over other types due to their low parasitic capacitance. MiM capacitors offer the highest Q-factor. Inductors can be realized from spiral inductors or from transmission lines. A transmission line offers lower Q but superior isolation from the substrate, which is an advantage for VCOs. An inductor implemented from a shorted stub also allows the use of a ground plane across the IC, enabling optimum distribution of the ground. A low-impedance ground connection is essential in mm-Wave circuits, especially in single-ended circuits.

In principle, all the passives that are routinely available in SiGe technologies can also be realized in CMOS. In practice, for cost reasons, RF options are often avoided in CMOS. Besides, the backend in CMOS is already several generations ahead of SiGe, which is often not an advantage for RF. For example, the strict metal density in CMOS may require tiling inside inductors and transmission lines. In such cases, the performance of these elements is difficult to predict.

IV. CIRCUIT EXAMPLES

By way of example, it is interesting to compare the performance reported for a similar 0.6-10GHz UWB receiver front-end in 65nm and 45nm generations [8]. Despite the substantial increase of f_T (by approximately 50%), the receiver in 45nm technology shows a 2.5dB higher Noise Figure and 9dB lower gain at more than double the power dissipation. This is a clear indication of reduced RF performance beyond the 65nm generation.

The 60GHz Doppler Radar transceiver front-end IC, recently realized in a 0.25um SiGe BiCMOS technology [9], is used as second example. The block diagram of this IC is shown in Fig. 3 [10].

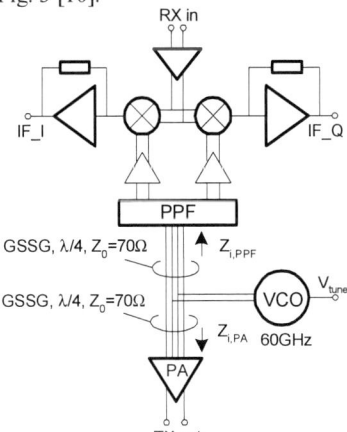

Fig. 3: 60GHz Doppler Radar Transceiver.

The IC layout is shown in Fig. 4. Extensive use of transmission lines for signal distribution and impedance matching via shorted stub inductors can clearly be seen.

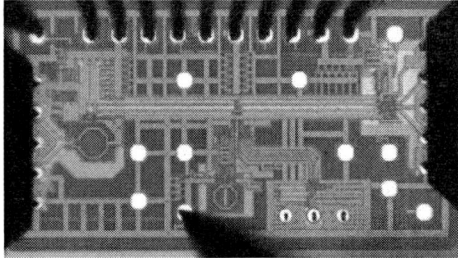

Fig. 4: Chip Photo of the 60GHz Doppler Radar Transceiver.

It is interesting to analyse the feasibility of this IC in CMOS. Although many functional blocks operating at 60GHz have been realized in recent history, to the authors' knowledge, a 60GHz transceiver of comparable RF integration has not yet been published. Note that the SiGe technology, used to integrate the IC of [10], provides an f_T/f_{max} of only 130/140GHz, respectively. Thus, the technology used is already outdated; technologies with improved RF performance are meanwhile widely available. In the following, the feasibility of several functional blocks in CMOS is analysed.

A. VCO

At frequencies above 10GHz, the varactor usually dominates the losses of the tank. For example, the quality factor of the varactor is only about 3 at 60GHz in the SiGe technology of [9]. A varactorless, tuneable LC-VCO topology is shown in Fig. 5 [11].

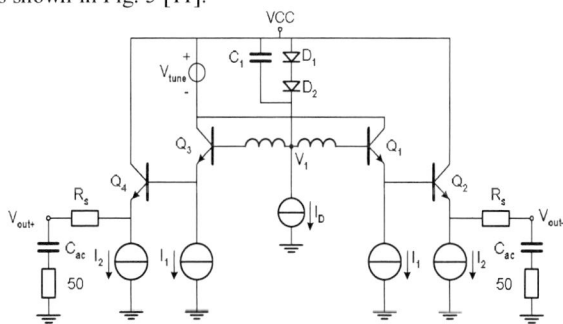

Fig. 5: Varactorless LC-VCO topology.

Undamping of the LC-tank is realized via resistively-loaded double emitter followers. Each emitter follower provides a 90° phase shift between the emitter and base currents. Thus, the resistive load of the output emitter followers translates to a negative input resistance of the first emitter followers, realizing the negative resistance needed to start and sustain oscillation. The double-emitter follower is designed to drive a 50Ω load. This makes distribution of the output signal via on-chip transmission lines straightforward. Tuning is realized via the bias-dependent collector-base junction capacitance of the first emitter followers.

A simple copy of this circuit in CMOS, replacing NPN by NMOS transistors, is not very attractive. The NMOS transistors provide a poor voltage gain from gate to source, leading to a low output signal amplitude. Moreover, the input capacitance is only weakly bias dependent, and thus, the oscillator would show a small tuning range. A more attractive CMOS VCO circuit is the cross-coupled differential pair, although this requires a varactor for frequency tuning. With a layout optimized for low gate series resistance, f_{cross} is in CMOS superior to f_{cross} in bipolar. Thus, mm-Wave oscillators are feasible in CMOS [7].

B. Quadrature signal generation: Polyphase filter

Quadrature signals are needed in zero-IF receivers, for example to obtain the sign of the Doppler frequency shift in radar. To generate a quadrature VCO signal, the use of a polyphase filter is attractive if good passives (R and C) are available [12]. This is usually the case in SiGe, but not in standard CMOS. In CMOS, a widely used architecture for the generation of quadrature signals is via two coupled oscillators. This has however not yet been demonstrated at 60GHz. An alternative approach is to use a 60GHz LC-ring-oscillator, injection-locked to a 20GHz quadrature signal [13]. This has been demonstrated successfully, although at least -3dBm signal power was needed at 20GHz to obtain a >1GHz lock range. Overall, the generation of 60GHz quadrature signals in CMOS is far more complex than in SiGe.

C. Signal Distribution

A drawback of a VCO based on a cross-coupled differential pair is that a dedicated output signal buffer is needed. A 50Ω buffer at mm-Wave frequencies is not straightforward in CMOS and usually shows a significant attenuation. Thus, although a 60GHz VCO is feasible in CMOS, buffering and distribution of its output signal is complex. This problem is believed to be the main bottleneck in CMOS, and limits the integration density at mm-Wave frequencies.

Signal distribution via on-chip transmission lines is widely applied at mm-Wave frequencies. Coplanar lines in the top metal layer on top of a metal ground plane are well suited to realize 50Ω characteristic impedance and low attenuation and yield excellent isolation from its surrounding, making it possible to include transmission lines as library elements. Clear examples of mm-Wave SiGe transceiver ICs making widespread use of on-chip transmission lines are the 12.5Gb/s, 20-input 20-output cross-connect IC for optical networking [14], the 24GHz automotive radar receiver IC [15] and the 60GHz radar transceiver IC shown in Fig. 4. Although the proposed transmission line configurations are also feasible in CMOS, highly integrated mm-Wave transceivers in CMOS are not available. This is believed to be due to the 50Ω buffering challenge, as explained before.

V. CONCLUSIONS

Current SiGe technologies provide already sufficient RF performance for fully integrated transceiver front-ends at 60GHz and 79GHz, while there is still headroom for further RF performance improvements in next generations. Although CMOS technologies offer comparable RF performance of the naked devices, substantial performance degradation occurs from naked device to circuit level. In CMOS, the impact of interconnect is large due to the high impedance level. A major bottleneck is RF signal distribution. While the use of on-chip transmission lines is OK in SiGe, in CMOS this is highly complex and has prohibited fully integrated transceivers at 60GHz and above. Moreover, with future generations of CMOS, the gap between intrinsic device performance and circuit performance will increase, leading to an effective decrease in RF performance for newer CMOS generations. First evidence of reduced RF performance in 45nm CMOS over 65nm CMOS has already been demonstrated for UWB front-ends. Thus, for mm-Wave applications, it is expected that SiGe technology will remain the best choice.

ACKNOWLEDGMENT

The authors wish to acknowledge the support of Domine Leenaerts (NXP Semiconductors Research) for permission to use material from his ISSCC 2008 workshop. We also like to thank X. Huang for his 60GHz mixer design work.

REFERENCES

[1] [Online]. http://www.infineon.com/radar

[2] R. Schneider et al., "KOKON – Automotive High Frequency Technology at 77/79GHz,", in Proc. EUMW, pp. 247-250, Oct. 2007.

[3] I. Gresham et al., "Ultra-Wideband Radar Sensors for Short-Range Vehicular Applications," IEEE Trans. Microwave Theory Tech., vol. 52, No. 9, pp. 2105-2122, Sept. 2004.

[4] Hyung Kyu Lim, "The 2nd Wave of the Digital Consumer Revolution: Challenges and Opportunities," ISSCC Dig. Tech. Papers, pp. 18-23, Feb. 2008.

[5] International Technology Roadmap for Semiconductors 2006 Update. [Online]. Available: http://www.itrs.com

[6] P. Wambacq et al., "Advanced Planar Bulk and Multigate CMOS Technology: Analog-Circuit Benchmarking up to mm-Wave Frequencies," ISSCC Dig. Tech. Papers, pp. 528-529, Feb. 2008.

[7] H. Veenstra et al., "10-40GHz design in SiGe-BiCMOS and Si-CMOS – linking technology and circuits to maximize performance," in Proc. European Microwave Week, 2005.

[8] R. van de Beek et al., "A 0.6-to-10GHz Receiver Front-End in 45nm CMOS," ISSCC Dig. Tech. Papers, pp. 128-129, Feb. 2008.

[9] P. Deixler et al., "QUBiC4X: An f_T/f_{max}=130/140GHz SiGe:C-BiCMOS Manufacturing Technology with Elite Passives for Emerging Microwave Applications," in Proc. IEEE BCTM, pp. 233-236, 2004.

[10] H. Veenstra et al., "60GHz Doppler Radar Transceiver in a 0.25µm SiGe BiCMOS technology," submitted for ESSCIRC 2008.

[11] H. Veenstra, E. v.d. Heijden, "Varactorless, tuneable LC-VCO for microwave frequencies in a 0.25µm SiGe technology," in Proc. IEEE BCTM, pp. 54-57, 2007.

[12] M. Notten, H. Veenstra, "60GHz Quadrature Signal Generation with a Single-phase VCO and Polyphase Filter in a 0.25µm SiGe BiCMOS technology," submitted for BCTM 2008.

[13] W.L. Chan et al. "A 56-to-65GHz Injection-Locked Frequency Tripler with Quadrature Outputs in 90nm CMOS," ISSCC Dig. Tech. Papers, pp. 480-481, Feb. 2008.

[14] H. Veenstra et al., "A 20-Input 20-Output 12.5Gb/s SiGe Crosspoint Switch for Optical Networking with <2ps RMS jitter," ISSCC Dig. Tech. Papers, pp. 174-175, Feb. 2003.

[15] H. Veenstra et al., "A SiGe-BiCMOS UWB Receiver for 24GHz Short-Range Automotive Radar Applications," in Proc. International Microwave Symposium, pp. 1791-1794, June 2007.

978-2-8748-7007-1/08 $25.00 © 2008 EuMA

Perspectives of (sub-) 32nm CMOS for Analog/RF and mm-wave Applications

M. Dehan[1], B. Parvais[1], A. Mercha[1], V. Subramanian[2], G. Groeseneken[1,2], W. Sansen* and S. Decoutere[1]

IMEC
[1]*Kapeldreef 75, Leuven, Belgium*
* *K.U.Leuven, Department ESAT-INSYS, 3001, Leuven, Belgium*
morin.dehan@imec.be

Abstract— **New process modules and device architectures for (sub-) 32nm CMOS lead to both opportunities and challenges for analog/RF and mm-wave circuit design. A survey will be given describing the advanced process modules and competing architectures (planar bulk CMOS versus FinFETS), and their impact on analog/RF performance. FinFETs will be shown to be better suited for analog baseband design and to have acceptable RF performance in the 1-10 GHz range, while planar bulk CMOS outperforms the FinFETs for sub-circuits above 10 GHz.**

I. INTRODUCTION

The international effort through the ITRS [1] presented a roadmap on the best estimates of introduction time, at the production level, of successive generations of leading technology nodes. It has become necessary to introduce revolutionary changes in the materials and process modules, and further scaling below 32nm might require alternative device architectures like FinFETs. It should be noted that the analog/RF and mm-wave requirements have limited weight in the determination of the CMOS technology choices, which is almost exclusive the prerogative of digital CMOS. As an example, Fig. 1 shows that FinFETs are capable of following the SRAM trend line for cell size scaling, and this might have much more impact for future device architecture.

II. SCALING ISSUES AND TECHNOLOGY TRENDS

A. Lithographical scaling and supporting device architectures

Although CMOS scaling has been for a long time driven by lithographical scaling capability, the reduction of the pattern sizes by itself is not enough. The appropriate measures have to be taken with respect to the device architecture to support the decreasing gate lengths, like the introduction of halo implants, high-k dielectrics and metal gate electrodes, or even for the most aggressively scaled devices a multi-gate architecture (Fig. 2).

The equation governing the lithography scaling is:

$$resolution = k_1 . \frac{\lambda}{NA} \qquad (1)$$

where k_1 is a technology constant, λ the wave length and NA the numerical aperture. Research options towards 32nm half pitch are summarized in table I, and comprise options working on all three components of (1): hyper-NA steppers (193nm immersion), wave-length scaling (Extreme UV lithography: EUVL), double patterning (k_1).

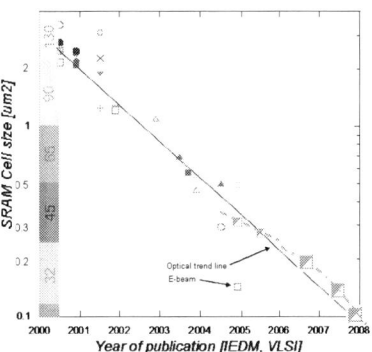

Fig. 1 SRAM trend line for cell size scaling. Half-open squares indicate FinFET SRAM cells.

EUVL, having much shorter wave length, is investigated as the next generation lithography technique with significant wave length reduction. The very short wave length has the advantage that it does no longer call for advanced correction techniques associated with the 193nm wavelength, such as optical proximity correction, attenuated phase shift mask technology, etc... EUVL being a reflection based optics system relaxes the efforts in lens optics, but a major effort is needed for photo resists and mask infrastructure. Double exposure techniques require additional efforts in overlay requirements, and add integration complexity.

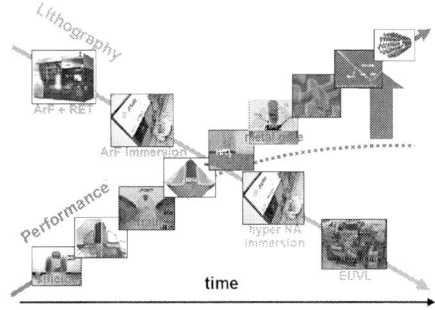

Fig. 2 Lithographical scaling versus performance scaling.

978-2-8748-7007-1/08 $25.00 © 2008 EuMA

TABLE I
LITHOGRAPHY SCALING

Parameter	NA	k_1		λ
Technique	193nm ArF immersion	Double patterning		EUVL
Wavelength	193 nm	193 nm		13.5 nm
NA	1.65	1.35-1.40		0.25
k_1	0.275	0.20		0.6
Challenges	Lens complexity Liquid with n_f>1.8	Overlay requirements Process integration		Optics lifetime Resist and mask infrastruct.

B. Gate leakage and short channel effects

Fig. 3 shows the gate leakage current density versus equivalent oxide thickness (EOT) for different gate dielectrics. Gate oxide thickness reduction using SiO_2 is no longer an option for downscaled CMOS, due to the high gate tunnelling currents. Heavy nitridation of the gate oxide (SiON) improves the leakage current until the 65nm CMOS node. Using a metal for the gate electrode material relaxes the need for ultra thin gate oxides a bit, as the absence of gate depletion allows the gate oxide to be a few Angstrom thicker. For 45nm and beyond, high-k dielectrics are being considered. Due to the high dielectric constant, good control over the channel can be obtained for relatively thick oxides. The physically thicker dielectric results in a lower electric field across the dielectric, and hence in reduced tunnelling current, but the gate leakage current increases again very strongly below a EOT of 1.5nm, due to direct tunnelling. Using materials with even higher dielectric constant, like Ta_2O_5 and $SrTiO_2$, might cause reliability problems, because the breakdown strength E_{bd} is reduced, as a direct result of the physics behind the dipole nature of the high-k dielectrics [2]. This relation between E_{bd} and dielectric constant k is shown in Fig. 4.

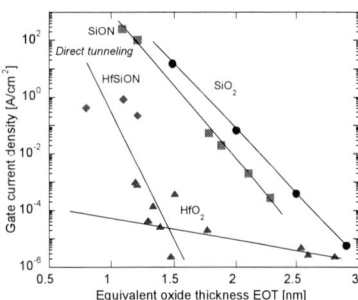

Fig. 3 Measured gate leakage current density versus equivalent oxide thickness, for different gate dielectrics.

Drain induced barrier lowering is already being suppressed by the use of increasingly higher doses of halo implants in the 90nm and 65nm node, with a detrimental effect on the voltage gain of the transistors due to barrier height modulation. The barrier that appears at the drain side because of the halo implants affects the output conductance (G_{ds}) also for long channel devices. Fully depleted technologies (e.g. FinFETs)

have demonstrated superior control of the short-channel effects, which enables the design of high-gain analog circuits.

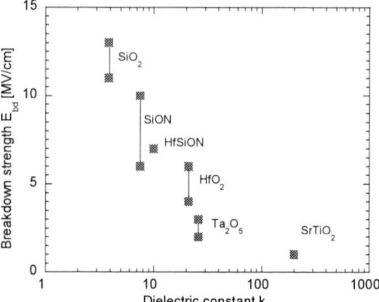

Fig. 4 Breakdown strength versus dielectric constant. After [2].

C. Gate work function engineering and mobility boosters

The introduction of metal gates (to reduce the gate depletion) for planar bulk CMOS requires a material with a correct work function for NMOS and PMOS. The choice of the material (Fig. 5) depends on the device architecture chosen, and the device specifications targeted (low operating power, low standby power, high performance). Alternative techniques to differentiate the work function of a metal through implants to make one material suitable for both NMOS and PMOS are still further investigated. Co-integration of two different types of gate electrode material is increasing the process complexity. Recently, attention was given to capping the high-k dielectric with different capping material for NMOS and PMOS to tune the work function.

Fully silicided gates are an interesting alternative to deposited metal gate materials. Symmetric threshold voltages for NMOS and PMOS can be obtained through the implantation of n-type and p-type dopant impurities into the poly gate prior to silicidation, or the use of different silicide thicknesses, resulting in different stoechiometry.

Fully depleted devices such as FinFETs need a mid-gap material. They have the advantage that symmetric threshold voltages are obtained with just one type of gate electrode material, but obtaining different flavors of transistors with respect to multiple threshold voltages is difficult.

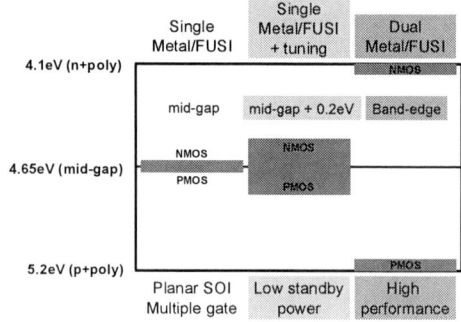

Fig. 5 Work function engineering for different applications and device architectures. .

978-2-8748-7007-1/08 $25.00 © 2008 EuMA

Many of the performance limitations (e.g. G_m, f_T,..) can be overcome if a higher hole and electron mobility is obtained. Stressor layers have become common in downscaled CMOS. NMOS and PMOS transistors however require respectively compressive and tensile stress, hence two masking steps. External stressor layers are compatible with other techniques like selective SiGe or SiC growth on source/drain regions, where the combined effect further boosts the performance. An example is given in Fig. 6, where the Ion for a given Ioff is plotted versus the gate current density, for different options in the gate stack and strain engineering. Significant improvement in drive current for a given level of Jg can be achieved with combining several performance booster techniques.

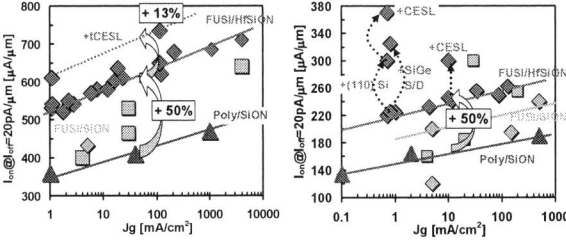

Fig. 6 Drive current for a given Ioff, as a function of the gate current density. Combined effect of introduction of high-k/metal gate options and strain engineering. Left : NMOS, Right : PMOS.

III. ANALOG AND RF PERFORMANCE

A. *Mobility and saturation velocity*

To understand the impact of the advanced gate stacks and device architectures on the device performance, we will first address the mobility and saturation velocity. In fig. 7, the low field mobility and saturation velocity are shown for transistors with identical gate length, but different high-k dielectrics. Although the low field mobility is affected by the differences in the gate dielectrics, it is not so pronounced for the saturation velocity, as the latter is a lateral field effect.

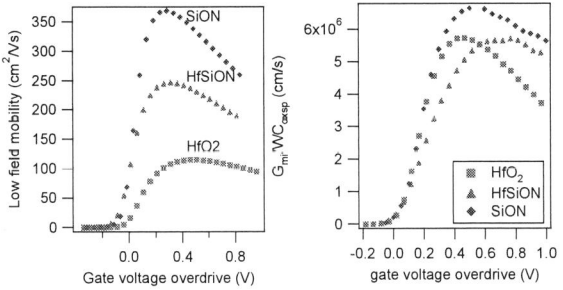

Fig. 7 Low field mobility and saturation velocity for different gate stacks.

Similarly, the low field mobility and v_{sat} is shown in Fig. 8, for a planar bulk NMOS versus n-channel FinFET transistor. Although most of the current in the FinFET flows along the sidewall where the mobility is lower, the peak mobility is higher and it is reached at lower current because the channel is

undoped (lower electric field and less scattering). This however does not translate in higher saturation velocity.

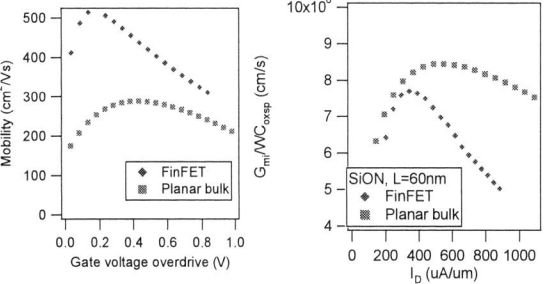

Fig. 8 Low field mobility and saturation velocity for planar bulk CMOS and FinFETs.

B. *Voltage gain, matching performance and 1/f noise*

The output conductance of fully depleted FinFETs is significantly improved for narrow gate lengths, and long channel devices do not suffer from drain barrier modulation by halos [3]. But the high fin access resistance degrades the transconductance. As shown in Fig. 9, the planar bulk CMOS transistors and FinFETs have a complementary G_m-A_V trade-off. There is a region in the G_m-A_V plot at high G_m, which can only be reached by planar bulk CMOS, albeit with a voltage gain that is rather low. At the other hand, there is a region with high voltage gain which can only be reached by FinFET transistors, albeit with lower achievable transconductance. Therefore, FinFET transistor are interesting for analog high-gain circuits, while planar bulk CMOS are the only choice for mm-wave applications.

The threshold voltage matching coefficients of FinFETs are relatively low (Fig. 10) in absence of channel doping. A complication however is the dependence of the matching coefficient on the width of the fins [4]. For very narrow fins, probably due to the fin roughness, the matching performance degrades compared to wider fins, which is problematic as the FinFETs are attractive as alternative device architectures below 32nm, where the fins need to be very narrow to support the associated gate lengths of these nodes.

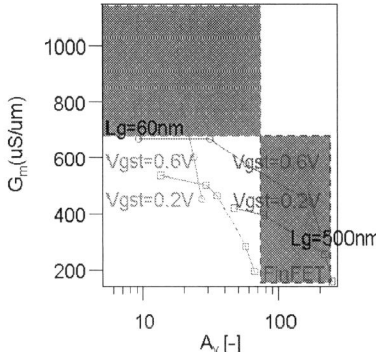

Fig. 9 Gm versus voltage gain for different gate lengths and gate overdrive voltages, comparing planar bulk CMOS with FinFETs.

978-2-8748-7007-1/08 $25.00 © 2008 EuMA

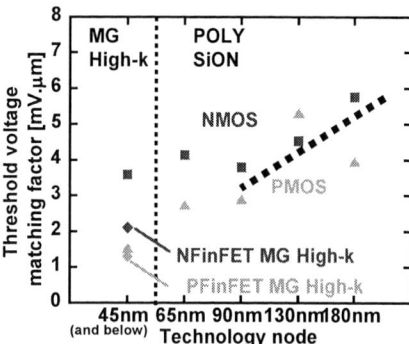

Fig. 10 Threshold matching factor trend chart, comparing different technology generations, gate stacks and device architectures.

The classic scaling of the 1/f noise with the gate oxide thickness has first been affected with the introduction of heavily nitrided gate oxides. Due to the high trapping density, the 1/f noise in high-k dielectrics is even higher than in SiON, although a lot of improvement can be obtained through process optimization. A trend chart is given in Fig. 11, where the reference line represents the data from ITRS roadmap, uncorrected for the introduction of high-k dielectrics. While initial results on 1/f noise were about 2 decades higher than the reference, progress in dielectric annealing [5], improved metal gate deposition, and the use of Dy-based or La-based capping oxides has significantly reduced the 1/f noise. However, most of these techniques suffer today from limited scaling capability, such that there is no solution yet for dielectrics with EOT around 1nm and below.

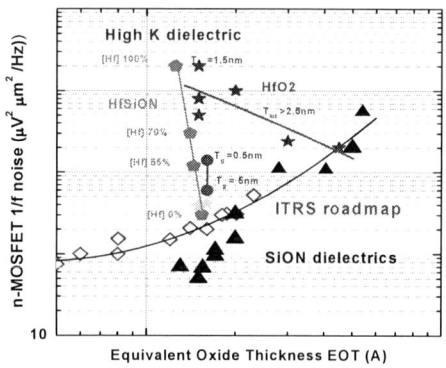

Fig. 11 Input referred voltage noise spectral density versus equivalent oxide thickness for different optimization techniques relative to the ITRS reference.

C. RF performance

The cut-off frequency of planar bulk CMOS devices seems to scale in line with the gate length scaling. However, if the control over the short channel effects would require a change in architecture towards multi-gate devices like FinFETs, a drastic loss in RF performance could be a consequence of this architecture change. As shown in Fig. 12, FinFET devices demonstrate today much lower f_{TS}, due to the fin access resistance and the high fringing capacitance between the fins and the gate due to the 3D architecture. Today, the highest values for f_T and f_{max} reported are about 140 GHz. These limitations seem to be quite fundamental, and if under pressure of digital scaling an architecture change towards multiple gate devices like FinFETs should be needed, it implies that the RF performance would no longer scale.

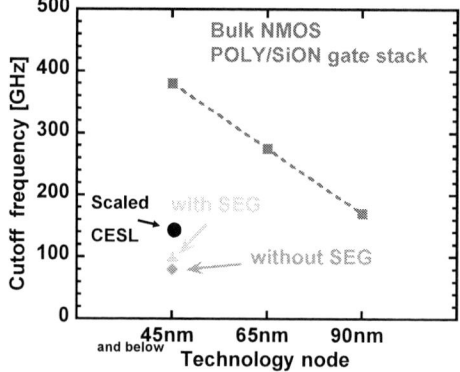

Fig. 12 Scaling of the cutoff frequency with technology node. FinFET devices reach significantly lower cutoff frequencies compared to planar bulk CMOS.

IV. CONCLUSIONS

A survey of the scaling limitations, and the opportunities offered with advanced process modules and architectures has been given, with respect to analog/RF and mm-wave performance. With the present state-of-the-art, the FinFET transistors outperform the bulk transistors with respect to gain and matching performance, while maintaining acceptable RF performance for applications in the 1-10 GHz range. Planar bulk CMOS outperforms the FinFETs above 10 GHz and is the only choice today for mm-wave applications.

ACKNOWLEDGMENT

The authors acknowledge the European Commission for supporting the IST project NANO-RF (IST-027150), and the IMEC colleagues of the PLANAR/EMERALD programs.

REFERENCES

[1] http://public.itrs.net
[2] J. McPherson et al., "Trends in the ultimate breakdown strength of high-dielectric constant materials", IEEE Trans. Electron Devices, Vol. 50, No. 8, pp. 1771-1778, August 2003.
[3] V. Subramanian et al., "Device and circuit-level analog performance trade-offs: a comparative study of planar bulk FETs versus FinFETS", Proceedings IEDM 2005, pp. 919-922, 2005.
[4] C. Gustin et al., "Stochastic matching properties of FinFETs", submitted to Electron Device Letters, 2006.
[5] Z.M. Rittersma et al., "Mixed Signal and noise properties of NMOSFETs with HfSiON/TaN gate stack", Proceedings ESSDERC 2005, pp. 105-108, 2005.

978-2-8748-7007-1/08 $25.00 © 2008 EuMA

Development of Ultrahigh-Speed InP/GaAsSb/InP DHBTs: Are Terahertz Bandwidth Transistors Realistic?

C.R. Bolognesi[1], H. Liu, O. Ostinelli, Y. Zeng

Institut für Feldtheorie und Höchstfrequenztechnik (IfH), THz Electronics Group,
ETH-Zürich, Gloriastrasse 35, Zürich CH-8092, Switzerland
[1]colombo@ieee.org

Abstract— In response to the continually increasing appetite for bandwidth, most transistor technologies have recently made great strides towards higher cutoff frequencies: Silicon MOSFETs, SiGe HBTs, InP –based HEMTs and a variety of InP –based HBTs all show cutoff frequencies fT and/or fMAX exceeding 300 GHz, and in some cases approaching 800 GHz. Proponents of various technologies have stated that the development of THz bandwidth devices is an attainable milestone for their technology of choice. Such ambitious goals naturally raise the question of whether such performances are in fact realistic given the well-known trends relating breakdown voltages and cutoff frequencies. Can the contending technologies be scaled in a way enabling THz cutoff frequencies while maintaining the well-behaved characteristics of less aggressively scaled previous generations? The present Invited Paper focuses on our efforts to push InP/GaAsSb DHBTs toward THz bandwidths.

I. INTRODUCTION

In response to the steadily increasing appetite for bandwidth and to the competition for market share between different semiconductor technologies, great strides have been made toward higher cutoff frequencies in most technologies: silicon MOSFETs, SiGe HBTs, and InP –based HEMTs and DHBTs have all shown cutoff frequencies exceeding 300 GHz. Indeed, according to the ITRS technology roadmap, several technologies target cutoff frequencies nearing 1 THz [1]. The question should be asked: Is this a realistic goal? Can the contending technologies be sufficiently scaled while maintaining well-behaved device characteristics, or is this just a manifestation of a "THz-hype" of the same ilk as its "Nano" close relative? By "well-behaved characteristics," we intend device characteristics resembling so-called "textbook descriptions" which are to be contrasted to otherwise freakish manifestations of transistor action. For example, consideration of trend lines in f_T *vs.* breakdown voltage (*BV*) plots show that most FETs and many bipolar transistors extrapolate to $f_T = 1$ THz with a $BV < E_G/q$! The principal exception to this trend remains InP –based HBTs, and InP/GaAsSb DHBTs in particular.

The present invited contribution reviews the progress we have made toward the development of THz bandwidth InP/GaAsSb DHBTs while pointing out the various technological and physical challenges we faced, along with the solutions implemented. It will be shown that the InP/GaAsSb DHBT scaled extremely well up to this point: transistor characteristics remain "normal" and "transistor-like" despite aggressive scaling. InP/GaAsSb/InP NpN DHBTs are attractive for wideband applications because high-performance devices can be fabricated based on relatively simple epitaxial layer stacks [2]. The simplicity follows from the staggered band alignment between InP and $GaAs_{0.51}Sb_{0.49}$: electrons can be injected directly from the p+ base into the InP collector without any grading layers [2, 3]. This simplifies epitaxial growth and processing, and it minimizes thermal resistance because a full InP collector can be used. Such structures enabled a cutoff frequency $f_T > 603$ GHz at room temperature with a breakdown voltage $BV_{CEO} = 4.2$ V, for a record $f_T \times BV_{CEO}$ product of 2.53 THz-V. The DHBT performance improves with cooling to reach $f_T > 700$ GHz with $BV_{CEO} = 4.4$ V at 5 K. To the best of our knowledge, this represents one of the best f_T ever reported (and conservatively so) for a DHBT of any kind. The $f_T \times BV_{CEO} > 3.10$ THz-V at 5 K is also unprecedented [4].

II. BASIC PHYSICAL OPERATION OF INP/GAASSB DHBTS

As already mentioned, InP/GaAsSb DHBTs are attractive for high-speed applications because they can be fabricated using relatively simple epitaxial layers which enhance device manufacturability. The technology has been profitably adopted by *Agilent Technologies* where it now enjoys commercialization since early 2005 despite a relatively short history —our development of InP/GaAsSb DHBTs only started in 1997, indicating a rapid transition from laboratory to industry. To the best of our knowledge, *Agilent* remains the only organization to have commercialized this technology at this point.

The basic InP/GaAsSb DHBT structure comprises an InP (or AlInAs) emitter, a GaAsSb base layer, and an InP collector. A representative equilibrium band diagram is shown in Fig. 1. Clearly, abrupt heterojunctions can be used since there is no electron blocking possible at the base/collector heterojunction: rather, electrons are ballistically injected into the InP collector, thus reducing the collector signal delay in comparison to type-I collectors of a similar thickness [5]. Such a structure enables the realization of 300-400 GHz transistors exhibiting an open-base $BV_{CEO} > 6$ V. For such transistors, the base transit time $_B$ represents a significant fraction of the total transit time, because the GaAsSb alloy in the base layer features a relatively low minority electron mobility when compared to GaInAs bases. The lower electron mobility obtained in GaAsSb is not a material quality issue, but rather a consequence of the low mobility in GaSb (3,000 cm²/Vs compared to 30,000 cm²/Vs for InAs) : this is then purely a bandstructure effect. This is essentially the price to be paid for the ease of injection of electrons into the collector. It became clear that the development of InP/GaAsSb DHBTs would greatly benefit from an aiding base built-in electric field to help speed-up electron transport across the base layer.

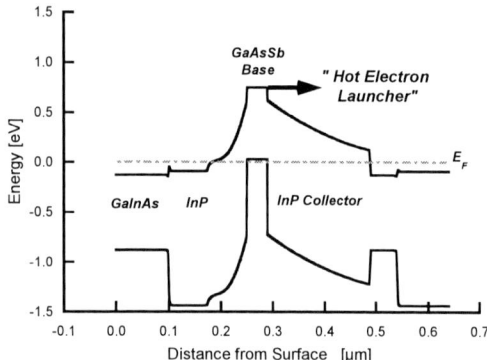

Fig. 1. Equilibrium band diagram for a uniform base InP/GaAsSb DHBT. Such structures enable 300-400 GHz devices with BVCEO > 6 V.

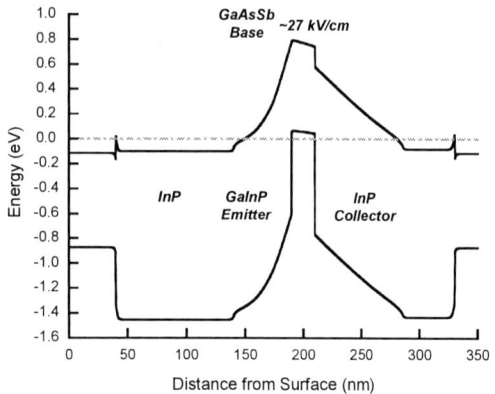

Fig. 2. Equilibrium band diagram for a graded base GaInP/GaAsSb DHBT. Such structures enable 600+ GHz devices with BVCEO > 4 V.

The question of exactly how to achieve this base grading in InP/GaAsSb-based DHBTs is far from trivial. One potential approach would be to introduce Al to the base material and form a quaternary graded (Al,Ga)AsSb base layer by ramping the Al- content from zero at the B/C side to some small value

on the emitter side. The advantage of this technique is that the base material would remain lattice-matched to the InP substrate. Unfortunately, an Al- containing base might result in reliability problems for the DHBTs, and can be expected to increase base contact resistances due to the rapid oxidation of surface layers prior to base metallization. Another approach involves the use of a (Ga,In)AsSb base layer and was pursued by a group at the University of Illinois [6]. Finally, we have chosen to simply achieve the required grading by ramping the As/Sb content of our ternary GaAs$_x$Sb$_{1-x}$ base layer. The advantage is that the base can be grown with zero net strain with respect to the InP substrate. On the negative side, it was initially not clear how to controllably achieve the required compositional change. Also, as far as we know, no other group had yet achieved base compositional grading by ramping the group-V element ratio across the base. Usually, compositional grading achieved by changing the group-III element composition across the base since growth occurs with V/III flux ratios >> 1. Under appropriate growth conditions to be reported separately, it turns out that a good grading can done through a variation of group V element ratio in the base alloy.

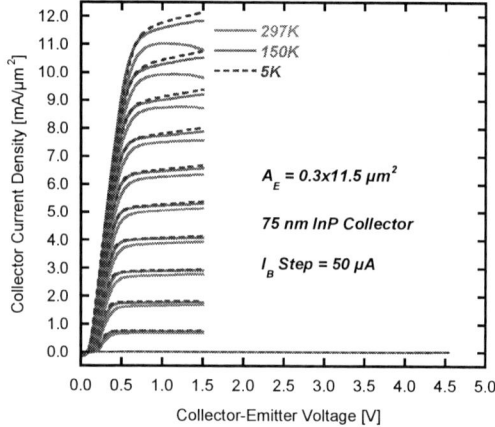

Fig. 3. Common-emitter characteristics at chuck temperatures of 297, 150 and 5 K. Despite the very thin collector, the transistor maintains a "textbook-like" behavior.

An example of such a structure is shown in Fig. 2 which depicts the band diagram of a transistor exhibiting an As grading from $x = 0.4$ on the collector side to $x = 0.6$ on the emitter side. Because lattice-matching is achieved for $x = 0.49$, the base structure is virtually strain-free. It is important to note that the structure incorporates a graded GaInP emitter which eliminates the conduction band discontinuity at the E/B junction, but that the Sb-rich B/C heterojunction enhances the band discontinuity at the electron launcher in the collector. The band diagram very nearly approximates the situation for an ideal heterojunction bipolar transistor. The zero conduction band discontinuity at the E/B interface leads to very low turn-on voltages which prove advantageous in limiting power dissipation, while the absence of collector grading minimizes the collector thermal resistance.

III. DEVICE CHARACTERIZATION

The common-emitter I-V characteristics of a transistor with a 20 nm graded base and a 75 nm InP collector are shown in Fig. 3 at chuck temperatures of 297, 150 and 5 K. Remarkably, the devices show very little temperature variation and minimal signs of heating even at current densities in excess of 10 mA/μm². Common-emitter breakdown voltages BV_{CEO} increase from 4.2 to 4.4 V with cooling.

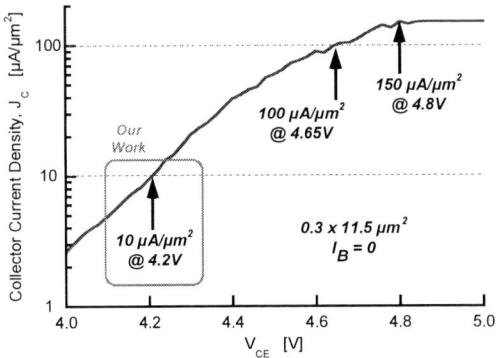

Fig. 4. Open base collector-emitter leakage current characteristic for the transistor shown in Fig. 3. Using higher 10× or 15× breakdown criteria can be results in inflated $f_T \times BV_{CEO}$ figures-of-merit by as much as 10-15%.

Care has to be exercised when comparing quoted breakdown voltages from the literature since various authors use inconsistent criteria. In some cases, the breakdown criteria is taken as a proportion of the current for peak f_T, and therefore slides higher as the device is scaled (i.e. as the real breakdown voltage decreases): this is clearly inadequate because it tends to mask the physical evolution of breakdown behavior as technologies scale. Throughout our work, we use a current density of 10 μA/μm² to determine the BV_{CEO}: the impact of the definition can be seen in Fig. 4 above.

A. Room Temperature RF Characterization

Fig. 5 shows the measured current and power gains up to 110 GHz using an HP 8510XF VNA and an LRRM off-wafer calibration standard for a 0.3 × 11.5 μm² DHBT with a 20 nm graded base and a 75 nm InP collector. The extracted current gain cutoff frequency f_T is 603 GHz, a value that can be confirmed through Gummel's f_T extraction method [4].

The small-signal equivalent circuit model corresponding to this performance is shown in Fig. 6 below. The model reveals that the relatively depressed maximum frequency of oscillation cutoff is due to the relatively broad emitter size and its associated capacitances.

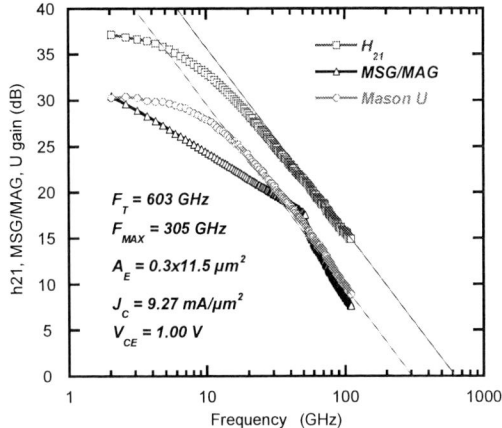

Fig. 5. Dynamic characterization at room temperature [4].

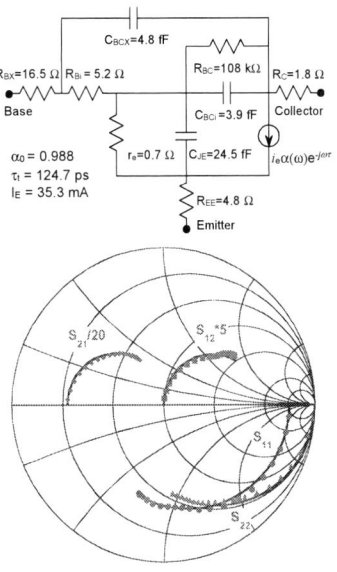

Fig. 6. Small-signal equivalent circuit model showing the excellent agreement between modeled and measured S-parameters. The model shows that higher f_{MAX} values can be achieved by scaling the emitter mesa size.

B. Cryogenic RF Characterization

The dynamic performance of InP/GaAsSb DHBTs has been shown to improve at cryogenic temperatures. Fig. 7 shows the f_T extraction using Gummel's method at chuck temperatures of 297, 150 and 5 K.

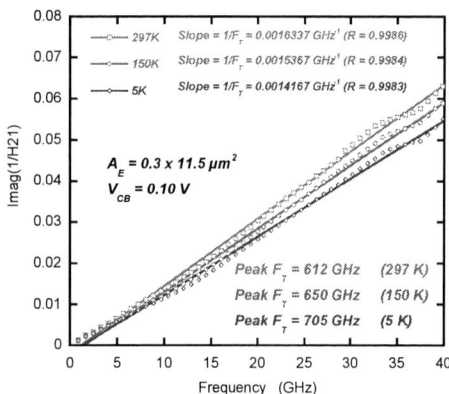

Fig. 6. Determination of current gain cutoff frequency f_T as a function of temperature through Gummel's method in a variable temperature probe station equipped with a 40 GHz vector network analyzer [4].

IV. CONCLUSIONS

InP/GaAsSb DHBTs presently exhibit the most favorable transistor scaling characteristics: these enable $f_T \times BV_{CEO}$ products higher than 2.5 THz-V at 300 K and 3.1 THz-V at 5 K. Such figures-of-merit are unmatched, and they do suggest that GaAsSb technology will prove to be a leading contender for the realization of well-behaved THz bandwidth transistors.

To date, InP/GaAsSb tend to display a cutoff frequency ordering $f_T > f_{MAX}$. As we have seen above in the case of the transistors described here, the extraction of a small-signal equivalent circuit model indicates this is largely due to insufficient scaling of the emitter mesa width. As such, this represents no fundamental problem.

It has been rather emphatically stated by some [7] that "*transistors having $f_{MAX} \ll f_T$ are of extremely limited utility in circuits.*" Clearly, a high f_{MAX} is required for analog applications —as stated above, this is only a question of process optimization for GaAsSb. In the meantime, one should recognize that the analytical approximations and linearizations used to estimate digital circuit performance trends as done in [7] necessarily represent oversimplifications of the problem at hand. In this context, one should be aware that an exhaustive study of HBT optimization for ultrahigh-speed digital circuit performance was carried by Ruiz-Palmero *et al.* through the physical simulation of multi-transistor circuits using a calibrated hydrodynamic device model [8]: the study relies neither on circuit approximations or linearizations, nor on SPICE-like device models. Rather, each transistor in the circuit was *physically simulated* and the

various technologies (InP/GaInAs SHBTs, and InP/GaAsSb and InP/GaInAs DHBTs) optimized to maximize digital circuit data rates in [8]. The study indicated that shortest gate delays of 1.29 ps and the fastest data rates of 300 Gb/s could only be achieved with InP/GaAsSb DHBTs characterized by $f_T = 1.1$ THz and $f_{MAX} = 570$ GHz. The results stand in marked contrast to the rule-of-thumb calling for very high f_{MAX} values (and/or requiring balanced $f_T \sim f_{MAX}$ cutoff frequencies) to maximize digital circuit performance, and they thus deserve further consideration. Strikingly, the 300 Gb/s of [8] were achieved for signal swings $V_{PP} = 520$ mV, much wider than in GaInAs –based HBTs. The result was directly attributable to the more favorable breakdown properties of the InP collectors used in InP/GaAsSb DHBTs.

ACKNOWLEDGEMENT

The authors wish to acknowledge the assistance of Dr. E. Gini and Mr. M. Ebnöther for their outstanding support of our MOCVD activities at ETH-Zürich. Mr. H. Benedickter and Mr. A. Alt are also acknowledged for their help with microwave measurements at the IfH.

REFERENCES

[1] F. Schwierz, "RF Transistors: Performance trends versus ITRS targets," Proc. 6th Int. Caribbean Conf. on Devices, Circuits and Systems, Mexico, pp. 195-200, April 26-28 2006.

[2] M.W. Dvorak C.R. Bolognesi, O.J. Pitts, S.P. Watkins; "300 GHz InP/GaAsSb/InP double HBTs with high current capability and BV_{CEO}>6 V," *IEEE Electron Device Letters*, vol. 22, pp. 361-363, 2001.

[3] X.G. Xu, J. Hu, S.P. Watkins, N. Matine, M.W. Dvorak, and C.R. Bolognesi, "Metalorganic vapor phase epitaxy of high-quality GaAs$_{0.5}$Sb$_{0.5}$ and its application to heterostructure bipolar transistors," *Applied Physics Letters*, vol. 74, pp. 9726-978, 1999.

[4] H.G. Liu, O. Ostinelli, Y. Zeng, and C.R. Bolognesi, "600 GHz InP/GaAsSb/InP DHBTs grown by MOCVD with a Ga(As,Sb) graded-base and $f_T \times BV_{CEO} > 2.5$ THz-V at room temperature," Tech. Dig. of Int. Electron Device Meeting (IEDM), pp. 667-670, Washington DC, December 10-12, 2007

[5] H.G. Liu, N.G. Tao, S.P. Watkins and C.R. Bolognesi, "Extraction of the average collector velocity in high-speed "Type-II" InP-GaAsSb-InP DHBTs," *IEEE Electron. Dev. Lett.*, vol. 25, pp. 769-771, 2004.

[6] W. Snodgrass, B.R. Wu, W. Hafez, K.Y. Cheng, and M. Feng, "Performance enhancement of composition-graded-base type-II InP/GaAsSb double-heterojunction bipolar transistors with f_T>500 GHz," *Appl. Phys. Lett.*, vol. 88, pp. 222101-3, 2006.

[7] M. Rodwell, E. Lind, Z. Griffith, S. R. Bank, A. M. Crook, U. Singisetti, M. Wistey, G. Burek, A.C. Gossard, "Frequency limits of InP -based integrated circuits," Proc. 19th IPRM, pp. 9-13, Matsue, Japan, May 14-18, 2007.

[8] J.M. Ruiz-Palmero, U. Hammer, H. Jäckel, H. Liu, and C.R. Bolognesi, "Comparative technology assessment of future InP HBT ultrahigh-speed digital circuits," *Solid-State Electronics*, vol. 51, pp. 842-859, 2007.

Technology Features and Design Tools for mm-wave Applications in CMOS

John J. Pekarik

IBM Semiconductor Research and Development Center
rue Jean Monnet, 38926 Crolles Cedex, France
pekarik@us.ibm.com

Abstract—**Leading-edge CMOS provides transistor performance that, by traditional measures, is more than adequate for implementing millimetre-wave transceivers. Analogue and RF design is supported by device-level libraries and extraction tools. Simple extensions to these will allow rapid design cycles for millimetre-wave applications.**

I. INTRODUCTION

Advances in CMOS technology driven by high-performance digital applications, provide advantages to the mm-wave designer that might not be apparent on first consideration. The most obvious advantage, performance, quantified by f_T, f_{MAX} or NF_{MIN} for example, has dramatically increased with geometry scaling and technology enhancements. CMOS technologies have been used to demonstrate circuits functioning at frequencies in and above the K-band. They are, by virtue of nanometer-scale design rules, able to implement staggering amounts of digital logic in a given area, thereby enabling the on-chip integration of sophisticated control logic for performance tuning and/or digital signal processing. Furthermore, the worldwide manufacturing capacity of silicon technologies driven by consumer applications like gaming and personal electronic appliances assures low-cost. This will certainly provide an impetus for the evolution of mm-wave consumer applications.

The consumer market demands rapid development cycles which contrast with traditional mm-wave design methods employing custom device designs of transistors and passive devices implemented on testchips and verified on hardware as "S-parameter" blocks. Mixed-signal design methods combining library-based digital design with schematic-based analog design have been successfully extended to radio-frequency applications below 10 GHz. This paper will focus on the device-level performance demonstrated by CMOS technologies and the support of these devices with models and design tools to enable schematic-based circuit design possible in the mm-wave regime.

II. TRANSISTOR PERFORMANCE

CMOS transistors follow the well-known Moore's law of scaling, thus leading to always-increasing functional integration with concurrent shrinking of dimensions. For MOS devices, $f_T \propto 1/L_G^\alpha$, where $(\alpha \sim 1)$ and, as a first order approximation, is independent of the gate oxide thickness. Data gathered from recent publications show good conformity

with the ITRS road-map as well as the simple inverse-scaling law, as depicted in Fig. 1.

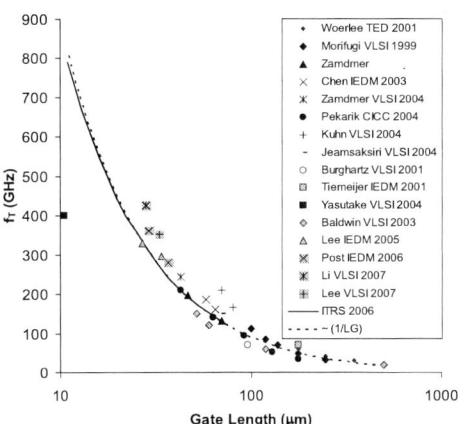

Fig. 1 Published values for f_T compared to the ITRS values and a simple $1/L_G$ scaling law.

For CMOS, through the 90nm node, the scaling of gate length and gate oxide thickness was roughly proportional. Further scaling of oxide thickness is limited by tunneling current and therefore, since transconductance (g_m) is inversely proportional to the oxide thickness, any scaling of g_m must be accomplished through an increase in carrier mobility.

Mechanical strain, which distorts the semiconductor crystal lattice, also distorts the energy band structure resulting in higher or lower carrier mobility depending on the carrier type (electrons or holes), whether the strain is compressive or tensile, whether the strain is uniaxial or biaxial, and the magnitude of the strain. Biaxial tensile strain in Silicon and compressive or tensile biaxial strain in SiGe can be induced with epitaxial growth on SiGe buffer layers yielding enhanced electron and hole mobilities and enhanced NFET and PFET performance [1]. Uniaxial strain can be induced by depositing stressed films on top of the completed FET prior to contacts [2] and if the stressed film is deposited prior to the final anneals, the stress is preserved in the device structure even after removal of the film [3]. These techniques are very effective in imparting tensile strain in the channel which thereby improves NFET performance. Compressive uniaxial strain to improve PFET performance has been achieved by selective epitaxial growth of SiGe in the source/drain region adjacent to the gate [4]. Combining these techniques can lead to dramatic simultaneous improvement in both NFET and PFET performance [5]

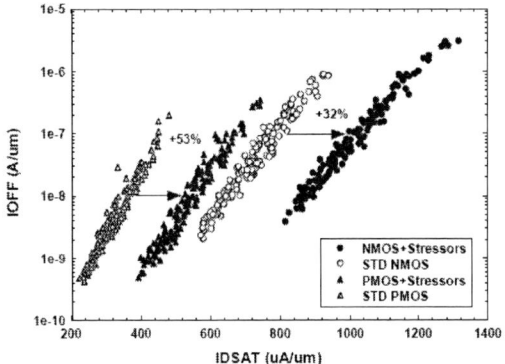

Fig. 2 Increase in FET drive current due to mobility enhancement techniques.

In order to avoid the high tunneling currents resulting from thin gate oxide, dielectric materials with higher effective permittivity (high-k) are being introduced. For an equivalent or lower effective oxide thickness these materials can be physically thicker and have dramatically lower tunneling current [6]. The introduction of metal gate electrodes addresses the problem of polysilicon depletion in the high vertical electric fields now inherent in scaled CMOS and the problem of high gate resistance from low polysilicon sheet resistance and contact resistivity. Coupling high-k dielectric materials with the appropriate choice of metal gates leads to FETs with high drive current [7] and improve high-frequency performance [8]. These technology elements are to be employed in high-performance 45nm digital CMOS [9] and result in excellent high-frequency performance shown below.

Equation (1) presents one way of calculating the transition frequency f_T [10] for deep submicron technologies:

$$f_T \approx \frac{g_m}{2\pi \cdot C_{gin}\sqrt{1+2\frac{C_{Miller}}{C_{gin}}}} = \frac{f_c}{\sqrt{1+2\frac{C_{Miller}}{C_{gin}}}} \quad (1)$$

With, g_m, the gate transconductance,

$$C_{gin} = C_{gsi} + C_{overlap} + C_{fringing} \quad (2)$$

$$C_{Miller} = C_{gdi} + C_{overlap} + C_{fringing} \quad (3)$$

$$f_c = \frac{g_m}{2 \cdot \pi \cdot C_{gin}} \quad (4)$$
, the intrinsic cut-off frequency

Where:
- C_{gsi}, C_{gdi} = the equivalent capacitance induced by the source / drain field effect into the channel

- $C_{overlap}$ = the equivalent capacitance given by the LDD (low doped drain / source regions) diffusion under the gate
- $C_{fringing}$ = the parasitic capacitance depending on the gate height and on the contact to gate distance .

The circuit designer has control of $C_{fringing}$ through the gate-to-contact spacing. Fig. 4 shows a SEM cross-section of a FET structure at the 45nm node with a schematic illustration of the contribution of the contacts to the gate capacitance. Increasing contact spacing can have dramatic affect on transistor performance by lowering this term. Fig 5 shows f_T vs. inverse gate length for two structures, one having the minimum allowed contact-to-gate spacing and one having a relaxed spacing. A third relationship on the plot shows f_T for transistors designed with the minimum contact spacing where the parasitic impedances, including the gate-to-contact capacitance contributions, were removed using deembedding techniques. This represents the performance of the intrinsic transistor.

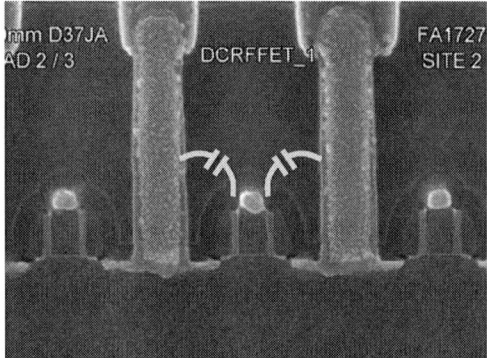

Fig. 4 Cross-sectional view of a MOS transistor from the Low Power CMOS of the 45nm node with the capacitance from gate-to-contact illustrated schematically.

The f_T quickly approaches intrinsic performance by increasing the gate spacing. However, increased contact spacing contributes to higher series resistance in the source and drain regions which appears in the simple expressions for f_{MAX} and NF_{MIN} shown in equations (5) and (6). Device models must accurately represent these effects to allow designers to optimize performance including power gain and noise figure.

$$f_{MAX} \approx \frac{g_m}{2\pi C_{gin}} \cdot \frac{1}{2\sqrt{(R_g + R_s + R_i) \cdot \left(g_d + g_m\frac{C_{Miller}}{C_{gin}}\right)}}$$

$$= \frac{f_c}{2\sqrt{(R_g + R_s + R_i) \cdot \left(g_d + g_m\frac{C_{Miller}}{C_{gin}}\right)}} \quad (5)$$

978-2-8748-7007-1/08 $25.00 © 2008 EuMA   112

$$NF_{MIN} \approx 1 + f/f_T \sqrt{g_m \cdot (Rg + Rs)} \qquad (6)$$

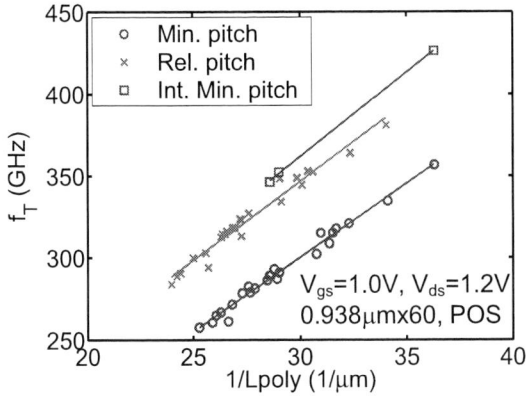

Fig. 5 f_T for NFETs from the 45nm node with minimum and relaxed gate to contact spacing along with data extracted from FETs having the minimum spacing for which the parasitic impedances of the wiring were de-embedded.

III. PASSIVE DEVICES IN CMOS WIRING

Cu interconnects in CMOS technology integrate metal levels at multiple pitch and thickness of Cu to satisfy requirements of high density wiring and current carrying capacity. One consequence with technology scaling is the increased number of metal levels with every new generation. As a result, the last level of metal is typically kept ~10μm away from the substrate. In addition to metal levels available in a digital CMOS, RF-specific integration schemes in sub-100nm CMOS nodes include a thick (~3μm) final Cu metal level. Cu interconnects are fabricated using damascene wiring and chemical-mechanical polishing to ensure that each metal level remains strictly planar. The planarity of the metal level in conjunction with well defined sidewalls is a significant factor aiding the formation of high quality factor passives.

A. Lumped-element Passive Devices

With added masks and processing steps, high-performance MIM capacitors and thin-film resistors can be integrated into the standard BEOL process flow. Fringing capacitors and inductors can be realized using the standard wiring levels. Each of these devices can be optimized for high-frequency performance by minimizing parasitic impedances through judicious layout.

Adjusting substrate resistivity is one means of influencing passive device performance [11] with technology parameters. Inductor peak Q as a function of wafer resistivity is shown in Fig. 6 for bulk CMOS technologies (peak Q for these inductors occurs between 8GHz and 25GHz depending on geometry and substrate resistivity). Initially, as the substrate resistivity is decreased from very high values, current flow in the substrate increases, resulting in larger power loss. Then as the spiral-substrate capacitive reactance becomes the dominant factor limiting the current, any further decrease in substrate resistance results in an improvement of Q. At even lower resistivities, eddy current generation in the substrate

begins to cause additional power loss leading to the ultimate fall-off in Q. The maximum Q and the onset of these mechanisms are influenced by the inductor size.

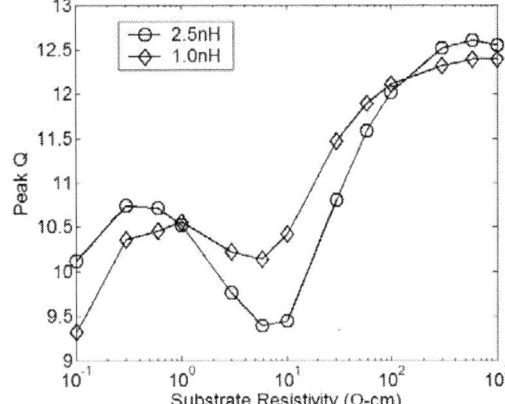

Fig. 6 Simulated effect of substrate resistivity on spiral inductor quality factor.

In general, resistivities of higher than 50 Ω-cm are unsuitable for bulk technologies due to latchup concerns. SOI technologies provide a buried oxide layer as a means of isolating the active devices from the bulk silicon substrate. While this layer is too thin to materially affect the parasitic capacitance between the spiral and the substrate, the isolation provided by the buried oxide allows the use of very high resistivity substrates without influencing active device characteristics allowing for less power loss and higher Q [12].

B. Distributed Passive Devices

Millimetre-wave applications will demand circuits which use not only resistors, capacitors, and inductors but also distributed elements. Transmission lines (t-lines) are the basic building blocks and are used as interconnects, matching elements, and serve as the foundation for various types of distributed devices.

Fig. 7 shows the t-line configurations in CMOS technologies. These include microstrip and coplanar lines for single and differential operation. These lines feature well-designed current return paths, such that a net zero current is maintained in a given cross section. With good shielding, line impedance can be well controlled and substrate losses and cross talk kept to a minimum. Microstrip lines up to 110GHz have been have been demonstrated in a 130nm BEOL featuring 3μm thick Al-Cu metal (Fig. 8). These t-lines exhibit a loss of ~0.8dB/mm. Properly shielded t-lines with 90° and 45° bends show no impact of the bends. Others forms of discontinuities (joints or splits) are also expected to minimally impact t-line performance since the cross-section dimensions are negligible compared to the wavelengths of interest. As technology scales, the t-line performance is not expected to change dramatically, since a thick top layer of metal will be retained and the BEOL "stack-up" height will remain the same.

978-2-8748-7007-1/08 $25.00 © 2008 EuMA 113

Numerous distributed elements have been implemented for 60GHz application using the above t-lines, including inductive and capacitive stubs, RF chokes, single-stub tuners for matching, and branch-line couplers. Inductive and capacitive components are typically created by varying the length of the short-circuit and open circuit stubs. At 60GHz, $\lambda/4$ =600μm, so these components can be easily integrated in silicon. Design of many of these components relies on t-line parameterized cells and corresponding models accurate to 100GHz. Models describe the skin and proximity effects and, for the coplanar structures, include the silicon substrate effects like substrate power loss, slow wave effect and frequency dependence of capacitance. The coupled line models also cover both the odd and even modes of signal propagation.

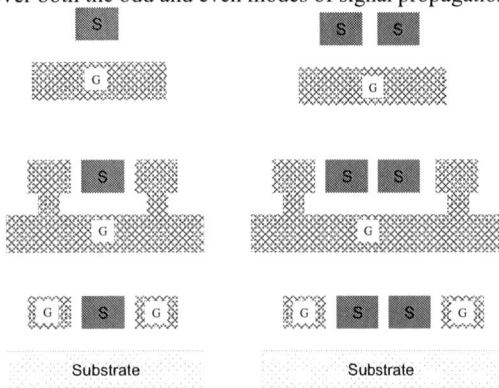

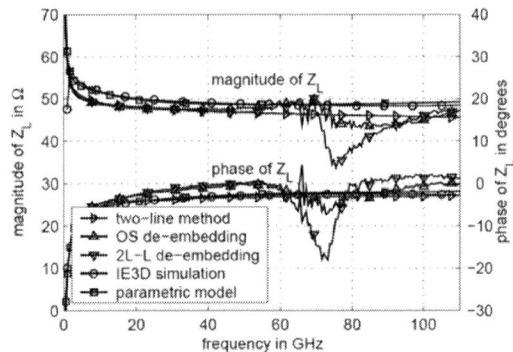

Fig. 7 Microstrip and coplanar-waveguide t-line configurations.

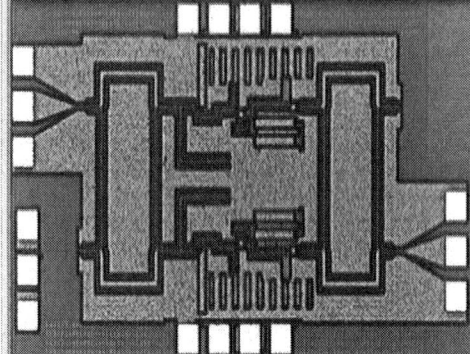

Fig. 8 Characteristic impedance of an on-chip microstrip t-line with passivation[13].

Microstrip branch-line, Lange couplers and CPW-based ring hybrid couplers showing good performance at 60 and 77GHz [14] have been demonstrated. The typical performance for these couplers are ~1dB insertion loss, ~15dB return loss, and 15-20dB isolation. Fig. 9 shows a balanced low-noise amplifier chip micro-graph for 77GHz (fabricated in a 130nm SiGe BiCMOS technology) featuring such distributed passives [15]. Coplanar tapers and branch line coupler are clearly seen in the micrograph.

IV. CONCLUSIONS

Transistor performance is now sufficient to support design

of millimetre-wave circuits. With the support of accurate models and integrated design tools, RF/AMS design techniques utilizing schematic-based elements representing "lumped" devices, interconnects and distributed devices can be employed.

Fig. 9 Chip micrograph showing microstip t-lines, branch-line couplers and CPW tapers in a 130nm BEOL.

REFERENCES

[1] Rim, K., et al., "Characteristics and device design of sub-100 nm strained Si N- and PMOSFETs", VLSI Technology, 2002. Digest of Technical Papers. 2002 Symp. on, 11-13 June 2002 Page(s):98 – 99

[2] Ito, S., et al., "Mechanical stress effect of etch-stop nitride and its impact on deep submicron transistor design", Electron Devices Meeting, 2000. IEDM Technical Digest. International, 10-13 Dec. 2000 Page(s):247 – 250

[3] Ota, K., et al., "Novel locally strained channel technique for high performance 55nm CMOS", Electron Devices Meeting, 2002. IEDM '02. Digest. International, 8-11 Dec. 2002 Page(s):27 – 30

[4] Ghani, T., et al., "A 90nm high volume manufacturing logic technology featuring novel 45nm gate length strained silicon CMOS transistors", Electron Devices Meeting, 2003. IEDM '03 Technical Digest. IEEE International, 8-10 Dec. 2003 Page(s):11.6.1 - 11.6.3

[5] Horstmann, M., et al., "Integration and optimization of embedded-sige, compressive and tensile stressed liner films, and stress memorization in advanced SOI CMOS technologies", Electron Devices Meeting, 2005. IEDM Technical Digest. IEEE International, 5-7 Dec. 2005 Page(s):233 - 236

[6] E. Guesev, et al., IBM J. Res. & Dev. Vol. 50 No. 4/5 July/Sept. 2006

[7] R. Chau, et al., "High-k/Metal–Gate Stack and Its MOSFET Characteristics", IEEE Electron Dev. Lett., Vol. 25, June 2004 p 408

[8] S. Nuttinck, "Ultrathin-Body SOI Devices as a CMOS Technology Downscaling Option: RF Perspective", IEEE Trans. On Electron Devicex, Vol. 53, No. 5, May 2006 p1193

[9] http://www-03.ibm.com/press/us/en/pressrelease/20980.wss, http://www.intel.com/pressroom/archive/releases/20070128comp.htm

[10] G. Dambrine, et al., "What are the Limiting Parameters of Deep-Submicron MOSFETs for High Frequency Applications?", IEEE Electron Device Letters, Vol. 24, No. 3, March 2003

[11] X. Zhu et al.," Micromachined on-chip inductor performance analysis," 16th Intl. Conf. on Micro Electro Mechanical Systems, p. 165, 2003.

[12] J. Kim et al., "Design and manufacturability aspect of SOI CMOS RFICs", Proc. Custom Integrated Circuits Conf., pp. 541 – 548, 2004.

[13] D. Goren et al., "On-chip interconnect-aware design and modeling methodology based on high bandwidth transmission line devices", Proc. Design Automation Conf., pp. 724 – 727, 2003.

[14] M. Chirala and B. Floyd, "Millimeter-Wave Lange and Ring-Hybrid Couplers in a Silicon Technology for E-band Applications", Microwave Symposium Digest, 11-16 June 2006, pp. 1547-1550.

[15] B. A. Floyd et al., "V-band and W-band SiGe bipolar low-noise amplifiers and voltage-controlled oscillators," RF Integrated Circuit Symposium Dig. Papers, pp. 295-298, 2004.

Proceedings of the 3rd European Microwave Integrated Circuits Conference

Design Aspects of 65-nm CMOS MMICs

Mikko Kärkkäinen, Mikko Varonen, Dan Sandström, Tero Tikka, Saska Lindfors, and Kari A. I. Halonen

TKK Helsinki University of Technology
SMARAD2 / Department of Micro and Nanosciences, Espoo, Finland

mmkarkka@ecdl.tkk.fi

Abstract—We present design aspects and techniques for millimeter-wave circuits implemented in 65-nm CMOS. Different transmission line topologies are discussed and measurement results for a conventional coplanar waveguide and slow-wave coplanar waveguide implemented in 65-nm CMOS are shown. The attenuation of the on-chip transmission lines can be reduced by using slow-wave coplanar waveguides. A 1-stage cascode amplifier in 65-nm CMOS employing inductors as matching elements is presented. On-chip interconnections of the amplifier are implemented and modeled using coplanar waveguides. The ground plane of the coplanar waveguide provides a good ground reference for the entire circuit.

I. INTRODUCTION

There are many emerging millimetre wave applications, which demand for low unit cost manufacturing solutions. The complementary metal oxide semiconductor (CMOS) technology has received a lot of interest since it enables mass production and integration of both digital and analogue functions on the same microchip. The device scaling of CMOS technologies improves the performance of the transistors in terms of a higher unity gain frequency (f_T) and maximum frequency of oscillation (f_{MAX}). We have already demonstrated 40 GHz and 60 GHz amplifiers and a V-band balanced mixer in 65-nm baseline CMOS achieving state-of-the-art performance [1][2]. The continuing scaling of bulk CMOS process typically introduces some challenges to the designer. These include lower supply voltage, stringent metal density requirements and thinner dielectric layers above the substrate leading to higher substrate losses of passives. In this paper, we discuss design aspects for millimetre wave circuits implemented in 65-nm CMOS.

II. TRANSMISSION LINES IN NANOSCALE CMOS

A. Design Considerations and Simulations

Thinner dielectric layers above the substrate of a nanoscale CMOS process and stringent metal density requirements set limitations for implementing transmission lines on silicon. A way to realize a conventional coplanar waveguide in nanoscale CMOS is presented in Fig. 1. The top metal layer is used for the centre conductor. Dummy metal is not allowed around the centre conductor or in between the ground planes of the CPW at any metal level. On the other hand, the metal density requirement has to be fulfilled which means that there has to be enough metal at all metal levels. This is accomplished by strapping all the other metal layers together with vias to form the ground plane for the CPW. The width of the centre conductor W and the distance between the center

conductor and the ground plane S can be used for realizing different characteristic impedances for the CPW. A wider centre conductor leads to lower conductor losses. In principle, the maximum width of the centre conductor is limited by the layout design rules of the chosen process. The metal density requirements set the limits for the maximum distance between the centre conductor and the ground plane for the CPW.

Because of the thin dielectric layers of a nanoscale CMOS process, the lossy silicon substrate is very close to the CPW, which causes increased substrate loss. One way to minimize the effect of the conductive substrate is to use the microstrip structure instead of the CPW. The microstrip line is realized between the top metal and lower metal planes. Ideally, this isolates the effect of the lossy silicon substrate. The removal of dummy metal from both underneath and the vicinity of the centre conductor can create a metal density problem. A way to realize a microstrip line in a CMOS technology is shown in Fig. 2 [3]. Drawing ground planes similar to a CPW-line fulfills the metal density requirements. These ground planes are then connected together using lower metal levels. The wide ground plane on the lower metal level must have longitudinal slots, which do not interfere with longitudinal ground currents of the microstrip line. When the height H of the dielectric material is rather low and when the top ground planes are located far from the center conductor, the signal propagates mostly in microstrip mode. In a nanoscale CMOS technology the more stringent metal density requirements render the design of millimeter wave circuits even more problematic. A rather large change in the width of the centre

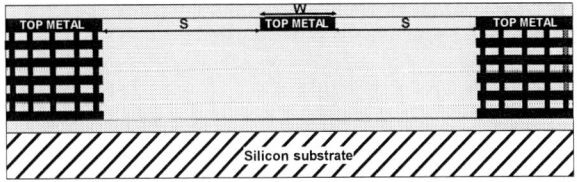

Fig. 1 Simplified cross-section of the conventional coplanar waveguide.

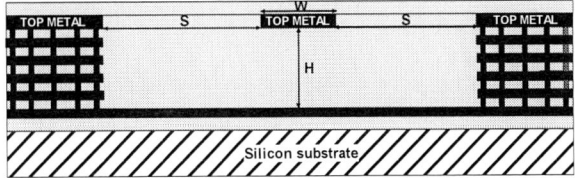

Fig. 2 Simplified cross-section of the microstrip line with sidewalls.

This work was funded by the Finnish Funding Agency for Technology and Innovation and supported by the Academy of Finland under UWI project.

978-2-8748-7007-1/08 $25.00 © 2008 EuMA

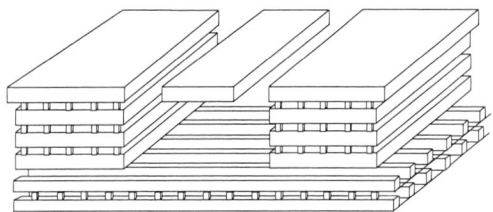

Fig. 3 Simplified cross-section of the slow-wave coplanar waveguide. Two lowest metal layers are strapped together with vias to form the floating shield strips.

conductor is needed to achieve a significant change in characteristic impedance, because of the low height of the dielectric layers. Thus, a wide range of impedances is difficult to realize, because the width of the centre conductor is limited by metallic losses in the narrow case and by the design rules in the wide case.

As discussed above, in the conventional coplanar waveguide the electromagnetic field penetrates into the silicon substrate, which increases losses. A metal shield structure can be drawn using the lowest metal levels to prevent the electromagnetic fields from penetrating into the lossy silicon substrate. An efficient way to realize the shield is the slow-wave structure employing floating shield strips [4]. The simplified cross section of the slow-wave coplanar waveguide implemented in this work is shown in Fig. 3. Two lowest metal layers are strapped together with vias to form the floating shield strips.

Electromagnetic simulations (Ansoft HFSS) were performed for the three different transmission line topologies presented above i.e. the conventional CPW, microstrip line and slow-wave CPW. For the transmission lines, a width W of a 12 µm was used for the centre conductor and the distance between the centre conductor and the ground plane S was 9 µm. The microstrip line was constructed by connecting sidewalls to the bottom metal ground plane and the height H was set to 2.4 µm. In the slow-wave CPW the width of the shield strip and the spacing between the strips was set to 1 µm.

The simulated attenuation α and phase constant β were calculated using equations found in [5]. The simulated attenuations per unit length are shown in Fig. 4. The Q-factor of a transmission line resonator can be calculated from

$$Q = \frac{\beta}{2\alpha}. \tag{1}$$

The simulated Q-factors are shown in Fig. 5. The conventional CPW has the highest attenuation and lowest Q-factor. The attenuation per unit length reduces significantly when using a microstrip line. The slow-wave CPW has the lowest attenuation and highest resonator Q-factor.

B. Measurement Results

Test structures were realized on a 65-nm CMOS for characterizing a conventional CPW and a slow-wave CPW. A test structure for the conventional CPW is shown in Fig. 6. A width of 12 µm was used for the centre conductor and the

distance between the centre conductor and the ground plane was 9 µm. The slow-wave CPW is constructed by strapping two lowest metal layers together to form the floating shield strips. The shield is designed using minimum design rules in order to suppress the induced current flow in the direction of the propagating RF-signal. This minimizes the ohmic losses and maximizes the reactive energy storage per unit length. The smallest allowable shield strip spacing minimizes the exposure of the overlying CPW to the conductive substrate [4].

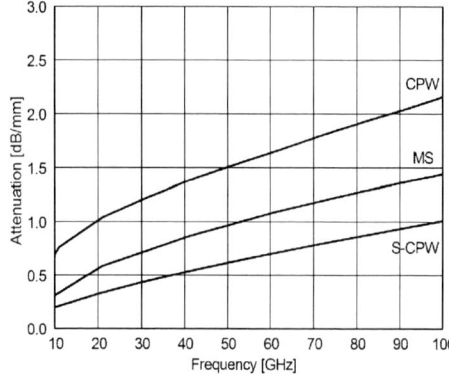

Fig. 4 EM-simulated attenuation per unit length dB/mm of the conventional coplanar waveuide (CPW), microstrip line (MS) and slow-wave coplanar waveguide (S-CPW).

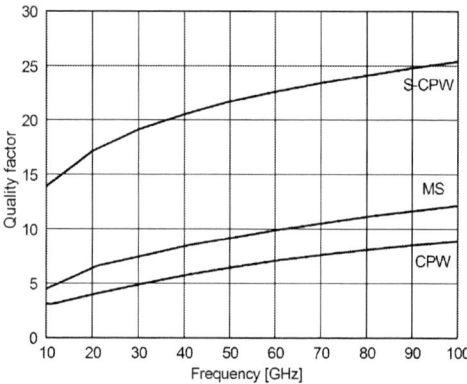

Fig. 5 EM-simulated Q-factor of the conventional coplanar waveuide (CPW), microstrip line (MS) and slow-wave coplanar waveguide (S-CPW).

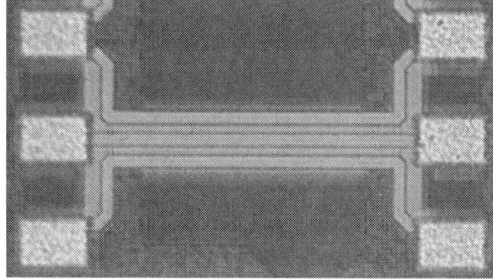

Fig. 6 Micrograph of a CPW test structure implemented in 65-nm CMOS.

978-2-8748-7007-1/08 $25.00 © 2008 EuMA

The measured characteristic impedance of the conventional CPW is around 47 Ω. Because of the shield strips, the relative dielectric constant for the slow-wave CPW is higher and the resulting characteristic impedance for the slow-wave version is lower (around 35 Ω).

The measured attenuation per unit length and quality factor for both conventional and slow-wave CPW are shown in Fig. 7 and Fig. 8, respectively. Even though the direct comparison of the CPW structures having different impedances is difficult, the transmission line attenuation of the slow-wave coplanar waveguide is significantly lower when compared to the conventional coplanar waveguide.

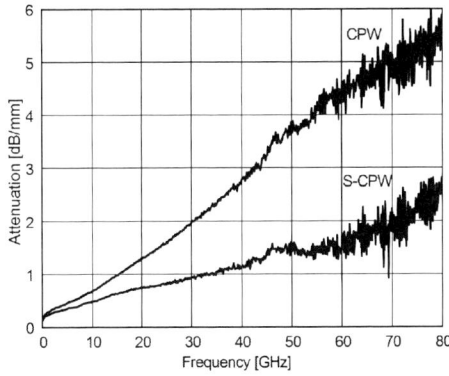

Fig. 7 Attenuation per unit length of the conventional and slow-wave CPW.

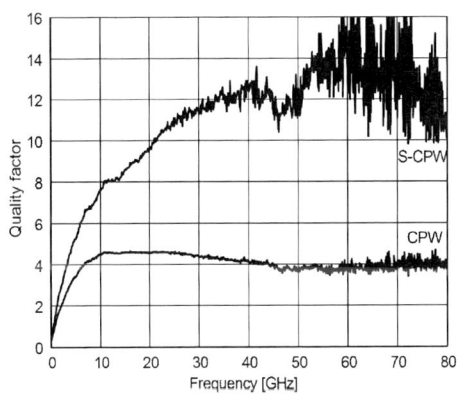

Fig. 8 Quality factor of the conventional and slow-wave CPW.

III. ACTIVE TEST STRUCTURES IN 65-NM CMOS

A. 30-GHz Amplifier

At millimeter waves, the transmission line matching networks provide a well-defined ground for the circuit. Another approach to perform on-chip matching is to use lumped elements such as inductors. At millimetre waves, the use of inductors becomes challenging, since the return currents of the circuit may not be explicit, which may result in inaccurate frequency response.

A principle layout/schematic and micrograph of a 1-stage cascode amplifier employing inductors as matching elements is presented in Fig. 9 and Fig. 10, respectively. On-chip

interconnections are implemented and modelled using coplanar waveguides. The coplanar waveguide provides a ground reference for the circuit.

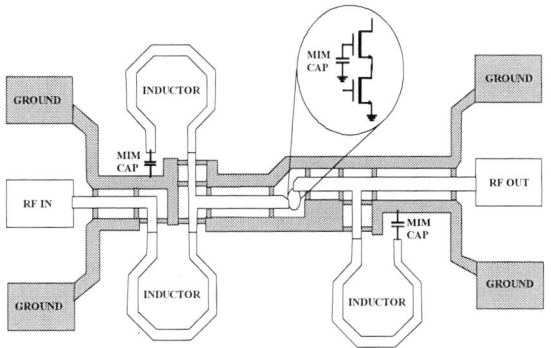

Fig. 9 Principle layout of the 30-GHz amplifier. Inductors are used for matching the cascode transistor. Coplanar waveguides are used for interconnections. The ground plane of the coplanar waveguide is used for ground reference for the circuit. Lower metal layers are used for connecting the ground planes of the CPW together around the discontinuities.

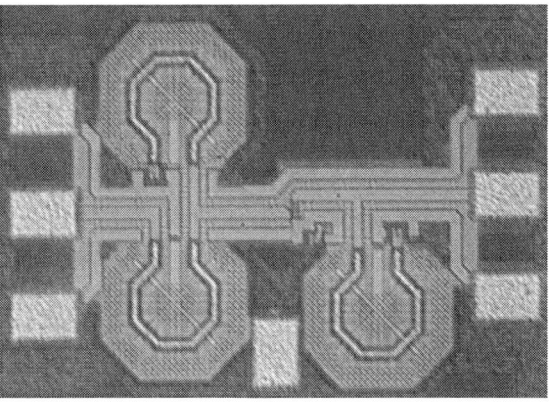

Fig. 10 Micrograph of the 30-GHz amplifier in 65-nm CMOS. Chip-area including pads is 0.54 mm x 0.36 mm.

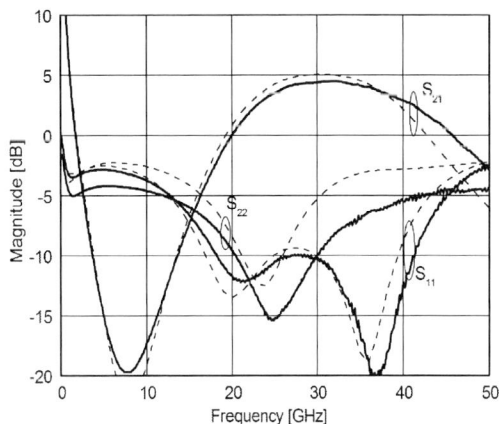

Fig. 11 Measured and simulated S-parameters of the 30-GHz amplifier implemented in 65-nm CMOS.

978-2-8748-7007-1/08 $25.00 © 2008 EuMA

The input of the amplifier is matched to 50 Ω using a series and short-circuited shunt inductor. The short-circuit is implemented using a metal-insulator-metal capacitor (2 pF). The low frequency stability is ensured using resistor-capacitor networks. A short-circuited shunt inductor is used for matching the output of the amplifier to 50 Ω.

The simulated and measured S-parameters of the amplifier are shown in Fig. 11. Because of the use of CPWs as interconnections and the ground plane of the CPW as a ground reference for the circuit, there is a good agreement between measured and simulated response. The measured small-signal gain is 4.5 dB at 32 GHz.

B. Transistor test structure in 65-nm CMOS

A CPW test structure, shown in Fig. 12, was developed for characterizing a common-source NMOS-transistor up to at least 60 GHz. The transistor data was de-embedded using open and short de-embedding [6]. The measured maximum stable gain is 9.4 dB at 60 GHz.

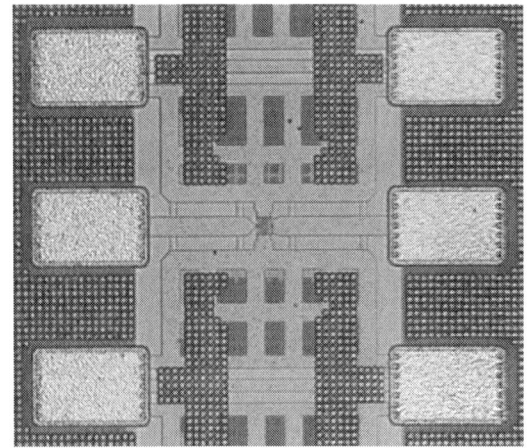

Fig. 12 A coplanar waveguide transistor test structure in 65-nm CMOS.

Although the design target was at 60 GHz the measurements were performed up to 110 GHz. The measured scattering-parameters of the test structure are shown in Fig. 13. As can be seen, a gain peak of 4.7 dB occurs at 97 GHz, which was not expected in the original design. The resonance is caused by the parasitic capacitance of the pad and by the length of the input and output CPWs. This can be simulated by using CPW and pad model for the test structure as presented in Fig. 14. Parasitic capacitances and resistances were extracted from the transistor layout. At millimeter wave frequencies, the effect of parasitic inductances becomes significant. Thus, small valued series inductors were used to model the access parasitic of the transistor. As can be seen from Fig. 13, a good agreement between simulations and measurements is achieved.

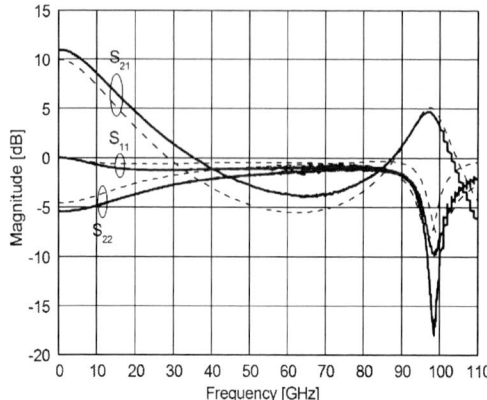

Fig. 13 Measured and simulated S-parameters of the transistor test structure.

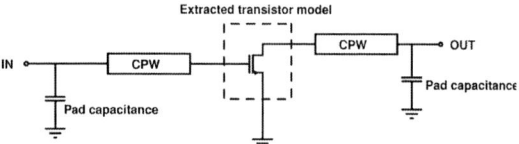

Fig. 14 The schematic for simulating the transistor test structure.

IV. CONCLUSIONS

In this paper we discussed and presented design aspects for implementing transmission lines in 65-nm CMOS. The attenuation of the on-chip transmission lines can be reduced by using slow-wave coplanar waveguides. A 30-GHz amplifier employing inductors and coplanar waveguides was presented. As the CPW provides a good ground reference for the circuit a good agreement between simulated and measured response is achieved.

ACKNOWLEDGMENT

We want to thank Hannu Hakojärvi, Millimetre Wave Laboratory of Finland - MilliLab, for on-wafer measurements.

REFERENCES

[1] M. Varonen, M. Kärkkäinen, and K. A. I. Halonen, "Millimeter-wave amplifiers in 65-nm CMOS," in *Proc. of the European Solid-State Circuit Conf.*, Munich, Germany. Sep. 2007, pp. 280-283.

[2] M. Varonen, M. Kärkkäinen, and K. A. I. Halonen, "V-band balanced resistive mixer in 65-nm CMOS," in *Proc. of the European Solid-State Circuit Conf.*, Munich, Germany. Sep. 2007, pp. 360-363.

[3] Y. Jin, M. A. T. Sanduleanu, E. Alarcon Rivero, and J. R. Long, "A millimeter-ave power amplifier with 25dB power gain and +8dBm saturated output power," in *Proc. of the European Solid-State Circuit Conf.*, Munich, Germany. Sep. 2007, pp. 276-279.

[4] T. S. D. Cheung, and J. Long, "Shielded passive devices for silicon-based monolithic microwave and millimeter-wave integrated circuits," *IEEE Journal of Solid-State Circuits*, vol. 41, pp. 1183-1200, May 2006.

[5] W. R. Eisenstadt, and Y. Eo, "S-parameter-based IC interconnect transmission line characterization," in *IEEE Trans. on Comp., Hybrids, and Manufacturing Tech., vol. 15, pp. 483-490, Aug. 1992.*

[6] M. C. A. M. Koolen, J. A. M. Geelen, M. P. J. G. Versleijen, "An improved de-embedding technique for on-wafer high-frequency characterization," in *Proc. IEEE Bipolar Circuits and Techn. Meeting*, Minneapolis, MN, Sept. 1991, pp. 188-191.

978-2-8748-7007-1/08 $25.00 © 2008 EuMA

Characterization and Modeling of Impact Ionization Effects on Small and Large Signal Characteristics of AlGaAs/GaInAs/GaAs PHEMTs

Charles Teyssandier[#], Fabien De Groote[*], Raphaël Sommet[*], Jean-Pierre Teyssier[*], Christophe Chang[#], Eric Leclerc[#], Bernard Carnez[#], Raymond Quéré[*]

[#]*United Monolithic Semiconductors SAS*
Route départementale 128, BP 46, 91401 ORSAY (France)

charles.teyssandier@ums-gaas.com
christophe.chang@ums-gaas.com

[*]*XLIM*

IUT GEII, 7 Rue Jules Valles 19100 BRIVE (France)
degroote@brive.unilim.fr

Abstract— This paper presents an analysis of the impact ionization phenomenon encountered in AlGaAs/GaInAs/GaAs PHEMTs. Two characterizations techniques have been used. At first, pulsed S-parameter measurements in the Impact Ionization (II) region have been required to identify the cut-off frequency of the phenomenon. Then, using these measurements, we propose a new small-signal model taking into account the frequency transition between quasi-static characteristics measured in pulsed conditions and microwave characteristics. Finally, this model has been tested in large signal conditions and the simulation results were checked through Load Pull Time domain Measurements (LPTM) to assert the validity of the model.

I. INTRODUCTION

The ever-growing demand in power as well as in higher frequency bands leads to the development of new power amplifiers with increased output power at higher frequencies. This development pushes the transistors towards their working limits. Thus, designers need improved models that are able to accurately describe the characteristics of the transistors in extreme operating conditions due to high compression or/and high output mismatch. Breakdown phenomena are among the most limitative effects in microwave transistors. A thorough analysis of their behavior allows a better understanding of the limits of the operating regions and thus a better use of the transistors. There are two main breakdown phenomena: the first one is related to the impact ionization and is called "on state" breakdown; the second, is the "off state" breakdown and appears when the transistor is pinched-off. This paper investigates the phenomenon of impact ionization in GaAs PHEMTs.

From pulsed I-V measurements and pulsed S-parameter measurements in the breakdown regions a cut-off frequency for the phenomenon is highlighted. This leads to a modification of the small and large signal equivalent circuit. Even if only one time constant for the Impact Ionization (II) is considered in the proposed model, the comparison between measured and simulated S-parameters and between measured and modeled load cycles validates the proposed model. Up to now, only static (pulsed) breakdown phenomena were considered for the design of amplifiers leading to incorrect load cycles prediction.

Modeling is performed on PPH25X technology from UMS. Main features of this AlGaAs/GaInAs/GaAs Pseudomorphic HEMT technology are: a double-side doped structure, a centred 0.25μm long T-shaped gate, a selective double recess for high breakdown voltage and via-holes under the source pads.

II. BEHAVIOR OF IMPACT IONIZATION

A. On-State Breakdown Mechanism

Impact Ionization phenomenon is the result of the collision of high-energy carriers with atoms of the crystal lattice. Incident electrons transfer a part of their energy to impacted particles. When this energy is sufficient, there is a succession cascade of electron-hole pairs, which leads to the breakdown effect.

In field effect transistors, the mechanism of II occurs in the channel, close to the drain where the electric field is the maximum (between the gate and drain). Pulsed I-V characteristics shown on Fig. 1 reveal the two types of breakdown phenomena in GaAs PHEMTs.

B. Small Signal Characterization

In order to investigate the small-signal behavior, the S-parameters of a 12x125 μm GaAs PHEMT were measured in pulsed mode in order to limit thermal effects. The measurements were performed at a quiescent point: Ids0 = 165 mA/mm and Vds0 = 8V, from which the pulses were issued, and at room temperature. The measured frequency range extends from 0.5 GHz to 20 GHz by 0.25 GHz steps for maximum accuracy on measurements. Y-parameters were calculated from these measurements. The main influence of II

978-2-8748-7007-1/08 $25.00 © 2008 EuMA

was observed on the Y_{22} parameters shown in Fig. 3. The frequency evolution of $Re\left(Y_{22}\left(f\right)\right)$ and $Im\left(Y_{22}\left(f\right)\right)$ are presented versus the pulsed bias point chosen such as Vds remains constant (= 7V) and Vgs varies from -1.2V (pinch-off voltage) to 1V (region where impact ionization is present).

It can be noticed that variations of both real and imaginary parts of Y_{22} are more important at low frequencies [1].

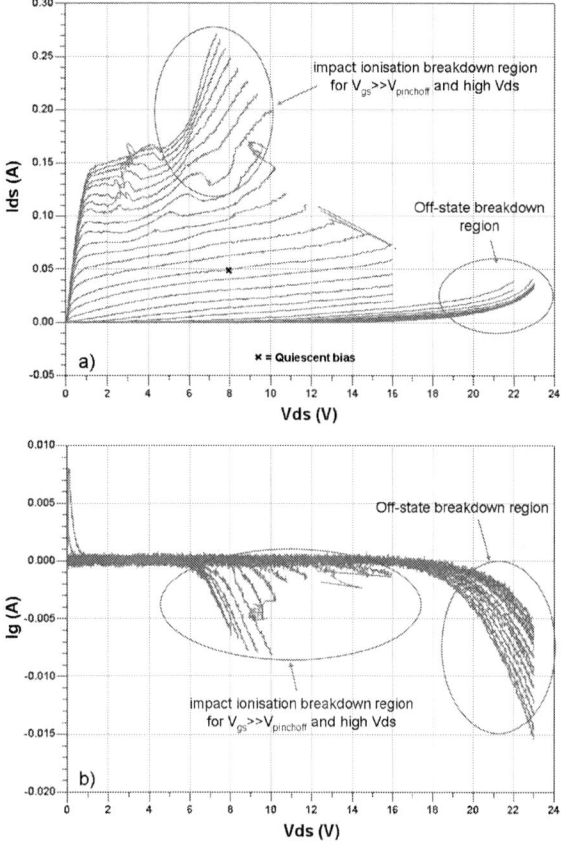

Fig. 1 Pulsed I(V) characteristics of the GaAs 4x75 μm PHEMT a) Drain current versus Vds b) gate current versus Vds

C. Evidence Of The frequency Dependence Of Impact Ionization

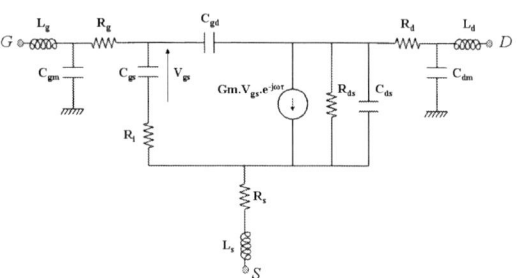

Fig. 2 Small-signal equivalent-circuit model of the PHEMT without impact ionization model.

The classical small-signal equivalent circuit model used in this paper is shown in Fig. 2.

After deembedding of the measured Y parameters, the RF output conductance (Gd_{rf}) and the output capacitance (C_{ds}), can be calculated through:

$$Gd_{rf} = Re\left(Y_{12}\right) + Re\left(Y_{22}\right)$$

$$\text{and } C_{ds} = \frac{1}{\omega}\cdot\left(Im\left(Y_{12}\right) + Im\left(Y_{22}\right)\right)$$

With $Re\left(Y_{12}\right) = 0$ because there is no resistance between drain and gate in our intrinsic model.

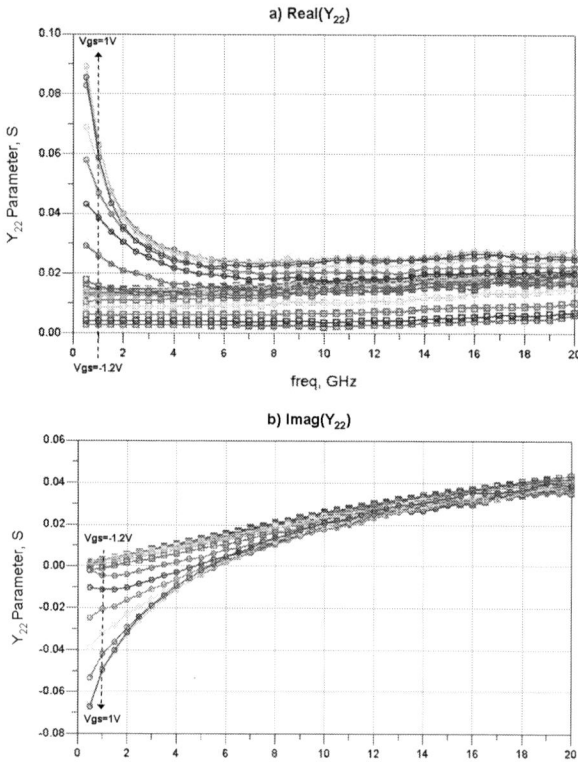

Fig. 3 Measured frequency dependence of intrinsic Y22 parameter of GaAs PHEMT from low to strong impact-ionization condition.

On the other hand, the static output conductance, called Gd_{dc}, is extracted from pulsed I(V)characteristics. In fact, to compare Gd_{dc} and Gd_{rf} in II conditions, Gd_{dc} is derived from drain current characteristics measured in pulsed conditions for a constant gate voltage (Vgs = 1V). The result is shown in Fig-4. The quasi static (f = 0 Hz) output conductance at Vds = 6.5V in the zone of impact ionization, is equal to 76 mS. From Y-parameters, in the same bias conditions, it can be observed that the output conductance decreases from 80 mS at 0.5 GHz to 25 mS at frequencies above 6 GHz. Moreover for gate bias voltages that do not put the transistors in II regions, the output conductance remains constant versus frequency. This suggests that II is a low frequency mechanism that does not exist at working frequencies greater than 4 GHz. This is the reason why a new II model is proposed in the following paragraph.

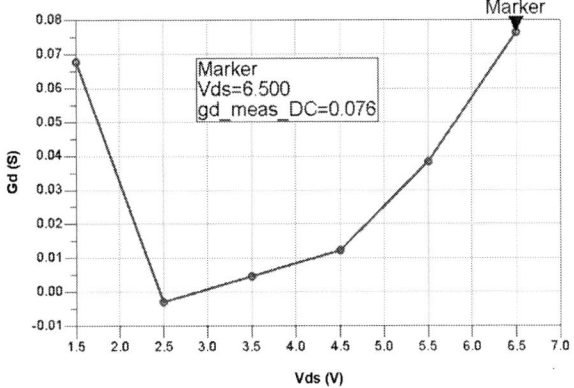

Fig. 4 Static output conductance Gd_{dc} of 12x125 GaAs PHEMT versus Vds at Vgs=1V

III. MODELING OF THE HIGH FREQUENCY BEHAVIOUR OF IMPACT IONIZATION

A. Modeling

As the main effect of II is observed on Y_{22}, the drain current is split in two sources in order to separate the II current, as shown in Fig-5.

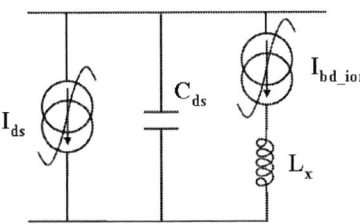

Fig. 5 Large-signal equivalent circuit model of GaAs PHEMT with low-pass filter for impact ionization effect.

The frequency dependence of the II mechanism is taken into account through the introduction of the self-inductance L_x in series with the breakdown current source I_{bd_ion}. On Fig. 6 the new small signal equivalent circuit of the large one presented in Fig. 5 is derived to calculate a new Y_{22}, G_x is the conductance due to the II and G_d is the conductance corresponding to the classical drain current without II.

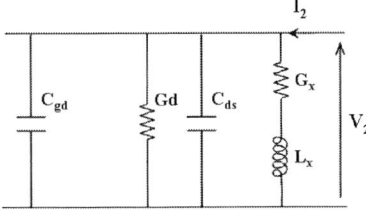

Fig. 6 Small-signal circuit to calculate Y22

$$Y_{22} = Gd + jC_{ds}\omega + jC_{gd}\omega + \frac{1}{R_x + jL_x\omega}$$

where $R_x = \frac{1}{G_x}$

So $Re(Y_{22}) = Gd + \frac{R_x}{R_x^2 + L_x^2\omega^2}$

$Im(Y_{22}) = C_{gd}\omega + C_{ds}\omega - \frac{1}{L_x\omega + \frac{R_x^2}{L_x\omega}}$

At low frequencies the output conductance is equal to $Re(Y_{22}) = G_d + G_x$ whereas at high frequencies we get: $Re(Y_{22}) = G_d$ and the cut-off frequency is given by:

$$Fc = \frac{R_x}{2.\pi.L_x}$$

B Small signal simulation results at Vgsi = 1V and Vdsi = 7V

On Fig. 7, the Y_{22} parameter simulation of the proposed model is compared to the measurements for a 4x75µm PHEMT. This characteristic is plotted for a frequency varying from 0.5GHz to 20GHz and with low-pass filters which cut impact ionization, the agreement is quite good.

The measurements were performed at room temperature. Y-parameters measurements presented in this paper were carried out at different temperatures. The comparison was made with an electrothermal model with same modelling structure for impact-ionization. The behaviour of $Re(Y_{22})$ versus temperature confirms that II effects appear for higher drain voltage when temperature increases [2].

C. Large Signal validation

The kink effect is a rise in drain courant and is related to both trap effects and II phenomenon. In [3], the authors show with short time scale for pulsed IV the kink can be hardly seen. In practice, it is difficult to have pulse durations that are close enough to the time constant of the II phenomenon.

In order to validate the proposed model in large signal conditions, Time Domain Load Pull Measurements were performed using a Large Signal Network Analyzer (LSNA) device [4]. To perform such validation the load cycle must enter in the II regions. This can be achieved by measuring the transistor under highly mismatched loading conditions.

The measurement frequency was chosen in order to satisfy the following conditions:
a- the frequency must be higher than the cut-off frequency of the II process
b- the frequency must be low enough in order to be able to measure a large number of harmonics for an accurate description of the highly nonlinear load cycle.

978-2-8748-7007-1/08 $25.00 © 2008 EuMA   121

Thus the large signal measurements were performed in CW mode at two frequencies: 2 GHz and 4 GHz. The results presented in the following are those obtained at 4 GHz.

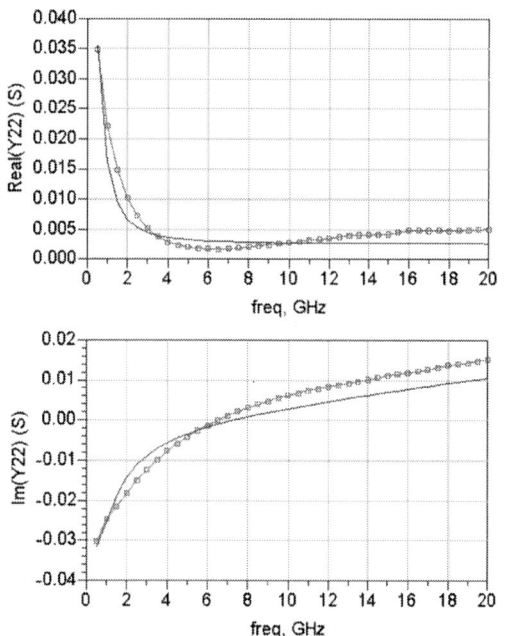

Fig. 7 Comparison between measurements (‑▫‑) and the simulation (⎯⎯) for 4x75µm PPH25X transistor

The results of the LSNA measurements are shown in Fig-8, plotted together with the static I-V characteristics taking the II phenomenon into account. It is clear from these curves that a large discrepancy between measured and modelled load cycles appears when the II current is taken into account at high frequencies. On the contrary with the new proposed model the correct agreement between the model and the measurements confirms the frequency limitation of the II process.

IV. CONCLUSIONS

The frequency dependence of the Impact Ionization process has been put into evidence using pulsed I-V and pulsed S-parameters measurements. The characteristics in the II regions have been obtained through these measurements. A new nonlinear model has been proposed which separates the main current and the II current taking into account the cut-off frequency. This model has been validated through time domain load pull measurements at 4 GHz. This model is currently used by designers to investigate the behaviour of transistors under mismatch conditions.

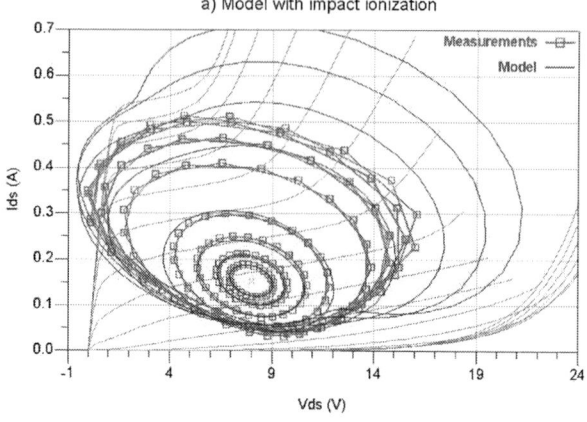

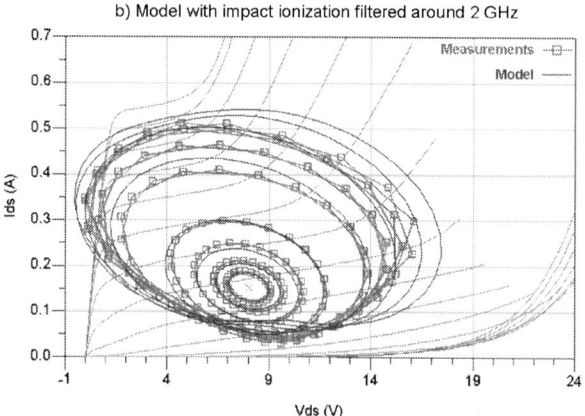

Fig. 8 Comparison between measurements (@4Ghz fundamental frequency) and simulations of the extrinsic output load cycles at load impedance (Z = 8-j30) from low level to 3dB compression: a) with Impact Ionization effects included and b) with Impact Ionization effects deleted by the filter.

REFERENCES

[1] M. Isler, K. Schünemann,."Impact-Ionization Effects on the High-Frequency Behovior of HFETs," *IEEE Transactions on Microwave Theory and Techniques*, vol. 52, pp. 858–863, March 2004.

[2] C. Groves, R. Ghin, J. P. R. David and G. J. Rees, "*Temperature Dependence of Impact Ionization in GaAs*," *IEEE Transactions on Electron Devices*, vol. 50, pp. 2027–2031, Oct. 2003.

[3] M.H. Somerville and al, "A Physical Model for the Kink Effect in InAlAs/InGaAs HEMT's," *IEEE Transactions on Electron Devices*, vol.47, pp.922-930, May 2000.

[4] J-P. Teyssier, D. Barataud, C. Charbonniaud, F. De Groote, J. Verspecht, J-M. Nébus, R. Quéré, "A Transistor Measurement Setup for Microwave High Power Amplifiers Design," APMC 2005, Suzhou (Chine), Dec. 2005.

Proceedings of the 3rd European Microwave Integrated Circuits Conference

A Methodology to Characterize the Low-Frequency Noise of InP Based Transistors

A. A. Lisboa de Souza [#1], J. C. Nallatamby [#2], M. Prigent [#3]

Xlim -Dep. C^2S^2 - Université de Limoges
7, Rue Jules Valles, Brive 19100 France
[1]antonio.desouza@brive.unilim.fr
[2]jcn@brive.unilim.fr
[3]prigent@brive.unilim.fr

Abstract—**This paper describes a methodology to measure the low-frequency noise of InP-based transistors. These transistors have demonstrated transition frequencies (f_t) greater than $200GHz$, generally achieved at current densities in excess of $200kA/cm^2$. Depending on the DC current gain, this may represent base currents of some mA. For the first time, curves of S_{ib}, S_{ic} and S_{ibic} for base currents of up to $3mA$ are demonstrated, in excellent agreement with those obtained from one-port measurements. This is only possible with an accurate experimental characterisation of the small-signal paramaters of the transistor, which are frequency-dependent due to self-heating.**

I. INTRODUCTION

InP based transistors are good candidates for milimeter wave applications such as W-band VCOs [1]. Maximum oscillation frequencies in excess of $400GHz$ have been demonstrated [2]. Such performance is generally achieved at very high current densities, which in general implies base currents of some mA. Under such conditions the impact of self-heating is considerable, and the impedances presented to the device should be carefully chosen to prevent low-frequency oscillations.

These conditions generally prevent the use of transimpedance amplifiers to characterize directly the low-frequency equivalent short-circuit current noise sources of such devices (illusttrated in Fig. 1), while the use of voltage amplifiers is still possible and allows a wide range of biasing resistors to be used [3].

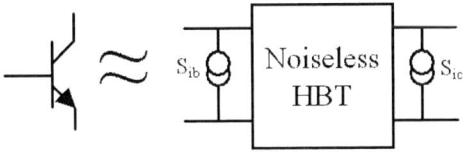

Fig. 1. Noisy bipolar transistor and equivalent model comprised of two possibly correlated equivalent short-circuit current noise sources.

However, since the noise is measured as voltage fluctuations, to deduce the current noise sources it is necessary to characterize the low-frequency small-signal parameters of the device.

This paper describes a methodology to measure the low-frequency equivalent short-circuit current noise sources (S_{ib}, S_{ic} and their correlation S_{ibic}) for base currents of up to $3mA$, which relies on the experimental characterization of the small-signal behavior of the transistor. The devices used in our experiments were fabricated at Alcatel Thales III-V Lab in their $1\mu m$ InP HBT process on epi-wafers grown by Picogiga International.

II. EXPERIMENTAL CHARACTERIZATION OF THE SMALL-SIGNAL H PARAMETERS

Since the noise at the ports of the transistor will be measured as voltage fluctuations and we are interested in extracting S_{ib}, S_{ic} and S_{ibic}, it is essential to characterize accurately the small-signal parameters of the device. Self-heating, which is a low-frequency mechanism, induces a frequency dispersion on such parameters within the frequency band of interest [4], which in this work goes from 100 to $100kHz$.

We have chosen to characterize the small-signal hybrid (h) parameters since they are somewhat meaningful with regards to the operation of a bipolar transistor, although other parameters such as Y or Z parameters give strictly the same results. The small-signal equivalent "h" circuit of the transistor is shown in fig. 2.

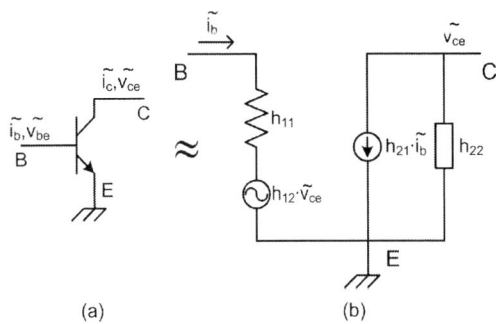

Fig. 2. Bipolar transistor (a) and equivalent circuit in terms of the hybrid parameters (b).

978-2-8748-7007-1/08 $25.00 © 2008 EuMA

To avoid the need of a short-circuit probe within the whole frequency band of interest to measure $h21$ directly, the 4 parameters are characterized through a series of impedance and voltage gain measurements under the biasing conditions to be used during noise measurements (see fig. 3). Moreover, to avoid disturbing the circuit, a (high impedance) AC current source is used to apply the stimulus, while the induced AC voltages are measured by high impedance probes.

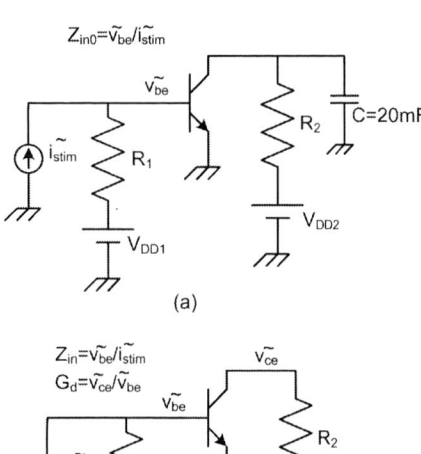

(a)

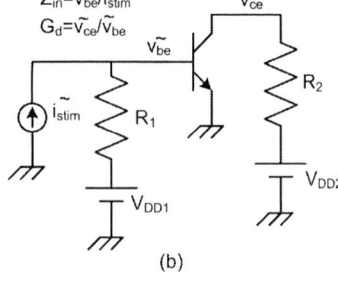

(b)

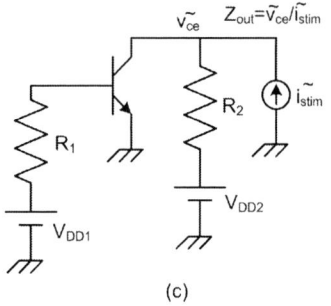

(c)

Fig. 3. Characterization of the small-signal h parameters.

There are 4 vector measurements over the system transistor + biasing circuitry to be performed: the input impedance under a shorted collector, Z_{in_0}, defined as the ratio of the voltage at the base to the stimulus current, is measured with the stimulus source applied to the base, while the shunting capacitor is connected to the collector. This is the only measurement for which the shunting capacitor is needed:

$$Z_{in_0} = \frac{\tilde{v}_b}{\tilde{i}} \text{ , with } \tilde{v_{ce}} = 0 \tag{1}$$

The input impedance, Z_{in}, defined as before, but without connecting the capacitor:

$$Z_{in} = \frac{\tilde{v}_b}{\tilde{i}} \tag{2}$$

The direct voltage gain, G_d, defined as the ratio of the voltage at the collector to that at the base (stimulus source applied to the base):

$$G_d = \frac{\tilde{v_{ce}}}{\tilde{v_b}} \tag{3}$$

Finally, the output impedance, Z_{out}, defined as the ratio of the voltage at the collector to the stimulus current, is measured with the stimulus source applied to the collector:

$$Z_{out} = \frac{\tilde{v_{ce}}}{\tilde{i}} \tag{4}$$

By using straightforward manipulations, it can be shown that the h parameters of the transistor relate to those 4 measurements as:

$$h11 = \frac{R_1 Z_{in_0}}{R_1 - Z_{in_0}} \tag{5}$$

$$h12 = \frac{(Z_{in} - Z_{in_0})R_1}{Z_{in}G_d(R_1 - Z_{in_0})} \tag{6}$$

$$h21 = \frac{Z_{in}G_d R_1}{Z_{out}(Z_{in_0} - R_1)} \tag{7}$$

$$h22 = \frac{Z_{out}Z_{in_0} + R_1 R_2 - Z_{in}R_2 - Z_{out}R_1}{Z_{out}R_2(R1 - Z_{in_0})} \tag{8}$$

To give an idea of the impact of self-heating on the small-signal behavior of the device, figures 4 and 5 present the h parameters characterized for a InP transistor featuring 4 emitter fingers of $15\times0.7\mu m$ each, biased at $I_{c_0} = 30$ and $60mA$ respectively ($V_{ce_0} = 2V$). As can be seen, both the h11 and h12 parameters are very sensitive to self-heating.

III. EXTRACTION OF S_{ib}, S_{ic} AND S_{ibic} FROM VOLTAGE FLUCTUATIONS

We proceed on measuring the voltage fluctuations simultaneously at the base and collector as shown in figure 6.

For practicity, as already mentioned the device is biased with the same resistors (namely, R_1 and R_2) used during the characterization of the small-signal parameters. The equivalent circuit for noise analysis is shown in Fig. 7.

$S_{i_{r1}}$ and $S_{i_{r2}}$ represent the current noise power of the biasing resistors R_1 and R_2 respectively. Using simple calculations, it can be shown that S_{ib}, S_{ic} and S_{ibic} relate to the voltage noise measured (S_{vb}, S_{bc} and S_{vbvc}), the biasing resistors (with their corresponding noise sources) and the complex h parameters as:

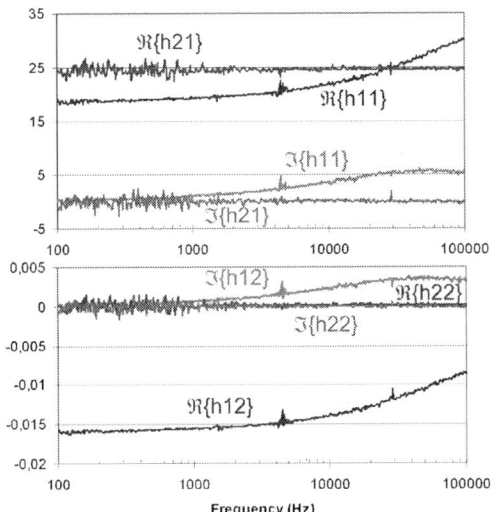

Fig. 4. Frequency dispersion of the h parameters for $I_{c_0} = 30mA$.

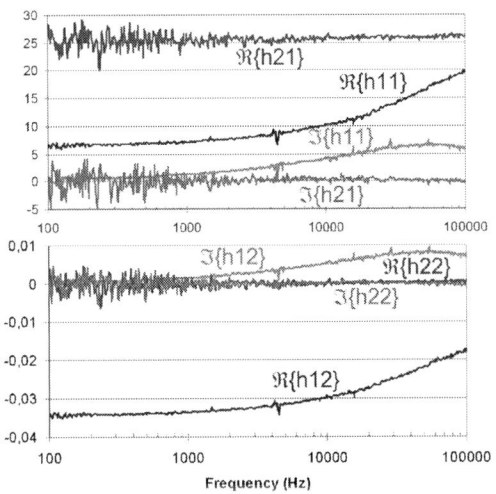

Fig. 5. Frequency dispersion of the h parameters for $I_{c_0} = 60mA$.

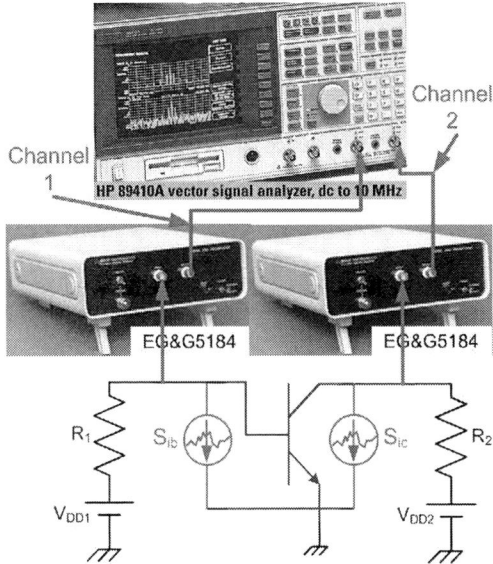

Fig. 6. Setup used to measure the low-frequency noise. The voltage fluctuations are measured simultaneously at the base and collector ports with low-noise Voltage Amplifiers (VA).

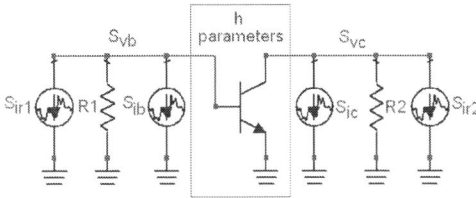

Fig. 7. Equivalent circuit for noise analysis.

$$S_{i_b} = |K_1|^2 \cdot S_{v_b} + |K_2|^2 \cdot S_{v_c} + K_1 \cdot K_2^* \cdot S_{v_b v_c} + \\ K_1^* \cdot K_2 \cdot S_{v_b v_c}^* - S_{i_{r1}} \quad (9)$$

$$S_{i_c} = |K_3|^2 \cdot S_{v_b} + |K_4|^2 \cdot S_{v_c} + K_3 \cdot K_4^* \cdot S_{v_b v_c} + \\ K_4^* \cdot K_3 \cdot S_{v_b v_c}^* - S_{i_{r2}} \quad (10)$$

$$S_{i_b i_c} = K_1 \cdot K_3^* \cdot S_{v_b} + K_1 \cdot K_4^* \cdot S_{v_b v_c} + \\ K_2 \cdot K_3^* \cdot S_{v_b v_c}^* + K_2 \cdot K_4^* \cdot S_{v_c} \quad (11)$$

, for which:

$$K_1 = \frac{R_2 \cdot h11 + R_2 \cdot R_1}{R_2 \cdot h11 \cdot R_1} \quad (12)$$

$$K_2 = \frac{-R_1 \cdot h12 \cdot R_2}{R_2 \cdot h11 \cdot R_1} \quad (13)$$

$$K_3 = \frac{R_2 \cdot h21 \cdot R_1}{R_2 \cdot h11 \cdot R_1} \quad (14)$$

$$K_4 = \frac{h22 h11 R_1 R_2 - h21 h12 R_2 R_1 + h11 R_1}{R_2 \cdot h11 \cdot R_1} \quad (15)$$

The curves in Fig. 8 were extracted with the procedure described above for a transistor featuring 4 emitter fingers of $15 \times 0.7 \mu m$ each, biased at $I_{c_0} = 30$ and $60mA$ ($V_{ce_0} = 2V$), leading to base currents of $I_{b_0} = 1.5$ and $3mA$ respectively.

Also included in the same plot are the curves obtained from one-port measurements (while the other port is shunted to ground with the help of a capacitor). As can be seen, an excellent agreement is found.

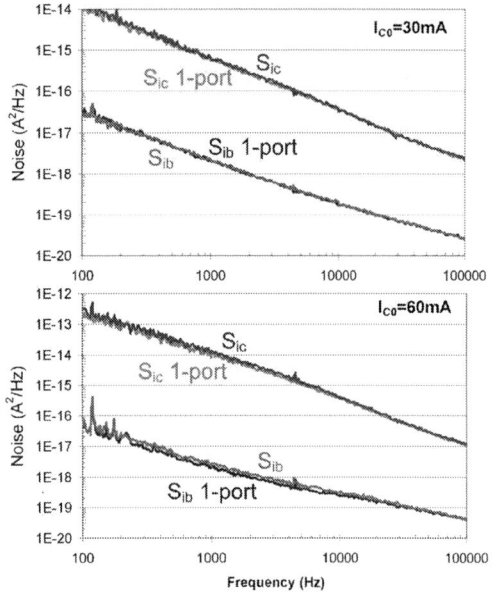

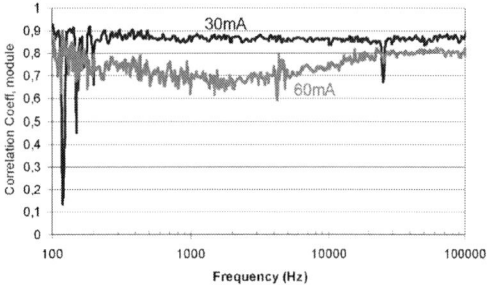

Fig. 8.   Curves of S_{ib} and S_{ic} extracted from two- and one-port measurements.

Figure 9 shows the module of the correlation coefficient between S_{ib} and S_{ic} obtained for the two biasing points cited. As can be seen, the noise sources are highly correlated.

Fig. 9.   Correlation coefficient between S_{ib} and S_{ic} for collector currents of 30 and $60mA$

IV. CONCLUSIONS

This paper presents a methodology capable of characterizing the low-frequency equivalent short-circuit current noise sources (along with their correlation) of bipolar transistors biased under very high current densities. The methodology has been applied to measure S_{ib}, S_{ic} and S_{ibic} of InP transistors under current densities requiring base currents of up to $3mA$.

For the first time under such base current values, curves of S_{ib}, S_{ic} and S_{ibic} in excellent agreement with those obtained from one-port measurements (S_{ib} and S_{ic}) are demonstrated.

ACKNOWLEDGMENT

This work was supported by the scholarship provided by CAPES (Ministry of Education, Brazil), and is part of the ATTHENA research project supported par ANR (France) [5].

REFERENCES

[1] Y. Baeyens, R. Pullella, C. Dorschky, J.-P. Mattia, R. Kopf, H.-S. Tsai, G. Georgiou, R. Hamm, Y. Wang, Q. Lee, and Y.-K. Chen, "Compact differential inp-based hbt vco's with a wide tuning range at w-band," in *Microwave Symposium Digest., 2000 IEEE MTT-S International*, vol. 1, 11-16 June 2000, pp. 349–352vol.1.

[2] J. Chingwei Li, T. Hussain, D. Hitko, P. Asbeck, and M. Sokolich, "Characterization and modeling of thermal effects in sub-micron inp dhbts," in *Compound Semiconductor Integrated Circuit Symposium, 2005. CSIC '05. IEEE*, 30 Oct.-2 Nov. 2005, p. 4pp.

[3] A. A. L. de Souza, J. C. Nallatamby, and M. Prigent, "Low-frequency noise measurements of bipolar devices under high dc current density: Whether transimpedance or voltage amplifiers," *Proceedings of the European Microwave Integrated Circuits Conference*, September 2006.

[4] O. Mueller, "Internal thermal feedback in four-poles especially in transistors," *Proceedings of the IEEE*, vol. 52, no. 8, pp. 924–930, Aug. 1964.

[5] "Atthena project." [Online]. Available: http://www.rmnt.org/com/archives/J3N2006-images/docs/ANR-ATTHENA.pdf

Revised RF Extraction Methods for Deep Submicron MOSFETs

J. C. Tinoco and J.-P. Raskin

Microwave Laboratory, Université catholique de Louvain

Place du Levant, 3, Maxwell Building, B-1348 Louvain-la-Neuve, Belgium

julio.tinoco@uclouvain.be

jean-pierre.raskin@uclouvain.be

Abstract— **Adequate modelling of MOS transistors for RF applications requires the accurate extraction of the extrinsic series resistances. In this paper, we fairly compare several RF extraction methods based on simulation results provided by an accurate foundry compact model of advanced RF MOSFETs. We present the relative sensitivity of each published RF characterization method to the measurement noise floor of Vectorial Network Analyzer. Additionally, the Bracale's method demonstrates to be less sensitive to the measurement noise but the extracted resistance values suffer from the mobility degradation due to the transversal electric field and the asymmetry of the device under test. Based on these theoretical and experimental results we propose a revised extraction procedure suitable for deep submicron transistors.**

I. INTRODUCTION

The MOSFET modelling and characterization at high frequency are based on the measurement of scattering (S-) parameters with a Vectorial Network Analyzer (VNA) and the definition of a lumped small-signal equivalent circuit as shown in Fig. 1. Two categories of lumped elements are generally considered:

- The intrinsic elements: g_m, g_d, C_{gs}, C_{gd} and C_{ds} which are dependent on the transistor bias conditions and scale with the transistor dimensions;
- The extrinsic series resistances: R_{se}, R_{de} and R_{ge} which are bias independent.

An adequate description of MOSFET behaviour requires the accurate extraction of the bias dependent intrinsic elements of the model. These intrinsic parameters cannot be directly extracted from the measured S-parameters. Indeed, the first step of the extraction procedure is the de-embedding of the extrinsic series resistances from the equivalent circuit.

Different characterization methods based on either DC or RF measurements have been proposed in the literature to extract the extrinsic series resistances of MOSFETs [1-9]. Contrary to the static extraction procedures, the RF techniques [7-10] allow to extract the gate resistance and separately the values of source and drain resistances as well.

In this paper, based on simulation results provided by an accurate foundry compact model of advanced RF MOSFETs, we fairly compare those RF extraction procedures considering the impact of VNA S-parameters measurement noise. Furthermore, the extraction techniques were applied on measured S-parameters on floating-body Silicon-on-Insulator (SOI) MOSFET with channel length of 0.13 μm. From those

simulation and experimental results, a new extraction procedure is proposed to overcome the main limitations of the current RF methods.

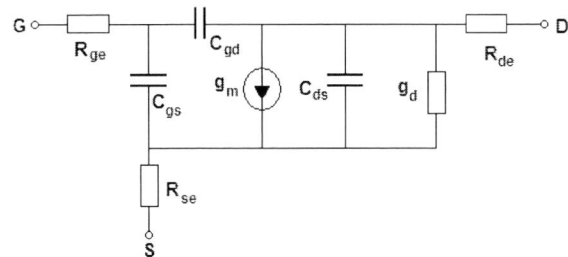

Fig. 1. Small-signal equivalent circuit of a MOSFET.

II. RF EXTRACTION METHODS

The philosophy of those extraction methods is mainly to bias the RF transistors at certain DC bias conditions under which the equivalent circuit can be simplified and then the number of unknowns reduced, facilitating its analysis. The main RF methods are: Lovelace [6], Torres-Torres [7], Raskin [8] and Bracale [9].

Recently, we have shown that Lovelace, Torres-Torres and Raskin's methods are able to determine accurately the series resistances when noise free S-parameters are considered [10]. However, the presence of noise on S-parameters strongly limits the efficiency of these extraction techniques, and in some cases they become non applicable. Table I summarizes the results previously obtained from ideal and noisy S-parameters.

TABLE I.　　SUMMARY OF EXTRACTED VALUES.

Extraction Methods	Extrinsic series resistances without noise			Extrinsic series resistances with noise		
	R_{se}	R_{de}	R_{ge}	R_{se}	R_{de}	R_{ge}
Lovelace	2.68	2.8	5	--	--	--
Torres-Torres	2.68	2.84	5	3.8	3.5	4.8
Raskin	2.79	2.88	5.03	3.19	4.65	5.8
Bracale	7.84	7.05	2.43	7.36	6.57	3.34
Considered values for Simulations	**2.87**	**2.87**	**5**	**2.87**	**2.87**	**5**

978-2-8748-7007-1/08 $25.00 © 2008 EuMA　　127

Fig. 2 shows the $Re(Z_{12})$ obtained from ideal and noisy simulated S-parameters, as it can be seen the noise produces a very big dispersion affecting the extraction procedure.

Some smoothing techniques were used, looking for reducing the noise impact on the S-parameters, and then to recover the extraction feasibility. However, no important improvement was observed for Lovelace and Torres-Torres' methods. In the case of the Raskin´s method, the improvement was evident but the number of measured frequency points has to be increased quite a bit.

Additionally, we demonstrated that the Bracale´s method is not adequate to extract accurately the series resistances of deep submicron MOSFETs as presented in Table I and explained hereafter. But on the other hand as illustrated in Table I, the extracted values with either ideal or noisy S-parameters are quite similar which indicates that this technique is less sensitive to measurement noise [10].

According to those results the Bracale´s method appears as a better candidate for developing a robust extraction procedure when the measurements noise becomes a major concern which is the case for nanometer scale transistors.

In this context, we have to deeply analyze the present limitations of the published Bracale's method [9]. Two major assumptions are made by Bracale *et al.* in [9]: the constant carrier mobility versus gate voltage overdrive and the perfect symmetry of the transistor when $V_{DS} = 0$ V. These two hypotheses are not valid anymore for deep-submicron devices [11, 12]. Therefore, we propose in this paper a revised Bracale´s method where the mobility degradation and MOSFET asymmetry are taken into account.

A. New Extraction Procedure

For the Bracale's extraction method the RF MOSFET is biased in inversion regime and at zero drain voltage ($V_{DS} = 0$ V, $V_{GS} > V_T$) [9]. Under this bias condition the device is considered symmetric ($C_{gs} = C_{gd} = C$), g_m tends to zero and g_d increases with V_{GS}. The real part of the impedance (Z-) parameters are function of the extrinsic series resistances, and are given by [9]:

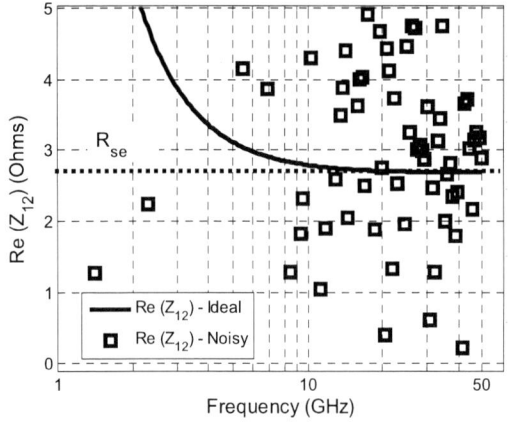

Fig. 2. Real part of Z_{12} with and wihout noise for silmulated S-parameters.

$$Re(Z_{22} - Z_{12}) = R_{de} + \frac{1}{2K(V_{GS} - V_T)} \quad (1)$$

$$Re(Z_{12}) = R_{se} + \frac{1}{2K(V_{GS} - V_T)} \quad (2)$$

$$Re(Z_{11} - Z_{12}) = R_{ge} - \frac{1}{4K(V_{GS} - V_T)} \quad (3)$$

with $K = \mu(W/L)C_{ox}$

where μ and C_{ox} are, respectively, the carrier mobility and the normalized gate oxide capacitance, and W, L are the transistor width and length, respectively.

Considering $C_{gs} \neq C_{gd}$, the real part of the Z-parameters can be expressed as:

$$Re(Z_{22} - Z_{12}) = R_{de} + \frac{1}{g_d(\alpha^{-1} + 1)\beta} \quad (4)$$

$$Re(Z_{12}) = R_{se} + \frac{1}{g_d(\alpha + 1)\beta} \quad (5)$$

$$Re(Z_{11} - Z_{12}) = R_{ge} - \frac{1}{g_d(\alpha + 1 + \alpha^{-1} + 1)\beta} \quad (6)$$

$$Re(Z_{22}) = R_{de} + R_{se} + \frac{1}{g_d\beta} \quad (7)$$

where $\beta = \left[1 + \frac{\omega^2}{g_d^2}\left(C_{gd} + C_{sd} - \frac{C_{gd}}{1+\alpha}\right)^2\right]$ and $\alpha = C_{gs}/C_{gd}$ which is defined as the symmetry coefficient.

If $C_{gd} + C_{sd} - C_{gd}/1 + \alpha << \omega/g_d$, the term $\beta = 1$, and for perfectly symmetric device $\alpha = 1$, thus (4)-(6) will be equal to (1)-(3). It is important to notice that in (7) is independent on α.

The output conductance can be obtained as:

$$g_d = \frac{dI_{DS}}{dV_{DS}} = \frac{W}{L}\mu_{eff}C_{OX}(V_{GS} - V_T) \quad (8)$$

Considering a single mobility degradation coefficient (θ), the effective mobility can be expressed as:

$$\mu_{eff} = \frac{\mu_0}{1 + \theta(V_{GS} - V_T)} \quad (9)$$

Substituting (9) in (8):

$$g_d = \frac{W}{L}\mu_0 C_{OX}\frac{(V_{GS} - V_T)}{1 + \theta(V_{GS} - V_T)} \quad (10)$$

Replacing (10) in (4)-(7), and considering a perfectly symmetrical device:

$$\text{Re}(Z_{22} - Z_{12}) = R_{de} + \frac{L}{2W\mu_0 C_{OX}} \frac{1}{V_{GS} - V_T} + \frac{L}{2W\mu_0 C_{OX}} \theta \quad (11)$$

$$\text{Re}(Z_{12}) = R_{se} + \frac{L}{2W\mu_0 C_{OX}} \frac{1}{V_{GS} - V_T} + \frac{L}{2W\mu_0 C_{OX}} \theta \quad (12)$$

$$\text{Re}(Z_{11} - Z_{12}) = R_{ge} - \frac{L}{4W\mu_0 C_{OX}} \frac{1}{V_{GS} - V_T} - \frac{L}{4W\mu_0 C_{OX}} \theta \quad (13)$$

$$\text{Re}(Z_{22}) = R_{de} + R_{se} + \frac{L}{W\mu_0 C_{OX}} \frac{1}{V_{GS} - V_T} + \frac{L}{W\mu_0 C_{OX}} \theta \quad (14)$$

The original Bracale´s method extracts the series resistances from the intercept of the linear regression of the corresponding Z-parameter vs. $1/(V_{GS}-V_T)$. According to (11)-(13), the intercepts are respectively:

$$b = R_{de} + \frac{L}{2W\mu_0 C_{OX}} \theta \quad (15)$$

$$b = R_{se} + \frac{L}{2W\mu_0 C_{OX}} \theta \quad (16)$$

$$b = R_{ge} - \frac{L}{4W\mu_0 C_{OX}} \theta \quad (17)$$

$$b = R_{de} + R_{se} + \frac{L}{W\mu_0 C_{OX}} \theta \quad (18)$$

Where $L/(W\mu_0 C_{ox})$ corresponds to the slope of (11)-(14).

Thus, the classical Bracale´s method overestimates R_{de} and R_{se}, while R_{ge} is underestimated.

According to (15)-(18), knowing θ, the real series resistances can be determined. From (10) we obtain:

$$\frac{1}{g_d} = \frac{L}{W\mu_0 C_{OX}} \frac{1}{(V_{GS} - V_T)} + \frac{L}{W\mu_0 C_{OX}} \theta \quad (19)$$

Therefore, θ can be extracted from a linear regression of $1/g_d$ vs. $1/(V_{GS}-V_T)$, as the intercept divided by the slope.

III. RESULTS AND DISCUSSIONS

The revised Bracale´s method is applied on ideal and noisy S-parameters, as mentioned above and described in [10]. Furthermore, measured S-parameters of a PD floating-body SOI MOSFET with channel length of 0.13 μm are used.

Fig. 3 shows the plot of $1/g_d$ vs. $1/(V_{GS}-V_T)$, used to determine the mobility degradation factor θ. Fig. 4 presents the linear regression of the $Re(Z_{12})$ vs. $1/(V_{GS}-V_T)$, obtained from ideal S-parameters. Table II summarizes the results obtained with this new extraction procedure.

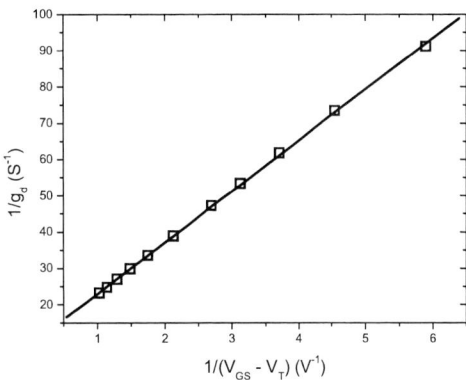

Fig. 3. θ determination from DC measurements.

As can be seen from Table II, the values obtained from the ideal and noisy S-parameters are very similar, reinforcing that contrary to the other RF techniques, this one is less sensitive to noise.

Furthermore, it can be seen in Table II that the source and drain resistances are considerably different, while the simulations were made using identical values of R_{se} and R_{de} (Table I). This difference comes from the inherent MOSFET asymmetry (even at $V_{DS} = 0$ V) which implies that the intrinsic capacitances C_{gd} and C_{gs} are not exactly equal.

Based on the ELDO simulations, Fig. 5 shows the variation of the symmetry coefficient as a function of V_{GS}.

As can be seen from Fig. 5, the symmetry coefficient is voltage dependent. According to (4)-(6), the real part of the Z-parameters are dependent on α. Thus, when a linear regression is made respect to $1/(V_{GS}-V_T)$, the voltage dependence of α will produce a modification of the straight line function parameters obtained and thus a modification on the final series resistances extracted.

The asymmetry involved in the MOSFET will produce alteration on the values extracted of the source and drain resistances. On the other hand, for the gate case in (6) appears $\alpha + \alpha^{-1}$ due to their opposite trend, the impact of the asymmetry is reduced allowing to obtain adequately R_{ge}.

TABLE II.   SUMMARY OF EXTRACTED SERIES RESISTANCES FROM REVISED BRACALE´S METHOD.

Resistances	Ideal S-parameters	Noisy S-parameters
R_{de}	2.34	2.57
R_{se}	3.63	3.6
R_{ge}	4.63	5.2
R_{ds}	5.97	6.2

In addition, from (7) it is possible to see that $Re(Z_{22})$ is not dependent on α, for this reason it is possible to determine adequately the total source-drain resistance $(R_{ds} = R_{de} + R_{se})$.

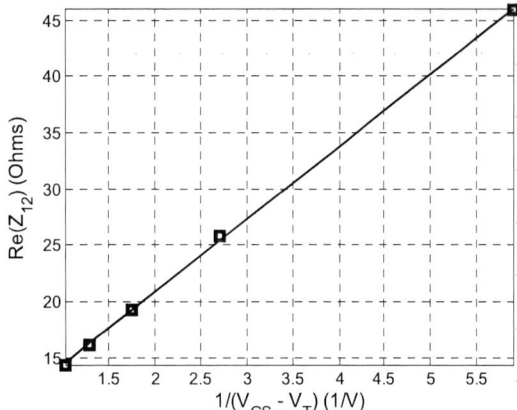

Fig. 4. Linear regression of the $Re(Z_{12})$ vs. $1/(V_{GS}-V_T)$ from simulated S-parameters.

Additionally, measured S-parameters confirm the high immunity of the revised Bracale's extraction technique against S-parameters measurement noise and the possibility to extract properly R_{ge} and R_{ds}.

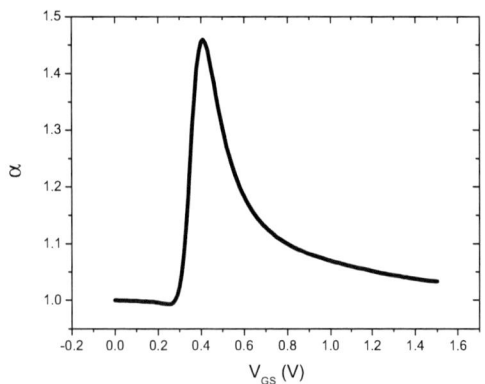

Fig. 5. Symmetry coefficient $\alpha = C_{gs}/C_{gd}$ vs. V_{GS}.

IV. CONCLUSIONS

Series resistances were extracted from simulated and measured S-parameters using the different RF techniques.

For the case of Lovelace, Torres-Torres and Raskin´s methods, the noise involved in the S-parameters measurements does not allow to obtain proper results. The use of smoothing techniques does not lead to significant improvements for Lovelace and Torres-Torres´ methods, indicating that they are very sensitive to the noise. For the case of Raskin´s method, smoothing techniques improve the extraction, however, the measured frequency points have to be increased quite a bit.

Original Bracele´s method does not allow accurate extraction of the series resistances due to the carrier mobility degradation for deep-submicron MOSFETs. This phenomenon was taken into account and a new procedure was established, where the mobility degradation coefficient is not neglected. This new method is less sensitive to S-parameters measurement noise and is valid to correctly extract the gate resistance as well as the total source-drain resistance of deep submicron advanced MOSFETs.

Finally, the authors are working on extension of the presented extraction method in order to take into account the inherent asymmetry of the nanometer scale MOSFETs in order to extract separately the source and drain resistances.

ACKNOWLEDGMENT

The authors would like to thank Dr. B. Parvais, Dr. M. Dehan, Dr. A. Mercha, Dr. S. Decoutere and Mr. V. Subramanian from IMEC, Leuven, Belgium for the fruitful discussions and ST-Microelectronics for providing the compact model of their 0.13 µm PD SOI MOSFET.

REFERENCES

[1] S P. I. Suciu and R. L Johnston, "Experimental derivation of the source and drain resistance on MOS transistors", *IEEE Transactions on Electron Devices*, vol. 27, no. 9, pp. 1846, 1980.

[2] J. Chern, P. Chang, R. Motta and N. Godinho, "A new method to determine MOSFET channel length", *IEEE Electron Device Letters*, vol. 1, no. 9, pp. 170, 1980.

[3] G. J. Hu, C. Chang and Y.-T. Chia, "Gate-voltage-dependet effective channel length and series resistance of LDD MOSFETs", *IEEE Transactions on Electron Devices*, vol. 34, no. 12, pp. 2469, 1987.

[4] F. H. De La Moneda, H. N. Kotecha and M. Shatzkes, "Measurement of MOSFET constants", *IEEE Electron Device Letters,* vol. 3, no. 1, pp. 170, 1982.

[5] S. Chung and J.-S- Lee, "A new approach to determine the drain-and-source series resistance of LDD MOSFET´s", *IEEE Transactions on Electron Devices*, vol. 40, no. 9, pp. 1709, 1993.

[6] D. Lovelace, J. Costa and N. Camilleri, "Extracting small-signal model parameters of silicon MOSFET transistors", *IEEE MTT-S*, pp. 865, 1994.

[7] R. Torres-Torres, R. S. Murphy-Arteaga and S. Decoutere, "MOSFET bias dependent series resitance extraction from RF measurements", *Electronics Letters*, vol. 39, no. 20, pp. 1476, 2003.

[8] J.-P. Raskin, R. Gillon, J. Chen, D. Vanhoenacker-Janvier and J.-P. Colinge, "Acurate SOI MOSFET characterizations at microwave frequencies for device performance optimization and analog modeling", *IEEE Transactions on Electron Devices*, vol. 45, no. 5, pp. 1017, 1998.

[9] A. Bracale, V. Ferlet-Cavrois, N. Fel, D. Pasquet, J. L. Gauthier, J. L. Pelloie and J Du Port de Poncharra, "A New Approach for SOI Devices Small-Signal Parameters Extraction", *Analog Integrated Circuits and Signal Processing*, vol. 25, pp.157, 2000.

[10] J. C. Tinoco and J.-P. Raskin, "RF-extraction methods for MOSFET series resistances: a fair comparison", *Proc. IEEE 7th International Caribbean Conference on Devices, Circuits and Systems*, Cancun, Mexico, April 28-30, 2008.

[11] D. Esseni and A. Abramo, "Modeling of electron mobility degradation by remote coulomb scattering in ultrathin oxide MOSFETs", *IEEE Transaction on Electron Devices*, vol. 56, no. 7, pp. 1665, 2003.

[12] T. Y. Chan, A. T. Wu, P. K. Ko, C. Hu and R. Razouk, "Asymmetrical characteristics in LDD and minimum-overlap MOSFETs", *IEEE Electron Device Letters*, vol. 7 no. 1, pp. 16, 1986.

Proceedings of the 3rd European Microwave Integrated Circuits Conference

Robust Packaged Diode Modelling with a Table-Based Approach

A. Rodríguez-Testera, O. Mojón, M. Fernández-Barciela and E. Sánchez

[#]*Dpto. Teoría de la Señal y Comunicaciones, Universidad de Vigo*
ETSI Telecomunicación, c/ Maxwell 36310 Vigo, Spain

Abstract— **A table-based nonlinear approach was used to predict the performance of a commercially packaged Schottky diode. Excellent results have been obtained which improve those of the analytical model provided by the manufacturer. Both models were extensively validated under DC, small and large-signal one-tone and two-tone excitations. Measurements for table-based model extraction and models validation were obtained by using a multi-tone nonlinear measurement system based on a LSNA. Time domain waveforms have once more been demonstrated as a key tool to compare large-signal models.**

I. INTRODUCTION

In general, diode nonlinear models used for circuit design with commercial packaged devices are of an analytical nature (e.g. SPICE diode model [1], one of the most widely used). On the other hand, a table-based approach was proposed by Root et al. [2], which accurately predicts on-wafer MODFET diodes performance under DC, and large-signal one-tone excitations. The model also showed good small-signal prediction for an on-wafer varactor diode.

Advantages of table-based approaches are their independence of the used technology, improved accuracy and direct extraction, thus avoiding parameters optimization. But even if table-based approaches has successfully been used to model transistors in MMIC designs [3-4], there is always the open question about how robust and practical these models are for modelling commercial packaged devices. Another caution sometimes addressed has been concerning their ability to predict multi-tone behaviour, since nonlinear functions higher-order derivatives are not constrained, in this case, by the expected behaviour of the analytical functions derivatives.

In this paper we generate a table-based nonlinear model of a commercially packaged Schottky diode and compare its performance with an analytical one (provided by the manufacturer) under DC, small and large-signal single-tone and two-tone excitations. All the measurements for model extraction and validation have been performed by using a multi-tone nonlinear measurement system based on the Large Signal Network Analyzer (Maury-NMDG LSNA).

II. TABLE-BASED NONLINEAR DIODE MODEL

Fig. 1 shows the Schottky diode nonlinear model topology. The nonlinear current and charge sources represent the table-based intrinsic model, while the rest of the elements (bias independent) model the parasitic network (which includes

package parasitics). C_{IN} and C_{OUT} are two extra capacitances which improve fitting to measured S-parameters.

To model the intrinsic Schottky diode we follow Root approach [2], but in a purely quasi-static formulation, since in this work we did not find it necessary to account for a relaxation time.

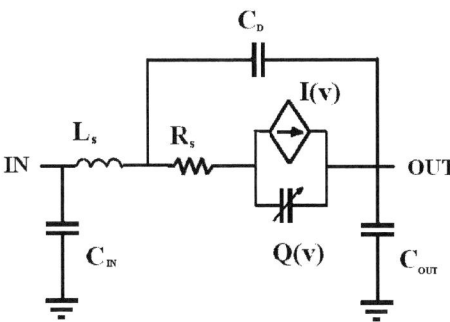

Fig. 1 Diode nonlinear model topology.

Table-based nonlinear current function is directly extracted from measured DC I-V data. Table-based nonlinear charge function is obtained from small-signal S-parameter measurements versus bias by using an integration procedure similar to [2] but based, in this case, on smoothing B-splines (MATLAB algorithms). Both intrinsic nonlinear model functions are defined as a function of intrinsic voltages and they will be spline interpolated during simulation time.

TABLE I

Parameter	Value	Units
C_{IN}	30	fF
C_{OUT}	200	fF
C_D	200	fF
L_S	2	nH
R_S	6	Ω

Model defined in this way does not require any type of function optimizations during extraction procedure, since there are no involved fitting parameters, i.e. ideality factors, saturation currents, and so on. On the other hand,

978-2-8748-7007-1/08 $25.00 © 2008 EuMA

measurements required for model extraction are just as simple to perform as those needed to extract an analytical model, just a more dense grid vs. voltage should be collected.

If the diode is to be operated under high input power levels, then it is important to complete model formulation of the intrinsic current function to consider reverse breakdown. We have included soft reverse breakdown modelling by extrapolating the table-based current function with an analytical approach that takes into account the manufacturer diode breakdown information.

This model formulation is valid for different type of diodes (varactor and MODFET diodes, as proved in previous works [2]) and technologies.

III. DEVICE AND MEASUREMENTS

The device used through out the discussion in this paper is an Agilent HSMS-2860 Schottky diode. It was mounted on a text fixture -designed by the authors- in a thru configuration, and thus operates as a two port device. Corresponding standard calibration impedances for TRL calibration were fabricated in the same substrate. Fig. 2 shows the test fixture and the calibration impedances.

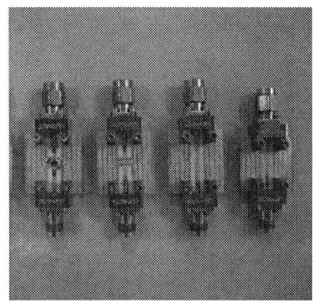

Fig. 2 . Test fixture and TRL calibration impedances.

Measurements for model extraction and validation purposes have been performed by using a multi-tone nonlinear measurement system based on the Large Signal Network Analyzer (Maury-NMDG LSNA). LSNA bandwidth is from 0.6 GHz to 50 GHz.

Parasitic elements values for this diode are shown in table I, and were obtained by conventional methods. The capacitances were extracted at low frequencies in reverse bias conditions. The inductance and series resistance were extracted taking into account one highly forward current bias point and another one near 0 V. A final optimization procedure can be performed by forcing a good fitting between the simulated (by using a small-signal equivalent circuit diode model) and measured S-parameters at different bias points. From the previous procedure, the inductance and series resistance values obtained were close to those provided by the manufacturer.

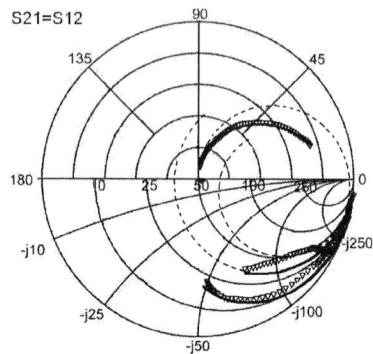

Fig. 3. Measured and simulated S-parameters. Bias point V=0 V (I=3nA). Frequency (0.6-6 GHz). Measurements (triangle), table-based model (line) and analytical model (dots).

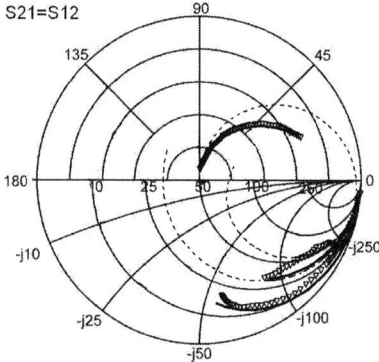

Fig. 4. Measured and simulated S-parameters. Bias point V=-1.5 V (I=-44nA). Frequency (0.6-6 GHz). Measurements (triangle), table-based model (line) and analytical model (dots).

Model nonlinear charge functions were extracted from small-signal measurements at 2 GHz vs. bias, in the range -1.5 V to 1.7 V. This range is the same we also used for extracting the current function from DC measurements. Since the device is mounted as a two port device, we will deal with four S-parameters instead of just one, S_{11}. We have found that, in this mode, the intrinsic charge, easily obtained from the imaginary part of Y_{21} or Y_{12} through an integration procedure, can be more accurately extracted when dealing with packaged diodes.

IV. RESULTS AND DISCUSSION

In this section we compare the table-based diode model predictions and the corresponding measurements, with those shown by the analytical SPICE diode model provided by the manufacturer. We did not try to improve the SPICE model to better fit our measurements, we just took the model as provided by the manufacturer. On the other hand, we are familiar with the extra difficulties that modeling packaged devices involve, in comparison to the use of on-die samples, and so the accuracy we should expect.

Figs. 3 and 4 show measured and simulated small-signal S-parameters at two different bias points in the range 0.6 GHz to 6 GHz. As can be clearly seen, improved results are obtained with the table-based approach.

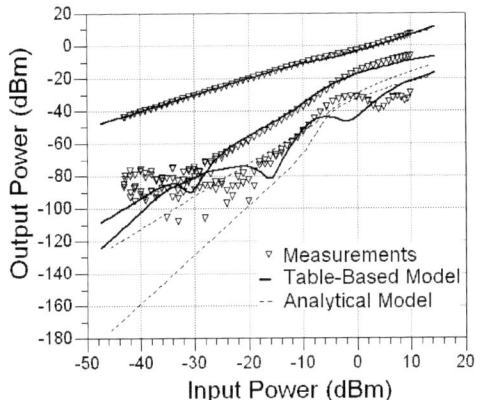

Fig. 5. Measured and modelled large-signal behaviour. Fundamental frequency=5.8 GHz, 2nd and 3rd harmonics. Bias point V=0 V (I=3 nA).

Figs. 5 and 6 show measured and simulated performance under single-tone large-signal excitation at two different bias points. Fundamental frequency is 5.8 GHz in both figures.

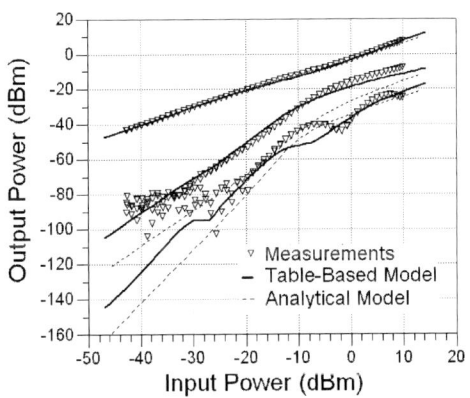

Fig. 6. Measured and modelled large-signal behaviour. Fundamental frequency=5.8 GHz, 2nd and 3rd harmonics. Bias point V=0.1375 V (I=3 uA).

Figs. 7 and 8 show measured and simulated DC and waveforms performance under single-tone large-signal excitations at two different bias points and different input power levels. Fundamental frequency is 5.8 GHz. As can be concluded from these figures, waveform comparison is a more robust and accurate way to validate a nonlinear model. From figs. 5 and 6 we could initially consider that both models have a similar large-signal behavior, but from the waveforms comparison it is clear that the predictions of the table-based model are much more accurate.

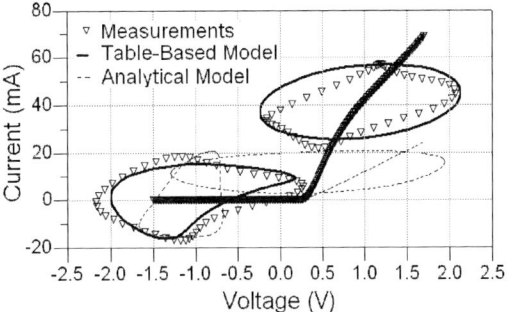

Fig. 7. Measured and modelled DC current, and time domain waveforms: Fund. Freq. =5.8 GHz, Pin=10 dBm. Bias point 1: V=-1 V (I=-33 nA). Bias point 2: V=1 V (I=35.8 mA).

Figs. 9 and 10 show measured and simulated large-signal behavior under two-tone excitations. Central frequency is 5.8 GHz, tone spacing 50 KHz. Two different bias points are considered. Again excellent results have been obtained with the table-based approach.

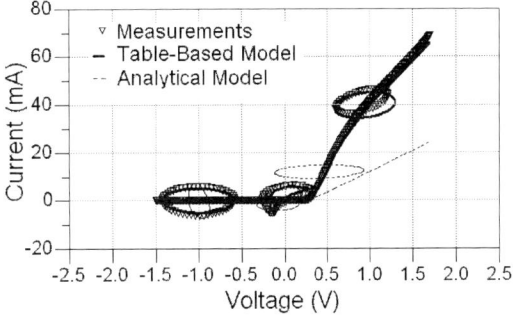

Fig. 8. Measured and modelled DC current, and time domain waveforms: Fund. Freq. =5.8 GHz, Pin=0 dBm .Bias point 1: V=-1 V (I=-33 nA). Bias point 2: V=0 V(I=3 nA). Bias point 3: V=1 V (I=35.8 mA).

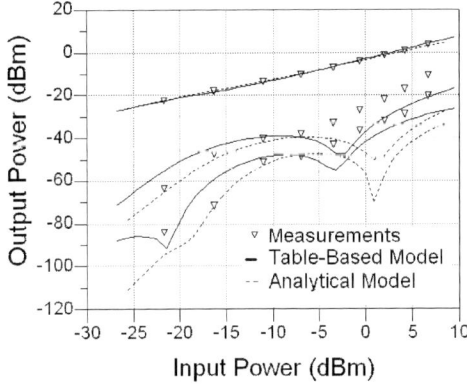

Fig. 9. Measured and modelled two-tone large signal behaviour. f_0=5.8 GHz ± 0.05 MHz. In plot are shown f_1, IMD3 ($2f_1$-f_2) and IMD5 ($3f_1$-$2f_2$). Bias point V=0.1375V (I=3 uA).

978-2-8748-7007-1/08 $25.00 © 2008 EuMA   133

Fig. 11 shows the behaviour of the proposed diode model under DC conditions, including high reverse voltages, in comparison with the measurements (covering a smaller voltage range) and the manufacturer analytical diode model.

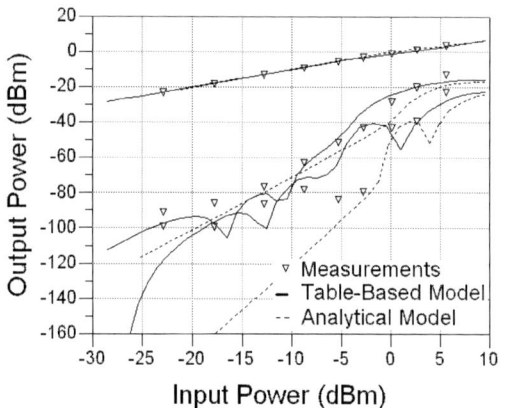

Fig. 10. Measured and modelled two-tone large-signal behaviour. f_0=5.8 GHz $\pm$ 0.05 MHz. In plot are shown, f_2, IMD3 ($2f_2$-f_1) and IMD5 ($3f_2$-$2f_1$). V=-1V (I=-44 nA).

V. CONCLUSIONS

This paper has demonstrated the ability of a simple table-based nonlinear approach to predict the performance of a commercial packaged Schottky diode through a comparison with extensive large-signal measurements. The model improves the results show by the diode manufacturer, which uses a SPICE diode model. Both models were extensively validated under DC, small and large-signal single-tone and two-tone excitations. Measurements for table-based model extraction and models validation were obtained by using a multi-tone nonlinear measurement system based on a LSNA. Excellent results were obtained with the table-based approach, especially in the time domain waveform prediction.

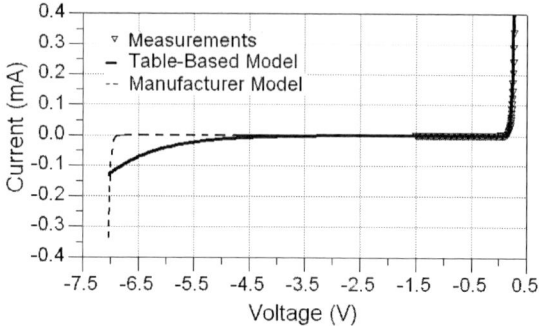

Fig 11. Measured and modelled (using the manufacturer and table-based approaches) DC current showing the reverse breakdown modelling.

ACKNOWLEDGMENT

This work was supported by the Spanish Ministerio de Educación y Ciencia (TEC2005-08377-C03-02/FEDER, HP2006-0120), and Universidad de Vigo.

REFERENCES

[1] L. W. Nagel and D. 0. Pederson, "Simulation program with integrated circuit emphasis (SPICE)," ERL Memo No. ERL-M520. University of California, Berkeley, May 1975.

[2] David E. Root, Marco Pirola, Siqi Fan, and Alex Cognata, "Measurement-Based Large-Signal Diode Model, Automated Data Acquisition System, and Verification with On-Wafer Power and Harmonic Measurements," *IEEE IMS MTT-S Digest*, pp. 261–264, 1993.

[3] M. Fernández-Barciela, P. J. Tasker, Y. Campos, M. Demmler, H. Massler, E. Sanchez, C. Currás, and M. Schlechtweg, "A simplified broad-band large-signal nonquasi-static table-based FET model," *IEEE Trans. Microwave Theory Tech.*, vol. 48, pp. 395-405, March 2000.

[4] Y. Campos-Roca, L. Verweyen, M Neumann, M. Fernandez-Barciela, M.C. Curras-Francos, E. Sanchez, A. Hulsmann, M. Schlechtweg, "Coplanar pHEMT MMIC frequency multipliers for 76-GHz automotive radar," *IEEE Microwave and Guided WaveLetters.*, vol. 9, pp. 242-244, June 1999.

Proceedings of the 3rd European Microwave Integrated Circuits Conference

X-Parameter Measurement and Simulation of a GSM Handset Amplifier

Jason M. Horn[#1], Jan Verspecht[*2], Daniel Gunyan[#3], Loren Betts[#4], David E. Root[#5], and Joakim Eriksson[†6]

[#]*Agilent Technologies, Inc., Santa Rosa, CA, 95403, USA*
[1]jason_horn@aglent.com, [3]daniel.gunyan@agilent.com, [4]loren_betts@agilent.com,
[5]david_root@agilent.com

[*]*Jan Verspecht, b.v.b.a., Opwijk, B-1745, Belgium*
[2]contact@jan_verspecht.com

[†]*Sony-Ericsson Mobile Communications AB, Nya Vattentornet, Lund, SE-221 88, Sweden*
[6]joakim.eriksson@sonyericsson.com

Abstract— **X-parameters, also referred to as the parameters of the Poly-Harmonic Distortion (PHD) nonlinear behavioral model, have been introduced as the natural extension of S-parameters to nonlinear devices under large-signal drive [1]-[3]. This paper describes a new approach to X-parameter characterization and nonlinear simulation - including large-signal experimental model validation - of a commercially available GSM amplifier. A specially configured Nonlinear Vector Network Analyzer (NVNA) and procedure for measuring, for the first time, X-parameters under pulsed bias conditions is presented. The measured pulsed bias X-parameters are then used with the PHD framework to enable accurate nonlinear simulation of device behavior, including harmonics (magnitude and phase) under pulsed bias large-signal conditions with mismatch. Independent NVNA measurements validate the predictions of the X-parameter simulations of output match under drive, and show the inadequacy of "Hot S22" techniques to predict such device performance.**

I. INTRODUCTION.

X-parameters have been previously introduced as the parameters of the Poly-Harmonic Distortion (PHD) nonlinear behavioral model [1]-[3], and can be viewed as the natural extension of S-parameters to nonlinear devices. A similar formalism is presented in [4]. While S-parameters are the linearization of the spectral map representing the behavior of a device at a DC operating point, X-parameters are the linearization of the spectral map about a Large Signal Operating Point determined by DC bias conditions and also one or more large RF input tones. X-parameters derived from measurement or simulation can be used in the PHD framework for very accurate simulation under large-signal drive in the presence of small to moderate mismatch, and can predict distortion through cascaded chains of nonlinear components. DC parameters can also be included to accurately predict bias-network interaction and PAE.

Like S-parameters, X-parameters can be easily and unambiguously identified from a set of simple experiments. The hardware required for X-parameter extraction is somewhat more complex, however, since small and large tones must be injected at multiple ports simultaneously, and cross-frequency phase coherence of measurements must be maintained. Due in part to the increased hardware complexity, previous X-parameter measurements have only been measured

at fixed DC bias with CW inputs. This paper introduces a system capable of taking the large-signal measurements necessary to extract X-parameters under pulsed bias operating conditions. The system is used to extract X-parameters of a commercially available Skyworks GSM amplifier (SKY77329) [5] under pulsed bias stimulus closely resembling real-world operating conditions, and the measured X-parameters are inserted into the PHD framework for simulation of the device under complex modulated stimulus in Agilent ADS.

II. DEVICE UNDER TEST

The motivation for measuring X-parameters under pulsed bias conditions was to model the behavior of a GSM amplifier under conditions as close as possible to actual GSM applications. The model needs to be able to handle large signal inputs, accurately predict harmonics, and also predict the affects of mismatch at the fundamental and harmonic frequencies. The specific IC considered was the Skyworks SKY77329 Power Amplifier Module, which is commercially available and used as the PA block in the Sony Ericsson W810 quad-band GSM/EDGE mobile phone.

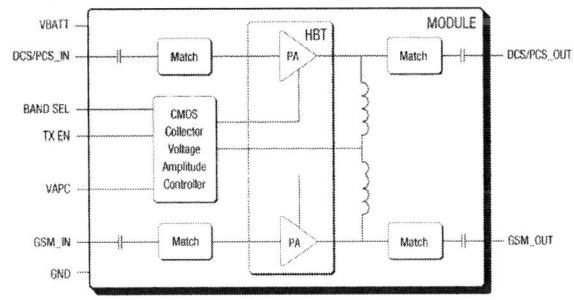

Fig. 1. Block diagram of SKY77329 Power Amplifier Module given on Skyworks Datasheet

The SKY77329 Power Amplifier Module is a multi-chip module (MCM) containing two RF amplifiers together with analog as well as digital control circuitry. The MCM contains a low-band amplifier operating around 900MHz and a high-band amplifier operating around 1800MHz. Each amplifier

978-2-8748-7007-1/08 $25.00 © 2008 EuMA

has its own RF input pins ("GSM_IN" and "DCS/PCS_IN") and RF output pins ("GSM_OUT" and "DCS/PCS_OUT"). These pins are AC coupled. Note that all of the other pins are common to the two amplifiers. The DC power is applied to the "VBATT" pin. The output amplitude is controlled by the voltage applied to the "VAPC" pin. The digital "BAND SEL" pin is used to control whether the low-band or the high-band amplifier is being used. The digital "TX EN" pin controls whether or not the amplifier is turned on.

III. Measurement Setup

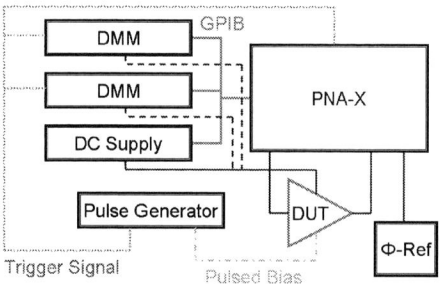

Fig. 2.   Block diagram of pulsed bias X-parameter measurement setup.

In order to take the large-signal measurements necessary for X-parameter extraction, a Nonlinear Vector Network Analyzer (NVNA) was configured using the Agilent PNA-X performance network analyzer. By combining the PNA-X with an external phase reference generator and appropriate instrument control and processing software, fully calibrated large-signal measurements were made possible with the high dynamic range, speed, and the feature set of the PNA-X, including triggering and pulsed capabilities. The built-in second source, combiner, and switching capabilities of the PNA-X were particularly useful since X-parameter measurements require multiple simultaneous input tones at harmonically related frequencies and multiple ports.

An Agilent E3631A DC power supply was used to sweep the device bias conditions and Agilent 34411A digital multimeters were used to measure DC current drawn by the device, enabling DC dependence and bias network interaction in simulation as well as PAE prediction. These instruments were controlled from the software running on the PNA-X through the built-in GPIB controller. An Agilent 81110A Pulse-/Pattern Generator was used to provide pulsed bias conditions mimicking actual GSM timing and to provide a trigger signal for measurement synchronization.

In order to model the device behavior under GSM-like operating conditions, the first output of the pulse generator was connected to the enable pin of the DUT and set to a 1/8 duty cycle pulse with a period of 4.615 μs. The second output was set to rise 50 μs after the first for measurement triggering purposes. The multimeters were set to use aperture mode integration, with an aperture of 300 μs to ensure that the measurement was taken entirely during the pulse. The PNA-X was also configured in trigger mode, and the IF bandwidth

was set to 3 kHz to meet the timing constraints. Measurement time was also a consideration, since the model was being extracted over a large range of operating conditions (9,720 Large Signal Operating Points per band). With this system, the total time required for measurement and extraction was about 5 hours per band.

IV. Enabling Simulation

The desired end result of these measurements was to produce a working nonlinear simulation model that resembles the DUT as much as possible in structure (including all relevant pins) and functionality, accurately representing the device behavior across its range of operation. Once the X-parameters have been measured, the next step taken toward accomplishing this goal was to generate PHD models for each band of the DUT. This was done using an auto-configuration script that creates and configures a model in Agilent ADS with the appropriate number of ports and harmonics based on the X-parameter measurements. Pins that were not swept in measurement, such as band-select and transmit-enable, were handled through external switching circuitry. This circuitry was combined with the high-band and low-band PHD models into a complete representation of the IC that includes full functionality of all pins and accurate representation of the device behavior across the operating range.

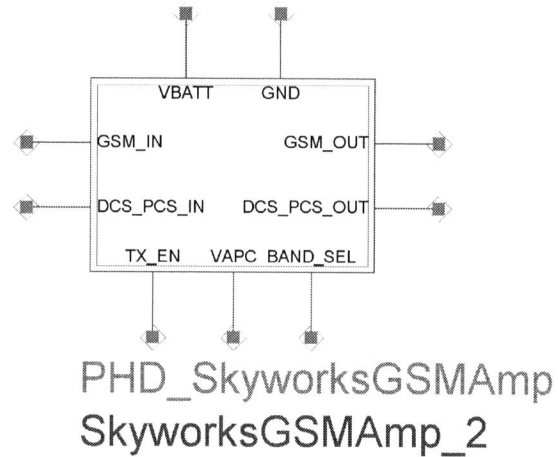

Fig. 3.   ADS block representing the SKY77329 IC with full functionality of all relevant input/output pins, enabling pulsed bias simulation.

V. Results

The measured X-parameters were used in simulation under a variety of conditions, and accurately reproduce the component behavior. The IC was characterized across a wide range of bias conditions, so we chose a few specific points of interest at which Skyworks had provided data on the IC and compared the simulation results to the specified behavior.

Output power was predicted as a function of VAPC, and results agreed very well with the provided data.

Output Power vs. VAPC

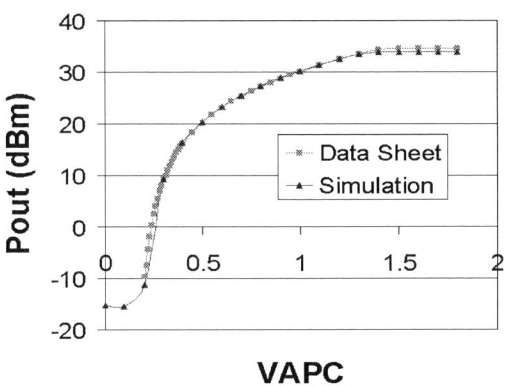

Fig. 4. Pout vs. VAPC comparison of simulation results and data provided with the IC.

Simulations sweeps were also run to verify the consistency of the measurements. Since the X-parameters for each large signal operating point are extracted from independent measurements, the smooth curves and consistent behavior of the DUT with respect to load changes, even on sensitive terms like the third harmonic phase, verify measurement consistency and repeatability.

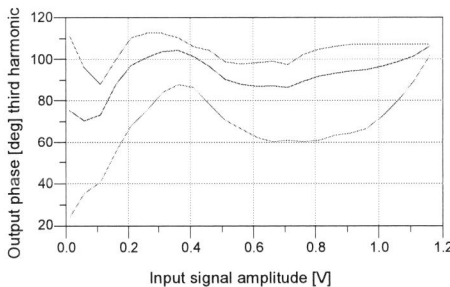

Fig. 5. Third harmonic phase vs. input power at fixed VAPC and VBATT. 25 ohm load shown in red, 50 in blue, and 100 in magenta.

To verify the accuracy of the X-parameters and illustrate their importance in characterizing a device like this, additional measurements were taken with tones injected at both the input and output ports of the DUT. The magnitude of the input tone at the output port was set such that the device saw an effective reflection coefficient of magnitude 0.5, and the phase was swept to cover a circular region on the smith chart. The measured b-waves of the device were then compared to the X-

parameter predicted results, as well as traditional Hot S22 predicted results for comparison. The actual measured b-waves are very close to the predictions from the X-parameter measurements. Hot S22 predictions are not close to the actual measurements.

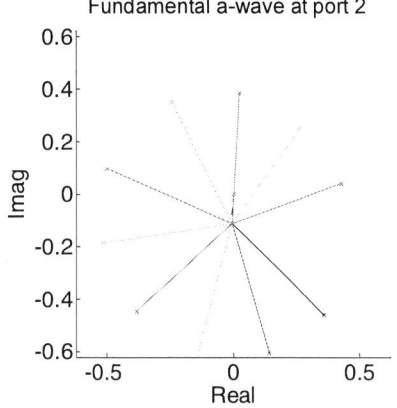

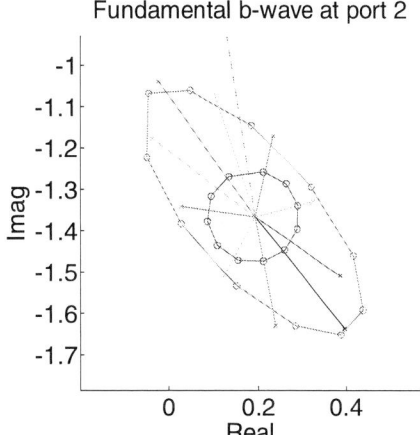

Fig. 6. **Top:** Measured complex amplitudes of injected tones at fundamental frequency at output DUT port, $a_{2,1}$ (colored "X"s at end of radial solid lines). **Bottom:** measured complex amplitudes of fundamental responses at output port, $b_{2,1}$ (colored "X"s at end of radial solid lines), X-parameter predictions (red circles) and classic Hot-S22 predictions (blue circles).

The inclusion of PHD blocks for both the upper and lower bands and additional switching circuitry in a single block in ADS allows correct simulation under pulsed bias conditions. The Envelope simulator can be used to accomplish this and much more with the resulting simulation block, including simulation with complex modulated stimulus. This allows simulation of the PA block under GSM-like conditions including the interaction effects with surrounding circuitry.

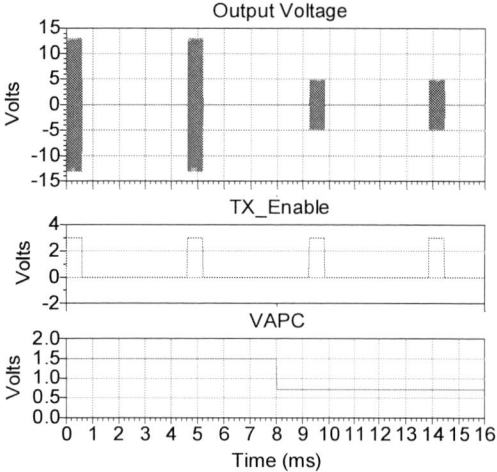

Fig. 7.   Envelope simulation results under pulsed bias conditions.

The envelope simulation results include time domain waveforms at input and output ports of the DUT. In other words, the full magnitude and phase information at the fundamental and harmonics, as well as DC information, is predicted within each pulse. This model inherits the properties given in [1]-[3] of the PHD model, but since the X-parameters are measured under pulsed bias conditions closely resembling the actual application, the resulting model is more faithful to the device in pulsed bias applications.

VI. Conclusion

X-parameters, which have previously been shown to enable accurate simulation of nonlinear devices under large-signal stimulus in a mismatched environment, were measured on a commercially available GSM handset amplifier under pulsed bias conditions that closely mimic real-world operating conditions. The measurements were done efficiently on state-of-the-art equipment, enabling the rapid extraction necessary to meet the demands of the short product development cycle in the mobile wireless industry. The measured X-parameters were then packaged into a virtual representation of the IC with functionality at all relevant pins that accurately reproduces the component behavior in simulation. In particular, the output match under large-signal operating conditions is demonstrated to be in excellent agreement with the simulation model based on automated X-parameter measurements. This has practical implications for designers, who can directly use the model as a design tool to help integrate the IC into a handset, and for IC manufacturers, who can distribute such a model as an electronic datasheet for their product.

Acknowledgement

The authors thank Agilent management for support.

References

[1]   D. E. Root et al., "Broad-Band Poly-Harmonic Distortion (PHD) Behavioral Models From Fast Automated Simulations and Large-Signal Vectorial Network Measurements," *IEEE Trans. MTT,* vol. 53, no. 11, pp. 3656-3664, November 2005

[2]   J. Verspecht and D. Root, "Polyharmonic Distortion Modeling," *IEEE Microwave Magazine,* vol. 7, no. 3, pp. 44-57, June 2006

[3]   J. Verspecht et al., "Multi-tone, Multi-port, and Dynamic Memory Enhancements to PHD Nonlinear Behavioral Models from Large-signal Measurements and Simulations," 2007 *IEEE/MTT-S IMS Digest,* pp. 969-972, June 2007

[4]   A. Soury et al., "Behavioral Modeling of RF and Microwave Circuit Blocs for Hierarchical Simulation of Modern Transceivers," 2005 *IEEE/MTT-S IMS Digest,* pp. 975-978, June 2005

[5]   Webpage as published on the internet at 2007-12-06: http://www.skyworksinc.com/products_detailpop2.asp?pid=99

An Integrated IQ Demodulator with Integrated Low-Power Multi-Gigabit BPSK / ASK Analog Signal Processor in 90nm CMOS

David A. Yeh[#1], Saikat Sarkar[#2], Stephane Pinel[#3], Padmanava Sen[#4], Joy Laskar[#5]

[#]*Georgia Electronic Design Center, School of Electrical and Computer Engineering, Georgia Institute of Technology, Atlanta, Georgia, 30308 U.S.A*

[1]davidyeh@ece.gatech.edu
[2]saikat@ece.gatech.edu
[3]pinel@ece.gatech.edu
[4]psen@ece.gatech.edu
[5]joy.laskar@ece.gatech.edu

Abstract— In this paper, two integrated low-power broadband 90nm-CMOS analog solutions are demonstrated to demodulate the multi-gigabit BPSK/ASK signal up to 3Gbps. In the coherent BPSK mode, a transmission speed over 2.5Gbps is achievable with a carrier frequency range of 8.5-9.5GHz for a total DC power consumption of 73mW with more than 20dB conversion gain. A minimum sensitivity of -35dBm is demonstrated for this demodulator with 33dB dynamic range. In the non-coherent ASK mode, a transmission speed over 3Gbps is achieved for a carrier frequency range of 6-9GHz at DC power consumption of 32mW. A minimum sensitivity of -26dBm is demonstrated for the demodulator with 21dB dynamic range. This is the best trade-off in terms of data rate and power consumption of CMOS demodulators reported at around 10GHz carrier frequencies.

I. INTRODUCTION

The rising demand of high-speed wireless applications has driven the technology development of the next-generation communication systems. Commercial applications such as high-definition multimedia streaming, downloading kiosk and point-to-point backhaul links all require multi-gigabit data transmission. Federal Communications Commission (FCC) has allocated several frequency spectra to accommodate the operation of these new devices.

A multi-gigabit transceiver implemented in the popular digital fashion such as orthogonal frequency-division multiplexing (OFDM) or multi-input multi-output (MIMO) present an unreasonable power budget in the system design. This is due to the power-hungry analog-to-digital converters (ADC) and one or multiple digital signal processors (DSP) all operating at gigahertzes. In contrast, the single carrier modulation and direct-sequence spread spectrum (DSSS) allows for low-power signal processing to recover the multi-gigabit data signal, operating at hundredths or thousandths of that power consumption.

There are tremendous potentials for such analog signal processors in the booming wireless market. Fig. 1 shows one example of the potential applications: an ultra-wideband (UWB) implementation. The analog signal processor in this direct-conversion architecture detects the incoming modulated signal at UWB frequency of 3.1-10.6GHz [1].

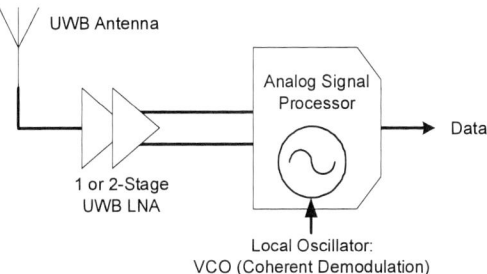

Fig. 1 Example of the analog demodulator in the direct-conversion application for an UWB receiver

With the increase in the complexity of the modulation scheme, there are significant and rising challenges in the analog demodulation techniques. In this paper, an analog signal processor for CMOS multi-gigabit receiver, capable of both coherent binary phase shift keying (BPSK) and amplitude shift keying (ASK) is presented. In addition, a non-coherent ASK demodulation technique directly detects the modulated data signal.

Section II of this paper describes the design aspects of the analog signal processor for coherent BPSK/ASK and the non-coherent ASK operation. Measurement results and the performance of the fabricated analog demodulator chip are presented in Section III. All integrated circuits are designed using STMicro's 90nm CMOS design kit.

II. ANALOG SIGNAL PROCESSOR

Fig. 2 shows the block diagram of the proposed integrated analog signal processor with the coherent BPSK/ASK part on the top and the non-coherent ASK part on the bottom. The two parts share the same differential input.

A. Coherent BPSK/ASK Demodulation

The differential BPSK/ASK-modulated signal is first down-converted into baseband signal by the two mixers with quadrature local oscillator (LO)) signals. However, the core of the coherent BPSK/ASK operation is the analog signal processor, which processes continuous-time inputs from the baseband I/Q paths, and demodulates coherently the data signal.

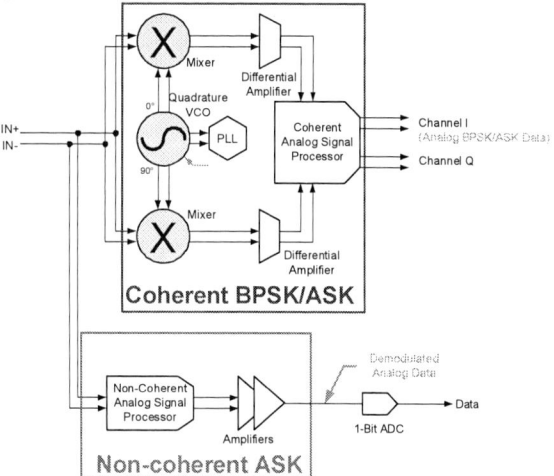

Fig. 2 Block diagram of the proposed dual-mode analog signal processor

A 1-bit ADC is used before the final data output to provide the digital output waveform at extremely low-power consumption.

1) Mixer: The schematic of the mixers is shown Fig. 3. It is based on the double-balanced Gilbert-Cell architecture. Table I shows a summary of the simulation results. The conversion gain is 5.7dB for a DC power consumption of only 5mW. The capacitors at the output are used to filter any undesirable high-frequency products, 2× LO leakage and LO feed-through. This mixer design has a bandwidth over 3GHz and it is capable of operating with a broad range of LO input frequencies (6-12GHz).

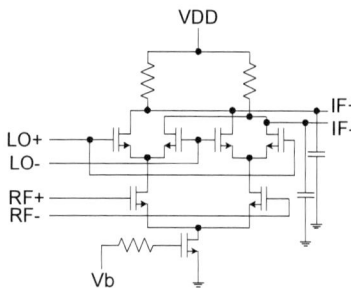

Fig. 3 Circuit schematics of the mixer

TABLE I
SIMULATION PERFORMANCE OF MIXER

Conversion Gain	5.7 dB
DC Power Consumption	5 mW
Input P_{1dB}	-2 dBm
Baseband Bandwidth	3 GHz

2) Differential Amplifier: Fig. 4 shows the schematic of a differential amplifier. It consists of a differential pair with a modified load to avoid the mismatch in the IC fabrication process [2]. The common-mode voltage of the differential branches appears at node A and this voltage is used to bias the PMOS load transistors. The differential amplifier has a gain of 12 dB and a bandwidth of 4GHz while consuming only 9.4mW

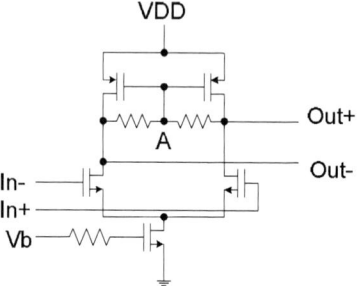

Fig. 4 Circuit schematics of the differential amplifier

3) Quadrature VCO: Fig. 5 shows the schematic of the quadrature VCO. The circuit employs a parallel-coupled architecture [3]. The NMOS and PMOS cross-coupled pairs provide the negative resistance for oscillation and the varactor pairs are used for frequency tuning of the LC tank. The differential buffer at the output of the QVCO is also used to control the output power.

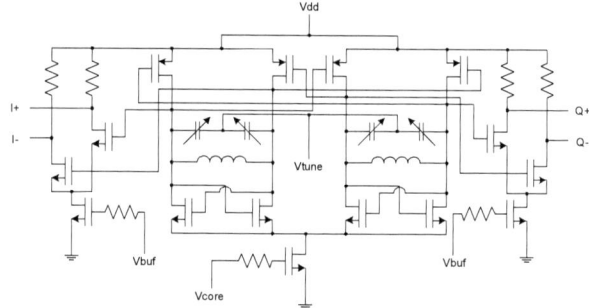

Fig. 5 Circuit schematics of the quadrature VCO

Table II shows the simulation performance of the quadrature VCO. A measured VCO tuning curve is shown in Fig. 6. It indicates K_{VCO}=1.81GHz/V, which is close to the simulation result of 2 GHz/V.

TABLE II
SIMULATION PERFORMANCE OF QUADRATURE VCO

Output Power	1 dBm
DC Power Consumption	22 mW
K_{VCO}	2 GHz/V
Phase Noise @ 1MHz offset	-105dBc
Tuning Frequency Range	8GHz~10GHz

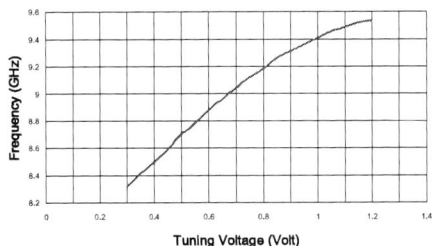

Fig. 6 Measured tuning curve of the quadrature VCO

B. Non-Coherent ASK Demodulation

As shown in the bottom part of Fig. 2, a non-coherent ASK demodulator has been added to directly detect the incoming modulated signal. The output of the ASK demodulator is the recovered data with the following two amplifiers and the differential-to-single-ended converter providing high-gain and low-pass filtering in the signal path. Due to the non-coherent nature of this demodulator, the jitter performance is worse than its coherent counterpart. However, a clock-data recovery (CDR) circuit can be employed to further condition the recovered data. 1-bit ADC is used before the final data output to provide the digital output waveform.

III. MEASUREMENT RESULTS

The size of the integrated chip is 1.08mm×0.98mm (including all I/O pads) as shown in Fig. 7(a). The coherent BPSK/ASK demodulator occupies 0.72mm×0.63mm and the non-coherent ASK demodulator takes up an area of 0.90mm×0.23mm. For measurement purpose, the chip is mounted into a fabricated FR-4 module with wire-bonding provides access to all pads (as shown in Fig. 7(b)).

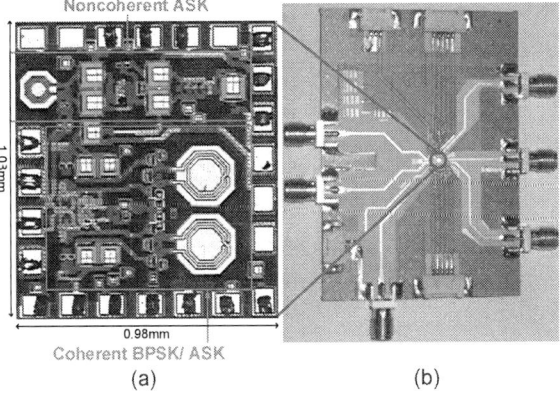

(a)   (b)

Fig. 7 (a) Photo of the integrated CMOS IQ demodulator with analog signal processor chip for multi-gigabit receiver; (b) photo of the module fabricated for testing purpose

The measurement setup is shown in Fig. 8. The pulse pattern generator outputs a multi-gigabit pseudo-random bit stream that modulates the LO (9GHz) through the off-the-shelf I-Q mixer component. By varying the DC-level offset of the pattern generator, one can construct either BPSK or ASK signal. The output of the mixer is attenuated before the 180°-hybrid converts the single-ended signal to differential inputs of the IQ demodulator. The data output are measured by a signal integrity analyser to plot eye-diagrams and extract bit-error rate (BER).

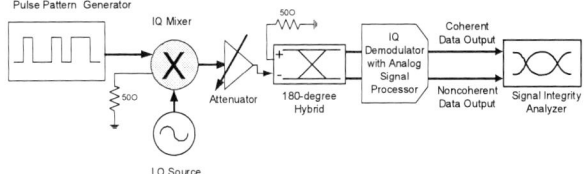

Fig. 8 Measurement setup for multi-gigabit demodulation

Table III provides a measurement summary of the coherent analog signal processor for BPSK demodulation in comparison to its simulation results. The minimum sensitivity, dynamic range and the total DC power consumption (including quadrature VCO, mixers, amplifiers and the analog signal processor) between the simulation and the measurement match very closely. The minimum sensitivity is 10dB better when operating at 1.5Gbps. With an input power beyond -31dBm, a carrier frequency variation of ±140MHz does not impact the performance of the coherent demodulation. The lower measured speed might be due to the unaccounted parasitic effects.

TABLE III
PERFORMANCE SUMMARY FOR IQ DEMODULATOR WITH INTEGRATED SIGNAL PROCESSOR

	Simulation	Measurement
DC Power Consumption	80W from 1.8V	73mW from 1.8V
Dynamic Range	30dB	33dB
Maximum Speed	3Gbps	≥2.5Gbps
Minimum Sensitivity	-36dBm	-35dBm
Synchronization Range	220MHz	280MHz

The sensitivity and dynamic range will be increased significantly by using a front-end variable-gain low-noise amplifier.

Fig. 9 shows the measured eye diagram and the corresponding bathtub curve for the coherent BPSK demodulation at a transmission speed of 2.0Gbps, carrier frequency of 8.9GHz, signal level of -37dBm. The bathtub curve indicates an extrapolated BER of 0.5×10^{-12}.

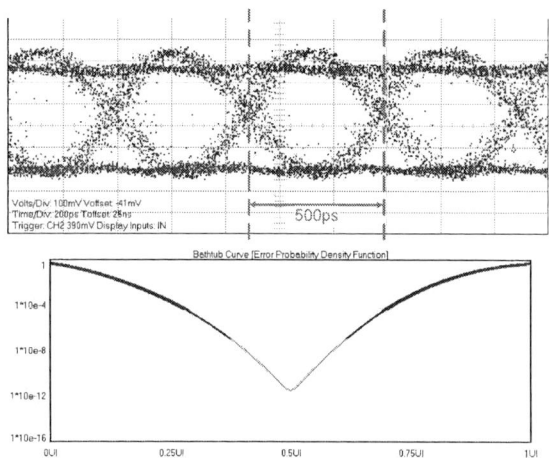

Fig. 9 Output eye diagram and the corresponding bathtub curve of the analog signal processor operating in the coherent BPSK mode at 2Gbps

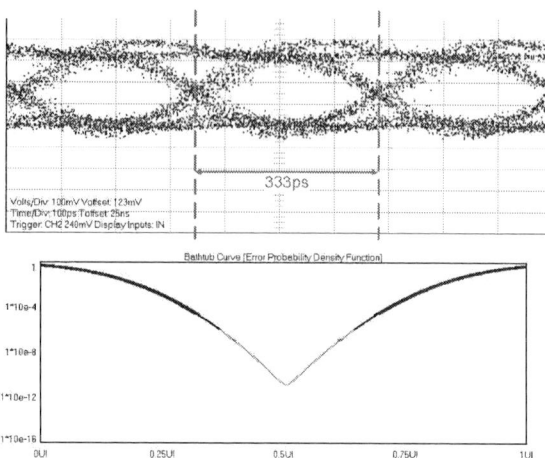

Fig. 10 Output eye diagram and the corresponding bathtub curve of the analog signal processor operating in the non-coherent ASK mode at 3Gbps

Table IV provides a measurement summary of the non-coherent ASK demodulation in comparison to its simulation results. The minimum sensitivity, dynamic range and DC power consumption between the simulation and the measurement match very closely. The DC power consumption figure includes the non-coherent analog signal processor, the differential amplifiers and the differential-to-single-ended converter (the last two parts account for 21.3mW). The measured minimum sensitivity is 6dB better when operating at 1.5Gbps.

TABLE IV
PERFORMANCE SUMMARY FOR NON-COHERENT ASK DEMODULATION

	Simulation	Measurement
DC Power Consumption	34mW from 1.8V	32mW from 1.8V
Dynamic Range	23dB	21dB
Maximum Speed	3.5Gbps	≥3Gbps
Minimum Sensitivity	-26dBm	-26dBm
Operating Freq. Range	6GHz~10GHz	6GHz~9GHz

Fig. 10 shows the measured eye diagram and the corresponding bathtub curve for the non-coherent ASK demodulation at a transmission speed of 3Gbps, carrier frequency of 8.2GHz, signal power level of -11dBm. The bathtub curve indicates an extrapolated BER of 0.5×10^{-10}.

IV. CONCLUSIONS

In this paper, an integrated IQ demodulator with integrated signal processor has been presented providing broadband solutions for multi-gigabit CMOS single-carrier UWB receiver. The coherent BPSK/ASK demodulation shows a maximum speed over 2.5GBps with a carrier frequency of 9GHz for a total DC power consumption of 73mW. The non-coherent ASK demodulation directly detects a modulated carrier range of 6~9GHz with a maximum speed over 3Gbps for a total power consumption of 32mW. By increasing the bandwidth in the signal path and the complexity of the demodulation approach, the transmission speed is extended to over 3.5Gbps data rate in each channel in QPSK mode for ultra high-speed applications.

ACKNOWLEDGMENT

Authors would like to acknowledge DARPA for their support in this work. This material is based upon work supported by DARPA and SPAWAR under Award No. N66001-06-1-2033. Any opinions, findings, and conclusions or recommendations expressed in this publication are those of the author(s) and do not necessarily reflect the views of the SPAWAR.

REFERENCES

[1] Y. Zheng, K-W Wong, M. Annamalai Asaru, D. Shen, W.H. Zhao, Y.J. The, P. Andrew, F. Lin, W.G. Yeoh, R. Singh, "A 0.18um CMOS Dual-Band UWB Transceiver", *IEEE Int. Solid-state Circuits Conf. Dig. Tech. Papers*, 2007, pp. 114-115.

[2] B. Razavi, *Design of Integrated Circuits for Optical Communications*, 1st ed., New York, U.S.A: McGraw-Hill, 2003.

[3] H-K Chen, D-C Chang, Y-Z Juang and S-S Lu "A Low Phase-Noise 9-GHz CMOS Quadrature-VCO using Novel Source-Follower Coupling Technique", *IEEE/MTT-S Int. Microwave Symp.*, p851-854, 2007.

Performance Analysis of Balanced and Unbalanced Feed-Forward Equalizer Structures for Multi-Gigabit Applications in 0.18μm CMOS Process

H. S. Kim[1], D. Bhatta, K. H. Lee , C. Scholz, E. Gebara, and J. Laskar

Georgia Electronic Design Center (GEDC), School of Electrical and Computer Engineering,

Georgia Institute of Technology,

85 5th Street NW, Atlanta, GA USA

[1]hskim@ece.gatech.edu

Abstract—**In this paper, a comparative study of two different structures for the Feed Forward Equalizer (FFE) is presented to emphasize the effect of structural differences on the performance of the passive delay line based FFEs with large number of taps. Both FFEs are designed for compensation of Inter Symbol Interference (ISI) in multi-Gb/s data link. The two test structures use the same building blocks but differ in the implementation. Both of them have nine taps with passive delay cells and are designed in 0.18μm CMOS technology.**

I. INTRODUCTION

Signaling rates through serial links have exceeded past the gigabit-per-second well into multi-gigabit-per-second regime. The existing channels were not designed for such high data rates and thus have severe bandwidth limitations. Low bandwidth channels severely degrade the rise time and fall time of signal and cause significant temporal spreading of signal pulse. This creates what is known as Inter Symbol Interference (ISI). ISI reduces the eye opening and increases the bit error rate. ISI in a channel depends on many frequency dependent physical factors like skin effect, dispersion in the substrate, substrate loss, reflection, etc. Thus, ISI is highly frequency dependent and typically worsens with increasing frequency.

The high cost involved in upgrading the entire existing networks with high speed channels leads to considerable efforts in developing techniques that allow higher data transmission through the existing infrastructures. Equalization in the electrical domain is a cost effective solution for significantly increasing the operational bandwidth of the existing bandwidth-limited channels [1]. An equalizer cancels the distortion introduced by the channel by implementing a channel specific transfer function. It compensates for the various linear dispersion effects in the channel by canceling the ISI components in the output signal and reduces the pulse broadening as shown in Fig. 1.

Equalization may be performed in either digital or analog domain. With the current available technology, it is very difficult to sample the signal at multi-gigabit-per-second data rate and perform clock recovery for digital equalization. Therefore, the subsequent discussions will address equalization by means of analog signal processing. In the

analog domain, better performance can be achieved by using predictive channel equalization or a Decision Feedback Equalizer (DFE) in addition to a Feed Forward Equalizer (FFE) [2]. Much of the signal degradation in channels with linear dispersion, however, can be compensated by using the FFE.

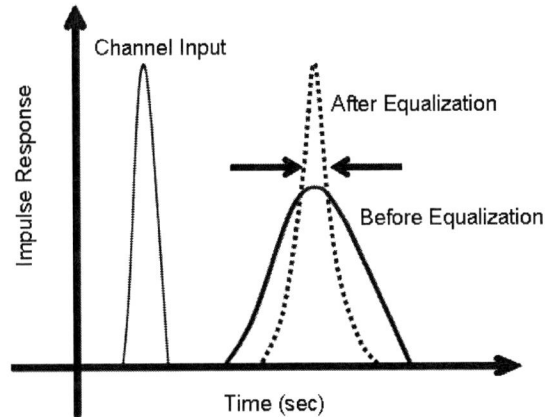

Fig. 1. Pulse broadening due to ISI and the effect of equalization.

Tapped delay line based FFEs can use either passive or active delay cells. Although to the best of our knowledge, there is no specific data available on the optimal number of taps, in general, more number of taps in the FFE provides better compensation for longer channels. However, it becomes very difficult to implement a good on-chip layout with more than a few taps using passive delay cells. The inductors used in passive delay cells occupy large area and cause difficulty in signal routing. The signal and bias lines become long, creating significant loss and parasitic cross coupling in the circuit. The bandwidth is also severely degraded due to the high amount of parasitic capacitive loading on the signal path. Moreover, it is very challenging to maintain linearity in gain and constant group delay over large bandwidth since the long lines have large parasitic components that can introduce spurious poles and zeroes in the characteristic response. The solution to these problems lies in the implementation of structures that can reduce the parasitic effects in FFEs with the large number of passive delay cells.

This paper highlights the effect of using different configurations on the FFE performance through the study of two functionally equivalent but differently structured FFEs. In Section II, an overview of the two test structures is given, along with a short description of the building blocks used. Section III provides results of the performance of the two structures with two fabricated layouts. Section IV concludes with a summary and comparative evaluation of the two FFE structures.

II. OVERVIEW OF THE TEST STRUCTURES

The current FFE structures under investigation are designed with target 3dB bandwidth specifications of 3GHz and a gain of 0dB for Gigabit Passive Optical Network (GPON) and coherent detection applications [2, 3]. Both FFE designs are based on tapped delay line Finite Impulse Response (FIR) filter structure. Each structure has nine taps with a uniform tap spacing of 80ps. The two structures differ from each other only in that the delay cells in one of the structures are all on the input side while those of the other structure are evenly distributed between the input and the output. For the ease of description henceforth, the two structures will be referred to as the unbalanced (delay cells only at the input) and the balanced (delay cells evenly distributed between input and output) structure.

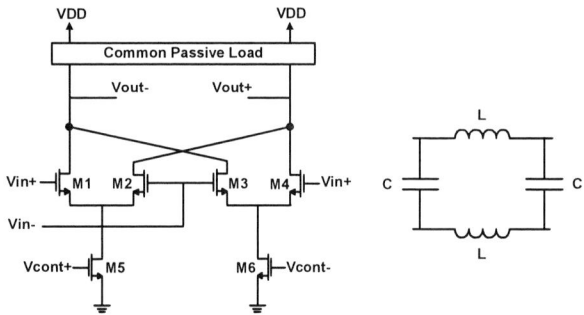

Fig. 2. VGA schematic and the passive delay cell structure.

The main building blocks of the FFE are Variable Gain Amplifiers (VGA) and delay cells which are shown in Fig. 2. The VGA block is implemented with a Gilbert cell based structure for linearity of operation. A common passive load is used for current summing at high speed data. To mitigate voltage head room issue, Vcont +/- is connected to a folded PMOS differential pair as a current steering block. The current steering block is degenerated with active device to achieve high linear performance [4]. The delay cell is implemented with a series inductor (L) and a shunt capacitor (C). This artificial transmission line conducted by LC ladder achieves the delay defined as $\tau = \sqrt{LC}$. Cut-off frequency is defined as $\omega = 2/\sqrt{LC}$ where the input impedance is purely reactive. The

delay of such a cell is fairly constant over a wide band of frequency. The passive delay cell can also provide fairly constant impedance characteristics over the frequency band of interest. Previous publications have shown active delay cells as an alternate technique for delay line implementation. Active delay cells provide benefits in terms of loss and die size at the cost of higher power consumption. In this study, however, the focus is on the passive delay cell based FFEs only.

A. Unbalanced Structure

The unbalanced structure is a direct form implementation of the FFE transfer function. The output of the FFE is derived by the summation of the output of different delay taps multiplied by appropriate weight as shown in Fig. 3. The nine taps are obtained from a long chain of eight unit delay cells each providing a delay of 80ps. The input delay line is terminated in a matched impedance to suppress multiple reflections.

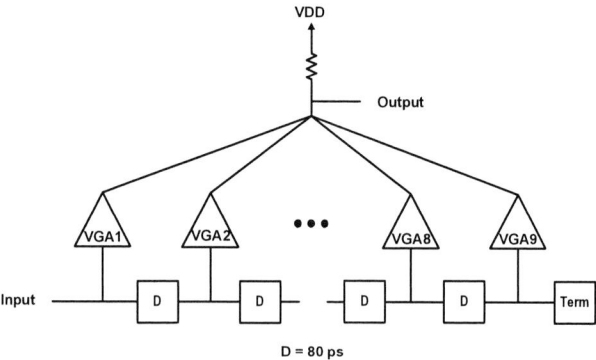

Fig. 3. Unbalanced FFE functional block diagram.

The delay cells, however, consume large area, so the signal routing from the VGA outputs to the summing node becomes very difficult. The long and unequal signal lines from the VGA outputs to the summing node can cause significant amount of random phase error over different process and temperature corners in the output of each tap and thus can degrade the overall accuracy of equalization. To address this problem at least partially, the unbalanced structure is folded at a point between the fourth and the fifth taps to reduce the length of the signal paths from the VGA outputs to the summing node. The bandwidth is degraded due to the high capacitive loading from the drain capacitances of all nine VGAs at a single node. Moreover, the signal in the input delay line suffers considerable loss due to the high series resistance of the inductor in the passive delay cells. This can cause large amount of deviation in the effective tap gains from the actual set value. Such errors can be minimized by taking into account the delay line characteristics when calculating the tap coefficients and by introducing booster amplifiers to compensate the losses in the long delay line. Lastly, the bias current consumed by all nine VGAs flow across the summing node, causing a significant DC voltage drop and reducing the voltage headroom. The voltage headroom in this structure is limited by a voltage drop across the summing node plus two stacks of transistors in the VGA.

B. Balanced Structure

The balanced structure is implemented as a travelling wave structure. A total of sixteen differential delay cells of 40ps each are distributed evenly between the input and the output paths as shown in Fig. 4. The signal is tapped from the nine delay points and added at the output with the VGA. The summation is achieved at different points in the output delay line, which consists of eight delay cells of 40ps each. Both the input and output delay cells are terminated in matching load to suppress multiple reflections. The signal at the output node is the linear superposition of the signal components arriving through different paths at the output. The overall delay in each path is the sum of the corresponding delays in the input and the output paths.

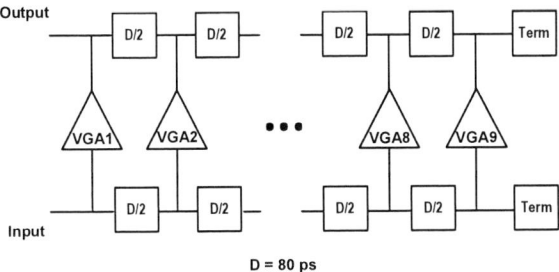

Fig. 4. Balanced FFE functional block diagram.

Since the output summation is not done at any single point, this structure can be implemented with shorter and equal length of signal paths from the VGA outputs to the summing nodes; it reduces the parasitic associated with the signal and bias lines, leading to a considerable improvement in bandwidth and voltage headroom. The distributed summing also reduces the effective overall capacitance at the output node, leading to an increased bandwidth. Also, the voltage headroom is only consumed by two stacks of transistors in the VGA because there are multiple summing nodes at the output delay line. This means each VGA can utilize more current for better overall gain of the FFE without the headroom issue.

In addition to bandwidth and voltage headroom improvements, the balanced FFE offers reflection immunity over the unbalanced FFE [5]. Using reflections coefficients at the termination of the delay lines, impulse response from the FFE can be analyzed. The first reflection from the unbalanced structure occurs at twice of the equalizer span and is proportional to the reflection coefficient at the termination. Because it is outside the FFE span, it can not be compensated. The first reflection in the balanced structure is twice the strength of that of the unbalanced structure. However, this first reflection is in the span of the FFE, it can be corrected by adapting the tap coefficients. The second reflection is too

small to affect the equalizer transfer function for both of the FFE structures [5].

III. RESULTS

Both structures are designed and fabricated in 0.18μm TSMC CMOS process. Fig. 5 shows microphotographs of the fabricated FFEs. Most of the area is occupied by inductors in L-C delay lines in both structures. The size of the unbalanced structure is 1.6x1.4 mm^2 and that of the balanced structure is 2.5x1.5 mm^2. The unbalanced structure has a smaller area compared to the balanced structure due to half the number of inductors used.

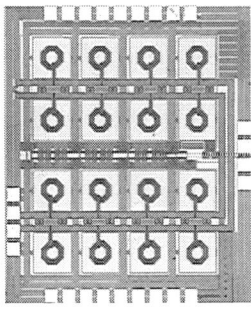

a)

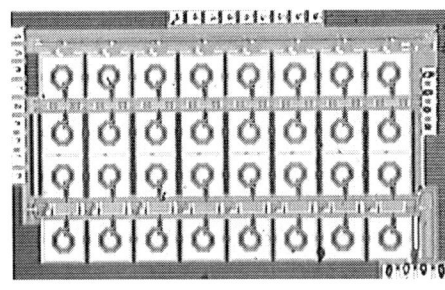

b)

Fig. 5. Microphotographs of the a) unbalanced and b) balanced FFEs.

Fig. 6 shows a close match between the measured and simulated transfer characteristic of the balanced FFE for the 1st tap and the 9th tap. Fig. 7 compares the measured transfer characteristic of the balanced and unbalanced FFE structures at the 1st and the 9th taps. At both of the taps, the balanced structure has a greater bandwidth than the unbalanced structure. The first tap has a bandwidth of 5.6GHz in the balanced structure compared to a bandwidth of 3.5GHz in the unbalanced structure. At the 9th tap, the balanced structure has a bandwidth of 2.3GHz while the unbalanced structure has a bandwidth of 0.8GHz.

The measured s-parameter datasets are used to extract the group delay information. Fig. 8 shows the extracted output at each tap of the balanced FFE for a step input with a rise time of 100ps. A delay of 70 to 90ps can be observed between the

978-2-8748-7007-1/08 $25.00 © 2008 EuMA   145

output responses of each adjacent tap. Table I lists the extracted tap delay for both of the FFE structure.

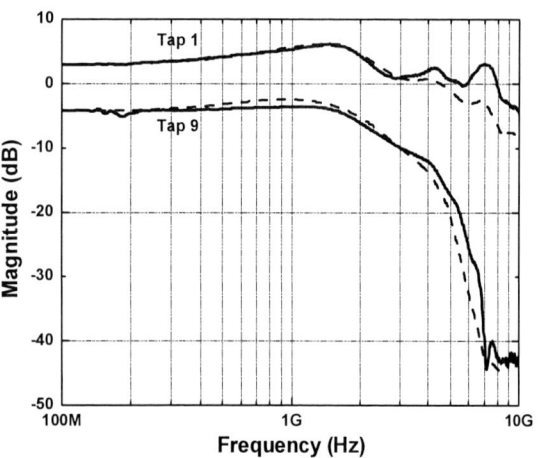

Fig. 6. Simulated (dashed) and measured (solid) forward transfer characteristic of the balanced FFE structure.

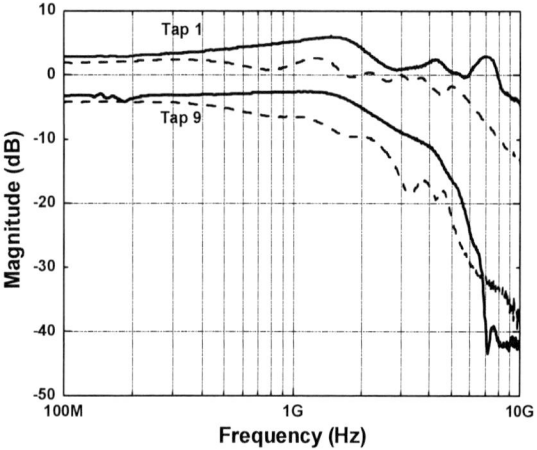

Fig. 7. Comparison of the measured forward transmission gain of the balanced (solid) and unbalanced (dashed) FFE structures.

IV. CONCLUSIONS

Comparative study of two different nine-tap passive delay line based FFE structures has been presented in this paper. Both of the FFE structures are functionally equivalent, use the same building blocks, and fabricated using 0.18μm CMOS technology. The two FFEs differ only in the configuration of the delay paths and the relative placement of the cells. The balanced structure has greater bandwidth compared to that of the unbalanced structure and is less effected by the reflection at the delay line termination. Hence, it can be concluded that the structural configuration of the passive delay line based FFEs has significant effect on the performance of the system.

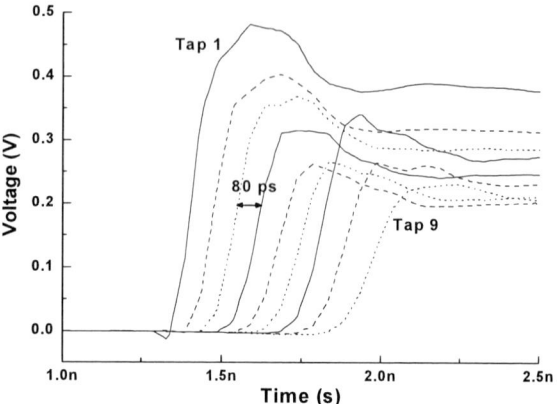

Fig. 8. Extracted output response at each tap of the balanced structure for a step input.

TABLE I

EXTRACTED TAP DELAY

	Balanced FFE (ps)	Unbalanced FFE (ps)
Tap (1, 2)	70	72
Tap (2, 3)	72	84
Tap (3, 4)	72	47
Tap (4, 5)	78	74
Tap (5, 6)	66	74
Tap (6, 7)	56	76
Tap (7, 8)	88	70
Tap (8, 9)	101	78

ACKNOWLEDGMENT

We would like to thank Andrea Romano from Pirelli Labs and Dr. Aldo Righetti from Corecom for their technical support.

REFERENCES

[1] H. Kim, J. de Ginestous, F. Bien, S. Chandramouli, C. Scholz, E. Gebara, J. Laskar, "Electronic dispersion compensator for a giga-bit passive optical network system," *2007 IEEE International Microwave Symposium*, June, 2007.

[2] Q. Xing-zhi, P. Bauwelinck, Y. Yi, D. Verhulst, J. Vandewege, B. De Vos, P. Solina, "Development of GPON upstream physical-media-dependent prototypes," in IEEE Journal of Light Technology, Vol. 22, Issue 11, pp. 2498-2508, Nov. 2004.

[3] M. Abrahams, and A. Maislos, "Insights on delivering an IP triple play over GE-PON and GPON," in *Optical Fiber Communication Conference*, pp8, March, 2006.

[4] F. Krummenacher, N. Joehl, "A 4 MHz CMOS Continuous-Time Filter with On-Chip Automatic Tuning," *IEEE J. Solid-State Circuits*, Vol. 23, pp. 750-758, June 1988

[5] S. Pavan, S. Shivappa, "Nonidealities in Traveling Wave and Transversal FIR Filters Operating at Microwave Frequencies," *2006 IEEE Trans. On Circuits and Systems*, Vol. 53, No.1, pp. 177-193, January 2006.

978-2-8748-7007-1/08 $25.00 © 2008 EuMA

Proceedings of the 3rd European Microwave Integrated Circuits Conference

A 20GS/s 8-Bit Current Steering DAC in 0.25μm SiGe BiCMOS Technology

Samiran Halder[1], Hans Gustat[1], Christoph Scheytt[1], Andreas Thiede[2]

[1]*Department of Circuit Design, IHP Microelectronics*
25, Im Technologiepark, 15236 Frankfurt (Oder), Germany
halder@ihp-microelectronics.com
gustat@ihp-microelectronics.com

[2]*Chair for High-Frequency Electronics,*
University of Paderborn
Warburger Strasse.100, 33098 Paderborn, Germany
[2]thiede@uni-paderborn.de

Abstract— **This paper presents the design of an 8-bit 20GS/s DAC. The DAC is implemented with a modified current steering architecture where unlike the conventional binary weighted architecture a R-2R ladder DAC architecture is used as the LSB sub-DAC. In simulation the 8-bit DAC shows 7.83 ENOB for 9GHz of input sinusoidal at 20GHz of sampling rate with the power dissipation of 2.5W. The measurement results of the 4-bit LSB sub-DAC show that the sub-DAC can work up to 30GHz with a power dissipation of 455mW.**

I. INTRODUCTION

In RF systems, the analog-digital interface is pushed towards the antenna, as the complex signal processing can be handled more efficiently in the digital domain. The direct digital synthesis (DDS) technique becomes more and more popular in the mobile communication arena due to the simple control procedure rather than an analog domain phase locked loop (PLL) based signal synthesis [1, 2]. The front end D/A converter (DAC) is a critical component in those systems. In high speed data links e.g. optical, satellite communication systems, medium resolution (4-8 Bits) DACs with sampling rate of up to 20 GHz are going to be used [3]. Another upcoming application of high-speed medium resolution DACs can be found in ultra wideband (UWB) communication systems. Different kinds of pulse forms are used, e.g. Gaussian and its derivatives. A DAC based direct waveform synthesis (DWS) is presented in [4]. The key requirement for this application is medium resolution with sampling rate more than ~16GHz and low power. In this paper an attempt has been made to serve the aforementioned applications by developing a low-power, medium resolution (8-bit) segmented current steering DAC with 20GS/s sampling rate.

II. DAC ARCHITECTURE

For very high sampling rate, current steering DAC architecture is an obvious choice [5, 6]. This architecture comes in several variants. Those are namely binary weighted, unary weighted or segmented architectures. The segmented current steering DAC architecture is most commonly used for medium to high accuracy DAC design as it provides possibly the better trade-off among the parameters e.g. speed, power, area and accuracy.

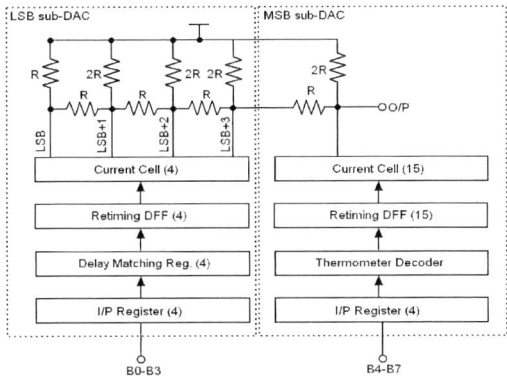

Fig. 1. Modified segmented current steering architecture of the 8-bit DAC

Fig.1. shows the block diagram of 8-bit modified segmented current steering architecture which is used in the current design. The percentage of segmentation is dictated by the static accuracy (INL and DNL) and the chip area. In the context of multi-GHz DAC design the length of the clock path is a very important issue. With an increasing segmented part the number of unit current cells of the MSB sub-DAC increases exponentially, and so does the length of the clock path. As a result, the delays among the current cells become unequal, which decreases the spurious free dynamic range (SFDR). A compromise is found in 50% of segmentation for the 8-bit DAC implementation i.e. the four LSB B0-B3 are implemented with the binary weighting and MSB B4-B7 are designed with unary weighted sub-DAC.

Unlike the conventional segmented current steering DAC, the LSB sub-DAC is having the same weight for all of the current cells and the binary weighting operation is accomplished with a modified R-2R resistive ladder. Its main advantage is a symmetrical and modular structure with a very predictive (and fast) dynamic behavior, which provides a great benefit in high resolution DAC design. Furthermore, it can be designed to match directly with the external load impedance of 50Ω. Although this R-2R resistive ladder shows different delays to the output for the different current cells,

978-2-8748-7007-1/08 $25.00 © 2008 EuMA   147

this delay variation is generally small compared to the sampling time period.

III. DAC IMPLEMENTATION

In a high-speed current steering DAC the main design constraints are imposed by the dynamic characteristics of the current switching. These dynamics in turn are influenced by the following main parameters: current mismatch, output impedance of the current switches and the timing alignment of the switches. The output impedance of the current switch directly dictates the linearity requirement of the current steering DAC. The current switch is implemented with a simple differential pair and an emitter degenerated cascode current source is used as the tail current source.

The proper timing alignment of the current switches is another factor of great impact to the dynamic performance of the DAC. Unequal delay among the current switches would result in higher current glitches at the output and the spurious free dynamic range (SFDR) of the DAC would be reduced. Thus, special attention has been paid to make the signal pathes from the retiming DFFs to the current switches as short as possible and equal in length, reducing the timing skew for the current switches.

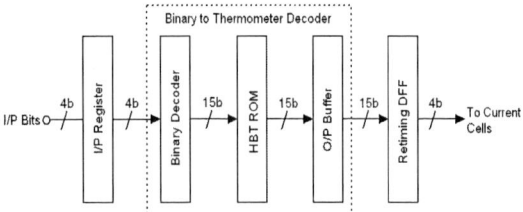

Fig. 2. Block diagram of proposed 4-bit thermometer decoder

In multi-GHz DACs the thermometer decoder is usually critical in terms of speed and power. The conventional implementation [5] requires two row and column decoders and additional logic unit to accomplish the thermometer decoding. The long chain of logic gates restricts the speed of operation. To enhance the speed of the thermometer decoder a new approach is proposed in this paper. The block diagram of the new thermometer decoder is presented in Fig. 2. In this technique the main thermometer decoding is performed with an HBT ROM based structure and the address decoding of this ROM is accomplished with a binary decoder. The 4-bit binary decoding needs a complex combinatorial operation. This imposes speed limitation due to the gate delay of the combinatorial logic. In this proposed architecture a novel method is adopted to design high-speed combinational logic. In [9] it has been shown that the wired OR/NOR function can be advantageously merged with the conventional ECL D-latch. This 4-input OR/NOR D-latch along with the conventional ECL D-latch construct the OR/NOR master-slave D-flipflop, which is used as a building block for the binary decoder. The HBT ROM has a pseudo-differential architecture. In Fig.3. one of the pseudo-differential paths of the HBT ROM is presented. The wired OR function is implemented with the parallel combination of emitter followers.

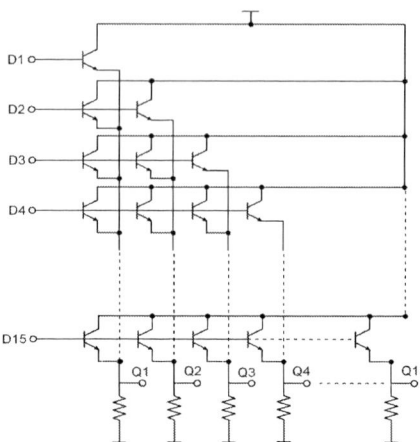

Fig. 3. Simplified schematic of one of the pseudo-differential paths of the HBT ROM

For any particular input data pattern only one of the outputs of the binary decoder goes high and all other outputs are low. A worst case scenario occurs when the input D1 goes high and all other inputs D2, D3, ..., D15 are low. In this case only one of the emitter followers tries to pull the output voltage (Q1) high and rest of the fourteen emitter followers push the output voltage down. With a sufficiently high input voltage (D1) the output goes high but the logic swing gets reduced. Therefore the logic levels are restored with a single differential stage output buffer.

For ultra high-speed DAC design the characterization comes as a great challenge, since providing synchronous test patterns at this speed is a tough task. One of the well accepted procedures is the built-in-self-test technique (BIST) [4], where an on-chip digital block provides the input bit pattern. In this design a binary up counter is included as the pattern generator. This counter is designed with a modified synchronous counter architecture. The block diagram of this synchronous counter is shown in Fig.4. A cross coupled master-slave ECL DFF is used to generate the MSB bit. The main improvement is done in the generation of (MSB-1)-bit. A modified master-slave DFF is used accomplish this operation, where the master D-latch is clocked with the MSB-bit and the slave latch is clocked with the input clock (CLK). A 6-bit ripple counter is used to generate the 6-LSB bits and (MSB-1)-bit is used as the clock for the ripple counter. At the output an 8-bit register is used to synchronize the output bits with the input clock.

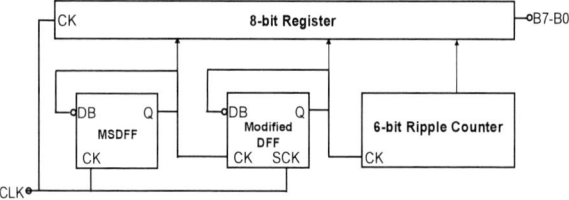

Fig. 4. Simplified block diagram of modified 8-bit synchronous counter

IV. RESULTS AND DISCUSSIONS

The 8-bit DAC is implemented in a 0.25 μm SiGe BiCMOS technology [9]. The minimum emitter size of the HBT is $0.21 \times 0.84 \mu m^2$ and the f_T, f_{MAX} are both 190GHz.

Fig. 5. Layout of 8-bit DAC

In the time critical signal paths, minimum size HBTs have been used for high speed and minimum parasitic capacitance load.

In Fig. 5. the layout of the 8-bit DAC is presented along with the 8-bit synchronous counter. This chip has two different power supplies. The main analog part i.e. the unit current cells and the R-2R ladder network works with 4.5V, whereas the digital parts of the DAC work with 2.5V of power supply. The full chip consumes 2.5W of power. At high signal speed the passive interconnects play a critical role for the static and dynamic behavior. In the DAC design relatively long interconnects (e. g. data, clock, output lines) are implemented with 50Ω microstrip transmission lines. These microstrip transmission lines are simulated in 2.5D electromagnetic simulator (ADS Momentum) and equivalent π-models for these transmission lines are generated. In the transient simulation these models were incorporated to achieve more realistic results.

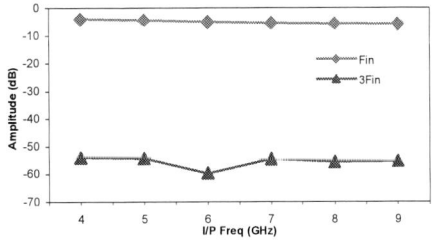

Fig. 6. Simulated fundamental and 3rd order frequency components for different input frequencies

The accuracy of this 8-bit DAC has been analyzed in the frequency domain. Because in the simulation the delay between the differential signals was always equal, the second order harmonic will show very low amplitude. Hence, for the accuracy calculation this second order harmonic is neglected and the relation of fundamental and third order harmonic is approximated as the total harmonic distortion (THD). The THD is used as the figure of merit to calculate the accuracy of the DAC in terms of effective number of bits (ENOB). In reality, noise, clock jitter and mismatch will reduce this simulated ENOB value. In Fig. 6. the amplitudes of the

fundamental and third order harmonics have been presented for different input frequencies. In this simulation full scale sinusoidal digital patterns are used as the input and the clock frequency (F_c) is 20GHz. From Fig. 6. it can be seen that for the full input frequency range the fundamental frequency has an almost amplitude and so does the third harmonic. Thus the DAC has an almost constant linearity over the full frequency range from 4GHz to 9GHz. It shows the lowest ENOB of 7.83 bit for 9GHz input.. Table I summarizes simulation results of the 8bit 20GS/s DAC.

TABLE I
SUMMARIZED SIMULATION RESULTS FOR 8-20GS/S DAC

Process	0.25μm SiGe BiCMOS (f_T=f_{MAX}=190GHz)
Simulated ENOB	7.83 bit
Conversion rate	20 GHz
Output resolution bandwidth	9GHz
Supply voltage	2.5V/4.5 V
Power dissipation	2.5W
Die area with pads	6 mm^2

The 4-bit binary sub-DAC of the full 8-bit segmented DAC has already been implemented and tested. This sub-DAC was tested on-wafer with a 40GHz probe station. For critical inputs and outputs 40GHz coaxial cables were used. A low phase noise sinusoidal signal from an Agilent E8257D with option UNX was used as input clock. Since the output load of the DAC is matched with the external 50Ω load, it was possible to connect the outputs directly to the Tektronix 6154 oscilloscope through DC blockers. The input bit pattern is generated by an Agilent 81250 parallel bit-error rate tester, which was configured as a bit sequence generator. Unfortunately the module used for the characterization can only generate bit rates ≤ 3.35GHz. Thus the DAC could not be tested at the highest input data rate. By measuring the static and dynamic characteristics at lower data rate the parameters have been extrapolated for the higher data rate. In Fig. 7. the measured INL and DNL of the 4-bit DAC is plotted. It achieved INL and DNL of 0.49LSB and 0.57LSB respectively.

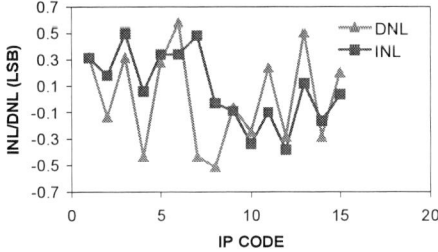

Fig. 7. INL/DNL plot of 4-bit sub- DAC

Fig. 8a and 8b represent reconstructed DAC outputs for different input bit patterns. Fig. 8a shows the one of the differential outputs of the DAC for an input pattern corresponding to a sinusoidal function. With a data rate of

978-2-8748-7007-1/08 $25.00 © 2008 EuMA

2.8GHz and a period of five patterns, the constructed sinusoidal has a frequency of 560 MHz. The DAC clock is 30GHz. In the output of the DAC a strong feedthrough from the clock input probe was observed.

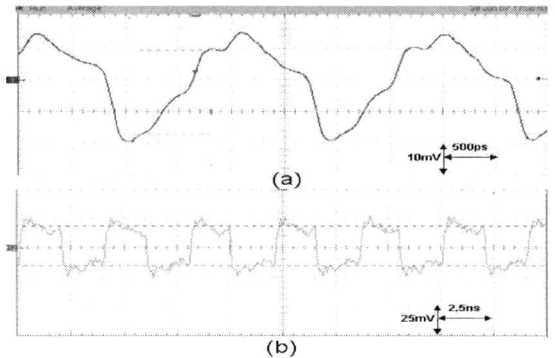

Fig. 8. (a) Sinusoidal and (b) Step reconstruction of the 4-bit sub-DAC

A full-swing step response of the DAC is presented in Fig. 8b with an input data rate of 500MHz and a clock rate of 15GHz. Due to the lower cutoff frequency of the DC blocker (1GHz) the flat tops have some non-zero slope. For the rise time measurement a reconstructed saw-tooth signal has been used. From this measurement the output bandwidth of the sub-DAC is calculated to be 3.85 GHz.

A common figure of merit (FOM) for a DAC relating sampling rate, power and resolution is expressed as,

$$FOM = \frac{Power}{2^N \cdot SamplingRate} \qquad (1)$$

where. N is the resolution of the DAC. For the 4-bit LSB sub-DAC it is 0.95pJ. The best FOM is found in [4], where special CML structures were used to reduce the power. In [10] 0.12μm technology was used with reduced supply voltage. The current work has a comparable FOM at higher sampling rate and higher supply voltage.

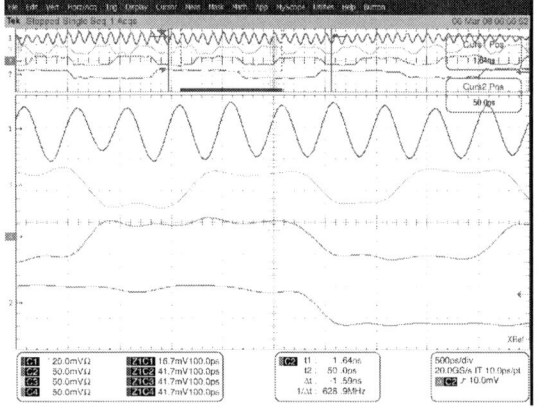

Fig. 9. Four MSB outputs of the 20GHz 8-bit synchronous counter

In a separate test-chip the modified 8-bit synchronous counter which would be used as the bit-pattern generator of the

8-bit DAC (as shown in Fig. 4) is implemented and characterized. This counter is also measured on-wafer. It includes an on-chip clock generator. A low jitter sinusoidal is used as the clock source, which is internally converted into clock signal. This counter is consumes 500mW of power for 20GHz of clock input. 50Ω output buffer are included for the characterization. The output bits of the synchronous counter are directly connected to the oscilloscope. In Fig. 9. the 4-MSB bits of the synchronous counter are presented. The measured MSB-3 signal already looks almost sinusoidal because of the bandwidth limitation of the measurement system and the output buffer. By measuring the exact frequency of the respective bits the operating frequency of the counter is calculated. It is found to be fully functional up to 20GHz.

V. CONCLUSIONS

In this paper the design of an 8-bit 20GS/s DAC has been presented. This DAC is implemented with a modified segmented current steering architecture where the LSB sub-DAC is implemented with a novel R-2R ladder. In the design of the binary weighted current steering DAC a new HBT ROM based thermometer decoder architecture is proposed. In simulation the 8-bit DAC shows 7.83ENOB with 9GHz of single-tone input at the sampling rate of 20GHz. The 4-bit LSB sub-DAC is already tested which is found to be functional up to 30GHz of sampling rate. The 4-bit sub-DAC achieves 0.49/ 0.57 LSB INL and DNL respectively with 3.85 GHz of output bandwidth. The sub-DAC shows a FOM of 0.95pJ, which is comparable with the state-of-the-art SiGe high-speed DACs. Additionally a novel 8-bit synchronous counter is presented which can be used as the input bit pattern generator of the 8-bit DAC.

ACKNOWLEDGMENT

The work was supported by the IHP. The authors like to thank the IHP technology team for providing the 0.25 μm SiGe BiCMOS technology SG25H1.

REFERENCES

[1] M. J. Flanagan, G. A. Zimmerman, " Spur-Reduced Digital Sinusoid Synthesis ", IEEE Transaction on Communication, Vol. 43, pp. 2254-2262, 1995.

[2] D. C. Larson, "High Speed Direct Digital Synthesis Techniques and Applications", Proc. GaAs IC Symposium, pp. 209-212, 1998.

[3] M. El Said, J. Sitch, M. Elmasry, "A 0.5 /spl mu/m SiGe pre-equalizer for 10 Gb/s single-mode fiber optic links", Proc. ISSCC, pp.224-225, 595, 2005.

[4] D. Baranauskas, D. Zelenin, " A 0.36W 6b upto 20GS/s DAC for UWB Wave Formation", Proc. ISSCC, pp. 580-581, 675, 2006.

[5] Van den Bosch, A. et al, " A 10-bit 1-GSample/s Nyquist current-steering CMOS D/A converter", IEEE Journal of Solid State Circuits, Vol. 36, pp. 315-324, 2001.

[6] B. Schafferer, R. Adams, "A 3V CMOS 400mW 14b 1.4GS/s DAC for Multi-Carrier Applications" Proc. ISSCC, pp.360-361, 532, 2004.

[7] M. Rodwell, "High Speed Integrated Circuit Technology, Towards 100GHz Logic", World Scientific, 2001, ISBN 981-02-4638-2

[8] H. Gustat, J. Borngraber, "NOR/OR register based ECL circuits for maximum data rate", Proc. BCTM, pp 90-93, 2005.

[9] B. Heinemann et al., "Novel Collector Design for High-Speed SiGe:C HBTs", Proc. IEDM, pp. 775-778, 2002.

[10] W. Cheng et al. "A 3b 40GS/s ADC-DAC in 0.12 μ m SiGe", Proc. ISSCC 2004, pp. 262-263, 2004.

Proceedings of the 3rd European Microwave Integrated Circuits Conference

A 1.2V Programmable ADC for a Multi-Mode Transceiver in 0.13μm CMOS

Dong-Young Chang, Sreekar Javvadi, Carlos Muñoz, Jason Abele, Ali Hadiashar, Mike Lugin, Rick Quintal, and Geoffrey Dawe

BitWave Semiconductor
900 Chelmsford Street,Lowell, MA 01851, USA
{dychang,sjavvadi,cmunoz,jason,ali,mlugin,rick,geoff}@bitwave.com

Abstract— **A flexible ADC for a multi-band and multi-mode transceiver is presented. It can be reconfigured between pipeline and ΔΣ architectures for power optimum operation in WLAN/WiMAX and GSM/CDMA2000/WCDMA, respectively. A maximum dynamic range of 76dB with less than 1mW at 1.2V supply was achieved using voltage scaling and closed-loop opamp design techniques. The prototype was built in 0.13μm CMOS and the active die area is 0.43mm².**

I. INTRODUCTION

The general performance specifications of an ADC used in a wireless receiver can vary significantly across the myriad of communication standards. For example, GSM in cellular requires the ADC to digitize a signal with as little as 200 kHz BW at a resolution of 12 or higher ENOB. On the other hand, WLAN or WiMAX systems work over a much broader signal BW, typically around 5 to 10 MHz, and require 8 to 9 bits of resolution. In practice, the power-and-area optimal choice for narrow-band/high-resolution applications would be ΔΣ modulators and for wide-band/low-resolution cases would be Nyquist-rate converters such as pipeline or algorithmic ADCs [1]. Portable systems that address multiple communication standards, therefore, have to either incorporate multiple ADCs, each with a different resolution, or choose a single ADC with its performance dictated by a mode with the highest resolution and speed requirement. Based on the BW-resolution trade-offs between standards, several flexible multi-mode ADCs have been presented in the past with limited architectural tunability [2, 3].

This work describes a programmable ADC that can be reconfigured between ΔΣ and pipeline architectures and thereby provides a power-optimal condition for different wireless standards that have high signal BW spreads. By carefully choosing each ADC architecture, extra circuitry coming from the reconfigurable implementation is negligible. A low-power opamp design methodology is also proposed to further emphasize power conscious design in this work.

II. ADC ARCHITECTURE

The block diagram in Fig. 1 shows a programmable ADC that can be reconfigured between a 2nd-order ΔΣ converter and a 9-bit pipeline converter. The ΔΣ modulator employs a feed-forward low-distortion architecture with a 3-bit quantizer.

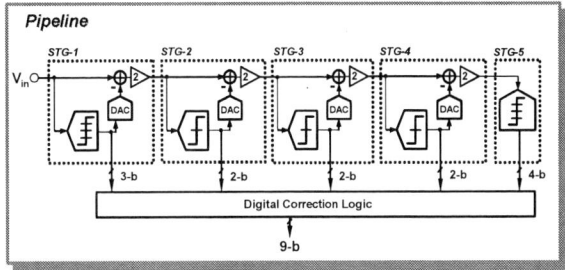

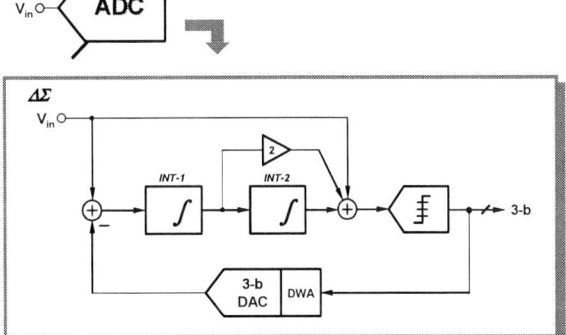

Fig.1 ADC architecture

Data-Weighted-Averaging (DWA) is used to suppress in-band distortion caused by the 3-bit switched-capacitor DAC mismatch. A total of two opamps are used to cover both ΔΣ and pipeline operation.

There are three important reasons for choosing this ΔΣ modulator architecture: First, the architecture ensures that only quantization noise passes through the integrators (INT-1 and INT-2) which allows all the internal voltage swings to be small with a multi-bit quantizer for low-voltage operation. Second, the architecture allows each integrator coefficient to be an integer power of 2 to remove any extra capacitors and switches which add more complexity, power, and area when reconfiguring into pipeline operation. Third, since most multi-mode transceivers need to be equipped with an analog baseband filter with strong out-of-band rejection to meet the toughest interferer cases, a modulator of 2nd order will suffice

978-2-8748-7007-1/08 $25.00 © 2008 EuMA   151

to meet the highest sensitivity GSM specification. The pipeline configuration employs a 5-stage (3-2-2-2-4) architecture with each stage comprising of a flash-ADC and an MDAC (a sub-DAC plus interstage gain amplifier). A front-end S/H amplifier can be removed by careful matching of sampling time constant between MDAC and flash-ADC in STG-1. When the ADC is working as a $\Delta\Sigma$ modulator, STG-2, STG-4, and STG-5 (partially) are disabled to avoid unnecessary power consumption.

In this work, the ADC is configured as a $\Delta\Sigma$ modulator in GSM, CDMA2000, and WCDMA cases with sampling rates of 26MHz, 39.3216MHz, and 61.44MHz, respectively. The pipeline configuration is used in WLAN and WiMAX modes with sampling rates of 40MHz and 33.6MHz, respectively. Within each ADC configuration, extensive bias tunability of individual circuit block through software is also implemented.

III. VOLTAGE SCALING

During pipeline operation, the internal voltage swing of each interstage gain amplifier output should be minimized as well, just like in the $\Delta\Sigma$ configuration, to operate under low supply conditions. The voltage scaling technique conceptual diagram of the pipeline ADC is depicted in Fig. 2 with corresponding residue plots of the first two stages. The ADC input range is defined by its reference voltage V_{REF}. By scaling down the reference voltage by a factor of two from STG-2 thru the last stage, all the internal opamp swings can be reduced by a factor of two with the dynamic range of the ADC remaining unchanged. To complete the scaling, the interstage gain of STG-1 needs to be reduced from 4 to 2. Note that in actual implementation it is important not to use multiple reference voltages within the ADC to avoid signal distortion. The ADC would normally use $0.5V_{REF}$ only at all stages, but STG-1 still needs a reference level of V_{REF} to handle an incoming analog input which can be achieved with a capacitor ratio of 2 that is accurately realized in most CMOS processes.

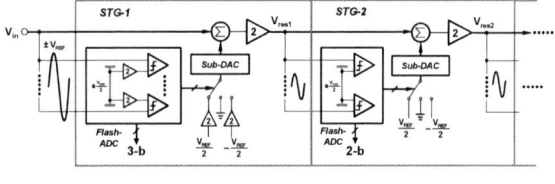

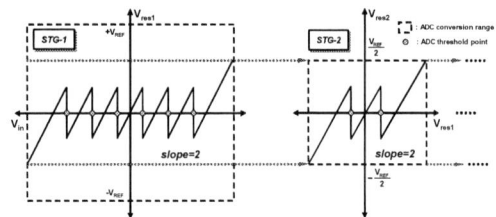

Fig. 2 Voltage scaling technique in the pipeline configuration

IV. OPAMP DESIGN

A high opamp open-loop gain is one of the most important parameters in the ADC design because it can affect noise shaping and signal distortion. As shown in Fig. 3, a single-stage folded-cascode amplifier with gain boosting is used for its low-power and stable operation. When the opamp is running under a closed-loop configuration, the transfer function contains a real pole defining the 3-dB BW ($\omega_{3\text{-}dB}$) of the overall amplifier and a zero defining the unity-gain BW of the gain boosting amplifier. The pole and zero are located closely enough to each other to be treated as a doublet. It is a well-known fact that in order to ensure stable operation, the zero should be placed between the real pole and the non-dominant high-frequency pole. Instead of using the conventional ad hoc design approach, a simple closed-loop analysis using a general form of the transfer function enables the discovery of the optimum damping ratio to achieve the fastest settling.

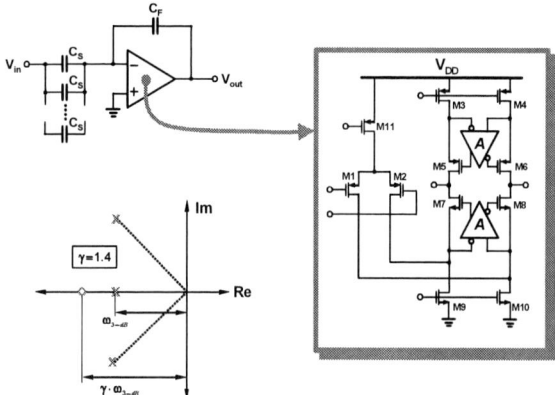

Fig. 3 Low-power opamp

For the analysis, a simplified small signal model shown in Fig. 4 can be used. β is a feedback factor for the opamp configuration shown in Fig.3, A is the open-loop gain of the gain boosting amplifier, and g_m is the transconductance of the MOS transistor. A closed-loop transfer function of the small-signal model shows that

$$T(s) = \frac{-\dfrac{g_{m1}g_{m2}\omega_{pa}}{C_1 C_L} \cdot \left((A_o + 1) + \dfrac{s}{\omega_{pa}}\right)}{s^3 + s^2\left(\omega_{pa} + \dfrac{g_{m2}}{C_1}\right) + s\left(\dfrac{g_{m2}(A_o+1)\omega_{pa}}{C_1} - \dfrac{g_{m1}g_{m2}\beta}{C_1 C_L}\right) + \dfrac{\omega_{pa}g_{m1}g_{m2}\beta(A_o+1)}{C_1 C_L}} \quad (1)$$

where, ω_{pa} is the unity-gain bandwidth of the gain boosting amplifier, C_1 is a parasitic capacitor and C_L is a load capacitor. Its general system transfer function can be written as

$$H_g(s) = \frac{\dfrac{1}{\beta} \cdot \left(\dfrac{s}{\omega_z} + 1\right)}{\left(\dfrac{s}{\omega_{cl}} + 1\right) \cdot \left(\dfrac{s^2}{\omega_n^2} + \dfrac{2\xi s}{\omega_n} + 1\right)} \quad (2)$$

978-2-8748-7007-1/08 $25.00 © 2008 EuMA   152

where, ω_z is a zero frequency, ω_{cl} is a closed-loop pole, ω_n is a natural frequency, and ξ is a damping ratio. By equating (1) with (2), actual design parameters can be derived as follows.

$$g_{m1} = \frac{\omega_{cl}\,\omega_n^2 C_L}{\omega_z \beta \cdot (\omega_{cl} + 2\xi\omega_n)} \tag{3}$$

$$g_{m2} = \xi\omega_n C_1 \cdot (\alpha + 2) \tag{4}$$

$$\gamma = \frac{1}{2\xi^2(\alpha+2)} \cdot \left[(1+2\alpha\xi^2) + \sqrt{(1+2\alpha\xi^2)^2 + 4\xi^2(\alpha+2)} \right] \tag{5}$$

$$\omega_z = \omega_{pa} \cdot (A_o + 1) \longrightarrow \omega_{pa} = \frac{\gamma\xi\omega_n}{A_o + 1} \tag{6}$$

where, $\alpha=1$. Since the optimum settling condition can be easily derived from the step response of the general system transfer function, all the opamp design parameters can be calculated. According to the analysis, a slight under-damped condition ($\xi=0.9$) helps settling behaviour. This implies that rather than having the pole and zero located at the same frequency to cancel one another, placing the zero at about 1.4 times higher frequency than the real pole ($\gamma=1.4$) makes the opamp have the fastest operation without any power penalty.

V. Measured Results

The architectural flexibility is shown in the measured output spectra in Fig. 5 for the $\Delta\Sigma$ configuration in GSM and WCDMA modes and the pipeline configuration in WLAN and WiMAX-10M. A plot of input power level versus SNDR performance for five different standards in Fig. 6 indicates a dynamic range for each setting. For all cases, the input signal range is 1.6Vp-p differential at 1.2V supply. The overall performance of the ADC across standards is summarized in Fig. 7 with each mode being optimally programmed by software. By reconfiguring the ADC architecture, the sampling clock frequencies remain under 100MHz across standards, keeping the overall power contribution of the digital circuitry small. The calculated FOMs are in the range of 0.32 and 1pJ/conv. The active die area is 0.43mm². The ADC can be reconfigured between the two different architectures and still be optimized for power, area, and performance without interrupting or burdening one another. Fig. 8 shows the die photo of the ADC implemented using a 0.13-um 1P8M CMOS process.

Acknowledgment

The authors would like to thank T. James, J. Kilpatrick, A. Abdelgany, M. D'Amato, E. Org, R. Cyr, and M. Farese for the support and encouragement. They are also grateful to J. Silva and P. K. Hanumolu at Oregon State University for useful technical discussions.

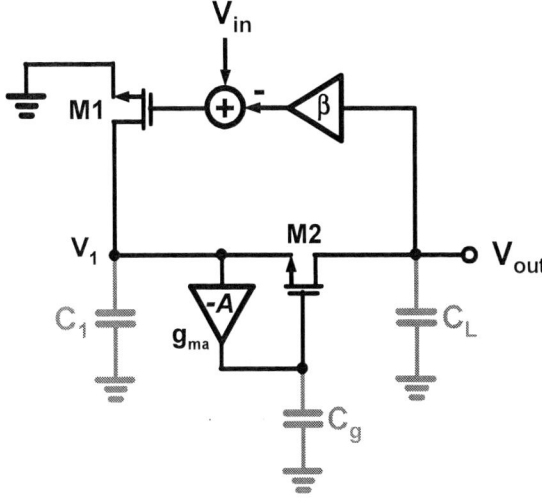

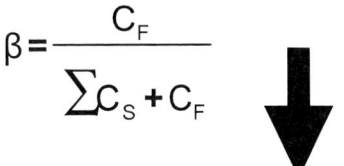

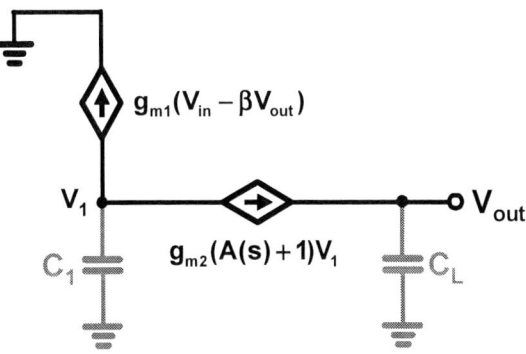

Fig. 4 Simplified small-signal model

References

[1] S. S. Mehta et al., "An 802.11g WLAN SoC", *IEEE J. Solid-State Circuits.*, vol. 40, no. 12, pp. 2483-2491, Dec. 2005.

[2] T. Burger et al., "A 13.5-mW 185-Msample/s $\Delta\Sigma$ modulator for UMTS/GSM dual-standard IF reception", *IEEE J. Solid-State Circuits.*, vol. 36, no. 12, pp. 1868-1878, Dec. 2001.

[3] S. Ouzounov et al., "A 1.2V 121-mode CT $\Delta\Sigma$ modulator for wireless receivers in 90nm CMOS", *IEEE ISSCC Dig. Tech. Papers*, pp. 242-243, Feb. 2007.

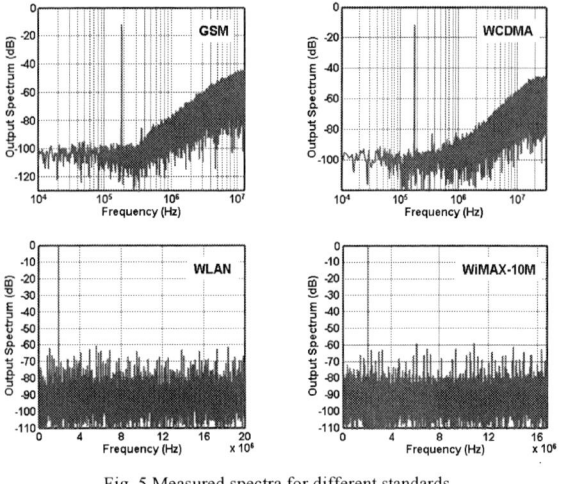

Fig. 5 Measured spectra for different standards

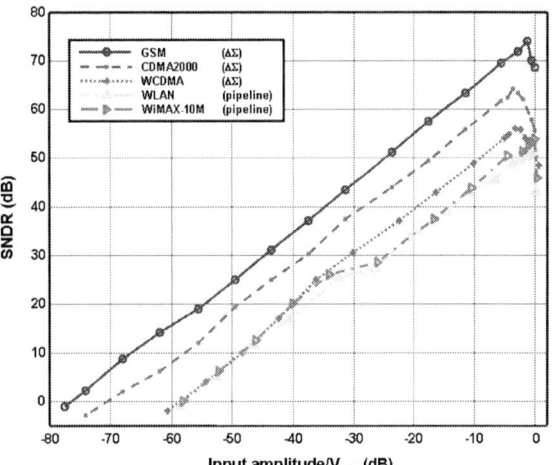

Fig. 6 Measured SNDR curves

Fig. 8 Die micrograph

	GSM	CDMA2000	WCDMA	WLAN	WiMAX-10M
Input BW	0.27MHz	0.6144MHz	1.92MHz	10MHz	5MHz
Sampling frequency	26MHz	39.3216MHz	61.44MHz	40MHz	33.6MHz
Peak SNDR	74dB	64dB	56dB	52dB	54dB
DR	76dB	67dB	56dB	57dB	58dB
Input Range	1.6Vp-p diff.				
Configuration	$\Delta\Sigma$	$\Delta\Sigma$	$\Delta\Sigma$	pipeline	pipeline
Power dissipation	0.9mW	1.2mW	2mW	5mW	4.2mW
FOM	320fJ/conv	530fJ/conv	1pJ/conv	430fJ/conv	650fJ/conv
Power supply	1.2V				
Process	0.13-um 1P 8M CMOS				
Active area	0.43mm^2				

Fig. 7 Performance summary

Proceedings of the 3rd European Microwave Integrated Circuits Conference

A 52GHz Millimeter-Wave PLL Synthesizer for 60GHz WPAN Radio

Ja-Yol Lee, Haecheon Kim, Hyun-Kyu Yu

Digital RF SoC Design Team, SoC R&D Group, ETRI
161, Gajeong, Yuseong, Daejeon, South Korea
[1]ljylna@etri.re.kr

Abstract— **In this paper, we present the design of 52-GHz frequency synthesizer for 60GHz WPAN application. The PLL consists of 26GHz PLL and 52GHz frequency doubler, generating two channels of output carriers with 2.08GHz step by using high-speed four-modulus divider. The proposed PLL represents phase noise of – 89dBc/Hz from 26.2GHz carrier and – 81dBc/Hz from 52.4GHz carrier, at 1MHz offset, respectively. Also, its integrated RMS phase noise from 1MHz to 100MHz is measured as 7.42° Output frequency tuning range from the PLL is 50 to 53-GHz. The synthesizer including frequency doubler consumes 160mA at 2.5V supply voltage and occupies 1.2 × 1.0 mm² chip area.**

I. INTRODUCTION

Multiple 60-GHz WPAN radios have been developed as wireless communication service of several-gigabit rate by using Si-based process technology [1][2].

In a 60GHz millimeter-wave transceiver, the frequency synthesizer is a key building block. It is very difficult to design the PLL-based programmable synthesizer directly at 60GHz band without tripler or doubler [3][4]. A 14.25-16GHz programmable PLL was presented as a frequency source for 60GHz direct-conversion receiver [3]. Its doubled output frequency becomes 28.5 to 32GHz. A 15.3-18GHz programmable PLL was developed as a local oscillator for 60GHz dual-conversion super-heterodyne transceiver [4]. Its tripled output frequency becomes 46 to 54GHz. Also, various fixed PLLs have been developed as millimeter-wave frequency sources [4]-[7], but they are not suitable for 60GHz WPAN radio covering 57 to 64GHz range.

In this paper, we have developed 52GHz PLL-based synthesizer for 60GHz dual-conversion super-heterodyne receiver. The synthesizer is divided into 26GHz programmable PLL and 52GHz frequency doubler. The 26GHz PLL consists of PFD, charge pump, loop filter, LC VCO, and four-modulus divider. The synthesizer shows a 50-53GHz locking range, and generates two channels of 50.32 and 52.4GHz when 262MHz reference is used. The PLL achieves phase noises of – 89dBc/Hz from 26.2GHz and – 81dBc/Hz from 52.4GHz, at 1MHz offset, respectively. The synthesizer represents spurious noise level of – 42dBc/Hz, and consumes 160mA at 2.5V.

In section II, a 60-GHz dual-conversion super-heterodyne receiver is briefly introduced. In section III, the 52GHz PLL including doubler is described in detail. The experimental results are presented in section IV, and finally, conclusion is drawn in section V.

II. 60GHz DUAL-CONVERSION RECEIVER

Fig. 1 shows a dual-conversion super-heterodyne receiver for 60GHz WPAN radio including RF and IF PLL. The 60-GHz dual-conversion receiver consists of RF LNA, RF Mixer, RF PLL, VGA, IF Mixer and IF PLL. In the 60GHz receiver, the LNA amplifies the RF signals between 57GHz and 64GHz, which are mixed down to 10GHz IF signal by the RF mixer. Then, the amplitude of the 10-GHz IF signal is controlled by the variable-gain amplifier (VGA). And then, it is fed to the double-balanced IF mixer which performs the IF-to-base-band signal conversion. The LO module is configured with 52GHz RF PLL and 10GHz IF PLL. The RF PLL should generate three channels between 48.24GHz and 52.4GHz in step of 2.08GHz or between 48.576GHz and 52.8GHz by 2.112GHz step. The IF PLL provides only 10GHz or 10.032GHz fixed carrier, depending on the reference clock.

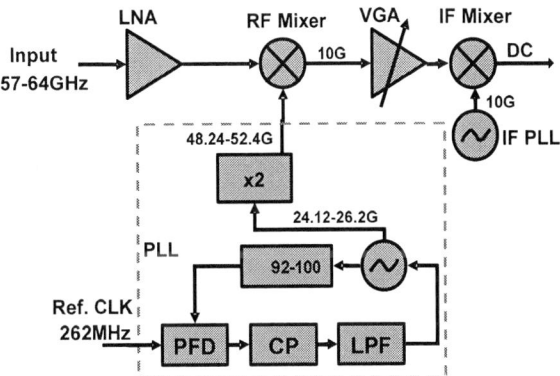

Fig. 1 60-GHz dual-conversion superheterodyne receiver

III. 52GHz PLL DESIGN

Fig. 2 represents the 52-GHz PLL-based frequency synthesizer consisting of a 26-GHz PLL and a frequency doubler. The 26-GHz PLL is composed of PFD/charge pump, 2nd-order loop filter, VCO, and four-modulus divider. The 2nd-order loop filter, configured with 12.4pF, 2.7kΩ, and 130pF, is optimized to have a settling time of 800ns and a spur level of 42dBc. The synthesizer can generate 3 channels between 48.24GHz and 52.4GHz by 2.08GHz step through the frequency doubler when 262-MHz reference clock is input. Both the precharged DFF-based PFD and the current-steering differential charge pump are used to achieve small turn-on

978-2-8748-7007-1/08 $25.00 © 2008 EuMA

time and fast operation [9]. In Fig. 2, a four-modulus divider is designed to provide the divide ratios of ÷ 20 to 25 for covering three channels.

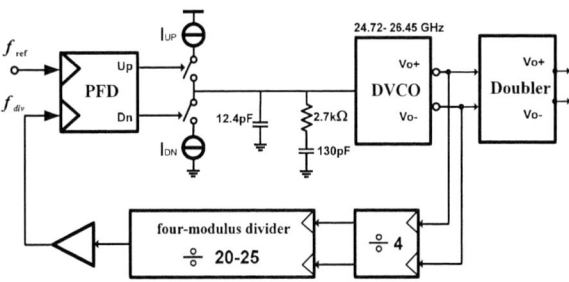

Fig. 2. 52GHz PLL architecture

A. Four-Modulus Divider

In millimeter-wave PLL, it is very difficult to design the high-speed multi-modulus divider for synthesizing multiple channels because there is digital-signal delay time. The major issues in designing the multi-modulus divider are high-speed operation, low noise, and low power consumption [3][4]. Fig. 3 represents the block diagram of the four-modulus divider used in the synthesizer. The four-modulus divider, which has four divide ratios controlled by mode control (MC) signal, is designed to simplify the hardware required for frequency synthesis. Therefore, it consists of a divide-by 4/5 prescaler, a divide-by-5 CMOS divider, and a control logic unit. The divide ratio of the four-modulus divider can be set to be ÷ 20, ÷ 23, ÷ 24, and ÷ 25, as shown in the table of Fig. 3.

For example, if both C1 and C0 are zero, the divide ratio becomes a divide-by-24. The operating timing waveform of the four-modulus divider is illustrated in Fig. 3 when it performs a divide-by-24 operation. As shown in Fig. 3, when MC is low, the divide-by 4/5 prescaler divides the input clock signal by ÷ 5. When MC is high, the 4/5 dual-modulus divider performs a divide-by-4 operation. In Fig. 3, the 4/5 prescaler divides the input signal of fvco/4 by ÷ 4 for one Po+ cycle and by ÷ 5 for four Po+ cycles while the followed divide-by-5 divider swallows five Po+ cycles. Thus, a total divide ratio (TDR) is calculated as TDR= (÷ 4) × 1 cycle + (÷ 5) × 4 cycles = ÷ 24.

A high-speed multi-modulus divider should be designed to contribute low noise to the sensitive VCO circuit in mixed-mode environment. Especially, the 4/5 dual-modulus prescaler in Fig. 3 should be designed to achieve both high speed and low noise. Therefore, the 4/5 dual-modulus divider of Fig. 4 is designed using a capacitive-degeneration ECL-like D-F/F which represents high conductance-zero frequency point and less possibility to miss the holding data at high frequency. In the ECL-like D-F/F, the emitter of the cross-coupled latch is not only capacitively degenerated, but also each clocking transistors(T_{r10}, T_{r11}, T_{r13}, T_{r14}) are separately connected to each latch transistors for enhancing speed. The 4/5 prescaler consists of one D-F/F and two NOR D-F/Fs. In Fig. 4, the outputs of both the second and the third D-F/Fs are fedback into the first NOR D-F/F as the control signals for generating

proper division ratio. Here, the MC signal is given to the third NOR D-F/F for modulating division ratio.

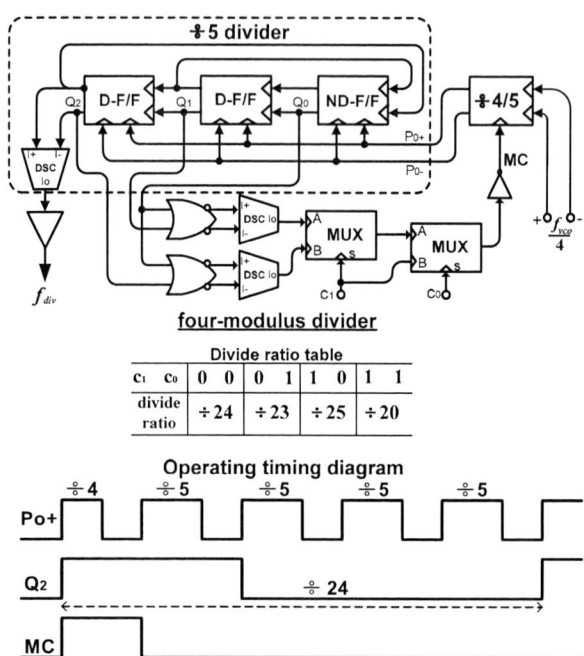

Divide ratio table							
c1 c0	0	0	0	1	1	0	1	1
divide ratio	÷ 24		÷ 23		÷ 25		÷ 20	

Fig. 3 Four-modulus divider circuit and its operating timing diagram

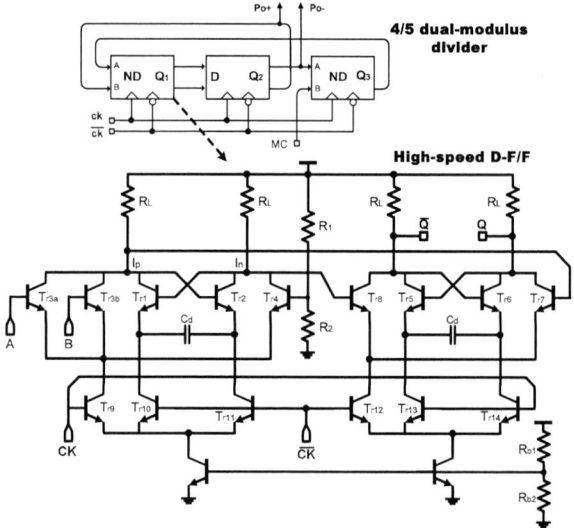

Fig. 4 Divide-by-4/5 dual-modulus divider and high-speed D-F/F

B. 26GHz LC-tank VCO

Fig. 5 shows the circuit diagram of the 26-GHz LC-tank VCO used in the synthesizer. It is a basic balanced differential oscillator that uses a cross-coupled differential pairs. In the LC VCO circuit, the cross-coupled pair generates negative conductance to compensate the LC-tank loss. For compensating the loss of inductor and achieving enough

978-2-8748-7007-1/08 $25.00 © 2008 EuMA   156

oscillation condition, a cross-coupled pair having much larger negative conductance is required, and thus the feedback capacitor C_f is inserted into the positive feedback path of the cross-coupled pair as shown in Fig. 5. The feedback capacitor C_f has a role to block DC flow and cross couple RF signal power. Also, C_f prevents the forward bias of the base-collector junction of the oscillation transistor, which results in high negative conductance as well as high oscillation signal amplitude. The high signal swing reduces phase noise of VCO. That is, the feedback capacitor pulls the negative conductance up to higher frequency band and larger amplitude. Both the input negative resistance and effective input capacitance of the VCO can be estimated as (1) and (2).

$$R_{in} = \frac{2(r_b + r_e)^2 + 2\left[\frac{1}{\omega C_f} + \left(\frac{\omega_T}{\omega}\right)\left(\frac{1}{g_m} + r_e\right)\right]^2}{(r_b + r_e) - \left(\frac{\omega_T}{\omega}\right)^2\left(\frac{1}{\omega_T C_f} + \frac{1}{g_m} + r_e\right)} \quad (1)$$

$$C_{in} = \frac{\left(\frac{\omega_T}{\omega^2}\right)\left(\frac{1}{\omega_T C_f} + \frac{1}{g_m} + r_b + 2r_e\right)}{2(r_b + r_e)^2 + 2\left[\frac{1}{\omega C_f} + \left(\frac{\omega_T}{\omega}\right)\left(\frac{1}{g_m} + r_e\right)\right]^2} \quad (2)$$

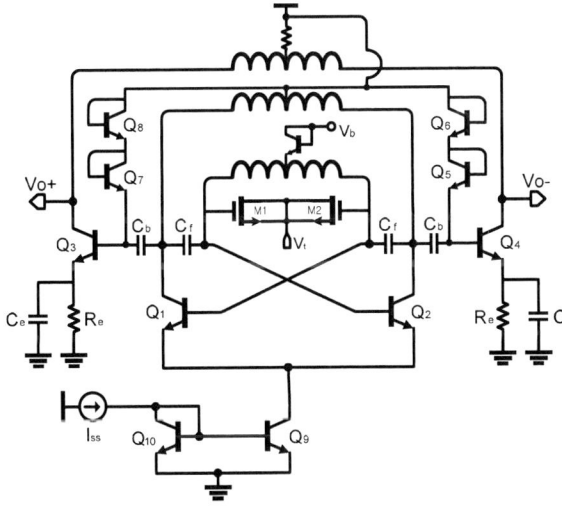

Fig. 5 Capacitive-feedback LC VCO

C. Frequency doubler

A 52-GHz frequency doubler is presented as shown in Fig. 6. In the doubler circuit, the collector nodes of the differential amplifier (Q1 and Q2) are put together for extracting the even-mode signal. Also, another even-mode signal with different phase is extracted from the combined emitter node of the differential amplifier. The common-base amplifier Q3 is used for amplifying the even-mode signal extracted from the

emitter node. Both Cm and Rm are used to control the amplitude and phase difference between the emitter extracted signal and the collector extracted signal [10]. The common-emitter amplifiers of Q4 and Q5 are used to amplify the extracted even-mode differential signals.

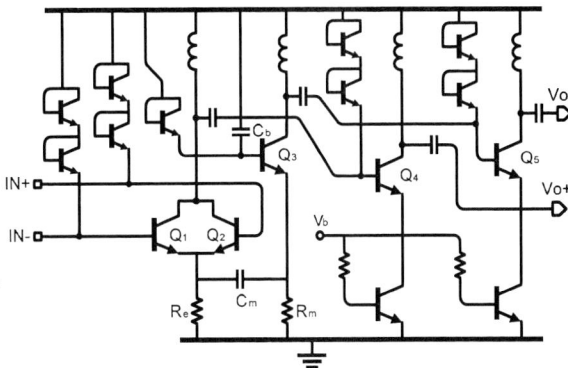

Fig. 6 Frequency doubler

IV. MEASURED RESULTS

Fig. 7 shows the measured tuning range of the VCO. Its tuning range is measured from 24.72GHz to 26.45 GHz. Fig. 8 shows the locked signal of 52.4 GHz when 262 MHz is input to PFD and ÷ 100 is selected. The spur level is measured as −42dBc. Fig. 9 represents its measured phase noises, which are − 89dBc/Hz from 26.2GHz and − 81dBc/Hz from 52.4GHz, respectively, at 1MHz offset. Its integrated RMS phase noise from 1MHz to 100MHz is estimated as 7.42°. The phase noise of the 52.4-GHz doubled carrier is degraded by about 7dB at offset frequency under 10MHz, compared with that of 26.2GHz fundamental carrier. This is close to the expected degradation of 6dB. Above 10MHz offset, the phase noise degradation of the doubled frequency carrier increases due to the noise floor of the measurement system, configured with Agilent 11970V-band harmonic mixer and E4440A spectrum analyzer.

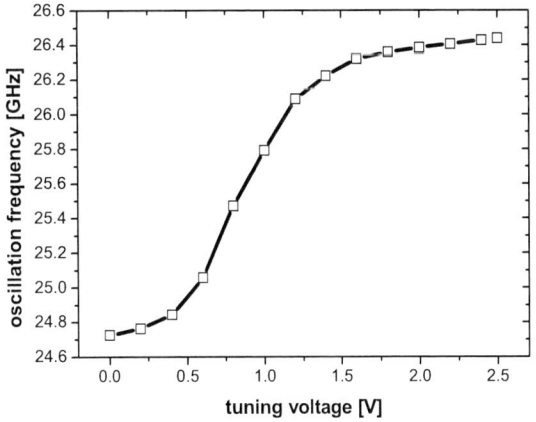

Fig. 7 Measured frequency tuning range

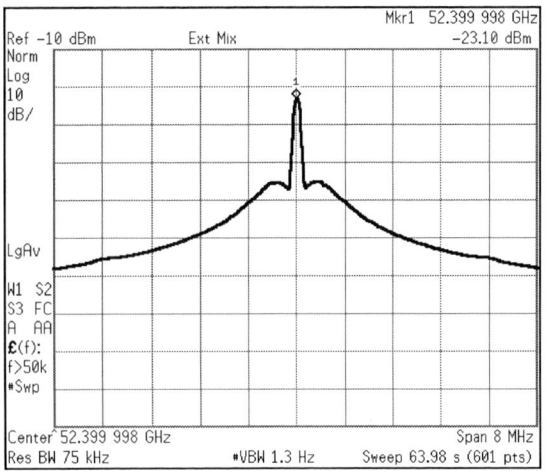

Fig. 8. Measured output spectrum of the PLL

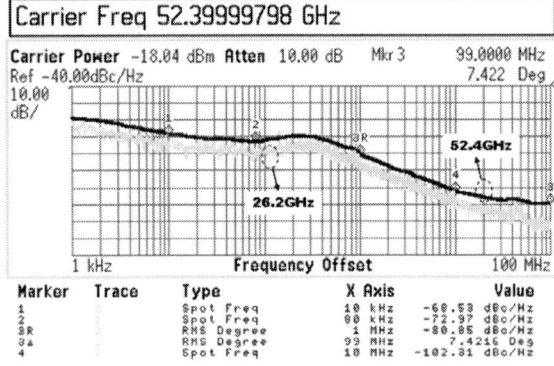

Fig. 9. Measured phase noise of the PLL

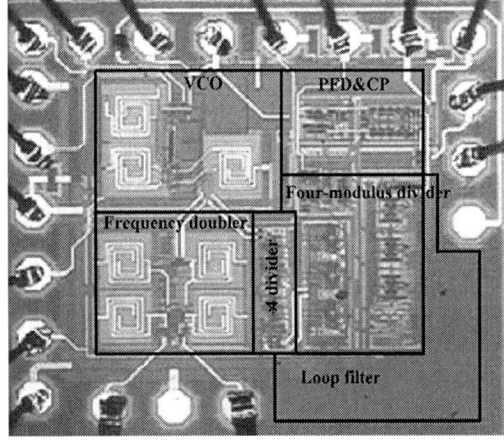

Fig. 10. Chip photograph of the PLL

Table I summarizes the measured results, and Table II compares this PLL with the reported other works. This PLL generates two-channel carriers between 50.32 and 52.4 GHz by 2.08 GHz step or between 50.688 and 52.8GHz by

2.112GHz step. The PLL consumes 160mA at 2.5V and 1.2mm² chip area. Fig. 10 shows the chip photograph of the PLL fabricated using 0.25µm SiGe BiCMOS process.

TABLE I
SUMMARY OF MEASURED RESULTS

Process	0.25 µm SiGe BiCMOS
Reference clock	262MHz
PLL frequency	50-53GHz
In-Band Phase Noise	- 80dBc/Hz from 26.2GHz carrier
@ 100kHz offset	- 73dBc/Hz from 52.4GHz carrier
Phase Noise	- 89dBc/Hz from 26.2GHz carrier
@1MHz offset	- 81dBc/Hz from 52.4GHz carrier
Phase Noise	- 110dBc/Hz from 26.2GHz carrier
@10MHz offset	- 102dBc/Hz from 52.4GHz carrier
RMS Jitter	7.42° (from 1MHz to 100MHz)
Reference Spur	- 42 dBc
Power consumption	160mA at 2.5V
Chip Area	1.2 × 1.0 mm²

TABLE II
COMPARISON OF THIS WORK TO THE REPORTED PLLS

Ref	[4]	[5]	[6]	[7]	[8]	This
Type	Tripled	Fund.	Fund.	Fund.	Fund.	Doubled
P_{dc}[mW]	258	650	57	78	80	400
Divider	Variable ÷56.5-64	Fixed ÷1024	Fixed ÷1024	Fixed ÷1024	Variable ÷256/258	Variable ÷80-100
PN $_{1MHz}$	-90	-72	-72	-80	-85	-81
f_o [GHz]	46-54	55-58	46-51	61-63	58-60	50-53

V. CONCLUSIONS

In this paper, a two-channel 52GHz PLL for 60-GHz dual-conversion receiver is presented. The PLL provides the output carriers of 50GHz to 53GHz, showing phase noise of - 89dBc/Hz from 26.2GHz fundamental carrier and - 81dBc/Hz from 52.4GHz doubled carrier, respectively, at 1MHz offset. The PLL achieves moderate phase noise compared to the previously reported PLLs.

REFERENCES

[1] S. K. Reynolds et al., "A silicon 60-GHz receiver and transmitter chipset for broadband communications," IEEE JSSC, vol. 41, no. 12, pp. 2820-2831, Dec. 2006.
[2] B. Razavi, "A 60-GHz CMOS receiver front-end," IEEE JSSC, vol. 41, no. 1, pp. 17-22, Jan. 2006.
[3] Ja-Yol Lee et al., "A 15-GHz 7-channel SiGe:C PLL for 60-GHz WPAN application," IEEE RFIC Sym. pp. 537-540, 2007.
[4] B. A. Floyd et al., "A 15 to 18-GHz programmable sub-integer frequency synthesizer for a 60-GHz transceiver," IEEE RFIC Sym. pp. 529-532, 2007.
[5] W. Winkler et al., "A fully integrated BiCMOS PLL for 60 GHz wireless Applications," IEEE ISSCC Dig., pp. 406-407, 2005.
[6] C. Cao, K.K. Oh, "A 50-GHz phase-locked loop in 130-nm CMOS," IEEE CICC, pp. 21-24, 2006.
[7] H. Hoshino et al., "A 60-GHz phase-locked loop with inductor-less prescaler in 90-nm CMOS," IEEE ESSCIRC Dig., pp. 472-475, 2007.
[8] C. Lee et al., "A 58-to-60.4GHz frequency synthesizer in 90nm CMOS," IEEE ISSCC Dig., pp. 196-197, 2007.
[9] Ja-Yol Lee et al., "A 9.1-to-11.5-GHz four-band PLL for X-band satellite & optical communication applications," IEEE RFIC Sym. pp. 533-536, 2007.
[10] F. Gruson et al., "A frequency doubler with high conversion gain and good fundamental suppression," IEEE MTT-s Dig., pp. 175-178, 2004.

Proceedings of the 3rd European Microwave Integrated Circuits Conference

Digitally-Assisted Linearization of Wideband Direct Conversion Receivers

[1]Edward A. Keehr, [1]Ali Hajimiri

[1]*Department of Electrical Engineering, California Institute of Technology*
1200 E. California Blvd. M/C 136-93, Pasadena, CA 91125, United States of America
[1]keehr@caltech.edu
[1]hajimiri@caltech.edu

Abstract— **A SAW-less direct-conversion receiver is presented which utilizes a mixed-signal feedforward path to regenerate and equalize IM3 products, thus accomplishing system-level linearization. The receiver system performance is dominated by a custom integrated front end realized in 130nm CMOS and achieves an uncorrected out-of-band IIP3 of -7.1dBm under the worst-case UMTS FDD Region 1 blocking specifications. IM3 equalization results in an effective IIP3 of +5.3dBm and reduces total input-referred error by over 23dB.**

I. INTRODUCTION

Improving the linearity of RF receivers has long been a challenging circuit design problem. As supply voltages drop with process scaling and as market demands continue to press for the elimination of off-chip components, such as SAW filters, standard techniques for meeting receiver linearity specifications will become inadequate. Direct conversion receivers are of considerable interest in this regard, as they constitute the dominant architecture in modern communication systems. Much progress has recently been made in improving the IIP2 of such circuits; however, a great need still exists to improve upon the IIP3 performance. This paper discusses the concept and details behind a system-level approach to satisfy this need.

II. SYSTEM-LEVEL CONCEPTS

A. System-Level Linearization via Equalization

In order to improve the system IIP3 for a SAW-less direct-conversion receiver, the original main path of the receiver is augmented with an alternate receiver path, as shown in Fig. 1 below.

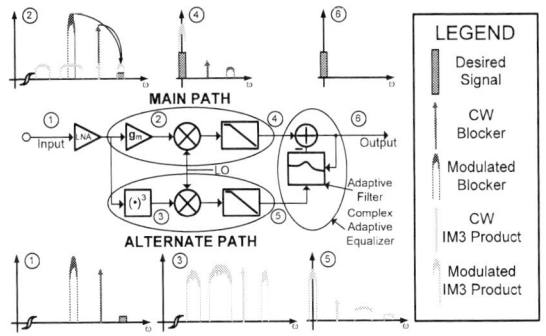

Fig. 1 IM3 equalization concept.

The alternate path generates IM3 products in the analog domain at RF and downconverts them to baseband using the same LO frequency as the main path receiver. The IM3 products are then digitized and are used as inputs to an equalizer which cancels the baseband IM3 products in the main path. Hence, a linearization via equalization is accomplished. Generating the IM3 products after the LNA at RF is a crucial feature of this architecture, since it is at this point in the system that the blocker magnitudes are at their largest.

This solution has the advantages of being both power-efficient and robust. Since the alternate path must pass only IM3 products, the dynamic ranges of its constituent blocks can be over 10dB less than in the main path, allowing for significant power savings in its overall design. As problematic blocker conditions occur roughly less than 10% of the time, the time-averaged power dissipation of the alternate path is further reduced by powering it on only when needed. The adaptive nature of the equalization guarantees robustness in the presence of changes in temperature, LO frequency, and blocker characteristics.

The idea of adaptively cancelling IM3 products has recently been described in a system-level study [1], where the alternate path resides completely within digital baseband. For IM3 cancellation to occur in this scheme, the ADCs must digitize all possible IM3-producing blockers. For example, in FDD UMTS Region 1 this requires digitizing frequency bands from 1670-1850MHz, 2015-2075MHz, and 1920-1980MHz (TX band), rendering this scheme unattractive from a power efficiency perspective. In order to overcome this issue, the mixed-signal approach proposed in this work passes only the problematic IM3 products through the alternate path ADCs. Hence, the alternate path ADC and digital baseband sampling rate requirements are no greater than those of the original main path.

B. Receiver System Architecture

The receiver system architecture described in this paper is shown below in Fig. 2. In order to provide a quantitative design objective, the design targets the FDD UMTS Region 1 standard, as it is well known for its linearity challenges. It contains a custom-designed integrated front-end in 130nm CMOS, analog baseband circuitry on PCB, and a digital back end implemented on an FPGA platform.

978-2-8748-7007-1/08 $25.00 © 2008 EuMA   159

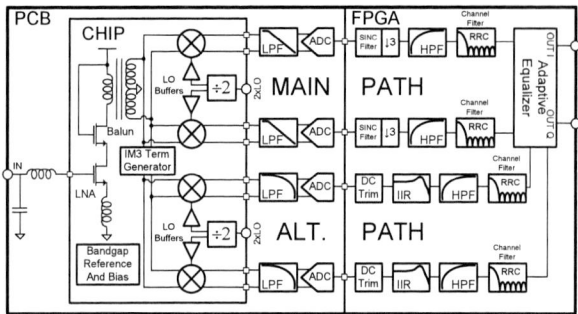

Fig. 2 Experimental UMTS receiver architecture.

Although it is possible to perform the adaptive equalization in analog circuitry, shifting as much of the signal processing burden as possible to the digital domain affords several advantages. For example, the behaviour of digital circuitry is relatively insensitive to process variations, and the continued scaling of digital processes has made baseband digital blocks power-competitive with equivalent blocks in the analog domain.

Considerations specific to modern cellular receivers also factor into the choice of predominantly digital equalization. For example, such receivers often implement adjacent channel rejection filters in the digital domain. To ensure that large adjacent channel signals do not interfere in the adaptive equalization, it is best to place the adaptive equalizer after these filters in the digital domain as well.

III. ANALOG CIRCUIT DESCRIPTIONS

A. Main Path Circuitry

As shown in Fig. 2, the LNA is an inductively degenerated cascode architecture which is terminated by a transformer balun. This balun serves as the LNA load inductor and subsumes the single-ended-to-differential conversion previously handled by the interstage SAW filter. The balun is followed by high-IIP2 mixers [2] which are folded in order to function under the 1.2V supply voltage. This choice obviates any IM2 equalization and allows the design effort to focus solely on IIP3 improvement. A frequency divide-by-two circuit which generates I and Q LO phases is also included on chip in order to minimize LO coupling into the substrate and chip ESD guard ring.

The main path baseband circuitry is implemented with discrete, commercially available components. The baseband low-pass filter is a third order Chebyshev architecture and is followed by an 8-bit pipelined ADC running at 50MHz.

B. Alternate Path Circuitry

The alternate path is a low-power, low-area variant of the main path, with the primary difference being the inclusion of an IM3 term generator. In order to conserve area, the alternate path mixer dispenses with the tuning inductor present in the main path mixer. Another I/Q generating frequency divider is used for the alternate path, as routing LO signal from the main path frequency divider increases the risk of LO coupling to other blocks in the system.

Like the main path, the alternate path baseband circuitry is implemented with discrete components. The alternate path baseband low-pass filter is a single real-pole embedded into the ADC buffer, which drives an 8-bit pipelined ADC running at 16.66MHz. Along with the free real pole at the output of the mixer, this level of postfiltering is sufficient to attenuate any IM3 products not involved in baseband equalization to below the alternate path noise floor.

IV. IM3 TERM GENERATOR

The IM3 term generator is an unconventional block, and as such, it merits additional consideration in its design. A pure cubing circuit is desired, as it passes negligible linear terms. If present, these terms will be treated as error by the adaptive equalizer, potentially degrading the receiver small signal gain, IM3 correction ratio, or both. Unlike other approaches that utilize the third-order Taylor series coefficient of the MOSFET [3], the design concept used for this IM3 term generator, shown below in Fig. 3, exploits the stronger second-order term by distributing the cubing between a conventional MOS squaring circuit and a Gilbert multiplier. The circuit schematic itself is shown in Fig. 4.

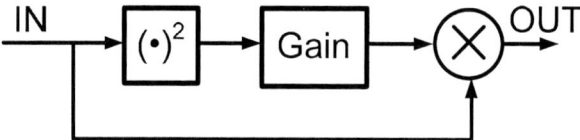

Fig. 3 Multistage cubing concept.

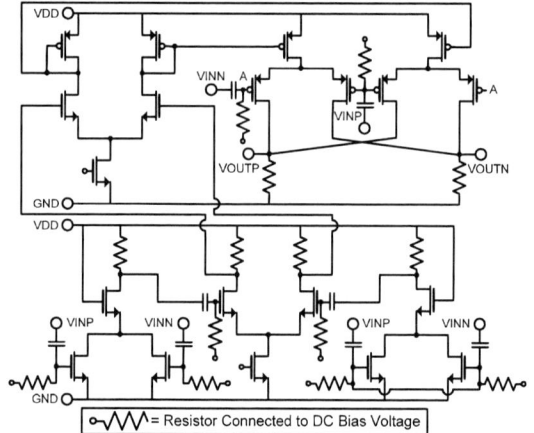

Fig. 4 IM3 term generator schematic.

As it produces a single-ended output, the squaring circuit must be followed by an active balun. To improve the generator CMRR, the negative terminal of the balun is tied to a replica squaring circuit whose gate terminals are shorted. This branch only generates common mode signal, which is then rejected by the CMRR of the balun.

Perhaps the most important aspect of the proposed IM3 term generator is the multistage nature of the cubing. Only the beat frequency terms of the squaring are required to be retained in order to complete the cubing operation. Hence, the

978-2-8748-7007-1/08 $25.00 © 2008 EuMA   160

bandwidth of the inter-multiplication circuitry can be substantially smaller than the RF frequencies of the blocker signals, as depicted in Fig. 5 below. In this case, the gain-bandwidth principle can be used to the designer's advantage, as substantial gain can be applied for less power than if the full IM2 spectrum up to 4GHz were retained in between nonlinear operations.

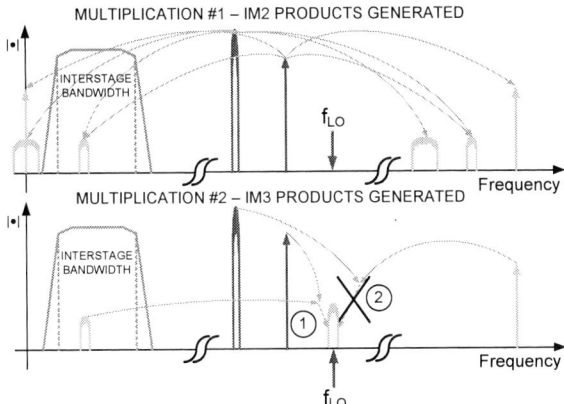

Fig. 5 Multistage cubing: frequency domain considerations.

V. DIGITAL EQUALIZATION

The path equalization implemented in this project is performed in the digital domain and is partitioned into fixed and adaptive portions. This choice stems from the fact that the minimal analog postfilters of the main and alternate path were found to be different in both type and order. Adaptive equalization of such a known IIR path difference is computationally inefficient. Therefore, the fixed equalization consists of a three-multiplier IIR filter in the alternate path. The remaining difference between the two paths is a complex DC gain and a small random mismatch in the baseband transfer function. This difference is broadband in the frequency domain and by the duality principle corresponds to a small number of FIR taps required in the adaptive equalizer.

The normalized LMS (NLMS) algorithm was chosen for the adaptive equalization scheme due to its simplicity and convergence speed. The division associated with this algorithm can be log2-quantized, allowing the use of a barrel shifter as a divider. Although a complex LMS-based algorithm such as NLMS can equalize the phase skew between the main and alternate paths, in general the presence of I-Q mismatch on either path limits the available IM3 cancellation. As shown in Fig. 6 below, an additional degree of freedom in the design was added to overcome this issue by feeding back the complex corrected signal back to independent I and Q taps on each of the incoming alternate path signals.

Yet another consideration is that the performance of the adaptive equalizer is limited in the presence of DC offset. To solve this problem, the proposed design includes high-pass filters in the digital domain and DC offset trimming circuitry in the alternate path. Periodic trimming must be performed

prior to the alternate path high-pass filters, or the step response incurred when enabling the digital portion of the alternate path will result in an exponential error transient at the output, prolonging equalizer convergence.

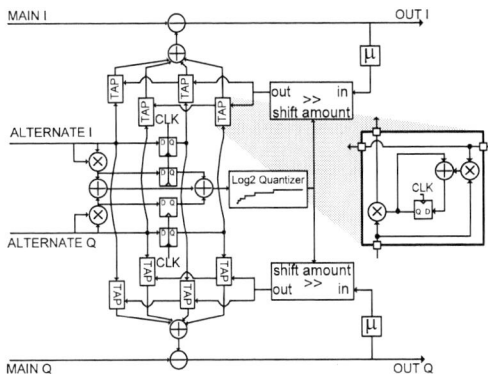

Fig. 6 Enhanced NLMS adaptive equalizer block diagram.

VI. MEASUREMENT AND PERFORMANCE

The emphasis of this project is to meet the IIP3 requirements implicitly posed by the UMTS out-of-band blocking test, which must be performed while the TX path is operating at maximum output power [4]. For the duplexer [5], the worst case specified IMD condition, with values referred to the LNA input, is -26dBm TX leakage at 1.98GHz, a -34dBm CW blocker at 2.05GHz, when the receiver LO frequency is set to 2.12GHz. Therefore, under experiment, the receiver is subject to a modified two-tone test, where one of the signals is QPSK-modulated and set to UMTS standards. Accounting for the 2dB loss of the duplexer and the 3dB increase in noise margin allowed under blocking conditions, the maximum allowed total input-referred error is -98dBm [6]. The results of this test are shown in Fig. 7 below and show that in this case under equalization, the input-referred error is -101dBm. (note: here f_{CW}=2.05125GHz, f_{LO}=2.1225GHz) Also shown in Fig. 7 are the results of the same test using the canonical NLMS algorithm. A main path quadrature mismatch of about 3° is partially responsible for the higher input-referred IM3 products.

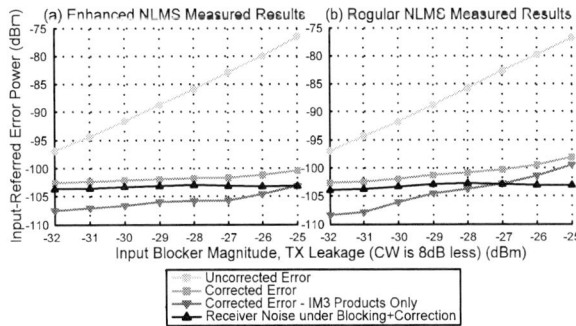

Fig. 7 Measured modified two-tone performance of the receiver, swept over amplitude.

The total input-referred error accounts for gain loss, thermal noise, IM2, and other IM products. Removing the effect of main path thermal noise and IM2 products yields a lumped input-referred quantity consisting of all other error sources. From this quantity, which is treated as residual IM3 error, a slope-of-3 line is extrapolated from the worst-case input blocker magnitude to obtain an effective IIP3 of +5.3dBm. This is an improvement of 12.4dB from the uncorrected IIP3 of -7.1dBm. This test was also performed at all 12 UMTS RX frequencies, with the results of this experiment shown below in Fig. 8. Note that in Figs. 7 and 8, the calculated thermal noise of the 50Ω input impedance is removed to isolate the performance of the receiver circuitry.

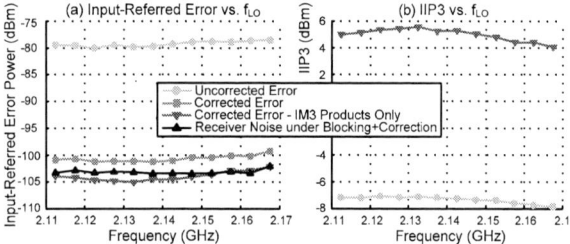

Fig. 8 Measured modified two-tone performance of the receiver, swept over LO frequency. At LNA input TX power is -26dBm, CW power is -34dBm.

Convergence behavior of the adaptive equalizer is shown in Fig. 9 below for the case where the TX leakage amplitude is -25dBm and the CW blocker amplitude is -33dBm at the LNA input at f_{LO}=2.1225GHz.

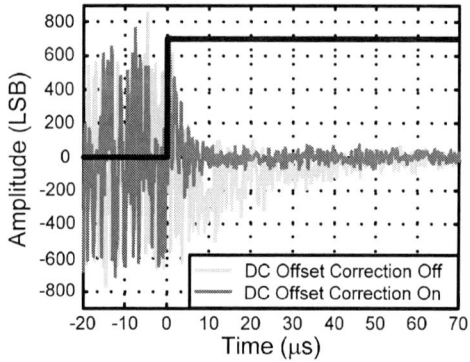

Fig. 9 Measured convergence of the adaptive equalization algorithm.

The measured performance summary of the receiver is shown below in Fig. 10. Fig. 11 shows a die photograph of the integrated RF CMOS front end. Digital power and area numbers are pre-layout estimates derived from a 90nm CMOS standard cell library with estimated wiring parasitics.

VII. CONCLUSION

This paper describes a UMTS receiver with an integrated front end that cancels IM3 products using a novel mixed-signal feedforward loop. Issues regarding the generation of reference IM3 products, path I-Q mismatches, and DC offset are considered, shown to be relevant, and overcome.

Parameter Measured at f_{LO}=2.1225GHz	Result
Total Active Analog Die Area	1.6mm × 1.5mm
Active Alternate Path Analog Die Area	0.5mm × 0.4mm
Analog Die Technology Node	130nm CMOS
Estimated Digital Alternate Path Area	0.42mm× 0.42mm
Digital Die Technology Node	90nm CMOS
Analog Die LNA+Main Path DC Gain	30.5dB
Complete LNA+Main Path DC Gain to ADC Input	70.2dB
Return Loss (S11) 2.11GHz-2.17GHz	<-13dB
IIP2@1.98GHz	+58dBm
Uncorrected IIP3 @1.98GHz/2.05125GHz	-7.1dBm
Effective IIP3@1.98GHz/2.05125GHz	+5.3dBm
ICP1@1.98GHz	-19dBm
Analog Die LNA+Main Path NF	5.1dB
Complete LNA+Main Path NF	5.5dB
Analog Die Supply Voltage	1.2V/2.7V
Analog Die LNA+Main Path Current	28mA (1.2V)
Analog Die Alternate Path Current	6.7mA (1.2V)
Estimated Digital Alternate Path Current	5.6mA (1.0V)
Baseband Signal Measurement Bandwidth	10kHz-1.92MHz

Fig. 10 Measured performance summary of receiver.

Fig. 11 RF CMOS front-end analog die.

ACKNOWLEDGEMENTS

The authors thank F. Bohn for the frequency divider IP and H. Mani and J. Yoo for testing assistance. This project was supported by an NDSEG fellowship and the Lee Center for Advanced Networking.

REFERENCES

[1] M. Valkama, A.S.H. Ghadam, L. Antilla and M. Renfors, "Advanced digital signal processing techniques for compensation of nonlinear distortion in wideband multicarrier radio receivers," *IEEE Trans. Microwave Theory and Techniques*, vol. 54, pp. 2356-66 June 2006.

[2] A. Liscidini, M. Brandolini, D. Sanzogni and R. Castello, "A 0.13 mm CMOS front-end, for DCS1800/UMTS/802.11b-g with multiband positive feedback low-noise amplifier," *IEEE Journal Solid State Circuits*, vol. 41, pp. 981-9, Apr. 2006.

[3] F. Shearer and L. MacEachern, "A Precision CMOS Analog Cubing Circuit," *IEEE NEWCAS*, pp. 281-284, June 2004.

[4] "UE Radio Transmission and Reception (FDD)," Tech. Specification Group, 3GPP, (TSG) RAN WG4, TS 25.101, v8.1.0, December 2007.

[5] muRata Corp Part Number DFYK61G95LBJCA, http://www.murata.com. Data sheet available at http://smartdata.usbid.com/datasheets/usbid/dsid/103495.pdf.

[6] A. Springer, L. Maurer and R. Weigel, "RF system concepts for highly integrated RFICs for W-CDMA mobile radio terminals," *IEEE Trans. Microwave Theory and Techniques*, vol. 50, pp. 254-267, Jan. 2002.

Wideband CMOS Receivers exploiting Simultaneous Output Balancing and Noise/Distortion Canceling

S.C. Blaakmeer[1], E.A.M. Klumperink[1], D.M.W. Leenaerts[2], B. Nauta[1]

[1] *University of Twente, CTIT Institute, IC Design group, P.O. Box 217, 7500 AE Enschede, The Netherland*
s.c.blaakmeer@utwente.nl, e.a.m.klumperink@utwente.nl, b.nauta@utwente.nl
[2] *NXP Semiconductors, Research, Eindhoven, The Netherlands*

Abstract— This paper deals with the problem of realizing wideband receiver front-ends in downscaled CMOS technologies, which are highly wanted for multi-standard radio receivers and cognitive radio applications. Instead of using many narrowband inductor based receivers, we prefer the use of one wideband receiver with sufficient bandwidth to cover all popular frequency bands up to 6GHz or even 10GHz. To relax RF filter requirements, high linearity is required, while high gain and low noise are important for good sensitivity. Downscaled CMOS technologies feature high speed transistors, but also decreasing supply voltages and increasing transistor non-idealities, which makes it increasingly difficult to achieve high gain and good linearity. It will be shown that a combination of a common-gate (CG) stage and an admittance-scaled common-source (CS) stage has attractive properties for implementing a wideband receiver with active balun, while simultaneously canceling the noise and distortion of the CG-stage. Example applications in a wideband Balun-LNA and combined Balun-LNA-Mixer will be shown.

I. INTRODUCTION

Wideband receivers are required for instance in upcoming Software-Defined Radio and Cognitive Radio architectures and for Ultra Wideband Communication in the 3-10GHz bands. There are many mobile wireless communication standards that use the frequency spectrum from a few hundred MHz up to 6 GHz and they are increasingly integrated in one device. Traditionally receivers with narrowband inductor based Low Noise Amplifiers (LNAs) are used, but this becomes more and more impractical if many radio interfaces are to be integrated. Moreover, on-chip inductors do not scale much with technology downscaling, so relatively to other components they become more expensive and therefore we prefer to avoid their use. Single-ended input LNAs are preferred to save I/O pins and because antennas and RF filters usually produce single ended signals. On the other hand, differential signaling in the receive chain is preferred in order to reduce second order distortion and to reject power supply and substrate noise. To avoid the use of an external broadband balun and its accompanying losses which add directly to the noise figure, it is advantageous to integrate a balun on chip.

In this paper we will review recently proposed circuits to realize wideband linear front-ends with no or only few inductors in CMOS [1]-[14]. The main focus is on LNAs, but we will also briefly discuss wideband I/Q down-converters. In section II we will discuss the relevance of high linearity in such receivers. In section III we will present an overview of recently proposed wideband receiver front-ends. We will

discuss why a Common Gate (CG)-stage is problematic as inductor-less wideband LNA. In section IV we show that combining a CG-stage with a Common Source (CS)-stage allows for achieving more gain. Furthermore, it can implement a wideband active balun in a very compact way, while simultaneously canceling the noise and distortion contribution of the CG-transistor. If the CS-stage is carefully optimized, both the linearity and noise of the resulting combined Balun-LNA can be good. Finally section V discusses a way of increasing the gain, while maintaining a high bandwidth, by avoiding making voltage gain at RF.

II. LINEARITY REQUIREMENTS FOR WIDEBAND RECEIVERS

Like a narrowband zero-IF, a wideband receiver is sensitive to the 2nd order intermodulation product generated by an AM modulated carrier via AM detection. However, a wideband receiver may also suffer from 2nd order intermodulation generated by interferers that have a sum or difference frequency equal to the wanted RF-input signal. The response to a modulated carrier can be suppressed by AC-coupling between the LNA-output and mixer-input and by driving and designing the mixer in a well-balanced way [15]. However, the intermodulation product generated at a frequency equal to the frequency of the wanted signal cannot be separated from the signal. Especially standards that operate on large bandwidths, like DVB-H (470–862 MHz) [16] or WiMedia UWB (3.1–10.6 GHz) [17], have a high probability that a combination of interferers renders an in-band intermodulation product. A receiver designed for these standards should have an LNA with sufficiently high IIP2 (and IIP3) in order to handle strong interferers like WLAN (IEEE 802.11a/b/g) and the GSM standards. The required intercept points depend strongly on the assumed interferer scenario and the assumed amount of pre-filtering of the interfering signals. For a WiMedia UWB receiver the required IIP2 is above +20 dBm and IIP3 above -9 dBm as derived in [18]. For a DVB-H receiver, the required IIP2 is in order of +22dBm using a GSM/WLAN interferer scenario.

III. WIDEBAND RECEIVERS IN LITERATURE

Table I shows an overview of recently published wideband LNAs and down-converters in CMOS with no or only a few inductors, published at the most important solid-state circuit conferences. Different types of techniques have been proposed, which will be briefly discussed below. Distributed amplifiers are not discussed as they heavily rely on inductors or transmission lines.

With the increasing f_T of MOS transistor, multi-GHz negative feedback amplifiers are becoming feasible and some interesting results have been achieved recently [3][7][8]. Still, several trade-offs exist between impedance matching, gain, noise and linearity. Until now, relatively modest IIP3 has been achieved which also varies strongly with frequency (typically in the range of -15dBm to -4dBm). IIP2 is often not reported, despite of its importance for wideband receivers. Furthermore, these circuits don't include balun functionality.

A Common Gate amplifier can achieve wideband impedance matching and gain with good linearity, but is it difficult to achieve a noise figure below 4dB. Moreover, at low supply voltage, there is not much voltage headroom to realize high voltage gain. Furthermore, a high load resistance, required for high gain, leads to bandwidth limitations. Therefore, CG-stages are often used in combination with inductive broad-banding to increase the bandwidth. However, we would like to avoid such inductors and investigated other ways to achieve a high gain. We will discuss now two techniques to increase the gain, while using standard transistors at 1.2V supply, and without the use of inductors. In section IV we explain how one can use a parallel CG- and CS-stage to realize a noise/distortion canceling LNA which also acts as balun, as proposed originally in [1] and later exploited and extended in [2][4][5][6][10][11][12][13]. In section V we will propose a technique to avoid making voltage gain at RF, but do this only after the down-conversion to IF.

IV. SIMULTANEOUS BALANCING AND NOISE/DISTORTION CANCELING

In the sections below we will briefly derive the conditions for simultaneous balancing, noise canceling and distortion canceling. We will neglect capacitive effects for simplicity. A more detailed discussion on high frequency limitations and robustness for component variations can be found in [1] [13].

A. Balancing (balun operation)

The Common Gate stage in Figure 1, biased with a current source, has a straightforward relation between its voltage gain ($A_{v,CG}$) and its input impedance ($R_{in,CG}$). The signal current (i_{Rcg}) flowing through the load resistor R_{CG} has to be equal to the signal current flowing at the input (i_{in}), as there is no alternative path to ground. Thus,

$$i_{in} = i_{Rcg} = \frac{v_{out,CG}}{R_{CG}} = \frac{v_{in} \cdot A_{v,CG}}{R_{CG}} \qquad (1)$$

As a result, the input impedance of the CG-stage can be expressed as:

$$R_{in,CG} = \frac{v_{in}}{i_{in}} = \frac{R_{CG}}{A_{v,CG}} \qquad (2)$$

For an ideal transistor, having infinite output resistance, this is obvious. In that case the input impedance can be written as $R_{in,CG} = 1/g_m$ and the gain equals $A_{v,CG} = g_m \cdot R_{CG}$. However, (1) and (2) are equally valid when the finite output resistance and the body-effect of a real transistor are taken into account.

TABLE I.

RECENT WIDEBAND LNAs AND DOWN-CONVERTERS IN CMOS WITH NO OR ONLY A FEW COILS

Ref	Bandwidth [GHz]	Gain A_V [dB]	NF [dB]	IIP2 [dBm]	IIP3 [dBm]	Pcore [mW]	Process V_{supply}	# coils area[mm²]	Functionality – Z-matching Technique
Bruccoleri et al JSSC 2004 [1]	0.2 – 2.0	10 – 14	< 2.4	+12	0	35	0.25μm 2.5V	0 0.075	LNA – Transimpedance +CS Noise Canceling
Cherazi et al CICC 2005 [2]	0.9 – 5	18 – 19	< 3.5	+4 (sim)	+1 (sim)	12	0.18μm 1.8V	4 ~0.4	Balun-LNA – CG+CS Noise Canceling
Zhan et al ISSCC 2006 [3]	0.5 – 8.2	22 – 25	< 2.6	?	-4 / -16	42	90nm 2.7V	0 0.025	LNA – Transimpedance Negative Feedback
Bagheri et al ISSCC06 [4] [6]	0.8 – 6	3–36 with IF-AMP	< 5.5	?	-3.5	29	90nm 2.5V	2 0.5	Balun-LNA+I/Q Mixer – CG+CS Stage
Blaakmeer et al RFIC 2006 [5]	2.7 – 4.5	18 – 19.6	< 5	?	-8	12.6	90nm 1.2V	1 0.2	LNA – CG+trafo/CS Noise Cancelling
Borremans et al ISSCC06 [7]	DC – 6	15 – 17.4	< 3.5	?	-15 / - 8	9.8	90nm 1.2V	0 0.002	LNA – Transimpedance with Active Feedback
Blaakmeer et al ESSC06 [12] [13]	0.2 – 5.2	13 – 15.6	< 3.5	+20	0	14	65nm 1.2V	0 0.009	Balun-LNA – CG+CS Noise Canceling
Ramzan et al ISSCC2007 [8]	1 – 7	15 – 17	< 3.5	?	-4.1	25	0.13μm 1.4V	0 0.019	LNA – Transimpedance with Active Feedback
Lee et al ISSCC07 [9]	2 – 8	23 with IF-AMP	< 4.5	+18	-7	31	90nm 2.5V	1 0.09	LNA+trafo-Balun+ I/Q Mixer – Negative FB.
Liao et al JSSC 2007 [10]	1.2 – 11.9	12.7 – 15.7	< 5	+10 / +20	-6.2	20	0.18μm 1.8V	5 0.59	LNA – CG Noise Canceling
Chen et al RFIC2007 [11]	0.8 – 2.1	14.5 – 17.5	< 2.6	?	0 / +16	17.4	0.13μm 1.5V	0 0.01	LNA – nMOS+pMOS CG Noise Canceling
Blaakmeer et al ISSCC08 [14]	0.5 – 7	18 no IF-AMP	< 5.5	+20	-3	16	65nm 1.2V	0 <0.01	Balun-LNA+I/Q Mixer – Noise Canceling

978-2-8748-7007-1/08 $25.00 © 2008 EuMA

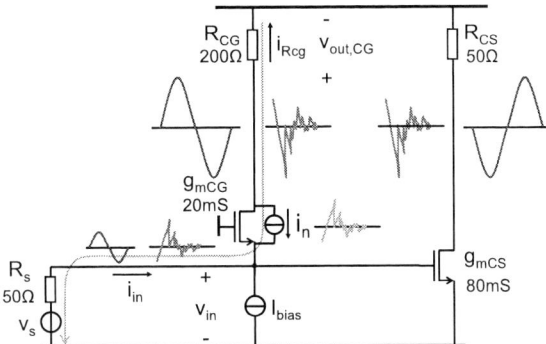

Fig. 1 The Balun-LNA, a combination of a Common Gate (CG) and admittance scaled Common Source (CS) stage to realize simultaneous output balancing, noise and distortion canceling

For an impedance match at the input, the input impedance of the CG-stage ($R_{in,CG}$) should equal the source resistance (R_S), thus the gain of the CG stage becomes:

$$A_{v,CG} = \frac{R_{CG}}{R_{in,CG}} = \frac{R_{CG}}{R_S} \qquad (3)$$

To create a balun, the gain of the CS-stage in Figure 1 should be equal, but have opposite sign, thus:

$$A_{v,CS} = -A_{v,CG} = -\frac{R_{CG}}{R_S} \qquad (4)$$

B. Noise Canceling

The noise generated by the CG-transistor in Figure 1 can be represented by a current source (i_n). This current generates both a voltage at the input-node ($v_{n,in} = \alpha_1 \cdot i_n \cdot R_S$) and a fully correlated anti-phase voltage at the CG-output ($v_{n,CG} = -\alpha_1 \cdot i_n \cdot R_{CG}$). The factor α_1 equals the voltage division between the input resistance ($R_{in,CG}$) and the source resistance (R_S), which equals 1/2 in case of impedance matching:

$$\alpha_1 = \frac{R_{in,CG}}{R_{in,CG} + R_S} \qquad (5)$$

The noise at the CS-output equals the CG-output noise ($v_{n,CS} = v_{n,in} \cdot A_{v,CS} = v_{n,CG}$), when the CS-gain $A_{v,CS}$ satisfies (4). Thus, the noise contribution of the CG-transistor can be canceled, as it becomes a purely common-mode signal at the differential output. This proofs that simultaneously balancing of the output signal and noise canceling is obtained. As the noise of the CG-transistor is cancelled, the CS-stage mainly determines the noise. By admittance scaling this noise contribution can be reduced at the cost of power consumption.

C. Distortion Canceling

As derived in [1], not only the noise of the impedance matching device, but also its distortion, due to the non-linearity of the transconductance, is canceled. We will show that also non-linearity of the output conductance of the CG-transistor is canceled.

The source signal (v_s) causes a non-linear drain-source current (i_{ds}) which is converted into a non-linear voltage at the input (v_{in}) via the (linear) source resistor R_S. The non-linear input voltage (v_{in}) can be written as a Taylor expansion of the signal source voltage (v_s):

$$v_{in} = \alpha_1 \cdot v_s + \alpha_2 \cdot v_s^2 + \alpha_3 \cdot v_s^3 + \alpha_4 \cdot v_s^4 + \ldots = \alpha_1 \cdot v_s + v_{NL} \qquad (6)$$

where the α's represent Taylor coefficients and v_{NL} contains all unwanted nonlinear terms and the first Taylor coefficient (α_1) is defined in (5).

The output voltage of the CG-stage can be written as:

$$v_{out,CG} = i_{in} \cdot R_{CG} = \frac{v_s - v_{in}}{R_S} \cdot R_{CG} = \left((1 - \alpha_1) \cdot v_s - v_{NL} \right) \frac{R_{CG}}{R_S} \qquad (7)$$

where (6) is used. The output voltage of the CS-stage can be written using (4):

$$v_{out,CS} = -v_{in} \frac{R_{CG}}{R_S} = -\left(\alpha_1 \cdot v_s + v_{NL} \right) \frac{R_{CG}}{R_S} \qquad (8)$$

The difference in sign of the wanted signal v_s and unwanted signal v_{NL} in (7) and (8) can be exploited: after subtraction only the linear signal remains:

$$v_{out,diff} = v_{out,CG} - v_{out,CS} = v_s \cdot \frac{R_{CG}}{R_S} \qquad (9)$$

In conclusion, all noise and distortion currents generated by the CG-transistor can be canceled, irrespective whether produced due to non-linearity of the transconductance or non-linearity of the output conductance. The gain required in the CS-stage to cancel the distortion products of the CG-transistor equals the gain required to obtain output balancing, leading to the conclusion that *simultaneous balancing and cancelation of unwanted noise and distortion currents of the CG transistor is possible*. As the distortion due to the CG-transistor is canceled, while R_{CG} is normally quite linear, the CS-stage will determine the overall linearity of the complete LNA. The linearity of the CS-stage has been analyzed in detail in [13]. It appears possible to realize very good IIP2 values above +20dBm, if the CS-stage is carefully optimized. The simultaneous noise canceling and distortion canceling idea has recently also been exploited to achieve high IIP3 [11].

V. BALUN-LNA WITH I/Q DOWN-CONVERTER

Although parallel operating CG and CS stages reduce the required voltage gain of the CG stage by a factor two, achieving a high bandwidth when driving a significant capacitive load is problematic. For 50Ω matching and 12dB voltage gain, a drain resistance of more than 200Ω is needed, which limits the load capacitance to 80fF for 10GHz -3dB bandwidth. To obtain more bandwidth, we propose to avoid creating voltage gain at RF, but do this at IF. Fig. 2 shows the principle: a CG-CS stage is stacked with current commutating mixer. The mixer transistors are in saturation and present a low impedance to the CG-CS stage output, therefore the bandwidth at these nodes is high. At IF, where much less bandwidth is required, the mixer output current is converted to voltage. The drain impedance is Z for the CG-stage and Z/4 for the CS-stage to realize simultaneous balancing and noise/distortion canceling at IF. By using LO square-wave signals with 25% duty-cycle, one CG-CS transconductance-stage can supply the required signal current for both a differential I- and Q- output. This results in a very power efficient down-converter. The IF-filter averages the current

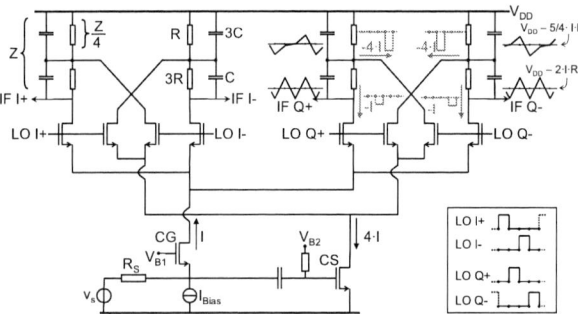

Fig. 2 A Balun-LNA with I/Q down-converter exploiting the CG-CS stage of Fig.1.

pulses through the load (see Fig. 2). As the CG/CS-bias currents do not flow continuously though the loads, the DC-drop across the loads is reduced, allowing an increased load impedance compared to the Balun-LNA of Fig. 1. This makes it possible to realize high voltage gain at IF, which remains high up to very high LO-frequencies.

Fig. 3 shows the gain, Noise Figure and S_{11} of a 65nm chip realizing the circuit of Fig. 2. Clearly it realized a very flat gain and noise figure up to 7GHz (the frequency range in simulation is actually higher, but is limited on the chip by the upper operating frequency of the LO-drivers) [14],[19].

VI. CONCLUSIONS

In this paper we reviewed recently proposed CMOS circuit techniques to realize wideband receivers. It turns out that a combination of a common gate and common source stage is an attractive option. It can implement an active balun, while it is also possible to exploit the simultaneous noise and distortion canceling property that reduces the noise and distortion of the common gate stage to negligible values. As this circuit has a gain equal to the sum of the gains of a CG and CS stage, it can realize an overall voltage gain close to 20dB even at a low supply voltage of 1.2V. It was also shown that the Balun-LNA can also be used as an RF transconductor, where a current commutation mixer I/Q mixer can be directly stacked on top of it to realize flat, very wideband gain.

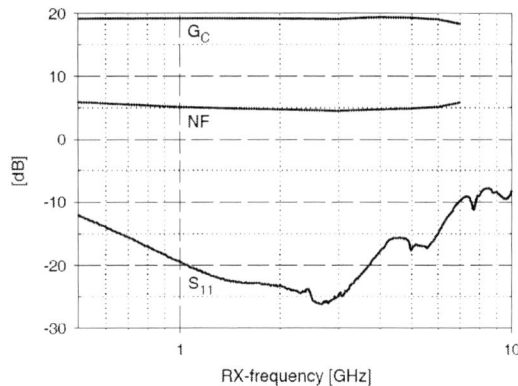

Fig. 3 Gain, Noise figure and S_{11} of a 65nm CMOS IC shown in Fig. 2. Note that the high and flat gain is achieved at only 1.2V supply.

REFERENCES

[1] F. Bruccoleri, E.A.M. Klumperink, and B. Nauta, "Wide-Band CMOS Low-Noise Amplifier Exploiting Thermal Noise Canceling," IEEE J. Solid-State Circuits, vol. 39, pp. 275-282, February 2004.

[2] S. Chehrazi, A. Mirzaei, R. Bagheri, and A. Abidi, "A 6.5 GHz Wideband CMOS Low Noise Amplifier for Multi-Band Use," Proc. IEEE Custom Integrated Circuits Conf. (CICC), pp. 801-804, Sept. 2005.

[3] J.-H.C. Zhan and S.S. Taylor, "A 5GHz Resistive-Feedback CMOS LNA for Low-Cost Multi-Standard Applications," ISSCC Dig. Tech.Papers, pp. 200-201, Feb. 2006.

[4] R. Bagheri et al, "An 800MHz to 5GHz Software-Defined Radio Receiver in 90nm CMOS", ISSCC Dig. Tech.Papers, 2006 pp. 1932 – 1941, Feb. 2006.

[5] S.C. Blaakmeer, E.A.M Klumperink, D.M.W. Leenaerts, and B. Nauta, "A wideband Noise-Canceling CMOS LNA exploiting a transformer," Dig. of papers RFIC Symposium, pp. 137-140, June 2006.

[6] R. Bagheri, A. Mirzaei, S. Chehrazi, M.E. Heidari, M. Lee, M. Mikhemar, W. Tang, and A.A. Abidi, "An 800-MHz–6-GHz Software-Defined Wireless Receiver in 90-nm CMOS," IEEE J. Solid-State Circuits, vol. 41, pp. 2860-2876, December 2006.

[7] J. Borremans, P. Wambacq, and D. Linten, "An ESD-protected DC-to-6GHz 9.7mW LNA in 90nm digital CMOS," presented at Digest of Technical Papers - IEEE International Solid-State Circuits Conference, 2007.

[8] R. Ramzan, S. Andersson, J. Dabrowski, and C. Svensson, "A 1.4V 25mW inductorless wideband LNA in 0.13μm CMOS," presented at Digest of Technical Papers - IEEE International Solid-State Circuits Conference, 2007.

[9] S. Lee, et al, "A broadband receive chain in 65nm CMOS," presented at Digest of Technical Papers - IEEE International Solid-State Circuits Conference, pp. 418-419, 2007.

[10] C. F. Liao and S. I. Liu, "A broadband noise-canceling CMOS LNA for 3.1-10.6-GHz UWB receivers," IEEE Journal of Solid-State Circuits, vol. 42, pp. 329-339, 2007.

[11] W. H. Chen, G. Liu, B. Zdravko, and A. M. Niknejad, "A highly linear broadband CMOS LNA employing noise and distortion cancellation," presented at Digest of Papers - IEEE Radio Frequency Integrated Circuits Symposium, pp. 61-64, 2007.

[12] S.C. Blaakmeer, E.A.M Klumperink, D.M.W. Leenaerts, and B. Nauta, "An Inductorless Wideband Balun-LNA in 65nm CMOS with balanced output," Proceedings of the 33rd European Solid-State Circuits Conference (München, Germany), pp. 364 - 367, Sept. 2007.

[13] S.C. Blaakmeer, E.A.M Klumperink, D.M.W. Leenaerts, and B. Nauta, "Wideband Balun-LNA with Simultaneous Output Balancing, Noise-Canceling and Distortion-Canceling", IEEE J. Solid-State Circuits, vol. 43, pp. 1341-1350, June 2008.

[14] S.C. Blaakmeer, E.A.M Klumperink, D.M.W. Leenaerts, and B. Nauta, "A Wideband Balun LNA I/Q-Mixer combination in 65nm CMOS", Digest of Technical Papers - IEEE International Solid-State Circuits Conference, pp. 418-419, 2007.

[15] D. Manstretta, M. Brandolini, and F. Svelto, "Second-order intermodulation mechanisms in CMOS downconverters," IEEE J. Solid-State Circuits, vol. 38, pp. 394-406, March 2003.

[16] ETSI. (2005, Nov.) Digital Video Broadcasting (DVB); DVB-H Implementation Guidelines. [Online], ETSI Document Number: TR 102 377, Version 1.2.1, Reference: RTR/JTC-DVB-175. Available: http://www.etsi.org/

[17] ECMA International. (2005, Dec.) High Rate Ultra Wideband PHY and MAC Standard. [Online], Available: http://www.ecma-international.org/publications/files/ECMA-ST/ECMA-368.pdf

[18] R. Roovers, et al, "An interference-robust receiver for ultra-wideband radio in SiGe BiCMOS technology," IEEE J. Solid-State Circuits, vol. 40, pp. 2563-2572, Dec. 2005.

[19] S.C. Blaakmeer, E.A.M Klumperink, D.M.W. Leenaerts, and B. Nauta, "The BLIXER, a Wideband Balun-LNA-I/Q-Mixer Topology," accepted for publication in IEEE Journal of Solid-State Circuits, Dec. 2008.

SiGe BiCMOS High Linearity Mixers for Base-Station Applications

Tero Tikka[#1], Ville Saari[#2], Jarkko Jussila[*3], Kari Halonen[#4], and Jussi Ryynänen[#5]

[#]*Department of Micro and Nanosciences / SMARAD2, TKK Helsinki University of Technology*
P.O. Box 3000, FI-02015 TKK, Finland
[1]`tero.tikka@ecdl.tkk.fi`

[*]*STMicroelectronics R&D Oy Finland*
Itämerenkatu 9, FI-00180 Helsinki, Finland

Abstract— This paper describes design issues related to high linearity SiGe BiCMOS active mixers, which are primarily targeted for WCDMA base-station applications. The effect of different mixer components to overall mixer dynamic range is described, and the measurement results from four different implementations are given to support this discussion. The different mixers are implemented using the same process as part of a complete receiver and thus the interface to baseband filter has been taken into account in the performance analysis. Since one of the goals in the mixer design has been to maximize the mixer bandwidth, the frequency limitations of different alternatives are discussed throughout the paper.

I. INTRODUCTION

A down-conversion mixer is considered as one of the most challenging design tasks in the integrated receivers, especially when targeting to high linearity. Currently, the most popular solution for an integrated BiCMOS mixer is based on the active Gilbert-cell topology. The modern BiCMOS processes still have sufficient supply voltage and the bipolar transistors offer low flicker noise, which enable active solutions with sufficient performance. The basic Gilbert-cell topologies offer gain, low NF and moderate IIP3, which are suitable for most wireless applications. However, if very high linearity for the mixer is required, linearization has to be applied to improve the linearity of the mixer. In base-station (BS) applications there is more freedom for selecting a proper pre-select filter compared to mobile applications. By choosing a high performance pre-select filter, the interfering signals from, for example, the Tx-band can be decreased and therefore the linearity requirement for the receiver can be reduced to some extent [1]. However, the good quality pre-select filters increase the cost of BS.

The mixers, presented in this paper as design examples, are targeted for BS direct-conversion receivers (DCRs) where the high dynamic range requirement sets demands for linearity together with low NF [1]-[3]. Therefore, this paper concentrates on the circuits, which improve the linearity of Gilbert-cell mixers while maintaining sufficient noise performance. In general, the high linearity was achieved by optimizing all sub-parts in the mixer as well as optimizing the interface to the following baseband filter. All the designs have been fabricated using same $0.25\text{-}\mu m$ SiGe BiCMOS process with a 2.5 V supply, thus giving a fare comparison between all implemented mixer structures.

A simplified schematic of a double balanced Gilbert-cell type mixer with a current-mode interface to baseband filter is shown in Fig. 1. The mixer itself can be divided into three sub-parts, an RF input stage, an LO switching stage, and a mixer load, which will be discussed in more detailed in the following sections of this paper. In general, the RF input stage acts as a transconductor, which in basic configuration defines noise and linearity parameters of the mixer. The LO switching transistors convert the signal down near DC in DCRs and the load defines the gain together with the g_m-stage. In addition, the load resistors R1 determine the common-mode level of the output. When pursuing high linearity performance none of the mixer sub-parts can be neglected and the performance optimization has to be applied to them all.

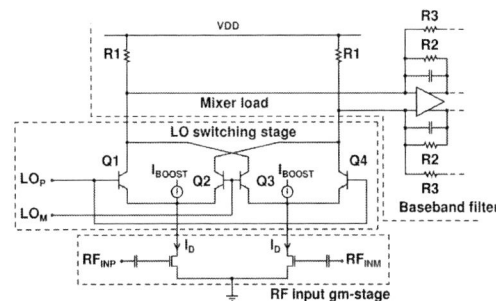

Fig. 1 Simplified schematic of a double balanced Gilbert cell mixer including the interface to baseband filter

In BiCMOS processes the obvious choice for the LO switches are bipolar transistors due their better flicker-noise properties. For the RF input stage the choice has to be made based on the performance requirements and obviously the used process defines what kinds of transistors are available. In designed mixers, MOS based structures have been chosen for the RF input stage due their better linearity although the noise properties are slightly worse than with bipolar transistors.

II. MIXER INPUT GM-STAGE LINEARIZATION

In basic Gilbert-cell type mixers, where the current through the input stage equals to switch stage, the input stage typically dominates the linearity performance. Thus, the obvious choice to start improving linearity is to concentrate on this stage. Different linearization methods of the Gilbert-cell input stage can roughly be divided into two categories; methods based on purely improving the linearity by parameter optimization, and topologies based on some feedback or feedforward technique such as degeneration or IM3 cancellation techniques.

A. Inductive Degeneration

An inductively degenerated input stage shown in Fig. 2 (a) offers very good linearity and rather good noise performance but at the cost of reduced gain and increased chip area [1]. Compared to resistive degeneration, the degeneration inductor does not cause DC voltage drop leaving adequate voltage headroom for all mixer subparts. The chip area penalty can be alleviated by using mutual inductance for I- and Q-mixers, as shown in Fig. 2 (b) [2]. In the example circuits, by using a shared inductor the total chip area of the I/Q-mixers is decreased approximately by 40% without any significant performance degradation compared to mixers with separate inductors [1].

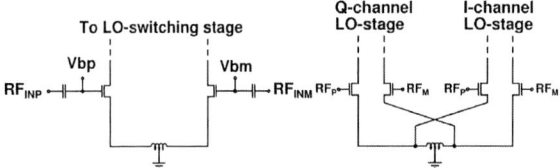

Fig. 2 Simplified schematics (a) of an inductively degenerated RF input gm-stage and (b) input stage using mutual inductor for both I- and Q-channel

An additional drawback of using inductive degeneration is highly frequency dependent performance, which limits the usability of inductive degeneration in wideband applications. The most critical problem is the gain response, which is inversely proportional to frequency. One possible solution to widen the frequency bandwidth is to add shunt resistors in parallel with the degeneration inductance [3]. Thus, the degeneration impedance is reduced leading to a more flat gain response at high frequencies. In the example design the shunt resistors, shown in Fig. 3, are implemented using NMOS-transistors and they can be switched on or off to optimize the performance depending the operation frequency.

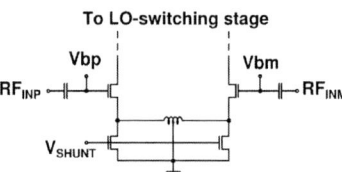

Fig. 3 A simplified schematic of a multibias dual-gate RF input gm-stage

B. Multibias Dual-Input Topology

One of the implemented mixers is based on the principle of IMD reduction using a multibias dual-gate transistor. The simplified schematic of the designed mixer input stage is shown in Fig. 4 [3], [4]. Compared to inductively degenerated gm-stage the chip area of the mixer is 85% smaller. The basic idea is to have separate biases for two differently sized transistors. Thus, when the output currents are summed

together the phase of the fundamental signal is equal and the intermodulation components have equal amplitude, but are in opposite phase.

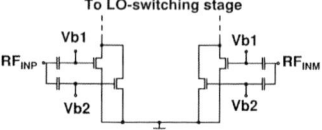

Fig. 4 A simplified schematic of a multibias dual-gate RF input gm-stage

According to the simulations, both the conversion gain and NF are sufficient over a very wide bandwidth. Thus it does not suffer from similar gain drooping as inductively degenerated input stage. In theory a very high linearity improvement can be expected from this topology. However, in example designs when the mixer is connected to LNA as part of the whole receiver, the simulated IIP3 of the whole RF-front-end could not reach the linearity of the receiver with inductive degeneration used in mixers.

III. GM-STAGE CURRENT BOOSTING

As mentioned earlier the input stage IIP3 can be improved by increasing the current through the input transistors. Additional boost current sources give the possibility to increase the current through the g_m-stage while current through the LO switching stage remain constant. The boost current sources are usually implemented with PMOS current mirrors, which have to be sized properly in order to minimize any additional parasitic capacitance and noise.

Despite the given freedom to optimize mixer currents, there is a potential hazard in g_m-boosting. Since the boost transistors need to be small in order minimize the parasitic capacitance, the current tuning range is limited. In addition, it is crucial to ensure that tuning of the boost current and the tuning of the input gm-current have similar tuning range. Otherwise the current through the LO switching stage is altered and therefore, the output common-mode level is affected.

IV. LO SWITCHING STAGE

The IIP3 of the switching stage depends on the current through the switching transistors, as shown in Fig. 5. The simulation was performed using an ideal g_m-stage, which does not affect the linearity and provides DC current for the LO switching stage. In addition, a low-impedance load explained in Chapter V was used to avoid large signals from modulating the V_{CE} of the switching transistors.

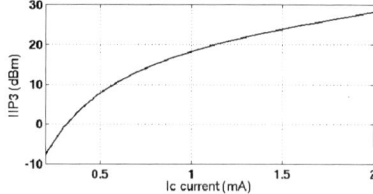

Fig. 5 Simulated IIP3 of the LO switching stage as a function of bias current

978-2-8748-7007-1/08 $25.00 © 2008 EuMA   168

As shown, the current through the switching stage must be high enough in order not to limit mixer linearity and in more general to maximize the overall linearity of the mixers. However, at the same time the current should be kept as small as possible to avoid NF degradation. The linearity of the LO switching stage is sensitive to the parasitic capacitance at the emitters of the switching transistors. Therefore, the layout has to be designed very carefully in order to minimize any additional parasitic capacitance. In addition, smaller transistors are faster and more linear but the cost is increased noise. Therefore the sizing must be optimized for the best performance.

V. MIXER-BASEBAND FILTER INTERFACE

The signal current from LO switches, in Gilbert-cell type mixers, can directly be driven into a baseband filter or it can be converted into voltage form at the mixer output. Depending on whether the output signal of the mixer is current or voltage, it has effects on the performance requirement of the mixer and the following baseband filter. In a voltage-mode operation, the output current from the LO switches is typically converted into voltage using resistive loads at the mixer output. In that case, the output signal modulates the collector-emitter voltage of the mixer switching transistors (Q_1-Q_4 in Fig. 1), which degrades the linearity of the mixer as can be seen from the comparison in Section VII. The amount of the linearity degradation, due to voltage-mode operation, depends on the available supply voltage and the value of the load resistors, i.e. the voltage gain of the mixer. Considering the following baseband filter, the load of the mixer can be simply used as the first filtering stage by adding capacitors in parallel with the load resistors. Since the formed RC pole is passive, it attenuates linearly out-of-band interfering signals before the active devices in the filter. Thus, the passive RC pole at the filter input increases considerably the out-of-band linearity of the baseband filter compared to an active filter implementation without a passive pole at the mixer output [5], [6]. Decreasing the pole frequency is obviously beneficial from the filter out-of-band linearity point of view. However, if the pole frequency is less than or approximately equal to the baseband bandwidth of the wanted signal, the pole attenuates also the high-frequency components of the wanted baseband signal, which increases the noise contribution from the following baseband stages. In an odd-order low-pass filter, it is possible to implement the real pole of the filter prototype as a separate passive RC structure.

In a current-mode mixer-baseband interface, the mixer output current is driven into a low-impedance node. Hence, in an ideal case, there is no voltage signal at the mixer output, which improves the linearity of the mixer. The low-impedance node can simply be implemented by connecting the inputs of an operational amplifier (opamp) to the mixer output [7]. The mixers discussed in this paper have a current-mode output, as shown in Fig. 1. A simplified model of the current-mode interface, where *gm* represents a mixer, is shown in Fig. 6 (a). The linearity of this baseband structure can easily be compared to a corresponding structure with the voltage-mode

input, shown in Fig. 6 (b). For this comparison, the output signal of the amplifier is approximated using a Taylor series

$$V_{out}(t) = k_1 V_{in}(t) + k_3 V_{in}(t)^3, \qquad (1)$$

where the linear gain is k_1 and k_3 is the third-order nonlinear coefficient. The ratio of the third-order nonlinear coefficients of the opamp-R amplifier with a current-mode input (k_3) and with a voltage-mode input (k_3') can be calculated to be

$$\frac{k_3}{k_3'} \approx \frac{1 + (R_2 / R_1)}{1 + (R_2 / R_1')}, \qquad (2)$$

when the linear gains of the circuits are assumed to be equal

$$\frac{k_1}{k_1'} \approx \frac{-gmR_2}{-(R_2 / R_1')} = 1. \qquad (3)$$

In the case of the current-mode input, the third-order nonlinear coefficient (k_3) can be diminished without affecting the linear gain (k_1) of the amplifier by increasing the mixer biasing resistance (R_1) connected between the opamp input and signal ground. Hence, in principle, a higher linearity can be achieved with the current-mode interface. In practice, there may be additional feedback resistors, such as R_3 shown in Fig. 1, which limit the achievable improvement in the filter linearity. In an IC implementation of the voltage-mode interface, the resistor R_1' that follows the passive RC pole at the mixer output must be realized with a voltage-current converter, which easily degrades the overall linearity, especially at low supply voltages. As stated before, the voltage-mode interface typically includes also a passive pole, which on the other hand significantly increases the out-of-band linearity. However, in the designed mixers, the current-mode interface was selected because it significantly increased the mixer linearity although high current consumption was needed in the baseband filter to meet its high linearity requirement.

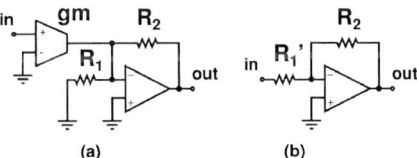

Fig. 6 Opamp-R amplifier with a) current-mode and b) voltage-mode input

VI. IIP2 PERFORMANCE

The IIP2 performance of the mixer is a very important issue in direct-conversion receivers and it can be improved either by some calibration scheme or by using special circuit topologies [8], [9]. For example in [9] a LC-folded cascode has been used to improve both IIP2 and IIP3. Folding increases also the voltage headroom of mixer sub-stages and therefore is beneficial for overall mixer linearity. However, in BiCMOS process good quality pnp transistors are rarely available. Thus PMOS transistors for input g_m-stage must be used. In addition, the folding with resonator effects on the mixer frequency response. The choice between different methods depends on the application and requirements. In narrowband systems,

978-2-8748-7007-1/08 $25.00 © 2008 EuMA

good results have been achieved with circuit topologies, which reduce the second harmonics and therefore improve the IIP2 performance. However, in the designs presented in this paper, IIP2 calibration has been used to investigate the wideband properties of calibration. According the receiver measurements, a +44 dBm IIP2 was achieved within the band of interest [1].

VII. EXPERIMENTAL RESULTS

All the presented mixers are part of fully integrated receivers. Thus, no explicit characterization for individual mixers can be done. However, based on the good matching between measurement and simulations results of the complete receiver, the simulation results of mixer itself should be valid. Table 1 presents collected simulation results of different mixers presented in this paper. All the mixers in Table I have a current consumption of approximately 7 mA, which gives good basis for comparison. In addition, to validate the design choices, the results from voltage-mode interface are included in Table I. In Table I, the design D1 describes a situation, where the DC level of the mixer output is increased to 2V by using 3.3 V supply for the mixer. In design D2, the supply is decreased to the nominal 2.5V supply resulting 1.2V common mode voltage at mixer output. As can be seen, this change alone degrades the linearity of mixer by 9 dBs. In design D3, an inductive degeneration is added to the input stage, which now increases linearity to original value. However, this reduces the gain by 4 dBs. Finally, designs D4-D7 show the results of the different implemented mixers. As can be seen the IIP3 of mixers, with the low impedance load, increases significantly compared to mixers with voltage mode interface. When comparing the implemented designs it can be observed that the sharing of the inductor has insignificant effect on the overall performance. In addition, as seen the IIP3 of multibias mixer is slightly lower compared to other implementations. In Fig. 7 is shown the measured frequency response of the inductively degenerated mixer referred to the multibias mixers response, which is almost flat. Even though the Q-value reduction was implemented for the degeneration inductors, shown in Fig. 4, the mixer with degenerated input stage has at least 7 dBs of variation in the band of 1.6 to 4GHz. Thus, compared to inductively degenerated mixer, the multibias mixer is a tempting alternative when taking also into account the low chip area required by the multibias mixer.

TABLE I
SIMULATION RESULTS OF DIFFERENT MIXER TOPOLOGIES @ 2GHZ

Configuration	Gain (dB)	IIP3 (dBm)	NF (dB)
D1, 1 kΩ load, 2 V CM-level	18.5	+8	N.A.
D2, 1 kΩ load 1.2 V CM level	18.5	-1	N.A.
D3, 1 kΩ load, inductive degen.	14.5	+8	N.A.
D4, Low Ω load, inductive degen. [1]	18.5	+15	12
D5, Low Ω load, mutual degen. [2]	19	+15	12.5
D6, Low Ω load, shunted degen. [3]	17/18	+15	11/12
D7, Low Ω load, multibias [3]	19	+10	10

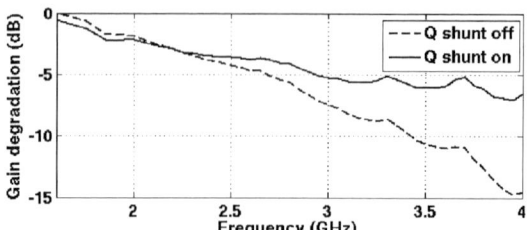

Fig. 7 Measured gain response of inductively degenerated mixer.

VIII. CONCLUSIONS

In this paper four different implementations of a down-conversion mixer for a WCDMA base-station receiver has been presented. All presented mixers have been optimized as part of the receiver and the design choices between the different topologies have been discussed in this paper. In addition, it has been shown that each sub-part in active mixers plays a significant role when targeting to high dynamic range. Furthermore, the effect of mixer-baseband interface to receiver performance has been studied.

ACKNOWLEDGMENT

The authors would like to thank Nokia Networks and Finnish Funding Agency for Technology and Innovation (TEKES) for financial funding. For technical assistance we would like to thank Mikko Hotti.

REFERENCES

[1] J. Ryynänen, M. Hotti, V. Saari, J. Jussila, A. Malinen, L. Sumanen, T. Tikka, K. Halonen, "WCDMA multicarrier receiver for base-station applications", *IEEE Journal of Solid-State Circuits*, vol. 41, no. 7, pp. 1542–1550, July 2006.

[2] T. Tikka, J. Mustola, V. Saari, J. Ryynänen, M. Hotti, J. Jussila, K. Halonen, "A 1-to-4 channel receiver for WCDMA base-station applications", in *IEEE Radio Frequency Integrated Circuits (RFIC) 2006 Symposium Digest of Papers*, pp. 23–26, June 2006.

[3] T. Tikka, J. Ryynänen, K. Halonen, "Multiband receiver for base-station applications", in *IEEE Proc. Radio and Wireless Symposium (RWS)*, pp. 871–874, January 2008.

[4] C.-F. Au-Yeung and K.-K. M. Cheng, "IMD reduction in CMOS double-balanced mixer using multibias dual-gate transistors," *IEEE Trans. Microwave Theory and Tech.*, vol. 54, no. 1, pp. 4-9, January 2006.

[5] T. Hollman, S. Lindfors, T. Salo, M. Länsirinne, K. Halonen, "A 2.7V CMOS dual-mode baseband filter for GSM and WCDMA," in *IEEE Proc. International Symposium on Circuits and Systems (ISCAS)*, pp. 316–319, May 2001.

[6] J. Jussila, J. Ryynänen, K. Kivekäs, L. Sumanen, A. Pärssinen, K. A. I. Halonen, "A 22-mA 3.0-dB NF direct conversion receiver for 3G WCDMA," *IEEE Journal of Solid-State Circuits*, vol. 36, no. 12, pp. 2025–2029, December 2001.

[7] P. M. Stroet, R. Mohindra, S. Hahn, A. Schuur, E. Riou, "A zero-IF single-chip transceiver for up to 22Mb/s QPSK 802.11b wireless LAN," in *IEEE International Solid-State Circuits Conference (ISSCC) Digest of Technical Papers*, pp. 204–205, February 2001.

[8] M. Brandolini, P. Rossi, D. Sanzogni and F. Svelto, "A +78 dBm IIP2 CMOS direct conversion mixer for fully integrated UMTS receiver," *IEEE Journal of Solid-State Circuits*, vol. 39, no. 1, pp. 223-229, January 2004.

[9] P. Sivonen, J. Tervaluoto, N. Mikkola and A. Pärssinen, "A 1.2-V RF front-end with on-chip VCO for PCS 1900 direct conversion receiver in 0.13-μ CMOS," *IEEE Journal of Solid-State Circuits*, vol. 41, no. 2, pp. 384-394, February 2006.

978-2-8748-7007-1/08 $25.00 © 2008 EuMA

Proceedings of the 3rd European Microwave Integrated Circuits Conference

5.5 GHz Low Voltage and High Linearity RF CMOS Mixer Design

Senhg-Feng Lu[1] and Jyh-Chyurn Guo[2]

Institute of Electronics, National Chiao-Tung University, 1001 Ta-Hsueh Rd, Hsinchu, Taiwan, R.O.C.

[1]sflu@seed.net.tw
[2]jcguo@mail.nctu.edu.tw

Abstract—**A CMOS mixer was design with a new circuit scheme to realize low voltage and high linearity simultaneously. A double balanced Gilbert cell was adopted as the basic topology and TSMC 0.18μm 1P6M CMOS process was employed for the on-chip RF circuit fabrication. The proposed new circuit scheme consists of LC-tanks as a capacitively coupled resonator for low voltage and multi-stage parallel RC networks for linearity improvement. Furthermore, multi-gated structure is applied at the RF input as a transconductance amplifier to enhance conversion gain and linearity. The new circuit scheme enables a successful low voltage operation at 1-V for 0.18μm technology. The measured circuit performance demonstrates superior linearity with IIP3 of 11 dBm and P_{1dB} of 2.2 dBm. The conversion gain can be maintained at 8.1 dB in a wide frequencies of 5GHz to 6.8GHz.**

I. INTRODUCTION

Mixer is one of the most important elements in RF front-end constituting the modern communication system. The well known Gilbert cell as shown in Fig.1 is the most generally used mixer architecture due to its advantages in port-to-port isolation and spurious output rejection. However, the traditional Gilbert cell mixers generally suffer a limitation in voltage scaling due to the cascade structure with three stacked transistors in series with a resistor. To realize low voltage operation, capacitively coupled resonating elements, e.g. LC tanks were proposed and demonstrated with the applications in RF circuits [1]. However, poor linearity featured by low IIP3 (-6dBm) appears as a critical weakness and demands a significant improvement. In this paper, a new circuit scheme adopting multi-stage parallel RC networks is proposed and implemented to achieve superior linearity.

II. CIRCUIT DESIGN PRINCIPLE

A 5.5 GHz down-conversion mixer aimed for high linearity at very low voltage to 1-V was designed with a new circuit scheme. The full circuit is composed of eight circuit blocks as shown in Fig.2. They are arranged with three on-chip and five off-chip elements, respectively. The three on-chip elements incorporate a pair of multiple-gate RF amplifiers [2], local switches (LO), and load IF circuits, denoted by the solid-line box. The off-chip circuits represented by dash-line box cover five elements, such as parallel LC-tanks, bypass capacitors, RF baluns, LO baluns, and a measuring circuit. QFN package is adopted to integrate the on-chip and off-chip circuits together. Fig. 3 displays the full circuit schematics in which the circuit topology corresponding to each functional block in Fig.2 is clearly defined.

To realize a low voltage design, capacitively coupled LC tanks first proposed by Manku et al. [1] are employed in this down-conversion mixer design. As shown in Fig.3, a pair of LC tanks are deployed at drain terminal of RF transconductance stage for 1-V supply voltage. One more pair of LC tanks are allocated at the source terminal of LO switches and ended to the ground. Ideally for inductors free from series resistances, the LC tanks can enable a short path at DC and an open path at resonance. In practice, the existence of series resistances in general inductors leads to a very low impedance at DC state whereas a very high impedance under RF operation. In this way, the DC voltage drop across the stacked structure with RF and LO stages in Fig.3 can be reduced to nearly a single stage to minimize the power dissipation. As for RF operation at resonance, the LC-tanks becomes a nearly open path and the bypass capacitor can pass the RF signal across RF and LO stages.

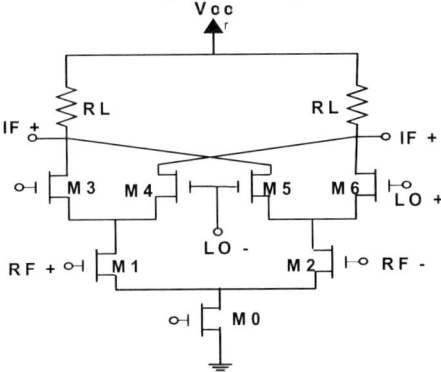

Fig. 1 CMOS Gilbert Cell mixer architecture

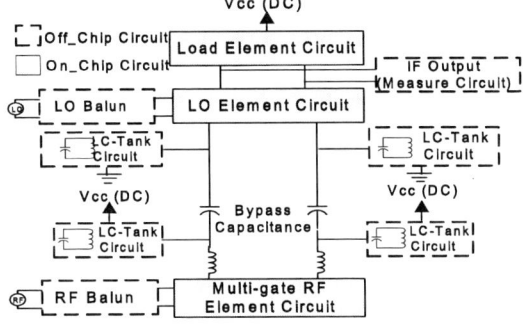

Fig. 2 CMOS RF mixer circuit block diagram

978-2-8748-7007-1/08 $25.00 © 2008 EuMA

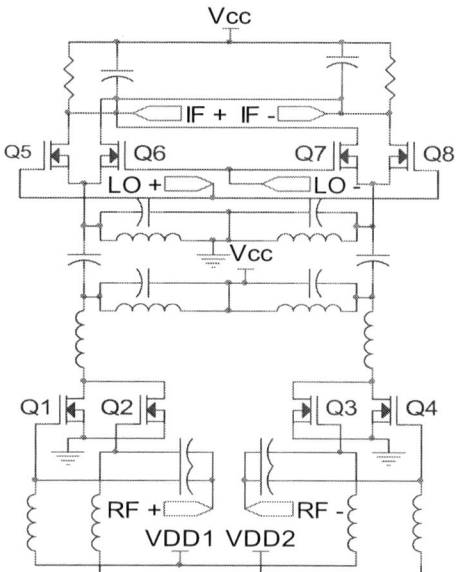

Fig. 3 The circuit schematics of a double balanced RF CMOS mixer with RF stage, LO switches, LO bias, IF load, LC-tanks, bypass capacitor, and baluns

Regarding a major target for high linearity, multi-gate transistors [2] were designed in the RF transconductance amplifiers for verification. The third-order nonlinear term cancellation realized by gate bias tuning on the multi-gated structure can help improve linearity. Besides, a new design with multi-stage parallel RC networks was implemented at the IF output to further enhance linearity. Simulation as shown in Fig.4 indicates that the multi-stage parallel RC networks adopted at IF stage, acting as a high-pass filter can effectively suppress the higher order harmonic components and push out the third order intercept point. Through this mechanism, the linearity defined by IIP3 can be significantly improved. As to design for higher conversion gain (CG), inductive degeneration was implemented by an on-chip inductor at RF output [3]. Note that an inductor of 3.799 nH was used in this design to optimize the output matching and improve CG.

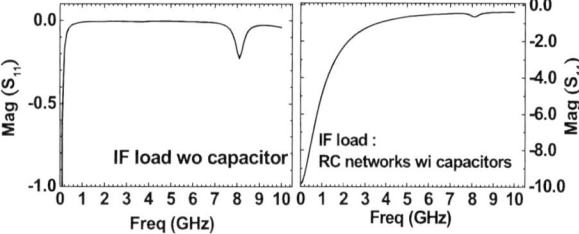

Fig 4 (a) IF stage with simple resistor network without capacitor (b) IF stage with parallel RC networks – the parallerl RC serves as a high-pass filter to effectively filter out higher order harmonics and improve lineariy

Eventually, the full circuit has to be mounted on PCB for measurement. The critical points to be considered are the parasitic inductance and resistance existing with the bond wires and the parasitic capacitance originated from the bonding pads. Fig. 5 presents the bond-wire package model. The series inductance is approximately 1nH/mm. For a bond wire with a length generally exceeding that of the on-chip inductor, it will impose a significant influence on the circuit performance. Due to the fact, the package model of bond-wire must be taken into account in the full circuit simulation. Fig. 6 illustrates the chip layout of the double balanced CMOS mixer in this design and Fig.7 specifies the pin assignment for bonding pads on board .

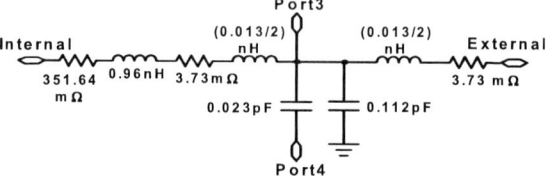

Fig. 5 A package model for the bonding wires and pads

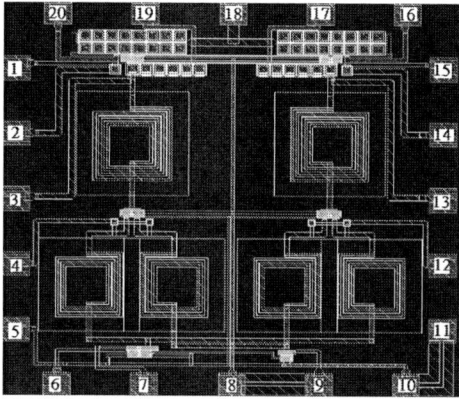

Fig. 6 Chip layout of a double balanced Gilbert mixer

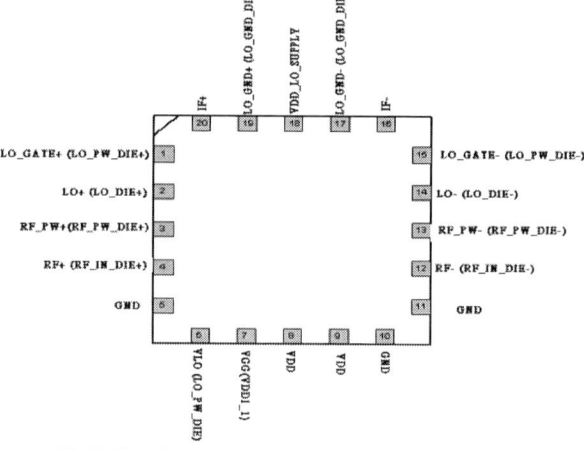

Fig. 7 Pin assignment for mixer chip on board bonding pads

III. RESULTS AND DISCUSSION

A. Simulation Results

0.18um MOSFET model was employed for on-Si-chip circuit simulation and SPIL QFN package model was integrated with core MOSFET model for a full circuit simulation. Fig.8(a) and (b) present the CG, P_{1dB}, and IIP3 under varying RF power, simulated using SS corner model, under 1-V and 25°C operation. Table 1 summarizes the full circuit simulation results using typical and corner models (TT, SS, FF) for MOSFETs as well as QFN package model for bonding wires and pads. The operation condition for the low voltage mixer is a supply voltage at 1-V and temperature at 25°C. The key performance parameters include conversion gain (CG), gain compression (P_{1dB}), the third-order intercept point (IIP3), and power consumption.

The simulation predicted a good performance of high CG, high linearity, and low power consumption. The power consumption at 1-V can be push to 2~3.7mW. The high linearity is featured by IIP3 of 8.2~12 dBm and 1dB output power compression (OP_{1dB}) of 4.89~6.22 dBm. CG can achieve 18.7~24.2. The high linearity is realized through the multi-stage RC networks at IF load.

TABLE I
Simulation results for a mixer on QFN chip operating under 1.0V and 25 °C
(TT : Typical, SS : Slow, FF : Fast)

Performance Parameters	SS	TT	FF
Conversion Gain (dB)	18.7	24.2	19.8
OP_{1dB} (dBm)	6.22	5.8	4.89
IIP3 (dBm)	12	10	8.2
Power Consumption (mW)	2.03	2.74	3.71
RF/LO : 5.5/5.499GHz, LO power : 2.5 dBm			

B. Measurement Results

The measured P_{1dB} and IIP3 are demonstrated in Fig. 9(a) and (b). The superior linearity with IIP3 of 11 dBm and P_{1dB} of 2.2 dBm proves the success of harmonic suppression through the high-pass filter realized by multi-stage parallel RC networks at IF load. IIP3 as high as 11 dBm at LO power of 2.5dBm matches very well with the simulation result of 12 dBm shown in Table 1. Fig. 10 (a) and (b) present the measured CG vs. input power and RF frequency in which CG above 8.1 dB can be maintained over a wide range of input power to 0dBm and frequencies in 5~6.8GHz. The degradation compared with simulation suggests a deviation from the desired optimal matching incorporating the circuit elements on PCB, such as LC tanks and balun. Improvement can be achieved by an extensive calibration on the package model and that for on-board circuit elements. Table 2 summarizes the measured performance parameters, such as CG, P_{1dB}, IIP3, and power consumption. The measured power consumption appears higher than simulation prediction. Process variation induced drift in resistances and IR drop is considered a potential reason responsible for the increased DC power. Fortunately, the power consumption keeps low at around 4 mW attributed to the sufficiently low voltage to 1.0V.

TABLE 2
Measurement Results under supply voltage at 1.0V, Temp = 25°C

Performance Parameters	Measured Results
Conversion Gain	8.1 dB @0dBm
P_{1dB}	2.2 dBm
IIP3	11 dBm
Power Consumption	4.1mW
RF/LO : 5.501/5.5GHz, LO power : 2.5 dBm	

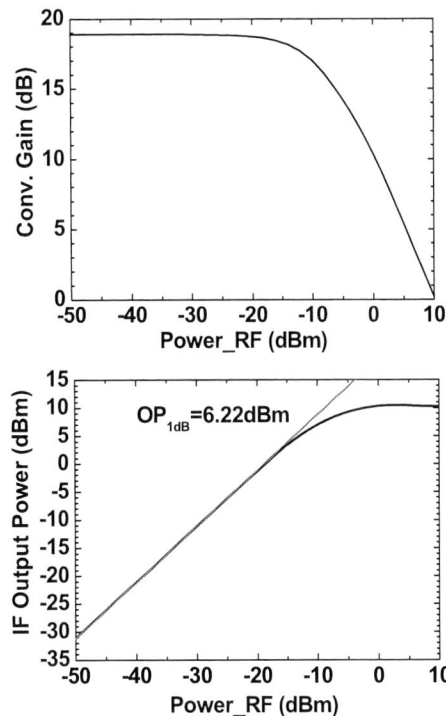

Fig. 8 (a) Simulated conversion gain and IIP3 under 1.0V and 25°, SS corner model was used for simulation

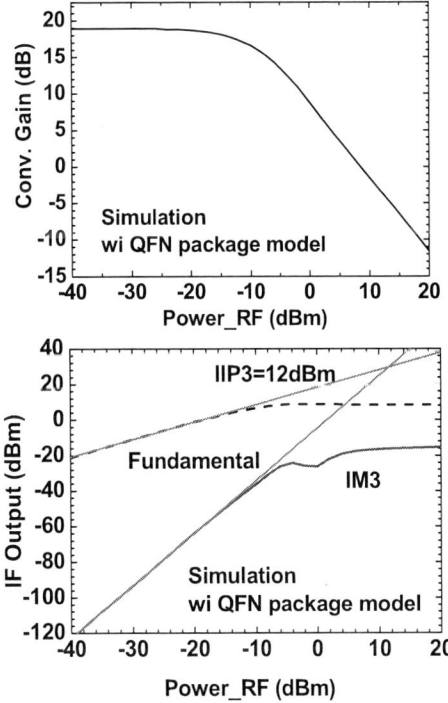

Fig. 8 (b) Simulated conversion gain and IIP3 under 1.0V and 25°C. SS corner model for MOSFET and QFN package model for bonding wires and pads were adopted for the full circuit simulation.

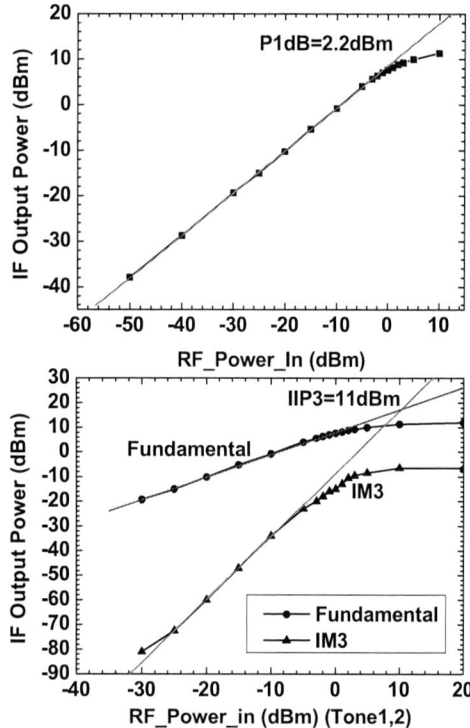

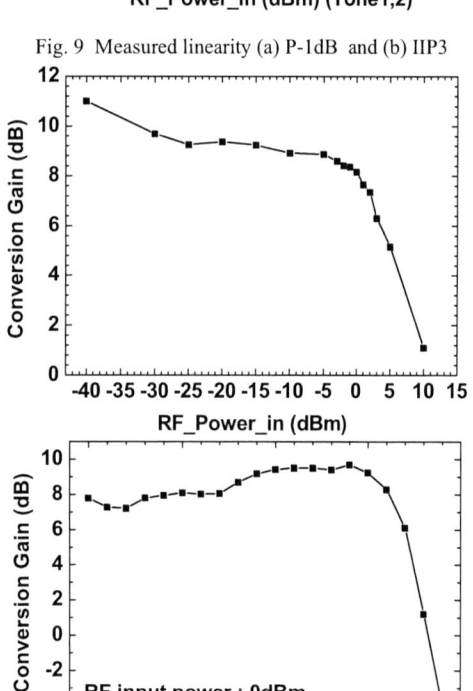

Fig. 9 Measured linearity (a) P-1dB and (b) IIP3

Fig. 10 (a) Measured conversion gain vs. RF input power
(b) Measured conversion gain vs. RF input frequency

IV. CONCLUSION

A 5.5 GHz down-conversion mixer has been fabricated in 0.18 um RF CMOS technology. The new circuit scheme enables a successful low voltage operation at 1.0V and low power consumption of around 4mW. This low voltage mixer demonstrates superior linearity with IIP3 of 11 dBm and P_{1dB} of 2.2 dBm. The conversion gain can be maintained at 8.1 dB over a broadband operation in 5GHz to 6.8 GHz.

The superior linearity in terms of high IIP3 and P_{1dB} proves the advantage realized by mutli-stage parallel RC networks at IF output. The success of low voltage operation validates a new design adopting LC-tanks at RF and LO stages. The major challenge remained with this work is a certain deviation between the whole chip simulation and measurement result. It is because that simulation accuracy is acceptable for on chip balun and LC tank design; however, a fully qualified simulation tool is lacking for balun and LC tank design, which is on PCB through the method of SMD. The PCB layout can be improved by considering the characteristic wavelength of microstrip lines. Regarding balun circuit design, a replacement of conventional design using passive components by active components [4,5] can further improve the circuit performance and reduce the chip area.

ACKNOWLEDGMENT

This work was supported in part by NSC under Grants NSC95-2221-E009-289 and NSC96-2221-E009-186. Also, the authors acknowledge the support from NDL CiC for test chip fabrication and RF Lab. for measurement.

REFERENCES

[1] T. Manku, G. Beck, and E.J. Shin, "A low-voltage design technique for RF integrated circuits," *IEEE Trans. Circuits and Systems II*, vol. 45, pp. 1408-1413, Oct. 1998.

[2] B. Kim, J. Ko, and K. Lee, "A New Linearization Technique for MOSFET RF Amplifier Using Multiple Gated Transistors," *IEEE Microwave and Guide Wave Letters*, Vol. 10, No.9, pp.371-373, 2000.

[3] B. Razzavi, RF Microelectronics, New Jersey, Prentice-Hall, 1998.

[4] I. J. Lin, C. Zelley. *0.* Boric-Lubecke, P. Goddl and R. Yan, "A silicon MMIC Active "A silicon MMIC active balun/buffer amplifier with high linearity and low residual phase noise," in 2000 *IEEE MTT-S Intemational Microwave Symposium Din.,* vol. 3. pp.1289-1292

[5] H. Koizumi, S. Nagata, K. Tateokq K. Kanazawal and D. Ueda, "A GaAs Single balanced mixer MMIC with built-in active balun for personal communication systems," in *IEEE Microwave and MiIlimeter- Wave Monolithic Circuifs Symposium,* May.15-16 1995 pp. 77-80.

978-2-8748-7007-1/08 $25.00 © 2008 EuMA

High Linearity Down-Conversion CMOS Mixers

Danilo Manstretta

Dipartimento di Elettronica, Università degli Studi di Pavia

via Ferrata 1, 27100, Pavia, Italy

danilo.manstretta@unipv.it

Abstract— **This paper gives a quantitative analysis of the main mechanisms setting fundamental limits to the linearity performances of CMOS direct down-conversion mixers. An advanced low voltage solution is proposed for 3G cell-phones in a 90nm CMOS technology that achieves: 3nV/√Hz average input referred noise in the band from 10kHz to 1.92MHz, a flicker noise corner of 300kHz, 9dBm IIP3 and 75dBm minimum IIP2 while drawing 5.4mA from a 1.2V supply.**

I. INTRODUCTION

As the quest for an all-purpose software-defined radio continues, it becomes more and more evident that the linearity performances that the RF front-end must satisfy are extremely demanding [1]. Furthermore, compatibility with deep submicron CMOS processes requires analog and RF circuits to work at ever lower supply voltages. Several advanced techniques are being studied to improve on the linearity of the RF front-end, both at the circuit and the architecture level. Only combining advanced circuit design together with advanced architecture solutions very high performance, low cost, flexible receivers may one day become possible.

In this paper we concentrate on the optimal performance that can be achieved in a CMOS direct down-conversion mixer. Direct conversion wireless receivers lend themselves to highly integrated low-cost solutions, minimizing the number of external components required. The path toward single-chip CMOS solutions has been successfully followed in several applications, from Bluetooth to WLAN. On the other hand, commercial solutions for more demanding applications such as 3G cell-phones still require external RF filters to meet the stringent linearity requirements. In wireless receivers the LNA usually dominates the receiver noise, while the down-conversion mixer usually represents the bottleneck in terms of linearity to noise ratio (spurious-free dynamic range). In direct-conversion CMOS solutions, the mixer must also meet stringent flicker noise and IIP2 requirements. In this paper we tackle the issue of high linearity down-conversion mixers exploring the fundamental mechanisms that limit the achievable performances of the basic active current-switching mixer. An advanced topology that is suitable for low-voltage operation is proposed that can meet the stringent UMTS requirements.

II. NOISE AND DISTORTION IN ACTIVE MIXERS

Noise and intermodulation distortion considerations will be carried out referring to the CMOS current commutating mixer reported in Fig. 1. In order to better understand the performance limitations of this circuit, in terms of noise and third order distortion, we will consider separately its three main sections: the transconductance stage at the RF input, the current commutating stage driven by the local oscillator (LO) signal and the IF output load. Second order distortion limitations will be dealt with separately.

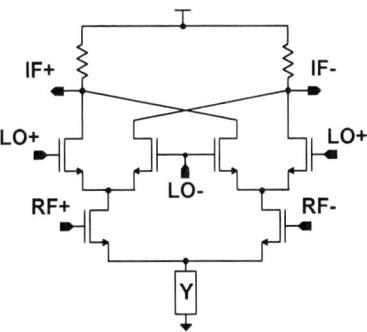

Fig. 1 Conventional CMOS current commutating mixer.

A. Transconductance stage and IF load

The input transconductor is usually implemented using a MOS differential pair. The input-referred thermal noise power spectral density is given by $8KT\gamma/g_m$, while flicker noise is up-converted at the mixer output and falls out-of-band. Input-referred third-order intercept point as a function of the overdrive voltage for an NMOS pair in a 90nm CMOS technology is reported in Fig. 2. The pseudo-differential pair shows 3-4dB improvement in IIP3 compared to fully-differential for the same noise. Since device transconductance is dictated by noise considerations, high overdrive voltages can give very good linearity but at the price of very large power consumption. For a 20mS transconductance and 20dBm IIP3, 12mA would be required in a pseudo-differential transconductor. The load is usually implemented using poly-silicon resistors that exhibit high linearity and low flicker noise.

B. Current-commutating stage

The current commutating stage represents the core of the mixer: it is here that frequency translation is carried out. The two cross-coupled pairs are operated in saturation and work as current buffers, providing low impedance to the input transconductor and ensuring isolation between IF and RF terminals. When driven by a large LO signal the transfer function is well approximated by a square wave toggling between +1 and -1. This approximation allows to determine with reasonable precision the down-conversion gain and the noise transfer functions [2]. The ratio between the output current at ω_{IF} and the input current at ω_{RF} ($\omega_{LO}\pm\omega_{IF}$) is approximately $2/\pi$ or -4dB. The output current noise at ω_{IF} is given by the superposition of the transconductor noise at $(2n+1)\omega_{LO}\pm\omega_{IF}$, resulting into an intrinsic thermal noise figure degradation of about 4dB. On the other hand, flicker noise from the transconductor is up-converted around $(2n+1)\omega_{LO}$ and can therefore be neglected.

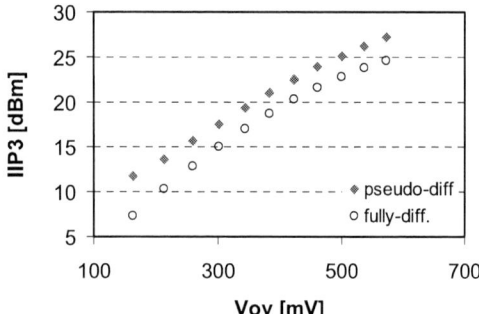

Fig. 2 Differential transconductors IIP3.

Noise and distortion are added by the transistors making up the switching pairs through two mechanisms:

1) *Finite Switching Time:* If an LO signal with a finite slope is used, a finite time T_{SW} is required to commutate the bias current of the switching pairs. When all transistors of the switching pairs are active at the same time, their noise is fully transferred to the output. The total thermal noise power in the output current increases linearly with the switching time T_{SW}. Flicker noise in the switching pair devices is also transferred to the output with a gain proportional to T_{SW}. This noise transfer mechanism is also referred to as *direct mechanism* [2]. During the switching time T_{SW}, third order intermodulation distortion is generated in the switching pair since current partition between two non-linear devices results in a non-linear transfer function [3]. IIP3 for an NMOS switching pair using minimum channel length devices in a 90nm CMOS technology and several device sizes has been simulated at 20MHz for a bias current of 2mA, a linear 20mS transconductor and a sinusoidal LO. Results are reported in Fig. 3. IIP3 is relatively high can be improved increasing the LO amplitude and devices size, therefore reducing T_{SW}. This limitation is often overlooked since other mechanisms are usually more important at high frequencies.

2) *Finite bandwidth*: Flicker noise performances can be improved using non-minimum channel length devices and a low bias current density [4]. This approach can lower T_{SW} but also results in a sub-optimum device f_T. As the device f_T moves closer to the operating frequency, bandwidth limitations at the switching pair common source nodes introduce an additional flicker noise transfer mechanism, commonly referred to as *indirect mechanism* [2] and

especially important in direct down-conversion mixers having stringent flicker noise requirements. Bandwidth limitations are also the main cause of third-order intermodulation distortion at RF [3]. Third-degree nonlinearity in the switching pair devices induce a current in the common source capacitance which appears, after down-conversion, at the output. In Fig. 3.b the IIP3 for a switching frequency of 2GHz is reported as a function of the LO amplitude for different device sizes, with and without a 20fF parasitic capacitance at the common-source. For a given device current density and LO amplitude (hence switching time), IIP3 is proportional to the bias current and is inversely proportional to the total common-source capacitance. At RF, the common-source capacitance, driven by the large rectified sinusoid proportional to the LO amplitude, limits the achievable IIP3. This introduces an important consideration for the design of the RF transconductor, whose output capacitance can seriously degrade the switching pairs IIP3.

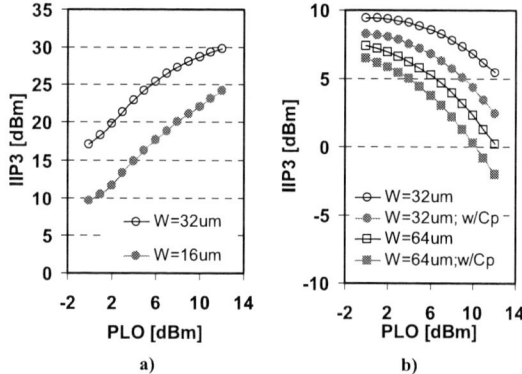

Fig. 3 Switching pairs IIP3 at different frequencies: a) 20MHz and b) 2GHz.

C. Second-order intermodulation distortion

The main sources of second order intermodulation distortion in mixers are self-mixing, transconductor nonlinearity accompanied by mismatch in the switching pairs and/or load resistors, and switching pairs nonlinearity and mismatches. Coupling of the RF signal into the local oscillator terminals can lead to second order distortion due to direct multiplication (self-mixing). The resulting mixer IIP2 is equal to the LO peak-to-peak amplitude driving the switching pair gates, multiplied by the RF-to-LO isolation [5] and is largely independent of other design details. IIP2 can be improved using large LO amplitudes and reducing RF-to-LO coupling. A sound layout practice is to prevent metal lines carrying RF and LO signals from crossing each other, or, if necessary, keep crossing orthogonal and carefully balance the parasitic capacitances between RF and LO lines. Using these simple precautions IIP2 in excess of 78dBm [6] has been demonstrated.

Low-frequency second-order intermodulation distortion products are generated in the RF transconductor. A pseudo-differential transconductor has better IIP3 performances compared to fully differential but generates large common-mode second-order intermodulation distortion products. Assuming perfectly matched devices, a fully differential transconductor does not produce any second-order intermodulation component. In fact, common mode components are not transmitted at pair output if the current source has zero conductance ($Y(0)=0$ in Fig. 1), while differential components are not generated if devices are matched. Even assuming mismatches and a real current source, with typical biasing currents and device dimensions, fully differential topologies have an IIP2 that is roughly 25-30 dB higher compared to pseudo-differential [5].

In a perfectly matched switching stage differential components of the low-frequency second-order intermodulation products generated by the transconductor would be fully up-converted at mixer output, while common-mode components would remain common-mode. Due to mismatches between devices, a finite differential DC-to-DC gain and a common-mode-to-differential conversion result [5]. Two distinct mechanisms have been identified: at low frequencies, the dominant mechanism is duty-cycle distortion of the time-varying current transfer function; at RF frequencies, bandwidth limitations determine an additional mechanism. Duty-cycle distortion is proportional to the ratio between V_{OFF} and the LO amplitude. With an LO amplitude of 600mV 0-pk, 2mA bias current and typical device sizes, a gain of about –50dB results. Figure 4 shows an equivalent representation of a single switching pair, with mismatches represented by an equivalent offset voltage V_{OFF} in series with the gate, driven by a large sinusoidal RF signal: a single source follower is driven by a rectified sinusoid superimposed to a square wave toggling between 0 and V_{OFF}, capturing the effect of device mismatches during commutations [4].

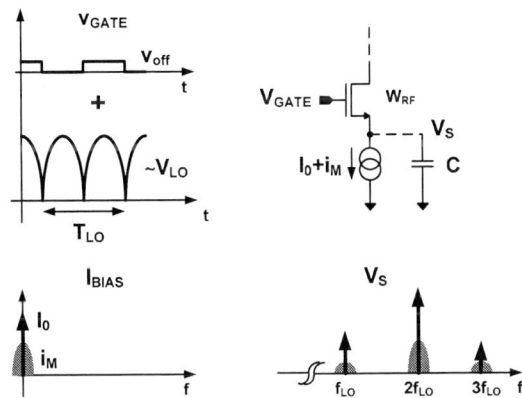

Fig. 4 Switching stage equivalent model for the analysis of second-order intermodulation distortion at high frequencies.

The source voltage is a low-pass filtered version of the applied waveforms and its spectrum presents even and odd harmonics of the local oscillator. The low-frequency intermodulation current i_M generated by the transconductor is superimposed to the biasing current and modulates the low-pass time constant. Sidebands around f_{LO} frequency and its

harmonics arise in the source voltage, inducing a current in the tail C which is then down-converted at mixer output by the switching pair commutation. The resulting IIP2 is reported in Fig. 5.a for 2mA bias current when driven by a 0.6V amplitude sinusoidal LO at 2GHz. Even at high overdrive voltages the achievable IIP2 is not adequate for most the stringent standards. Common mode components are also converted into differential by mismatches in the load resistors. The use of highly linear poly-silicon resistors, with carefully inter-digitized layout, maintains mismatches below 0.1% making this a non dominant contribution.

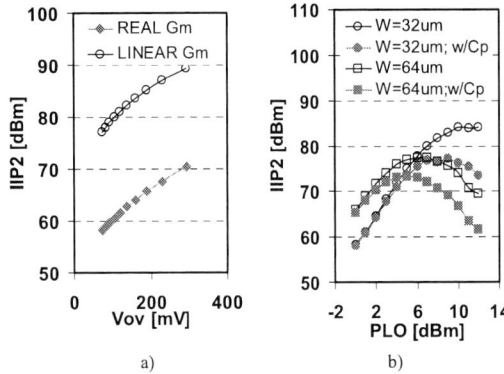

Fig. 5 Switching pair IIP2 simulations in 90nm CMOS: a) IIP2 at 6dBm LO with linear and pseudo-differential transconductor; b) IIP2 as a function of LO amplitude for different device size with and without a 20fF parasitic tail capacitance.

An intrinsic limit to switching stage IIP2 is due to non-linearity and mismatches in the switching stage. The parasitic capacitor, loading the source switching pair, is at the origin of the stage inter-modulation distortion at RF. Approximate expressions for IIP2 due to the described mechanism can be found in [4]. For a given current density, IIP2 increases more than proportionally with the bias current (smaller mismatches are associated with larger devices) and decreases as the frequency of operation is increased due to the lower impedance of the source capacitance. The results of extensive simulations carried out on a mixer implemented in a 90nm CMOS technology using a fixed bias current of 2mA and 1.2V supply are shown in Fig. 5. Minimum channel length devices are used to minimize parasitic capacitances. In Fig. 5.a a pseudo-differential transconductor is used and the resulting IIP2 as a function of the overdrive voltage is compared to the intrinsic IIP2 of the switching pairs. Notice that the non-linearity in the transconductor dominate the non-linearity of the switching pairs by about 20dB. Higher IIP2 values at high overdrive voltages are associated with relatively low transconductance values (ranging from 10mS to 4mS in Fig. 5.a), hence meeting typical noise constraints would lead to high current consumption. Furthermore, in a direct down-conversion mixer, flicker noise requirements usually dictate the use of non-minimum channel length devices with associated larger tail node capacitances. In Fig. 5.b the IIP2 of the switching pairs driven by a linear 20mS transconductor is shown for different device sizes. The strong dependence on the LO amplitude is explained by the fact that several contributions, arising from non-linear up-conversion around LO harmonics, are down-converted by the switching pair action with different amplitudes and phases depending on the LO amplitude. At the optimal drive level, a smaller device, with lower associated capacitances, results in a higher IIP2. However if a fixed 20fF parasitic capacitance is added at the common source, IIP2 degradation is more pronounced for smaller devices. Minimization of the capacitive loading at the sensitive common source node is key to achieve high IIP2 and low noise.

III. HIGH LINEARITY DIRECT DOWN-CONVERSION MIXER

Compatibility with scaled CMOS technology makes operation at low supply voltage a key requirement. Designing high dynamic range CMOS direct down-conversion mixers at low voltage with reasonable power consumption proves quite challenging. High IIP3 can be achieved at low voltage using a pseudo-differential transconductor. On the other hand, as shown in the previous section, common-mode low-frequency second order distortion products generated by the pseudo-differential transconductor severely limit the achievable IIP2. Preventing injection of the common-mode intermodulation components into the switching pairs would eliminate the input transconductor contribution to IIP2, minimizing the common-mode to differential conversion caused by mismatches in the switching pairs and in the load resistors. AC coupling between transconductor and switching pairs is a possible solution but comes at the price of an almost double power consumption. An alternative solution is proposed in Fig. 6.

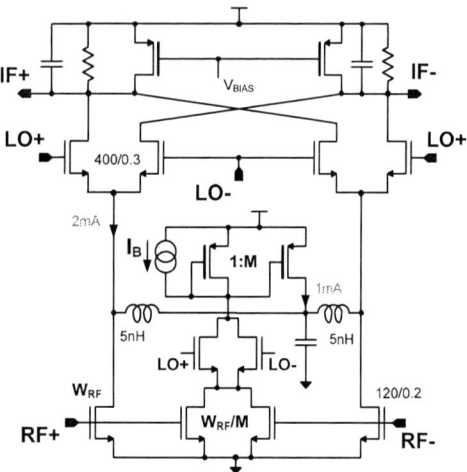

Fig. 6 High linearity CMOS mixer with IM2 cancellation

A scaled-down replica of the pseudo-differential transconductor is used to generate a common mode second-order distortion product which is then amplified in the current domain using a PMOS current mirror and injected at the common sources of the switching pairs, cancelling the

contribution arising from the transconductor. A 10nH center-tapped differential inductor is used to tune-out the switching pair common-source capacitance, improving 1/f noise and intrinsic IIP2 and IIP3 performances of the switching pairs [6]. Injecting the second-order distortion cancellation current at the center-tap of the inductor avoids any noise penalty.

The circuit was optimized in a 90nm CMOS technology at 1.2V supply to satisfy UMTS requirements with sizeable noise margin. The proposed circuit can work at lower supply voltages compared to [6] and avoids bandwidth limitations in the second-order distortion suppression loop inherent to feedback implementations [7]. A summary of the circuit performances is reported in Table I.

TABLE I
MIXER PERFORMANCE SUMMARY

Down-Conversion gain	10dB
Input Referred Noise (Averaged 10kHz - 1.92MHz)	3 nV/√Hz
1/f noise corner	300 kHz
IIP3	9 dBm
Minimum IIP2	75 dBm
Current Consumption	5.4mA
Voltage Supply	1.2V

IV. CONCLUSIONS

In-depth analysis of the fundamental mechanisms underlying noise and intermodulation distortion has enabled the design of a high dynamic range low-voltage CMOS direct down-conversion mixer. The need for an efficient, wideband, high performance solution still poses an exciting challenge.

ACKNOWLEDGMENT

This work has been carried out in the framework of the Italian national research program FIRB (contract nr. RBIP063L4L).

REFERENCES

[1] A. A. Abidi, "The Path to the Software-Defined Radio Receiver", *IEEE Journal of Solid-State Circuits*, vol. 42, no. 5, May 2007, pp. 954-966

[2] H. Darabi, A. Abidi, "Noise in RF-CMOS Mixers: A Simple Physical Model", *IEEE Journal of Solid State Circuits*, vol. 35, pp. 15-25, Jan. 2000

[3] M. Terrovitis and R. G. Meyer, "Intermodulation distortion in current commutating CMOS mixers," *IEEE J. Solid-State Circuits*, vol. 35, no. 10, pp. 1461–1473, Oct. 2000.

[4] D. Manstretta, R. Castello, and F. Svelto, "Low 1/f noise CMOS active mixers for direct conversion," *IEEE Trans. Circuits Syst. II*, vol. 48, pp.846–850, Sept. 2001.

[5] D. Manstretta, M. Brandolini, F. Svelto, "Second-order inter-modulation mechanisms in CMOS downconverters," IEEE Journal of Solid-State Circuits, vol.38, no.3, pp. 394- 406, Mar 2003

[6] M. Brandolini, P. Rossi, D. Sanzogni, and F. Svelto, "A +78dBm IIP2 CMOS Direct Downconversion Mixer for Fully Integrated UMTS Receivers", *IEEE Journal of Solid State Circuits*, vol. 41, pp. 552-559, Mar. 2006

[7] M. Brandolini, M. Sosio, and F. Svelto, "A 750 mV, 15 kHz 1/f noise corner, 51-dBm IIP2, direct-conversion front-end for GSM in 90-nm CMOS," in *Proc. IEEE Int. Solid-State Circuits Conf.*, Feb. 2006, pp.374–376.

Advanced Modeling of MISHFET Devices and their Performance in Current-Mode Class-D Power Amplifiers

J. Cumana, C. Lautensack, M. Eickelkamp, J. Goliasch, A. Noculak, A. Vescan and R. H. Jansen

Chair of Electromagnetic Theory (ITHE)

RWTH Aachen University

Kopernikusstr. 16, D-52074 Aachen, Germany

jansen@ithe.rwth-aachen.de

Abstract—**GaN HFETs and MISHFETs are promising power devices for RF and microwave power applications. However, the performance of devices can be compromised under some operating conditions. From the device development point of view, device optimization is necessary to obtain the best possible performance. For device modeling and design purposes, the device needs to be characterized and modeled accurately in order to foresee how the device will behave under realistic operating conditions.**

In this paper, an improved EEHEMT1-based model for GaN MISHFETs, will be introduced. This model is capable of describing the knee region of the device's output characteristics, dispersion effects as well as gate diode behavior accurately. The models will be incorporated in a switched-mode amplifier topology and evaluations will be made to determine the suitability of MISHFETs in these amplifiers.

I. INTRODUCTION

Today a number of impressive GaN HFET transistors with output power as high as 180W are commercially available. Making use of their excellent properties like high electron mobility of $1500\,\frac{cm}{Vs}$ and very high breakdown field of $2.7 \cdot 10^6\,\frac{V}{cm}$, these devices are able to deliver high RF power in communication bands like WiMAX and LTE [1]. Their large bandwidth and high operation voltage in addition to higher impedances along with their high frequency operation capability [2] will allow them to compete against well-established technologies.

Further developements on these devices were made by isolating the gate electrode leading to Metal-Insulator-Semiconductor-HFETs (MISHFETs) [3]. These are designed to further increase performance over regular GaN HFETs in terms of drain current, linearity and gate leakage current as first presented in [4]. Single chip amplifers using this kind of devices delivering over 100 W at 2.14 GHz [5] and MISHFETs for frequencies up to $f_t = 180\,GHz$ have been reported [6]. The technology sets new demands on modeling and circuit design. Many reports in technology aspects are available [7], only few adress modeling and application to new circuit topologies [9].

In this paper, the HFET-model used at ITHE will be applied to MISHFETs and the accuracy will be evaluated. In addition

enhancements will be presented to improve the accuracy of MISHFET models. Section II of this paper will therefore address modeling issues of GaN HFETs and MISHFETs taking into account additional effects as knee walkout and dispersion effects. A way of describing the gate diode characteristics and breakdown behavior of MISHFETs is also presented. In Section III a highly efficient Class-D topology incorporating GaN MISHFETs and HFETs will be compared. Advantages of MISHFETs used in Switched-mode amplifers will be pointed out.

II. DEVICE MODELING

A. Basic MISHFET-Model

The MISHFETs and HFETs investigated in this work were fabricated at the ITHE GaN-Lab. All devices had the dimensions $w_G = 2x250\,\mu m$, $l_{GS} = 1.5\,\mu m$, $l_G = 1\,\mu m$, $l_{GD} = 7.5\,\mu m$. The Si_xN_y-dielectric of the MISHFETs was grown to a thickness of $6.4\,nm$. As this paper focusses on modeling strategies and switch mode performance of the devices, a more detailed description of the general MISHFET characteristics (e.g. g_m, f_T, f_{max}) and their production processes will not be given at this point but will be published elsewhere [11].

The model used at the ITHE for describing HFETs is implemented in Agilent's Advanced Design System (ADS) and is based upon a SDD-replication of Agilent's EEHEMT1 model [10]. Compared to the lumped EEHEMT1 model included in ADS, SDD models allow modification of the model equations if necessary. However, such modifications have only been made if they lead to a significant improvement of the model quality as in case of the optimized description of the knee region of HFETs presented earlier [12]. Basis of the extraction of the model parameters are pulsed S-parameter measurements with $f = 1 - 20$ GHz and the quiescent voltages $V_{GSQ} = V_{DSQ} = 0V$. From these S-parameters a linear model for every bias point was generated using the tool MDL-GRED/LINMIC of CST[13]. Afterwards optimized multibias parameters of the nonlinear EEHEMT1 model were extracted. Based upon these parameters the enhanced ITHE model was

978-2-8748-7007-1/08 $25.00 © 2008 EuMA

created including improved descriptions of MHz- and GHz-dispersion effects, knee walkout, gatelag and drainlag, thermal effects, gate source diode and breakdown.

The same approach was for the first time used for modeling MISHFETs and checked for compatibility issues. When extracting the parameters of the linear single bias models and the basic multibias EEHEMT1 model, both descriptions showed to be well suited for modeling MISHFETs. The observed DC and RF error values were comparable to the HFETs model quality. Also MHz- and GHz-dispersion frequencies were measured to be in the same order of magnitude and could be described with the existing approach. The optimization of the curve shape of the drain current in the knee region used for modeling HFETs also showed to be well suited for describing MISHFETs as depicted in figure 1. This seems to be even more important for MISHFETs because we assume the dependence of knee voltage on V_{GS} originates from dislocations or traps located at the surface of the device. Due to the additional dielectric layer and the not yet optimized passivation processes their ammount should be larger for MISHFETs.

Though most of the HFET modeling approach showed to be well suited for describing MISHFETs, the gate source diode and breakdown characteristics deserve further attention.

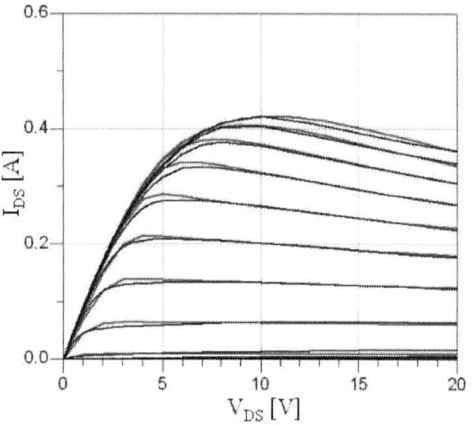

Figure 1. Simulated (red) and measured (black) IV-characteristics of the investigated MISHFET

B. Diode Characteristics

The MISHFET's gate-source diode characteristic differs from usual HFET gate-source diodes with Schottky characteristic. When applying positive gate-source voltages smaller than $V_{GS} = 2V$ to the devices investigated the gate current I_{GS} does not depend on V_{GS} exponentially but linearly. The reason for this is the ohmic behavior of the gate isolation layer. At $V_{GS} > 2V$, which is higher than the usual operation condition, the gate current starts increasing exponentially.

This behavior cannot be described by the diode description included in Agilent's EEHEMT1 model because Schottky behavior is assumed. Therefore a simple but effective approach

used and validated at the ITHE, was applied. By adding a large parallel resistor to the gate source diode, the device characteristics for $V_{GS} > 0V$ can be described very accurately as shown in figure 2.

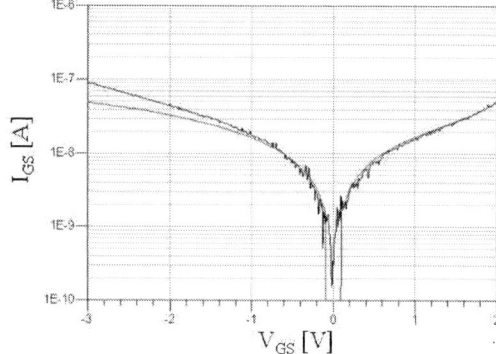

Figure 2. Simulated (red) and measured (black) diode-characteristic of the investigated MISHFET

C. Breakdown Characteristics

When increasing the drain-gate voltage of a GaN-HFET, at a certain voltage V_{br} the drain-gate current I_{DG} starts increasing exponentially, which can damage the device. This well-known effect is called gate-drain breakdown, with V_{br} being the breakdown voltage. For modeling of the gate-drain breakdown, V_{br} is defined as the gate-drain voltage, at which I_{DG} exceeds a predefined threshold (e.g. 1 mA/mm). As mentioned earlier, the investigated MISHFETs show a gate leakage current several orders of magnitude lower than comparable HFETs due to their enhanced gate isolation. Therefore, when performing breakdown measurements, I_{DG} of the MISHFET does not reach the threshold used for HFETs without the gate isolation being damaged.

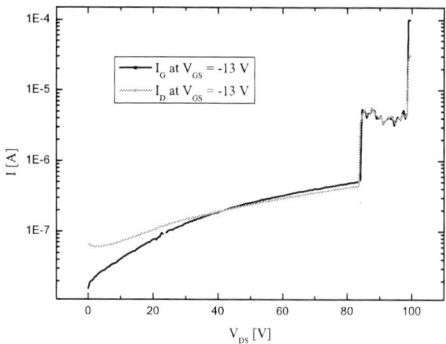

Figure 3. Measured breakdown characteristic of the MISHFET

Figure 3 shows the result of a breakdown measurement during which the MISHFET dielectric was damaged. The

978-2-8748-7007-1/08 $25.00 © 2008 EuMA   180

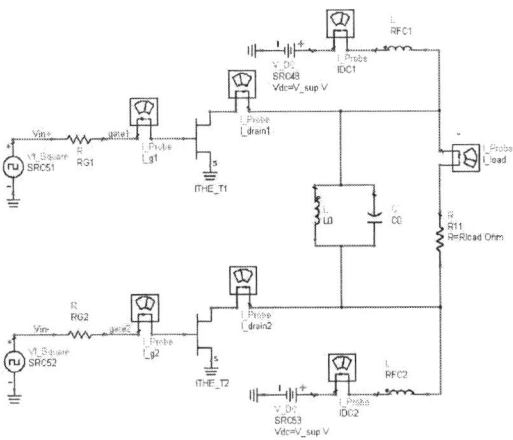

Figure 4. Basic topology of a current-mode class-D power amplifier

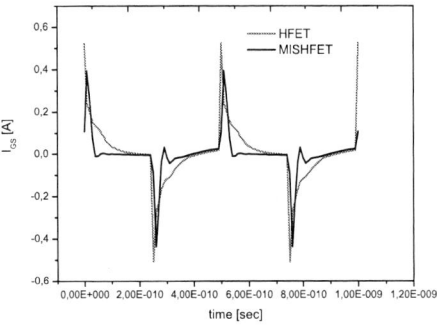

Figure 5. Gate currents of HFET and MISHFET

stepwise increase of I_{DG} at $V_{DG} = V_{DS} - V_{GS} = 97V$ and $V_{DG} = 110V$ is due to the formation of conducting channels through the gate isolation. This can either lead to a permanent loss of the isolating benefits without the device being destroyed or to a destruction of the device. In any case drain-gate-voltages that cause damage of the gate dielectric must not be used in device simulation. Therefore the authors decided to set the breakdown voltage of the model to $V_{br} = 80V$ and to avoid using voltages that high for the simulations presented in section III.

III. CIRCUIT SIMULATION

A. Current-Mode Class-D Power Amplifier

In the last section, the modeling approaches for the HFET and MISHFET were presented, taking into account the most relevant characteristics of both device technologies. To further evaluate differences between MISHFETs and HFETs concernig their applicability in class-D current-mode power amplifiers, simulations were performed using the ITHE models in ADS. The power amplifier used is shown in figure 4. It has a basic current-mode class-D topology, in which two devices are driven 180° out-of-phase. In this case, 2GHz rectangular signals were used to drive the gate of the transistors. Due to the balanced configuration of the amplifier, the output only needs to be short circuited at odd harmonics. This is done by means of a resonator which is tuned to the fundamental frequency and absorbs the parasitic output capacitance of the devices. The shape of the resulting voltage and current waveforms at the drain side of the devices is a half-wave rectified sinusoidal and square wave [14]. This amplifier class that has been successfully used at low frequencies, reapears again in the RF scenario showing promising results at microwave frequencies [15].

B. Comparison: HFET vs. MISHFET

The gate capacitance is one of the most important aspects to be considered when switching transistor devices. This parasitic capacitance must be charged beyond the threshold voltage to achieve turn-on. The gate driver must provide a high enough output current to charge the gate capacitance within the time required by the system design[17]. Figure 5 displays the simulated gate currents for the case of a class-D power amplifier using HFETs and MISHFETs respectively. The 1V-amplitude rectangular signals driving the gates, originate high current spikes at the input of the devices. The figure shows that the amplitude of the peak current and the time required to charge the gate capacitance of the MISHFET is lower than that of the HFET device. The dielectric material under the gate has contributed to a reduction of the gate capacitance and therefore less current is required to switch the device from the off state to the ohmic region .

As mentioned above, one of the basic principles of a current-mode class-D power amplifier, is to manipulate the harmonics of the output impedance in order to achieve an appropriate shaping of voltage and currents at the drain side. In this way overlaping is reduced. If ideally no overlap is present, the loadline will resemble a kind of "L" in the I-V plane. Figure 6 depicts the loadlines of both devices for $15V$ drain supply voltage. It is readily seen that while the shape described by the MISHFET's loadline approaches the theoretical case, the HFET's trace crosses hyperbolas of higher power. Additional simulations show that further increasing the supply voltage, deteriorates the loadline of the HFET even more, signalizing an increased level of overlaping and as a consequence higher power dissipation in the devices. This process of loadline degeneration starts in the MISHFET at higher drain supply voltages, indicating that this device technology is able to keep its level of efficiency up to higher operating voltages. Figure 6 also shows another remarkable result. For the same drain supply voltage, the MISHFET device achieves higher peak RF voltage and current, making it capable of delivering higher output power.

In order to compare the maximum possible performance of both device technologies, the parameters of the switched-mode amplifier were tuned. The amplitude of the drive signals, supplied voltage V_{SUP} and load impedance R_{LOAD} were

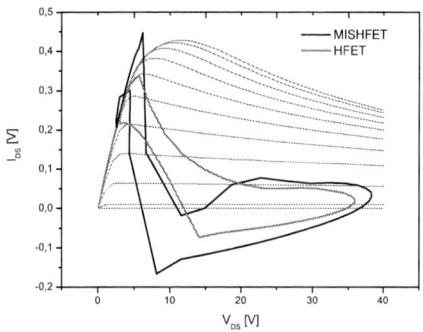

Figure 6. Load line at drain side of HFET and MISHFET devices

Table I
SIMULATION RESULTS OF HFET AND MISHFET CLASS-D AMPLIFIER

	V_{SUP}(V)	R_{LOAD}(Ω)	PAE (%)	η(%)	P_{OUT}(W)
HFET	13	230	62.0	58.6	2.0
MISHFET	15	220	67.0	63.0	2.8

adjusted for each amplifier, in such way that maximum power added efficiency (PAE) was obtained. The results are listed in table I.

The peak voltage of the square input signal was $3.25V$ with a DC component of $-1.75V$ for the MISHFET, and $1.6V$ with a DC component of $-1.4V$ for the HFET. The peak value of gate current I_{GS} was about $430mA$ in both transistor technologies. Table I clearly shows that higher values of power-added efficiency (PAE) and drain efficiency (η) are achieved by the amplifier using MISHFETs. The MISHFET class-D power amplifier exhibits a noticeably 40% higher output power in comparison to the HFET class-D amplifier. These results not only demonstrate that MISHFET devices are capable of working properly in current-mode class-D amplifiers, but also show that considerably better performance can be expected from them compared to HFETs. It is worth mentioning that the HFETs used are from a more mature technology compared to the MISHFETs. Additional improvements in MISHFET technology, like the use of high-k materials as dielectrics, have been made recently [16]. They show superior values of transconductance and drain saturation current, promising even better results in the switched-mode amplifier terrain.

IV. CONCLUSION

In this paper improvements were proposed to properly describe MISHFET devices based on the HFET large signal model used at the ITHE. The approach shown here allows characterizing dispersion effects and knee walkout as well as to provide an exact description of the gate diode behavior of GaN MISHFETs. This leads to an accurate way of modeling this kind of devices.

The design of current-mode class-D power amplifers is presented, incorporating the derived models of GaN HFETs

and MISHFETs into simulations. Comparison of both depicts the superiority of the latter in terms of output power and power added efficiency stimulating further investigation of MISHFETs used in switched-mode amplifers.

ACKNOWLEDGMENT

The authors would like to thank BMBF "Bundesministerium für Bildung und Forschung" Germany for the financial support given (Project No. 01BU0609).

REFERENCES

[1] A. Bindra, M. Valentine and K. Vic, *Next-generation wireless propels GaN power transistors, CMOS RFICs and passive,* RF Design Magazine, January, pp. 16-21, 2006.

[2] T. Palacios, A. Chakraborty, S. Rajan, C. Poblenz, S. Keller, S. P. DenBaars, J. S. Speck and U. K. Mishra, *High-Power AlGaN/GaN HEMTs for Ka-Band Applications,* IEEE Electron Device Letters, Vol. 26, No. 11, pp. 781- 783, November 2005

[3] M. A. Khan, X. Hu, G. Sumin, A. Lunev, J. Yang, R. Gaska and M. S. Shur, *AlGaN/GaN metal oxide semiconductor heterostructure field effect transistor,* IEEE Electron Device Letters, Vol. 21, No. 2, pp. 63-65, February 2000

[4] M. A. Khan, G. Simin, J. Yang, J. Zhang, A. Koudymov, M. S. Shur, R. Gaska, X. Hu and A. Tarakji, *Insulating Gate III-N Heterostructure Field-Effect Transistors for High-Power Microwave and Switching Applications,* IEEE Transactions on Microwave Theory and Techniques, Vol. 51, No. 2, February 2003.

[5] M. Kanamura, T. Kikkawa, T. Iwai, K. Imanishi, T. Kubo and K. Joshin, *An over 100 W n-GaN/n-AlGaN/GaN MISH-HEMT Power Amplifier for Wireless Base Station Applications,* Electron Device Meeting, 2005, IEDM Technical Digest. IEEE International, pp. 572- 575, 2005.

[6] M. Higashiwaki, T. Matsui and T. Mimura, *30-nm-gate AlGaN/GaN MIS-HFETs with 180 GHz f_T,* 64th Device Research Conference, pp. 149-150, June 2006.

[7] V. Adivarahan, M. Gaevski, W.H. Sun, H. Fatima, A. Koudrymov, S. Sygi, G. Simin, J. Yang, M. Afir Khan, A. Tarakji, M.S. Shur and R. Gaska, *Submicron Gate Si_3N_4/AlGaN/GaN-Metal-Insulator-Semiconductor Heterostructure Field-Effect Transistors,* IEEE Electron Device Letters, Vol. 24, No. 9, pp. 541- 543, September 2003

[8] M. Marso, G. Heidelberger, K.M. Indlekofer, J. Bernat, A. Fox, P. Kordos and H. Lüth, *Origin of Improved RF Performance of AlGaN/GaN MOSHFETs Compared to HFETs,* IEEE Transactions on Electron Devices, Vol. 53, No. 7, pp. 1517- 1523, July 2006

[9] A. Fox, M. Marso, G. Heidelberger and P. Kordos, *RF characterization and modeling of AlGaN/GaN based HFETs and MOSHFETs,* International Conference on Advanced Semiconductor Devices and Microsystems (ASDAM), 2006, pp.109-112, October 2006

[10] Agilent Technologies: ADS-Documentation, Nonlinear Devices [online] http://eesof.tm.agilent.com/docs/adsdoc2001/manuals/pdf.html. Page Mill Road Palo Alto, CA 94304 U.S.A.

[11] M. Eickelkamp, C. Lautensack, A. Noculak, H. Kalisch, R.H. Jansen and A. Vescan, *Investigation of passivation strategies for AlGaN/GaN-MISHFETs on Silicon Substrates,* WOCSDICE Leuven, Belgium, 2008

[12] C. Lautensack, S. Chalermwitsukul , R. H. Jansen, *Modification of EEHEMT1 Model for Accurate Description of GaN HEMT Output Characteristic,* APMC 2007

[13] CST - Computer Simulation Technology GmbH, Darmstadt, Germany [online] http://www.cst.com

[14] B. Berglung, J. Johansson, T. Lejon, *High efficiency power amplifiers.* In: Ericsson Review, No. 3, 2006.

[15] H. Kobayashi, J. M. Hinrichs, P. M. Asbeck, *Current-Mode Class-D Power Amplifiers for High-Efficiency RF Applications.* In: IEEE Trans. on Microwave Theory and Techniques, vol. 49, Dec. 2001, pp. 2480 2485.

[16] D. Gregusova, R. Stocklas, K. Cico, T. Lalinsky, P. Kordos, *AlGaN/GaN metal-oxide-semiconductor heterostructure field-effect transistors with 4 nm thick Al_2O_3 gate oxide.* In: Semicond. Sci. Technol. 22, 2007, pp. 947 951.

[17] Abid Hussain, *Driving Power MOSFETs in High-Current, Switch Mode Regulators.* In: Microchip Technology Inc., AN786, 2002.

978-2-8748-7007-1/08 $25.00 © 2008 EuMA

Decrease in Slow Current Transients and Current Collapse in GaN-based FETs with a Filed Plate

A. Nakajima, K. Itagaki and K. Horio

Faculty of Systems Engineering, Shibaura Institute of Technology
307 Fukasaku, Minuma-ku, Saitama 337-8570, Japan
horio@sic.shibaura-it.ac.jp

Abstract—**Two-dimensional transient analyses of AlGaN/GaN HEMTs and GaN MESFETs are performed in which a deep donor and a deep acceptor are considered in a semi-insulating buffer layer. Quasi-pulsed *I-V* curves are derived from the transient characteristics. It is studied how the existence of field plate affects buffer-related lag phenomena and current collapse. It is shown that in both FETs, the lag phenomena and current collapse could be reduced by introducing a field plate, because electron injection into the buffer layer is weakened by it, and trapping effects are reduced. The dependence on insulator-thickness under the field plate is also studied, suggesting that there is an optimum thickness of insulator to minimize the current collapse and drain lag.**

I. INTRODUCTION

AlGaN/GaN HEMTs are now receiving great attention because of their potential applications to high power and high temperature microwave devices [1]. However, slow current transients are often observed even if the gate voltage or the drain voltage is changed abruptly [2]. This is called gate lag or drain lag, and is problematic for circuit applications. The slow transients mean that dc *I-V* curves and RF *I-V* curves become quite different, resulting in lower RF power available than that expected from the dc operation. This is called power slump or particularly current collapse in the GaN device field. These are serious problems, and many experimental works are reported [1-5], but few theoretical works are made [5-7]. The lags and current collapse can be reduced by surface passivation [3] and by using a field plate [4]. These are considered due to the decrease in surface-state effects. In this work, we have made two-dimensional transient analyses of field-plate AlGaN/GaN HEMTs and GaN MESFETs with a semi-insulating buffer layer in which deep levels are considered, and found that buffer-related lag phenomena and current collapse could also be reduced by introducing a field plate.

II. PHYSICAL MODEL

Fig.1 shows (a) AlGaN/GaN HEMT and (b) GaN MESFET structures analyzed here. The gate length L_G and the field-plate length L_{FP} are typically set to 0.3 μm and 1μm, respectively. Polarization charges of 10^{13} cm^{-2} are set at the heterojunction interface, and the surface polarization charges are assumed to be compensated by surface-state charges, as in [8]. As a model for a semi-insulating buffer layer, we use a

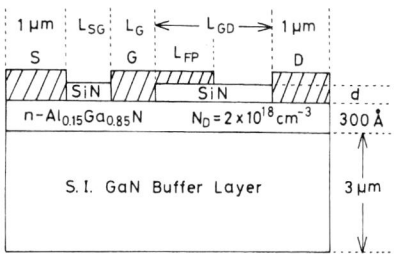

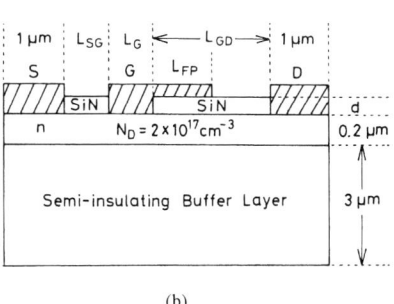

Fig. 1 Device structures analyzed in this study. (a) AlGaN/GaN HEMT, (b) GaN MESFET.

three-level compensation model which includes a shallow donor, a deep donor and a deep acceptor [6,7]. Some representative experiments show that two levels ($E_C - 1.75$ eV, $E_C - 2.85$ eV) are associated with current collapse in GaN-based FETs [2], and hence we use energy levels of $E_C - 2.85$ eV ($E_V + 0.6$ eV) for the deep acceptor and of $E_C - 1.75$ eV for the deep donor. Other experiments show shallower energy levels for deep donors in GaN [9], and hence we vary the deep donor's energy level E_{DD} as a parameter. Mainly, we present the data for $E_C - E_{DD} = 1.0$ eV and 0.5 eV here. The deep-donor density N_{DD} and the deep-acceptor density N_{DA} are typically set to 2×10^{17} cm^{-3} and 10^{17} cm^{-3}, respectively. The shallow-donor density N_{Di} is set to 10^{15} cm^{-3}. When $N_{DD} > N_{DA}$, the deep donors donate electrons to the deep acceptors, and hence the ionized deep-donor density N_{DD}^+ becomes nearly equal to N_{DA} under equilibrium, and it acts as an electron trap [6,7].

Basic equations to be solved are Poisson's equation including ionized deep-level terms, continuity equations for

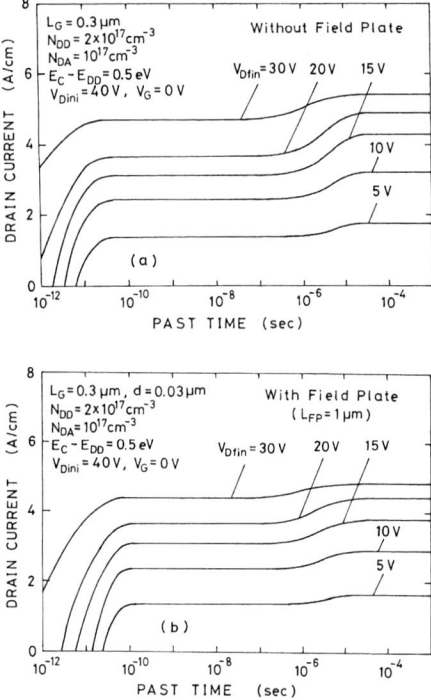

Fig. 2 Calculated drain-current responses of AlGaN/GaN MESFETs when V_D is changed abruptly from 40 V to V_{Dfin}, while V_G is kept constant at 0 V. $N_{DD} = 2 \times 10^{17}$ cm^{-3}, $N_{DA} = 10^{17}$ cm^{-3} and $E_C - E_{DD} = 0.5$ eV. (a) Without a field plate, (b) with 1 μm-length field plate.

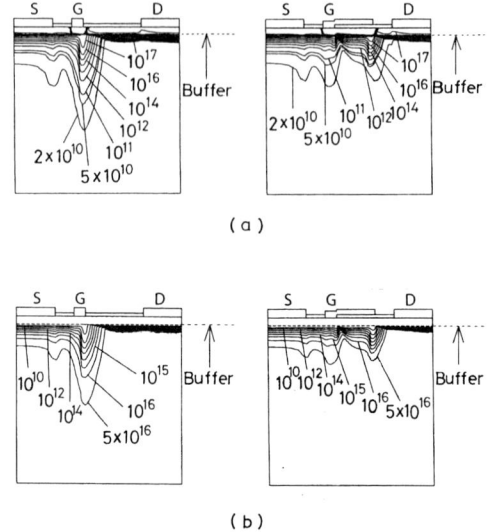

Fig. 3 (a) Electron density profiles and (b) ionized deep-donor density N_{DD}^+ profiles at $V_G = 0$ V and $V_D = 40$ V. $d = 0.03$ μm. $N_{DD} = 2 \times 10^{17}$ cm^{-3}, $N_{DA} = 10^{17}$ cm^{-3} and $E_C - E_{DD} = 0.5$ eV. The left is for the case without a field plate, and the right is for the field-plate structure ($L_{FP} = 1$ μm).

electrons and holes which include carrier loss rates via the deep levels and rate equations for the deep levels. These equations are put into discrete forms and are solved numerically. We have calculated the drain-current responses when the drain voltage V_D and/or the gate voltage V_G are changed abruptly.

III. DRAIN CURRENT RESPONSES

Fig.2 shows an example of calculated drain-current responses of AlGaN/GaN HEMTs when V_D is lowered abruptly from 40 V to V_{Dfin}, where V_G is kept constant at 0 V. Fig.2(a) is for a structure without a field plate ($L_{FP} = 0$) and Fig.2(b) is for a case with a field plate ($L_{FP} = 1$ μm). Here the thickness of SiN passivation layer d is 0.03 μm. In both cases, the drain currents remain at low values for some periods and begin to increase slowly, showing drain-lag behavior. It is understood that the drain currents begin to increase when the deep donors in the buffer layer begin to emit electrons [7]. It is seen that the change of drain current is smaller for the case with a field plate, indicating that the drain lag is smaller for the field-plate structure. We will discuss below why this happens.

Fig.3 shows (a) electron density profiles and (b) ionized deep-donor density N_{DD}^+ profiles at $V_G = 0$ V and $V_D = 40$ V. The left is for the structure without a field plate, and the right is for the field-plate structure. In Fig.3(a), we see that for the structure without a field plate, electrons are injected deeper into the buffer layer under the gate, particularly under the drain edge of the gate region. These electrons are captured by the deep donors, and hence N_{DD}^+ decreases there as seen in Fig.3(b). As mentioned before, when V_D is lowered abruptly, the drain current remains at a low value for some periods and begins to increase slowly as the deep donors begin to emit electrons (and N_{DD}^+ increases), showing drain lag. In the case of field-plate structure, as seen in Fig.3(a), electrons are injected into the buffer layer under the drain edge of field plate as well as under the gate. But the overall injection depth is not so deep as compared to the case without a field plate. This is because the electric field at the drain edge of the gate becomes weaker by introducing a field plate. Hence, the change of N_{DD}^+ by capturing electrons is smaller for the field-plate structure as seen in Fig.3(b). Therefore, the drain lag becomes smaller for the field-plate structure.

IV. PULSED I-V CURVES AND CURRENT COLLAPSE

Next, we have calculated a case when V_G is also changed from an off point. V_G is changed from threshold voltage V_{th} to 0 V, and V_D is changed from 40 V to V_{Don} (on-state drain voltage). The characteristics become similar to those in Fig.2, although some transients arise when only V_G is changed (gate lag). From these turn-on characteristics, we obtain a quasi-pulsed I-V curve.

978-2-8748-7007-1/08 $25.00 © 2008 EuMA   184

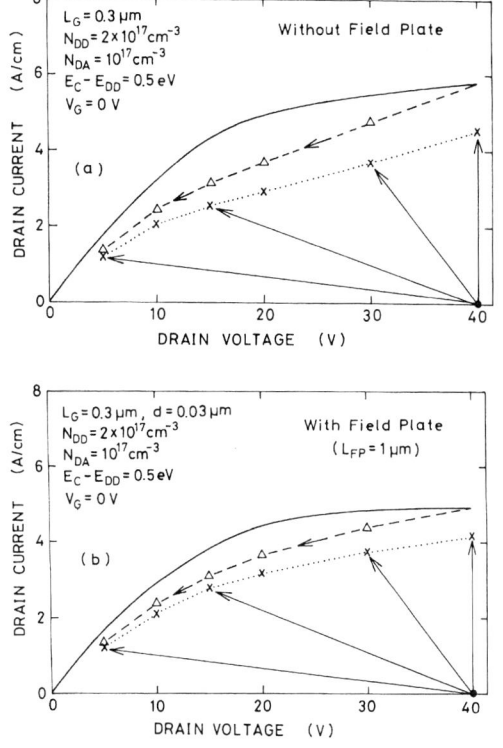

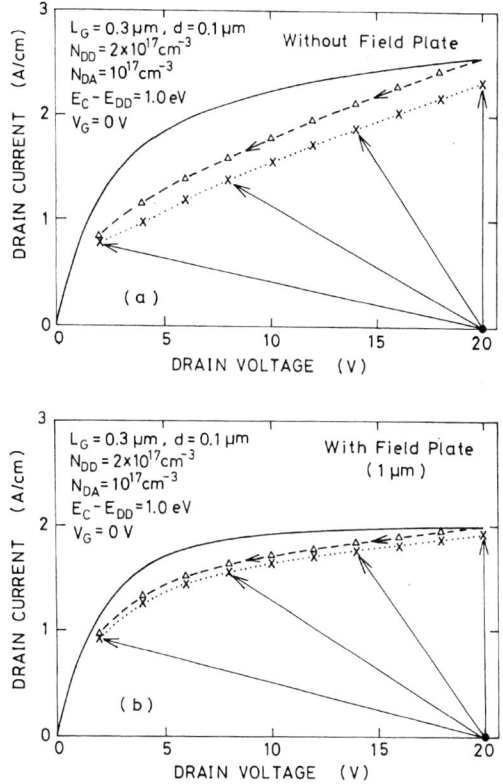

Fig. 4 Steady-state I-V curves ($V_G = 0$ V; solid lines) and quasi-pulsed I-V curves ($\triangle$, x) of AlGaN/GaN HEMTs. (a) Without a field plate, (b) with 1 μm-length field plate. ($\triangle$): Only V_D is changed from 40V ($t = 10^{-8}$ s), (x): V_D is lowered from 40 V and V_G is changed from V_{th} to 0 V ($t = 10^{-8}$ s).

Fig. 5 Steady-state I-V curves ($V_G = 0$ V; solid lines) and quasi-pulsed I-V curves ($\triangle$, x) of GaN MESFETs. (a) Without a field plate, (b) with 1 μm-length field plate. ($\triangle$): Only V_D is changed from 20V ($t = 10^{-8}$ s), (x): V_D is lowered from 20 V and V_G is changed from V_{th} to 0 V ($t = 10^{-8}$ s).

In Figs.4 and 5, we plot by (x) the drain current at $t = 10^{-8}$ s after V_G is switched on for AlGaN/GaN HEMT and GaN MESFET, respectively. Fig.4(a) and Fig.5(a) are for the structures without a field plate and Fig.4(b) and Fig.5(b) are for the field-plate structures ($L_{FP} = 1$ μm). These curves are regarded as quasi-pulsed I-V curves with pulse width of 10^{-8} s. They stay rather lower than the steady-state I-V curves (solid lines), indicating current collapse behavior. In Figs.4 and 5, we also plot another pulsed I-V curve ($\triangle$), which is obtained from figures like Fig.2 (where only V_D is changed), indicating drain-lag behavior. From Fig.4 and Fig.5, we can definitely say that the lag phenomena (drain lag, gate lag) and current collapse become smaller for the field-plate structure.

We have next studied dependence of lag phenomena and current collapse on SiN thickness d. Fig.6 and Fig.7 show drain-current reduction rate $\Delta I_D/I_D$ (ΔI_D : current reduction, I_D : steady-state current) due to current collapse, drain lag or gate lag, with d as a parameter. Fig.6 is for AlGaN/GaN HEMTs, and Fig.7 is for GaN MESFETs. Here $d = 0$ ($L_{FP} = 1$ μm) corresponds to a case of $L_G = 1.3$μm without a field plate. When d is thick, the current collapse and lag phenomena are relatively large because the field plate does not almost affect the characteristics. As d becomes thinner, the current collapse

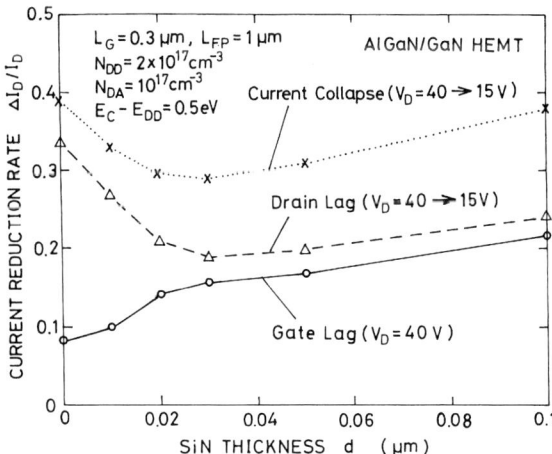

Fig. 6 Current reduction rate $\Delta I_D/I_D$ due to current collapse, drain lag or gate lag for AlGaN/GaN HEMTs, with SiN thickness d as a parameter. $L_{FP} = 1$μm.

978-2-8748-7007-1/08 $25.00 © 2008 EuMA

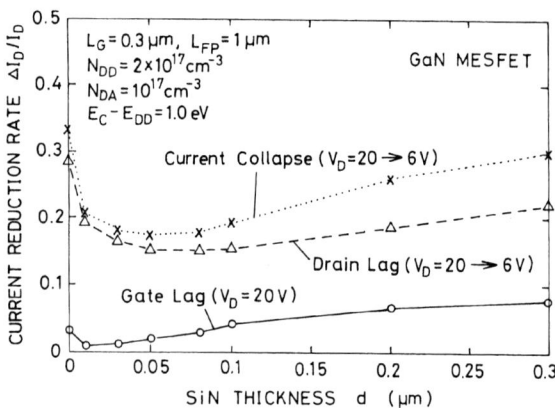

Fig.7 Current reduction rate $\Delta I_D/I_D$ due to current collapse, drain lag or gate lag for GaN MESFETs, with SiN thickness d as a parameter. $L_{FP} = 1\mu m$.

and lag phenomena become smaller because the trapping effects are reduced, although the rates of current collapse and drain lag increase for very thin d, where electron injection under the field plate becomes significant. When $d = 0$ ($L_G = 1.3\ \mu m$), that is, without a field plate, the current collapse becomes rather noticeable. (In a previous work [10], we showed that the current collapse was not so dependent on the gate length L_G in the structure without a field plate.) From Fig.6 and Fig.7, we can say that there is an optimum thickness of SiN to minimize the buffer-related current collapse and drain lag in AlGaN/GaN HEMTs and GaN MESFETs.

V. CONCLUSION

Two-dimensional transient analyses of the field-plate AlGaN/GaN HEMTs and GaN MESFETs with a semi-insulating buffer layer have been performed in which a deep donor and a deep acceptor are considered in the buffer layer. Quasi-pulsed *I-V* curves have been derived from the transient characteristics. It has been shown that the drain lag is reduced by introducing a field plate because the trapping effects become smaller. It has also been shown that the gate lag and current collapse are also reduced in the field-plate structure. It is suggested that there is an optimum thickness of SiN passivation layer to minimize the buffer-related current collapse and drain lag in GaN HEMTs and MESFETs.

REFERENCES

[1] U. K. Mishra, L. Shen, T. E. Kazior and Y.-F. Wu, "GaN-based RF power devices and amplifiers", *Proc. IEEE*, vol.96, pp.287-305, 2008.

[2] S. C. Binari, P. B. Klein and T. E. Kazior, "Trapping effects in GaN and SiC Microwave FETs", *Proc. IEEE*, vol.90, pp.1048-1058, 2002.

[3] G. Koley, V. Tilak, L. F. Eastman and M. G. Spencer, "Slow transients observed in AlGaN/GaN HFETs: Effects of SiN$_x$ passivation and UV illumination", *IEEE Trans. Electron Devices*, vol.50, pp.886-893, Apr. 2003.

[4] A. Koudymov, V. Adivarahan, J, Yang, G. Simon and M. A. Khan, "Mechanism of current collapse removal in field-plated nitride HFETs", *IEEE Electron Device Lett.* vol.26, pp.704-706, 2005.

[5] J. Tirado, J. L. Sanchez-Rojas and J. I. Izpura, "Trapping Effects in the transient response of AlGaN/GaN HEMT devices", *IEEE Trans. Electron Devices*, vol.54, pp.410-417, 2007.

[6] K. Horio, K. Yonemoto, H. Takayanagi and H. Nakano, "Physics-based simulation of buffer-trapping effects on slow current transients and current collapse in GaN field effect transistors" *J. Appl. Phys.*, vol.98, no.12, pp.124502 1-7, Dec. 2005.

[7] K. Horio and A. Nakajima, "Physical mechanism of buffer-related current transients and current slump in AlGaN/GaN high electron mobility transistors", *Jpn. J. Appl. Phys.*, vol.47, no.5, pp.3428-3433 2008.

[8] S. Karmalkar and U. K. Mishra, "Enhancement of breakdown voltage in AlGaN/GaN high electron mobility transistors using a field plate", *IEEE Trans. Electron Devices*, vol.48, pp.1515-1521, Aug. 2001.

[9] H. Morkoc, *Nitride Semiconductors and Devices*, Springer-Verlag, 1999.

[10] K. Itagaki, N. Kobayashi and K. Horio, "Analysis of buffer-related lag phenomena and current collapse in GaN FETs", *phys. stat. soli. (c)* vol.4, pp.2666-2669, 2007.

Increased reliability of AlGaN/GaN HEMTs versus temperature using deuterium.

G. Astre[#1], J.G. Tartarin[#2], J. Chevallier[+3], S. Delage[°4].

[#] University of Toulouse, LAAS CNRS & Paul Sabatier University, 7 av. Colonel Roche, 31.077 Toulouse cedex 4, France
[1]gastre@laas.fr
[2]tartarin@laas.fr

[+] GEMaC, 1 place Aristide Briand, 92195 Meudon, France
[3]jacques.chevallier@cnrs-bellevue.fr

[°] TIGER, Alcatel-Thales III-V Lab, Route de Nozay, 91460 Marcoussis, France
[4]sylvain.delage@3-5lab.fr

Abstract— Low frequency noise (LFN) is a reliable diagnostic tool to evaluate and locate the defects of a technology. In this study, LFN is used to assess effects of deuterium (H^+ ions) in diffusion condition on the robustness of 0.25 *2*75 µm² gate area AlGaN/GaN high electron mobility transistors (HEMT) grown on Si substrate. H^+ Ions are diffused from the above AlGaN/GaN layer through the AlGaN/GaN interface and GaN layer, notably under the gated channel where the defects are located. Two batches of devices are stressed under high temperature condition at 400°C during 5 minutes (step 1) and 500°C during 15 minutes (step 2). The first batch is composed with 8 deuterated transistors while the second batch is composed with 8 non deuterated transistors. Static measurements and low frequency noise spectral density measurements of the drain current (S_{ID}) are examined after each step of temperature. The first step does not reveal any degradation, while the second step highlights significative differences between the deuterated and non deuterated devices: LFN of deuterated devices remains constant, whereas LFN of non deuterated devices increases (GR superimposed with 1/f flicker noise). The deuteration of the devices can open the way to robust temperature devices, as AlGaN/GaN HEMT are dedicated to applications at high power and high temperature.

I. INTRODUCTION

The development of wideband gap technology is convenient for integrated circuits design such as VCO and LNA. AlGaN/GaN high electron mobility transistors stand as promising candidates to substitute GaAs and SiGe technologies for power applications at high frequency. They take advantages from a higher breakdown voltage, higher power density and higher thermal conductivity that are suitable for power applications while keeping good carrier mobility for high frequency applications. Recent studies revealed also very good low noise performances for this technology [1] [2]. Thus, GaN devices can pretend to be used in transceivers for power amplification, for low noise amplification and for frequency synthesis (voltage controlled oscillation) over a wideband of frequencies ranging from L up to Ka band.

LFN measurements provide valuable information about noise source location and failure prediction [3], and can also assess devices robustness. In a previous work, deuterium has been used as a probe in diffusion condition to locate the defects density at the AlGaN/GaN interface as well as in the GaN 2DEG using SIMS (Secondary Ions Mass Spectrometry) measurements [4]. This result corroborates interpretations from LFN measurements for devices grown on SiC substrate [5] [6]. In this work, we focus on the effects of deuterium on AlGaN/GaN HEMT devices under thermal stress in order to improve the devices robustness. LFN and static measurements are performed at two time and temperature steps to evidence the different behaviours between the deuterated devices and the non-deuterated devices.

II. ALGAN/GAN HEMT ON SI

HEMT devices are grown by MOCVD techniques on silicon substrate using the AlGaN/GaN HEMT TIGER usual process. Transistors feature 0.25*2*75 µm² gate area and 24% alumina content.

For this study, deuterium is brought in a half of the wafer before gate deposition, while the second half is finalized without deuterium. After this process step, deuterium is located in the AlGaN layer (from source to drain) and diffuses toward the GaN layer by means of a RF plasma (3W during 30 minutes at a temperature of 300 K) (figure 1).

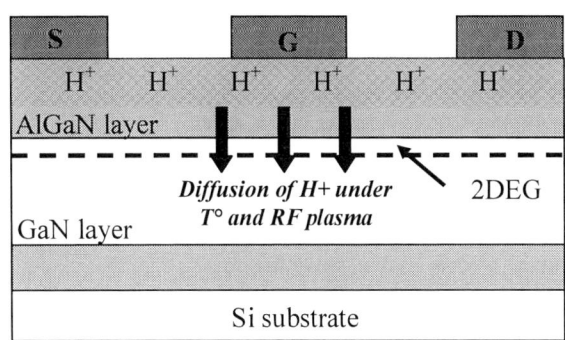

Fig. 1 Deuterium diffusion in AlGaN/GaN structure

III. INITIAL MEASUREMENTS

Initial static and LFN measurements are performed on each batch of samples (15 deuterated devices for "batch 1" and 58 non-deuterated devices for "batch 2"). No significant difference is found between the deuterated and non deuterated samples before the application of the temperature stress, neither on the DC characteristics (I_{dss}, V_T, R_{on} ...), nor on the LFN spectra. A slightly higher drain current on batch 1 can be attributed to an increase of the charge density due to the diffusion of deuterium. The two batches have equivalent R_{on} of 34,5 Ω (meaning value) and an I_{dss} of 100 mA and 105 mA respectively for the meaning value of batch 2 and batch 1(figure 2).

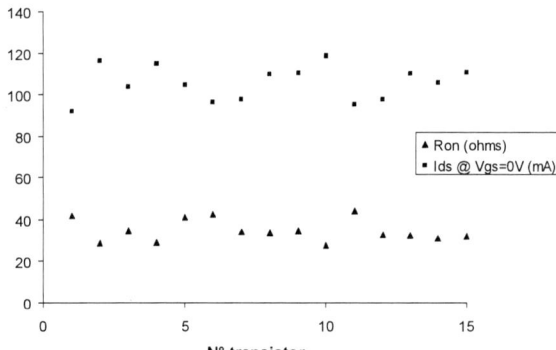

Fig. 2 DC characteristics repartition of devices from batch 1

S_{ID} spectra, measured at V_{GS}=-3 V and V_{DS}=10 V feature a $1/f^{\gamma}$ behaviour (with γ=1,27) (figure 3).

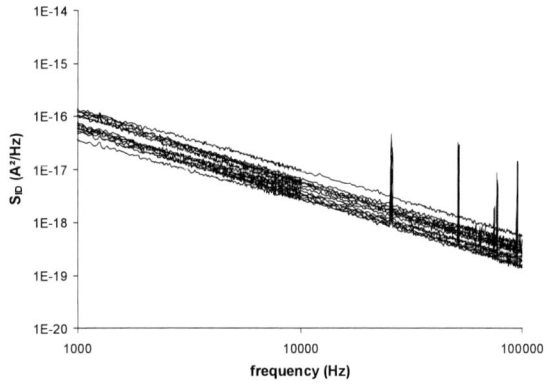

Fig. 3 S_{ID} spectra repartition of batch 1 (15 devices)

We can conclude about the initial good homogeneity of the two batches. Next, two batches of 8 transistors, representative of each initial batch are chosen for the study.

IV. THERMAL STRESS

A. Stress conditions

First, the two batches are thermally stressed at 400°C during 5 minutes in an air oven (step 1), above the classic deuterium diffusion temperature (300°C). Then, temperature is elevated at 500°C during 15 minutes (step 2). All devices are measured at T_0 (initial), T_1 (step 1) and T_2 (step 2).

B. Results and discussion

Static measurements reveal a decrease of drain and gate leakage currents after each step. The decrease of leakage current can be attributed to the improvement of surface states induced by thermal annealing.

After step 1, constant or improved (slightly lower) spectra are measured for both batch 1 and batch 2: it thus cannot be attributed to deuterium effects. This is probably due to AlGaN/GaN interface improvement during the burn-in phase.

Step 2 represents harder stress conditions applied to the samples of each batch. However, no degradation is measured on DC characteristics and LFN spectra for the deuterated structures (figure 4).

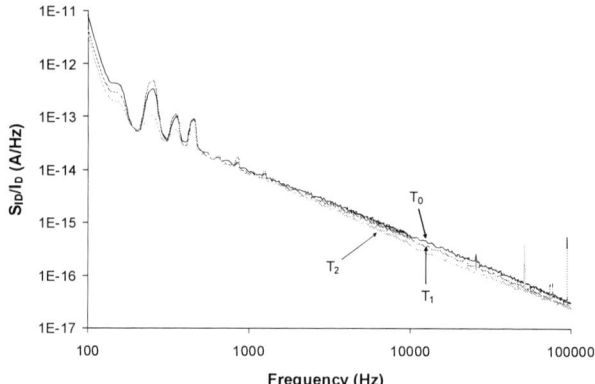

Fig. 4 S_{ID} spectra at T_1, T_2 and T_3 of deuterated device

Samples from batch 2 systematically have their LFN spectra degraded after this step 2, with different degradation signatures: the flicker noise $1/f^{\gamma}$ source increases for some devices, while LFN spectra of some other devices have changes on their frequency index γ. Whatever the variation of the flicker noise source, all the devices from batch 2 suffer from the apparition of a trapping-detrapping noise source (G-R center) (figure 5). These defects for devices of batch 2 are attributed to the lower crystal quality under the gate (where the defects are located). Temperature plays here a strong role in the activation of new noise sources.

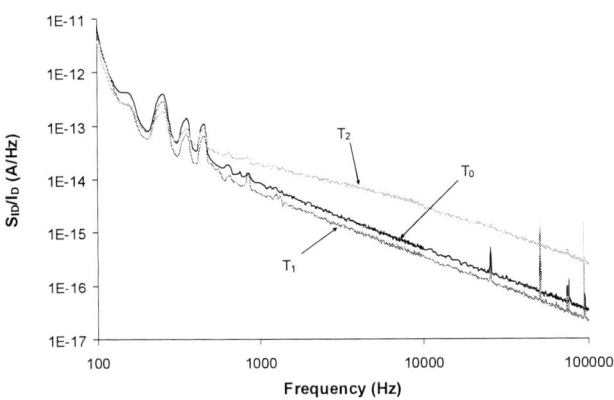

Fig. 5 S_{ID} spectra at T_1, T_2 and T_3 of non deuterated device

The devices from batch 2 seem to be sensitive to different degradation mechanisms and their LFN spectra systematically degrades whatever the cause. The investigation on the comprehension of the degradation mechanisms is not in the scope of this study as we propose here the evidence of deuteration effect on the devices robustness improvement. In previous works, elevated temperature stresses (at less than 240°C) on HEMT have shown reductions on the drain current, on the transconductance gain and an increase on the R_{on} resistance with no change on the pinch-off voltage [7]. These changes have been attributed to the crystal quality (misfit dislocations). Devices from this study keep their I_{DSS} biasing constant (same remark for the quiescent point of LFN measurements). Neither the mobility nor the number of the carriers seems to be affected by the deuteration process during the early reliability stresses, but this technological step impacts strongly the LFN signature of the devices.

Recent studies [3] [5] [6] have already underlined traps location at the AlGaN/GaN interface and in GaN bulk (2DEG) under the gated channel. Temperature activates defects for devices without deuterium in the noisy part of the device (batch 2), and the devices degrades. Transistors with deuterium in the channel under the gate (batch 1) do not suffer from any degradation: H^+ seems to passivate the defects, and the devices are thus more reliable.

V. CONCLUSIONS

Conspicuous effect of deuterium on devices robustness seems to offer a new solution to improve the reliability of GaN

devices. We have evidenced the effect of deuterium diffusion in the noisy part of the devices on the low frequency noise spectra. Deuteration can open the way to more temperature-robust GaN technologies, but other investigations must be driven using different stress and bias conditions. Degradation processes are not clear yet in AlGaN/GaN devices. Results proposed in the literature do not yet propose a simplified and uniformed model for all the events issued from the many papers about reliability or robustness tests. Different kinds of investigations are still on going to identify the degradation mechanisms.

REFERENCES

[1] Lan, X. Wojtowicz, M. Smorchkova, I. Coffie, R. Tsai, R. Heying, B. Truong, M. Fong, F. Kintis, M. Namba, C. Oki, A. Wong, T., "A Q-band low phase noise monolithic AlGaN/GaN HEMT VCO", IEEE Microwave and Wireless Components Letters, Vol. 16 (2006), pp 425-427.

[2] G. Soubercaze-Pun, J.G. Tartarin, L.Bary, J.Rayssac, E.Morvan, B. Grimbert, S.L. Delage, J-C. De Jaeger, J. Graffeuil 'Design of X-band GaN oscillator: from the low frequency noise device characterization and large signal modeling to circuit design', IEEE MTT Symposium 2006, San Francisco (U.S.A), pp.747-750.

[3] J.G. Tartarin, G. Soubercaze-Pun, A Rennane, L. Bary, R. Plana, J.C. De Jaeger, M. Germain, S. Delage, J. Graffeuil, « Using low frequency noise characterization of AlGaN/GaN HEMT as a tool for technology assessment and failure prediction", Noise in Devices and circuits II, Proc. of SPIE, Vol. 5470, Masapolmas, Spain, MAY 2004, pp 296-306.

[4] J. Mimila Arroyo, "Effects of deuterium diffusion on the electrical properties of AlGaN/GaN heterostructures", Material Research Society Symposium, MRS 2005, San Fransisco, USA, Symp.E.

[5] J.G Tartarin, G. Soubercaze-Pun, L.Bary, C.Chambon, S.Gribaldo, O. Llopis, L. Escotte, R. Plana, S. Delage, C. Gacquière, J. Graffeuil, "Low frequency and linear high frequency noise performances of AlGaN/GaN grown on SiC substrate", 13th GAAS Symposium, Paris, 2005, pp 277-280.

[6] J.G. Tartarin, G. Soubercaze-Pun, J.L. Grondin, L. Bary, J. Mimila-Arroyo, J. Chavallier, "Generation-Recombination In AlGAN/GaN HEMT On SiC Substrate, Evidenced By Low Frequency Noise Measurements And SIMS Characterisation", International Conference on Noise and Fluctuations, 2007, Tokyo, Japan.

[7] Y.C. Chou, D. Leung, I. Smorchkova, M. Wojtowicz, R. Grundbacher, L. Callejo, Q. Kan, R. Lai, P.H. Liu, D. Eng, A. Oki, "Degradation of AlGAN/GaN HEMTs under elevated temperature lifetesting", Microelectronics Reliability 44 (2004), Elsevier, pp. 1033-1038.

Proceedings of the 3rd European Microwave Integrated Circuits Conference

X-Band GaN SPDT MMIC with over 25 Watt Linear Power Handling

Jochem Janssen[1], Marc van Heijningen[1], Keith P. Hilton[2], Jessica O. Maclean[2], David J. Wallis[2], Jeff Powell[2], Michael Uren[2], Trevor Martin[2] and Frank van Vliet[1]

[1]TNO Defence, Security and Safety
Oude Waalsdorperweg 63, Den Haag, The Netherlands
Jochem.Janssen@tno.nl
[2]QinetiQ Ltd,
Malvern, Worcestershire WR14 3BE, United Kingdom

Abstract— **Single Pole Double Throw (SPDT) switches are becoming more and more key components in phased-array radar transmit/receive modules. An SPDT switch must be able to handle the output power of a high power amplifier and must provide enough isolation to protect the low noise amplifier in the receive chain when the T/R module is transmitting. Therefore Gallium Nitride technology seems to become a key technology for high power SPDT switch design. The technology shows good performance on microwave frequencies and is able to handle high power. An X-band SPDT switch, with a linear power handling of over 25 W, has been designed, measured and evaluated. The circuit is designed in the coplanar waveguide AlGaN/GaN technology established at QinetiQ.**

I. INTRODUCTION

Gallium Nitride technology is an upcoming technology in the microwave field. Robust key components can be designed with good performance at microwave frequencies [1]. A typical circuit which can be designed in GaN technology is a single pole double throw (SPDT) switch. Due to development of GaN technology a relative large circulator can be substituted with a solid-state SPDT switch (see Figure 1).

The SPDT switch, as a substitute for a circulator, determines the operating mode of the T/R module. It switches the operation of the module in transmit or receive mode. An SPDT switch has to be able to handle the transmitted power by the high power amplifier and must be able to provide enough isolation to protect the low noise amplifier against power overload when the T/R module is transmitting.

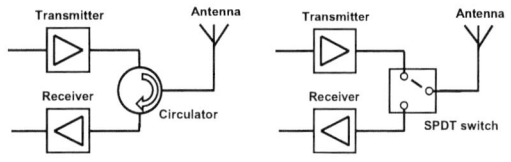

Fig. 1 Block schematic of a T/R module with circulator and a T/R module with an SPDT switch.

The following paragraphs present the establishment of the technology, and the modelling, design and evaluation of an X-band SPDT switch MMIC with high power handling in AlGaN/GaN technology. The targets for this design were 20 W power handling, 1 dB insertion loss and -40 dB isolation. The return losses had to be lower than -10 dB. The circuit is designed in the scope of the EDA project Korrigan [2].

II. ALGAN/GAN TECHNOLOGY

The circuits were implemented using the coplanar waveguide GaN MMIC technology established at QinetiQ. The epitaxial growth was carried out in a Thomas-Swan CCS MOVPE reactor and used a layer structure consisting of 25 nm of undoped $Al_{0.25}Ga_{0.75}N$ on a 1.9 μm thick Fe doped insulating GaN layer grown on 2" 4H SI SiC. The process flow was fairly conventional, using TiAlPtAu alloyed Ohmic contacts and a NiAu 0.25 μm T gate. Devices were passivated with a PECVD $SiN_x/SiO_2/SiN_x$ multilayer dielectric stack which doubled as the MIM capacitor dielectric and had a breakdown voltage >200 V. Thin film resistors were implemented using NiCr with a sheet resistance of 27 Ohm/sq. Inductors and coplanar transmission lines were implemented in 3 μm plated gold with air-bridged underpasses using a 0.8 μm evaporated gold feed metal layer. The GaN HFETs had a contact resistance of 0.47 Ohm.mm, gm of 230 mS/mm, Idss0 of 1030 mA/mm, pinch-off voltage of -5.15 V, an on-resistance of 3.55 Ohm.mm, f_T of 40 GHz and a gate breakdown voltage >100 V.

III. SWITCH FET MODELLING

The used FETs are 8x125 μm devices in switch configuration (Figure 2). In this configuration the drain and source are used as input and output, and the gate to switch the drain-source channel. At the start of the design there was no switch FET model available and there were no switch FET test-structures to measure. Therefore the used switch FETs are modelled with ADS Momentum.

978-2-8748-7007-1/08 $25.00 © 2008 EuMA

The reference device for the modelling is an 8x75 µm FET in common source configuration, which is measured, and simulated in Momentum. With Momentum and the measurement data of the common source configuration device, the FET layout is changed into switch configuration. Hereafter the results from the 8x75 µm device are scaled up to a 8x125 µm switch FET.

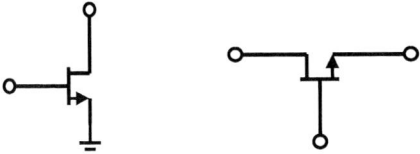

Fig. 2 Schematic. Left: FET in common source configuration. Right: FET in switch configuration.

A switch can be 'on' or 'off', so two states can be distinguished [3]; the on-state (short circuit) and the off-state (open circuit). When the switch is in on-state a layer is added to the layout, which matches with the measured drain-source on-resistance (R_{ds}). When the switch is in off-state a gap between the drain and source is introduced, which represents the series drain-source capacitance of the measured off-capacitance (C_{ds}). Figure 3 and Figure 4 show the Momentum layout of a single gate finger in both states.

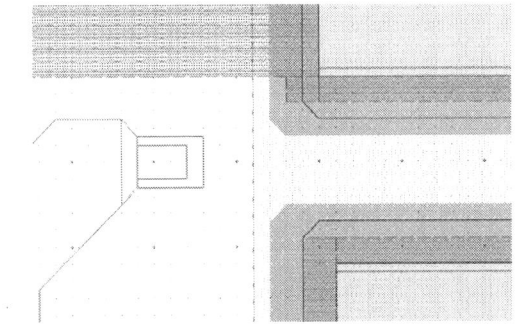

Fig. 3 Momentum model of a single gate finger junction, on-state case.

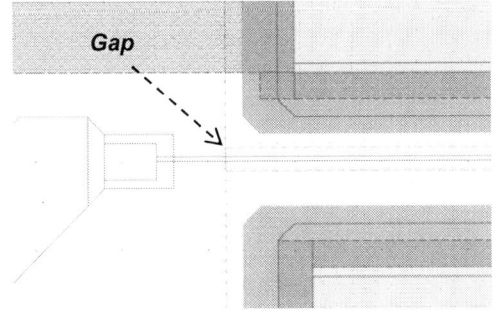

Fig. 4 Momentum model of a single gate finger junction, off-state case.

A comparison of the simulation results of R_{ds} and C_{ds} in Momentum with the measurement results are presented in Figure 5 and Figure 6. The measured off-state is at -10 V gate voltage. The measured on-state is at 0 V gate voltage. Measurement and simulation show a good match at X-band. So the Momentum simulation can be used for the 8x125 µm switch FET simulations.

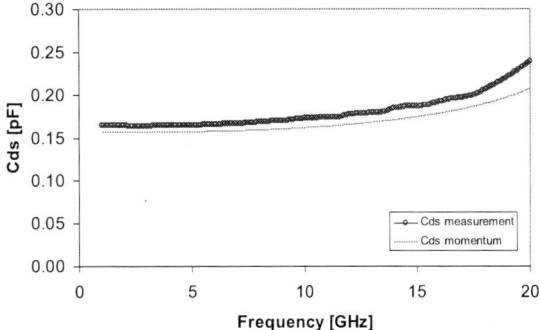

Fig. 5 Comparison of C_{ds} of the Momentum based model with measurement results of the 8x75 µm common source FET.

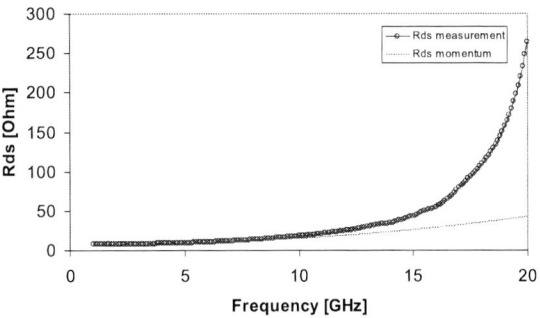

Fig. 6 Comparison of R_{ds} of the Momentum based model with measurement results of the 8x75µm common source FET.

The simulation results of the on-state and off-state of the Momentum based 8x125 µm switch FET model are presented in Figure 7 and Figure 8. In the on-state the gate voltage is 0 V, while in the off-state the gate voltage is tuned to the measurement results of -10 V.

The Momentum based models are only valid for small-signal simulations, so no large signal information was available for simulations. During measurement of the SPDT switch the gate voltage for off-state operation is tuned lower to -20 V, which gives a better result on power handling.

978-2-8748-7007-1/08 $25.00 © 2008 EuMA

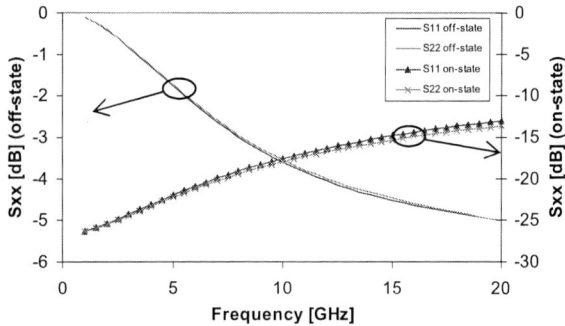

Fig. 7 Simulated return loss (S_{11} and S_{22}) of the Momentum based model of 8x125 µm FET in both states.

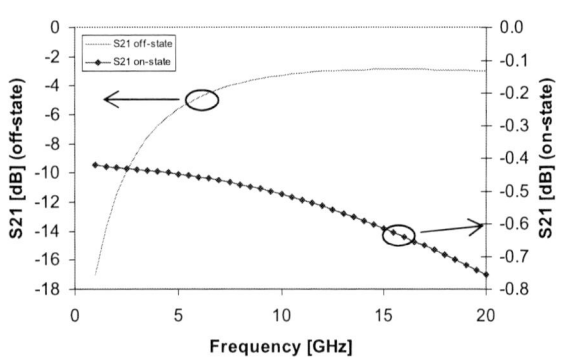

Fig. 8 Simulated S21 of the Momentum based model in both states.

IV. SPDT MMIC DESIGN

The topology consists of two stages of series-parallel switch pairs. Inductors are placed in parallel with the series switches, to create a resonance at 8 GHz and 11 GHz. This resonance gives an isolation improvement by the resonance of the off-state capacitance with the inductors in parallel. The ports of the SPDT switch are matched to 50 Ω at X-band with an LC matching network. 1.5 kΩ gate resistors are applied to improve the insertion loss per FET. A simplified schematic of the topology is shown in Figure 9 and a photograph in Figure 10.

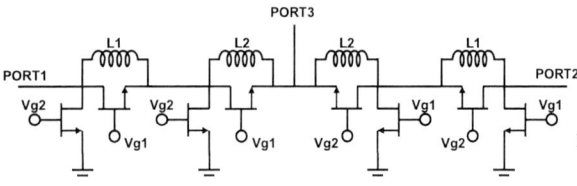

Fig. 9 Topology of the designed SPDT switch.

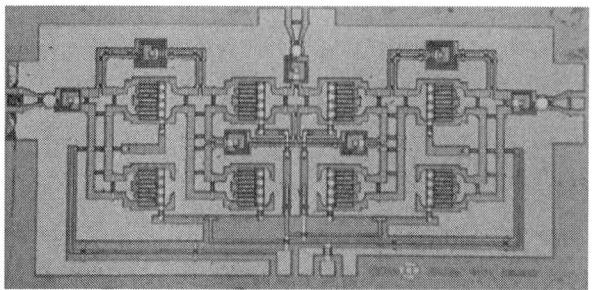

Fig. 10 Chip photograph, size 4x1.8 mm.

V. MEASUREMENT RESULTS

S-parameters measurement results, with the simulation results included are presented in Figure 11 to Figure 14. In Figure 11 and Figure 12 the insertion loss and isolation are presented. Figure 13 and Figure 14 show the matching performance of the circuit. The bias voltage of the switch is 0 V (on-state FETs) and -20 V (off-state FETs).

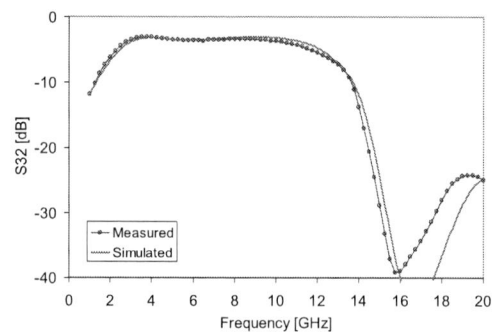

Fig. 11 Measured and simulated insertion loss.

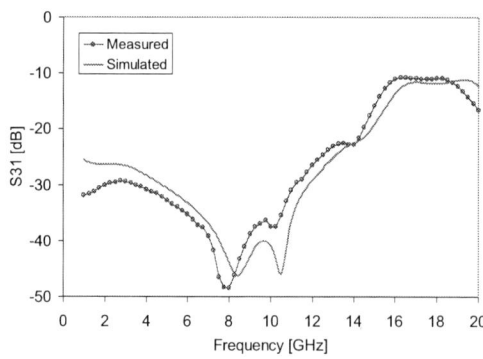

Fig. 12 Measured and simulated isolation.

978-2-8748-7007-1/08 $25.00 © 2008 EuMA

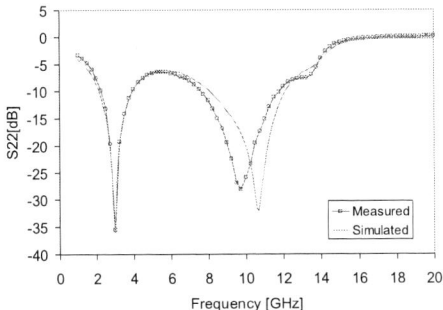

Fig. 13 Measured and simulated port return-loss (Front-end electronics side).

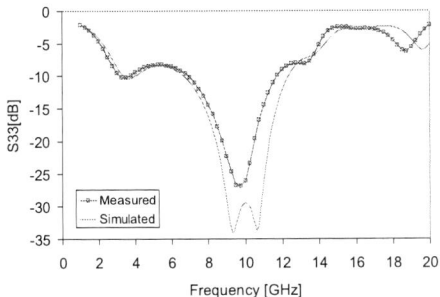

Fig. 14 Measured and simulated port return-loss (antenna side).

A good match is achieved between small-signal measurement and simulation. The achieved isolation is very good (35 dB in X-band). The input and output return losses are better than 10 dB in X-band. The insertion loss is high (3.5 dB), due to a relative low off-state isolation of the devices.

Power sweep measurement results are presented in Figure 15 and Figure 16. Figure 15 presents the pulsed power sweeps in on-state at different duty-cycles. A pulse-period of 100 µs is used. A power handling of 44 dBm is measured. Figure 16 presents the pulsed power sweep in off-state versus gate voltage. The isolation is uniform over the measured dynamic range.

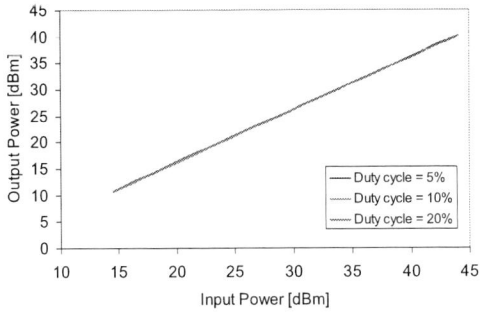

Fig. 16 Measured power sweep with -20 V off-state gate voltage

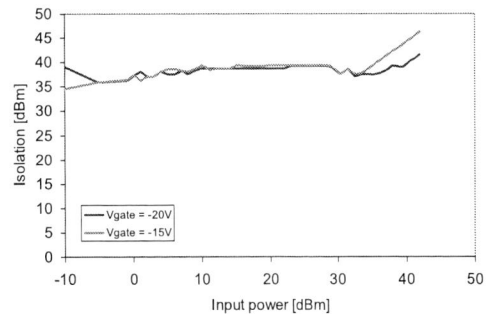

Fig. 15 Measured large signal isolation versus input power.

When presenting higher input power to the SPDT switch the input capacitor breaks down at 45 dBm input power. The input capacitors are not DC-decoupling capacitors, but capacitors to ground as part of the matching network, which makes these sensitive for high voltage swings. In the second iteration this is improved by applying only series DC-decoupling input capacitors.

VI. CONCLUSIONS

An X-band SPDT switch, with over 25 W linear power handling is designed in gallium nitride (GaN), coplanar waveguide technology. The employed resonance inductors in parallel to the series FETs improve the isolation of the switch.

Analysis and measurement are carried out successfully. The switch is designed with Momentum based switch FET models, characterised from common source FET. The measurement results of the SPDT switch are matching with the Momentum based simulations results. Measurement shows that input capacitors are the power limitations of the presented SPDT switch.

ACKNOWLEDGMENT

The results in this paper have been made in the scope of the European Defence Agency (EDA) funded program, KORRIGAN.

REFERENCES

[1] Mishra, U.K., et al., " GaN-Based RF Power Devices and Amplifiers ", Invited paper, Proceedings of the IEEE, Vol 96, No 2, February 2008.
[2] Gauthier, G. and Reptin, F., "KORRIGAN: Development of GaN HEMT Technology in Europe", CS MANTECH Conference, 2006, Vancouver, Canada.
[3] Devlin L. "Design of integrated switches and phase shifters". *July 2003*, www.plextek.co.uk/papers/swdes2.pdf.

High Power Microstrip GaN-HEMT Switches for Microwave Applications

V. Alleva[#1], A. Bettidi[#2], A. Cetronio[#2], W. Ciccognani[*3], M. De Dominicis[#1], M. Ferrari[*3], E. Giovine[*4], C. Lanzieri[#2], E. Limiti[*3], A. Megna[#1], M. Peroni[#2], P. Romanini[#2]

[#1]Elettronica SpA, Rome 00131, Italy

[#2]SELEX Sistemi Integrati SpA, Rome 00131, Italy

[1]marco.dedominicis@elt.it

[2]acetronio@selex-si.com

[*3]Electronic Department, University of Rome "Tor Vergata", Rome 00133, Italy

[*4]CNR-IFN, Rome 00156, Italy

[3]limiti@uniroma2.it

Abstract— **In this paper the design, fabrication and test of X-Band and 2-18GHz wideband high power SPDT MMIC switches in microstrip GaN technology are presented. Such switches have demonstrated state-of-the-art performances. In particular the X-Band switch exhibits 1dB insertion loss, better than 37dB isolation and a power handling capability at 9 GHz of better than 39dBm at 1dB insertion loss compression point; the wideband switch has an insertion loss lower than 2.2dB, better than 25dB isolation and a power handling capability of better than 38dBm in the entire bandwidth.**

I. INTRODUCTION

AlGaN/GaN HEMT is a pacing technology mainly targeted for high power applications at microwave and millimetre-wave frequencies, owing to its high critical breakdown field, its one-order of magnitude higher power density, its saturation drift velocity and the availability of a high thermal conductivity SiC substrate [1]. Apart from their exceptional power performance [2]-[3], GaN HEMTs are promising candidates for robust low-noise applications due to their low-noise performance combined to their high power handling capability, providing major advantages in terms of linearity and robustness [4].

Although some possibility for RF switching devices has been investigated, there have been few studies to date on the use of GaN HEMTs for high power microwave and RF control applications [5], [6]. For this applications low-loss/high power RF switches are requested, especially for multifunction antenna systems where a single aperture is used to serve multiple applications by simply switching between transmit and receive channels of the component.

The status on RF power switches to date is that traditional p-i-n diode based T/R switches encounter additional losses as a result of their intrinsic dc power consumption; the drive towards GaAs FET-based switches is hampered by the relatively low breakdown voltage of these components which require multi-stage design configurations with active device series connections to divide the maximum voltage of the input signal; new semiconductor technologies based on wide-bandgap materials such as Galliun Nitride (GaN), because of their higher breakdown voltages, promise to extend the power

level of such microwave circuits by at least a factor of five and to appreciably reduce the overall chip size and cost.

This contribution illustrates that GaN-HEMT technology is sufficiently mature for power switch applications and as such in a good position for dedicated work to further optimize "switch transistor" technology and topology to improve current performance status. To this goal the present work illustrates the design, fabrication and RF performance of X-Band and 2-18GHz wideband MMIC SPDT switches, based on a Microstrip (MS) AlGaN/GaN HEMT on SiC technology, demonstrating the good level of maturity reached in terms of fabrication yields and performance.

II. DEVICE/CIRCUIT FABRICATION PROCESS

The SPDT switches reported below have been fabricated with the current SELEX Sistemi Integrati GaN-HEMT Microstrip (MS) MMIC technology. The process is based on an epi-layer structure of GaN/AlGaN/GaN deposited on semi insulating SiC substrates by either MOCVD or MBE techniques. The mask levels for MMIC fabrication are based on a mix and match procedure utilising both I-Line Stepper and Electron Beam Lithography (EBL) processes. The latter is used only for the fabrication of the high resolution quarter micron Gate dimensions necessary for the HEMT devices. Drain and Source electrodes of the devices are made by ohmic contact formation of a Ti/Al/Ni/Au metallisation to the GaN/AlGaN epi-layer via a high temperature alloying cycle. Wafer passivation for surface protection is carried out by SiN plasma-enhanced chemical vapour deposition (PE-CVD), while the active device isolation is achieved via Fluorine ion implantation. The SiN passivation film deposition has been optimised in order to minimize the carrier trap concentration at the interface with the semiconductor in order to minimise the detrimental drain dispersion phenomenon of the switch transistors.

After active device formation the MMIC fabrication process comprises: the deposition of NiCr thin-film resistors, Metal-Insulator-Metal (MIM) capacitors and electro-plated inductors and interconnect transmission lines, with air-bridges where necessary.

The fabrication process is concluded with back-side wafer processing for the fabrication of through substrate via-hole interconnects. Said process comprises: wafer thinning down to circa 70μm, via-hole etching by means of an Inductively Coupled Plasma (ICP) dry etch process and finally back-side (substrate and via-hole) metallisation with a 10μm thick electro-plated Au film deposited on an appropriate barrier metal layer.

III. X-BAND POWER SWITCH DESIGN AND PERFORMANCE

To design the microstrip X-band SPDT switch, S-parameters measurements have been performed on two devices with gate periphery equal to 600μm (6x100μm) and 300μm (3x100μm). In particular S-parameter measurements have been performed at different gate biases in cold FET conditions and a model for on-state and off-state as a function of the gate width has been extracted.

The design goal was to achieve an X-Band switch with bandwidth 8GHz to 11GHz, insertion loss better than 1dB, isolation higher than 35dB and power handling at 1dB compression (P1dB) greater than 5Watt. The objective was to demonstrate that GaN-based switches can simultaneously achieve reasonably high power handling capability without sacrificing the isolation and the insertion loss. In order to satisfy these requirements, a circuit configuration, shown in Fig. 1, consisting of one series and three shunt transistors has been adopted.

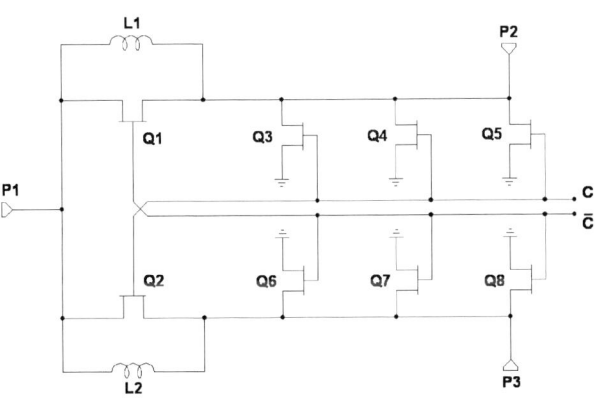

Fig. 1 Circuit topology of the X-band AlGaN/GaN SPDT power switch.

To increase the isolation performance of the series devices (Q1 and Q2), an inductive (L1) compensation has been implemented, as proposed in [7]. When the transistor Q1 is in the ON state (i.e. ohmic mode), the inductance has no effect since it is in parallel with a low impedance and switch insertion loss is not affected in the frequency band of interest. Vice versa, when Q1 is OFF (i.e. pinch off), it behaves as a large resistor (several thousands ohms) in parallel with a large capacitance; this parasitic capacitance reduces the isolation at high frequencies. To overcome this problem, the inductance L1 has been chosen to resonate at the upper side of the frequency band of interest so to increase the isolation of the switch.

A micrograph of the fabricated SPDT switch, with overall dimensions 2.4x1.9 mm, is illustrated in Fig. 2.

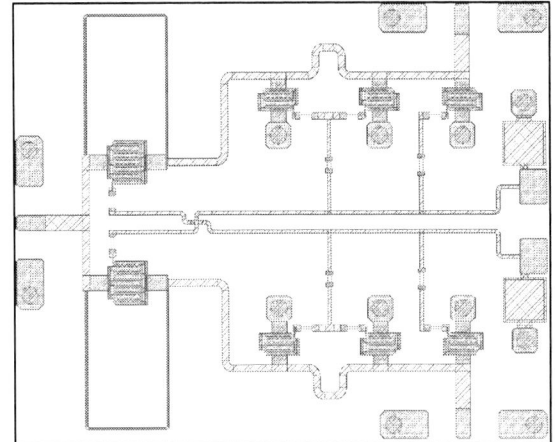

Fig. 2 Micrograph of the X-band MMIC SPDT switch.

Small-signal performances of the X-Band switch has been evaluated by means of on-wafer measurements utilizing a vector network analyzer and related air-coplanar probes with orthogonal CPW SOLT calibration. Comparisons between measured and simulated results for insertion loss, isolation and related port matching are presented respectively in Fig. 3 and Fig. 4. As shown, there is excellent agreement between measured and simulated results. In particular the insertion loss equals 1dB and isolation is better than 37 dB in the whole bandwidth. The input and output port matching are better than 13dB.

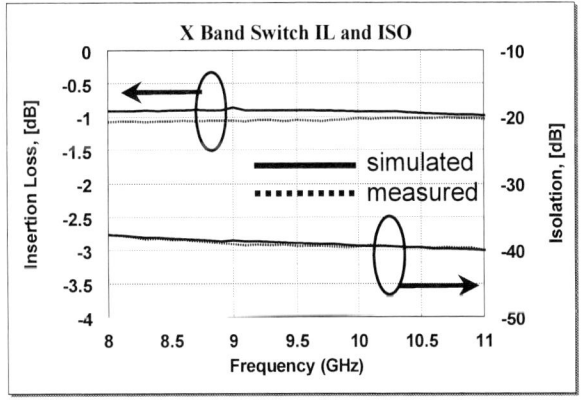

Fig. 3 Measured and simulated insertion loss and isolation for X-band switch. Control voltage: VC=0V for insertion loss measurement and VC=-20V for isolation measurement.

To characterize the power performance, the switch has been mounted in a test-jig with coaxial input and output ports. The Pin-Pout characteristics of the switch have been measured between 8 and 11GHz, in pulsed operation (duty-cycle 25% and pulse-width 100usec). As shown in Fig. 5, a very high power handling has been measured at 9GHz, where an input power equal to 39dBm causes no compression phenomena.

978-2-8748-7007-1/08 $25.00 © 2008 EuMA

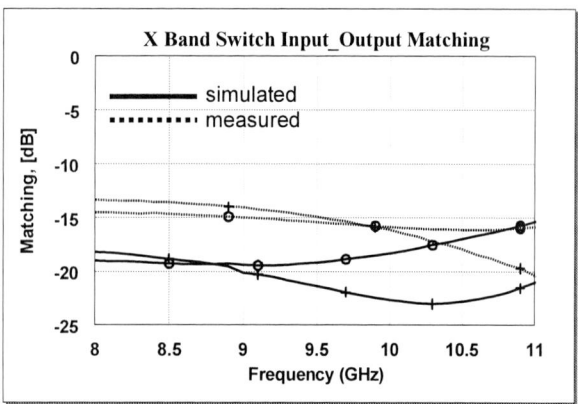

Fig. 4 Measured and simulated port matching for X-band switch (circles: S11, crosses: S22).

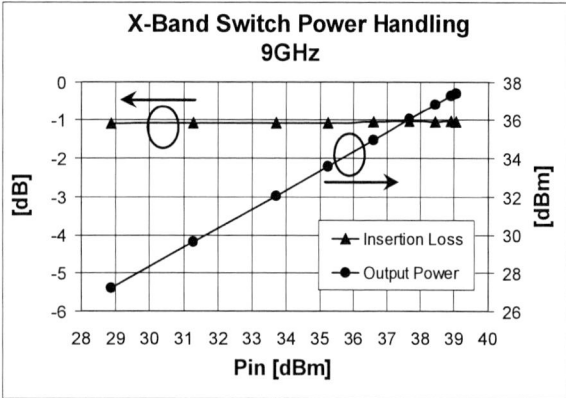

Fig. 5 Measured insertion loss and output power as a function of input power for X-band MMIC switch at 9GHz (VC=-20V).

IV. WIDEBAND POWER SWITCH DESIGN AND PERFORMANCE

The wideband SPDT power switch is based on the same transistor topologies and models outlined for the X-Band version above. The switch has been designed with a smaller series transistor, with gate periphery equal to 400µm (4x100µm), to achieve low insertion loss in the entire bandwidth.

In this case, the design goal was to achieve a wideband switch with bandwidth 2GHz to 18GHz, insertion loss better than 2dB, isolation higher than 25dB and power handling at 1dB compression (P1dB) greater than 5Watt.

As illustrated in Fig. 6, a three-FET circuital topology has been adopted, with the first device in series with the RF signal and the other two in shunt configuration. In this schematic the series HEMTs Q1 and Q3-Q4 are controlled by two control voltages $\overline{C}$ and C, one at zero gate bias and the other at a negative bias larger than the transistor pinch-off voltage.

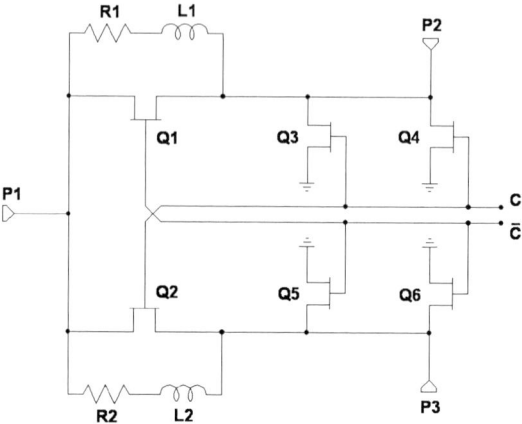

Fig. 6 Circuit topology of the wideband SPDT power switch

Similar to the X-Band circuit, the wideband design has been performed using the compensation concept [7]: in particular a resistor R1 has been added in series with the inductance L1. When Q1 is in the ON state, this series elements don't have relevant effects because in parallel with a low impedance and the switch insertion loss is not affected in the frequency band of interest. When Q1 is OFF (i.e. pinch off), the capacitance of the device resonates with the inductance L1. This resonance still reduces the capacitive effect near the resonance frequency and consequently increases the switch isolation; differently from the X-band design, the resistance has the effect to reduce the quality factor of the resonance thus to improve the positive effect introduced by the inductance for a wide band of frequencies.

A micrograph of the fabricated wideband SPDT switch, with overall dimensions 2.0x1.7 mm, is illustrated in Fig. 7.

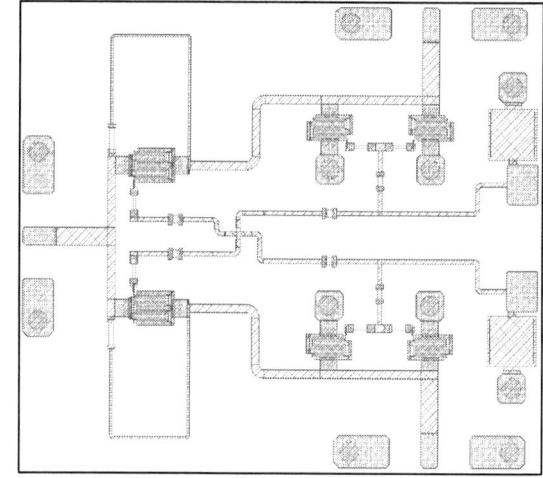

Fig. 7 Micrograph of the 2-18 GHz MMIC SPDT Switch

Small-signal performances of the wideband switch has been evaluated by on-wafer measurements, in a similar manner to the X-band switch reported above. Comparison of measured and simulated insertion loss and isolation are illustrated in Fig. 8.

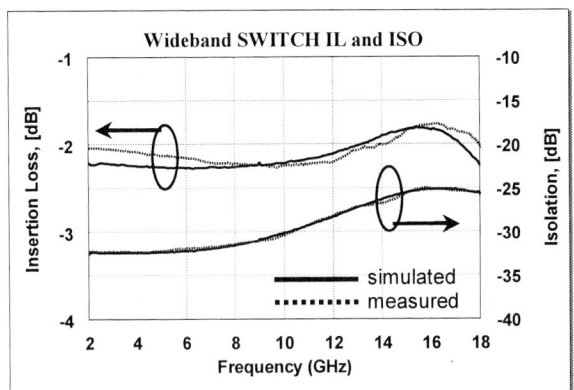

Fig. 8 Measured and simulated insertion loss and isolation of wideband switch. Control voltages: V_C=0V for insertion loss measurement and V_C=-20V for isolation measurement.

As for the X-band circuit, the wideband switch exhibits excellent correlation between measured and simulated performances. In particular, measured insertion loss is lower than 2.2dB in the overall bandwidth and isolation is higher than 25dB. The input and output port matching are better than 11dB.

With a control voltage equal to -20V, the switch has an insertion loss compression of 1dB for input power higher than 38dBm in the entire bandwidth [8]. Fig. 9 shows the measured insertion loss and output power as a function of input power at 12GHz.

V. CONCLUSION

In this contribution we have illustrated the design, fabrication and RF performance evaluation of high power X-Band and wideband 2-18GHz switches. GaN-HEMT technology is demonstrating therefore a good level of maturity for microwave power switch applications and as such in a good position for dedicated development for optimising "switch transistor" technology and topology to specific applications.

The two MMIC SPDT switches, based on MS AlGaN/GaN HEMT on SiC substrate technology, have demonstrated state-of-the-art linear performances. In particular the X-band power switch has demonstrated an insertion loss equal to 1dB, an isolation higher than 37dB and a power handling capability, at 9GHz, better than 39dBm at the 1dB insertion loss compression point; the wideband switch has shown an insertion loss lower than 2.2dB, an isolation higher than 25dB and a power handling capability better than 38dBm at the 1dB insertion loss compression point in the entire bandwidth.

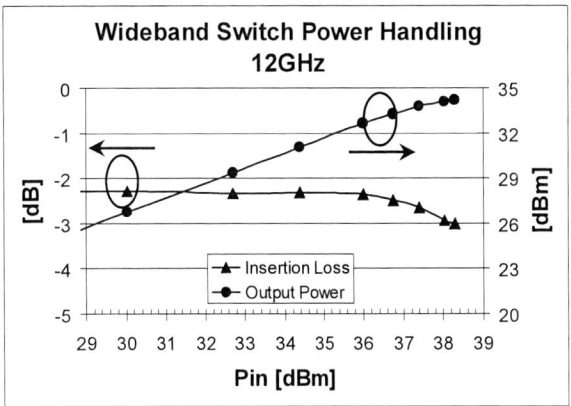

Fig. 9 Measured insertion loss and output power as a function of input power for X-band MMIC switch at 12GHz (VC=-20V).

ACKNOWLEDGMENT

This research activity has been carried out in the KORRIGAN RPT N° 102.052 Project funded within the EUROPA framework in the CEPA 2 priority area. The authors wish to acknowledge the consortium members for their contribution and the European Defence Agency for its financial support.

REFERENCES

[1] R. J. Trew, "SiC and GaN – Is There One Winner for Microwave Power Applications ?", *Proceedings of the IEEE,* , vol. 90, no. 6, pp. 1032–1047, June 2002.

[2] D. E. Meharry, R. J. Lender, K. Chu, L. L. Gunter and K. E. Beech, "Multi-Watt Wideband MMICs in GaN and GaAs", in *IEEE MTT-S International Microwave Symposium*, 3-8 June, 2007, pp. 631-634.

[3] M. Micovic, A. Kurdoghlian, H. P. Moyer, P. Hashimoto, A. Schmitz, I. Milosavjevic, P. J. Willadesn, W.-S. Wong, J. Duvall, M. Hu, M. J. Delaney, D. H. Chow, "Ka-Band MMIC Power Amplifier in GaN HFET Technology", in *IEEE MTT-S International Microwave Symposium*, 6-11 June 2004, vo. 3, pp. 1653-1656.

[4] D. Krausse, R. Quay, R. Kiefer, A. Tessmann, H. Massler, A. Leuther, T. Merkle, S. Muller, C. Schworer, M. Mikulla, M. Schlechtweg, and G. Weimann, "Robust GaN HEMT low-noise amplifier MMICs for X-band applications", *Gallium Arsenide applications Symposium*, October 2004, pp. 71-74.

[5] V. Kaper, R. Thompson, T. Prunty, J.R. Shealy, "Monolithic AlGaN/GaN HEMT SPDT switch", *Gallium Arsenide applications Symposium*, October 2004, pp. 83-86.

[6] H. Ishida, Y. Hirose, T. Murata, A. Kanda, Y. Ikeda, T. Matsuno, K. Inoue, Y. Uemoto, T. Tanaka, T. Egawa, and D. Ueda, "A high power Tx/Rx switch IC using AlGaN/GaN HFETs", *International Electron Device Meeting*, December 2003, pp. 23.6.1-23.6.4.

[7] A. Ezzeddine, R. Pengelly, H. Badawi, "A High Isolation DC to 18 GHz Packaged MMIC SPDT Switch", *European Microwave Conference*, October 1988, pp. 1028 - 1033.

[8] A. Bettidi, A. Cetronio, M. De Dominicis, G. Giolo, C. Lanzieri, A. Manna, M. Peroni, C. Proietti, P. Romanini, "High Power GaN-HEMT Microwave Switches for X-Band and Wideband Applications", *accepted to be presented at 2008 IEEE Radio Frequency Integrated Circuits (RFIC) Symposium.*

InAs/In$_{1-x}$Ga$_x$As Composite Channel High Electron Mobility Transistors for High Speed Applications

Edward Yi Chang[1], Chien-I Kuo[1], Heng-Tung Hsu[2], Chia-Yuan Chang[1]

[1]*Department of Materials Science and Engineering, National Chiao-Tung University*
[1]edc@mail.nctu.edu.tw
[2]*Department of Communications Engineering, Yuan Ze University*

Abstract—**80-nm InAs channel HEMTs with different lattice matched sub-channels, In$_{0.53}$Ga$_{0.47}$As and In$_{0.7}$Ga$_{0.3}$As, have been fabricated. The device with InAs/ In$_{0.7}$Ga$_{0.3}$As composite channel exhibits high drain current density (1101 mA/mm), and high transconductance (1605 mS/mm) at drain bias V$_{DS}$ = 0.8 V. The high current gain cutoff frequency (f_t) of 360 GHz and maximum oscillation frequency (f_{max}) of 380 GHz of the device with InAs/ In$_{0.7}$Ga$_{0.3}$As were obtained at V$_{DS}$ = 0.7 V in comparison to the InAs/In$_{0.53}$Ga$_{0.47}$ As channel HEMTs with f_t = 310 and f_{max} = 330 GHz. This is due to the high electron mobility and electron confinement in the InAs/ In$_{0.7}$Ga$_{0.3}$As channel. In addition, a low gate delay time 0.84 psec was obtained at V$_{DS}$ = 0.5 V. The excellent performance of the InAs channel HEMTs demonstrated in this study shows great potential for high speed and very low power logic applications with the optimal design of In$_{0.7}$Ga$_{0.3}$As/InAs/In$_{0.7}$Ga$_{0.3}$As composite channel.**

I. INTRODUCTION

Low DC power consumption devices and circuits have always been desired for all kinds of communications systems especially the satellite communications. With the breakthrough in semiconductor hetero-structure technologies, planar III-V compound semiconductor field effect transistors (FETs) or high electron mobility teansistors (HEMTs) have been identified as one of the most attractive devices for such applications. Besides, in order to extend Moore's law well into the next decade, III-V based heterostructure devices is also a potential candidate FETs for low-power logic applications beyond Si CMOS technology in 22 nm node era [1-3].

The excellent RF-performance has been demonstrated by InAlAs/InGaAs HEMTs on InP substrate or GaAs substrate [4-6]. In fact, higher electron mobility and velocity can be realized by reducing of gate length (L_g) or increasing of the indium content in the InGaAs channel. Due to the high electron mobility (~ 20000 cm^2/V • s), high drift velocity, reasonable energy band gap (0.36 eV), and large conduction band offset in InAs, InAs-channel HEMT is considered very promising for high speed and low power logic applications.

In this study, channel structures of HEMTs with different indium concentration as upper and lower lattice-matched sub-channels, including In$_{0.53}$Ga$_{0.47}$As and In$_{0.7}$Ga$_{0.3}$As are fabricated and evaluated for the RF and digital performances.

II. EXPERIMENT

The epitaxial structure of the InAs-channel with different In$_{1-x}$Ga$_x$As cladding layer as the sub-channel (In$_{0.53}$Ga$_{0.47}$As and In$_{0.7}$Ga$_{0.3}$As) HEMTs were grown by molecular beam epitaxy (MBE) on InP substrate, respectively. These sub-channels were applied to minimize the interface roughness scattering and further enhance the electron confinement in the thin strained InAs layer and improve the electron transport properties [7]. In addition, both of these two types of devices used the InP etching stop layer to improve the selectivity of the wet chemical recess etch and provide semiconductor surface passivation on each side of the gate to reduce the trapping effect on the InAlAs surface [8]. With the use of the InP etching stop layer, the lateral recess length was easy to control. A high uniform threshold voltage (V$_{th}$) was achieved and RF performance was improved using the InP etch layer [9].

For the device fabrication, the active area of the device was isolated by wet etch. Source and drain Ohmic metal were formed with 240-nm-thick Au/Ge/Ni/Au and alloyed by rapid thermal annealing at 250 oC for 30 sec. As a result of the highly Si doped cap, a low Ohmic contact resistance (Rc) and an sheet resistance (Rsh) were obtained by using the transmission line model method. A tri-layer resist system of ZEP-520/PMGI/ZEP520 was used for the E-Beam lithography. The gate-recess was performed precisely by wet chemical etching using succinic acid-based solution because the side-recess spacing played an important role in balancing short-channel effect and frequency response [10]. The Ti/Pt/Au gate metal was formed by evaporation and lift off. The gate length of the T-shaped gate was 80 nm. Fig. 1 shows the cross-sectional scanning electron microscopy (SEM) image of the T-shaped gate resist profile. Finally, devices were passivated using a 100-nm-thick PECVD (plasma enhanced chemical vapor deposition) silicon nitride film.

III. RESULT AND DISSCUSION

Fig. 2 (a) and 2 (b) shows the current-voltage characteristics of 2 × 50 µm devices of InAs/In$_{0.53}$Ga$_{0.47}$As composite channel HEMTs. As observed from the figure, this device can be well pinched off with a threshold voltage of -0.8 V. The fabricated InAs MHEMT shows a drain current density of 834 mA/mm and transconductance, g$_{m,max}$ of 1450 mS/mm at V$_{DS}$ = 0.8 V. On the other hand, the DC characteristics of the InAs/In$_{0.7}$Ga$_{0.3}$As composite channel

978-2-8748-7007-1/08 $25.00 © 2008 EuMA

HEMTs are shown in the Fig. 3 (a) and (b). The drain current density of 1101 mA/mm and $g_{m,max}$ of 1605 mS/mm are observed. The high transconductance was due to higher carrier concentration and superior electron transport properties in the InAs/In$_{0.7}$Ga$_{0.3}$As channel. The result indicates that the InAs-channel HEMTs can be operated at low drain bias condition to reduce dc power dissipation and is suitable for low power applications.

The S-parameters of the 2 × 50 μm device were measured from 5 to 80 GHz using on-wafer probing system with HP8510XF network analyzer. A standard Load-Reflection-Reflection-Match (LRRM) calibration method was used to calibrate the measurement system. The parasitic effects (mainly capacitive) due to the probing pads have been carefully removed from the measured S-parameters using the same method as in [11] and the equivalent circuit model in [12]. Fig. 4 shows the frequency dependence of the current gain H$_{21}$, the power gain MAG/MSG, and the Mason's unilateral gain U of the InAs/In$_{0.53}$Ga$_{0.47}$As device measured at V$_{DS}$ = 0.7 V and V$_{gs}$ = -0.4 V. The f_T and f_{max} are 310 and 330 GHz, respectively by extrapolating H$_{21}$ and MAG/MSG by least square fitting with a -20 dB/dec slope. Comparison to the InAs/In$_{0.7}$Ga$_{0.3}$As channel device, a very high current gain cut-off frequency f_T of 360 GHz and the maximum oscillation frequency f_{max} of 380 GHz were obtained and the results are shown in Fig. 5. The excellent RF performance of InAs/In$_{0.7}$Ga$_{0.3}$As composite channel device is mainly caused by high transconductance.

To characterize such device for high-speed logic applications, the gate delay time (CV/I) or intrinsic speed was also evaluated. C is the total gate capacitance including gate-to-source capacitance (C$_{gs}$) and gate-to-drain capacitance (C$_{gd}$) which extracted from the high frequency S-parameter. V is the applied drain voltage and equal to the power supply voltage (V$_{CC}$). I is the on-state current with the bias point. According to the definition in [13-14], the gate delay time of both devices were calculated to be 0.87 (InAs/In$_{0.53}$Ga$_{0.47}$As) and 0.84 (InAs/In$_{0.7}$Ga$_{0.3}$As) psec at the same drain voltage V$_{DS}$ = 0.5 V, respectively. Fig. 6 shows the comparison of delay time for different types of devices (InAs HEMTs, advanced Si-MOSFETs and InSb HEMTs) as a function of gate length [2, 15]. In this plot, it can be seen that the 80 nm InAs HEMTs exhibits excellent intrinsic speed than the state of-the-art Si NMOSFETs at the same gate length. It indicates the InAs is a material with great potential for low-voltage high-speed III-V-based logic circuit applications. Overall, the InAs/In$_{0.7}$Ga$_{0.3}$As HEMTs exhibit very well intrinsic speed, DC and RF performances owing to the appropriate epi-structure design, and short gate length.

IV. CONCLUSIONS

In this study, high performance 80 nm InAs/In$_{1-x}$Ga$_x$As composite-channel HEMTs with different indium content sub-channel has been fabricated and characterized. The high current gain cutoff frequency (f_T) of 360 GHz and maximum oscillation frequency (f_{max}) of 380 GHz were obtained at V$_{DS}$ = 0.7 V with a calculated low gate delay time (0.84 psec at

V$_{CC}$ = 0.5 V) due to the high electron mobility in the InAs/In$_{0.7}$Ga$_{0.3}$As channel. Overall, the device with In$_{0.7}$Ga$_{0.3}$As sub-channel displays better characteristics than with In$_{0.53}$Ga$_{0.47}$As sub-channel because of the carrier confinement improvement and superior transport property. Finally, the device exhibits the sate-of-the-art performance and the device technology demonstrated is suitable for high-frequency milimeter wave and high-speed logic applications.

ACKNOWLEDGMENT

The authors would like to acknowledge the assistance and support from the National Science Council and the Ministry of Economic Affairs, Taiwan, R.O.C., under the contracts NSC 96-2752-E-009-001-PAE and 95-EC-17-A-05-S1-020. Part of this work was supported by the "Nanotechnology Support Project" of the Ministry of Education, Culture, Sports, Science and Technology (MEXT), Japan.

REFERENCES

[1] D. H. Kim, J. A. Alamo, J. H. Lee, and K. S. Seo, "Performance Evaluation of 50 nm In$_{0.7}$Ga$_{0.3}$As HEMTs for beyond-CMOS logic applications," in *IEDM Tech. Dig.*, 2005, pp.767-770.

[2] R. Chau, S. Datta, and A, Majumdar, "Opportunities and Chanllenges of III-V Nanoelectronics for future high-speed,low-power logic applications," in *Proc. IEEE CSIC Dig.*, pp. 17-20, 2005.

[3] C. I. Kuo, H. T. Hsu, and E. Y. Chang, "InAs-Channel Based Quantum Well Transistor," Electrochemical and Solid-state Lett., 11 (7) pp. H193-H196 2008.

[4] Y. Yamashita, A. Endoh, K. Shinohra, K. Hikosaka, T. Matsui, S. Hiyamizu, and T. Mimura, "Pseudomorphic In$_{0.52}$Al$_{0.48}$As/ In$_{0.7}$Ga$_{0.3}$As HEMTs with an ultrahigh f_t of 562 GHz," *IEEE Electron Device Lett.*, vol. 23, pp. 573-575, Oct. 2002.

[5] K. Shinohra, Y. Yamashita, A. Endoh, I. Watanabe, K. Hikosaka, T. Matsui, T. Mimura, and S. Hiyamizu, "547 GHz f_t In$_{0.7}$Ga$_{0.3}$As/ In$_{0.52}$Al$_{0.48}$As HEMTs With Reduced Source and Drain Resistance," *IEEE Electron Device Lett.*, vol. 25, pp. 241-243, May 2004.

[6] S. J. Yeon, M. Park, J. Choi, and K. Seo, "610 GHz InAlAs/In$_{0.75}$GaAs Metamorphic HEMTs with an ultra-short 15-nm-gate," in *IEDM Tech. Dig.*, pp.613-616 2007.

[7] T. Akazaki, K. Arai, T. Enoki, and Y. Ishii, "Improved InAlAs/InGaAs HEMT Characteristics by Inserting an InAs Layer into the InGaAs Channel," *IEEE Electron Device Lett.*, vol. 13, pp. 325-327, June. 1992.

[8] G. Meneghesso, D. Buttari, E. Perin, C. Canali, and E. Zanoni, "Improvement of DC, low frequency and reliability properties of InAlAs–InGaAs InP-based HEMTs by means of an InP etch stop layer," in *IEDM Tech. Dig.*, Dec. 1998, pp. 227–230.

[9] T. Suemitsu, H. Yokoyama, T. Ishii, T. Enoki, G. Meneghesso, and E. Zanoni, "30-nm two-step recess gate InP-based InAlAs/InGaAs HEMTs," *IEEE Trans. Electron Devices*, vol. 49, no. 10, pp. 1694–1700, Oct. 2002.

[10] T. Suemitsu, H. Yokoyama, Y. Umeda, T. Enoki, and Y. Ishii, "High-performance 0.1-μm gate enhancement-mode InAlAs/InGaAs HEMTs using two-step recessed gate technology," *IEEE Trans. Electron Devices*, vol. 46, no. 6, pp. 1074-1080, Jun. 1999.

[11] Y. Yamashita, A. Endoh, K. Shinohara, M. Higashiwaki, K. Hikosaka, T. Mimura, S. Hiyamizu, and T. Matsui "Ultra-short 25nm Gate Lattice Match InAlAs/InGaAs HEMTs Within the Range of 400GHz Cut Off Frequency," *IEEE Electron Device Lett.*, vol. 22, pp. 367-369 August 2001

[12] G. Dambrine, A. Cappy, F. Heliodore, and E. Playez, "A new method for determining the FET small signal equivalent circuit," *IEEE Trans. Microwave Theory Tech.*, vol. 36, pp.1151-1159, 1988

[13] R. Chau, S. Datta, M. Docyz, B. Doyle, B. Jin, J. Kavalieros, A. Majumdar, M. Metz, and M. Radosavljevic, "Benchmarking Nanotechnology for High-Performance and Low-Power Logic Transistor Applications", *IEEE Trans. Nanotechnology*, Vol. 4, No. 2, pp. 153–158, 2005.

[14] J. Guo, A. Javey, H. Dai, and M. Lundstrom, "Performance analysis and design optimization of near ballistic carbon nano tube field effect transistors," in *IEDM Tech. Dig.*, 2004, pp.703-706.

[15] T. Ashley, A. R. Barnes, L. Buckle, S. Datta, A. B. Dean, M. T. Emeny, M. Fearn, D. G. Hayes, K. P. Hilton, R. Jefferies, T. Martin, K. J. Nash, T. J. Philips, W. H. A. Tang, and R. Chau, "InSb-based quantum well transistors for high speed, low power applications," in Proc. CS-Mantech Conf., Apr. 2005.

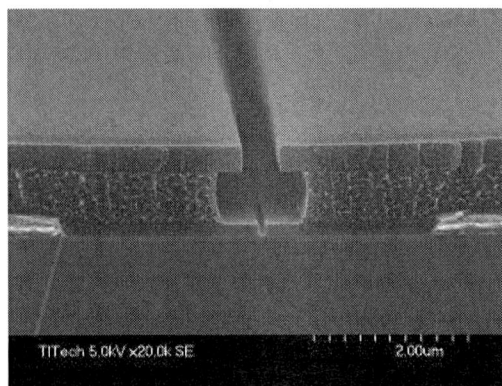

Fig. 1 Cross-sectional SEM image of the 80nm T-shaped gate resist profile.

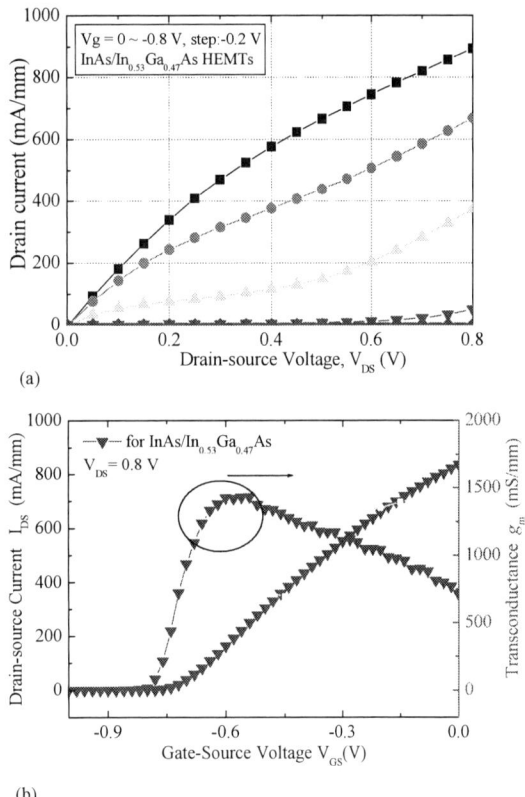

(a)

(b)

Fig. 2. Current-Voltage characteristics of the InAs/In$_{0.53}$Ga$_{0.47}$As 0.08 × 100 µm^2 HEMTs. (a) Drain-source current versus drain-source voltage curves (b) Transconductance versus gate-source voltage curves.

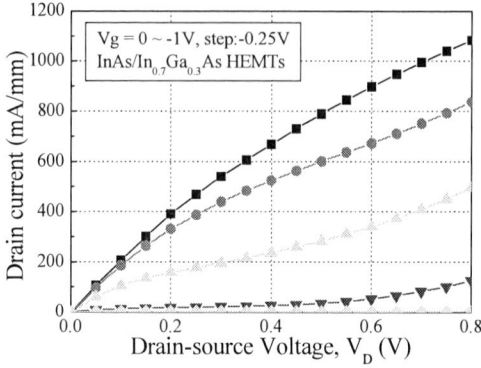

(a)

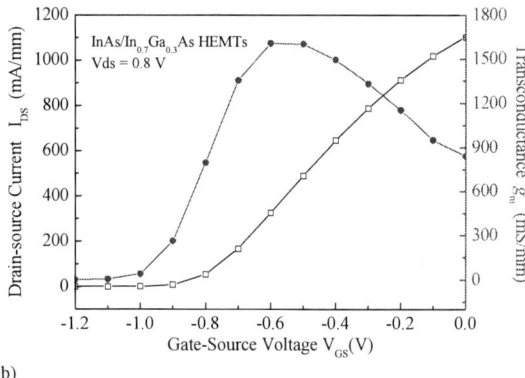

(b)

Fig. 3 Current-Voltage characteristics of InAs/In$_{0.7}$Ga$_{0.3}$As 0.08 × 100 µm^2 HEMTs. (a) Drain-source current versus drain-source voltage curves (b) Transconductance versus gate-source voltage curves

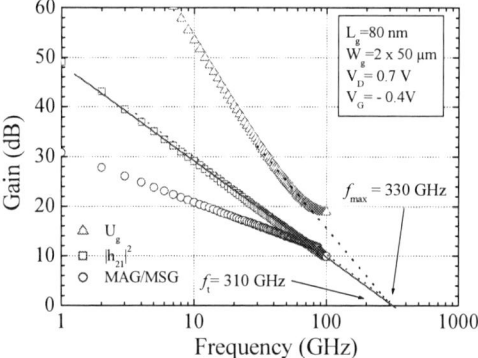

Fig. 4 Typical current gain |h$_{21}$|, MAG/MSG and U_g as a function of frequency of a 0.08 µm × 100 µm InAs/In$_{0.53}$Ga$_{0.47}$As HEMT. The V$_{DS}$ is 0.7 V, and the V$_G$ is – 0.4 V.

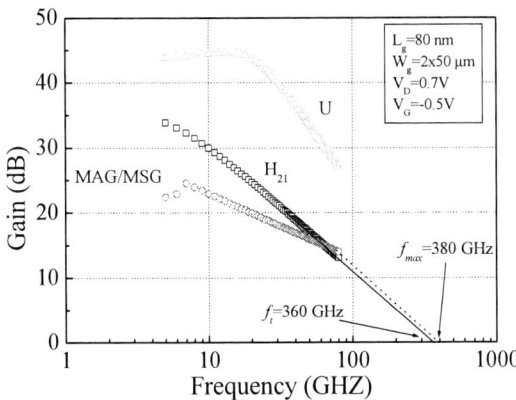

Fig. 5 Frequency dependence of the current gain H_{21}, the power gain MAG/MSG, and the unilateral gain U of the InAs/In$_{0.7}$Ga$_{0.3}$As composite channel HEMTs. The frequency range was from 5 to 80 GHz, and the device was biased at $V_{DS} = 0.7$V and $V_{gs} = -0.5$V.

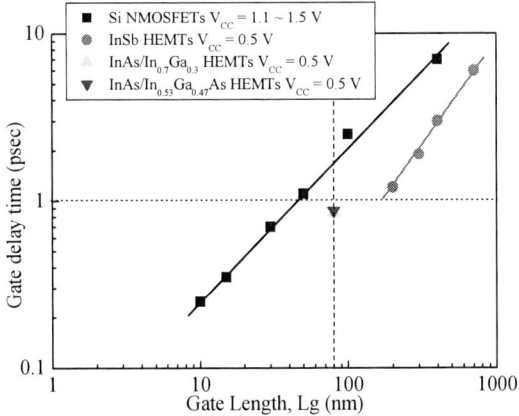

Fig. 6 Gate delay of InAs, InSb HEMTs and Si NMOSFETs as a function of gate length.

Integrated Schottky Structures for Applications Above 100 GHz

Byron Alderman[1], Hosh Sanghera[1], Bertrand Thomas[1], David Matheson[1], Alain Maestrini[2], Hui Wang[2], Jeanne Treuttel[2], Jose V. Siles[3], Steve Davies[4], Tapani Narhi[5]

[1]*Rutherford Appleton Laboratory, Chilton, Oxfordshire, OX11 0QX, UK*
[2]*Observatoire de Paris, LERMA, 75014 Paris, France*
[3]*Universidad Politécnica de Madrid, Ciudad Universitaria s/n, 28040 Madrid, Spain*
[4]*University of Bath, Bath, BA2 7AY, UK*
[5]*ESA/ESTEC, Keplerlaan 1, 2200 AG Noordwijk ZH, The Netherlands*
[1]B.Alderman@rl.ac.uk

Abstract— Recent developments in the fabrication of GaAs integrated Schottky structures for applications above 100 GHz are presented. Two approaches are discussed; the fabrication of integrated circuits using a GaAs foundry service, coupled with the research based post-processing of these structures, and the fabrication of discrete and integrated Schottky structures using a bespoke research laboratory.

I. INTRODUCTION

Low capacitance GaAs Schottky diode technology is required for millimetre and sub-millimetre wave heterodyne receivers. Schottky diodes operate at both ambient and cryogenic temperatures and are uniquely able to cover the frequency range from DC to above 1 THz. Schottky diode technology has been evolving for many years and has traditionally been driven by the demands of radio astronomy and remote sensing of the atmosphere. Ground based applications, e.g. security imaging, are now increasing in importance. For these applications, Schottky based technology offers an attractive alternative to detectors and sources that require cryogenic cooling [1].

Despite the growing demand for Schottky devices operating above 100 GHz, there remains limited availability within Europe and there is currently no space-qualified process available. The European Space Agency (ESA) has initiated a programme to investigate the use of the GaAs foundry service from United Monolithic Semiconductors (UMS) to fill the current gap between demand and availability (ESA AO/1-5084/06/NL/GLC). This programme aims to investigate the performance limitations of this GaAs foundry service and to explore ways of post-processing GaAs wafers to enhance device performance, for example, to reduce the dielectric loading around the anode to reduce the parasitic capacitance and the effect of dielectric loading. Using this approach, integrated Schottky structures have been designed for operation at frequencies upto 380 GHz.

Schottky diode fabrication is a relatively simple process which can be established in a research environment using optical lithography with simple manual alignment, deposition and etching tools. Structures in which the Schottky contact is integrated with an embedding network can also be fabricated in such an environment. In fact, a small research laboratory,

with its inherent flexibility, can be very effective in optimising Schottky structures. Whereas a GaAs foundry can be considered as a fixed process with a small number of wafers procured with a single reticule design repeated across a wafer, in a research environment it is often the case that relatively small samples are processed using contact lithography, rather than a stepper, and that each sample iterates to an optimum set of process conditions. A foundry therefore offers a stable and reliable fabrication process whereas a research laboratory offers the ability to develop novel structures.

As operating wavelengths of Schottky structures move into the sub-millimetre wave range, novel integration and substrate transfer or removal techniques are required [2]. Here we report on the design, fabrication and test of devices fabricated at a GaAs foundry to which additional post-processing has been applied in order to improve their performance at higher frequencies. We also report on the fabrication and test of discrete and integrated Schottky structures designed and fabricated in a bespoke research laboratory which are being developed specifically to operate in the sub-millimetre wave range. In both cases, technology demonstrators are being targeted at the 200 to 400 GHz range.

II. GAAS FOUNDRY DEVICES

The use of a GaAs foundry service to fabricate high frequency Schottky structures has the potential to supply a large number of identical devices without the effort of developing the fabrication technology, which is often very expensive and time consuming. However, circuits operating above 100 GHz often require non-standard processing to reduce the parasitic capacitance and dielectric loading of waveguide cavities. These techniques are expensive to develop and there is little incentive for them to be available within a foundry service, given the current level of demand. For these reasons we have investigated the electromagnetic advantages and corresponding effort of post-processing foundry devices. Our results indicate that provided the quality of the Schottky contact can be retained for anodes that are smaller than is defined by typical design rules, the structure of the circuit metallisation can be designed to be suitable to at least 380 GHz for discrete diodes. Furthermore, provided that appropriate post-processing can be performed, integrated structures could operate over the same frequency range.

A. Circuit Designs

A range of integrated Schottky circuits have been designed using a combination of 3D finite element analysis (HFSS) and non-linear circuit simulators (ADS). These circuits include:

- Sub-harmonic mixer at 183 GHz
- Sub-harmonic mixer at 380 GHz
- Frequency doubler at 190 GHz
- Frequency tripler at 90 GHz

These designs have been made with the assumption that certain post-processing will be performed on the GaAs wafer delivered by the foundry. For example, the sub-harmonic mixer at 380 GHz has been designed as a membrane structure; that is, the substrate of the circuit will be entirely removed leaving an epi-layer of thickness 4 μm. This ultra-thin circuit will be placed in a waveguide cavity, supported only by beam-leads, thin gold foils that extend beyond the edge of the circuit.

The design of the frequency doubler at 190 GHz was made assuming a substrate transfer technique. This approach is similar to the membrane process but instead of leaving the epi-structure suspended in air, the GaAs substrate is removed and replaced by quartz. This transferred substrate offers a lower dielectric constant and improved thermal conductivity, as compared to GaAs.

B. Post-processing

Epi-structures, with diodes and passive matching elements have been successfully transferred from a GaAs substrate to quartz. The process to do this involves an epitaxial lift-off followed by transfer to quartz with an epoxy adhesion layer. The thickness of the epoxy layer was approximately 1 μm and the quartz was 50 μm thick. Fig. 1 shows an optical image of this circuit. The total length of the circuit is 1.8 mm. The diodes are situated in a vertical line towards the right hand side of this image and are in an anti-series configuration. Fig. 2 is an SEM micrograph of this structure, with the quartz substrate visible towards the bottom of the image.

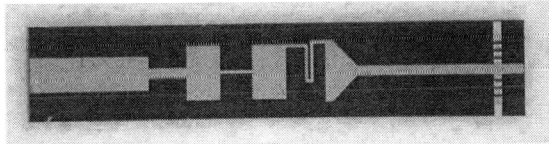

Fig. 1 Optical image of a 190 GHz doubler circuit (total length 1.8 mm).

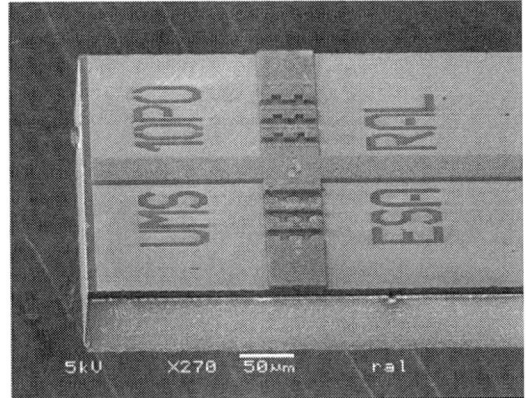

Fig. 2 SEM micrograph of a doubler circuit that has been transferred to a quartz substrate.

C. Results

RF results have been demonstrated on the 190 GHz frequency doubler with a transferred quartz substrate. The results presented in Fig. 3 match simulated performance in both bandwidth and expected output power. The multiplication efficiency is recorded to be of the order 6%. This is quite low for a traditional frequency doubler in this range, however, this has been achieved with an epi-structures which is optimised for mixer applications [3].

Significant RF results are yet to be reported for the mixer and frequency tripler demonstrators.

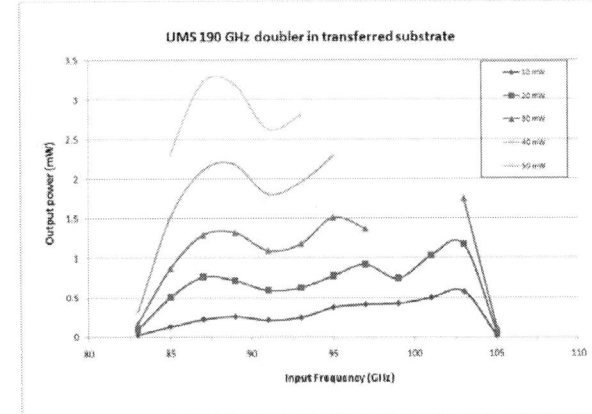

Fig. 3 RF results of a 190 GHz doubler using a transferred substrate process.

III. BESPOKE DEVICE FABRICATION

The ability to have full control over the design and fabrication of integrated Schottky structures offers significant advantages to the design of Schottky circuits operating above 100 GHz. The ability to fabricate air-bridged structures, significantly reduces the parasitic capacitance presented in parallel to the Schottky junction, improving bandwidth and performance. Techniques can be developed to fabricate very small anodes, typically sub-micron, and to ensure that where a pair of diodes are required, for example in the standard anti-parallel configuration of a sub-harmonic mixer, that these anodes have identical electrical performance. Furthermore, with full control of the process, it is possible to start investigating the use of substrate transfer or membrane techniques to gain control of the ways in which the substrate operates in the circuit, i.e., the ability to replace the GaAs of the substrate with a material with a high thermal conductivity, low dielectric constant, or to remove the substrate entirely.

A. The Development of a Stable Schottky Process

The mechanical and electrical yield of these devices is excellent given the limitations of contact lithography used. Extraction of key DC parameters, namely the series resistance and ideality, provides a good indication of the quality of Schottky diodes. Results presented in Table I give the averaged measured values of these parameters for series of 100 diodes of circular anodes of diameters from 1.1 to 2 μm. Associated standard deviations are given in parenthesis following the mean values. The ideality factors were calculated from measurements at currents of 10 and 100 μA, whereas the series resistance was extracted at a forward current of 1 mA. The anode diameters are quoted to an accuracy of 7% and were measured immediately prior to the deposition which forms the Schottky diode [4].

TABLE I
EXTRACTED DIODE PARAMETERS: AVERAGE AND STANDARD DEVIATIONS

Anode diameter (μm)	Series Resistance (Ω)	Diode Ideality
1.1	12.98 (0.45)	1.164 (0.002)
1.4	11.46 (0.38)	1.161 (0.002)
1.7	10.13 (0.48)	1.158 (0.002)
2.0	9.16 (0.29)	1.156 (0.002)

B. Discrete Flip-Chip Devices

The technique of flip-chip soldering discrete diodes to a circuit, typically gold-on-quartz, is commonly applied to millimetre wave circuits operating to approximately 400 GHz. Fig. 4 shows a schematic of a typical flip-chip device which has dimensions 120 x 40 x 15 μm³, (length x width x thickness). An SEM micrograph of the air-bridge detail of one of these structures is shown in Fig. 5. In this image, the length of the air-bridge is 20 μm and the channel over which this passes is 4 μm.

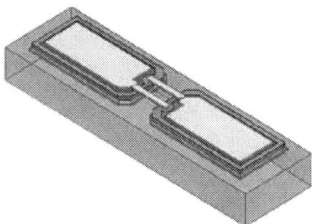

Fig. 4 A schematic of a flip-chip anti-parallel pair of Schottky diodes.

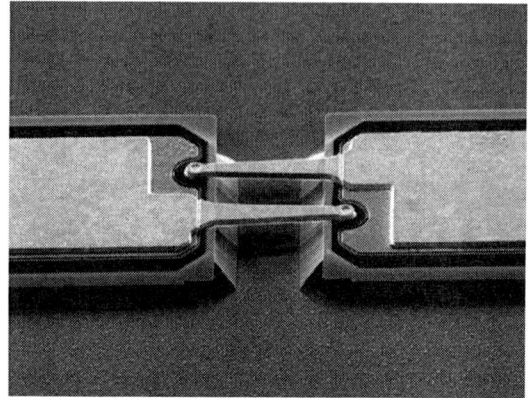

Fig. 5 An anti-parallel pair of air-bridged Schottky diodes.

Discrete flip-chip diodes from RAL have been measured in a range of fixed-tuned sub-harmonic mixers from 160 to 380 GHz. Only at 183 GHz have the circuits been specifically designed for these devices, where a double side band (DSB) mixer noise temperature below 500 K was recorded. A summary of these results is shown in Fig. 6.

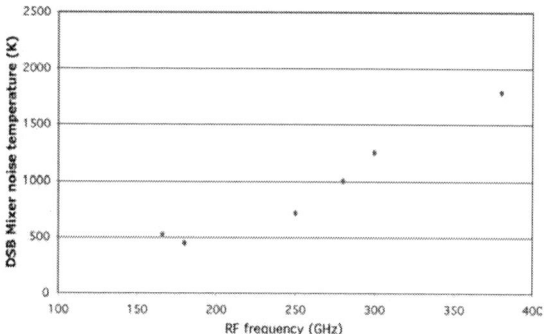

Fig. 6 A summary of mixer results using discrete diodes from RAL.

C. Integrated Structures

Beyond the frequency at which flip-chip diodes can be accurately placed without degrading the circuit performance, integrated Schottky structures are required. In the work presented here, initial demonstration is given of the circuit of a sub-harmonic mixer designed for 183 GHz. The approach taken in this demonstrator was to fabricate the entire circuit on 50 μm thick GaAs substrate. Alternative techniques are required at higher frequencies but this remains an important stepping-stone to higher frequency demonstrators. Fig. 6 shows an optical image of this integrated structure, housed in a waveguide cavity, with an SEM micrograph of an array of these devices shown in Fig. 7, together with a detailed view of the air-bridge structure.

978-2-8748-7007-1/08 $25.00 © 2008 EuMA

Fig. 6 An optical image of a diode/filter structure in a waveguide cavity.

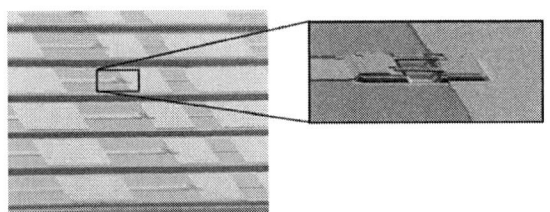

Fig. 7 An SEM micrograph of an integrated diode/filter structure of a 183 GHz sub-harmonic mixer.

Best performance for this device has realised a DSB mixer noise temperature of 550 K at 168 GHz. The offset in frequency from the design target of 183 GHz is due to the doping density of the GaAs being double that for which it was designed. A new structure on the intended epi-material is currently being processed.

IV. CONCLUSIONS

Integrated Schottky structures have been fabricated using both a modified foundry process from UMS and a bespoke research fabrication laboratory at the Rutherford Appleton Laboratory, UK. Both approaches have been shown to be suitable for the fabrication of integrated Schottky structures at frequencies near 200 GHz. Furthermore, simulations have indicated that with a combination of foundry based processing and additional post-processing in a research environment, integrated structures can be fabricated that are suitable for applications to 380 GHz.

ACKNOWLEDGMENT

The authors wish to thank E. Leclerc at United Monolithic Semiconductors for his substantial contribution to this research.

REFERENCES

[1] S.A. Maas, Nonlinear Microwave Circuits: Artech House, 1998.
[2] E. Schlecht, G. Chattopadhyay, A. Maestrini, A. Fung, S. Martin, D. Pukala, J. Bruston and I. Mehdi, "200, 400 and 800 GHz Schottky Diode 'Substrateless' Multipliers: Design and Results," *in Proc. IEEE MTT-S International,* pp. 1649-1652, Phoenix, AZ, May 2001
[3] D. Porterfield, T. Crowe, W. Bishop, D. Kurtz, E. Grossman, "A High-Pulsed-Power Frequency Doubler to 190 GHz", Proc. of the 30th Intl. Conf. on Infrared and mm-Waves, pp. 78-79, September 19-23, 2005, Williamsburg, Virginia, USA.
[4] B. Alderman, H. Sanghera, C. Price, B. Thomas, and D. N. Matheson, "Fabrication of reproducible air-bridged Schottky diodes for use at frequencies near 200 GHz," *Infrared and Millimetre Waves Conference 2007.*

978-2-8748-7007-1/08 $25.00 © 2008 EuMA

Heterostructure Barrier Varactor Quintuplers for Terahertz Applications

Jan Stake[#1], Tomas Bryllert[*2], Arne Øistein Olsen[*3], Josip Vukusic[#4]

[#]*Department of Microtechnology and Nanoscience, Chalmers University of Technology*
SE-412 96 Göteborg, Sweden
[1]jan.stake@chalmers.se
[4]josip.vukusic@chalmers.se

[*]*Wasa Millimeter Wave AB*
SE-42341 Torslanda, Sweden
[2]bryllert@wmmw.se
[3]olsen@wmmw.se

Abstract— **We present progress and status of Heterostructure Barrier Varactor quintupler sources for 170 GHz and 210 GHz (G-band). The source modules feature an ultra-compact waveguide block design, and a microstrip matching circuit on high-thermal-conductivity AlN to improve the power handling capability. Furthermore, we present progress on design and fabrication of integrated HBV circuits for terahertz applications.**

I. INTRODUCTION

There is a need for compact, highly functional and reliable receivers and transmitters at terahertz frequencies [1-3]. High power sources are needed as local oscillators for heterodyne receivers (arrays, SSB mixers) and transmitters for standoff imaging systems. In this work, the main objectives are to demonstrate compact sources for submillimeter wave receivers and as transmitters in THz-radars [4].

Because of the inherent difficulty to generate power at these frequencies, the output power from a lower frequency source can be multiplied [5-7] to higher frequencies using a nonlinear device such as the Heterostructure Barrier Varactor (HBV) diode [8]. The HBV has a symmetric capacitance-voltage (C-V) characteristic, operates unbiased and only generates odd harmonics of the pump signal. These specific properties simplifies the design of high order multipliers (×3, ×5, ×7, etc.) [9, 10]. Moreover, since cascading the epitaxial growth scales the voltage handling capability of the HBV, this device is well suited for high power generation [11-12].

The progress on HBV multiplier includes multi diode quasi-optical circuits [13, 14], NLTLs [15] and highly efficient single diode waveguide circuits [16]. For instance, an HBV quintupler (×5) with a state-of-the-art conversion efficiency of 11% has been demonstrated at 100 GHz [17, 18], HBV triplers (×3) have been shown to provide 0.2 W at 113 GHz [11], 10 mW and at least 10% efficiency has been demonstrated at short millimeter wavelengths [19-21], and 1 mW has been reported for a HBV tripler up to 450 GHz [22]. In terms of output power, the best results have been achieved using a filter circuit on AlN instead of quartz due to a better heat sink for the flip-chip mounted diode.

In this paper, we present design and fabrication of high power G-band HBV quintuplers (×5), optimized for a pump power of 1-2 Watt (Q-band).

II. HBV QUINTUPLERS

This work, utilizes a new ultra compact waveguide block design, see Fig. 1. The HBV diode is flip-chip soldered onto a microstrip circuit that contains the impedance matching elements and waveguide probes. The microstrip circuit is then mounted in a waveguide block with waveguide input/output interfaces (input WR-22). One of the ambitions of the work was to make a design that was reliable and reproducible – therefore care was taken to minimize the number of manual steps in the fabrication and mounting. No DC electrical connection between the microstrip circuit and the waveguide block was therefore used.

Fig. 1 Photograph of the WR-05 quintupler block. The input interface is a WR-22 waveguide

978-2-8748-7007-1/08 $25.00 © 2008 EuMA

A. HBV diodes

An HBV is a symmetric device composed of a high bandgap semiconductor (barrier) that is surrounded by moderately doped low band-gap semiconductors (modulation layers). We also use a pseudomorphic (3 nm) AlAs layer in the centre of the barrier in order to increase the effective potential barrier, which leads to a very low leakage current. The epimaterial is a generic HBV design with three barriers [23, 24] and designed to exhibit a low thermal resistance [11, 25, 26], see Fig. 2. The epitaxial material was grown on InP by molecular beam epitaxy (EPI930) at the Nanofabrication Laboratory at Chalmers. Standard III–V processing techniques were used to fabricate the HBV devices.

Fig. 2 SEM image of a high-power HBV diode chip. Four-mesas are series connected in order to increase the power handling capability

B. Filter circuit design

The microstrip circuit was fabricated on an AlN substrate (thermal conductivity ~ 170 W/mK) to improve the power handling capability. No DC connection between the waveguide block and the circuit was used since simplicity was one of the design objectives. This also means that open waveguide probes were used both at the input and at the output side. Fig. 3 shows the opened block with the microstrip mounted and ready for final assembly.

The optimum embedding impedances were extracted from harmonic balance simulations using the Chalmers HBV device model [27]. This model self-consistently calculates the interdependent electrical and thermal properties of the device. Three-dimensional FEM modeling was applied to calculate the thermal resistance and electrical series resistance used in the model.

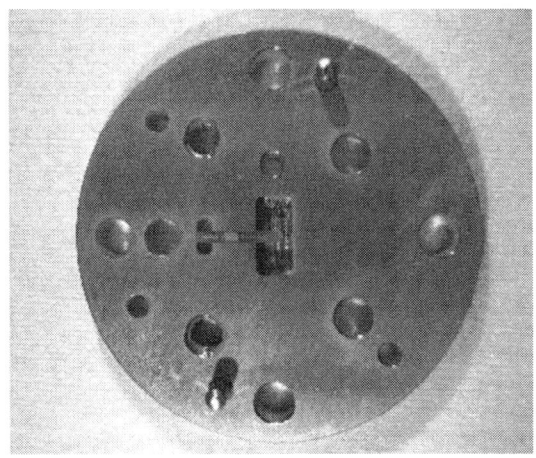

Fig. 3 One half of the waveguide multiplier block with a microstrip circuit on AlN substrate mounted

III. RESULTS

The input signal to the multiplier was provided by a HP83650B frequency synthesizer followed by a Spacek power amplifier. A waveguide isolator was used between the power amplifier and the HBV multiplier (DUT). The output power was measured using an Erickson PM2 power meter. In figure 4 the output power is plotted as a function of input power showing a maximum output power of more than 20 mW for the 210 GHz quintupler. This result was achieved with a 500 μm^2 HBV diode from the same batch as used for the high power W-band tripler in [11]. The test was limited by available input (pump) power at 40 GHz and higher efficiency is expected at higher power levels. The output spectrum was checked with a Fourier Transform Spectrometer, which confirmed that the output signal only contains the fifth harmonic.

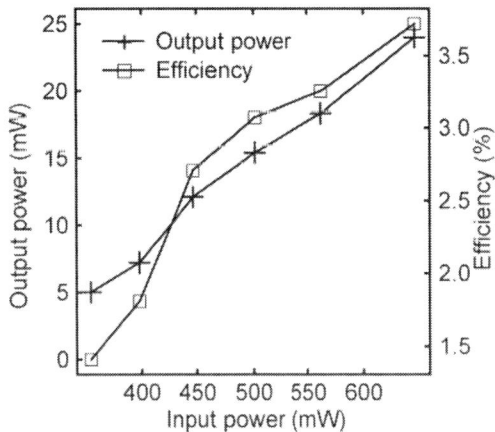

Fig. 4 Output power and efficiency at an output frequency of 202 GHz for an HBV quintupler

To achieve a high conversion efficiency, optimum embedding impedances at the pump frequency, the idler and the output frequency must be presented to the HBV simultaneously [9]. However, this is a formidable task to achieve at these frequencies, especially, for hybrid circuits and flip-chip mounted diodes. Further improvements of the 210 GHz quintupler are expected with a MMIC fabrication approach. With better fabrication tolerances, simplified circuit modelling and a reduced thermal resistance, the circuit can handle even more input RF power, provide better efficiency and bandwidth. In fig. 5., a monolithically integrated 94 GHz HBV tripler prior separation into individual chips is shown. RF measurement results of this high power demonstrator are pending.

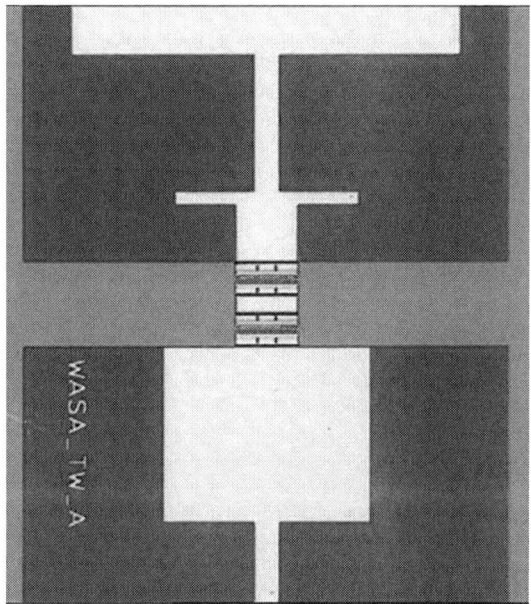

Fig. 5 Photograph of a 94 GHz InP-HBV-MMIC tripler

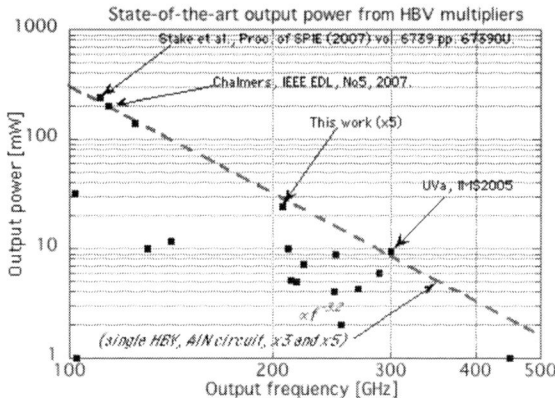

Fig. 6 Reported state-of-the-art HBV multipliers

IV. CONCLUSIONS

World-record output power performance of more than 20 mW at 200 GHz for an HBV quintupler has been demonstrated. A new and novel design of the waveguide block has been presented that makes the machining of the block simple and repeatable. In fig. 6, the reported output power for single HBV diode multipliers are shown (triplers and quintuplers). Today, the output power for HBV multipliers are comparable to state-of-the-art Schottky doublers [28, 29] at short millimeter wave frequencies. Finally, the HBV multipliers can be further improved by optimizing circuits, devices and using monolithic integration techniques (MMICs).

ACKNOWLEDGMENT

The authors would like to thank Carl-Magnus Kihlman, Chalmers, for fabricating the waveguide blocks and Mahdad Sadeghi, Nanofabrication laboratory, Chalmers, for the growth of epi-material.

This work was supported in part by the European Space Agency, in part by the Swedish Foundation for Strategic Research (SSF), in part by the Swedish National Space Board, and in part by the Swedish Defence Material Administration (FMV), the Swedish Emergency Management Agency (KBM) and the Swedish Agency for Innovation Systems (VINNOVA) through the FOI projects NanoComp and Radar eyes.

REFERENCES

[1] P. H. Siegel, "Terahertz technology," *IEEE Transactions on Microwave Theory and Techniques,* vol. 50, pp. 910-928, 2002.

[2] P. H. Siegel, "THz instruments for space," *IEEE Transactions on Antennas and Propagation,* vol. 55, pp. 2957-2965, 2007.

[3] M. Tonouchi, "Cutting-edge terahertz technology," *Nature photonics,* vol. 1, pp. 97-105, 2007.

[4] R. Appleby and R. N. Anderton, "Millimeter-Wave and Submillimeter-Wave Imaging for Security and Surveillance," *Proc. IEEE,* vol. 95, pp. 1683-1690, 2007.

[5] P. Penfield and R. P. Rafuse, *Varactor applications.* Cambridge: The MIT Press, 1962.

[6] A. Räisänen, "Frequency multipliers for millimeter and submillimeter wavelengths," *Proc. IEEE,* vol. 80, pp. 1842-1852, Apr 26 1992.

[7] G. Chattopadhyay, E. Schlecht, J. Ward, J. Gill, H. Javadi, F. Maiwald, and I. Mehdi, "An All-Solid-State Broad-band Frequency Multiplier Chain at 1500 GHz," *IEEE Transactions on Microwave Theory and Techniques,* vol. 52, pp. 1538-1547, 2004.

[8] E. L. Kollberg and A. Rydberg, "Quantum-barrier-varactor diode for high-efficiency millimetre-wave multipliers," *Electronics Letters,* vol. 25, pp. 1696-1698, 1989.

[9] L. Dillner, J. Stake, and E. L. Kollberg, "Analysis of symmetric varactor frequency multipliers," *Microwave and Optical Technology Letters,* vol. 15, pp. 26-29, May 1997.

[10] J. Stake, S. H. Jones, L. Dillner, S. Hollung, and E. L. Kollberg, "Heterostructure-barrier-varactor design," *IEEE Transactions on Microwave Theory and Techniques,* vol. 48, pp. 677-682, 2000.

[11] J. Vukusic, T. Bryllert, T. A. Emadi, M. Sadeghi, and J. Stake, "A 0.2-W heterostructure barrier varactor frequency tripler at 113 GHz," *IEEE Electron Device Letters,* vol. 28, pp. 340-342, May 2007.

[12] A. Rahal, R. G. Bosisio, C. Rogers, J. Ovey, M. Sawan, and M. Missous, "A W-Band Medium Power Multi-Stack Quantum Barrier Varactor Frequency Tripler," *IEEE Microwave Guided Wave Lett.,* vol. 5, pp. 368-370, 1995.

[13] J. B. Hacker et al.. "A high-power W-band quasi-optical frequency tripler," in *IEEE International Microwave Symposium,* vol. 3 pp. 1859-1862, 2003.

[14] S. Hollung, J. Stake, L. Dillner, and E. L. Kollberg, "A 141-GHz Integrated Quasi-Optical Slot Antenna Tripler," in *IEEE AP-S International Symposium and USNC/URSI National Radio Science Meeting*, Orlando, FL, USA, 1999, pp. 2394-2397.

[15] S. Hollung, J. Stake, L. Dillner, M. Ingvarson, and E. L. Kollberg, "A Distributed Heterostructure Barrier Varactor Frequency Tripler," *IEEE Microwave and Guided Wave Letters*, vol. 10, pp. 24-26, 2000.

[16] J. Vukusic, B. Alderman, T. A. Emadi, M. Sadeghi, A. Ø. Olsen, T. Bryllert, and J. Stake, "HBV tripler with 21% efficiency at 102 GHz " *Electronics Letters*, vol. 42, pp. 355-356, 2006.

[17] T. Bryllert, A. Ø. Olsen, J. Vukusic, T. A. Emadi, M. Ingvarson, J. Stake, and D. Lippens, "11% efficiency 100 GHz InP-based heterostructure barrier varactor quintupler," *Electronics Letters*, vol. 41, p. 30, 2005.

[18] A. Ø. Olsen, M. Ingvarson, B. Alderman, and J. Stake, "A 100-GHz HBV Frequency Quintupler Using Microstrip Elements," *IEEE Microwave and Wireless Components Letters*, vol. 14, pp. 493-495, 2004.

[19] T. David, S. Arscott, J.-M. Munier, T. Akalin, P. Mounaix, G. Beaudin, and D. Lippens, "Monolithic integrated circuits incorporating InP-based heterostructure barrier varactors," *IEEE Microwave and Wireless Components Letters*, vol. 12, pp. 281-283, 2002.

[20] Q. Xiao, J. L. Hesler, T. W. Crowe, I. Robert M. Weikle, and Y. Duan, "High Efficiency Heterostructure-Barrier-Varactor Frequency Triplers Using AlN Substrates," in *IMS 2005*, Long Beach, CA, 2005.

[21] R. Meola, J. Freyer, and M. Claassen, "Improved frequency tripler with integrated single-barrier varactor," *Electronics Letters*, vol. 36, pp. 803-804, 2000.

[22] M. Saglam, B. Schumann, K. Duwe, C. Domoto, A. Megej, M. Rodriguez-Gironés, J. Müller, R. Judaschke, and H. L. Hartnagel, "High-performance 450-GHz GaAs-based heterostructure barrier varactor tripler," *IEEE Electron Device Letters*, vol. 24, pp. 138- 140, 2003.

[23] L. Dillner, W. Strupinski, S. Hollung, C. Mann, J. Stake, M. Beardsley, and E. L. Kollberg, "Frequency Multiplier Measurements on Heterostructure Barrier Varactors on a Copper Substrate," *IEEE Electron Device Letters*, vol. 21, pp. 206-208, 2000.

[24] T. A. Emadi, T. Bryllert, M. Sadeghi, J. Vukusic, and J. Stake, "Optimum barrier thickness study for the InGaAs/InAlAs/AlAs heterostructure barrier varactor diodes," *Applied Physics Letters*, vol. 90, pp. 012108-3, 2007.

[25] M. Ingvarson, B. Alderman, A. Ø. Olsen, J. Vukusic, and J. Stake, "Thermal constraint for heterostructure barrier varactor diodes," *IEEE Electron Device Letters*, vol. 25, pp. 713-715, 2004.

[26] J. Stake, L. Dillner, S. H. Jones, C. M. Mann, J. Thornton, J. R. Jones, W. L. Bishop, and E. L. Kollberg, "Effects of self-heating on planar heterostructure barrier varactor diodes," *IEEE Transactions on Electron Devices*, vol. 45, pp. 2298-2303, 1998.

[27] M. Ingvarson, J. Vukusic, A. Ø. Olsen, T. A. Emadi, and J. Stake, "An electro-thermal HBV model," in *IMS 2005*, Long Beach, CA, 2005, pp. 1151-1153.

[28] J. Ward, E. Schlecht, G. Chattopadhyay, A. Maestrini, J. Gill, F. Maiwald, H. Javadi, and I. Mehdi, "Capability of THz sources based on Schottky diode frequency multiplier chains," in *IEEE International Microwave Symposium*. vol. 3 Fort Worth, Texas, 2004, pp. 1587-1590.

[29] T. Crowe, D. Porterfield, J. Hesler, W. Bishop, D. Kurtz, and H. Hui, "Terahertz technology for imaging and spectroscopy," in *Terahertz for military and security applications IV*, 2006.

Metamorphic MMICs for Operation Beyond 200 GHz

A. Tessmann, I. Kallfass, A. Leuther, H. Massler, M. Schlechtweg, O. Ambacher

Fraunhofer Institute for Applied Solid State Physics (IAF), Tullastr. 72, D-79108 Freiburg, Germany

e-mail: Axel.Tessmann@iaf.fraunhofer.de

Abstract— **In this paper, we present the development of advanced millimeter-wave and submillimeter-wave monolithic integrated circuits for use in active and passive high-resolution imaging systems operating beyond 200 GHz. A 210 GHz subharmonically pumped dual-gate field-effect transistor (FET) mixer has been successfully realized using our 100 nm InAlAs/InGaAs based depletion-type metamorphic high electron mobility transistor (mHEMT) technology in combination with grounded coplanar circuit topology (GCPW). Furthermore, a G-band low-noise amplifier MMIC demonstrating a linear gain of more than 16 dB between 180 and 220 GHz and a state-of-the-art noise figure of 4.8 dB was fabricated using a gate length of 50 nm. Finally, a submillimeter-wave monolithic integrated circuit (S-MMIC) could be realized based on an advanced 35 nm mHEMT technology, offering a small-signal gain of more than 15 dB between 270 and 310 GHz.**

I. INTRODUCTION

Millimeter- and submillimeter-wave semiconductor technologies with cut-off frequencies of more than 500 GHz have enabled the realization of both stationary and mobile imaging systems offering high resolution and compact size. These advanced sensors find manifold applications in safety and security systems, e. g. for the stand-off control of persons and the non-invasive control of baggage and letters to detect concealed weapons, explosives or drugs. Further applications are in avionics as landing aids or runway control, in reconnaissance and surveillance as well as in medical imaging, quality control and astronomy [1-3]. Modern millimeter-wave sensors enable stand-off detection and investigation through barriers as well as the behavior analysis of potential individual perpetrators in crowds. Atmospheric windows in the millimeter- and submillimeter-wave frequency range are located around 94, 140, 220 and 340 GHz. An essential benefit for the majority of these applications is that numerous materials (fabric, wood, plastic) as well as fog, smog and dust are transparent at these frequencies, and the absorption is sensitive to changes in the dielectric constant. As image resolution improves with increasing frequency the G-band (140–220 GHz) and H-band (220–325 GHz) are of great interest for high-resolution sensor systems [4,5].

In this paper, we report on the development of coplanar G-band and H-band MMICs for use in high-resolution imaging systems. The millimeter-wave components are based on advanced metamorphic HEMT technologies, featuring gate lengths of 100 nm, 50 nm and 35 nm. Compared to InP substrates, mHEMT technology on GaAs substrates is less expensive, taking advantage of the high crystal quality, greater mechanical strength and the large diameter of available GaAs wafers. The utilized grounded coplanar waveguide (GCPW) technology is very attractive at millimeter- and submillimeter-wave frequencies, due to the high isolation between adjacent lines, the low source inductance of the active devices, the very compact transmission line dimensions, and the suppression of parasitic substrate modes.

II. METAMORPHIC HEMT TECHNOLOGY

For the fabrication of the 100 nm and 50 nm gate length circuits, we used an $In_{0.52}Al_{0.48}As/In_{1-x}Ga_xAs/In_{0.53}Ga_{0.47}As$ composite channel structure which was grown by molecular beam epitaxy (MBE) on 4-inch semi-insulating GaAs wafers. The 50 nm gate length technology features an indium content of 80 % in the main channel, resulting in high carrier mobility and velocity, while the 100 nm technology has a channel indium content of 65 % and thus offers a higher breakdown voltage, as well as increased reliability. For the third mHEMT technology the gate length was reduced to 35 nm and a single InGaAs channel with an In content of 80 % was used. These modifications result in an f_T of more than 500 GHz and a very high maximum transconductance of $g_{m,\,max} = 2500$ mS/mm.

To adapt the lattice constant a metamorphic buffer was grown with a linear $In_yAl_{0.48}Ga_{0.52-y}As$ transition in composition. The transistor mesa was wet-chemically etched, and GeAu was deposited for ohmic contacts. The gate recess was etched using a succinic acid based solution. The active devices consist of T-shaped Pt-Ti-Pt-Au gates, which were defined by e-beam lithography. Finally, the devices were passivated with 250 nm chemical vapor deposited (CVD) silicon nitride for good reliability and robustness. The process further includes $50\,\Omega$/sq NiCr thin film resistors, 0.225 fF/μm^2 metal-insulator-metal (MIM) capacitors and two levels of metal interconnects with a 2.7 μm thick plated Au layer air bridge technology. A detailed description of the IAF mHEMT technology can be found in [6]. The electrical DC- and RF-parameters of the three technologies are summarized in Table I. With an indium content of 65 % in the main channel the 100 nm gate length technology achieved an average

978-2-8748-7007-1/08 $25.00 © 2008 EuMA

TABLE I

ELECTRICAL DC- AND RF-PARAMETERS OF THE IAF METAMORPHIC HEMT TECHNOLOGIES

	$l_g = 100$ nm	$l_g = 50$ nm	$l_g = 35$ nm
R_c	0.07 Ω·mm	0.05 Ω·mm	0.03 Ω·mm
R_s	0.23 Ω·mm	0.15 Ω·mm	0.1 Ω·mm
$I_{D, max}$	900 mA/mm	1200 mA/mm	1600 mA/mm
V_{th}	-0.3 V	-0.25 V	-0.3 V
$BV_{off\text{-}state}$	4.0 V	2.2 V	2.0 V
$BV_{on\text{-}state}$	3.0 V	1.6 V	1.5 V
$g_{m, max}$	1300 mS/mm	1800 mS/mm	2500 mS/mm
f_T	220 GHz	380 GHz	515 GHz
f_{max}	300 GHz	500 GHz	> 700 GHz
$MTTF$	3×10^7 h	2.7×10^6 h	n.a.

extrinsic transit frequency of $f_T = 220$ GHz and a maximum oscillation frequency of $f_{max} = 300$ GHz, while for the 50 nm technology an f_T of 380 GHz and an f_{max} of 500 GHz was measured for a 2×30 μm common source device at a drain voltage of $V_{ds} = 1$ V. The 35 nm technology achieved a maximum extrinsic transit frequency of $f_T = 515$ GHz (as shown in Fig. 1) and a maximum oscillation frequency of $f_{max} > 700$ GHz, which were extrapolated based on the current unity gain and Mason's unilateral gain for a 2×10 μm common source device, measured on-wafer.

Another important issue is the reliability of the mHEMTs due to the high dislocation density in the metamorphic buffer. Therefore, the long-term stability of the metamorphic devices was determined by accelerated lifetime tests, which were performed in air. Based on a 10 % $g_{m,max}$ degradation failure criterion, a median time to failure (MTTF) of well above 10^6 h at a channel temperature of 125 °C was achieved [5].

III. MILLIMETER-WAVE MONOLITHIC ICS

The availability of mature InP-based pseudomorphic and GaAs-based metamorpic HEMT technologies with cutoff frequencies up to 600 GHz allows the realization of active electronic circuits covering the entire millimeter-wave frequency range. Using our 100 nm mHEMT technology we demonstrate the first active, subharmonic down-conversion mixer operating beyond 200 GHz [7]. Besides a lower conversion loss, the advantages of active FET mixers over conventional diode mixers comprise a potentially lower noise figure, lower local oscillator (LO) power requirements, and most important, the on-chip integration with other circuit components like LNAs to form multi-funtional single-chip receivers. Figure 2 shows the chip photograph of the realized G-band dual-gate FET mixer MMIC. Its total chip size is 1×1.5 mm^2. The mixer's conversion gain characteristic was evaluated on-wafer with all power levels normalized to the

probe tip reference plane. At its LO input, the MMIC was driven by an Agilent E8257D signal generator whose output signal was multiplied in an HP source module and amplified with an in-house W-band HPA module, resulting in a maximum of 10 dBm LO-power at the chip level. The RF signal was obtained from an Oleson G-band frequency extension module fed by an Agilent E8257 signal generator. The down-converted IF component was measured with a spectrum analyzer. The applied bias voltages were $V_D = 1.5$ V, $V_{G1} = 0$ V and $V_{G2} = -0.7$ V. Under active operation, the mixer consumed 36 mW of DC power with a total drain current of 24 mA.

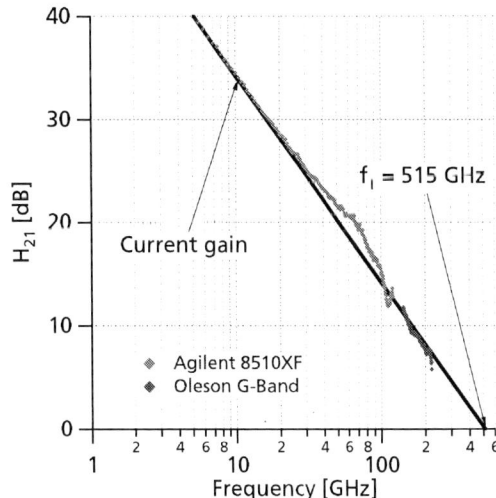

Fig. 1. Measured current gain of a 2×10 μm mHEMT ($l_g = 35$ nm) and extrapolated transit frequency f_T of 515 GHz.

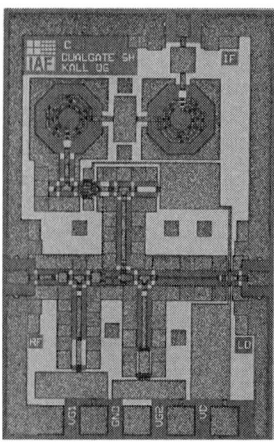

Fig. 2. Chip photograph of subharmonic dual-gate FET mixer MMIC realized in IAF 100 nm mHEMT technology.

Figure 3 shows the measured conversion gain performance of the subharmonic dual-gate mixer MMIC. The mixer was driven with 10 dBm LO power and the IF was measured at 400 MHz. The evaluation of the conversion gain was based on the RF power measurement using a G-band power sensor (ELVA-1 DPM 06/05), which in turn was calibrated with an Erickson calorimeter. The mixer MMIC achieved a maximum conversion gain of -4.7 dB at 214 GHz. Over the full measured bandwidth of 165 to 220 GHz, the conversion gain stayed above –10 dB. Figure 3 also includes the comparison to the simulated conversion gain, obtained from a harmonic balance analysis within the Agilent ADS design environment. The frequency settings during simulation took into account four and two harmonics of the LO and RF signals, respectively. In-house nonlinear FET models were employed to predict the frequency translating effect. Good agreement between measured and predicted conversion gain characteristics was obtained.

In addition to the subharmonic mixer circuit, a G-band four-stage grounded coplanar waveguide (GCPW) low-noise amplifier MMIC was designed in 50 nm mHEMT technology to achieve an excellent noise figure in combination with high gain at 210 GHz. The grounded coplanar waveguide technology was adopted to prevent the excitation of parasitic substrate modes, which can substantially degrade the gain characteristic, especially at very high frequencies. A chip photograph of the realized four-stage low-noise amplifier MMIC is shown in Fig. 4. To achieve reasonable gain up to 220 GHz, the gate width of the utilized common-source devices was chosen to $w_g = 2 \times 10$ μm.

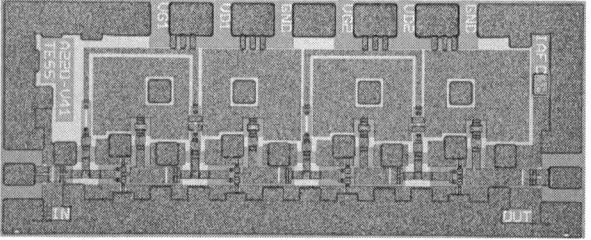

Fig. 4. Chip photograph of the four-stage 210 GHz low-noise amplifier MMIC. The chip size is 0.65×1.5 mm^2.

The on-wafer measured small-signal gain of the MMIC is shown in Fig. 5. The LNA circuit achieved a linear gain of more than 16 dB between 180 and 220 GHz, when applying a drain voltage of $V_d = 0.8$ V, a gate voltage of $V_g = -0.15$ V and a drain current of $I_d = 30$ mA (375 mA/mm). Furthermore, noise figure measurements were performed at room temperature ($T = 293$ K) from 180 to 206 GHz by using a commercial G-band noise diode. The measured average noise figure of the low-noise amplifier was only 4.8 dB over the characterized frequency range.

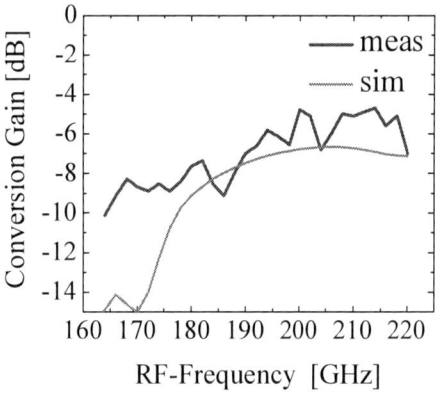

Fig. 3. Measured and simulated conversion gain versus RF frequency of the subharmonic dual-gate FET mixer MMIC at 10 dBm LO-power.

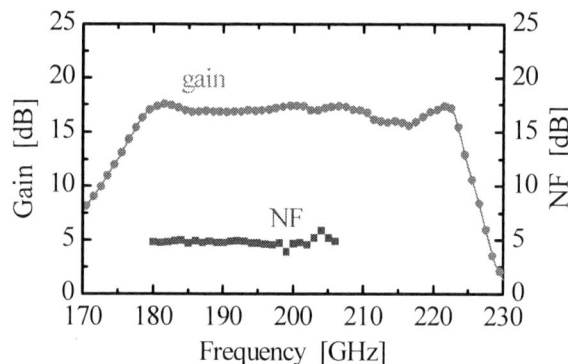

Fig. 5. On-wafer measured gain and noise figure (NF) of the four-stage 210 GHz low-noise amplifier MMIC.

978-2-8748-7007-1/08 $25.00 © 2008 EuMA

IV. SUBMILLIMETER-WAVE MONOLITHIC ICs

Based on our advanced 35 nm mHEMT technology, we could successfully realize first submillimeter-wave monolithic integrated circuits (S-MMICs) demonstrating measured gain at 300 GHz, which is the threshold of the submillimeter-wave frequency regime. Fig. 6 shows an SEM photograph of a coplanar 270 GHz single-stage LNA S-MMIC, featuring a total chip-size of only 0.15 mm^2. This circuit achieves a small-signal gain of more than 7 dB at 270 GHz and approximately 2.5 dB at 320 GHz, as shown in Fig. 7. Additionally, a four-stage 300 GHz amplifier S-MMIC was designed, offering sufficient gain and bandwidth to be used in high-resolution transceivers and imagers. The monolithic circuit consists of four 2×10 μm common source devices. Figure 8 shows the on-wafer measured S-parameters of the coplanar 300 GHz LNA, demonstrating a linear gain of more than 15 dB between 270 and 310 GHz.

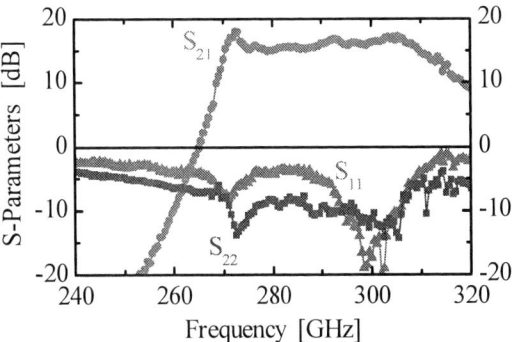

Fig. 8. On-wafer measured S-parameters of a four-stage 300 GHz LNA S-MMIC ($l_g = 35$ nm).

V. CONCLUSIONS

Metamorphic HEMT technology has been demonstrated to be highly suitable for the development of advanced millimeter-wave and submillimeter-wave circuits. A subharmonically pumped dual-gate FET mixer was realized achieving a conversion gain of -4.7 dB at 214 GHz, while a G-band LNA circuit demonstrated a small-signal gain of more than 16 dB between 180 and 220 GHz together with a record noise figure of only 4.8 dB. Furthermore, S-MMIC amplifier circuits were successfully designed, fabricated and measured, demonstrating a linear gain of more than 15 dB between 270 and 310 GHz.

REFERENCES

[1] W. R. Deal, X. B. Mei, V. Radisic, W. Yoshida, P.H. Liu, J. Uyeda, M. Barsky, T. Gaier, A. Fung, R. Lai, "Demonstration of a S-MMIC LNA with 16-dB Gain at 340-GHz," *2007 IEEE CSIC Symposium Digest*, pp. 75-78, Oct. 2007.

[2] D. M. Sheen, D. L. McMakin, and T. E. Hall, "Three-Dimensional Millimeter-Wave Imaging for Concealed Weapon Detection," *2001 IEEE Transactions on Microwave Theory and Techniques*, vol. 49, no. 9, pp. 1581-1592, Sept. 2001.

[3] H. Essen, A. Wahlen, R. Sommer, W. Johannes, R. Brauns, M. Schlechtweg, A. Tessmann, "High-bandwidth 220 GHz experimental radar," 2007 *IEE Electronics Letters*, Vol. 43, No. 20, pp. 1114-1116, 27th Sept. 2007.

[4] D. Streit, R. Lai, A. Oki, A. Gutierrez-Aitken, "InP HEMT and HBT applications beyond 200 GHz," *14th International Conference on Indium Phosphide and Related Materials (IPRM' 02)*, pp. 11-14, May 2002.

[5] A. Tessmann, "220-GHz Metamorphic HEMT Amplifier MMICs for High-Resolution Imaging Applications," *2005 IEEE Journal of Solid-State Circuits*, vol. 40, no. 10, pp. 2070-2076, Oct. 2005.

[6] A. Leuther et al., "50 nm MHEMT Technology for G- and H-Band MMICs," *19ᵗʰ International Conference on Indium Phosphide and Related Materials (IPRM' 07)*, pp. 24-27, May 2007.

[7] I. Kallfass, H. Massler, A. Leuther, "A 210 GHz, Subharmonically-Pumped Active FET Mixer MMIC for Radar Imaging Applications," *2007 IEEE CSIC Symposium Digest*, pp. 71-74, Oct. 2007.

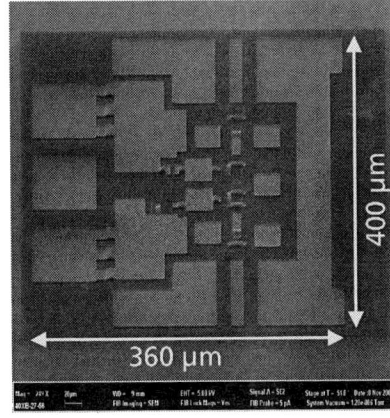

Fig. 6. SEM photograph of a 270 GHz single-stage LNA S-MMIC ($l_g = 35$ nm). The chip-size is 0.36 x 0.4 mm^2.

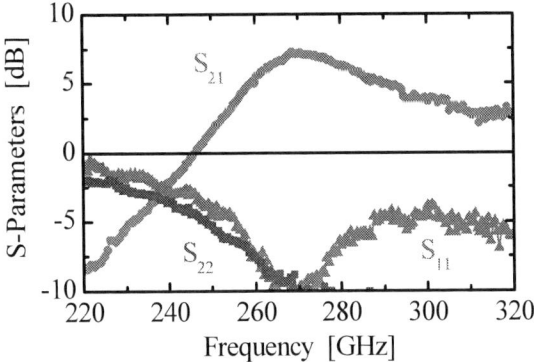

Fig. 7. On-wafer measured S-parameters of a single-stage 270 GHz LNA S-MMIC ($l_g = 35$ nm).

Industrial MHEMT Technologies for 80 - 220 GHz Applications

Derek Smith[1], Gilles Dambrine[2], Jean-Claude Orlhac[3]

[1]OMMIC

2 chemin du Moulin, Limeil Brevannes, France

[1]`derek.smith@ommic.com`

[2]IEMN , Villeneuve D'Ascq, France

[2]EADS-Astrium, Toulouse, France

Abstract— **This paper describes the industrial development methodology of a family of MHEMT technologies with gate lengths from 130 down to 50 nm. State of the art MMIC LNA performance for the commercial 70 nm technology at 90 and 150 GHz is presented.**

I. INTRODUCTION

The last several years has seen new and vital applications developing that require the fabrication of semiconductor devices at higher and higher frequencies. Two typical examples are for Earth Observation where, for example, man's impact on the natural equilibrium of nature has to be monitored so as to be controlled and the use of passive imaging for stand-off control of aircraft passengers, at sports events etc enabling the detection of unwanted objects.

Passive imaging typically uses millimetre wave frequencies such as ~ 94 GHz, 140 GHz, 220 GHz which represent Electromagnetic Radiation Windows. Systems are becoming available at 94 GHz using active devices and there is a definite push towards the higher frequencies for reasons of resolution, size, associated information etc.

These passive systems need low noise amplifiers with excellent noise figures that are produced on an industrial line so as to provide the product stability, reliability and cost structure that will allow the full scale deployment of the above mentioned systems and in particular for the passive imaging for security and control.

This paper presents the methodology and results of work in OMMIC to develop the MMIC technologies with the required performance from an industrial line.

II. DEVELOPMENT METHODOLOGY

The OMMIC technologies are all based on a highly modular process flow with a minimum number of processing changes between the different technologies. As an example Table I shows the basic modules for the D01PH PHEMT process, the D01MH high indium content 130 nm MHEMT process and the D007IH very high indium content 70 nm MHEMT process.

TABLE I
CHANGE TO PROCESS FLOW FOR EACH GENERATION OF MHEMT TECHNOLOGY

Mask Layer	Process Step Description	Change necessary compared to previous technology ? (Starting point : 130 nm PHEMT)		
		130 nm	70 nm	50 nm
	Epitaxy	YES	YES	YES
1 – LI	Layer Isolation	YES	NO	NO
2 – OH	Ohmic Contact	YES	NO	NO
3 – GM	Gate Metallisation and Lithography	YES	YES	YES
4 – BE	Capacitor Bottom Electrode	NO	NO	NO
5 – CG	Contact to Gate (Si_3N_4 via)	NO	NO	NO
6 – TE	Capacitor Top Electrode	NO	NO	NO
7 – CO	Contact Opening ($SiO_2 - Si_3N_4$ via)	NO	NO	NO
8 – IN	Interconnect Metal	NO	NO	NO
9 – CB	Contact Bonding (Si_3N_4 via)	NO	NO	NO
10 – VH	Back Side Via Hole	YES	NO	NO
11 – CS	Chip Separation	NO	NO	NO

As can be seen in Table I the D01MH was based on the D01PH process with a new epitaxial structure and optimisation of the gate structure. All other processing modules are left unchanged except the Via Hole where the MHEMT buffer layer needs to be etched through. The D007IH technology is itself based on the D01MH process with a new epitaxial structure and a new gate topology (double mushroom with BCB passivation). OMMIC is now developing a 50 nm MHEMT based on the 70 nm technology.

This approach has many advantages in terms of shorter development time as well as easier NPI (New Process Introduction) onto the production line. The work involved in the device modelling, libraries, reliability studies etc becomes dramatically reduced with a large amount of re-use of information. The Design of Experiments can be planned based on a well-understood basic process.

III. 0.13 MICRON MHEMT

The first MHEMT technology developed at OMMIC was based on a fully released 130 nm PHEMT technology. This new MHEMT technology was designed to be a general-purpose technology with reasonable breakdown voltages and a good noise figure [1]. The process has an asymmetric mushroom gate of 130 nm (see Fig 1). The choice of the indium content is a compromise between the RF performance (given by electron mobility [2]) and the Breakdown Voltage (given by the band gap width [3]) – a compromise value of 40 % has been chosen with a composite channel and single doping plane.

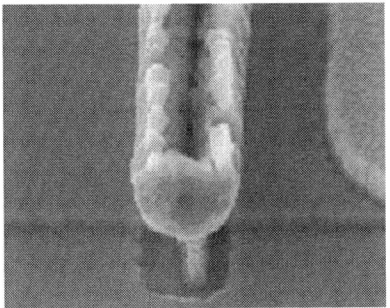

Fig. 1 SEM of the 130nm Gate

A comparison of the performance obtained from a 130 nm PHEMT and a 130 nm MHEMT technology is presented in Fig. 2 and Fig. 3.

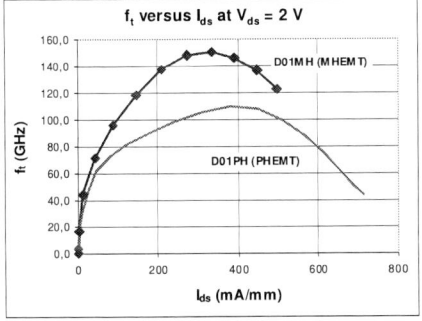

Fig. 2 Comparison of the f_t for a 130 nm PHEMT and MHEMT Process

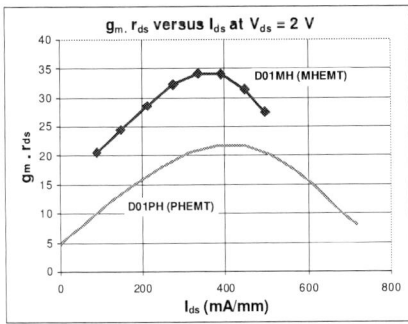

Fig. 3 Comparison of the $g_m.r_{ds}$ for a 130 nm PHEMT and MHEMT Process

It can be seen that a considerably higher f_t is attainable with the MHEMT process (150 GHz) compared to the PHEMT (105 GHz) with the same Gate Length. The measured f_{max} is 250 GHz. Similarly the gm.rds (low frequency gain) is far higher for the MHEMT device (34 compared to 22). Optimum performance is also obtained at a lower value of drain current for the MHEMT process. The principal advantage of the MHEMT technology is shown in Fig. 4 where an excellent Noise Figure of 0.83 dB is associated with a gain of 11.7 dB at 30 GHz (measured on-wafer on a 8x30 μm FET). This improved performance is achieved for the same level of process complexity (130 nm gate) as the PHEMT process.

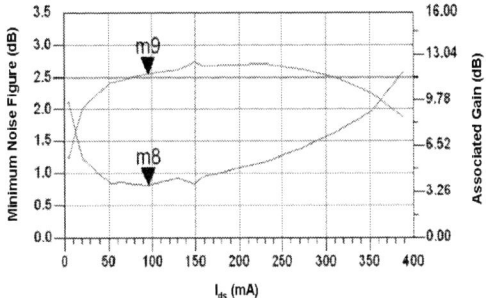

Fig. 4 Minimum Noise Figure and Associated Gain at 30 GHz

A summary of the main parameters of the 130 nm MHEMT and 130 nm PHEMT technologies is presented in Table II.

TABLE II
MAIN PARAMETERS OF THE PHEMT AND MHEMT TECHNOLOGIES

Parameter	Units	130 nm PHEMT	130 nm MHEMT
R_s	Ω.mm	0.15	0.2
I_{dsmax} @ V_{gs} = + 0.7 V	mA/mm	650	600
V_t @ I_{ds} = 1 mA/mm	V	- 0.85	- 0.7
V_{BGD} @ I_g = 1 mA/mm	V	- 11	- 11
$g_{m (max)}$	mS/mm	750	900
$f_{t (max)}$	GHz	105	150
MSG at 30 GHz	dB	15	16
NF_{min} at 30 GHz	dB	1.1	0.83
G_{ass} at 30 GHz	dB	7.5	11.7

IV. 0.07 MICRON MHEMT

The new requirements at higher frequencies as described above has led OMMIC to develop a new MHEMT device with higher ft and a lower noise figure. The choice of the gate length is 70 nm (a compromise between performance and yield at the time of the process development). The intended applications require a low noise figure and low power consumption – this removes the previous limitation on the band gap width (required for higher break down voltages) and allows the use of an indium content of 70 %. A composite channel is used with a 53 % GaInAs sub-channel so as to improve the on state breakdown voltage.

978-2-8748-7007-1/08 $25.00 © 2008 EuMA

For such small gate lengths, the static fringing capacitance of the gate becomes the main limitation to the performance for a given epitaxial structure [4]. This has led OMMIC to develop a double mushroom gate topology as shown in Fig. 5. The epitaxial structure itself has to be seen as a 3 dimensional problem and for very short gate lengths attention has to be given to short channel effects [5] and leads to the use of a very thin barrier.

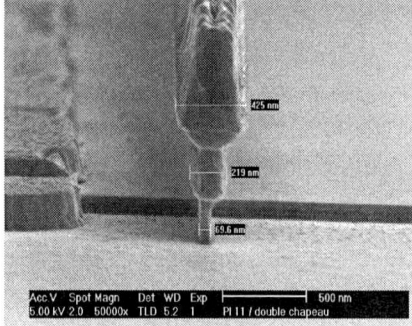

Fig. 5 Double Mushroom Structure of the 70 nm Technology

The measured ft and fmax are of the order of 300 GHz. The extrinsic DC gm is very high at 1.6 S/mm. The on wafer measured noise figure of a 4x15 µm transistor is shown in Fig. 6. The minimum noise figure is 0.55 dB and the associated gain 12.5 dB at 30 GHz.

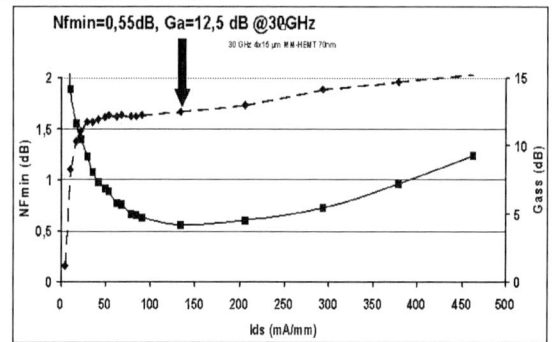

Fig. 6 Minimum Noise Figure and Associated Gain at 30 GHz

The main characteristics of the process are detailed in TableIII.

TABLE III
MAIN PARAMETERS OF THE 70 NANOMETRE MHEMT TECHNOLOGY

Parameter	Units	70 nm MHEMT
R_s	Ω.mm	0.2
I_{dsmax} @ V_{gs} = + 0.2 V	mA/mm	700
V_t @ I_{ds} = 1 mA/mm	V	- 0.55
V_{BGD} @ I_g = 1 mA/mm	V	- 2.5
$g_{m (max)}$	mS/mm	1600
$f_{t (max)}$	GHz	300
MSG at 30 GHz	dB	17
NF_{min} at 30 GHz	dB	0.55
G_{ass} at 30 GHz	dB	12.8

This process was used to fabricate Low Noise Amplifiers at 80 – 100 GHz and 140 – 160 GHz as part of an ESA project for Earth Observation Satellites. The 80 – 100 GHz MMIC is a 3 stage co-planar design. As can be seen in Fig. 7 the noise figure is typically 2.5 dB with 20 dB of associated gain over the operating bandwidth. It can also be seen in Fig. 8 that the noise figure is quite independent of the drain current and can be used at very low power consumption. This is a state of the art performance using an industrial MMIC process.

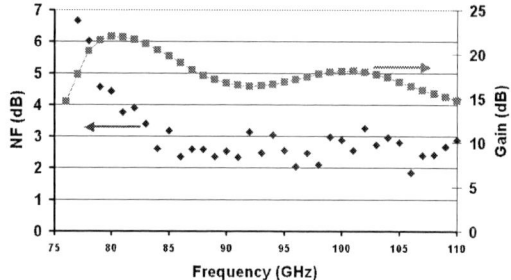

Fig. 7 On Wafer Noise Figure vs Frequency for 3 stage LNA

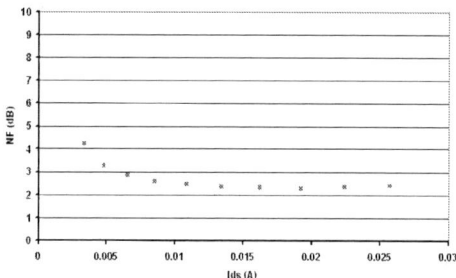

Fig. 8 On Wafer Noise Figure vs Drain Current for 3 stage LNA at 90 GHz

The 140 – 160 GHz MMIC is a 4 stage co-planar design and the on wafer measured S-parameters, noise figure and associated gain are presented in Fig.9 and Fig. 10. The on-wafer noise figure is typically 4.5 dB with 20 dB of associated gain. The gain is flat with an excellent input match over the 140 – 160 GHz frequency band.

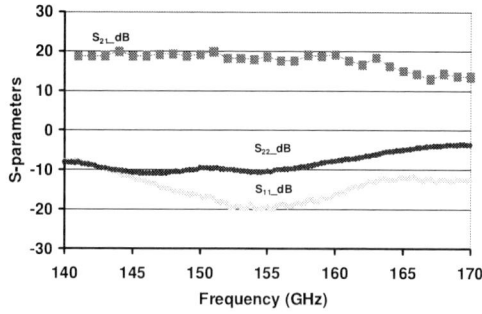

Fig. 9 S-Parameters vs Frequency for 4 stage LNA

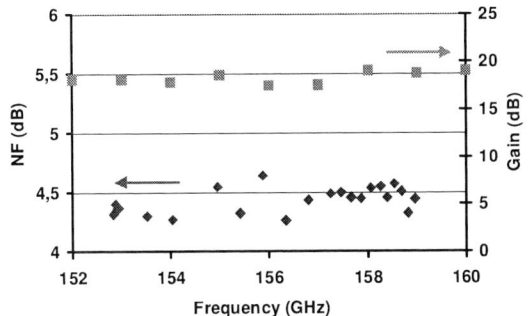

Fig. 10 Noise Figure vs Frequency for 4 stage LNA

V. 0.05 MICRON MHEMT

The 70 nm MHEMT technology has the required performance so as to enable the production of LNAs at 90 or 140 GHz - as is shown in this paper - and has the potential to allow the production of LNAs up to 180 GHz and beyond.

However in order to achieve very low noise figures at 220 GHz and beyond OMMIC has begun the development of its next generation process (D005IH) based on a 50 nm gate with an increased Indium content of 80 %. A SEM photograph showing the double mushroom gate in detail is presented in Fig 11. Since the development of the 70 nm process enough has been learned about the fabrication and control of the very short gate lengths to allow the 50 nm now to be considered as a next generation industrial process.

It is intended to release a production technology with an f_t greater than 400 GHz. Preliminary results on the 50 nm gate structure prior to the optimisation of the epitaxial wafers already demonstrate an f_t of greater than 350 GHz.

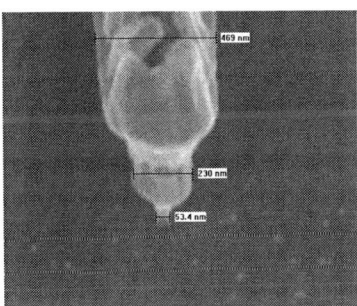

Fig. 11 SEM photograph of the gate

VI. CONCLUSIONS

This paper has presented the evolution of OMMIC's processes from a 130 nm PHEMT to a 50 nm MHEMT Transistor based technology. Results have been presented for LNA MMICs operating at 80-100 GHz and 140-160 GHz that are state of the art from a commercial foundry. The most recent developments at 50 nm are introduced that will lead to similar results but at 220 GHz and beyond while still being 'industrial'. The methodology used for the development of each technology based on a modular process flow dramatically improves the development and release time for the technology and ensures its 'industrial' character.

ACKNOWLEDGMENT

The authors would like to thank the Process Development and Epitaxial Growth Groups of OMMIC who have developed the technologies mentioned in this article. The measurements were performed at the IEMN and the circuits designed by EADS-Astrium, France.

Part of this work was carried out during the "MMIC technology for future atmospheric sounders" ESA study, contract N° 16264/02/NL/MM.

REFERENCES

[1] Release for production of a 150GHz, 125nm gate 40% In Metamorphic GaAs HEMTMMIC process; J.Bellaïche; P.Baudet; S.Demichel; M.Renvoisé; H.Maher; J.F.Pautrat; MG.Périchaud; S.Lafont; CS MANTECH 2006 Pages : 24-27.

[2] Characteristics of strained In0.65Ga0.35As/In 0.52Al0.48As HEMT with optimized transport parameters; Ng, G.I.; Hong, W.-P.; Pavlidis, D.; Tutt, M.; Bhattacharya, P.K.; Electron Device Letters, IEEE Volume 9, Issue 9, Sept. 1988 Page(s):439 – 441.

[3] Enhancement-mode Al0.66In0.34As/Ga0.67 In0.33As metamorphic HEMT, modeling and measurements; Boudrissa, M.; Delos, E.; Gaquiere, C.; Rousseau, M.; Cordier, Y.; Theron, D.; Jaeger, J.C.; Electron Devices, IEEE Transactions on Volume 48, Issue 6, June 2001 Page(s):1037 – 1044

[4] 50-nm Self-Aligned and "Standard" T-gate InP pHEMT Comparison: The Influence of Parasitics on Performance at the 50-nm Node; Moran, D.A.J.; McLelland, H.; Elgaid, K.; Whyte, G.; Stanley, C.R.; Thayne, I.; Electron Devices, IEEE Transactions on Volume 53, Issue 12, Dec. 2006 Page(s):2920 – 2925

[5] Short-Channel Effect Limitations on High-Frequency Operation of AlGaN/GaN HEMTs for T-Gate Devices; Jessen, G.H.; Fitch, R.C.; Gillespie, J.K.; Via, G.; Crespo, A.; Langley, D.; Denninghoff, D.J.; Trejo, M.; Heller, E.R.;Electron Devices, IEEE Transactions on Volume 54, Issue 10, Oct. 2007 Page(s):2589 - 2597

Proceedings of the 3rd European Microwave Integrated Circuits Conference

Characterization and Numerical Simulations of High Power Field-Plated pHEMTs

A. Chini [1#], M. Esposto[#], G. Verzellesi[#], S. Lavanga[*], C. Lanzieri[*], A. Cetronio[*]

[#]*Department of Information Engineering University of Modena and Reggio Emilia*
Via Vignolese 905 – 41100 Modena - ITALY
[1]chini.alessandro@unimore.it
[*]*SELEX Sistemi Integrati S.p.A.*
Via Tiburtina Km. 12,400 – 00131 Rome - ITALY

Abstract— This paper presents the results obtained both by experimental measurements and numerical simulations carried out on state-of-the-art Field-Plated GaAs-based pHEMTs. The effect of field-plate length on DC and RF operation of pHEMTs will be discussed showing that the adoption of an optimal field-plate structure can significantly boost the device RF power performance, resulting in power density up to 2W/mm measured under continuous wave RF signals at 2GHz. The physical origin of the DC-to-RF dispersion in the fabricated devices has been associated with a hole-trap located at 0.65eV from the valence band as obtained from current-DLTS measurements. The experimental results will also be supported and validated by numerical simulations. It will be shown that the beneficial effects arising from the adoption of the field-plate structure lie in its control on the trapped charge population responsible for the DC-to-RF dispersion mechanism.

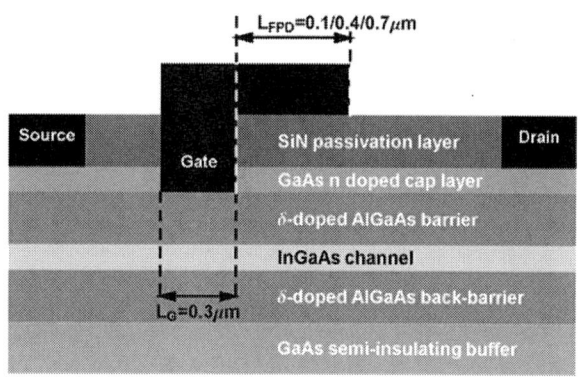

Fig.1 Schematic, not-to-scale, cross-section of fabricated pHEMTs.

I. INTRODUCTION

Achieving high power densities and high efficiencies is nowadays the main challenge in the development of any RF power device since its performance improvements are some of the main keys in the development of modern wireless communication systems and active phased-array radars. One of the main limitation in reaching state-of-the-art operation in compound-semiconductor devices is represented by the well know DC-to-RF dispersion mechanism that consists of a reduction of the maximum drain current that can flow through the device when the latter is subjected to large RF signal drives. Among the various solutions that have been proposed, the adoption of a field-plate structure has been proven to be an effective way in order to reduce DC-to-RF dispersion both for GaAs-based [1,2] and GaN-based [3] devices.

Aim of this paper is to investigate the benefits induced by the adoption of a field-plate structure in GaAs-based pHEMTs and suggest a possible interpretation for the field-plate action in reducing the DC-to-RF dispersion through the use of numerical simulations.

II. DEVICE FABRICATION

The field-plated pHEMT devices studied in this paper were fabricated in the GaAs Foundry of Selex Sistemi Integrati following a standard process flow except for the definition of the field-plate and gate contact.

As can be seen in figure 1, the epitaxial structure consists of a standard double heterojunction InGaAs/AlGaAs channel capped with a lightly n-doped GaAs cap layer. After ohmic contacts formation, device isolation was carried out by means of ion implantation. At this point, a 50nm SiN layer was deposited on the whole wafer by using a PECVD process. This SiN layer represents the dielectric layer that physically sustains the field plate. Afterwards, a first lithography step was carried out in order to open a window through the SiN layer. SiN layer was then dry etched and the GaAs layer was removed by wet etching. A second lithography step was then carried out in order to pattern both gate foot and overhang towards the drain contact in order to form a gamma-gate contact that acts as a field-plate terminal, see figure 1. On the same wafer devices with three different field-plate extensions (L_{FPD}) of 0.1μm, 0.4μm and 0.7μm were fabricated. Gate foot length was estimated to be 0.3μm while various gate widths (W_G) were fabricated ranging from 4x100μm, 8x75μm and 10x100μm. Nominal gate-source (L_{GS}) and gate-drain (L_{GD}) spacings are 0.2μm and 1.5μm, respectively.

III. DEVICE CHARACTERIZATION

Fabricated devices were subjected to electrical characterization by means of DC, 80ns pulsed I-V and RF measurements. Due to the presence of DC to RF dispersion in the fabricated devices current Deep Level Transient

978-2-8748-7007-1/08 $25.00 © 2008 EuMA

Spectroscopy (I-DLTS) measurements were also carried out in order to identify the responsible trap level(s). The experimental results obtained will now be described in the following subsections.

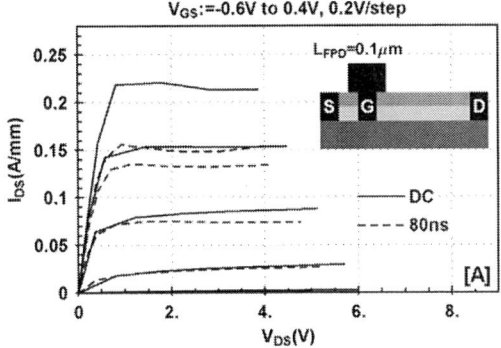

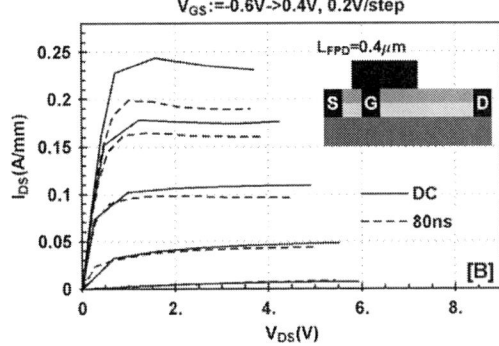

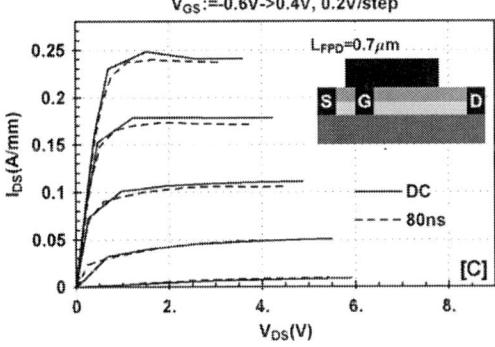

Fig. 2 Comparison of DC and 80ns pulsed I-V characteristics for devices with L_{FPD} of 0.1μm [A], 0.4μm [B] and 0.7μm [C]. The amount of DC-to-RF dispersion is larger for the device with L_{FPD}=0.1μm and it decreases at the increasing of L_{FPD}.

A. DC and pulsed I-V characterization

Device DC characteristics were first measured and, as it can be seen in figure 2, all the devices tested showed pinch-off voltages of approximately -0.4V and maximum drain currents at V_{GS}=+0.4V in the 0.22A/mm - 0.24A/mm range regardless of the field-plate extension. On the other hand, device off-state breakdown voltages increased at the increasing of L_{FPD}.

Measured off-state breakdown voltage were 25V, 38V and 44V for L_{FPD} of 0.1μm, 0.4μm an 0.7μm respectively. The increase of breakdown voltage with the increasing of L_{FPD} has been discussed in a previous work and will not be addressed here [5].

Further to the increase of the breakdown voltage, the adoption of a larger L_{FPD} yielded also other benefits in device performance. When comparing the DC I-V characteristics with 80ns pulsed I-V characteristics for the device with the smallest L_{FPD}, see figure 2A, it is straightforward to notice that pulsed characteristics are lower than DC ones. This means that DC-to-RF dispersion is affecting the device and that its RF performance will be limited both in terms of output power and power added efficiency. However, by comparing the results obtained in figures 2B and 2C, we can observe that, at the increasing of L_{FPD}, the amount of dispersion decreases, thus suggesting that devices with the largest L_{FPD} will behave the best when subjected to large signal RF drive.

B. RF Power Measurements

After DC and pulsed I-V characterization the fabricated devices were tested with continuous wave RF power measurements carried out at 2GHz. The measurement setup used in this work is based on an Agilent PNA Vector Network Analyzer, a PAF Dragon Load-Pull System, and two manual tuners for load and source matching. Due to the differences in breakdown voltages we tested all the devices at the same operating drain voltage in order to evaluate the improvement in large signal performance due to the field-plate structure. Three devices with a gate width of 1mm for each of the three L_{FPD} available on the processed wafer were used for this comparison. The devices were biased at V_{DS}=10V and matched for maximum output power. The obtained results are summarized in table I. The conclusion of this comparison is that pulsed I-V characteristics can predict the large signal behaviour since the devices with the largest pulsed currents, i.e. with the largest L_{FPD}, are those yielding the best power performance. In fact, devices with L_{FPD}=0.7mm yielded up to 1.6dBm more power than those with L_{FPD}=0.1μm.

TABLE I
COMPARISON OF CW RF PERFORMANCES AT 2GHZ

| L_{FPD} (μm) | V_{DS}=10V | | |
	Gain (dB)	P_{OUT} (dBm) @ 3dB compression	Peak PAE (%)
0.1	20.9	27.6	51.5
0.1	20.6	27.8	51.1
0.1	20.7	27.6	49.8
0.4	20.7	28.2	53.8
0.4	20.3	28.2	53.8
0.4	20.9	28.0	52.8
0.7	18.5	29.4	59.0
0.7	18.2	29.3	58.5
0.7	18.7	29.2	58.0

978-2-8748-7007-1/08 $25.00 © 2008 EuMA

Since the devices with $L_{FPD}=0.7\mu m$ were the most promising one in order to achieve high output power levels we measured them also at higher operating voltages. The highest power density was obtained from devices with $W_G=8x75\mu m$. Biased at $V_{DS}=21V$, they yielded power densities up to 2W/mm with peak PAEs of 65%, see figure 3.

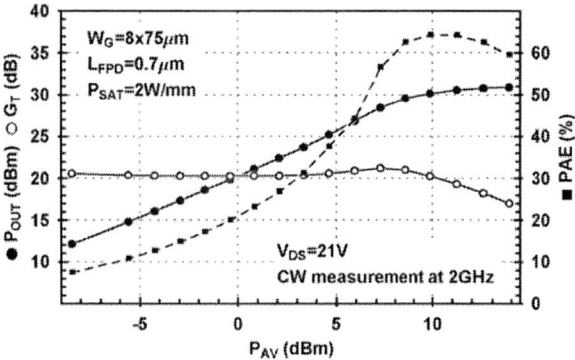

Fig. 3 Uncooled Room Temperature CW RF power measuremen t at 2GHz on a 8x75μm wide device with $L_{FPD}=0.7\mu m$. The device yielded a maximum output power density of 2W/mm with a peak PAE of 65%.

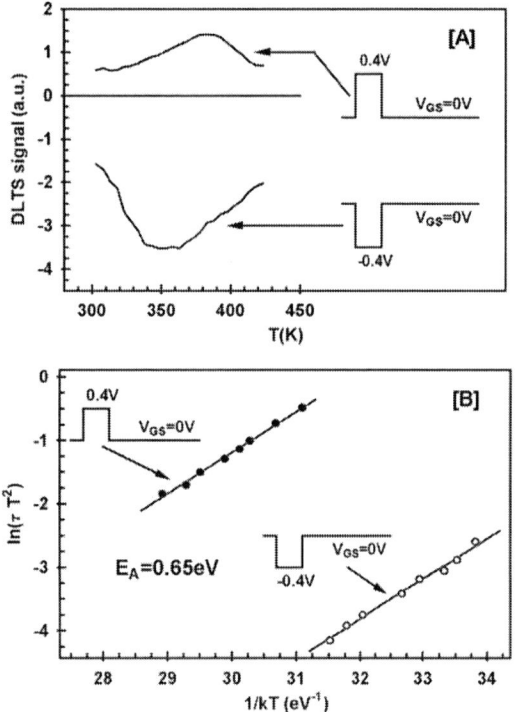

Fig. 4 Current-DLTS signals obtained at a drain voltage of 2V by applying positive and negative gate pulses whose period was 100μs and pulse width 10μs [A]. From the measured data an Arrhenius plot can be obtained yielding an activation energy of 0.65eV for both the traps capture and emission process [B].

C. Current-DLTS Measurements

From results shown in section A, it is clear that some trapping phenomena must occur at least in the devices with the smallest L_{FPD}, since we observed DC-to-RF dispersion and reduced output power for these ones. DC-to-RF dispersion has been widely studied in the past on GaAs-based devices and it has been proposed that it can be explained by means of hole traps located at the device surface [4]. In our case, since all the devices were subjected to the same processing steps, the cause of the dispersion has to be present regardless of the L_{FPD} adopted. It must therefore be assumed that the field-plate structure somehow mitigates or controls trap dynamics, thus reducing the amount of current dispersion. However, in order to carry out numerical simulations aimed at gaining insight into the physical mechanism involved, a careful experimental trap characterization has to be carried out. Since the amount of dispersion is larger for the devices with $L_{FPD}=0.1\mu m$, we chose these devices to carry out current-DLTS measurements.

Figure 4A shows DLTS spectra obtained by applying both positive and negative gate voltage pulses at a drain voltage of 2V. The underlying capture and emission transients are characterized by time constant in the 1μs-10μs range. From the Arrhenius plot depicted in figure 4B we extracted an activation energy of 0.65eV both for the emission and the capture transient. The fact that both emission and capture transient yield the same activation energy has been previously observed in [4], where it was explained by means of a hole-trap located at the surface of the device. In our case, we expect that the trap involved in the current transient phenomena will be located at the GaAs/SiN interface.

IV. NUMERICAL SIMULATIONS

The device structure was thus simulated by means of the commercial DESSIS-ISE (Synopsis Inc.) simulator. The structure of the simulated devices is exactly the same as the one depicted in figure 1 with the introduction of a trap level at the GaAs/SiN interface located at 0.65eV from the GaAs valence band with a concentration of $2x10^{12}cm^{-2}$. Numerical simulations were carried out for all three values of L_{FPD} and the effect of the field-plate extension on drain current dispersion was evaluated by simulating gate pulses from -0.8V to 0.2V, with drain voltages fixed at 4V and a rise time of 10ns. As can be seen in figure 5A, numerical simulations predict the time constant involved in the experimentally observed drain current transients, as well as the fact that the devices with the largest L_{FPD} are subjected to little or no current dispersion compared to those with smaller field-plate extension. Some insights on the field-plate action in reducing the transient phenomena can be gained by observing figure 5B, where the concentrations of empty hole traps at t=10ns are depicted for the three L_{FPD} studied. When t=10ns, the gate voltage has reached its steady state value of +0.2eV but trap population has not yet changed significantly from the starting condition of t=0s when the gate voltage was -0.8eV, i.e. below pinch-off. From figure 5B, it is straightforward to notice that the device with the shortest L_{FPD} has the largest amount of empty hole traps, i.e. the highest surface negative charge. This translate into a smaller 2DEG concentration in the device access region and thus into smaller initial drain current. When time increases, holes are captured from the hole traps, thus reducing the amount of empty traps. This leads to a reduction of the surface negative charge, an increase in the 2DEG concentration and an increase of the drain current towards its steady state value. The decrease in empty traps at increasing L_{FPD} is at the origin of the power performance improvement observed for increasing L_{FPD}.

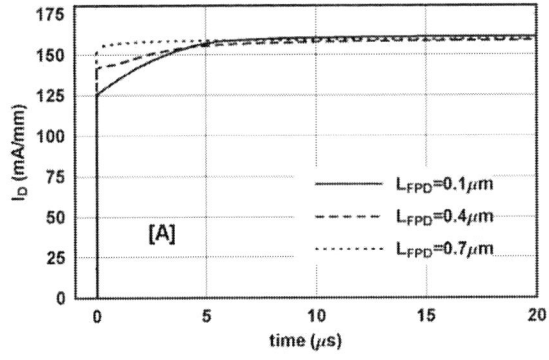

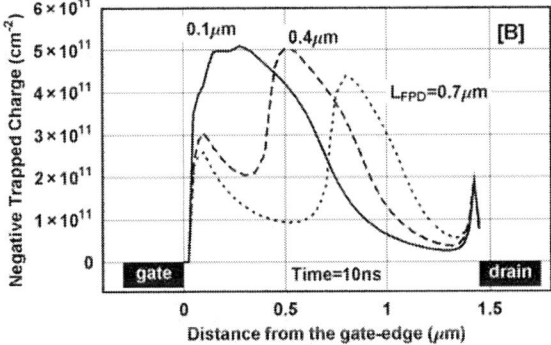

Fig. 5 [A] Simulated drain current transients obtained by applying a positive gate voltage pulse from -0.8V to +0.2V at a drain voltage of 4V. Numerical simulations are in agreement with the experimental results where devices with $L_{FPD}=0.1\mu m$ are showing the largest amount of DC to RF dispersion. [B] Amount of empty surface hole traps at t=10ns. The amount of empty traps (i.e. the amount of negative trapped charge) is larger for $L_{FPD}=0.1\mu m$ compared to the population profiles obtained with $L_{FPD}=0.4$ and $0.7\mu m$.

V. CONCLUSIONS

A detailed analysis, based on both experimental measurements and numerical simulations, of the effect of field-plate length in GaAs pHEMTs has been presented. It has been shown that power performance is improving at the increasing of field-plate length resulting in up to a 1.6dBm increase in output power. This improvement is related to the reduction of trapping phenomena that induce DC-to-RF current dispersion. The origin of DC-to-RF dispersion in the fabricated devices has been associated with a hole trap located at 0.65eV from the valence band edge, as obtained from

current-DLTS measurements carried out on the fabricated devices. Finally, the effect of field-plate extension on reducing drain current dispersion has been explained by means of numerical simulations that take into account the presence of surface traps located at the GaAs/SiN interface. To the best of authors knowledge this is the first time that the often observed improvement in device performance due to the field plate structure has been explained by means of numerical simulations.

ACKNOWLEDGMENT

This work was partially supported by the Italian Ministry for University and Research (MIUR) under the PRIN 2005 project "High breakdown voltage FETs for high power and efficiency applications".

REFERENCES

[1] K. Asano, Y. Miyoshi, K. Ishikura, Y. Nashimoto, M. Kuzuhara and M. Minuta, "Novel high power AlGaAs/GaAs HFET with a field-modulating plate operated at 35V drain voltage", International Electron Devices Meeting (IEDM) 1998, 6-9 December, 1998, pp. 59-62.

[2] A. Wakejima, K. Ota, and K. Matsunaga, "A GaAs-based field-modulating plate HFET with improved WCDMA peak-output-power characteristics", *IEEE Transactions on Electron Devices*, Vol. 50, No.9, 2003, pp. 1983-1987.

[3] A. Chini, D. Buttari, R. Coffie, S. Heikman, S. Keller and U. K. Mishra "12W/mm power density alGaN/GaN HEMTs on sapphire substrate", *IEE Electronics Letters*, Vol. 40, Issue 1, 2004, pp. 73-74.

[4] G. Verzellesi, A. F. Basile, A. Cavallini, A. Castaldini, A. Chini and C. Canali, "Light Sensitività of Current DLTS and Its Implications on the Physics of DC-to-RF Dispersion in AlGaAs-GaAs HFETs", *IEEE Transactions on Electron Devices*, Vol. 52, No.4, 2005, pp. 594-602.

[5] A. Chini, S. Lavanga, M. Peroni, C. Lanzieri, A. Cetronio, V. Teppati, V. Camarchia, G. Ghione and G. Verzellesi, "Fabrication, Characterization and Numerical Simulation of High Breakdown Voltage pHEMTs", *The 1st European Microwave Integrated Circuits Conference*, 10-13 Sept. 2006.

Proceedings of the 3rd European Microwave Integrated Circuits Conference

A Computational Load-Pull Method for TCAD Optimization of RF-Power Transistors in Bias-Modulation Applications

O. Bengtsson[#1], L. Vestling[*2], J. Olsson[*3]

[#]University of Gävle, Gävle, SE-801 76, Sweden

[1]bob@hig.se

[*]Uppsala University, The Ångström Laboratory, Solid State Electronics,

P.O. Box 534, Uppsala, SE-751 21, Sweden

[2]lars.vestling@angstrom.se

[3]jorgen.olsson@angstrom.uu.se

Abstract— **In this paper a method for TCAD evaluation of RF-Power transistors for high-efficiency operation using drain bias-modulation is presented. The method is based on large signal time-domain transient computational load-pull. With the method, intrinsic device parasitics and mechanisms affecting device efficiency under drain bias modulation can be investigated and optimized for the application making it very useful for RFIC design. A case study has been done on a CMOS compatible LDMOS. For verification under dynamic operation two-tone signals with varying envelope has been simulated. The results show a possible 15% increase in the efficiency of a modulated signal for the studied device at the expense of increased phase distortion observable also in the time-domain waveforms generated. Since the method is based on TCAD it is also useful in the investigation of e.g. dynamic breakdown during high envelope under bias-modulation operation.**

I. INTRODUCTION

Investigation and modeling of nonlinear distortion in RF-power transistors have for a number of years been of great interest to the research community. New telecom standards with wider bandwidth and higher peak-to-average ratio have placed higher demands on the linearity of the power amplifiers. Linearization techniques have had to be implemented to meet system requirements at the expense of efficiency. To overcome this much effort is now spent on improving overall efficiency by using high efficient switch mode amplifiers and / or implementing efficiency enhancement architectures like envelope tracking, envelope elimination and restoration, EER, or Doherty configuration [1]. In envelope tracking systems the amplifier works in a linear mode but efficiency is increased by modulating the amplifier bias based on the signals envelope, Fig 1. In EER the signal is limited to constant amplitude and the amplifier works in switch mode operation. The amplitude modulation is restored by modulating the bias of the amplifier, Fig. 2. These methods are successfully used in low-power amplifiers for cellular handheld applications, [2]-[5]. In order to optimize devices for these high-efficiency modes of operations it is of great importance to understand what parasitics have an affect and how. Normally device evaluation for bias-modulation

applications is conducted using circuit simulation tools and available non-linear models. With the models high efficiency bias modulated amplifiers have been designed in standard circuit simulators then fabricated and evaluated [6].

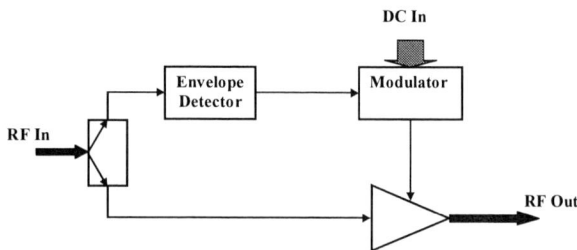

Fig. 1 Schematic envelope tracking system.

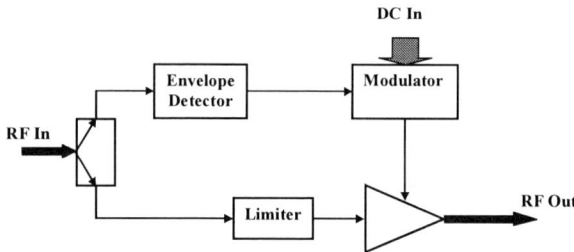

Fig. 2 Schematic EER system for improved efficiency.

For device evaluation this procedure is time consuming and very lengthy from device design to feedback. It also requires a good device model which is seldom available. Technology CAD or TCAD is normally used for device optimization pre-fabrication. With TCAD the fundamental semiconductor equations are solved for a finite element, FEM, physical model of the device. Commercial tools like Atlas from Silvaco and Dessis from Synopsys readily provide DC, small-signal and transient electrical solutions for 1D-3D structures. The ability to store solutions during electrical simulations enables

978-2-8748-7007-1/08 $25.00 © 2008 EuMA

the study of transport and breakdown mechanisms in the structure. Since TCAD is based on FEM it can be time consuming for good accuracy but with improving computer performance and computation algorithms even large signal simulations for RF-power devices are now feasible on ordinary personal computers. In this work the presented method was evaluated for a CMOS compatible medium voltage LDMOS for RF-Power applications at 1 GHz, [7]. The device was biased in low class AB and evaluated for envelope tracking.

II. METHOD OUTLINE

Initially a CW large-signal time-domain, LSTD, computational load-pull simulation was conducted for a constant drain supply voltage of 12 V in the class-AB mode [8], [9]. CW voltage simulations were then conducted for different input voltages with the fundamental optimum load-impedance created by a parallel resonance circuit. From this swept v_{IN} response the 1 dB compression point, P_{1dB}, was found and the device was re-tuned for optimum efficiency in this point. These simulations were used as reference for the bias-modulated simulations.

A. Single-tone CW Simulations

The bias-modulator was modeled based on the DC-IV curve simulation. Low end of drain-supply voltage, V_{DM}, was set by the knee voltage, V_{KNEE}, and high-end of the drain supply voltage, V_{D-MAX}, normally set by the breakdown voltage, was in this case set to 12 V since the device was designed for 12 V operation. V_{D-MAX}, was modeled to be reached at an input peak voltage equivalent to the 1 dB compression point. A linear model was used for the modulator (1).

$$V_{DM}(t) = V_{KNEE} + \frac{\hat{v}_{IN}(t)}{\hat{v}_{P1dB}}(V_{D-MAX} - V_{KNEE}) \quad (1)$$

From this equation the drain supply voltage was calculated for each individual CW input peak voltage simulated. This produced an estimated "static" response expected from the drain bias-modulation, Fig 3.

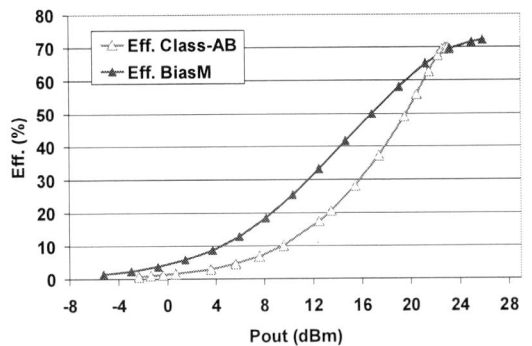

Fig. 3 Drain-efficiency versus output power for CW simulated LDMOS at class-AB with V_D=12 V and class-AB with bias modulation.

The efficiency shows a great increase of 10-15% in the midlevel regions with an output power of 8 dBm to 16 dBm. The increased output power in compression for the bias-modulated signal is due to a non-limited drain-supply voltage. Since the breakdown was not reached in this study the drain supply was allowed to increase beyond 12 V. The drain bias-modulation also affects the phase-response. Compared to the ideal case the phase response versus output power, related to the AM-PM conversion, shows a quite different characteristic for the drain bias modulated case, Fig. 4.

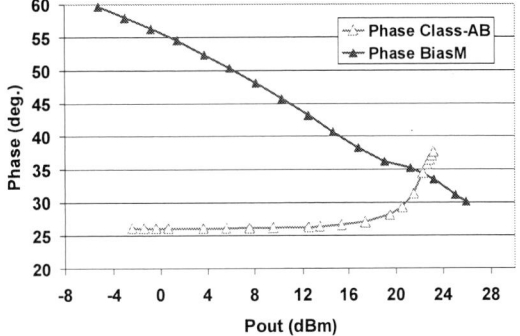

Fig. 4 Output power and phase distortion in reference to ideal phase shift for CW simulated LDMOS at class-AB with V_D=12 V and class-AB with bias modulation.

The phase response of the normal class-AB shows the expected almost constant phase until compression is reached. The drain bias-modulated case however shows an almost linear phase response with a 30° difference in phase between the lowest amplitude and compression. This is expected to add severe phase distortion.

B. Two-tone Varying Envelope Simulations

The above result is based on a modeled modulator and CW signals with constant envelope. To verify the results during dynamic operation a two-tone signal with varying envelope was used. A novel LSTD computation load-pull setup was used for the drain bias-modulation, Fig 5.

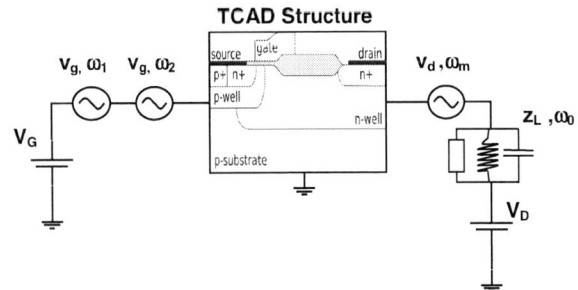

Fig. 5 Schematic outline of computational load-pull LSTD setup for drain bias-modulation simulation.

978-2-8748-7007-1/08 $25.00 © 2008 EuMA

The two input signals creates a modulation peak envelope voltage, PEV, equal to two times the individual tone peak voltage (2), [1].

$$\hat{v}_S = 2\hat{v}_{IN}\cos(\omega_m t)\cos(\omega_c t) \qquad (2)$$

Where $\hat{v}_S$ is the momentary input signal voltage, $\hat{v}_{IN}$ is the single tone peak voltage, ω_m is the modulation frequency equal to half the tone spacing and ω_c is the carrier frequency or the mean frequency of the two signal frequencies ω_1 and ω_2. The load-impedance was created using the same resonance structure as for the constant V_D case. In series with the load a voltage source was introduced to create the bias-modulation. The bias modulation generator was set to the modulation frequency ω_m with the peak amplitude given by the second part of (1). The constant part of the drain voltage V_D was set to the knee voltage. With proper timing of the signals in the LSTD simulation a full period bias-modulated simulation was conducted, Fig 6.

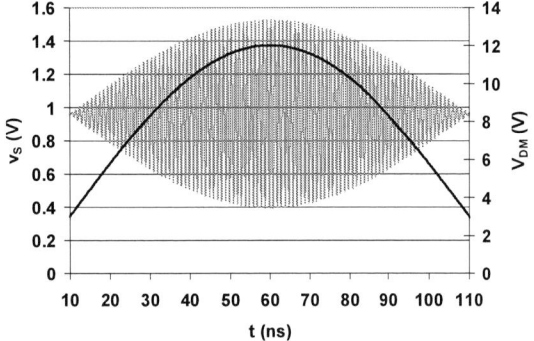

Fig. 6 Two-tone input signal, v_S (grey), at 1.00 GHz and 1.01 GHz and modulated drain bias V_{DM} (black). Input peak- voltage PEv$_S$= -5 dBV.

From Fig. 4 it was expected that the drain-modulated signal produce more phase shift at less output power. A study of the drain current, i_D, for the dynamic signal also show this, Fig. 7.

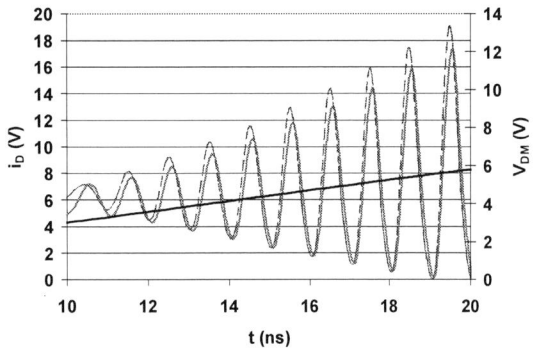

Fig. 7 Drain current for the class-AB case (blue-dashed) and the bias-modulated case (red-solid) in the low envelope region of the signal.

The peak current of the bias-modulated signal is much smaller and lags severely. At higher envelope the signal has caught up with the constant V_D signal both in amplitude and phase, Fig. 8, which was also expected from Fig. 4.

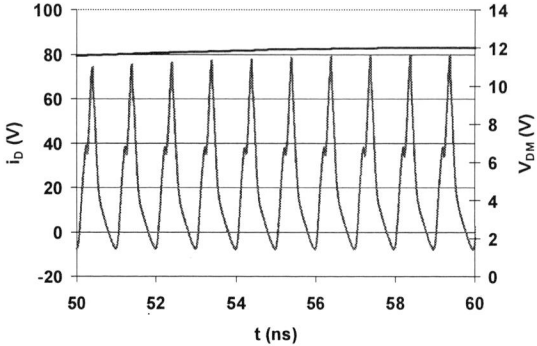

Fig. 8 Drain current for the class-AB case (blue-dashed) and the bias modulated case (red-solid) in the high envelope region of the signal.

C. Efficiency Comparison

In order to compare the possible efficiency enhancement using bias modulation under dynamic operation the normal class-AB and the drain bias-modulated case were simulated using the same two-tone signals at three different amplitudes. The LSTD results were analyzed using FFT as in ordinary computational load-pull. For the calculation of efficiency the power provided at the modulation frequency was however added as supply power, Fig 9.

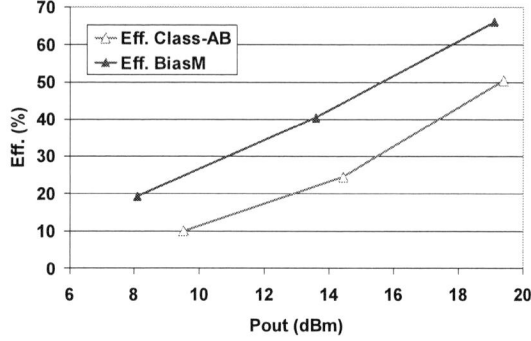

Fig. 9 Drain-efficiency versus average output power for a two-tone simulated LDMOS at class-AB with V_D=12 V and class-AB with bias modulation. Input signals at 1.00 GHz and 1.01 GHz.

The results show a possible efficiency enhancement of about 15% in dynamic operation for the two-tone signal. This however comes at an expense of increased intermodulation distortion, IMD, and phase distortion. IM3 has increased almost 20 dB in the low-power region by introducing the drain bias-modulation, Fig. 10.

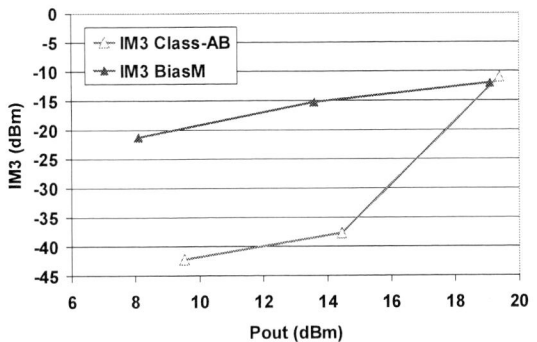

Fig. 10 Third order IMD versus average output power for a two-tone simulated LDMOS at class-AB with V_D=12 V and class-AB with bias modulation. Input signals at 1.00 GHz and 1.01 GHz.

The phase distortion for the dynamic range covered is about 5° for class-AB but 10° for bias-modulation, Fig. 11.

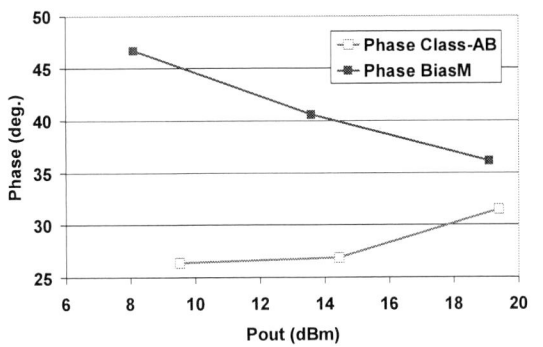

Fig. 11 Phase distortion in reference to the ideal phase shift versus average output power for a two-tone simulated LDMOS at class-AB with V_D=12 V and class-AB with bias modulation. Input signals at 1.00 GHz and 1.01 GHz.

Due to the lengthy computations of two-tone LSTD simulations an adaptive accuracy was used during the TCAD simulations. Higher order IMD at lower power were therefore not considered.

III. CONCLUSIONS

A computational load-pull method for the investigation of bias-modulation of RF-power devices based on large-signal time-domain simulations has been developed. The usefulness of the method has been shown on a case study of a CMOS compatible LDMOS producing expected power performance and efficiency improvement of about 15% in the backed-off region for a two-tone signal but also validating phase distortion from the simulated time-domain waveform under dynamic signal operation. The case study shows a much increased phase distortion of 10° for the studied dynamic range of the bias-modulated device compared to 5° for the standard class-AB case. With this method it is possible to investigate the intrinsic device mechanisms at TCAD level under bias-modulation conditions like envelope tracking and EER at an early stage in the device design process. It is also possible to investigate e.g. the effect of dynamic breakdown during high envelope or bias modulation effect on intermodulation and phase distortion. The contribution from the device can be studied independent of the modulator enabling focus on power device improvement.

REFERENCES

[1] S. C. Cripps "RF Power Amplifiers for Wireless Communications," Artech House, 1999.

[2] G. Hannington, P-F. Chen, P. M. Asbeck, and L. E. Larson, "High-Efficiency Power Amplifier Using Dynamic Power-Supply Voltage for CDMA Applications," *IEEE Trans. Microwave Theory & Tech.*, vol. 47, no. 8, pp. 1471-1476, August 1999.

[3] J. Staudinger, B. Gilsdorf, D. Neuman, G. Norris, G. Sadowniczak, R. Sherman, and T. Quach, "High Efficiency CDMA RF Power Amplifier Using Dynamic Envelope Tracking technique," *2000 IEEE MTT-S Int. Microwave Symp. Dig.*, vol. 2, pp. 872-876, June 2000.

[4] M. Ranjan, K. H. Koo, G. Hannington, C. Fallesen, and P. Asbeck, "Microwave Power Amplifiers with Digitally-Controlled Power Supply Voltage for High Efficiency and High Linearity," *2000 IEEE MTT-S Int. Microwave Symp. Dig.*, vol. 1, pp. 493-496, June 2000.

[5] B. Sahu, and G. A. Rincón-Mora, "A High-Efficiency Linear RF Power Amplifier With a Aower-Tracking Dynamically Adaptive Buck-Boost Supply," *IEEE Trans. Microwave Theory & Tech.*, vol. 52, no. 1, pp. 112-120, January 2004.

[6] K. Y. Kim, J. H. Kim, S. M. Park, and C. S. Park, "Parasitic Capacitance Optimization of GaAs HBT Class E Power Amplifier for High Efficiency CDMA EER Transmitter." *2007 IEEE RFIC Symp. Dig.*, pp 733-726, June 2007.

[7] O. Bengtsson, A. Litwin, and J. Olsson, "Small Signal and Power Evaluation of Novel BiCMOS Compatible, Short Channel LDMOS Technology," *IEEE Trans. Microwave Theory Tech.*, vol. 51, pp. 1052-1056, Mar. 2003.

[8] G. H. Loechelt, and P. A. Blakey, "A Computational Load-Pull System for Evaluating RF and Microwave Power Amplifier Technologies," *2000 IEEE MTT-S Int. Microwave Symp. Dig.*, vol. 1 , pp. 465-468, June 2000.

[9] [9] R. Jonsson, Q. Wahab, S. Rudner, and C. Svensson, "Computational load pull simulations of SiC microwave power transistors," *Solid-State Electronics*, vol. 47, pp. 1921-1926, 2003.

Thermal Model Extraction of GaN HEMTs for Large-Signal Modeling

Samir Dahmani, Endalkachew S. Mengistu, and Günter Kompa

University of Kassel, Department of High Frequency Engineering,
Wilhelmshöher Allee 73, D-34121 Kassel, Germany, Phone: +49-561 8046528, Fax: +49-561 8046529
dahmani@uni-kassel.de

Abstract—Self-heating has a large effect on electrical performance of RF power devices. In the past, several methods were developed to estimate the average channel temperature of FETs. Some of these are based on approximate closed form expressions. These techniques give acceptable results under limited conditions and only for specific device layouts. In this proposed work, we present an accurate method for the extraction of the thermal profile of large-size AlGaN/GaN HEMTs using both Finite Element Method (FEM) simulation and measurement techniques. The thermal investigation of the complete structure of the device permits an accurate calculation of the distributed device temperature taking into account the temperature dependence of the thermal conductivities. This analysis also helps device designers tuning physical and geometrical parameters of the structure. The thermal resistances and the thermal time constants under each finger and of the whole FET structure are calculated from static and transient FEM thermal simulations, respectively. Alternatively, the thermal time constant is also determined from drain current transient measurements. Using this procedure, we obtained detailed thermal profile for AlGaN/GaN HEMT and implemented the resulting thermal subcircuit in the large-signal model of GaN HEMTs. The simulated static and transient I(V) characteristics of the large-signal model are in good agreement with the measured data.

I. INTRODUCTION

With improved power density and output power of AlGaN/GaN HEMTs, much effort is made to investigate the effect of temperature on their performance. Accurate knowledge of the channel temperature of these devices greatly improves their large-signal model. Theoretically, an increase in device temperature leads to a decrease in electron mobility, saturation electric field, and drift velocity, which decreases the current and gain, and affects impedance matching [1,2].

In measurement based modeling techniques of high power transistors, the temperature is considered uniform throughout the chanel of the device. This implies that the contribution of each finger to total output power is the same. However, with increased dissipated power, the channel under lateral fingers are at lower temperature than those under the central fingers. Therefore, outer fingers contribute more to the output power than the inner ones due to this non-uniform temperature distribution. This non-uniform self-heating effect, neglected in earlier works, is considerd in present paper.

Hence, the objective of the proposed work is to offer an accurate electro-thermal modeling procedure. In section (2), thermal analysis and FEM simulation of the GaN HEMT device is considered. First, the steady state thermal analysis is considered, from which the equivalent thermal resistance of the channel is deduced. Second, transient thermal simulation is conducted to extract the thermal time constants for thermal model implementation. Some selected simulation results are also presented. As an alternative method, transient measurement technique is presented for determining the thermal time constant in section (3). Finally in section (4), the thermal model is integrated into a large-signal look-up table based model in Agilent ADS® [3]. The simulated static and transient I(V) characteristics will then be compared with static and transient DC measurement results to check the validity of the thermal model.

The AlGaN/GaN multi-finger HEMT structure considered here is based on available data in reference [4]. Structures have gate-to-gate pitch of 50 μm, source-to-drain separation of 2.5 μm, and gate width of 250 or 400 μm. Four layer AlGaN/GaN/AlGaN/SiC simplified structure has been used for thermal investigation.

II. THERMAL ANALYSIS AND MODELING

The GaN HEMT device physics reveals that its electrical behavior is highly dependent on temperature variations. An inaccurate estimation of the channel temperature will lead to a significant shift in the drain-source current magnitude. Under large signal operation, as temperature increases, the electron mobility decreases, and consequently the drain-source current, given by (1), is decreased.

$$I_{ds} = q\, n_s(T)\, v(T)\, d_{2DEG}\, L_{ch} \qquad (1)$$

The dissipated power density is calculated and partitioned on the different fingers. This power density is included in the thermal procedure via the heat generation term Q of the heat transfer equation described by,

$$\nabla \cdot \big(\kappa(T)\,\nabla T\big) + Q - \rho_m C_p \frac{\delta T}{\delta t} = 0 \qquad (2)$$

This equation is implemented taking into account all heat transfer mechanisms and boundary conditions of the device. Conductive, convective, and radiative terms, represented by the three terms in (3), respectively, are included in the heat generation term (Q) using the energy conservation principle.

$$Q = -k\frac{dT}{dx} + h(T_{m1} - T_{m2}) + \varepsilon_{em} K_{S.B}(T_{m1}^4 - T_{surface}^4) \qquad (3)$$

An important parameter is the thermal conductivity of each layer. This term is to be considered delicately due to its high effect on device temperature. It has been shown that it is temperature dependent. Empirical fitting expressions used here are available in references [5, 6]. The above procedure is simulated, physical quantities are described over domain of integration and heat generation terms related to dissipated power and reference temperature $T_0 = 300°K$ are applied on appropriate boundaries. The resulting matrix formulation includes the thermal profile of the complete structure, from which the thermal resistance can be determined by taking the ratio of the temperature rise to dissipated power [7].

$$R_{Th} = \frac{\Delta T}{P_{diss}} \qquad (4)$$

Two important analyses will be presented, the steady state and the transient thermal simulations, from which thermal resistances and time constants are determined successively.

A. Steady State Thermal Simulation

In the steady state thermal analysis of the 3.2 mm AlGaN/GaN HEMT the thermal profile of the structure is determined. The non-uniform distribution of temperature is obtained, from which the thermal resistance of each finger is determined separately. The total thermal resistance is calculated by means of the thermal resistance of individual fingers using averaging or integration over each finger area.

In Fig. 1, the non-uniform temperature distribution is shown. The temperature rise in lateral fingers shows less increase due to heat dissipation to lateral edges. The temperature rise is reported to be around 58.39° under the central finger. Temperature profile at the channel-to-gate surface and the interface between the SiC substrate and the AlGaN layer are evaluated, from which the thermal resistance can be defined at any point. The thermal resistances of the 8 individual fingers are illustrated in Fig. 2. These latter are used in the determination of the total thermal resistance of the GaN HEMT under discussion. Details can be found in [8].

B. Transient Thermal Simulation

The main idea is to quantify the thermal behavior of the GaN HEMT device in the presence of transient DC power dissipation, from which the thermal time constant can be deduced for individual fingers and for total structure. This thermal time constant is used in the thermal model implementation of the look-up table based electrical model. Moreover, this procedure helps us in enhancing the thermal model through the knowledge of the transient behavior of individual fingers.

Fig. 3 represents simulation results for the 2 x 400 µm structure. The obtained results have an exponential trend. After exponential fitting, the resulting thermal time constant is 36.62 µsec. Analyzing data, as illustrated in inset of Fig. 3, we noticed that the temperature rise is very fast below 1 µsec and relatively very slow above. The effect of such variations cannot be described at circuit level by a simple RC thermal model implementation commonly used for slow thermal time constant.

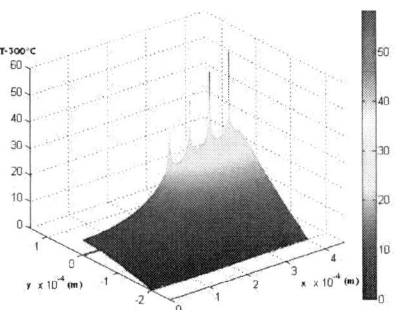

Fig. 1. Thermal profile of a 3.2 mm (8 x 400 µm) HEMT structure, due to symmetry only 4 fingers are presented. The temperature gradient is higher at the channel under the gate.

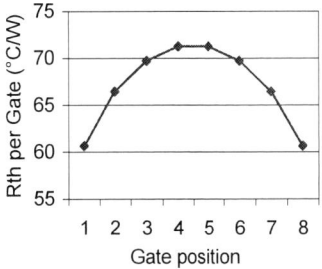

Fig. 2. Thermal resistance for Individual fingers of a 3.2 mm device, the total thermal resistance is the equivalent thermal resistance of the 8 fingers.

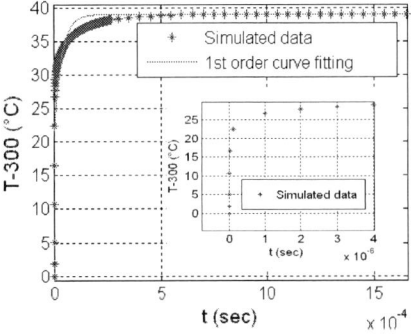

Fig. 3. Transient thermal simulation and first order curve fitting for the 2 x 400 µm HEMT device. Inset shows fast transient occurs for t < 1 µsec.

Similarly, in Fig. 4a, simulated data and corresponding curve fitting are shown for individual fingers in 8 x 400 µm device. The slow thermal time constants are 83.44 µsec, 83.60 µsec, 85.26 µsec and 91.80 µsec from central to lateral finger, respectively.

A second order exponential curve fitting is given in Fig. 4b, it accurately approximates the simulated data. The resulting time constants are 23.71 nsec and 87.51 µsec for fast and slow transient thermal profile, respectively. This second order approximation can be interpreted as the real thermal behavior of the AlGaN/GaN HEMT device. Fast and slow transient thermal variations are for epitaxy and SiC layers, respectively.

978-2-8748-7007-1/08 $25.00 © 2008 EuMA

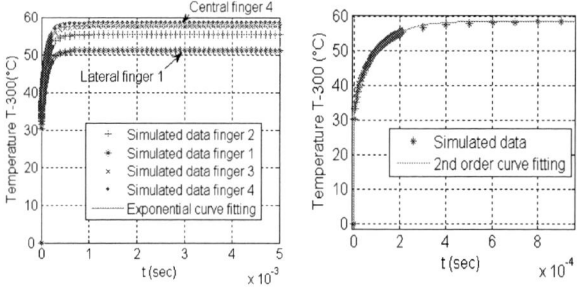
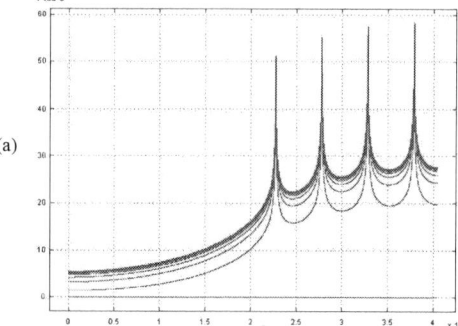

Fig. 4. Transient Simulated data and exponential curve fitting for 3.2 mm device (a) for individual fingers, four fingers are shown, (b) second order curve fitting for the central finger.

Fig. 5 depicts the epitaxy and SiC layers transient thermal behavior and can be used to validate this second order approximation. First, in Fig. 5a, the saturation temperature of the SiC substrate under fingers level can be determined; under the central finger it is 34°C. From Fig. 5b, the temperature rise on the central finger is about 58.89°C. Comparing these data with the second order fitting, using Matlab optimization tool box, we noticed that the two extracted slow and fast thermal time constants match the transient temperature of SiC and epilayer, respectively. This gives an enhanced implementation of the thermal model. Thermal resistances are obtained through calculation of the ratio of temperature rise in each layer to the dissipated power. And, the thermal capacitances for each layer are the ratio of time constants to thermal resistances.

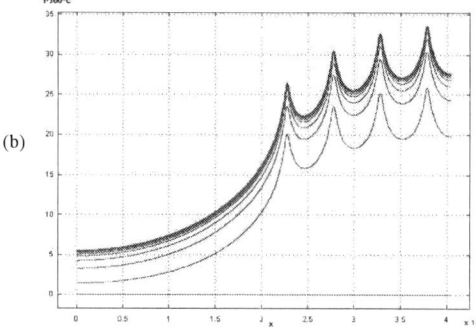

Fig. 5. Transient thermal profile (a) at the interface between the SiC substrate and epitaxial layer, (b) at the surface under the fingers. Δt = 50 µsec.

III. TRANSIENT DRAIN CURRENT MEASUREMENTS

Drain current transient measurements can be used to determine thermal and charge trap related time constants of FETs. This second method, as contrasted to the FEM thermal simulation of the HEMT structure discussed in the previous section, obtains the thermal constant indirectly by observing the change in drain current as function of device temperature.

In estimating the thermal time constants, trap related effects should be excluded and vice versa. Since charge trap related effects are function of the average gate- and drain-voltages, transient measurements from a bias point $(V_{GS0}, V_{DS0}) = (0V, 0V)$ have negligible trapping effects. Transient drain current measurements from such bias point can be used to obtain the thermal time constants. This is similar to the pulsed DC measurement of a FET where the I(V) characteristics from the bias point (0V, 0V) is taken as dispersion free and hence as reference in the dispersive drain current model derivation [9].

Transient measurement results for different size GaN HEMTs, using DiVA D265EP pulse system, are presented below. The drain current transient shown in Fig. 6a is for the 2 x 250 µm GaN HEMT (on wafer) with "pulsed-to" point v_{GS}, $v_{DS} = (0V, 4V)$. Since the "pulsed-to" point is along $v_{GS} = 0V$ and the drain-voltage remains low, the decrease in the drain current as function of time is due to self-heating.

As discussed in previous section, the drain current transient also shows a fast initial fall (related to the fast temperature increase of the thin epitaxy layers) and then followed by a slow decrease due to thicker SiC substrate.

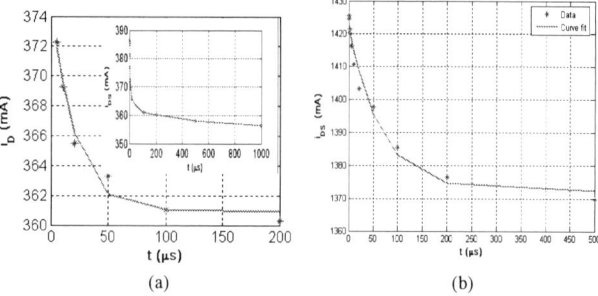

Fig. 6. Drain current transients: measurement and curve fitting (a) 0.5 mm GaN HEMT. Bias: $V_{G0}, V_{D0} = (0V, 0V)$, pulsed to $v_{GS}, v_{DS} = (0V, 4V)$. Inset for $0 < t < 1ms$. (b) 3.2 mm GaN HEMT. Bias: $(V_{GS0}, V_{DS0}) = (0V, 0V)$, pulsed-to point $(v_{GS}, v_{DS}) = (-1V, 4V)$.

A first order curve fit gives a thermal time constant of about 20 µs for a 2 x 250 µm GaN HEMT, and about 30 µs for a 2 x 400 µm device. The drain current transient to the pulsed-to point (v_{GS}, v_{DS}) for 3.2 mm HEMT is shown in Fig. 6b. Again multiple thermal time constants are required to fit the transient, as discussed earlier, but a first order fit gives a time constant about 65 µs.

Table I below summarises the results for thermal resistance and thermal time constant for the GaN HEMT devices under investigation. First, the thermal resistance is compared for different device sizes using FEM simulation and closed form expression given in [10]. The results show generally good agreement, whereas the technique in [10] overestimates the thermal resistance for FETs with small number of fingers.

978-2-8748-7007-1/08 $25.00 © 2008 EuMA

TABLE I
THERMAL RESISTANCE AND TIME CONSTANT FOR INVESTIGATED HEMTS.

	2 x 250 μm	2 x 400 μm	8 x 400 μm
R_{th} (°C/W) [10]	-	-	6.4
Rth simulated	-	19.6	8.35
τ_{th} (μs) measured	20	30	65
τ_{th} (μs) simulated	29	36	85

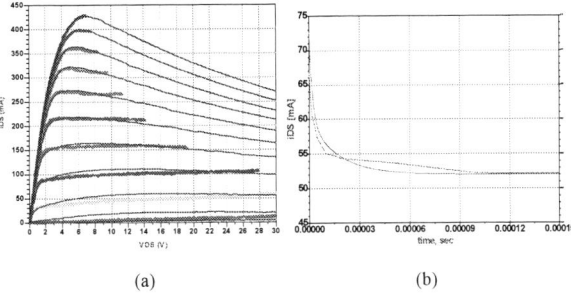

(a)   (b)

Fig. 7. Model simulation results (dots) and measurements (bold) of a 0.5 mm GaN HEMT. (a) Static DC. V_{GS} = +1V (top), 0.5V step. (b) Drain current transients. Bias: V_{G0}, V_{D0} = (-3V, 5V) and pulsed-to v_{GS}, v_{DS} = (-3V, 30V).

Also, the thermal time constants from the two methods are compared which are in good agreement. The transient measurement setup does not permit the evaluation of the fast thermal response, as the minimum sample time is 200 ns. Moreover, the maximum transient time that can be captured is limited to 1 ms. These two constraints limit the system use for multiple time constants implementation and for larger devices characterization, respectively. The FEM simulation approach gives both time constants and with no power limits, making it promising for enhanced thermal model implementation and larger HEMTs investigation.

IV. HEMT LARGE-SIGNAL MODEL SIMULATION

For FETs with substantial trap and thermal related effects, an efficient technique for deriving their dispersive drain current models is to use pulsed DC I(V) measurements from different quiescent bias conditions and at different ambient temperatures. The method is based on a few sets of pulsed DC I(V) measurements, a good estimation of the thermal resistance of the device and time constants related to trap and thermal effects. These steps enable us to predict the drain current variation as function of average drain and gate-voltages, self-heating and/or external temperature. Such a dispersive table-based large-signal model has been reported earlier [3].

In testing the thermal resistance extraction procedure described in the previous section, the static DC simulation of the large-signal model is the most relevant. Under static DC simulation of the large-signal model, the device channel temperature must be correctly determined from the dissipated power at a given bias point and the thermal resistance of the FET structure. The good match between simulation results and measured data shown in Fig. 7a is a verification of the validity of the extracted thermal resistance.

The large-signal model is also tested under transient simulations. The plot in Fig. 7b compares measurement with the large-signal model simulation. Here the bias and pulsed-to points are (V_{GS0}, V_{DS0}) = (-3V, 5V) and (v_{GS}, v_{DS}) = (-3V, 30V), respectively. In this case, self-heating dominates and the model compares well with measurement.

V. CONCLUSION

In this paper an investigation of electro-thermal modeling for FETs has been presented. First, steady state thermal profile of multi-finger GaN HEMT structure is investigated, from which the thermal resistances are deduced. The thermal resistances under each gate are separately calculated and from which the total thermal resistance is determined. Transient thermal analysis was also presented, which enables to determine the fast and slow thermal time constants of the individual fingers and complete structure. The results show that, for FETs with large number of fingers, the channel temperature under different gates are substantially unequal.

An alternative technique for the determination of the thermal time constant from transient drain current measurement has also been presented, which gave comparable results to the FEM method. The simulation procedure has shown advantages in describing both fast and slow time constants. Moreover, it is not limited in application as the transient measurement system. Consequently, it provides us with a valuable tool to enhance thermal model implementation using multiple time constants for larger devices. Finally, these additional inputs for a thermal model of the dispersive large signal model of the HEMT resulted in improved static DC and transient simulation of the implemented large-signal model.

REFERENCES

[1] S. Nuttinck et al., "Direct On-Wafer Non-Invasive Thermal Monitoring of AlGaN/GaN Power HFETs Under Microwave Large Signal Conditions," Gallium Arsenide Applications Symposium, 12th GAAS, pp. 79-82, Amsterdam, Oct. 2004.

[2] D. Denis, C. M. Snowden, and I. C. Hunter, "Design of Power FETs Based on Coupled Electro-Thermal-Electromagnetic Modeling," IEEE Trans., *MTT-S Int. Microw. Symp. Dig.*, pp. 461-464, Jun. 2005.

[3] E. S. Mengistu and G. Kompa, "A Large-Signal Model of GaN HEMTs for Linear Power Amplifier Design," in *Proc. 1st European Microwave Integrated Circuits Conference*, pp. 292-295, Manchester, UK, Sep. 2006.

[4] R. Lossy et al., "Large Area AlGaN/GaN HEMTS Grown on Insulating Silicon Carbide Substrates," Phys. Stat. Sol. (a) 194, No. 2, pp. 460-463, 2002.

[5] Y. Chang, Y. Zhang, and Y. Zhang, "A Thermal Model for Static Current Characteristics of AlGaN/GaN HEMTs Including Self-Heating Effect," J. Appl. Phys., Vol. 99, 044501-1/5, 2006.

[6] E. R. Heller and A. Crespo, "Electro-Thermal Modeling of Multifinger AlGaN/GaN HEMT Device Operation Including Thermal Substrate Effects," Journal of Microelectronics Reliability, doi:10.1016 / 2007.01.090.

[7] K. A. Filippov and A. A. Balandin, "Self-Heating Effects in GaN/AlGaN Heterostructure Field-Effect Transistors and Device Structure Optimization," Nanotech, Vol. 3, pp. 333-336, 2003, www.nsti.org, ISBN 0-9728422-2-5.

[8] S. Dahmani, E. S. Mengistu, and G. Kompa, "Electro-Thermal Modeling of Large-Size GaN HEMTS," *German Microwave Conference*, March,10-12, Hamburg, 2008.

[9] A. K. Jastrzebski, "Characterization and Modeling of Temperature and Dispersion Effects in Power MESFETs," *24th European Microwave Conference*, pp. 1319-1324, 1994.

[10] A. M. Darwish et al., "Thermal Resistance Calculation of AlGaN/GaN on SiC Devices," *IEEE Trans. Microwave Theory & Tech.*, Vol. 52, pp. 2611-2619, Nov. 2004.

978-2-8748-7007-1/08 $25.00 © 2008 EuMA

Proceedings of the 3rd European Microwave Integrated Circuits Conference

Modeling of Radiation, Conductor, and Dielectric Losses in SIW Components by the BI-RME Method

Maurizio Bozzi[#1], Luca Perregrini[#2], Ke Wu[*3]

[#] *University of Pavia, Department of Electronics*
Via Ferrata 1, 27100, Pavia, Italy
[1]maurizio.bozzi@unipv.it
[2]luca.perregrini@unipv.it
[*] *Poly-Grames Research Center, École Polytechnique de Montréal*
2500, Chemin Polytechnique, Montréal, Québec, Canada, H3T 1J4
[3]ke.wu@polymtl.ca

Abstract — This paper describes the modeling of the different types of losses in substrate integrated waveguide (SIW). In particular, radiation leakage, conductor losses, and dielectric losses are considered. The modeling is based on the Boundary Integral-Resonant Mode Expansion (BI-RME) method. This method permits a fast and accurate determination of the wideband frequency response of SIW interconnects and components, by providing the admittance matrix of the circuit in the form of a pole expansion in the frequency domain. The effect of conductor and dielectric losses are included in the definition of the admittance matrix, whereas radiation leakage is accounted by defining fictitious side ports, terminated with matched loads.

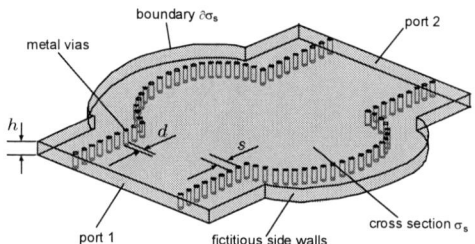

Fig. 1.   An arbitrary substrate integrated waveguide circuit.

I. INTRODUCTION

The substrate integrated waveguide (SIW) technology represents a new and very promising candidate for circuits and components operating in the microwave, millimeter-wave and terahertz region. SIWs belong to the family of substrate integrated circuits (SICs) and are fabricated by using two rows of conducting cylinders and/or slots embedded in a dielectric substrate that connect two parallel metal plates [1,2] (Fig. 1). In this way, the non-planar rectangular waveguide can be made in planar form compatible with existing planar processing techniques, preserving most of the advantages of the conventional metallic waveguide (e.g., high quality-factor, high power-handling capability, etc.). The major advantage of SIW technology is that it permits the fabrication of a complete circuit in planar form (including planar circuitry, transitions with microstrip lines and coplanar waveguides, rectangular waveguides, active components and antennas) using a standard printed circuit board (PCB) or other planar processing techniques.

In the design of SIW components, one of the major issues is related to the minimization of losses. There are three mechanisms of loss in the SIW structures [3]. Due to their similarity to rectangular waveguides, SIW structures exhibit conductor losses due to the finite conductivity of metallic walls and dielectric losses due to the loss tangent of dielectric medium. Moreover, the presence of gaps along the side walls is subject to radiation loss, due to a possible leakage through the gaps. There are design rules which permit to minimize the

radiation leakage, based on the ratio between the longitudinal spacing s and the diameter d of the metal vias is sufficiently small ($s/d<2.5$) [3]. Nevertheless, the evaluation of the different contributions of loss is particularly important in the design of SIW components, especially in the design of structures with unconventional shapes at millimeter-wave frequencies.

An efficient technique for the modeling of SIW interconnects and components is based on the Boundary Integral-Resonant Mode Expansion (BI-RME) method, which was applied in [4] to the analysis of lossless SIW components. This method permits a fast and accurate determination of the wideband frequency response of SIW circuits, by providing their admittance matrix in the form of a pole expansion in the frequency domain. The BI-RME method also allows automatically deriving multimodal equivalent circuit models of SIW discontinuities [4]. The proposed method can be adopted to obtain libraries of parametric models of SIW discontinuities, which can be included in computer aided design tools for an efficient design of complex SIW circuits.

This paper extends the capabilities of the BI-RME method to the modeling of radiation leakage, conductor loss, and dielectric loss in SIW circuits. The effect of conductor and dielectric losses are included directly in the definition of the admittance matrix, in conjunction with a perturbation approach. Conversely, radiation leakage is accounted for by defining additional fictitious ports at the sides of the SIW interconnect, and by terminating these ports with matched loads.

978-2-8748-7007-1/08 $25.00 © 2008 EuMA   230

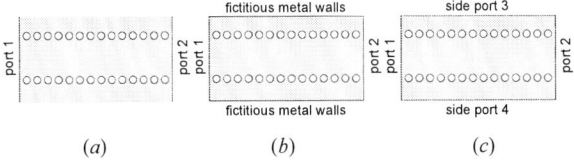

Fig. 2. Geometry of an SIW interconnect: (a) real geometry; (b) geometry normally considered in the BI-RME analysis; (c) geometry considered in the BI-RME analysis with radiation losses, with the definition of additional side ports.

II. MODELING OF DIELECTRIC AND CONDUCTOR LOSSES

The effect of dielectric and conductor losses is included directly in the representation of the admittance matrix of the SIW circuit, by using a perturbation approach.

Let us consider a planar waveguide component of arbitrary shape (Fig. 1), connected to the exterior through rectangular ports, with metal conductivity σ_c and filled with a dielectric medium with the following dielectric permittivity

$$\varepsilon = \varepsilon_0 \varepsilon_r + \frac{\sigma_d}{j\omega} \qquad (1)$$

where ε_0 is the dielectric permittivity of vacuum, ε_r is the relative dielectric permittivity of the medium, σ_d is the conductivity of the medium.

From [5], the ij-th term of the generalized admittance matrix (GAM) relating modal currents and voltages of the terminal rectangular waveguide modes can be expressed as

$$Y_{ij}(k) = \frac{A_{ij}}{j\eta k} + \frac{jk}{\eta} \sum_{p=1}^{P} \frac{C_{ip} C_{ip}}{k_p^2 + jkk_p/Q_p - k^2} \qquad (2)$$

where $k = \omega\sqrt{\varepsilon\mu_0}$ is the wave-number and $\eta = \sqrt{\mu_0/\varepsilon}$ is the characteristic impedance. The term A_{ij} is related to the low-frequency behavior of the admittance matrix, k_p is the resonance wave-number of the p-th mode of the cavity obtained by short-circuiting the ports, C_{pi} is related to the coupling between the p-th cavity mode and the i-th port mode, and Q_p represents the quality factor of the p-th cavity mode.

The convergence of the series in (2) can be accelerated by extracting a quasi-static term (namely the limit k→0) [6]. By taking into account (1), it finally results that

$$Y_{ij}(k) = \frac{A_{ij}}{j\eta_0 k_0} + \sigma_d B_{ij} + \frac{jk_0\varepsilon_r}{\eta_0} B_{ij} +$$
$$+ \frac{k_0^2 \varepsilon_r^{3/2}}{\eta_0} \sum_{p=1}^{P} \frac{C_{ip} C_{ip}}{k_p Q_p \left(k_p^2 + jk_0 k_p \varepsilon_r^{1/2}/Q_p - k_0^2 \varepsilon_r\right)} + \qquad (3)$$
$$+ \frac{jk_0^3 \varepsilon_r^2}{\eta_0} \sum_{p=1}^{P} \frac{C_{ip} C_{ip}}{k_p^2 \left(k_p^2 + jk_0 k_p \varepsilon_r^{1/2}/Q_p - k_0^2 \varepsilon_r\right)}$$

where $k_0 = \omega/c$ and $\eta_0 = 120\pi\ \Omega$. In (3), the frequency independent terms A_{ij} and B_{ij} are calculated as in [7], and k_p and C_{ip} are computed as in [6].

The quality factor Q_p depends on both conductor and dielectric losses

$$\frac{1}{Q_p} = \frac{1}{Q_p^{(c)}} + \frac{1}{Q_p^{(d)}} \qquad (4)$$

where $Q_p^{(c)}$ is the quality factor related to conductor losses and $Q_p^{(d)}$ is related to dielectric losses. In particular [5]

$$\frac{1}{Q_p^{(c)}} = \frac{R_s}{\omega_p \mu_0} \int_S |\mathbf{H}_p|^2 dS = \frac{\delta_p}{d} + \frac{\delta_p}{2} \int_{\partial\sigma_S} |\mathbf{h}_p|^2 d\ell \qquad (5)$$

$$\frac{1}{Q_p^{(d)}} = \vartheta_e = \frac{\sigma_d}{\omega_p \varepsilon_0 \varepsilon_r} \qquad (6)$$

where R_s is the surface resistance and δ_p is the skin depth at resonance frequency ω_p of the p-th cavity mode, $\mathbf{H}_p$ is the magnetic modal vector of the p-th cavity mode, $\mathbf{h}_p$ is the magnetic modal vector of the p-th mode of a waveguide with cross-section σ_S, and $\partial\sigma_S$ is the boundary of σ_S and does not includes the ports.

In order to determine the GAM relating modal currents and voltages of the SIW port modes, the mode spectrum of the terminal SIW is calculated as in [8], and the transformation of the GAM described in [4] is applied.

III. MODELING OF RADIATION LOSSES

The effect of radiation losses, due to the leakage through the gaps, is modeled by defining additional ports at the sides of the SIW (Fig. 2). This solution allows overcoming an intrinsic limitation of the BI-RME method, which is usually applied to closed structures.

More specifically, additional rectangular ports are defines at the two sides of the SIW circuit, and these ports are terminated with matched loads. In order to simulate the radiation condition, a sufficient number of modes should be defined at each port, and the cutoff frequency of these modes should be smaller than the operation frequency. For this reason, it is not possible to consider a single unit cell of the SIW structure, and a sufficiently large number of cells need to be included in the analysis.

Once the admittance matrix of the 4-port circuit has been derived, the corresponding scattering matrix is computed, and the sub-matrix connecting the real ports of the SIW circuit is extracted. The reduced matrix accounts for the effect of radiation leakage.

IV. VALIDATION EXAMPLES

The first example refers to an SIW interconnect, firstly presented in [2]. This structure was fabricated to evaluate the conductor and dielectric losses, and therefore the dimensions were chosen in such a way, that radiation losses are negligible compared with the conductor and dielectric losses. The geometry of the SIW is shown in Fig. 3a (dimensions in mm: w=3.97, d=0.635, s=1.016, h=0.254; dielectric substrate with ε_r=9.9, tanδ=0.0002, σ_c=5·10^7 S/m). In the application of the

978-2-8748-7007-1/08 $25.00 © 2008 EuMA

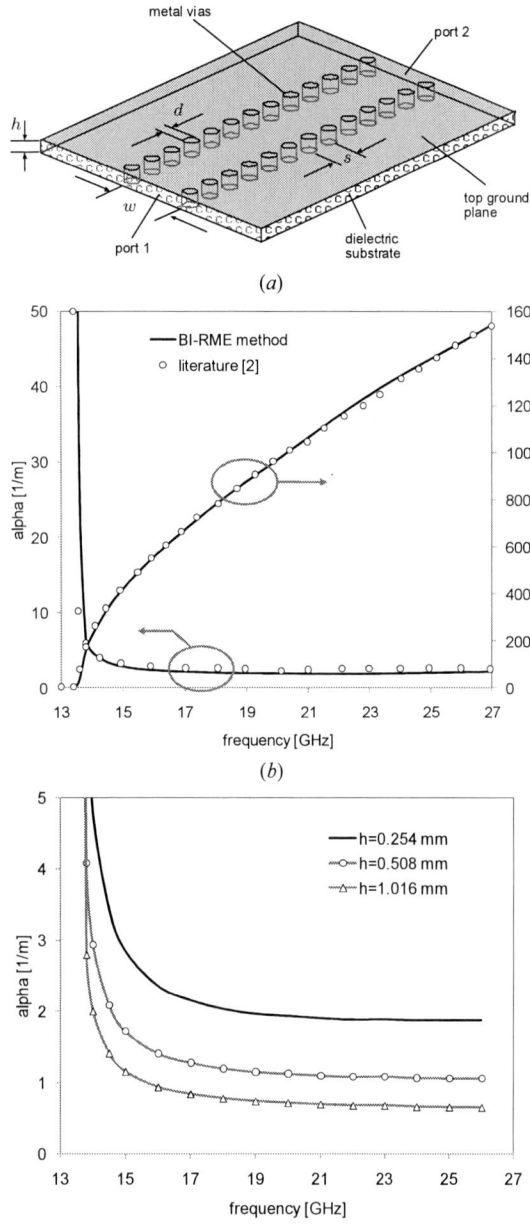

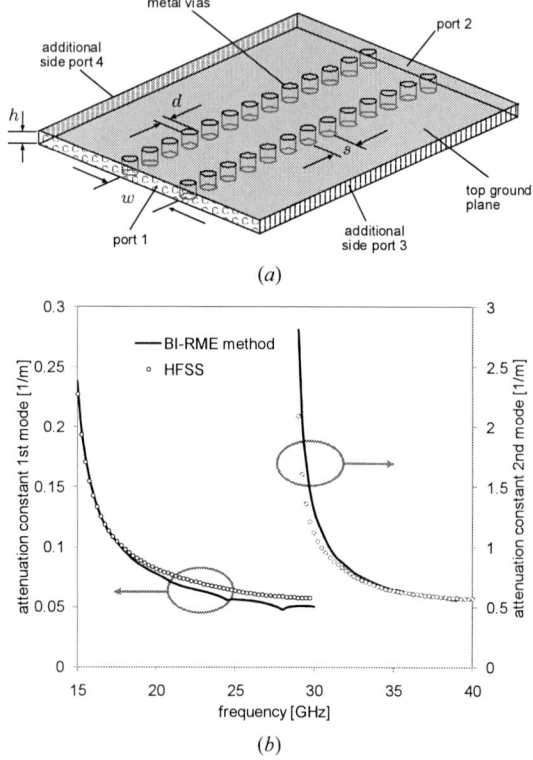

Fig. 4. Modeling of a radiation loss in an SIW interconnect (solid lines: BI-RME method; circles: data from HFSS [2]): (*a*) geometry of the SIW structure; (*b*) attenuation constant of the first two modes.

Fig. 3. Modeling of a lossy SIW interconnect: (*a*) geometry of the SIW structure; (*b*) attenuation and propagation constants of the fundamental mode (*h*=0.254 mm); (*c*) attenuation constant of the fundamental mode for different values of the substrate thickness *h*.

BI-RME method, a waveguide section is analyzed, in order to obtain the GAM in the form (3), and then an eigenvalue problem is formulated, along the lines of [8]. For each frequency of interest, the real and imaginary parts of the first eigenvalue yield the value of the attenuation and propagation constants of the fundamental SIW mode, respectively. Fig. 3*b*

shows the attenuation and propagation constants; the results obtained by using the BI-RME method are compared with simulated and measured data from [2]. In Fig. 3*c*, the effect of the substrate thickness *h* on the losses is considered: the attenuation constants calculated with three different thickness values are compared, showing how the SIW losses can be mitigated by using thicker substrates.

The second example refers to an SIW interconnect fabricated with ideal dielectric material and perfectly conducting metal, where the only source of loss is the radiation leakage. The geometry of this structure is shown in Fig. 4*a* (dimensions in mm: *w*=7.2, *d*=0.8, *s*=2, *h*=0.508; dielectric substrate with ε_r=2.33). The attenuation constants of the first and second SIW modes are shown in Fig. 4*b* and compared with the results obtained with the commercial software HFSS. It is noted that, in the BI-RME analysis, 10 modes were defined for each side port, even is few of them are above their cutoff frequency in the band of interest.

The third example refers to an SIW filter, comprising three centered posts with different diameter, shown in Fig. 5*a* (dimensions in mm: *w*=21.06, *d*=2, *s*=4, d_1=2, d_2=0.5, s_1=14, *h*=1; dielectric substrate with ε_r=2, σ_d=0.001 S/m, σ_c=4·10^7 S/m). The scattering parameters obtained by using the BI-RME method are compared with those obtained by

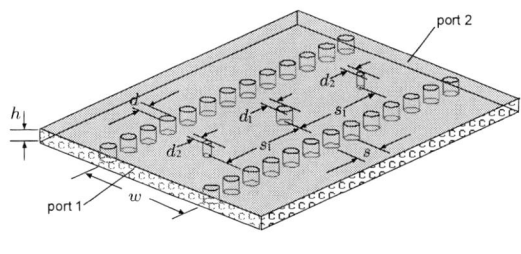

(a)

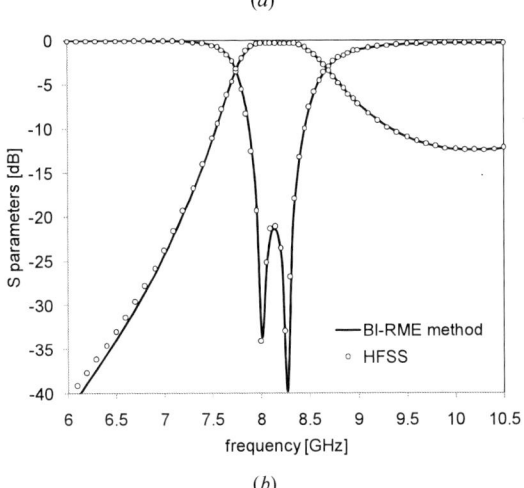

(b)

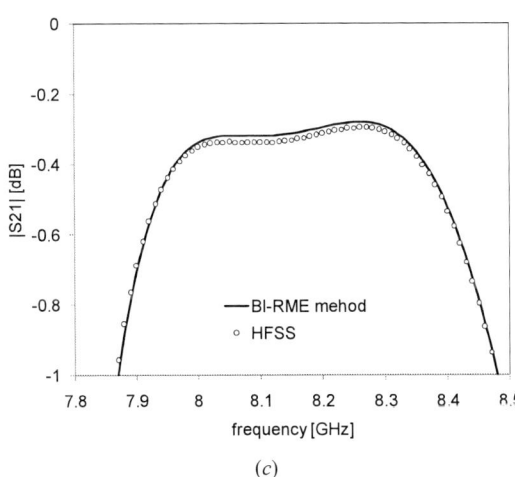

(c)

Fig. 5. Modeling of a three-pole SIW filter: (a) geometry of the filter; (b) scattering parameters versus frequency (solid lines: BI-RME method; circles: HFSS); (c) detail of the insertion loss in the pass-band.

using the commercial code HFSS (Fig. 5b). In order to better appreciate the accuracy of the proposed method, a detail of the insertion loss in the pass-band is shown in Fig. 5c. The required computing time was 23 sec for the wideband analysis in the entire frequency band when using the BI-RME method, and 30 sec per frequency point in the case of HFSS. It is noted

that the BI-RME method provides a substantial reduction in computing time, thus allowing a powerful tool for the efficient design of complex SIW circuits.

V. CONCLUSION

The BI-RME method is extended in this paper to the modeling of lossy SIW components. Three different mechanisms of loss have been considered, namely the radiation leakage, the conductor loss, and the dielectric loss.

The BI-RME method yields the generalized admittance matrix of the SIW component in the form of a pole expansion in the frequency domain, and permits to account for both conductor and dielectric losses. Moreover, the effect of radiation leakage is taken into account by defining additional ports at the sides of the SIW structures, and by terminating these ports with matched loads. The proposed method has been validated through the analysis of the attenuation and propagation constants of SIW interconnects and of the frequency response of an SIW filter.

It is finally noted that the BI-RME method allows for a very fast modeling of the SIW component and it also permits to automatically derive equivalent circuit models of lossy SIW discontinuities, which can be used for the fast design of complex SIW circuits.

REFERENCES

[1] F. Xu and K. Wu, "Guided-Wave and Leakage Characteristics of Substrate Integrated Waveguide," *IEEE Trans. on Microwave Theory and Techniques*, Vol. MTT-53, No. 1, pp. 66-73, Jan. 2005.

[2] D. Deslandes and Ke Wu, "Accurate Modeling, Wave Mechanisms, and Design Considerations of a Substrate Integrated Waveguide," *IEEE Trans. on Microwave Theory and Techniques*, Vol. MTT-54, No. 6, pp. 2516-2526, June 2006.

[3] M. Bozzi, M. Pasian, L. Perregrini, and K. Wu, "On the Losses in Substrate Integrated Waveguides," *37th European Microwave Conference 2007*, Munich, Germany, 2007.

[4] M. Bozzi, L. Perregrini, and K. Wu, "Direct Determination of Multi-mode Equivalent Circuit Models for Discontinuities in Substrate Integrated Waveguide Technology," *2006 IEEE MTTS International Microwave Symposium*, San Francisco, CA, 2006.

[5] K. Kurokawa, *An Introduction to the Theory of Microwave Circuits*, Academic Press, 1969.

[6] G. Conciauro, P. Arcioni, M. Bressan, L. Perregrini, "Wideband Modeling of Arbitrarily Shaped H-Plane Waveguide Junctions by the 'Boundary Integral-Resonant Mode Expansion' Method," *IEEE Trans. on Microwave Theory & Techniques*, Vol. MTT-44, No. 7, pp. 1057-1066, July 1996.

[7] P. Arcioni, M. Bressan, G. Conciauro, and L. Perregrini, "Generalized Y–Matrix of Arbitrary H–Plane Waveguide Junctions by the BI–RME Method," *International Microwave Symposium Digest* (IMS1997), pp. 211-214, 1997.

[8] Y. Cassivi, L. Perregrini, P. Arcioni, M. Bressan, K. Wu, and G. Conciauro, "Dispersion Characteristics of Substrate integrated rectangular waveguide," *IEEE Microwave & Wireless Components Letters*, Vol. 12, No. 9, pp. 333–335, Sept. 2002.

Proceedings of the 3rd European Microwave Integrated Circuits Conference

Tapped integrated inductors: Modelling and Application in Multi-Band RF Circuits

Morin Dehan [#1], Jonathan Borremans [#*], Piet Wambacq [#*] Stefaan Decoutere [#]

[#]*IMEC*
Kapeldreef 75, Heverlee, Belgium
[1]`morin.dehan@imec.be`
[*]*Vrije Universiteit Brussel, Department ETRO*
Pleinlaan, Brussels, Belgium

Abstract— **As CMOS scales down and sees is cost per mm2 increasing, area-aware RF design solutions are called for. Integrated inductors with multiple taps allow for low-area multi-band RF circuit design. This work reports on the design of these inductors and provides modelling and extraction procedures demonstrated on 4-port measurements. Additionally, the application of such inductors is demonstrated on a low-area switchable dual-band VCO in 90 nm CMOS with an area of only 0.04 mm2. Dual-band operation around 3.5 and 10 GHz is achieved, with a high FOM of 182 dB. This performance demonstrates the opportunities using tapped inductors for high-performance area-aware RF design.**

I. INTRODUCTION

With silicon technologies scaling down to nanometer dimensions, mask costs per square mm increase. In order to repay some of that expense, an area-aware approach towards RF design is appropriate. RF circuit design typically relies plentiful on inductors for input matching, resonant loads of amplifiers, resonators for a VCO, etc. However, integrated inductors with a high quality factor (Q) occupy a large area. Unfortunately, inductors do not scale along with the technology. Hence, minimizing the number of on-chip inductors is favourable. Another approach to improve the area efficiency of scaled CMOS is to increase the level of integration and the amount of flexibility of ICs. For example, multi-standard and multi-band wireless transceivers emerge. These rely on yet more inductors to achieve wideband or multi-band operation. Tapped inductors – also referred to as multi-tap inductors (M-inductor) – allow for inductor reuse. Indeed, an inductor with additional taps, together with a set of capacitors, can form a high-order resonant tank. The achieved multiple resonances can be used for wideband or multi-band operation in various circuits. Another way of seeing this is as a set of inductors – traditionally realised as separate inductors – conveniently realized within a small area. For example, the single-ended example of a resonant load in Fig. 1 can be used to achieve a wideband or dual-band resonant impedance.

For RF and microwave models of integrated inductive components, e.g. Inductors, designers typically rely on either full-wave electromagnetic simulators or lumped element models. For the latter, the values of the components are determined either from relatively simple and compact equation or from tables of extracted parameters. Choosing

between either kinds of model is mostly the results of a compromise between accuracy and speed.

In this work we elaborate on the design and modelling of such tapped inductors. A lumped element model has been used. We present a modelling procedure for adequate wideband models, and demonstrate them on a 4-port inductor, measured up to 20 GHz. Next, we discuss applications of the tapped inductors in multi-band RF circuits. We demonstrate a 3.5-and-10 GHz dual-band VCO with an excellent phase-noise Fig.-of-merit of 182 dB, showing that tapped inductors are sound with multi-band RF design.

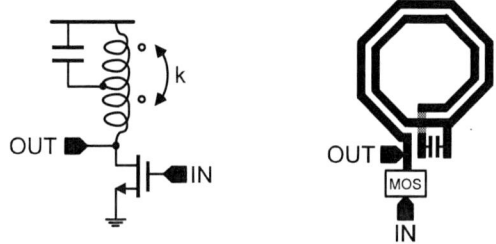

Fig. 1 An example amplifier with dual-band load (left), and possible compact layout using a tapped inductor (right)

II. TAPPED INDUCTOR LAYOUT AND MODELLING

A. Topology Underscope

We propose a modelling strategy by means of a 4-port inductor example. An M-inductor topology used as test case is presented in Fig. 2. It is a fully symmetrical square multi-turns inductor used further in this work as the tank of a VCO (Section IV). It only differs from a traditional symmetrical layout by the two extra terminals connected to internal nodes of the spiral, yielding a 4-port device. The whole structure is designed in the top metal layer of an advanced digital metal stack with 9 layers of metal.

A patterned ground shield (PGS) made of the two lowest metal layers has been used to reduce the substrate losses [1]. The key geometrical parameters of the spiral are summarized in Table I.

978-2-8748-7007-1/08 $25.00 © 2008 EuMA

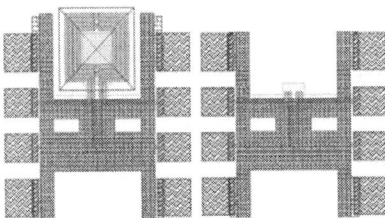

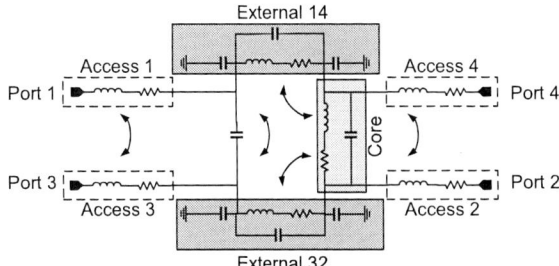

Fig. 2 Top view of the multitap inductor test structure layout (left), and its corresponding open dummy (right). All access lines are surrounded by dummies to fulfil the metal density requirement. Dummy generation was suppress around the spiral to reduce parasitic capacitances.

Fig. 3 Proposed model for the 4-port multitap inductor.

TABLE I
GEOMETRICAL PARAMETERS OF THE MULTITAP INDUCTOR

Number of Turns	4	Conductor width	7.5μm
Inner Diameter	80 μm	Spacing between conductors	2.2μm
Distance to PGS	4.7 μm	Metal Thickness	.81 μm

B. Modelling

To model this component, we followed the methodology proposed by Long in [1]. First, we evaluate the self and mutual inductances of all conductors which compose the spiral, by using the equations proposed by Greenhouse in [3]:

$$L_i = \frac{\mu_0}{2\pi} l \left[\ln\left(\frac{2l_i}{w+t}\right) + 0.5 + \frac{w+t}{3\,l_i} \right] \quad (1)$$

$$M_{ij} = \frac{\mu_0}{2\pi} l \times \left(\ln\left(l/GMD + \sqrt{1 + l^2/GMD^2}\right) - \quad (2) \right.$$
$$\sqrt{1 + GMD^2/l^2} + GMD/l$$

Where L_i is the self inductance, μ_0 is the vacuum permeability; l, w, and t are the length, the width and the thickness of the conductor respectively. M_{ij} is the mutual inductance between 2 parallel conductors of length l, and GMD is the geometric mean distance between these 2 conductors.

Then, based on the work of Sakurai [4], the coupling capacitances between each conductor and the capacitances between the conductors and the ground plane are evaluated. The resistance associated to each straight wire which composes the spiral, is derived from the sheet resistance. From this preliminary calculation, we can construct the electrical model of the spiral by modelling each straight conductor of the spiral by a traditional π-equivalent circuit, similar to the one proposed by [5], and by connecting all these conductors together to obtain the M-inductor.

The circuit created in this way contains a huge number of nodes. Moreover, the complexity of the equivalent circuit will depend on the number of conductors that compose the M-inductor.

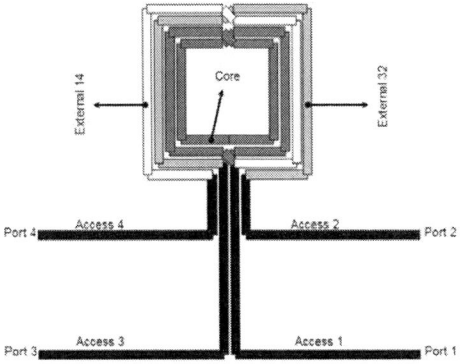

Fig. 4 Schematic view of the M-inductor

This raises the level of this model to an unnecessary complexity. To make the model simpler, we reduce it to obtain the relatively compact one presented in Fig. 3. The M-inductor is divided into 7 different parts: 4 access lines, 2 external inductors, and a core inductor (Fig. 4). Each part of the layout corresponds directly to some elements of the model. They are first modelled as a simple π-equivalent circuit, neglecting the presence of the other parts. The mutual inductance between two parts, obtained by summing the mutual inductances between the conductors of these 2 parts, is finally added to the equivalent circuit.

C. Experimental Validation

The inductor and its associated open deembedding structure (Fig. 2) were measured using a four-port PNA-X network analyzer from Agilent using two G-S-S-G RF probes from Picoprobe. The data consisted in 4 x 4 S-parameter matrices ranging from 10 MHz to 20 GHz which were converted into Y-parameters using general conversion formulas. They are defined using a two-step calibration procedure. First, a full 4-port SOLT (Short Open Load Thru) calibration procedure using a commercially available alumina substrate defines the reference planes at the probe tips. Next, an on-wafer de-embedding was performed by subtracting from the Y-parameters of the spiral, those of a corresponding open dummy.

978-2-8748-7007-1/08 $25.00 © 2008 EuMA

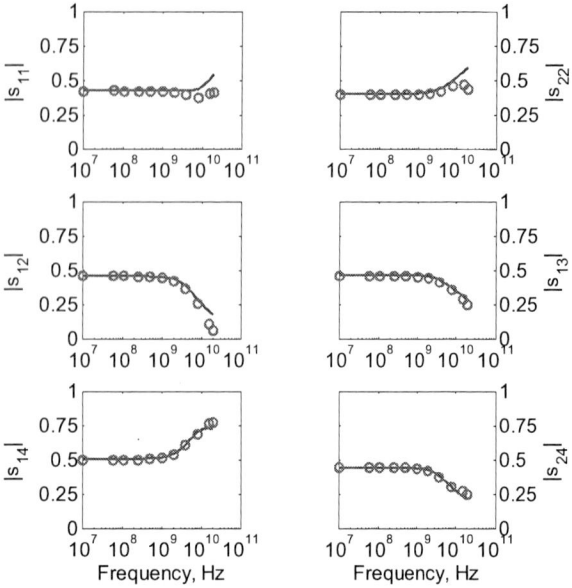

Fig. 5 S-parameters comparison between measurement (symbols) and from the model (solid line)

A comparison between measured and simulated S-parameters is shown in Fig. 5. Only 6 parameters are displayed as, thanks to the symmetry of the components, there are only 6 parameters out of 16 which are different from each other. Good agreement is obtained between our model and the experiment. Some minor discrepancy appears at the higher end of the spectrum (f>10GHz). We attribute it mostly to the accuracy of the deembedding technique that we used. While more elaborate deembedding structure may be more adequate for higher frequencies, this choice was solely related to limited available silicon area.

Since the M-inductor will be used as part of the LC tank of a VCO, a high-impedance configuration, we also compare the measured and simulated differential impedances of the spiral. The differential impedances are calculated from the 4-port Z-parameters, and not obtained from differential measurements. We define the two input impedances:

$$Z_{diff\,13} = Z_{11} + Z_{33} - Z_{13} - Z_{31} \qquad (3)$$

$$Z_{diff\,24} = Z_{22} + Z_{44} - Z_{24} - Z_{42} \qquad (4)$$

They correspond to the differential impedance between the ports 1 and 3 (Zdiff 13), and between the ports 2 and 4 (Zdiff 24). When differential impedance is calculated between 2 ports, we define mathematically that the 2 other ports are connected to perfect open circuits.

We calculate also the differential trans-impedance:

$$Z_{diff\,1324} = Z_{12} + Z_{34} - Z_{32} - Z_{14} \qquad (5)$$

Comparison between measurement and simulation are shown in Fig. 6. Good agreement is observed between the measured and simulated impedances over the whole frequency range. This builds enough confidence in the model to use it to design differential circuits like a VCO. Note that the simple open deembedding accurately fits the modelling purpose up to 20 GHz for the most critical parameters for VCO design (Eq. 3-5).

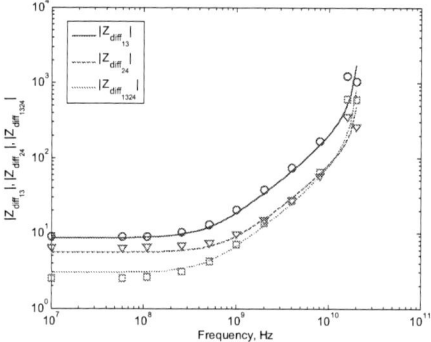

Fig. 6 Magnitude of the differential impedance between ports 1-3 and ports 2-4. Good agreement is observed between the model (lines) and the measurements (symbols).

III. APPLICATION EXAMPLE: A SINGLE-INDUCTOR DUAL-BAND VCO

As explained in the introduction, tapped inductors can find applications in both wideband and multi-band RF circuits. An example is in a multi-band VCO. It has been shown that a VCO using a high-order tank can achieve oscillation at different distinct frequencies. For example, a fourth-order tank supporting two resonances may be formed with four inductors and two capacitors as in [6]. An elegant compact second order tank implementation can be realized completing the inductor from Fig. 4 with varactors to form the tank in Fig. 7a. In first order, these are three inductors, loaded by two capacitors. The inductors are coupled, but note that the coupling is low due to the specific layout. When used in a differential fashion, the half-circuit of the tank is shown in Fig. 7b. It has two sets of complex conjugate poles at ωH and ωL, and therefore two resonance frequencies. Thus it allows for two possible oscillation frequencies when used in an oscillator. A robust approach to excite the desired frequency of oscillation in a stable fashion is by using feedback as depicted in Fig. 8. Positive or negative feedback pushes the low- or high-frequency poles into the right half plane, causing instability [7]. In fact, it can be shown that oscillation occurs when

$$1 \pm G_m \Re(Z_{12}) < 0 \qquad (6)$$

where G_m is the trans conductance of the activated G_m-cell, and Z_{12} the trans-impedance of the Z-matrix of the 2-port differential half-circuit of the tank. Note that the part of this parameter related to the M-inductor is accurately modeled as can be seen in Fig. 6. For the capacitors, a model from the foundry's design kit is used. This method of mode selection allows for stable operation, without parasitic hopping of the VCO between different oscillation modes. A different way of interpreting the design of this oscillator is to see it as an amplifier, where passive feedback is employed to cause instability at the desired frequency.

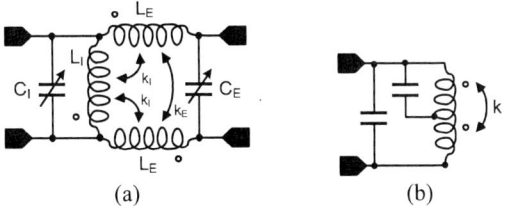

(a)   (b)

Fig. 7 (a) Fourth-order resonator tank with two possible oscillation modes (b) differential-mode half-circuit of the tank

Relying on this architecture, a 3.5-and-10 GHz VCO has been designed in a 90 nm digital CMOS using the same inductor from Section II. The tunable capacitors are achieved as MOS-varactors.

Fig. 9 shows a chip micrograph of the oscillator. The wafer-probed measured tuning range (Fig. 10) shows operation from 3.1 to 4 GHz and from 8.8 to 11 GHz. In the lower band (4 GHz) a phase-noise of -122 dBc/Hz has been measured, for the higher band (10 GHz) -116 dBc/Hz at 2.5 MHz offset after divide-by-two operation. The core oscillator draws between 2 and 8 mA depending of the mode of operation. The phase-noise Fig.-of-merit has been calculated:

$$FoM = 10\log_{10}\left[\frac{1}{P_{DC}[mW] \cdot PN}\left(\frac{f_0}{\Delta f}\right)^2\right] \qquad (6)$$

Using only a digital CMOS process, a high FOM of 182 dB, averaged over the frequency of operation, has been achieved on a very low area (only 0.04 mm^2).

IV. CONCLUSIONS

We have proposed a model for tapped inductors that demonstrates to achieve adequate performance up to very high frequencies. Further, we have elaborated on how these high-Q devices can be employed for multi-band RF circuit design, and illustrated this on a dual-band VCO example. Using only a digital technology, with a thin top layer, a high FOM has been achieved for a very low circuit area. This successfully demonstrates that inductor reuse by means of tapped inductors is an appropriate approach towards area-aware RF design.

ACKNOWLEDGMENT

The authors acknowledge the EMIC lab. from the UCL, Belgium, for the 4-ports RF measurements.

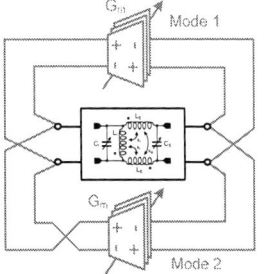

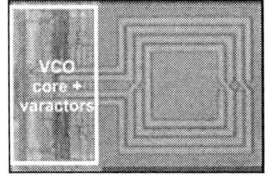

Fig. 8 Operation principle on the dual-band VCO

Fig. 9 Chip photograph of the dual-band VCO

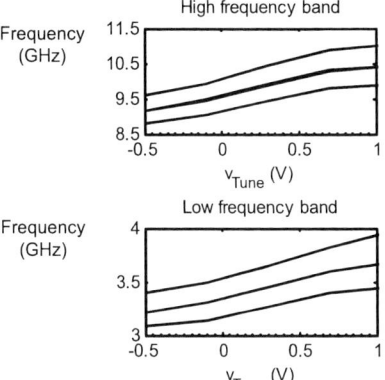

Fig. 10 Measured tuning range of the dual-band VCO in high- and low-frequency operation.

REFERENCES

[1] C. T. S. D. Cheung, J. R. Long, "Shielded passive devices for silicon-based monolithic microwave and millimeter-wave integrated circuits », JSSC, Vol. 41, No. 5, May 2006, pp. 1183-1200.

[2] J.R.Long, and M.A.Copeland, "The modeling, characterization, and design of monolithic inductors for silicon RF IC's," JSSC, Vol. 32, pp 357 - 369, 1997

[3] H.Greenhouse, "Design of Planar Rectangular Microelectronic Inductors," IEEE Trans. On Parts, Hybrids, and Packaging, Vol 10,pp.101-109, 1974

[4] T.Sakurai, "Closed Form expressions for interconnection delay, coupling and crosstalk in VLSI's," IEEE TED, Vol. 30, pp.183-185, 1983

[5] C.P.Yue, an d S.S.Wong, "Physical modeling of spiral inductors on silicon," IEEE TED, Vol. 47, pp.560-568, 2000

[6] N. T. Tchamov, S. S. Broussey, I. S. Uzunov, K. K. Rantala, "Dual-Band LC VCO Architecture With a Fourth-Order Resonator". IEEE TCAS II, Vol. 52, No. 3, March 2007, pp. 277-281.

[7] A. Bevilacqua, F. P. Pavan, C. Sandner, A. Gerosa, A. Neviani, "A 3.4-7 GHz Transformer-Based Dual-mode Wideband VCO", *Proceedings of ESSCIRC*, pp. 440-443, Sept. 2006

Proceedings of the 3rd European Microwave Integrated Circuits Conference

6-24 GHz Mixer Using 0.25µm Enhancement Mode PHEMT Technology in a Low Cost Chip Scale Package

Sushil Kumar, Julie Kessler & Henrik Morkner

Avago Technologies, 350 W. Trimble Road, San Jose, CA 95131

Abstract— **This paper discusses development of a 6-24GHz mixer in a novel chip scale package. The mixer and package was fabricated together using Avago's enhancement mode (E-mode) PHEMT technology. This chip scale package is high performance, low cost and it totally eliminates all the assembly steps (such as die attach, bond wire etc) required to package a singulated die in a package.**

The mixer has been tested at two different stages of fabrication, first Un-Capped (like without top-lid in case of conventional package) and after final GaAs-Capped (with top-lid on). The measured conversion loss of un-capped mixer is ~9dB upto 22GHz @LO=+16dBm. Conversion loss of capped wafer is marginally lower than uncapped mixer upto 22GHz. The IIP3 of uncapped mixer mixer is about +19dBm and capped mixer IIP3 is about 1-2dB lower than Capped mixer in most of the band.

Rest of the performances of (Capped and Un-capped) mixers are very similar. L-R Isolation ~35dB, L-I Isolation ~40dB. IF test frequency is 2GHz.

To the best of author's knowledge this is the first reported chip scale packaged Mixer.

I. INTRODUCTION

Communication system that uses 6GHz and above are very attractive as these provide wide bandwidth for achieving high data rate and large capacity. With the emergence of new unlicensed and licensed bands several new application are under investigation and in Implementation. The success of these systems relies on low cost, miniaturized high performance components and this is the motivation behind development of this mixer. One of the key components of Tx/Rx chain is a mixer. Typical application of a mixer is shown in following block diagram.

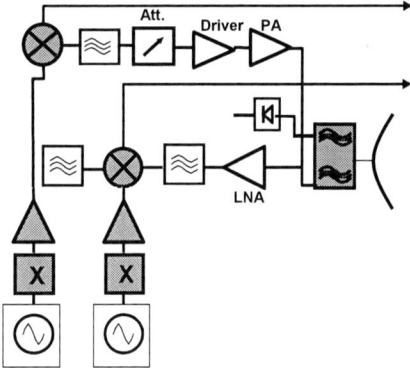

Fig. 1 Block Diagram of a transceiver

II. DEVICE CHARACTERISTICS

Avago Technologies E-mode process parameters are given in table (1). Unlike depletion mode device, an enhancement mode device needs positive supply only. Compared to D-mode, the channel of E-mode device is very thin and therefore E-mode channel capacitance is almost twice of D-mode process. Due to E-mode's high channel capacitance, circuit design is a little challenging.

TABLE I
FET CHARACTERISTICS OF E-MODE AND D-MODE PROCESSES

Parameters	E-mode
Lg (Gate Length)	0.25µm
MIM Cap	0.4 fF/µm²
TFR Resistor	--
Mesa Resistor	213Ω/sq
Gm (mS/mm) @Vds=3V	580
Vgs @peak Gm (V)	+0.7
Ids @peak Gm (mA/mm)	171
Imax (mA/mm)	330 @Vgs=+1V
Bvgd @1mA/mm (V)	-17
Vto @1mA/mm (V)	+0.97
Vth @1mA/mm (V)	+0.25
Ft (GHz) @Vds=2V	55

III. PACKAGE DESIGN

Avago Technologies has developed a GaAsCap packaging technology for building high value, mass production RFIC/mmW MMIC components in a true chip-scale, wafer level package. There are a number of value propositions associated with this technology;

- High performance at frequencies from DC-mmW
- Eliminates one assembly step of die attach and wire bonding as complete package is processed with IC.
- Virtually eliminates parasitics associated with plastic, lead frames and bond wires & provides true air cavity
- Enables ultra-thin IC substrates to dissipate heat and whole package is thinner than current solutions so much better for thermal dissipation for PAs.
- Finished GaAsCap wafer can be RF probed as accurately as 'on wafer' probing and so suitable for large scale manufacturing test and does not need any

978-2-8748-7007-1/08 $25.00 © 2008 EuMA

PCB or custom test fixture/contactor board for testing.

- Estimated cost is $0.10/mm^2) even at mmW frequencies.

A chip scale (GaAsCap) wafer is composed of a pair of bonded GaAs wafers, in which all the I/O's are routed through Vias to the backside of the device wafer. The combination of the gasket and the cap wafer provides an air cavity, structural and protection for the device wafer. When the wafer is sawn between the gaskets, it literally becomes thousands of individually packaged parts. Photograph in fig. (2) is a finished GaAsCap wafer.

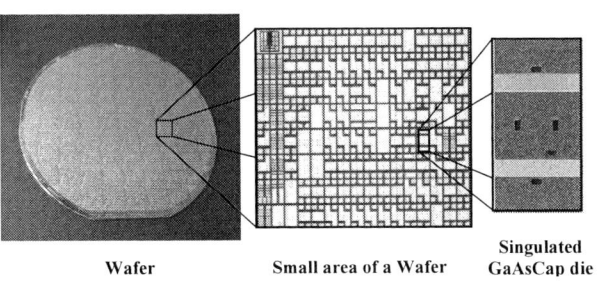

| Wafer | Small area of a Wafer | Singulated GaAsCap die |

Fig. 2 Photograph of a finished GaAsCap wafer.

Figure (3) illustrates a cross-sectional view of a GaAsCap wafer. The base wafer is a standard processed GaAs wafer. The backside vias serve as I/O and ground pads for the package. The backside metal is plated thick enough to ensure adequate coverage in the bottom of the vias without inhibiting the use of standard solder pastes in assembly. The cap wafer provides an air cavity, protection for the devices, and enough structural support to allow thinning of the device wafer to 1.5 mils thickness.

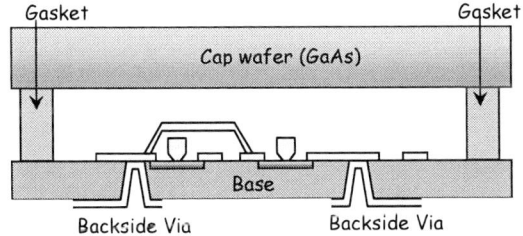

Fig. 3 Cross-sectional view of a GaAsCap wafer.

IV. CIRCUIT DESIGN

The designed mixer is based on single balance design. S-D connected FET has been used as diode. It has a LO balun that feeds diode and mixes with RF/IF frequency to generate desired IF/RF frequency as an down/up converter. A diplexer has been used for RF/IF. Several Momentum simulations have been run to optimize the performance of the LO balun. This balun provides amplitude balance ±0.5dB and phase balance ±0.5° from 5-25GHz. Use of 180° hybrids ensures excellent L-R isolation and eliminates or minimizes the need of a band pass filter to filter out LO power at R-port. Design of LO

balun is based on Marchand technique. To reduce the chip size the balun has been folded into rectangular spiral shape. In addition to Balun entire layout has also been simulated using ADS Momentum to take into account all coupling among close proximity traces and all sorts of junction discontinuities. The insertion loss of low pass section (IF) of diplexer is <0.5dB from DC-3GHz and the insertion loss of high pass section (RF) is <1dB from 5-25GHz. Such low loss diplexer and excellent amplitude and phase matched Balun are key to this mixer low conversion loss and very wide band performance. Simplified schematic of Mixer is shown in fig.4.

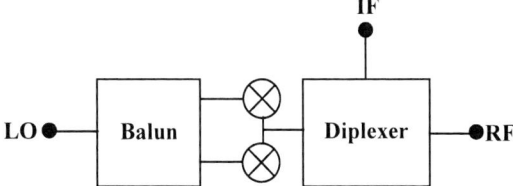

Fig. 4 Simplified Schematic of Mixer.

The photograph of mixer chips developed is shown in fig. (5a,b). The GaAsCap package footprint is 2mm x 2mm.

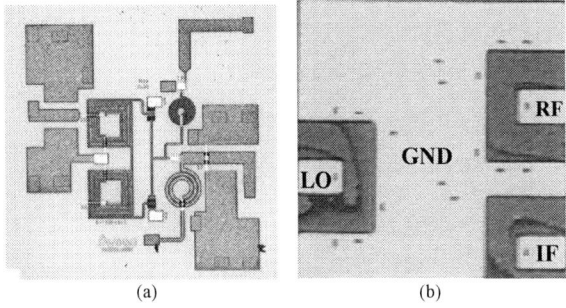

| (a) | (b) |

Fig. 5 (a) Top View of Un-capped Package. (b) Bottom view of Un-Capped/GaAs-Capped Package.

V. MEASURED PERFORMANCE

The Un-Capped and GaAs-capped mixers has been soldered on high frequency Roger PCB board and characterized in connectorized PCB media as shown in fig.6(a,b,c). Measured performance of mixer includes all losses such as connectors, PCB trace. This loss is of the order of 0.2-1.5dB from DC-25GHz.

The measured mixers (UnCapped & GaAs-Capped) performance is shown from fig.(7)- fig.(12) as an Up-Converter. The down conversion performance is better or similar to Up-conversion. The LO (frequency) = (RF-IF)/2 GHz and IF=2GHz.

Fig.(7) shows the measured uncapped mixer C.L. at Plo=+14 to +20dB. It shows 8.5-10dB conversion loss from 5-24GHz @ Plo=+16dBm.

978-2-8748-7007-1/08 $25.00 © 2008 EuMA

(a)

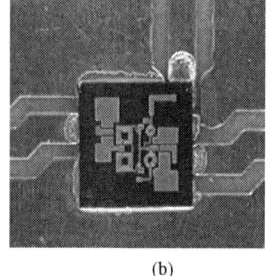

(b)

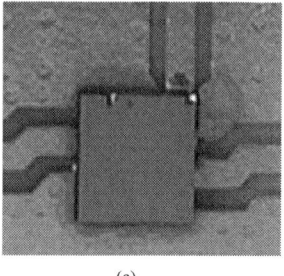

(c)

Fig. 6 (a) Photograph of Un-Capped Pkg in a Test Fixture. (b) Top View of Un-capped Package in fixture. (c) Top View of GaAs-capped Package in fixture.

Fig.(8) shows C.L. of GaAs-Capped mixer at Plo=+14 to +20dB. GaAs-Capped mixer has slightly higher loss than Un-Capped mixer.

Fig.(9) & Fig.(10) shows the measured IIP3 of Un-Capped & GaAs-Capped mixer in fixture. It can be seen from this measurement that IIP3 fluctuates significantly with frequency. This behavior attributes of RF/IF circuit impedance variation. This impedance variation produces different load impedance at harmonics termination of mixer. Also, the IIP3 of GaAs-Capped mixer is lower than Un-Capped mixer. The reason for GaAs-Capped mixer performance degradation compared to Un-Capped mixer is due to GaAs-Cap lid close proximity affect.

The Power leakage and Isolation behaviors of Un-Capped & GaAs-Capped mixer are quite similar so they are shown as a one mixer measurement in fig.(11) & (12).

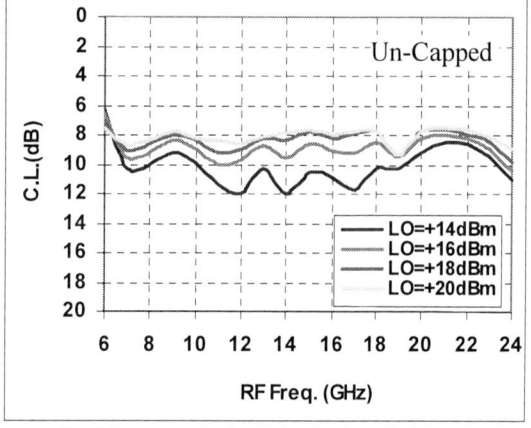

Fig. 7 Up-Conversion Loss of an Un-Capped mixer

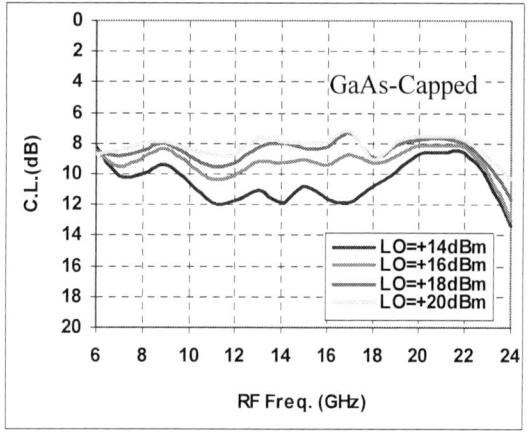

Fig. 8 Up-Conversion Loss of an GaAs-Capped mixer

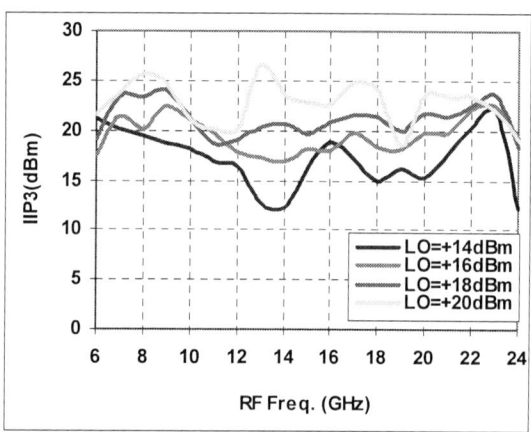

Fig. 9 Up-Conversion IIP3 of an Un-Capped mixer

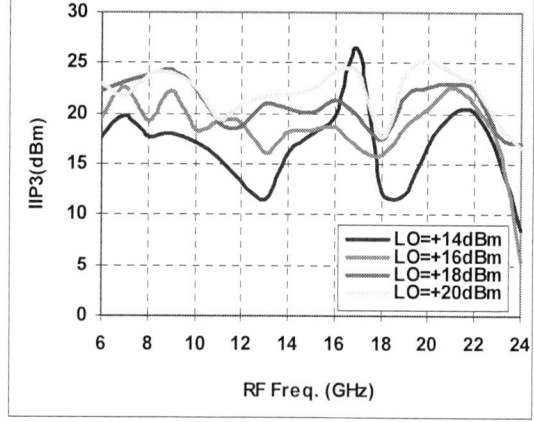

Fig. 10 Up-Conversion IIP3 of an GaAs-Capped mixer

978-2-8748-7007-1/08 $25.00 © 2008 EuMA

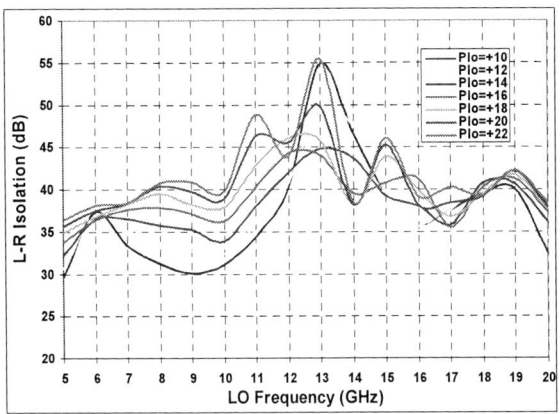

Fig. 11 L-R Isolation of an Un-Capped/GaAs-Capped mixer

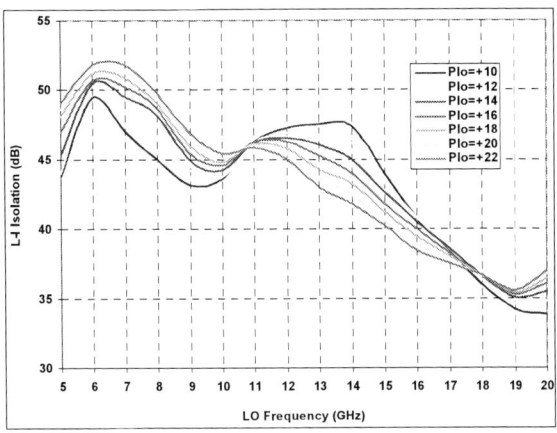

Fig. 12 L-I Isolation of an Un-Capped/GaAs-Capped mixer

VI. CONCLUSIONS

A low cost, broadband high performance mixer in a novel package has been developed. This package fabrication technique made it possible to integrate the package as a MMIC fab process step and thus totally eliminated and die-to-package assembly complexity, time and cost.

ACKNOWLEDGMENT

Authors are thankful to Avago Technologies Fab team for fabrication of the designed mixer. Authors are grateful to Hue B. Tran for doing all testing.

REFERENCES

[1] S. Kumar, M. Vice, H. Morkner, W. Lam, "Enhancement mode GaAs PHEMT LNA with Linearity Control (IP3) and phased matched Mitigated Bypass Switch and Differential Active Mixer," 2003 IEEE MTT-S Digest, pp. 1577-1580, June 2003.

[2] H. Morkner, S. Kumar, M. Vice, " A 18-45GHz Double Balanced Mixer with Integrated LO Amplifier and unique suspended Broadside coupled Balun, IEEE GaAs Integrated Circuit Symposium 2003, 25th Annual Technical Digest, USA, Nov. 2003, pp 1577-1580.

[3] Trantella C.J., "Ultra small MMIc Mixers for K and Ka band Communications ." 2000 IEEE MTT-S digest pp 647-650.

35-65GHz MMIC QPSK Modulator

V. F. Fusco, C. Wang, T. Pochiraju

ECIT, Queen's University Belfast
Belfast, BT3 9DT,
`v.fusco@ecit.qub.ac.uk`

Abstract—**A novel method for directly producing QPSK modulation from baseband IQ signals is presented. The key feature of the architecture is the absence of mixers and as a result unwanted mixing products. A broadband MMIC modulator chip is demonstrated at V band to showcase the performance of the modulator. It completely covers the world wide frequency band allocation for the forthcoming 60GHz indoor wireless communication system. The power consumption of the circuit is less than 50mW and its 1dB compression point is 17dBm. The overall insertion loss of the circuit is 9.5±2dB over the frequency band 35-65GHz with phase state errors below ±15° . Up to 2Gbps bit rate signaling rates should be achievable using the circuit.**

I. INTRODUCTION

In order to improve spectral efficiency and information capacity, phase modulation techniques such as BPSK or QPSK are used. QPSK involves phase-only modulation and provides a good balance between information capacity, ease of implementation and bit error rate. QPSK modulation permits twice the amount of information to be carried within the same bandwidth as BPSK with little additional complexity. Moreover there has been a definite drive to achieve direct modulation in order to reduce RF front-end complexity through removal of IF stages, [2-5], [7]. Although some transceiver chips have recently been presented at millimeter wave frequencies, the presence of mixers in these topologies adds complexity and cost.

The use of the 60 GHz band is rapidly gaining worldwide interest in the high data rate indoor wireless area [6]. A primary reason for this interest is the fact that at least 5 GHz of spectrum is available in this frequency band worldwide. The bandwidth of the QPSK modulator determines the maximum data rates that can be transmitted. The V band modulator presented here allows operation over more than 25GHz bandwidth.

Traditional passive QPSK modulators generally employ a pair of balanced BPSK modulators which are then typically power combined using a Wilkinson or Lange coupler [1], [3]. This adds to the insertion loss of the system and can contribute to phase and amplitude errors. In the proposed topology, Fig.1, the final serial output is taken from the second modulating switch, this reduces insertion loss.

II. PROPOSED QPSK MODULATOR

The block diagram of the basic modulator is shown in Fig. 1. Assuming ideal behavior for the coupler and phase shifters

the four phase states generated for various switch positions are shown in Table 1.

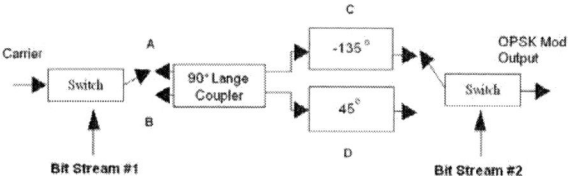

Fig. 1 Block diagram of designed QPSK modulator

TABLE I

Generation of QPSK Phase States

Bit #2 (Switch #2)	Bit #1 (Switch #1)	Path Chosen	Relative Output Phase	Simulation Port Set
0	0	BC	0°	S21
0	1	AC	90°	S65
1	1	AD	180°	S87
1	0	BD	270°	S43

The bit streams operating the switches are generated from the data stream as illustrated in Fig. 2 for the Gray sequence (00-01-11-10). The operation of the circuit can be explained in terms of these bit streams and the information in Table 1. The QPSK modulated output is collected serially from the second switch. At the same time, the S parameters of the modulator were acquired by setting different simulation ports for various switch positions.

In Fig. 2, a pulse generator was used to output two pulse streams at the same frequency and duty cycles but with a 90° relative delay. This results in all four phase states being periodically generated. This implies that each bit stream is of exactly half the frequency of the original data stream. Since each of these bit streams controls a switch, the switches only need operate at half the desired data rate.

A QPSK modulator needs to be evaluated in terms of bandwidth, insertion loss and encoded symbol error. Each of these metrics is linked to the phase and/or amplitude performance of the modulator. For the architecture described here the Lange coupler and phase shifter have the greatest bearing on these characteristics.

978-2-8748-7007-1/08 $25.00 © 2008 EuMA

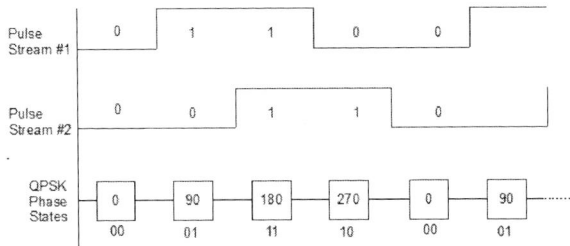

Fig. 2 Generation of Pulse streams and corresponding phase states

A. Broad band 90° coupler

In MMIC implementations at V band, Lange couplers can be used to provide a broadband 90° phase split. This coupler is manufactured on 100um thick GaAs substrate; the fingers are 10um wide with 12um gaps. It acquires an amplitude mismatch of ±0.5dB and a phase error of < 3° over the frequency range 45GHz to 80GHz centered at 72GHz. See Fig.3 and Fig.4.

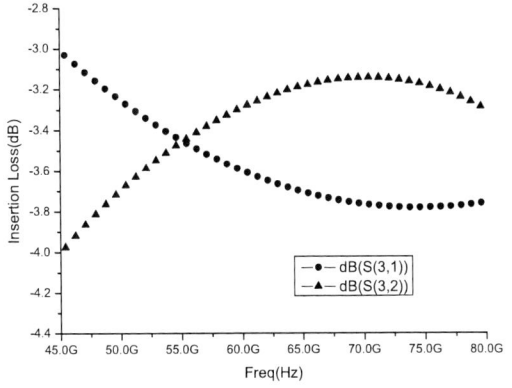

Fig. 3 The insertion loss for coupled port and through port

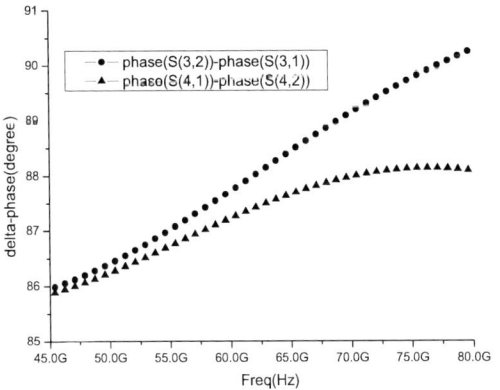

Fig. 4 The delta phase between coupled port and through port

B. Broad band 180° phase shifter

Boire et al [5] demonstrated a wideband 180° phase shifter. The basic configuration of this phase shifter consists of a shorted 3 dB quadrature coupler [8] and a π-type network. The coupler has its direct and coupled ports terminated with ideal short circuits, and the π-network is constructed with a series transmission line between two shunt short circuit transmission lines. The two networks are similar in all aspects, except that the transmission phase difference between the two is 180°. The relationship is independent of the electrical length of the two networks, and thus independent of frequency. Please see Fig. 9 for V band 180° MMIC phase shifter implementation. An unfolded Lange coupler, with direct and coupled ports terminated with short circuits, is more suitable to get 180° phase ahead of π-network with shunt short circuit transmission lines. From Fig. 5, the phase error is ±3° from 40GHz to 72GHz.

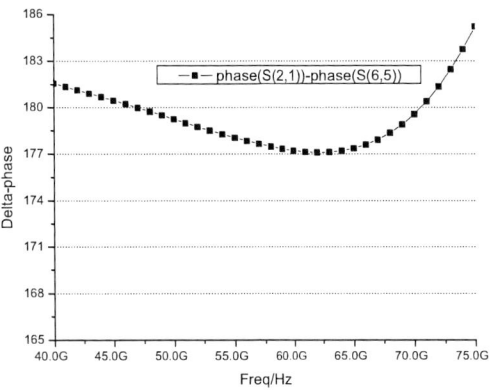

Fig. 5 The phase difference between short-type and π-type networks

One drawback of the Boire 180° bit phase shifter is the large differential insertion loss between the two networks.

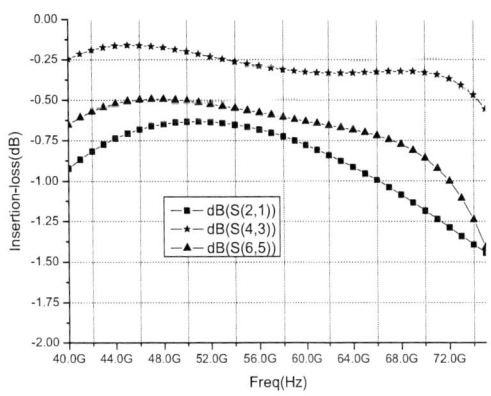

Fig. 6 The insertion loss difference among short-type, short-resist-type and π-type networks

978-2-8748-7007-1/08 $25.00 © 2008 EuMA   243

In Fig.6, dB(S(4,3)) stands for the insertion loss of the π - network with shunt short circuit transmission lines, and the insertion loss for shorted 3 dB Lange coupler is dB(S(21)). There is a big amplitude mismatch between them. Based on this observation, a couple of small resistors are added to the π -network between the shunt transmission line and via-hole ground which will increase the insertion loss and keep the phase unchanged. The revised circuit insertion loss dB(S(6,5)) is nearly the same as dB(S(2,1)), which ensures excellent amplitude match. The phase shifter yields an amplitude mismatch of less than 0.25dB over the band 45-65GHz with an insertion loss of less than 1dB.

C. Absorptive switch

Though the insertion loss of absorptive switches is a little higher than reflective ones, they yield better VSWR performance, especially for the isolated path. On comparison, the modulator based on absorptive switches showed much better phase performance. Figure 7 shows the measured return loss of the absorptive switch for the input, isolated and through ports. These are below -10dB over a very broad frequency band.

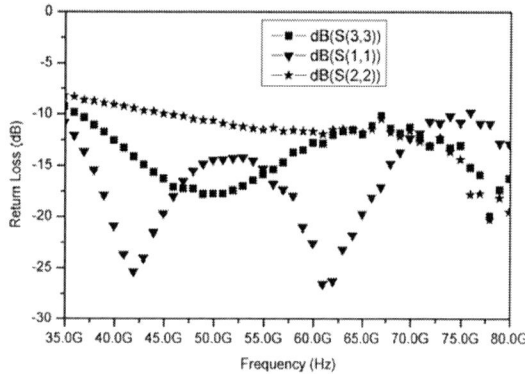

Fig.7 The return losses for input port 1, isolated port 2 and through port 3 of the absorptive switch

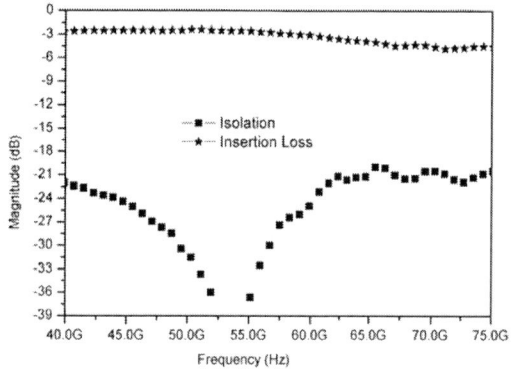

Fig.8 The insertion loss and isolation of the absorptive switch

As seen in Fig.8, the insertion loss is below 3dB and isolation is above 20dB for this absorptive switch from 40GHz to 65GHz.

III. MEASUREMENT RESULTS

The MMIC QPSK modulator was manufactured through standard 0.15 um pHEMT technology, (UMS PH15), [9]. All of passive circuit parts were optimized through EM simulation.

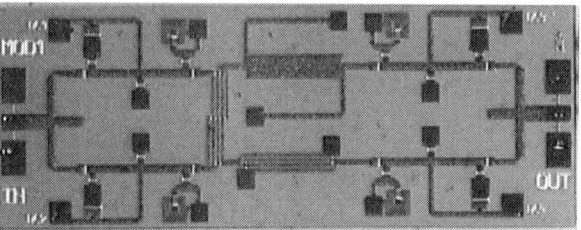

Fig. 9 The photograph of MMIC QPSK modulator

Fig. 9 give out the circuit photograph, the chip size is 2.7mm*1mm*0.1mm. Nonlinear diode models were used to simulate the switch. These diode components are designed for low junction capacitance as well as low series resistance for switch design and exhibit calculated cut-off frequencies higher than 300 GHz ($Fc=1/(2\pi RsCj[0V])$). The on and off-states switch voltages are respectively are -1V and +1.2V. The total current is about 50mA when the control voltage is 1.2V. The phase performance is shown in Fig.10. The simulation and measurement results have good accord. The phase errors are ±15° from 35GHz to 65GHz for different switch states.

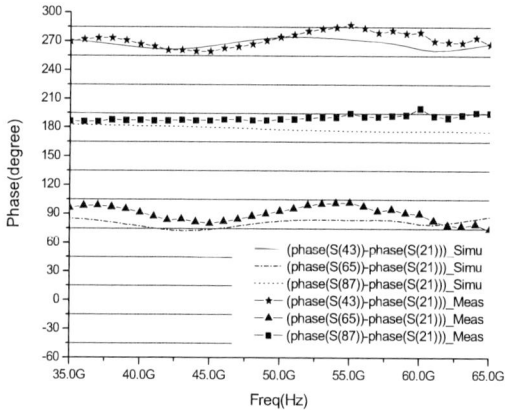

Fig. 10 The phase performances vs. Carrier frequency for QPSK modulator

Fig. 11 gives out the constellation Diagrams for QPSK modulator respectively for 46GHz, 56GHz, 61GHz and 66GHz. The modulator has better amplitude and phase performances between 56GHz and 66GHz, which makes it suitable for 60GHz indoor wireless systems.

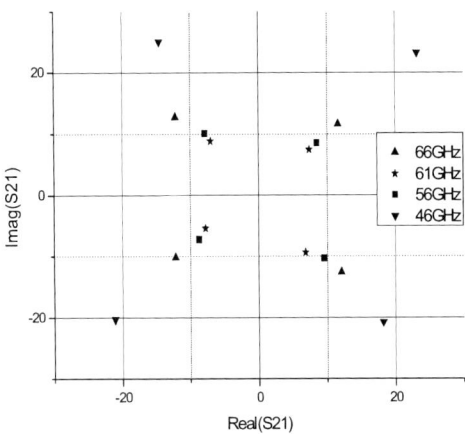

Fig. 11 The constellation Diagrams for QPSK modulator

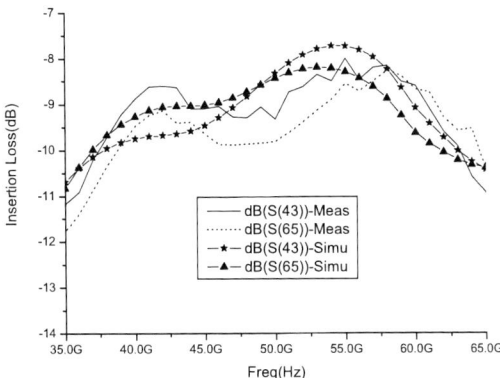

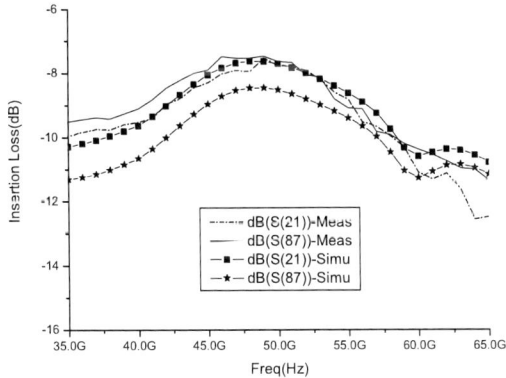

Fig. 12 The insertion loss vs. Carrier frequency for QPSK modulator

Figure 12 shows the insertion loss for the four phase states. Port setting was shown in Table 1. Again, the measured results agree with the simulation results. The insertion loss difference between the former two states and the latter two states was introduced by the amplitude mismatch of Lange coupler. The insertion loss suffered in the circuit is determined by the losses in the coupler (about 3.5dB), phase shifter loss (1dB) and the switch loss (3dB/switch). The loss in the absorptive switches increases with carrier frequency suggesting that reducing the insertion loss at higher frequencies will increase the bandwidth of the modulator. The overall insertion loss is better than 9.5±2dB over the frequency band 35-65GHz.

IV. CONCLUSIONS

This paper demonstrates a V band MMIC QPSK modulator. With very broad frequency band and very fast switching speed it is thought be capable of achieving up to 2Gbps bit rate at 60GHz frequency band. The low power consumption and absence of active components like mixers could be highly beneficial too with respect to cost and complexity of implementation.

ACKNOWLEDGMENT

The authors would like to acknowledge the support of the QUB SoCaM initiative and technical staff.

REFERENCES

[1] S. Yi, A. P. Freundorfer and D. Sawatzky, "A QPSK direct digital modulator in GaAs HBT at 28 GHz," *Canadian Conf. Electrical and Computer Engineering*, pp. 1882-1885, May 2005.

[2] M. El-Gabaly, B. R. Jackson and C. E. Saavedra, "An L-band Direct-Digital QPSK Modulator in CMOS," *Intl. Sym. Signals, Systems and Elec.*, pp. 563-566, August 2007.

[3] S. Kumar and G. Wells, "A 2.75-4.75 GHz QPSK Modulator with Low Amplitude and Phase Errors," *Electronics Letters*, vol. 26, no. 14, pp. 961-962, July 1990.

[4] J. G. Proakis and M. Salehi, "Communication systems engineering," *Prentice Hall*, 2nd Edition, pp. 357, 2002.

[5] D.Boire C. Internadio and J. E. Begenford, "A 4.5-18GHz phase shifter," *IEEE MTT-S International Microwave Symposium Digest.*, January 1985.

[6] B. Bosco, S. Franson, R. Emrick, "A 60 GHz transceiver with multi-gigabit data rate capability," *Radio and Wireless Conference,* Sept. 2004.

[7] I. Magrisso and A. Madjar, "A millimeter wave direct QPSK modulator MMIC using PIN technology and a novel approach to self error-correction," *32nd European Microwave Conf.*, pp. 1-4, October 2002.

[8] K.C. Gupta, R. Garg, I. J. Bahl, "RFIC and MMIC design and technology," Artech House, June 1979.

[9] http://ums.openkast.com/home/home_page.php

Wideband CMOS *LC* VCOs for IEEE 802.15.4a Applications

C. Stagni[1], A. Italia[2], and G. Palmisano[3]

Dipartimento di Ingegneria Elettrica Elettronica e dei Sistemi, Facoltà di Ingegneria, Università di Catania
V.le. A. Doria 6, 95125, Catania, Italy
[1]ceciliastagni@gmail.com
[2]aitalia@diees.unict.it
[3]gpalmisano@diees.unict.it

Abstract— Two wideband *LC* VCOs, customized for IEEE 802.15.4a applications, are presented in this paper. Both circuits have been designed and implemented in a 90-nm CMOS technology. The first VCO has been designed for the 3.1-5 GHz UWB band. It exhibits a tuning range of 40% from 3.2 GHz to 4.8 GHz with a tuning voltage of 1.2 V. The measured phase noise at 1-MHz offset from a 4-GHz carrier is −114 dBc/Hz. The 4-GHz VCO current consumption is only 1.5 mA from a 1.2-V supply voltage. The second VCO represents an improved design intended for the 6-10 GHz frequency operating range. By exploiting differentially tuned varactors, the proposed 8-GHz VCO is able to guarantee a 45% tuning range and a phase noise at 1-MHz offset frequency lower than −102 dBc/Hz. The 8-GHz VCO draws 2.6 mA from a 1.2-V supply voltage.

I. INTRODUCTION

In the near future an outstanding spread of low-data rate wireless personal area networks (LR-WPANs) is expected to comply with the growing demand for industrial, vehicular, residential and medical applications. The IEEE 802.15 Task Group 4a (TG4a) is defining a new standard [1] for these low-cost, very low-power, short-range wireless applications, intended as an alternate PHY layer for the existing IEEE 802.15.4. TG4a recognizes ultra wideband (UWB) communications as the future technology for the LR-WPANs. The draft version of the IEEE 802.15.4a standard [1] defines an UWB physical layer, which operates in the 3.1-5 GHz and 6-10 GHz frequency bands. The signal 3-dB bandwidth is approximately 500 MHz. A wider 3-dB bandwidth can be provided by several optional channels (5, 8, 12, 16). The protocol exploits an impulse radio based signaling scheme. The data modulation technique is a combination of both PPM and BPSK.

The development of wideband low-power, low-cost radio front-end for new generation of LR-WPAN systems represents a key-point for the spread of mass-market equipment. To this purpose, a proper choice of receiver architecture is of primary importance. The direct conversion and quadrature analog correlation represent suitable architectures since they exhibit optimal performance in terms of bit error rate (BER) and energy per useful bit [2]. However, a strong design effort must be carried out to implement low-power wideband RF circuits.

Among the RF front-end building blocks, the VCO design is very challenging since it is extremely difficult to obtain a

wide tuning range along with low power consumption. The CMOS technology scaling is further complicating the design of wideband VCOs. Indeed, the reduced minimum device length results in a lower supply voltage, which limits the tuning control voltage and hence the achievable tuning range. Moreover, shorter length devices exhibit lower transconductance gain, which involves a higher bias current to fulfil start-up condition.

This paper presents two wideband *LC* VCOs for IEEE 802.15.4a applications. The first VCO has been designed for the 3.1-5 GHz frequency band while the second VCO covers the 6-10 GHz channels. Both VCOs demonstrate excellent tuning range with very low power consumption. Thanks to an accurate design of the *LC* tank, very good phase noise performance has been also achieved all over the tuning range.

II. CIRCUITS DESCRIPTION AND DESIGN

The simplified schematic of the proposed VCOs is shown in Fig. 1. Although complementary cross-coupled topology offers better phase noise performance for a given bias current, a differential cross-coupled NMOS topology has been preferred since it ensures lower parasitic capacitances and allows low voltage operation. The *LC* tank consists of an integrated differential spiral inductor (*L*), N+POLY/N-WELL MOS varactors (C_V), and two switched metal-insulator-metal (MIM) capacitors (C_{sw1}, C_{sw2}). The switched capacitors,

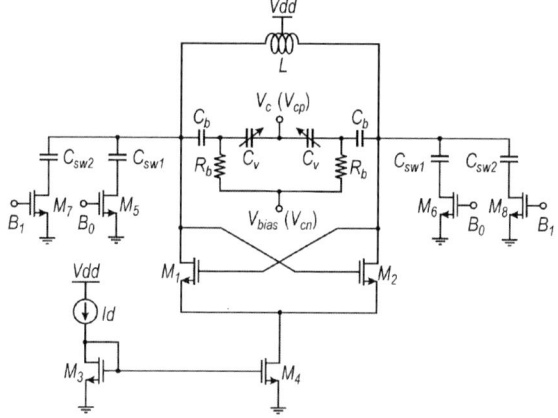

Fig. 1 Simplified schematic of the proposed VCOs

controlled by a two-bit channel select digital word (B_0, B_1), provide coarse tuning steps allowing the channel selection. The N+POLY/N-WELL MOS varactors provide continuous frequency tuning. Varactors are ac-coupled to the tank by means of MIM capacitors (C_b). A single-ended frequency tuning scheme has been adopted for the 4-GHz VCO. The varactor gate terminal has been properly biased to maximize its capacitance tuning range. The 8-GHz VCO exploits differentially-tuned varactors controlled by the differential voltage ($V_{cp}-V_{cn}$). This approach widens the capacitance tuning range, which results 30% higher compared to a single-ended frequency tuning scheme.

A. LC Tank Design

The design of wideband VCOs involves a difficult tradeoff between phase noise, tuning range and power consumption performance and an accurate choice of the LC tank plays a key-role to achieve the optimum design. Obtaining wide tuning range along with low phase noise is one of the main issues in the design of wideband LC VCOs. Indeed, high-gain varactor, required to achieve a wide tuning range, degrades phase noise performance due to the AM-to-FM conversion mechanism [3]. To overcome this drawback, the targeted tuning range can be split into sub-bands by exploiting a switched capacitor array. The size of the switched capacitor array must be careful designed. An excessive number of switched capacitors could result in worst phase noise and tuning range performance. Indeed, the finite switch resistance R_{ON} reduces the overall tank quality factor, while the switch parasitic capacitances limit the achievable tuning range. Both VCOs use an array of two switched-capacitors. As shown in Fig. 2, the targeted frequency range of 8-GHz VCO has been split in four overlapping sub-bands, which cover the standard higher band channels. On the other hand, the 4-GHz VCO covers the required frequency range by exploiting three values of the digital channel select word.

The dimensions of both MOS switches and capacitors $C_{sw1,2}$ must be carefully set since these devices heavily affect the overall tank quality factor and the tuning range. The quality factor of a switched capacitor can be expressed by the following equation

$$Q_{sw} = \left(\omega_o R_{ON} C_{sw}\right)^{-1} \qquad (1)$$

where ω_o is the oscillation frequency and C_{sw} is the capacitance value. By defining C_{ON} and C_{OFF} as the equivalent switched capacitor on-state and off-state capacitances respectively, the C_{ON}/C_{OFF} ratio can be written as follow

$$\frac{C_{ON}}{C_{OFF}} \cong \frac{C_{sw}}{\dfrac{C_{sw} \cdot C_{swoff}}{C_{sw} + C_{swoff}}} = 1 + \frac{C_{sw}}{C_{swoff}} \qquad (2)$$

where C_{swoff} is the switch parasitic capacitance. Equations (1) and (2) show the critical tradeoff between tuning range and phase noise. To achieve a wideband tuning range, small parasitic capacitance C_{swoff} should be provided to maximize the C_{ON}/C_{OFF} ratio. However, this results in small MOS

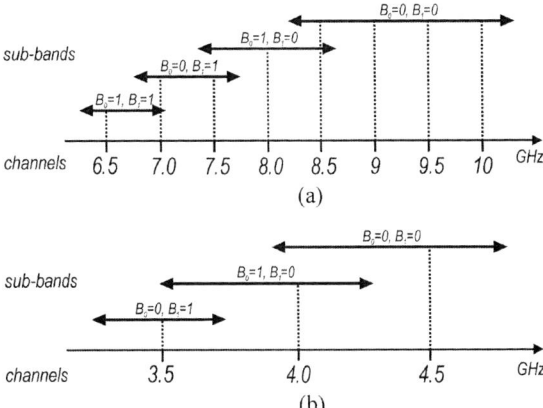

Fig. 2 Sub-bands of the proposed VCOs (a) 8-GHz VCO (b) 4-GHz VCO

switches, which could affect the overall tank quality factor because of an excessive R_{ON} resistance. Therefore, minimum length MOS devices must be used to reduce C_{swoff}. The width must be chosen to tradeoff phase noise and tuning range performance.

The varactor geometrical parameters have been accurately chosen in order to optimize its quality factor Q_{VAR}. Indeed at high operating frequencies, Q_{VAR} could significantly decrease the overall tank quality factor. Typical Q_{VAR} values are 70 and 30 at 4 GHz and 8 GHz respectively. To this purpose, minimal-length and minimal-width varactors have been employed.

As fas as concerned the choice of the tank inductor, a design strategy has been adopted to achieve simultaneous optimization of phase noise and tuning range with the constraint of minimum power consumption. Initially, the inductance selection scheme assumes the phase noise as main design goal. Afterwards, tuning range and power consumption constraints will be considered to set the final tank inductor. This is an iterative design strategy that allows the VCO target specification to be addressed.

Several works [4]-[5] demonstrate that exists an optimum value of oscillator amplitude, which minimizes the phase noise. Indeed, the phase noise reaches a minimum value at the boundary between the current-limited regime and the voltage-limited regime [5] From a different point of view, this means that for a given bias current at the operating frequency the phase noise can be optimized with a proper choice of the tank inductor. In particular, for a fixed overall tank quality factor exists an optimum inductance value, which minimizes the phase noise. This dependence can be exploited to set the optimum value of both the inductance L and the quality factor Q_L of the tank inductor. Fig. 3 shows phase noise at 1-MHz offset from a 4-GHz carrier versus tank inductance for three values of Q_L. The phase noise is given with Q_L as design parameter, but it takes into account the overall tank quality factor, which is lower because of the degradation effect of switched capacitor array and varactors. As shown in Fig. 3, the optimization of the phase noise involves a proper choice of both inductance and quality factor. Indeed, a higher quality factor of the tank inductor results in a better phase noise only if its inductance value is appropriate. In wideband design, the frequency dependence of both L and Q_L must be taken into account to design the optimum inductor. Indeed, it must provide the optimum pair (L, Q_L) that simultaneously minimizes and keeps symmetric the phase noise all over the range of operating frequencies. The variation of varactor quality factor within the tuning range should be also taken into account to ensure small phase noise variation. To clarify the proposed design approach, Table I can be consulted. It reports the simulated phase noise performance of the 4-GHz VCO with three differential inductors. The L_B can be easily considered the optimum inductor. Indeed, it provides smaller phase noise variation ensuring better performance at the boundaries of the desired tuning range in comparison with L_A and L_C.

The tank inductor selection, performed considering the optimization of phase noise described above, defines univocally the varactor size and hence the achievable tuning range, which could be smaller than the targeted one. A lower inductance value could be selected to increase the varactor size thus obtaining the desired tuning range at disadvantage of phase noise. However, simultaneous optimization can be achieved if a higher tail current is employed. Indeed, the

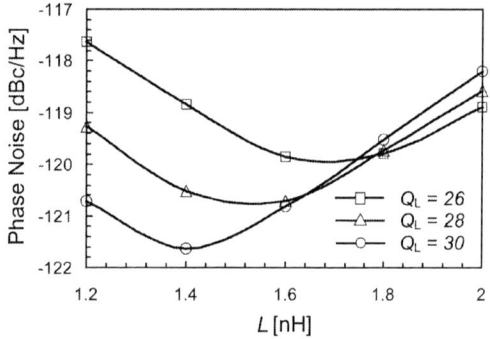

Fig. 3 Phase noise versus tank inductance (1-MHz offset frequency, 4 GHz carrier)

TABLE I
EXAMPLE OF PHASE NOISE OPTIMIZATION

Parameter	Operating frequency [GHz]	Inductance [nH]	Quality factor	Phase noise (Δf=1 MHz) [dBc/Hz]
Inductor L_A	3.5	1.4	24.3	−116.6
	4	1.41	26.8	−117.8
	4.5	1.43	28.5	−118.4
Inductor L_B	3.5	1.79	25.8	−117.4
	4	1.8	28.1	−119
	4.5	1.83	29.3	−117.4
Inductor L_C	3.5	2.2	26.4	−119.8
	4	2.23	28.3	−118.4
	4.5	2.27	28.7	−116

maintenance of the optimum oscillation amplitude and hence the optimization of phase noise can be accomplished with a lower inductance value increasing the VCO current. On the other hand, if a higher tuning range than the targeted one is obtained, simultaneous optimization can be provided with a lower current consumption. Therefore, a fundamental trade-off exists among phase noise, tuning range and power consumption. Phase noise and tuning range specifications can be simultaneously obtained at disadvantage of power consumption. The extensive use of simulations helps the optimization process to converge to the targeted VCO design for all process corners.

B. VCO Core Design

For an LC VCO, the start-up condition is given by the well known inequality $g_m \geq 1/R_P$. The equivalent tank resistance R_P exhibits a frequency dependent behaviour that cannot be ignored in wideband oscillator design. The worst-case scenario is at the low-end of the tuning range, since R_P reaches its lowest value. The channel width and the multiplicity of the cross-coupled NMOS devices have been set to guarantee robust start-up condition. Minimum-length transistors have been preferred to keep low the parasitic capacitances in parallel to the tank. Finally, the tail current generator is generally considered the most significant source of flicker noise [3] in differential LC VCO. Its reduction is mandatory in every oscillator design since it upconverts into $1/f^3$ phase noise through the AM-to-FM conversion mechanism [3]. To this aim, long-channel length devices with high aspect ratio (W/L) should be adopted.

III. EXPERIMENTAL AND SIMULATION RESULTS

The VCOs have been fabricated in a 90-nm CMOS technology by STMicroelectronics. The technology features seven metal layers and integrated inductors are implemented by stacking the three topmost metal layers. Fig. 4 shows the die microphotographs of the fabricated circuits. Both VCOs exploit a differential patterned ground shield inductor. The 4-GHz VCO tank inductor has a differential inductance value of 1.5 nH and a quality factor ranging from 24 to 28 over the all frequency range. The 8-GHz VCO uses a horseshoe 450 pH tank inductor, which exhibits a quality factor ranging from 16 to 22.

Fig. 5 shows the measured and the simulated tuning range of 4-GHz VCO. The circuit achieves a wide tuning range of 40% from 3.2 GHz to 4.8 GHz with a tuning voltage ranging from 0 to 1.2 V. The measured phase noise at 1-MHz offset frequency is −114 dBc/Hz at 4 GHz (B_0=1, B_1=0, channel 2), −112 dBc/Hz at 3.5 GHz (B_0=0, B_1=1, channel 1), and −112.2 dBc/Hz at 4.5 GHz (B_0=0, B_1=0, channel 3). Thanks to the proposed selection strategy of the tank inductor, the phase noise variation within the band of interest is lower than 2 dB. Moreover, a symmetric phase noise performance has been obtained. The circuit draws only 1.5 mA from a 1.2-V supply voltage. The 8-GHz VCO is being tested and the expected tuning range is shown in Fig. 6. The circuit provides a 45% tuning range from 6.4 GHz to 10.1 GHz. The simulated phase

978-2-8748-7007-1/08 $25.00 © 2008 EuMA

noise at 1-MHz offset frequency is lower than −102 dBc/Hz for all the standard higher band channels. The 8-GHz VCO draws 2.6 mA from a 1.2-V supply voltage.

Table II summarizes the performance of the proposed circuits and reports the wideband VCO state-of-the-art. The VCOs demonstrate excellent tuning range performance and extremely low current consumption. Moreover, the FOM [6] and PFTN [5] performance of measured 4-GHz VCO are comparable with the state-of-the-art. It is worth mentioning that such a result has been obtained with a 90-nm CMOS technology, which limits the available tuning control voltage and suffers from lower gain and phase noise performance compared to longer channel-length technologies.

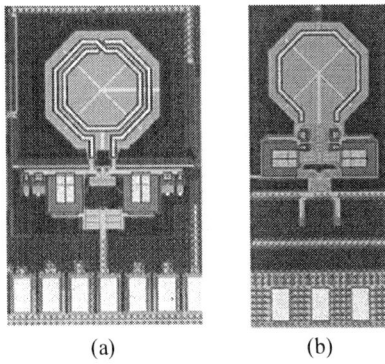

(a)   (b)

Fig. 4 Die microphotographs (a) 4-GHz VCO (b) 8-GHz VCO

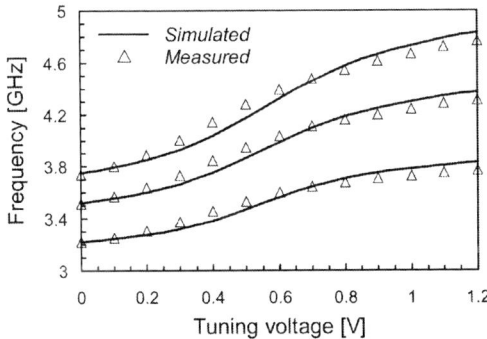

Fig. 5 Measured and simulated tuning range of 4-GHz VCO

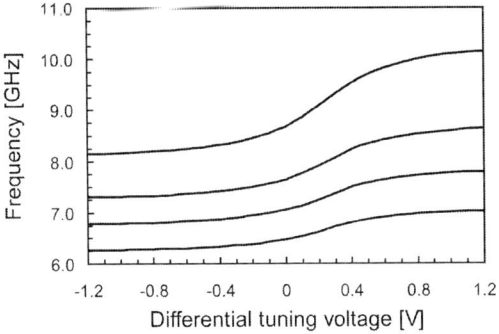

Fig. 6 Simulated tuning range of 8-GHz VCO

TABLE II
COMPARISON WITH THE WIDEBAND VCO STATE-OF-THE-ART

Process	Center Freq [GHz]	P$_{DC}$ [mW]	Tuning range	FOM [dB]	PFTN [dB]	Ref.
130-nm SOI CMOS	4.33	2-3	58.7% (1.4 V)	−186.6	10.3	[6]
180-nm CMOS	1.8	2.6-10	73% (1.5 V)	−184.8	8.5	[7]
130-nm CMOS	4.15	2.7-9.2	50.6% (1.2 V)	−179.3	3	[8]
180-nm CMOS	7.65	14.4	34.5% (1.8 V)	−177.2	−7.28	[9]
90-nm CMOS	**4**	**1.8**	**40% (1.2 V)**	**−183.5**	**1.8**	**This work (measured)**
90-nm CMOS	**8.25**	**3.12**	**45% (1.2 V)**	**−177.4**	**−3.4**	**This work (simulated)**

IV. CONCLUSIONS

Two wideband *LC* VCOs for IEEE 802.15.4a applications have been implemented in a 90-nm CMOS technology. The first VCO, designed for the 3.1-5 GHz lower band, provides a 40% tuning range and exhibits a measured phase noise at 1-MHz offset frequency lower than −112 dBc/Hz. The circuit draws 1.5 mA from a 1.2-V supply voltage. The second VCO represents an improved design, which covers the 6-10 GHz operating band. By exploiting differentially tuned varactors, the 8-GHz VCO ensures a 45% tuning range while drawing only 2.6 mA from a 1.2-V supply voltage.

REFERENCES

[1] *Standard draft proposal*, IEEE 802.15.4a [Online]. Available: www.ieee802.org/15/pub/TG4a.html.

[2] J. Ryckaert, *et al.*, "A CMOS ultra-wideband receiver for low data-rate communication," *IEEE J. Solid-State Circuits*, vol. 42, pp. 2515-2527, Nov. 2007.

[3] S. Levantino, C. Samori, A. Bonfanti, S. L. J. Gierkink, A. L. Lacaita, and V. Boccuzzi, "Frequency dependence on bias current in 5-GHz CMOS VCOs: impact on tuning range and flicker noise upconversion," *IEEE J. Solid-State Circuits*, vol. 37, pp. 1003-1011, Aug. 2000.

[4] C. Samori, A. L. Lacaita, A. Zanchi, S. Levantino, and G. Cali, "Phase noise degradation at high oscillation amplitudes in LC-tuned VCO's," *IEEE J. Solid-State Circuits*, vol. 35, pp. 96-99, January 2000.

[5] D. Ham, and A. Hajimiri, "Concepts and methods in optimization of integrated LC VCOs," *IEEE J. Solid-State Circuits*, vol. 36, pp. 896-909, June 2001.

[6] N. H. W. Fong, J.-O. Plouchart, N. Zadmer, D. Liu, L. F. Wagner, C. Plett, and N. G. Tarr, "Design of wide-band CMOS VCO for multiband wireless LAN applications," *IEEE J. Solid-State circuits*, vol. 38, pp. 1333-1341, Aug. 2003.

[7] A. D. Berny, A. M. Niknejad, and R. G. Meyer, "A 1.8-GHz LC VCO with 1.3 GHz tuning range and digital amplitude calibration," *IEEE J. Solid-State Circuits*, vol. 40, pp. 909-917, April 2005.

[8] D. Hauspie, E.-C. Park, and J. Cranincks, "Wideband VCO with simultaneous switching of frequency band, active core, and varactor size," *IEEE J. Solid-State circuits*, vol. 42, pp. 1333-1341, July 2007.

[9] G.-Y. Tak, S.-B. Hyun, T. Y. Kang, B. C. Choi, and S. S. Park, "A 6.3-9 GHz CMOS fast settling PLL for MB-OFDM UWB applications," *IEEE J. Solid-State circuits*, vol. 40, pp. 1671-1679, Aug. 2005.

Proceedings of the 3rd European Microwave Integrated Circuits Conference

A 60-GHz LNA and a 92-GHz Low-Power Distributed Amplifier in CMOS with Above-IC

C. Pavageau[1], O. Dupuis, M. Dehan, B. Parvais, G. Carchon[2], and W. De Raedt[3]

IMEC, Kapeldreef 75, B-3001 Leuven, Belgium

[1]Christophe.Pavageau@imec.be

[2]Geert.Carchon@imec.be

[3]Walter.DeRaedt@imec.be

Abstract— This paper demonstrates a broadband LNA for 60-GHz WPAN and a 92-GHz low-power Distributed Amplifier (DA) in an advanced CMOS technology. A post-processed technology (Above-IC), used for packaging and bonding pads redistribution, provides ultra-low-loss on-chip passives in a cost-effective solution. In the WPAN bandwidth (57-64 GHz), the LNA has a 13.4 dB peak gain, a NF between 5.6-6.7 dB and a gain flatness of 1.7 dB. A 3-dB bandwidth of 11 GHz is achieved. The DA shows a 6.5 dB gain and a 3-dB BW of 92 GHz, giving a 195 GHz gain-bandwidth product (GBW), for a dc power consumption of 43 mW. Thank to the asset of Above-IC, this DA performs a ratio of the GBW and power consumption of 4.31 GHz/mW, which is by far the best reported tradeoff among similar architecture in CMOS, at least 2.8 times higher than others.

I. INTRODUCTION

In the past year, the fast increase of the high-frequency performance of CMOS technologies, together with the ever growing demand of the high-speed communication market have driven important efforts on building transceiver circuits at millimeter-wave frequencies with digital CMOS technologies, due to their inherent low cost, high volume and high integration capabilities. But digital process offer low resistivity substrates, thin dielectric and thin metallic layers in the back-end-of-line (BEOL), leading to high losses for on-chip passive components and as a consequence, counterbalancing benefits from the improvement of transistor's high-frequency performance. Microstrip line (MS), implemented with standard digital BEOL layers, experiments losses in the range of 1 dB/mm at 30 GHz [1].

Several options are available to decrease on-chip component losses, such as thick copper layers in the BEOL or high-resistivity (HR) substrates [2]. For the design of DA in CMOS, it has been successfully shown that employing CPW lines on HR substrates instead of MS lines lead to a twofold increase of the GBW product [1]. However, HR substrates as well as thick copper layers are not standard technology options.

A more attractive and cost-effective solution for on-chip transmission lines is to realize them above the passivation, with a thin-film post-processed technology [3]-[4]. This technology does not add extra-cost because it is used for packaging and bonding pads redistribution during chips manufacturing. The concept, known as Above-IC technology, is depicted in Fig. 1. In this case, multiple benzo-cyclobutene

dielectric (BCB) and thick copper layers are deposited on top of the passivation. Hence, transmission lines, featuring ultra-low-loss (0.3 dB/mm at 100 GHz [3]) can be realized thus alleviating common limitations of CMOS for millimeter-wave. As metal layers are added above the passivation, one may also reduce the number of BEOL layer, hereby further reducing costs.

Above-IC advantages for 5-GHz VCO has been already demonstrated in 90 nm CMOS, showing a reduction of the power consumption by a factor 17 compared to a similar design with a BEOL inductors. Benefits for LNA design have also been experienced at 17 GHz [5] and 40 GHz [6], with drastic gain and noise figure improvements as well as large power consumption reduction.

This paper demonstrates the potential for millimeter-wave applications of the aforementioned Above-IC technology combined with an advanced CMOS process, through the design and measurements of a broadband 60-GHz LNA and an efficient low-power DA. Together with a reduced number of layers in the BEOL, this concept provides a low-cost solution for high quality on-chip passives on CMOS.

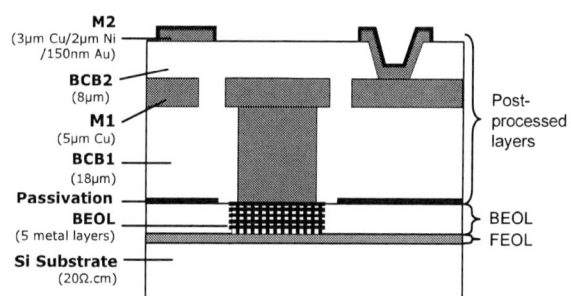

Fig. 1 Schematic of the thin-film Above-IC interconnect scheme on top of the BEOL layers.

II. DEVICE TECHNOLOGY

The designs take benefits of an advanced CMOS process from IMEC, providing transistor with a gate length shrinked from 130 nm to 45 nm and composed of poly-Si/SiON (EOT=1.6 nm). The extrapolated peak f_t and f_{max} are 250 GHz and 400 GHz. The BEOL features five thin copper layers with thicknesses of 0.3 μm for Metal-1 and 0.5 μm for the others, TaN resistors and ONO MIM capacitors (1.1fF/μm²).

978-2-8748-7007-1/08 $25.00 © 2008 EuMA   250

III. CIRCUIT DESCRIPTION AND MEASUREMENTS

For both the LNA and the DA, on-wafer S-parameters measurements were performed using an Agilent PNA operating up to 110 GHz. RF pads were not calibrated out, so that measurements include their effects.

A. Broadband 60-GHz LNA

The broadband LNA is designed for 60-GHz WPAN (IEEE 802.15.3) allowing an unlicensed frequency bandwidth of 7 GHz, from 57 GHz to 64 GHz.

The LNA is implemented as a single-ended dual-stage in cascode configuration (Fig. 2). Biasing and matching networks are combined to reduce the overall size. The input, inter-stage and output matching networks are implemented with a combination of MS lines in Above-IC and BEOL. According to the design methodology in [7], the size of the input transistor is first chosen to match the real part of the optimum noise impedance to 50 Ω. Then, inductive degeneration is used to achieve simultaneous conjugate matching for the optimum noise and input impedances. For the first stage, NMOS transistors of 18x2 μm are used while for the second stage, the transistor size is 50x2 μm. The circuit micrograph is shown in Fig. 3. The chip area (pads excluded) is 0.36 mm².

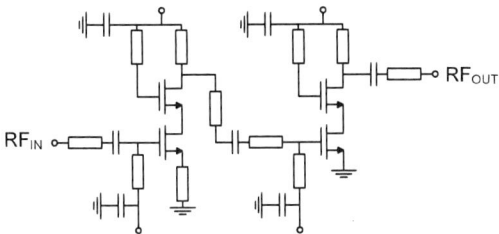

Fig. 2 Circuit schematic of the broadband 60-GHz LNA.

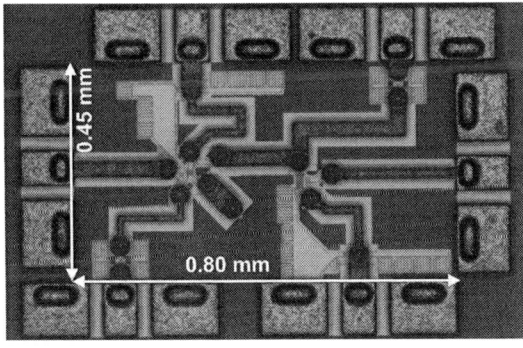

Fig. 3 Micrograph of the broadband 60-GHz LNA.

The measurement of the LNA noise figure has been realized with the use of a V-band noise source, a down-converter mixer and a frequency multiplier (x4) combined with a signal generator (E8257C) to generate the LO signal. The signal was down-converted to 50MHz and treated by a N8973A noise figure analyzer. Fig. 4 shows the measured S-parameters and the noise figure (NF). From 57 to 64 GHz, the LNA exhibits a maximum gain of 13.4 dB at 61 GHz, a gain flatness of 1.7 dB, input return losses lower than -14 dB and a noise figure (NF) between 5.6 and 6.7 dB. The dual-stage cascode design leads to isolation better than -45 dB. The output return losses are lower than -3 dB. It is worth to mention that this strong output mismatch reduces drastically the gain of the LNA. It is attributed to the use of preliminary compact models extracted at an early stage of development of the CMOS process.

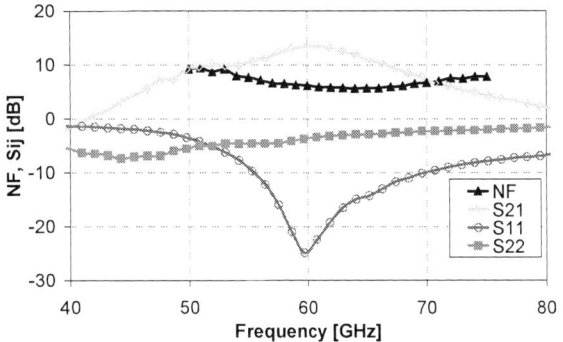

Fig. 4 Measured S-parameters and NF of the 60-GHz LNA.

TABLE I
PERFORMANCE SUMMARY OF THE BROADBAND 60-GHz LNA

Parameter	
Frequency [GHz]	57 − 64
min/max Gain [dB]	11.7/13.4
min/max NF [dB]	5.6/6.7
3-dB bandwidth [GHz]	11
V_{dd} [V]	1.8
P_{DC} [mW]	95
Area [mm²]	0.36

B. Low Power Distributed Amplifier

The distributed amplifier (DA) architecture has been widely used with III-V technologies to achieve multi-decade flat gain for instrumentation and broadband optical communication systems. DAs are composed of parallel gain stages, whose input and output capacitances are combined with inductive components, therefore realizing artificial transmission lines matched to the system impedance. In theory, infinite gain-bandwidth product (GBW) is possible by continuously adding stages, but a practical limitation arises due to passive element losses. In digital CMOS BEOL, the excessive high losses of transmission line deeply decrease the maximum achievable GBW product [1]. Therefore, the Above-IC technology has a significant asset for the design of DAs by bringing ultra-low-loss MS line (0.3dB/mm at 100 GHz).

The DA schematic is depicted in Fig. 5. As the circuit was intended for technology benchmarking, a conventional architecture was used, with cascode stages to circumvent the

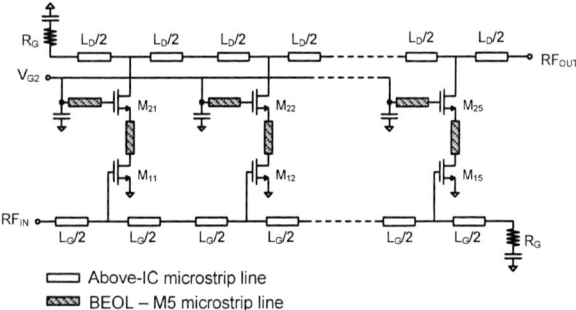

Fig. 5 Circuit schematic of the 5-stages DA with Above-IC.

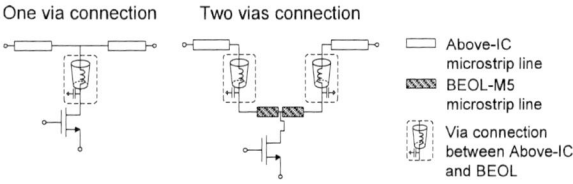

Fig. 6 Schematic diagram of the transistor connection to the WLP microstrip line with one or two vias.

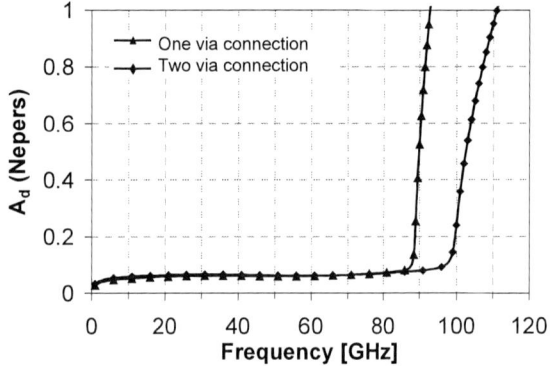

Fig. 7 Simulated losses per section A_d of the artificial drain line with one or two vias connection.

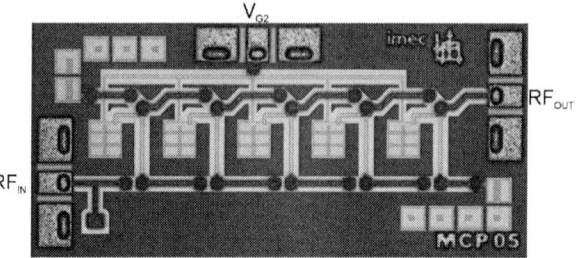

Fig. 8 Micrograph of the low-power DA (0.7x1.58 mm²).

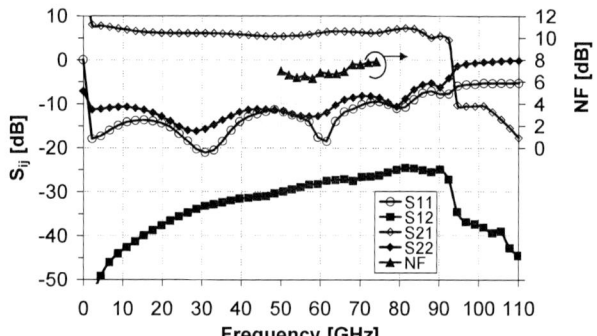

Fig. 9 Measured S-parameters and Noise Figure NF.

TABLE II
STATE-OF-THE-ART OF DA IN CMOS WITH CASCODE ARCHITECTURE.

Reference	[1]	[9]	[10]	[12]	*this work*
Technology	SOI	bulk	SOI	SOI	bulk
Physical gate length [nm]	115	-	46	-	45
f_t [GHz]	63	-	196	147	250
Number of stages	4	8	9	4	5
S_{21} [dB]	7 ±0.5	**9.8** ±1.2	**11** ±1.2	**9.7** ±1.6	**6.5** ±1.5
BW [dB]	**43**	**44**	**90**	**59**	**90**
S_{11}/S_{22} [dB]	<-9 /<-6	<-8 /<-14	<-7 /<-5	<-5 /<-12	<-5 /<-5
NF [dB]	-	2-7 <40GHz	4.8-6.2 <18GHz	3.2-3.8 <40GHz	6.3-8.1 50-75GHz
P_{DC} [mW]	75	103	210	132	**43**
Area [mm²]	1.8	1.5	1.3	0.3	1.1
Figure-of-Merit					
GBW [GHz]	101	136	**319**	180	195
GBW/P_{DC} [GHz/mW]	1.34	1.32	1.52	1.37	**4.31**

Miller effect of the MOSFET and loss-compensation technique [8] to get excellent gain flatness. A detailed description of the architecture can be found in [1].

Artificial lines are synthesized with MS line sections in Above-IC of 10 µm width, leading to an unloaded impedance characteristic of 84 Ω. These lines are connected to the underlying BEOL and transistors by means of 30 µm-width vias, modelled by an inductor of 20 pH with a parallel capacitance of 11 fF. Two options are possible to connect transistors (Fig. 6). The first uses one via but in this case, the via inductance adds in series with the transistor access and in parallel with the capacitance of MS line sections. Another option uses two vias with in between a short line in BEOL to connect the transistor. As opposed to the first option, via ind-

tances and capacitances are merged respectively in series and parallel with those of Above-IC line sections to realize the artificial line. Fig. 7 shows the simulation of the attenuation per section A_d of the output artificial line with both options. In order to match the 50-Ω impedance condition, the one via option requires Above-IC lines 1.5 times longer than the two vias option, leading to a cutoff frequency 10 GHz lower.

The circuit microphotograph is shown in Fig. 8. The DA exhibit a gain of 6.5±1.5 dB from 2 to 90 GHz and a 3-dB cutoff frequency of 92 GHz (Fig. 9), resulting in a GBW of 195 GHz. The power consumption is 43 mW from a 1.2 V supply voltage.

The increase of gain at lower frequencies is due to an under-estimation of 30% of the resistive termination. Up to 80 GHz, S_{11} and S_{22} are better than -10 dB and -8 dB, respectively. The measured NF is 6.3-8.1 dB over 50-75 GHz. Linearity measurements were performed at 10 GHz, showing a OP_{1dB} of 5.5 dBm and a IMD3 of 52.5 dBc resulting in an equivalent 16 dBm OIP3.

Table II shows a comparison with state-of-the-art DA in CMOS having similar cascode architecture to the one we used. The ratio of the GBW to the power consumption was chosen as a figure-of-merit (FOM). It has the advantage of being relatively independent of the number of stages and transistor width. The DA in SOI [10] benefits from transistor having the closest features to the CMOS technology we used, both in terms of transistor gate length and high frequency performance [11]. This design achieves a comparable BW of 90 GHz, a better GBW product of 320 GHz and a FOM of 1.52 GHz/mW. However, the DA with Above-IC achieves a FOM of 4.31 GHz/mW which is 2.8 times higher, showing a significantly more efficient design in term of power consumption. This huge improvement is due to the ultra-low-loss transmission lines in Above-IC.

IV. CONCLUSIONS

A 60-GHz broadband LNA and a 92-GHz efficiently low-power designed DA were realized in an advanced CMOS technology combined with a post-processed Above-IC technology.

The LNA obtains a peak gain of 13.4 dB, a minimum NF of 5.6 dB together with excellent gain and NF flatness of 1.7 and 1.1 dB, respectively, in the WPAN bandwidth of 57-64 GHz.

The DA achieves an average gain of 6.5 dB and a 3-dB cutoff frequency of 92 GHz. Thank to the asset of Above-IC ultra-low-loss passives, this DA performs a ratio of the GBW product to the power consumption of 4.31 GHz/mW, which outperforms by at least a factor 2.8 other published results in CMOS with comparable architecture. This result demonstrates the suitability of CMOS technologies combined with Above-IC for implementing efficiently low-power designs at millimeter-wave frequencies.

In essence, the Above-IC technology does not add extra-cost because it is used for packaging and bonding pads redistribution during chip manufacturing. Thereby, it is a competitive solution to provide ultra-low-loss on-chip passives for millimeter-wave applications at 60-GHz and above, compared to other solutions such as thick copper layers in the BEOL or high-resistivity substrates which are non-standard and expensive options of CMOS technologies.

ACKNOWLEDGMENT

The authors acknowledge the European Commission for supporting the IST project NANO-RF (IST-027150).

REFERENCES

[1] C. Pavageau, M. Si Moussa, J-P. Raskin, D. Vanhoenaker-Janvier, N. Fel, J. Russat, L. Picheta, F. Danneville, "A 7-dB 43-GHz CMOS distributed amplifier on high resistivity SOI substrates", Accepted for publication in *IEEE Trans. Microwave Theory & Tech.*, vol 56, no.3, March 2008.

[2] F. Gianesello, et al, "1.8 dB insertion loss 200 GHz CPW band pass filter integrated in HR SOI CMOS Technology", in *IEEE Int. Microwave Symp.*, pp. 453–456, June 2007.

[3] G. Carchon, et al., "Thin-Film as enabling passive Integration for RF SoC and SiP," in *ISSCC Dig. Tech. Papers*, pp. 398–399, Feb. 2005.

[4] G. J. Carchon, W. De Raedt, E. Beyne, "Wafer-level packaging technology for high-Q on-chip inductors and transmission lines", *IEEE Trans. Microwave Theory & Tech.*, vol. 52, no 4, pp. 1244–1251, 2004.

[5] D. Linten, et al., "Implementation of 6kV ESD Protection for a 17GHz LNA in 130nm SiGeC BiCMOS", in *IEEE Int. Conf. Semiconductors Electronics*, pp. A7–A12, 2006.

[6] S. Pruvost, et al., "40GHz Low Noise Receiver Circuits using BCB Above-Silicon Technology Optimized for Millimeter-wave Applications", in *IEEE RFIC Symp. Dig.*, pp. 137–140, 2007.

[7] S.P. Voinigescu, et al., "A scalable high-frequency noise model for bipolar transistors with application to optimal transistor sizing for low-noise amplifier design", in *IEEE J. Solid-State Circuits*, Sept. 1997.

[8] S. Deibele, J.B. Beyer, "Attenuation compensation in distributed amplifier design," *IEEE Trans. Microwave Theory and Tech.*, vol. 37, no. 9, pp. 1425–1433, 1989.

[9] K. Moez, M. Elmasry, "A 10dB 44GHz Loss-Compensated CMOS Distributed Amplifier," in *ISSCC Dig. Tech. Papers*, pp. 548–621, Feb. 2007.

[10] J. Kim, et al., "A 12dBm 320GHz GBW Distributed Amplifier in a 0.12µm SOI CMOS," in *ISSCC Dig. Tech. Papers*, pp. 478–479, Feb. 2004.

[11] J. Kim, et al., "Design and manufacturability aspect of SOI CMOS RFICs", in *IEEE Custom Integrated Circuit Conf.*, pp. 541–548, 2004.

[12] F. Ellinger, "60-GHz SOI CMOS Travelling-Wave Amplifier With NF Below 3.8 dB From 0.1 to 40 GHz," *IEEE J. Solid-State Circuits*, Feb. 2005.

978-2-8748-7007-1/08 $25.00 © 2008 EuMA

20 GHz Power Amplifier Design in 130 nm CMOS

Mattias Ferndahl [#1], Ted Johansson [*], and Herbert Zirath [#]

[#] *Chalmers University of Technology, Department of Micro Technology and Nano Science, Microwave Electr. Lab.,*
SE-412 96 Göteborg, Sweden
[1] mattias.ferndahl@chalmers.se

[*] *Infineon Technologies Nordic AB SE-164 81 Kista, Sweden*
Present address: Huawei Technologies Sweden AB, P.O. Box 94, SE-164 94 KISTA, Sweden

Abstract— **Five different 20 GHz power amplifiers in 130 nm CMOS technology have been designed and characterized. The power amplifiers explore single versus cascode configuration, smaller versus larger transistor sizes, as well as the combination of two amplifiers using power splitters/combiners. A maximum output power of 63 mW at 20 GHz was achieved. Transistor level characterization using load pull measurements on 1 mm gate width transistors yielded 148 mW output power. These numbers are, to the authors' knowledge, the highest reported for CMOS above 10 GHz. Transistor modeling and layout for power amplifiers are also discussed.**

I. INTRODUCTION

Due to the down-scaling and resulting high transit frequencies, CMOS is becoming a realistic alternative to III-V technologies for microwave applications, also above today's commercial applications in the lower GHz-region. Several successful design examples well up into the V-band have been published [1]–[4]. While the short channel lengths have enabled excellent small-signal high-frequency performance, it has also led to lower breakdown voltages and thus a smaller allowed RF voltage swing. As a result the maximum output power is reduced and the design of CMOS Power Amplifiers (PA) becomes more challenging. Surprisingly little has however been reported regarding CMOS PAs at frequencies above 10 GHz. The results by Vasylyev et al [5] represents highest power, with 17.8 dBm output at 17 GHz, while Yao et al [2] with 6.4 dBm at 60 GHz represents highest power combined with high frequency. Investigations on device level have also been carried out by Scholvin et al [6] and Ferndahl et al [7], [8].

This paper reports a study of K-band power amplification using a 130 nm CMOS technology. Results from load pull and S-parameter measurements on several devices are presented. These show, for CMOS at K-band, very high output power densities.

The transistor measurements were also used to add RF-extensions to the intrinsic transistor models, provided by the foundry. These extended models were utilized in the design of five different PAs. The PA designs investigate both device size and single transistor versus cascode PAs. It is shown that both the output power and the gain is improved for the cascode PAs.

The paper is organized as follows. Section II briefly explains the modeling of the transistor and passive elements. In Section III power transistor measurements and their layout are presented and discussed. The circuit design is covered in Section IV and results in Section V. Comparisons and conclusions are made in Section VI and Section VII, respectively.

II. MODELING

The available transistor model for the used process is based on the BSIM4 model, with some extensions of library cells to handle the layout parasitics up to the polysilicon/metal 1 layers. To include the effects of the metal interconnects up to the top metal layer, series resistances and inductances were added to all three terminals (gate, drain, source), see Fig. 1. The parameter values were found through a multi-bias optimization using measured extrinsic data and simulated intrinsic data. The measured data was de-embedded using a statistical equivalent-circuit-based de-embedding procedure [9]. The pad capacitance was extracted from measurements of test structures. Additional transmission lines models were developed using process information such as substrate thickness and conductivity. These passive models were also validated using 2.5D EM simulations.

III. TRANSISTOR LEVEL MEASUREMENTS AND LAYOUT

To evaluate 130 nm CMOS as a technology for power amplification at K-band, load pull measurements were performed on 1 mm gate width, minimum gate length devices. A set of different transistor layouts, having 1 mm total gate width, with varying unit gate finger width from 5μm and up, were

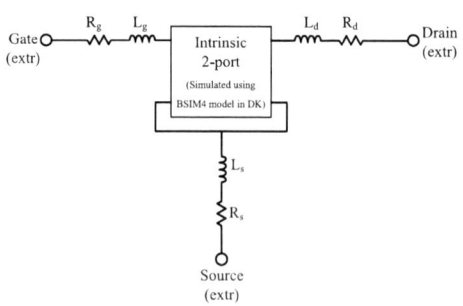

Fig. 1. Intrinsic BSIM3/4 model with RF extension.

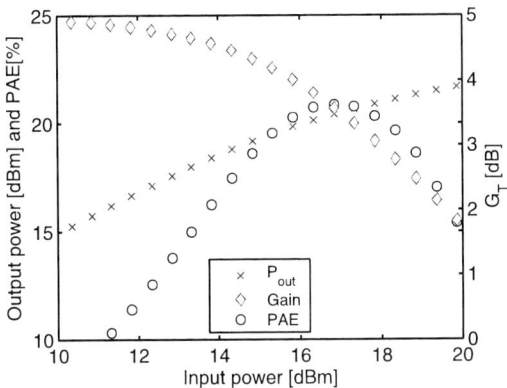

Fig. 2. Transducer gain, output power and PAE from load pull characterization of a 1 mm gate width 130 nm CMOS at 23 GHz. Source and load impedance were optimized for maximum output power ($Z_S = 9 - j0.13$, $Z_L = 9.2 - j0.16$).

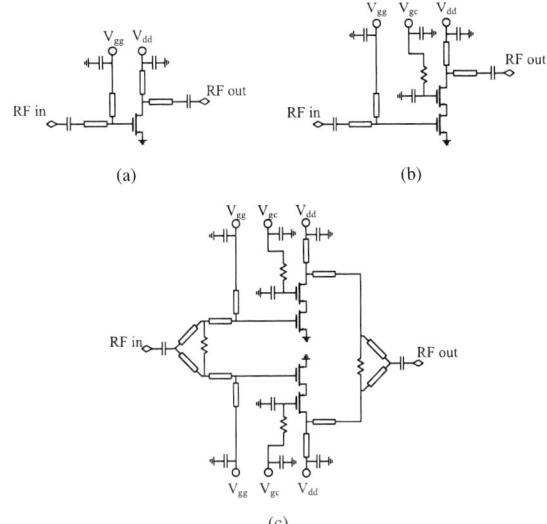

Fig. 3. Schematic of (a) the single transistor PAs, (b) the cascode PAs, and (c) two combined cascode PAs.

characterized. The layout of large transistor cells for power amplification at high frequencies, about a fifth of f_T, is problematic from two aspects. First, the finger gate width should be kept short enough to have sufficient gain, since the high gate resistance from wide gate fingers will deteriorate the high frequency performance. The second issue is how to layout a multi-finger-transistor to avoid problems with electromigration and reliability due to the large drain/source currents, this while keeping the parasitic capacitances and resistances as low as possible. A unit gate finger width of 10 μm gave satisfactory performance in terms of both gain and metal current density related reliability. The resulting maximum output power, gain, and power added efficiency (PAE) for the 100 fingers x 10 μm device are shown in Fig. 2. The maximum achieved output power density at 3 dB compression was 148 mW/mm. The gain of the 1 mm device was however too low (5 dB) and smaller transistor sizes (200 μm and 500 μm) were chosen for the PA designs.

Also I/O devices with higher allowed V_{dd} were studied, but the gain of these devices at microwave frequencies is too low due to the thick gate oxide and longer gate length. This makes minimum gate length devices the only viable option.

To further increase the output power, cascode amplifiers were also designed. This allows for increased supply voltage and thus higher output power as $P_{out} \propto V_{dd}^2$. This option and the two transistor sizes are evaluated in the next section using five different PA designs.

IV. CIRCUIT DESIGN

Five different PAs were designed and characterized: single transistor PA and cascode PA with both 200 μm and 500 μm transistors, and also a PA with two 200 μm cascode PAs which were combined on-chip using Wilkinson combiners. Fig. 3 gives an overview of the schematics. All PAs were single stage. Additional gain can be added by using one or several driver stages in front of the PAs, but will not effect the

maximum output power, which was the main interest in this study. Microstrip transmission lines were used as matching elements with the signal line in the top metal layer and ground plane in the bottom metal layer. All transmission lines were done with 50 Ω characteristic impedance, except for the shorted stubs to the drain, which were made 20 μm wide, corresponding to an impedance of 27 Ω. The wider drain line helps to accommodate the large drain current, so that problems with electromigration and/or Joule heating are avoided.

The power amplifiers were fabricated in a 130 nm RF CMOS process with four Cu layers, two thick top Al metal layers and MIM capacitors. The designs were made using Agilent's ADS simulator and Cadence layout environment. Chip photos of the power amplifiers can be found in Fig. 4.

V. RESULTS

The PAs were measured on wafer with 1.5 V drain bias for the single transistor PAs and 3.0 V drain and cascode bias for the cascode PAs. From simulation it could be seen that the voltage swing is almost equally partitioned between the input common-source transistor and the common-gate transistor of the cascode. The peak-to-peak voltages between transistor terminals, for example the drain and source of the input transistor, were all well below 3V, i.e.less than twice the maximum allowed supply voltage of 1.5 V. The gain versus output power at 20 GHz for all five power amplifiers are presented in Fig. 5 and a summary, with comparison to previously published work, in Table I and Fig. 8.

A comparison with simulated and measured S-parameter data of the the combined 200 μm cascode amplifiers shows rather good agreement, see Fig. 6, although a slight frequency shift towards lower frequencies is observed. The transducer gain and PAE of both the combined 200 μm cascode PA

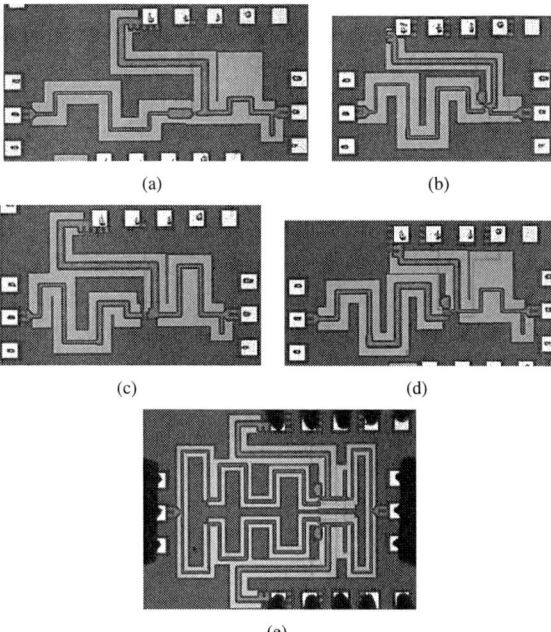

(a)   (b)

(c)   (d)

(e)

Fig. 4.   Chip photo of the five different power amplifiers.

and the 200 μm cascode PA show very good agreement between simulation and measurement, see Fig. 7. That both the small signal and the large signal measurements are in agreement with simulation shows that the models used are valid both for the passive matching networks, including the Wilkinson combiner, and the transistors. Furthermore, in Fig. 7 the simulated transducer gain without the RF-extension of the transistor model is shown. This clearly shows that these parasitics must be included, to have accurate simulations. No re-modeling or fitting to measured data was used for this comparison.

VI. DISCUSSION

As seen from Table I and Fig. 5 the gain and output power are rather poor for both the 200 μm and the 500 μm single transistor PA. The output power and gain are however increased with three dB, or more, when cascode PAs were used. This since a larger voltage supply can be used as the voltage is partitioned between the input and the output transistor of the cascode cell. Also the output conductance decreases and thus the available gain.

By combining two 200 μm cascode PAs the output power is increased further but with a small penalty in gain as the power splitter and combiner introduces losses. The PA with best performance, in terms of output power, was the PA with two on-chip combined 200 μm cascode PAs, and, in terms of gain, the 200 μm cascode PA.

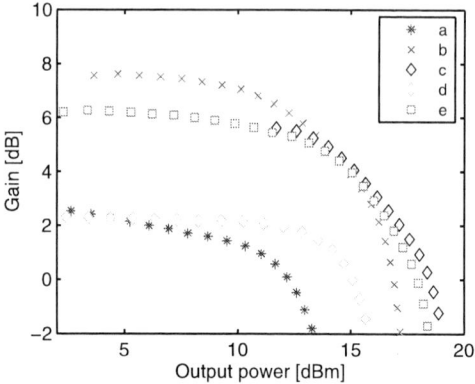

Fig. 5.   Transducer gain versus output power for the single transistor 200 μm PA (a), the 200 μm cascode PA (b), the two combined 200 μm cascode PAs (c), the single transistor 500 μm PA (d), and the 500 μm cascode PA (e). All at 20 GHz.

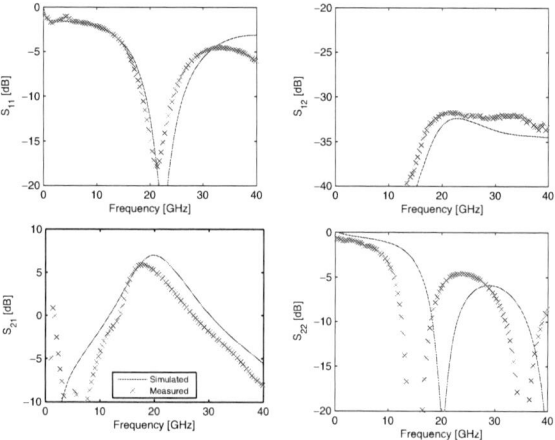

Fig. 6.   Measured and simulated S-parameters of the combined 200 μm cascode PA.

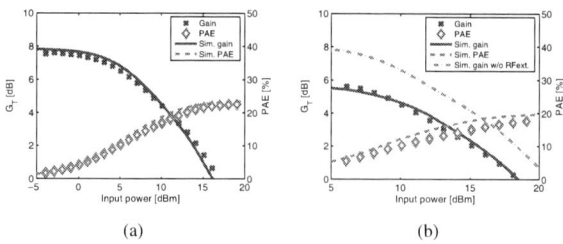

(a)   (b)

Fig. 7.   Measured and simulated transducer gain and PAE of the (a) 200 μm cascode PA and (b) the combined 200 μm cascode PA.

TABLE I

SUMMARY OF RESULTS AND COMPARISON OF 10 TO 40 GHz CMOS POWER AMPLIFIERS.

Power amplifier	$Freq.$ [GHz]	G_T/stage [dB]	P_{1dB} [dBm]	P_{sat} [dBm]	PAE [%]	V_{supply} [V]	Size mm^2	Tech.	Stages
This work									
(a) Single 200 μm	20	2.8	7.8	12	36	1.5	0.91	130 nm	1
(b) Cascode 200 μm	20	7.5	11.6	16.9	22	3.0	0.7	130 nm	1
(c) Two cascode 200 μm comb.	20	5.6	14.6	18.3	17	3.0	1.13	130 nm	1
(d) Single500 μm	20	2.5	13.6	15	22	1.5	0.72	130 nm	1
(e) Cascode 500 μm	20	6.2	13	18	16	3.0	0.72	130 nm	1
Previously published work									
[5]	17	7.2	15	17.1	15.6	1.5	0.9	130 nm	2
[12]	18	6.5	8	10.9	23.5	1.5	0.78	130 nm	4
[10]	24	3.5	11	14.5	6.5	2.8	1.3	180 nm	2
[13]	36-51	3.6	7.5	10.6	10	1.5	0.72	90 nm	4

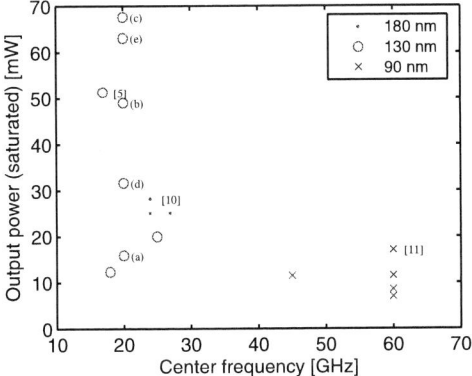

Fig. 8. Output power versus frequency for published power amplifiers in CMOS between 10 an 70 GHz. This paper: (a)-(e).

VII. CONCLUSIONS

A set of different power amplifiers in 130 nm CMOS technology operating at 20 GHz have been presented. Both transistor size and configuration as well as the combining of two power amplifiers using Wilkinson splitter/combiners were studied. A maximum output power of 63 mW at 20 GHz for two on-wafer combined cascode PA and more than 100 mW from load pull characterization of a single transistor were achieved. Both these values are, to the authors' knowledge, the highest reported above 10 GHz for CMOS. Furthermore transistor modeling and the layout of transistors for power amplifiers was discussed.

ACKNOWLEDGMENT

This work was supported by VINNOVA, Sweden through the European MEDEA+ project HiMission.

REFERENCES

[1] M. Ferndahl, B. M. Motlagh, A. Masud, I. Angelov, H. O. Vickes, and H. Zirath, "CMOS Devices and Circuits for Microwave and Millimetre Wave Applications," in *Proc. 35th European Microwave Conf.*, 2005, p. 105.

[2] T. Yao, M. Q. Gordon, K. K. W. Tang, K. H. K. Yau, M. T. Yang, P. Schvan, and S. P. Voinigescu, "Algorithmic Design of CMOS LNAs and PAs for 60-GHz Radio," *IEEE J. Solid-State Circuits*, vol. 42, no. 5, pp. 1044–1057, 2007.

[3] B. Razavi, "CMOS transceivers at 60 GHz and beyond," in *IEEE Int. Symp. on Circuits and Systems, Proc.*, New Orleans, LA, United States, 2007, Proc. - IEEE Int. Symp. on Circuits and Systems, pp. 1983–1986.

[4] B. Heydari, M. Bohsali, E. Adabi, and A. M. Niknejad, "Millimeter-wave devices and circuit blocks up to 104 GHz in 90 nm CMOS," *IEEE J. Solid-State Circuits*, vol. 42, no. 12, pp. 2893–2903, 2007.

[5] A. V. Vasylyev, P. Weger, W. Bakalski, and W. Simbuerger, "17-GHz 50-60 mW power amplifiers in 0.13μm standard CMOS," *IEEE Microwave and Wireless Components Letters*, vol. 16, no. 1, pp. 37–39, 2006.

[6] J. Scholvin, D.R. Greenberg, and J. A. del Alamo, "Fundamental Power and Frequency Limits of Deeply-Scaled CMOS for RF Power Applications," in *Int. Electron Devices Meeting Tech. Dig.*, 2006, pp. 1–4.

[7] M. Ferndahl, H.-O. Vickes, H. Zirath, I. Angelov, F. Ingvarson, and A. Litwin, "90-nm CMOS for microwave power applications," *IEEE Microwave and Wireless Components Letters*, vol. 13, no. 12, pp. 523–525, 2003.

[8] M. Ferndahl, H, Nemati, B. Parvais, H. Zirath, and S. Decoutere "Deep submicron CMOS for millimeter wave power applications," *IEEE Microwave and Wireless Components Letters*, vol. 18, no. 5, May, 2008

[9] M. Ferndahl, C. Fager, K. Andersson, P. Linnér, H. O. Vickes, and H. Zirath, "A general statistical equivalent-circuit-based de-embedding procedure for high-frequency measurements," *IEEE Trans. Microwave Theory and Tech.*, vol. SUBMITTED, 2008.

[10] A. Komijani, A. Natarajan, and A. Hajimiri, "A 24-GHz, +14.5-dBm fully integrated power amplifier in 0.18-μm CMOS," *IEEE J. Solid-State Circuits*, vol. 40, no. 9, pp. 1901, 2005.

[11] D. Chowdhury, P. Reynaert, and A. Niknejad, "A 60GHz 1V +12.3dBm Transformer-Coupled Wideband PA in 90nm CMOS," in *IEEE Int. Solid-State Circuits Conf. Dig. Tech. Papers*, 2008.

[12] Cao Changhua, Xu Haifeng, Su Yu, and K. K. O, "An 18-GHz, 10.9-dBm fully-integrated power amplifier with 23.5PAE in 130-nm CMOS," in *Proc. 31st European Solid-State Circuit Conf.*, 2005, pp. 137–140.

[13] J.-H. Tsai, Y.-L. Lee, T.-W. Huang, C.-M. Yu, and J. G. J. Chern, "A 90-nm CMOS broadband and miniature Q-band balanced medium power amplifier," in *IEEE MTT-S Int. Microwave Symp. Dig.*, 2007, pp. 1129–1132.

Advanced Components for Applications in S-Band and X-Band Radars

Timothy Boles

Tyco Electronics – Wireless Systems Segment
100 Chelmsford St., Lowell, Massachusetts, 01851, USA
bolest@tycoelectronics.com

Abstract— **Performance issues in S-Band and X-Band radar applications, have been investigated in parallel paths. The first approach continued with the basic GaAs based MESFET and pHEMT devices with the addition of field plate structures to enhance the transistor source-to-drain breakdown, enabling operation at higher voltages and producing significant improvements in device operation. The second direction questioned the basic material properties underlying the device structures. This methodology has led to the investigation of a number of pHEMT and HEMT designs based on SiC, GaN on SiC, and GaN on silicon devices for both S-Band and X-Band radar applications.**

I. INTRODUCTION

The performance requirements of S-Band and X-Band radar systems have fueled various developments in both compound semiconductor device design and the fundamental underlying material technologies since the early 1990's. These developments were driven by the fact that the capabilities, even with projected performance improvements, of the existing workhorse devices, silicon bipolar transistors for high pulsed power applications and GaAs MESFET's for higher frequency needs, had reached the point of diminishing returns.

In terms of S-Band power for radar applications, high voltage silicon bipolar devices have and continue to deliver >50 watts of peak transmitted power into the commercial marketplace since the late 1970's. The issues with this technical approach are centered on the inability to deliver these pulsed power levels with more than 6 dB to 7 dB gain. In addition, these high power silicon BJT's, due to the relatively large parasitic output capacitance, cannot instantaneously cover the entire bandwidth at S-Band radar frequencies. Lastly, these devices are hybrid matched, discrete components with no chance of being integrated into a true multi-stage MMIC.

For operation at X-Band, high voltage, silicon based technologies, either bipolar or MOSFET design geometries, have failed to demonstrate acceptable frequency response above 4.0 GHz, much less than at significantly higher frequencies. The standard solution for X-Band radars has classically been GaAs based, utilizing either MESFET or pHEMT device structures. While achieving the frequency response with these III-V compound semiconductors was readily realizable, the difficulty with this approach has been the relatively low voltage operation that severely restricted the

device peak power handling.

To address the performance issues in these radar applications, two paths have been investigated in parallel. The first approach continued with the basic GaAs based MESFET and pHEMT devices with the addition of field plate structures to enhance the transistor source-to-drain breakdown, enabling operation at higher voltages and producing significant improvements in peak power, power gain, and operating bandwidth. The second direction questioned the basic material properties underlying the device structures. This methodology resulted in the elimination of homo-junction silicon as a transistor material base as a contender for improved S-Band performance; and has led to a number of pHEMT and HEMT designs based on SiC, GaN on SiC, and GaN on silicon devices for both S-Band and X-Band radar applications.

This paper will discuss improvements that have been investigated both from a field plated MESFET approach and material based solutions utilizing GaN on silicon.

II. DISCUSSION

In order to examine the effect of modifications of device structure on the high frequency performance of existing GaAs material technology, a high voltage, implanted MESFET transistor was developed. This device design is an extension of the M/A-COM MSAG structure, an acronym for <u>M</u>ulti-function <u>S</u>elf-<u>A</u>ligned <u>G</u>ate, which is a 10 volt power FET technology, based on a refractory self-aligned metal gate, ion-implanted, planar GaAs MESFET process. An additional benefit of the basic MSAG process approach is the ability to include, in the same MMIC circuit, FETs individually optimized by means of selective process adjustments for power, low noise, switching, and digital functionality.

To extend this basic FET structure to higher operating voltages and increased output power, the active layer ion-implantation parameters for the power MSAG process have been further optimized and a gate connected field plate has been added. The basic design of a high voltage power FET, as shown in Fig. 1, is a straightforward extension of the gate overlay metallization which is already part of the standard MSAG device. This "gamma" gate field plate is used to reduce gate resistance and enhance microwave performance. These design improvements to the standard MSAG structure, designated HVMSAG, has produced BV_{gdo} breakdown voltages greater than 50 volts enabling operation at supply voltages up to 28 volts. The prototype HVMFET design was followed by a device optimization effort which focused on best performance, i.e. power and efficiency, at S-band, and for MMIC applications.

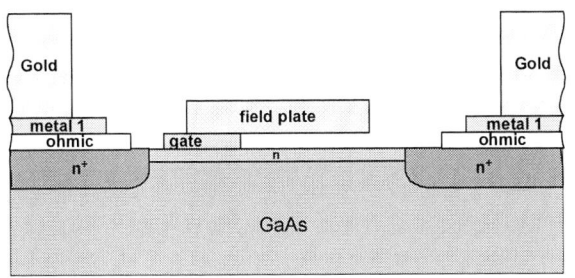

Fig. 1 Crossection of HVMFET Showing Gate Connected Field Plate (GCFP). Region under the Field Plate is Filled with Passivation Dielectric.

This effort consisted of empirically evaluating device-design factors such as gate length, field-plate dimension, source-drain spacing, ion implantation energies, implant doses, etc. A mask with a full-factorial layout of FETs, covering all critical dimension variants, was used in the study. Process factors such as implant energy and dose were varied. A total of over 400 device variants were produced, tested dc, and small-signal RF. Based on these results, the most promising candidate FETs were pursued and evaluated with load-pull measurements. From this study, an HVFET was chosen that best satisfied all the criteria for power performance in an S-band MMIC at 24V to 28V drain bias and subjected to full RF characterization.

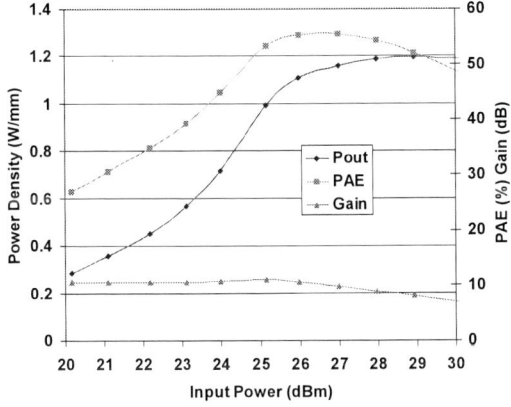

Fig. 2 Large Unit Cell FET S-Band Performance @ 24 Volt Bias

In Fig. 2, the results of a full RF characterization of a large periphery unit cell FET are presented. It can be seen that power densities of 1.2W/mm at a drain bias of 24V, with approximately 10dB of associated gain, and at a power added efficiency of 55% were able to be achieved at S-Band. When applied to a full-up MMIC, this ability to sustain operation at higher dc bias voltages, has resulted in peak RF power levels of approximately 60 watts with over 44% power added efficiency, as can be seen in Fig. 3. This high frequency

performance is able to be delivered at S-Band radar frequencies.

Further improvements in the performance of the HVMSAG structure were attempted through the addition of a source connected field plate (SCFP). A crossection of this structure is presented in Fig. 4. Performance enhancements were anticipated due a significant reduction in the parasitic terminal capacitances and a further reduction in peak fields in the drain region adjacent to the gate associated with the basic FET structure.

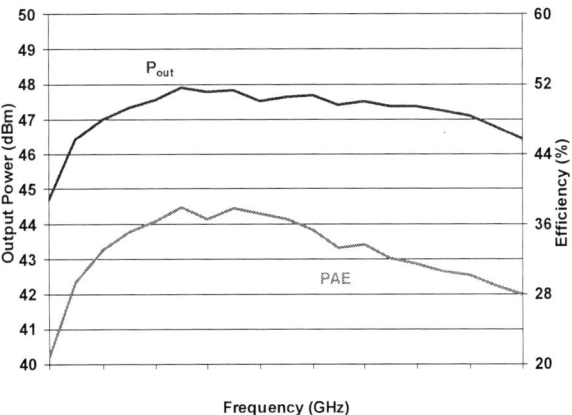

Fig. 3 Output Power and Power Added Efficiency at S-Band for HVMSAG Two Stage MMIC

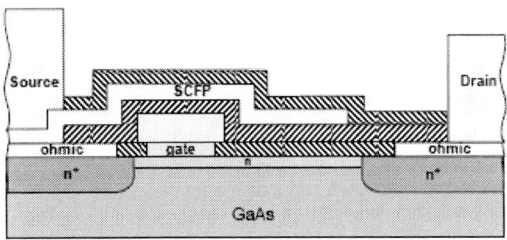

Fig. 4 Crossection of HVMSAG FET showing Source Connected Field Plate (SCFP) with passivation dielectric shown.

Again, this effort was investigated by using a similar, rigorous empirical technique as was employed on the GCFP HVMAG power FETs. In this case, individual mask layers were added to the standard GCFP design to generate an array of 256 SCFP FETs representing the full factorial combination of several values for the primary critical dimensions associated with the field plates: the GCFP width; the SCFP overlap spacing; the separation between the drain implants and SCFP, etc. As before with the GCFP, these device variants were measured for dc, small signal s-parameters, and pulsed IV characteristics tested to provide criteria for the most likely candidates for more complete, large signal RF evaluation.

As a further control point to examine the effects of field plates unambiguously, the SCFP FET most similar in channel design to the HVMSAG GCFP FET was selected for comparison. These devices share the same channel implant structure, gate length, and field plate extension past the gate center line. As can be seen in Fig. 4, the SCFP FET in this comparison has a conventional T-gate, a structure with minimal overhang past the gate Schottky contact with equal dimension on both source and drain sides, without what would be considered a GCFP. A third device, identical to the SCFP FET, but omitting the field plate, essentially a standard "T" gate FET is included in the comparison to clearly identify the effects of each field plate structure.

A fourth design was created combining a GCFP and SCFP into one FET was also included in this large signal RF evaluation. This approach was aimed at trying to obtain the best high frequency performance characteristics of the GCFP with the SCFP without a major compromise in adding parasitic reactance.

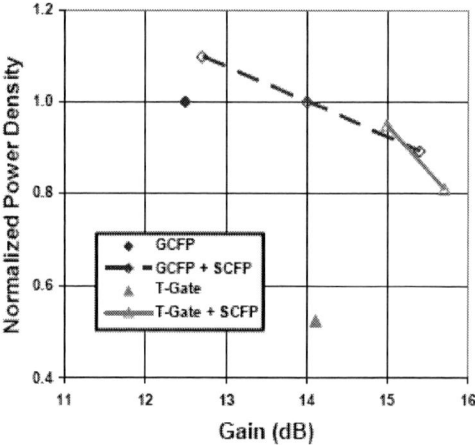

Fig. 5 Saturated Output Power and Gain Comparison for Four Different MSAG FET Configurations

The results of large signal RF testing at S-Band frequencies obtained from each of these varied FET structures are summarized in Fig. 5. The standard HVMSAG (GCFP) which is the benchmark for all the design variants and was used to normalize the FETs relative to power density exhibited the lowest gain of the devices. As expected, due to it's significantly lower V_{dgo} breakdown voltage, the standard "T" gate configured FET can be seen to deliver the lowest power density but with an approximately 2 dB increase in RF gain when compared to the standard GCFP. The "T" gate in combination with the SCFP approach produced the highest large signal gain relative the GCFP standard, as much as a 4 dB improvement, but at a penalty of an approximate 10% reduction in power density. The last variant, the combined GCFP and the SCFP approach, produced a range of performance that from an overall perspective matched the power density of the standard HVMSAG FET while achieving RF gain at S-Band that overlapped that seen with the SCFP. This work has demonstrated that optimum large and small signal RF performance at S-Band frequencies can only be achieved with proper design of both gate-connected and source-connected field plate structures.

In terms of X-Band radar applications, MSAG switch FETs with a typical R_{on} of 1.6 Ω•mm have been developed to address the need for lower passive circuit insertion loss. In addition, a MSAG FET with high-drive and high gain capabilities was developed to address requirements for higher TOI. Both of these MSAG FETs are fully compatible with other FETs in the MSAG suite making it possible to design and fabricate higher-performance, low-cost multi-function control circuits up through X-Band frequencies.

While significant improvements have been observed with the HVMSAG structure in RF performance in S-Band radar applications when compared to the prior art silicon bipolar transistors, this GaAs high voltage MESFET technology is limited in frequency response primarily in S-Band. Even the predicted improvements by adding a source connected field plate to this design structure are expected to be seen in terms of output power, power added efficiency, stability, bandwidth, device f_T, and overall frequency response, practical operation beyond C-Band is not anticipated. To address the higher frequency X-Band radar needs, other material systems need to be considered.

In Table I below, a comparison between the basic physical and electrical properties of various semiconductor materials is presented. It can be seen that GaN is a wide bandgap (WBG) material ($E_g = 3.4$ eV) that has received much attention for high power RF applications. GaN's band gap and breakdown electric field strength allow high voltage operation; while the high saturation velocity indicates that high frequency operation at least through X-Band should be readily achievable.

First, from a basic material characteristic perspective, GaN has a high frequency dielectric constant of 5.3 versus almost 12 for either silicon or GaAs. This lower dielectric constant will translate into lower intrinsic inter-terminal capacitances which will provide a significant reduction in degenerate RF feedback for GaN based transistors as compared to similar silicon or GaAs based structures.

Second, GaN, as a material, has an inherent breakdown field strength that is roughly 10 times that of silicon or GaAs. Thus, for equivalent silicon or GaAs devices a GaN HEMT transistor requires only one tenth of the spacing in the drain region to support the field required to achieve the same BV_{dgo} breakdown. Another way of looking at this is that for similar device dimensions, a GaN device will have a drain-to-source breakdown 10 times that of a silicon or GaAs structure. This higher field strength can be used to translate the required GaN transistor into a physically smaller active area which again will result in a significant reduction in terminal capacitive feedback increasing the intrinsic device bandwidth. Alternately, a portion of the 10x improvement in field strength can be used to increase the BV_{gdo} breakdown allowing the

TABLE I
COMPARISON OF PROPERTIES OF SEMICONDUCTOR MATERIALS

PROPERTY	SI	GAAS	SIC	GAN
Band Gap (eV)	1.1	1.4	3.2	3.4
Breakdown Electric Field (V/cm)	$3x10^5$	$4-5x10^5$	$2-4x10^6$	$5x10^6$
Electron Mobility (cm^2/Vs)	1100	6000	370	1350
Saturation Velocity (cm/s)	$1.0x10^7$	$2.0x10^7$	$2.0x10^7$	$2.7x10^7$
Dielectric Constant	11.8	12.9 (static) 10.9 (high freq)	9.7 (static) 6.7 (high freq)	9.5 (static) 5.3 (high freq)
Thermal Conductivity (W/cm C)	1.5	0.5	4.9	1.3
Expansion Coefficient	$3.6x10^{-6}$	$6.0x10^{-6}$	$2.4x10^{-6}$	$3.2x10^{-6}$
FET Technology	LDMOS	HFET	MESFET	HFET
Power Density (W/mm)	0.8	1-1.5	2-4	>>2

transistor structures to operate at higher supply voltages, improving the device linearity and performance.

From a design perspective, power density improvements of 2x to 20x when compared to silicon or GaAs based transistors been reported for GaN HEMT structures. This will physically translate into a much smaller device for the GaN small signal/medium power applications. Again, since electrode capacitance is an area dependent function, this increase in power density will result in significantly reduced junction capacitances. This capacitance reduction will increase the fundamental device f_T and bandwidth. Also, as in the case of the improvement of field strength for GaN, this higher a portion of this power density improvement can be used to provide additional drive back off in a Class A or Class AB operation in order to increase overall linearity while still increasing the intrinsic device bandwidth.

Lastly, it has been reported that GaN HEMT devices have demonstrated both improved noise figure and 1/f noise performance when compared to a GaAs pHEMT. The ability to handle high power densities and the observed relatively high thermal conductivity compared to GaAs logically calls attention to the possible construction of an X-Band low noise amplifier receiver that does not require a limiter protection function for survivability.

The device structure that has received the most attention has been a HEMT structure based on the AlGaN/GaN system grown on various substrates, i.e. SiC, sapphire, and silicon. Homoepitaxial growth is not possible as GaN substrates are not available. Silicon carbide has received the main focus due to its close lattice match to GaN and its excellent thermal properties making it a good substrate candidate for power devices. The main disadvantages associated with SiC stem from mechanical issues such as wafer size and crystal defect density. The alternate approaches have been to grow the heterostructure on Sapphire or Silicon. Sapphire has a very poor thermal conductivity making it a bad candidate for power devices while silicon has a very large lattice mis-match to GaN making it generally difficult to grow high quality films. It can be seen from Table I, that the material system of GaN on silicon substrates offers the best compromise between in thermal conductivity for high power performance and mechanical substrate properties for high quality GaN films and low cost manufacturability. This GaN on silicon material

system is the approach being pursued at Tyco Electronics for front-end X-Band radar LNA applications.

III. CONCLUSION

The needs of S-Band radar applications in a true matched MMIC are being actively addressed by new device design geometries based on mature GaAs MESFET HVMSAG processes with excellent success. For the higher frequency X-Band radar system requirements, it is necessary to pursue new semiconductor materials and technology in order to realize both the needed power handling, bandwidth, frequency response, and device ruggedness.

ACKNOWLEDGEMENT

The author would like to thank the Engineering and Technology staffs in both the Wireless Systems Segment Roanoke and Walker facilities for their inputs and insights into the technology capabilities for the HVMSAG and GaN on Si FET structures.

REFERENCES

[1] M. J. Drinkwine, T. Winslow, D. Miller, D. Conway, and B. Raymond, "An Ion-Implanted GaAs MESFET Process for 28V S-Band MMIC Applications", 2006 CS MANTECH Conference Technical Digest, April, 2006, pp. 187-190.

[2] M.L. Balzan, M.J. Drinkwine, and T.A. Winslow, "GaAs MESFET with Source-Connected Field Plate for High Voltage MMICs", 2008 CS MANTECH Conference, April, 2008.

[3] M. J. Drinkwine, H. Singh, and M. Ashman, "Low-Cost, High-Performance Multifunction X-band Control MMICs Using Ion-Implanted FET Technology", 2007 CS MANTECH Conference Technical Digest, 2007.

Low-cost S-band Multi-function MMIC

A.P. de Hek, M. Rodenburg and F.E. van Vliet

TNO Defence, Security and Safety

Oude Waalsdorperweg 63, 2509 JG The Hague, The Netherlands

peter.dehek@tno.nl

marien.rodenburg@tno.nl

frank.vanvliet@tno.nl

Abstract— **This paper discusses the design and performance of a four port S-band multi-function MMIC. The multi-function chip consist of a 6-bit phase shifter, 6-bit attenuator, transmit receive switches, a low noise amplifier, a medium power amplifier and integrated LVCMOS control logic. It is shown that excellent results have been obtained for the presented multi-function chip which has been realised in the low cost high volume PP50-11 0.5 μm PHEMT process of WIN Semiconductors.**

I. INTRODUCTION

Internationally there is a trend towards the development of phased array radar systems, which combine both search and tracking capabilities in one radar system. S-band is the preferred choice as frequency band for such systems. Phased array radar systems can consist of several thousands of modules. Therefore, it is essential to reduce the costs of these modules as much as possible to make such systems economically attractive. The component count and the price of the used components need to decrease. This can be achieved through integration of the necessary functionality for beam steering and the low-noise and driver amplifiers necessary for transmit and receive onto a single multi-function MMIC.

The presented S-band multi-function chip is to the author's knowledge the first low-cost solution which is also commercially available. In addition, as far as we know the results of only one prototype S-band multi-function chip have been published [1]. Our multi-function chip has a greatly improved frequency band, gain, noise figure, attenuation and phase shifter accuracy when compared to the results presented in [1].

In the next sections, the design, layout and measurement results of the S-band multi-function chip will be discussed. As first step, the design will be discussed in section II. In section III an overview of the layout is given. Finally, in section IV the obtained measurement results are compared to the simulation results.

II. DESIGN MULTI-FUNCTION MMIC

For the multi-function chip a four port topology was selected, see figure 1. This topology was selected to have the maximum flexibility for the use of the chip in various applications ranging from phased array radar to telecommunication and instrumentation applications.

The multi-function chip consists of an SPDT switch at its input and output. A 2-stage low-noise amplifier at the input

and a driver amplifier at the output of the chip are used. In between the amplifiers a 6-bits phase shifter and a 6-bit attenuator have been placed.

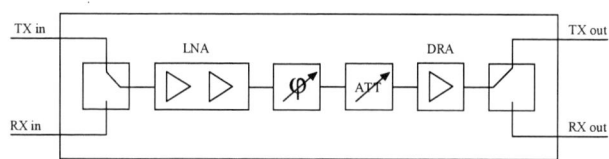

Fig. 1 Block diagram of the multi-function chip

An essential step in every chip design is the choice of the technology in which the chip will be realised. Factors that play a role in the technology selection are cost, yield, power performance and low noise performance. For the multi-function chip the PP50-11 process of WIN Semiconductors has been selected. This is a 0.5 μm PHEMT process, which is fabricated on 6" wafers. This process in our experience is very repeatable and has a high yield for relatively large MMICs (> 25 mm^2). Therefore this technology is very suited for the development of complex low-cost MMICs. A major advantage of the selected technology for the current application is the high gate-source breakdown voltage of more than 10 V. This high breakdown voltage allows for the selection of an off-voltage for the switches of -5 V which greatly enhances the compression level of the switch FETs used in the SPDT switches, phase shifter and attenuator. Since the selected technology is a power technology the noise figure is higher than can be achieved with other technologies. Nevertheless it is the authors believe that this minor disadvantage is more than compensated for by the high power performance of this technology, which improves the multi-chip performance for all other specifications. In addition, it should be realised that it is difficult to realise noise figures below 1 dB with multi-function chips as the one discussed. Therefore, for really low noise figures always a low-noise amplifier in front of the multi-function chip need to be used. In the remainder of this section the design of the individual building blocks will be discussed.

A. Single Pole Double Throw switches

The topology depicted in figure 2 is used for the SPDT switches. In each branch one series switch and two parallel switches are used. In the on-state the series switch is turned on and the parallel switches are turned off. In the parallel branch the transistors are switched the other way around to improve the isolation as much as possible.

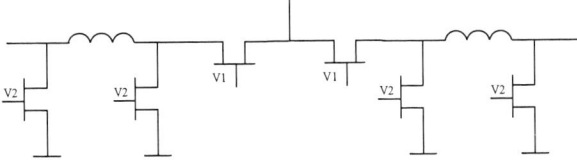

Fig. 2 Topology SPDT switch

The used -5 V in the off-state enables a high compression level of the switches. As a result no transistors need to be stacked to achieve the desired compression level of more than 25 dBm. In the on-state 0 V is used. This results in a low loss in the series switches. The depicted switch has a loss of better than 0.6 dB and an isolation of better than 50 dB. The inductors in between the parallel switches form an artificial 50 Ω transmission line. With the help of this line a matching of better than 25 dB has been achieved. The common input has a matching of better than 22 dB.

B. Amplifier design

For the multi-function chip an overall gain of more than 20 dB was specified. It turns out that three amplifier stages are needed to realise this gain.

To keep the noise figure as low as possible it was decided to design a two-stage noise amplifier (LNA), which is put directly after the SPDT switch. The targeted noise figure for the multi-function-chip is smaller than 2.5 dB. This number can be achieved when a two-stage LNA is placed inside the common leg. An alternative could have been placing the LNA in front of the SPDT switch. In this way the noise figure is further reduced with the loss of the switch. A disadvantage of this approach is that also for transmit additional gain either at the input or at the output of the chip need to be added. This results in an increased chip layout and the DC dissipation of the multi-function chip is unnecessarily increased with the dissipation of the added amplifier stage(s).

The LNA is classical in the sense that the first stage is realised with a transistor, which has an additional source inductance for optimal noise matching. In the second stage a feedback amplifier is used to make the output of the amplifier as good as possible 50 Ω. In this way a matching of better than 15 dB has been achieved. The LNA has a gain of 26 dB and a noise figure of 1.5 dB

As driver amplifier a one-stage feedback amplifier has been designed. Feedback has been used to flatten the gain as much as possible over the specified frequency band. A gain of 12 dB with a matching better than 15 dB has been realised. The one dB compression level of this amplifier is 16.5 dBm.

Both amplifiers are biased at a drain voltage of 5 V and a gate voltage of -0.95 V. The gate voltage has been realised with a gate bias circuit, which gives a 100% compensation for threshold variations [2].

The necessary DC dissipation at the one dB compression point is 475 mW for the LNA, 300 mW for the driver amplifier and 80 mW for the gate bias circuit.

C. 6-bit phase shifter design

The design of the 6-bit phase shifter is discussed in [3]. The smallest phase step is 5.625°. In between the phase shifter and the attenuator no buffer amplifier for the minimisation of the effect of the load and source variations between the different states has been placed. It turns out that the matching of both phase shifter and attenuator is so good that no significant impact is seen when they are connected directly together. A matching of better than 15 dB, an RMS phase error of less than 2° and an amplitude variation of less than 1 dB have been achieved.

D. 6-bit attenuator design

A 6 bit attenuator is designed for the multi-function chip. The smallest attenuation step is 0.45 dB and the largest is 14.4 dB. The desired attenuation is realized by switching between a through connection and an attenuator, see figure 3.

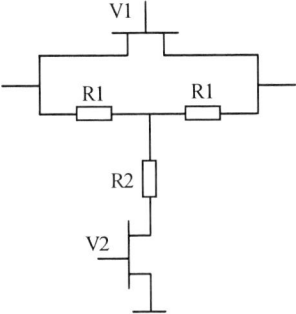

Fig. 3 Topology attenuator bits

For all attenuator bits a T section is used except for the 7.2 dB bit where a pi-section was used. To improve the matching for all bits except for the 0.45 dB and 0.9 dB bit resistor R2 is switched on and off with an additional transistor.

In simulation a matching of better than 15 dB and an RMS amplitude error of less than 0.2 dB and maximum phase error of better than 2.5° has been realised.

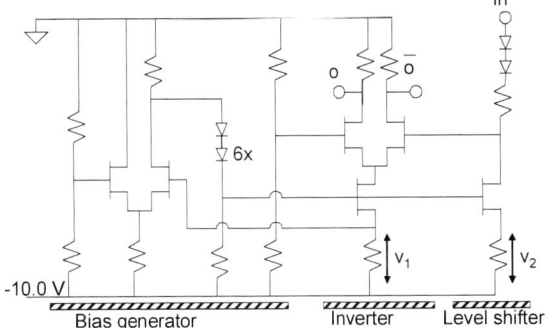

Fig. 4 Schematic of LVCMOS compatible level shifter inverter

E. LVCMOS control logic

The switch voltages are generated with the LVCMOS compatible level shifter inverter depicted in figure 4. The design of this control logic is discussed in detail in [4]. For the current design the off-voltage has been lowered from -3 to -5 V. For a proper working the supply voltage of this circuit had to be changed to -10 V.

III. LAYOUT

In this section the layout of the multi-function chip will be discussed. The used PP50-11 process has the following features 0.5 μm depletion mode power PHEMT transistors, TaN and GaAs resistors, MIM capacitors and viaholes through 100 μm thick substrates.

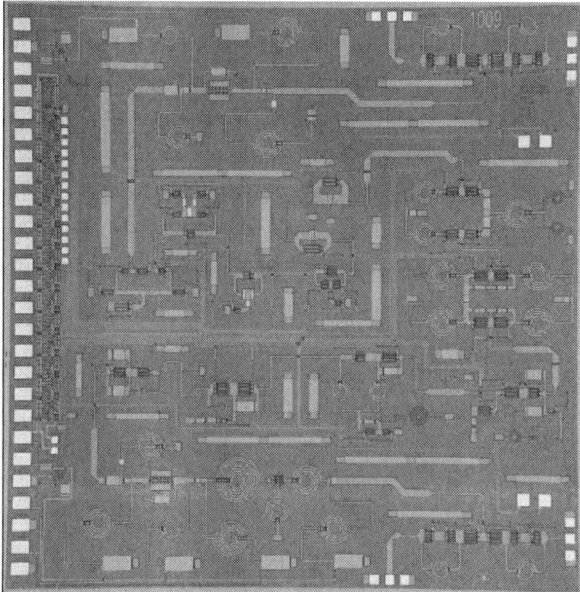

Fig. 5 Photograph of layout of the multi-function chip (5.5x5.5 mm^2)

A photograph of the realised multi-function chip has been depicted in figure 5. The control logic is visible along the left edge of the chip. The SPDT switches are located at the right bottom and upper corner. The LNA and the 6-bit phase shifter are visible at the bottom of the chip. The 6-bit attenuator is visible at the top of the chip. Between all vital parts of the multi-function chip screens consisting of metal lines with viaholes are placed. These screens help too improve the isolation between the various parts of the multi-function chip.

IV. MEASUREMENT RESULTS

The multi-function chip input matching and the small-signal gain are depicted in respectively figure 6 and 7. All measurement results have been obtained at an ambient temperature of 25 °C in the reference state of the attenuator and phase shifter. The depicted results show a good agreement between the measured and simulated performance. The RMS phase error is depicted in figure 8. Again an excellent

agreement between simulation and measurement results has been obtained.

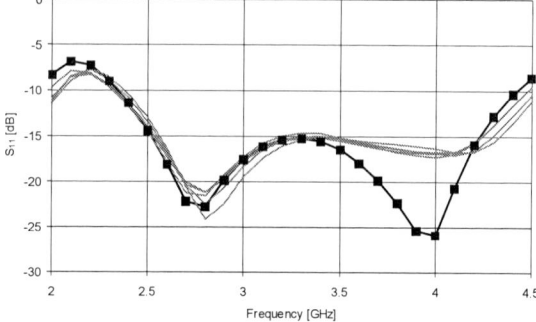

Fig. 6 Input matching multi-function chip in reference state (simulation black squares)

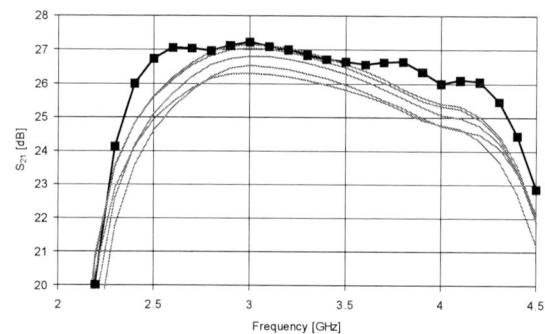

Fig. 7 Gain multi-function chip in reference state (simulation black squares)

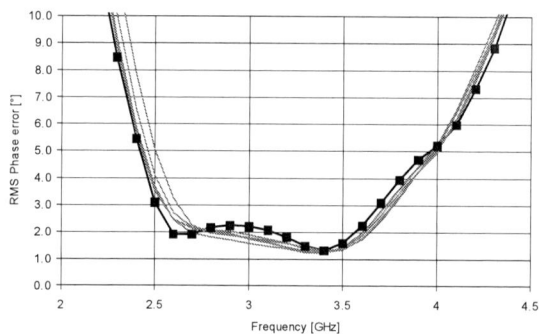

Fig. 8 RMS phase error multi-function chip (simulation black squares)

The RMS attenuation error is depicted in figure 9. The measurement results show that for the 3 – 4.5 GHz frequency band an RMS attenuation error of better than 0.2 dB has been obtained. Between 2.5 – 3 GHz the attenuation error is higher than simulated. Nevertheless an attenuation error smaller than 0.4 dB has been measured over mentioned frequency band. The measured noise figure is depicted in figure 10. The depicted results show that between 2.5 – 4 GHz a noise figure of approximately 3.5 dB has been measured. The measured noise figure is 1.5 dB higher than expected on the basis of simulations.

The measured output power at a fixed input power of -10 dBm is depicted in figure 11. Again an excellent agreement between measurement and simulation results can be observed. The used input power will result in a compression level of approximately 2 dB. At this compression level an output power between 14 and 15 dBm has been measured.

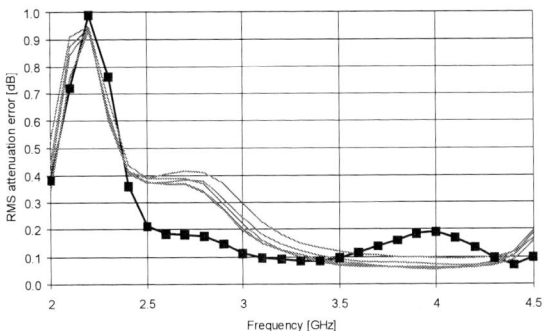

Fig. 9 RMS attenuation error multi-function chip (simulation black squares)

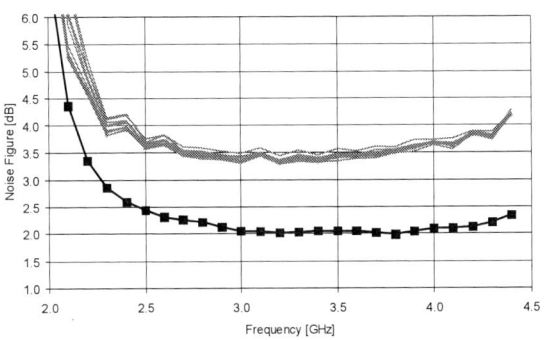

Fig. 10 Noise Figure multi-function-chip in reference state (simulation black squares)

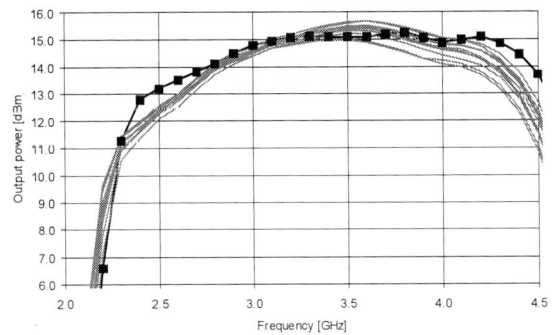

Fig. 11 Measured output power multi-function chip in reference state at an input power of -10 dBm (simulation black squares)

V. CONCLUSIONS

In this paper the design and measurement results of an S-band multi-function MMIC are discussed. An excellent agreement between measurement and simulations is demonstrated. The multi-function MMIC is realised in the low-cost high volume PP50-11 process of WIN Semiconductors. The obtained measurement results at 25 °C are summarized in table 1.

TABLE I
PERFORMANCE S-BAND MULTI-FUNCTION CHIP

	Measurement
Frequency [GHz]	2.8 - 3.5
Gain reference state [dB]	26.5 ± 0.5
Matching [dB]	< -13
Noise Figure [dB]	< 3.6
Phase control	360°/6-bit
RMS Phase error [°]	< 2.5
Attenuation control	28.35 dB/6-bit
RMS attenuation error [dB]	< 0.45
Pout-1dB [dBm]	> 13
Chip size [mm²]	5.5 x 5.5

REFERENCES

[1] N. Billström, H. Berg, K. Gabrielson, E. Hemmendorff, M. Herz, "T/R "core-chips" for S-, C- and X-band radar systems", European Microwave Conference Proceedings, October 2004.

[2] A.P. de Hek and E.B. Busking, "On-chip active gate bias circuit for MMIC amplifier applications with 100% threshold voltage variation compensation", Eumic conference proceedings, September 2006.

[3] A.P. de Hek, M. Rodenburg and F.E. van Vliet, "A Cost-effective High-Power S-band 6-bit Phase Shifter with Integrated LVCMOS Control Logic", Eumic conference proceedings, October 2007.

[4] M. van Wanum, G. van der Bent, M. Rodenburg, A.P. de Hek, "Generic robust LVCMOS compatible control logic for GaAs HEMT switches", Eumic conference proceedings, September 2006.

Proceedings of the 3rd European Microwave Integrated Circuits Conference

Silicon-Germanium for Phased Array Radars

Håkan Berg[€1], Heiko Thiesies[€2], Erik Hemmendorff[€3], Georgios Sidiropoulos[£4], Jonas Hedman[€5]

[€]*Microwave & Antennas, Saab Microwave Systems, Saab AB*
SE-412 89 Göteborg, Sweden
[1]`hakan.j.berg@saabgroup.com`
[2]`heiko.thiesies@saabgroup.com`
[3]`erik.hemmendorff@saabgroup.com`
[5]`jonas.hedman@saabgroup.com`

[£]*Ericsson AB*
SE-417 56 Göteborg, Sweden
[4]`georgios.sidiropoulos@ericsson.com`

Abstract— **Phase and amplitude controlling ICs realized in a low cost standard silicon process are demonstrated. The design of several ICs at S-, C- and X-band has shown that silicon germanium is a strong contender to gallium arsenide where lowest noise figure is not vital. This applies also to the T/R-modules suited for military AESA-radars. The circuits presented in this paper are manufactured by austriamicrosystems in their 0.35μm SiGe-BiCMOS process with an f_T of 70 GHz.**

I. INTRODUCTION

If one could use only one semiconductor technology for all functions in a radar T/R-module the obvious choice is of course gallium arsenide. It is a good compromise between maturity, output power, noise figure, integration and accessibility. On the other hand for the low noise amplifiers there exists strong contenders e.g. gallium nitride or indium phosphide and the same is true for high power amplifiers where both silicon carbide and exclusive III-V compounds can outclass the performance of gallium arsenide. Even for the core functionality, amplitude and phase control, there is a strong contender in silicon germanium. Although it does not outperform gallium arsenide when classic microwave parameters are compared there are benefits from choosing a classic silicon based technology. The obvious benefit is cost; silicon is more or less by definition a low cost technology, this is true already for quite low volumes. There is more to the cost issue however than simply the die prize. If a BiCMOS process is chosen integrated digital control functionality may replace an otherwise additional digital IC in the T/R-module. The overall module cost can thereby be reduced. There are however other benefits which are of equal importance; the maturity guarantees a high yield and other features such as dedicated ESD-protected periphery cells and verified digital gates.

What is often overlooked is that the high level of integration does not only apply to digital circuits. Of maybe equal importance is the possibility to design microwave circuits with a high density. This allows for a designer to design more complex microwave circuits without the penalty of parasitics that follows from a large circuit area.

Saab Microwave Systems have a large experience in designing MMICs for T/R-modules; both LNAs, so-called core-chips and power amplifiers. Lately however there has

been a shift of interest towards standard low cost technologies throughout the whole radar business. Saab has been no exception in this respect and has therefore investigated the possibility to use silicon ICs where possible. The results have been promising when it comes to performance. At the same time extensive tests has shown very good results when environmental durability is concerned.

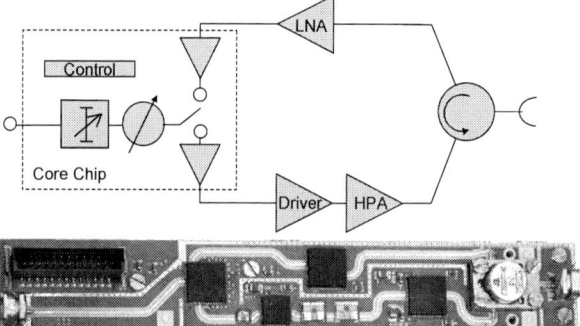

Fig. 1 Example of low cost T/R-module based on packaged ICs together with basic architecture.

If the architecture in Fig. 1, with LNA and power amplifiers outside the core-chip, is used for the T/R-module the noise figure and output power is less important for the core-chip itself. This makes it possible to use silicon germanium to realize the core functionality. This has been considered for phased array radar applications for some time [1]-[3].

II. DESIGN CONSIDERATIONS

Compared to when designing core-chips in gallium arsenide there are differences to do the same in silicon. Below are listed a few of these without any order of precedence:

- MOSFETs are poor switches at high frequencies. If a MOSFET-switch is designed to have the same on-resistance as a pHEMT-switch the off-capacitance is substantially larger. This results in higher loss if a dual pole switch is to be designed. At the same time this off-capacitance is very lossy compared to the corresponding one in gallium arsenide.

978-2-8748-7007-1/08 $25.00 © 2008 EuMA

- Silicon can be considered a lossy dielectric or even a resistive material depending on frequency of interest. Using silicon as substrate results in that microstrip can not be used. Differential circuits is therefore the obvious choice. Even CPW is an alternative although not used in designs presented in this paper. The losses in for example inductors however are still comparable to those of inductors supplied in gallium arsenide design kits.

- The design flow supported by silicon foundries always include LVS, layout versus schematic, DRC, design rule check, and parasitic extraction. This contributes to the possibility to design complex circuits with a high degree of confidence.

- When noise figure, linearity and parasitics are considered gallium arsenide transistors outclasses those in silicon and silicon germanium.

- The physical size of comparable devices is typically much smaller in silicon than in gallium arsenide. This allows for more complex designs which can enhance the performance of the total IC.

- The DC-behavior of silicon devices is usually well modeled. One can therefore allow for internal bias circuits which is necessary if for example cascode amplifiers are to be used. At the same time it is possible to integrate more complex analog circuits e.g. operational amplifiers and temperature sensors etc.

- The high yield compared to gallium arsenide ICs together with the access to CMOS circuits means that it is possible to use digital control circuits internally on chip.

- The small size of microwave devices and the access to digital circuits means that it is possible to design core-chips with a large extra number of possible phase and amplitude states. The states which are best suited for the application are then selected. This means that if an attenuator linear in decibel is to be designed on-chip one can instead design an attenuator with arbitrary attenuation steps as long as it can cover the required attenuation with sufficient resolution. This can reduce the complexity of the chip and circuit architecture.

- The maturity of silicon processes and its good environmental protection relaxes the requirements on the environmental protection of the next level in T/R-module hierarchy.

- Most silicon foundries support only time-domain simulators e.g. Spectre. This can be a problem since the next system level is usually designed using microwave design tools.

The conclusion is that it is not a good idea to move a design directly from a gallium arsenide process to a silicon germanium process. The biggest issue however is the poor performance of the switches. One must therefore spend much effort to first of all redesign the switched applications such as phase shifters and attenuators.

To begin, the switched attenuators are replaced and the amplitude control can instead be realized using variable gain amplifiers, VGAs. One commonly used approach is to use an active mixer of multiplier type where an analog control signal is applied instead of an LO-signal. There are however more ways to implement a VGA; the main idea is still to avoid switched attenuators. This is usually done in analog types of VGAs.

To realize the phase shifters without switches is a bit more complicated. There is always the possibility to use a vector modulator; something that might be inevitable if broadband phase shifters are to be designed. In that case only one 90 degree phase shifter needs to be realized together with VGAs. In this work however phase shifters based on switched LC-networks are used. The goal in these cases has been to minimize the effect of the switches on the total performance. One way of doing so is to use phase shifter architectures where the switches can be placed in parallel to capacitors. This way the large parasitic capacitance of the MOSFET can be absorbed by the capacitor. This would then result in a capacitor with a more or less poor Q-factor. The parasitic capacitance itself however does not cause any degradation of the performance. Phase shifter architectures where this effect is used are reported in [1].

The third application for switches is the T/R-functionality when transmit-receive functionality is required. In this case even more complex architectures have been implemented and are patent pending. The idea is to make the T/R-switching by active three-ports where the transmission between the desired ports is defined by the active elements. This allows for the IC architecture to be more of a T-like structure as in Fig. 1 instead of the loop type that is commonly used in gallium arsenide designs [5]. There are a few disadvantages with the loop type architecture that can be avoided by doing this. The largest disadvantage is that any leakage in the switch will degrade the resolution and if the states are plotted one can see a figure looking more or less like curtains in the wind. The leakage can have its origin in several mechanisms; the switch, circuits placed close together on the chip which is inevitable since the most sensitive nodes shall come together in the switch. The physical distance is also a problem at module level where everything comes even closer in relative distance. The high requirement on isolation in the switch is as mentioned above not compliable with the poor switch performance using MOSFETs. The next issue that needs to be addressed is the fact that differential circuits are to prefer in silicon RFICs. This means that baluns are needed at input and output. Lattice baluns, coupled inductors or active baluns could be used depending on the requirements on common-mode suppression, noise figure and bandwidth. In this work only passive baluns are used. First order lattice baluns are broadband and well matched with low losses. They do however have rather poor common mode suppression over a larger frequency band. This can be fixed by using the second order type [4] which however consumes quite a bit of die area. For narrow band applications tuned coupled inductors can be used. In the ICs in this work both first order lattice baluns as well as the tuned coupled inductor type of baluns have been used.

III. DESIGNED CORE-CHIPS IN SILICON

There have been a few core-chips designed and manufactured in silicon germanium at Saab Microwave as part of an ongoing development towards low cost AESA radars. So far core-chips at S- and C-band have been demonstrated. Substantial work however has been spent aiming towards X-band and multiband applications. Most of that work has not been aimed directly towards T/R-modules and is therefore omitted in this paper.

A. S-band core-chip (Score)

A single leg core-chip for S-band has been designed and manufactured. The layout and performance are optimized to fit into a leaded plastic package to enable a low cost module based on commercial board technology. It has a bandwidth of 3.0-3.5 GHz and is well suited for airborne surveillance radar. It has an extreme available resolution covering 45dB and 360°; a total of 21 bits corresponding to 2 million states. Since this resolution is usually not needed one can choose the state that best corresponds to the required state. This approach with an excess number of states requires a serial interface which has been implemented. Added value from the high level of integration is that also a temperature sensor is implemented. This is one example of the possibility to reduce the number of components at module level. The IC has a total die size of 3.8×2.8 mm

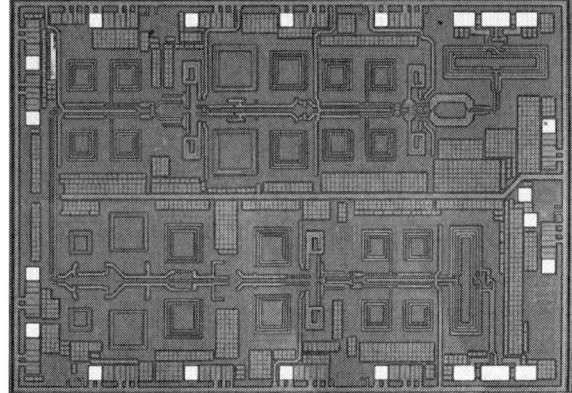

Fig. 2 Microphotograph of the S-band core-chip (Score)

The IC above has its input at the top and output at the bottom. ESD-protected pads can be seen at the edges of the IC. All internal bias voltages and current references are created internally from a single +5V supply.

In Fig. 3 one can see that the 2 million available states present an extreme resolution in phase and gain for gain settings between +10 and -35dB. One can also see that there is some difference between the three colored curves showing three different phase states. This implies that there is some coupling between in- and output on the chip. The magnitude of this coupling can be calculated to be in the order of -50 dB. This is in the same order of magnitude as the coupling between two GSG-probes at the distance in question.

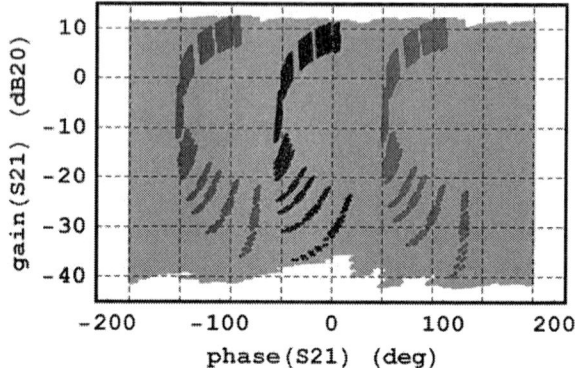

Fig. 3 All phase and gain states for Score. In blue, black and red three amplitude control sweeps with different phase states are highlighted. In grey the 2 million states are shown.

B. C-band core-chip (SiGeMINI)

As part of an AESA study program at Saab Microwave Systems a silicon germanium core-chip [1] at C-band was designed in parallel to the work with gallium arsenide ones [5]. The SiGe core chip consists of a single leg topology with phase and amplitude control. It was most of all considered a demonstrator of the potential of low cost silicon in high frequency applications with performance suitable for radar. The amplitude control has been divided into two parts; One 5-bit fine amplitude control that can be used to compensate for differences in gain originating from temperature and process variation in both IC and module, etc. There is also an analog amplitude control with high dynamic that can for example be used for tapering the antenna. The analog control signal is accessible directly since this is a demonstrator not directly aiming at a product. For the same reason no internal voltage references has been used. These could instead be controlled from the outside in order to increase the testability during the measurements. Extra test circuits were included on chip which made the total chip area 5.04×3.4 mm. The IC is designed to fit inside a 7×7 mm QFN package.

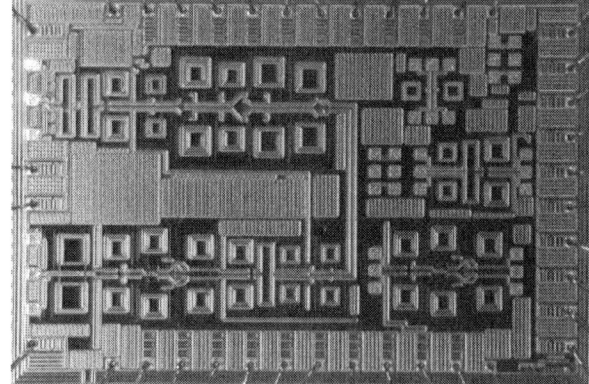

Fig. 4 Microphotograph of the C-band core-chip (SiGeMINI).

Below the measured phase and gain states are shown where the analog amplitude control is not activated. One can see that there is a good resolution in phase and amplitude over gain settings between +10 and +2 dB. The phase states shown cover only half a circle; 0 to 180 degrees since the amplitude control also switches the polarity of the signal.

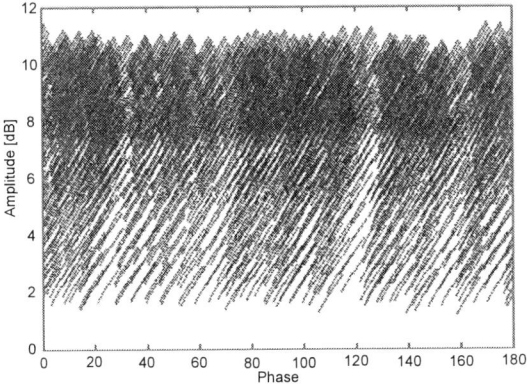

Fig. 5 Phase and gain states for SiGeMINI when the analog amplitude control is disabled.

One can see that even though the analog amplitude control is not activated the IC has more than sufficient resolution. This reduced amplitude control is however too small if antenna tapering shall be covered by the core-chip. This would however be fulfilled using the analog amplitude control.

C. C-band core-chip with T/R-function (Sleipner)

Using the sub circuits from SiGeMINI a core-chip including T/R-functionality was designed. It has a T-like architecture as mentioned above. The switching is done by switching the gain on and off in the three legs.

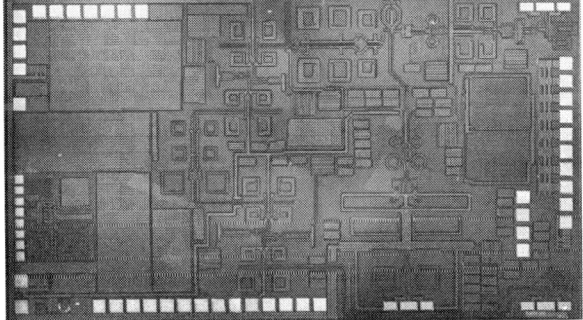

Fig. 6 Microphotograph of the C-band core-chip (Sleipner).

This design was done as a "drop-in" replacement to a gallium arsenide core-chip so the size and pad locations are everything but optimal. RF-port locations together with the conductive substrate caused some leakage between in- and output. This resulted in that the dependence between phase and amplitude states were larger than desired. The states however cover the phase state diagram for gain settings between +11 and -17 dB with an extremely high resolution. The best suited states can then be chosen using a table in a ROM implemented on-chip.

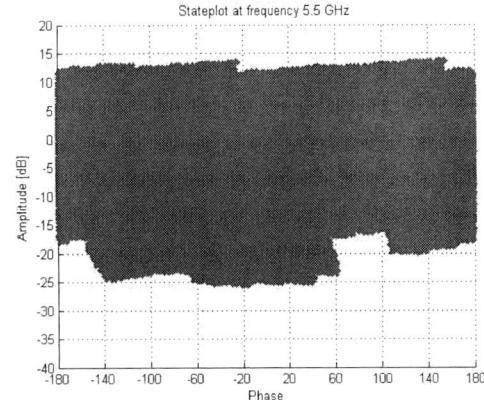

Fig. 7 Phase and gain states for Sleipner.

D. Summary and comparison

TABLE I
SUMMARIZED RESULTS

Parameter	Gemini-S (GaAs) [5]	Gemini-C (GaAs) [5]	Score (SiGe)	SiGeMINI (SiGe)	Sleipner (SiGe)
Frequency band	3-3.5 GHz	5-6 GHz	3-3.5 GHz	5-6 GHz	5-6 GHz
Max gain Rx/Tx	11dB/24dB	12 dB	10dB	10dB	16dB/13dB
Noise Figure (Rx)	<10 dB	<8 dB	<8 dB	<12 dB	<7 dB
Input TOI (Rx)	>16 dBm	>16 dBm	>15 dBm	>10 dBm	>5 dBm
P1dB (Tx)	+20 dBm	+16 dBm	+8 dBm	+10 dBm	+13 dBm
Gain control	7 bit, 40 dB	6 bit, 20 dB	11 bit, 45 dB	5 bit + analog, 45dB	6 bit, 31.5dB
Phase control	6 bit, 360°	7 bit, 360°	10 bit, 360°	9 bit, 360°	6 bit, 360°
Interface	Parallel	Parallel	Serial	Serial /parallel	Serial
Supply voltages	+3.5V, -5V	+3.5V, -0.5V, -5V	+5V	+5V, +4V, +2V	+5V

IV. CONCLUSION

Three core-chips manufactured in a standard silicon germanium process are presented. It has shown to be an attractive alternative for some circuit applications of which phase and amplitude control in AESA type radars is one.

ACKNOWLEDGEMENT

The authors wish to thank the Swedish ministry of defense for supporting this work.

REFERENCES

[1] H. Thiesies and H.Berg, "A Phase and Amplitude Control Front End Chip in SiGe for Phased-Array C-band Radar Applications", in *Proc. IEEE MTT-S.*, 2005.

[2] R. Tayrani, "Broad-Band SiGe MMICs for Phased-Array Radar Applications", *IEEE Journal of Solid State Circuits*, vol. 38, pp. 1462-1470, Sep. 2003.

[3] M. A. Mitchell, "An X-band SiGe Single-MMIC Transmit/Recieve Module for Radar Applications", in *Proc. Radar 2007*, pp. 664-669.

[4] D. Kuylenstierna and P. Linnér, "Design of broad-band lumped-element baluns with inherent impedance transformation", *IEEE Trans. Microwave Theory and Technique*, vol. 52, pp. 2174-2186, Dec. 2004.

[5] N. Billström, et. Al., "T/R "Core Chips" for S-, C- and, X-Band Radar Systems", in *Proc. EuMC*, 2004, pp. 1029-1032.

Thales Components and Technologies for T/R Modules

Yves MANCUSO

Thales Systèmes Aéroportés, 2 av. Gay-Lussac, 78851 Elancourt, France

+33(0)1 34 81 91 27

[1] yves.mancuso@fr.thalesgroup.com

Abstract — **This paper presents new developments and perspectives in Phased Arrays Radars and Electronic Warfare for the next generations of T/R modules (medium/long term), in order to decrease the mass production cost, while increasing the level of performance and reliability. In terms of physical architecture, even if the brick one is more current at mid-term, the tile concept is investigated for conformal and/or multifunction phased array antennas : a 3 dimension module will lead to a drastic reduction of size and weight of the antenna.**

MMICs are always the key components, with evolutions towards multifunction chips, new processes like GaN, SiGe, MEMS power switches.

Concerning the packaging, a technological roadmap indicates the different capabilities : thick film multilayer ceramic circuits, co-fired ceramics based on LTCC or HTCC processes, surface-mounted packages on printed circuits boards, and 3D architectures.

The interconnection domain is also now more and more important in order to be compatible with the level of integration required for the microwave modules : fuzz buttons, flex, sub-miniature connectors.

All these, technologies mastered by Thales are dual for Airborne and Space, Military and Civilian applications.

I. INTRODUCTION

Electronically Scanned Antenna introduces a technological breakthrough in Airborne Radar. First the beam agility makes search and tracking completely independent, opens the way to new detection strategies and improves dramatically the tactical situation acquisition. It enables also simultaneous mode implementation. By suppressing the mechanical parts, it increases the reliability and makes easier the platform installation. Active Electronically Scanned Antenna (AESA) offers even more potentialities with an improved power budget thanks to loss reduction and beam-shaping capabilities, which enables to adapt the radiating pattern to the environment.

Recently, Active ESA solutions are being developed in X band fighter radars (F22, Rafale ,…) or lower S and L band AEW radars. Large amount of fundings are being spent in these technologies.

The cost reduction of the current AESA is the main concern for these developments. Mid-term and long-term solutions at components, modules, packaging, global architecture level will be discussed.

Fig. 1. Example of X band Active ESA RBE2 for Rafale

AESA have an intrinsically wide bandwidth and high gain and using them for the benefit of different sensors of an aircraft is a way to cut the global cost of a system. Moreover the lack of room where to place the different antennas and the need to have a 4 π steradian coverage is an even stronger driver to share the front end of the sensors. It imposes new technologies like tile modules to reduce their thickness and in some cases conformal antennas to match to the platform shape.

Considering the very different requirements and constraints for transmission and reception, it will be beneficial to separate (when possible) the transmit antenna from the receive antenna. Ultimately, in Network Centric operation concept and for covert operation transmission and reception, functions will be distributed on different platforms changing the requirements and the architectures of the involved antennas.

II. ON GOING DEVELOPMENTS IN X BAND AESA

The feasibility and manufacturability of an X band fighter radar AESA is no longer a technical issue, and has been demonstrated in several development or programs, and especially within the Rafale flights since 5 years. The challenge is now more technological and industrial in order to decrease the cost of these antennas composed of thousands of Transmit/Receive Modules (TRMs).

The antenna cost issue is illustrated by the average cost distribution of the main AESA radar parts :

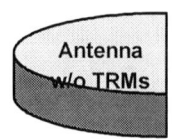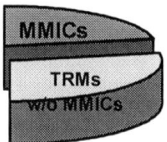

Fig. 2. AESA cost Breakdowm

A. Components

1) MMICs / GaAs Cost

GaAs is generally known to be very expensive but Fig. 2 shows that the GaAs cost represents a relatively moderate part of the total cost (less than 25%). This cost depends on the GaAs market price (on which the radar designer has almost no impact) and the total GaAs surface used in the TRM.

Reduction of this surface is based on the integration density reached either by the technology or the design, for example with multifunctions chips or corechips.

HBT technology currently used for HPA MMICs allows to optimize the required surface by easily modulating the base (low current), versus drain modulation (high current) for PHEMT process. So the PAE (Power Added Efficiency) has to be considered at module level and not only at MMIC level.

With the UMS HB20PX process, the better state-of-the- art has been achieved providing in X band 10W with more than 40% PAE.

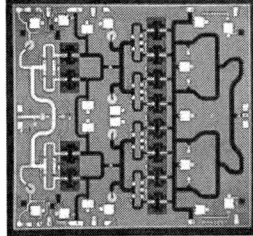

Fig. 3. X band GaAs HBT 10 W 40% PAE HPA MMIC (18mm2)

2) GaN developments

GaN semiconductors provide many potential advantages for power microwave applications, for example increased input power level capability. Modules used in radar applications could benefit of this new technology in both their Transmit and Receive path : power handling capability of GaN should allow the reception amplifier to face high power electromagnetic aggression.

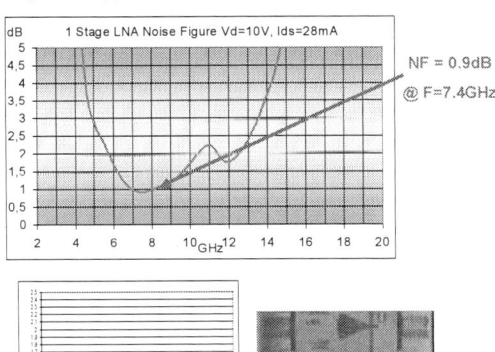

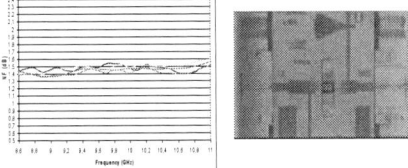

Fig. 4. First results for hybrid X band GaN LNAs noise figures.

An example of THALES / UMS result is an X-Band MMIC HPA built in GaN microstrip technology, with an average output power of 40-50 W and a power added efficiency of 33-37%. As a peak result for the transistors used, a power density of 7W/mm was reached @ Vd=32V.

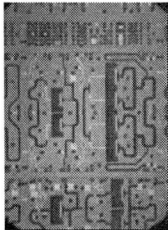

Fig. 5. X band GaN MMIC HPA

3) Corechips

Multifunction core chips integrates in a single chip all the low power functions (phase shifter, gain control, switches, amplifiers) reducing the GaAs surface needed and the wiring cost .Hereafter is an example of an Electronic Warfare corechip.

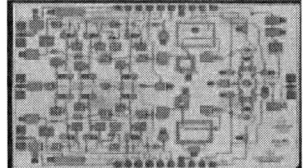

Fig. 6. Wide band GaAs corechip

For longer term, the latest evolution of SiGe technology towards microwaves frequencies could be a breakthrough to cut the cost. Analogous to the realisation of single-chip television or radio front-end, the "ultimate goal" might here be described as a "single" core chip for the T/R module.

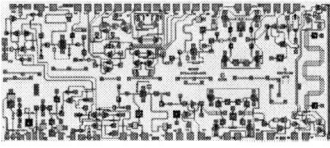

Core chip surface 20 mm²
(GaAs)

2000 µm x 2200 µm
SiGe phase shifter

Fig. 7. A GaAs multifunction for an X-band TR module and a first demonstration for a complete SiGe phase shifter including the controls.

4) MEMS

There is an increasing interest of aerospace industry towards MEMS devices, which already found several applications in other sectors. The potential benefits from the use of these devices are good electrical performances, power savings, and small size. Silicon technology is used for the development of MEMS RF switches, especially power switches in X band (10 W power handling).

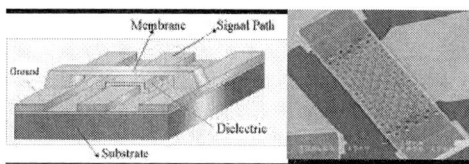

Fig.8. Structure of a power RF MEMS switch

B. Packaging And Modules

The way to connect the components, to protect them, to distribute signals and power supplies, to evacuate calories is a major contributor to the global cost.

The first TRMs were pluggable units that could be replaced individually for maintenance purpose. Considering the intrinsic TRM reliability it appears that the connectors dramatically increase the cost... and severely degrade the global reliability. New designs use TRMs directly soldered on planks which assemble tens of them on a single printed circuit board.

Ongoing developments tend to suppress the TRM package and to connect all the RF components, the control circuits, the power supplies on a Multifunction Carrier which includes all RF, low power signals and power distribution. The radiating elements could also be integrated in the board.

These low cost solutions are based on organic packages, the RF and DC interfaces being realised through BGA or lead frame. The main advantage of these organic packages is to offer TCE matched compatibility with printed circuit boards, which allows quite large packages.

Fig.9. Examples of a RF organic packages

To summarize, we can draw the following technological roadmap indicating the different evolutions :
- thick film multilayer ceramic circuits ;
- co-fired ceramics based on LTCC or HTCC processes ;
- surface-mounted packages on printed circuits boards like organic or ceramic ;
- 3D module architectures with high density integration.

C. Signal Distribution

Today RF distribution is made with RF lines. RF optical distribution is a promising technology for mid term.

For longer term, there is a strong need for more flexible and efficient phased-array beam management, particularly on the receiving end. This is calling for a more intelligent technique to combine the signals of different modules, and pushes towards a digital beam forming type of processing using distributed Radar receivers.

Chip sets for communication products are orders of magnitude cheaper than for military transceivers. Without doubt, the requirements based on the military needs and the mass nature contribute part to this. However, the combination of RF electronics with digital control and processing on a single Si die is a lesson that can be learnt from the telecommunication world.

The "single-chip-radar-receiver" mentioned above will open the path to the "Digital TRM" concept and the full beam forming which will solve the RF distribution problems.

III. SHARING THE RESOURCES

Sharing the resources (the antennas) between the different sensors of an aircraft is a way to cut the global cost of the system. But the lack of room where to place these antennas (the "windows") is certainly a stronger reason to share the front end of the sensors.

Sharing the apertures, which is also a way to reduce the aircraft RCS, is then a new requirement for future systems. It will lead :
- to increase the AESA bandwidth (from some tens percents to multi octave) ;
- to reduce the antenna thickness (to a few centimeters) in order to be able to install easily these antennas on the aircraft body. In some case conformal arrays will be required.

New concepts such as "tile TRM" antenna are then being developed, where the module footprint has to be compliant with the radiating element size, that is to say about 10x10 mm² for a wide band antenna (6-18 GHz). In order to fulfil this requirement, a 3D module construction is investigated.

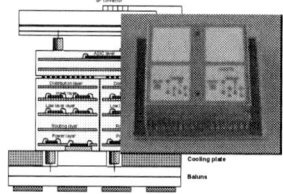

Fig. 10. Examples of a 3D architecture

This approach requires new technologies for the manufacturing of the high integrated module sub-levels, the internal interconnection between two sub-levels and the external connection of the T/R module. The module includes different layers ; for example :

- a control ASIC chip,
- low level RF components (phase-shifter, LNA, attenuator …),
- RF power components (HPA, driver…).

The several RF levels are achieved on soft substrates. The power components are attached on a layer directly in contact with the cooling system of the antenna. Interconnections between the RF layers are performed using μBGA, bumps or Fuzz buttons transitions. A particular attention is given to the thermal aspects, due to the CW operation, and the miniaturisation of the interconnections because of the high integration.

Different module architectures are investigated to realise the assembly. They differ each other from technologies and interconnections required to realise the T/R module.

IV. CONCLUSIONS

We have presented a survey of the evolutions of Airborne Phased Array Radars. Current developments are mainly focused on the cost reduction of the X band Active Electronically Scanned Antenna. Beyond the continuous technologies improvements at the MMIC level (GaAs HBT, GaN, SiGe, MEMS), several breakthroughs are foreseen at mid or long term.

Surface mounted packages on printed boards, organic BGAs, plastic packages are mid term solutions preceding longer term 3D architectures.

Signal distribution is another challenge where optical RF is being developed before the implementation of full digital TRMs and generalized Digital Beam Forming.

Sharing the apertures on a platform (Radar, ESM/ECM, Communications...) is a way to not only decrease the global cost, but to extend the coverage of the sensors and their performances. It needs broadband antennas (up to multi-octave) and low thickness antennas (few centimeters) in order, if required, to conform them on the platform frame.

Tile TRM concept and 3D architectures are already in labs.

ACKNOWLEDGMENT

The author thanks colleagues at THALES for various technical contributions.

He also wish to acknowledge the French MoD for supporting many actions.

Proceedings of the 3rd European Microwave Integrated Circuits Conference

GaN MMIC based T/R-Module Front-End for X-Band Applications

P. Schuh*, H. Sledzik*, R. Reber*, A. Fleckenstein*, R. Leberer*, M. Oppermann*,
R. Quay[†], F. van Raay[†], M. Seelmann-Eggebert[†], R. Kiefer[†] and M. Mikulla[†]

*EADS Deutschland GmbH, Defence Electronics, Wörthstr. 85, 89077 Ulm, Germany
[†]Fraunhofer Institute of Applied Solid-State Physics, Tullastr. 72, 79108 Freiburg, Germany

patrick.schuh@ieee.org

Abstract—Amplifiers for a next generation of T/R-modules in future active array antennas are realized as monolithically integrated circuits (MMIC) on the bases of novel AlGaN/GaN HEMT structures. Both, low noise and power amplifiers are designed for X-band frequencies.

The MMICs are designed, simulated and fabricated using a novel via-hole microstrip technology. Output power levels of 6.8 W (38 dBm) for the driver amplifier (DA) and 20 W (43 dBm) for the high power amplifier (HPA) are measured. The measured noise figure of the low noise amplifier (LNA) is in the range of 1.5 dB.

A T/R-module front-end with mounted GaN MMICs is designed based on a multi-layer LTCC technology.

I. INTRODUCTION

Active electronically scanned array (AESA) radars are increasingly being favoured over conventional mechanically-scanned systems. The achievable radar range of such an AESA radar is mainly determined by the output power and the noise figure of the antenna. T/R-modules are key elements in active phased array antennas for radar and electronic warfare applications [1]. Inside these T/R-modules two main building blocks are the amplifier chain in the transmit path and the low noise amplifier in the receive path.

In most of today's T/R-Modules GaAs MMIC amplifiers are used with typical output power levels in the range of 5 W to 10 W [2], [3]. Higher output power levels, broader bandwidth, increased power added efficiency (PAE) values, and higher operating voltages are advantages for performance improvement to meet future requirements. For these applications the use of amplifiers based on AlGaN/GaN is a very promising approach [4]–[7]. Another important parameter is the PAE of the amplifier. Due to the higher breakdown voltage of the GaN device compared to a GaAs device, the supply voltage can be significantly increased. This leads to an additional increase in efficiency on system level, because of lower losses in the power supply.

Besides the noise figure performance also the robustness against high input power overdrive is a key issue for the receive path in a T/R-module. The AlGaN/GaN technology with its high breakdown voltage is very good suited for robust low noise applications [8].

To satisfy these future needs, a T/R-module front-end composed of novel GaN MMICs and multi-layer packaging technology [9] is designed.

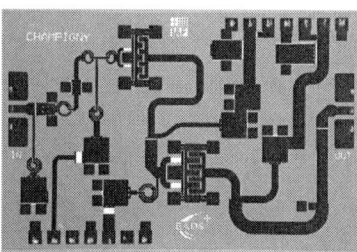

Fig. 1. Photo of a GaN DA MMIC chip. Chip size : 3 mm ×2 mm

In this paper the design and the achieved performance of single GaN MMIC amplifiers and of a whole T/R-module front-end based on these GaN MMICs are presented.

The MMIC and front-end design, simulation and measurements are done at EADS Defense Electronics, Ulm. The wafer and MMIC fabrication is done at the Fraunhofer Institute of Applied Solid-State Physics, Freiburg.

II. GaN MMIC TECHNOLOGY AND MODELING

The AlGaN/GaN HEMT MMIC technology is based on multi-wafer growth of single heterojunction devices on 3-inch s.i. SiC substrates by MOCVD. The 3-inch HEMT technology uses electron-beam-defined gates with 0.25 μm gate-length including fieldplates for high-power operation.

The two-terminal breakdown voltages of the power HEMTs are $BV > 100$ V. Typical output power densities are beyond $5 \frac{W}{mm}$ at 10 GHz with an associated drain efficiency $\eta > 50\%$ at $V_{DS} = 28$ V for high efficiency operation. For low-noise operation, the AlGaN/GaN HEMTs yield a minimum noise figure of $NF < 0.8$ dB at 10 GHz at $V_{DS} = 10$ V.

After the frontside processing the full 3-inch SiC wafer is thinned to 100 μm thickness and a via-hole backside process is applied.

For the design a library of passive microstrip components is available, including all technology specific elements like MIM capacitors, resistors, and inductances. The large-signal GaN HEMT modeling is based on an in-house two-dimensional voltage-lag model to appropriately describe thermal effects and low-frequency dispersion, and their impact on gain and PAE [10].

978-2-8748-7007-1/08 $25.00 © 2008 EuMA

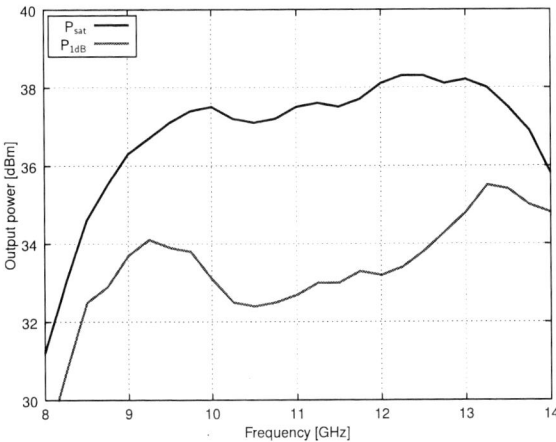

Fig. 2. Measured output power versus frequency of the GaN MMIC DA at 1 dB gain compression and at saturation with $V_{DS} = 30V$ and $V_{GS} = -4.9V$.

III. DRIVER AMPLIFIER MMIC

The MMIC driver amplifier is designed as a two stage amplifier with one transistor $8 \times 60\,\mu m$ gate-finger width in the first stage and one transistor $8 \times 125\,\mu m$ gate-finger width in the second stage (see Fig. 1). The amplifier is designed for $50\,\Omega$ impedance at the input and output port and for operation in linear mode, not using the whole available output power of the transistor. One design objective was to provide enough input power for two HPAs in parallel configuration in the frequency band from 8 GHz to 12 GHz. Although this amplifier will be used for operation in linear mode, the output matching is designed using harmonic balance simulation.

The maximum measured output power is higher than 38 dBm while operating in saturation mode with up to 5 dB gain compression. This output power leads to a power density of $7.4\,\frac{W}{mm}$ at transistor level. The achieved output power is sufficient for driving one or two high power amplifiers in the frequency range between 8.5 GHz and 14 GHz (Fig. 2), while operating the driver amplifier below the 1 dB compression point.

IV. HIGH POWER AMPLIFIER MMIC

The MMIC high power amplifier is designed as a two stage amplifier based on 4 transistors with $8 \times 125\,\mu m$ gate-finger width in the second stage and 2 transistors with the same size in the first stage (see Fig. 3). The amplifier is designed for $50\,\Omega$ impedance at the input and output port. The output combiner is optimized for maximum output power in the frequency range from 8.5 GHz to 11 GHz. Previous intensive load-pull simulations have been performed to find the optimum load impedance for maximum output power.

For electrical stabilization of the second stage RC-networks close to all transistor inputs are used. Odd mode stability analysis has been performed for the parallel transistor structure. The first stage transistors are stabilized by the gate bias circuits. The interstage network and the output combiner

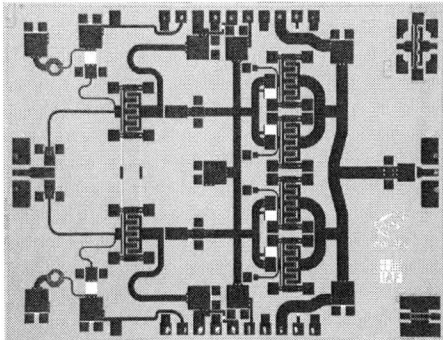

Fig. 3. Photo of a GaN HPA MMIC chip. Chip size : 4 mm × 3 mm

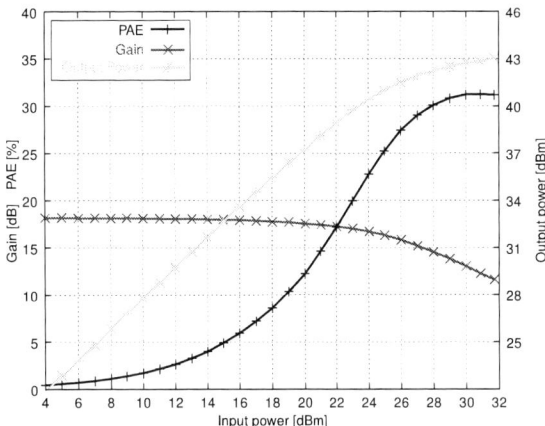

Fig. 4. Measured power performance of the GaN MMIC HPA at 11 GHz with $V_{DS} = 30V$ and $V_{GS} = -5V$. $P_{max} = 43$ dBm $PAE_{max} = 31\%$

are optimized by electromagnetic simulations of distributed elements. An improved PAE performance is obtained by application of second harmonic short-networks at all gates of the second stage transistors.

The output power, gain and PAE measurement results in pulsed mode are shown in Fig. 4. The maximum measured output power is 20W, when biased for class-AB operation. This output power leads to a power density of $5.7\,\frac{W}{mm}$ at transistor level. In this case the small signal gain is about 18 dB and the associated PAE value is 31%. The small signal gain is higher than 15 dB over the whole frequency range from 8.5 GHz to 11 GHz (see Fig. 5). More then 14 W output power is measured over a frequency range from 8.75 GHz to 11.5 GHz.

V. LOW NOISE AMPLIFIER MMIC

For the designed MMIC low noise amplifier the same transistor size of $8 \times 30\,\mu m$ is used for the first and second stage (see Fig. 6). For this first iteration design the transistor layout wasn't optimized for low noise operation. It is similar to the power transistor layout. The transistor size of the first stage is a tradeoff between low noise performance and large

978-2-8748-7007-1/08 $25.00 © 2008 EuMA

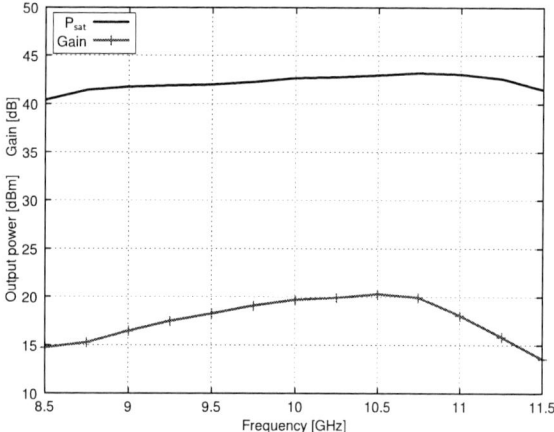

Fig. 5. Measured saturated output power and small signal gain versus frequency of the GaN MMIC HPA. Bias Point: $V_{DS} = 30V$ and $V_{GS} = -5.0V$

signal robustness.

To allow a simultaneous noise match and input match over a sufficiently broad bandwidth an additional source inductance is used. The design model is based on small signal and noise parameter measurements of the transistor, without any inductive feedback.

For decoupling purpose, but also to increase the robustness [11], a resistor is used in the gate bias network.

In the frequency range from 8 GHz to 12 GHz a gain between 15 dB and 17 dB is simulated. The expected noise figure is about 1.75 dB to 1.8 dB.

In Fig. 7 the measured small signal gain and the noise figure of the GaN MMIC LNA is shown. The gain is a little bit lower than simulated, especially around 12 GHz a distinct slope is measured. The obtained noise figure is better than expected over the whole frequency band. The minimum noise figure is only about 1.45 dB. For frequencies up to 11 GHz the noise figure of this GaN LNA is only 0.5 dB worse than the normally used GaAs LNA [12].

Besides the good noise performance the behavior of the LNA with large input power signals is very important for this type of LNA. The saturated output power of the LNA is about 24 dBm (see Fig. 8). First robustness tests have been performed up to 4 W input power. At this input power level the LNA is about 35 dB in compression. The output power is decreasing, because the bias of the first stage transistor is shifted toward deep pinch-off.

VI. T/R-MODULE FRONT-END

For the front-end a multi-layer low temperature cofired ceramic (LTCC) substrate is used. After obtaining first experience with an amplifier chain on LTCC with GaN MMICs [13] some modifications are realized.

One big challenge is the thermal situation with these GaN MMICs having a very high power density compared to today's state-of-the-art GaAs MMICs. A first approach was to use AlN

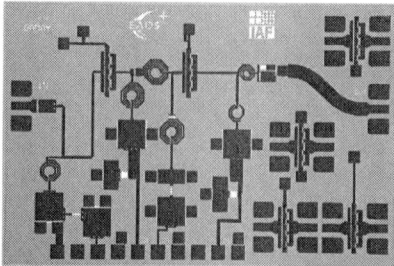

Fig. 6. Photo of a GaN LNA MMIC chip. Chip size : 3 mm × 2 mm

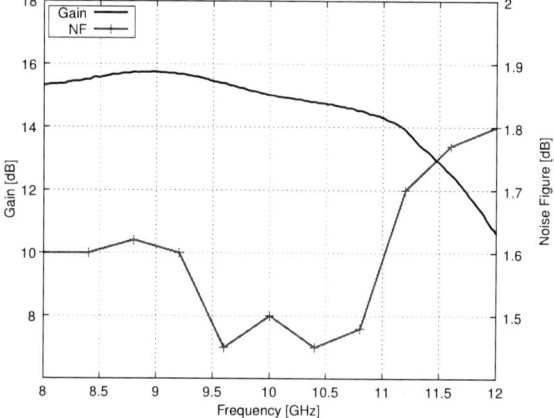

Fig. 7. Measured gain and noise figure versus frequency of the GaN MMIC LNA with $V_{DS} = 15V$ and $V_{GS} = -3.5V$. $NF_{min} = 1.45$ dBm

high temperature cofired ceramic (HTCC) instead of LTCC because of its significantly higher thermal conductivity. In this case the DA and the two HPAs were mounted directly on the HTCC substrate in cavities. The thermal situation for the DA was improved but the situation for the HPAs were dramatically declined.

The favored solution today is using the mature and proven LTCC technology for T/R-Modules used in series production and soldering the DA similar to the HPAs on a CuMo heatsink.

The T/R-Module front-end consists of a whole transmit path with one DA and two HPAs in parallel, a circulator, a receive path with LNA and limiter and a GaAs switch combining both pathes. It can be used as front-end for a software defined radar or can be combined with any core chip using CMOS, SiGe or GaAs technology.

Besides the RF relevant components, like power splitter and combiner, also the whole DC control electronics is integrated on the substrate. The control electronics provides the pulsed 30V drain voltage for the GaN high power MMICs, the LNA bias voltages and the T/R-switch control. In Fig. 9 the completely assembled front-end is shown.

For the transmit operation mode the output power, gain and PAE measurement results in pulsed mode are shown in Fig. 10. The maximum measured output power is 32W.

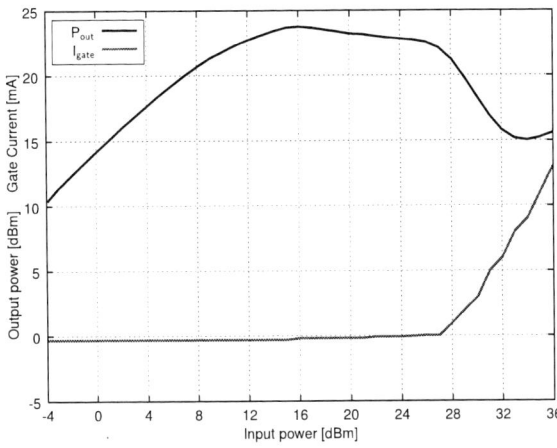

Fig. 8. Measured output power and gate current versus input power of the GaN MMIC LNA. Bias Point: $V_{DS} = 15V$ and $V_{GS} = -3.5V$

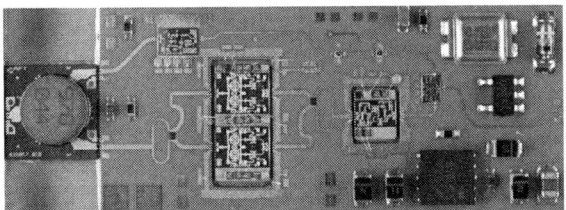

Fig. 9. Photo of a T/R-Module front-end with GaN MMIC chips.

In this case the small signal gain is about 31 dB and the associated front-end PAE value is 24%, also taking the disipation power of the control electronics into account. More then 20 W output power is measured over a bandwidth of 1.6 GHz.

In the receive operation mode a small signal gain in the range of 13 dB is measured. The associated noise figure is about 2.2 dB at room temperature.

VII. CONCLUSION

A whole GaN MMIC amplifier chip set for a X-band T/R-module front-end were designed, simulated, fabricated and measured. Output power levels up to 20 W for the transmit path HPA and a noise figure of 1.45 dB for the receive path LNA are achieved. A T/R-module front-end with three types of amplifiers (DA, HPA and LNA) integrated on multi-layer LTCC substrates is successfully demonstrated. With such a front-end, T/R-Modules with more then 20 W transmit output power and a receive noise figure below 3 dB can be realized.

ACKNOWLEDGMENT

This work was partly funded by the German "Bundeswehr Technical Center for Information Technology and Electronics" (WTD81), Greding and the German Federal Ministry of Defense (BMVg), Bonn. This financial support is gratefully acknowledged.

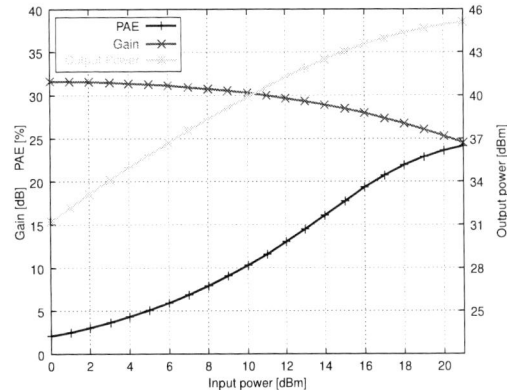

Fig. 10. Measured power performance of the GaN T/R-Module front-end at 11 GHz with $V_{DS} = 30V$ and $V_{GS} = -5V$. $P_{max} = 45$ dBm $PAE_{max} = 24\%$

REFERENCES

[1] W. Holpp and C. Worning, "New electronically scanned array radars for airborne applications," in *Asia Pacific Microwave Conference Proceedings*, (Bangkok), Dec. 2007.

[2] H. Hommel and H.-P. Feldle, "Current status of airborne active phased array (AESA) radar systems and future trends," in *European Radar Conference Digest*, (Amsterdam), pp. 12–123, 2004.

[3] B. A. Kopp, M. Borkowski, and G. Jerinic, "Transmit/receive modules," *IEEE Transactions on Microwave Theory and Techniques*, vol. 50, pp. 827–834, Mar. 2002.

[4] T. Edwards, "Semiconductor technology trends for phased array antenna power amplifiers," in *European Radar Conference Proceedings*, (Manchester), pp. 269–272, Oct. 2006.

[5] P. Schuh, R. Leberer, H. Sledzik, M. Oppermann, B. Adelseck, H. Brugger, R. Behtash, H. Leier, R. Quay, and R. Kiefer, "20W GaN HPAs for next generation X-Band T/R-Modules," in *IEEE Microwave Theory and Techniques Symposium Digest*, (San Francisco), pp. 726–729, June 2006.

[6] D. M. Fanning, L. C. Witkowski, C. Lee, D. C. Dumka, H. Q. Tserng, P. Saunier, W. Gaiewski, E. L. Piner, K. J. Linthicum, and J. W. Johnson, "25 W X-Band GaN on SI MMIC," in *Compound Semiconductor Manufacturing Technology Conference Digest*, 2005.

[7] S. T. Sheppard, R. P. Smith, W. L. Pribble, Z. Ring, T. Smith, S. T. Allen, J. Milligan, and J. W. Palmour, "High power hybrid and MMIC amplifiers using wide-bandgap semiconductor devices on semi-isulating SiC substrates," in *Device Research Conference Digest*, pp. 175–178, June 2002.

[8] R. Schwindt, V. Kumar, O. Aktas, J.-W. Lee, and I. Adesida, "Al-GaN/GaN HEMT-based fully monolithic X-band low noise amplifier," *Physica Status Solidi (c)*, vol. 2, pp. 2631–2634, May 2005.

[9] R. Rieger, B. Schweizer, H. Dreher, R. Reber, M. Adolph, and H.-P. Feldle, "Highly integrated cost-effective standard X-band T/R module using LTCC housing concept for automated production," in *European Conference on Synthetic Aperture Radar Digest*, pp. 303–306, 2002.

[10] M. Seelmann-Eggebert, T. Merkle, F. van Raay, R. Quay, and M. Schlechtweg, "A systematic state-space approach to large-signal transistor modeling," *IEEE Transactions on Microwave Theory and Techniques*, vol. 55, pp. 195–206, Jan. 2007.

[11] M. Rudolph, R. Behtash, R. Doerner, K. Hirche, J. Würfl, W. Heinrich, and G. Tränkle, "Analysis of the survivability of GaN low-noise amplifiers," *IEEE Transactions on Microwave Theory and Techniques*, vol. 55, pp. 37–43, Jan. 2007.

[12] H.-P. Feldle and R. Reber, "Monolithic low noise amplifier for X-Band applications," in *Gallium Arsenide Application Symposium Proceedings*, (Amsterdam), pp. 205–209, 1998.

[13] P. Schuh, R. Leberer, H. Sledzik, M. Oppermann, B. Adelseck, H. Brugger, R. Quay, M. Mikulla, and G. Weimann, "Advanced high power amplifier chain for X-Band T/R-Modules based on GaN MMICs," in *European Microwave Integrated Circuit Conference Proceedings*, (Manchester), pp. 241–244, Oct. 2006.

978-2-8748-7007-1/08 $25.00 © 2008 EuMA

Current Trends and Challenges in III-V HBT Compact Modeling

Matthias Rudolph

Ferdinand-Braun-Institut für Höchstfrequenztechnik (FBH)
Gustav-Kirchhoff-Str. 4, D-12489 Berlin, Germany
rudolph@fbh-berlin.de

Abstract—This paper gives an overview on recent achievements in the modeling of GaAs or InP based HBTs. The emphasis lies on the description of weakly nonlinear behavior, and on advanced descriptions for $1/f$ and shot noise for nonlinear simulation. Although compact HBT modeling already reached a high level of accuracy, certain limitations remain that will also be addressed.

I. INTRODUCTION

GaAs-based HBTs have been for quite a while the devices of choice for power amplification in the lower GHz range, especially for use in mobile devices. The vast majority of GaAs-HBT MMICs fabricated today therefore are power amplifiers (PAs) designed to deliver around 1 W in the range $1-2$ GHz, powered from a battery. InP based HBTs are capable to achieve highest cutoff frequencies while maintaining reasonably high breakdown voltages. E.g., this technology is well suited to design modulator drivers for multi-Gbps optical links that require high voltage swings and highest switching speeds. Recently developed GaAs HBT technologies target base station amplifiers rather than mobile PAs, with tenfold bias supply voltage and output power: beyond 10 W at 2 GHz from a 28 V supply. This results in much higher output impedances, that allow for easier and, especially, broadband output matching.

Since III-V HBTs have come to market more than two decades after their unipolar counterparts, compact models are not yet as well established. This can be seen from the fact that simply the number of models that are available in circuit simulators is much lower than the number of MESFET or HEMT models. There are mainly two modern HBT models commonly available. One is the FBH model that will be used as the reference model in this paper. The other one that follows the same philosophy although with different formulas, is the AgilentHBT model [1]. On the other hand, designers are often not sure if they can trust the model they get, if it is not well understood what effects are to be accounted for.

It is the aim of this paper to review the state-of-the-art in III-V HBT modeling. The status of the models will be discussed, as well as some recent advancements, and finally some unsolved issues will be addressed. In order to keep this article readable while addressing all relevant issues, the different issues will be touched rather than providing an in-depth discussion. For details please refer to the literature cited in each section.

Fig. 1. Micro-photograph of single-finger HBT from FBH Foundry. The $20\,\mu m$ thick emitter airbridge serves as a heat spreader, that is capable of equalizing the temperature distribution in multi-finger devices at 28 V operation. The airbridge has been removed in the foreground in order to show the inner device.

A. HBT Model Requirements

To begin with, three of the expectations shall be recalled that led to the development of HBTs. These are

- the capability to be operated at high power densities,
- high linearity, and
- low $1/f$ noise.

HBTs are vertical devices, see Fig. 1. The collector current flows through the bulk device, in contrast to HEMT devices where the drain current flows within a narrow horizontal channel close to a heterointerface. Therefore, higher power densities can be achieved in HBTs since the whole bulk of the device carries the current. Lower $1/f$ noise can be expected because the current does not flow along an interface. Besides of power amplifiers, HBTs are also well suited for applications like linear amplifiers and low phase-noise oscillators.

In order to fully exploit the capabilities of HBTs, it can be concluded that a model must at least account for self-heating at high dissipated powers, predict weak nonlinearity well, and is required to provide a noise description that is suitable for oscillator simulation.

That HBTs in general show significant self-heating has already become common knowledge, and will therefore not be discussed again in this paper. Instead, we will review the state-of-the-art regarding transit-time and noise modeling, and

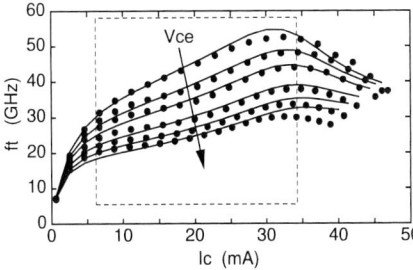

Fig. 2. Typical bias dependence of transit frequency, for $3 \times 30\,\mu m^2$ HBT, Symbols: Measurement, lines: model. $V_{ce} = 1 \ldots 5\,V$.

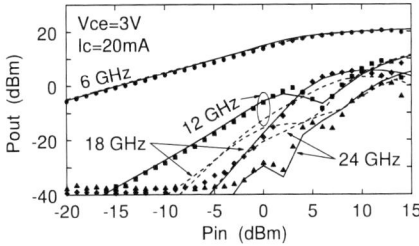

Fig. 3. Power-spectral measurement and simulation of $3 \times 30\,\mu m^2$ HBT, Symbols: Measurement, solid lines: FBH model, broken lines: electro-thermal model neglecting bias-dependence of f_t.

name the current challenges that remain to be solved.

II. TRANSIT TIMES AND WEAK NONLINEARITY

It is common to all types of bipolar transistors that the transit frequency first increases with current, and, finally degrades again at high current densities due to the base push-out effect. What is special about III-V HBT is the significant bias dependence of f_t at moderate currents, see the dashed box in Fig. 2. This bias dependence is mainly due to a variation of collector transit time τ_c. A change in base-collector capacitance C_{bc} that decreases with current is also observed. The reason for this effect is the strong impact of current and voltage on the collector depletion region, and thus on τ_c and C_{bc} [2]. Dedicated HBT models therefore provide special formulas for the collector charge Q_c in this area, with $C_{bc} = \partial Q_c/\partial V_{bc}$, and $\tau_c = \partial Q_c/\partial I_c$. An other part of the total transit time is modeled through the base-emitter charge Q_e, $\tau_e = \partial Q_e/\partial I_e$. Basically, it is possible to partition the total charge with some degree of freedom between the two junctions [3]. Thus, common large-signal models do not provide real time delays, but approximate time-delay effects through RC time constants. This approximation affects the linear current gain α:

$$\alpha = \underbrace{\frac{\alpha_0\,e^{-j\omega\tau}}{1 + j\omega/\omega_\alpha}}_{\text{small signal}} \approx \underbrace{\frac{\alpha_0 - j\omega\tau_c}{1 + j\omega\tau_e}}_{\text{linearized large-signal}} \tag{1}$$

with $\tau = \tau_c$ and $\omega_\alpha = 1/\tau_e$.

Accurate modeling of the collector charge is not only necessary in order to predict the bias dependence of f_t. This

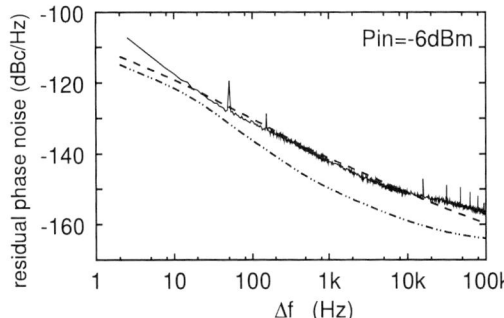

Fig. 4. Double sideband residual phase noise versus frequency-offset Δf of a $3 \times 30\,\mu m^2$ HBT, with $-6\,dBm$ input power at 3.5 GHz, $V_{cc} = 3\,V$, $I_c = 30\,mA$. Measurements (—) compared to simulation: model implementation using lowpass sources ($-\cdots-$), and cyclostationary sources (- -). From [8].

bias dependence has also been shown to be the main cause for weak nonlinearity in class-A operation of the HBT [4], [5]. Fig. 3 highlights the effect of proper collector charge modeling on weak nonlinearity prediction. While a full-featured model (solid lines) yields correct values for the harmonics, a reduced model (dashed lines) fails for all harmonics except the fundamental. This reduced model did not account for a current dependence in Q_c, as it is the case in traditional bipolar models, but its values were tuned in order to get the best fit for the bias point under investigation.

The transit-time model therefore can be regarded as equally important for model accuracy as the thermal model.

III. NOISE MODELING

Significant advancements have been achieved recently in the modeling of the HBT noise properties. With respect to $1/f$ noise, the main issue is how the low-frequency noise is upconverted in nonlinear operation. At high frequencies, it is now well accepted that the correlation of the shot-noise sources must not be neglected.

A. $1/f$ Noise

In the nonlinear regime, the $1/f$ noise is upconverted and, e.g., can be observed in an oscillator close to the carrier as phase noise. This upconversion process, however, is not intrinsically accounted for by the nonlinear simulator. But, as has been shown recently, it significantly depends on the way the noise sources are defined.

It is a precondition that the model provides at least two noise sources that allow for accurate simulation of the $1/f$ noise in the linear domain as a function of frequency, current, and source impedance. But in order to predict $1/f$ upconversion, it is required to rely on cyclostationary sources. This means that the noise level is not just a function of the bias current, but is defined as a function of the instantaneous large-signal current. Cyclostationarity means, in this respect, that the driving current is a time periodic signal. Such a source will generate noise sidebands at all harmonics of the large-signal current.

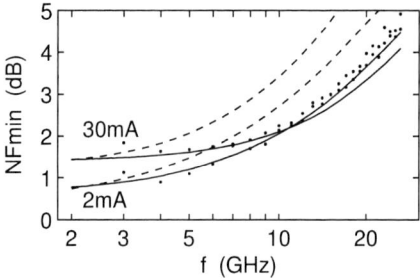

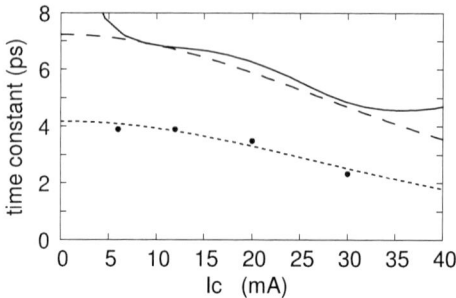

Fig. 5. Minimum noise factor $NF_{\min}$ of $3\times30\,\mu m^2$ HBT at $V_{CE} = 1$ V. Measured values (•), simulated with conventional large-signal noise model (− −), and simulated with proposed new noise model (—). From [13].

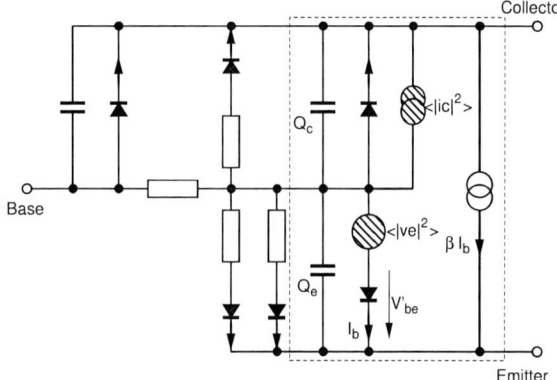

Fig. 6. Schematic of intrinsic HBT with shot noise sources. From [13].

Fig. 7. Time-constants for $3\times30\,\mu m^2$ HBT as a function of collector current at $V_{CE} = 4$ V: Correlation time constant τ determined from noise measurements (•), total intrinsic time constant of current gain α determined from S-parameter measurements (—), sum of base and collector transit-time $\tau_b + \tau_c$, approximated by formulas of the FBH large-signal model (- - -), 65% of calculated τ_c estimated from FBH large-signal model formulas ($\cdots$). From [13].

This kind of noise source is easily implemented as a subcircuit, and has been shown to be required for reliable oscillator phase noise simulation [7]–[9].

Fig. 4 shows measurement and simulation of residual phase noise [8]. The HBT is measured at 3.5 GHz in a 50-Ohm environment, with class-a bias and an available source power of −6 dBm. The graph shows the noise sideband close to the carrier as a function of offset frequency. The traditional noise implementation (dash-dotted line) clearly fails to predict the residual phase noise of the HBT. The dashed line, on the other hand, was simulated relying on a cyclostationary implementation. The improvement in accuracy due to this new approach is obvious, although both simulation approaches yield identical results in the small-signal regime.

B. Shot Noise

It is well known that the two shot noise sources of the HBT are correlated, and that this correlation can be approximated by the intrinsic transit-time [10]–[12]. The drawback so far was that this intrinsic time constant is not easily predicted. Thus, it had to be extracted from noise measurement in the small signal case, even if the full small-signal model is known. The large-signal case is even more involved, since the correlation time constant is a function of bias. This is the reason that large-signal models so far commonly simply neglect the correlation

completely.

Fig. 5 shows the minimum noise figure $NF_{\min}$ of a $3\times30\,\mu m^2$ HBT in two bias points. The impact of the correlation is clearly observed, since the curves obtained with the correlated model (solid lines) closely follow the measured data (bullets), while neglecting the correlation renders the result highly inaccurate as frequency increases (dashed lines).

The results shown in Fig. 5 were obtained from a recently proposed large-signal shot noise model [13]. The basic assumption of this new approach is that it is possible to approximate the noise performance with two noncorrelated noise sources. These sources need to be properly placed, as shown in Fig. 6. The basic configuration resembles the small-signal model introduced by v.d. Ziel [14] with the important difference that in this large-signal model only the time constant τ impacts the noise correlation, i.e., only the time delay associated with the collector charge is visible in the correlation of the noise sources.

Fig. 7 gives an idea of the time constants of a HBT. The straight line shows the total transit time $\tau_t = \tau + 1/\omega_\alpha$ of the small-signal model as extracted from S-parameters. The bullets represent the correlation time constants determined from noise measurement. It can be seen that the collector transit time model (long dashes) dominates the transit times at moderate currents. It is also shown that 65% of this time constant (short dashes) yields a good fit of the noise correlation. This property is exploited in the new noise model by modeling the different parts of the transit time through the different charges of the model.

IV. CHALLENGES

Although models for III-V HBTs are mature and allow for reliable simulation of nonlinear circuits, there are – and will be – always some challenges remaining. The three most important will be addressed in the following.

978-2-8748-7007-1/08 $25.00 © 2008 EuMA

A. Large Power Devices

Modeling large power devices is a challenging task, since two important conditions fulfilled for smaller devices may not hold anymore. This holds for all types of transistors, and HBTs are no exception.

The first assumption is that the device behaves as a homogeneous lumped element. However, large devices are prone to show distributed effects, either electrically due to large feeding structures, or thermally due to unequal heating.

The second assumption is that it is basically possible to measure the whole range of relevant bias points. As power levels increase, the safe measurement region decreases. It could be limited by the device's power handling capability. But also the measurement setup can set constraints, e.g. due to the maximum current ratings of the bias tees.

These issues do not call for a new formulation of the intrinsic device, nor would this be a solution. Since these effects change from device to device, there can not be one model that fits all. These challenges are to be faced in daily work and often require additional information from detailed analysis of the passive feeding structures or thermal interaction [6].

B. Operation in Deep Saturation

While the model for the active region of the HBT is mature, not much effort has been spent so far to the description of the saturation region. The model for the forward-biased base-collector junction is basically still close to the very basic Gummel Poon model. This might be an issue in the design of saturated amplifiers.

C. Operation Beyond f_t

It is possible to fabricate HBTs that provide maximum oscillation frequencies f_{max} much higher than the transit frequency f_t, and, hence, to design oscillators well beyond f_t [15]. The model, however, is limited towards higher frequencies due to its transit-time model: it approximates time delay effects through CR time constants, which corresponds to the first-order Taylor series expansion and requires $\omega\tau \leq 1$. As f_t corresponds roughly to the intrinsic time-delay of the device, it is evident that accuracy of a standard model rapidly deteriorates as frequency approaches f_t. In order to extend model validity to and beyond f_t, it will be required to develop sophisticated nonquasistatic time-delay models [16], [17], but a general comprehensive model has not been proposed so far.

V. Conclusions

HBT modeling converged in the last years, and the key issues have been solved. Thus, models are available that allow for reliable circuit simulation, including HBT self heating, prediction of weak nonlinearities, and a sophisticated description of capacitances and transit-time effects.

Also regarding noise modeling for nonlinear simulation, significant advancements have been achieved recently. Concerning the $1/f$ model, it has been shown that cyclostationary noise sources yield a significant improvement when simulating phase noise. A new model topology of the shot-noise model was also discussed that allows for a good approximation of the correlation of the noise sources. This advanced noise model is available as Verilog-A code, but not yet in a model provided by simulator vendors.

Finally, some limitations of the state-of-the-art models were addressed.

In conclusion, it can be stated that HBT models can be considered to be mature and at least as accurate as their HEMT counterparts.

References

[1] M. Rudolph, *Introduction to modeling HBTs*, Boston, London: Artech House, 2006.

[2] L.H. Camnitz, N. Moll, "An Analysis of the Behaviour of Microwave Heterostructure Bipolar Transistors," in *Compound Semiconductor Transistors — Physics and Technology*, S. Tiwari, Ed. New York: IEEE Press, 1993, pp. 21 – 45.

[3] M. Rudolph, R. Doerner, "Consistent Modeling of Capacitances and Transit Times of GaAs-Based HBTs," *IEEE Trans. Electron Dev.*, vol. 52, 1969 – 1975, Sept. 2005.

[4] M. Iwamoto, P.M. Asbeck, Th.S. Low, C.P. Hutchinson, J.B. Scott, A. Cognata, X. Quin, L.H. Camnitz, D.C. D'Avanzo, "Linearity Characteristics of GaAs HBTs and the Influence of Collector Design" *IEEE Trans. Microwave Theory Tech.*, vol. 48, pp. 2377 – 2386, Dec. 2000.

[5] M. Rudolph, R. Doerner, "Large-Signal HBT Model Requirements to Predict Nonlinear Behaviour," in: *IEEE MTT-S Int. Microwave Symp. Dig.*, 2004, 43 – 46.

[6] M. Rudolph, R. Doerner, "Large-Signal Modeling of High-Voltage GaAs Power HBTs," in: *IEEE MTT-S Int. Microwave Symp. Dig.*, 2005, 457 – 460.

[7] J.-Ch. Nallatamby, M. Prigent, M. Camiade, A. Sion, C. Gourdon, J.J. Obregon, "An Advanced Low-Frequency Noise Model of GaInP/GaAs HBT for Accurate Prediction of Phase Noise in Oscillators," *IEEE Trans. Microwave Theory Tech.*, vol. 53, pp. 1601 – 1612, May 2005.

[8] M. Rudolph, F. Lenk, O. Llopis, W. Heinrich, "On the Simulation of Low-Frequency Noise Upconversion in InGaP/GaAs HBTs," *IEEE Trans. Microwave Theory Tech.*, vol. 54, 2954 – 2961, Juli 2006.

[9] P. A. Traverso, C. Florian, M. Borgarino, F. Filicori, "An Empirical Bipolar Device Nonlinear Noise Modeling Approach for Large-Signal Microwave Circuit Analysis," *IEEE Trans. Microwave Theory Tech.*, vol. 54, pp. 4341–4352, Dec. 2006.

[10] A. v.d. Ziel, *Noise in Solid State Devices and Circuits*. J. Wiley & Sons, 1986, pp. 109–119.

[11] M. Rudolph, R. Doerner, L. Klapproth, P. Heymann, "An HBT noise model valid up to transit frequency," *IEEE Electron Device Lett.*, vol. 20, no. 1, pp. 24 – 26, Jan. 1999.

[12] J. Herricht, P. Sakalas, M. Schröter, P. Zampardi, Y. Zimmermann, F. Korndörfer, A. Simukovic, "Verification of π-Equivalent Circuit based Microwave Noise Model on A$_{III}$B$_V$ HBTs with Emphasis on HICUM, in *IEEE MTT-S Intl. Microwave Symp. Dig*, Long Beach, CA, 2005.

[13] M. Rudolph, F. Korndörfer, P. Heymann W. Heinrich, "Compact Large-Signal Shot-Noise Model for HBTs," *IEEE Trans. Microwave Theory Tech.*, vol. 56, 7 –14, Jan. 2008.

[14] A. v.d. Ziel, "Theory of Shot Noise in Junction Diodes and Junction Transistors," *Proc. IRE*, pp. 1639 – 1646, Nov. 1955.

[15] J. Hilsenbeck, F. Lenk, W. Heinrich, J. Würfl, "Low Phase Noise MMIC VCOs for Ka-Band Applications with Improved GaInP/GaAs-HBT Technology", *IEEE GaAs IC Symp. Dig.*, 2003, pp. 223 – 226.

[16] S.V. Cherepko, J.C.M. Hwang, "Implementation of Nonquasi-Static Effects in Compact Bipolar Transistor Models," *IEEE Trans. Microwave Theory Tech.*, vol. 51, no. 12, pp. 2531 – 2537, Dec. 2003.

[17] M. Rudolph, "Limitations of Current Compact Transit-Time Models for III-V-Based HBTs," *IEEE MTT-S Int. Microwave Symp. Dig.*, 2008, WE3C-05.

A Nonlinear Drain Resistance pHEMT model for Millimeter-wave High Power Amplifiers

Akira Inoue[#1], Hirotaka Amasuga[*2], Seiki Goto[*3], and Moriyasu Miyazaki[#4]

[#]*Information Technology R&D Center, Mitsubishi Electric Corporation*

5-1-1 Ofuna, Kamakura, Kanagawa 247-8501, Japan
[1]Inoue.Akira@cw.MitsubishiElectric.co.jp

[*] *High Frequency & Optical Semiconductor Works, Mitsubishi Electric Corporation*

4-1 Mizuhara, Itami, Hyogo 664-8641, Japan

Abstract— **A high power pHEMT with longer drain-gate separation can operate at higher voltage. However, it shows large output power loss at millimeter-wave in addition to the conventional parasitic power dissipation. The nonlinear drain resistance Rd of the pHEMT is found to cause the large power loss, although it behaves as a conventional resistor at low frequency. The nonlinearity of the Rd is modeled and shows good agreement with the measurement. Comparisons of pHEMTs with different nonlinear Rd also support the model.**

I. INTRODUCTION

GaAs pHEMTs are popular in microwave and millimeter-wave applications such as satellite communications, cell phones, wireless LAN, and automobile radar systems. Generally, efficiency and power of the transistors decrease at higher frequencies. Losses from parasitic resistances bring lower output power and gain which result in poor efficiency. Power losses of the transmission lines are also believed to be significant.

Recently, high operating voltage high power pHEMTs were developed for the millimeter-wave high power amplifiers [1,2]. To enhance power density, such devices have longer gate-drain length Lrd to achieve high breakdown voltage. However, measured performances of such high power pHEMTs are much poorer than the simulation that includes parasitic losses. The nonlinear drain resistance is found to play important role in this phenomenon

In this paper, the power loss problem of the high power pHEMTs is reviewed [2,3]. A nonlinear drain resistance pHEMT model is proposed to describe the power loss problem accurately. The physical meaning of the power loss is explained by a simple theory. Experiments and simulations at several frequencies support the model. To ensure the correctness of the model, pHEMTs with different nonlinear drain resistances are fabricated and compared.

II. POWER LOSS PROBLEM

Recent high power pHEMTs are driven at high drain voltage to achieve higher power density. Therefore, the length between gate and drain (Lrd) becomes longer to enhance breakdown voltage. Figure 1 shows the cross section of the pHEMT. Epitaxial structure consists of a n-AlGaAs/i-InGaAs/n-AlGaAs double-hetero structure, double doped electron supply layers, an i-AlGaAs barrier layer, and an i-GaAs buried layer.

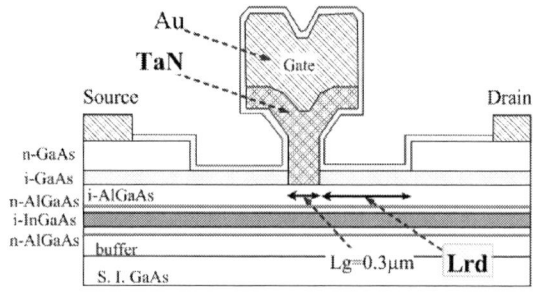

Fig. 1. Cross section of the pHEMT

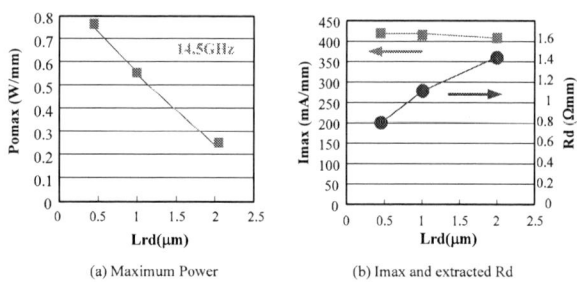

(a) Maximum Power   (b) Imax and extracted Rd

Fig. 2. Pomax, Imax, and extracted Rd vs. Lrd

Although the breakdown voltage was enhanced with longer Lrd, the output power was not increased to be expected. Figure 2 shows the maximum output power Pomax at 14.5GHz of pHEMTs with the Lrd of 0.45, 1.0, and 2.05μm. Although Imax is nearly same, the Pomax decreases with longer Lrd. The drain resistance Rd is less than 1.4Ω/mm that is not enough to explain the power decrease.

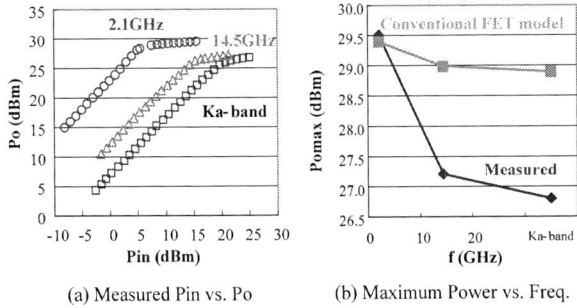

(a) Measured Pin vs. Po   (b) Maximum Power vs. Freq.

Fig.3. Frequency dependence of the output power of the pHEMT with 1μm Lrd.

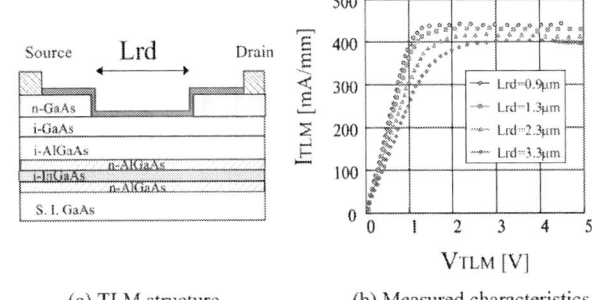

(a) TLM structure   (b) Measured characteristics

Fig. 5. TLM structure and IV characteristics.

The large signal performances of a 1μm Lrd PHEMT at several frequencies are shown in Fig.3. Higher the frequency, the maximum output power Pomax becomes smaller. Simulated results with conventional FET model that includes parasitics are shown in Fig. 3 (b). The extracted Rd is 1.1Ωmm. Measured and modeled S-parameters and small signal parameters are shown in Fig. 4. At 2.1GHz, the model shows good agreement with the measured Pomax. However, the model cannot describe the large decrease of the Pomax at 14.5GHz and Ka-band. Although the model shows slight decrease of the Pomax with respect to the frequency, it is not enough to describe the large power loss at high frequency. Since it is difficult to describe these phenomena with conventional model, we call them as the power loss problem at millimeter-wave.

To model the 1μm Lrd pHEMT, we used half of 1.3μm Lrd TLM, because 0.35μm or less portion of the gate-drain region is considered as the intrinsic part. The TLM current I_{TLM} is modeled by the following equation,

$$I_{TLM} = I_{TLM0} \tanh(\alpha_1 V_{TLM} + \alpha_2 V_{TLM}^2)(1 + \lambda V_{TLM}), - (1)$$

where α_1, α_2, λ are fitting parameters. Modeled I_{TLM} (solid line in Fig. 6) shows good agreement with measured one (cross marks).

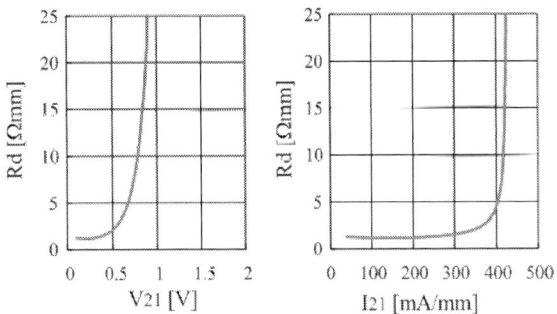

Fig. 6. Measured and modeled I-V characteristics of the TLM

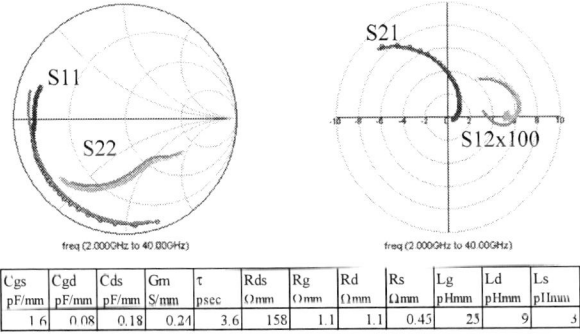

Fig.4. Measured (circles) and modeled (solid line) S-parameters, and small signal parameters (table).

Cgs pF/mm	Cgd pF/mm	Cds pF/mm	Gm S/mm	τ psec	Rds Ωmm	Rg Ωmm	Rd Ωmm	Rs Ωmm	Lg pHmm	Ld pHmm	Ls pHmm
1.6	0.08	0.18	0.24	3.6	158	1.1	1.1	0.45	23	9	3

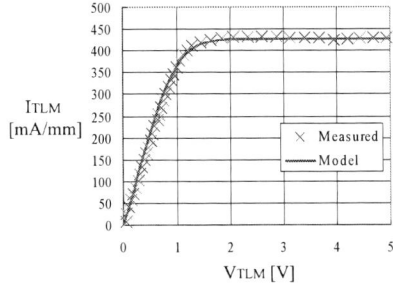

Fig. 7. Calculated Rd of the pHEMT

III. NONLINEAR DRAIN RESISTANCE MODEL

The region between gate and drain is physically a nonlinear resistance that arises from electron velocity saturation [4]. The nonlinearity of the Rd was measured by TLM (Transmission Line Model) structures. IV characteristics of various Lrd structures are measured with 1μsec wide pulses (Fig.5).

Since Rd is the half of the TLM, Rd is calculated by setting $V_{TLM}=2V_{21}$, $I_{TLM}=I_{21}$, where V_{21} and I_{21} denote voltage and current through the gate-drain region. Results are shown in Fig. 7. Rd rapidly increases when it becomes higher than 1V. However, the 425mA/mm maximum current of the Rd was considered to be large enough to influence nothing to the device, since the Imax of the pHEMT is 400mA/mm. Figure 8 shows measured pulsed IV characteristics of the pHEMT and simulated results with both linear and nonlinear Rd models (solid lines), where Vgs is set from -1.2V to 1.2V. Both models show nearly same results except the knee of the maximum current. Rd maintains the low value with the Ids of less than 360mA/mm. Therfore, IV characteristics become similar. S-parameters are also same at the operating bias, because Rd remains linear value at that bias. The nonlinear Rd has nearly no influence on the DC and small signal characteristics.

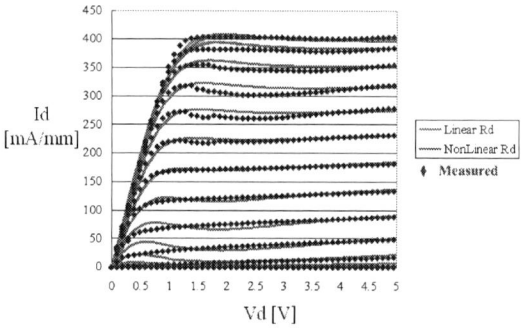

Fig. 8. Measured and simulated pulsed IV of the pHEMT

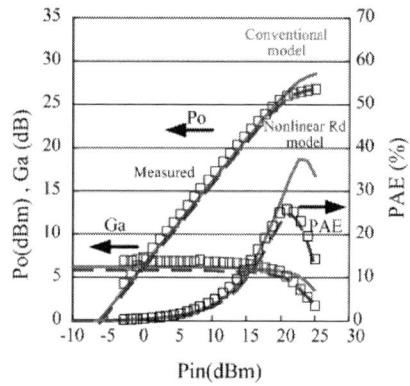

Fig. 9. Measured and simulated Pin vs. Po at Ka-band.

However, the models show different results under large signal operation. Figures 9 and 10 show measured results (square), linear Rd model (solid line), and nonlinear Rd model (dotted line), normalized to the gate width of 1mm. At 2.1 GHz, both models show quite similar results to the measured ones. On the other hands, at Ka-band, the conventional model shows much higher output power of 28.9 dBm, while the proposed model exhibits 26.8 dBm that is the same as the measured power. At Ka band, the proposed model also demonstrates nearly the same PAE as the measured one. From these results, we can conclude that in order to achieve

accurate millimeter wave band power simulations, it is necessary to precisely model the drain resistance nonlinearity.

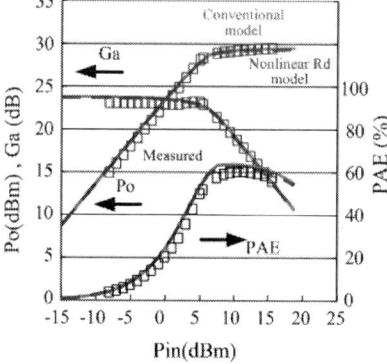

Fig. 10. Measured and simulated Pin vs. Po at 2.1GHz.

To understand the physical meaning of the large signal and high frequency effects of the nonlinear Rd, we calculated the Rd by using a simple equivalent circuit shown in Fig. 11.

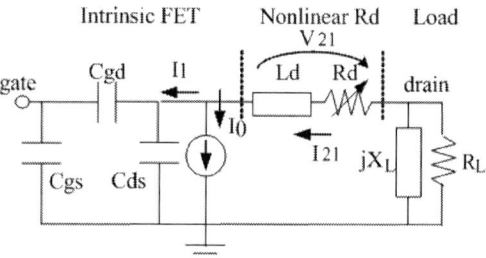

Fig.11. Simplified equivalent circuit of the pHEMT.

The impedance associated with Ld and Rd is small compared with R_L and the device capacitances, then

$$X_L \cong \frac{1}{\omega C_1} \quad , \text{where} \quad C_1 = Cds + \frac{CgdCgs}{Cgd + Cgs} . \quad - (2)$$

Cancelling imaginary components by jX_L, the current I_{21} through Rd is,

$$I_{21} = I_0 e^{j\omega t} + I_1 \cong I_0(1 + j\omega C_1 R_L)e^{j\omega t} . \quad - (3)$$

Therefore, the voltage V_{21} of the gate-drain region is,

$$V_{21} = I_{21}(Rd + j\omega Ld)$$
$$= I_0 \sqrt{(1 + \omega^2 C_1^2 R_L^2)(Rd^2 + \omega^2 Ld^2)}e^{j(\omega t + \varphi)} . \quad - (4)$$

Equation (4) shows that as the frequency and Lrd increases, ω, Rd, and Ld become larger, therefore, V_{21} increases. Higher V_{21} increases Rd. This results in larger power loss. The Rd increase only occurs with large I_0 that corresponds to the large signal operation. This equation describes the power loss problem that shows smaller maximum power with longer Lrd at higher frequency. Physically, this means that the current of charging and discharging combined capacitance C_1 exceeds the Imax of the FET and goes beyond the critical current where the Rd blows up.

IV. pHEMT WITH IMPROVED RD

The nonlinear Rd model predicts the reduction of millimeter-wave power loss by enhancing the maximum current of the gate-drain region. To verify this, stepped recess structure pHEMTs in Fig. 12 is proposed. We made two pHEMTs with the buried layer thickness Tb=150nm and 50nm. Epitaxial layers are same except the thickness Tb. Thick buried layer reduces the surface effects and increases the maximum currents of the gate-drain region. TLMs with both structures are also fabricated and measured.

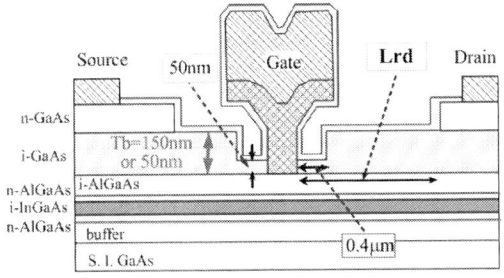

Fig.12. Cross section of the stepped recess pHEMT.

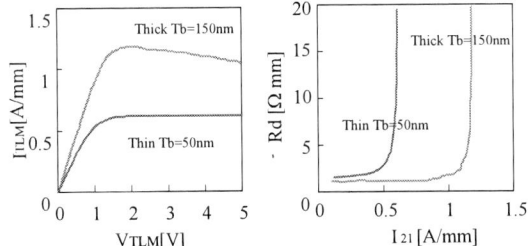

Fig.13. Measured IV of the TLMs and calculated Rd.

Fig. 13 shows the measured TLM IV-characteristics and the Rd calculated from it. Proposed thick buried layer exhibits nearly twice larger maximum currents compared with the thin one. Although Rd is nearly same in the low current region, thick Tb device keeps low Rd even at the I_{TLM} of 1A/mm.

Since the intrinsic parts of both pHEMTs are same, DC IV characteristics of both pHEMTs are quite similar as shown in Fig. 14.

Maximum output power Pomax at Ka-band is shown in Fig. 15. Several Lrd pHEMTs of both structures are measured (circles) and simulated with the nonlinear Rd model (triangles). The Pomax of the thick Tb pHEMT is significantly higher than that of thin one especially with longer Lrd. The thick buried layer overcomes the power loss problem from the nonlinear Rd. Since the model well describes the results, correctness of the nonlinear Rd model is proved.

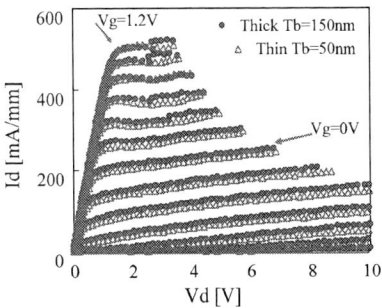

Fig.14. Measured pulsed IV of the both pHEMTs.

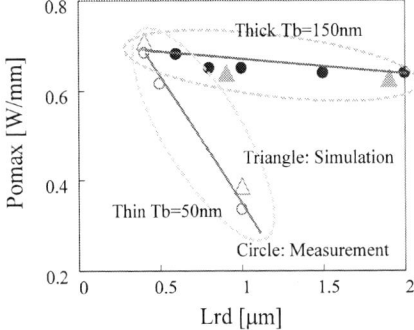

Fig.15. Measured and simulated Pomax vs. Lrd.

V. CONCLUSIONS

The millimeter-wave power loss problem of high power pHEMTs are well described by the proposed nonlinear drain resistance model. The problem is overcome by a stepped recess structure pHEMT which enhances the maximum current of the gate-drain region. High power and high frequency FETs can be widely modeled by the proposed nonlinear drain resistance.

ACKNOWLEDGMENT

Part of this work was supported by "The research and development for expansion of radio spectrum resources" of the Ministry of Internal Affairs and Communications, Japan.

REFERENCES

[1] H. Amasuga et al, "A High Power Density TaN/Au T-gate pHEMT with High Humidity Resistance for Ka-Band Applications," IEEE IMS2005 Digest, June 2005.

[2] M. F. Wong, J. A. del Alamo, A. Inoue, T. Hisaka and K. Hayashi, "Impact of Drain Recess Length on the RF Power Performance of GaAs PHEMTs," 6th Topical Workshop on Heterostructure Microelectronics Abstracts, pp.28-29, August 2005.

[3] A. Inoue, H. Amasuga, S. Goto, T. Kunii, M.F. Wong, J.A. del Alamo, T. Oku and T. Ishikawa, "A Nonlinear Drain Resistance Model for a High Power Millimeter-wave PHEMT," IEEE MTT-S Int. Microwave Symp. Dig, 2005.

[4] D. R. Greenberg, and Jesus A. del Alamo, "Velocity Saturation in the Extrinsic Device: A Fundamental Limit in HFET's," IEEE Trans. Electron Devices, Vol.41, No.8, pp.1334-1339, Aug.1994.

Proceedings of the 3rd European Microwave Integrated Circuits Conference

Widebandgap Semiconductor HFET Models for Microwave CAD

R.J. Trew, W. Kuang, Y. Liu, and G.L. Bilbro

ECE Department, Box 7911, North Carolina State University, Raleigh, NC, 27695-7911, USA

trew@ncsu.edu

Abstract— **Physics-based device models integrated into harmonic-balance microwave CAD simulators add flexibility and the ability to investigate both device and circuit design parameters before fabrication and prototyping. Accurate formulation of these models requires that relevant physical phenomena affecting the performance of these devices be identified and suitable models developed. In this work it is shown that inclusion of space-charge induced source resistance, RF channel breakdown, and gate tunnel leakage and surface conduction in AlGaN/GaN HFETs produce a simulator that produces excellent agreement between simulated and measured data for amplifiers fabricated with these devices. This type of simulator is very useful for advanced optimization investigations.**

I. INTRODUCTION

Large-signal device models are extensively used in microwave CAD software programs and permit components such as oscillators, power amplifiers, mixers, and other nonlinear components to be designed efficiently. Two basic modeling approaches are used, with the vast majority of models in use based upon equivalent circuit techniques. The other approach is generally termed 'physics-based' models, and these models are based upon solution of the semiconductor device equations applied to specific device geometry. While the equivalent circuit based models require extensive experimental characterization to determine the model element values, the physics-based models require knowledge of the charge transport characteristics, breakdown fields, and device doping and dimension data. The equivalent circuit models are generally only applicable to the parameter space over which they have been calibrated and are not ideally formulated for device design or investigations of anomalous effects observed in experimental data. Also, it is difficult to include certain physical phenomena that are known to affect device performance without extensive reformulation of the model. Physics-based device models are much easier to employ for these type investigations.

AlGaN/GaN HFETs are a challenge to model under large-signal RF operation due to a variety of physical effects that occur. A specific problem that is difficult to model is the 'reliability' problem experienced by these devices where the dc channel current and RF output power are observed to continually decrease as a function of operation time [1]. In this work a physics-based model that includes pertinent physical phenomena that have a first-order effect upon device performance is discussed. The model accurately describes the RF performance of AlGaN/GaN HFET microwave power amplifiers, including drain and gate leakage currents, and the

time dependence of dc and RF performance. Therefore, the model is useful for investigation of reliability issues.

II. SATURATION MECHANISMS

Microwave HFETs saturate under large-signal excitation by means of several distinct physical phenomena. It is known, for example, that the forward and reverse conduction characteristics of the gate diode are of fundamental importance, and therefore, accurate models for the forward and reverse conduction characteristics are required. The dc and dynamic load lines for a class A nitride power amplifier are shown in Fig. 1.

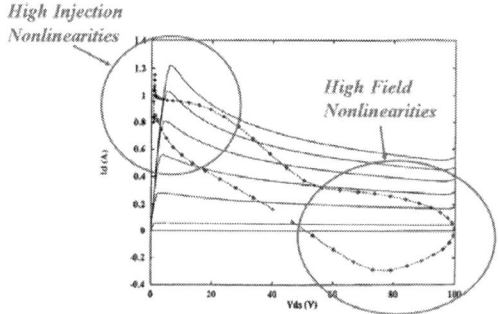

Fig. 1 Dynamic load line under maximum PAE conditions for a class A AlGaN/GaN HFET power amplifier (Vds=40 V).

As indicated in Fig. 1, the dynamic load line clips at both the high voltage, low current, and low voltage, high current extremes. During the high current portion of the RF cycle the channel current density approaches and can exceed the threshold for space-charge limited current flow. This results in the initiation of a nonlinear channel resistance in the gate-source region [2]. This nonlinear source resistance has been demonstrated in both experimental and theoretical work, and inclusion of the nonlinear resistance in HFET models results in excellent agreement between simulated and measured data.

During the high voltage portion of the RF cycle the gate electrode and the area under the gate edge in the conducting channel are subjected to very high electric fields. Two high field effects can occur: (1) electron tunneling from the gate to deep level states in the surface region of the AlGaN surface layer in the HFET, and (2) avalanche ionization within the conducting channel. Both effects can occur, and which one dominates is determined by the one with the lowest threshold. In GaAs and InP-based MESFETs and HEMTs gate

978-2-8748-7007-1/08 $25.00 © 2008 EuMA   286

breakdown at the gate edge generally dominates, with channel breakdown occurring for certain submicron gate length devices. In AlGaN/GaN HFETs avalanche breakdown generally occurs within the conducting channel, with the gate electrode demonstrating tunnel leakage and surface conduction. This is shown in Fig. 2, which shows the measured drain and gate leakage currents as a function of voltage for a selection of devices across a wafer.

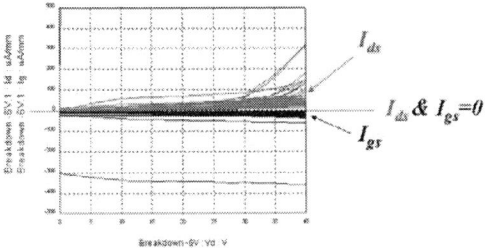

Fig. 2 Ids and Igs vs voltage at pinch-off for an AlGaN/GaN HFET.

As indicated the gate current (Igs) is small, but negative for all values of applied drain voltage indicating that tunnel leakage and gate conduction occurs. However, rapid increase due to breakdown is not observed. The channel current (Ids) shows only small leakage at pinch-off until some of the devices begin to show rapid increase in current, which is due to avalanche ionization in the channel. For this wafer some devices show channel breakdown at 30 V, with most showing breakdown at 50 V. This behavior is typical for nitride HFETs and indicates that accurate models for channel breakdown and gate tunnel leakage are required.

A novel gate tunnel leakage and surface conduction model has been developed [3]. The surface conduction mechanism is illustrated in Fig. 3, and is based upon a trap-to-trap electron hopping transport mechanism.

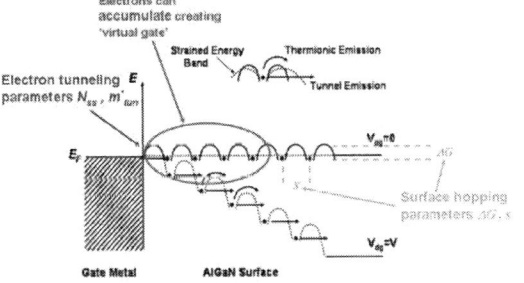

Fig. 3 Gate tunnel leakage and surface conduction model.

In this model the critical parameters are the trap spacing, s, and activation energy, ΔG. These parameters are used to describe the electron velocity in the trap-to-trap conduction mechanism, where the electron velocity is expressed as

$$v = \frac{2skT}{h}\exp\left(-\frac{\Delta G}{kT}\right)\sinh\left(\frac{qEs}{2kT}\right) \qquad (1)$$

Using values s=0.1 nm and ΔG=0.25 eV produces the simulated time-dependent gate leakage current and drain current shown in Fig. 4. As indicated, excellent agreement is obtained between the simulated and measured data. The ΔG value determined from the measured gate current data is consistent with a nitrogen vacancy defect that has been shown to exist in the nitride semiconductors [4]. The time dependence to the gate leakage current occurs due to space charge feedback from the electrons that leak to the semiconductor next to the gate. As the electrons accumulate, the space charge suppresses the electric field at the gate edge, thereby reducing the gate tunnel current. The time-dependent decrease in drain current results from a reduction in the 2DEG density in the conducting channel due to space-charge from the electrons on the AlGaN surface.

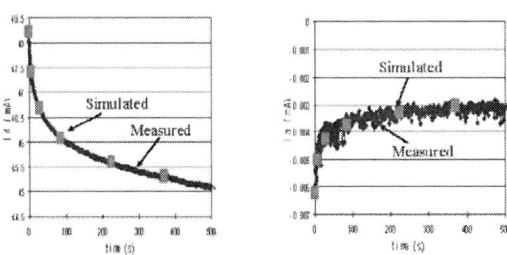

Fig. 4 Drain current and gate leakage current as a function of time for a device biased at pinch-off.

III. LARGE-SIGNAL HFET MODEL

The physics-based HFET model suitable for integration into a harmonic balance simulator is shown in Fig. 5.

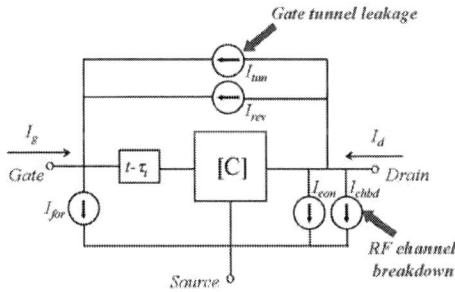

Fig. 5 HFET model including gate leakage and RF channel breakdown

The gate tunnel leakage and surface conduction mechanism is included and represented by a current generator between the gate and drain, and the RF channel breakdown is represented with a current generator connected between the drain and source. The nonlinear resistance is not shown in Fig. 5, but is included as a current-dependent resistance in the source. The source resistance is linear until the threshold for space-charge limited current is approached, which typically occurs before the peak in power-added efficiency is achieved. The nonlinear,

978-2-8748-7007-1/08 $25.00 © 2008 EuMA

time-dependent source resistance can be effectively modeled with the expression [2].

$$r_{ss} + \Delta r_{ss} = \frac{r_{ss}}{1 - \dfrac{I}{I_{sc}}} \qquad (2)$$

where r_{ss} is the dc magnitude of the gate-source region resistance, Δr_{ss} is the space-charge induced increase in r_{ss}, I_{sc} is the threshold for the onset of space-charge limited current, and I is the channel current.

The magnitude of the breakdown voltage that controls the drain-to-source current generator is critical in obtaining good agreement between the simulated and measured data. RF channel breakdown controls the onset of gain saturation and nonlinear performance due to waveform clipping. Unfortunately, avalanche ionization in a 2DEG is not well understood and is difficult to calculate due to lack of information regarding high field material parameters. For the work reported here, the breakdown voltage is used as a parameter that is adjusted to achieve agreement between the simulated and measured data. In this manner, the simulator can be used in a parameter extract mode to determine the RF breakdown. Generally it is observed that the RF breakdown voltage is less than the measured dc breakdown voltage, and the RF breakdown voltage increases with drain bias. That is, the same device biased at a higher drain voltage will demonstrate an increased RF breakdown voltage compared to a device biased at lower drain voltage. This phenomenon is currently being investigated in more detail.

IV. MICROWAVE PERFORMANCE

. The model has successfully simulated a variety of microwave AlGaN/GaN HFET amplifiers ranging from S-band to Ka-Band. In all cases the onset of the space-charge induced source resistance is observed and inclusion of this phenomenon results in excellent agreement between the simulated and measured dc and RF performance. The threshold channel current density for the onset of the space-charge limited current can be used as a extractable fitting parameter or calculated self-consistently from device physics. The threshold will vary with the amount of charge ionization occurring within the channel since holes generated in the ionization process drift towards the source region where they recombine with 2DEG electrons, thereby reducing the channel charge density. Therefore, the channel ionization and space-charge limited current processes are inter-dependent.

The simulated performance of an X-band AlGaN/GaN HFET power amplifier, compared with the measured data, is shown in Figs. 6-8.

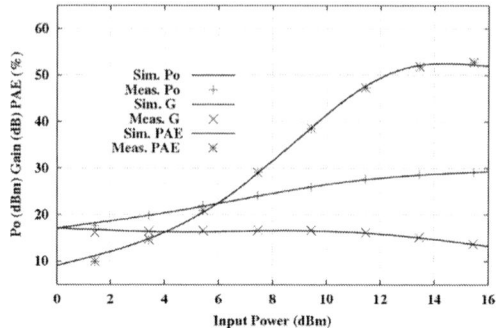

Fig. 6 Simulated and measured RF performance for a 10 GHz class A-B AlGaN/GaN HFET amplifier (Vds=25V)

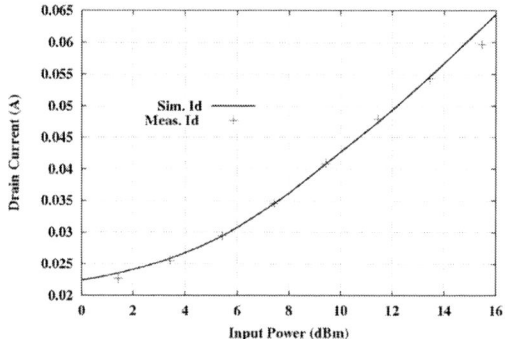

Fig. 7 Simulated and measured dc drain current for a 10 GHz class A-B AlGaN/GaN HFET amplifier (Vds=25V)

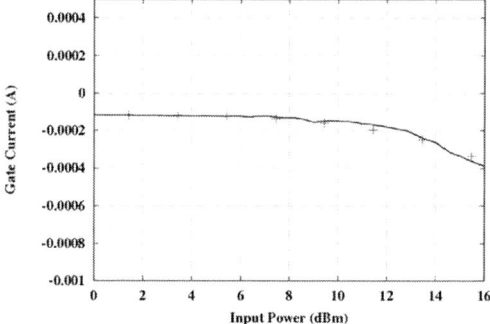

Fig. 8 Simulated and measured dc gate current for a 10 GHz class A-B AlGaN/GaN HFET amplifier (Vds=25V)

The solid lines are the simulated results and the discrete points are the measured data. The HFET has a gate length of Lg=0.2μm and a gate width of W=500μm. The drain was biased at Vds=25V with the gate biased to produce class A-B operation. The simulated and measured dc drain and gate currents as a function of RF input power are shown in Figs. 7 and 8, respectively. Note that the gate current is small, but finite and negative for all input power drive levels, indicating that tunnel leakage is occurring without any RF drive. This occurs since the dc bias produces an electric field at the gate edge with magnitude sufficient to produce electron tunneling. As the RF drive increases, the magnitude of the tunneling current increases, as indicated.

The resistance of the gate-source region thoughout the RF cycle is shown in Fig. 9. The solid line indicates the condition at the onset of space-charge limited current, and the dotted line indicates the condition when the amplifier is tuned for maximum power-added efficiency.

As indicated in Fig. 9, the source region resistance under no drive is about $r_{ss}=1.25$ Ω. During the low voltage, high current portion of the RF cycle the source region resistance becomes time-dependent and increases as the channel current density approaches the space-charge limited regime. As the amplifier is driven to maximum power-added efficiency, the source region resistance demonstrates significant increase, with this particular device showing an increase from the r_{ss} =1.25 Ω value to a magnitude just under r_{ss} =6 Ω.

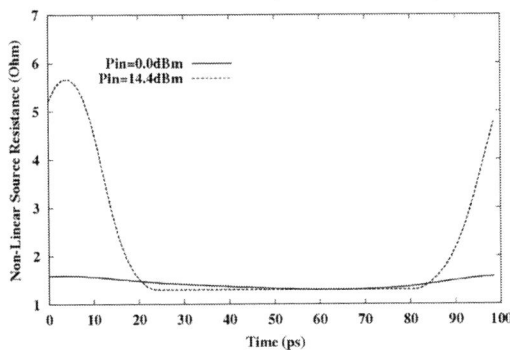

Fig. 9 Time-dependent source resistance for a 10 GHz class A-B AlGaN/GaN HFET amplifier at the onset of space-charge limited current (solid line) and at maximum power-added efficiency (dotted line)

Although all devices investigated to date have demonstrated the existence of the nonlinear source resistance, the magnitude of the phenomenon varies with device structure, operating conditions, and frequency of operation. Optimized S-Band devices often demonstrate an increase in r_{ss} by about a factor of two, whereas Ka-Band devices can show an increase in r_{ss} of an order of magnitude. Generally, as the device is scaled to smaller dimensions for higher frequency operation, the space-charge resistance demonstrates greater increases as the device is driven into saturation. The effect can be minimized by careful design.

The nonlinear source resistance can have a significant effect upon the linearity of the device [5]. Since the transconductance and gain of the amplifier are very sensitive to the magnitude of the source resistance even small variations can produce significant degradation in linearity. Practical utilization of these devices for communications applications where linearity is a priority will require design optimization to minimize the nonlinear source resistance phenomenon.

V. CONCLUSIONS

Physics-based device models can add significant flexibility to investigate phenomena affecting device and circuit performance, as well as serve as a tool to investigate new and optimized device designs before prototyping is necessary. The AlGaN/GaN HFETs present challenging modeling issues since these devices operate at very high electric fields and very high current densities, sufficient to produce a variety of performance limiting effects. It is demonstrated in this work that a physics-based model for AlGaN/GaN HFETs that includes the major performance limiting phenomena experienced by these devices has been developed. These effects include the onset of space-charge limited current transport in the gate-source region, RF breakdown in the 2DEG conducting channel, and gate tunnel leakage and surface conduction by a trap-to-trap hopping mechanism between the gate and drain. Techniques to include these effects in a physics-based device model integrated into a harmonic balance simulator have been developed. The simulator demonstrates excellent agreement between the simulated and measured dc and RF data.

ACKNOWLEDGMENT

This work was supported in part by the U.S. Army Research Office under Grant DAAD19-03-1-0148 and by the Office of Naval Research under Multiuniversity Research Initiative Grant N00014-05-1-0419.

REFERENCES

[1] K.S. Boutros, P. Rowell, and B. Brar, "A study of output power stability of GaN HEMTs on SiC substrates," *42nd Annual International Reliability Physics Symposium*, Phoenix, AZ, 2004, pp. 577-578.

[2] R.J. Trew, Y. Liu, G.L. Bilbro, W.W. Kuang, R. Vetury, and J.B. Shealy, "Nonlinear Source Resistance in High Voltage Microwave AlGaN/GaN HFET's," *IEEE Trans. Microwave Theory and Tech.*, vol. 54, pp. 2061-2067, May 2006.

[3] R.J. Trew, Y. Liu, W. Kuang, and G.L. Bilbro, "Reliability Modeling of High Voltage AlGaN/GaN and GaAs Field-Effect Transistors," Proc. of SPIE, Vol. 6894 1H1-7, 2008.

[4] Z.Q. Fang, D.C. Look, W. Kim, Z. Fan, A. Botchkarev, and H. Morkoc, "Deep Centers in n-GaN Grown by Reactive Molecular Beam Epitaxy," *Appl. Phys. Lett.*, vol. 72, No. 18, pp. 2277-2279, May 1998.

[5] Y. Liu, R.J. Trew, and G.L. Bilbro, "RF Linearity and Nonlinear Source Resistance in AlGaN/GaN HFET's," *IEEE International Microwave Symposium (IMS)*, Honolulu, HI, June 3-8, 2007.

A Non-Quasi-Static SOI MOSFET Model

Darren R. Burke and Thomas J. Brazil

School of Electrical, Electronic and Mechanical Engineering, University College Dublin
Belfield, Dublin 4, Ireland
darren_burke@yahoo.com
tom.brazil@ucd.ie

Abstract— **The static and large-signal behaviour of a new model for a submicron partially-depleted (PD) body-tied (BT) silicon-on-insulator (SOI) MOSFET was recently shown to give excellent agreement with measurements. Here, we complete the model validation with a detailed study of its small-signal capabilities up to a frequency of 50 GHz. Additionally, a new direct procedure is described enabling the extraction of a full parasitic network without the need for any on-wafer de-embedding structures.**

I. INTRODUCTION

The many advantages associated with silicon-on-insulator (SOI) MOSFETs compared to bulk devices have been known for some time [1], resulting in SOI technology attaining increasing importance in high-speed wireless communications systems. The conventional quasi-static (QS) SPICE models, such as BIM3v3 [2], which assume an instantaneous charging of the inversion layer, generate questionable results when used to describe the performance of high-speed circuits. Hence, the BSIMSOI model, which is based on the BSIM3 model, also becomes of limited value when used at RF frequencies. The work presented in [3] demonstrates an extraction technique for a non-quasi-static (NQS) SOI model up to a frequency of 40 GHz. This model, however, is only applicable for small-signal applications. The authors of [4] have accounted for NQS effects by using the channel segmentation approach [5]. As shown, the predicted small-signal results in [4] agree well with experiment, though no evidence is provided for the large-signal capabilities of the model. In [6], we introduced a new compact DC and large-signal partially-depleted (PD) body-tied (BT) SOI MOSFET model. In this letter, the model validation is completed through a detailed study of its small-signal capabilities up to a frequency of 50 GHz. Further, bias-dependent capacitances are added to the model to improve its performance, while resistances which model the gate tunneling currents are also included. Finally, a straightforward extraction technique that does not require any on-wafer de-embedding structures is used to determine the extrinsic parasitic component values.

II. MODEL IMPROVEMENTS AND PARASITIC NETWORK

NQS effects are incorporated into the SOI model by using a novel formulation of the channel segmentation approach. A key advantage of the segmentation method compared to other NQS modelling approaches is that it automatically gives full consistency between results from large-signal and small-signal

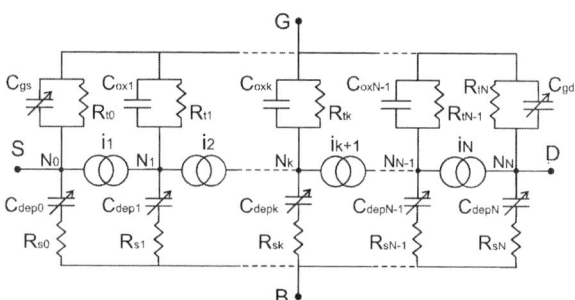

Fig. 1 Intrinsic SOI model with N channel segments.

models. A circuit diagram of the new segmented model, which is based on that developed in [6], is shown in Fig. 1. The intrinsic device consists of N channel segments, with each segment having a current source, i_k, an oxide capacitance, C_{oxk}, a gate tunneling resistor, R_{tk}, and a series connection of a depletion capacitance, C_{depk}, and a substrate resistance, R_{sk}.

The small-signal and large-signal prediction capabilities of the model can be improved if the two oxide capacitances at nodes $k = 0$ and N in Fig. 1 are replaced with nonlinear bias-dependent capacitances. In this work, these capacitances are modelled using the following novel expression:

$$C_{gx}(t) = (1/N)\left\{C_x + \alpha_1 \tanh\left[\beta_1\left(v_{GB}(t) - V_1\right)\right]\right\} \cdot \left\{1 - \alpha_2 + \alpha_2 \tanh\left[-\beta_2\left(v_{DB}(t) - V_2\right)\right]\left(0.5\left(1 + \tanh\left[\beta_1\left(v_{GB}(t) - \alpha_3\right)\right]\right)\right)\right\}$$

(1)

where x denotes either the source or drain node, and C_x, α_1, α_2, α_3, β_1, β_2, V_1, and V_2 are fitting parameters. The inclusion of these bias-dependent capacitances also means that even if only one channel segment is used, good matching can be achieved over the entire bias plane, as long as the frequency is kept low. Note, also, that C_{gs} is guaranteed to equal C_{gd} at $V_{ds} = 0$ V since the same coefficients are used in both expressions.

Multi-bias *S*-parameter measurements are required to obtain the values of the fitting parameters for these bias-dependent capacitances. However, as the number of channel segments increases, the impact of these capacitances decreases due to the factor N in (1). Thus, only a small number of *S*-parameter measurements are required if a large value of N is

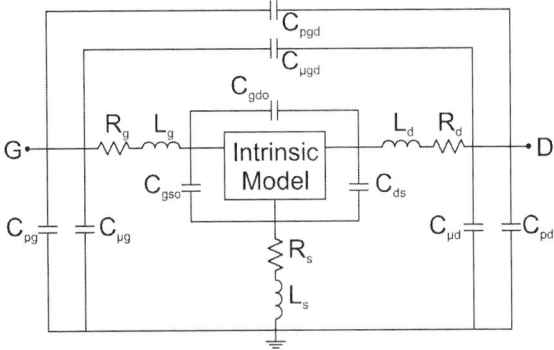

Fig. 2. Complete parasitic network for the SOI wafer used in this work.

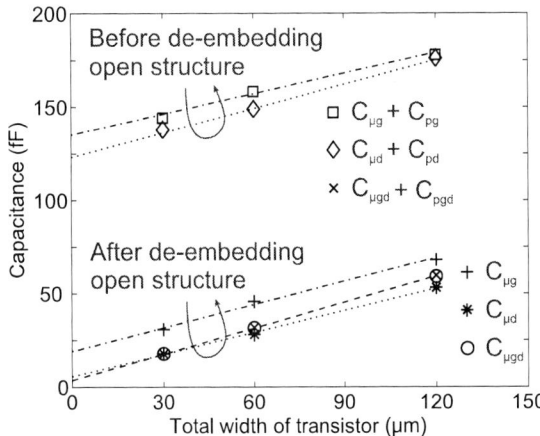

Fig. 3. Access capacitances as a function of the finger width before and after de-embedding the open structure. Note that the value of $C_{pgd} \approx 0$ fF.

used, since the C_{gx} capacitances only need to be approximately estimated.

The dimensions of modern devices are continuously scaled down to smaller sizes to increase speed and packing density, which results in a significant increase in the magnitude of the tunneling current through the gate oxide. In this work, the tunneling current has been modelled empirically by placing resistors in parallel with the C_{oxk} capacitors in Fig. 1. These resistors are set equal to

$$R_{tk} = \frac{\dfrac{R_c}{Wh}\left[1 + \dfrac{EV_{DS}}{1 + (V_{DS}/F)^2}\right]}{(1 + G \exp\{HV_{GB}\})} \qquad (2)$$

where W is the channel width, h is the length of each channel segment, and R_c, E, F, G, H, and J are empirically chosen constants based on experimental measurements. The bracketed term in the numerator of (2) accounts for the saturating decrease of the gate tunneling current with increased drain bias, while the denominator models the exponential increase in tunneling current for increasing gate voltages.

III. PARASITIC NETWORK EXRACTION

Before any S-parameter measurements were taken, an LRRM calibration of the system was performed. Following this, the measurement reference planes are shifted up to the probe tips, i.e., to the extremities of the parasitic network shown in Fig. 2. In order to validate the proposed new parasitic network extraction procedure, it is necessary to initially de-embed the S-parameters of an open de-embedding structure from the measured S-parameters. Assuming this structure to be purely capacitive, which is only really valid for relatively low frequencies, this will subtract the C_{pg}, C_{pd}, and C_{pgd} pad capacitances from Fig. 2. Note that the average values of these capacitances over the frequency range 0.5 to 50 GHz were

extracted to be, respectively, 114.6 fF, 119.2 fF, and ~ 0 fF.

Next, the linear regression routine used in [3] to extract the values of the access capacitances, $C_{\mu g}$, $C_{\mu d}$, and $C_{\mu gd}$, was performed. Three different transistors were used to extract the access elements, all having 30 gate fingers, a 0.13 µm channel length, and active zone widths equal to 1, 2, and 4 µm. The values of the extracted access capacitances, as shown in Fig. 3, were $C_{\mu g} = 19$ fF, $C_{\mu d} = 5.5$ fF, and $C_{\mu gd} = 4.2$ fF.

The technique used in [3] to extract the access capacitances can also be used to extract the parallel combination of the access and pad capacitances before the open structure is de-embedded, since both are independent of the bias and device width. This approach would prove to be useful in cases where an open de-embedding structure is not situated on the wafer. Further, the assumption that the equivalent circuit of the open structure is purely capacitive would be no longer required. The results of the extracted parallel combinations of the capacitances are shown in Fig. 3. The extracted values for $C_{pg}+C_{\mu g}$, $C_{pd}+C_{\mu d}$, and $C_{pgd}+C_{\mu gd}$ are, respectively, 135 fF, 123.1 fF, and 3.5 fF. Comparing these results with the previous two sets of results indicates that the method used in [3] to extract the access capacitances can also be used to extract the parallel combination of the pad and access capacitances. Note that a TRL calibration was performed in [3], which positioned the reference planes close to the intrinsic device. Thus, the effects of the pad parasitics were minimal.

After de-embedding the pad and access capacitances, the series impedances are extracted next using the 'Cold FET Extraction Procedure' proposed in [7]. The extracted element values are $R_g = 11.0$ Ω, $R_d = 1.0$ Ω, $R_s = 0.89$ Ω, $L_d = 11.8$ nH, and $L_s = 4.1$ nH. Note that the gate inductance was assumed to be zero as the extracted value was ≈ 0. The extracted resistances and inductances are then de-embedded.

978-2-8748-7007-1/08 $25.00 © 2008 EuMA

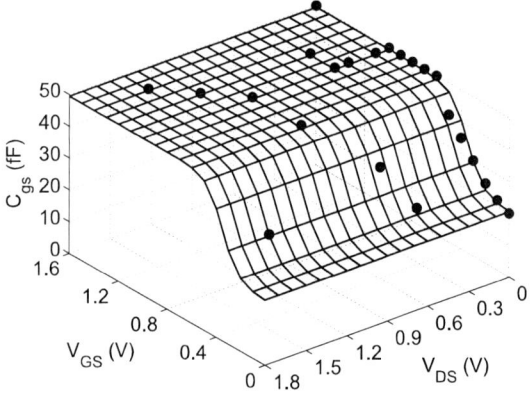

(a)

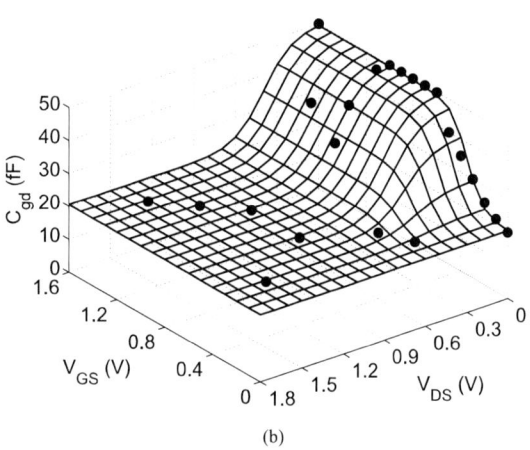

(b)

Fig. 4. Comparison of measured (dots) and modelled (lines) (a) C_{gs} and (b) C_{gd} versus V_{GS} and V_{DS}.

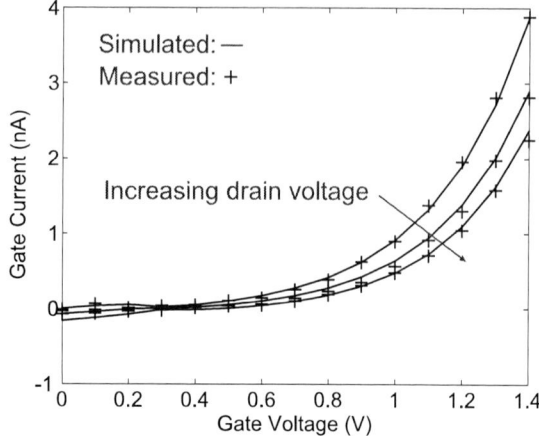

Fig. 5. Comparison of measured (crosses) and simulated (lines) gate tunneling current. Drain voltages range from 0 V to 0.8 V in steps of 0.4 V.

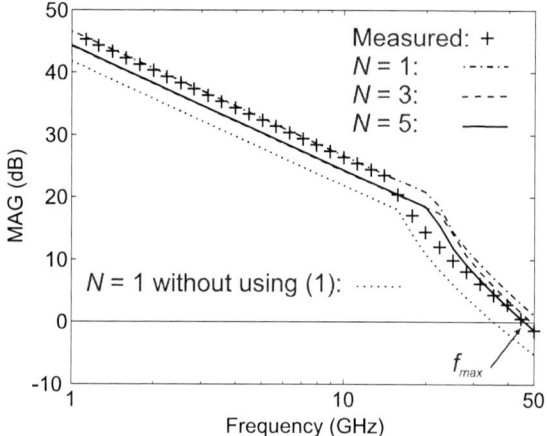

Fig. 6. Comparison of measured (crosses) and simulated (lines) MAG curves for varying N.

Finally, to extract C_{gso} and C_{gdo}, which model the overlap and fringing capacitances, and C_{ds}, which models the drain-source coupling capacitance, we bias the device at $V_{GB} = V_{DS} = 0$ V. These capacitances can then be given by $C_{gs0} = Im\{Y_{11}+Y_{12}\}/\omega = 56$ fF, $C_{gd0} = Im\{-Y_{12}\}/\omega = 56$ fF, and $C_{ds} = Im\{Y_{22}+Y_{12}\}/\omega = 37$ fF, where ω is the angular frequency.

IV. MODEL VERIFICATION

Measurements were performed on an n-type PD BT SOI MOSFET which had a channel length of 0.13 μm and width of 30x4 μm. The device had a front and back gate oxide thickness of, respectively, 2.0 nm and 400 nm, and a silicon thickness of 150 nm. The body was connected to the source.

In order to obtain the model parameter values for the nonlinear bias-dependent capacitance expression, multi-bias S-parameter measurements are required. The measured C_{gs} and C_{gd} capacitance values can then be extracted using the standard pi-model for the intrinsic device in Fig. 2. Following that, the parameter values in (1), with N set equal to 1, were optimized to fit (1) to the measured results. Fig. 4 demonstrates the nonlinear multi-bias capabilities of (1). As shown, the bias-dependent capacitance model reproduces the measured C_{gs} and C_{gd} capacitances with very good accuracy.

Fig. 5 shows the comparison between the measured gate tunneling current and the simulated gate tunneling current using (2). As shown, (2) models the gate current over a wide bias range with very good accuracy.

Fig. 6 compares the measured and simulated maximum available gains (MAG) for $N = 1$, 3, and 5 at $V_{GB} = V_{DS} = 1.2$ V. As shown, the QS model ($N = 1$) gives the least accurate results at high frequencies. As N increases, however, the accuracy of the predicted maximum frequency of oscillation, f_{max}, improves. Also shown in Fig. 6 is the MAG curve for the QS model without using (1) to replace the C_{ox0} and C_{oxN}

capacitances in the intrinsic model. The improved accuracy from using (1) is clearly evident.

Fig. 7(a) compares the measured and simulated, using $N = 5$, S-parameters for $V_{GB} = V_{DS} = 1.2$ V, while Fig. 7(b) shows the comparison at $V_{GB} = 1$ V and $V_{DS} = 0.5$ V. As seen, good agreement is observed in both sets of plots, thereby verifying the small-signal capabilities of the new PD BT SOI model.

V. CONCLUSIONS

The verification of a new physics-based compact microwave model suitable for PD BT SOI MOSFETs has been completed here. The small-signal capabilities of the NQS model have been demonstrated up to a frequency of 50 GHz, while the inclusion of empirical gate tunneling resistances has extended the capabilities of the model further. In addition, a technique to extract a full parasitic network has been verified that does not require any on-wafer de-embedding structures.

ACKNOWLEDGMENT

The authors wish to thank Prof. D. Vanhoenacker-Janvier of the Laboratoire d'Hyperfréquences, Université catholique de Louvain, Belgium, for providing the measurement results.

REFERENCES

[1] J.-P. Colinge, *Silicon-on-Insulator Technology: Material to VLSI*, 2nd edition, Boston: Springer, 1997.

[2] W. Liu, *MOSFET Models for SPICE Simulation including BSIM3v3 and BSIM4*, New York: John Wiley & Sons, Inc., 2001.

[3] J.-P. Raskin, R. Gillon, J. Chen, D. Vanhoenacker-Janvier, and J.-P. Colinge, "Accurate SOI MOSFET characterization at microwave frequencies for device performance optimization and analog modeling," *IEEE Trans. Electron Devices*, vol. 45, no. 5, pp. 1017–1025, May 1998.

[4] B. Iniguez *et al.*, "Deep-submicron DC-to-RF SOI MOSFET macro-model," *IEEE Trans. Electron Devices*, vol. 48, no. 9, pp. 1981–1988, Sep. 2001.

[5] Y. Tsividis, *Operation and Modeling of the MOS Transistor*, 2nd edition, New York: McGraw-Hill, 1999.

[6] D. R. Burke, M. El Kaamouchi, D. Vanhoenacker-Janvier, and T. J. Brazil, "DC and large-signal microwave MOSFET model applicable to partially-depleted, body-contacted SOI technology," in *IEEE MTT-S Int. Microwave Symp. Dig.*, 2007, pp. 585-588.

[7] A. Bracale *et al.*, "A new approach for SOI devices small-signal parameters extraction," in *Analog Integrated Circuits and Signal Processing*, Norwell, MA: Kluwer, 2000, pp. 157-169.

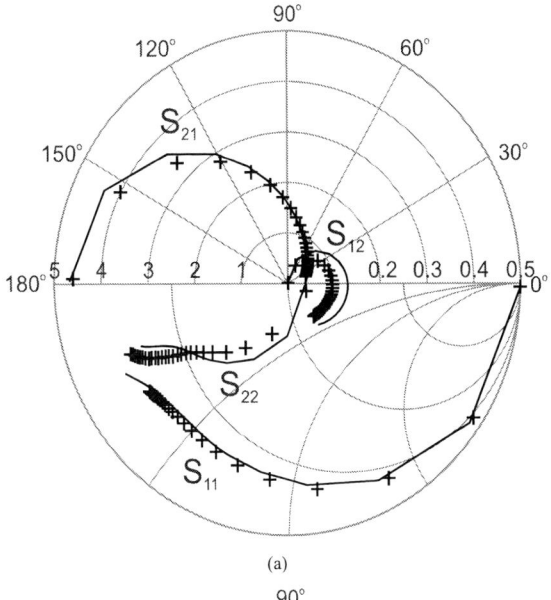

(a)

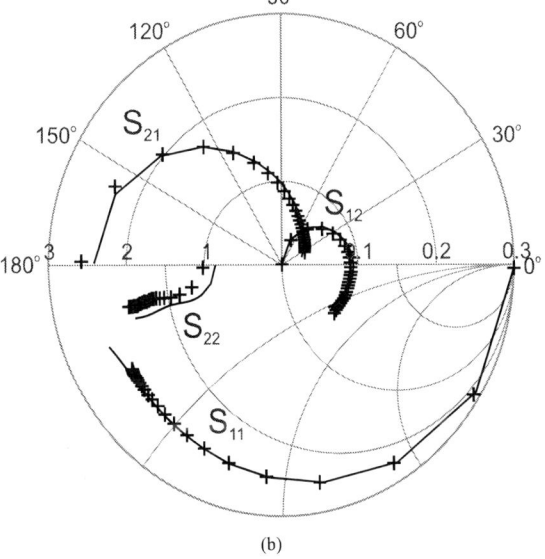

(b)

Fig. 7. Comparison of measured (crosses) and simulated (lines) S-parameters up to 50 GHz, in steps of 2 GHz, using $N = 5$ at (a) $V_{GB} = V_{DS} = 1.2$ V, and (b) $V_{GB} = 1.0$ V and $V_{DS} = 0.5$ V.

Accurate Nonlinear Electron Device Modelling for Cold FET Mixer Design

Valeria Di Giacomo[#1], Alberto Santarelli[*], Antonio Raffo[#], Pier Andrea Traverso[*], Dominique Schreurs[°], Julio Lonac[+], Davide Resca[*], Giorgio Vannini[#], Fabio Filicori[*], Maurizio Pagani[§]

[#] *University of Ferrara – Department of Engineering*
Via Saragat 1, 44100 – Ferrara, Italy
[1]valeria.digiacomo@unife.it

[*] *University of Bologna – Department of Electronics Information and Systems*
Viale Risorgimento 2, 40136 – Bologna, Italy

[°] *Electronic Engineering Department, Katholieke Universiteit Leuven*
B-3001 Leuven, Belgium

[+] *MEC s.r.l*
Viale Pepoli 3/2, 40123 Bologna, Italy

[§] *Ericsson Lab Italy S.p.A.*
Via Cadorna, 73, 20090 Vimodrone (MI), Italy

Abstract— **A nonlinear empirical model is here adopted to model the cold-FET behaviour of a GaAs PHEMT, in the framework of a resistive mixer application. The model, purely mathematical and technology independent, is suitably identified in the device operative region of interest and is validated in large-signal conditions by exploiting a measurements setup based on LS-VNA.**

I. INTRODUCTION

Microwave communication systems require a higher and higher integration level: all the functional blocks tend to be embedded on the same MMIC. This leads to an increasing interest in FET-based mixers, which, unlike diodes, are fully technological compatible with the other system components, besides providing good performances. In this context, particular attention is paid to cold FET mixers, due to their high linearity and low phase-noise [1].

This kind of application requires nonlinear electron device models capable of accurately predicting the weak, local non linearity around the V_{DS} = 0 V bias condition, while the LO signal swings within the maximum rating range of the gate voltage. For this reason, readily available foundry models, mostly thought for "hot" operating conditions, are unsuitable. Unfortunately, also in the literature few models can be found satisfying these requirements [2]-[4]. Conventional equivalent circuit models, in fact, usually do not suit well this kind of applications, since the analytical functions they are based on, depending on a relatively small number of parameters, provide reasonably good global fitting of the nonlinear behaviour at the cost of local prediction accuracy. Therefore, Look-Up-Table-based models can be a good choice, since they can be tailored to an optimal description of the local nonlinearities: in such a way, their extraction can be carried out to a limited extent in the device operating region of interest, in order to optimize the model performances for the specific application.

In this paper, the empirical, purely mathematical, Nonlinear Discrete Convolution (NDC) model [5] is validated for electron devices in cold FET configuration, in the framework of a resistive mixer application with f_{LO} = 4.8 GHz, f_{RF} = 4.2 GHz, down-converting at f_{IF} = 600 MHz. The model is suitably identified in the particular region of interest, providing very good accuracy in both predictions of current and voltage waveforms and estimation of output power at the IF frequency for several levels of the LO power.

II. MODEL DESCRIPTION AND IDENTIFICATION

The NDC model [5] describes the device dynamics (including nonquasi-static effects) in terms of elementary delay operators through a non-linearly controlled discrete-time convolution. The model can be derived, without any constrictions on the physical device structure, by considering that the instantaneous device currents depend on both present and past voltage values, and by adopting a purely mathematical functional description for the current/voltage relationship, i. e. for a single-port device:

$$i(t) = \Psi \left| v(t-\tau), \vartheta_C \right|_{\tau=0}^{\infty}, \qquad (1)$$

where Ψ is a nonlinear functional, ϑ_C is the device case temperature and $i(t)$ and $v(t)$ are the current and voltage at the device port, respectively. By separately dealing with parasitic and low-frequency dispersive effects, which feature a slow-dynamics, a short duration of the memory effects with respect to the typical operating frequencies of the device can be assumed. The hypothesis is well verified by means of both numerical simulations and experimental validation. A finite memory time T_M, suitably chosen, can be thus adopted for the electron device, leading to:

$$i(t) = \tilde{\Psi} \left| v(t-\tau), P_0, V_0, \vartheta_C \right|_{\tau=0}^{T_M} =$$
$$= \tilde{\Psi} \left| v(t), v(t-\tau) - v(t), P_0, V_0, \vartheta_C \right|_{\tau=0}^{T_M}, \qquad (2)$$

where V_0 and P_0 are the mean voltage values and the dissipated power, respectively, explicitly introduced in order to account for dispersive effects. Equation (2), in fact, allows great model simplification: the current-voltage functional description can be linearized with respect to purely dynamic voltage deviations between present and past voltages ($v(t-\tau)-v(t)$), leading to a convolution integral between voltage deviations and an impulse response function nonlinearly controlled by the instantaneous port voltages. By approximating the integral with a discrete summation of delayed terms, the final model equation is:

$$i(t) = F_{LF}[v(t), V_0, P_0, \vartheta_C] + \sum_{p=1}^{N_D} g_p[v(t)] \cdot [v(t - p\Delta\tau) - v(t)] \ (3)$$

where the g_p functions are voltage-controlled impulse responses and N_D is a suitable number of intervals in which the memory time is divided ($T_M = N_D \cdot \Delta\tau$). The linear dependence on the voltage dynamic deviations is justified by considering that under short-memory conditions the dynamic deviations are small also in the presence of large voltage signals.

The F_{LF} function in (3) represents the low frequency device behaviour: a suitable dispersive model should be adopted in conjunction to the NDC model, accounting dispersive phenomena due to traps and thermal effects, by means of the current dependence on mean voltage values (V_0) and dissipated power (P_0), respectively [5]. However, dispersive phenomena affect cold-biased devices to a negligible extent, thus the F_{LF} function is here replaced by the simple DC I/V characteristics.

The nonlinear model (1) is directly identifiable on the basis of DC I/V characteristics and small-signal S parameters and can be easily implemented in commercial circuit simulators by simply using the tools normally available for the construction of user-defined models (i.e., non-linear purely algebraic functions and delay operators).

The F_{LF} and the g_p functions can be suitably identified starting from empirical data, according to the procedure outlined in [5], which allows a systematic and unambiguous link between conventional measurements and prediction of the large-signal electron device behaviour. In particular, the model has been here extracted on the basis of DC and S-parameter data, measured in the operating region of interest for resistive mixer device configuration: V_{GS} ranging from off-state to forward condition and V_{DS} in the neighbourhood of the cold FET bias.

III. EXPERIMENTAL VALIDATION

The NDC model has been extracted for a GaAs 0.25-μm PHEMT with 140-μm width. To this aim, DC and S-parameter measurements were carried out for V_{GS} ranging from -3 V to 0.6 V (step 0.1 V) and V_{DS} between -0.8 V and 0.8 V (step 0.05 V); the S parameters were measured from 4 GHz up to 46 GHz exploiting an on-wafer TRL calibration. A memory time T_M=2.7 ps, divided into N_D=3 time slots, was chosen, according to the considerations reported in [5]. The static trans- and output I/V characteristics of the electron

device are reported in Fig. 1, where the chosen quiescent condition for the current application, V_{GS}=-1.2 V and V_{DS}=0 V, is also outlined for reference.

A first model validation was conducted in small signal conditions: the S parameters in the nominal bias point are plotted in Fig. 2, showing a very good model prediction.

In order to fully validate the model, a large-signal characterization was carried out on the device in cold FET configuration, by means of on-wafer waveform measurements under two-tone excitation, [6]. The measurement setup, Fig. 3, exploits a large-signal VNA, which acquires the spectral components in phase and amplitude (up to 20 GHz) of the incident and scattered voltage waves at the DUT ports. Two 50-Ω sources apply the signals at the gate port (LO signal) and at the drain port (RF signal). The measurement setup configuration adopted does not allow to tune source and load impedances, thus the device characterization has been carried out with termination impedances quite far from the real ones in mixer applications. This characterization is however significant in order to estimate the model prediction capabilities for cold FETs in large-signal operations.

In our test, the device is biased at V_{GS}=-1.2 V and V_{DS}=0 V and the input signal frequencies are 4.8 GHz for the LO generator and 4.2 GHz for the RF one. Different power levels are provided by the LO source, while the RF source power is constant.

The predicted and measured output power at the conversion IF frequency, i.e., 600 MHz, is plotted in Fig. 4a, for the different levels of the LO input power, showing a very good model performance for the specific application. The conversion loss, obtained as the ratio between the IF output power and the RF input power, is plotted in Fig. 4b: good agreement is achieved also in this test. An important requirement in mixer design is often the rejection of the LO and RF components at the output port (single or double balance): to this aim, it is important for the model at disposal to well predict also the unwanted output components. The predicted and measured output powers at the LO frequency is reported in Fig. 4c and the globally good model performance can be evaluated from Fig. 4d, where the whole spectrum of the reflected output waveform is displayed in correspondence of the maximum input power level.

Finally, the model performances are evaluated in the time domain by comparing predictions and measurements of the output voltages and currents in Fig. 5, again for the maximum LO source power level: the whole waveforms are well predicted, even though high nonlinearity is here involved.

The presented large signal validation outlines the very good capabilities of the NDC model in predicting the electrical FET behaviour under cold operation. This makes the model a good candidate for the design of a resistive mixer.

Thus, a double balanced cold FET mixer has been designed by adopting a single device topology with a circulator connected at the drain port in order to separate the input RF from the output IF signal. The FET has been biased under almost pinched-off channel bias conditions (V_{GS}=-1.2 V). Two matching networks were synthesized in order to maximize the output power at the IF and to have conjugate match at the gate port (LO input). Finally, a low-pass filter guarantees the mixer balance. The obtained performances are reported in Table I.

TABLE I
MAIN MIXER FIGURE OF MERIT

$f_{LO} = 4.8$ GHz $f_{RF} = 4.2$ GHz $f_{IF} = 600$ MHz $P_{av}^{LO} = 24$ dBm $P_{av}^{RF} = -10$ dBm	Conv. Loss ($P_{in}^{RF} / P_{out}^{IF}$)	LO Rejection ($P_{out}^{IF} / P_{out}^{LO}$)	RF Rejection ($P_{out}^{IF} / P_{out}^{RF}$)
	4.45 dB	18 dB	32 dB

IV. CONCLUSIONS

The empirical, behavioural NDC model has been adopted here to model the cold-FET behaviour of a PHEMT. The model has been validated in large-signal conditions, showing very good performances in predicting some of the main figures of merit in mixer applications. This leads us to conclude that the NDC model can be a valuable aid in resistive mixer design.

ACKNOWLEDGMENT

This work was funded in part by MIUR (Italian Ministry of Instruction, University and Research) and was performed in the framework of the former NoE TARGET.

REFERENCES

[1] S. Maas, *The RF and Microwave Circuit Design Cookbook*, Artech House, 1998.

[2] E.W. Lin, W.H. Ku, "Device considerations and modeling for the design of an InP-based MODFET millimeter-wave resistive mixer with superior conversion efficiency", *IEEE Trans. on Microwave Theory and Techniques*, Vol. 43, Issue 8, pp. 1951-1959, Aug. 1995.

[3] J. A. Pla, W. Struble, "Nonlinear model for predicting intermodulation distortion in GaAs FET RF switch devices," *IEEE MTT-S Dig.*, pp. 641-644., June 1993.

[4] I. Angelov, H. Zirath, N. Rorsman, "A new empirical nonlinear model for HEMT devices," *IEEE MTT-S Dig.*, pp. 1583-1586, June 1992.

[5] F. Filicori, A. Santarelli, P. A. Traverso, A. Raffo, G. Vannini, M. Pagani, "Non-linear RF device modelling in the presence of low-frequency dispersive phenomena", *Wiley Int. Journal of RF and Microwave Computer-Aided Engineering*, Vol. 16, Issue 1, pp. 81-94, Jan. 2006.

[6] D. Schreurs, J. Verspecht, B. Nauwelaers, A. Barel, M. van Rossum, "Waveform Measurements on a HEMT Resistive Mixer", *47th ARFTG Conference Digest-Spring*, Vol. 29, pp. 129-135, June 1996.

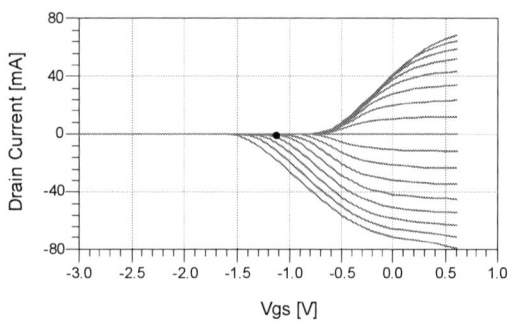
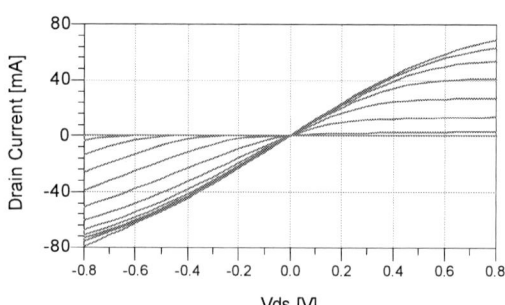

Fig. 1: PHEMT DC I/V characteristics. The black point on the trans-characteristic marks the nominal bias condition chosen for the cold FET model validation: V_{GS}=-1.2 V, V_{DS}=0 V

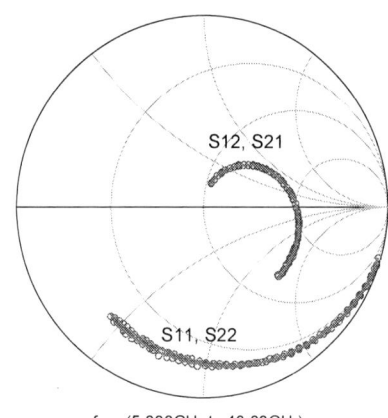

freq (5.000GHz to 46.00GHz)

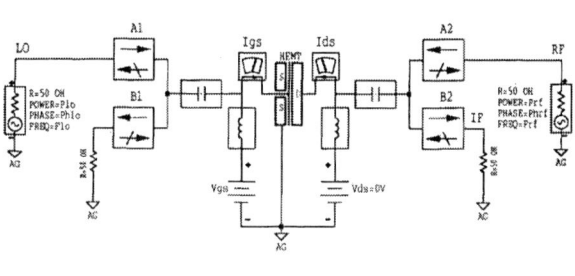

Fig. 2: PHEMT S parameters in V_{GS}=-1.2 V, V_{DS}=0 V, for frequency ranging between 5 and 46 GHz. Model predictions (red line) compared with measured data (blue circles).

Fig. 3: Measurement setup configuration adopted for the large-signal characterization of the cold FET

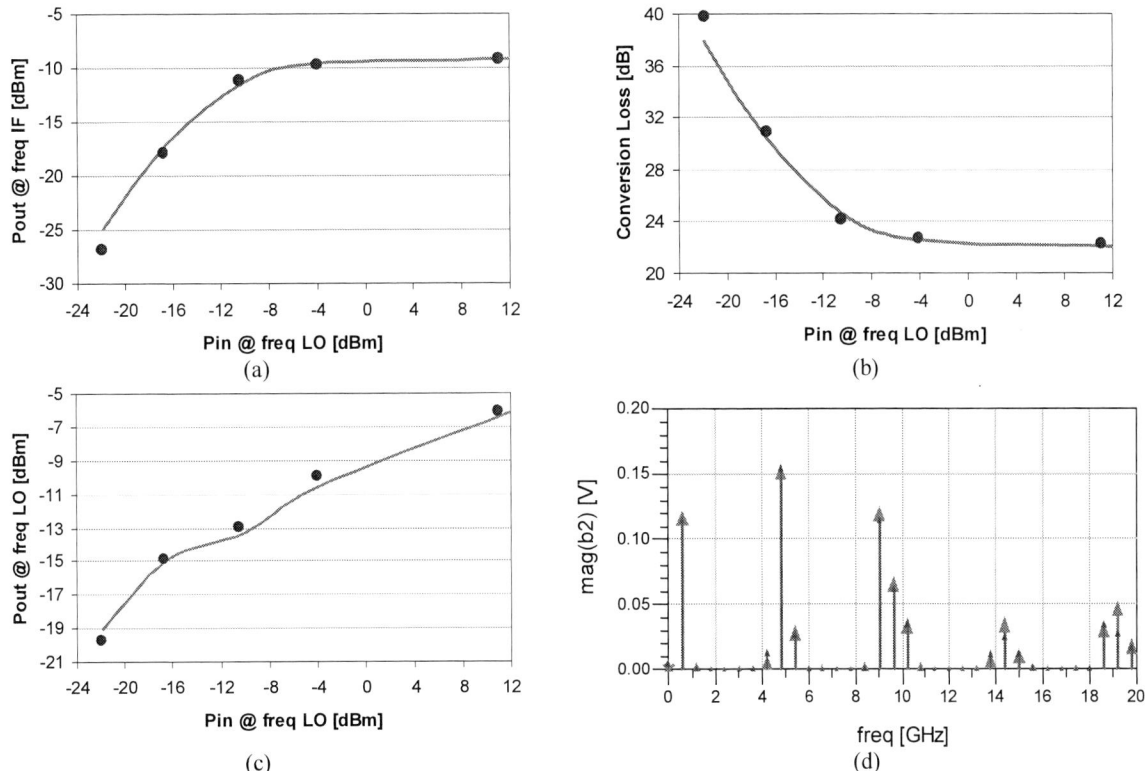

Fig. 4: LS validation (source and load terminations equal to 50Ω). (a): Output power at the conversion IF frequency (600 MHz) versus LO input power. (b): Conversion Loss. (c): Output power at the LO frequency (4.8 GHz). Model predictions (red line) and measurements (blue circles). (d): Spectrum of the output reflected waveform. Predictions (red bold line) and measurements (blue thin line).

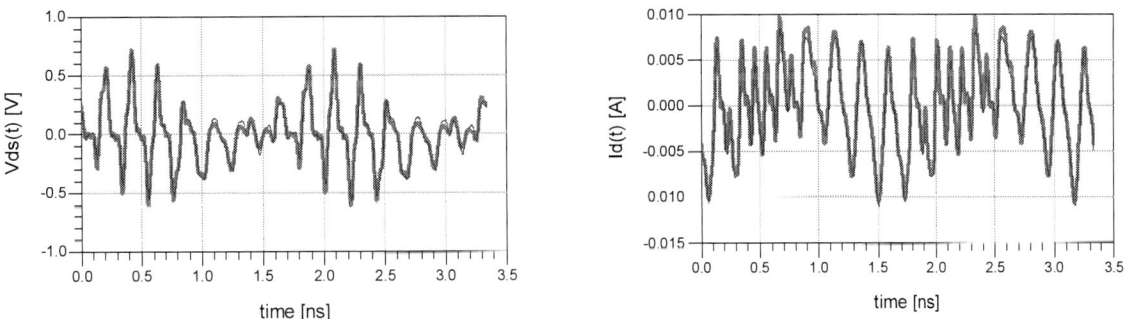

Fig. 5: Voltage and current waveforms at the output device port for the maximum level of input power. Model predictions (red bold line) and measurements (blue thin line). Source and load terminations equal to 50Ω.

978-2-8748-7007-1/08 $25.00 © 2008 EuMA

Gate Width Dependence of Noise Parameters and Scalable Noise Model for HEMTs

Yu Zhu, Cejun Wei, Oleksiy Klimashov, Binghui Li, Cindy Zhang, and Yevgeniy Tkachenko

Skyworks Solution Inc.
20 Sylvan Road, Woburn, MA 01801, USA
yu.zhu@skyworksinc.com

Abstract— **Explicit expressions describing the gate width dependences of HEMTs noise parameters have been obtained experimentally. The minimum noise figure and optimum source admittance are proportional to gate width, and noise resistance is inversely proportional to gate width. A scalable noise model is then developed, which accurately predicates noise parameters in a broad gate width range. The scalable noise model can be attached to any nonlinear signal model to predicate both noise and nonlinear signal responses.**

I. INTRODUCTION

HEMTs with different gate widths are used in MMIC design, and gate width can be optimized to achieve desired linearity and noise performances. HEMTs amplifiers with both low noise and high linearity performances are important for microwave receiver applications [1]–[2]. A scalable HEMT device model is thus required to accurately predicate nonlinear signal as well as noise performances.

Noise model has conventionally been realized by implementing noise sources inside a nonlinear signal model [3]–[4]. Therefore, the modeled noise response depends on both the noise sources and the equivalent circuit parameters of the signal model. It is usually difficult to accurately predicate small signal response in a broad bias range with a nonlinear signal model. Any error introduced in signal equivalent circuit model corrupts the accuracy of noise response prediction. In addition, noise model extraction requires extensive modeling effects. The signal equivalent circuit needs to be carefully de-embedded to determine the noise sources. The issues mentioned above become more remarkable when a scalable noise model needs to be generated.

A novel approach has recently been introduced [5]. Instead of being built inside the signal model, the noise model is built independently and can be attached to any nonlinear signal model. The noise model accuracy is not affected by signal model, and the noise model extraction is significantly simplified.

While the noise parameters have been expressed as functions of noise sources and signal equivalent circuit parameters [3]–[4], there are no explicit expressions describing the dependences of noise parameters on gate width. Such expressions will be guidance in selecting gate width in circuit design.

In this paper, the scaling rules describing the gate width dependences of the noise parameters are first revealed experimentally, which are then utilized to develop a scalable

noise model by extending the novel approach introduced in [5]. It is noteworthy that the simple scaling rules obtained cannot be used directly in conventional noise model, since the noise scaling effect have been achieved by scaling the signal equivalent circuit and /or the noise sources so far [6]–[7].

II. GATE WIDTH DEPENDENCE OF HEMTs NOISE PARAMETERS

The noise measurements were performed on the depletion mode AlGaAs/InGaAs pseudomorphic HEMTs with gate width of 200, 300, 400, 500, 600, 700, and 800μm, respectively, and the unit gate width is 50μm. The noise measuring frequency range is 0.8-6GHz, and the drain current range is 0.1-1.0 Idss.

Figure 1 shows the measured minimum noise figure, NFmin, versus drain current for different gate width. NFmin increases with current for all of devices in the current range of 0.1 − 1.0 Idss. Increasing gate width is an efficient way to achieve low NFmin at high current range. In order to find the relation among the noise behaviors for different gate width, the measured NFmin are plotted against current density as shown in Fig. 2. The NFmin behaviors for different gate width become similar to each other. NFmin is then normalized with a reference device with 400μm gate width. The normalized NFmin are plotted against gate width for different current density shown in Fig. 3. The normalized NFmin for different current density are very close, and a slight increase in NFmin with increasing gate width has been observed. The scaling rule for NFmin is thus expressed as linear function of gate width as

$$NF_{\min}(W,J) = NF_{\min}(W_0,J)[1+C(W-W_0)], \quad (1)$$

where, W is the gate width, W0 the gate width of reference device, J the current density, and C the only fitting parameter, which is independent of bias and frequency. The deviations from linear function observed in Fig. 3 are believed to be due to the measurement error.

Figure 4 shows the current dependence of noise resistance, Rn, for different gate width. Overall, Rn decreases with increasing gate width, and the sweet spot, where Rn reaches it minimum, moves to high current range with increasing gate width. Figure 5 shows the measured Rn versus current density. Rn behaviours become similar to each other, and the sweet spots appear at the same current density for different gate width. Figure 6 shows the normalized Rn versus the inverse of

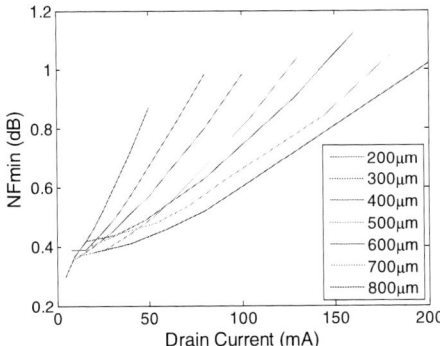

Fig. 1 Measured NFmin versus current for HEMTs with different gate width at Vd=5V, and frequency=3GHz

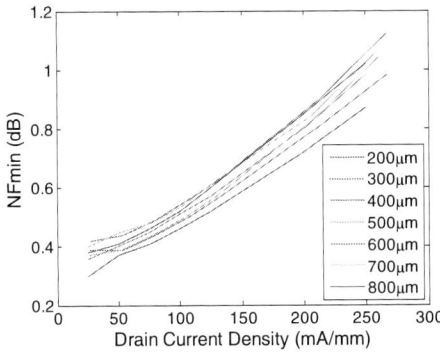

Fig. 2 Measured NFmin versus current density for HEMTs with different gate width at Vd=5V, and frequency=3GHz

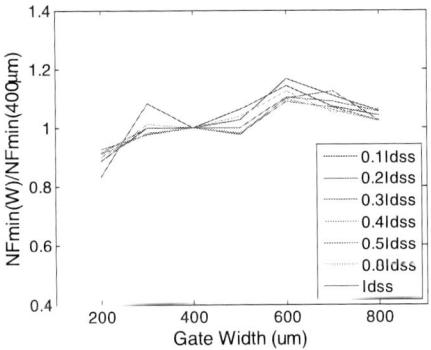

Fig. 3 Gate width dependence of normalized NFmin for different current density at Vd=5V, and frequency=3GHz

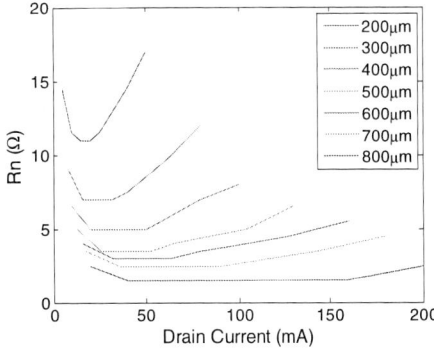

Fig. 4 Measured noise resistance versus current for HEMTs with different gate width at Vd=5V, and frequency=3GHz

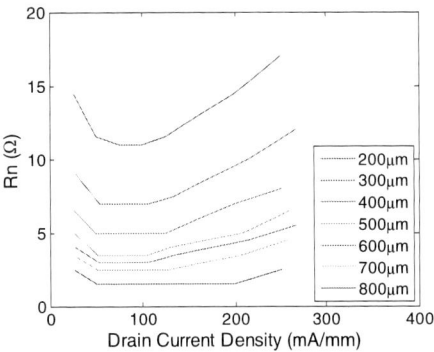

Fig. 5 Measured noise resistance versus current density for HEMTs with different gate width at Vd=5V, and frequency=3GHz

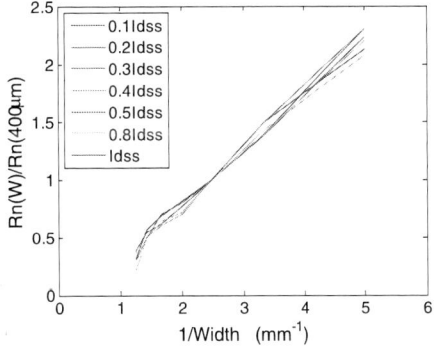

Fig. 6 Gate width dependence of normalized noise resistance for different current density at Vd=5V, and frequency=3GHz

the gate width. The curves for different current density are almost overlapped. It is interesting to note that the slope is exactly reference gate width W_0, and the Rn scaling rule can thus be well represented by a linear function of 1/W as,

$$R_n(W,J) = R_n(W_0,J)\frac{W_0}{W},\qquad(2)$$

and no extra fitting parameter is needed.

The optimal source admittance can be expressed as Yopt = Gopt+jBopt, where Gopt and Bopt are optimal source conductance and susceptance, respectively. The current dependences of Gopt are shown in Figs. 7. Gopt increases with increasing gate width as expected. When plotted in terms of current density, the Gopt behaviors become similar for different gate width as shown in Fig. 8. Figure 9 shows the normalized Gopt versus gate width for different current

978-2-8748-7007-1/08 $25.00 © 2008 EuMA

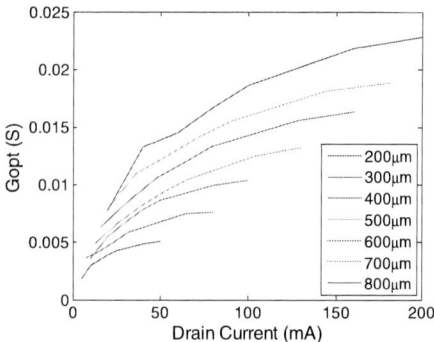

Fig. 7 Measured Gopt versus current for HEMTs with different gate width at Vd= 5V, and frequency=3GHz

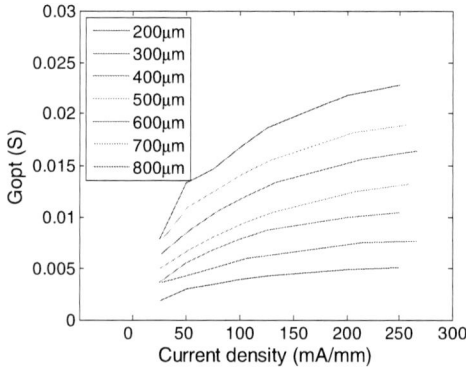

Fig. 8 Measured Gopt versus current density for HEMTs with different gate width at Vd=5V, and frequency=3GHz

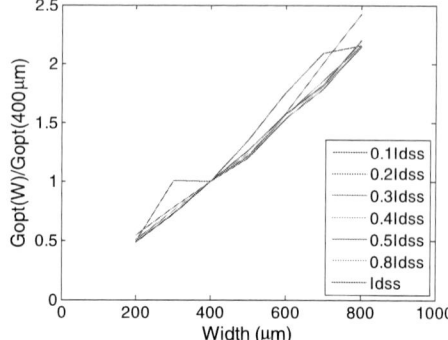

Fig. 9 Gate width dependence of normalized Gopt for different current density at Vd=5V, and frequency=3GHz

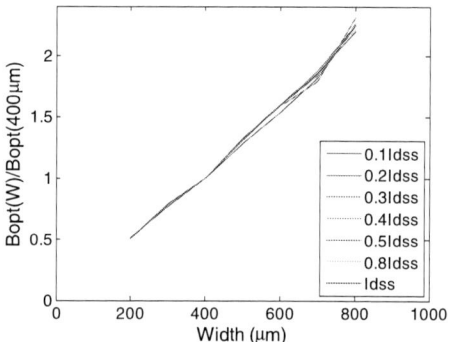

Fig. 10 Gate width dependence of normalized Bopt for different current density at Vd=5V, and frequency=3GHz

density. The same manipulation has also been applied to the measured Bopt, and the gate width dependence of normalized Bopt is shown in Fig. 10. The scaling rules for Gopt and Bopt can be expressed as linear functions of W as

$$G_{opt}(W, J) = G_{opt}(W_0, J)\frac{W}{W_0}, \qquad (3)$$

and

$$B_{opt}(W, J) = G_{opt}(W_0, J)\frac{W}{W_0}, \qquad (4)$$

respectively. It has also been found that the gate width dependence of Γopt, the optimal reflection coefficient, is not as straightforward as that of Yopt. It is thus much easy and accurate to use Yopt instead of Γopt in scaling calculation.

III. MODEL IMPLEMENTATION

The noise model proposed in [5] is reviewed briefly here. The noisy nonlinear model is realized with a cascade of a noise model and a noise free nonlinear signal model. The noise model behaves as a 0 Ohm resistor with arbitrary noise

performance. The noise performance is described via a correlation matrix, which can be calculated directly from measured noise parameters. Since no signal model parameter and noise sources are involved in noise calculation, a scalable noise model can be generated easily by directly utilize the scaling rules obtained.

A scalable noise model has been developed with the user complied model in ADS. Instead of using multiple sets of measured noise data for different gate width, only one set of measured data of the reference device is used in the model. When attached to a nonlinear signal model, the noise parameters of the reference device are first picked up according to the device bias conditions. Since the noise model is built as a nonlinear element, it gets the drain voltage and drain current values dynamically from the signal model. The desired noise parameters corresponding to a specific gate width are then calculated using the scaling rules. The measured and simulated noise parameters are shown in Fig. 11 (a), (b), and (c), respectively. Very good agreements have been achieved between the measurement and simulation.

While the frequency and bias dependence of noise parameters are represented by one set of measured data here, it is, of course, possible to implement formula to replace the measured data.

978-2-8748-7007-1/08 $25.00 © 2008 EuMA   300

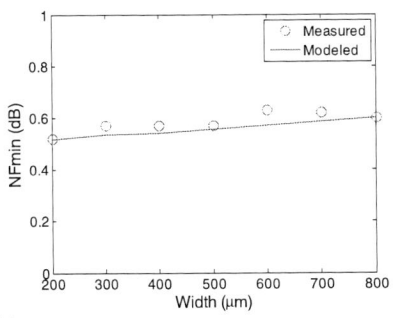

(a)

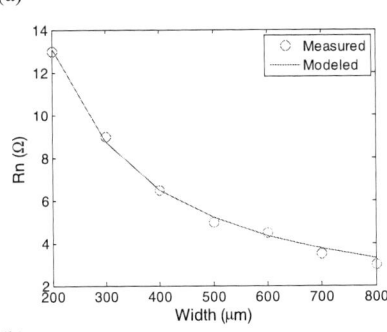

(b)

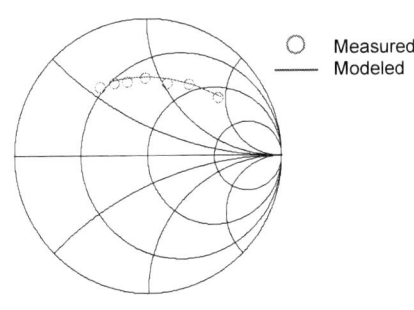

(c)

Fig. 11 Comparisons between measured and modeled noise parameters at Vd=5V, Jd=125mA/mm, and frequency=3GHz, (a) minimum noise figure, (b) noise resistance, (c) optimal reflection coefficient

IV. CONCLUSIONS

The gate width dependences of the four noise parameters have been experimentally revealed. The scaling rules are simple and accurate, and only one fitting parameter is needed. Based the scaling rules and one set of measured noise data, a scalable noise model has been developed as a user complier model in ADS. Good agreement between measured and modeled noise performances has been achieved in a broad gate width range. The scalable noise model can be attached to any nonlinear signal models to predicate both noise and nonlinear signal performances.

REFERENCES

[1] K. W. Kobayashi, Y. Chen, I. Smorchkova, R. Tsai, M. Wojtowicz, and A. Oki, "A 2 watt, sub-dB noise figure GaN MMIC LNA-PA amplifier with multi-octave bandwidth from 0.2-8 GHz," *2007 IEEE MTT-S Int. Microwave Symp. Dig.*, vol. 1, pp. 619-622, June 2007.

[2] E. Byk, P. Quentin, M. Camiade, S. Tranchant, "Plastic packaged high linearity low noise amplifier for 12-30GHz multi-band telecom applications," *2006 IEEE MTT-S Int. Microwave Symp. Dig.*, vol. 3, pp. 1903-1906, June 2006.

[3] W. M. Pospieszalski, "Modeling of noise parameters of MESFETs and MODFETs and their frequency and temperature, dependence," *IEEE Trans. Microwave Theory & Technique*, vol. 37, pp. 1340-1350, September 1992.

[4] B. Klepser, C. Bergamaschi, M. Schefer, C. G. Diskus, W. Patrick, and W. Bachtold, "Analytical bias dependent noise model for InP HEMT's," *IEEE Trans. Electron Devices*, vol. 42, pp. 1882-1889, November 1995.

[5] Y. Zhu, C. Wei, O. Klimashov, C. Zhang, and Y. Tkachenko, "Nonlinear Signal Model with Linear Noise Response for Low Noise Circuits Design," *2007 Asia Pacific Microwave Conference*, December 11-14, 2007.

[6] B. Hughes, "Designing FET's for broad noise," *IEEE Trans. Microwave Theory & Technique*, vol. 41, pp. 190-198, February 1993.

[7] B. Klepser, C. Bergamaschi, M. Schefer, S. Diskus, W. Patrick, and W. Bachtold, "Analytical bias dependent noise model for InP HEMT's," *IEEE Trans. Electron Devices*, vol. 42, No. 11, pp. 1882-1889, November 1995.

A 70GHz VCO with 8GHz Tuning Range in 0.13um CMOS Technology

Zongru Liu, Efstratios Skafidas and Rob Evans

National ICT Australia,

Department of Electrical and Electronic Engineering

University of Melbourne,

Parkville Victoria 3010, Australia

[1]z.liu@ee.unimelb.edu

[2]stan.skafidas@nicta.com.au

[3]rob.evans@nicta.com.au

Abstract— **A 70GHz VCO with 8GHz tuning range is implemented on 0.13um CMOS. It has an output power of -4dBm and a phase noise of -107dBc/Hz at 10 MHz carrier offset. From 0 to 70 degree Celsius the output power varies from -4dBm to -8dBm and exhibits a maximum frequency deviation of 200MHz over this range. The VCO has the highest figure of merit (-169.8dBc/Hz) of any VCO fabricated on bulk CMOS operating above 60GHz.**

I. INTRODUCTION

Millimeter wave oscillator design on CMOS is required by applications such as wireless personal area network (WPAN) and automotive radar. Furthermore, these applications require large tuning range, in excess of 4GHz for automotive and 7GHz for WPAN, moderate output power, low phase noise and low variation with temperature changes.

Three candidate architectures are usually considered for millimeter wave application on CMOS: 1) Fundamental, 2) VCO with frequency doubler, 3) Push-push VCO [4]. The fundamental architecture has narrow tuning range and also requires a wide band divider in the phase locked loop (PLL) which can be difficult to build at millimeter wave frequencies on CMOS. Architectures based on frequency doubling or push-push topology are a better choice because they can double the tuning range of a fundamental architecture. However, the frequency doubling architecture requires additional circuits such as a doubler/multiplier and filters which can consume considerable space. Insufficiently filtered harmonics generated by the doubler modulate the fundamental frequency and increase phase noise. The push-push architecture combines frequency generation and frequency doubling in one circuit. Furthermore since the varactors operate at a lower frequency, compared to a fundamental based architecture, they have a higher quality factor resulting in high Q tank circuits. Furthermore as the internal and external tank frequencies are separated the push-push VCO experiences less load pulling. Hence the push-push architecture is well suited for high frequency low phase noise applications.

In this paper, a 70GHz CMOS push-push voltage controlled oscillator (VCO) implemented on standard 0.13um bulk CMOS technology is presented. This VCO can be tuned from 65.8GHz to 73.6GHz, has an output power of -4dBm and has a phase noise of -107dBc/Hz at 10 MHz offset from the carrier (66GHz). The device output, when measured versus temperature from 0 to 70 degree Celsius varies from -4dBm to -8dBm and exhibits a maximum frequency deviation of 200MHz from its nominal frequency. The VCO presented in this paper has the best figure of merit (-169.8dBc/Hz) of any millimeter wave voltage controlled oscillator implemented on bulk CMOS operating above 60GHz.

II. VCO CORE STRUCTURE

A push-push voltage controlled oscillator is implemented in this paper. The core of the differential cross-coupled LC oscillator with push-push output is shown in Fig 1. Each arm has an LC tank circuit that determines the frequency of oscillation. Frequency dependent signals at the drain are then cross-coupled to the other transistor's gate which creates a negative impedance -1/gm. This negative impedance must exceed the losses of the resonator to ensure sustained oscillation. It has been reported in the literature that the push-push oscillator output frequency can be much higher than the transistors f_{max} [2]. In our design this allows us to dispense of the cascode transistor which consumes voltage headroom.

Most of the traditional designs include a current source to set the bias current and provide high impedance. Due to the oscillator transistor's mixing effect caused by nonlinearity, the low frequency noise of the current source is up converted to high frequency around the even harmonics and then down converted to the fundamental frequencies [5]. However the removal of the current source makes the circuit susceptible to power supply noise. Care must be exercised to ensure that power and ground plane are of sufficiently low noise. In our design, the current source is omitted to suppress this phase noise contribution. In the push-push oscillator the fundamental and odd harmonics cancel and power is delivered to the load at the even harmonics. The suppression of even-mode oscillation is crucial for the quality of the desired signal at $2f_0$.

978-2-8748-7007-1/08 $25.00 © 2008 EuMA

III. VCO ACTIVE DEVICE CONSIDERATION

In voltage controlled oscillator design active device sizes and layout must be carefully considered as they influence oscillator operating frequency and phase noise. In CMOS designs the MOSFET exhibit flicker noise that when upconverted, increases the phase noise of the oscillator especially in the $1/f^3$ region.

In [9] the flicker noise was shown to have an input referred voltage noise $\left(\overline{V_n^2}\right)$ equal to:

$$\overline{V_n^2} = \frac{M}{C_{ox}W_gL_g}\frac{1}{f} \quad (1)$$

where M is a process dependant constant, C_{ox} is the oxide capacitance, W_g is the gate width and L_g is the gate length. From (1) it can be seen that by increasing gate area decreases the flicker noise. Unfortunately increased gate area increases device capacitance which reduces the output frequency of the device. In this design the transistors (M_1, M_2), shown in Fig 1, have been optimised to produce an oscillator with reduced phase noise. Through a simulation study it was found the transistor sizes that produced the smallest phase noise where transistors for this push-push oscillator where ones with 50 fingers and a total width of 50um biased at 0.6mA/uM gate width.

Another important consideration is the careful selection of the varactors. Ideally a VCO would have as large a tuning range as possible. Unfortunately in a varactor based implementation there is a trade off between tuning range and phase noise. The gate leakage current of the varactor increases the VCO's phase noise and the parasitic capacitance reduces the oscillation frequency [6]. To achieve maximum possible tuning range with acceptably low phase noise, carefully designed inversion mode MOS varactors are employed. The NMOS varactors C are laid out as a multi-finger structure to reduce gate resistance and enhance the resonator's Q factor. In this design, the gate length of the NMOS varactors used in this design is 260nm and their total width being 100um in comprised of 2um fingers order to achieve a C_{max} to C_{min} ratio of 3.

The inductors (L_1, L_2) were implemented as transmission lines In order to reduce edge and fringe capacitances the inductors were implemented on the top metal layer to achieve the highest Q possible. These RF transmission lines are 100um long, 25um wide and have an equivalent inductance value of approximately 50pH. Not shown in the schematic, following the push-push output is a buffer amplifier.

IV. MEASUREMENT RESULTS

The fully integrated VCO is measured on-wafer using a Suss-Microtech probe station with 110GHz coplanar probes and an Anritsu 40GHz spectrum analyzer and a 65GHz signal generator. Fig 2 shows the output frequency as the tuning voltage varies from 0 to 2V. The output frequency varies from 65.8GHz to 73.6GHz. After calibrating the cable and pads loss, the output power at 70GHz is -4dBm. The core power consumption is 32mW from a 1.6V supply.

Phase noise was measured by heterodyning the VCO's signal with a signal generator to 2.5 GHz intermediate frequency. Fig 3 shows the measured phase noise is -92 dBc/Hz at 1 MHz offset (66GHz) and -107 dBc/Hz at 10 MHz offset (66GHz).

Temperature variation was also tested from 0 to 70 degree Celsius. Fig 4 shows the output frequency versus temperature. The frequency shift is about 200MHz in this range. This is mainly due to the relatively small temperature dependence of the passive network in the VCO core. Table 1 compares this work with recently published oscillators above 60GHz. This work has the highest figure of merit (FOM), -169.8dBc/Hz when frequency tuning range (FTR) is taken into account. Fig 5 shows the die micro-graph with a size of 0.50x1.045 mm^2 including output buffer and testing pads.

ACKNOWLEDGMENT

NICTA is funded by the Australian Government as represented by the Department of Broadband, Communications and the Digital Economy and the Australian Research Council through the ICT Centre of Excellence program.

REFERENCES

[1] Luiz M. Franca-Neto, Ralph E. Bishop, Brad A. Bloechel, "64 GHz and 100GHz VCOs in 90nm CMOS Using Optimum Pumping Method," *ISSCC Dig. Tech. Papers*, pp.444-445, Feb., 2004

[2] R. C. Liu et al., "A 63GHzVCO Using a Standard 0.25um CMOS Process," *ISSCC Dig. Tech. Papers*, pp.446-447, Feb., 2004

[3] W. Winkler., "60GHz and 76GHz Oscillators in 0.25um SiGe:C BiCMOS," *ISSCC Dig. Tech. Papers*, pp.454-455, Feb., 2003

[4] K. Kobayashi, et al., "A 108-GHz InP-HBT Monolithic Push-Push VCO with Low Phase Noise and Wide Tuning Bandwidth," *IEEE J. Solid State Circuits*, vol. 34, no.9, Sept. 1999, pp.1225 – 1232.

[5] P. Andreani, H. Sjoland, "Tail Current Noise Suppression in RF CMOS VCOs," *IEEE J. Solid State Circuits*, vol. 37, no.3, Mar. 2002, pp.342 – 348.

[6] Chihun Lee and Shen-luan Liu, "A 58 to 60.4GHz Frequency Synthesizer in 90 nm CMOS," *ISSCC Dig. Tech. Papers*, pp.196-197, Feb., 2007

[7] P. C. Huang, M. D. Tsai, H. Wang, C. H. Chen, C. S. Chang., "A 144GHz VCO in 0.13um CMOS Technology," *ISSCC Dig. Tech. Papers*, pp.404-405, Feb., 2005

[8] B. Heydari, M. Bohsali, E. Adabi, A. M. Niknejad., "Low-Power mm-wave Components up to 104 GHz in 90nm CMOS," *ISSCC Dig. Tech. Papers*, pp.200-201, Feb., 2007

[9] Y. Tsividis, "Operation and Modelling of the MOS transistor", 2nd Ed. Boston, MA, Wiley, 1981.

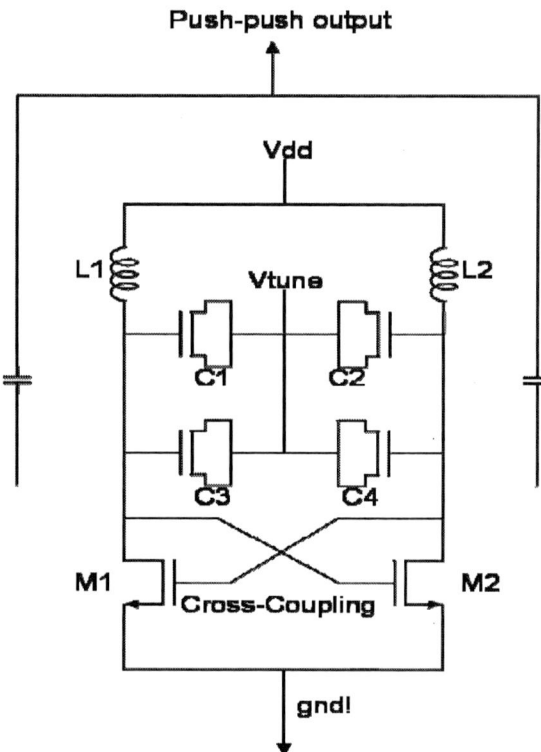

Fig. 1 Schematic of VCO core

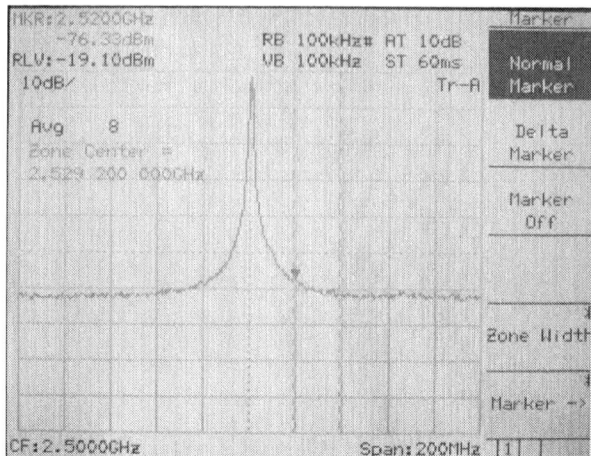

Fig. 3 Measured phase noise at IF

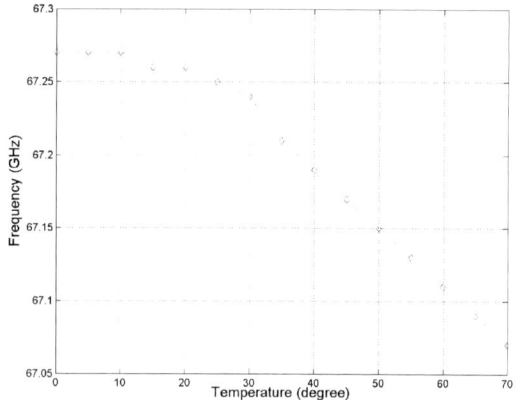

Fig. 4 Frequency variation as a function of ambient temperature

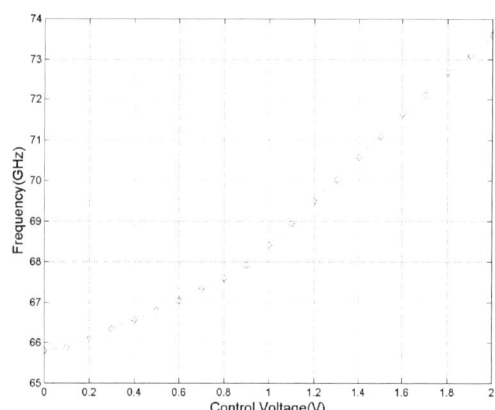

Fig. 2 Measured tuning range of the fabricated voltage controlled oscillator

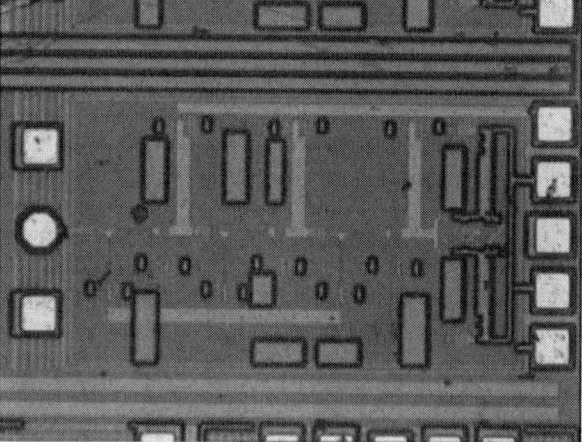

Fig. 5 Micrograph of fabricated 70 GHz push-push voltage controlled oscillator

TABLE I

COMPARISON OF OSCILLATORS ABOVE 60GHz

Process	Freq (GHz)	Power Supply (V)	Core Power Diss(mW)	Phase Noise dBc/Hz	Offset (MHz)	Power Output (dBm)	Tuning Range (GHz)	Temp Variation (degree C)	FOM Tune dBc/Hz	Ref
0.25um SiGe BiCMOS	60	-3	73.8	-87	1	-17	9.8	N/A	-168.1	ISSCC 03 [3]
0.25um SiGe BiCMOS	76	-4	128	-91	1	-7	N/A	N/A	N/A	ISSCC 03 [3]
0.25um CMOS	63	1.8	118.8	-85	1	-4	4.5	N/A	-157.3	ISSCC 04 [2]
90nm CMOS	64	1	20	-110	10	-65	N/A	3GHz (-50 to 110)	N/A	ISSCC 04 [1]
90nm CMOS	100	1	30	-110	10	-65	N/A	5GHz (-50 to 110)	N/A	ISSCC 04 [1]
0.13um CMOS	114	1.2	8.4	-107.6	10	-22.5	2.4	N/A	-165.9	ISSCC 05 [7]
90nm CMOS	104	1	6	-70(sim)	1	-8	3.12	N/A	-152.1	ISSCC 07[8]
0.13um CMOS	70	1.6	32	-107	10	-3	7.8	200MHz (0 to 70)	-169.8	This Work

$$FOM(tune) \quad = \quad Phn \quad - \quad 20 \ \log \ \left(\frac{f_0}{\Delta f} \cdot \frac{FTR(\%)}{10} \right) + 10 \ \log \ \left(\frac{P_{diss}}{1 \, mW} \right)$$

A Low-voltage, Fully-integrated (1.5-6)GHz Low-Noise Amplifier in E-mode pHEMT Technology for Multiband, Multimode Applications

Zulfa Hasan-Abrar[#1], Yut H. Chow[#2], Yong W. Eng[#3]

[#]Avago Technologies Malaysia Sdn. Bhd., Wireless Semiconductor Division, Phase III,Bayan Lepas FIZ, 11900 Bayan Lepas, Penang, Malaysia.

[1]zulfa.hasan-abrar@avagotech.com
[2]yut-hoong.chow@avagotech.com
[3]yong-wah.eng@avagotech.com

Abstract— This paper describes the design and implementation of a fully-integrated MMIC low-voltage, low-noise amplifier (LNA) for use in multimode, multiband receivers using 0.25um enhancement-mode GaAs pHEMT technology. The LNA has two cascaded gain stages and is fully usable down to 0.8V supply voltage and 5mA total current drain. Power supply inductors, bypass capacitor and interstage matching are integrated on the die. An external inductor can be added to improve input match and gain. At 1.4V supply, it achieves broadband (1.5-6)GHz gain of 17.5dB and typical noise figure of 1.5dB while consuming 18mA of total current. Gain variation is typically less than 1.5dB. Input IP3 is better than -4dBm across the band. The complete chip occupies an area of 1.1mm².

I. INTRODUCTION

Recent years have seen the proliferation of RF standards, all of which aims to address the apparent need for communications of one form or another. Examples of these are the UMTS, WLAN, WMAN, GPS and UWB standards. The demand for higher data rates in particular, have pushed communication devices to operate at higher frequencies and with wider modulation bandwidths. Commercially, it is highly desirable to integrate all these functions and standard capabilities into a single, ubiquitous and portable communications device. However, these standards all operate on different frequency bands, utilize different communication protocols and require different levels of performance. Various hardware architectures and circuit techniques have been developed that aim to fulfill these requirements while keeping the power consumption within reasonable levels. A true multimode, multi-band radio will be capable of transmit and receive functions of all these different protocols and all the different frequencies

In this paper we show some examples of different multimode, multiband RF front-end architectures. We then describe a fully integrated LNA MMIC that will enable the near-term realization of this multimode, multiband receiver. This LNA is realized in a 0.25um enhancement-mode (E-mode) pHEMT process that enables smaller die area while maintaining the desirable low-noise properties of depletion-mode pHEMTs. The LNA exhibits 17.5dB of gain across the

(1.5-6)GHz band with a typical noise figure of 1.5dB while consuming only 18mA from a 1.4V supply. Unlike cascode amplifiers, this LNA can be used down to 0.8V supply and still maintain broadband performance. The complete chip occupies an area of (1.0x1.1)mm.

II. EXAMPLES OF RF FRONT-END ARCHITECTURES FOR VARIOUS RF STANDARDS

Fig 1 below shows typical architectures of W-CDMA transceivers used in cell phone handsets. It is separated either by frequency bands or by technology [1].

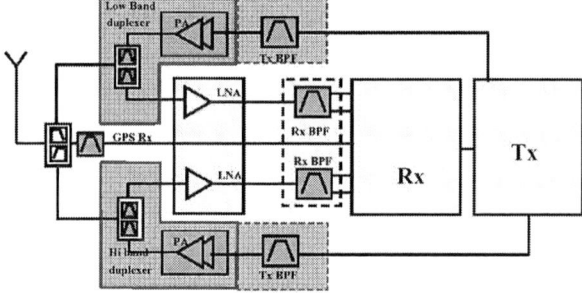

Fig. 1a Typical W-CDMA architecture based on frequency band separation.

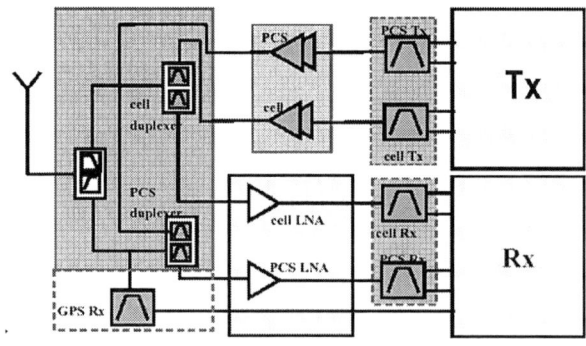

Fig. 1b.   Typical W-CDMA architecture based on technology separation.

For the new IEEE 802.16d/e (WMAN) standard that accommodates multiple-input and single-output (MISO) capabilities, the architecture is shown in Fig 2 below. MIMO architectures can be extended based on this.

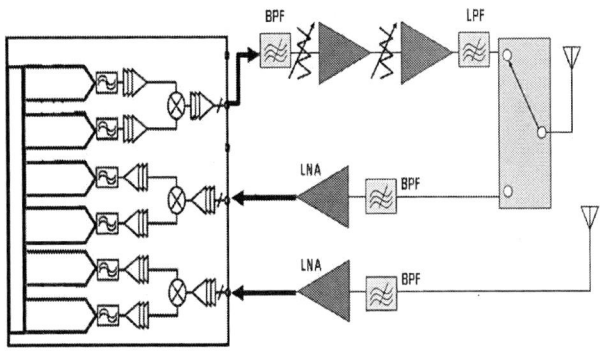

Fig. 2.   MISO architecture for WMAN application

These examples show that the LNA can be co-located on the front-end section of the radio, thus the LNA can be designed as a single device with proper filtering for the various frequency band filters. This is termed a "concurrent LNA" [2], an example of which is shown in Fig 3 for WLAN applications covering the IEEE 802.11 "g" and "a" bands at 2.4GHz and (5-6)GHz respectively. A broadband LNA that covers the multiple standards in the front-end can therefore potentially be suitable as an ideal concurrent LNA solution.

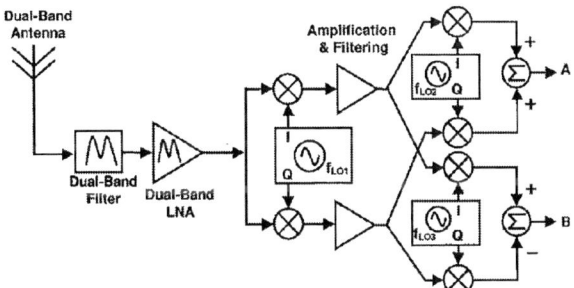

Fig. 3.   Concurrent LNA architecture for dual-band WLAN application.

III. LNA MMIC DESIGN

The complete LNA is integrated into a single, E-mode GaAs MMIC chip with all the input, output and interstage matching. RF bypass on the supply rail is also included.

A. Enhancement-mode pHEMT Technology

The LNA is designed using Avago Technologies' enhancement-mode (E-mode) pHEMT process. This is a 0.25um minimum feature size process built on a GaAs-AlGaAs epitaxial layer using molecular beam epitaxy. Low-voltage operation is improved by a threshold voltage of about 0.3V, significantly lower than many SiGe processes. There is no penalty to the transition frequency (fT) of an E-mode process compared to a depletion-mode process, thus the traditional advantages of GaAs such as high electron mobility and high transconductance are preserved. Fig 4 below shows on-wafer probed Fmin vs Vds performance at different bias currents for a single device, demonstrating that excellent NF performance is possible even for large supply voltage variations.

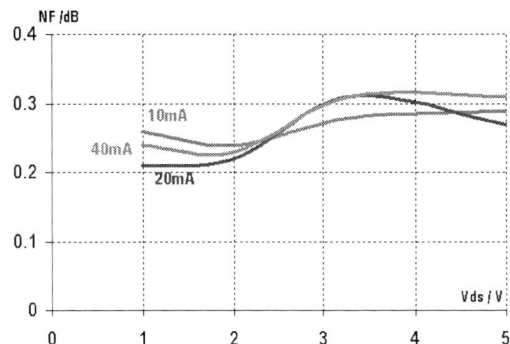

Fig. 4.   Fmin vs Vds curve at different bias current.

B. Circuit Level Design

Fig 5 below shows the complete LNA MMIC schematic diagram. It is a 2-gain stage design with all input, output, and interstage matching integrated onto the chip. Unlike more conventional LNAs, a cascode is not employed. This, together with the E-mode technology, enables operation down to very low supply voltages.

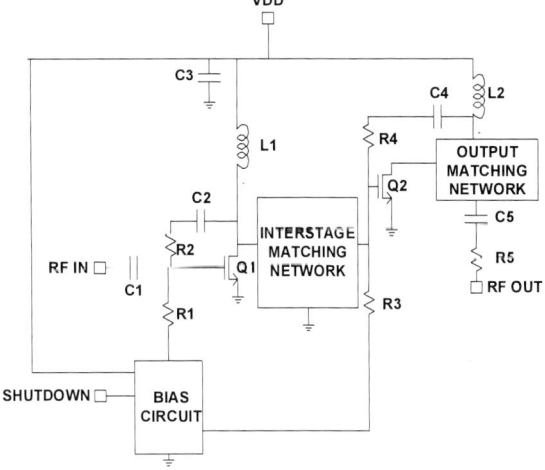

Fig. 5.   Schematic diagram of LNA MMIC

Broadband operation is detemined by the choice of pHEMT device periphery, input impedance matching, interstage matching design and considerations for MMIC parasitics. A

978-2-8748-7007-1/08 $25.00 © 2008 EuMA   307

large device periphery will enable easier matching, but would degrade gain and affect linearity for a given drain current. Since complete LNA integration into the MMIC is desired, this precludes use of complicated passive input matching. Lossy input matching will also degrade noise figure, as will increased lossy feedback.

Interstage matching is critical to obtain gain flatness. A tee-match network was chosen because it presents a high-pass response to offset the low-pass response of the first gain stage. At frequencies above 3GHz, parasitics become non-negligible, even at the MMIC level. Bondwires, device capacitances and coupling between on-chip spiral inductors have to be considered as part of the matching design elements. Fig 6 below shows a 2.5D electromagnetic simulation result of the complete MMIC that includes side-current coupling and thick metallization effects of the inductors. Current density and coupling effects can be clearly observed.

Large-signal performance is dependent on the device size and the current through the device in the final gain stage. Ultimately, the linearity is determined by the supply voltage since this determines the clipping point at the output load.

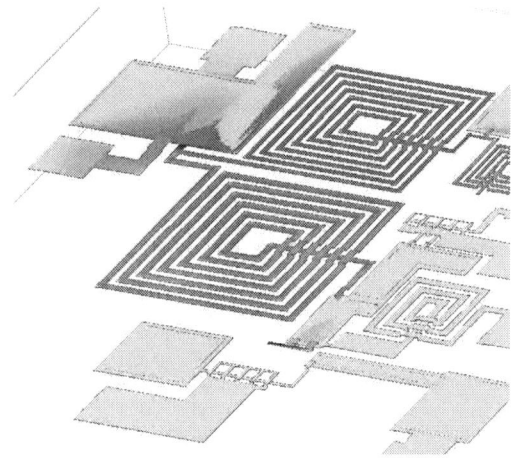

Fig. 6.   EM simulation of the complete LNA MMIC.

The complete integrated LNA MMIC occupies an area of (1.0x1.1)mm, a photo of which is shown in Fig 7.

IV. MEASUREMENT DATA

The complete MMIC was mounted on a 10mil thick Rogers4350 printed circuit test board. Fig 8 shows the testboard used in the LNA MMIC measurements. In the test circuit, only a single 0.1uF bypass capacitor is used in the supply line.

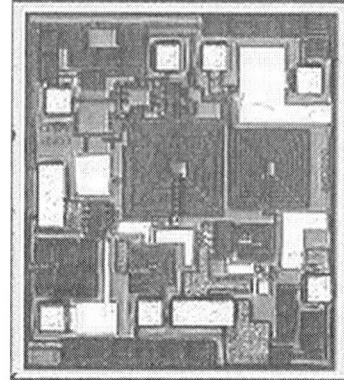

Fig. 7.   Photograph of MMIC LNA chip.

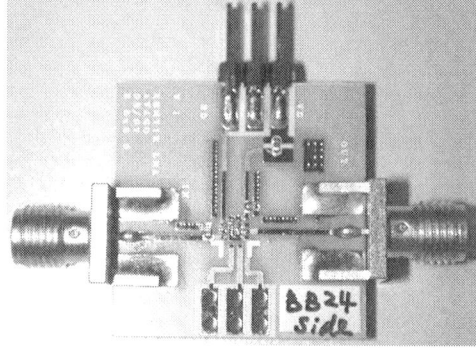

Fig. 8.   Photograph of testboard with MMIC LNA chip.

Fig 9 compares the input match characteristics with and without an external input shunt inductor. While the performance of the MMIC alone is acceptable, it was found that the external inductor improves on the already broadband input characteristics by improving the gain response. This is demonstrated in Fig 10 with a 1.4V, 18mA bias where the flat gain characteristics extends to beyond 6GHz with only 1.5dB gain variation.

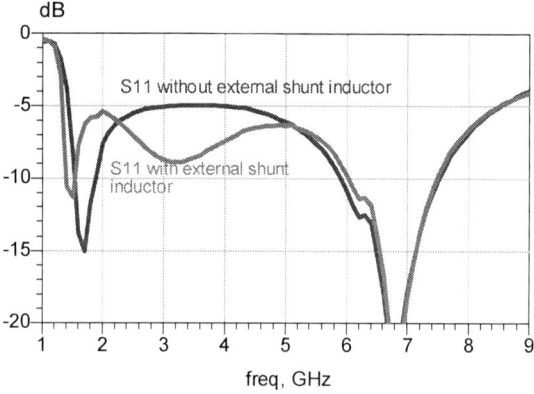

Fig. 9.   Comparison of S11 with and without external shunt inductor.

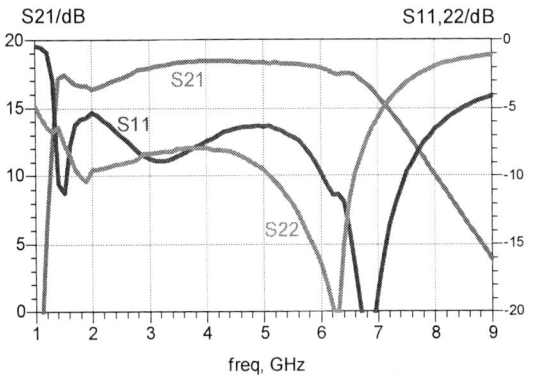

Fig. 10. Small-signal response of LNA showing less than 1.5dB variation from (1.5-6.5)GHz.

Fig 11 and 12 details the broadband noise figure and third order intercept points under the same bias conditions.

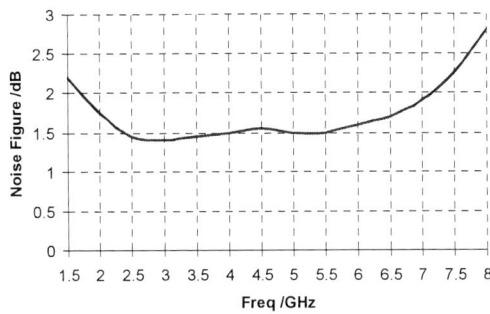

Fig. 11. Broadband Noise Figure performance

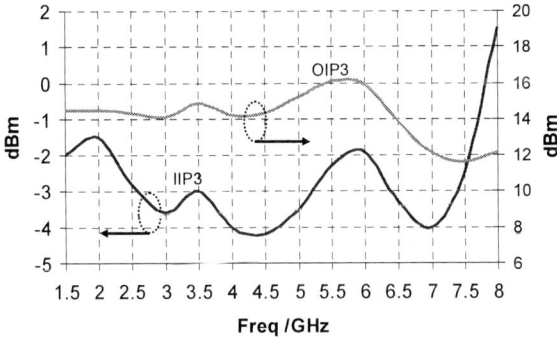

Fig. 12. Broadband IP3 performance

IP3 performance is also very well behaved, it does not exhibit significant fluctuations across the band. This demonstrates that the MMIC performance is truly broadband across both small-signal and large-signal parameters.

The LNA performance at different supply voltages and currents are detailed in Fig 13. Notice that broadband performance is maintained, even at a 4mW (0.8v, 5mA) total

power dissipation. While gain has reduced because the bias current has dropped, gain flatness is well maintained with less than 3dB variation across the pass band.

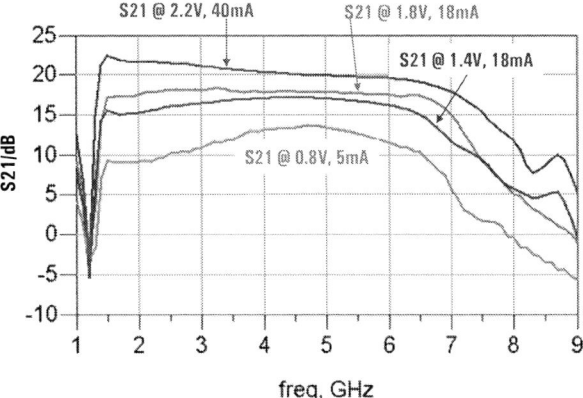

Fig. 13. Broadband Gain performance under different bias conditions.

V. CONCLUSION

The design and implementation of a broadband, low-voltage LNA MMIC suitable for multimode, multiband RF applications has been described. The LNA is fabricated on a 0.25um E-mode pHEMT process. It exhibits broadband performance of (1.5-6)GHz from 2.2V, 40mA bias down to 0.8V, 5mA. Both large-signal and small-signal parameters show consistent performance across the band. At 1.4V and 18mA, the LNA achieves 17.5dB of gain and typical noise figure of 1.5dB with worst case IIP3 of -4dBm. The complete MMIC occupies 1.1mm^2 of die area and actual application requires only a single power supply bypass capacitor.

ACKNOWLEDGMENT

The authors wish to acknowledge the support of the management team at Avago Technologies WSD R&D in Penang during the development of this MMIC.

REFERENCES

[1] Chris Chung, "RF Front-end Technology," *Proc Korea Communications Expo 2006.*, Seoul, May 2006.
[2] H. Hashemi, and A. Hajimiri, "Concurrent Multiband Low-Noise Amplifiers—Theory, Design, and Applications," *2003 IEEE MTT.*, vol. 50, No 1, , pp. 288-301, Jan 2002.

Space Mapping Algorithm with Improved Convergence Properties for Microwave Optimization

Slawomir Koziel[#1], John W. Bandler[*2]

[#]*School of Science and Engineering, Reykjavik University*
Kringlan 1, IS-103 Reykjavik, Iceland
[1]koziel@ru.is

[*]*Department of Electrical and Computer Engineering, McMaster University*
1280 Main Street West, Hamilton, Ontario, Canada L8S 4K1
[2]bandler@mcmaster.ca

Abstract—**A space mapping algorithm with improved convergence properties for microwave design optimization is presented. In contrast to the previously published technique, a new convergence control method can be applied to surrogate models using non-extractable parameters, in particular, the output space mapping—one of the most useful approaches to date. We demonstrate that the new algorithm allows for faster convergence without compromising the quality of the final design.**

I. INTRODUCTION

Space mapping (SM) [1]-[4] has been successfully used in microwave design for over a decade. It allows efficient optimization of computationally expensive "fine" models based on full-wave electromagnetic simulators through iterative optimization and updating of so-called "coarse" models, less accurate but cheaper to evaluate. Coarse models are often circuit equivalents of the microwave structures in question. When set up properly, SM algorithms can provide satisfactory results after only a few evaluations of the fine model.

One of the open problems in space mapping is the right choice of coarse model and surrogate model type. Both are problem dependent and substantially influence the performance of the algorithm [5]. Preliminary results regarding an automated choice of the surrogate model can be found in [6].

Another issue is that the convergence of SM algorithms is not guaranteed in general [7]. One consequence is that, in many cases, the SM algorithm quickly finds an acceptable solution, however, it then falls into oscillation with respect to the design variables and/or objective function value and it is unclear when to terminate the optimization process. This issue is related to the coarse/surrogate model choice because the right selection may alleviate the convergence problems to some extent [5].

An initial investigation regarding the convergence control of an SM algorithm by adjusting the impact the available fine model data has on the current surrogate model have been presented in [8]. It is shown that the convergence rate of the algorithm can be improved by a proper choice of parameter extraction weights so that the subsequent iteration points have less and less impact on the surrogate model. This method, however, cannot be used in the case of non-extractable parameters. In particular, it does not apply to output SM which is one of the most useful approaches that allow good local alignment between the fine model and the surrogate [2].

In this paper we present an extended convergence control method which can be used with any type of SM surrogate model containing both extractable and non-extractable parameters. We present two examples of microwave design problems that demonstrate the usefulness of the proposed approach and show its advantages over both the standard SM and the initial convergence control method [8].

II. SPACE MAPPING OPTIMIZATION ALGORITHM

Let $\boldsymbol{R}_f$ denote the response vector of a fine model of the device of interest. Our goal is to solve

$$\boldsymbol{x}_f^* = \arg\min_x U\left(\boldsymbol{R}_f(\boldsymbol{x})\right) \tag{1}$$

where U is a given objective function, e.g., minimax. The fine model is computationally expensive, which makes a direct optimization of $U(\boldsymbol{R}_f(\boldsymbol{x}))$ prohibitive. We assume that the coarse model $\boldsymbol{R}_c$ is also available, which is a simplified and less accurate representation of the fine model, but, it is much faster to evaluate.

To solve (1), we consider an optimization algorithm that generates a sequence of points $\boldsymbol{x}^{(i)}$, $i = 0, 1, 2, \ldots$ as follows:

$$\boldsymbol{x}^{(i+1)} = \arg\min_x U\left(\boldsymbol{R}_s^{(i)}(\boldsymbol{x})\right) \tag{2}$$

where $\boldsymbol{R}_s^{(i)}$ is a family of surrogate models. Surrogate models are constructed from $\boldsymbol{R}_c$ so that that the misalignment between $\boldsymbol{R}_s^{(i)}$ and the fine model is minimized for each i. Let $\overline{\boldsymbol{R}}_s$ be a generic SM surrogate model, i.e., $\boldsymbol{R}_c$ composed with suitable mappings. At iteration i the surrogate model $\boldsymbol{R}_s^{(i)}$ is defined as

$$\boldsymbol{R}_s^{(i)}(\boldsymbol{x}) = \overline{\boldsymbol{R}}_s(\boldsymbol{x}, \boldsymbol{p}^{(i)}) \tag{3}$$

where

$$\boldsymbol{p}^{(i)} = \arg\min_p \sum_{k=0}^{i} w_{i,k} \| \boldsymbol{R}_f(\boldsymbol{x}^{(k)}) - \overline{\boldsymbol{R}}_s(\boldsymbol{x}^{(k)}, \boldsymbol{p}) \| \tag{4}$$

is a vector of model parameters and $w_{i,k}$ are weighting factors. The standard weighting scheme is $w_{i,k} = 1$ for all i and $k = k_{min}$, $k_{min}+1, \ldots, i-1, i$. The common choice is $k_{min} = 0$, i.e., all available points are used.

A variety of SM surrogate models is available [1]-[3], e.g., the input SM [1], where the surrogate model takes the form

$$\overline{\boldsymbol{R}}_s(\boldsymbol{x}, \boldsymbol{p}) = \overline{\boldsymbol{R}}_s(\boldsymbol{x}, \boldsymbol{B}, \boldsymbol{c}) = \boldsymbol{R}_c(\boldsymbol{B} \cdot \boldsymbol{x} + \boldsymbol{c}) \tag{5}$$

Here, parameters $\boldsymbol{B}$ and $\boldsymbol{c}$ have to be obtained by solving a (non-linear) parameter extraction problem (4).

Another popular approach is the so-called output SM

$$\overline{\boldsymbol{R}}_s(\boldsymbol{x}, \boldsymbol{p}) = \overline{\boldsymbol{R}}_s(\boldsymbol{x}, \boldsymbol{d}) = \boldsymbol{R}_c(\boldsymbol{x}) + \boldsymbol{d} \tag{6}$$

978-2-8748-7007-1/08 $25.00 © 2008 EuMA   310

where d is a correction term accounting for the difference between the fine and coarse model responses at iteration i, so that $d^{(i)} = R_f(x^{(i)}) - R_c(x^{(i)})$. Model (6) ensures perfect alignment between the surrogate and the fine model at $x^{(i)}$. It is often used on top of other space mapping types. Note that parameter d is not extracted but calculated after the explicitly extractable space mapping parameters are known.

III. CONTROLLING CONVERGENCE OF SM ALGORITHMS

A. General Strategy of Convergence Control

It is intuitively clear and also confirmed by theoretical results [2], [7], that convergence of the SM algorithm, i.e., the relation between $\|x^{(i+2)} - x^{(i+1)}\|$ and $\|x^{(i+1)} - x^{(i)}\|$ is mainly determined by the relation between the subsequent surrogate models, $R_s^{(i)}$, and $R_s^{(i+1)}$ at any given iteration i. If the models are more similar to each other, which is determined by the closeness of the relevant SM parameters, the ratio $\|x^{(i+2)} - x^{(i+1)}\|/\|x^{(i+1)} - x^{(i)}\|$ is likely to be smaller, and the convergence rate is likely to improve. The values of the SM parameters are determined by the relative impact of previous iteration points on the parameter extraction process (4). This impact can be manipulated as shown below.

B. Extractable Space Mapping Parameters

In the case of extractable parameters, as in (5), the convergence control can be realized by a proper choice of weights w_{ik} in (4). In particular, by choosing $w_{ik+1} < w_{ik}$ for all k, one can expect that the influence of subsequent iteration points on the values of the SM parameters will decrease. In an extreme case, $w_{i0} = 1$, and $w_{ik} = 0$ for $k = 1, 2, \ldots$, we would have $p^{(k)} = p^{(0)}$ for $k = 1, 2, \ldots$, i.e., all the surrogate models $R_s^{(k)}$, $k = 0, 1, \ldots$, would be the same and the SM algorithm would converge in one iteration. This is not a good choice since the algorithm could not find a satisfactory solution when it is forced to stop after the first iteration.

The practical scheme proposed in [8] is the following:

$$w_{ik} = \alpha^k \qquad k = 0, 1, \ldots, i \qquad (7)$$

where α is a constant, typically smaller than one.

C. Non-Extractable Space Mapping Parameters

Non-extractable parameters are not dependent on the weighting factors w_{ik} so, in this case, we need another method to implement convergence control.

Let us denote by r all extractable surrogate model parameters so that we have $\bar{R}_s(x, p) = \bar{R}_s(x, r, d) = R_s(x, r) + d$ (here, R_s is the coarse model composed with extractable parameters) with $d^{(i)} = R_f(x^{(i)}) - R_s(x^{(i)}, r^{(i)})$. Now, instead of $d^{(i)}$ we will use the modified term $\bar{d}^{(i)}$ defined as $\bar{d}^{(0)} = d^{(0)}$ and

$$\bar{d}^{(i)} = (\beta)^i \cdot d^{(i)} + (1 - (\beta)^i) \cdot \bar{d}^{(i-1)} \qquad (8)$$

for $k = 1, 2, \ldots$, with $0 < \beta \leq 1$.

D. Practical Guidelines

The parameters α and β should be chosen small enough to ensure convergence of the SM algorithm but also large enough to allow the algorithm to find an acceptable solution. It may not be possible to satisfy these two conditions for some coarse models and SM types. Therefore, convergence control could be successfully employed if the algorithm either (i) converges and finds a satisfactory solution but one wants to

obtain a better convergence rate, or (ii) does not converge but seems to be able to find a satisfactory solution at least for some $x^{(i)}$. In other cases, the only choice is to employ a better coarse model or to try a different SM type [6]-[8]. As it is not known beforehand whether the algorithm falls into one of the two cases above, it is advisable to enable (7) and/or (8) after the algorithm finds a solution satisfying the design specifications (or a solution sufficiently close to satisfying the design specifications) for the first time.

IV. EXAMPLES

A. Capacitively-Coupled Dual-Behavior Resonator Filter

Consider a second-order capacitively-coupled dual-behavior resonator (CCDBR) microstrip filter [9] shown in Fig. 1. The design parameters are $x = [L_1 \, L_2 \, L_3 \, S]^T$. The fine model R_f is simulated in FEKO [10]. The coarse model R_c is the circuit model implemented in Agilent ADS [11] and shown in Fig. 2. The design specifications are $|S_{21}| \geq -3$ dB for 3.8 GHz $\leq \omega \leq$ 4.2 GHz, and $|S_{21}| \leq -20$ dB for 2.0 GHz $\leq \omega \leq$ 3.2 GHz and for 4.8 GHz $\leq \omega \leq$ 6.0 GHz.

We use the input and output SM surrogate model $R_s(x) = R_c(x + c^{(i)}) + d^{(i)}$. The starting point is the coarse model optimal solution $x^{(0)} = [2.415 \, 6.093 \, 1.167 \, 0.082]^T$ mm. (specification error +7.8 dB). Fig. 3 shows the coarse and fine model responses at $x^{(0)}$. It should be noted that the SM algorithm using only the input SM is not able to yield a solution satisfying the design specifications (the specification error of the best solution found in 15 iterations is +0.3 dB).

We optimize our filter three times, using the SM algorithm with the weighting scheme (7) and the modified d term (8) with (i) $\alpha = \beta = 1.0$ (which is equivalent to the standard SM algorithm, i.e, the algorithm (2)-(4) with $w_{ik} = 1$ for $k = 0, 1, \ldots, i$), (ii) $\alpha = 0.5$ and $\beta = 1.0$ (i.e., the SM algorithm [8] with weighting scheme (7) and the standard d term, and (iii) $\alpha = \beta = 0.5$ (i.e., the new algorithm presented in Section III) The termination condition is $\| x^{(i)} - x^{(i-1)} \| < 10^{-3}$.

Table I shows the optimization results. The convergence plots and the evolution of the specification error for all three cases are shown in Figs. 4 and Fig. 5, respectively. Fig. 6 shows the fine model response at the final solution $x^{(7)} = [3.181 \, 4.895 \, 1.255 \, 0.0503]^T$ mm found by the SM algorithm using the scheme (7) and d term (8) for $\alpha = \beta = 0.5$.

It is seen that the performance of the SM algorithm using the weighting scheme (7) and the d term (8) is better than the performance of both the standard SM algorithm and the algorithm [8], especially with respect to convergence properties. Neither the standard SM algorithm nor the SM algorithm [8] converges: there is no indication when the process should be terminated. Also, these two algorithms exhibit an oscillatory behaviour with respect to specification error. Note that although the specification error value at the best solution found by the standard algorithm is close to the one obtained by the new algorithm, the specification error at the final iteration is rather poor. The new algorithm is free from these problems and exhibits a clear convergence pattern as well as a consistent behavior with respect to the specification error value.

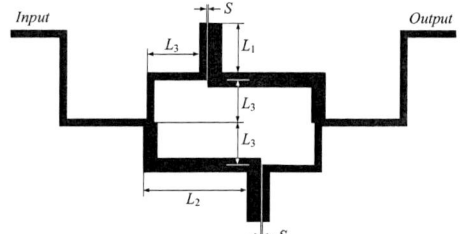

Fig. 1. Geometry of the CCDBR microstrip filter [9].

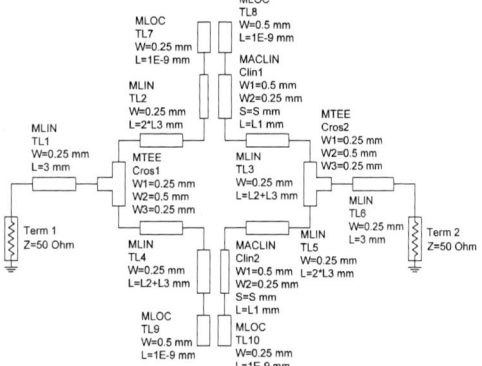

Fig. 2. Coarse model of the CCDBR microstrip filter (Agilent ADS).

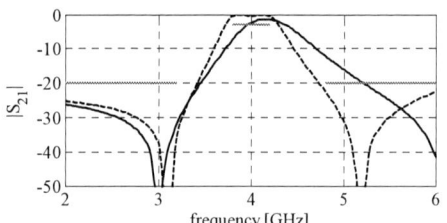

Fig. 3. CCDBR filter: coarse (dashed line) and fine (solid line) model response at the starting point $x^{(0)}$.

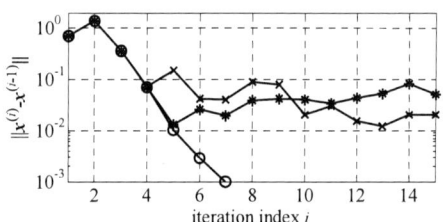

Fig. 4. CCDBR filter: convergence plots for the standard SM algorithm (*), the SM algorithm [8] (×), and the new SM algorithm using the weighting scheme (7), and the d term (8) (o).

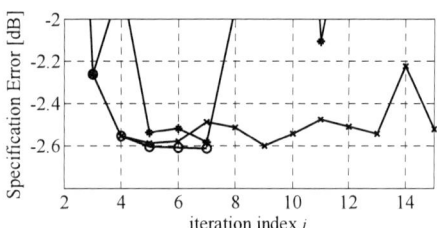

Fig. 5. CCDBR filter: specification error versus iteration index for the standard SM algorithm (*), the SM algorithm [8] (×), and the new SM algorithm using the weighting scheme (7), and the d term (8) (o).

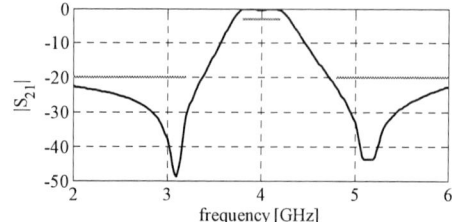

Fig. 6. CCDBR filter: fine model response at the final solution found by the SM algorithm using the weighting scheme (7) and the d term (8).

TABLE I
CCDBR FILTER: SPACE MAPPING OPTIMIZATION RESULTS

Algorithm	Specification Error		Number of Fine Model Evaluations	Convergence*
	At Final Iteration	At Best Solution		
$\alpha = 1.0$, $\beta = 1.0$ (Standard SM)	–0.6 dB	–2.5 dB	16	Algorithm did not converge
$\alpha = 0.5$, $\beta = 1.0$ (Algorithm [8])	–2.5 dB	–2.6 dB	16	Algorithm did not converge
$\alpha = 0.5$, $\beta = 0.5$ (New algorithm)	–2.6 dB	–2.6 dB	8	Convergence in 7 iterations

*Convergence according to the termination condition $\| x^{(i)} - x^{(i-1)} \| < 10^{-3}$

B. Compact Bandpass Filter Using Microstrip Resonators with Open Stub Inverter

Consider the bandpass microstrip filter with open stub inverter [12] shown in Fig. 7. The design parameters are $x = [L_1\ L_2\ L_3\ S_1\ S_2\ W_1]^T$. The fine model is simulated in FEKO [10]. The coarse model is implemented in Agilent ADS [11] (Fig. 8). The design specifications are $|S_{21}| \geq -3$ dB for 1.95 GHz $\leq \omega \leq$ 2.05 GHz, and $|S_{21}| \leq -20$ dB for 1.5 GHz $\leq \omega \leq$ 1.8 GHz and for 2.2 GHz $\leq \omega \leq$ 2.5 GHz.

We use the following surrogate model: $R_s(x) = R_{c.f}(x, F^{(i)}) + d^{(i)}$, which corresponds to frequency space mapping [1] and output space mapping, where $R_{c.f}$ is R_c evaluated at frequencies different from the original frequency sweep, according to the mapping $\omega \rightarrow f_1 + f_2 \cdot \omega$, with $F = [f_1\ f_2]^T$. The initial design is the coarse model optimal solution $x^{(0)} = [25.0\ 5.0\ 1.221\ 0.652\ 0.187\ 0.1]^T$ mm (specification error +16 dB). As before, we optimize the filter using the SM algorithm with the weighting scheme (7) and the modified d term (8) with (i) $\alpha = \beta = 1.0$, (ii) $\alpha = 0.5$ and $\beta = 1.0$, and (iii) $\alpha = \beta = 0.5$.

The optimization results are shown in Table II. The convergence plots and the evolution of the specification error are shown in Figs. 9 and 10, respectively. Fig. 11 shows the fine model response at the final solution $x^{(8)} = [23.762\ 5.00\ 1.00\ 0.704\ 0.166\ 0.100]^T$ mm found by the SM algorithm using the weighting scheme (7) and d term (8) for $\alpha = \beta = 0.5$; its performance is better than the performance of both the standard SM algorithm and the algorithm [8], especially with respect to convergence properties. As before, neither the standard SM algorithm nor the SM algorithm [8] converges, and these two algorithms exhibit an oscillatory behaviour with respect to specification error. The new algorithm has a clear convergence pattern as well as consistent behavior with respect to the specification error value.

978-2-8748-7007-1/08 $25.00 © 2008 EuMA

V. CONCLUSIONS

An efficient technique for controlling the convergence properties of space mapping optimization algorithms is presented that can be used with all types of space mapping surrogate models containing both extractable and non-extractable parameters. The performance of the method is demonstrated using microwave design examples.

ACKNOWLEDGEMENT

This work was supported in part by the Natural Sciences and Engineering Research Council of Canada under Grant RGPIN7239-06 and Grant STPGP336760-06, and by Bandler Corporation. J.W. Bandler is also with Bandler Corporation, P.O. Box 8083, Dundas, ON, Canada L9H 5E7.

REFERENCES

[1] J.W. Bandler, Q.S. Cheng, S.A. Dakroury, A.S. Mohamed, M.H. Bakr, K. Madsen, and J. Søndergaard, "Space mapping: the state of the art," *IEEE Trans. Microwave Theory Tech.*, vol. 52, no. 1, pp. 337-361, Jan. 2004.

[2] S. Koziel, J.W. Bandler, and K. Madsen, "A space mapping framework for engineering optimization: theory and implementation," *IEEE Trans. Microwave Theory Tech.*, vol. 54, no. 10, pp. 3721-3730, Oct. 2006.

[3] D. Echeverria and P.W. Hemker, "Space mapping and defect correction," *CMAM The International Mathematical Journal Computational Methods in Applied Mathematics*, vol. 5, no. 2, pp. 107-136, 2005.

[4] S. Amari, C. LeDrew, and W. Menzel, "Space-mapping optimization of planar coupled-resonator microwave filters," *IEEE Trans. Microwave Theory Tech.*, vol. 54, no. 5, pp. 2153-2159, May 2006.

[5] S. Koziel and J.W. Bandler, "Coarse and surrogate model assessment for engineering design optimization with space mapping," *IEEE MTT-S Int. Microwave Symp. Dig*, Honolulu, HI, 2007, pp. 107-110.

[6] S. Koziel and J.W. Bandler, "Space-mapping optimization with adaptive surrogate model," *IEEE Trans. Microwave Theory Tech.*, vol. 55, no. 3, pp. 541-547, March 2007.

[7] S. Koziel, J.W. Bandler, and K. Madsen, "Quality assessment of coarse models and surrogates for space mapping optimization," to appear, *Optimization and Engineering*, 2008.

[8] S. Koziel and J.W. Bandler, "Controlling convergence of space-mapping algorithms for engineering optimization," *Int. Symp. Signals, Systems and Electronics, URSI ISSSE 2007*, Montreal, Canada, 2007, pp. 21-23.

[9] A. Manchec, C. Quendo, J.-F. Favennec, E. Rius, and C. Person, "Synthesis of capacitive-coupled dual-behavior resonator (CCDBR) filters," *IEEE Trans. Microwave Theory Tech.*, vol. 54, no. 6, pp. 2346-2355, June 2006.

[10] FEKO® *User's Manual*, Suite 4.2, June 2004, EM Software & Systems-S.A. (Pty) Ltd, 32 Techno Lane, Technopark, Stellenbosch, 7600, South Africa.

[11] Agilent ADS, Version 2005C, Agilent Technologies, 1400 Fountaingrove Parkway, Santa Rosa, CA 95403-1799, 2005.

[12] J.R. Lee, J.H. Cho, and S.W. Yun, "New compact bandpass filter using microstrip λ/4 resonators with open stub inverter," *IEEE Microwave and Guided Wave Letters*, vol. 10, no. 12, pp. 526-527, Dec. 2000.

TABLE II
BANDPASS FILTER WITH OPEN STUB INVERTER: SM OPTIMIZATION RESULTS

Algorithm	Specification Error		Number of Fine Model Evaluations	Convergence*
	At Final Iteration	At Best Solution		
$\alpha = 1.0$, $\beta = 1.0$ (Standard SM)	–0.6 dB	–1.4 dB	16	Algorithm did not converge
$\alpha = 0.5$, $\beta = 1.0$ (Algorithm [8])	–0.5 dB	–1.3 dB	16	Algorithm did not converge
$\alpha = 0.5$, $\beta = 0.5$ (New algorithm)	–1.7 dB	–1.7 dB	9	Convergence in 8 iterations

*Convergence according to the termination condition $\| x^{(i)} - x^{(i-1)} \| < 10^{-3}$

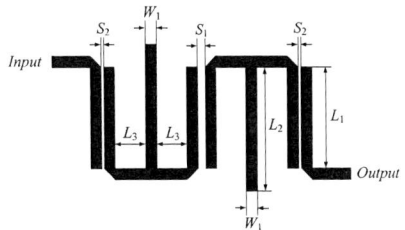

Fig. 7. Geometry of the bandpass filter with open stub inverter [12].

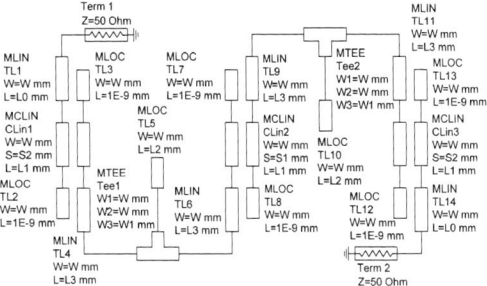

Fig. 8. Coarse model of the bandpass filter with open stub inverter (Agilent ADS).

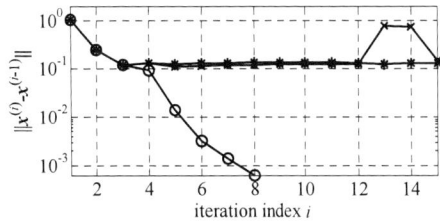

Fig. 9. Bandpass filter with open stub inverter: convergence plots for the standard SM algorithm (*), the SM algorithm [8] (×), and the new SM algorithm using the weighting scheme (7), and the *d* term (8) (o).

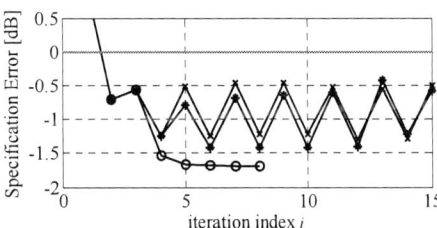

Fig. 10. Bandpass filter with open stub inverter: specification error versus iteration index for the standard SM algorithm (*), the SM algorithm [8] (×), and the new SM algorithm using the weighting scheme (7), and the *d* term (8) (o).

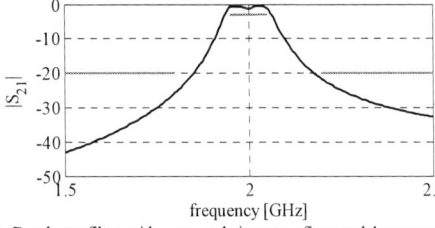

Fig. 11. Bandpass filter with open stub inverter: fine model response at the final solution found by the SM algorithm using the weighting scheme (7) and *d* term (8).

Full W-Band High-Gain LNA in mHEMT MMIC Technology

Walter Ciccognani, Franco Giannini, Ernesto Limiti, Patrick E. Longhi

Electronic Engineering Department, University of Rome Tor Vergata
Via del Politecnico,1 – 00133 Roma, ITALY
longhi@ing.uniroma2.it

Abstract— In this contribution the possible applications, technology, design and measurements of a W-Band high gain LNA are given. The main features of the four stage LNA can be summarised as following: a 25dB average gain with ±2dB ripple from 70 to 105GHz, where gain is higher than 21dB on the entire 70-110GHz range. LNA predicted noise figure is 2.7dB between 80 and 95GHz and less than 3.2dB up to 108GHz while the chip's power consumption is 35mW. The technology used is a 70nm GaAs mHEMT process from OMMIC.

I. INTRODUCTION

By analyzing the satellite communication service applications, the current trend shows a steady and growing demand for resources in terms of allocated available bandwidth. The bands traditionally used by lower frequencies until the Ka, are now very crowded, given the presence of many operating links with GEO and LEO satellites. This framework is expected to deteriorate in the coming years, affecting both civil and military communications and data exchange: systems increase in traditional bands creates a corresponding increase in communications interference, at the expense of the quality of service. In addition, requests for safe communication needs are growing: a particularly important requirement in military systems (e.g. data on military missions in sensitive areas), but also in civilian systems involved in the exchange of sensitive information (e.g. medical data in telemedicine applications). Therefore the exploration of higher frequency bands becomes necessary in order to alleviate and overcome the abovementioned issues. Ka-Band is the frontier for existing services and applications. On the other hand, frequency bands immediately above (Q 35-50 GHz and V 50-75 GHz), are rather difficult to use for terrestrial or satellite communications due to the absorption of atmospheric gases.

In this scenario, W-Band (75-110 GHz) represents a new opportunity. The availability of large bandwidth and smaller absorption, allows the rise of advanced innovative services featuring high standards in terms of exchanged data volume, interactivity, quality and number of users. Given this frequency range, no telecommunications system, terrestrial or satellite, has been developed at present and at the same time there is rare experimental data on W-Band propagation.

Further possible targeted applications include radar systems, with the development of millimetre wavelength radar or radiometric payloads, which can be installed not only on satellites, but also on vehicles for the exploration of the solar system planets and the moon. In the field of observation, the use of high frequencies, along with large antennas located on Earth and in space, will take the radioastronomy scientific community a step forward, allowing investigation with a resolution, cover and space, up to now unavailable. This possibility will encourage the development of large antennas in the millimetre frequency range: the European radioastronomy community has forecast for the coming decade the capability for existing antennas to operate from the centimetre wavelength region well into the millimetre one. The availability of high performance front-end active subsystems, as the LNA object of this contribution, will spur the technological development of large antennas using these receivers.

In this context a low-power, high gain, low noise amplifier has been developed as a test vehicle for the available technology, the proposed design procedures and the characterisation phase.

Some interesting results on the design and realisation of W-Band LNAs have been produced lately. In [1] a 15dB gain LNA with NF equal to 3.2dB between 90 and 98GHz has been realised using a 0.1μm InP technology. More recently a single stage LNA with 7.3dB gain and 2.5dB NF in the 75-95GHz band using a 50nm InP technology has been reported in [2]. Reference [3] has reported the realisation of a 20.5dB gain and 3.9 noise figure at 94 GHz using a 0.2μm gate length InAs/AlSb metamorphic HEMT process. Finally, if we extend our search to HBTs, we can consider a low-noise SiGe-amplifier (LNA) for the frequency range from 75 up to 85GHz integrated in a 0.18μm BiCMOS technology. The LNA shows a 6.2dB noise figure at 77GHz and simultaneously extremely high gain, adjustable from nearly 0dB up to 33dB at 77GHz [4]. All the proposed circuits are realised either on emerging technologies (InP, InAs/AlSb) that hopefully will deliver high reliability and convenience in the future but are at present still difficult to access and manage or on other technologies (SiGe), that are currently producing interesting results in the field of high-power and high-integration circuits that seem to be still unprepared to deliver optimum noise performance. In this paper, instead, an LNA realised on commercial GaAs mHEMT process is presented, proving that state-of-the-art performance can be achieved in the W-Band without having to necessarily employ cutting edge technologies.

II. TECHNOLOGY

The MMIC LNA has been realised making use of OMMIC's foundry facilities and in particular the preliminary and under development D007IH process. The active device is a 70nm channel length mHEMT. A high Indium content has been added to the HEMT's channel to improve the active device's noise performance and frequency limits. The GaAs substrate is 100μm thick and preliminary CAD models for passive elements and microstrip transmission lines are available for the design. Fig. 1 depicts the active device's (25x2μm) NF (solid line, right axis) and maximum stable/available gain (dotted line, left axis) as a function of frequency in the 0-160GHz frequency range when the source is terminated on an ideal ground. At 96GHz, i.e. the frequency at which the stability factor becomes unitary, the maximum available gain is 9.2dB and noise figure equals 1.9dB.

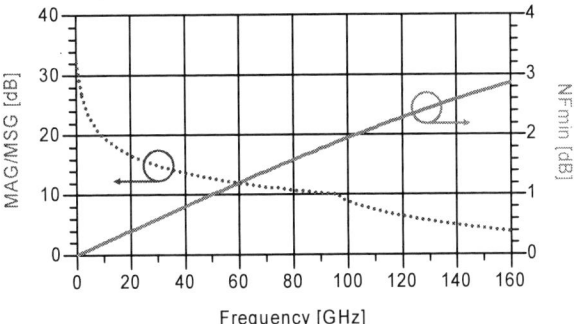

Fig. 1. Selected 25x2μm FET gain (dotted line, left axis) and noise performance (solid line, right axis) as a function of frequency at the selected bias point V_{DS}=1.0V, I_D=7mA.

III. DESIGN METHODOLOGY

A. Multi-stage LNA design issues

The adopted design flow was oriented towards obtaining high gain and limited ripple over a bandwidth as broad as possible, associated to low noise behaviour and prescribed port impedance.

A 4-stage topology was adopted to obtain an average gain around 25dB in the W-Band range. The four stages are very similar to each other since a minimum noise measure technique was adopted to design the single stage: the first stage's gain is not very high, approximately 6dB, and therefore unable to conceal the subsequent stages' noise contribution. Therefore, to obtain low noise behaviour from the overall amplifier not only the noise figure of the single stage has to be considered but its available gain too, as clearly appears from an analysis of the cascaded stages Friis' formula and its effects. As a result, the four stages were all designed to fulfil the minimum noise measure condition, thus obtaining the minimum noise figure of the overall amplifier. The latter effect can be explained by considering that the noise figure of an infinite cascade of identical stages is equal to the noise measure of the single stage plus one. Thanks to the previous considerations and for the sake of simplicity, the LNA was realised by cascading four practically identical stages therefore obtaining a simplification of the design procedures.

Resulting die size is 2.8x1.8mm² and its microphotograph is depicted in Fig. 2 (left).

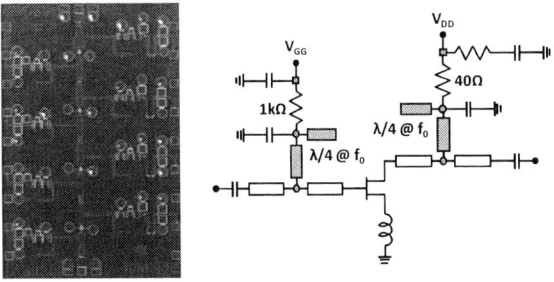

Fig. 2. (left) MMIC microphotograph. Chip size is 2.8 x1.8mm² and (right) Single stage circuit schematic

B. Single-stage LNA design

The first step of the design flow was the selection of the device geometry. At millimetre wavelength "wide" FETs, composed by more than four fingers and unit finger width larger than 30μm, are very difficult to match to 50Ω since input and output optimum terminations are featured by very low impedance levels, close to a short circuit. Moreover, multiple finger devices exhibit a decrease in cut-off frequency f_T dictated by parasitic effects, practically masking the beneficial effects of a 70 nm process technology. Therefore a 2 finger FET with 25μm unit finger width was selected. For this geometry the matching to 50Ω is not too complicated and the f_T is high enough for the targeted frequency band. Typical low noise bias conditions were selected for each active device: V_{DS}=1.0V and I_D=7mA, resulting in a reduced LNA dc power consumption. As stated before this aspect is quite critical in radioastronomy application where the front-end module has to be cooled down to about 15-25K to significantly reduce the active devices' noise contribution.

Since the active device is conditionally stable in the operating bandwidth, a suitable reactive feedback was added between the FET's common terminal (source) and ground to obtain in-band stability as seen in Fig. 2 (right). The inductive behaviour was obtained by adding a thin microstrip line acting as a series inductor in the considered frequency range.

This technique is typically used in LNA design since it eases the IM/NF trade-off that has to be performed at the active device's input. The procedure becomes particularly helpful when the device is conditionally stable and therefore an adequate simultaneous I/O match cannot be realised without the risk of triggering oscillation phenomena. The drawback of this method is the reduction of the active device's maximum available gain. Therefore the amount of feedback (i.e. the reactance value) must be carefully selected so that the active device can be more simply matched at its terminals and to avoid an excessive active device's gain degradation [5].

Out-of-band stability was achieved by inserting resistive elements in the FET's bias networks as seen in Fig. 2 (right). The bias networks were designed to be "transparent" in the operating bandwidth, therefore not adding extra noise contribution. This effect was obtained by joining a quarter-

978-2-8748-7007-1/08 $25.00 © 2008 EuMA

wave open stub at the end of a quarter-wave transmission line used for bias injection. The latter two elements are represented by the grey transmission lines in Fig. 2 (right). In the operating frequency range the bias network's impedance is high enough to conceal its lossy elements noise contribution. On the other hand, at lower frequencies, the line's electrical lengths are small enough so that the resistive elements shunt the FET's input and output ports therefore stabilising the active device at lower frequencies.

Since the four stages are practically identical, the distance between two consecutive FETs has been set to 0.75mm. So transmission line length between two adjacent FETs has not been used as a free design variable for matching. Another design choice consists in reducing as much as possible the number of discontinuities in the microstrip line that produce heavy parasitic phenomena in the operating bandwidth. Accordingly, only the transmission line width W and the distance between the bias injection and the FETs' gate or drain terminals have been optimized as seen in Fig. 2 (left). Moreover, the value of the DC-block capacitors between each stage was select to alleviate the matching difficulties.

The entire LNA layout was simulated electromagnetically by means of MOMENTUM, Agilent's EM solver.

IV. CHARACTERISATION

The MMIC was mounted on a test-jig for dc biasing purposes. Each LNA bias pad was connected to ground through two shunt 10pF and 1nF off-chip capacitors. Series resistors, 1kΩ for the gate line and 10Ω for the drain line, were inserted between the external connectors and the off-chip capacitors to add extra resistance on the dc bias path. The test-jig photo and the microphoto of the decoupling capacitors and drain bias line are shown in the following Fig. 3 (left) and Fig. 3 (right) respectively.

Fig. 3.   (left) Test-jig photograph and (right) MMIC off-chip RF by-pass network and drain bias lines

The MMIC was characterised in terms of S-parameters between 70 and 110GHz using the HP8510XF vector network analyzer. The RF coplanar ground-signal-ground probe tips pitch is 150μm.

To obtain the nominal bias conditions ($V_{DS} \approx 1.0V$ and $I_D = 7mA$) the chip was biased at $V_{DD} = 1.3V$ since each stage has approximately 40Ω resistance on the drain bias line. The gate voltage V_{GG} was adjusted around -0.1V to obtain the nominal drain current on each stage. The overall LNA total power consumption is therefore quite low and in the order of 35mW. The predicted output 1dBcp is around 2dBm.

Fig. 4 depicts the LNA measured linear parameters. The average gain is 25dB, with 2dB ripple from 70 to 105GHz and the $|S_{21}|$ is greater than 21dB on the entire 70-110GHz

measured frequency band. The measured $|S_{22}|$ is better than -10dB from 72 to 90GHz while the $|S_{11}|$ is better than -5dB from 70 to 88GHz.

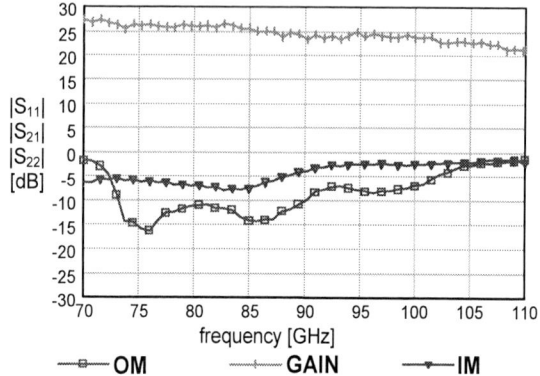

Fig. 4.   LNA measured linear parameters

Finally the expected LNA's noise behaviour is presented as the corresponding characterisation phase is still in progress. The LNA's simulated noise figure at room temperature is approximately 2.7dB between 80 and 95GHz and less than 3.2dB until 108GHz as shown in Fig. 5.

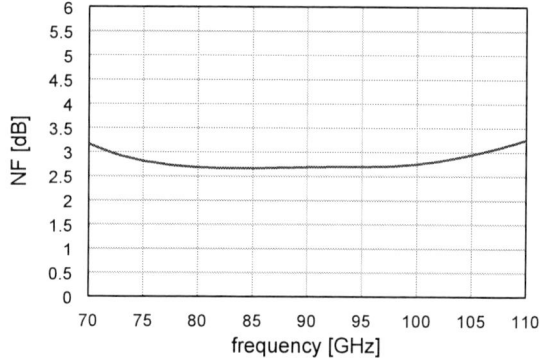

Fig. 5.   LNA expected noise figure at room temperature

It is to note that FET noise characterisation has not been actually carried out in the LNA design bandwidth: device noise parameters have been extracted from on wafer characterisations up to 60GHz and the model has been fitted afterwards at higher frequencies. The small-signal device equivalent circuit model has been however validated up to the operating frequency, thus providing confidence also on the resulting noise model. Moreover, the entire LNA layout was electromagnetically simulated by means of a 2.5D commercial CAD software. This type of EM-simulation should more accurately assess the resistive losses of the passive microstrip structure as compared to circuit simulations and consequently the LNA's NF.

TABLE I

COMPARISON OF SOME W-BAND LNAS IN GALLIUM ARSENIDE HEMT TECHNOLOGY REPORTED IN LITERATURE

REF	ACTIVE DEVICE	TL structure	L_{gate} [µm]	n° of FETs	AVG-GAIN [dB]	GAIN/FETs [dB]	f_{MIN} [GHz]	f_{MAX} [GHz]	BW [GHz]	RIPPLE [dB]	NF [dB]	YEAR
[6]	InAlAs/InGaAs m-HEMT	CBCPW	0.1	4	21.5	5.4	75	100	25	±1	3.0	2006
[7]	InAlAs/InGaAs m-HEMT	CPW	0.05	4	21.5	5.4	70	105	35	±1,5	2.5	2005
[8]	AlGaAs/InGaAs /GaAs p-HEMT	CPW	0.15	6	19.0	3.2	70	105	35	±2.0	N/A	2000
[9]	InAlGaAs m-HEMT	CPW	0.07	3	19.0	6.3	75	100	25	±1.0	3.0	2003
THIS WORK	m-HEMT	microstrip	0.07	4	25.0	6.3	70	105	35	±2.0	2.7(*)	2008

(*) = simulated

TABLE I compares some W-Band LNAs in HEMT technology reported in literature. The LNA presented in this contribution exhibits the highest W-Band average gain (25dB) reported for LNAs using a similar technology and one of the highest ratio between LNA gain and number of FETs in the circuit (6.3dB) over a considerable frequency region (70-105GHz). The columns f_{MIN} and f_{MAX} indicate the frequency band over which the parameters average gain, ripple and noise figure have been assessed. The expected noise behaviour is aligned with the ones obtained in similar technologies Future activity will be focused on two main guidelines. The first concerns the LNA's room temperature noise characterisation. The second regards the LNA's packaging in a test-jig for cryogenic measurement at 25K. The test-jig, already partially developed, consists in an aluminium carrier using a special glue to compensate the different geometric temperature variations of GaAs and Aluminium. WR-10 waveguide connectors are mounted for the RF I/O ports while a single micro D 9-pin connector is used for the 8 DC bias line connections. The cryogenic probe station has been developed in house and measurements up to 110GHz will be performed using the ANRITSU ME7808C Broadband VNA (40MHz to 110GHz).

V. CONCLUSIONS

The possible applications, the design procedures and test of a W-Band low dc bias power LNA on GaAs substrate have been presented and discussed. The measured LNA's main features can be summarised as follows: the average gain is 25dB, with ±2dB ripple from 70 to 105GHz and the gain is greater than 21dB on the entire 70-110GHz frequency range. The LNA's expected noise figure, which will be presented in the final contribution, is approximately 2.7dB between 80 and 95GHz and less than 3.2dB up to 108GHz while the chip's power consumption is around 35mW. The measured $|S_{22}|$ is better than -10dB from 72 to 90GHz and the $|S_{11}|$ is better than -5dB from 60 to 88GHz. The technology used is a 70nm GaAs mHEMT process by OMMIC.

ACKNOWLEDGMENT

The authors wish to acknowledge the microwave electronics group from the DEIS-University of Bologna for their much appreciated assistance and valuable support during the characterisation phase, and of course, for the use of their test equipment.

This work was funded by the PHAROS-RadioNet Project (FP6 – R113CT 2003 5058187) whose support is gratefully acknowledged. Visit the website www.radionet-eu.org.

REFERENCES

[1] J.W. Archer, R. Lai and R. Gough, "Ultra low-noise Indium-Phosphide MMIC amplifiers for 85-115GHz," IEEE Trans. on Microwave Theory and Techniques, Vol. 49, no. 11, Nov. 2001, pp. 2080-2085.

[2] K. Elgaid, H. McLelland, C.R. Stanley, I. G. Thayne, "Low Noise W-Band MMIC Amplifier using 50nm InP Technology for Millimeterwave receivers Applications," Proc. of the 2005 IPRM Conf., 15-15 May 2005, Glasgow (United Kingdom), pp 523-525.

[3] J.B. Hacker et al., "An Ultra-Low Power InAs/AlSb HEMT W-Band Low-Noise Amplifier," Microwave Symposium Digest, 2005 IEEE International MTT-S, pp. 1029-1032.

[4] R. Reuter, Y. Yin, "A 77 GHz (W-band) SiGe LNA with a 6.2 dB Noise Figure and Gain Adjustable to 33 dB," Proc. of the BCTM 2006, 8-10 Oct. 2006, Maastricht (The Netherlands).

[5] W. Ciccognani, F. Giannini, E. Limiti, P.E. Longhi, "Determining Optimum Load Impedance for Noisy Active 2-Port Networks," Proc. of the EuMC 2007, 28-30 Oct. 2007, Munich (Germany).

[6] A. Tessmann, M. Kuri, M. Riessle, H. Massler, M. Zink, W. Reinert, W. Bronner, A. Leuther, "A Compact W-Band Dual-Channel Receiver Module," Microwave Symposium Digest, 2006 IEEE International MTT-S, pp. 85-88.

[7] M. Schlechtweg, A. Tessmann, A. Leuther, C. Schworer, H. Massler, "Advanced mm-Wave ICs and Applications," Proc. of the IEEE RFIT 2005, 30 Nov – 02 Dec 2005, Singapore.

[8] A. Tessmann, W. H. Haydl, M. Neumann, and J. Rüdiger, "W-Band Cascode Amplifier Modules for Passive Imaging Applications," IEEE Microwave and Guided Wave Letters, Vol. 10, no. 5, May 2000, pp 189-191.

[9] C. Schworer, A. Tessmann, A. Leuther, H. Massler, W. Reinert and M. Schlechtweg, "Low-Noise W-Band Amplifiers for Radiometer Applications Using a 70nm Metamorphic HEMT Technology," Proc of the 11th GaAs Symp, 6-10 October 2003, Munich (Germany), pp. 373-376.

Proceedings of the 3rd European Microwave Integrated Circuits Conference

An Electrothermal Model of High Power HBTs for High Efficiency L/S Band Amplifiers

A.Xiong [*1], R.Sommet [*], O.Jardel [*], T.Gasseling [†2], A.A.Lisboa de Souza [*], R.Quéré [*] and S.Rochette [‡3]

XLIM - UMR CNRS n°6172, University of Limoges
7, rue Jules Vallès, 19100 Brive La Gaillarde FRANCE
[1]alain.xiong@xlim.fr

†*AMCAD Engineering*
ESTER technopole B.P.6915, 87069 Limoges FRANCE
[2]gasseling@amcad-engineering.com

‡*THALES ALENIA SPACE*
26, avenue J.F. Champollion B.P. 1187, 31037 Toulouse Cedex 1, FRANCE
[3]stephane.rochette@thalesaleniaspace.com

Abstract—This paper deals with an electrothermal model of high power heterojunction bipolar transistor (HBT) intended for CAD. The first section describes the model topology and sets the implemented equations that allow to take into account of the physical phenomena. The model also integrates scaling rules function of emitter length (W) and number of fingers (N). For the thermal aspect, low frequency impedance measurement approach has been led. Model simulations have been compared to quasi-isothermal I-V measurements, pulsed S parameters measurements, and load-pull multi-harmonics measurements.

I. INTRODUCTION

HBTs have become of great interest for high power and high frequency applications. Their modeling is still a permanent challenge in term of robustness, accuracy and convergence speed for circuit designers. In spite of these criteria and although many HBT models have been investigated, only few are suitable to design power amplifiers that contain several tens of HBT. The model we propose here is dedicated to designers. It relies on physics-based equations and takes into account of most phenomena such as leakage currents, self heating, charges transit time and redistribution time, and also breakdown effects. This development has been performed within the framework of an L/S band SSPA demonstrator for the European Spatial Agency.

II. MODEL TOPOLOGY

The aim was to develop a transistor model able to describe the electrothermal behavior both in static mode and in dynamic mode. So the modeling process has been split into three stages. The first one is devoted to the description of the transistor currents in static regime and leads to the convective model. The second stage concerns the microwave behavior described by capacitances, inductances, transit times. The last stage consists in adding the thermal dependence in the HBT model.

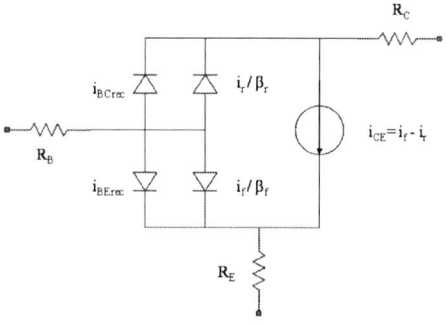

Fig. 1.   Schematic of the DC model

A. DC Model

For the model, a typical π topology as shown in Fig. 1 has been adopted. Diodes crossed by the forward current i_f and the reverse current i_r control the current source i_{CE} with a direct current gain β_f and a reverse current gain β_r. Diodes crossed by i_{BErec} and i_{BCrec} have been added to take into account of the leakage currents due to recombination effects. Classical equations are used :

$$i_f = I_{SF}(T)\left(exp^{\frac{V_{BE}}{N_f \cdot V_t}} - 1\right) \quad (1)$$

$$i_{BErec} = I_{SE}(T)\left(exp^{\frac{V_{BE}}{N_E \cdot V_t}} - 1\right) \quad (2)$$

$$i_r = I_{SR}(T)\left(exp^{\frac{V_{BC}}{N_r \cdot V_t}} - 1\right) \quad (3)$$

$$i_{BCrec} = I_{SC}(T)\left(exp^{\frac{V_{BC}}{N_C \cdot V_t}} - 1\right) \quad (4)$$

The temperature dependency is added in the previous equations to take into account of the thermal effects.

$$I_{SX}(T) = I_{SX}.exp^{\left(-\frac{T_{SX}}{T}\right)} \quad (5)$$

978-2-8748-7007-1/08 $25.00 © 2008 EuMA

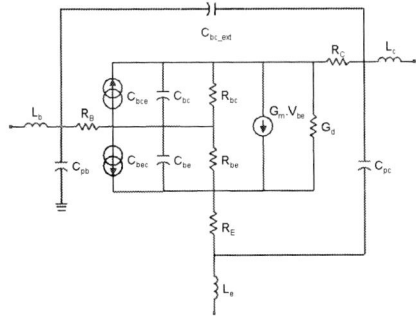

Fig. 2. Schematic of the AC small signal circuit

$$\beta_f = \frac{\beta_{f0}}{1 - a + a.R^b}, \quad R = \frac{T + T_0}{T_{nom} + T_0} \quad (6)$$

T_0 is equal to $273^o K$ and T_{nom} represents the baseplate temperature. The determination of the DC parameters, i.e. the diodes emission coefficients N_F, N_E, N_R, N_C, the saturation currents I_{SF}, I_{SE}, I_{SR}, I_{SC}, the temperature parameters T_{SF}, T_{SE}, T_{SR}, T_{SC}, the reverse current gain β_r and the forward current gain β_f with its parameters β_{f0}, a and b, leads to the set up of the convective model.

B. AC Small Signal Model

The model is presented on Fig. 2. The main parasitics are resistors R_B, R_C, R_E defined earlier in DC model, capacitances due to access coupling C_{pc}, C_{bc_ext}, contact capacitance C_{pb}, and access inductances L_b, L_c, L_e. To define the intrinsic capacitances of the small signal model, a non quasi static (NQS) charge repartition has been adopted. With this approach, charges depend both on base-emitter voltage and base-collector voltage [1]. NQS effects also allow to take into account of charge redistribution time due to command voltages V_{BE} and V_{BC}. This phenomenon can be modelled by adding two additional trans-capacitances C_{bce} and C_{bec} in parallel with classical capacitances C_{bc} and C_{be}. To integrate the punch-through effect not included in classical relation [2], junction capacitance equation has been modified.

$$C_j = \frac{C_{j0}}{\left(1 - \frac{v}{\Phi_D}\right)^z} + C_p \quad (7)$$

where C_{j0} is the zero bias capacitance, Φ_D the diffusion voltage. z and C_p are model parameters.

1) Q_{be} charge: The base-emitter charge results from the junction charge Q_{bej}, the depletion charge Q_{bed} and transcapacitance charge Q_{bec} which represents charge redistribution effects due to V_{BC} variation in the base.

$$Q_{bec} = Q_{bcd}.F_c \quad (8)$$

F_c is a model parameter and Q_{bcd} will be define later in Q_{bc} charge calculation. Moreover, by including Q_{bek} charges in Q_{bec} which allows to take into account of Kirk effect [3]. The total base-emitter charge is formulated as :

$$Q_{be} = Q_{bej} + Q_{bed} + Q_{bek} + Q_{bcd}.F_c \quad (9)$$

The junction charge Q_{bej} is given by integration of Eq. (7). Q_{bed} depend on the transit time at low current density.

$$Q_{bed} = \tau_f.i_f.(1 - F_d) \quad (10)$$

$$\tau_f = \tau_{f0}.(1 - \frac{V_{bc}}{V_{bc_{inv}}}).(1 - \frac{i_f}{i_{finv}}).(1 + \frac{\Delta T}{K_{\tau_t}}) \quad (11)$$

The last term of Eq. (11) expresses the linear thermal dependence of transit time with temperature. The expression of Q_{bek} is simplified from HICUM model [4][5]:

$$Q_{bek} = \tau_k.i_f.(1 - F_k) \quad (12)$$

$$\tau_k = \tau_{k0}.w^2 \quad (13)$$

$$w = \frac{1 - \frac{i_{fk}}{i_f} + \sqrt{(1 - \frac{i_{fk}}{i_f})^2 + A_{\tau_k}}}{1 + \sqrt{1 + A_{\tau_k}}} \quad (14)$$

$$i_{fk} = i_{fk0}.(1 - \frac{V_{bc}}{V_{bc_{invk}}}) \quad (15)$$

τ_k is the transit time at high collector current densities, w is the normalized injection width in the collector, i_{fk} the critical current to onset high-current effects.

2) Q_{bc} charge: The base-collector charge results from the junction charge Q_{bcj}, the depletion charge Q_{bcd} and the transcapacitance charge Q_{bce} to take into account of redistribution effects due to V_{BE} voltage. Q_{bcj} is given by integration of Eq. (7). Q_{bcd} is modelled by :

$$Q_{bcd} = \tau_r.i_r \quad (16)$$

where τ_r is the reverse transit time. And Q_{bce} is defined as :

$$Q_{bce} = Q_{bed}.F_d + Q_{bek}.\frac{F_k}{1 - F_k} \quad (17)$$

For AC small signal model, the parameters are C_{j0}, Φ_D, z, C_p, the transit time parameters, τ_{f0}, K_{τ_t}, τ_{k0}, τ_r, and model parameters $V_{bc_{inv}}$, i_{finv}, $V_{bc_{invk}}$, i_{fk0}, A_{τ_k}, F_c, F_d, F_k.

C. Large Signal Model

The large signal model [6] can be synthesised as shown in Fig. 3. The thermal part of the circuit and the breakdown model I_{Br} will be detailed in the next sections.

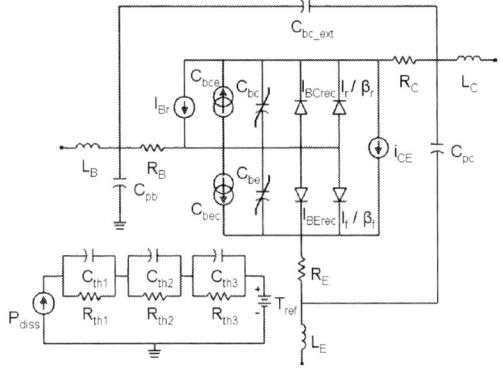

Fig. 3. Large signal circuit with breakdown effect

978-2-8748-7007-1/08 $25.00 © 2008 EuMA

TABLE I
PARAMETERS SCALING RULES

Currents	Extrinsic elements
$I_s = I_{s_u}.W.N$	$L_b = L_{b_u}.\frac{W}{N} + L_{b_0}$
Capacitances	$L_c = L_{c_u}.N + L_{c_0}$
$C_{j0} = C_{j0u}.W.N$	$L_e = \frac{L_{e_u}}{W^c.N^d}$
Parasitic resistor	$C_{pc} = C_{pc_u}.W.N + C_{pc_0}$
$R_i = \frac{R_{i_u}}{W.N}$	$C_{pb} = C_{pb_u}.N + C_{pb_0}$

D. Scaling Rules

Because the model is dedicated to designers, it is interesting to integrate scaling rules. However, the initial model must be precise and must present good convergence capabilities. In our model, lateral scaling parameters are function of the emitter length W and the number of fingers (emitters) N. Unitary parameters have been defined as shown in table I. Intrinsic parameters like currents and capacitances can be easily scaled by applying linear dependence. This rule commonly adopted in literature gives generally good results. Scaling rules for extrinsic elements have been extracted from measurement observations realized on different transistor sizes.

E. Parameters extraction method

Many parameters extraction methods exist [2]. Our model is directly derivated from Gummel-Poon model. We have used a software developed in our lab based on simulated diffusion optimization algorithms. The optimization is made both on the IV input and output data [7].

III. THERMAL MODELING

Low frequency impedance measurements have been performed to determine the thermal time constants [8]. The thermal model is composed of multiple parallel RC cells connected to a current source representing the total dissipated power (Fig. 3). A trade-off between accuracy and complexity has led to a circuit of three RC cells. Comparison between thermal impedance measurements and the model is proposed on Fig. 4. The global thermal resistance has also been validated by 3D finite elements simulation.

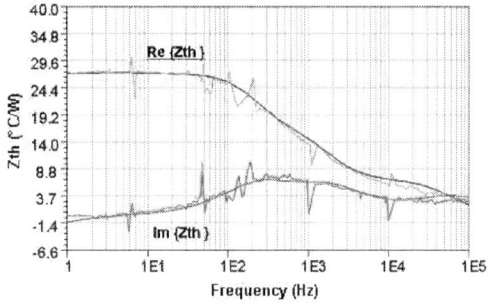

Fig. 4.   Zth simulation and measurements

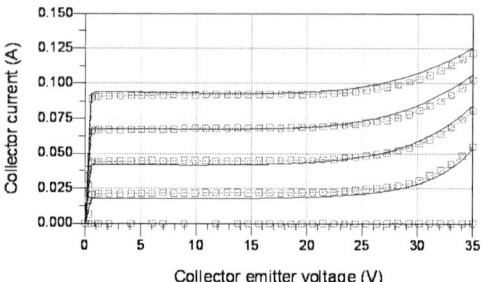

Fig. 5.   Breakdown effects simulations and measurements

IV. BREAKDOWN MODEL

Breakdown effect comes from electrons multiplication due to impact ionisation at high electric field. This phenomenon limits the base-collector voltage excursion and can be taken into account simply by including in the calculation of coefficient α_f (classically obtained from β_f), a parameter M which represents the V_{CB} dependence [9].

$$\alpha_f^* = \alpha_f.M, \quad M = \frac{1}{1 - (\frac{V_{CB}}{BV_{CB0}})^n} \quad (18)$$

BV_{CB0} is the voltage to onset breakdown effect and n a model parameter. In the equivalent circuit, it is represented by a current source I_{Br} between base and collector as shown in Fig. 3. The model behaviour can be adjusted thanks to pulsed IV measurements as illustrated in Fig. 5.

V. MODEL VALIDATION

The first study has been performed on a $3 \times 2 \times 110$ GaAs HBT from the UMS technology (3 bicells transistor, $(2 \times 110)\mu m^2$ emitter surface). Then, scaling rules have been applied to characterize a $10 \times 2 \times 110$ HBT. All the results presented in this section concern this device. The first characterizations rely on pulsed I-V and pulsed [S] parameters measurements performed with a 600ns pulse width and a signal recurrence of $6\mu s$ to limit self heating effects and overshoot in the pulse shape. Within the pulse, a 350ns RF signal has been applied to perform [S] parameters measurements. The transistor was thermally controled. As the quiescent bias point was closed to V_{be0}=0V and V_{ce0}=0V, the junction temperature has been assumed to be equal to the controled temperature. The model has been implemented in Agilent ADS software with SDD.

Pulsed I-V measurements and simulations have been compared for $25°C$ and $75°C$. These comparisons validate the convective model and the thermal dependency of the model parameters. The results are shown in Fig. 6.

To validate the small signal behaviour, pulsed quasi isothermal [S] parameters for several bias points have been compared in the frequency range from 1 to 10 GHz at $25°C$ and $75°C$. Comparisons are presented in Fig. 7.

Load-pull CW measurements have been performed with active loop condition [10] on a transistor operating in C

978-2-8748-7007-1/08 $25.00 © 2008 EuMA

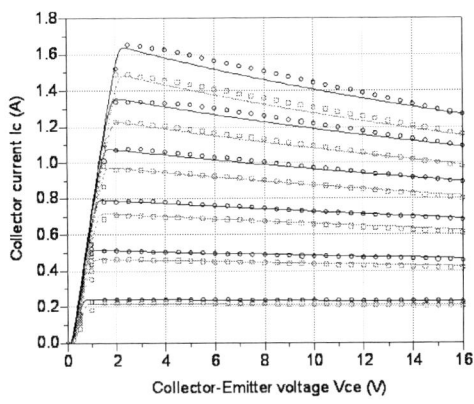

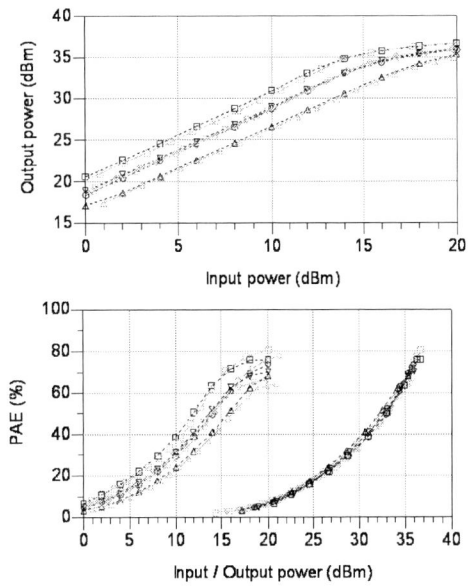

Fig. 6. $Ic = f(V_{ce})$ simul. — $25°C$/ - - $75°C$, meas. $\circ 25°C$ / $\square 75°C$

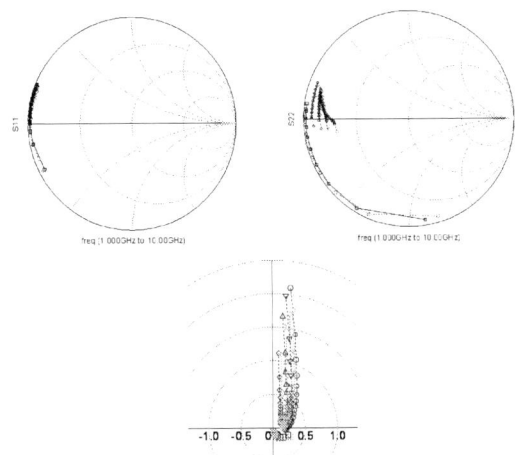

Fig. 8. Ouput power, PAE — simulation / - - measurement

Fig. 7. S parameters — simulations and - - measurements at $25°C$

class. The study has been led with 3 harmonics optimization in L band for 1.257GHz($\square$), 1.5GHz(∇) and 1.6GHz($\triangle$). Comparison between output power, power added efficiency (PAE) exhibit a good agreement in Fig. 8. The 3rd harmonic optimization has allowed to reach more than 70% of PAE for a 3.5 watts of output power.

VI. CONCLUSIONS

We have investigated HBT modelling to develop a scalable compact electrothermal model able to take into account of a maximum of physical phenomena with good convergence capabilities. The method has been described step by step. The model has been successfully implemented in ADS and Microwave Office, and validated by a full characterization campaign on UMS technology GaAs HBT devices.

ACKNOWLEDGMENT

The authors would like to thank European Spatial Agency, Thales Alenia Space, United Monolithic Semiconductor and AMCAD engineering, their partners in the L/S band SSPA project.

REFERENCES

[1] J. G. Fossum and S. Veeraraghavan, "Partitioned-charge-based modeling of bipolar transistor for non-quasi-static circuit simulation," *IEEE Trans. on ED*, vol. 7, no. 12, pp. 652–654, December 1986.

[2] M. Schroter and A. Chakravorty, *HICUM, A Geometry Scalable Physics-Based Compact Bipolar Transistor model*, 2005, users Manual http://www.iee.et.tu-dresden.de/iee/eb/eb_homee.html.

[3] C. T. Kirk, "A theory of transistor cutoff frequency falloff at high current densities," *IEEE Trans. on ED*, vol. 9, pp. 914–920, 1962.

[4] M. Schroter, S. Lehmann, S. Frégonèse, and T. Zimmer, "A computationally efficient physics-based compact bipolar transitor model for circuit design — part i: Model formulation," *IEEE Trans. on ED*, vol. 53, no. 2, February 2006.

[5] S. Frégonèse, S. Lehmann, T. Zimmer, M. Schroter, D. Céli, B. Ardouin, H. Beckrich, H. Brenner, and W. Krauss, "A computationally efficient physics-based compact bipolar transistor model for circuit design – part ii: Parameter extraction and experimental results," *IEEE Trans. on ED*, vol. 53, no. 2, February 2006.

[6] O. Jardel, R. Quéré, S. Heckmann, H. Bousbia, D. Barataud, E. Chartier, and D. Floriot, "an electrothermal model for gainp/gaas power hbts with enhanced convergence capabilities," *IEEE Trans. on MTT*, vol. 53, no. 2, February 2006.

[7] J. P. Teyssier, J. P. Viaud, J. J. Raoux, and R. Quéré, "Fully integrated nonlinear modelling and characterization system of microwave transistors with on-wafer pulsed measurements," in *IEEE MTT-S International*, vol. 3, May 1995, pp. 1033–1036.

[8] A. A. Lisboa de Souza, J. C. Nallatamby, M. Prigent, and R. Quéré, "Dynamic impact of self-heating on input impedance of bipolar transistors," *Electronics Letters*, vol. 42, no. 13, June 2006.

[9] S. Heckmann, R. Sommet, J. M. Nebus, J. C. Jacquet, D. Floriot, P. Auxemery, and R. Quere, "Characterization and modeling of bias dependent breakdown and self-heating in gainp/gaas power hbt to improve high power amplifier design," *IEEE Trans. on MTT*, vol. 50, no. 12, pp. 2811–2819, 2002.

[10] F. Blache, J. M. Nebus, P. Bouysse, and L. Jallet, "A novel computerized multiharmonic active load-pull system for the optimization of high efficiency operating classes in power transistors," *IEEE Trans. on MTT*, vol. 3, pp. 1033–1036, December 1995.

Regenerative Frequency Divider SiGe-RFIC with Octave Bandwidth and Low Phase Noise

Thomas Wallin[∈1], Johan Hellen[∈2], Håkan Berg[∈3], Sven-Erik Elfgren[∈4]

∈ *Microwave & Antennas, Saab Microwave Systems, Saab AB*
SE-412 89 Göteborg, Sweden

[1]thomas.wallin@saabgroup.com
[2]johan.hellen@saabgroup.com
[3]hakan.j.berg@saabgroup.com
[4]sven-erik.elfgren@saabgroup.com

Abstract— **Using traditional RFIC design architecture a broadband regenerative frequency divider is designed. By integrating the frequency divider in a single IC the size is reduced and the bandwidth is increased without compromising the phase noise performance. A one octave bandwidth, 3.2-6.4 GHz, is achieved with a phase noise floor below -157dBc/Hz. For a narrower frequency band, 4.0-5.6 GHz, a phase noise floor below -167dBc/Hz is measured. This is directly comparable to regenerative frequency dividers designed using conventional discrete components or commercially available dividers. At the same time the bandwidth is exceeded only by digital dividers, they however have a typical phase noise floor of -150 dBc/Hz at these frequencies. Both $f_0/2$ and $3f_0/2$ are available at the output while the input frequency, f_0, is suppressed to at least -20dBc. The IC is manufactured by austriamicrosystems in their 0.35μm SiGe-BiCMOS process with an fT of 70 GHz and is packaged in a 5x5 mm QFN plastic package.**

I. INTRODUCTION

Saab Microwave Systems is producing radar systems for air- sea- and ground based applications. The radar systems are based on Doppler technique where low phase noise is a crucial parameter, especially when it comes to detecting small targets in a clutter environment.

A common technique to generate the necessary microwave frequencies in radars today is to have a reference frequency from a quartz or SAW device. By the use of multipliers and mixers all necessary frequencies are generated from this reference.

An attractive alternative is to use a cavity oscillator or an optoelectric oscillator at X-band as a reference. Such oscillators exhibit better phase noise performance than conventional sources when compared at the same frequencies. If such a high frequency reference is used frequency dividers must be used instead of multipliers. That means that several frequency dividers working at different frequencies are needed. Therefore it would be desirable to integrate them on ICs to keep the overall size down. In order to keep the number of individual components and thereby the cost down the same IC must cover several frequencies. At the same time the frequency dividers must have a low residual phase noise in order to make use of the oscillator's good phase noise performance. To cope with the bandwidth an alternative is to use digital frequency dividers which can have very large bandwidth. But they have not so far showed acceptable phase

noise performance and produce a high harmonic content due to the inherent square wave output. The requirement on low phase noise makes regenerative frequency dividers an interesting component. Regenerative frequency dividers was first described by Miller in 1939 [1] and further investigated in [2]-[4]. A general block diagram of a regenerative divider is shown below. It can be explained as a phase locked oscillator where the gain around the loop must equal unity with a phase shift of a number of wavelengths. The difference between this circuit and conventional oscillators is that the input frequency, f_0, is a prerequisite for the oscillation to occur. Without the signal at the LO port there is no gain through the mixer and thereby not through the loop. Both frequencies; $f_0/2$ and $3f_0/2$ are generated which is useful in many cases. However, $3f_0/2$ has to be suppressed in the loop in order not to increase the residual phase noise of the divider [5].

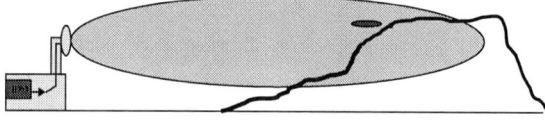

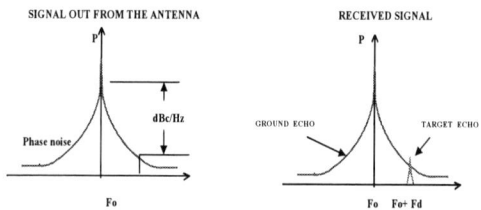

Fig. 1   Importance of low phase noise in Doppler radars.

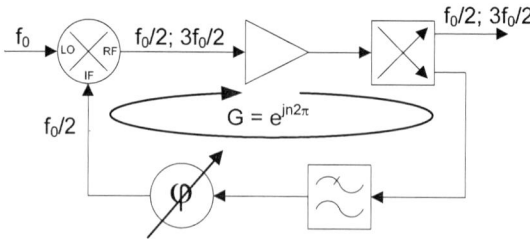

Fig. 2   General block diagram of a regenerative frequency divider.

II. DESIGN CONSIDERATIONS

The bandwidth of a frequency divider is defined by the loop gain, G, shown in Fig. 2. The loop gain shall be unity with a phase of a multiple of wavelengths. Since it is relatively easy to design broadband mixers and amplifiers the bandwidth is defined by where the phase condition is fulfilled. If the feedback loop should be a wavelength or more it is not possible to fulfil the phase condition over a large frequency band. The obvious solution is therefore to choose the number of wavelengths to zero. That on the other hand implies that the physical size must be very small compared to the wavelength. In order to achieve such a small size it is more or less inevitable to integrate the frequency divider on-chip if a broadband divider is to be designed [6].

The choice of semiconductor process fell upon a SiGe-HBT process. There are a few reasons for this; the low phase noise corner frequency of SiGe-HBTs, the possibility to have a high level of integration and thereby making the circuits small and the cost issue where silicon germanium is a very cost effective solution.

In the table below the most important design goals are shown. Although it is desirable to have a large bandwidth the phase noise is the most important parameter. The possibility to access both frequencies; $f_0/2$ and $3f_0/2$ complicates the design at chip level but simplifies the design at system level. It is therefore important that both frequencies can be accessed.

TABLE I
DESIGN GOALS

Parameter	Value
Input frequency, f_0	5.0 GHz ±1.0 GHz
Ouput frequencies	$f_0/2$ and $3f_0/2$
Phase noise	<-155 dBc @1kHz offset relative $f_0/2$
Output power	0 dBm
Input power	5 dBm

III. CIRCUIT DESIGN

Since the objective was to design a broadband frequency divider the idea was to design a very broadband core and let the bandwidth be defined solely by the loop filter. This way it is easy to design frequency dividers for different frequency bands using the same core design. The loop filter is necessary to make half the input frequency, $f_0/2$, dominate over $3f_0/2$ in the loop in order to minimize phase noise [5].

Since circuits with zero phase shift shall be achieved over a large frequency band standard analog building blocks are to prefer over impedance matched ones. The matching networks usually introduce phase shift which is not affordable in this design.

The core consists of a resistively loaded Gilbert cell mixer together with an emitter follower as a feedback buffer. The necessary loop gain is achieved by the mixer why the buffer only needs to decouple the output of the mixer from the filter.

The loop filter is simply an LC-tank in parallel to the output of the feedback amplifier. There are multiple reasons to use an LC-tank. First of all the filter must not add any extra phase shift, at least not at centre frequency. This more or less rules out low pass filters. Since the signal does not need to pass through an inductor its influence within the frequency band of operation is minimized. The second reason has to do with the fact that it is possible to bias the circuits through a parallel resonator while a series resonator would require bias nets. An additional reason is the size; only one centre-tapped inductor is necessary. Since the inductor will be part of a resonator tank its parasitic capacitors does not influence the performance why it is possible to have high coupling between the turns and thus minimize the size. The LC-tank only needs to make sure that the condition for oscillation through the loop is fulfilled at $f_0/2$ while it is not at $3f_0/2$.

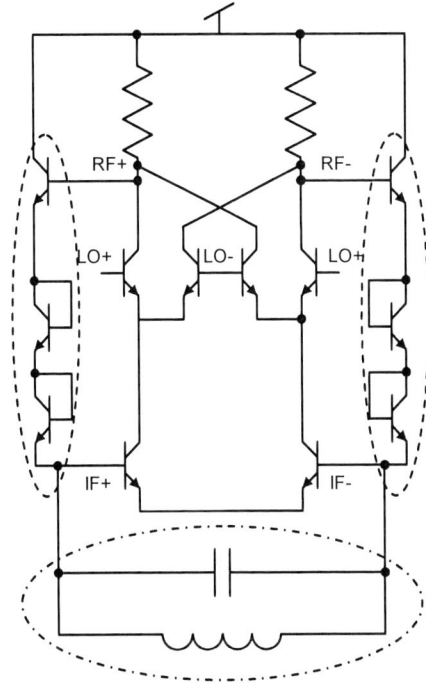

Fig. 3. Schematics of the frequency divider. Marked in dashed lines are the feedback amplifiers and in dash-dotted line is the loop filter.

In order to keep a low phase noise the noise figure of the mixer is crucial. In order to minimize this; the mixer has to be noise matched at the input. The impedance to which the mixer should be matched is mainly defined by the impedance of the LC-tank at resonance. Since that impedance is high and resistive the noise matching is easily done by just scaling the mixer so that the optimum source impedance matches the impedance of the LC-tank at resonance. Below the layout of the frequency divider is shown with a differential in and output. The output is connected through an emitter follower in order to not load the mixer. It is matched to the output by simply adding series resistors at the output.

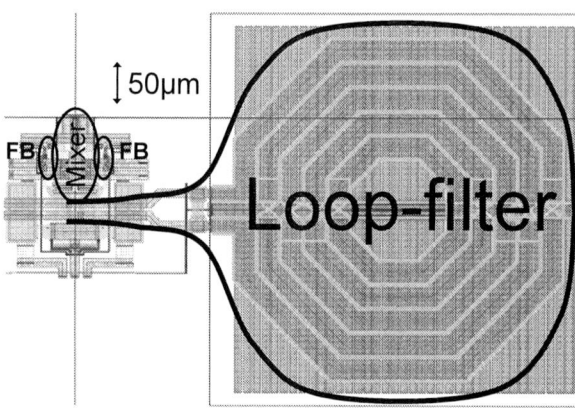

Fig. 4   Layout of of the frequency divider. Shown are the mixer, feedback amplifiers, FB, and loop-filter.

IV. IC ARCHITECTURE

A problem that arose during the design work is that the LO-voltage becomes deformed. This happens since the two nodes are not loaded symmetrically. The voltage over the IF-port has exactly half the frequency compared to that of the LO-port, resulting in that every switching of IF-polarity occurs when the LO-voltage has the same polarity. This results in one of the two nodes of the LO-port to be nearly constant while the other is carrying the signal. This is not desirable for differential circuits and can be solved by the connection of two frequency dividers in anti-parallel. This way the IF-signal will switch at both polarities of the LO-voltage resulting in symmetry.

There is a lot more to be won by connecting two dividers in anti-parallel. The two output signals will be identical unless apart from a constant 90 degree phase difference at $f_0/2$ and 270 degrees at $3f_0/2$. This feature can be used in order to filter out the two frequency components using a 90 degree hybrid to combine the two outputs. This method is similar to the way an image rejecting mixer works and produces $f_0/2$ at one output of the hybrid and $3f_0/2$ at the other. Since the outputs of the two dividers are combined it results in a 3 dB reduction of the phase noise.

Alternatively the 90 degree phase difference can be utilized directly to feed image rejecting mixers where two LO-signals with a 90 degree phase difference are used. This can substantially reduce the complexity of the frequency generation at top level.

Since all circuits are differential the input and output ports utilize simple first order lattice baluns. The two outputs should be identical if a perfect 90 degree difference is to be produced. In this design however the baluns are optimized for the two frequency bands, $f_0/2$ and $3f_0/2$. This is done to make sure that the phase noise can be properly measured in both frequency bands.

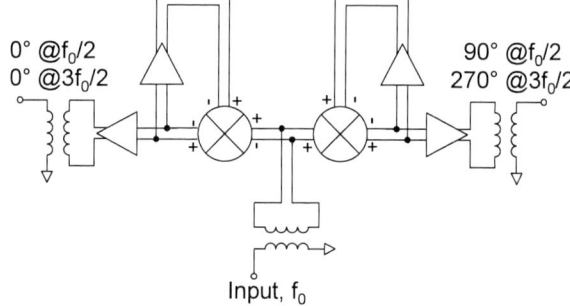

Fig. 5.   Architecture of the complete IC with two frequency dividers and baluns at in- and outputs.

Below the layout of the whole IC is shown. At the left there is the input balun, in the centre of the IC are the two frequency dividers and at the right the two output baluns are shown. The rest of the IC is covered by ESD-protection periphery cells, bias circuits and metal fill. Between the baluns and frequency dividers differential 100Ω transmission lines are used. The layout is optimized to fit in a QFN 5x5 mm plastic package and the die size is 3.2x3.2 mm. The size is set by the number of I/O pads.

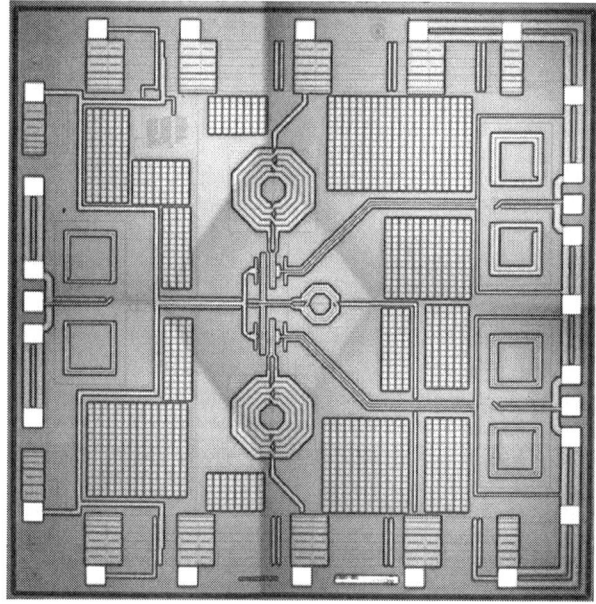

Fig. 6.   Microphotograph of the complete IC incorporating bias circuits, baluns and ESD-protection.

V. Measured Results

The measurements have been done using a 90 mA current consumption from a 5.5V source, i.e. each frequency divider consumes 45mA. The output spectrum has been measured over a wide frequency band. Below the spectrums for three different input frequencies are shown. They show a good suppression of the input frequency, f_0 over a wide frequency band. It is more than 20 and 10 dB below the output power at $f_0/2$ and $3f_0/2$ respectively. From the spectrums one can see that the divider got a stable bandwidth over one octave. This shall be compared to the bandwidth usually presented for regenerative frequency dividers up to 5%. Relative bandwidths up to 50% has been reported using self-oscillating mixers [7] although no phase noise measurements are presented.

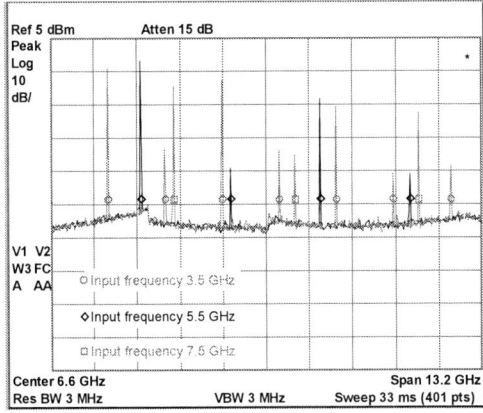

Fig. 7 Output spectrum for three different input frequencies; 3.5 GHz (red), 5.5 GHz (black) and 7.5 GHz (green).

The phase noise measurements have been done for frequencies between of 3.2 and 6.4 GHz using an Agilent E5500 phase noise measurement system. These are shown in Fig. 8. The measured phase noise over a frequency band of 24%, Fig. 9, is well comparable to earlier reported regenerative frequency dividers and commercially available ones [8].

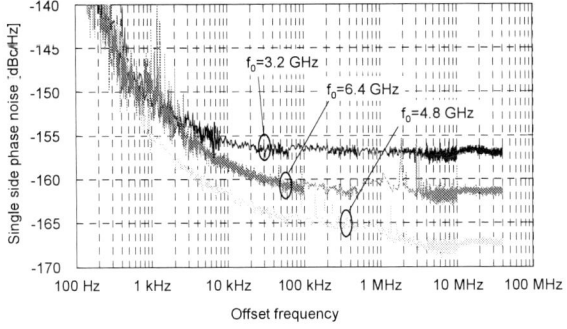

Fig. 8. Phase noise relative to the carrier frequency of $f_0/2$. The measurement is done for input frequencies, f_0, of 3.2, 4.8 and 6.4 GHz.

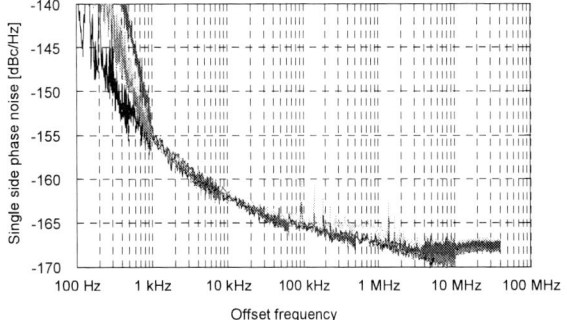

Fig. 9. Phase noise relative to the carrier frequency of $f_0/2$. The measurement is done for input frequencies, f_0, of 4.4 (blue), 4.8 (green), 5.2 (black) and 5.6 GHz (red).

VI. Conclusion

A regenerative frequency divider integrated in a low cost SiGe-RFIC is demonstrated. It shows excellent residual phase noise performance over an extremely wide frequency band, 24%. In a frequency band of one octave, 3.2-6.4GHz it shows phase noise well below digital dividers. The frequency divider is packaged in a standard QFN plastic package allowing a big part of the frequency generation to be realized using low cost PCB technologies. This together with its performance makes it a key component if high performance frequency generators with several frequency dividers are to be designed.

Acknowledgement

The authors wish to thank the Swedish ministry of defence for supporting this work. The authors also wish to thank Heiko Thiesies for carefully reading and reviewing this paper and Per Westling for great help with phase noise measurements.

References

[1] R. L. Miller, "Fractional-frequency generators utilizing regenerative modulation", in *Proc. IRE*. vol. 27, pp. 446-457, July 1939.

[2] R. G. Harrison, "Theory of Regenerative Frequency Dividers Using Double-Balanced Mixers", in *Proc. MTT-S*, 1989, pp. 459-462.

[3] R. H. Derksen, Volker Lück, and H. M. Rein, "Stability Ranges of Regenerative Frequency Dividers Employing Double Balanced Mixers in Large-Signal Operation", *IEEE Transactions on Microwave Theory and Techniques*, vol. 39, pp. 1759-1762, Oct. 1991.

[4] E. Rubiola, M. Oliver, and J. Groslambert, "Phase Noise of Regenerative Frequency Dividers" in *Proc. 5th European Freq. and Time Forum*, 1991, pp. 115-122.

[5] Ferre-Pikal, E.S. and Walls, F.L (1997), "Low PM Noise Regenerative Dividers", in *Proc. IEEE Int. Freq. Cont. Symp.*, 1997, pp. 478-48.

[6] L. Landén, C. Fager, H. Zirath, Herbert, "Regenerative GaAs MMIC Frequency Dividers for 28 and 14 GHz", in *Proc. EuMC*, 2000, vol. 1, pp. 184-186.

[7] I. Kipnis, "20 GHz Frequency-Dividers Silicon Bipolar MMIC", *Electronics Letters*, vol. 23, Issue 20, pp. 1085-1087, 1987

[8] M. Mossammaparast, C. McNeilage, P. Stockwell, J.H. Searls, M.E. Suddaby, "Low Phase Noise Division From X-band to 640MHz", in *Proc. Freq. Control Symp. and PDA Exh.*, 2002, pp. 685-689

New Macromodeling Approach to Phase Noise Analysis of Locked Oscillators

M.M. Gourary[1], S.G. Rusakov[1], S.L. Ulyanov[1], M.M. Zharov[1], B.J. Mulvaney[2], K.K. Gullapalli[2]

[1]IPPM, Russian Academy of Sciences, 3 Sovetskaya str., Moscow, Russia
gourary@ippm.ru

[2] Freescale Semiconductor Inc., 7700 W. Parmer Lane, Austin, Texas, USA
brian.mulvaney@freescale.com

Abstract—A new oscillator macromodel in the form of phase differential equation is proposed. The comparison of new macromodel with the Adler equation and with the macromodel based on Floquet theory is presented. It is shown that the proposed approach allows to perform phase noise analysis of any oscillator circuit with arbitrary periodic injection waveform. The approach can be easily implemented into a circuit simulator.

I. INTRODUCTION

The effective way to reduce the oscillator phase noise is the application of injection locking mode [1]. So RF designers need phase noise analysis of locked oscillators to be implemented into a circuit simulator. The efficient technique to perform phase noise analysis is based on the usage of oscillator phase macromodel. At the present time two main approaches are used to analyze phase behavior of oscillator: the Adler equation and the Floquet-based phase equation.

The Adler equation [2] for LC oscillator under small sinusoidal injection with frequency ω_{inj} has the form

$$\frac{d\theta}{dt} = \omega_0 - \omega_{inj} - \frac{\omega_0}{2Q} \cdot \frac{V_{inj}}{V_{osc}} \sin(\theta) \qquad (1)$$

where $\theta(t)$ is the slow-varying phase of oscillator waveform, ω_0 is the fundamental of free-running oscillator, Q is the quality factor of LC tank, V_{inj}, V_{osc} are magnitudes of injection signal and capacitance voltage respectively. Advantages of the approach: the injection locking condition can be easily obtained by steady-state solution of (1) with $d\theta/dt = const$; the stability and phase noise of locked oscillator can be obtained by analyzing small deviations from the steady-state solution [3].

Equation (1) can be applied to LC oscillator only. The similar equation can be obtained for some other classes of oscillators and injection-locked frequency dividers [4]. But in any case this approach cannot be implemented into a circuit simulator to provide the analysis of an arbitrary oscillator circuit defined by its netlist.

The phase differential equation based on Floquet theory [5] is presented in the form

$$\frac{d\alpha}{dt} = v^T(t + \alpha)b(t) \qquad (2)$$

where *v(t)* is a perturbation projection vector (PPV), *b(t)* is an external excitation, α is unknown phase variable.

In comparison with (1) the important advantage of this phase equation is its applicability to an arbitrary oscillator circuit. Here locking conditions cannot be simply obtained by the steady-state condition $d\alpha/dt = const$ because the solution of (2) corresponding with constant frequency does not exist. The problem can be solved by averaging (2) before the determination of the steady-state solution [6]. Such approach allows to obtain the locking condition but it cannot be applied to the noise analysis.

In this paper we present a new oscillator phase differential equation that can be applied to arbitrary oscillator circuit. The equation has the steady-state solution under an arbitrary locking excitation and allows to perform the noise evaluation by the standard small-signal analysis near DC operating point.

The paper is organized as follows. In Section II basic expressions of the harmonic balance oscillator analysis are presented. The phase differential equation of the macromodel is derived in Section III. The features of new macromodel are given in Section IV in comparison with other known macromodels. Section V presents the application of the macromodel to the phase noise analysis of locked oscillators. Simulation examples are given in Section VI.

II. BACKGROUND

In the context of Harmonic Balance (HB) analysis the oscillator is described by the algebraic system in the frequency domain [7]

$$I(X) + j\omega K \cdot Q(X) = 0 \qquad (3)$$

Here X, I, Q are the vectors of harmonics of nodal voltages, currents and charges respectively. Vectors X, I, Q contain components X_{kl}, I_{kl}, Q_{kl}, where k is harmonic index, l is nodal index. ω is the unknown fundamental frequency, K is a diagonal matrix of harmonic indexes

$$K = diag[\ldots, -kE_N\ldots, -E_N, 0E_N, E_N, \ldots, kE_N, \ldots], \qquad (4)$$

E_N is unit matrix of the circuit size N. The unambiguous solution of (3) can be obtained by fixing the phase of the first harmonic of the reference node voltage

$$Re(X_{1,r}) = 0 \tag{5}$$

The solution of (3, 5) can be determined by Newton iterations with solving the augmented linear system

$$J \cdot \Delta X + jKQ \cdot \Delta\omega = R, \ Re(\Delta X_{1,r}) = 0 \tag{6}$$

where R is the residual vector, J is the Jacobian matrix

$$J = G + j\omega K \cdot C \tag{7}$$

Here $G = \partial I(X)/\partial X$, $C = \partial I(X)/\partial X$ are block Toeplitz matrices of harmonics of nodal conductances and capacitances. The Jacobian matrix (7) at the solution point ($J_0 = G + j\omega_0 K \cdot C$) is singular matrix, and so there exists the eigenvector U corresponding to zero eigenvalue and the eigenvector V of transposed Jacobian such that

$$J_0 \cdot U = 0, \ J_0^T \cdot V = V^T \cdot J_0 = 0 \tag{8}$$

The eigenvector V is the HB representation of PPV (2) that can be arbitrary normalized. The traditional normalization is defined by the condition $V^T C U = 1$ that is equivalent to

$$jV^T KQ = 1/\omega_0 \tag{9}$$

The equivalence is resulted from the equality $j\omega_0 K \cdot Q = CU$ because both sides represent capacitance currents in the frequency domain.

The solution of (3) can be defined for arbitrary phase shift ϕ by the replacement (5) with $Re(X_{1,r}(\phi)\exp(j\phi)) = 0$. For the given condition the solution, currents and charges are shifted by ϕ. This corresponds with multiplying of the vectors by a diagonal matrix $\exp(j\phi K)$ where K is defined by (4)

$$X(\phi) = e^{j\phi K} X, \ I(\phi) = e^{j\phi K} I, \ Q(\phi) = e^{j\phi K} Q, \tag{10}$$

The conductance matrix with the shift ϕ is evaluated by

$$G(\phi) = \frac{\partial I(X(\phi))}{\partial X(\phi)} = \frac{\partial I(X(\phi))}{\partial X}\left(\frac{\partial X(\phi)}{\partial X}\right)^{-1}. \tag{11}$$

Similarly the capacitance matrix $C(\phi)$ can be evaluated. So from (10, 11) one can obtain $G(\phi) = e^{j\phi K} G e^{-j\phi K}$, $C(\phi) = e^{j\phi K} C e^{-j\phi K}$ resulting in the following expressions for shifted Jacobian

$$J_0(\phi) = e^{j\phi K} \cdot J_0 \cdot c^{-j\phi K}, \ J_0^T(\phi) = e^{-j\phi K} \cdot J_0 \cdot e^{j\phi K} \tag{12}$$

Shifted values of eigenvectors can be obtained by

$$U(\phi) = e^{j\phi K} U, \ V(\phi) = e^{-j\phi K} V \tag{13}$$

because we have from (8, 12, 13)

$$J_0(\phi) \cdot U(\phi) = 0, \ V^T(\phi) \cdot J_0(\phi) = 0 \tag{14}$$

III. DERIVATION OF THE MACROMODEL EQUATION

In this section we consider the oscillator behavior at the excitation presented as harmonics of oscillator fundamental

with time-dependent slowly varying phasors $B_{kl}(t)$ and slowly varying DC vector $B_0(t)$

$$b(t) = \frac{1}{2}B(t)\exp(jK\omega_0 t) + B_0(t) \tag{15}$$

We assume that the oscillator behavior under such excitation is represented as the phase-shifted unperturbed solution $X(\phi)$ (10) with small time-varying amplitude deviations $\Delta X(t)$

$$X(t) = X(\phi(t)) + \Delta X(t) \tag{16}$$

Then we can write the linearized HB system for such excitation in the form of (6) representing the frequency deviation $\Delta\omega$ as the corresponding instantaneous value $\frac{d\phi(t)}{dt}$

$$J_0(\phi) \cdot \Delta X(t) + jKQ \cdot \frac{d\phi(t)}{dt} = \frac{1}{2}B(t) + B_0(t) \tag{17}$$

After left multiplying (17) by $V^T(\phi)$ and taking into account (9, 13, 14) we obtain the equation independent on amplitude variations $\Delta X(t)$

$$\frac{1}{\omega_0}\frac{d\phi}{dt} = \frac{1}{2}V^T \exp(-j\phi(t)K) \cdot B(t) + V_0^T \cdot B_0(t), \tag{18}$$

Taking into account complex conjugate properties of vectors B, V ($B_{k,l} = B^*_{-k,l}$, $V_{k,l} = V^*_{-k,l}$) the phase equation (18) can be written in the form

$$\frac{1}{\omega_0}\frac{d\phi}{dt} = W(B(t), \phi(t)) + V_0^T \cdot B_0(t) \tag{19}$$

where

$$W(B(t), \phi(t)) = Re\left(\sum_{k>0} V_k^T \cdot B_k(t) e^{-jk\phi(t)}\right) \tag{20}$$

Here V_k, B_k are vectors of kth harmonics of all nodes.

In some cases it is more convenient to represent the excitation by the sum of excitations (15) with magnitudes $B^{(n)}$ ($n=1,2,...$). Then the phase model is described by the equation

$$\frac{1}{\omega_0}\frac{d}{dt}(\phi) = \sum_n (W(B^{(n)}(t), \phi(t)) + V_0^T \cdot B_0^{(n)}(t)) \tag{21}$$

IV. VALIDATION

A. *DC excitation*

When the oscillator is excited by the slow varying DC signal applied to l-th node the equation (19) is presented as

$$\frac{d\phi}{dt} = \omega_0 V_{0,l} \cdot B_{0,l}(t) \tag{22}$$

This expression corresponds with the traditional VCO representation as an ideal integrator in PLL macromodels [8].

B. *Injection locking conditions*

The periodic excitation with the frequency $\omega_0 + \Delta\omega$ can be considered as the excitation with the frequency ω_0 and slow

978-2-8748-7007-1/08 $25.00 © 2008 EuMA

linearly varying phase $\Delta\omega \cdot t$ of each harmonic $B_{kl}(t) = B_{kl} \cdot \exp(jk\Delta\omega \cdot t)$. Substituting this expression into (20) the equation (19) can be written as follows

$$\frac{d\phi}{dt} = \omega_0 W(B, \phi(t) - \Delta\omega \cdot t) \qquad (23)$$

It is seen that the steady-state solution of (23) is $\phi(t) = \Delta\omega \cdot t + \phi_0$, where ϕ_0 satisfies the equation

$$W(B, \phi_0) = \Delta\omega / \omega_0 \qquad (24)$$

that coincides with the injection locking phase algebraic equation derived in [9]. Thus the macromodel (19, 20) correctly defines the injection locking solution.

C. Adler Equation

The Adler equation (1) is obtained for the sinusoidal excitation with frequency ω_{inj} at one input node. In such case $\Delta\omega = \omega_0 - \omega_{inj}$, and (23) can be presented as

$$\frac{d\phi}{dt} = \omega_0 |V_{1,l} \cdot B_{1,l}| \cos(\psi + (\omega_0 - \omega_{inj})t - \phi(t)) \qquad (25)$$

Here ψ is the phase of $V_{1,l} \cdot B_{1,l}$. After introducing new variable $\theta(t)$ (the full phase of the oscillator) defined by

$$\theta(t) = \psi + (\omega_0 - \omega_{inj})t - \phi(t) - \pi \cdot 3/4 \ , \qquad (26)$$

the equation (25) can be transformed to

$$\frac{d\theta}{dt} = \omega_0 - \omega_{inj} - \omega_0(|V_{1,l}||B_{1,l}|)\sin(\theta) \qquad (27)$$

Hence (27) coincides with Adler equation (1) if $A_{inj} = |B_{1,l}|$ and $V_{1,l}$ is taken in accordance with LC oscillator characteristics (see [9, 10]).

D. Floquet-based Phase Model

When the excitation is applied to l-th node only the Floquet-based phase model (2) is defined as

$$\dot{\alpha} = v_l(t + \alpha)b_l(t) \qquad (28)$$

Let the excitation be a slow-varying m-th harmonic

$$b_l(t) = \frac{1}{2}(B_{m,l}(t)\exp(jm\omega_0 t) + B_{-m,l}(t)\exp(jm\omega_0 t)) \qquad (29)$$

PPV $v_l(t)$ can be represented as Fourier series with components $V_{k,l}$. Then

$$v_l(t + \alpha) = \sum_k V_{k,l} \cdot \exp(jk\omega_0(t + \alpha)) \qquad (30)$$

After multiplication (29) by (30) we can obtain

$$v_l(t + \alpha) \cdot b_l(t) = \frac{1}{2}\sum_k V_{k,l} B_{m,l}(t)e^{j(k+m)\omega_0 t + k\omega_0\alpha} \qquad (31)$$

The terms in the sum in (31) with $k + m \neq 0$ are highly

oscillatory sinusoids that poorly correspond to the slow-varying phase waveform. Such terms can be considered as a small addition to amplitude variations and should be removed from the phase analysis. Then we obtain from (31)

$$v_l(t + \alpha) \cdot b_l(t) \approx Re(V_{m,l} B_{m,l}(t)e^{jm\omega_0\alpha}) . \qquad (32)$$

After substituting (31) and $\phi = \omega_0\alpha$ into (2) one can obtain

$$\frac{1}{\omega_0}\frac{d\phi}{dt} = Re(V_{m,l} B_{m,l}(t)e^{jm\phi}) \qquad (33)$$

Thus the proposed phase model (19, 20) can be derived from the Floquet-based phase model (2) by neglecting highly oscillatory terms. This result is obtained for the single harmonic, single node excitation but its extension to the general case is straightforward.

V. PHASE NOISE OF LOCKED OSCILLATOR

Let the oscillator be excited by the following signals:
- sufficiently large locking signal with oscillation frequency ω_0, and harmonics magnitudes B_{kl} ($k > 0$),
- small sinusoidal signal applied to n-th node with frequency $m\omega_0 + \delta\omega$, and magnitude b_{mn}.

Then (21) can be represented as follows

$$\frac{1}{\omega_0}\frac{d\phi}{dt} = W(B, \phi(t)) + Re(V_{mn} b_{mn} e^{j(\delta\omega \cdot t - m\phi(t))}) \qquad (34)$$

We can write the solution of (34) as a small deviation from the injection locking phase: $\phi(t) = \phi_0 + \eta(t)$ where ϕ_0 is defined by (24). Then (34) takes the form

$$\frac{1}{\omega_0}\frac{d\eta}{dt} = W(B, \phi_0 + \delta) + Re(V_{mn} b_{mn} e^{(\delta\omega \cdot t - m\phi_0 - m\eta)}) \quad (35)$$

We can linearize rhs of (35) with respect to $\eta(t)$. It can be shown that $\eta(t)$ in the second term (35) can be neglected because it corresponds with second order terms in Taylor expansion (at small V_{mn}). After linearizing (35) we obtain

$$\frac{1}{\omega_0}\frac{d\eta}{dt} - \dot{W}(B, \phi_0)\eta(t) = Re(V_{mn} b_{mn} e^{j(\delta\omega \cdot t - m\phi_0)}) \qquad (36)$$

where $\dot{W}$ is determined by the differentiation of (20)

$$\dot{W} = \frac{dW}{d\phi} = \sum_{k>0} Re\left(-jk \sum_{k>0} V_k^T \cdot B_k(t)e^{-jk\phi}\right) \qquad (37)$$

Equation (36) is the first order differential equation with sinusoidal excitation that provides the transfer factor from input excitation to the oscillator phase (η / b_{mn})

$$|H_{mn}^{lock}|^2 = \omega_0^2 |V_{mn}|^2 / (\omega_0^2 \dot{W}(B, \phi_0)^2 + \delta\omega^2) \qquad (38)$$

Note that the transfer factor of the free running oscillator corresponds to zero injection magnitudes ($B=0$) in (38)

$$\left|H_{mn}^{free}\right|^2 = \omega_0^2 \left|V_{mn}\right|^2 / \delta\omega^2 \qquad (39)$$

After comparison (38) with (39) one can conclude

$$\left|H_{mn}^{lock}\right|^2 = \left|H_{mn}^{free}\right|^2 \frac{\delta\omega^2}{(\omega_p^2 + \delta\omega^2)} \qquad (40)$$

where $\omega_p = \omega_0 \dot{W}(B, \phi_0)$ \qquad (41)

The phase noise in the oscillator is defined by squares of transfer factors multiplied by PSD of noise sources. Thus the phase noise in locked oscillators can be determined by the phase noise in free running oscillators by analogy with (40)

$$\left|L_\phi^{lock}\right| = \left|L_\phi^{free}\right| \frac{\delta\omega^2}{(\omega_p^2 + \delta\omega^2)} \qquad (42)$$

Such expression can be found in the literature (see [4]) but with the definition of ω_p for special cases of oscillator circuits. Our approach allows to evaluate ω_p (41) for an arbitrary oscillator, and it can be easily implemented into a circuit simulator.

Note that for the single harmonic, single node sinusoidal injection the corner frequency (41) is reduced to $\omega_p = kA\omega_0 \left|V_{kl}\right|$ where A is the injection magnitude.

VI. EXAMPLES

The proposed approach was implemented in the circuit simulator and numerical experiments were performed with a number of harmonic and ring oscillator circuits.

Simulation results for BiCMOS differential ring oscillator [1], and for 3-stages CMOS ring oscillator are presented in Fig. 1, 2 respectively. The results are shown for examples with white noise sources, and for examples with flicker noise sources. In each case the phase noise was evaluated both for free running oscillator and locked oscillator.

VII. CONCLUSIONS

To provide noise simulation of locked oscillators the new phase macromodel in the form of differential equation is proposed. The macromodel allows to analyze the behavior of the oscillator phase when the excitation is defined as a set of slow-varying harmonics of the fundamental. It is shown that the macromodel coincides with the Adler equation for the case of LC oscillators. The new macromodel can be obtained from the well-known macromodel based on Floquet theory by neglecting highly oscillatory terms. This distinction provides the existence of the steady-state solution of the phase equation for locked oscillators and allows to perform phase noise analysis of an arbitrary oscillator circuit under arbitrary periodic injection waveform.

REFERENCES

[1] P. Kinget, R. Melville, D. Long, V. Gopinathan, "An injection-locking scheme for precision quadrature generation," *IEEE J. of Solid-State Circuits*, vol. 37, no. 7, pp. 845 - 851, July 2002

[2] R. Adler. A study of locking phenomena in oscillators. *Proc. of the IRE and Waves and Electrons*, 34: 351-357, June 1946.

[3] H. Chang et al, "Phase Noise in Externally Injection-Locked Oscillator Arrays," *IEEE Trans. Microwave Theory Tech.*, vol. 45, No. 11, pp. 2035-2042, Nov. 1997.

[4] S. Verma, H. Rategh, T. Lee, "A Unified Model for Injection-Locked Frequency Dividers," *IEEE J. of Solid-State Circuits*, vol. 38, no. 6, June 2003.

[5] A. Demir, A. Mehrotra, and J. Roychowdhury, "Phase Noise in Oscillators: A Unifying Theory and Numerical Methods for Characterization," *IEEE Trans. on Circuits and Systems - I*, vol. 47, pp. 655-674, May 2000.

[6] X. Lai and J. Roychowdhury, "Automated Oscillator Macromodelling Techniques for Capturing Amplitude Variations and Injection Locking," *Proc. ICCAD conf.*, pp. 687-694, 2004.

[7] K.S. Kundert, J. White, A. Sangiovanni-Vincentelli, *Steady-State Methods for Simulating Analog and Microwave Circuits*, Kluwer Academic Publishers, Boston, 1990.

[8] A. Demir, "Computing Timing Jitter From Phase Noise Spectra for Oscillators and Phase-Locked Loops With White and 1/f Noise," *IEEE Trans. on Circuits and Systems - I*, vol. 53, No. 9, Sep. 2006.

[9] M.M. Gourary, S.G. Rusakov, S.L. Ulyanov, M.M. Zharov, B.J. Mulvaney, K.K. Gullapalli, "Analysis of Oscillator Injection Locking by Harmonic Balance Method," *Design, Automation and Test in Europe Conf.*, pp. 318-323, 2008.

[10] X. Lai and J. Roychowdhury, "Analytical Equations For Predicting Injection Locking in LC and Ring Oscillators," *IEEE Custom Integrated Circuits Conf.*, pp. 461-464, 2005.

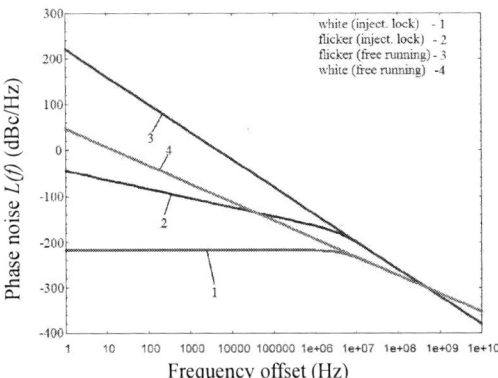

Fig. 1. Phase noise in BiCMOS Differential Ring Oscillator

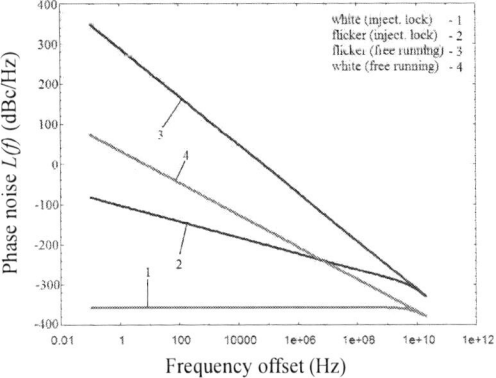

Fig. 2. Phase noise in 3-stages CMOS Ring Oscillator

Proceedings of the 3rd European Microwave Integrated Circuits Conference

Performance of Unstuck Γ Gate AlGaN/GaN HEMTs on (001) Silicon Substrate at 10 GHz

J-C. Gerbedoen [#1], A. Soltani[1], N. Defrance[1], M. Rousseau[1], C. Gaquière[1], J-C. De jaeger[1],

S. Joblot[2,3], Y.Cordier[2]

[1]*IEMN, UMR-CNRS 8520, USTL, Avenue Poincaré, 59652 Villeneuve d'Ascq Cedex, France*
[#]`Jean-Claude.Gerbedoen@iemn.univ-lille1.fr`
[2]*CRHEA-CNRS, rue Bernard Grégory, Sophia Antipolis, 06560 Valbonne, France*
[3]*ST-Microelectronics, 850 rue Jean Monnet, 38926 Crolles, France*

Abstract - **This paper shows the capability of AlGaN/GaN High Electron Mobility Transistors (HEMTs) on (001) oriented silicon substrate with 300nm gate length using unstuck Γ gate for low cost device microwave power applications. The total gate periphery of 300µm, exhibits a maximum DC drain current density of 600mA/mm at V_{DS}=7V with an extrinsic trans-conductance ($g_{m\ max}$) around 200mS/mm. An extrinsic current gain cutoff frequency (f_T) of 37GHz and a maximum oscillation frequency (f_{max}) of 55GHz are deduced from S_{ij}-parameters measurements. At 10GHz, an output power density of 2.9W/mm associated to a power added efficiency (PAE) of 20% and a linear gain of 7dB are obtained at V_{DS}=30V and V_{GS}=-2V.**

I. INTRODUCTION

The AlGaN/GaN High Electron Mobility Transistors (HEMTs) on SiC substrate represent the most promising devices for millimeter-wave power applications [1]. In this contribution, devices are fabricated on a (001) oriented silicon substrate presenting the advantages to be the cheapest due to the availability of substrate on a large scale and finally the possibility of co-integration with MOS technology. Up to now, many efforts were realized on the Si (111) substrate and many publications have shown good power performances. For these devices the state of the art is 7W/mm at 10GHz [2] and 5.1W/mm at 18GHz [3]. Nowadays, there is also a challenge to fabricate devices on Si (001) orientation substrate to reduce more and more the material cost and to be compatible with Silicon technology. Recently, devices fabricated on Si (001) substrate with 100nm Si_3N_4 Γ gate technology have demonstrated a microwave power performance of 1W/mm at 2.15GHz [4]. This performance can be enhanced by improving material quality [5] as well as device technology, as reported by Palacios with associated to Ge-spacer technology [6].

The present paper describes the cutoff frequency capabilities and microwave power obtained on transistor fabricated on this kind of substrate as well as a fabrication process aimed to take advantage of the good material properties. The Γ gate structure is optimised with the help of a 2D-energy-balance model developed at our laboratory [7]. This study opens a new field of interest for GaN-based devices on low cost silicon substrate for telecommunication and general public power application.

II. MATERIAL GROWTH

The heterostructures used in the present study are grown at CRHEA with the collaboration of ST-Microelectronics on a High Resistive (HR) (001) silicon substrate (ρ>5kΩ.cm) with a misorientation of 6° toward the (110) direction. Ammonia (NH_3) is used as the nitrogen precursor during the growth by molecular beam epitaxy (MBE). The epitaxial structure consists of a 1.2µm thick of AlN/GaN stress accommodation layer, a 0.8µm thick of GaN buffer, a 1nm thick of AlN spacer layer to improve the carrier confinement in the 2D electron gas [8], a 25nm thick of $Al_{0.24}Ga_{0.76}N$ barrier and 1nm thick of unintentionally doped GaN cap layer. In comparison with the first results published [4] the threading dislocation density has been divided by a factor 2 and is estimated at 1×10^{10} cm^{-2} by plane-view TEM (Transmission Electronic Microscopy) [5]. This improvement is mainly due to the optimization of the AlN/GaN stress accommodation layer. Finally, the use of this layer allows growing a AlGaN/GaN structure free of cracks by accommodating the tensile strain induced by the difference in thermal expansion coefficient between GaN and Si during the cool down from 800°C to room temperature.

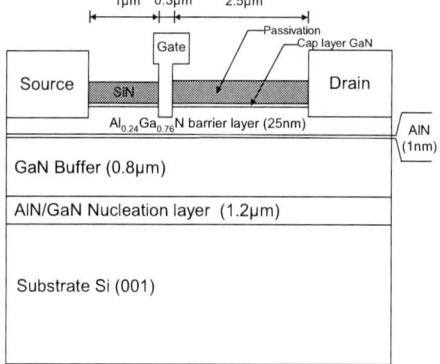

Fig. 1 Structure of AlGaN/GaN HEMT on silicon substrate oriented (001).

III. DEVICE SIMULATION

A scheme of the studied structure is presented in fig. 1. In a first time, in order to optimise the gate shape, a 2D-energy-balance simulation was carried out. The model is based on the conservation equations deduced from Boltzmann's equation

978-2-8748-7007-1/08 $25.00 © 2008 EuMA   330

and coupled to Poisson's equation. The optimisation of the Γ gate extension towards the drain is performed by means of the study of the electric field and carrier energy described by the model.

The figure 2 represents the distribution of the electric field in the 2DEG channel (E_x) for different Field-Plate lengths (L_{FP}) ranging from 0µm to 1.2µm. On a general way, the peak electric field reduction is already important for small field plate length. When using very long field plate, the gain is reduced when the Field Plate length reaches 0.4µm. This reduction of the peak electric field can explain an improvement of the breakdown voltage for mainly two reasons: first, a spreading of the electric field leads to a spreading of the average energy of electrons, reducing impact ionisation. The other reason is a drop of the electric field at the gate exit improving also the gate breakdown by reducing the tunnelling current across the Schottky contact. Furthermore the theoretical study showed that the Γ gate extension involves an increase of the parasitic capacitances. Consequently, the best compromise taking into account the possibility to increase the breakdown behaviour as well as good performance at the operating frequency (10GHz) was 0.4µm.

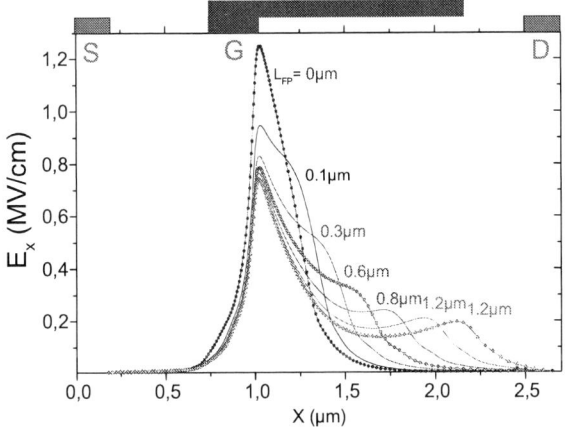

Fig.2 Distribution of the electric field in the channel for the Γ gate at V_{DS}=30V and V_{GS}− 2V.

IV. DEVICE PROCESSING

The fabrication of AlGaN/GaN HEMT on (001) oriented silicon substrate with 300nm gate length using a recess starts with the deposition of a Si_3N_4 film passivation around 80nm thick by plasma enhanced chemical vapour deposition (PECVD) at 340°C. The source and drain ohmic contacts are defined by electron beam lithography on the Si_3N_4 layer and then etched by highly anisotropic CHF_3/CF_4 plasma to open Si_3N_4. More than half of the barrier layer is etched by digital etching due to the presence of an AlN layer in the structure involving ohmic contact formation difficulties. Digital etching technique consists in an oxidation of the semiconductor surface by oxygen plasma in Reactive Ion Etching (RIE) at 50W, 300mTorr since 40sec followed by a desoxidation by HCl solution, to selectively remove this oxide layer [9]. Low

power is used to minimise as possible the degradation of the 2DEG channel [10]. This process provide an etch linearity (~0.6nm per cycle), a good reproducibility and very good roughness surface (RMS is about 0.8nm after process). A 13nm depth recess is performed using 22 cycles of this two-steps etching process to enhance the contact resistance. An annealing at 400°C for 15min under nitrogen is carried out to repair damage due to the etching [10]. The contacts are formed using a Rapid Thermal Annealing (RTA) of evaporated Ti/Al/Ni/Au (12/200/40/100 nm) metallization at 870°C for 30sec under nitrogen atmosphere. Transmission Line Method (TLM) measurement shows that ohmic contact recess strongly improves the contact resistance, from about 4Ω.mm for unrecessed ohmics to 0.7-0.8Ω.mm after recess depth optimization. The mesa isolation is defined by He+ ion multiple implantations based on different doses and energies through the Si_3N_4 passivation. To realize the footprint defined by electron beam lithography, the Si_3N_4 layer is etched using the same anisotropic plasma than for ohmic contact, permitting to reduce the resist undercut. This step is very important for the gate foot definition. The gate foot is then slightly recessed using digital etching technic in the barrier layer. A 10nm depth gate recess is carried out corresponding to 17 digital etching cycles. Then, the wafer is annealed at 400°C for 15min under nitrogen for the same reason than previously. A 40nm thick Germanium (Ge) sacrificial layer is then deposited by evaporated technique on the Si_3N_4 layer. Second e-beam lithography is used to open the Ge layer at the gate foot place. After, a third lithography, using a bilayer resist, making the lift-off easier, permits to define the Γ gate on the Ge and Si_3N_4 opening with about 400nm extension of the top gate at the drain side corresponding to the optimum value obtained from simulation (fig. 2). The gate metallization used is based on evaporated Pt/Ti/Pt/Au (25/25/25/200 nm). Then, the sacrificial layer is removed by a wet chemical etching solution based on H_2O_2. The gate obtained is then unstuck from the Si_3N_4 layer permitting an important decrease of parasitic capacitive elements under the gate top with an air gap of 40nm [6].

The device features are 2×150µm gate width with drain-gate spacing from 1.5 to 3.5µm, a gate-source spacing of 1µm and a gate length (L_G) of 300nm. Hall effect measurement shows a sheet carrier density of $1.04×10^{13}cm^{-2}$, an electron mobility of 1718cm² V⁻¹s⁻¹ and a sheet square resistance of about 352Ω/sq at room temperature.

V. MEASUREMENT RESULTS

A. Static characteristics

The devices selected for measurement have a drain-gate distance L_{GD}=2.5µm. The DC characteristics are measured with programmable alimentation multi-slots like HP4142 or Agilent 5270. The figure 3 shows that a drain current density I_{DSS}=600mA/mm is obtained at V_{DS}=7V. The pinch-off voltage is around -5.5V.

978-2-8748-7007-1/08 $25.00 © 2008 EuMA

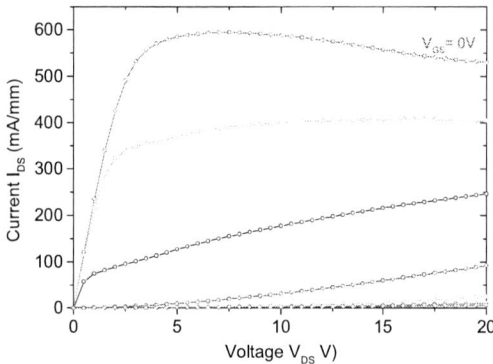

Fig. 3 Static I_{DS}-V_{DS} characteristics of a 2x150x0.3μm² AlGaN/GaN HEMT. Vgs swept from -6V to 0V by step of 1V.

An extrinsic peak transconductance ($g_{m\ max}$) of about 200mS/mm is measured at a gate bias of -2.2V for a drain bias of 20V. The figure 4 shows the Schottky characteristics with and without annealing at 350°C for 15min under nitrogen atmosphere. Improvement of the Schottky contact behavior is noted after annealing under reverse and forward bias condition, showing a decrease of the trapping phenomena under the gate.

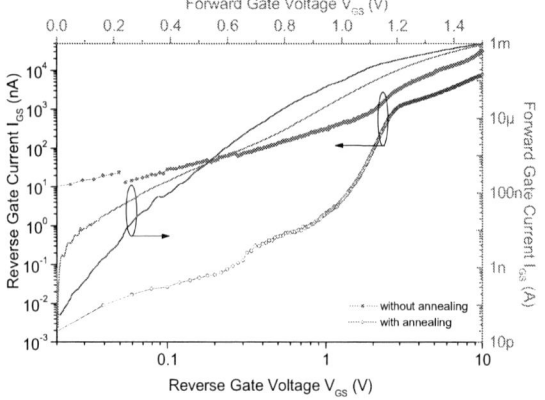

Fig. 4 I_{GS}-V_{GS} measurement with and without annealing post metallisation at 350°C.

Without annealing, the reverse gate current leakage is around 200μA/mm at V_{GS}=-10V and drops to 54μA/mm for the annealing Schottky contact. Under forward gate bias conditions, the Schottky barrier height (Φ_b) is 0.4eV associated to an ideality factor (η) of 4 before annealing. After annealing, these values are 0.7eV and 2 respectively. The breakdown voltage measured under transistor configuration is about 200V at pinch-off.

B. RF Measurements

Small signal microwave measurements were also carried out on this transistor under probes. The calibration procedure is performed on wafer, in the probes plane, using a TRM method with a 40GHz Power Network Analyzer (PNA) type E8363B. S-parameters are measured in the 250MHz to 40GHz frequency range. The value of the unity current gain cutoff frequency (f_T) is determined by the extrapolation of the current gain $|h_{21}|$. The maximum oscillation frequency (f_{max}) is determined from the Mason's gain.

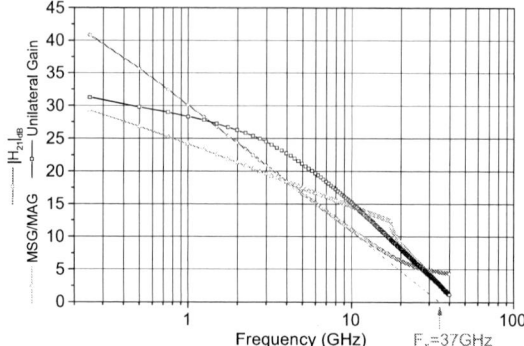

Fig. 5 Measured current gain versus frequency and maximum available gain versus frequency for a 2×150×0.3μm².

The fig.5 shows the current gain (h_{21}) and the maximum available gain (MAG) versus frequency. The deduced extrinsic f_T and f_{max} values are 37GHz and 55GHz respectively at V_{DS}=20V and V_{GS}=-2V. These measurements are obtained without access lines deembedding. This good RF performance is attributed to the optimized device processing (Γ gate extension and mushroom spacer technology) and also to the high material quality (HR buffer even in microwave conditions).

C. Pulsed Measurements

Pulsed measurements under probes are also carried out in order to determine the trapping phenomenon influence and the thermal effects.

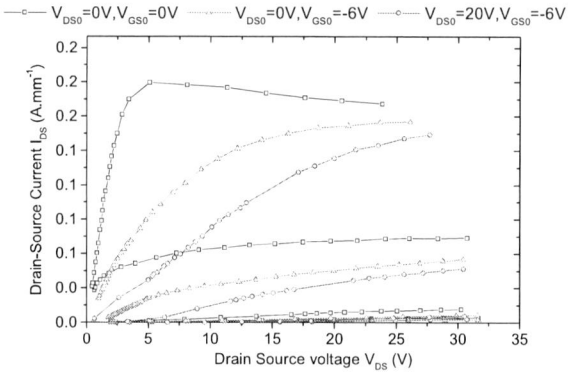

Fig. 6 DC Pulsed I_{DS}-V_{DS} measurement of a 2x150x0.3μm² at three quiescent points with V_{GS} from 0V to -6V step -2V.

The figure 6 shows the DC pulse $I_{DS}(V_{GS},V_{DS})$ characteristics for different quiescent bias points. The results at V_{DS0}=0V and V_{GS0}=0V give reference characteristics where the thermal effects are avoided. In this configuration, the maximum I_{DS} current at V_{GS}=0V is 209mA corresponding to 693mA/mm, a little bit higher than the DC measurement (fig.3). At V_{DS0}=0V and V_{GS0}=-6V, corresponding to the pinch-off voltage, the influence of the gate lag effects involves

978-2-8748-7007-1/08 $25.00 © 2008 EuMA

a reduction of the drain current. This phenomenon is classically explained by the activation of traps mainly located near the surface and under the gate. From the quiescent bias point V_{DS0}=20V and V_{GS0}=-6V, the drain lag effects are observed. This phenomenon is due to the activation of traps levels within the buffer.

The pulse duration used is 400ns with a duty cycle of 0.4%. At V_{DS0}=20V, the current drop is 11% and 25% due to respectively the gate and drain lag effect showing that the trapping phenomena is not too important for this new device and that good microwave power performance can be expected. The drain lag effect involves a drop of the maximum drain current associated to an increase of the access resistance. This phenomenon is usually observed on different nitride based transistor whatever the substrate nature. Nevertheless, an increase of the knee voltage is observed. It can be improved by a pretreatment before Si_3N_4 deposition.

VI. LARGE SIGNAL CHARACTERIZATION

At last, under probe large signal power measurement is performed at 10GHz. It is based on an active load-pull setup permitting to get the optimum load impedance especially for small transistor. At the bias point V_{DS}=30V and V_{GS}=-2V corresponding to class A operation, selected devices show good microwave power performance (fig. 7).

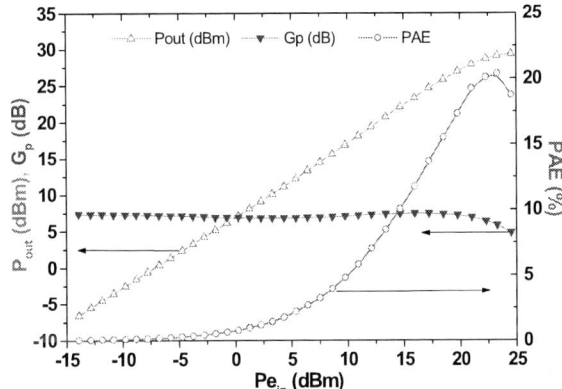

Fig. 7 Power density P_{out} (dBm) vs P_{inj} (dBm) (Red line); Power gain G_P (dB) (Blue line) and Power added efficiency PAE (%) (Black line)

An output power density of 2.9W/mm associated to a maximum Power Added Efficiency (PAE) of 20% and a linear gain of about 7dB are obtained in class deep AB (V_{DS}=30V; V_{GS}=-2V). Measurement was also carried out at V_{DS}=30V and V_{GS}=-3V corresponding to class AB. In this case, the linear gain increases to 15dB but for both biases the same power gain, around 5dB, is obtained at maximum output power. In class AB, the output power is 2.65W/mm associated to a Power Added Efficiency of 22%. It is noted that the optimum load impedance is not critical for this transistor. Furthermore, the microwave power performance can be enhanced increasing the V_{DS} bias point. But in this process the Schottky contact is not enough electrically robust. This result constitutes the large signal state of art at 10GHz and is very

promising for next III-N based power transistors on (001) silicon substrate.

VII. CONCLUSIONS

Unstuck Γ gate AlGaN/GaN HEMTs grown on (001) Silicon substrate were fabricated. A record output power density as high as 2.9W/mm was obtained at 10GHz with a Γ gate extension of 400nm. The transistors exhibit extrinsic f_T and f_{max} values of respectively 37GHz and 55GHz at V_{DS}=20V and V_{GS}=-2V. These results show the capability of such transistors for low cost microwave power applications such as telecommunication base stations. Furthermore, this new way is compatible with the Silicon technology for circuit fabrication. Compared to previous results [4], the improvement is due to better epitaxial layer structure on Si (001) combined with optimized technological processing steps using a Field Plate.

ACKNOWLEDGMENT

The authors would like to thank the DGA (Direction Générale pour l'Armement) for the financial support under contract n°06-137606/DGA/D4S/MRIS. The work on material growth is supported by the French Office "MINEFI" in the frame of the project "Nano2008".

REFERENCES

[1] Y.-F. Wu, A. Saxler, M. Moore, R. P. Smith, S. Sheppard, P. M. Chavarkar, T. Wisleder, U. K. Mishra, and P. Parikh, "30 W/mm GaN HEMTs by Field Plate Optimization", *IEEE Electron Device Letters*, **25**(3) (2004) 117-119.

[2] D.C. Dumka, C.Lee, H. Q. Tserg, P. Saunier, and M. Kumar, "AlGaN/GaN HEMTs on Si substrate with 7 W/mm output power density at 10 GHz", *Electronics Letters*, **40**(16) (2004) 1023-1024.

[3] D. Ducatteau, A. Minko, V.Hoel, E. Morvan, E. Delos, B. Grimbert, H. Lahreche, P. Bove, C. Gaquière, J-C. De Jaeger and S. Delage, "Output Power Density of 5.1 W/mm at 18 GHz with an AlGaN/GaN HEMT on (111) Si Substrate", *IEEE Electron Device Letters*, **27**(1) (2006) 7-9.

[4] S. Boulay, S. Touati, A.A. Sar, V. Hoel, C. Gaquière, J-C. De Jaeger, S. Joblot, Y. Cordier, F. Semond and J. Massies, "AlGaN/GaN HEMTs on a (001) oriented Silicon Substrate Based on 100 nm SiN Recessed Gate Technology for Microwave Power Application", *IEEE Transactions on Electron Devices*, **54** (2007) 2843-2848.

[5] S. Joblot, Y. Cordier, F. Semond, S.Chenot, P. Vennéguès, O. Tottereau, P. Lorenzini, and J. Massies, "AlGaN/GaN HEMTs grown on silicon (001) substrates by molecular beam epitaxy", *Superlattices Microstruct.*, **40** (2006) 295-299.

[6] T. Palacios, E. Snow, Y. Pei, A. Chakraborty, S. Keller, S.P. DenBaars and U. K. Mishra, "Ge-Spacer Technology in AlGaN/GaN HEMTs for mm-Wave Applications", *Electron Devices Meeting, (2005). IEDM Technical Digest. IEEE International.*

[7] M. Rousseau, J.D. Delemer, J.C. De Jaeger and F. Dessenne, "Two-dimensional hydrodynamic model including inertia effects in carrier momentum for power millimetre-wave semiconductor devices", *Solid-State Electronics*, **47**(8) (2003) 1297-1309.

[8] S. Joblot, Y. Cordier, F. Semond, P. Lorenzini, S. Chenot and J. Massies, "AlGaN/GaN HEMTs on (001) silicon substrates", *Electronics Letters*, **42** (2006) 117-118.

[9] D. Buttari, S. Heikman, S. Keller, and U.K. Mishra, "Digital etching for highly reproducible low damage gate recessing on AlGaN/GaN HEMTs", *IEEE Lester Eastman conference*, (2002) 461-469.

[10] J. Lee, D. Liu, H. Kim, and W. Lu, "Post annealing effect on device performance of AlGaN/GaN HFETs", *Solid State Electronics*, **48** (2004) 1855-1859.

Proceedings of the 3rd European Microwave Integrated Circuits Conference

A Reflection-Type Biphase Modulator with Balanced Loads

Walter Ciccognani, Franco Di Paolo, Mauro Ferrari[1], Franco Giannini, Ernesto Limiti

Dipartimento di Ingegneria Elettronica, Università degli Studi di Roma Tor Vergata

Via del Politecnico 1, 00133 Roma - ITALY

[1]m.ferrari@ing.uniroma2.it

Abstract— **In this contribution a new balanced biphase modulator topology is presented. The new biphase modulator allows the reduction of the resulting chip area occupation by using 3 couplers only, in contrast with traditional reflection-type balanced biphase modulators with 4 couplers, and exhibiting comparable performances. A 45-65 GHz test vehicle adopting the proposed topology has been designed and realised in monolithic form in 70 nm mHEMT technology, resulting in a 1.8 x 1 mm^2 die size. The novel topology and the resulting measured performances of the test vehicle are described in the following.**

I. INTRODUCTION

A biphase modulator is a subsystem capable of simultaneously modifying amplitude, in a continuous mode, and phase over two discrete phase states (0/180 degrees) of an input RF signal according to a suitable control. A biphase modulator (BM) has a series of specific applications: it can be adopted to perform a BPSK modulation in direct modulation schemes [1] or as a component of a vector modulator (VM), simply using two biphase modulators combined adding a 90 degree hybrid and a power combiner. The latter arrangement is capable of simultaneously and continuously changing amplitude and phase over a 360 degree range. The use of VM leads to the elimination of up-converting chains (mixer, filters, etc..) to realise complex modulation scheme, acting directly at RF. Besides complex modulation schemes, other VM applications consists in FMCW radars used in ACAS (Automotive Collision-Avoidance Systems), in phased-array antennas and in the feed-forward linearization technique, where VMs can be used instead of the attenuator - phase shifter cascade in the nulling loop, thus providing enhanced performances [2]. In any case, the basic and enabling building block of a VM is indeed the BM, for which a novel design methodology is presented in the following section.

II. TRADITIONAL BIPHASE MODULATOR TOPOLOGIES

A large number of BMs and VMs types have been presented in literature. The simplest VM or BM are obtained cascading attenuators and phase shifters as in [3] and [4]. This choice has some drawbacks as the difficulty to realise a high-performance variable attenuator with constant phase and a phase shifter with constant insertion loss. The most attractive and common approach instead resides in the use of reflection-type biphase modulator [5]. This topology is schematically depicted in Fig. 1. Adopting ideal couplers, the BM scattering matrix is given by (1).

$$[S] = \begin{bmatrix} 0 & -j\Gamma(V_G) \\ -j\Gamma(V_G) & 0 \end{bmatrix} \tag{1}$$

where $\Gamma(V_G)$ is the reflection coefficient of the load, typically realized via a cold-mode FET controlled by its gate voltage V_G.

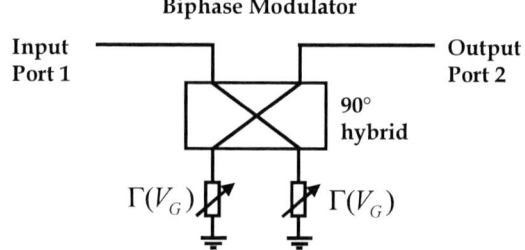

Fig. 1 A reflection-type BM topology

To implement the biphase transfer function defined by (2), an ideal BM should fulfill condition (3), i.e. symmetry with respect to its control voltage:

$$\begin{cases} |S_{21}(V_G)| = k|V_G| \\ \angle S_{21}(V_G) = \dfrac{\pi}{2} + \dfrac{\pi}{2} sign(V_G) \end{cases} \tag{2}$$

$$S_{21}(V_G) = -S_{21}(-V_G) \ , \quad \forall V_G \tag{3}$$

FET's parasitics, whose effects are especially evident at high frequencies, do not allow acceptable performance adopting such simple connection, given the lack of symmetry in the device behavior, as evidenced in Fig. 2 where a typical cold-FET reflection coefficient is plotted as a function of the gate control voltage @ 55 GHz. To correct this problem, two main strategies are possible: the first one, as reported in [6], is based on parasitics compensation, therefore being suitable for narrowband applications only; the second one is the balancing

978-2-8748-7007-1/08 $25.00 © 2008 EuMA

technique, actually the most used for broadband applications [7]-[10].

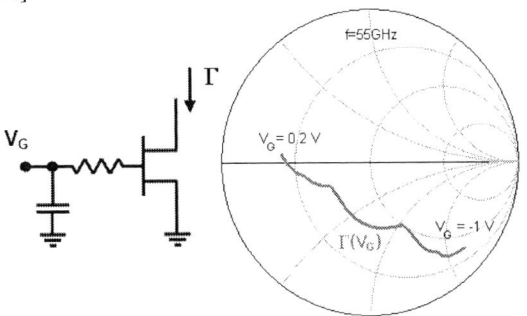

Fig. 2 Simple load realized with a cold-FET and $\Gamma(V_G)$ @ 55GHz

In the latter methodology, two simple BMs, controlled by complementary voltages, are combined in a push-pull structure using 90 degree hybrids as depicted in Fig. 3:

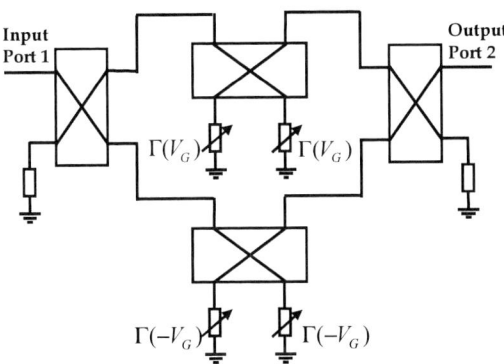

Fig. 3 Balanced reflection-type BM

where the total transfer coefficient of the balanced BM is given by

$$S_{21}^{TOT}(V_G) = \frac{1}{2}\left[S_{21}(V_G) - S_{21}(-V_G)\right] \quad (4)$$

Such connection exhibits two main drawbacks: the first is the need of four couplers and the second is the difficult access to the control terminals of the structure.

Due to the former drawback, to realize a VM the total number of couplers for the whole structure (composed by two BM and the input coupler) becomes nine, while for the latter, access to all control terminals is not possible if the use of underpasses or bridges is avoided, as highly recommended at very high frequencies.

III. PROPOSED VM TOPOLOGY

The idea underlying the proposed BM is based on the observation that the transmission coefficient S_{21} and reflection coefficient Γ of the simple BM are identical with the exception of a 90 deg phase term (as in (1)). Instead of balancing the simple BM (as in Fig. 3), the same result may be therefore attained by directly balancing the loads. This novel principle (that has been called *Balanced Load*) is described by expression (5) and is represented by the connection in Fig. 4 (left), where the resulting reflection coefficient is plotted as a function of the gate control voltage (right).

$$\Gamma^{TOT}(V_G) = \frac{1}{2}\left[\Gamma(V_G) - \Gamma(-V_G)\right] \quad (5)$$

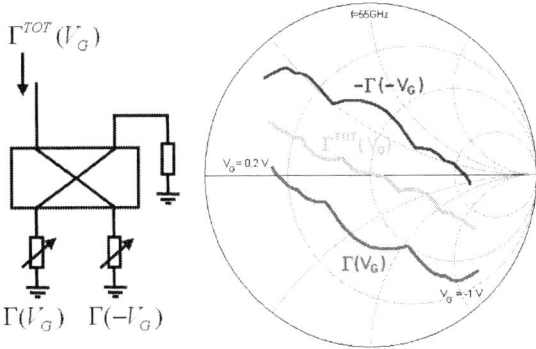

Fig. 4 Balanced load and $\Gamma_{TOT}(V_G)$ representation @ 55GHz

The resulting reflection-type BM adopting balanced loads is therefore schematically depicted in Fig. 5 where only three 90 degree hybrids are used, while its transfer function is given by:

$$S_{21}^{TOT}(V_G) = -j\Gamma^{TOT}(V_G) \quad (6)$$

Novel Balanced Biphase Modulator

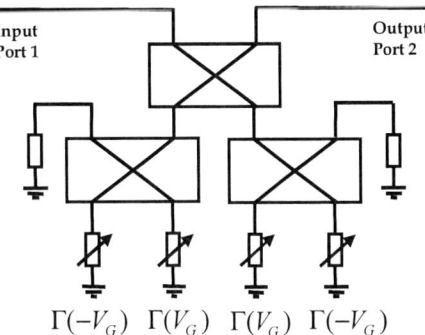

Fig. 5 Reflection-type biphase vector modulator with balanced loads

978-2-8748-7007-1/08 $25.00 © 2008 EuMA

Expression (6) is fully equivalent to (4) and both verify condition (3): nevertheless, in the proposed Balanced Load approach three couplers only are needed and control signals are easily accessed on a single side of the structure, thus easing the overall VM routing. Please note, in the previous discussion, that V_G and $-V_G$ are two complementary voltages, eventually different in magnitude. As an example, when V_G=-1 V, $-V_G$=0.2 ; on the contrary, when V_G=0.2 V, $-V_G$=-1.

IV. TEST VEHICLE DESIGN

To validate the proposed topology, a test vehicle has been designed and realised in monolithic form making use of a 70 nm GaAs mHEMT technology from OMMIC and selecting, as a load, a 2x20 μm device. The resulting chip, depicted in Fig. 6, occupies 1.8 x 1 mm^2 and has been designed to operate in the 45÷65 GHz band.

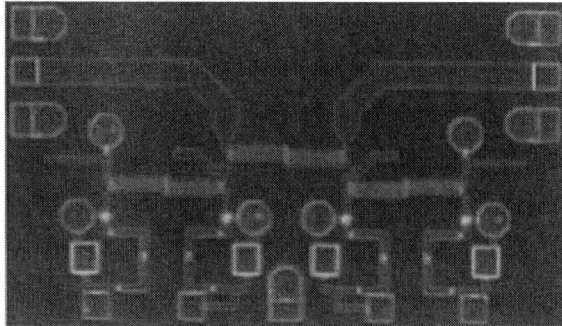

Fig. 6 Test vehicle BM

The design of the whole BM has been performed making a combined use of both circuit and 2D electromagnetic simulation tools, the latter both to validate passive models (tee junctions, steps, tapers, etc..) and to take into account potential asymmetries and EM interactions arising from coupling between different sections of the circuit.

After a 2D EM simulation of the critical parts and passive elements, the final design has been therefore obtained after a subsequent co-simulation of the entire BM.

As evidenced in Fig. 6, the BM needs two complementary voltages (V_G and $-V_G$) to control amplitude and phase of each BM (clearly visible in the lover portions of the circuit in Fig. 6 in the following order: $-V_G$ V_G V_G $-V_G$).

By complementary varying such controls from -1.2V to 0.2 V for $-V_G$, and from 0.3V to -0.9V for V_G , values resulted by a calibration, a possible symbol constellation of 27 states exhibiting a 180 deg coverage has been obtained and reported in Fig. 7 where the control voltages are indicated as ($-V_G,V_G$).

For all 27 states, the relative S_{21} phase (covering the same range) is plotted in Fig. 8 as a function of frequency and the relative S_{21} amplitude for a fixed phase state is plotted in Fig. 9. In these plots, the minimum insertion loss state of the BM has been chosen as reference state.

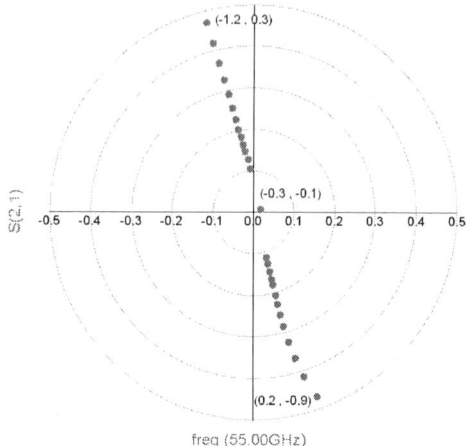

Fig. 7 BM constellation @ 55 GHz

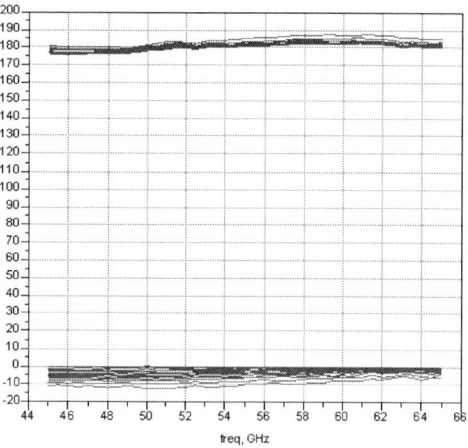

Fig. 8 Relative phase [deg] for the BM S_{21} by varying control voltages.

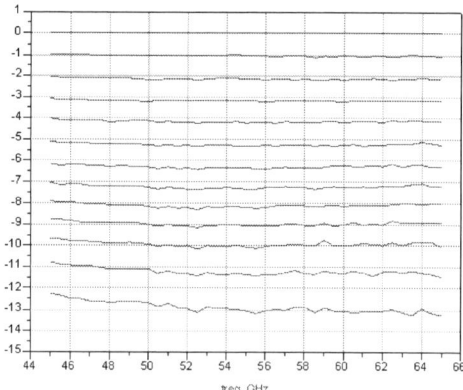

Fig. 9 Relative amplitude [dB] of the BM S_{21} by varying control voltages (for a fixed phase state)

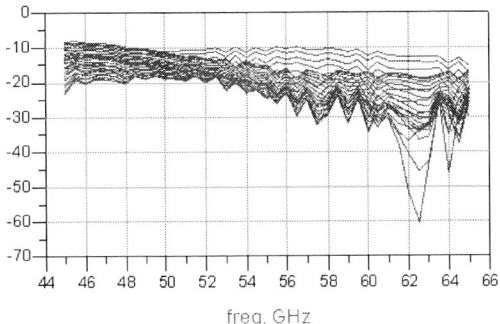

Fig. 10 BM Return in loss in dB at port 1

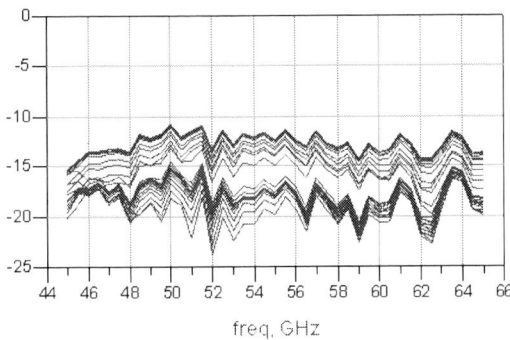

Fig. 11 BM Return loss in dB at port 2

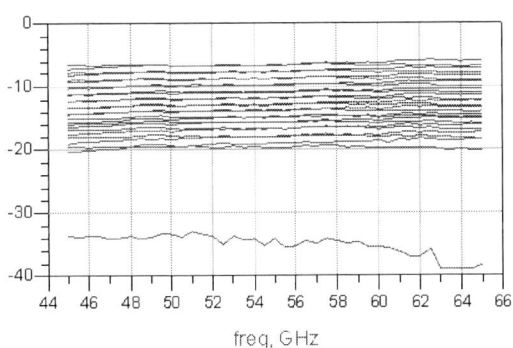

Fig. 12 BM Insertion-State S_{21} magnitude in dB

For the above states of the BM, input and output return loss are presented in Fig. 10 and in Fig. 11, where better than 10 dB input and output matching is demonstrated.

The insertion loss of the BM, presented in Fig. 12, is always better than 6.5 dB all over the frequency range of interest.

V. CONCLUSIONS

A novel BM topology has been presented, based on the novel "Balanced Loads" approach here introduced. Such new topology allows the reduction of the overall BM complexity by reducing the number of necessary couplers from four to three, as compared to traditional balanced BMs. A 45-65 GHz band test vehicle of the new topology has been designed and realized in MMIC form making a combined use of 2D EM and circuit simulators. The resulting chip, offering good performances as a BM, exhibits an input/output return loss better than about 10 dB, an insertion loss always better than 6.5 dB and a good flatness in term of amplitude and phase response, all over the 45÷65 GHz design bandwidth.

ACKNOWLEDGMENT

This work was funded by the PHAROS-RadioNet Project (FP6 – R113CT 2003 5058187) whose support is gratefully acknowledged. Visit the website www.radionet-eu.org.

REFERENCES

[1]   S. Kumar, "Directly modulated VSAT transmitter", *Microwave Journal*, April 1990,pp.255-264.

[2]   L.Silverman, C.Del Plato, "Vector Modulator Enhances Feed-Forward Cancellation", *Microwaves & RF*, March 1998.

[3]   I. D. Lucyszyn, "RFIC and MMIC Robertson, S. design and technology," *The Istitution of Electrical Engineers*, London, 2001, pp. 481-483.

[4]   L G.B. Norris, D.C. Boire, G. St. Onge, C. Wutke, C. Barratt, W. Coughlin, J. Chickanosky, "A fully monolithic 4÷18 GHz digital vector modulator", *IEEE MTT-S International Microwave Symposium Digest*, Volume 2, 1990,pp. 789-792.

[5]   L.M.Devlin, B.J.Minnis, "A versatile vector modulator design for MMIC", *IEEE MTT-S International Microwave Symposium Digest,*, 1990, pp.519-522.

[6]   I M. Chongcheawchamnan, S. Bunnjaweh, D. Kpogla, D. Lee, I.D. Robertson, "Microwave I-Q vector modulator using a simple technique for compensation of FET parasitics", *IEEE Trans. Microwave Theory Tech.*, Volume 50, Issue 6, June 2002, pp. 1642-1646.

[7]   I. D. Lucy A.E. Ashtiani, S. Nam, A. d'Espona, S. Lucyszyn, and LD. Robertson, "Direct multilevel carrier modulation using millimeter-wave balanced vector modulators", *IEEE Trans. Microwave Theory Tech.*, Volume 46, December 1998.,pp. 2611-2619.

[8]   A.E. Ashtiani, T. Gokdemir, A. Vilches, Z. Hu, I. D. Robertson, S.P.Marsh, "Monolithic GaAs/InGaP HBT balanced vector modulators for millimeter-wave wireless systems", *IEEE Radio Frequency Integrated Circuits Symposium*, 2000, pp.187-190

[9]   S. Nam, N. Shala, K. S. Ang, A. E. Ashtiani, T. Gokdemir, I. D. Robertson, S. P. Marsh, "Monolithic millimeter-wave balanced bi-phase amplitude modulator in GaAs/InGaP HBT technology," IEEE MTT-S Microwave Symposium Digest, 1999, pp.243-246.

[10] A.E. Ashtiani, T. Gokdemir, S. Nam, I.D. Robertson, "Compact 38 GHz MMIC balanced vector modulators employing GaAs/InGaP HBTs," *IEE Electronics Letters*, Volume 35, NO. 10, 13th May 1999, pp.817-818.

Design Manufacturing and Packaging of a 5-bit K-Band MEMS Phase Shifter

S. Bastioli[#1], F. Di Maggio[§2], P. Farinelli[#3], F. Giacomozzi[*4], B. Margesin[*5], A. Ocera[#6], I. Pomona[§7], M. Russo[§8], R. Sorrentino[#9]

[#]University of Perugia, Dept. of Electronic and Information Engineering, Via G. Duranti, 93, 06125, Italy
[1]bastioli@diei.unipg.it,
[3]farinelli@diei.unipg.it,
[6]ocera@diei.unipg.it,
[9]sorrentino@diei.unipg.it

[§]Selex Communications Spa, Via Sidney Sonnino, 6 - 95045 Misterbianco (CT) , Italy
[2]francesco.dimaggio@selex-comms.com,
[8]massimo.russo@selex-comms.com,
ignazio.pomona@selex-comms.com

* MEMSRaD Research Unit, FBK-irst, Via Somarive 14, 38050 Trento, Italy
[4]giaco@fbk.eu,
[5]margesin@fbk.eu

Abstract — This work presents the design, manufacturing and packaging of a novel K-band 5-bit MEMS phase shifter for applications in Satellite COTM (Communication On The Move) Terminals. The first 4 bits are realized by using a switched line topology whereas the less significant bit consists of a loaded line section. The device has been manufactured in microstrip technology on high resistivity silicon substrate by using the 8-masks FBK MEMS process. Excellent performances were measured for the MEMS switches as well as the single bits constituting the phase shifter. The phase shifter full wave simulations show excellent performance in the frequency band of interest 20.2-21.2 GHz. Return loss and insertion loss better than 17 dB and 2 dB and phase error minor than 2 degrees are obtained for all the $2^{\wedge 5}$ phase shifter states. The design and manufacturing of the low cost packaging solution is also presented.

I. INTRODUCTION

Microwave phase shifters are basic components used in a large variety of communication and radar systems, microwave instrumentation, and industrial applications [1-4]. As commercial and military systems increasingly move towards smaller and high performance antenna systems, MEMS technology can be applied to develop electronically reprogrammable phase shifters with low-power consumption, low loss, and excellent linearity.

In recent years a large number of high performance reconfigurable MEMS phase shifters has been presented working at different frequency bands. However the actual integration in telecommunication systems of such promising devices has not been fully demonstrated. This is not only due to the well known reliability issues connected with MEMS technology but also to the few high performance packaged components which have been presented up to now [5-7]. The

packaging is indeed one of the most critical parts of such devices due to the not easy compatibility of the standard packaging techniques with MEMS technology. In addition a big effort is needed for the design of wideband RF feed-through able to preserve the extremely high performance of MEMS devices.

This paper presents the design, manufacturing and packaging of a compact 5-bit K-band MEMS phase shifter to be used in a Phased Array Antenna for Satellite COTM (Communication On The Move) Terminals. Particular attention has been devoted to do the best electromagnetic and electromechanical design choices in order to realize at least 20 identical packaged devices to be mounted on a phased array demonstrator. The technology as well has been developed to guarantee a repeatable process with high yield.

The phase shifters have been manufactured on 200 µm thick high resistivity silicon substrate by using the 8-masks FBK MEMS process [8]. The electro-magnetic design of the 5-bit phase shifter is presented in Section 2 together with full-wave simulations. Section 3 shows the manufacturing and experimental results. The measured RF performance of the MEMS SPST (Single-Pole-Single-Throw) switches as well as the single bits constituting the phase shifter is presented. Section 4 describes the package design together with CST Studio EM full wave simulations. All relevant discontinuities were modelled and compensated in order to ensure high transmission characteristic in the 0-25 GHz frequency band.

II. PHASE SHIFTER DESIGN

The 5-bit MEMS phase shifter has been realized with hybrid architecture to operate in the bandwidth 20.2-21.2 GHz. For the first four bits, 180°, 90°, 45° and 22.5°, a switched line topology has been chosen. This topology indeed turned to be the best trade-off among large phase shift, low loss and reduced space occupation in the selected frequency band. Series ohmic MEMS switches have been used to realize the SPDT building blocks which allow to drive the signal into the desired path. An air-bridge typology similar to the one previously developed and presented in [9] has been adopted due to the very high performance in terms of both RF response and reliability (no contact degradation up to $10^{\wedge 9}$ low power hot switching, power handling better than 5W). On the other hand the fifth bit, 11.25°, is based on a loaded line topology, which is convenient for this bit since a small phase shift is required. It consists of a microstrip section loaded by 2 shunt capacitive MEMS switches. They are simultaneously activated in order to increase the loaded capacitance and obtain the desired phase shift. Since a small capacitance variation is needed, 0.63 μm thick stopping pillars have been placed in the microstrip line below the bridge central part. In this way the bridge never comes in intimate contact with the RF line and a C_{on}/C_{off} ratio around 5 can be realized. L-C matching sections have been added at the input and output ports in order to compensate for the packaging connections as described in section IV. The total space occupation of the device is 11.35 x 3.2 mm², a photo being shown in Fig.1.

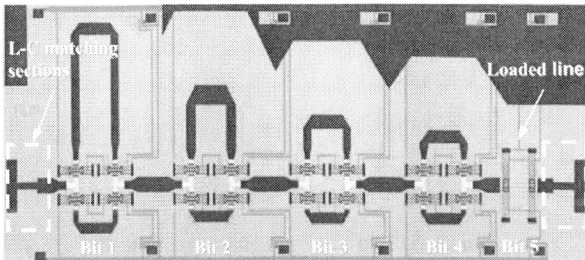

Fig. 1 Photo of the 5-bit MEMS phase shifter.

Each bit of the phase shifter was designed by using the full wave simulator ADS Momentum [10]. Afterwards the simulations of the single bits were circuitally cascaded in order to predict the performance of the whole structure. Simulated results show a return loss and an insertion loss better than 17 dB and 2 dB respectively for all $2^{\wedge 5}$ phase shifter states. Phase error is minor than 2 degrees in the 20.2-21.2 GHz frequency band. The simulations do not account for the switch contact resistance and consequently higher values of insertion loss are expected for all states.

III. MANUFACTURING & EXPERIMENTAL RESULTS

The MEMS phase shifter has been fabricated at Fondazione Bruno Kessler Laboratories, employing the well-established eight-mask surface micro-machining process. The process allows the electrodeposition of two gold layers of different thicknesses for highly complex movable bridges and microstrip lines. The air bridges are realized without the need of any planarization step by using 3 μm thick photoresist as sacrificial layer. The bridge release is done with a modified plasma ashing process in order to avoid sticking problems. The bias network is realized by depositing a high resistivity 0.63 μm thick poly-silicon layer covered with silicon oxide for isolating the DC from the RF lines. This layer is also used for realizing the stopping pillars used in the capacitive switches and the contact bumps for the ohmic switches. The process allows the monolithic manufacturing of both ohmic and capacitive switches. Low Temperature Oxide is used as dielectric for capacitive switches as well as MIM capacitors. A third gold layer is deposited for the realization of low resistance metal-to-metal electro-mechanic contacts for ohmic switches.

In order to verify the performance of the phase shifter building elements, both the ohmic and capacitive SPST switches have been realized in coplanar technology (Fig.3) and measured in the 5-30 GHz frequency band (Fig.4). Excellent performance was measured for both devices. The series ohmic switch shows an off-state isolation better 15 dB and an insertion loss better than 0.4 dB up to 30 GHz (contact resistance ~ 1.8 Ohm). On the other hand, an insertion loss better than 0.2 dB, an off-state capacitance of 55 fF and an on-state capacitance of 270 fF were measured for the shunt capacitive switch, resulting in a C_{on}/C_{off} ratio equal to 4.9. This value is in a very good agreement with the theoretical one since, due to the stopping pillars, the down state capacitance of such a switch is not sensitive to dielectric roughness or bridge shape factors.

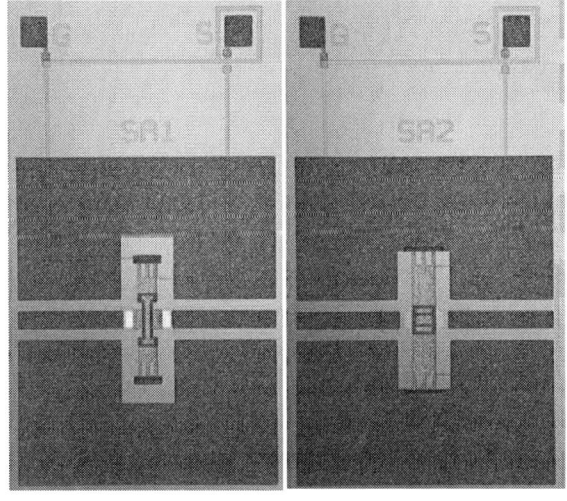

(a)　　　　　(b)

Fig. 3 Photo of the MEMS SPST switches (coplanar implementation) constituting the phase shifter (a) Series ohmic switch. (b) Shunt capacitive switch.

978-2-8748-7007-1/08 $25.00 © 2008 EuMA

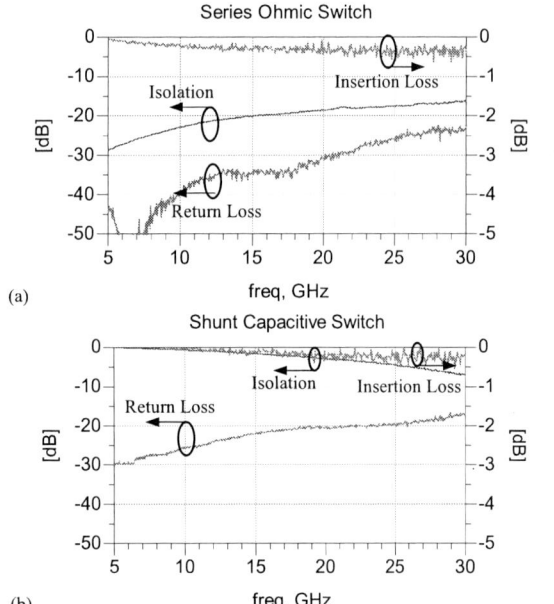

(a)

(b)

Fig. 4 Measured performance of the MEMS SPST switches. Return loss (red curve), Insertion loss (pink curve) and Isolation (blue curve). (a) Series ohmic switch. (b) Shunt capacitive switch.

The single bits constituting the phase shifter have been realized as well in order to analyze the performance and the phase error introduced by every single part of the circuit. Via-less coplanar to microstrip transitions have been designed enabling on-wafer probe tests. A 15-25 GHz TRL calibration kit has been manufactured on the same wafer in order to remove the contribution of such transitions during the measurements. The measured performance of the single bits are presented in Fig.5 and summarized in Table1. Return loss better than 20 dB and insertion loss better than 1.3 dB were measured for all bits in the entire frequency band of interest 20.2-21.2 GHz. The loss is dominated by the ohmic switch contact resistance, which is about 0.9 Ohm for every contact point (1.8 Ohm for every SPST switch). Note that Bit 5 is extremely low loss with respect to the others since it does not utilize series ohmic switches. The phase shift is linear over a very wide frequency band and the phase error is below 2.4 degrees in all bits (Fig. 5c).

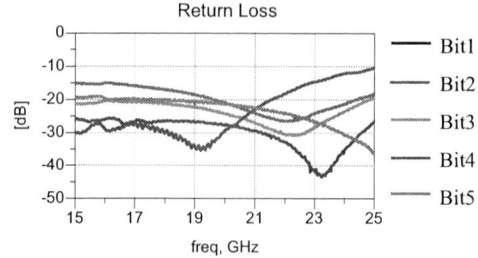

(a)

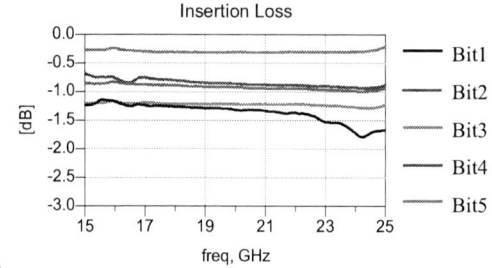

(b)

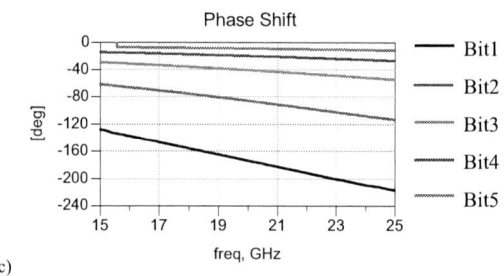

(c)

Fig. 5 Measured performances of the single bits. (a) Return Loss. (b) Insertion Loss. (c) Phase Shift.

IV. PACKAGE DESIGN

A low cost packaging solution has been developed in order to integrate the phase shifter into a phased array antenna by using automatic surface mounting equipment.

Laminate RT Duroid 5880 [11] was chosen as a substrate since it is a good trade-off between high RF performance and low cost. Full hermeticity is not guaranteed but this was not required for the specific device applications, which are in terrestrial communications. Five layers, 254 μm thick each, have been necessary to complete the 1-level MEMS package. Wire bonding techniques have been adopted to interconnect the phase shifter with the RF signal line in the package. Three parallel bonding wires (25 μm in diameter) have been used for each interconnection in order to reduce the total series inductance introduced by the wires. In addition L-C matching sections have been added in the silicon die as well as in the package side in order to obtain a good matching in a very wide frequency band.

A transition with two via-holes has been designed to bring the RF signal from the intermediate laminate layer into the

TABLE1

	Theoretical Phase Shift [degrees]	Measured Phase Shift [degrees]	Phase Error	Measured Insertion Loss [dB]	Measured Return Loss [dB]
Bit1	180	180,17	0,17	1,18	41,72
Bit2	90	89,52	-0,48	0,95	24,25
Bit3	45	42,58	-2,42	1,08	30,62
Bit4	22,5	20,97	-1,53	1,01	23,56
Bit5	11,25	9,56	-1,69	0,15	32,45

Table1. Single bit performance @ 20.7 GHz

external bottom surface thus allowing surface mounting solutions. Capacitive pads have been patterned between the two via-holes stages in order to compensate for the parasitic inductance generated from the vias.

The 3D layout of the transition is shown in Fig. 5 whereas Fig.6 presents the comparison between the simulated parameters with and without the compensation structures. The simulations were obtained by using the 3D simulator CST Studio [12]. Simulated results show an insertion loss and a return loss better than 0.5 dB and 25 dB, respectively, from DC up to 25 GHz. The total space occupation of the packaged phase shifter is 16.95 x 8.40 x 1.27 mm^3.

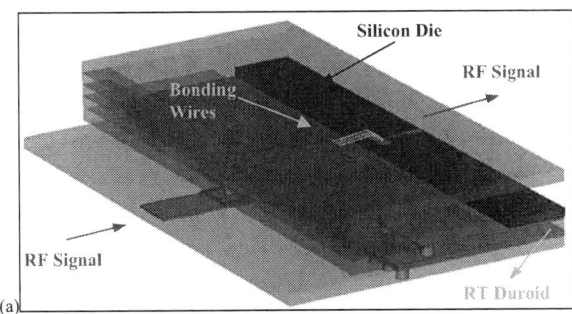

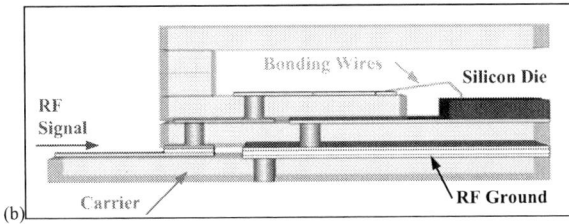

Fig. 5 3D-Layout of the 1-level plastic package transition. (a) Prospective view; (b) side view.

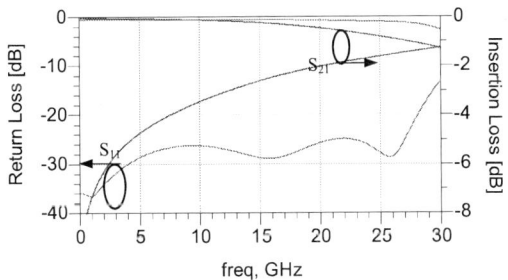

Fig. 6 CST Studio simulations of the package transition. Row transition in blue and compensated one in red.

VI. CONCLUSIONS

The design, manufacturing and packaging of a novel K-band 5-bit MEMS phase shifter has been presented. High RF performances were measured for the MEMS switches constituting the device, both ohmic and capacitive devices. The single bit showed as well excellent performance in the frequency band of interest 20.2-21.2 GHz. Return loss and insertion loss better than 20 dB and 1.3 dB and phase error

minor than 2.4 degrees were obtained for all 5 bits. The design of a low cost packaging solution has also been shown. Particular attention was focused on the optimization of the die-package transition in order to preserve the extremely high performance of the device. The measurements of the packaged device are about to be completed.

ACKNOWLEDGEMENT

The authors would like to acknowledge OPTO-I [13] who took care of the package manufacturing.

REFERENCES

[1] Schoebel, J.; Buck, T.; Reimann, M.; Ulm, M.; Schneider, M.; Jourdain, A.; Carchon, G.J.; Tilmans, H.A.C. "Design considerations and technology assessment of phased-array antenna systems with RF MEMS for automotive radar applications"; *IEEE Trans on Microwave Theory and Techniques*, Vol. 53, Part 2, June 2005 pp:1968 – 1975

[2] M. Kim, J.B. Hacker, R.E. Mihailovich, and J.F. DeNatale, "A DC-to-40 GHz four-bit RF MEMS true-time delay network," *IEEE Microwave and Wireless Components Lett.*, vol. 11, no. 2, pp. 56–58, Feb. 2001.

[3] G.L. Tan, R.E. Mihailovich, J.B. Hacker, J.F. DeNatale, and G.M.Rebeiz, "A very-low-loss 2-bit X-band RF MEMS phase shifter," presented at the 2002 IEEE Int. Microwave Symp., Seattle, WA

[4] J.S.Hayden , A.Malczewski, J.Kleber, C.L.Goldsmith, G.M. Rebeiz "2 and 4-Bit DC-18 GHz Microstrip MEMS Distributed Phase Shifters" 2001 IEEE

[5] Pillans, B.; Eshelman, S.; Malczewski, A.; Ehmke, J.; Goldsmith, C. "Ka-band RF MEMS phase shifters for phased array applications" *Radio Frequency Integrated Circuits (RFIC) Symposium, 2000. Digest of Papers. IEEE* pp:195 - 199 11-13 June 2000.

[6] Varian, K.; Walton, D. "A 2-bit RF MEMS phase shifter in a thick-film BGA ceramic package";*Microwave and Wireless Components Letters, IEEE* Vol. 12, Sep 2002 pp:321 – 323

[7] Kingsley, N.; Papapolymerou, J."Organic Wafer-Scale packaged miniature 4-bit RF MEMS phase shifter" *IEEE Trans on Microwave Theory and Techniques,*Vol. 54, Issue 3, March 2006 pp.1229 – 1236

[8] A. Ocera, P. Farinelli, F. Cherubini, P. Mezzanotte, R. Sorrentino, B. Margesin, F. Giacomozzi, "A MEMS-Reconfigurable Power Divider on High Resistivity Silicon Substrate", IEEE MTT-S International Microwave Symposium, Honolulu, 3-8 June 2007

[9] P.Farinelli, B.Margesin, F.Giacomozzi, G.Mannocchi, S. Catoni, R.Marcelli, P.Mezzanotte, L.Vietzorreck, F. Vitulli, R.Sorrentino, F. Deborgies, "A low contact-resistance winged-bridge RF-MEMS series switch for wide-band applications", Proceedings of the European Microwave Association Vol. 3, Sept. 2007; pp: 268–278

[10] ADS Agilent Momentum www.agilent.com

[11] Rogers Corporation www.rogerscorporation.com/mwu/pdf/5000data.pdf

[12] CST Studio www.cst.com

[13] OPTO-I Optoelettronica Italia www.optoi.com

Microwave Compact Passive Circuit Model of Isolated Interconnect over a Silicon Substrate with a Through-Silicon Via (TSV) Ground Supply Network

Wayne Woods, Guoan Wang, Hanyi Ding

IBM Microelectronics
Essex Junction, VT
whwoods@us.ibm.com
gawang@us.ibm.com
hding@us.ibm.com

Abstract— **As operating frequencies increase in state-of-the-art wireless designs, highly accurate modelling of critical interconnect paths routed over silicon is crucial for first-pass design success [1]. With this in mind, the interconnect stack of an IBM silicon germanium (SiGe) process incorporating a TSV ground supply network was modelled with model accuracy and efficiency as the goals. A unique modelling methodology for assigning the values of silicon skin-effect circuit model elements is discussed. The final model is verified with hardware measurements and found to accurately estimate the frequency-dependent resistance, capacitance, and inductance of a single line over silicon at microwave frequencies in a compact, efficient, pre-layout circuit model that includes the effects of process variation.**

I. INTRODUCTION

Microwave circuits are sensitive to the electrical parasitics of even small interconnect lines. Many signal paths in microwave circuits are shielded microstrip transmission lines [2]. Critical interconnect in high-frequency digital applications often take the form of coplanar waveguides. However, many lines are routed directly over the silicon substrate and are isolated with no nearby ground return lines either by design or necessity. Isolated lines routed directly over silicon generally have higher inductance than lines with nearby current return paths like microstrips and coplanar waveguides. Circuit designers sometimes use isolated interconnect as straight-line inductors with high Q values [3].

In microstrip lines, resistance increases due to skin-effect in the signal conductor as well as the ground plane below and is generally proportional to the square of the frequency. Also, the line inductance drops at the onset of skin-effect in the signal conductor. Full-bandwidth microwave compact passive models of microstrips include the effect of skin-effect on resistance and inductance [4]. In coplanar interconnect geometries, the effect of the silicon substrate on signal resistance is generally mitigated due to the much lower resistance ground return lines parallel to the signal line and can be ignored in many cases. However, the effect of the silicon substrate on signal capacitance must be considered when there are no crossing lines below the signal line to shield

it from the silicon substrate and the spacing to the ground return lines is large. Full-bandwidth microwave compact passive coplanar waveguide models include accurate modelling of the capacitance over silicon and the metal conductor skin-effect on inductance and resistance [5].

In isolated signal lines over silicon the signal resistance increases much more with frequency due to the skin-effect within the silicon substrate than due to the skin-effect in the conductor. This is caused by the return current in the silicon crowding towards the dielectric-to-silicon interface under the metal line with frequency. At low frequencies, the return current is spread throughout a large area of silicon, and the effect is small, but as frequency increases, the effect on signal line resistance becomes dramatic [6, 7]. As in the case of coplanar waveguides, the silicon substrate affects the capacitance of the signal line and causes a drop in signal capacitance near the relaxation frequency of the silicon. At low frequencies, the silicon substrate acts as a ground plane, but at higher frequencies, it acts like a lossy dielectric. A full-bandwidth microwave compact passive model of isolated lines over a thinned silicon substrate with a TSV ground supply network has been developed here that uses an efficient circuit model topology and modelling methodology to accurately model signal resistance, capacitance, and inductance from DC to microwave frequencies.

II. MODEL DEVELOPMENT

A. Model Topology

Fig. 1 shows a model topology which can be used to model a conductor over a silicon substrate [8]. Note that a perfect electrical conductor (PEC) is assumed to be on the backside of the silicon substrate. Several literature sources are available concerning the modelling of interconnect over silicon. One can readily find predictive equations in the literature for line inductance, line capacitance, and line conductance through the substrate for an isolated line over a silicon substrate [8, 9]. However, in practical applications, these equations can deviate from those values measured from hardware and predicted by commercial EM solvers such as HFSS [10] for a given technology. Fitting equations can then be employed to adjust the literature equations to a particular semiconductor process. The process modelled here has TSVs in the silicon substrate. Although, no TSVs are assumed directly below an isolated line over silicon, a silicon substrate with a TSV ground supply network is a different EM environment from the uniform silicon cross-sections assumed in the literature.

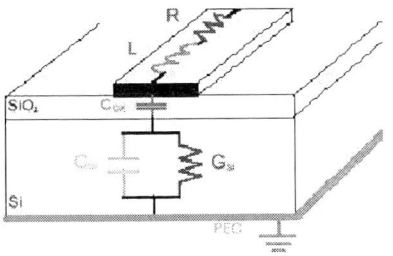

Fig. 1 Model topology of a line over silicon; resistance: R, inductance: L, capacitance to silicon,: C_{OX}, through silicon: C_{Si}, substrate conductance: G_{Si}

The inductance and resistance of a signal line over silicon are frequency dependent and a skin-effect network must be added in series to the inductor and resistor in the circuit topology shown in Fig. 1 to enable a full-bandwidth passive model. Fig. 2 shows a parallel inductor and resistor skin-effect section added into the signal path that models both the drop of inductance and increase in line resistance due to the onset of skin-effect in the metal signal line.

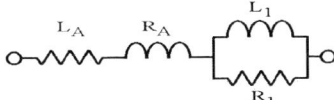

Fig. 2 Signal path circuit model of a metal line using a parallel inductor and resistor pair as a skin-effect network

The signal path circuit model in Fig. 2 models frequency-dependent resistance and inductance due to the onset of skin-effect in the metal line. This signal path circuit model is adequate for modelling the initial inductance and resistance change near the onset of skin-effect. Additional skin-effect sections must be added to make the resistance proportional to the square of frequency for a full-bandwidth microwave model [4, 5]. When modelling metal conductors that are not isolated, L_A represents the inductance of the line at very high frequency and R_A represents the resistance at DC. In the skin-effect network, L_1 represents the decrease of inductance and R_1 the increase of resistance due to the metal conductor skin-effect.

The effect of the silicon substrate skin-effect dominates line resistance of an isolated line over silicon at higher frequencies. So, a single skin-effect network is employed to model the initial drop in inductance and increase in resistance due to the skin-effect onset in the metal conductor. At higher frequencies, the skin-effect due to the silicon substrate affects resistance much more than that of the metal, but does not affect the inductance in the frequency range of the model.

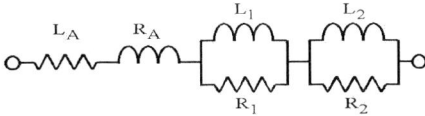

Fig. 3 Signal path circuit model of a conductor over silicon; one network (L_1, R_1) for skin-effect of the conductor and one for that of silicon (L_2, R_2)

A single skin-effect network can be used to model the resistance increase of the line due to the silicon without affecting the line inductance below frequencies much higher than the maximum frequency, (ω_{max}), enabled by the model. The final model topology is a repeating pi model consisting of the capacitance and silicon conductance components shown in Fig. 1 and the signal path network shown in Fig. 3. The number of pi-sections needed in the model is determined by the maximum frequency of the model (ω_{max}) and the maximum line length allowed. The line length modelled by each pi-section should not exceed $\lambda/10$. However, the number of sections should be kept to the minimum needed to reduce the total number of circuit elements required by the model. If the model is implemented in Verilog-A [11], the number of sections can be a function of length and frequency range.

B. Modelling Methodology

The general form of the equivalent resistance through the signal path circuit model in Fig. 3 is shown in Fig.4. In Fig. 4, ω_1 represents the frequency at the onset of skin-effect in the conductor and ω_{max} represents the maximum frequency range enabled by the model. The transition frequency of the silicon skin effect network, ω_2, is large and not shown in the frequency range depicted in Fig. 4.

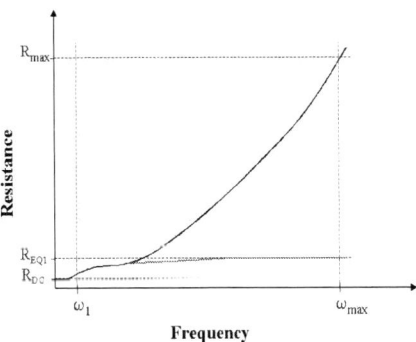

Fig. 4 General resistance versus frequency of circuit model in Fig. 4

To ensure that the transition frequency of the silicon skin-effect network is well above the maximum usable frequency of the model, ω_2 is set equal to $N*\omega_{max}$ where $N > 2$. The relationship of the two parallel skin-effect networks: $R_1=\omega_1*L_1$ and $R_2=\omega_2*L_2$ relates the transition frequencies to the circuit element values in Fig. 3. The value of L_1 is the difference between the low and high frequency inductance of the line and can be found in the literature e.g. [9]. R_A in Fig. 3 is set to the DC resistance of the conductor. Since, ω_2 is set to $N*\omega_{max}$, the value of L_2 is given by

978-2-8748-7007-1/08 $25.00 © 2008 EuMA 343

$$L_2 = \left(\frac{N+1}{N}\right)\left(\left(\frac{R_{max} - R_{DC}}{\omega_{max}}\right) - \left(\frac{\omega_{max}\,\omega_1}{\omega_{max}^2 + \omega_1^2}\right)L_1\right) \quad (1)$$

where R_{max} is given by

$$R_{max} = R_{Si}(\omega_{max}) + R_{metal}(\omega_{max}) \quad (2)$$

In Equation (2), R_{Si} is the return resistance through the silicon substrate and R_{metal} is the resistance of the metal in the signal path where $R_{metal} = R_{DC} + R_{metal\_skin\_effect}$. The equivalent resistance of the conductor skin-effect section (L_1 parallel to R_1) at frequency ω_1 is equal to $R_1/2$. The skin-effect resistance in a conductor rises proportionally to the square of the frequency. So, an estimate of the skin-effect resistance of the conductor at frequency ω_{max} is given by

$$R_{metal\_skin\_effect} = \frac{R_1}{2}\sqrt{\frac{\omega_{max}}{\omega_1}} \quad (3)$$

and, because: $R_1 = \omega_1 * L_1$, Equation (1) is then rewritten as:

$$L_2 = \left(\frac{N+1}{N}\right)\left(\frac{R_{Si}}{\omega_{max}} + \frac{1}{2}\sqrt{\frac{\omega_1}{\omega_{max}}}L_1 - \left(\frac{\omega_{max}\,\omega_1}{\omega_{max}^2 + \omega_1^2}\right)L_1\right) \quad (4)$$

where R_{Si} is the resistance of the silicon substrate ground return path at ω_{max}. Once L_2 is determined by Equation (4), R_2 is then fixed because $R_2 = \omega_2 * L_2$. Expressions for ω_1 and L_1 in Equation (4) can be found in the literature. The silicon skin-effect network in Fig. 3 causes a drop in inductance at ω_2, therefore, L_A must be set to $L_{infinity} - L_2$ so that the high-frequency inductance in the enabled frequency bandwidth is equal to $L_{inifinity}$.

III. MODEL VERIFICATION

The model was compared to simulations with HFSS and hardware measurements taken from test structures in IBM's 5PAE SiGe process using a thinned silicon substrate with a TSV ground supply network. The resistivity of the silicon substrate used here is 0.6 Ω-m and the substrate thickness is 145 μm. The value of R_{Si} in Equation (4) was determined using HFSS to be 65 mΩ/μm at ω_{max}=120 GHz. Also, the value of N in Equation (4) was set to 3. The line length was 300 μm in all test cases. Fig. 5 shows the model resistance versus that predicted by HFSS.

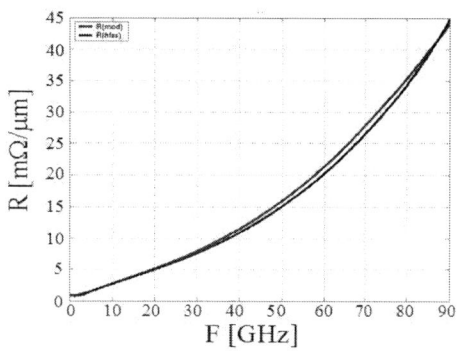

Fig. 5 Resistance versus frequency of HFSS (black) and the model (blue)

Fig. 5 shows good agreement of the model resistance versus HFSS up to 90GHz. Figs. 6-9 show data from the model compared to simulations, hardware measurements, and HFSS and Z2D simulations. Z2d is a 2-D EM field-solver developed by IBM.

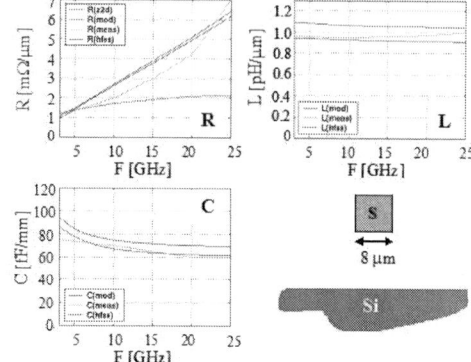

Fig. 6 Resistance, inductance and capacitance of model (blue), HFSS (black), measurement (green), and z2d (red) for line width = 8 μm

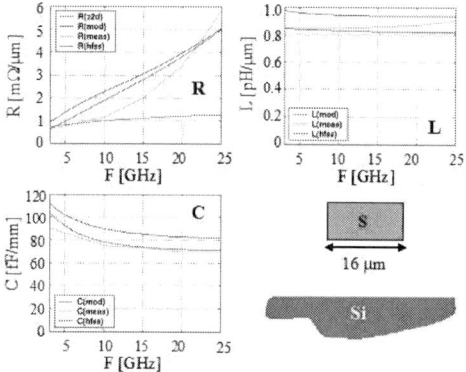

Fig. 7 Resistance, inductance and capacitance of model (blue), HFSS (black), measurement (green), and z2d (red) for line width = 16 μm

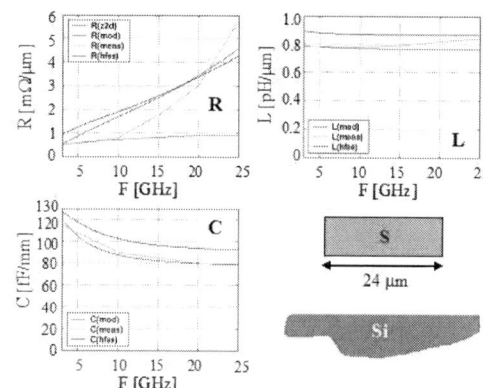

Fig. 8 Resistance, inductance and capacitance of model (blue), HFSS (black), measurement (green), and z2d (red) for line width = 24 μm

978-2-8748-7007-1/08 $25.00 © 2008 EuMA

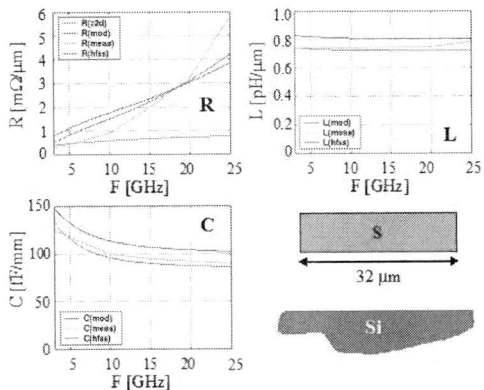

Fig. 9 Resistance, inductance and capacitance of model (blue), HFSS (black), measurement (green), and z2d (red) for line width = 32 µm

The resistance, capacitance, and inductance values of Figs. 5-9 were obtained by converting two-port S-parameter model, simulation, and measurement data using the formulation given in [12]. Measurements were made on a thinned silicon wafer with TSVs connecting the G-S-G probe ground pads to a metal ground plane on the backside of the silicon substrate. The HFSS simulation setup did not consider the TSVs. This contributes to the difference in inductance between the HFSS simulations and the measured results. The TSVs complicate the measurement de-embedding for isolated lines over silicon. The increase in the measured inductance with frequency is likely an artifact of the de-embedding in the TSV environment. The electro-magnetically complex TSV measurement set-up is a good representation of the actual isolated lines over the thinned silicon substrate with a TSV ground supply network.

IV. CONCLUSIONS

The model shows good agreement to the resistance values predicted by HFSS for all four of the line widths considered. The dominating effect of the silicon ground return resistance can be seen by comparing the red curve from Z2D that models only the conductor resistance and not the silicon return resistance. At 25 GHz the resistance due to the silicon skin-effect is a factor of two or three larger than the Z2D simulated resistance of only the metal conductor DC resistance and skin-effect. The model includes both the conductor and silicon skin-effects with only two skin-effect networks. The same value of R_{si} was used in all of the test case widths. Field solver study revealed a very slight width dependence of Rsi up to widths of 32 µm. For widths greater than 32 µm, the width dependence of R_{si} becomes more pronounced and should be included in a model. The resistance behavior of the measured test structures show an accelerating increase from 3 GHz to 25 GHz while the behavior predicted by the model and HFSS is essentially linear in Figs. 6-9. The match of the resistance behavior between the measured data and the model is clearly better when the silicon skin-effect is included. The modelling

methodology presented efficiently models the line resistance contribution due to the silicon skin-effect. The potential increase in accuracy from changing the model topology and adding more skin-effect modelling sections must be balanced against the need to reduce model complexity. The small number of circuit elements required by this model and the ease of model migration to different semiconductor processes using this methodology is highly desirable. Comparisons of this pre-layout model to the hardware measurements from a thinned silicon substrate with a TSV ground supply network show the model accurately predicts performance. Also, this is a pre-layout compact model that includes the effects of process variation so it can be used early in the schematic design phase and in Monte Carlo simulations making it very useful to high-frequency analog circuit designers.

ACKNOWLEDGMENT

The authors would like to thank Michael McPartlin for providing the measurement data.

REFERENCES

[1]   R. Singh, D. L. Harame, M. M. Oprysko, *Silicon Germanium Technology, Modelling, and Design*, IEEE Press and Wiley, New York 2004.

[2]   A. Valdes-Garcia, S. Reynolds, U. R. Pfeiffer, "A 60 GHz Class-E Power Amplifier in SiGe", *IEEE Asian Solid-State Circuits Conference*, pp. 199-202, Nov. 2006.

[3]   E. C. Yi-Jan, L. K. Wei-Min, L. Jongsoo, J. D. Cressler, J. Laskar, G. Freeman, "A low-Power Ka-band SiGe HBT VCO Using Line Inductors", *Radio Frequency Integrated Circuits Symposium,* Digest of Papers, pp. 587-590, June 2004.

[4]   T. Zwick, Y. Tretiakov, D. Goren, "On-Chip SiGe Transmission Line Measurements and Model Verification up to 110 GHz", *Microwave and Wireless Components Letters*, IEEE, vol. 15, issue 2, pp. 65-67, Feb. 2005.

[5]   D. Goren, S. Shlafman, B. Sheinman, W. Woods, J. Rascoe, "Silicon-Chip Single and Coupled Coplanar Transmission Line Measurements and Model Verification up to 50 GHz", *Signal Propagation in Interconnects*, IEEE Workshop, May 2007.

[6]   U. Arz, H. Grabinski, D. F. Williams, "Influence of the Substrate Resistivity on the Broadband Propagation Characteristics of Silicon Transmission Lines", *54th Automatic RF Technologies Group Conference Digest*, vol. 36, pp. 65-70, Dec. 2000.

[7]   H. Hasegawa, M. Furukawa, H. Yanai, "Properties of Microstrip Lines in Si-SiO$_2$ Systems", *Proceedings of the IEEE*, vol. 59, issue 2, pp. 297-299, Feb. 1971.

[8]   A. Weisshaar and H. Lan, "Accurate Closed-Form Expressions for the Frequency-Dependent Line Parameters of On-Chip Interconnects on Lossy Silicon Substrate", *IEEE MTT-S Digest*, vol. 3, pp. 1753-1756, 2001.

[9]   Y. Eo and W. R. Eisenstadt, "High-Speed VLSI Interconnect Modelling Based on S-Parameter Measurements", *IEEE Trans. Components, Hybrids, and Manufacturing Technology*, vol. 16, no. 5, pp. 555-562, August 1993.

[10] Ansoft Corporation, *High-Frequency Structure Simulator User's Guide*, 2008.

[11] Cadence Design Systems, Inc., *Verilog-A Language Reference*, 2008

[12] Y. Eo, W. R. Eisenstadt, "High-Speed VLSI Interconnect Modelling Based on S-Parameter Measurements", *IEEE Transactions on Components, Hybrids, and Manufacturing Technology,* vol. 16, No. 5, August 1993.

978-2-8748-7007-1/08 $25.00 © 2008 EuMA

Spectral Response Modelling of Heterojunction Phototransistors for Short Wavelength Transmission

Hassan A. Khan[1], Ali A. Rezazadeh[2], Suba C. Subramaniam[3]

Microwave and Communication Systems Research Group, School of Electrical & Electronic Engineering,
University of Manchester, Manchester, M60 1QD, UK

[1]hassan.khan@postgrad.manchester.ac.uk
[2]ali.rezazadeh@manchester.ac.uk
[3]suba.subramaniam@manchester.ac.uk

Abstract— An accurate spectral response model for phototransistors has been proposed. The model is based on the calculation of photogenerated carriers through absorption in base, collector and sub-collector regions of phototransistor. Absorption pattern in *AlGaAs/GaAs* heterojunction phototransistors has been analysed and discussed using the proposed model. Collection efficiency, being strictly a geometry and wavelength dependent parameter, is not considered unity unlike all the precedents and its importance is highlighted. With the aid of the absorption model, absolute responsivity of a phototransistor is predicted for the first time. Power absorption profile and quantum efficiency have also been modelled. The analysis is performed for short wavelength transmission and the measurements corroborate the simulated results.

I. INTRODUCTION

There has been an ever-growing interest in heterojunction phototransistors (HPTs), in lightwave communications, over p-i-n and avalanche photodiodes due to their internal gain and lack of excess noise resulting from avalanching [1]-[5]. For short wavelength applications, mostly p-i-n photodetectors have been employed but GaAs-based HPTs show superior performance to p-i-n photodiodes [4], [6], [7]. The parameters such as *responsivity*, *optical gain* and *noise equivalent power* (or specific detectivity) are important in the characterisation of photodetectors. *Responsivity* depends on several inherent factors including material absorption coefficient, refractive index, device structure, doping and temperature of operation along with the external factors such as bias voltage and the energy of incident radiation. *Optical gain* depends on coupling efficiency, collection efficiency and internal gain of the transistor [8]. *Noise equivalent power*, along with responsivity, takes into account the leakage currents in the phototransistor. Therefore, a thorough understanding of these parameters is vital in the design optimisation of future phototransistors.

Several models have already been developed for photodetectors and phototransistors [9]-[13]. In all these models, collection efficiency is taken as unity for the analysis which may not be the case for every detector. Collection efficiency significantly depends on vertical device dimensions, material properties, doping concentrations and incident photon energy. Its dependence on these parameters is discussed for the first time in this paper. Absorption coefficient is usually taken as a constant for all doping in a material which may not be the case especially at wavelengths closer to the threshold wavelength of the absorbing material [14]. In the present work the variation of absorption coefficient with doping is analysed and taken into consideration for the power absorption profile. In our earlier works [8] and [13], we presented a spectral response model for hetero-structures but this is only valid for relative measurement of spectral response and cannot be used for absolute responsivity measurements. In this paper, a modified formulation has been presented to accurately determine the absolute responsivity of HPTs. Moreover, optical characteristics of the phototransistor can be predicted with the aid of this modified model as well as the prior knowledge of device electrical characteristics.

II. ABSORPTION MODEL FOR PHOTOTRANSISTORS

Fig. 1 illustrates the vertical dimensions, energy bandgap relationship and the optical power absorption profile of an N-p-n *AlGaAs/GaAs* HPT. It is shown that some of the flux is reflected from the surface of phototransistor which can be minimised by using anti-reflective coating. However, the phototransistor understudy had no anti-reflective coating so the reflected flux is mentioned for completion. It is assumed that the absorption starts from the base region and no absorption occurs in emitter region. This was taken care of by illuminating the base region (of a large geometry HPT) directly with an external radiation to excite the photo-carriers. The absorption in the base layer (ϕ_{abs1}) is dependent on incident wavelength (λ) and doping concentration (N_A) which can be modelled by

$$\phi_{abs1}(\lambda, N_A) = (\phi_{inc} - \phi_{ref})\left[1 - e^{-(b-a)\alpha_b}\right] \quad (1)$$

Where α_b is the absorption coefficient of the base layer,

$\phi_{inc} = \dfrac{P_{in}\lambda}{hc}$ is the incident flux on the base layer and

$\phi_{ref} = \dfrac{P_{in}\lambda}{hc}R_f$ is the reflected flux from the base layer, P_{in} is

the input power, h is Planck constant, c is the speed of light and R_f is Fresnel reflection coefficient which is given as [15]

$$R_f(\lambda) = \left[\frac{(n_b - n_o)}{(n_b + n_o)}\right]^2 \quad (2)$$

978-2-8748-7007-1/08 $25.00 © 2008 EuMA

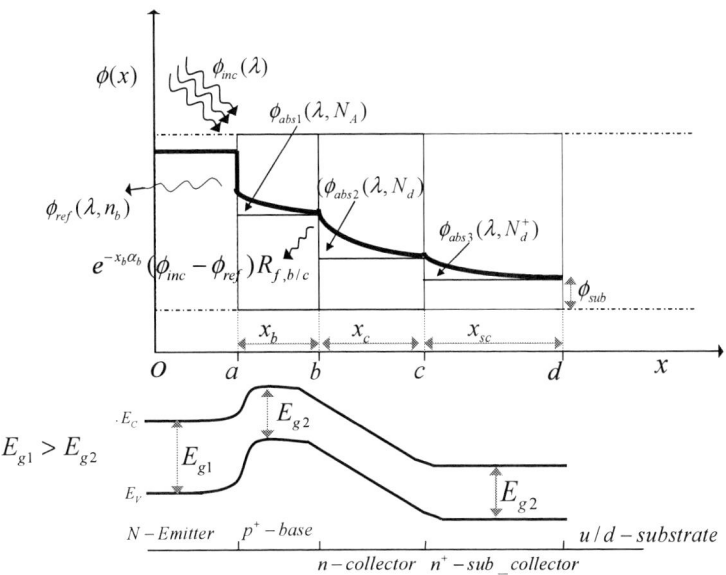

Fig. 1 Schematic of optical flux absorption and propagation in different layers of phototransistor along with the energy band diagram

Where n_o is refractive index of air which is 1 and n_b is refractive index of the base layer. It is clear from (1) that the flux absorbed is dependent on the width of the layer and not the layer-coordinates. $R_{f,b/c}$ in fig. 1 models the reflectivity between base and collector material. The base-collector for single-HPTs is a homojunction and therefore $R_{f,b/c}$ can be ignored for the flux modelling in the collector and sub-collector regions. The flux absorbed in these two regions can be represented by (3) and (4) respectively.

$$\phi_{abs2}(\lambda, N_d) = (\phi_{inc} - \phi_{ref}) \left[1 - e^{-(c-b)\alpha_c} \right] e^{-(b-a)\alpha_b} \quad (3)$$

$$\phi_{abs3}(\lambda, N_d^+) = (\phi_{inc} - \phi_{ref}) \left[1 - e^{-(d-c)\alpha_{sc}} \right] e^{-(b-a)\alpha_b - (c-b)\alpha_c} \quad (4)$$

Where α_c and α_{sc} are absorption coefficients, N_d and N_d^+ are doping concentrations for collector and sub-collector respectively. This detailed analysis of every layer is important in order to analyse the collection efficiency (η_c). The useful flux absorbed (ϕ_{absT}) in phototransistor can thus be given by

$$\phi_{absT} = \phi_{abs1}(\lambda, N_A) + \psi_{abs2}(\lambda, N_d) + \phi_{abs3}(\lambda, N_d^+) \quad (5)$$

The flux absorbed in the substrate ($\phi_{sub} = \phi_{inc} - \phi_{ref} - \phi_{absT}$) does not contribute towards the photoresponse but leads to a fall in the collection efficiency which is given as

$$\eta_c(\lambda) = \frac{\phi_{absT}}{\phi_{absT} + \phi_{sub}} \quad (6)$$

Another important parameter is the responsivity (R_{HPT}) of a phototransistor defined as a ratio of photogenerated current to incident optical power. The R_{HPT} and quantum efficiency (η_q) are given by (7) and (8) respectively [15].

$$R_{HPT} = \frac{\beta I_{ph}}{P_{in}} \quad (7)$$

$$\eta_q = \frac{hc I_{ph}}{\lambda q P_{in}} \quad (8)$$

Where β is the internal gain of the transistor and I_{ph} is the photogenerated current. For calculation of I_{ph}, the recombination should also be taken into account along with the flux absorption nonlinearities discussed above. For the subcollector, the primary photogenerated current can be written as [2]

$$I_{ph,sc} = q\phi_c A \left[1 - \frac{e^{-\alpha_{sc} x_{sc}}}{\alpha_{sc} L_p + 1} \right] \quad (9)$$

Where ϕ_c is the value of flux at the collector boundary (x = b) and L_p is hole diffusion length.

The spectral response can be predicted for phototransistors through the amount of flux absorbed in each section of the device and it can be written as

$$R_{spec} = \frac{\beta q}{P_{in}} \int_\lambda \int_0^{d-a} \alpha e^{-\alpha x} \, dx \, d\lambda \quad (10)$$

$$R_{spec} = \frac{\beta q}{P_{in}} \int_\lambda \left[\alpha_b \int_0^{b-a} e^{-\alpha_b x} dx + \alpha_c \int_0^{c-b} e^{-\alpha_c x} dx + \alpha_{sc} \int_0^{d-c} e^{-\alpha_{sc} x} dx \right] d\lambda \quad (11)$$

A change in the origin (in the lower-limit of integral) for every layer should be noticed in (11). Absorption is a distance-dependent parameter and so a change in the origin is necessary for accurate prediction of spectral response. This can also be written in terms of responsivity of each layer.

$$R_{spec} = \int_0^{x_b} R_b(\lambda)\,dx + \int_0^{x_c} R_c(\lambda)\,dx + \int_0^{x_{sc}} R_{sc}(\lambda)\,dx \quad (12)$$

The prediction of spectral responsivity can be utilised to model the optical characteristics of transistors. This can be achieved by calculating the amount of photo-generated carriers in the three active regions of transistor through the proposed model. These photo-carriers can then be linked to the base current of a transistor biased in forward-active mode. Thus, characteristic curves due to optical illumination can be predicted. MATLAB has been used to carry out this analysis.

III. EXPERIMENTAL DETAILS

The HPTs measured in this work are lattice matched *AlGaAs/GaAs* Npn transistors with graded B/E junction. The devices were fabricated using standard photolithographic techniques. The layer structure is given in table 1. B/C and B/E junction area is 0.23 mm² and 0.023 mm² respectively.

TABLE I
LAYER STRUCTURE FOR GRADED EMITTER/BASE HPT UNDER REVIEW

Material	Mole Fraction (%)	Thickness (µm)	Doping (cm^{-3})
n-GaAs	-	0.19	Si - 5x10^{18}
n-Al$_x$Ga$_{1-x}$As	30 – 0	0.02	Si - 5x10^{17}
n-Al$_x$Ga$_{1-x}$As (emitter)	30	0.15	Si - 5x10^{17}
n-Al$_x$Ga$_{1-x}$As	0 – 30	0.02	Si - 5x10^{17}
p-GaAs (base)	-	0.09	C - 2x10^{19}
n-GaAs (collector)	-	0.5	Si - 2x10^{16}
n-GaAs (sub-collector)	-	1.0	Si - 5x10^{18}
SI-GaAs	-	400	u/d

The *AlGaAs/GaAs* HPT was set on a 4 dc-probe station. Three probes were specifically connected to HPTs terminal points and the fourth one was connected to optical fibre to shine the incident optical radiation on the device base terminal. The probes were connected to Keithley Semiconductor Parameter Analyser where accurate measurements can be carried out.

IV. RESULTS AND DISCUSSIONS

Material properties, used for simulation, are taken from [14] and [16]. In Fig. 2, R and η_q begins to fall with λ sharply after 750 nm. Conventionally, the fall is associated closer to 875 nm which is the bandgap of *GaAs*. This can be associated to two possible reasons. Firstly, the bandgap varies with doping which lowers the threshold wavelength. This is incorporated in simulations by adding the effect of doping onto the absorption coefficient for *GaAs* [14]. Secondly, η_c falls significantly at around 750 nm which causes R and η_q to drop with λ. The simulated value of *11.5 A/W* for the responsivity at 635nm is very similar to the measured value of *11 A/W*.

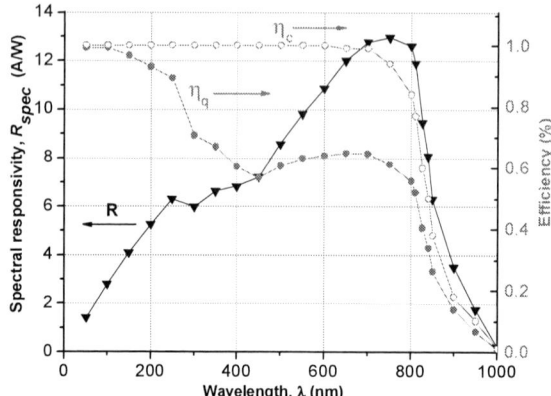

Fig. 2 Simulated spectral response for the *AlGaAs/GaAs* HPT, V_{CE}= 2.5 V

The effect of band gap variation and η_c lowering for phototransistors is incorporated for the first time in the analysis of phototransistors and can prove critical in device modelling and performance enhancement.

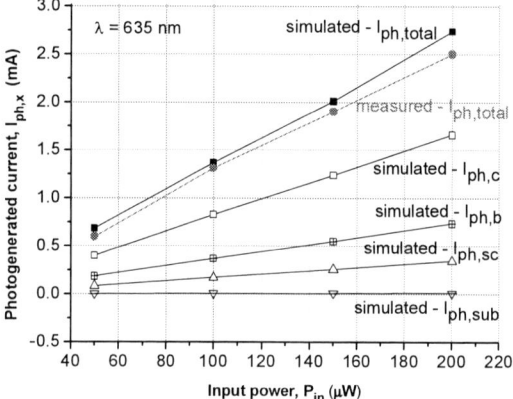

Fig. 3 Measured and simulated photogenerated currents with input optical power for various *AlGaAs/GaAs* HPT layers, V_{CE}= 2.5 V

Fig. 3 depicts comparison between the measured and simulated photogenerated current. Current contribution for every absorbing layer has also been shown. The power absorption profiles for the *AlGaAs/GaAs* HPT at 635 nm and 840 nm are shown in fig. 4. The absorption coefficient at 635 nm does not vary with the change in doping and so the optical flux is a decreasing exponential for the three layers (α = 3.42 µm^{-1}) [16]. However, the effect of change in absorption coefficient is clearly seen at 840 nm where it is no longer a single-exponential. In this case, the absorption is layer-dependent as the absorption coefficients for base, collector and sub-collector are 0.53 µm^{-1}, 1.35 µm^{-1} and 0.15 µm^{-1} respectively [14]. Referring to fig. 1, the flux propagation in different layers has been modelled by (13) and numerically shown in fig. 4.

$$\phi(x) = (\phi_{inc} - \phi_{ref}) \begin{cases} e^{-\alpha_b x} & ; 0 \le x \le x_b \\ e^{-\alpha_c x} - e^{-\alpha_b x_b - \alpha_c x} \\ & ; x_b < x \le x_b + x_c \\ e^{-\alpha_c x_c - \alpha_{sc} x} - e^{-\alpha_b x_b - \alpha_c x_c - \alpha_{sc} x} \\ & ; x_b + x_c < x \le x_b + x_c + x_{sc} \\ e^{-\alpha_c x_c - \alpha_{sc} x_{sc}} - e^{-\alpha_b x_b - \alpha_c x_c - \alpha_{sc} x_{sc}} \\ & ; x_b + x_c + x_{sc} < x \end{cases} \quad (13)$$

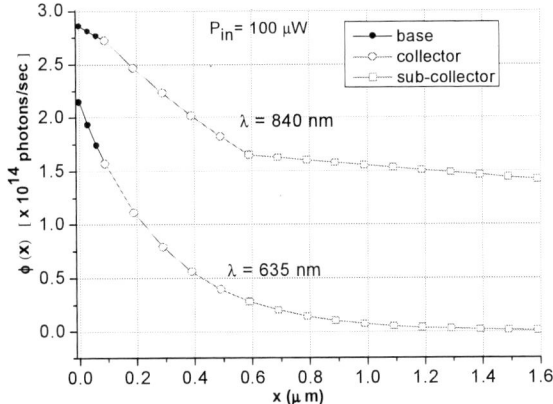

Fig. 4   Power absorption profile for the *AlGaAs/GaAs* HPT at different incident wavelengths

V. CONCLUSIONS

An accurate absorption model for the GaAs-based HPT has been developed and characterised. The collection efficiency of phototransistors should not be considered unity if the wavelength of incoming signal is close to the bandgap of the absorbing layer. Due to a shift in cut-off wavelengths, smaller values for responsivity are observed for the *AlGaAs/GaAs* HPT at 800nm. It is shown that for the accurate prediction of spectral response the effect of doping on the absorption coefficient should also be incorporated. The model developed can be applied to any other material systems (e.g. *InP/InGaAs* or *InGaP/GaAs*) providing a useful tool in predicting the optical characteristics of HPTs.

ACKNOWLEDGMENT

The authors would like to thank Tuyen Vo for valuable technical discussions and Sarmad Sohaib for his assistance in MATLAB. Hassan Khan would also like to express his gratitude to Higher Education Commission (HEC) Pakistan for funding this project.

REFERENCES

[1] J. C. Campbell, A. G. Dentai, C. A. Burrus, Jr., and J. F. Ferguson, "InP/InGaAs heterojunction phototransistors," *IEEE Journal of Quantum Electronics*, vol. QE-17, pp. 264-9, 1981.

[2] N. Chand, P. A. Houston, and P. N. Robson, "Gain of a heterojunction bipolar phototransistor," *IEEE Transactions on Electron Devices*, vol. ED-32, pp. 622-627, 1985.

[3] S. Chandrasekhar, M. K. Hoppe, A. G. Dentai, C. H. Joyner, and G. J. Qua, "Demonstration of enhanced performance of an InP/InGaAs heterojunction phototransistor with a base terminal," *IEEE Electron Device Letters*, vol. 12, pp. 550-2, 1991.

[4] R. Sridhara, S. M. Frimel, K. P. Roenker, N. Pan, and J. Elliott, "Performance enhancement of GaInP/GaAs heterojunction bipolar phototransistors using dc base bias," *Journal of Lightwave Technology*, vol. 16, pp. 1101-6, 1998.

[5] M. N. Abedin, T. F. Refaat, O. V. Sulima, and U. N. Singh, "AlGaAsSb-InGaAsSb HPTs with high optical gain and wide dynamic range," *IEEE Transactions on Electron Devices*, vol. 51, pp. 2013-2018, 2004.

[6] K. D. Pedrotti, R. L. Pierson, Jr., N. H. Sheng, R. B. Nubling, C. W. Farley, and M. F. Chang, "High-bandwidth OEIC receivers using heterojunction bipolar transistors Design and demonstration," *Journal of Lightwave Technology*, vol. 11, pp. 1601-14, 1993.

[7] H. Kamitsuna, "Ultra-wideband monolithic photoreceivers using HBT-compatible HPTs with novel base circuits, and simultaneously integrated with an HBT amplifier," *Journal of Lightwave Technology*, vol. 13, pp. 2301-7, 1995.

[8] S. A. Bashar and A. A. Rezazadeh, "Fabrication and spectral response analysis of *AlGaAs/GaAs* and InP/InGaAs HPTs with transparent ITO emitter contacts," *IEE Proceedings-Optoelectronics*, vol. 143, pp. 89-93, 1996.

[9] J. A. Gonzalez-Cuevas, T. F. Refaat, M. N. Abedin, and H. E. Elsayed-Ali, "Modeling of the temperature-dependent spectral response of In(1-x)Ga(x)Sb infrared photodetectors," *Optical Engineering*, vol. 45, pp. 044001, 2006.

[10] A. A. de Salles, A. S. Hackbart, and L. N. Spalding, "A simple model for the GaAs HBT high-frequency performance under optical illumination," *Microwave and Optical Technology Letters*, vol. 7, pp. 392-5, 1994.

[11] S. M. Frimel and K. P. Roenker, "Gummel-Poon model for Npn heterojunction bipolar phototransistors," *Journal of Applied Physics*, vol. 82, pp. 3581-92, 1997.

[12] S. W. Tan, W. T. Chen, M. Y. Chu, and W. S. Lour, "A new model for the phototransistor," presented at Extended Abstracts of the Fourth International Workshop on Junction Technology, 15-16 March 2004, Shanghai, China, 2004.

[13] S. A. Bashar and A. A. Rezazadeh, "Characterisation of transparent ITO emitter contact InP/InGaAs heterojunction phototransistors," presented at Proceedings of the 3rd IEEE International Workshop on High Performance Electron Devices for Microwave and Optoelectronic Applications, EDMO 95, 27 Nov. 1995, London, UK, 1995.

[14] H. C. J. Casey, D. D. Sell, and K. W. Wecht, "Concentration dependence of the absorption coefficient for n- and p-type GaAs between 1.3 and 1.6 ev," *Journal of Applied Physics*, vol. 46, pp. 250-257, 1975.

[15] S. M. Sze. "Semiconductor Devices. Physics and Technology", 2nd edition, John Wiley & Sons, USA 2001.

[16] "Properties of GaAs", INSPEC Publication, EMIS Data Review Series No. 2, 2nd Edition, 1990, pp. 513 - 528.

Proceedings of the 3rd European Microwave Integrated Circuits Conference

A Low-Power Ultra-Compact CMOS LNA with Shunt-Resonating Current-Reused Topology

Muh-Dey Wei[1], Sheng-Fuh Chang[2], Yu-Chun Liu

Department of Electrical Engineering, Department of Communications Engineering,
Center for Telecommunication Research, National Chung Cheng University,
168 University Rd., Min-Hsiung, Chia-Yi, Taiwan
[1]d95415006@ccu.edu.tw
[2]ieesfc@ccu.edu.tw

Abstract— **A low-power ultra-compact CMOS low-noise amplifier (LNA) in a shunt-resonating current-reused topology is presented. The common-source transistors are connected with a shunt-resonating inter-stage match network such that the bias current is shared to have low power consumption and RF signal is doubly amplified to have high gain and low noise figure. The implemented 0.18-μm CMOS LNA achieves 15.2-dB power gain and 3.0-dB noise figure, while only consuming 1.81 mW. Compared with previously published current-reused LNA, the proposed LNA has smallest chip size of 0.28 mm², excluding the I/O pads, and the highest FOM of 2.77.**

I. INTRODUCTION

SILICON-BASED CMOS technology has become a competitive technology for rapidly-developing wireless broadband communications due to its low-cost, high-level integration of radio transceiver with base-band processors. The low-noise amplifier (LNA) is one of the key components in an RF receiver front-end because its dominant noise contribution to the whole radio receiver. The design of LNA involves numerous aspects of requirements, including noise figure (NF), gain, linearity, input/output match, power consumption, and chip size. A number of LNA topologies [1]-[5], such as the cascaded, the cascode, and the g_m-boosted topologies, have been proposed to achieve the highest performance.

Among the variety of topologies, the current-reused topology has been shown the most promising solution [1]-[4]. As shown in Fig. 1(a), the current-reused topology is basically composed of two stages of common-source (CS) MOSFET transistors, in which the bias current is shared by these transistors such that the power consumption is only half of the cascaded topology. The RF signal is amplified by the first and the second CS transistors such that the power gain is as high as the cascade topology. The crucial design point in the current-reused topology is the inter-stage network, which directs the RF signal and simultaneously allows the same DC current flowing through two transistors. In other words, the inter-stage network plays dual functions of RF impedance match and DC bias with RF choking effect. Because the resistive choking, as used in Fig. 1(a), reduces the voltage headroom, the linearity is degraded. An inductor in Fig. 1(b) is used to replace the resistor as a high-impedance choke [2], but this inductor takes much larger chip area. In Fig. 1(c),

there is another modified current-reused topology [3]-[4], which uses the series-resonance inter-stage network to provide better gain shape. However, its chip size is much larger due to more inductors are used.

In this paper, we present an enhanced current-reused topology using the shunt-resonance inter-stage network, as shown in Fig. 1(d). This topology overcomes the drawbacks in of the previous current-reused LNAs and will be demonstrate in the following to have an extremely high FOM and the small chip size.

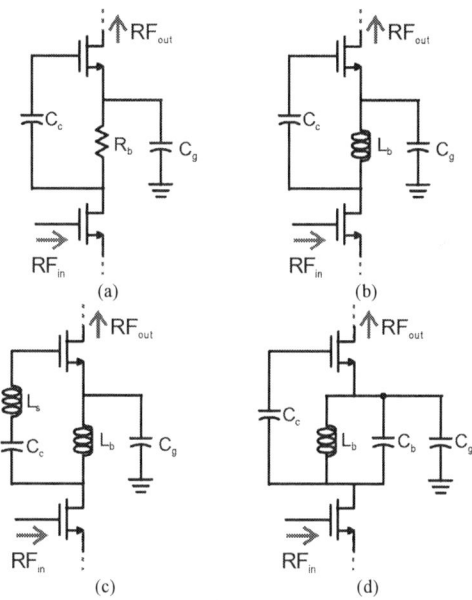

Fig. 1 Current-Reused Topology, (a) resistive blocking[1], (b) inductive blocking[2], (c)series-resonance coupling[3]-[4], (d) proposed in the paper.

II. CIRCUIT DESIGN

A. Proposed Shunt-Resonating Current-Reused Topology

Fig. 2 shows the proposed shunt-resonating current-reused LNA. The transistors M_1 and M_2 operate in the common-source (CS) amplification configuration and they share the

978-2-8748-7007-1/08 $25.00 © 2008 EuMA   350

same DC bias current flowing through L_b. The amplified RF signal from M_1 is impeded by the high reactance of the shunt combination of L_b and C_b and directed to the gate node of M_2 through the coupling capacitor C_2. The overall trans-conductance is equal to the multiplication of the individual trans-conductance of M_1 and M_2. Therefore, it has a comparable power gain to the cascaded CS topology. The capacitor C_3 is presented as the RF bypass for M_2. The shunt-peaking inductor L_1 is used to enhance the bandwidth and also acts as a reactive output load.

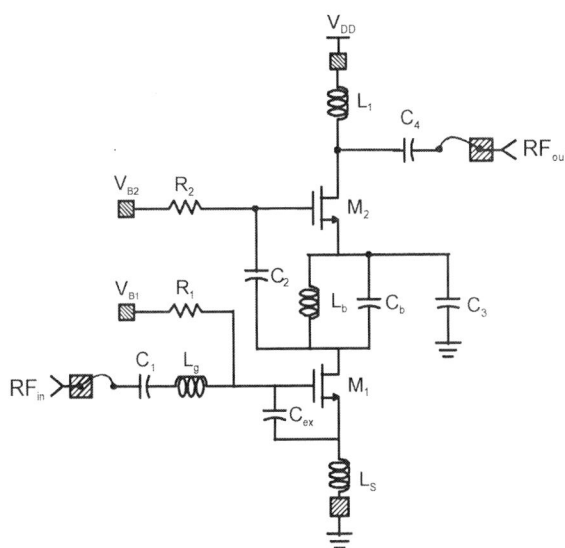

Fig. 2 Schematic of proposed LNA

The C_b shunted across L_b has another advantage of increasing of the effective inductance of L_b such that the required actual inductance of L_b can be reduced and, hence, its layout size is reduced. A practical spiral inductor L_b in CMOS process can be modelled as an ideal inductor in series with its parasitic resistance R_p. When shunted by a capacitor C_b, the equivalent shunt impedance Z_{in} is written as[6]:

$$Z_{in} = R_{ps} + j\omega L_{bs} \tag{1}$$

$$R_{ps} \sim \frac{R_p}{(1-\omega^2 L_b C_b)^2} = \frac{R_p}{\left(1-\left(\dfrac{\omega}{\omega_0}\right)^2\right)^2} \tag{2}$$

$$L_{bs} \approx \frac{L_b}{1-\omega^2 L_b C_b} = \frac{L_b}{1-\left(\dfrac{\omega}{\omega_0}\right)^2} \tag{3}$$

where $\omega_0 = 1/\sqrt{L_b C_b}$ is the resonant frequency of the L_b-C_b shunt network. According to the layout scaling rule, the size reduction factor of the inductor is given as:

$$R.F. \equiv \left(\frac{sqL_{bs}[mm^2] - sqL_b[mm^2]}{sqL_b[mm^2]}\right)\cdot 100(\%) \tag{4}$$

where sqL_{bs} and sqL_b are the practical layout size of CMOS spiral inductor. The effective inductance and the size reduction factor are presented in Fig. 3, where a spiral inductor of 3-nH shunted with a metal-insulator-metal (MIM) capacitor of 0.3-pF. The inductance has a 44% increase and the size reduction factor is 65% at 3.0-GHz.

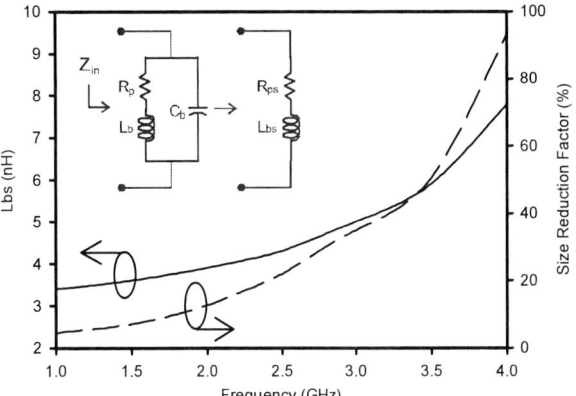

Fig. 3 Shunt-Resonance equivalent inductor and simulation result.

B. Simultaneous Noise and Input Matching Network

A capacitance C_{ex} is connected across the gate and source nodes of M_1 for easier noise and gain simultaneous match than the conventional source inductively-degenerated LNA [7]-[8]. From Fig. 2, the input impedance of the LNA with inductive source degeneration L_S can be written as:

$$Z_{in} = \frac{1}{j\omega C_1} + j\omega\left(L_S + L_g\right) + \frac{1}{j\omega C_t} + \frac{g_{m1}L_S}{C_t} \tag{5}$$

where C_1 is the DC-blocking capacitor, g_{m1} is the trans-conductance of M_1, and $C_t=C_{gs1}+C_{ex}$. Therefore, the minimum noise figure [8] is obtained as:

$$F_{min} = 1 + \frac{2}{\sqrt{5}}\frac{\omega}{\omega_{T1}}\sqrt{\gamma\delta\left(1-|c|^2\right)} \tag{6}$$

where $\gamma(\approx 0.67-1.33)$ and $\delta(\approx 1.33-4)$ are excess noise parameters of CMOS device, $c(\approx j0.4)$ is the correlation coefficient between the gate noise and drain noise. The transistor width is determined based on the minimum noise figure and then the reactive elements (L_s, L_g, and C_{ex}) are designed for input and output impedance match.

C. CMOS Implementation

By using 0.18-μm CMOS technology, the width of M_1 is determined as 90 μm and M_2 is selected as 48 μm, smaller than M_1 to reduce parasitic capacitances. The inductors L_s and L_g are 1 nH and 4.6 nH, respectively. The inter-stage shunt-resonating inductor L_b is 3.2 nH and the capacitor C_b is 0.24 pF. The output matching network inductor L_1 of 1.5 nH. All inductors are realized with the on-chip spiral inductor structure. All capacitors are implemented with the MIM-capacitor structure, where C_1= 2.1 pF, C_{ex}= 46 fF, C_2= 3.2 pF,

978-2-8748-7007-1/08 $25.00 © 2008 EuMA   351

$C_3 = 0.6$ pF, and $C_4 = 0.25$ pF. The resistances are $R_1 = R_2 = 5.7$ kΩ, which are realized with the high-resistance implant (HRI) poly-silicon layer.

III. MEASUREMENT RESULTS

The implemented CMOS LNA was measured in a chip-on-board manner with the vector network analyzer. The microphotograph of the chip is shown in Fig. 4, where the chip area including and excluding I/O pads are only 0.37 and 0.28 mm^2, respectively. The measured S-parameters are shown in Fig. 5. The maximal power gain is 15.2 dB at 3.0 GHz with a 3-dB bandwidth spans from 2.8 to 3.3 GHz. The reverse isolation is better than 22 dB. The input return loss is 12 dB and output return loss is 11 dB at 3.0 GHz. The measured noise figure reaches a minimum of 3.01 dB at 3.08 GHz, as depicted in Fig. 6.

The measured P_{1dB} of the LNA is -19.4 dBm from the measured relationship of output and input power levels. The input-referred third-order compression point (IIP$_3$) was measured by the two-tone excitation method, where two RF signals centered at 3.0 GHz with 10 MHz spacing are fed at input. The output spectrum is illustrated at the inlet of Fig. 8, which gives the IIP$_3$ as -10.8 dBm. The DC power consumption is 1.81 mW from the 1.0-V supply.

The performance is summarized in Table I, where published current-reused CMOS LNAs are also listed. Since the CMOS LNAs design are traded off among numerous, even conflicting, specifications. To compare their performances on the same basis, a modified definition of figure of merit from those in [9]-[11] is used. This FOM includes the chip size, where it is not counted in [9]-[11]:

$$\text{FOM} = \frac{Gain[dB] \cdot IIP_3[mW]}{P_{DC}[mW] \cdot \left(NF[dB] - 1 \right)} \cdot \frac{Freq.[GHz]}{Chip\,Size[mm^2]} . \quad (7)$$

As indicated from Table I, our proposed LNA has the highest FOM of 2.77.

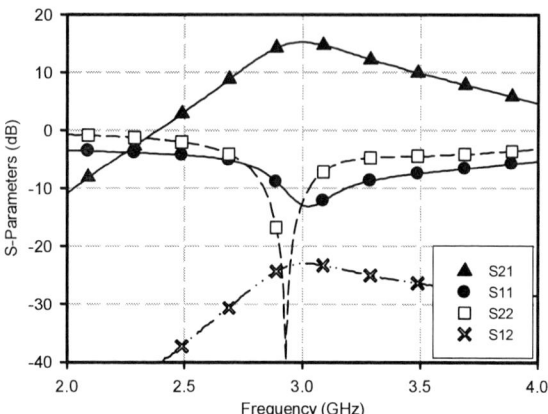

Fig. 5 Measured S-Parameters.

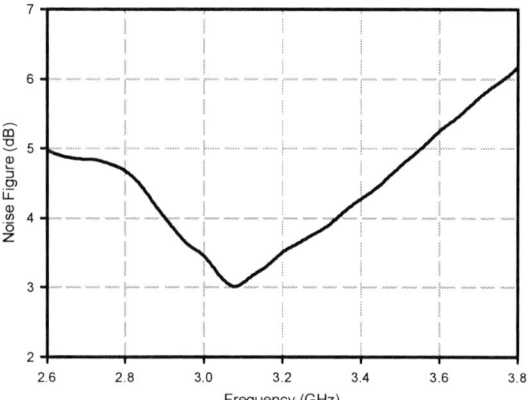

Fig. 6 Measured noise figure

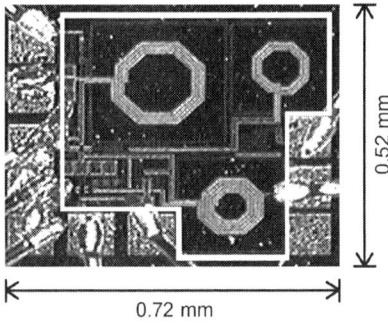

Fig. 4 Microphotograph.

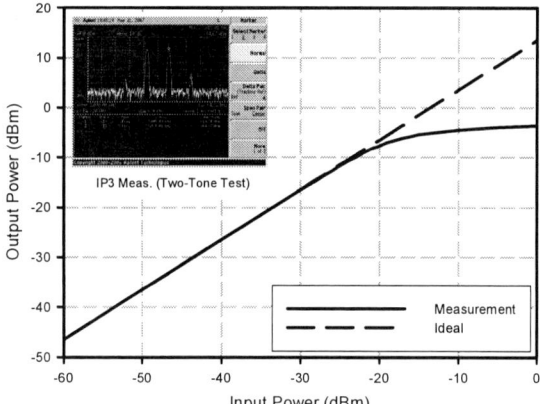

Fig. 7 Linearity – P_{1dB} and IP$_3$

TABLE I
COMPARISON OF LOW-POWER LNAs PERFORMANCE

Ref.	Tech. (μm)	S_{11} (dB)	S_{21} (dB)	Freq. (GHz)	Iso. (dB)	NF (dB)	IIP_3 (dBm)	P_{1dB} (dBm)	P_{DC} (mW)	V_{DD} (V)	Chip size including I/O pads (mm^2)	FOM
[5]	CMOS 0.18	-10	21	5.4	-	3	-23	-	2.7	1.8	0.43	0.24
[13]	CMOS 0.13	-13	13	2.4	> 20	3.6	-	-	6.5	1.2	0.72	-
[12]	CMOS 0.18	-12	9.2	4.5-5.5	-	4.5	-16	-27	0.9	0.6	0.95	0.38
[10]	CMOS 0.13	-13	9.1	3.0	> 33	4.7	-11	-25.0	0.4	0.6	2	0.72
[4]	CMOS 0.18	-15	12.5	5.7	-	3.7	-0.45	-11	14.4	1.8	1.4	2.37
[2]	CMOS 0.18	-11	16.4	5.7	-	3.5	-	-	3.2	1.0	0.58	-
This Work	CMOS 0.18	-11.1	15.2	2.8-3.3	> 22	3.01	-10.8	-19.4	1.81	1.0	**0.37**	**2.77**

IV. CONCLUSIONS

In this paper, we present a low-power ultra-compact CMOS LNA with a shunt-resonating current-reused topology, where the shunt resonating inter-stage network is proposed to connect two common-source MOSFET transistors. This inter-stage network, providing dual functions of impedance match and DC current sharing, takes much smaller chip size than other topologies. The implemented 0.18-μm CMOS LNA achieves 15.2-dB power gain and 3.01-dB noise figure, while only consuming 1.81 mW from 1.0 voltage supply. The chip size, excluding and including the I/O pads, is only 0.28 and 0.37 mm^2, respectively. The proposed LNA achieves the highest FOM, compared with recently published current-reused LNAs.

ACKNOWLEDGMENT

The authors would like to thank National Science Council (NSC) and Chip Implementation Center (CIC) for support of the work and chip implementation.

REFERENCES

[1] B. Razavi, RF Microelectronics, 1st ed. Upper Saddle River, NJ: Prentice-Hall, 1997.

[2] Y. S. Wang and L. H. Lu, "5.7 GHz low-power variable-gain LNA in 0.18 μm CMOS," Electronics Lett., vol. 41, pp. 66-68, 2005.

[3] C.-Y. Cha and S.-G. Lee, "A 5.2-GHz LNA in 0.35-μm CMOS utilizing inter-stage series resonance and optimizing the substrate resistance," IEEE J. of Solid-State Circuits, vol. 38, pp. 669-672, Apr. 2003.

[4] C.-H. Liao and H.-R. Chuang, "A 5.7-GHz 0.18-μm CMOS gain-controlled differential LNA with current reuse for WLAN receiver," IEEE Microw. Wireless Compon. Lett., vol. 13, pp. 526-528, Dec. 2003.

[5] J. S. Walling, S. Shekhar, and D. J. Allstot, "A g_m-Boosted Current-Reuse LNA in 0.18μm CMOS," in IEEE Radio Freq. Integr. Circuits Symp., 3-5 June, 2007, pp. 613-616.

[6] M. Shouxian, M. Jian-Guo, Y. K. Seng, and D. M. Anh, "A modified architecture used for input matching in CMOS low-noise amplifiers," IEEE Trans. on Circuits and Systems II: Express Briefs, vol. 52, pp. 784-788, Nov. 2005.

[7] P. Andreani and H. Sjoland, "Noise optimization of an inductively degenerated CMOS low noise amplifier," IEEE Trans. on Circuits and Systems II: Express Briefs, vol. 48, pp. 835-841, Sep. 2001.

[8] T.-K. Nguyen, C.-H. Kim, G.-J. Ihm, M.-S. Yang, and S.-G. Lee, "CMOS low-noise amplifier design optimization techniques," IEEE Trans. on Microw. Theory and Tech., vol. 52, pp. 1433-1442, May 2004.

[9] S. Asgaran, M. J. Deen, and C.-H. Chen, "A 4-mW monolithic CMOS LNA at 5.7GHz with the gate resistance used for input matching," IEEE Microw. Wireless Compon. Lett., vol. 16, pp. 188-190, Apr. 2006.

[10] H. Lee and S. Mohammadi, "A 3GHz subthreshold CMOS low noise amplifier," in IEEE Radio Freq. Integr. Circuits Symp., 11-13 June 2006.

[11] A. Amer, E. Hegazi, and H. Ragai, "A Low-Power Wideband CMOS LNA for WiMAX," IEEE Trans. on Circuits and Systems II: Express Briefs, vol. 54, pp. 4-8, Jan. 2007.

[12] H. H. Hsieh and L. H. Lu, "Design of Ultra-Low-Voltage RF Frontends With Complementary Current-Reused Architectures," IEEE Trans. on Microw. Theory and Tech., vol. 55, pp. 1445-1458, July 2007.

[13] M. E. Kaamouchi, M. S. Moussa, P. Delatte, G. Wybo, A. Bens, J.-P. Raskin, and D. Vanhoenacker-Janvier, "A 2.4-GHz Fully Integrated ESD-Protected Low-Noise Amplifier in 130-nm PD SOI CMOS Technology," IEEE Trans. on Microw. Theory and Tech., vol. 55, pp. 2822-2831, Dec. 2007.

Proceedings of the 3rd European Microwave Integrated Circuits Conference

Efficient Frequency Domain plus Spatial Expansion Method For Semiconductor Devices Modeling

Giorgio Leuzzi[1], Vincenzo Stornelli[2]

Dept. of Information and Electrical Engineering
Monteluco di Roio, 67100, L'Aquila, Italy.

[1]leuzzi@ing.univaq.it; [2]stornelli@ing.univaq.it

Abstract— **We here present a method that combines frequency-domain Fourier series expansion and space-domain polynomial expansion of the physical quantities inside the semiconductor, for an efficient numerical modelling of high-frequency active devices, based on the solution of the physical transport equations in the semiconductor. The frequency- and space-domain expansions drastically reduce the number of time and space sampling points where the equations are computed, greatly reducing the computational burden with respect to classical finite-differences approaches. Moreover, the frequency-domain technique eliminates the need for time-to-frequency transforms for a spectral solution, and allows easy inclusion of frequency-dependent parameters of the semiconductor especially important at very high frequencies (e.g. dielectric constant). Also the coupling with a EM program, for a global modeling simulator, becomes straightforward, due to the reduced interconnection nodes with the physical simulator. A demonstrator for PC implementing a quasi-2D model with a hydrodynamic formulation with the first three moments of Boltzmann's Transport Equation is given, and its results are compared with a standard finite-difference time-domain approach and measured results.**

I. INTRODUCTION

High accuracy and fast semiconductor device modelling is becoming increasingly important for the understanding of basic properties of advanced devices, for the prediction of their performances, and for the optimisation of their structure, with reduced fabrication effort. Moreover, as the operating frequency is pushed to ever higher frequencies, the layout-dependent coupling and propagation effects within the device must be taken into account by the model [1]. This can be obtained by coupling a numerical full-wave electromagnetic analysis of the passive embedding structure of the device to a numerical physics-based analysis of the transport phenomena in the active part of the device. Both electromagnetic and transport phenomena can be accurately modelled by means of partial differential equations in time and space domain, that is, Maxwell's and Boltzmann's equations respectively. A common numerical approach for their solution is the reduction to finite-difference equations, by means of discretisation of the time and space variables *t* and *(x,y,z)* [2]. This approach, while providing a very simple implementation scheme, has some drawbacks.

First of all, the time step in the discretised transport equations must be smaller than the relaxation time constants

in the equations themselves, in order for the solution algorithm to converge. Similarly, the space step in the discretised transport equations must be smaller than the Debye length in the semiconductor, that is much smaller than the typical distance over which the physical quantities in an actual device vary. Therefore, the number of time and space steps is unnecessarily large, making the algorithm inefficient. Secondly, a finite differences approach in the time domain leads to a transient solution from an arbitrary initial state. Thirdly, a time-domain solution of the transport equations in the active semiconductor region implies a time-domain solution also for the equations of the electromagnetic field in the large passive embedding structure external to the active semiconductor region. In order to overcome the drawbacks of the finite-difference discretisation in the time domain, frequency-domain approaches have been proposed; in particular, Harmonic Balance [3] and Spectral Balance [4] algorithms have been demonstrated. Additional advantages of the frequency-domain scheme include a more efficient handling of a two-tone excitation of the device, and of frequency-dispersive behaviour of the material. A disadvantage of this approach is the limitation to periodic excitations, a limitation usually not important for typical applications. A more serious disadvantage is the need to solve the problem simultaneously for all time instants, instead of the step-by-step procedure of the time-domain finite-difference approach. The consequent larger size of the problem partly reduces the benefit of the longer time step, while not offsetting the mentioned advantages. In this paper, also the space-domain finite-difference discretisation scheme is removed, by expanding the physical quantities in polynomials in the space variable; the unknowns are now the coefficients of the polynomials. The equations are written in a number of space points equal to the degree of the polynomials (plus one), i.e. in a very limited number of points. This approach reduces the size of the problem within the semiconductor; moreover, it reduces the number of space points where the transport equations must be connected to the passive embedding structure. For the verification of the feasibility of this approach, the DC and microwave characteristics of a high-frequency FET in both linear and nonlinear regime are computed, and compared with the results of a standard formulation and with measurements. The results are satisfactory.

978-2-8748-7007-1/08 $25.00 © 2008 EuMA

II. QUASI-TWO-DIMENSIONAL MODEL

The region between source and drain (see Fig. 1) is divided into three subregions, i.e. the part of the channel directly located under the gate metallisation (L_G), and the two regions between gate and source L_{SG} and between gate and drain L_{DG}. Carrier dynamics in all regions are described by the first three moments of Boltzmann's Transport Equation [5]; it is also assumed that the collision terms can be approximated by a relaxation time model [5]. Therefore, the equations in each region are the particle, momentum and energy conservation equations, Eq.(1), coupled to Poisson's equation, Eq.(2):

$$
\begin{cases}
\dfrac{\partial n_{L_i}(x,t)}{\partial t} + \dfrac{\partial \left(n_{L_i}(x,t) \cdot v_{L_i}(x,t) \right)}{\partial x} = 0 \\[2ex]
\dfrac{\partial v_{L_i}(x,t)}{\partial t} + v_{L_i}(x,t) \cdot \dfrac{\partial v_{L_i}(x,t)}{\partial x} = \dfrac{q\xi_{L_i}(x,t)}{m^*} + \\[2ex]
\quad - \dfrac{2}{3 n_{L_i}(x,t)m^*} \dfrac{\partial \left(n_{L_i}(x,t)w_{L_i}(x,t) \right)}{\partial x} - \dfrac{v_{L_i}(x,t)}{\tau_v} \\[2ex]
\dfrac{\partial w_{L_i}(x,t)}{\partial t} + v_{L_i}(x,t) \cdot \dfrac{\partial w_{L_i}(x,t)}{\partial x} = q v_{L_i}(x,t) \cdot \xi_{L_i}(x,t) + \\[2ex]
\quad - \dfrac{2}{3n} \dfrac{\partial \left(n_{L_i}(x,t)v_{L_i}(x,t)\left(w_{L_i}(x,t) \right) \right)}{\partial x} - \dfrac{\left(w_{L_i}(x,t) - w_0 \right)}{\tau_w}
\end{cases}
\tag{1}
$$

$$
\frac{\partial \xi_{L_i}(x,t)}{\partial x} = \frac{\left(n_{L_i}(x,t) - N_{cL_i}(x,t) \right)}{\varepsilon_o(\omega)}
\tag{2}
$$

where $n_{L_i}(x,t)$ is the electron density, $v_{L_i}(x,t)$ is the electron velocity, $\xi_{L_i}(x,t)$ is the longitudinal electric field, and $w_{L_i}(x,t)$ is the electron energy; the relaxation times τ_v and τ_w as well as the equilibrium energy w_0 are obtained by steady-state Monte Carlo analysis. $N_{cL_i}(x,t)$ is the equilibrium carrier density in the channel as provided by the channel charge-control model, as explained below. Eq.(1) and Eq.(2) are written in each of the three regions, and the solutions are matched at the boundaries for consistency. The excitation voltage appears as the boundary condition of the Poisson's equation at the source and drain terminals. The charge-control in the region under the gate metallization is obtained by solving the Poisson's equations (3-4) in the vertical direction z, together with Eq.(5) that computes the charge density from the wave function:

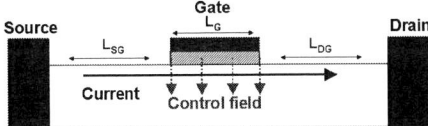

Fig. 1. Q2D formulation scheme

$$
-\frac{\hbar^2}{2m_0} \frac{d}{dz}\left[\frac{1}{m_z^*} \frac{d\psi_i(z)}{dz} \right] + qV(z)\psi_i(z) = E_i \psi_i(z)
\tag{3}
$$

$$
\frac{d}{dz}\left[\varepsilon(z) \frac{dV(z)}{dz} \right] = q\left[N_D^+(z) - n_g(z) \right]
\tag{4}
$$

$$
n_g(z) = \sum_i n_i(z)\left| \psi(z) \right|^2
\tag{5}
$$

where m_z^* is the effective mass; E_i and $\psi_i(z)$ are the energy level and the wave function of the i-th subband respectively (i-th Eigenvalue and Eigenfunction of Schrödinger's equation); $V(z)$ is the electrostatic potential in the semiconductor; $\varepsilon(z)$ is the position-dependent dielectric constant, and $N_D^+(z)$ the ionized donor density; $n_i(z)$ is a 3D density of states computed by means of a numerical approximation of the Fermi-Dirac statistic. Dopant ionization is also taken into account in the model. The charge-control in the two lateral regions between gate and source, and between gate and drain, is computed from the channel-to-surface voltage, obtained from the electromagnetic solver in the embedding region. As a simpler alternative, it can be obtained by a linear interpolation between the gate and source and gate and drain voltages. The system is solved by a classical finite-difference time-domain method [6], and the results are stored in a LUT, then fitted by a Lagrange polynomial giving the desired $N_{cL_G}(x,t)$ dependence on control voltage.

III. THE SPECTRAL BALANCE AND SPATIAL POLYNIMIALS

For a global analysis, the device is first divided into an embedding, passive linear region (substrate, passivation dielectric, access pads, etc.), and an active, non linear region, in our case the channel (this statement will be clarified fin figure 4b where the active device channel is depicted in filled grey slices). The two regions are connected at a suitable number of points, e.g. at the gate, drain and source metallisations, at the surface between gate and source and gate and drain, etc.; the embedding region is also connected to the external excitation, i.e. the input signal, and to the external load (Fig.2). The embedding region is analysed by means of a numerical electromagnetic field solver in the frequency domain, at DC, fundamental frequency and harmonic frequencies of the input signal. Any standard algorithm or commercial CAD tool can be used for the analysis. As a result, the embedding region is lumped into an equivalent representation; for instance, a Thévenin equivalent, formed by a complex N-port admittance matrix and N equivalent voltage sources, where N is the number of connection points between the embedding and the active regions. A Thévenin equivalent is computed for each frequency of analysis. The physical quantities of interest in the active region, i.e. the electron density, velocity and energy, and the electric field, are expanded in polynomials with respect to the space variable:

978-2-8748-7007-1/08 $25.00 © 2008 EuMA

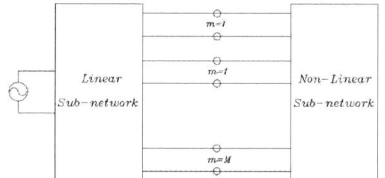

Fig.2. The partition between passive and active regions for global analysis

$$\begin{cases} n(x,t) = \sum_{m=0}^{M} a_m(t) \cdot x^m \\ v(x,t) = \sum_{m=0}^{M} b_m(t) \cdot x^m \end{cases} \qquad (6)$$

The choice of polynomials is justified by the actual behaviour of the physical quantities in the active region; however, it could be replaced by other basis functions, with some limitations, as will be clear from the following discussion.

The coefficients of the polynomials are dependent on time, because the space distributions of the physical quantities are driven by the time-dependent input signal; if the latter is time periodic, the time dependence of the coefficients is periodic as well, and can be written as a Fourier series:

$$\begin{cases} a_m(t) = \sum_{k=-K}^{K} A_m^k \cdot e^{jk\omega t} \\ b_m(t) = \sum_{k=-K}^{K} B_m^k \cdot e^{jk\omega t} \end{cases} \qquad (7)$$

By replacing (7) in (6) we get:

$$\begin{cases} n(x,t) = \sum_m \sum_k A_m^k \cdot e^{jk\omega t} \cdot x^m \\ v(x,t) = \sum_m \sum_k B_m^k \cdot e^{jk\omega t} \cdot x^m \end{cases} \qquad (8)$$

These expressions, and the alike for electron energy and electric field, must now be replaced in the equations (1) and (2). Let us consider for simplicity only the continuity equation, that we repeat here:

$$\frac{\partial n(x,t)}{\partial t} + \frac{\partial \left[n(x,t) \cdot v(x,t) \right]}{\partial x} = 0 \qquad (9)$$

First, we see that the derivative with respect to time of the electron density is immediately obtained by simply deriving the series in (8) term by term. Then, we consider the product between electron density and velocity; and similarly all the products present in (1) and (2). Its calculation is straightforward, when considering that the product of two Fourier series (in our case, with respect to the time variable) can be explicitly expressed again as a Fourier series:

$$x(t) \cdot y(t) = \sum_{k_x} X_{k_x} \cdot e^{jk_x \omega_0 t} \cdot \sum_{k_y} Y_{k_y} \cdot e^{jk_y \omega_0 t} =$$
$$= \sum_{k_z} Z_{k_z} \cdot e^{jk_z \omega_0 t} = z(t) \qquad (10)$$

where the coefficients of the product series are given by the convolution of the coefficients of the two factor series:

$$Z_{k_z} = \sum_{k_x} X_{k_x} \cdot Y_{k_z - k_x} \qquad (11)$$

Similarly, the product of two polynomials (in our case, with respect to the space variable) is again a polynomial:

$$p(x) \cdot q(x) = \sum_{h_p} P_{h_p} \cdot x^{h_p} \cdot \sum_{h_q} Q_{h_q} \cdot x^{h_q} =$$
$$= \sum_{h_r} R_{h_r} \cdot x^{h_r} = r(x) \qquad (12)$$

where the coefficients are expressed as:

$$R_{h_r} = \sum_{h_p} P_{h_p} \cdot Q_{h_r - h_q} \qquad (13)$$

Combining the two convolutions (10) and (12), products of series as in (8) are easily expressed again as series of the same type; their coefficients are computed by means of convolutions as (11) and (13). As a consequence, the second term in equation (9) is the derivative with respect to space of a series in the form of (8), and therefore it is immediately obtained. Eventually, by simple algebraic manipulations, the continuity equation (9) is written in the form:

$$\sum_m \sum_k C_m^k \cdot e^{jk\omega t} \cdot x^m = 0 \qquad (14)$$

and similarly all other equations in (1) and (2). For each equation, the unknowns are the $(M+1) \cdot (2K+1)$ real numbers in the coefficients, where M is the order of the polynomes in the space variable, and K is the order of the highest harmonic in the Fourier series. Therefore, as stated above, we have to write each equation in $(M+1)$ points in space, and in $(2K+1)$ points in time, in order to have a sufficient number of equations. This is a huge reduction with respect to a standard finite-difference approach. We remark that the expansion in Fourier series has already been implemented [4], but the multiplication between Fourier series has been dealt with by a different approach. The series are evaluated in a number of sampling points, then multiplied in the time domain at each of the sampling points; the products are Fourier-transformed, and the coefficients of the product series are therefore obtained. This approach is very convenient when nonlinear functions of the unknowns are present in the equations, i.e. when an unknown is the argument of an exponential function; this happens for example in the analysis of nonlinear circuits, and the unknowns are voltages or currents. However, when only multiplications are present, as in the transport equations (1), the approach proposed in this work is more convenient. As a disadvantage of the present method, it must be mentioned that the nonlinear system is composed by the equations (1) and (2), each in a form of a series as in (14), evaluated at $(M+1)$ points in space and in $(2K+1)$ points in time. The system is nonlinear, and must be solved by an iterative procedure. When the Newton-

978-2-8748-7007-1/08 $25.00 © 2008 EuMA   356

Raphson method is used, the Jacobian matrix is full, while with a finite-difference approach it is banded, and therefore less burdensome to invert. However, the large reduction of the sampling point normally offsets the disadvantage.

IV. RESULTS

The described formulation has been implemented in a computer code for a PC and coupled to a commercial EM simulator; a standard finite-difference time-domain formulation and a standard Harmonic Balance formulation have also been implemented, for verification and comparison. A pseudomorphic HEMT for high frequencies from Selex S.I has been analyzed. The HEMT layer structure and its layout are shown in Fig.3. S-parameters have been computed, and compared to measured ones (Fig.4). The measurements have been performed with a VNA with on-wafer probes and on-chip calibration. Moreover in Fig. 5, an example results of the electron velocity inside the active channel, calculated with the proposed method is shown, with an applied $V_{ds} = 3V$ and

$V_{gs} = -0.7 + 0.1 \, sin(\omega t)$ and compared with a standard time–domain solution. As apparent, the frequency domain simulator gives directly the steady state results skipping the transitory. The frequency domain plus spatial expansion result in more than five time shorter simulation time, on a 1.6GHz clock and 512 MB RAM PC.

V. CONCLUSION

In this work, an approach combining space-domain polynomial expansion and time-domain Fourier series expansion for the unknown physical variables in an active device has been described. An algorithm for a faster solution of the transport equations together with Poisson's equation has been described and implemented, and results have been found to be in complete accordance with standard finite-differences formulations, with great reduction in computing time.

REFERENCES

[1] Y. A. Hussein, El-Ghazaly, S. M. Goodnick, "An Efficient Electromagnetic-Physics-Based Numerical Technique for Modeling and Optimization of High-Frequency Multifinger Transistors," *IEEE Trans. Microwave Theory Tech.*, Vol. 51, No. 12, pp. 2334–2346, December 2003.

[2] S. Selberherr, *Analysis and Simulation of Semiconductor Devices*, Springer-Verlag, Wien, 1984.

[3] R.W.Dutton, B.Troyanovski, et al., "Device Simulation for RF Applications," in Proc. International Electron Devices Meeting, pp. 301-304, Dec. 1997.

[4] G. Leuzzi, V. Stornelli, "Global Modeling Analysis of HEMTs by the Spectral Balance Technique", *IEEE Trans. on Microwave Theory and Technique*, Vol. 55, No. 6, June 2007.

[5] P. Sandborn, A. Rao, P. Blakey, "An assesment of approximate nonstationary charge transport models used for GaAs device modeling," *IEEE Trans. Microwave Theory Tech.*, Vol. 36 No.7, pp. 817-829, July 1989.

[6] I.-H. Tan, G.L. Snider, and E.L. Hu, "A self-consistent solution of Schrödinger-Poisson equations using a nonuniform mesh", *J. of Applied Physics*, Vol. 68, pp. 4071,15 October 1990.

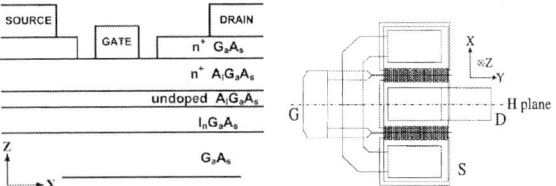

Fig. 3. The pHEMT cross-section and layout structure.

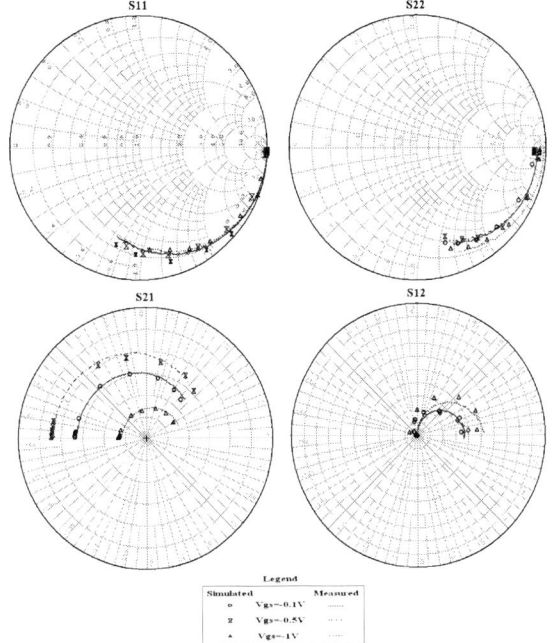

Fig. 4. Comparison between simulated and measured S-parameters for the considered pHEMT for V$_{DS}$=3V and three different gate bias voltage. The frequency range is 4-39 GHz in 7-GHz steps.

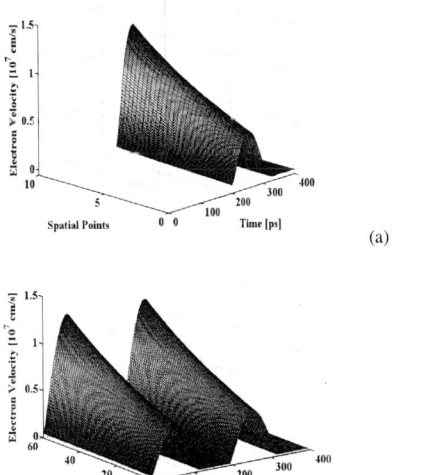

Fig. 5. Electron velocity as a function of space (from source to drain) and time, computed with the proposed algorithm (a) and with a standard timedomain algorithm (b).

978-2-8748-7007-1/08 $25.00 © 2008 EuMA

Ka-Band Wide-Bandwidth Voltage-Controlled Oscillators in InGaP-GaAs HBT Technology

Chau-Ching Chiong[#1], Hong-Yeh Chang[*], Ming-Tang Chen[#]

[#] *Academia Sinica, Institute of Astronomy and Astrophysics*

P.O. Box 23-141, Taipei 10617, Taiwan, R.O.C.

[1]ccchiong@asiaa.sinica.edu.tw

[*] *Department of Electrical Engineering, National Central University*

No. 300, Jhongda Rd., Jhongli 32001, Taiwan, R.O.C.

Abstract—Monolithic wide-bandwidth low-phase-noise voltage-controlled oscillators (VCOs) using 2-μm InGaP-GaAs heterojunction bipolar transistor (HBT) technology are presented in the paper. The tuning range of the VCOs are 28.0 to 34.0 GHz and 33.8 to 39.1 GHz, with a phase noise of -100.7 and -103.8 dBc/Hz at 1-MHz offset from the carrier. The overall dc power consumption of the differential-output VCOs is 85 mW with a supply voltage of -2.5 V. The VCOs feature wide tuning range and low phase noise at the same time, with a figure of merit (FOM) of -176 and -171 dB. These are the first Ka-band VCOs with wide tuning bandwidth using commercial GaAs HBT process.

I. INTRODUCTION

In modern radio astronomical receiving system, due to its requirement of ultra low noise performance and wide tuning range, power source has been long time dominated by the yttrium-iron-garnet (YIG) and the Gunn oscillators. However, with high power consumption (YIG) and mechanical tuning system (Gunn), they are not suitable for next generation large-array, multi-pixel applications. To solve this problem, an integrated, monolithic microwave integrated circuit (MMIC) voltage controlled oscillator (VCO) is potentially an attractive choice in these applications for its low-price, light-weightiness, easily-controlled and low power-consumption.

With the advance of MMIC technologies, the noise performance of MMIC VCOs is now approaching those of the YIG and the Gunn oscillators up to 6 GHz [1]. For higher frequency regime, MMIC VCOs with low phase noise and wide tuning range are relatively rare [2]. In Ka-band, some MMIC VCOs with low phase noise have been presented using GaAs-, InP- and SiGe-based heterojunction bipolar transistor (HBT) technologies [3]-[8] thanks to HBT's low flicker-noise ($1/f$), but their tuning range is quite narrow. VCOs with wide tuning range can be seen in the literatures. However, the power consumption in these circuits are high [9], [10].

In our previous attempt [11], we have successfully demonstrated wide-bandwidth GaAs HBT VCOs based on a single-ended double-tuned configuration working in Ku- and K-band. In this paper, two differential-output MMIC VCOs with double-tuned topology is designed for Ka-band operation by using a commercial 2-μm GaAs HBT technology provided by the WIN semiconductors Corp. The VCOs feature wide tuning range and low noise performance with low power

consumption (<100 mW), and thus are promising power source in Ka-band.

II. CIRCUIT TOPOLOGY AND DESIGN METHODOLOGY

The basic building block of the proposed double-tuned VCO is shown in Fig. 1. Common-base configuration is chosen for wide-band operation. Negative resistance is provided by feedback path via base inductor, L_b. The design is a "double-tuned" oscillator, in which there are two varactors in the circuit to obtain wide tuning bandwidth and better linearity in the curve of oscillating frequency versus tuning voltage [12], [13].

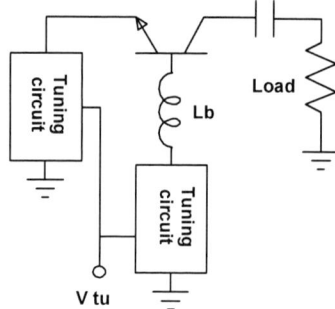

Fig. 1 Configuration of the VCO with CB and double-tuned topology

The schematic of the differential double-tuned VCO can be seen in Fig. 2. The emitter size for the core bipolar junction transistors (BJTs), Q1 and Q2, are 1×10 μm^2. It has a maximum unit current gain frequency (f_T) of 40 GHz, and a maximum unit power gain frequency (f_{max}) of 100 GHz. Some key design considerations in this topology can also be found in [11].

Harmonic Balance in AWR's Microwave Office [14] was used to predict the oscillation frequency and output power of the VCO. LC values at resonator (L_{e1}, L_{e2}, D_e) are chosen according to [9]. L_{bb} and D_b are chosen by monitoring circuit instability at desired frequency range in simulator. As L_b decreases, the corresponding frequency range with negative resistance will be shifted to higher frequency side. With proper selection of V_{bb} with regard to the tuning voltage (V_{tu}), D_b reduces the equivalent inductance viewed from the base of core BJT as V_{tu} goes up. In this way, the double-tuned topology avoids gain saturation in single-tuned VCO as V_{tu}

978-2-8748-7007-1/08 $25.00 © 2008 EuMA

goes up, and helps the circuit to achieve wide tuning bandwidth and good tuning linearity. A wide tuning range VCO is thus realized, though typical capacitance tuning ratio of the varactor in the process is small (~ 4). At the end, LC network at collector (L_c, C) is chosen by maximizing output power in simulation.

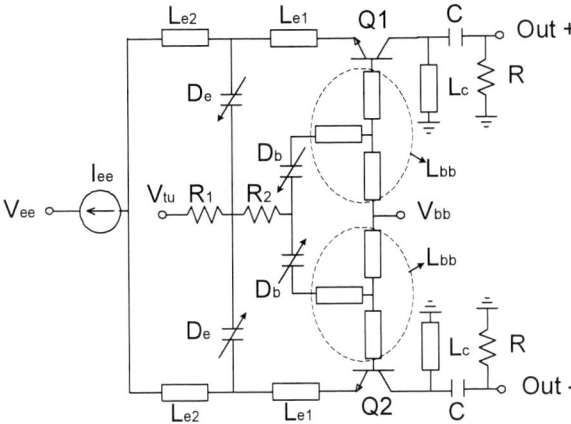

Fig. 2 Schematic of the differential, double-tuned VCO

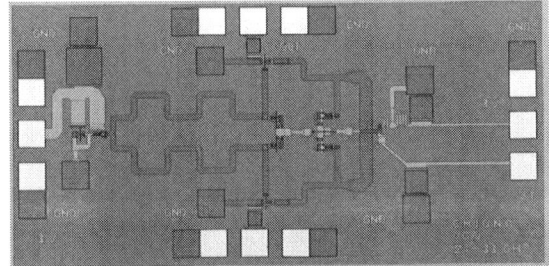

Fig. 3 Chip photo of the VCO A with a chip size of 2×1 mm^2

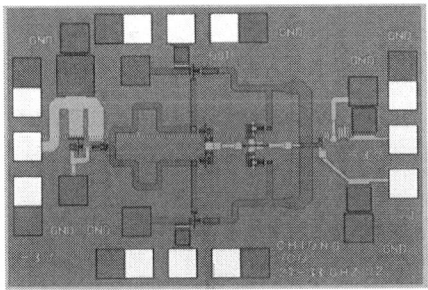

Fig. 4 Chip photo of the VCO B with a chip size of 1.5×1 mm^2

The metal-insulator-metal (MIM) capacitor, spiral inductor, and thin film resistor are available in the process. The microstrip lines are used as inductive element. The base-collector (BC) junction of HBT transistor is used as a varactor. The varactor cell D_e is formed with a 1×10 μm^2 BC junction, while D_b is formed with a 2×20 μm^2 BC junction. Two varactors are controlled by single tuning voltage for the purpose of simplifying the oscillator control. To further

suppress the noise and stabilize the oscillation of the VCO, a few in-band bypass capacitors are added in the bias circuit. All of the passive components, such as microstrip lines, dc-block and bypass capacitors, are all simulated with a full-wave EM simulator [15]. In the center of the differential circuit are the virtual ground nodes for RF signal. Differential topology thus helps to reduce the number of RF ground node and capacitor cell in the circuit. This simplifies various RF design issues (via holes etc.) at higher frequency. The chip photo of VCO A and VCO B are shown in Fig. 3 and Fig. 4 with a chip size of 2×1 mm^2 and 1.5×1 mm^2, respectively.

III. EXPERIMENTAL RESULTS

The measurements of the VCOs are performed via on-wafer probing with an Agilent E4448A spectrum analyzer. The dc power is supplied by Keithley 2400 SourceMeter. Although the design is differential, the measurement was performed singled-ended. The total dc current consumption, I_{ee}, of the VCO is 34 mA with V_{ee}/V_{bb}= -2.5/-0.55 V. The measured output frequency and output power of the two VCOs versus the tuning voltage (V_{tu}) are shown in Fig. 5, 6, 8, and 9. Only one oscillation mode was found within 0-50 GHz. The simulation results are also plotted in these figures for comparison. These figures show that Harmonic Balance simulator provides good estimate of oscillating frequency for VCO design, whereas predicted output power shows up to 10 dB difference between simulation and measurement.

The maximum output power of the VCOs is 3 and 0 dBm at the output GSG pad. A summary of the oscillating frequency and the tuning range of each VCO can be found in Table I. The tuning range of VCO A and VCO B are 19% and 15%, respectively. The spectrum of VCO A with $V_{tu} = 0$ V is shown in Fig. 7, with a phase noise of -100.7 dBc/Hz at 1-MHz offset frequency. The spectrum of VCO B with $V_{tu} = 2.5$ V can be found in Fig. 10, with a phase noise of -103.8 dBc/Hz at 1-MHz offset frequency.

The comparisons of the previously reported Ka-band VCOs and this work are summarized in Table I. The performance of a VCO can be evaluated by two figures of merit (FOM), which can be defined as in [16]

$$FOM_1 = L(\Delta f_{offset}) - 20\log\left(\frac{f_o}{\Delta f_{offset}}\right) + 10\log(P_{diss}), \quad (2)$$

$$FOM_2 = FOM_1 + 20\log\left(\frac{f_o}{\Delta f_{tune}}\right), \quad (3)$$

where $L(\Delta f_{offset})$ is the phase noise measured at offset frequency Δf_{offset}, f_0 is the oscillation frequency, P_{diss} is the dc power consumption (mW), and Δf_{tune} is the tuning range. Therefore, VCO A exhibits FOM$_1$/FOM$_2$ of -171/-157 dB, while VCO B exhibits FOM$_1$/FOM$_2$ of -176/-159 dB. A comparison of the FOM with previously reported MMIC Ka-band VCOs shows promising performance with our circuit. Our designs are the first Ka-band VCOs with wide tuning bandwidth using commercial GaAs HBT process, and the dc power consumption is only 1/2 to 1/4 of that in other Ka-band wide bandwidth VCOs in the literature [9], [10].

978-2-8748-7007-1/08 $25.00 © 2008 EuMA

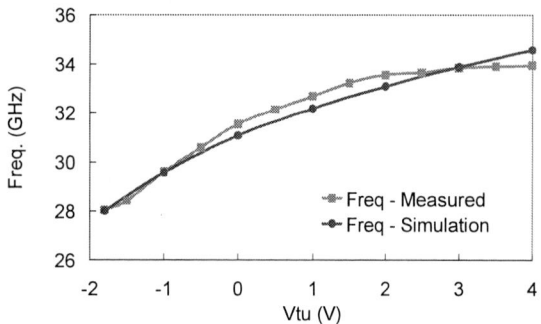

Fig 5 Measured and simulated output oscillating frequency of VCO A versus the tuning voltage, V_{tu}

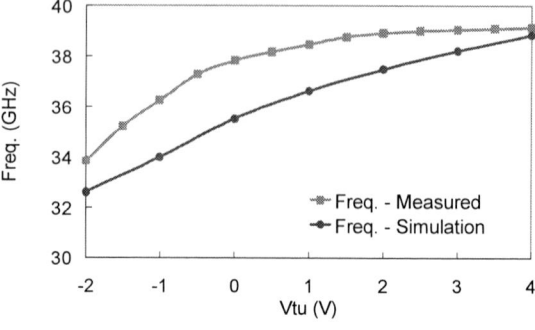

Fig 8 Measured and simulated output oscillating frequency of VCO B versus the tuning voltage, V_{tu}

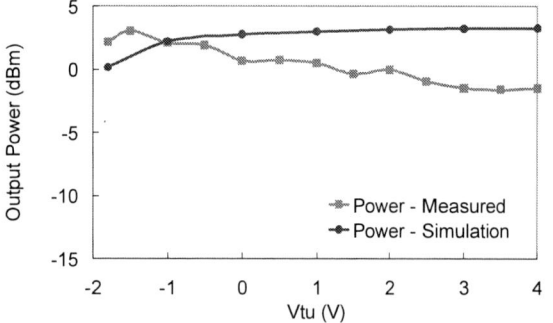

Fig. 6 Measured and simulated single-ended output power at GSG pad of VCO A versus the tuning voltage, V_{tu}

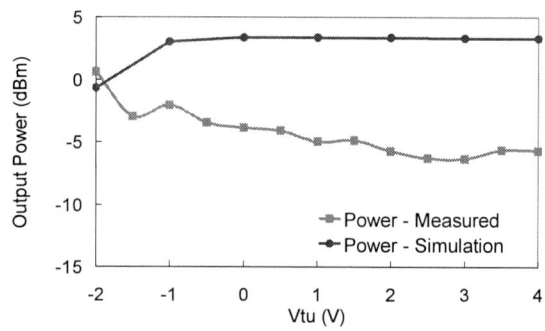

Fig. 9 Measured and simulated single-ended output power at GSG pad of VCO B versus the tuning voltage, V_{tu}

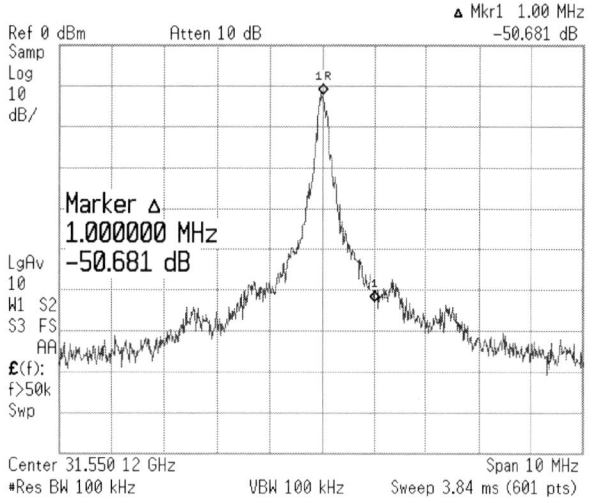

Fig. 7 Measured output spectrum of VCO A with V_{tu} = 0 V. Measured phase noise at 1-MHz offset is -100.7 dBc/Hz

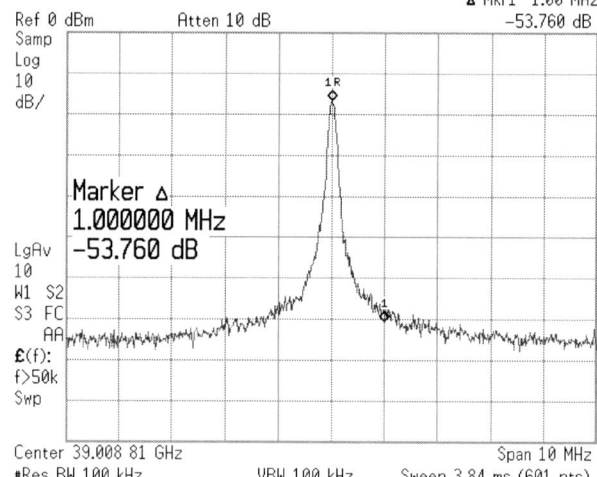

Fig. 10 Measured output spectrum of VCO B with V_{tu} = 2.5 V. Measured phase noise at 1-MHz offset is -103.8 dBc/Hz

978-2-8748-7007-1/08 $25.00 © 2008 EuMA

TABLE I
COMPARISONS OF PREVIOUSLY REPORTED KA-BAND VCOS AND THIS WORK

Ref.	Process	f_T/f_{max} (GHz/GHz)	Center Frequency (GHz)	Tuning Range (GHz)	Phase Noise @ 1MHz offset (dBc/Hz)	Output Power (dBm)	P_{diss} (mW)	FOM$_1$ / FOM$_2$
[3]	InGaP/GaAs HBT	50/70	38.5	1.1	-114	-3	125	-184/-153
[4]	2-µm InGaP/GaAs HBT	40/100	41.7	0.9	-105	-10	20	-185/-152
[5]	SiGe BiCMOS	47/-	24.7	1.4	-93	-37	22	-167/-142
[6]	1-µm GaInP/GaAs HBT	40/150	34.2	0.4	-108	4	420	-172/-134
[7]	0.12-µm SiGe HBT	50/130	32.2	1.8	-99	-14	2.64	-185/-160
[8]	InGaAs/InP HBT	100/60	26.5	1.8	-112	0	10.2	-190/-167
		100/60	33.5	2.2	-107	-1	14.3	-186/-162
[9]	0.35-µm SiGe HBT	70/75	41	11	-109	3.5	280	-177/-165
[10]	InP DHBT	140/170	43.5	13	-100 @ 10 MHz	-8.5	197	-150/-139
This work								
VCO A	2-µm InGaP/GaAs HBT	40/100	31.0	6.0	-101	3	85	-171/ -157
VCO B	2-µm InGaP/GaAs HBT	40/100	36.5	5.3	-104	0	85	-176/ -159

IV. CONCLUSIONS

The design and measurement of two wide-bandwidth differential VCOs using 2-µm GaAs HBT technology have been presented in this paper. Both VCOs demonstrate wide bandwidth performance by using a double-tuned common-base configuration. The center frequency of VCO A and B are 31.0 and 36.6 GHz, with tuning bandwidth of 19% and 15%. The VCOs feature phase noise of -101 and -104 dBc/Hz at 1-MHz offset from the carrier frequency. The maximum output power is about 3 dBm. The dc power consumption is 85 mW. A comparison of the FOM with previously reported VCOs shows promising performance with the proposed topology.

ACKNOWLEDGMENT

This work was supported in part by the ALMA-T research project in ASIAA, Taipei, Taiwan, R. O. C.

REFERENCES

[1] A. K. Poddar, U. L. Rohde, K. J. Schoepf, "Cost-Effective VCOs Replace Power-Hungry YIGs," *Microwaves and RF Journal*, Apr. 2006.

[2] M. Tsuru, K. Kawakami, K. Tajima, K. Miyamoto, M. Nakane, K. Itoh, M. Miyazaki, Y. Isota, "A Triple-Tuned Ultra-Wideband VCO," *IEEE Trans. on Microwave Theory and Techniques*, vol. 56, no. 2, pp.346-354, Feb. 2008.

[3] K. Choumei, T. Matsuzuka, S. Suzuki, S. Hamano, K. Kawakami, N. Ogawa, M. Komaru, and Y. Matsuda, "A Ka-Band Direct Oscillation HBT VCO MMIC with a Parallel Negative Resistor Circuit," in *Microwave Symposium Digest, IEEE MTT-S International*, 2005, pp. 1175-1178.

[4] C.-H. Lin, K.-H. Liang, H.-Y. Chang, Y.-J. Chan, C.-C. Chiong, and E. Bryerton, "A Q-band Low Phase Noise Voltage Controlled Oscillator Using Balanced π-Feedback in 2-µm GaAs HBT Process," in *IEEE Radio Frequency Integrated Circuits (RFIC) Symposium*, pp. 119-122, 2007.

[5] J.-H. Conan Zhan, J. S. Duster, and K. T. Kornegay, "A 25-GHz Emitter Degenerated LC VCO," *IEEE Journal of Solid-State Circuits*, vol. 39, no. 11, pp. 2062-2064, Nov. 2004.

[6] J. Hilsenbeck, F. Lenk, W. Heinrich, and J. Wuerfl, "Low Phase Noise MMIC VCOs for Ka-Band Applications with Improved GaInP/GaAs-HBT Technology," in *IEEE GaAs IC Symposium Digest*, 2003, pp. 223 -226.

[7] Y.-J. E. Chen, W.-M. L. Kuo, Z. Jin, J. Lee, Y. V. Tretiakov, J. D. Cressler, J. Laskar, and G. Freeman, "A Low-Power Ka-Band Voltage-Controlled Oscillator Implemented in 200-GHz SiGe HBT Technology," *IEEE Trans. on Microwave Theory and Techniques*, vol. 53, no. 5, pp. 1672-1681, May 2005.

[8] J. Lin, Y. K. Chen, D. A. Humphrey, R. A. Hamm, R. J. Malik, Al Tate, R. F. Kopf, and R. W. Ryan, "Ka-Band Monolithic InGaAs/InP HBT VCO's in CPW Structure," *IEEE Microwave and Guided Wave Letters*, vol. 5, no. 11, pp. 379-381, Nov. 1995.

[9] H. Li and H. M. Rein, "Millimeter-Wave VCOs with Wide Tuning Range and Low Phase Noise, Fully Integrated in a SiGe Bipolar Production Technology," *IEEE Journal of Solid-State Circuits*, vol. 38, no. 2, pp. 184-191, Feb. 2003.

[10] L. Zhang, R. Pullela, C. Winczewski, J. Chow, D. Mensa, S. Jaganathan, and R. Yu, "A 37~50 GHz InP HBT VCO IC for OC-768 Fiber Optic Communication Applications," in *IEEE Radio Frequency Integrated Circuits (RFIC) Symposium*, pp. 85-88, 2002.

[11] C. C. Chiong, H. Y. Chang, and M. T. Chen, "Wide-Bandwidth InGaP-GaAs HBT Voltage-Controlled Oscillators in K- and Ku-Band," in *Proceeding of Global Symposium on Millimeter Wave (GSMM)*, Nanjing, Apr. 2008, pp. 185-188.

[12] B. N. Scott, M. Wurtele, and B. B. Cregger, "A Family of Four Monolithic VCO MIC's Covering 2-18 GHz," in *IEEE Microwave and Millimeter-Wave Monolithic Circuits Symposium Digest*, May 1984, pp. 58-61.

[13] K. Tajima, Y. Imai, Y. Kanagawa, K. Itoh, "A 5 to 10 GHz Low Spurious Triple Tuned Type PLL Synthesizer Driven by Frequency Converted DDS Unit," in *Microwave Symposium Digest, IEEE MTT-S International*, 8-13 June 1997, vol. 3, pp. 1217-1220.

[14] Microwave Office/Analog Office User Guide, Version 7.5, Applied Wave Research Inc., June 2007.

[15] Sonnet User's Manual, Release 11, Sonnet Software Inc., North Syracuse, NY, Mar., 2007.

[16] A. Tasic, W. A. Serdijn, and J. R. Long, "Design of Multistandard Adaptive Voltage-Controlled-Oscillators," *IEEE Trans. on Microwave Theory and Techniques*, vol. 53, no. 2, pp. 556-563, Feb. 2005.

978-2-8748-7007-1/08 $25.00 © 2008 EuMA

Circuital Modelling of Shunt Capacitive RF MEMS Switches

Giancarlo Bartolucci[#1], Romolo Marcelli[*2], Simone Catoni[*3], Benno Margesin[+4],

Flavio Giacomozzi[+5], Andrea Lucibello[*6], Viviana Mulloni[+7], Paola Farinelli[°8]

[#]*Department of Electronic Engineering of the University of Roma "Tor Vergata"*
Via del Politecnico 1, 00133 Roma - ITALY
[1]bartolucci@eln.uniroma2.it

[*]*CNR-Istituto per la Microelettronica e Microsistemi*
Via del Fosso del Cavaliere 100, 00133 Roma – ITALY
[2]romolo.marcelli@cnr.it
[3]simone.catoni@isl-altran.it
[6]andrea.lucibello@imm.cnr.it

[+]*Fondazione Bruno Kessler – Centro per la Ricerca Scientifica e Tecnologica*
Via Sommarive 18, 38050 Povo (TN) – ITALY
[4]margesin@fbk.eu
[5]giaco@fbk.eu
[7]mulloni@fbk.eu

[°]*Dipartimento di Ingegneria Elettronica e dell'Informazione - Università degli Studi di Perugia*
Via G. Duranti 93, 06125 Perugia - ITALY
[8]farinelli@diei.unipg.it

Abstract— **A novel circuital model for a shunt connected RF MEMS coplanar switch, based on a fully analytical approach, is presented. The numerical simulations performed with the proposed new model are in good agreement with experimental measurements.**

I. INTRODUCTION

In many RF telecommunication systems the marriage of micromechanical techniques and electronic devices has led to the replacement of standard semiconductor switches with two-state MEMS components. The design approach is the same typically adopted for the synthesis of conventional high frequency integrated circuits, and it is based on the use of commercial CAD software packages for the circuital simulation of the whole network. For this aim, accurate and simple models are required for each MEMS device, especially if they can be considered as building blocks of more complicated configurations, like phase shifters and matrices. The aim of this paper is to present a lumped element equivalent circuit for a shunt connected coplanar micromechanical switch. An analytic approach is used to obtain the values of the inductors, resistors and capacitors. A comparison between experimental and simulated data confirms the validity of the proposed equivalent circuit.

II. LUMPED ELEMENT MODEL AND SWITCH REALIZATION

A coplanar MEMS switch in shunt connected configuration is typically composed by a moving bridge pulled down by a DC bias voltage. In the ON state the bridge is in the up position and the signal flows through the central conductor of the CPW. When the bridge is actuated it stops the signal on the main line, realizing a RF connection between the central conductor and the lateral ground planes (OFF state). To separate the DC and RF contributions, highly resistive pads laterally placed with respect to the central conductor of the CPW are used for the DC bias, even if in this position cause an increase in the threshold voltage for collapsing the bridge. When the bridge is up, the device is just a transmission line surmounted by a metal strip in correspondence with the dielectric layer deposited on the central conductor direction, for realising the needed capacitance in the DOWN state. In this case, an increase in the transmission loss is expected because of the small electrical mismatch due to this discontinuity. A technologically actuated switch has been studied for modelling the DOWN state. Such a device has been obtained by directly depositing the Cr/Au layers without using a sacrificial layer, i.e. simulating the ideal situation in which the actuation area is fully covered by the bridge when it is in the DOWN state (actuated). By using this solution we avoided, at this stage, residual air gap contributions due to: (i) the mechanical re-shaping and the non-perfect flatness of the bridge with respect to the dielectric film used for the capacitance during the actuation, and (ii) the dielectric film roughness. The first effect can be attributed to the technological tolerance due to a non-perfect alignment of the electrical pads used for the actuation. In fact, they are placed on the side of the central conductor of the CPW, to separate the DC and RF signals, and the reliability in the film deposition for both materials (usually low temperature oxide (LTO) or SiO_2 on top of polysilicon) will possibly give misalignment effects. The second one is a technological limitation related to the intrinsic resolution and reliability of the deposition techniques, which do not allow a surface roughness better than few tens of nm [1]. In order to develop an equivalent circuit for this component the same approach proposed in [2] for the series connected MEMS switch is here used for the switch in shunt configuration. More in detail, the structure can be considered composed by gaps between conductors and metallic strips with rectangular transversal sections. In Fig. 1 the schematic for the micromechanical bridge used for realising the shunt connected switch is shown.

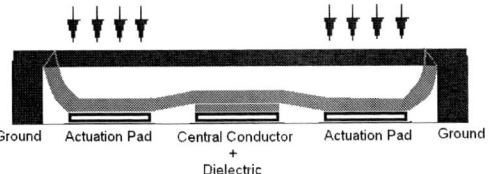

Fig. 1 Schematic of a micromechanical bridge used for realising the shunt connected switch.

The bridge at rest is diagrammed as well as the actuated bridge, putting to evidence the two states by using the black colour for the ON state and the grey one for the OFF state. Lateral pads are biased by the DC voltage in order to pull down the bridge, which in the OFF state touches the dielectric layer deposited onto the central conductor of the CPW.
In Fig. 2 the lumped element equivalent circuit model of the RF MEMS switch is presented. The RF signal is carried up to the bridge region by means of a uniform CPW input line.

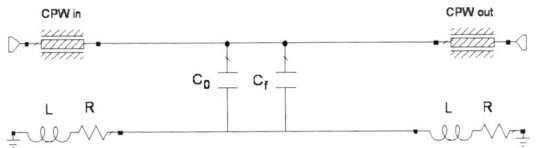

Fig. 2 The proposed lumped element equivalent circuit model for the MEMS switch.

The equivalent circuit of the bridge itself is composed by a number of capacitive, inductive and resistive contributions. In particular, the actuation area is modelled by defining a parallel plate capacitance C_0 depending on the size of the central conductor of the CPW and on the thickness of the dielectric layer. The capacitance C_f accounts for the fringing effects.
The metal strips exceeding the actuation area, going up to the CPW ground plane, have been modelled by using two RL branches, obtained following the analytic approach in [2, 3].
In particular:

$$R = \frac{l}{\sigma w t}\left\{ \frac{0.43093 x_w}{1+0.041(w/t)^{1.19}} + \frac{1.1147+1.2868 x_w}{1.2296+1.287 x_w^3} + \right.$$
$$\left. +0.0035(w/t-1)^{1.8}\right\} \tag{1}$$

$$x_w = (2 f \sigma \mu w t)^{1/2} \tag{2}$$

$$L = \frac{\mu_0 l}{2\pi}\left\{ ar\sinh\left(\frac{l}{w+t}\right) + \frac{l}{w+t} ar\sinh\left(\frac{w+t}{l}\right) + \frac{w+t}{3l} - \right.$$
$$\left. -\frac{1}{3}\left(\frac{l}{w+t}\right)^2\left[\left(1+\frac{(w+t)^2}{l^2}\right)^{3/2}-1\right]\right\} \tag{3}$$

where l is the length, w is the width and t is the thickness of the metal layer; σ is the conductivity, μ is the permeability and f is the working frequency.
The capacitor made by the central part of the shunt switch has been modelled accounting for two contributions: (i) zero order capacitance, and (ii) fringing fields.
The zero order capacitance is given by $C=\varepsilon_{eq}A/d_{tot}$, where ε_{eq} is the equivalent dielectric constant which accounts for the dielectric and, in the ON position, for the air too; A is the area of the capacitor and $d_{tot}=d+g_0$ is the distance between the two parallel plates of the capacitor. d is the thickness of the dielectric used in the shunt configuration, while g_0 is the air gap between the dielectric layer and the metal bridge when the switch is non actuated. The fringing fields, always present and related mainly to the ON condition, when the capacitance has a low value, which give a contribution in the order of 10% to 20%. The value of ε_{eq} can be derived in static conditions by assuming that the insertion of a dielectric layer between the plates of the capacitor made by the bridge and the central conductor of the coplanar waveguide corresponds to impose a charge re-distribution at the interface between the two media, thus originating a series of two capacitors. Under this assumption, the total capacitance will be:

$$C = \varepsilon_0 A \frac{\varepsilon_1 \varepsilon_2 (d+g_0)}{\varepsilon_1 d + \varepsilon_2 g_0} \frac{1}{d+g_0} = \frac{\varepsilon_{eq} A}{d+g_0} \tag{4}$$

In the present case, we can define $\varepsilon_1=1$ (Air) and $\varepsilon_2=\varepsilon$ for the dielectric, which turns out in:

$$\varepsilon_{eq} = \varepsilon_0 \frac{\varepsilon(d+g_0)}{d+\varepsilon g_0} \tag{5}$$

$g_0=2.8$ μm is the air gap for the realized structure, $d=0.3$ μm is the dielectric thickness of Si_3N_4, and $\varepsilon=7.6$. As it can be easily obtained from (5), when the bridge is in the UP position it will be $\varepsilon_{eq}\approx1$, and when the bridge is fully collapsed $\varepsilon_{eq}=\varepsilon$
The zero order capacitance has been calculated by using Eq.(4), for which $A=150\times88$ μm^2, and the fringing capacitance is obtained by means of the analytical approach proposed in [4] with the characteristic impedance Zc evaluated by using the theoretical results presented in [5].

III. MEASUREMENT RESULTS

In Fig. 3 the realised switch is shown. It is worth noting the placement of the lateral actuation pads, for separating the DC feeding from the RF path. In Fig. 4 and in Fig. 5 the experimental data of the S-parameters for the switch in the ON state and in the OFF one (technologically actuated device) are presented. The comparison between measured data and analytically obtained ones shows that the circuital approach satisfies the MEMS shunt modelling up to 40 GHz. Successively, a real device has been measured and compared with the reference one. As a result, see Fig. 6, the real shunt switch always shows a frequency of resonance for the OFF state, $F_{OFF,measured}$, higher than that of the reference switch, $F_{OFF,expected}$. The measured frequencies of resonance can be understood in terms of a residual air gap due to both, the unavoidable roughness which is intrinsic to the technological processes and to the non-perfect alignment between the dielectric layer used for the capacitance and the bridge. In other words, we can assume that the disagreement is due to the not full actuation of the switch, which can be modelled by introducing a residual air gap. On the other hand, the air gap contribution can be eliminated by using the so-called floating metal solution. In that case the dielectric used for the capacitor is metalized on the top, to obtain a metal-insulator-metal (MIM) capacitor independent of the shape of the bridge when the switch is actuated [6]. Parasitic contributions due to the real shape of the collapsed bridge are not expected within the microwave range

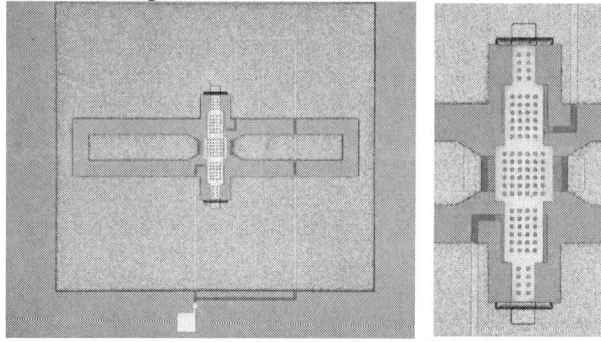

Fig. 3 Photo of the shunt connected RF MEMS switch (left), with a detailed view of the device (right).

In Fig. 8 an electromagnetic simulation performed by means. of a commercial software package (ADS, Agilent) is compared with experimental results. It has to be considered an almost exact solution of the problem, to be used as a reference, together with the experiment for the validation of the circuital modelling. It is worth noting that the EM approach is necessary for the design of the single switch, but it is not convenient for the simulation of a complete system involving several switches, unless they are so close each other to have an electromagnetic interaction. Specifically, matrices and phase shifters can take benefit from this approach. The possibility for a full circuital approach is demonstrated in the following Fig. 9, where the model based on the circuit of Fig.

2 and related formulas has been applied to the floating metal device.

IV. CONCLUSIONS

In this paper shunt capacitive micromechanical switches for RF applications have been characterized and modelled within the microwave range. An equivalent lumped circuit has been proposed to describe the electrical response in the ON and OFF states of the devices. Closed form expressions for the inductors, resistors and capacitors have been used to model the device following an analytical approach. In particular, the contribution of the residual air gap due to the not full actuation of the device has been studied, and this effect has been studied by introducing an additional contribution to the capacitor of the equivalent circuit. An alternative technological solution based on a metal layer deposited onto the dielectric used for the capacitive response, and having a floating potential, has been also considered. The equivalent circuit obtained for this structure, as well as that of the previous ones, agrees very well with the experimental findings, thus demonstrating the usefulness of this approach for the synthesis of complicated structures based on a huge number of RF MEMS switches.

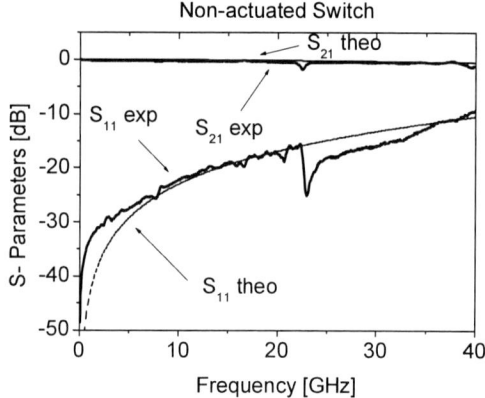

Fig. 4 Measured and simulated S-parametrs for the switch in the OFF position (bridge UP).

ACKNOWLEDGMENT

This activity has been partially supported by the ESA-ESTEC Program on "Microwave Electrostatic Micro-machined Devices for on Board Applications". F. Vitulli and G. Mannocchi from Thales-Alenia Space Italia are kindly acknowledged for the scientific management of the above activity, as well as F. Deborgies from ESA-ESTEC for his continuous support. R. Sorrentino, from University of Perugia and L. Vietzorreck from München Universität for contributions in the switch design.

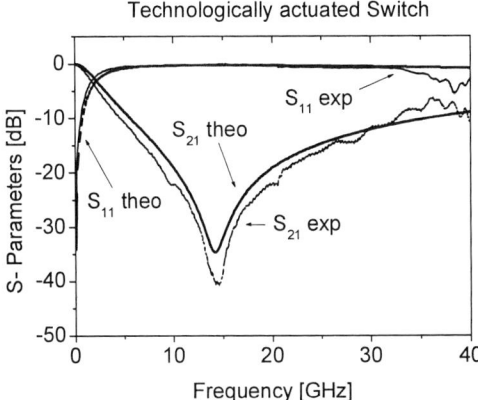

Fig. 5 Measured and simulated S-parameters for the switch in the ON position (bridge DOWN). The bridge is *technologically actuated*, i.e. the Au has been deposited directly onto the wafer to simulate an almost ideal actuation.

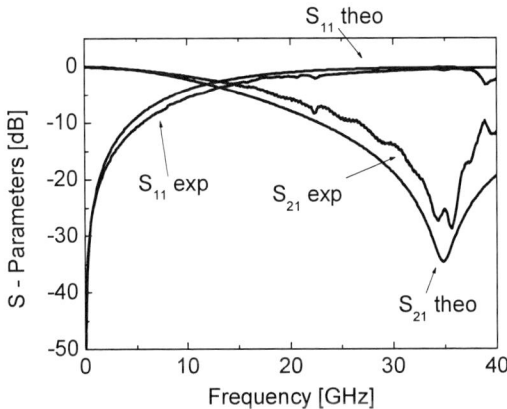

Fig. 6 Experimental (exp) and theoretical (theo) results for the really actuated RF MEMS switch. A frequency of resonance higher than expected is experienced due to the contribution of the residual air gap.

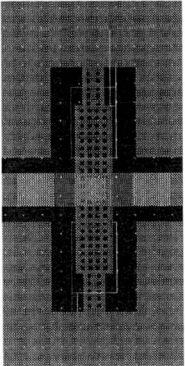

Fig. 7 Diagram of the shunt switch with the floating electrode

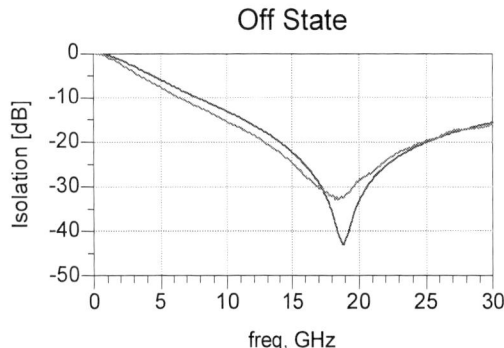

Fig 8 Electromagnetic simulation of the S-parameters for the floating metal configuration, compared to the experimental result in the OFF state (bridge in the DOWN position).

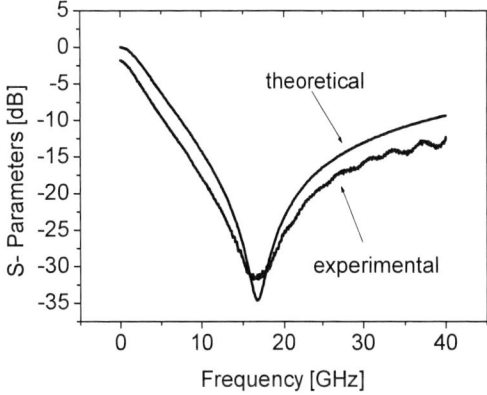

Fig. 9 Experimental response of the floating metal RF MEMS switch in the OFF state as compared to the theoretical response obtained by means of the circuital approach.

REFERENCES

[1] Dimitrios Peroulis, *Member, IEEE*, Sergio P. Pacheco, *Member, IEEE*, and Linda P. B. Katehi, "RF MEMS Switches With Enhanced Power-Handling Capabilities", *IEEE Trans. Microwave Theory Tech.*, Vol. 52, No. 1, January 2004, pp.50-68.

[2] R. Marcelli, G. Bartolucci, G. Minucci, B. Margesin, F. Giacomozzi, F. Vitulli: "Lumped element modelling of coplanar series RF MEMS switches", *Electronics Letters*, Vol. 40, No. 20, September 2004, pp. 1272-1273.

[3] E. Pettenpaul, H. Kapusta, A. Weisgerber, H. Mampe, J. Luginsland, and I. Wolff: "CAD Models of Lumped Elements on GaAs up to 18 GHz", IEEE Trans. On MTT, Vol. 36, No. 2, pp.294-304 (1988).

[4] I. Wolff and N. Knoppik: "Rectangular and Circular Microstrip Disk Capacitors and Resonators", *IEEE Trans. Microwave Theory Tech.*, Vol. MTT-22, No. 10, pp.857-864, October 1974.

[5] T. Edwards, *Foundations for Microstrip Circuit Design*, Second Edition, John Wiley and Sons, Chichester, 1992.

[6] X Rottenberg, R P Mertens, B Nauwelaers, W De Raedt and H A C Tilmans, "A distributed RF-MEMS capacitive series switch" *J. Micromech. Microeng.*, Vol. **15** S97-S102 (2005).

Alumina and LTCC Technology
for RF MEMS Switches
and True Time Delay Lines

Roberta Buttiglione[#1], Simone Catoni[*2], Giorgio De Angelis[*3], Massimiliano Dispenza[#4],

Annamaria Fiorello[#5], Kari Kautio[+6], Manu Ladhes[+7], Romolo Marcelli[*8], Kari Ronka[+9]

[#]*SELEX-SI*
Via Tiburtina km 12.400, 00131 Roma - ITALY
[1]rbuttiglione@selex-si.com
[4]mdispenza@selex-si.com
[5]afiorello@selex-si.com

[*]*CNR-Istituto per la Microelettronica e Microsistemi*
Via del Fosso del Cavaliere 100, 00133 Roma – ITALY
[2]simone.catoni@isl-altran.it
[3]giorgio.deangelis@imm.cnr.it
[8]romolo.marcelli@cnr.it

[+]*VTT Technical Research Centre of Finland*
Kaitoväylä 1, Oulu P.O. Box 1100, FI-90571 Oulu - Finland
[6]kari.kautio@vtt.fi
[7]manu.lahdes@vtt.fi
[9]kari.ronka@vtt.fi

Abstract— **The technology for the realization of RF MEMS ohmic series switches on LTCC substrates has been set up for their implementation in true time delay line configurations. Alumina and LTCC structures for validating the technological process on the individual switches have been obtained and characterized. Static delay lines on LTCC have been manufactured and tested to get the expected electrical performances for the final configuration.**

I. INTRODUCTION

In electronically scanned antennas the steering of the beam can be defined by changing the amount of phase shift imposed between adjacent T/R modules. If large instantaneous bandwidth signals are used, the phase shift would not be the same for all the frequency components of the spectrum, thus giving rise to the beam squint effect. By imposing a net time delay between adjacent T/R modules which would be intrinsically independent of frequency such a problem can be solved. On the other hand, long delays are required to be implemented for this scope, and it would be fruitful to develop a fabrication process for integrating RF-MEMS switches [1-3] directly onto LTCC multilayer substrates. Actually, the advantages of MEMS switches (wide bandwidth, low losses, linearity ...) could be conjugated with the ease of building-up long lines in a compact form, thanks to multilayer LTCC (Low Temperature Cofired Ceramic) technology [4,5], taking advantage from the possibility to realize buried transmission lines. In this paper, results about the technology for the realization of RF MEMS switches on alumina and directly onto an LTCC substrate, and their implementation in true time delay lines (TTDL) will be presented.

II. SWITCH TECHNOLOGY

A. RF MEMS on alumina

The initial process was developed on alumina substrates 254 µm thick, with the aim to finally transfer it onto LTCC HERAEUS CT707. The main problem in developing such a technology was actually encountered in the different stress characteristics of the bridge suspended membrane. To better investigate the problem, several runs were considered with a process flow including only the steps necessary to build up the membrane. Finally, a thicker alumina substrate (635 µm) was used, to reduce the thermal conductance of the substrate, thus minimizing the previously observed effect, as it can be seen on Fig. 1, where the exploited device is shown. Actually, it has to be accounted that the thermal conductivity of LTCC is 2 W/mK ca., i.e. one tenth lower than the standard alumina, and a proper thermal sink is necessary for the technology transfer to LTCC.

B. RF MEMS on LTCC

A set of analysis and tests have been performed on advanced LTCC substrate materials exhibiting either high dielectric constant (Ferro ULF-K140; Ferro ULF-K280; Heraeus CT765-4.3; ESL 41260) or ferroelectric properties (ESL40012). Objective of the test was to assess the feasibility of available material systems as well as the manufacturing process.

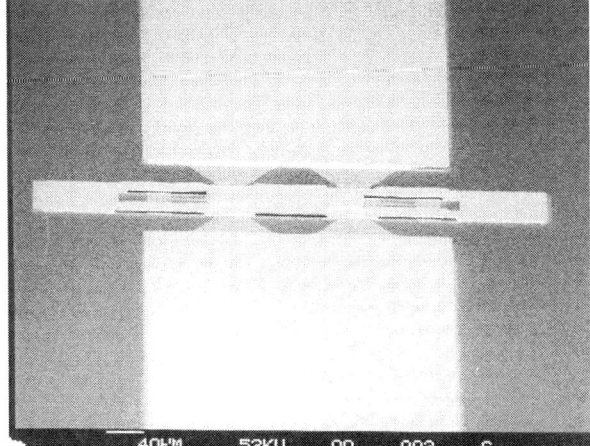

Fig. 1 RF MEMS ohmic series switch realized onto 635 μm thick alumina. SiO_2 was used as a sacrificial layer, removed by plasma etching

Surface treatment methods for LTCC planarization have been addressed to provide high quality platforms for thin-film processing. A large set of lapping/polishing conditions as well as different tape materials were tested. It turned out that the waviness of the substrates can be reduced by using flat setter plates during sintering process, but also lapping the substrates after the sintering. Surface roughness is reduced by polishing process, but carrying out only a polishing step without even a quick lapping step with fine particle abrasive slurry is not cost-effective, due to long processing times. Several commercially available materials were tested, and, when the ranking of the samples is done, it can be seen that the samples with Ferro and DuPont are the poorest. Better results have been obtained with Heraeus CT800 and Heralok with different top layers. Finally, the best samples were Heralok with AHT-01-005 top layer, and Heraeus CT707. The last one fullfilled the criterion of pore depth lower than 300 nm.

Accurate electromagnetic simulation requires precise knowledge of the electrical properties of the LTCC substrate and metallization. Ring resonator structures were designed and manufactured, consisting of surface (microstrip) and buried (stripline) configurations. By studying the resonance frequencies and widths of the resonance peaks, effective relative dielectric constant ($\varepsilon_{r,eff}$) and the attenuation constant (α) can be extracted at the resonance frequencies. From all of these data, it turns out $\varepsilon_r=7.2$ ca. from 17 to 70 GHz ca.

After that, a process was established to realized RF MEMS on planarized LTCC substrates, following the process given in Table I. Both technologically actuated switches, to be used as a reference for the expected electrical performances, and real switches, with suspended bridges, have been manufactured, having the general layout of Fig. 1, but different lateral dimensions. For the technologically actuated ones, the dimples height is equal to the gap spacing between bridge and substrate, so that switches are originated in down state. Work is in progress for testing the single switches on LTCC.

TABLE I
PROCESS FLOW FOR THE REALIZATION OF RF MEMS ON LTCC

Step	Material/Process
Buffer layer	1. Ta_2O_5 deposition
Windows for vias	2. Ta_2O_5 etching
Conductive pattern & Delay lines	3. NiCr/Pt/Au thin film deposition
	4. Pattern definition: Photolithography
	5. Au electroplating
	6. Au thin film ionic etching
Actuation lines, dimples, pull-down electrode, membrane anchors: Ta_xN_y/NiCr/Pt	7. Pattern definition: Photolithography
	8. NiCr/Pt thin film ionic etching
	9. NiCr/Pt/Au thin film deposition
Passivation layer	10. Pattern definition: Photolithography
	11. Ta_2O_5 deposition and lift-off
Sacrificial layer	12. SiO_2 deposition
Membrane anchors	13. Anchors windows definition: Photolithography
	14. SiO_2 etching to realize anchors windows
Dimples	15. Dimples windows definition: Photolithography
	12. SiO_2 etching to realize the dimples
Suspended Membrane	17. Au deposition
	18. Membrane structure definition: Photolithography
	19. Au Ion etching
Au sealing ring	20. Au electroplating: 8 um thick
Sacrificial layer removal and release of the membrane	21. Sacrificial layer etching

III. RF MEMS CHARACTERIZATION ON ALUMINA

In this section the results of the measurement campaign developed on two runs on alumina will be presented. DC tests, RF measurements and short time reliability cycles have been performed, accompanied with the characterization of the reference, technologically actuated switches. The electrical response of the switches in the UP and DOWN position are shown in Fig. 3 (isolation) and in Fig. 4 (transmission). Actually, the bridges have dimensions ranging from 180x500 μm^2 and 120x250 μm^2. In both runs a Ta_2O_5 passivation was introduced, and the contours of this patterned layer are visible

978-2-8748-7007-1/08 $25.00 © 2008 EuMA

in Fig. 2. Actuation voltages *Va* in the order of 20 to 50 volt have been experienced depending on the bridge dimensions, with de-actuation voltages *Vd* close to (1/2)*Va*. In particular, the SLTP17 (120X250 μm^2) and SLTP18 (120X300 μm^2) configurations, due to their reasonable actuation voltages and good cycling reliability, and because of their geometrical characteristics (shortest membranes) seem to be the best candidates for the future implementation of the dynamical TTDL module. From the analysis of Fig. 3, the isolation exhibits a maximum of -7.78 dB and a minimum of -13.46 dB for an operating frequency around 35 GHz.

Fig. 2 Contact region of the RF MEMS on LTCC with the Ta_2O_5 passivation visible from the contours of this patterned layer.

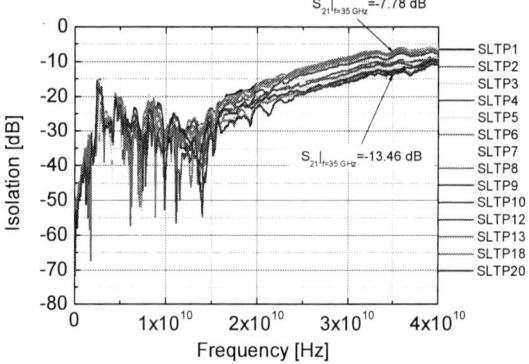

Fig. 3 Measured isolation for the RF MEMS switches on alumina. The performances are better at frequencies lower than 10 GHz, but the isolation is getting worst above F=15-20 GHz, probably due to the metal electrode used for the actuation. Actually, a realization by using a dielectric thin film solution could help in improving the isolation in the high frequency range.

IV. STATIC TTDLs ON LTCC

In this paragraph, results about static delay lines realized onto LTCC substrates will be presented. They have been minded as a reference for the expected performances of the entire structure including the RF MEMS switches. In particular, the influence of reflections by the open stubs, always present in a such device, have been studied. For obtaining good reflection characteristics, the length has to be in the order of 200 μm.

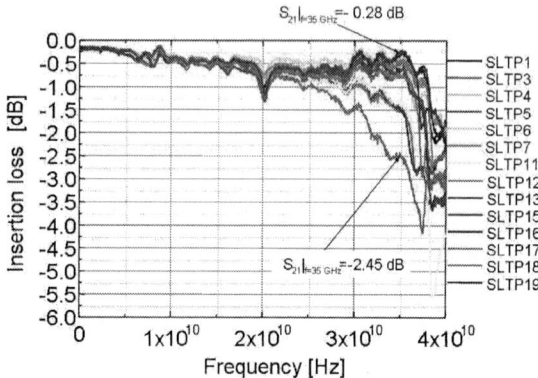

Fig. 4 Measured Insertion Loss for the RF MEMS switches on alumina in the DOWN state. These losses include also the contribution introduced by the via holes transitions and by the I/O lines in the order of 0.1 dB, so for example in the case of the SLTP4 switch (120x400 μm^2) the losses associated to the switch itself are in the order of 0.4 dB.

This implies that the suspended membrane, i.e. the MEMS bridge, has to be very close to the T-Junction transition of the module. Positioning the MEMS devices in such a way could introduce unwanted coupling effects. To decrease the contribution of the previously discussed phenomena, the shape and the dimensions of the T-Junction transitions have been modelled, and the shape of the MEMS devices has been accurately analyzed. In particular, a fixed-fixed beam arrangement with a total length of 400 μm (including the beam supports) has been imposed for the MEMS geometry. The actual T-Junction shape was chosen to have a simplified geometry, without tapering. The simulations of the lines have been performed by means of a 3D FEM based simulator (HFSS), in good agreement with the presented experimental results. In Fig. 5 and in Fig. 6 the schematic diagram of the TTDL module to be realized and the photo of the actual test structures are shown.

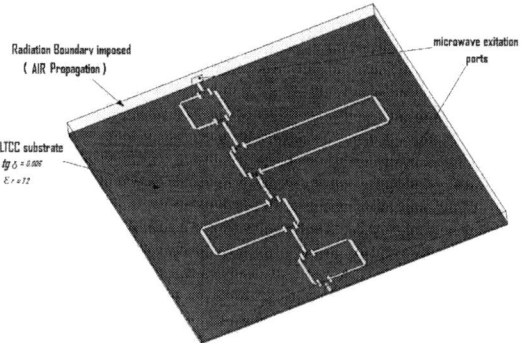

Fig. 5 Schematic diagram of the TTDL module, with the series switches.

The measurements performed on the realized structures exhibit additional losses with respect to the simulations, as evidenced from the comparison between Fig. 7 and Fig. 8.

These additional losses are mainly due to the not ideal resistivity ρ of the gold film used to manufacture the lines. As a result of the comparison between the expected ρ and that experimentally determined, it turns out a value three times higher with respect to the nominal value. As a consequence, 1.8 dB of additional losses in the case of the long path at 35 GHz have to be added to the expected transmission parameter. In Fig. 9 the delay time characterization for the longest path is also shown.

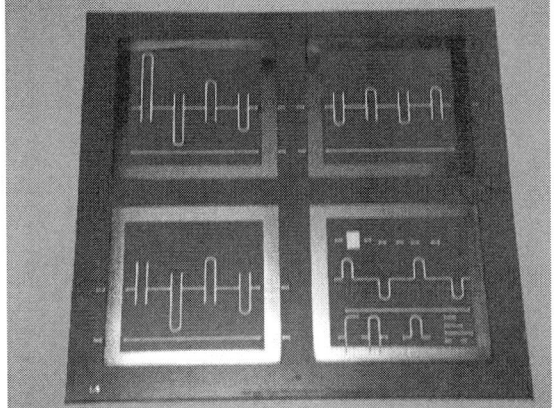

Fig. 6 Photo of some TTDL test structures realized on LTCC

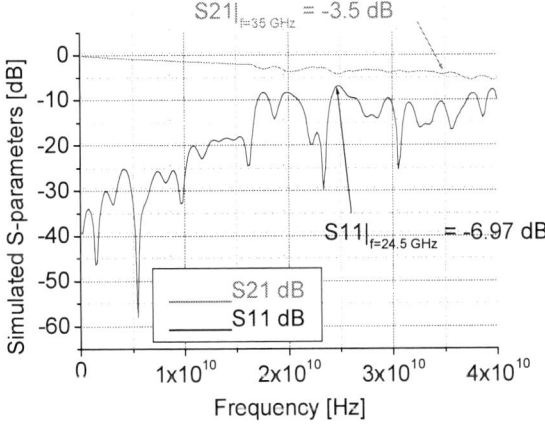

Fig. 7 3D simulations of the Scattering parameters of the longest path of the static delay line.

V. CONCLUSIONS

The design, realization and test of RF MEMS switches and TTDL structures onto alumina and LTCC substrates has been presented. The technology for single devices has been set up on both substrates, and problems related to the roughness and compatibility of processes have been solved. A preliminary theoretical and experimental investigation has been also performed on static delay lines on LTCC before their implementation by using the MEMS technology.

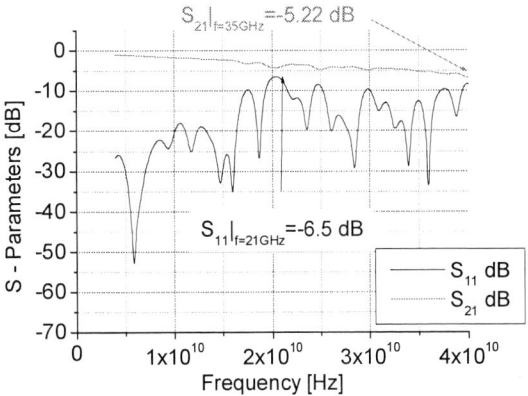

Fig. 8 Measurements of the S-parameters of the longest path of the static delay line.

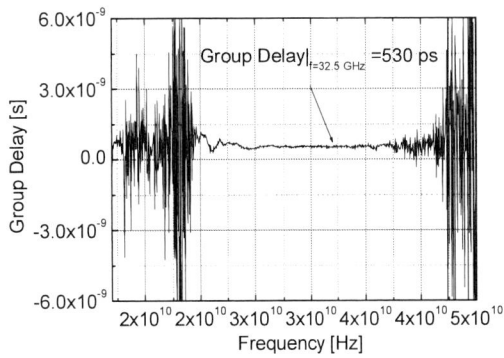

Fig. 9 Measurement of the Time Delay for the longest path of the test static delay line on LTCC.

ACKNOWLEDGMENT

This activity has been funded from EDA under the Contract TEMPO "Technologies for the Miniaturisation and the Packaging of True Time Delay Modules" Contract Number 04/102.049/024.

REFERENCES

[1] N. Scott Barker, and G. M. Rebeiz, "Distributed MEMS True-Time Delay Phase Shifters and Wide-Band Switches", *IEEE Trans. on Microwave Theory and Tech*, Vol. 46, No. 11, pp.1881-1890, Nov. 1998

[2] M. Kim, J. B. Hacker, R. E. Mihailovich, and J. F. DeNatale, "A DC-to-40 GHz Four-Bit RF MEMS True-Time Delay Network", *IEEE Microwave and Wireless Components Lett.*, Vol. 11, No. 2, pp.56-59, Feb. 2001

[3] J. B. Hacker, , R. E. Mihailovich, Moonil Kim, and J. F. DeNatale, "A Ka-Band 3-bit RF MEMS True-Time-Delay Network", *IEEE Trans. on Microwave Theory and Tech*, Vol. 51, No. 1, pp.305-308, Jan. 2003

[4] J. Lenkkeri et al., "Prospects and limits of LTCC technology", *Advancing Microelectronics*, Vol. 32, No. 3, pp. 9 – 12, 2005..

[5] J. Tuominen et al., "Surface planarisation of LTCC for high performance thin-film applications", *Proceed. of IMAPS Nordic Annual Conference*, Göteborg, Sweden, September 17 - 19, pp. 166 – 170, 2006.

Proceedings of the 3rd European Microwave Integrated Circuits Conference

The Fundamental Design Approach of The RF-DC Conversion Circuit for Optimizing Its Characteristics

Tsunayuki YAMAMOTO [#1], Kazuhiro FUJIMORI [#2], Minoru SANAGI [#3], Shigeji NOGI [#4]

[#] *The Graduate School of Natural Science and Technology, Okayama University*
3-1-1 Tsushima-Naka, Okayama, 700-8530, Japan
[1] yamamoto@micro.elec.okayama-u.ac.jp
[2] fujimori@elec.okayama-u.ac.jp
[3] sanagi@elec.okayama-u.ac.jp
[4] nogi@elec.okayama-u.ac.jp

Abstract— A Rectifying Antenna (Rectenna) is one of the most important components for a wireless power transmission. It has been developed for many applications such as Space Solar Power System (SSPS), Radio Frequency IDentification (RF-ID), and electric vehicle etc. The Rectenna consisting of RF-DC conversion circuits and receiving antennas needs to be designed for high conversion efficiency to achieve efficient power transmission. In this paper, we investigated the RF-DC conversion circuits by the theoretical analysis. And, we propose the theoretical formula for calculating the output DC voltage in giving the input voltage. The result that is calculated by the formula agreed with the simulator's result, so, we confirmed the validity of the formula we derived. The theory and the formula we derived are useful in order to know the basic operation of the RF-DC conversion circuit, and to design the circuit efficiently.

I. INTRODUCTION

Recently, microwave and millimeter-wave are used for not only the radio-communication but also the wireless power transmission. For example, Space Solar Power System (SSPS)[1][2][3], Radio Frequency IDentification (RF-ID), and electric vehicle[4][5], etc..

In the wireless power transmission, one of the most important components is rectifying antenna (rectenna). The rectenna has the circuit to convert the radio frequency to the direct current (RF-DC), and the receiving antennas. For realizing an efficient wireless power transmission system, the rectenna must be designed with high RF-DC conversion efficiency.

Various kinds of RF-DC conversion circuits have been investigated [6][7]. In our past studies, we paid an attention to the RF-DC conversion circuit with the resonance structure[8], and we studied the RF-DC conversion circuit with the high RF-DC conversion efficiency when the input power is small [9][10].

For designing the RF-DC conversion circuit with the high conversion efficiency, it is necessary to decide the various optimum parameters of the circuit, for example, by the theoretical analysis, by using the simulator, etc.. In the past report, the theoretical analyses of the RF-DC conversion circuits were done [11][12]. But, in these reports, the output DC voltage

has been treated as already-known. And, it is not possible to calculate the theoretical conversion efficiency without giving the output DC voltage to the equation. So, for calculating the theoretical conversion efficiency, it is necessary to measure the output DC voltage and to give the measurement result to the equation.

Other design method is using the simulator. In our past report, we designed the RF-DC conversion circuits by using the LE-FDTD (Lamped-Element Finite-Difference Time-Domain) metod [13][14]. The result that is calculated by LE-FDTD method agrees with the measurement result well. But, a lot of times is necessary to design the optimum RF-DC conversion circuit by using LE-FDTD method. Therefore, an efficient design is difficult.

In this paper, we treated the diode as a simple model considering only the direct current, and we propose the theoretical formula for calculating the output DC voltage in giving the input voltage. By using the theoretical formula we derived, it can instantaneously know the response when the input parameters change. So, the formula we derived is useful for efficient designing the RF-DC conversion circuit.

II. THE THEORETICAL FORMULA FOR DESIGNING THE RF-DC CONVERSION CIRCUIT

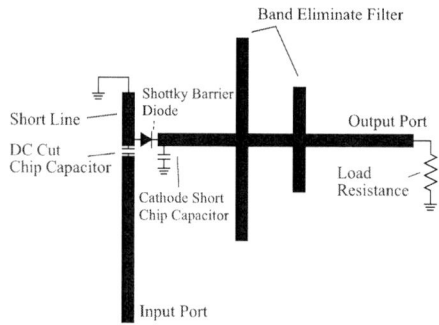

Fig. 1. The RF-DC Conversion Circuit We are Studying

978-2-8748-7007-1/08 $25.00 © 2008 EuMA   370

The Fig.1 shows the RF-DC conversion circuit we study (For 5.8[GHz]). This circuit is composed of the Shottky Barrier Diode (SBD), chip capacitor, band-eliminate filter, and load resistance. The input power is applied to the anode terminal of the diode, and the rectification is done. The rectified microwave passes through the band-eliminate filter, and only the DC output power occurs at the load resistance.

In our past reports[9][10], we know that there are strong correlation in the conversion efficiency and the voltage amplitude of the anode terminal of the diode. However, we don't understand in detail, how the change in the anode voltage infuluences the conversion efficiency. Therefore, if we can theoretically analyze the output DC voltage when the input parameters change, it seems that an efficient design of the RF-DC conversion circuit is possible. So, we derive the theoretical formula for calculating the output DC voltage in giving the input voltage.

In this section, deriving of the theoretical formula is described as follows.

First, it is necessary to think about the modeling of the diode. In high frequency, the affection of the reactive elements (junction capacitance, package parasitic inductance, etc.) is large. So, for exact analysis of the RF-DC conversion circuit, it is necessary to consider these elements. But, in this paper, the exact calculation is not our purpose. Our purpose is to know the basic operation of the RF-DC conversion circuit. So, we exclude these elements in this paper.

Fig.2 shows the simple RF-DC conversion circuit model considering only direct current.

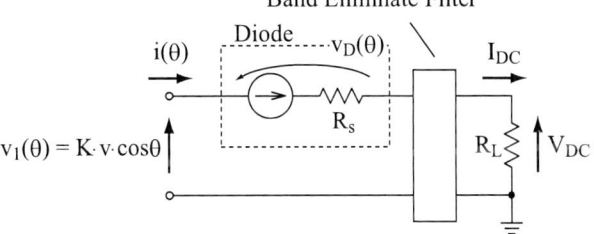

Fig. 2.   Simple RF-DC Conversion Circuit Model

In this model, the diode is composed of the non-linear current source and ohmic contact R_s. R_L is load resistance, V_{DC} is output DC voltage, $i(\theta)$ is alternate current, and I_{DC} is output direct current. The input voltage of the input port and the voltage of the diode anode-terminal are shown as

[The Input Voltage of the Input Port]

$$v(\theta) = v \cdot \cos\theta \qquad (1)$$

[The Voltage of the Anode Terminal]

$$v_1(\theta) = K \cdot v \cdot \cos\theta \qquad (2)$$

To make the problem simple, diode's built-in voltage 'V_{bi}' is excluded.

v in (1)(2) is the amplitude of the input voltage. K in (2) is constant, and if the traveling wave is inputted to the diode, $K = 1$, if the standing wave is inputted to the diode, $K = 0 \sim 2$.

First, the voltage of between the anode terminal and the cathode terminal '$v_D(\theta)$' can be expressed as

$$v_D(\theta) = K \cdot v \cdot \cos\theta - V_{DC} \qquad (3)$$

Therefore, $i(\theta)$ is calculated as

$$i(\theta) = \frac{K \cdot v \cdot \cos\theta - V_{DC}}{R_s} \quad (Diode\ ON) \qquad (4)$$
$$= 0 \quad (Diode\ OFF) \qquad (5)$$

The waveform of $i(\theta)$ is shown in Fig.3. In Fig.3, θ_1 is the phase angle where the diode is turned off.

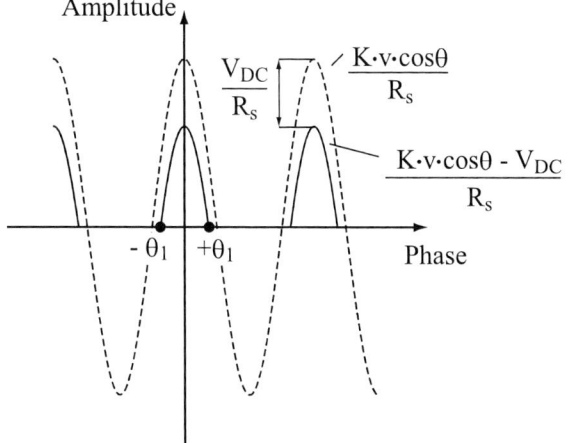

Fig. 3.   Waveform of $i(\theta)$

The current which flows in R_L is only the direct current. This direct current is calculated by using the Fourier Series Expansion to (4).

$$I_{DC} = \frac{1}{\pi \cdot R_s}(K \cdot v \cdot \sin\theta_1 - V_{DC} \cdot \theta_1) \qquad (6)$$

On the other hand, the direct current which flows in R_L can be expressed as

$$I_{DC} = \frac{V_{DC}}{R_L} \qquad (7)$$

Therefore, (8) is derived as

$$\frac{1}{\pi \cdot R_s}(K \cdot v \cdot \sin\theta_1 - V_{DC} \cdot \theta_1) = \frac{V_{DC}}{R_L} \qquad (8)$$

To pay an attention to $v_1(\theta)$, (9) is derived as

$$K \cdot v \cdot \cos\theta_1 - V_{DC} = 0 \qquad (9)$$

By (9) and ($0 \le \theta_1 < \frac{\pi}{2}$), next equation (10) and (11) are derived as

$$\sin \theta_1 = \sqrt{1 - \left(\frac{V_{DC}}{K \cdot v}\right)^2} \qquad (10)$$

$$\theta_1 = \cos^{-1}\left(\frac{V_{DC}}{K \cdot v}\right)$$

$$= \frac{\pi}{2} - \left\{\left(\frac{V_{DC}}{K \cdot v}\right) + \frac{1}{6}\left(\frac{V_{DC}}{K \cdot v}\right)^3 + \cdots\right\}$$

$$\simeq \frac{\pi}{2} - \left\{\left(\frac{V_{DC}}{K \cdot v}\right) + \frac{1}{6}\left(\frac{V_{DC}}{K \cdot v}\right)^3\right\} \qquad (11)$$

By substituting (10) and (11) in (8), (12) is calculated as

$$\sqrt{1 - \left(\frac{V_{DC}}{K \cdot v}\right)^2}$$
$$-\frac{V_{DC}}{K \cdot v}\left\{\frac{\pi}{2} - \left(\frac{V_{DC}}{K \cdot v}\right) - \frac{1}{6}\left(\frac{V_{DC}}{K \cdot v}\right)^3\right\} =$$
$$\frac{\pi R_s}{R_L} \cdot \frac{V_{DC}}{K \cdot v} \qquad (12)$$

When (12) is arranged, (13) is derived as

$$\frac{1}{36}\left(\frac{1}{K \cdot v}\right)^8 V_{DC}{}^8$$
$$+\frac{1}{3}\left(\frac{1}{K \cdot v}\right)^6 V_{DC}{}^6$$
$$-\frac{1}{3}A\left(\frac{1}{K \cdot v}\right)^5 V_{DC}{}^5$$
$$+\left(\frac{1}{K \cdot v}\right)^4 V_{DC}{}^4$$
$$-2A\left(\frac{1}{K \cdot v}\right)^3 V_{DC}{}^3$$
$$+\left(A^2 + 1\right)\left(\frac{1}{K \cdot v}\right)^2 V_{DC}{}^2$$
$$-1 = 0 \qquad (13)$$

where A in (13) is expressed as

$$A = \frac{\pi(2R_s + R_L)}{2R_L} \qquad (14)$$

(13) is the eight order equation of output DC voltage 'V_{DC}'. So, the V_{DC} can be obtained by solving (13).

In (13), eight solutions exists. In eight solutions, six complex solutions exist, and two real solutions exists. In two real solutions, the one is positive, and another one is negative. θ is ($0 \le \theta_1 < \frac{\pi}{2}$), so, the solution is positive. The solution that we require is only one.

By substituting R_s, R_L, K, and v in (13) (14), V_{DC} can be uniquely calculated in giving the input (K and v).

III. THE THEORETICAL OUTPUT DC VOLTAGE AND THE CONVERSION EFFICIENCY

By using (13), we calculated the output DC voltage and conversion efficiency corresponding to the load resistance. And, to determine the validity of the formula we derived, we compared the calculated results with the results of the simulator, which uses a Harmonic Balance method. In the simulator, the diode was treated as an ideal diode. So, the parasitic elements were excluded.

Now, conversion efficiency 'η' can be expressed as

$$\eta = \frac{2Z_c}{R_L}\left(\frac{V_{DC}}{v}\right)^2 \qquad (15)$$

where Z_c in (15) is characteristic impedance of the input port.

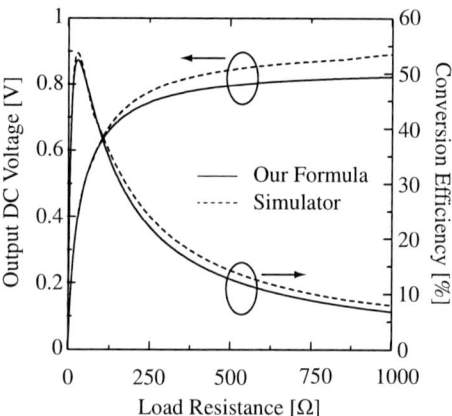

Fig. 4. The Output Voltage and The Conversion Efficiency

Fig.4 shows the comparison result of the output DC voltage and the RF-DC conversion efficiency when $R_s = 11$ (from M/A-COM MA4E2054-1141T Data-sheet), $K = 1$, $v = 1$, $Z_c = 50$, and $R_L = 1 \sim 1000$. ($v = 1$ means that the input power is 10 mW.)

In Fig.4, the calculated results by our formula agrees with the simulator's results very well. So, we confirmed the validity of our formula.

Next, we investigated the relationship of the anode voltage and the conversion efficiency. Fig.5 shows the calculated conversion efficiency by using our formula. Dotted line is $K = 0.8$, and the solid line is $K = 1.0$.

In Fig.5, the conversion efficiency when $K = 1.0$ is larger than the conversion efficiency when $K = 0.8$. Especially, the maximum conversion efficiency of both is greatly different. This result agrees with our past report [9][10]. In our past reports, we know that there are strong correlation in the conversion efficiency and the voltage amplitude of the anode terminal of the diode. From the calculating result of our formula, the validity of our measurement results was proven.

From this result, we designed the RF-DC conversion circuit that the anode voltage is larger. And, the conversion efficiency

978-2-8748-7007-1/08 $25.00 © 2008 EuMA

'70.7%' can be achieved up to the present. Fig.6 shows the conversion efficiency of the optimum RF-DC conversion circuit. (This circuit has two diode. Please refer to [9] [10] for details.)

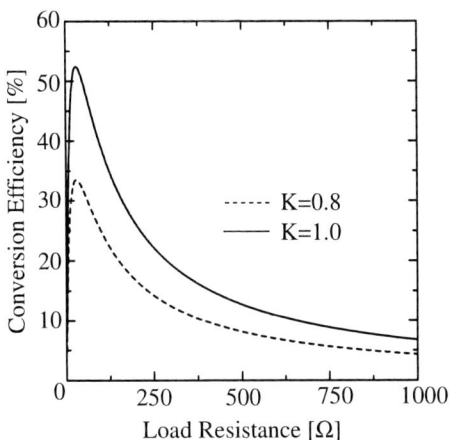

Fig. 5. The Difference of The Conversion Efficiency by The Difference of The Anode Voltage Amplitude

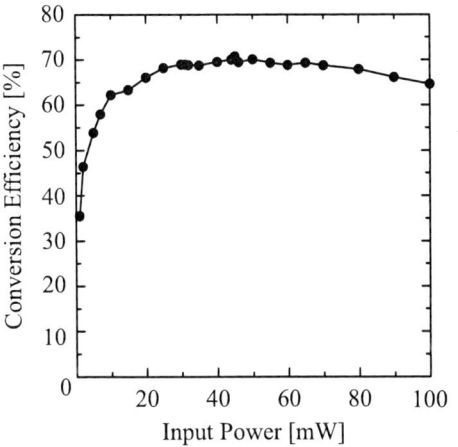

Fig. 6. The Conversion Efficiency of The RF-DC Conversion with The Optimum Design Parameters

Since built-in voltage of the diode, the reactive elements, etc. are not considered, the results of Fig.6 is different from our past reports [9][10]. But, by using (13), it is possible to know the response of the RF-DC conversion circuit when the input changes. And, it is possible to know the basic operation of the RF-DC conversion circuit. This formula is useful to design the more efficient RF-DC conversion circuits.

IV. CONCLUSIONS

In this paper, we treated the diode as simple model considering only the direct current, and we proposed the theoretical formula for calculating the output DC voltage in giving the input. The calculation result agreed with the simulator's result,

so, we confirmed the validity of our formula. The theory and formula we derived are useful in order to design the RF-DC conversion circuit with high conversion efficiency. By considering various parameters (built-in voltage, break-down voltage, etc.), the more and more exact calculation will be possible.

REFERENCES

[1] T.ITO, Y.FUJINO and M.FUJITA, "Fundamental Experiment of a Rectenna array for microwave Power Reception", IEICE Trans. Commun., vol.E76-B, no.12, pp.1508 - 1513, Dec. 1993.

[2] W.C Brown, "The history of power transmission by radio waves", IEEE Transactions on Microwave Theory and Techniques, vol.MTT-32, No.9, pp.1230 - 1242, Sep.1984.

[3] N.SHINOHARA, H.MATSUMOTO, K.HASHIMOTO, "Solar Power Station / Satellite (SPS) with Phase Controlled Magnetrons", IEICE Trans. Electron., vol.E86-C, No.8, pp.1550 - 1555, Aug. 2003.

[4] Y.FUJINO, T.ITO, M.FUJITA, N.KAYA, H.MATSUMOTO, K.KAWABATA, H.SAWADA, T.ONODERA, "A Driving Test of a Small DC Motor with a Rectenna Array", IEICE Trans. Commun., vol.E77-B, No.4, pp.526 - 528, Apr. 1994.

[5] N.SHINOHARA, K.MATSUMOTO, "Wireless Charging System by Microwave Power Transmission for Electric Motor Vehicles", IEICE Trans. Technical Report of IEICE SPS2006-18 (2007-02), pp.21 - 24, Feb. 2007.

[6] J.O.McSpadden, L.Fan and K.Chang, "A High Conversion Efficiency 5.8GHz Rectenna", IEEE MTT-S Digest, pp.547 - 550, 1997.

[7] Young-Ho Sun and K.Chang, "A High-Efficiency Dual-Frequency Rectenna for 2.45 and 5.8GHz Wireless Power Transmission", IEEE Trans. Microwave Theory Tech. vol.50, pp.1784 - 1789, July 2002.

[8] H.KITAYOSHI and K.SAWAYA, "A Study on Rectenna for Passive RFID-Tag", Proceedings of the 2006 IEICE General Conference, CBS-1-5, pp.S-9 - S-10, 2006.

[9] T.YAMAMOTO, K.FUMIMORI, M.SANAGI, S.NOGI, "The mW-class High Efficient RF-DC Conversion Circuit using the Resonance Structure", Proceedings of the 2007 International Symposium on Antennas and Propagation, 3B1-4, pp.660 - 663, Aug. 2007

[10] T.YAMAMOTO, K.FUJIMORI, M.SANAGI, S.NOGI, "The Design of mw-Class RF-DC Conversion Circuit using the Full-Wave Rectification", Proceedings of the 37the European Microwave Conference, pp.905 - 908, Oct. 2007.

[11] Tae-Whan Yoo and K.Chang, "Theoretical and Experimental Development of 10 and 35 GHz Rectennas", IEEE Transactions on Microwave Theory and Techniques, vol.40, No.6, June 1992.

[12] J.O.McSpadden, L.Fan and K.Chang, "Design and experiments of a high-conversion-efficiency 5.8GHz rectenna", IEEE Trans. Microwave Theory Tech, vol.45, pp.2053 - 2060, Dec. 1998.

[13] T.TAKAGAKI, T.YAMAMOTO, K.FUJIMORI, M.SANAGI, S.NOGI, "Efficient Design Apporoach of RF-DC Conversion Circuit Including Undesirable Radiation", Proceedings of the 2006 International Symposium on Antennas and Propagation, FB1-2, 1-4, Nov. 2006.

[14] T.TAKAGAKI, T.YAMAMOTO, K.FUJIMORI, M.SANAGI, S.NOGI, "Efficient Design Approach of mw-class RF-DC Conversion Rectenna Circuits by FDTD Analysis", Proceedings of Asia-Pacific Microwave Conference 2006, FROF-09, Dec. 2006.

Proceedings of the 3rd European Microwave Integrated Circuits Conference

The Impact of Technology Node Scaling on nMOS SPDT RF Switches

Tushar K. Thrivikraman[1], Wei-Min Lance Kuo, Jonathan P. Comeau, and John D. Cressler

School of Electrical and Computer Engineering, 777 Atlantic Drive, N.W.
Georgia Institute of Technology, Atlanta, GA 30332–0250 USA
Tel: (404) 894-5161 / Fax: (404) 894-4641
[1]tthrivi@ece.gatech.edu

Abstract—**This work presents a comparison of three single-pole double-throw (SPDT) nMOS RF switches implemented in commercially available Si-based 180, 130, and 90 nm technologies. In addition, a new series-shunt switch is presented that offers a means to improve switch isolation. Measured results of these RF switches demonstrates how technology node scaling impacts RF switch design and provides insight into the complicated trade-offs between insertion loss, isolation, and linearity.**

I. INTRODUCTION

The need for highly-integrated, low-cost circuit and system solutions has fueled aggressive technology scaling in the RF market. A rapidly growing area of IC design is presently focused on Si-based RF through mm-wave transceivers. Silicon-based RFIC's offer many benefits over III-V technologies, but traditionally have not been suitable for high-frequency circuit design due to their poor performance. Aggressive technology scaling, however, (i.e., 45 nm node [1]) and the addition of bandgap-engineered Silicon-Germanium (SiGe) technologies, are presently fueling a dramatic performance improvement in Si-manufacturing compatible platforms, yielding unity gain cutoff frequency (f_T) in excess of 500 GHz [2]. These advancements have enabled Si-based technologies to become major players even in traditional highly-demanding applications such as defense radar systems [3].

A crucial building block of any radar or wireless communication IC is the RF switch. The switch has many roles in a typical RF front-end, including transmit/receive, digitally controlled gain, and phase state selection. The design of low-loss, high-isolation, and high-linearity switches is a very actively researched topic [4], [5], and [6]. However, to date, little investigation has been done to quantify the effects of technology scaling on RF switch design and performance.

In the present work, we explore RF single-pole double throw (SPDT) nMOS switch designs in three different commercially-available Si-based technologies: a second-generation 180 nm SiGe HBT BiCMOS node (IBM 7HP), a third-generation 130 nm SiGe HBT BiCMOS (IBM 8HP), and a 90 nm RF CMOS technology (IBM 9RF). The focus of this work is to gain a better understanding of how technology node scaling affects design choices and tradeoffs, and how one best achieves optimum switch performance.

In addition to presenting measured SPDT switch performance results from three different Si-based technologies, a new strategy for increasing switch isolation is also presented.

Section II introduces the SPDT switch topology used for these designs and compares the technologies used for fabrication. Section III presents the measured results, followed by an analysis and comparison of the SPDT switches, and concludes with a summary.

II. DESIGN

All three SPDT switches were designed using the topology highlighted in Fig. 1. In order to improve insertion loss, an isolated triple-well nMOS device is used as the series switch. The isolated p-well is achieved by floating the deep n-well of the triple-well nMOS device, reducing parasitic losses by increasing the effective substrate resistance in the body of the device. The shunt nMOS devices increase isolation while only minimally degrading the insertion loss. Both the source and drain of the device were held at the same *dc* potential, and therefore only leakage current is dissipated, enabling these switches to consume virtually no power. When the digital signal, S, is high, M2 is on and M4 is off, completing the connection between RF_{in} and RF_{out2}.

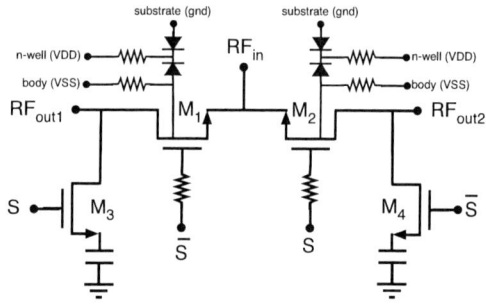

Fig. 1. Schematic of single-pole double-throw (SPDT) switch using triple-well nMOS devices.

In order to better understand the trade-offs of using a triple-well technology, a comparison of the parasitic capacitances between a standard nMOS device and an isolated triple-well device is shown in Fig. 2. The 10 kΩ resistors on the gate and body of the device isolate the gate and body nodes from *ac* ground, thus improving the insertion loss of the switch. However, switch isolation is degraded due to these floating nodes, allowing RF energy to leak from the source to the drain even when the switch is off.

978-2-8748-7007-1/08 $25.00 © 2008 EuMA   374

In order to improve isolation, these large resistors can be bypassed in the off-state, as shown in Fig. 3. When S is high, the bypass nMOS devices are off, and the body is isolated. When S is low, and the series switch is off, the bypass nMOS devices are on and the body is no longer isolated from ground, thus improving switch isolation. This technique can be further improved by customizing the layout of the device to reduce the source-to-drain capacitances, as discussed in [7].

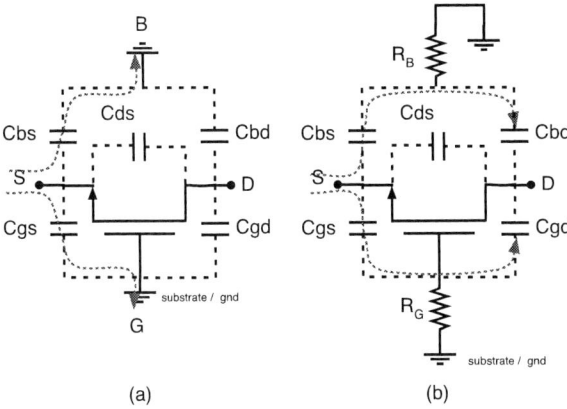

(a)   (b)

Fig. 2. Comparison of non-isolated switch (a) and triple-well isolated nMOS switch with floating body (b).

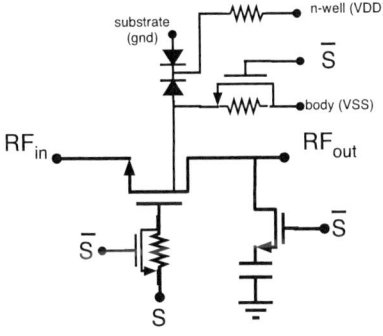

Fig. 3. Schematic of series-shunt switch with switched floating body and gate for improved isolation performance (*patent pending*).

The selection of device geometry results in a balance in trade-offs between the device resistance ("r_{on}") and parasitic device capacitances. A transistor with a large width will provide a smaller r_{on}, but will also increase overlap as well as source and drain to body and gate capacitances, thus increasing the switch insertion loss at high frequency. In order to minimize the insertion loss, the correct balance between the on-resistance and capacitances must be determined. The cross-generational RF switches discussed in the present work were sized to achieve optimum insertion loss at 10 GHz. For each lithography node, it is expected that the optimal width for a lower insertion loss will scale with gate length. As gate length

decreases, r_{on} decreases, and therefore a smaller device will reduce the parasitic capacitances. The optimal gate width for the 180 nm and 130 nm technologies was chosen as 200 μm, while the 90 nm technology was determined to be 160 μm. All devices were fabricated with 10 fingers, allowing for a very compact layout. V_{DD} for the 130 nm and 90 nm technologies was selected to be 1.2 V, while the 180 nm technology was 1.8 V.

In addition to the differences in gate length between the technologies, there are also other differences relevant to this investigation. Both 180 nm and 130 nm technologies were fabricated with 7 metal layers, including a 4 μm thick aluminum top metal. However, the triple-well nMOS device in the 180 nm platform contained a deep trench ring around the device, which aids in further isolating the p-well from the substrate. In addition, the 90 nm technology was fabricated with 9 metal layers. Fig. 4 shows a micrograph of the three fabricated switches, where the highlighted box shows the reference plane that was achieved via de-embedding. In addition to these circuits, two-port versions of each switch were also fabricated, with one output port of the switch terminated with a 50 Ω resistor for ease of measurement. The layout for the three switches was kept as uniform as possible to ensure valid comparisons.

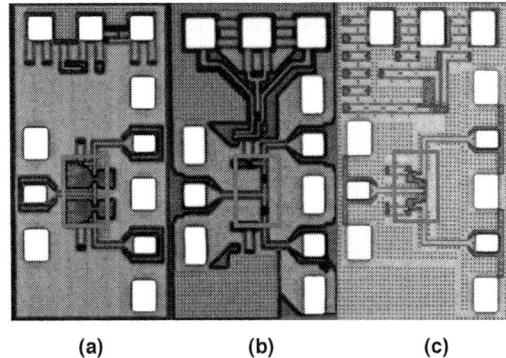

(a)   (b)   (c)

Fig. 4. Micrograph of SPDT switches in different technologies; (a) 180 nm, (b) 130 nm, (c) 90 nm.

III. MEASURED RESULTS

The switch characterization was performed in an RF-shielded room using an Agilent E8363B PNA for two-port on-wafer measurements. The three-port on wafer measurements were conducted by utilizing the Agilent PNA with a four-port extension module. 'Open' and 'short' deembedding structures were used to remove the pad parasitics from the two-port test structures. Both the two- and three-port device measurements showed good consistency. In addition, multiple die were measured and were also found to be in good agreement. Switch linearity characteristics were measured on the two-port devices using an Agilent PNA, E8267D PSG signal generator, and E4446A spectrum analyzer. All technologies were measured with the same instrument setup and calibration, reducing the overall measurement error when comparing these technologies.

As shown in Fig. 5, the 90 nm nMOS SPDT switch exhibits the lowest insertion loss (1.0 dB at 10 GHz). However, the insertion loss of the 180 nm switch (1.3 dB at 10 GHz) is unexpectedly lower than that of the 130 nm switch (1.5 dB at 10 GHz). Fig. 6 shows that all switches exhibit better than 15 dB of isolation; however, the 180 nm switch has slightly higher isolation than the 130 or 90 nm node switches. Fig. 7 shows the results of a single series-shunt switch in a 130 nm technology with our new enhanced isolation scheme, as discussed above (Fig. 3). The isolation of this new switch is improved by approximately 3 dB at 10 GHz, with only a slight increase in insertion loss (0.5 dB). Therefore, this technique can be used to improve the isolation performance of the triple-well isolated SPDT switches.

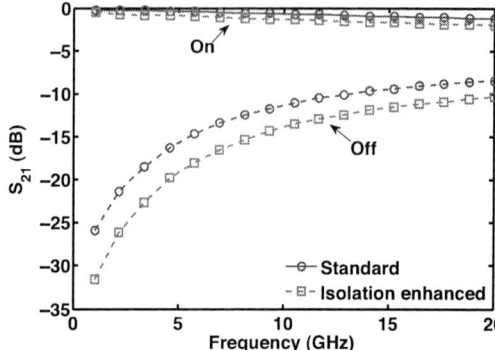

Fig. 7. Comparison between standard series-shunt switch and the new isolation-enhanced series-shunt switch in 130 nm technology.

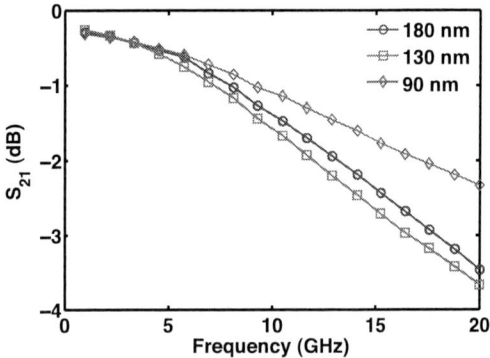

Fig. 5. S_{21} for the SPDT switches in the 'on' state across technology generations.

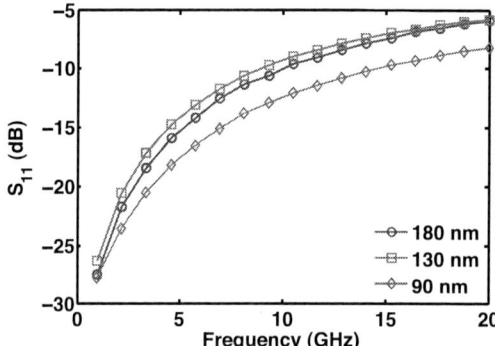

Fig. 8. S_{11} for SPDT switches in the 'on' state across technology generation.

dBm for both the 130 nm and 90 nm technologies.

Our data highlight the generational differences between the 180 nm, 130 nm, and 90 nm technology platforms. However, a clear trend across technology node is immediately not apparent. Therefore, further analysis is needed to determine the effects of lithography node on switch performance. To aid in understanding the apparent discrepancy between the 130 and 180 nm insertion loss values, measured and simulated results for the 180 nm and 130 nm switches are compared in Fig. 9 across 8 - 12 GHz. The post-layout extracted switches were simulated within the Cadence simulation environment using Agilent's GoldenGate simulation tool.

The measured data for the various switch designs exhibits a faster roll-off with frequency than simulations predict and thus presents a large simulation to measurement discrepancy. The modified simulations attempt to correct this correlation error by adding a 30 - 40 fF shunt capacitor between the source/drain to ground. The addition of this shunt capacitance correctly matches measured performance, and thus can be understood to be inaccurately capture by the present parasitics extraction tools. We believe that for the 130 nm switch, a sub-optimal device size was selected due to these unmodeled parasitics, leading to a selection of a larger than necessary device compared to the 180 nm or 90 nm nodes.

The sub-optimial device selection due to the additional

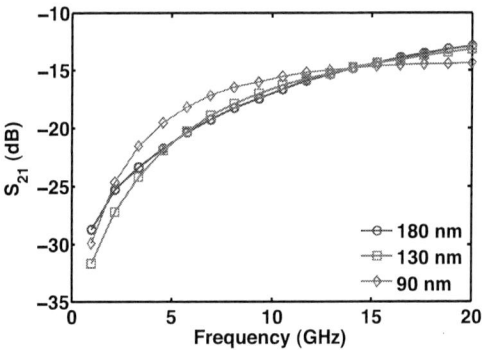

Fig. 6. S_{21} for the SPDT switches in the 'off' state across technology generations.

For these switch designs, even though no external matching elements were used, the return loss of the switches (Fig. 8) shows good matching below 10 GHz, with S_{11} measured to be less than -10 dB for all switch variants.

Linearity, as measured by the input third-order intercept (TOI), is a very important metric for high-frequency switches. As expected, the input TOI results (not shown) show degradation with technology node scaling. Here, the 180 nm technology exhibits the highest TOI of 27 dBm, compared to 25

978-2-8748-7007-1/08 $25.00 © 2008 EuMA

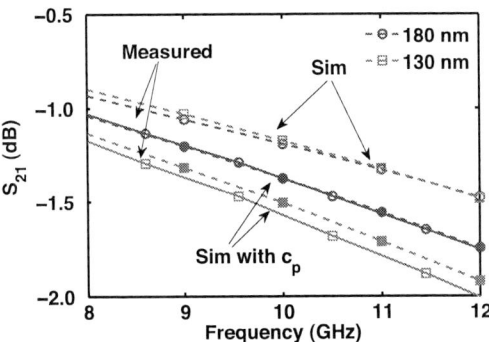

Fig. 9. Comparison of simulation and measured results for SPDT switches highlighting simulation to measurement correlation.

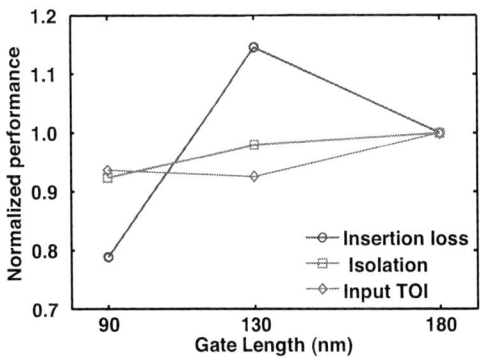

Fig. 11. Normalized insertion loss, isolation, and input TOI across technology generations.

parasitic capacitances is clearly shown in Fig. 10. In this figure, two gate widths (200 and 180 μm) of the 130 nm switch are plotted, with and without the additional capacitances. With c_p equal to 0, a 200 μm device shows slightly better insertion loss over a 180 μm device. However, with the addition of a 34 fF capacitance, a 180 μm device shows superior performance. Therefore, these unaccounted device parasitics have a large impact on device size selection and performance, and must be carefully modeled.

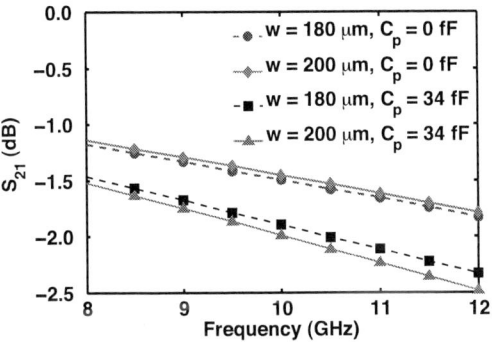

Fig. 10. Simulations of 200 and 180 μm 130 nm nMOS SPDTs, with and without source/drain to ground parasitic capacitance.

Fig. 11 plots normalized performance metrics (insertion loss, isolation, and input TOI) with respect to the 180 nm switch performance, as a function of gate length. For insertion loss, it is clear that the 130 nm node does not reflect the anticipated trend, for the reasons discussed above. However, for switch isolation, the trend does indicate that as gate length decreases, the isolation will slowly degrade. This is intuitive, since the parasitic source-to-drain capacitances increase as the gate length decreases, which is a major contributor to device isolation. In addition, input TOI also shows a decrease in performance with technology node scaling; however, the exact trend is not clear, and further investigation is needed to understand the linearity behavior of the switches, including bias dependence and gate width. This is presently underway.

IV. SUMMARY

In this work, we present single-pole double throw (SPDT) nMOS RF switches in three different Si-based technologies to highlight the performance differences across technology node. We show that performance trends do exist across technology generations, however, care should be taken in optimizing switch performance to provide maximum benefits for each technology node. For our designs, the 90 nm node shows the best insertion loss performance, but while exhibiting the worst isolation.

We expect these performance trends to continue as technology nodes continue to scale, and thus, for sub-90 nm nodes, techniques such as the switched triple-well isolation technique might offer benefits in improving switch isolation while still maintaing low insertion loss. Future work will focus on understanding how device size and bias can be optimized and how technology scaling impacts this optimization and circuit performance.

ACKNOWLEDGMENT

The authors are grateful for the support of the Georgia Tech Research Institute Fellows Council and Dr. D. Parekh for IRAD project support, M. Mitchell, the members of the SiGe Devices and Circuits Group, and the Georgia Electronic Design Center at Georgia Tech.

REFERENCES

[1] S. Lee et al., "Record RF performance of 45-nm SOI CMOS technology," IEEE IEDM, pp. 255–258, Dec. 2007.
[2] J. Yuan et al., "On the frequency limits of SiGe HBTs for TeraHertz applications," in Proc. IEEE BCTM, 2007, pp. 22–25.
[3] J. P. Comeau et al., "A Monolithic 5-Bit SiGe BiCMOS Receiver for X-Band Phased-Array Radar Systems," in Proc. IEEE BCTM, 2007.
[4] C.-C. Wu et al., "A 0.13 μm CMOS T/R switch design for ultrawideband wireless applications," IEEE ISCAS Symp. Dig., pp. 3758 –3761, May 2006.
[5] J. P. Comeau et al., "Design and layout techniques for the optimization of nMOS SPDT series-shunt switches in a 130nm SiGe BiCMOS technology," IEEE RFIC Symp. Dig., pp. 457–460, 3-5 June 2007.
[6] W.-M. L. Kuo et al., "Comparison of Shunt and Series/Shunt nMOS Single-Pole Double-Throw Switches for X-Band Phased Array T/R Modules," IEEE SiRF, pp. 249–252, Jan. 2007.
[7] Q. Li et al., "CMOS T/R switch design: Towards ultra-wideband and higher frequency," IEEE JSSC, vol. 42, no. 3, pp. 563–570, March 2007.

Design and Temperature Dependent Analysis of GaAs Multilayer Transmission Lines

J. Yuan, A. A. Rezazadeh, J. Lu, Q. Sun and V. T. Vo

The electromagnetic Centre for Microwave and mm-Wave Designs and Applications,
School of Electrical and Electronics Engineering, The University of Manchester, P.O. Box 88, Manchester, M60 1QD UK

Tel: +44 161 306 4823, Email: junyi.yuan@postgrad.manchester.ac.uk

Abstract— **In multilayer technology, the generated heat can not be dissipated effectively due to its multilayer structure. In this paper we designed, fabricated and characterised a group of GaAs multilayer coplanar waveguide (CPW) transmission lines and studied the effects of temperature on these components. Furthermore a de-embedding technique has been applied and the effect of de-embedding shown to be critical in removing the pads parasitics. The temperature dependents of the transmission lines are carefully analyzed from -25°C to 125°C and the results indicated that the characteristic impedances and effective dielectric constants of the transmission lines remained constant within the temperature range. These results are reported for the first time and provide an insight into the design optimization of multilayer circuits for compact 3D MMIC applications.**

I. INTRODUCTION

The newly developed CPW multilayer technology [1] shows great advantages comparing to conventional planar microstrip technology. Firstly the CPW structure makes devices and components to be grounded without the need of via-holes. This means it is not necessary to thin down the substrate reducing the number of processing steps and the cost of manufacturing. Secondly this CPW structure minimized the parasitics associated with the via-holes. Thirdly, since the signal is shielded from other components in the circuits the layout can be more compact while good isolation is maintained. The other advantage of CPW design is that it further increases the package density and thus provides commercial competitive wafer size. By far a large number of passive devices using CPW multilayer technology have been reported and good performances and bandwidth are realized with over 60% size reduction. [2-5].

However, multilayer passive components require to be integrated with the active devices to form practical 3D MMICs. In multilayer technology, metal conductors and dielectric layers are placed upon active devices, where the generated heat can not efficiently be dissipated from the circuits. Therefore the temperature in the multilayer circuits would be expected to be much higher than the standard microstrip or conventional CPW circuits. Thus study of the thermal characterisations of multilayer passive devices is essential for optimum design of 3D MMICs.

Thermal characterisations of active devices are widely investigated [6-8] because their small signal model parameters are normally temperature dependent. On the contrast, temperature characterisations of passive components have not

been studied since they are not expected to generate heat as those of the active devices.

In this paper we present the temperature dependence of various CPW multilayer transmission lines fabricated on semi-insulating GaAs substrates. The fabrication details of these components are given in our previous paper [1]. The structure simplicity of the transmission lines is extremely suitable for investigating the temperature dependences of the basic physical parameters of various conductor and dielectric layers employed in the construction of multilayer MMICs. On-wafer S-parameters together with the appropriate formulations were employed to extract characteristic impedances, effective dielectric constants and dissipation losses of the transmission lines over the temperature range from -25°C to 125°C. In addition the effects of de-embedding in removing the parasitics associated with the probing pads for better data extraction are also discussed.

II. EXPERIMENTAL

A. Characterisation of Transmission Lines

The fabricated CPW transmission lines have pads on each side, which are designed for RF probing. Normally these pads have characteristic impedances of 50Ω, but the transition between the probing pads and the transmission lines have different impedances and loss properties. Therefore a signal reflection exists between the two measured ports. When the half wavelength of measuring frequency equals to the length of the transmission line, the reflection starts to resonate and causes errors in the measured S-parameters. For that reason, pads de-embedding are necessary when high frequency transmission line is considered.

The transmission line can be modelled as an intrinsic DUT and two pads of parasitics. The parasitics of a pad can be represented as a serial impedance (Z_{pad}) and a shunt conductance (Y_{pad}). The de-embedding procedure also requires open and short test patterns to characterise Y_{pad} and Z_{pad} respectively. Open pad is modelled as a simple shunt conductance and short pad as a serial impedance. The concept of de-embedding procedure used here is similar to that given in [9]. However, various steps in the calculations have been modified in order to adopt the procedure for the components used in this work. As described in Figure 1, firstly the measured short and open pattern S-parameters are employed to characterise Y_{pad} and Z_{pad}, which are used to form ABCD matrix of the pads. Then the measured S-parameters of the transmission line are converted to the ABCD parameters. Finally the intrinsic ABCD parameters of the transmission lines can be obtained.

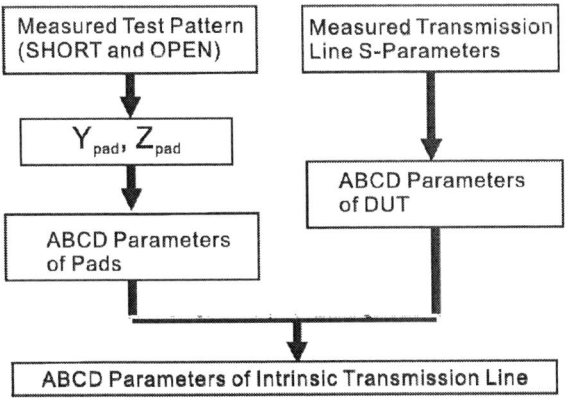

Figure 1   Transmission line de-embedding flow chart.

Following the de-embedding, the intrinsic transmission line parameters such as characteristic impedance, effective dielectric constant, $\varepsilon_{r,eff}$ and dissipation loss can be evaluated from the intrinsic S-parameters [10] using the following equations:

$$Z_0 = Z_{sys} \sqrt{\frac{(1+S_{11})^2 - S_{21}^2}{(1-S_{11})^2 - S_{21}^2}} \qquad (1)$$

$$\varepsilon_{r,eff} = \left(\frac{\beta c}{\omega}\right)^2 \qquad (2)$$

$$\text{Dissipation Loss} = 20 \times \log_{10} e^{-\alpha l} \ (dB) \qquad (3)$$

Where Z_{sys} is the measuring system impedance, l is the physical length of transmission line in millimetre, α and β are the real and imaginary part of the complex propagation constant respectively, c is the speed of light and ω is the angular frequency.

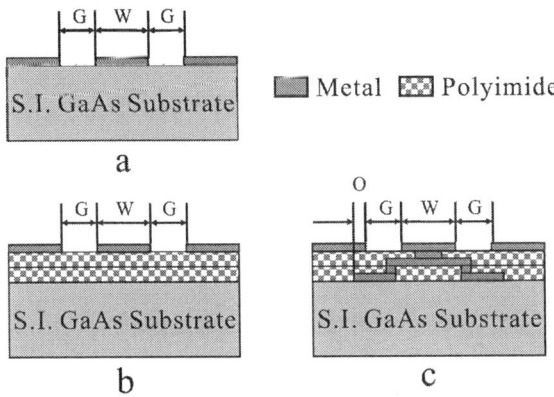

Figure 2   Cross-sectional views of various 3D transmission line structures: (a) conventional CPW, (b) planar on two dielectric layers and (d) V-shaped with overlap.

The cross-sectional views of a group of 3D CPW transmission lines studied are illustrated in Figure 2. These transmission lines are fabricated on a 0.5mm thick S.I. GaAs substrate using 3 metal layers and 2 dielectric layers. The dielectric and metal employed in the fabrication procedure are polyimide and Ti/Au respectively. All transmission lines have a physical length of 2mm.

The planar structure transmission line on the polyimide (Figure 2b) is selected to demonstrate the performance improvement by the de-embedding procedure used in this work. The RF on-wafer measurements are carried out using HP85107A Network Analyzer from 45 MHz to 40 GHz. The evaluated intrinsic characteristic impedances before and after de-embedding are shown in Figure 3.

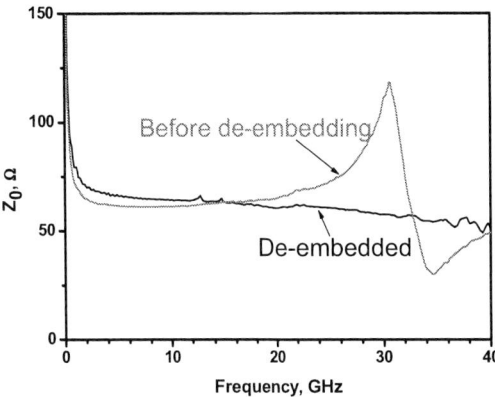

Figure 3   Comparison of characteristic impedances before and after de-embedding of the planar on two dielectric layers transmission line (Figure 2b).

From Figure 3 it is clear that the de-embedding procedure eliminates the resonance of Z_0 above 25GHz and extends the applicable data up to 40GHz. The very high impedances below 1GHz are mainly due to the dominating effects of series resistance and shunt conductance of the transmission line.

The extracted characteristic impedances, effective dielectric constants and dissipation losses for the three transmission lines at room temperature ($25^{\circ}C$) are shown in Figure 4. All the extracted parameters are shown up to 20GHz since some of the data at higher frequency contain oscillations.

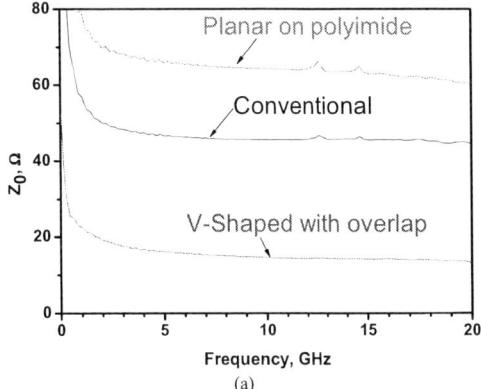

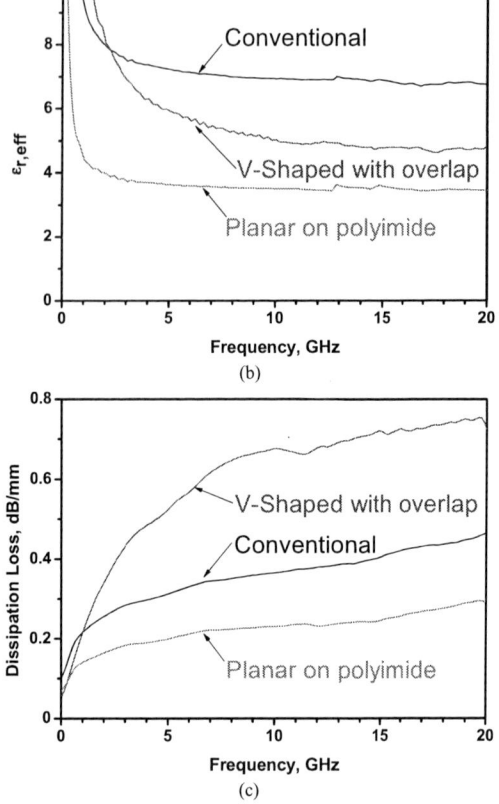

Figure 4 Measured results for a group of multilayer transmission lines: (a) characteristic impedances, (b) effective dielectric constants and (c) dissipation loss.

The unusual high values of effective dielectric constants below 2GHz shown in Figure 4b are due to the mathematical limitation of our extraction procedure using S-parameters, which makes the results unreliable at lower frequency.

From Figure 4a it can be seen that the conventional CPW line is designed to have a characteristic impedance of 50Ω. For the MMIC designs, high and low impedance transmission lines are considered to be very useful, especially in amplifier matching networks. Multilayer design provides a great flexibility for the designers since for example the characteristic impedance of the planar transmission line can be increased simply by raising the metal layer and applying a low permittivity polyimide layer (ε_r of 3.7) beneath the conductor layer. In this case the thickness of the two polyimide layers is 5μm in total and Z_0 changes from 48Ω to 62Ω i.e. an increase of about 30%. The effective dielectric constant of this transmission line is decreased as the electric fluxes penetrate through the low permittivity dielectric layer and the air instead of the S.I. GaAs substrate. Moreover, since air and polyimide are less lossy than the GaAs (ε_r of 12.9). Furthermore the dissipation loss of the planar transmission line on polyimide drops significantly to a lower value providing a low loss transmission line for the realization of the compact 3D MMICs.

Furthermore, a lower characteristic impedance transmission line can readily be achieved by employing a V-Shaped structure and by designing the lower metal layer to have an overlap with the top ground layer, as shown in Figure 2c. In this design the fabricated V-Shaped transmission line with overlap has an overlap size of 10μm. This structure allows most of the electric fluxes to penetrate through the polyimide layer of the overlap area. This results the increase in the signal to ground capacitance which leads to a drop of characteristic impedance from 48Ω to about 18Ω (Figure 4a). The drawback of this approach is that large amount of electric fluxes are concentrated at the overlap area which causes more current leaks through the relatively thin polyimide layer of the overlap area resulting an increase in dielectric loss. The high electric field of the overlap area also causes the well known current crowding effect resulting in an increase in conductor loss. Therefore, the dissipation loss of this type of transmission line is significantly higher than the conventional CPW line, as shown in Figure 4c

B. Temperature Dependent Characteristics

S-parameters of the transmission lines are measured by RF on-wafer probing from -25°C to 125°C with an accuracy of ± 2°C. The temperature control system employs Temptronic TP03200A Thermo Chuck and Cascade Micro Chamber.

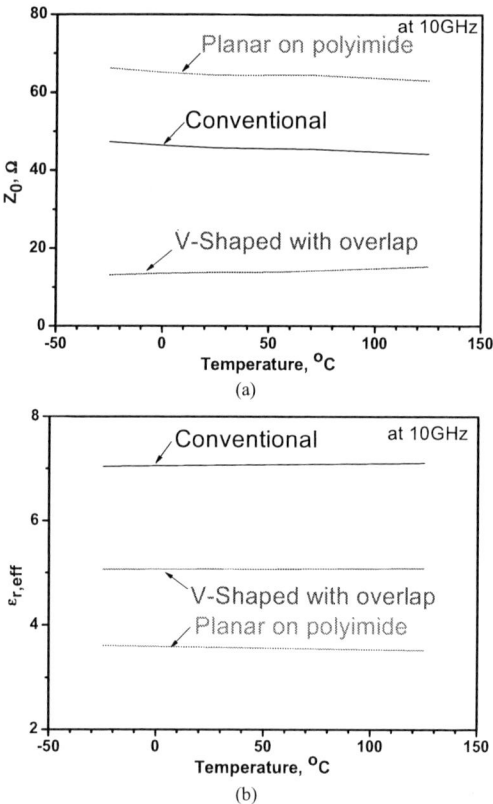

Figure 5 Transmission line parameters vs. temperature at 10GHz: (a) characteristic impedances and (b) effective dielectric constants.

The VNA calibration LRRM standard is carried out each time the temperature changes and the wafer is hold for 30 minutes for temperature stabilizing. The extracted characteristic impedances and effective dielectric constants of the transmission lines at 10GHz as a function of temperature are shown in Figure 5. The data illustrate that the values of Z_0 and $\varepsilon_{r\,eff}$ change in a very limited range and show no apparent trend when the temperature changes from -25°C to 125°C.

It is possible to qualitatively explain the observed variations of the Z_0 and $\varepsilon_{r.eff}$ with temperature by considering the characteristic impedance of a transmission line:

$$Z_0 = \sqrt{\frac{R + j\omega L}{G + j\omega C}} \qquad (4)$$

Where at high frequency this can be simplified to:

$$Z_0 = \sqrt{\frac{L}{C}} \qquad (5)$$

From the data of Figure 5a, the small variation of Z_0 with temperature means that the ratio of L to C does not change a great deal. The capacitance of a parallel capacitor and the inductance for a cylindrical coil of cross section area of A are commonly defined as:

$$C = \frac{\varepsilon_0 \varepsilon_r A}{d} \qquad (6)$$

$$L = \frac{\mu_0 \mu_r N^2 A}{l} \qquad (7)$$

Where d is the distance between two flat conductors, ε_r is the permittivity of the material between two conductors, μ_0 is the permeability of free space, μ_r is the relative permeability of the core material, N is number of the turns and l is the length of coil in meters.

According to (6) and (7), C and L are proportional to ε_r and μ_r respectively. From the results given in Figure 5 the effective dielectric constants remained rather constant indicating that C does not change with temperature, which also means L stays constant with temperature. Therefore, the mere variations of C and L with temperature observed confirmed that the permittivity of the polyimide and the permeability of the gold layer remained rather unchanged.

III. CONCLUSIONS

In this work, we presented a technique to characterise the effects of temperature on multilayer CPW transmission lines. The employed CPW multilayer transmission lines are fabricated on S.I. GaAs substrate using three metal layers and two polyimide layers fabrication technology. The results show that a variety of multilayer CPW transmission lines for the realization of compact 3D MMICs can easily be constructed by utilizing the flexibility of the multilayer technique. The applied de-embedding technique shows an effective way in removing pads parasitics. In addition, the temperature

dependences of the parameters of the transmission lines are conducted and analysed. The results indicated that the characteristic impedances and dielectric constants of the multilayer transmission lines remained relatively constant over the temperature range from -25 to 125^{0}C. These observations are very important in the design optimizations of compact 3D MMICs and their applications.

REFERENCES

[1] V. T. Vo, L. Krishnamurthy, Q. Sun, and A. A. Rezazadeh, "3-D low-loss coplanar waveguide transmission lines in multilayer MMICs," *Microwave Theory and Techniques, IEEE Transactions on*, vol. 54, pp. 2864-2871, 2006.

[2] V. T. Vo, L. Krishnamurthy, S. Qing, and A. A. Rezazadeh, "Miniature CPW Inductors for 3-D MMICs," in *Microwave Symposium Digest, 2006. IEEE MTT-S International*, 2006, pp. 1377-1380.

[3] L. Krishnamurthy, Q. Sun, V. T. Vo, G. Parkinson, D. K. Paul, K. Williams, and A. A. Rezazadeh, "A comparative study of active and passive GaAs microwave couplers," in *Gallium Arsenide and Other Semiconductor Application Symposium, 2005. EGAAS 2005. European*, 2005, pp. 353-356.

[4] L. Krishnamurthy, V. T. Vo, R. Sloan, K. Williams, and A. A. Rezazadeh, "Broadband CPW multilayer directional couplers on GaAs for MMIC applications," in *High Frequency Postgraduate Student Colloquium, 2004*, 2004, pp. 183-188.

[5] Q. Sun, J. Yuan, V. T. Vo, and A. A. Rezazadeh, "Design and Realization of Spiral Marchand Balun Using CPW Multilayer GaAs Technology," in *Microwave Conference, 2006. 36th European*, 2006, pp. 68-71.

[6] N. Rinaldi, "Small-signal operation of semiconductor devices including self-heating, with application to thermal characterization and instability analysis," *Electron Devices, IEEE Transactions on*, vol. 48, pp. 323-331, 2001.

[7] T. Egawa, Z. Guang-Yuan, H. Ishikawa, H. Umeno, and T. Jimbo, "Characterizations of recessed gate AlGaN/GaN HEMTs on sapphire," *Electron Devices, IEEE Transactions on*, vol. 48, pp. 603-608, 2001.

[8] S. Chitrashekaraiah, V. T. Vo, and A. A. Rezazadeh, "Linear Temperature Dependent Small Signal Model for InGaP/GaAs DHBTs Using IC-CAP," in *Microwave Symposium Digest, 2006. IEEE MTT-S International*, 2006, pp. 1093-1096.

[9] Ming-Hsiang CHO, Guo-Wei HUANG, Chia-Sung CHIU, Kun-Ming CHEN, An-Sam PENG, and Y.-M. TENG, "A Cascade Open-Short-Thru (COST) De-Embedding Method for Microwave On-Wafer Characterization and Automatic Measurement," *IEICE Transactions on Electronics*, vol. E88-C, pp. 845-850, May 2005.

[10] W. R. Eisenstadt and Y. Eo, "S-parameter-based IC interconnect transmission line characterization," *Components, Hybrids, and Manufacturing Technology, IEEE Transactions on [see also IEEE Trans. on Components, Packaging, and Manufacturing Technology, Part A, B, C]*, vol. 15, pp. 483-490, 1992.

DC-Contact RF MEMS Switches using Thin-Film Cantilevers

Hui Shen, Songbin Gong, N. Scott Barker

Charles L. Brown Dept. of ECE, University of Virginia

351 McCormick Rd. Charlottesville, VA, USA • 22904-4743

shenhui@virginia.edu songbin@virginia.edu Barker@virginia.edu

Abstract—**This paper describes the development of DC-contact RF-MEMS SPST, SP3T, and SP4T switches implemented with a thin-film cantilever. Using aluminium as the sacrificial layer in the fabrication process, flat cantilevers are realized with a measured actuation voltage of 50~70 V. The SPST switch is used as a building block to realize more complicated SP3T and SP4T switches for use in true-time delay phase shifters. The preliminary measurements of the SP3T and SP4T switches demonstrate isolation of 20 dB and insertion loss less than 2 dB up to 50 GHz.**

I. INTRODUCTION

Single-pole single-throw (SPST) DC-contact RF-MEMS switches have demonstrated excellent isolation with 25 dB at 50 GHz for an up-state capacitance of 2 fF [1]. These switches have been used successfully to develop a 4-bit switched-line phase shifter operating to 40 GHz by implementing a single-pole double-throw (SPDT) switch with the above mentioned SPST switch [2]. In order to implement more compact switched-line phase shifters it is necessary to use single-pole multi-throw switches. Unfortunately, these switches require long connecting lines when they are used in a multi-throw topology due to their physically large layout [3]. The long connecting lines (~250 µm) introduce significant inductance that severely limits the high frequency performance of the multi-throw switch resulting in an isolation of 20 dB at 20 GHz.

In order to increase the upper frequency limit of multi-throw switches, this research is focused on the development of in-line DC-contact switches using cantilevers in order to achieve more compact designs and ultimately higher frequency performance in the multi-throw switch and switched-line phase shifter designs. Previous work on cantilever based DC-contact switches has focused on relatively thick beams (>7 µm) formed by electroplating [4]. Such thick beams have the advantage of being very stiff and thus insensitive to residual stress from the fabrication process. However, this design also results in low isolation due to the need to place the actuation electrode near the tip of the cantilever.

Fig. 1: Initial DC-contact RF-MEMS switches fabricated with a photoresist sacrificial layer.

In order to achieve higher isolation, the height of the cantilever beam above the pull-down electrode must be increased. Unfortunately, for these thick beams, this greatly increases the associated pull-in voltage. Therefore, to maintain a reasonable pull-in voltage while increasing the height of the beam, the thickness must be reduced.

Initial attempts at fabricating cantilevers using a 0.8 µm thick beam made from gold evaporated on top of a photoresist sacrificial layer resulted in significant distortion, as seen in Fig. 1, due to the resulting stress gradient within the beams. Recent research by Stanec *et al.* has demonstrated that the origin of this residual stress is the large difference in coefficient of thermal expansion (CTE) between the polymer based photoresist and the gold film [5],[6]. Therefore, this research has investigated the use of metal sacrificial layers as a means to reduce the CTE mismatch and thus prevent any stress gradient from forming within the cantilevers.

II. SPST AND SP4T SWITCH DESIGN

In the initial designs, the up-state capacitance is chosen to be 5.4 fF in order to provide reasonable isolation and return loss up to 50 GHz for the SP3T and SP4T switches, while maintaining a beam height of 2.6 µm. The resulting spring constant of the cantilever is 7.3 N/m with a pull-down voltage of 53 V. The up-state capacitance is determined by a combination of direct coupling between the beam and the output line and indirect coupling through the pull-down

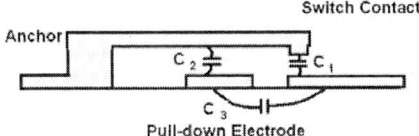

Fig. 2 Diagram of the coupling between the beam, contact pad, and pull-down electrode.

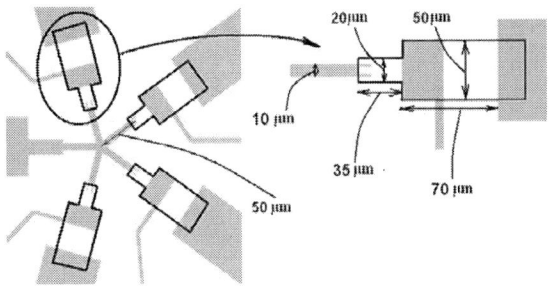

Fig. 3 Layout of the SP4T switch using in-line DC-contact cantilever-based RF-MEMS switches.

electrode as shown in Fig. 2. The up-state capacitance is approximately given by:

$$C_{up} \approx \frac{C_2 \cdot C_3}{C_2 + C_3} + C_1$$

The tip of the cantilever is designed to reduce the direct coupling through C_1 down to the level of coupling through C_2 and C_3. With this design, a SP4T switch can be realized that only requires 50 µm long stubs from the center to each cantilever, as shown in Fig. 3, with a return loss below -20 dB up to 50 GHz. The isolation of the SPST switch is 6 dB worse than the SP3T or SP4T due to the large reflection that occurs when the SPST is in the off-state.

III. Fabrication Process

The switches are fabricated on 110 µm 12,000 Ω·cm silicon with a 0.6 µm SiO$_2$ capping layer. The fabrication process is outlined in Fig. 5. In step (a), Al is deposited over the entire wafer and then patterned using a chlorine-based reactive ion etch. The Au circuit layer is then deposited in the gaps, resulting in a planarized surface. Next, the Al sacrificial layer is deposited (b) and patterned to open up the anchor hole for the cantilever beam. This is filled with Au to once again planarize the surface(c). Lastly, 0.8 µm of Au is evaporated to form the beam using a lift-off process. The wafer is then diced and lapped and then released using a critical point dryer (d). SEM images of the completed SP3T and SP4T switches are shown in Fig. 6. These images demonstrate the ability of the process to yield comparatively thin cantilevers with excellent flatness.

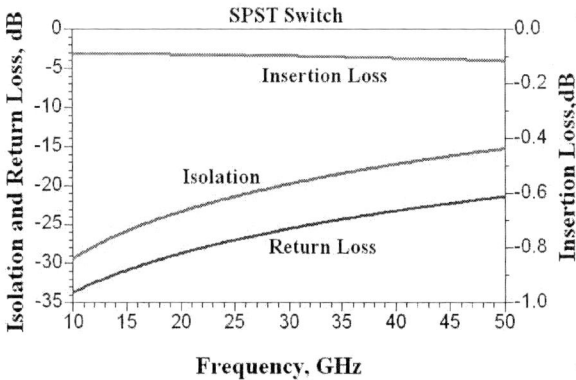

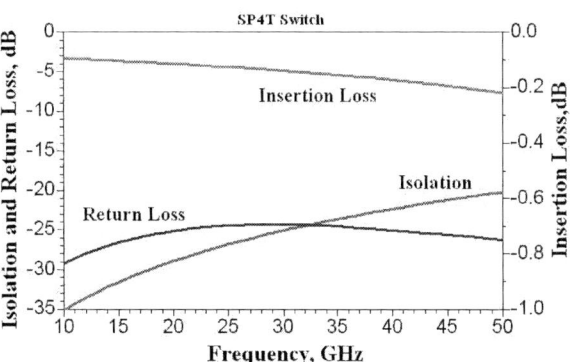

Fig. 4 Circuit simulation results of the SPST and SP4T switch using C_{up}=5.4 fF and R_{on}=1 Ω.

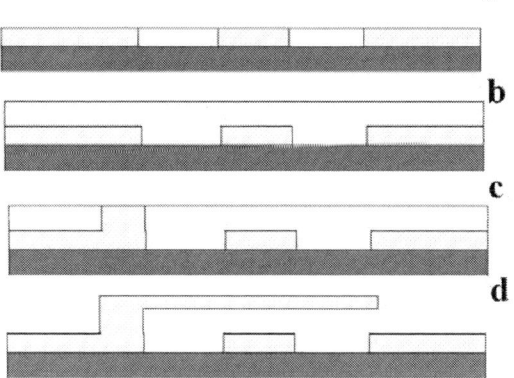

Fig. 5 Outline of the cantilever fabrication process: (a) circuit layer deposition and planarization; (b) sacrificial layer deposition; (c) anchor definition and fill; (d) beam deposition and release.

978-2-8748-7007-1/08 $25.00 © 2008 EuMA

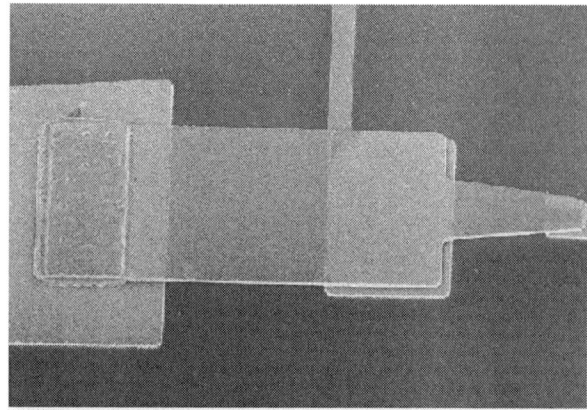

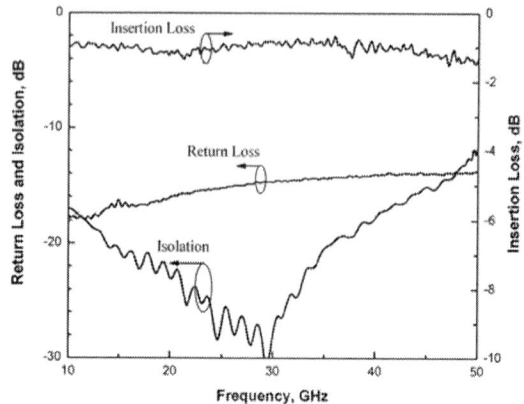

Fig. 7 Measured insertion loss, return loss, and isolation of the SPST switch.

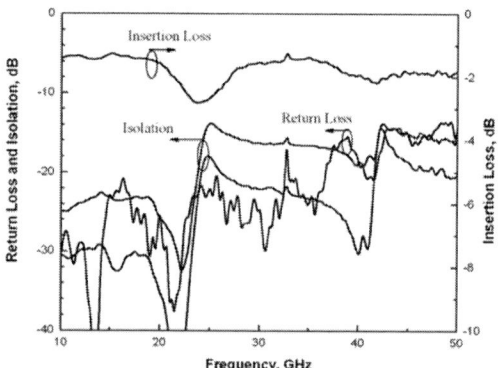

Fig. 8 Measured results for the SP3T switch.

Fig. 6 SEM images of the fabricated SPST, SP3T and SP4T switches.

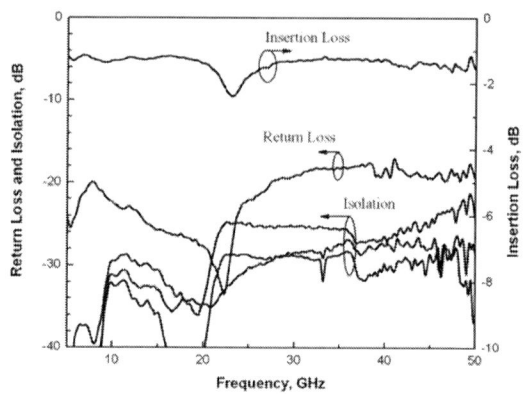

Fig. 9 Measured results for the SP4T switch.

IV. MEASUREMENTS

The fabricated devices are measured using an 8510C VNA with Picoprobe 67A-GSG-150P probes. A TRL calibration is used with standards fabricated on chip. The measured performance of the SPST switch is shown in Fig.7 and demonstrates an insertion loss better than 2 dB and isolation below 14 dB up to 50 GHz. The measurements of the SP3T and SP4T switches are shown in Fig. 8 and Fig. 9 respectively.

As designed, the SP3T and SP4T switches have better isolation (>20 dB) than the SPST switch.

As shown in Fig. 10, each port of the multi-throw switches is connected to a probing pad. Because the measurement reference planes are positioned 200 μm from the probe pads, the loss of long microstrip line connected to the switch is also included in the measurements. During the measurements, the

978-2-8748-7007-1/08 $25.00 © 2008 EuMA

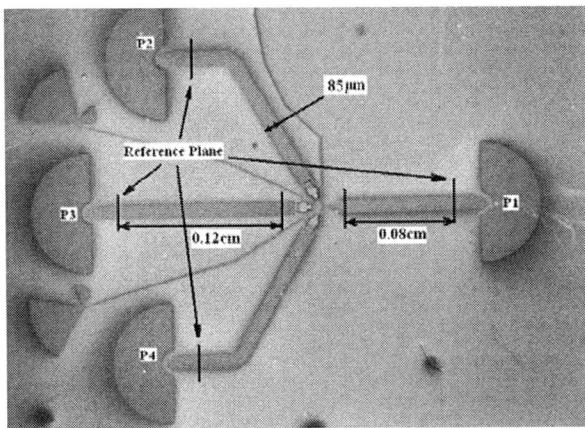

Fig. 10 SEM image of the SP3T switch and probing pads.

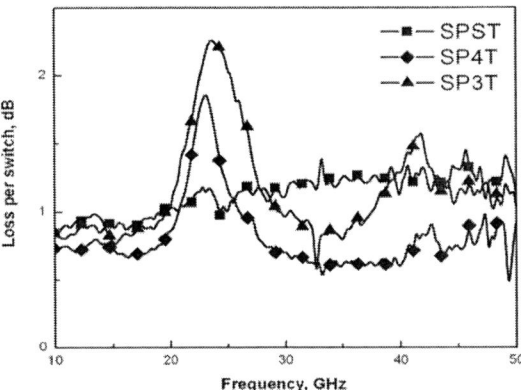

Fig. 11 The loss of SPST, SP3T, and SP4T switches after de-embedding the transmission-line loss.

unused ports in the SP3T and SP4T designs are left open due to the limited number of probes available. This also contributes to the discrepancy between the simulations and measurements.

The insertion loss of the switches is measured more accurately once the losses due to the transmission line sections are de-embedded from the measurements. The loss of each switch is calculated by the following formula:

$$Loss(dB) = -10\log(\frac{|S_{21}|^2}{1-|S_{11}|^2}) - \alpha \cdot L$$

where α (dB/cm) is the attenuation of the microstrip line determined by the TRL calibration using Multical, L is the length of the microstrip line de-embedded from the measurement results (0.2 cm for the SP3T and SP4T and 0.02 cm for the SPST). As seen in Fig. 11, the losses for each switch at 10 GHz are: 0.84 dB for SP3T; 0.73 dB for SP4T; 0.88 dB for SPST. Given these losses, the contact resistance of each switch is calculated: 8.8 Ω for SP3T; 7.7 Ω for SP4T; 9.2 Ω for SPST. These high contact resistances are likely due to insufficiently clean contact surfaces and poor dimple shape. In addition, for the SP3T and SP4T switches, significant radiation is observed around 24 GHz. Full-wave simulations indicate that this is due to the Au bias lines. This radiation is effectively eliminated through the use of resistive bias lines.

V. CONCLUSIONS

DC-contact RF-MEMS switches implemented with thin-film cantilevers are designed, fabricated and measured. Flat cantilevers are realized by using Al as the sacrificial layer in the fabrication process. The SPST switch is used as a building block to realize more complicated SP3T and SP4T switch designs that can be used in compact true-time delay phase shifters. The preliminary measured results of the SP3T and SP4T switches demonstrate isolation of 20 dB and insertion loss below 2 dB up to 50 GHz.

REFERENCES

[1] R.E. Mihailovich, M. Kim, J.B. Hacker, E.A. Sovero, J. Studer, J.A. Higgins, and J.F. Denatale, "MEMS relay for reconfigurable RF circuits" *IEEE Microwave Wireless Comp. Lett.*, Vol. 11, No. 2, pp. 53-55, Feb. 2001.

[2] M. Kim, J.B. Hacker, R.E. Mihailovich, and J.F. Denatale, "A DC-to-40 GHz Four-Bit RF MEMS True-Time Delay Network" *IEEE Microwave Wireless Comp. Lett.*, Vol. 11, No. 2, pp. 56-58, Feb. 2001.

[3] G.-L. Tan, R. E. Mihailovich, J. B. Hacker, J. F. DeNatale, and G. M. Rebeiz, "Low-Loss 2- and 4-bit TTD MEMS Phase Shifters Based on SP4T Switches", *IEEE Transactions On Microwave Theory And Techniques*, Vol. 51, No. 1, Jan., 2003

[4] P. M. Zavracky, N. E. McGruer, R. H. Morrison and D. Porter, "Microswitches and microrelays with a view toward microwave applications," *Int J. RF Microwave CAE*, Vol. 9, No. 4, pp. 338-347, 1999.

[5] J. R. Stanec, M. R. Begley, and N. S. Barker, "Mechanical Properties of Sacrificial Polymers used in RF-MEMS applications," *Journal of Micromechanics and Microengineering*, pp. 2086-2091, Oct. 2006.

[6] J. R. Stanec, C. H. Smith III, I. Chasiotis and N. S. Barker, "Realization of low-stress Au cantilever beams," *Journal of Micromechanics and Microengineering*, pp. N7-N10, Feb., 2007.

Proceedings of the 3rd European Microwave Integrated Circuits Conference

Impact of Diode Geometry on Local Oscillator Breakthrough in Sub-Harmonic Mixers

Venkata Gutta [#1], Anthony Fattorini [*2], Anthony E. Parker [#3], James T. Harvey [*4]

#*Department of Electronic Engineering, Macquarie University*
Sydney, NSW 2109, Australia
[1]vgutta@ieee.org, [3]tonyp@ieee.org

*Mimix Broadband Inc.
10795 Rockley Rd, Houston, TX 77099, USA
[2]tfattorini@mimixbroadband.com, [4]jharvey@mimixbroadband.com

Abstract— An investigation in to the asymmetry of the current-voltage characteristics and the local-oscillator breakthrough in anti-parallel diode sub-harmonic mixers is presented. Twenty nine bare anti-parallel diode pair circuits, have been used to identify those aspects of the diode geometry, that have a strong influence on the diode mismatch and consequently the local-oscillator breakthrough. The circuits were fabricated on a six-inch Gallium Arsenide high electron mobility transistor process.

I. INTRODUCTION

Mixers facilitate frequency conversion in radio systems. Their performance is critical to that of the overall radio system. Increased spurious response of mixers, such as the local oscillator (LO) breakthrough, requires extensive filtering in radios. Filtering increases the complexity as well as the cost of the radio.

Anti-parallel diode pair (APDP) mixers are pumped by a sub-harmonic of the LO-frequency, owing to their anti-symmetrical current-voltage characteristics. They provide an inherent rejection of the LO-breakthrough. Such mixers circumvent the problems associated with implementing fundamental frequency oscillators, capable of meeting low phase-noise requirements at millimeter-wave frequencies [1].

In practise, APDP mixers suffer from some unwanted LO-breakthrough, which has been attributed to diode mismatch. Diode mismatch manifests as an asymmetry in the anti-parallel diode current-voltage characteristics [1], [2], [3].

The aspects of the diode geometry and the fabrication process responsible for the asymmetry, which in turn may help explain LO-breakthrough are yet to be identified. This information is vital to mixer designers on two accounts. Firstly identifying whether the LO-breakthrough is caused by the process tolerances or the diode geometry will reveal if it can be improved without requiring control of the fabrication process parameters. Secondly, if geometrical aspects were found to be the limiting factors, then it will aid in identifying those physical features that need to be addressed for improved performance.

Such knowledge can also help the device designers choose device technologies or processes that can ameliorate the LO-

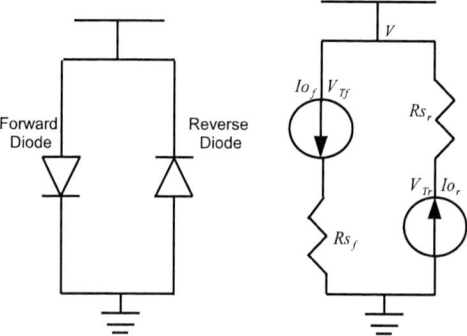

Fig. 1. The left hand side of the figure shows an anti-parallel arrangement of diodes often used as a sub-harmonic mixer. The mismatch between the diodes of an anti-parallel diode pair circuit, expressed in terms of the asymmetry between the intrinsic diode parameters is shown on the right hand side.

breakthrough issue. This investigation has focused on bare-APDP circuits fabricated on a GaAs low-noise high electron mobility transistor (HEMT) process. The equivalent circuit diagram of an APDP circuit can be seen in Fig. 1

The mismatch that exists between the diodes, in spite of their close proximity was studied by measuring the asymmetry of current-voltage characteristics and the LO-breakthrough in a number of bare-APDP circuits. These measurements were statistically analyzed to identify those aspects that have a significant impact on the LO-breakthrough.

Statistical analysis of the current asymmetry and the LO-breakthrough measurements, is useful in understanding the implications of the geometry and the process tolerances. The asymmetry in the current-voltage characteristics can be resolved in to asymmetries associated with each of the intrinsic diode parameters as shown in Fig. 1. The diode parameters include saturation current I_o, diode thermal voltage V_T and series resistance R_s. These parameters in turn provide a link to the geometry of the diode.

The current-voltage characteristic of an APDP circuit, accounting for the diode mismatch can be written as:

978-2-8748-7007-1/08 $25.00 © 2008 EuMA

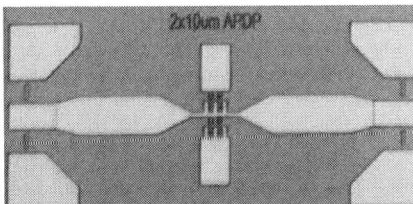

Fig. 2. The figure shows the layout of the APDP circuit used in the investigation. In the case of the diode located at the top, the gate contact is connected to the ground. In the other diode, the metal is connected to the ground. This creates an inherent physical asymmetry in the layout.

$$I(V) = I_{of} e^{\frac{V}{V_{Tf}}} - I_{or} e^{-\frac{V}{V_{Tr}}}, \qquad (1)$$

where I is the diode current, V is the applied bias, I_{of} and I_{or} represent the forward and the reverse-biased saturation currents, V_{Tf} and V_{Tr} represent the forward and the reverse-biased thermal voltages. It should be noted, that the formulation of V_T in this context includes the ideality factor.

A diode-parameter extraction methodology has been used to extract (V_{Tf}, I_{of}, R_{sf}) from the measured forward-biased and (V_{Tr}, I_{or}, R_{sr}) from the reverse-biased characteristics respectively. It utilized a semi-log plot of the measured dc current, $\log I$ against the applied bias V. The slope of a straight-line fit was used to extract the diode thermal voltage and the intercept provided the diode saturation current. The diode series resistance was determined by the ratio of the voltage deviation from the straight-line fit (at the diode current) to that of the diode current [4].

II. DIODE ASYMMETRY

Diode mismatch in an APDP mixer manifests as an asymmetry in the current-voltage characteristic. This in turn results in some LO-breakthrough. The impact of process variations and geometry on the LO-breakthrough, can be found by quantifying the diode mismatch in terms of the asymmetries of the individual diode parameters. The resulting asymmetries are plotted against the LO-breakthrough for all the sample circuits, to help identify those asymmetries that correlate with the LO-breakthrough.

The asymmetry of an intrinsic diode parameter, has been defined as the ratio of its magnitude during forward-bias operation to its magnitude under reverse-biased operation. The asymmetries of the individual diode parameters can be defined as:

$$I_o \text{ Asymmetry} = \frac{I_{of}}{I_{or}}, \qquad (2)$$

$$V_T \text{ Asymmetry} = \frac{V_{Tf}}{V_{Tr}}, \qquad (3)$$

$$R_s \text{ Asymmetry} = \frac{R_{sf}}{R_{sr}}. \qquad (4)$$

Accordingly extracted values of the intrinsic parameters (V_{Tf}, I_{of}, R_{sf}) and (V_{Tr}, I_{or}, R_{sr}) have been used to

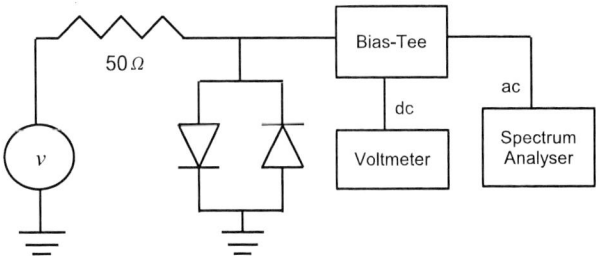

Fig. 3. A figure of test setup showing the anti-parallel diode pair circuit used to measure the LO-breakthrough and the dc-offset. An input signal at 18 GHz was used to excite the APDP circuit and the resulting 36 GHz signal and the dc-offset were measured. The bias-tee was used to isolate the ac and the dc paths. The dc-offset was measured at the voltmeter and the LO-breakthrough was measured using a 40 GHz spectrum analyzer.

compute the asymmetries of the diode parameters in twenty-nine APDP circuits.

III. FABRICATION PROCESS

Bare anti-parallel diode pair circuits, fabricated on WIN Semiconductor's $0.15\mu m$ GaAs HEMT process, were used in this investigation. E-beam lithography was employed in the fabrication of the six-inch wafer. The process has an f_T of 95 GHz. The diode structure utilized a gate, with dimensions $0.15\mu m$ and $10\mu m$ respectively. Each diode comprised of two gate fingers. The drain and the source metallizations of the HEMT structure were combined to form a metal-semiconductor junction with the gate.

IV. MEASUREMENT

The current-voltage characteristics, the LO-breakthrough and the dc-offset were measured from samples located on the same quarter-section of a six-inch GaAs wafer.

A. Current-Voltage Characteristics

An applied dc bias, ranging from -1.3 to +1.3 volts in 0.065 volt steps was applied to each APDP circuit. The current was measured with a precision of up to $1nA$, at a controlled temperature of 300 K. Elimination of data corresponding to faulty devices limited the usable sample population to 29 circuits.

B. LO-Breakthrough

The test setup used to measure the LO-breakthrough and the dc-offset is shown in Fig. 3. The LO-breakthrough and the dc-offset of the twenty nine samples were measured at various drive powers of 10, 12, 15 and 18 dBm respectively. A single-tone input signal with a frequency of 18 GHz was used to excite the APDP circuit and the resulting LO-breakthrough at 36 GHz was measured. A low-pass filter (not shown in Fig. 3) with an out-of-band rejection of 50 dB was used to minimize the input signal generator's second-harmonic at 36 GHz, so as to ensure measurement integrity. The dc-offset was measured at the dc input of the bias-tee, which was used to isolate the dc from the ac-path.

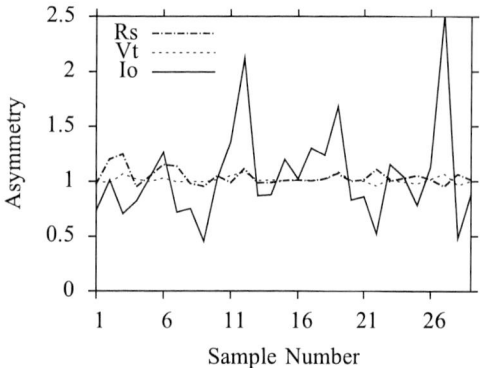

Fig. 4. Asymmetry of the intrinsic diode parameters extracted from the measured current-voltage characteristics of 29 samples of bare anti-parallel diode pair circuits is shown. Significant variation in I_o asymmetry is evident. Magnitude of R_s asymmetry takes a value, that is either greater than or equal to unity across the sample.

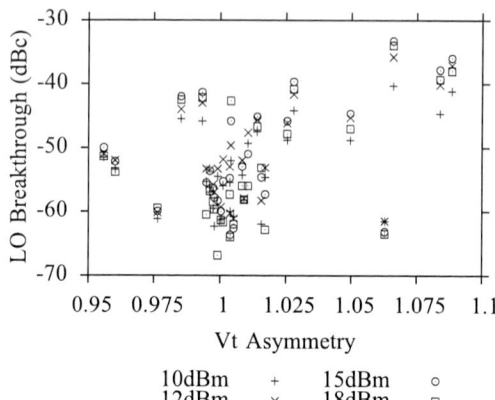

Fig. 6. The data does not show a correlation between the asymmetry of thermal voltages and the LO-breakthrough. The correlation coefficient at various LO-drive powers remains unchanged at around 0.4. The data seems to suggest, that asymmetry in the thermal voltages does not cause LO-breakthrough.

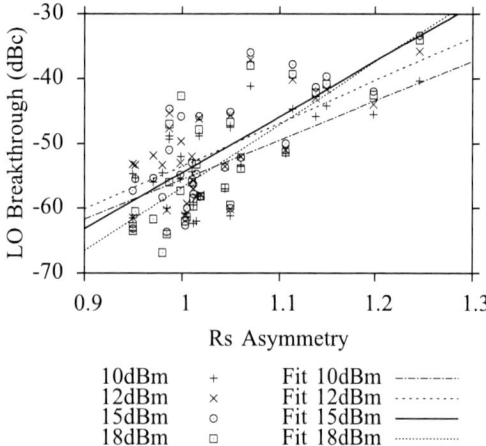

Fig. 5. LO-breakthrough shows an increase in magnitude with the increase in R_s asymmetry, for various LO drive powers. The correlation coefficient of the sample data increases from 0.67 to 0.8 with increasing drive power. The data shows a correlation between LO-breakthrough and R_s asymmetry.

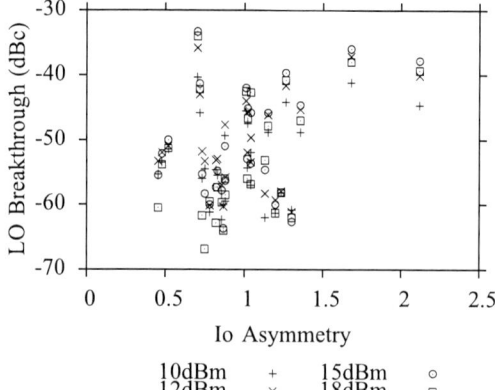

Fig. 7. The data does not show any correlation between the I_o asymmetry and LO-breakthrough. The magnitude of the correlation coefficient varies from 0.05 to 0.16 across the LO drive powers of 10, 12, 15, and 18 dBm respectively. The above figure suggests, that I_o asymmetry has no influence in the generation of LO-breakthrough.

V. RESULTS

Variation in the asymmetries of the intrinsic diode parameters I_o, V_t, R_s can be seen in Fig. 4. Asymmetry in I_o is significantly larger than the asymmetry in either V_T or R_s. Furthermore, I_o and V_T asymmetries appear to be random in regards to their skew. In clear contrast R_s asymmetry seems to be skewed in one direction across the sample.

A general trend of increasing LO-breakthrough with the R_s asymmetry is evident in Fig. 5. The correlation between the R_s asymmetry and the LO-breakthrough, particularly improves with the LO-drive power. No correlation can be observed between the LO-breakthrough and the V_t asymmetry in Fig 6. Absence of any correlation is also replicated between the LO-breakthrough and the I_o asymmetry in Fig 7.

Measured dc-offset strongly correlates with the LO-breakthrough in Fig. 8. This correlation is stronger at higher drive powers. A correlation that improves with the LO-drive power, also exists between the dc-offset and the R_s asymmetry. This can be seen in Fig. 9.

VI. DISCUSSION

The local oscillator breakthrough and the dc-offset are indicators of the diode mismatch in an APDP mixer. The diode mismatch may be a result of many different parameters, some of which may be process related and others more systematic in origin.

The magnitude of I_o, V_T and R_s asymmetries in Fig. 4 show, that the I_o asymmetry is significantly larger than the other two asymmetries. This holds true across the sample. Conventional thinking advocates, that any asymmetry can generate LO-breakthrough and larger the asymmetry, greater the LO-breakthrough.

978-2-8748-7007-1/08 $25.00 © 2008 EuMA

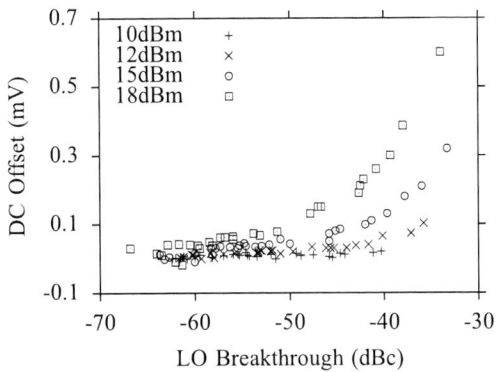

Fig. 8. DC-offset is a strong function of LO-breakthrough and this dependency increases with drive power. The positive dc-offset is a sign, that a systematic effect maybe at play. The positive dc-offset is a strong indication, that thermal effects owing to layout asymmetry maybe at play.

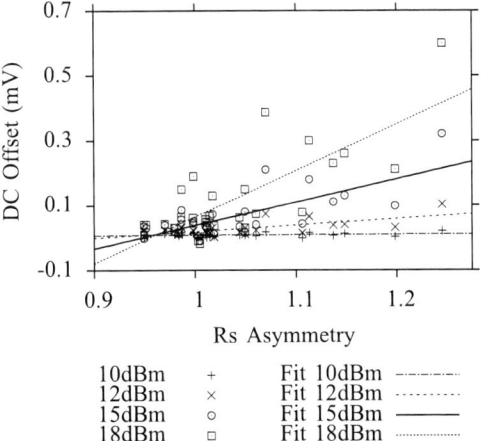

Fig. 9. DC-offset strongly correlates with the series resistance asymmetry. The dc-offset is pre-dominantly unidirectional and has a positive magnitude, which can be attributed to an inherent layout asymmetry. Thermal effects arising out of this layout asymmetry, maybe contaminating the series resistance.

It would be logical to expect an increased LO-breakthrough for large I_o asymmetry. Contrary to such expectations, the LO-breakthrough plotted against the I_o asymmetry in Fig. 7 does not show any correlation. In addition no clear correlation can be observed between the LO-breakthrough and the V_T asymmetry from Fig. 6. However, a strong correlation seems to exist between the R_s asymmetry and the LO-breakthrough.

A correlation also seems to exist between the LO-breakthrough and the dc-offset. It should follow, that a correlation must also exist between the R_s asymmetry and the dc-offset. This can in fact be seen from Fig. 9. All these results when analyzed together seem to indicate, that the LO-breakthrough is generated by the R_s asymmetry.

But the correlation between the LO-breakthrough and the R_s asymmetry is inconsistent with certain other aspects of the measured data. For instance, Figs. 8 and 9 show, that the dc-offset is unidirectional and has a positive magnitude, across

the sample. It should be noted, that R_s asymmetry has a skew in Fig. 4. Asymmetries of the other two diode parameters, I_o and V_t in Fig. 4 appear to have no skew. Since the series resistance is also subject to process variations, it would be logical to expect an R_s asymmetry with no skew.

This apparent conundrum can be explained, if it were to be postulated, that the diodes are subject to certain amount of heating owing to the power dissipation. Presence of heating makes it difficult to extract the series resistance accurately. If the diodes in the diode pair were to have different thermal resistances, then the R_s asymmetry may in fact be a manifestation of the asymmetry of the thermal resistances. But before such an alternative mechanism can be put forth, it is essential to consider the physical layout of the anti-parallel diode pair.

The layout of the APDP circuit can be seen in Fig. 2. In one of the diodes, the gate contact is connected to the ground. In the case of the other diode, the metal contact of the Schottky junction is connected to the ground. The disparity between the dimensions of the gate and the metal contact will result in slightly different levels of heat-sinking between the diodes. This will result in different thermal resistances between the diodes of an APDP circuit.

Based on this evidence, it can be asserted, that the differing thermal resistances of the diodes are the real reason behind the correlation observed between the R_s asymmetry and the LO-breakthrough. This assertion is further supported by the improvement in the correlation with the LO-drive power, seen in Fig. 5. Any difference in thermal resistance that might exist between the two diodes, will accentuate with the drive power. It is probable, that the correlation observed between the R_s asymmetry and the LO-breakthrough is actually an indication of the existence of a temperature differential between the diodes.

VII. CONCLUSION

The statistics on the asymmetry of the diode parameters and the LO-breakthrough suggest, that the series resistance asymmetry is the dominant cause of the LO-breakthrough. An identified mechanism is the difference in thermal resistances, owing to the inherent layout asymmetry. This mechanism is the subject of an ongoing investigation.

ACKNOWLEDGMENT

This work is supported by the Australian Research Council.

REFERENCES

[1] M. Cohn, J. Degenford, and B. Newman, "Harmonic mixing with an antiparallel diode pair," *Microwave Theory and Techniques, IEEE Transactions on*, vol. 23, no. 8, pp. 667–673, Aug 1975.
[2] R. Hicks and P. Khan, "Analysis of balanced subharmonically pumped mixers with unsymmetrical diodes," *Microwave Symposium Digest, MTT-S International*, vol. 81, no. 1, pp. 457–459, Jun 1981.
[3] K. Itoh, K. Kawakami, O. Ishida, and K. Mizuno, "Unbalance effects of an antiparallel diode pair on the virtual local leakage in an even harmonic mixer," *Microwave Symposium Digest, 1998 IEEE MTT-S International*, vol. 2, pp. 857–860 vol.2, Jun 1998.
[4] S. Maas, *Nonlinear Microwave and RF Circuits*. Boston: Artech House, 2003.

RF Noise Shielding Method and Modelling for Nanoscale MOSFET

Jyh-Chyurn Guo[1], Yi-Min Lin[2] and Yi-Hsiu Tsai[3]

Institute of Electronics, National Chiao-Tung University, 1001 Ta-Hsueh Rd, Hsinchu, Taiwan, R.O.C.

[1]jcguo@mail.nctu.edu.tw

[2]ymlin.ee93g@nctu.edu.tw , [3]goodluck.ee94g@nctu.edu.tw

Abstract—**RF noise shielding methods with different coverage areas (Pad and TML shielding) were implemented in two port test structures adopting 100-nm MOSFETs. Noise measurement reveals an effective suppression of NF_{min} but increase of NF_{50}, simultaneously from the shielding methods. The suppression of NF_{min} is contributed from the reduction of $Re(Y_{opt})$ while the noise resistance R_n is kept nearly the same. A lossy substrate model developed in our original work for a standard structure without shielding can be easily extended based on the layout and topology of the shielding schemes to predict the noise shielding effect and explain the mechanisms. The extended lossy substrate model indicates that the elimination of substrate loss represented by substrate RLC networks is the major mechanism contributing the reduction of NF_{min}. However, the increase of parasitic capacitance generated from the shielding structures is responsible for the degradation of f_T and NF_{50}. The results provide an important insight and guideline for low noise RF circuit design.**

I. INTRODUCTION

Noise coupling through Si substrate has been identified as a critical killer to mixed signal IC with digital and analog circuits on a single chip. To overcome this failure mechanism, many works have been done on substrate noise isolation techniques. Among the proposed methods, heavily doped guard ring (GR), triple well, and deep trench are most frequently used [1-3]. However, their noise isolation capability is generally limited to few GHz [3], and become ineffective in advanced RF CMOS circuits with operating frequency driven by nanoscale technology to well beyond 10 GHz. Besides, most of the characterization and analysis focused on the isolation between two features on the same chip, such as port-to-port, pad-to-pad, or device-to-device isolation in terms of $|S_{12}|$ but quite few literatures covered a systematic study of shielding effect on RF noise in miniaturized devices, which are vulnerable to lossy substrate effect. A ground shielded bond pad structure was proposed and fabricated in Si bipolar technology [4]. A significant improvement over pad-to-pad isolation ($|S_{12}|$) and suppression on LNA noise figure (NF) was demonstrated. The experimental results prove the ground shielding effect on isolation ($|S_{12}|$), gain ($|S_{21}|$), and noise (NF). However, a simple resistance model was assumed and implemented to simulate the substrate coupling effect. This simplified model may be valid in sufficiently low frequency ($\leqq 10$ GHz) but is no longer accurate to fit high frequency domain up to tens of GHz. The validity of a simple RC model and the limitation of

frequency have been investigated through a serious comparison between electroquasistatic (EQS) and electrodynamic (ED) models [5]. The results indicate an inductive like characteristics in noise propagation through the substrate and suggest that ED model is indispensable to realize an accurate simulation in high frequency up to several tens of GHz. Unfortunately, the EM analysis requires complicated computation and extensive memory, and is not suitable for circuit simulations. All the mentioned challenges trigger our motivation of this work.

In our previous work, a lossy substrate model in an equivalent circuit form has been developed to accurately predict the RF noise measured from sub-100 nm MOSFETs under high frequency up to 18 GHz [6-8]. The substrate RLC networks, for the first time proposed in our original model, incorporating inductive impedance together with RC networks can simulate the substrate noise coupling through the pad and transmission line (TML) with a broadband accuracy and scalability over different pad structures and TML topologies [9]. In this paper, we will demonstrate that the original lossy substrate model for standard structure without shielding can be easily extended for those with shielding to predict the influence on high frequency S-parameters and noise parameters. An interesting result with an opposite trend in minimum noise figure (NF_{min}) and 50Ω noise figure (NF_{50}) will be presented and discussed.

II. RF NOISE SHIELDING STRUCTURE DESIGN AND EXTENDED LOSSY SUBSTRATE MODEL

100 nm RF n-MOSFETs were fabricated in 130nm CMOS process as the core devices for this study. Multi-gate-finger structures with various finger widths and numbers, W/N=4μm/6, 2μm/12, 1μm/24 under a fixed total width, W_{tot}=WxN=24μm were designed to investigate the trade-off between gate resistance (R_g) and capacitances. The experimental results indicate that the smaller W and larger N can reduce R_g but increase parasitic capacitances at gate terminal. The former one can help suppress gate induced excess noise. Unfortunately, the later one generally degrades f_T due to increased gate capacitances and may overwhelm the advantage of smaller R_g. In this paper with limited pages, W/N=4μm/6 is selected for presentation due to the best high frequency performance represented by highest f_T. Note that W_{tot}=24μm is a relatively small dimension selected for

achieving lower current and low power, but taking a trade-off with lower g_m and higher noise resistance (R_n), and raised challenge to low noise design.

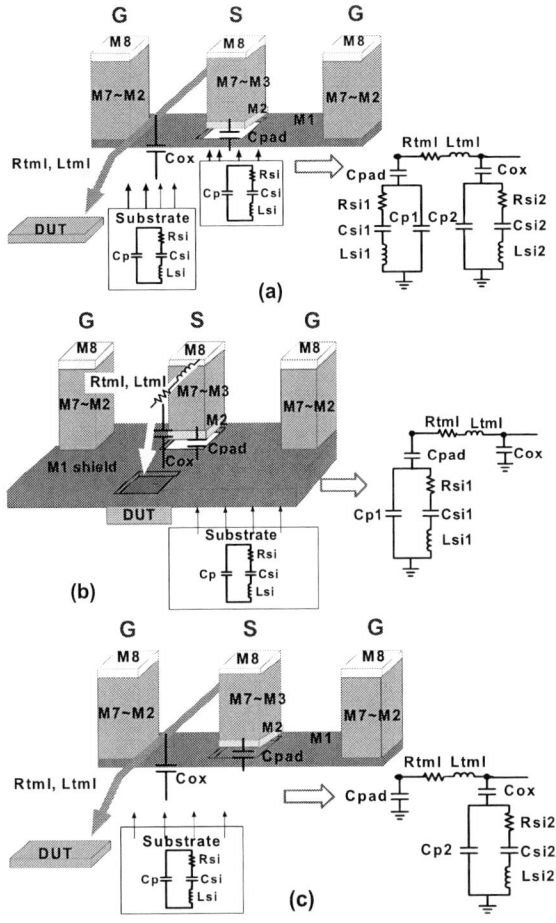

Fig.1 RF test structures adopting device under test (DUT), GSG pads, TML, and different shielding schemes (a) standard without shielding (b) TML shielding (c) pad shielding, and the corresponding equivalent circuit models under an ideal shielding condition.

To develop RF noise shielding methods in miniaturized devices for low noise RF CMOS design, two different shielding schemes were implemented using 0.13μm BEOL (Back-End-of-Line) process with 8 layers of Cu and FSG as IMD (Inter-Metal Dielectric). G-pads for grounding were constructed with stacked metals from the bottom (M1) to the top (M8). S-pads for signal supply were built from M2 to M8, i.e. stacked metals excluding M1. The preserved M1 is employed as the noise shielding plate with two different coverage areas deployed under the TML and pad, defined as TML shielding and pad shielding. Note that TML connecting DUT (device under test) to S-pad is composed of M8. Fig. 1 illustrates the 3D structures for DUT, GSG pads, TML, and the proposed shielding schemes. Fig.1(a) is a standard structure without shielding and the other two adopting TML and pad shielding are shown in Fig.1(b) and (c) respectively.

Following the test structures, equivalent circuit models adapted to two shielding schemes can be easily developed based on our original lossy substrate model in Fig.1(a). For an ideal shielding, the substrate loss can be eliminated and then the substrate RLC networks (R_{Si}, C_{Si}, L_{Si}, and C_P) under the TML or pad can be removed to leave a simple capacitor, as shown in Fig.1(b) and (c).

The definition of lossy substrate model parameters and extraction method can be referred to our original work [6-7]. The inductance L_{Si} introduced in the substrate networks of our model is a key parameter facilitating accurate simulation of substrate loss and noise propagation at very high frequency [7]. The referred ED model can explain the physics underlying L_{Si}, which involves contribution of magnetic vector potential in the electric field [5]. A perfect shielding can eliminate substrate loss and remove substrate networks under the pad and TML. Then, the original lossy substrate model is reduced to a simple capacitor, such as C_{pad} and C_{ox} corresponding to pad and TML shielding. Note that C_{pad} and C_{ox} can be calculated from layout and process parameters to serve as the initial values. This simplified equivalent circuit can reduce the parameter extraction flow. The model parameters extracted in this reduced flow assuming an ideal shielding, act as an initial model for further optimization to ensure accuracy over extremely high frequency.

TABLE I
RLC model parameters of the extended lossy substrate models for four test structures with different shielding schemes

W4N6	Pad RLC model parameters						
Shielding	C_{pad} (fF)	C_{p1} (fF)	C_{Si1} (fF)	L_{Si1} (pH)	R_{Si1} (Ω)	L_{tml} (pH)	C_c (fF)
X	60.54	84.17	234.2	10.44	230.9	46.71	1.50
TML (M1)	64.25	58.62	119.6	211.4	259.4	18.92	0.58
Pad (M1)	161.1	x	x	x	x	20.92	0.70
	C_{ox} (fF)	C_{p2} (fF)	C_{Si2} (fF)	L_{Si2} (pH)	R_{Si2} (Ω)	R_{tml} (Ω)	
X	21.63	1.106	34.94	65	429.7	0.2	
TML (M1)	29.75	21.61	45.2	248.2	207.5	0.19	
Pad (M1)	22.31	59.66	53.53	744.7	136.9	0.199	

Table 1 summarizes a full set of model parameters extracted through an optimal fitting to the measured S-parameters up to 50 GHz. The results indicate that pad shielding can fully eliminate substrate network under the pad but TML shielding cannot. It suggests that substrate loss is dominated by coupling through the pads in the specified GSG pad topology (M2~M8), and then pad shielding enables a more effective isolation against substrate loss compared with TML shielding. Note that pad shielding leads to a dramatic increase of C_{pad} by around 2.5 times and may degrade high frequency performance due to the added parasitic capacitance. Fig. 2 presents open pad S-parameters over a broadband of 50 GHz, and a good agreement between measurement and simulation using the optimized lossy substrate models adapted to various shielding schemes. The standard structure without shielding indicates a dramatic fall-off in mag(S_{11},S_{22}) with increasing frequency, which reveals a significant substrate loss. TML shielding has very minor effect and demonstrates similar results. On the other hand, pad shielding can recover mag(S_{11},S_{22}) to near a constant independent of frequency and approaching unity, but phase(S_{11},S_{22}) toward more negative. The former one indicates an effective suppression of substrate loss and the later one

978-2-8748-7007-1/08 $25.00 © 2008 EuMA

reveals an increase of capacitance consistently correlated to increase of C_{pad}.

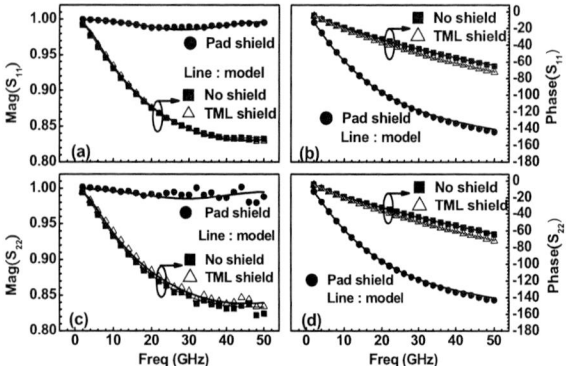

Fig.2 Open pad S parameters for three test structures with different shielding methods (no, TML and pad shielding). A comparison between measurement and simulation by extended lossy substrate models over wide frequency up to 50 GHz (a) mag(S_{11}) (b) phase(S_{11}) (c) mag(S_{22}) (d) phase(S_{22})

III. NOISE SHIELDING EFFECT ON HIGH FREQUENCY PERFORMANCE

The extended lossy substrate models proven for open pads adopting specified shielding schemes were integrated with intrinsic MOSFET at gate/drain (port-1/port-2) for a two-port network circuit simulation to identify the impact on high frequency and noise characteristics [6-7]. Note that an extensive calibration has been done on the intrinsic MOSFET models in terms of V_T, mobility, velocity saturation, CLM, DIBL, overlap and fringing capacitances to realize a good match with measured I-V and C-V characteristics for 100 nm nMOS (not shown for brevity).

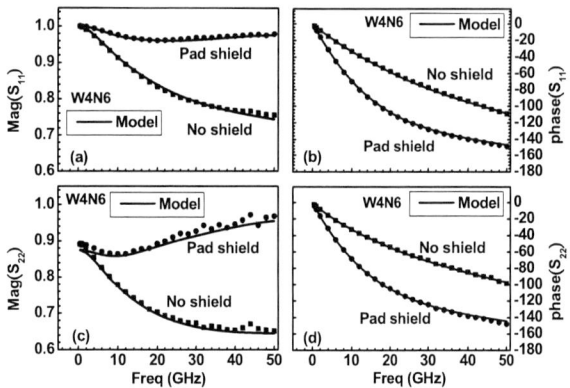

Fig. 3 100 nm nMOS (W/N=4μm/6) S parameters for two test structures (no and pad shielding). A comparison between measurement and simulation by extended lossy substrate models over wide frequency up to 50 GHz (a) mag(S_{11}) (b) phase(S_{11}) (c) mag(S_{22}) (d) phase(S_{22})

The high frequency accuracy is validated by a satisfactory fitting to the measured S-parameters up to 50 GHz, as shown in Fig.3 for a standard one without shielding and another one

with pad shielding. Again, the apparent drop of mag(S_{11},S_{22}) with increasing frequency due to substrate loss, revealed by the standard structure without shielding can be recovered in devices with pad shielding. However, the increase in negative phase suggests additional parasitic capacitances and a potential impact on high frequency performance. Fig. 4 indicates the cut-off frequency f_T corresponding to three test structures and the dramatic degradation suffered by those with pad shielding. The two-port network circuit simulation using the proven lossy substrate models can consistently predict the degradation, shown in Fig. 4(a). The impact considered due to parasitic capacitances introduced from shielding plate (M1) is proven by an analytical expression of f_T, given as $g_m/2\pi(C_{gg}^2 - C_{gd}^2)^{1/2}$ and a good match with that extracted from unit current gain, i.e. $|H_{21}|=1$, shown in Fig. 4(b).

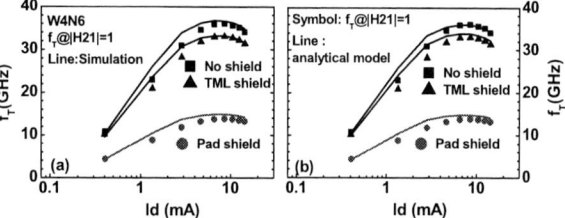

Fig. 4 100 nm nMOS (W/N=4μm/6) measured and simulated f_T for three test structures with different shielding schemes (a) f_T extracted from $|H_{21}|=1$ (b) f_T extracted at $|H_{21}|=1$ and calculated by analytical model $f_T=g_m/2\pi(C_{gg}^2-C_{gd}^2)^{1/2}$

IV. NOISE SHIELDING EFFECT ON RF NOISE PARAMETERS

The influence of shielding structures on RF noise is of our major focus. Four noise parameters (NF$_{min}$, R_n, Re(Y_{opt}), Im(Y_{opt})) were measured by ATN-NP5B under fixed V_{gs}(0.8V for max. g_m) and sweeping frequency to 18 GHz. Fig. 5 exhibits four noise parameters measured from 100 nm nMOS in two port test structure with various shielding schemes in Fig.1(a)~(c). The results indicate an effective NF$_{min}$ suppression of around 1.8/2.05 dB at 10/18 GHz realized by pad shielding but very minor effect from TML shielding. It can be understood from shielding effect on S-parameters demonstrated in Fig.2 that substrate loss can be effectively eliminated by pad shielding but not for TML shielding. The reduction of Re(Y_{opt}) in Fig.5(c) makes a major contribution to NF$_{min}$ suppression whereas R_n keeps nearly the same. The results infer an important insight that R_n is a parameter representing intrinsic device property (g_m, g_{do}, and R_g) independent of substrate loss and shielding effect. On the other hand, Re(Y_{opt}) closely reflects excess noises introduced from the lossy substrate and can be reduced through an effective shielding against the substrate coupling. Im(Y_{opt}) is one more important noise parameter, which performs an optimal matching to the source admittance at the input of DUT. The substantial increase of Im(Y_{opt}) (inductive mode) in test structures adopting pad shielding is realized through a tuner in ATN-NP5B to compensate for additional parasitic capacitance introduced by shielding plate. In this way, the degradation of f_T identified from DUT with shielding (Fig.4), due to added parasitic capacitance can be recovered through

978-2-8748-7007-1/08 $25.00 © 2008 EuMA

an optimal admittance compensation enabled by $\text{Im}(Y_{opt})$, and then NF_{min} can be achieved corresponding to the recovered f_T and an expression of noise parameters in a noisy two-port network given by (1)-(2).

$$F = F_{min} + \frac{R_n}{G_s} |Y_s - Y_{opt}|^2 \qquad (1)$$

$$F_{min} = F(Y_s = Y_{opt}), \ NF_{min} = 10 \times \log(F_{min}) \qquad (2)$$

Regarding NF_{50}, the noise figure normally used in the practice of RF circuit design reveals an interesting result in shielding effect. Fig. 6 demonstrates a significant increase of NF_{50} corresponding to pad shielding that is going a direction opposite to what NF_{min} behaves. The adverse effect on NF_{50} from shielding is considered due to lack of admittance matching for compensating excess capacitances introduced by shielding plate. The proposed mechanism is supported by a consistent correlation with the degradation of f_T shown in Fig. 6(a). The dramatic drop of f_T incurred by shielding can explain the increase of NF_{50} accelerated at higher frequency. The result provides an important guideline in RF circuit design that an appropriately selected inductor is indispensable to realize compensation for parasitic capacitances, which is particularly critical for low noise design incorporating shielding schemes.

Note that noise simulation based on an improved thermal noise model (a replacement of default noise model in BSIM3) and the proven lossy substrate model can accurately predict the measured noise parameters. The major features incorporated in the improved noise model are short channel effects (velocity saturation, CLM, and carrier heating), substrate resistance induced potential fluctuation effect in drain current noise, and gate resistance induced excess noises in both drain and gate current noises.

V. CONCLUSION

RF noise shielding methods have been implemented and demonstrated an effective suppression of NF_{min} in 100 nm MOSFETs. A lossy substrate model incorporating inductive impedances in the substrate network can accurately predict substrate loss effect over extremely high frequency up to 50 GHz and the impact on noise parameters. The extended lossy substrate model adapted to noise shielding schemes proves the noise reduction due to elimination of substrate loss through the removal of substrate RLC network from the original one without shielding. The adverse effect on NF_{50} from shielding reveals an impact from the introduced excess capacitances and suggests an appropriate compensation required for RF circuit design. The proposed noise shielding methods and lossy substrate models with proven broadband accuracy for various shielding schemes can facilitate low noise RF CMOS design.

ACKNOWLEDGMENT

This work was supported in part by NSC under Grant NSC96-2221-E009-186. Also, the authors acknowledge the support from NDL CiC for test chip fabrication and RF Lab. for measurement.

REFERENCES

[1] S. Bronckers, et al., in IEEE RFIC Symp. Proceedings, 2007, pp.753-756
[2] S. Wane, et al., in 2004 IEEE RFIC Symp. Digest, pp.179-182
[3] J. C. Guo, et al. in Symp. on VLSI Tech. Dig., June, Japan, 2003, p.39-40
[4] J. T. Colvin, et al. in IEEE BCTM Conference Proceeding, 1998, pp.109-112
[5] G. Manetas, et al. IEEE Trans. on Electromagnetic Compatibility, vol.49, no. pp.577-584, 2007
[6] Jyh-Chyurn Guo, et al., IEEE T-MTT, vol.54, pp.3975-3985 Nov. 2006
[7] Jyh-Chyurn Guo, et al., IEEE T-ED, vol. 53, pp.339-347, Feb. 2006
[8] J. C. Guo, et al., in 2006 RFIC Symp. Digest, pp.349-352
[9] J. C. Guo, et al. in 2007 RFIC Symp. Digest, pp.299-302

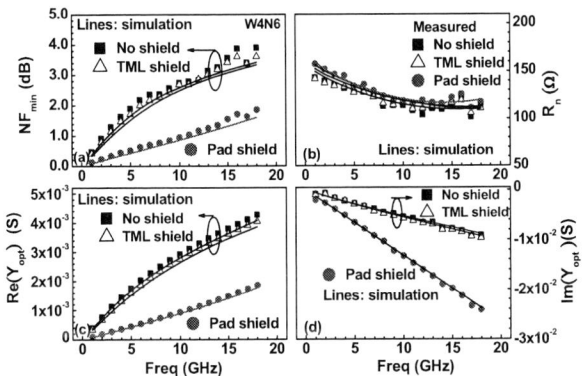

Fig. 5 100 nm nMOS (W/N=4µm/6) noise simulation and measurement for three test structures with different shielding schemes (no, TML, and pad shielding) (a) NF_{min} (b) R_n (c) $\text{Re}(Y_{opt})$ (d) $\text{Im}(Y_{opt})$

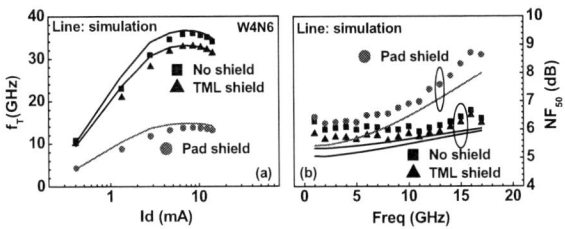

Fig. 6 100 nm nMOS (W/N=4µm/6) f_T before de-embedding and NF_{50} for three test structures with different shielding schemes (a) f_T (b) NF_{50}. Simulation (lines) can consistently predict measurement.

An Optimum Cascode Topology for High Gain Micro/Millimeter Wave CMOS Amplifier Design

Mohammad-Reza Nezhad-Ahmadi, Behzad Biglarbegian, Hassan Mirzaei, Safieddin Safavi-Naieini

Department of Electrical Engineering, University of Waterloo, Waterloo, Ontario, N2L 3G1, Canada

mrnezhad@maxwell.uwaterloo.ca
bbiglarb@maxwell.uwaterloo.ca
hmirzaei@maxwell.uwaterloo.ca
Safavi@maxwell.uwaterloo.ca

Abstract— **An optimum gate-loaded cascode topology for maximizing the gain of a stable CMOS amplifier in micro/millimeter wave band is presented. By adding a piece of transmission line in the gate of cascode transistor, choosing an appropriate matching circuit that includes biasing, and exploiting the proper transmission line structure the gain per stage of CMOS amplifier can be increased and an optimum design can be achieved. Based on this topology a 28GHz amplifier in 0.18μm CMOS technology has been designed and fabricated. Gain of about 16dB was measured for a two-stage cascode at 28GHz in a 0.18μm CMOS technology.**

I. INTRODUCTION

The demand for high speed point to point communications such as wireless local access networks (WLAN), anti-collision radars and wireless personal area networks has aroused the progression of applications at microwave and millimeter wave frequency bands. Extremely fast scaling of gate length in transistors has enabled CMOS technology to work at GHz range. Although the gain and noise performance of CMOS transistor are not as good as GaAs and SiGe counterparts, this technology will be one of the main candidates for future short range GHz wireless system such as sensors, RFIDs, and WPAN devices. Recently, different research groups have implemented various amplifiers on different technologies in various microwave [1]~[4] and millimeter wave bands [5]~[11]. Different topologies were utilized to design a small and efficient amplifier. At the end of microwave band several amplifier on CMOS technology using common-source (CS) [1]~[3] or Cascode [4] topology have been reported.

This paper presents an optimum amplifier topology on 180nm CMOS technology using a novel gate-loaded cascode transistor topology. Using this new topology one can design a more stable amplifier with the higher gain. In Section II the proposed cascode topology will be discussed. Section III describes the design of a two-stage cascode amplifier based on proposed building block in section II. In section IV the measurement and simulation results for 28GHz amplifier is demonstrated, while the conclusions are given in section V.

II. GATE-LOADED CASCODE TOPOLOGY

A. Topology and matching elements

Cascode topology has been the main candidate for most of CMOS LNA designs since cascode amplifiers have the benefit of low power consumption and minimum noise figure simultaneously as well as good isolation. In this topology two MOS transistors are connected from one's source to the other's drain. Normally the bias voltage of the transistors is applied from the gate of the top transistor. One of the main concerns for designing amplifiers is stability of the amplifier. Normally, designers try to design the transistor for having the maximum available gain. On the other hand, this maximum gain can produce instability. In our new method of design, one can attain a higher gain in the stable region.

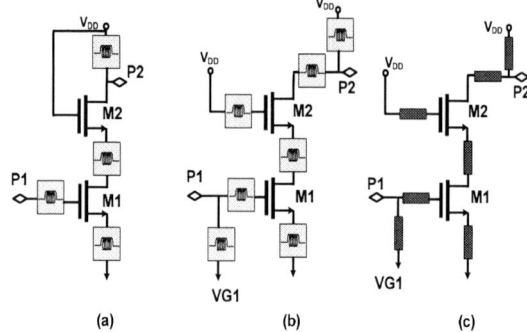

Figure 1: Cascode amplifier building block a) conventional topology b) proposed optimum topology with lumped matching elements c) proposed optimum with transmission line matching elements

Figure 1 depicts the conventional and proposed amplifier building block for cascode amplifier design. The matching elements can be implemented by either lumped elements or transmission lines. The proposed building block consists of a cascode topology with transmission line segments in source of the lower transistor, gate of the upper transistor and in between the two transistors. Also biasing of drain and gate has been provided through the transmission lines that are part of the matching circuit. As shown in Figures 1b and 1c, by adding a piece of transmission line in the gate of upper transistor, gain of the cascode can be increased compare to conventional cascode designs in Figure 1a that connect the gate of the second device directly to VDD. The proposed transmission line for the input and output matching circuits are designed in the Grounded Coplanar Wave Guide (CPWG) technology. Figure 2 illustrates three different transmission line structures in silicon technology. Table I shows the EM simulation result of three different structures. The CPWG presents lower attenuation constant (dB/mm) compare to microstrip and CPW structures and therefore is a better choice for the implementation of matching elements. It also significantly reduces the coupling of RF signal to other circuits on the same substrate through the low resistivity substrate.

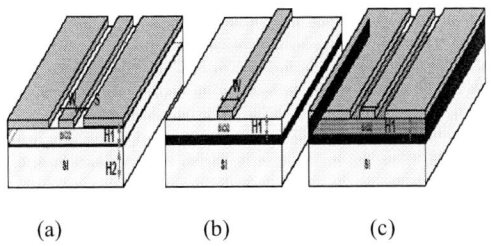

(a)   (b)   (c)

Figure 2: Different transmission line structures on silicon a) CPW b) Microstrip c) CPWG

TABLE I

ATTENUATION VERSUS FREQUENCY FOR A 50OHM LINE IN DIFFERENT TRANSMISSION LINE STRUCTURES

Transmission line type	Frequency		
	20 GHz	25 GHz	30 GHz
CPW W=16μm S=8μm	0.98	1.5	1.75
Microstrip W=16μm, H1=7μm 1.52	1.52	1.72	1.9
CPWG W=7.5μm, S=4μm, H1=7μm	0.4	0.5	0.56

III. AMPLIFIER DESIGN

A. Transistors

The MOS transistors for this amplifier are sized for maximum available gain. The device size is 80μm/0.18 μm for 28GHz amplifier in 0.18μm CMOS technology. Each Cascode stage consumes about 12mA. Hence the total dissipated power is about 36 mW.

B. Amplifier Design

Figure 3 shows the schematic of a two-stage cascode amplifier using the proposed cascode building block. All the CPWG transmission lines for biasing, interconnecting and matching are simulated and optimized in ADS. Momentum was utilized for accurate modeling the interaction between T-junctions, bends and transmission lines. For the transistor the provided model by foundry (TSMC) was used.

The transmission lines at the gate of the top transistors play a major role in the stability of the amplifier and by assigning the appropriate value for the electrical length and the characteristic impedance of the lines a stable amplifier with the maximum gain is achievable. Here in this design for a

two-stage cascode amplifier, a gain about 20 dB with an input matching better than 15 dB and the NF of about 4 dB at 28 GHz is simulated.

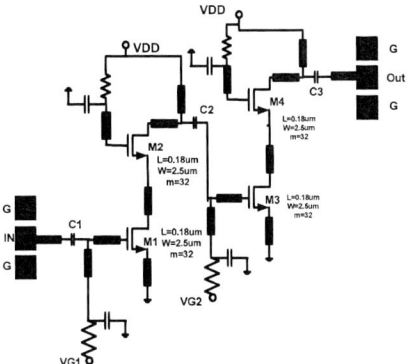

Figure 3: The designed two-stage cascade amplifier for the 28 GHz amplifier on CMOS 180nm technology

IV. MEASUREMENT RESULTS

A. 30GHz amplifier measurement results

Figure 4 shows the gain results of the fabricated amplifier. The amplifier was designed to have maximum gain at 28GHz. Measurement results shows more than 16 dB which is lower than the simulation results in 28 GHz. This variation can be resulted from the poor model of the transistor provided by the foundry above 20 GHz, as well as the dielectric losses of the substrate of the transmission lines. Figure 5 shows input/output matching and the isolation of the amplifier is shown in Figure 6. The simulation results have a good agreement with the measurements and the amplifier. It shows at least 15 dB return loss at both input and output and more than 50 dB isolation between the output and input port. The die micrograph of the fabricated amplifier is shown in Figure 6. A brief comparison between the aforementioned amplifier and the similar amplifiers is done in Table II.

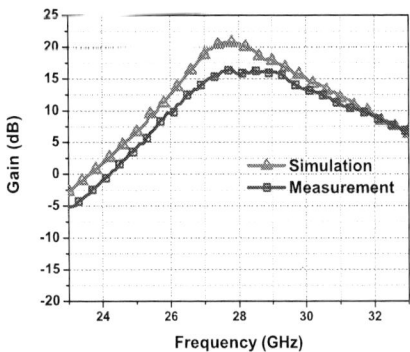

Figure 4: The measured and simulated results of the amplifier. The two stage cascode amplifier shows 16 dB gain at 28 GHz

TABLE II
COMPARISON OF THE RESULTS OF THE OPTIMUM AMPLIFIER WITH SIMILAR
AMPLIFIERS IN CMOS TECHNOLOGY

Ref	[1]	[2]	[3]	[4]	This work
Process	CMOS 0.18μm	CMOS 0.18μm	CMOS 0.18μm	CMOS 0.12μm	CMOS 0.18μm
Frequency (GHz)	24	21.8	25.7	22.5	28
Gain (dB)	13	15	8.9	5.5	16
Gain per Stage (dB)	6.5	5	2.9	5.5	8
Chip Size (mm²)	0.34	--	0.735	0.56	0.3
Input/ Output Return Loss	15/20	-/-	14/12	15/10	17/17
P_{DC}(mW)	14	24	54	5.4	36
Power Supply	1	1.5	1.8	1.2	1.5
Circuit Topology	2stages CS	3stages CG,CS	3stages CS	Cascode 1stage	Cascode 2stages

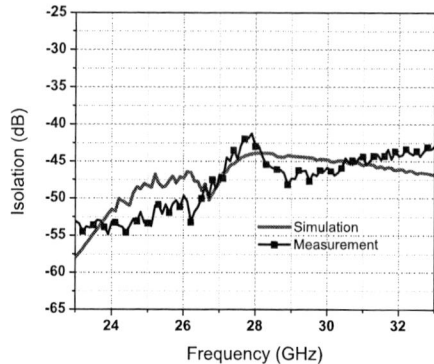

Figure 6: Isolation results of the simulation and measurement results of the 28-GHz amplifier

V. CONCOLUSION

An optimum cascode topology for micro/millimeter wave amplifier design was presented. By adding a piece of transmission line to the gate of the second cascode, choosing proper matching circuits that includes biasing of the transistors, and exploiting the proper transmission line structure the gain per stage of cascode CMOS amplifier can be increased. The transmission lines at the gate of the top transistors play a major role in the stability of the amplifier and by assigning the appropriate value for the electrical length and the characteristic impedance of the lines a stable amplifier with the maximum gain is achievable.

ACKNOWLEDGMENT

This work was supported by NSERC (National Science and Engineering Research Council, Canada), RIM (Research In Motion, Canada) and CITO (Communication Information Technology Ontario).

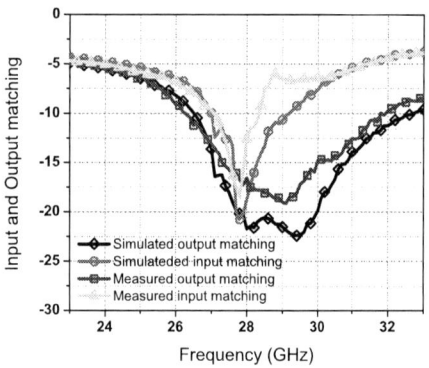

Figure 5: Input and output matching of the simulation and measurement results of the 28-GHz amplifier

REFERENCES

[1] D. Alldred, B. Cousins, and S. P. Voinigescu, "A 1.2 V, 60 GHz radio receiver with on-chip transformers and inductors in 90 nm CMOS," in Proc. IEEE Compound Semiconductor Integrated Circuits Symp., Nov. 2006, pp. 51–54. Radio and Wireless Symposium, 2007 IEEE 9-11 Jan. 2007 Page(s):487 - 490

[2] K. Yu, Y. Lu, D. Chang, V. Liang and M. Frank Chang; "K-Band Low-Noise Amplifiers Using 0.18 μm CMOS Technology", IEEE Microwave and Component Letters, vol. 14, No. 3., March 2004, pp. 106-108.

[3] S. Shin, M. Tsai, R. Liu, K. Lin and H. Wang; "A 24-GHz 3.9-dB NF Low-Noise Amplifier Using 0.18 μm CMOS Technology", IEEE Microwave and Component Letters, vol. 15, No. 7., July 2005, pp. 448-450.

[4] D.Pienkowski,G. Boeck, "22 GHz Amplifier using a 0.12 μm CMOS Technology", International Conference on Microwaves, Radar & Wireless Communications, MIKON 2006.

[5] T. Yao, M. Gordon, K. W. Tang, K. H. Yao, M. Yang, P.Schvan, S.P. Voinigecue, "Algorithmic Design of CMOS LNAs and PAs for 60-GHz Radios," IEEE J. Solid States Circuits, vol. 42, no. 5, pp. 1044-1057, May 2007.

[6] C.H. Doan, S.Emami, A.M. Niknejad, R.W. Brodersen, "Milimeter-wave CMOS design," IEEE J. Solid-State Circuits, vol.40, no.1, pp. 144-155, Jan.2005

[7] B.Razavi, "A 60-GHz CMOS Receiver Front-End.", IEEE J. of Solid-State Circuits, vol. 41, no. 1, pp. 17-22, Jan. 2006

[8] B. A. Floyd, S. K. Reynolds, U. R. Pfeiffer, T. Zwick, T. Beukema, and B. Gaucher, "SiGe bipolar transceiver circuits operating at 60 GHz," IEEE J. Solid-State Circuits, vol. 40, no. 1, pp. 156–157, Jan. 2005C.

[9] H. Wang, Y. H. Cho, C. S. Lin, H. Wang, C. H. Chen, D. C. Niu, J. Yeh, C. Y. Lee, and J. Chern, "A 60 GHz transmitter with integrated antenna in 0.18 μm SiGe BiCMOS technology," in IEEE Int. Solid-State Circuits Conf. (ISSCC) Dig. Tech. Papers, Feb. 2006, pp. 186–187.

[10] B. Floyd, S. Reynolds, U. Pfeiffer, T. Beukema, J. Grzyb, and C. Haymes, "A silicon 60 GHz receiver and transmitter chipset for broadband communications," in IEEE Int. Solid-State Circuits Conf. (ISSCC) Dig. Tech. Papers, Feb. 2006, pp. 649–658.

[11] D. Alldred, B. Cousins, and S. P. Voinigescu, "A 1.2 V, 60 GHz radio receiver with on-chip transformers and inductors in 90 nm CMOS," in Proc. IEEE Compound Semiconductor Integrated Circuits Symp., Nov. 2006, pp. 51–54. Radio and Wireless Symposium, 2007 IEEE 9-11 Jan. 2007 Page(s):487 - 490

Efficient Design Methodology of RF-MEMS based Tuner

David DUBUC, Chloé Bordas, Katia GRENIER

LAAS-CNRS, University of Toulouse
7 avenue du Colonel ROCHE, 31077 Toulouse, France
dubuc@laas.fr grenier@laas.fr

Abstract— This paper describes a design methodology specifically developed for RF-MEMS based tuner, which translates into optimal performances in terms of both impedance coverage of the Smith Chart and $|\Gamma_{MAX}|$ value. The design methodology is moreover associated with an RF-MEMS technology developed for medium power (in the watt range) applications and exhibiting a design robustness regarding the capacitive RF-MEMS contact quality. A tuner has been designed thanks to this efficient procedure, fabricated and measured. The circuit exhibits excellent measured RF-performances: a wide impedance coverage is reached from the first test with a value of $|\Gamma_{MAX}|$ of 0.82 at 14GHz.

I. INTRODUCTION

These last years, RF-MEMS technology has been intensively developed and several demonstrators (phase shifters, reconfigurable filters, impedance tuners, ...) have been successfully designed and present real performances improvement compared with classical technologies [1].

Nevertheless, concerning the RF-MEMS based tuner, no explicit design methodology has been proposed [2-10]. This translates into (1) several design-fabrication-measurement cycles to reach the desired performances and (2) non optimal designs, as all the degrees of freedom have not been properly and fully exploited.

This paper presents an explicit design methodology of RF-MEMS based tuner, which leads to an optimal impedance coverage of the Smith Chart. We would like to point out that the measured performances have been obtained from the first test, which validates the proposed methodology. Moreover the measured RF-performances are in a good agreement with the simulations and are in the state of the art for RF-MEMS based tuner integrated on Silicon.

The second part of this paper presents the RF-MEMS technology developed at LAAS-CNRS and the tuner topology investigated. The third part deals with the proposed design methodology, whereas the measurements are presented and discussed in the fourth part. The conclusion summaries this paper.

II. RF-MEMS BASED TUNER TECHNOLOGY AND TOPOLOGY

The RF-MEMS technology used for this paper has been developed at the LAAS-CNRS and is described in [11]. The RF-MEMS devices are associated to MIM capacitors (see fig. 1) in order to add the capacitor ratio of the resulting varactor

to the list of freedom parameters available for designers. We can then define the capacitor ratio of the varactor by properly designing the MIM capacitors value without redefining the movable part of the RF-MEMS devices (which has been optimised regarding the technological constraints). Moreover, this configuration brings some robustness to the total capacitor ratio, which is initially 30% dispersed because of the technological dispersion of the contact quality between the mobile membrane and the dielectric in the down position. Indeed we prove that the dispersion of the varactor (RF-MEMS associated with MIM capacitor) is reduced down to 5%.

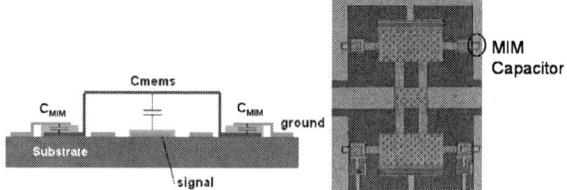

Fig. 1 Varactor (MEMS+MIM) topology and micro-photography

Thanks to this device, we propose to evaluate our design methodology (described in the next paragraph) on a single stub tuner implementing 6 RF-MEMS based varactors as presented in the figure 2.

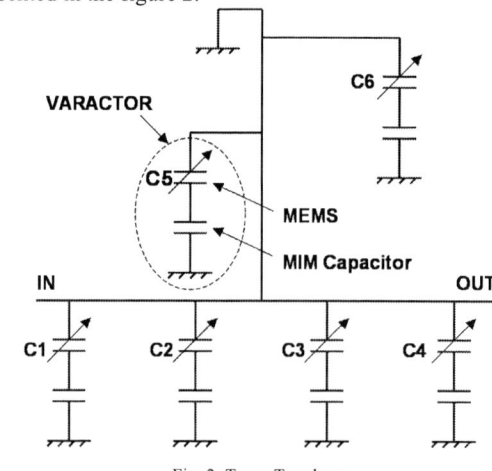

Fig. 2 Tuner Topology

III. DESIGN METHODOLOGY

Thanks to the proposed technology and topology, the parameters, which have to be optimised, are :

- the MIM capacitor value : C_{MIM} (we consider that the MEMS capacitor – without the MIM- are fixed by technological constraints)
- the characteristic impedance of the unloaded line (without the varactors) : Z_0
- the spacing s between the MEMS capacitor both for the input and the output lines and for the stub

These parameters have to be optimised, targeting :

- an impedance coverage
  1. as uniform as possible
  2. providing high values of $|\Gamma|$
  3. providing also low values of $|\Gamma|$
- Technologically feasible (this limits some dimensions).

The target 3 is fulfilled when the Characteristic impedance of the loaded line, with all the MEMS are all in the up position, named $Z_{c,up}$, is near 50Ω :

$$Z_{c,up} = 50\Omega \qquad \text{(design equation n°1)}$$

The first target (target 1) is difficult to analytically expressed. To circumvent this difficulty, we propose to consider that this target is reached if, for each transmission line (TL) , the phase difference between the two MEMS states is 90°. Indeed, when a phase difference of 90° is reached for a TL, an half wise rotation is observed in the Smith Chart then leading to "a best impedance coverage". To express this constraint, we introduce a parameter, which represents the two-states-difference of the normalised length of TL regarding the wavelength :

$$\delta = \frac{l}{\lambda_{down}} - \frac{l}{\lambda_{up}}$$

The impedance coverage will then be optimally uniform if :

$$\delta = 1/4 \qquad \text{(design equation n°2)}$$

which becomes our 2nd design equation on which this work was build. After some mathematical manipulations, the proposed figure of merit can be expressed as a function of the designed parameters:

$$\delta = \frac{f\sqrt{\varepsilon_{r0}}}{c} 2s \left(\sqrt{1 - R + R K_{up}} \right) - \sqrt{K_{up}}$$

where $K_{up} = (Z_0/Z_{c,up})^2$; R, s and ε_{r0} correspond to the capacitor ratio C_{down}/C_{up}, the spacing between varactors and the relative permittivity of the unloaded line respectively.

The second design equation then translates into an explicit expression of the capacitor ratio (then named R_{opt}), which permits to design the value of the MIM capacitors of the varactors :

$$R_{opt} = \frac{\left(\frac{\delta}{B} + \sqrt{K_{up}} \right)^2 - 1}{K_{up} - 1} \qquad B = \frac{f\sqrt{\varepsilon_{reff}}}{c} 2s$$

We finally deduce the optimal value of the MIM capacitor from this optimal capacitor ratio of the varactor and the up-state value of the MEMS devices (without MIM capacitor) :

$$C_{MIM}^{opt} \approx (R_{opt} - 1) \times C_{MEMS}^{up}$$

This last expression assumes that the MEMS capacitor ratio is large enough compared with the one of the resulting varactor.

Finally, the last target (target 2) is fulfilled when the down-state capacitor value of the varactor is sufficiently large to 'short circuit the signal' leading to the edge of the Smith Chart. As this value is already defined by the designed equation n°2, we will optimise the target 2 by tuning the s value which is on the other side constraint by the Bragg condition [12] and the technological feasibility. The s value will then be a parameter to optimise iteratively in order to reach the best compromise between "wide impedance coverage (i.e. designed equation 1 and 2) and "technological feasibility".

We have applied this procedure to our tuner, considering the existing technology and have reached, after some iterations, the values summarised in the table 1.

TABLE I

VALUES OF THE TUNER'S PARAMETERS USING THE PROPOSED METHODOLOGY

Transmission line Characteristic Impedance		63Ω
MEMS capacitor (theoretical)	up	70 fF
	down	4000 fF
MIM capacitor		500 fF
Total Capacitor	up	60 fF
	down	450 fF
Total Capacitor Ratio		7-8

IV. RF-PERFORMANCES AND DISCUSSIONS

The tuner has been fabricated (see microphotography in fig. 3) and on-wafer measured. We would like to emphase that, in the objective to integrate this tuner with active circuits, the developed technology is on silicon (2kΩ.cm) with a BCB interlayer of 20 μm.

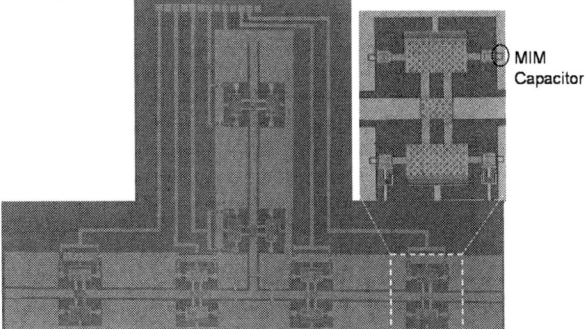

Fig. 3 micro-photography of the fabricated RF-MEMS single stub tuner

978-2-8748-7007-1/08 $25.00 © 2008 EuMA

The on-wafer S parameters have been measured from 400MHz to 30 GHz for the 2^6=64 possible states. The measured and simulated (with Agilent ADS) S11 parameter vs frequency, when all the MEMS devices are in the down position, are shown in fig. 4. This demonstrates the accuracy of the predicted model of the RF-MEMS varactors and TL over a wide frequency range.

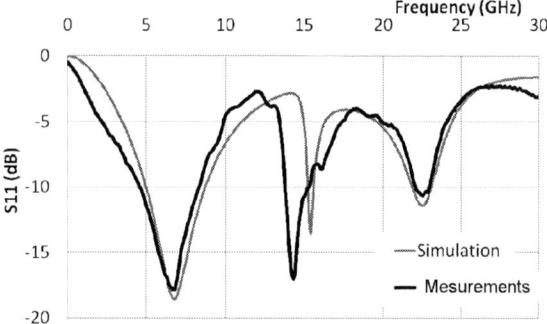

Fig. 4 Measured and simulated S11 parameter when all MEMS devcies are in the down position

The fig. 5 presents the measured and simulated impedance coverage at 12,4 GHz (64 points) with 50Ω input and output terminations. There is a good correspondence in the impedance coverage and especially a wide one as expected.

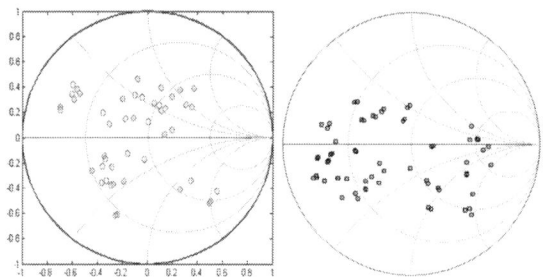

Fig. 5 Measured (rigth) and simulated (left) impedances coverage of the tuner at 12,4 GHz

This result then validates our proposed design methodology as a wide impedance coverage is reached from the first test. We would like to point out that the frequency shift from 20GHz (considered during the design phase) to 12,4GHz is due to a technological drift of the dielectric thickness of the MIM capacitors (which was taken into account for simulations presented in this paper) and then does not invalidate our design procedure.

Moreover, the circuit exhibits excellent RF-performances as 0.82 value at 14 GHz is reached for the $|\Gamma_{MAX}|$ parameter (see table 2), which is at the state of the art of silicon-integrated RF-MEMS based tuner.

TABLE I

MEASURED $|\Gamma_{MAX}|$ VALUES VS FREQUENCY OF THE TUNER

Frequency GHz	$|\Gamma_{MAX}|$
10	0,7
12,4	0,75
14	0,82
18	0,76
20	0,66
23	0,79

To conclude this discussion, we present the predicted impedance coverage at 20GHz of a 12 bits (3 stubs) tuner designed using our methodology. The high resulting impedance coverage, as well as the high value for $|\Gamma_{MAX}|$, bring an other validation of our procedure.

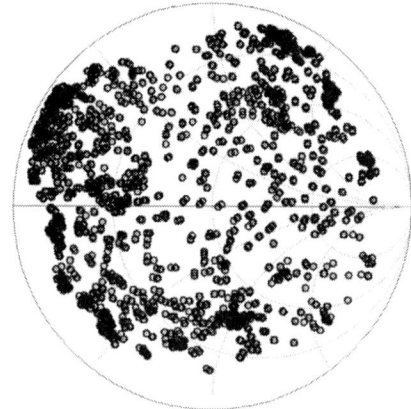

Fig. 6 Impedance coverage of a 12 bits (3 stubs) RF-MEMS tuner (simulations)

V. CONCLUSION

This paper has presented a design methodology of RF-MEMS based tuner that has been validated by measurements. The fabricated tuner exhibits high impedance coverage and high value of $|\Gamma_{MAX}|$ (0.82 at 12,4GHz). Moreover the good agreement between measured and simulated performances validates the proposed approach and predicted models. The proposed design methodology, which can be applied to a large range of RF-MEMS tuner topology, then provides an efficient procedure to reach optimal circuit from the first test.

ACKNOWLEDGMENT

These work has been supported by Thales Alenia Space, Toulouse, France. We would like to thank M. Paillard, J.L. Cazaux and O. Vendier from Thales Alenia Space, Toulouse, France for fruitful discussions.

978-2-8748-7007-1/08 $25.00 © 2008 EuMA

REFERENCES

[1] G. Rebeiz, "RF MEMS: Theory, Design, and Technology", Wiley 2003.

[2] T. Vähä-Heikkilä, G. M. Rebeiz, "A 20-50 GHz reconfigurable matching network for power amplifier applications", IEEE MTT-S Digest, 2004, pp. 717-720.

[3] Q. Shen, N. S. Baker, "RF-MEMS based tunable matching network", IEEE 2003, pp. 313-316.

[4] Q. Shen, N. S. Baker, "A reconfigurable RF MEMS based double slug impedance tuner", European Microwave Conference 2005, Paris, pp. 537-540.

[5] Q. Shen, N. S. Baker, "Reconfigurable matching with a 10-30 GHz distributed RF-MEMS tuner", APMC 2005.

[6] Q. Shen, N. S. Baker, "Distributed MEMS tunable matching network using minimal-contact RF-MEMS varactors", IEEE MTT, 2006, pp. 2646-2658.

[7] Y. Lu, L. P. B. Katehi, D. Peroulis, "High-power MEMS varactors and impedance tuners for millimeter-wave applications", IEEE MTT, November 2005, Vol. 53, No. 11, pp. 3672-3678.

[8] B. Lakshminarayanan, T. Weller, "Reconfigurable MEMS transmission lines with independent Z0- and β- tuning", IEEE MTT-S Digest, 2005.

[9] H.-T. Kim, S. Jung, K. Kang, J.-H. Park, Y.-K. Kim, Y. Kwon, "Low-loss analog and digital micromachined impedance tuners at the Ka-band", IEEE MTT, December 2001, Vol. 49, No. 12, pp. 2394-2400.

[10] T. Vähä-Heikkilä, J. Varis, J. Tuovinen, G. M. Rebeiz, "A reconfigurable 6-20 GHz RF MEMS impedance tuner", IEEE MTT-S Digest, 2004, pp. 729-732.

[11] C. Bordas, D. Dubuc, K. Grenier, "Technological optimization of RF-MEMS capacitive switches with enhanced Power handling – application to K-band MEMS-based tuner", APMC 2006, Yokohama, Japan.

[12] S. Barker, G. Rebeiz, "Distributed MEMS true-time delay phase shifters and wide-band switches" IEEE Trans. On MTT, Vol. 46, Issue 11, Part 2, Nov. 1998 pp:1881 - 1890

Proceedings of the 3rd European Microwave Integrated Circuits Conference

Experimental Study of Ground Plane Width Effect in Multilayer MCM CPW Lines

K. K. Samanta & I. D. Robertson

Institute of Microwave and Photonics, School of Electronics and Electrical Engineering,
University of Leeds, Leeds, LS2 9JT, UK

Abstract — **This paper describes the accurate characterization and detailed experimental study of ground width of multilayer CPW lines up to 110GHz fabricated using photoimageable thick film technology for MCM applications. Further, the effective dielectric constant and characteristic impedance are extracted from the measured S-parameters by conversion to ABCD parameters to investigate the practical effect of ground plane width on a multilayer CPW.**

I. INTRODUCTION

The rising need for higher packaging density and reduced mass, size and cost have been driving the trend towards 3-D multilayer microwave circuits. For designing multilayer multichip modules (MCMs) and monolithic microwave integrated circuits (MMICs) at millimeter-wave frequencies, coplanar waveguide (CPW) is considered as an alternative to microstrip lines [1], [2]. Its principal advantage is the location of the signal grounds as well as signal line on the same substrate surface. This eliminates the need for via holes and thus simplifies the fabrication process, allows easy integration of shunt and series circuit elements [1]-[4]. In practice, the conductor backing is used to provide mechanical strength to the substrate and also to provide a heat sink. In a conductor-backed multilayer CPW, as the dielectric constant (ε_r) for the parallel-plate mode is always higher than the effective dielectric constant for the fundamental CPW mode (ε_{eff}), the CPW mode is inherently faster and hence leaky [5-6]. Coplanar waveguide with finite ground plane widths (FGCPW) are thus often used to overcome this problem.

Multilayer FGCPW is also preferred for an MCM application, due to its advantage of increased circuit density and because novel circuit components can be implemented that are not possible in CPW. In addition, by proper design of the ground plane width, FGCPW can be used up to a very high frequency without significant leakage losses. Fig 1 shows the cross sectional view of a multilayer finite ground width conductor-backed coplanar waveguide (FG-CBCPW) used in this work. The test structures were fabricated using Hibridas photoimageable technology [7] on an Alumina base substrate (CoorsTek ADS-96R) using a conventional off-contact screen printer. Layers of photoimgeable Hibridas dielectric paste (HD1000) and low loss silver conducting paste (HC4700) are printed alternately using screen printing technique to form the required height of dielectric. Printing of each dielectric layers gives a height of about 10 μm where as

that for conducting layer is 5 μm. This technology has been shown suitable for MCM application [8-9] and a wide range of high performance miniaturized components has been developed up to 110 GHz [10-11].

In this paper, the performance of conductor-backed finite ground multilayer FGCPW is studied experimentally. The transmission line parameters are extracted from the measured S-parameters by conversion to ABCD parameters. The effect of the ground width on multilayer FGCPW line on a ceramic substrate is reported.

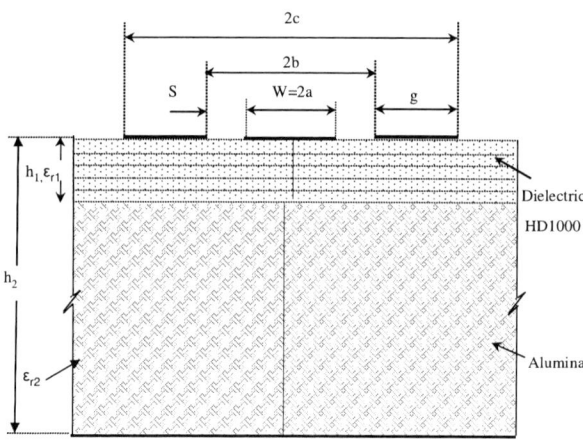

Fig. 1 The cross sectional view of a multilayer finite ground width and conductor backed coplanar waveguide (FG-CBCPW)

II. ANALYSIS

Fig 2 shows the layout of a multilayer finite ground width and conductor backed coplanar waveguide (FG-CBCPW). Both sides of the CPW lines are designed with tapered transition to facilitate on wafer measurements. The FGCPW was fabricated using multilayer photoimageable Hibridas dielectric paste (HD1000) of height, h_1, which was printed onto an alumina substrate of height (h_2-h_1). The relative dielectric constants of the two layers were ε_{r1} and ε_{r2} respectively. The low loss silver conducting paste (HC4700) was printed to form the top metal tracks.

978-2-8748-7007-1/08 $25.00 © 2008 EuMA   402

A wide range of test lines was characterized using measured s-parameters from an Agilent 8510XF system with Cascade Summit 10000 probe station. The effective dielectric constant (ε_{eff}) and characteristic impedance were extracted from the measured data by conversion to ABCD parameters.

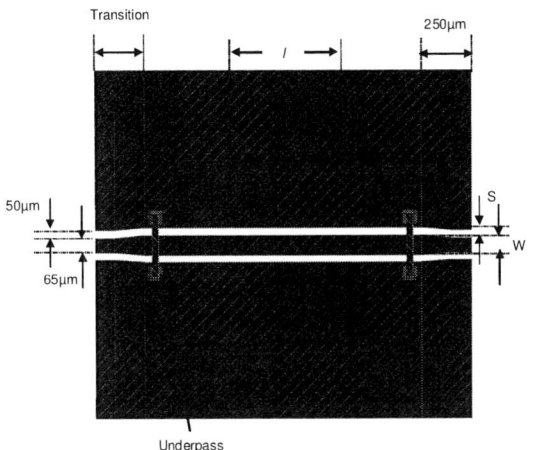

Fig. 2 Layout of a multilayer finite ground width and conductor backed coplanar waveguide (FG-CBCPW)

The ABCD matrix of a transmission line of length, ι, propagation constant, γ, and characteristic impedance, Z_{ch}, can be expressed as [13]:-

$$\begin{pmatrix} A & B \\ C & D \end{pmatrix} = \begin{pmatrix} \cosh \gamma\iota & jZ_{ch} \sinh \gamma\iota \\ jY_{ch} \sinh \gamma\iota & \cosh \gamma\iota \end{pmatrix} \quad \text{.......................(1)}$$

The ABCD parameters in terms of measured S-parameter can be represented as [13]

$$A = \frac{(1+S_{11})(1-S_{22}) + S_{12}S_{21}}{2S_{21}}$$

$$B = Z_o \frac{(1+S_{11})(1+S_{22}) - S_{12}S_{21}}{2S_{21}} \quad \text{...............(2)}$$

$$C = Y_o \frac{(1-S_{11})(1-S_{22}) - S_{12}S_{21}}{2S_{21}} \quad \text{...............(3)}$$

$$D = \frac{(1-S_{11})(1+S_{22}) + S_{12}S_{21}}{2S_{21}}$$

In case of measured s-parameter data from the VNA , Z_0 is the normalization resistance, 50 Ohm. Then Z_{ch} can be obtained from:-

$$\frac{B}{C} = Z_{ch}^2$$

$$or, \quad Z_{ch} = \sqrt{\frac{B}{C}} \quad \text{.......................................(4)}$$

The propagation constant, γ, can be found from:-

$$B.C = \sinh^2 \gamma\iota$$

$$\gamma = \frac{\sinh^{-1} \sqrt{B.C}}{l \times e^{-3}}$$

Where, ι is the length of the transmission line

$$\beta = \frac{2\pi}{\lambda_g} = \frac{2\pi}{\lambda_0 / \sqrt{\varepsilon_{eff}}}$$

The effective dielectric constant is then calculated from:-

$$\varepsilon_{eff} = \left(\frac{c_0}{2\pi f} imag(\gamma) \right)^2 \quad \text{.........................(5)}$$

Where, c_0 is the velocity in the free space

III. EFFECT OF GROUND PLANE WIDTH IN FGCPW TRANSMISSION LINES

The CPW lines were fabricated with different ground plane widths on a dielectric of several heights (20μm, 30 μm, 50μm and 60μm), coated onto an alumina base substrate. The FGCPW are fabricated for ground plane widths, g = 140 μm, 210μm, 280 μm, 350 μm, 560 μm, 700 μm and 1400 μm. The layers of Hibridas dielectric and silver conductor paste were

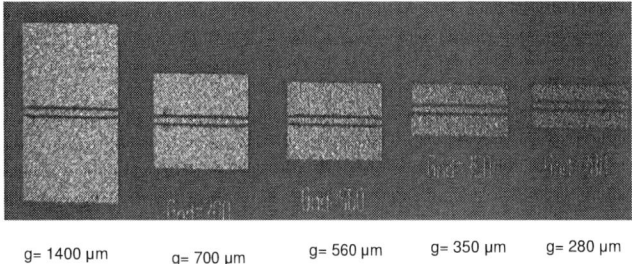

g= 1400 µm   g= 700 µm   g= 560 µm   g= 350 µm   g= 280 µm

Fig. 3 Microphotograph of some of the fabricated FGCPW lines with various ground plane widths (g) of 1400μm, 700μm, 560μm, 350μm and 280μm

printed alternately to achieve the required height of the dielectric. In this paper, the performance of CPW transmission lines for a dielectric layer of 50 μm above alumina substrate of ε_r=9.5 were characterized for $W = (2a) =$ 90 μm and $S = (b-a) = 50$ μm. The CPWs were designed and fabricated for different ground plane widths, which correspond to a ground plane width to centre conductor width (g/W) ratio of 1.5, 2.4, 3.1, 3.9, 6.2, 7.8 and 15. A microphotograph of some of the fabricated FGCPW structures is shown in Fig. 3.

A. Experimental Results:

The measured S_{11} for the CPWs for ground width (g) of 140 μm, 280 μm, 560 μm and 700 μm is depicted in Fig. 4, which remains below -10dB up to 80GHz. At low frequencies (<30GHz) line impedance is more closed to 50 Ohm for higher ground width.

The variation of insertion loss (S_{21}) with frequency for different ground width, 140μm to 700μm is shown in Fig.5. This shows that the attenuation at 80 GHz varies from 0.15 dB/mm to 0.25 dB/mm depending on the ground width.

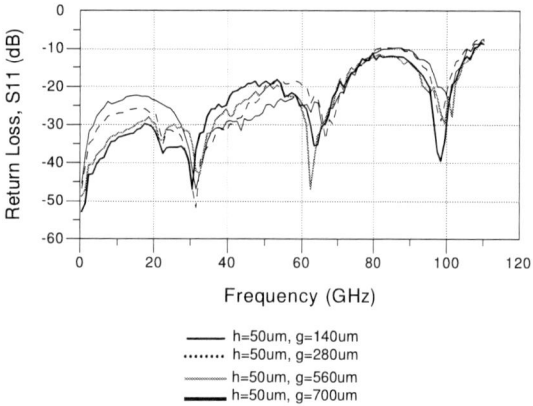

Fig. 4 The measured and the simulated S_{11} of the FGCPW for different ground plane widths

At low frequencies there is no measurable difference in attenuation with ground width if the normalised ground width (g/W) is more than 3. For narrow widths, the attenuation is dominated by conductor loss, whereas for wider ground width dielectric and radiation loss is higher. So, as the frequency increases, attenuation for the line with wide ground width increases faster than that for a narrow line. Due to lower effective dielectric constant the CPW structures are inherently leaky, but the power transfer to a parasitic mode is not significant if the higher order modes (coupled slotline and parallel plate waveguide mode) are not coupled with the fundamental CPW mode. A line with wider g is more dispersive than that with narrow width. The region of high dispersion is due to coupling to the parasitic mode, which is stronger when $h_2/2b$ is small. In Fig. 5, it is observed that for a normalised ground width > 6 (g> 560 μm) losses increases

drastically and for $2c > \lambda_d /2$ (half-wavelength in the dielectric) the resonance occurs.

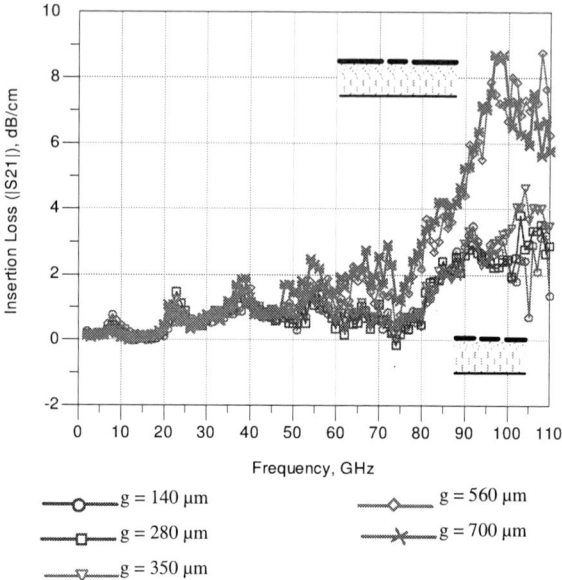

Fig. 5 The measured and the simulated S_{21} of the FGCPW for different ground plane widths

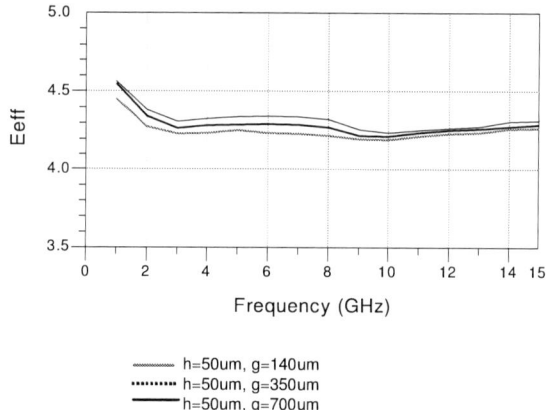

Fig. 6 Measured effective permittivity vs. frequency for ground plane widths of 140μm, 350um, and 700μm

B. Parameter Extraction:

The FG-CPW has been analyzed experimentally using measured s-parameters. The effective dielectric constant, $\varepsilon_{eff.}$ is calculated using the equation (5) for ground width 140μm, 350um and 700μm. For narrow ground plane width, the energy is more closely bounded with in the region of the substrate and consequently ε_{eff} is high in a FGCPW. With

978-2-8748-7007-1/08 $25.00 © 2008 EuMA

ground width the ε_{eff} converges to that of the conventional CPW. For a multilayer FGCPW the computed variation of ε_{eff} with frequency and ground plane width as a parameter is depicted in Fig. 6. The characteristic impedance (Z_0) is calculated using the equation (4) for different ground width from 140 μm to 700 μm. The variation of characteristic impedance (Z_0) with ground plane width is shown in Fig. 7. This shows a significant effect of ground width on Z_0. As the ground plane width increases, the Z_0 converges to that of the conventional CPW.

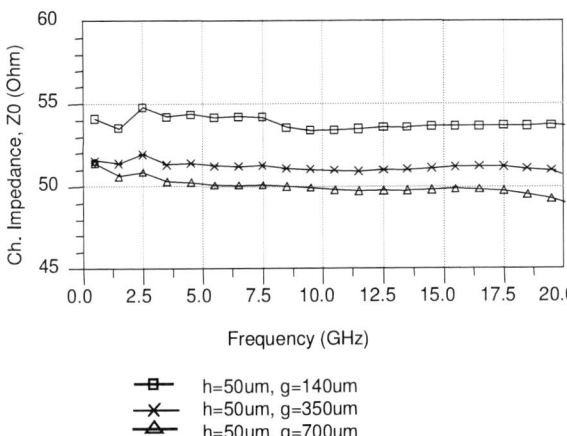

Fig. 7 The measured characteristic impedance with frequency for ground width of 140μm, 350μm and 700μm

IV. CONCLUSIONS

This paper has demonstrated that it is necessary to extract the effective dielectric constant and characteristic impedance of FGCPW, supported on a multilayer MCM substrate, in order to investigate the effect of ground plane width on the propagation parameters of a multilayer MCM. Using this approach, the significant effect of ground plane width on characteristic impedance has been demonstrated experimentally to enable miniaturized mm-wave MCM applications.

V. REFERENCES

[1] G. Ghione and C. U. Naldi, ''Coplanar Waveguides for MMIC Applications: Effect of Upper Shielding, Conductor Backing, Finite-Extent Ground Planes, and Line-to-Line Coupling,'' *IEEE Trans. Microwave Theory Tech.*, Vol. 35, No. 3, pp. 260—267, March 1987.

[2] R. N. Simons, *"Coplanar Waveguide Circuits, Components, and Systems"*, John Wiley & Sons, 2001, ISBNs: 0-471-16121-7 , pp.16-20, 114-118.

[3] Y. Noguchi and N. Okamoto, ''Analysis of Characteristics of the Coplanar Waveguide with Ground Planes of Finite Extent,'' *Trans. IECE., Japan,* Vol. 58-B,No. 12, pp. 679—680. Dec. 1975

[4] M. Cai, P. S. Kooi, M. S. Leong, and T. S. Yeo, ''Symmetrical Coplanar Waveguide with Finite Ground Plane,'' *Microwave Optical Tech. Lett.,* Vol. 6, No.3, pp. 218—220, March 1993.

[5] W.-T. Lo, C.-K. C. Tzuang, S. T. Peng, C.-C. Tien, C.-C. Chang, and J.-W. Huang, ''Resonant Phenomena in Conductor-Backed Coplanar Waveguides (CBCPW's),'' *IEEE Trans. Microwave Theory Tech.,* Vol. 41, No. 12, pp. 2099—2107, Dec. 1993.

[6] R. Jackson, ''Mode conversion at discontinuities in Finite-width conductor-backed coplanar waveguide*'', IEEE Trans. on Microwave Theory and Techn.,* vol. 37, no. 10, pp. 1582-1589,Oct. 1989.

[7] K. K. Samanta and I. D. Robertson, "Advanced Multilayer Thick-film Technology for Cost-Effective Millimetre-Wave Multi-Chip Modules", *IEEE Tenth High Frequency Postgraduate Student Colloquium,* UK, 5-6 September, 2005.

[8] K. K. Samanta, D. Stephens and I. D. Robertson, "Design and Development of a 60GHz Multi-Chip Module Receiver Employing Substrate Integrated Waveguides" *IEE Proc. on Microwaves, Antennas & Propagation*, October 2007, Vol.1, issue 5, p. 961-967.

[9] K. K. Samanta, D. Stephens and I. D. Robertson, "60 GHz Multi-Chip-Module Receiver with Substrate-Integrated-Waveguide Antenna and Filter", *Electronics Letters*, 8th June 2006 Vol. 42, Issue 12, p. 701-702.

[10] K. K. Samanta and I. D. Robertson, "Characterization of TFMS and CPW Lines and Interconnections up to 110GHz in Multilayer Photoimageable Thick Film Technology", *European Microwave Conference*, October, 2006.

[11] Stephens D. ; Young P. R. ; Robertson I. D., "Millimeter-Wave Substrate Integrated Waveguides and Filters in Photoimageable Thick-Film Technology", *IEEE Transactions on Microwave Theory and Techniques*, Issue 99, 2005, pp:1−7.

[12] K. K. Samanta, D. Stephens and I. D. Robertson, "Ultrawideband Characterisation of Photoimageable Thick Film Materials for Microwave and Millimetre-wave Design", *IEEE MTTS, International Microwave Symposium2005*, Long Beach, USA, 11-17 June, [CD ROM].

[13] D.M. Pozar, "Microwave Engineering", *John Wiley & Sons,* 2005, ISBN 0-471-44878-8.

A Low-Power 3-5-GHz UWB Down-Converter with Resistive-Feedback LNA in a 90-nm CMOS Process

Giuseppina Sapone[1], and Giuseppe Palmisano[2]

DIEES, Facoltà di Ingegneria, Università di Catania
Viale Andrea Doria 6, 95125 Catania, Italy
[1]gsapone@diees.unict.it
[3]gpalmisano@diees.unict.it

Abstract— **This paper presents the design and the measurement results of a 3-5-GHz down-converter fabricated in a 90-nm CMOS technology. The circuit consists of a single-ended resistive-feedback low-noise amplifier and two I/Q double-balanced mixers. A transformer-based on-chip single-ended-to-differential conversion allows gain and noise performance to be optimized at a very-low power. A post-layout stability analysis on the LNA is also reported, giving the guidelines to avoid possible oscillations introduced by the single-ended LNA. The down-converter achieves a 23-dB conversion gain, a noise figure of 3.4 dB, and an input third-order intercept point of −8 dBm, while drawing only 9 mA from a 1.2-V supply voltage.**

I. INTRODUCTION

Ultra-wideband (UWB) communication prospective evolved rapidly since its adoption in military applications in 1960s. In 2002, Federal Communications Commission allocated the 3.1-10.6-GHz range for UWB unlicensed use [1], encouraging the development of commercial mass-market applications. The transmission in such a huge portion of the frequency spectrum is allowed in minimum 500-MHz wide channels, though with restriction in signal emissions (power spectral density < -41.3 dBm / MHz) to enable coexistence with well-established narrow-band wireless systems. Even with these limits, the high achievable throughput capacities can be traded-off for distance, mainly splitting the possible wireless applications in high-data rate (HDR, up to 480 Mb) short-range (≤ 10 m) connectivity and low-data rate (LDR, 100 kb to 27 Mb) medium-range (≤ 100 m) communication within sensor networks, the latter with location-tracking capabilities [2], [3]. Both these technologies are part of the Wireless Personal Area Network (WPAN) family of applications. Communication in such a huge bandwidth leads designers to face topics like ultra-wideband input matching and gain performance within a low power consumption budget.

This paper describes the design and measurement results of a fully integrated zero-IF I/Q down-converter operating in the 3-5 GHz frequency range. Throughout the work, details are given on the possible causes of instability introduced by the adoption of the resistive-feedback single-ended low-noise amplifier (LNA) configuration. Though the circuit is designed for LDR systems, it exhibits competitive performance for both HDR and LDR UWB applications in the 3-5-GHz frequency range.

II. CIRCUIT DESIGN

The simplified schematic of the down-converter is depicted in Fig. 1. The circuit includes a single-ended wideband LNA with resonant load, followed by two double-balanced I/Q mixers. The on-chip transformer (T_L) acts as the LNA resonant load and performs the single-ended-to-differential (S/D) conversion of the RF signal at the output of the LNA to the mixers, as well.

During the last years, several techniques were being adopted to implement the wideband 50-Ω input matching of the LNA for UWB/WPAN applications, such as multi-section reactive networks, input common-gate (CG), shunt resistive feedback. Passive filtering is an effective technique to broaden the input matching, but it requires a large on-chip area that makes it more suitable for 3-10-GHz applications [4]. On the other hand, the use of an input CG seems to be a power-efficient and area-saving solution, at the cost of a slightly higher noise figure. However, if the CG transistor is adopted in a single-ended topology, it requires a wideband high-impedance source biasing to avoid a significant loss of the RF signal. This impedance can be implemented by using a very-large on-chip or external inductive component [5], or by means of a Tee-bias probe [6] (in on-wafer measurements). The differential topology solves this problem, but it proves to be unsuitable in a wideband low-noise / low-power design. Indeed, a differential LNA requires an external balun that inevitably introduces high losses due to the required wideband characteristic [7], thus degrading the overall receiver noise figure. Moreover, the target application poses severe limits on power consumption that are better fulfilled by adopting a single-ended amplifier. Therefore, it is a fundamental advantage to implement a LNA with single-ended input.

Resistive shunt-feedback is a simple area-saving solution that proves to be appropriate for the realization of the input matching in the 3-5-GHz UWB band. Differently from a narrow-band inductively-degenerated LNA, an AC-coupled resistive feedback between the output and the input of the cascode amplifier is added, as shown in Fig. 1, with the effect of lowering the quality factor of the narrow-band input network, formed by L_G - M_1 - L_S, thus enlarging the matching bandwidth of the LNA, as described in [8]. Previous papers presenting a resistive-feedback amplifier are mainly focused on the description and the advantages of the topology [8], [9].

However, an actual implementation leads the designer to deal with some crucial aspects. Indeed, although resistive feedback is an effective technique to achieve a wide matching bandwidth, proper design and layout together with accurate post-layout analyses are required to avoid instabilities, due to the connection of R_F and C_F. Indeed, a common-source transistor with capacitive and parasitic inductive feedback can be identified as a well-known oscillator topology [10]. The presence of the resistance R_F mitigates possible oscillations.

In a single-ended LNA topology, another source of instability is represented by the mainly inductive path that connects the gate of M_2 (point X) to the V_{DD} bias point. It can be modelled as shown in Fig. 2. This path also includes the bonding wire inductance, whose effect would not be considered in a differential design. The inductive parasitic is predominant and is divided in two sections, L_{G1} and L_{G2}. These two components model the sections from X to C_G, and from C_G to C_{VDD} (point Y). C_G is a capacitance of several pF with the aim of grounding the gate of M_2, thus rejecting part of the noise injected from the supply. C_{VDD} provides an ac-ground to the V_{DD} path, whereas L_{VDD} models the inductance component from Y to the power supply source, including the bonding wire.

A post-layout stability analysis of the LNA was performed up to 25 GHz, to consider the fifth harmonic of the input signal. Simulations of the stability factors K and B_1 [10] were carried-out, though only K is reported since its violations of the unconditional stability condition (K > 1 & B_1 > 0) are wider than the ones of B_1. Two fundamental sources of instability are evidenced. The first one is the series resonance between inductance $L_T = L_{G1} + L_{G2}$, which can be as large as several hundreds of pH, and its parasitic capacitance. This resonance leads to K < 1 for L_T > 200 pH, as shown in Fig. 3a. In this case the insertion of the series resistance R_G is beneficial since it lowers the quality factor of the resonance. There is a value of R_G above which is K > 1. Fig. 3b plots K with L_T = 400 pH, for several values of R_G.

The problem can be further alleviated by the insertion of C_G, which not only rejects part of the noise injected from V_{DD}, but also splits the inductive path in two sections, L_{G1} and L_{G2}. In

this manner, the section L_{G2} becomes almost uninfluential to the resonance and can cover large distances within a die, whereas L_{G1}, which is still critical for stability, can be minimized by connecting C_G as close as possible to X. As a drawback, C_G introduces instability due to the shunt resonance with L_{G1} and R_G. Fig. 4a plots K vs. frequency for several values of L_{G1} (R_G = 100 Ω, C_G = 5pF). In this configuration the stability condition is satisfied for L_{G1} < 60 pH. Fig. 4b plots K for L_{G2} ranging from 0 to 600 pH (L_1 = 50pH, R_G = 100 Ω, C_G = 5 pF) to verify the scarce influence of L_{G2} after the insertion of C_G.

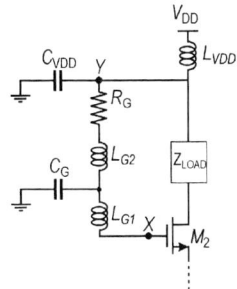

Fig. 2 Model of the parasitic path from the gate of M_2 to V_{DD}.

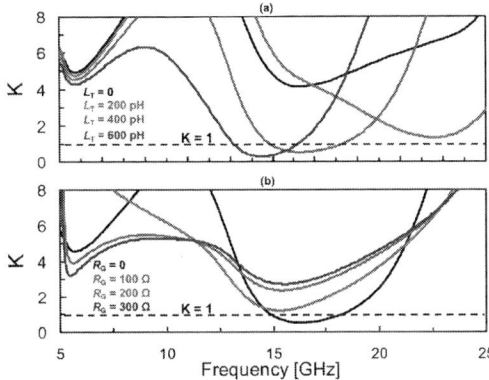

Fig. 3 K for several values of (a) L_T, (b) R_G.

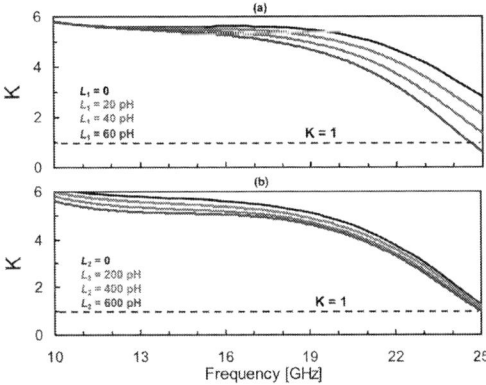

Fig. 4   K for several values of (a) L_1 (R_G = 100 Ω, C_G = 5 pF), (b) L_2 (L_1 = 50 pH, R_G = 100 Ω, C_G = 5 pF).

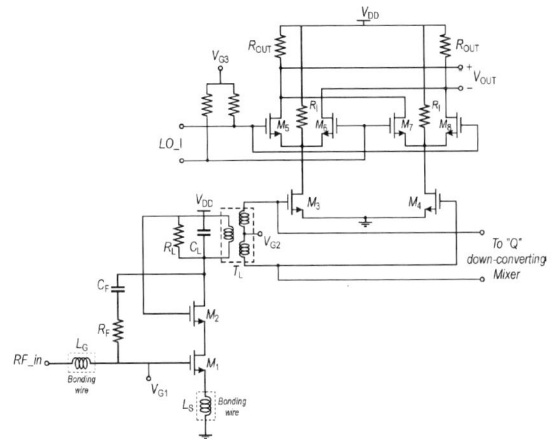

Fig. 1 Transistor-level schematic of the UWB down-converter.

978-2-8748-7007-1/08 $25.00 © 2008 EuMA

One of the main targets of the LNA design is to provide sufficiently high gain to minimize the overall receiver noise figure. In order to decrease the power consumption, the required gain was achieved, at a fixed current budget, by a proper reduction of the V_{GS} and a corresponding increment of the width of the input transistor M_1, still preserving the linearity performance. To this aim, the transistor sizing $(W_1 / L_1) = (W_2 / L_2) = (240/0.1)$ guarantees an excellent compromise among gain, noise and linearity.

A resonant RLC network (R_L, T_L, C_L) implements the wideband load of the LNA. A value of 200 Ω was chosen for the resistor R_L to guarantee the required bandwidth flatness. The transformer T_L has been entirely designed for the current application by means of the $2D\frac{1}{2}$ EM simulator Momentum. It adopts an interleaved symmetrical structure that guarantees a coupling coefficient as high as 0.8. Besides implementing the resonant load of the LNA, T_L accomplishes the S/D conversion of the RF signal to the mixers. The transformer T_L is also exploited to bias the voltage-to-current converters through the secondary winding center-tap.

T_L is the only on-chip inductive component in the circuit, since source and gate inductances ($L_G = 2.2$ nH, $L_S = 0.35$ nH) were implemented by means of bonding wires. In order to preserve the single-ended circuit performance from assembling tolerances, it was accounted for a ±20% variation on both L_G and L_S values.

The I/Q mixers adopt a Gilbert-quad based double-balanced topology, which guarantees high common-mode and substrate noise rejection. As shown in Fig. 1, the bias currents of the quad M_5 - M_8 and of the transconductor M_3 - M_4 are set independently, thanks to resistors R_I that subtract a fixed amount of current to the switching section. This technique allows the desired value of transconductance to be achieved, while limiting the bias current of the quad transistors, which need a lower voltage swing from the local oscillator (LO) to proper switch the signal current. Moreover, the resistive output load will experience a lower dc-voltage drop, allowing the use of higher-value resistors.

III. EXPERIMENTAL RESULTS

The die photo of the down-converter is presented in Fig. 5. The circuit was fabricated in a 90-nm CMOS technology. The whole die area is $1100\,\mu m \times 1300\,\mu m$, including the electrostatic discharge (ESD) protection pad-ring. It is fully compatible with all design rules of an advanced CMOS process, including the invasive density rules. As described in the previous section, the single-ended RF input makes both layout and assembling parasitic estimation crucial. Each pad, which includes ESD-protection diodes, offers to the signal path a shunt capacitance that was estimated to be as high as 100 fF. This is an issue of great concern, especially for high-frequency wideband circuits, whose input signal matching and noise figure performance are significantly degraded by pad capacitive contributions. For this reason, in most reported papers smaller non-standard input pads are adopted, thus considerably improving achievable performance. However, in this work we preferred to use standard pads, thus preserving

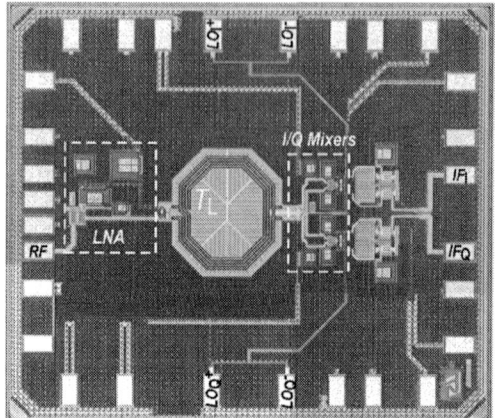

Fig. 5 Die photograph of the UWB down-converter.

the reliability of the fabricated circuit at the cost of a degradation of matching and noise performance. External baluns in the LO signal paths and on-chip output buffers are used only for testing purposes.

The chip was mounted on a standard FR4 substrate using die-on-board assembling. The down-converter was characterized providing three LO frequencies (f_{LO}) to the Gilbert quad separately, i.e. 3.5 GHz, 4 GHz, 4.5 GHz that are roughly the central frequencies of the 3-5-GHz band-plan, as specified in the IEEE 802.15.4a standard [3]. The input frequency (f_{RF}) was swept to cover each of the three 500-MHz channels. The mixers were tested feeding the Gilbert quad with a 350 mV-peak LO signal, which offered the best compromise between gain and noise performance.

Fig. 6 plots the input reflection coefficient (S_{11}) and the conversion gain of the whole down-converter in the band of interest. Maximum measured values of gain are 23.3 dB, 21.1 dB and 19.6 dB in lower, medium and upper channels, respectively. Fig. 7 depicts the double-sideband noise figure (DSB NF) in the same channels. The DSB NF values are as low as 3.4 dB in the two lower bands, whereas the value rises up to 5.5 dB in the upper band, accordingly to the lower measured gain. A single-tone linearity test revealed an input P_{1dB} of −18, −20, −16.2 dBm in the three bands, respectively. An input third-order intercept point (IIP_3) of −8 dBm was also measured using two 10-MHz-spaced tones centered at 4.1 GHz. The circuit draws 9 mA ($I_{LNA} = 5$ mA, $I_{MIXERS} = 3$ mA, $I_{BIAS} = 1$ mA) from a 1.2-V power supply.

Table I summarizes achieved performance, comparing it with state-of-the-art down-converters designed for both LDR and HDR 3-5-GHz UWB applications [5], [9], [11]-[13] and two receivers for a wider range [14], [15].

Where specified, the presented data have been drawn from measurements or extrapolations of the LNA circuit only; otherwise, the reported performance refers to the whole down-converter. Although more than one paper presents the RF receiver [5], [9], [12], including also baseband circuits, the noise depends almost totally on the RF front-end, allowing the comparison to be carried out correctly. The single-ended approach for the LNA is the most common choice to fulfil the

978-2-8748-7007-1/08 $25.00 © 2008 EuMA

TABLE I
COMPARISON WITH STATE-OF-THE-ART UWB DOWN-CONVERTERS

Reference	Technology	RF Frequency [GHz]	Single / Diff. (S/D) LNA	Gain [dB]	NF [dB]	IP_{1dB} [dBm]	IIP_3 [dBm]	P_{DC} [mW]	V_{DD} [V]
[9]	0.18-μm CMOS	3 - 5	D	16*	(3.9 - 4.2)*	—	−4.5	14.4*	1.8
[11]	0.18-μm CMOS	3 - 5	S	10	(3.15 - 4.8)*	−22	−13	46.8	1.8
[12]	0.18-μm CMOS	3 - 5	D	24	8	—	—	8.82	1.8
[13]	0.13-μm CMOS	3 - 5	D	37 / 4	3.6 - 4.1	—	−22 / 2	51	1.5
[5]	0.13-μm CMOS	3 - 5	S	32	6.5 - 8.4	−27.5	—	11.25	1.5
[14]	0.18-μm CMOS	3 - 8	S	17.4	(4.3 - 4.8)*	−10	—	24.3	1.8
[15]	65-nm CMOS	2 - 8	S	23	4.5	—	−7	27.84	1.2
This work	**90-nm CMOS**	**3 - 5**	**S**	**23**	**3.4 - 6**	**−20**	**−8**	**10.8**	**1.2**

* LNA only

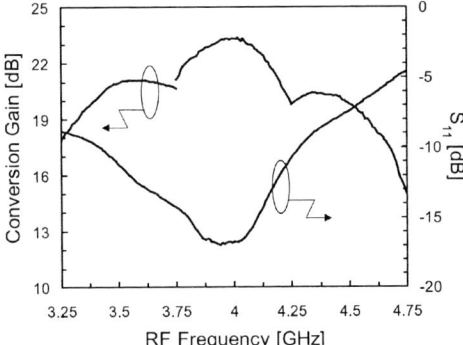

Fig. 6 S_{11} and conversion gain as a function of frequency.

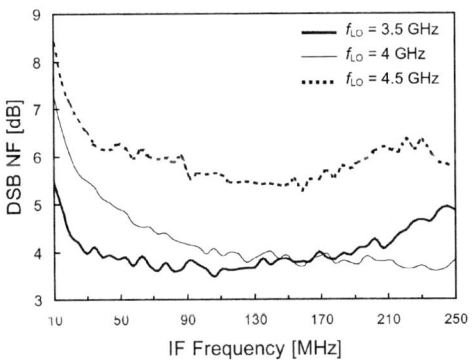

Fig. 7 DSB NF as a function of frequency.

requirements of the UWB application, thus avoiding the use of a lossy wideband balun at the RF input. The comparison with other works reveals that the presented down-converter exhibits excellent performance in terms of gain, *NF*, and linearity, within a very-low power budget.

IV. CONCLUSION

A 90-nm CMOS down-converter for 3-5-GHz UWB low-power applications has been presented. The LNA adopts a resistive-feedback topology with on-chip transformer load. As summarized in Table I, a 23-dB conversion gain, a

DSB NF of 3.4 dB, and an *IIP$_3$* of −8 dBm have been measured. The comparison with state-of-the-art UWB down-converters reveals that the presented circuit achieves excellent noise performance at a very low power budget (only 9 mA from 1.2-V supply). The achieved performance allows the proposed down-converter to suit both LDR and HDR UWB wireless systems in the 3-5-GHz band.

REFERENCES

[1] FCC, "Final Rule of the Federal Communications Commission, 47 CFR Part 15, Sec. 503," Federal Register, vol. 67, no. 95, May 2002.

[2] IEEE 802.15 WPAN High Rate Alternative PHY Task Group 3a (TG3a) [online]. Availble from: http://www.ieee802.org/15/pub/TG3a.html.

[3] *Wireless Medium Access Control (MAC) and Physical Layer (PHY) Specifications for Low-Rate Wireless Personal Area Networks (WPANs), Add Alternate PHYs,* IEEE standard 802.15.4a, 2007.

[4] A. Bevilacqua, and A. M. Niknejad, "An ultrawideband CMOS low-noise amplifier for 3.1–10.6-GHz wireless receivers," *IEEE J. Solid-State Circuit,* vol. 39, pp. 2259–2268, Dec. 2004.

[5] B. Razavi *et al.,* "A UWB CMOS transceiver," *IEEE J. Solid-State Circuit,* vol. 40, pp. 2555–2562, Dec. 2005.

[6] C.-F. Liao, and S.-I. Liu, "A broadband noise-canceling CMOS LNA for 3.1–10.6-GHz UWB receivers," *IEEE J. Solid-State Circuit,* vol. 42, pp. 329–339, Feb. 2007.

[7] A. Valdes-Garcia, C. Mishra, F. Bahmani, J. Silva-Martinez, and E. Sánchez-Sinencio, "An 11-Band 3–10 GHz receiver in SiGe BiCMOS for multiband OFDM UWB communication," *IEEE J. Solid-State Circuit,* vol. 42, pp. 935–948, Apr. 2007.

[8] C.-W. Kim, M.-S. Kang, P. T. Anh, H.-T Kim, and S.-G. Lee, "An ultra-wideband CMOS low noise amplifier for 3–5-GHz UWB system," *IEEE J. Solid-State Circuit,* vol. 40, pp. 544–547, Feb. 2005.

[9] S. Iida, "A 3.1 to 5GHz CMOS DSSS UWB transceiver for WPANs," in *IEEE ISSCC. Dig. Tech. Papers,* Feb. 2005, pp. 214-215.

[10] G. Gonzales, *Microwave Transistor Amplifiers-Analysis and Design,* Upper Saddle River, New Jersey. Prentice-Hall, 1997.

[11] J. Paek, B. Park, and S. Hong, "A 3-5 GHz RF receiver front-end for UWB wireless system," in *Proc. 36th EuMC,* Sep. 2006, pp. 1511-1514.

[12] J. Ryckaert *et al.,* "A 16mA UWB 3-to-5GHz 20Mpulses/s quadrature analog correlation receiver in 0.18μm CMOS," in *IEEE Int. Solid-State Circuit Conf. Dig. Tech. Papers,* Feb. 2006, pp. 368-377.

[13] C. Sandner *et al.,* "A WiMedia/MBOA-compliant CMOS RF transceiver for UWB," *IEEE J. Solid-State Circuit,* vol. 41, pp. 2787–2794, Dec. 2006.

[14] B. Gun Choi *et al.,* "A direct-conversion receiver for low-voltage low-power multi-band UWB with a novel single-level mixer," *in Proc. EuRAD,* Oct. 2005, pp. 255-258.

[15] S. Lee *et al.,* "A broadband receive chain in 65nm CMOS," in *IEEE ISSCC Dig. Tech. Papers,* pp. 418-419, Feb. 2007.

A Fully Integrated 60 GHz SiGe BiCMOS Mixer

Sang-Heung Lee, Ja-Yol Lee, and Haecheon Kim

Electronics and Telecommunications Research Institute (ETRI)
161 Gajeong-dong, Yuseong-gu, Daejeon, 305-700 Korea
shl@etri.re.kr

Abstract— **In this paper, a 60 GHz MMIC down-conversion mixer for 60 GHz WPAN is designed and fabricated on chip using 0.25 μm SiGe:C BiCMOS process technology. This 60 GHz mixer is fully integrated on chip, including active input balun and output balun circuits. The results of the fabricated mixer measured at RF 60 GHz show conversion gain of 10.7 dB, LO to IF isolation and RF to IF isolation of above 30 dB, and input P1dB of -17 dBm. Also, the results of the fabricated mixer measured between RF 57 and 63 GHz show conversion gain of 12.0 ~ 10.7 dB, LO to IF isolation and RF to IF isolation of above 28 dB, and input P1dB of -17 ~ -18 dBm. The chip size of the manufactured mixer is 1.3 mm x 0.8 mm.**

I. Introduction

Millimeter-wave integrated circuits are traditionally implemented using compound semiconductor such as gallium arsenide or indium phosphide [1-2]. Recent advances in SiGe BiCMOS and Si CMOS technologies, and continued progress in various III-V technologies, have made it possible to build low-cost radio receivers and transmitters operating at millimeter-wave frequencies [3-4], including the 57 to 64 GHz Industrial, Scientific, and Medical (ISM) band because the large amount of bandwidth available there allows high data rates. This enables high data-rate communications for wireless personal-area network (WPAN) or point-to-point applications, with possible data rates of at least 150 to 1000 Mb/s. SiGe bipolar technology has been used to fabricate low-noise amplifiers, voltage-controlled oscillators, mixer, and power amplifiers in the 60 and 77 GHz bands as well [5-6].

In this paper, the design and measured results of a double-balanced Gilbert mixer are discussed, which is used for a superheterodyne receiver with a RF input 57 ~ 63 and is fabricated by IHP 0.25 μm SiGe:C BiCMOS process technology.

II. 60 GHz SiGe Bicmos Mixer Design

The heterodyne receiver for 60 GHz WPAN is shown in Fig. 1, which consists of a RX-RF module, a RF front-end module, and a LO module. The RF front-end module covering unlicensed band 57 ~ 63 GHz is composed of an antenna, a band-selection filter, and a LNA. The LO module is configured with 60 GHz PLL and an I/Q generator. In the RX-RF module, double-balanced mixer is utilized for down converting RF signal to IF signal by mixing RF signal and LO signal.

Architecture and circuit of a 60 GHz down-conversion mixer are shown in Fig. 2 (a) and (b) respectively, which are

composed of active RF/LO input baluns, double-balanced Gilbert mixer, IF amplifier, and IF balun. The 57 ~ 63 GHz single-ended RF signal and 56 ~ 62 GHz single-ended LO input are converted to be differential to feed in the double balanced mixer by on-chip active baluns. The converted differential IF signals from mixer core circuits are amplified by 1-stage differential amplifier, and finally single-ended IF output signal is obtained through IF balun.

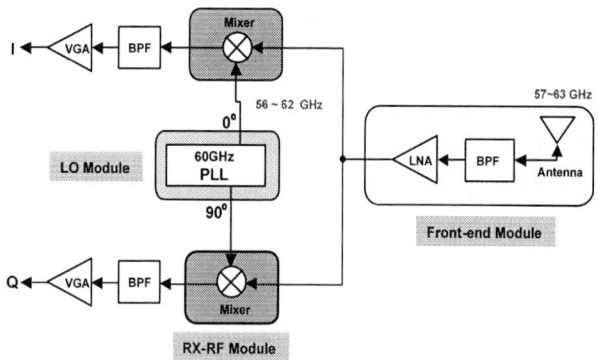

Fig. 1 Architecture of 60 GHz RF receiver.

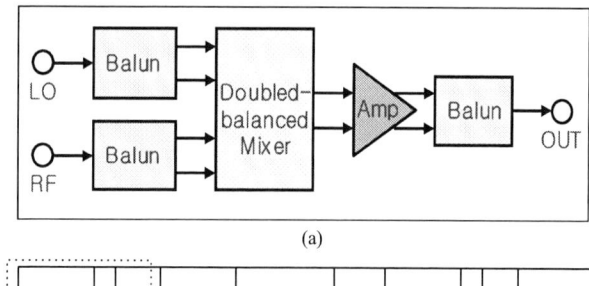

(a)

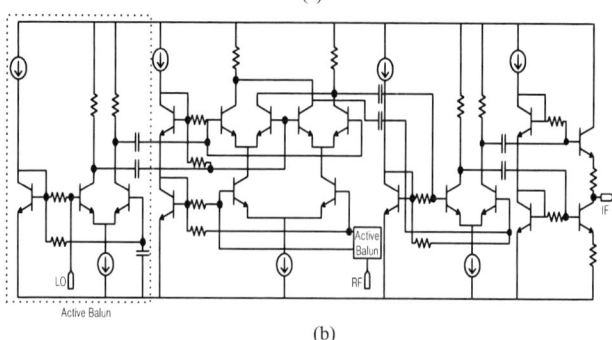

(b)

Fig. 2 60 GHz down-conversion mixer. (a) Architecture, (b) circuit.

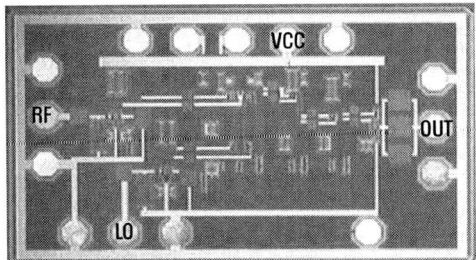

Fig. 3 Microphotograph of the fabricated mixer.

In an active balun (LO balun or RF balun) of the first stage of Fig. 2, the input is applied to the base of transistor and the output measured at the output between the two collectors, which are at the same dc potential. The gain of balun is determined by transconductances of transistors and the collector resistors. In the double-balanced Gilbert mixer of the second stage, when the switching transistors of mixer are driven fully differentially, theoretically, there is no LO leakage at the output. An active balun of the final stage is a push-pull balun, which is composed of a common-emitter with degeneration and the common-collector. The degeneration resistor controls the gain of the common emitter path for the same amount of gain for both inverting and non-inverting signals, resulting in maximum cancellation of LO leakage at the output of the balun. The resistor of common collector is included for IF output impedance matching. The gain of the balun is the sum of the two signal path gains [7-8]. These on-chip input and output baluns are designed and integrated together with this mixer core and IF amplifier and provide good isolation.

In design of 60 GHz down-conversion mixer, ADS and cadence design tools were used. Design of 60 GHz down-conversion was performed as the following: (1) circuit design and simulation in ADS environment, (2) circuit layout in cadence environment, and (3) extraction/verification in cadence environment.

III. Fabrication and Measurements

The photograph of manufactured mixer is shown in Fig. 3 and the entire size of this chip is 1.3 mm x 0.8 mm. In Fig. 3, the input/output signals and supply voltage on pad were indicated.

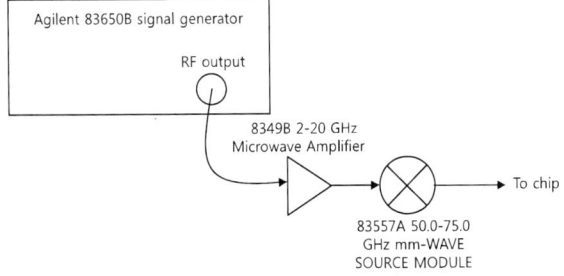

Fig. 4 60 GHz signal source generation setup.

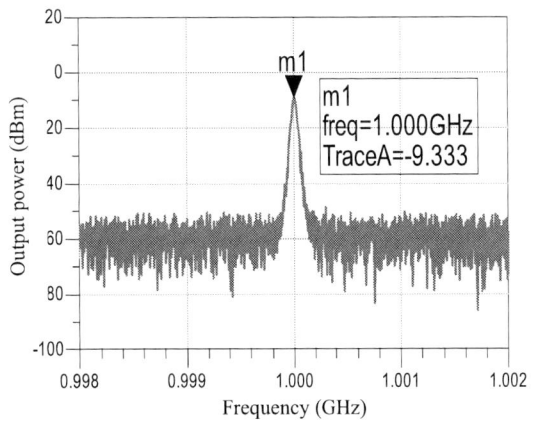

Fig. 5 Port-to-port isolation. S(2,1) and S(1, 2) mean LO to RF isolation, and RF to LO isolation, respectively.

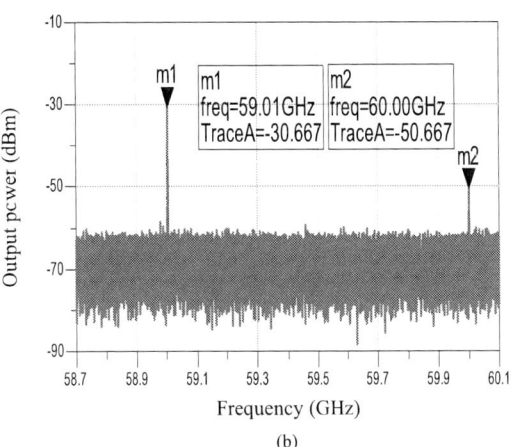

(a)

(b)

Fig. 6 Frequency spectrum measured at IF port. (a) IF signal, (b) LO & RF signals.

As shown in Fig. 4, signal source generation was done using Agilent 83650B signal generator, 8349B 2-20 GHz

978-2-8748-7007-1/08 $25.00 © 2008 EuMA   411

microwave amplifier, and 83557A 50.0-75.0 GHz mm-wave source module and output frequency spectrum was obtained using Agilent 8565E spectrum analyzer and 11974V 50-70 GHz preselected RF section.

Fig. 5, Fig. 6, Fig. 7. Fig. 8, Fig. 9, and Fig. 10 show the results measured at RF 60 GHz and LO 59 GHz. Fig 5 shows port-to-port isolation, and LO to RF isolation and RF to LO isolation are about -29 dB and -28 dB, respectively. Fig. 6(a) and 6(b) show the output characteristics of the fabricated mixer when applying a LO input power of 0 dBm at 59 GHz and a RF input power of -20 dBm at 60 GHz. From Fig. 6, the conversion gain is 10. 7 dB (from Fig. 6(a)) and LO to IF isolation and RF to IF isolation are above 30 dB (from Fig. 6(b)). Fig. 7 shows conversion gain, port-to-port isolation, and input P1dB when applying LO input power between -7.5 dBm and 5.0 dBm at 59 GHz and RF input power of -20 dBm at 60 GHz. From Fig. 7, conversion gain is between 2 and 13 dB, LO to RF isolation and RF to LO isolation are above 28 dB, input P1dB is between -17 dBm and -18 dBm. Fig. 8 and Fig. 9 show input return loss at RF 60 GHz and LO 59 GHz and output return loss at IF 1 GHz, respectively. Fig. 10 shows conversion gain and port-to-port isolation according to RF input power at RF 60 GHz when applying a LO input power of 0 dBm at 59 GHz, and LO to RF isolation and RF to LO isolation are above 30 dB and input P1dB is -17 dBm.

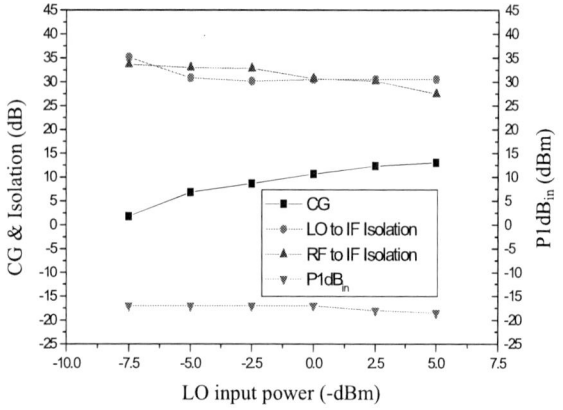

Fig. 7 Conversion gain, isolation and input P1dB according to LO input power.

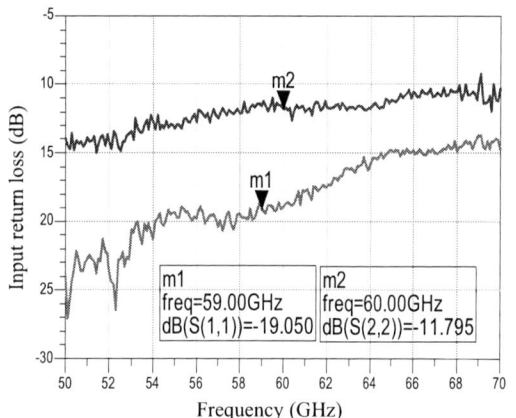

Fig. 8 Input return loss. S(1,1) and S(2,2) mean return loss for LO and return loss for RF, respectively.

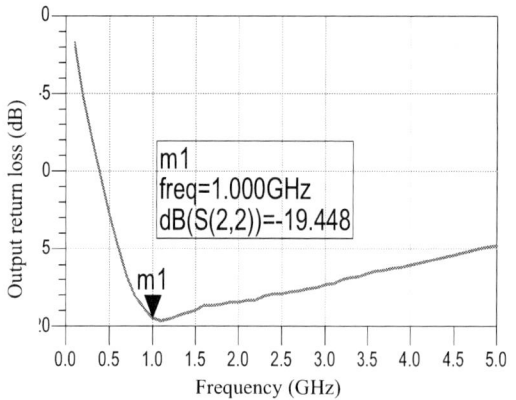

Fig. 9 Output return loss.

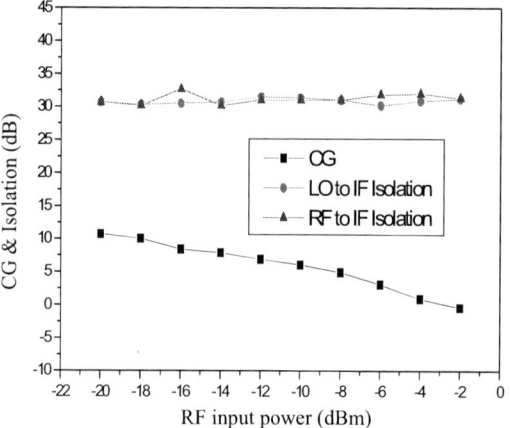

Fig. 10 Conversion gain and isolation.

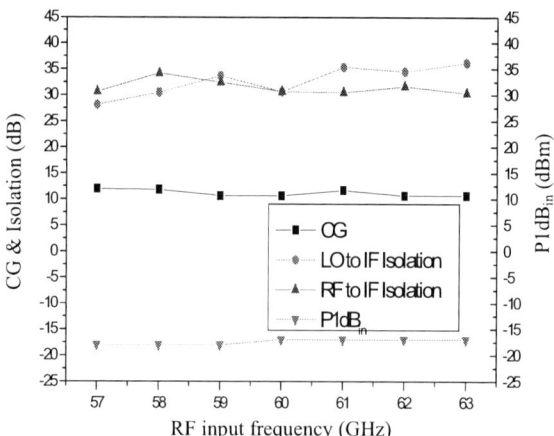

Fig. 11 Conversion gain, isolation, and input P1dB.

Fig. 11 shows conversion gain, port-to-port isolation and input P1dB measured between RF 57 GHz and 63 GHz (IF fixed to 1 GHz). The conversion gain is 12.0 ~ 10.7 dB, LO to RF isolation and LO to IF isolation are above 28 dB, and input P1dB is -17 ~ -18 dBm.

TABLE I

PERFORMANCE SUMMARY FOR RF 60 GHz

Parameters	Measured values
RF frequency [GHz]	60
LO frequency [GHz]	59
Conversion gain [dB]	10.7
LO to RF isolation [dB]	28 >
RF to LO isolation [dB]	28 >
LO to IF isolation [dB]	30 >
RF to IF isolation [dB]	30 >
LO input return loss [dB]	-19
RF input return loss [dB]	-12
IF output return loss [dB]	-19
P1dB$_{in}$ [dBm]	-17
IC [mA] @ VCC=2.5V	15

TABLE II

PERFORMANCE SUMMARY FOR RF 57 ~ 63 GHz

Parameters	Measured values
RF frequency [GHz]	57~63
LO frequency [GHz]	56~62
Conversion gain [dB]	12.0~10.7
LO to RF isolation [dB]	27 >
RF to LO isolation [dB]	25 >
LO to IF isolation [dB]	> 28
RF to IF isolation [dB]	> 30
LO input return loss [dB]	< -17
RF input return loss [dB]	< -12
IF output return loss [dB]	< -20
P1dB$_{in}$ [dBm]	-17 ~ -18
IC [mA] @ VCC=2.5V	15

TABLE III

COMPARISON OF PUBLISHED SI-BASED MIXERS

Parameters	[3]	[5]	this
RF frequency [GHz]	61.5	61.5	60
Conversion gain [dB]	16	9	10.7
P1dB$_{in}$ [dBm]	-17	-7	-17
Current consumption [mA]	115 @ 2.7V	19 @ 2.7 V	15 @ 2.5V
Chip size [mm^2]	1.9 x 1.65	1.24 x 1.8	1.3 x 0.8
Architecture	Double-balanced	Single-balanced	Double-balanced

The performances of the mixer measured at RF 60 GHz and between RF 57 GHz and 63 GHz were summarized in Table I and Table II, respectively. Also, the results of the 60 GHz down-conversion mixer measured at RF 60 GHz was compared with published Si-based results, and they were summarized in Table III. This mixer showed lower current (power) consumption, compared with Si-based results.

IV. CONCLUSIONS

In this paper, a 60 GHz MMIC down-conversion mixer for 60 GHz WPAN was designed and fabricated on chip using 0.25 μm SiGe:C BiCMOS process technology. This 60 GHz mixer was fully integrated on chip, including active input balun circuits (LO, RF) and output balun circuit (IF). The results of the fabricated mixer measured at RF 60 GHz showed conversion gain of 10.7 dB, LO to RF isolation and LO to IF isolation of above 30 dB, and input P1dB of -17 dBm. Also, the results of the fabricated mixer measured between RF 57 and 63 GHz showed conversion gain of 12.0 ~ 10.7 dB, LO to RF isolation and LO to IF isolation of above 28 dB, and input P1dB of -17 ~ -18 dBm. In comparison with other Si-based results of 60 GHz band, this mixer showed lower current (power) consumption. The chip size of the manufactured mixer is 1.3 mm x 0.8 mm.

REFERENCES

[1] A. Fujihara et al., "High performance 60-GHz coplanar MMIC LNA using InP heterojunction FET's with AlAs/InAs superlattice layer," *IEEE MTT-S International Microwave Symposium Digest*, pp. 21-24, Jun. 2000.

[2] K. Nishikawa, et al., "Compact LNA and VCO 3-D MMIC's using commercial GaAs PHEMT technology for V-band single-chip TRX MMIC," *IEEE MTT-S International Microwave Symposium Digest*, pp. 11717-1720, Jun. 2002.

[3] B. Floyd, et al., "SiGe bipolar transceiver circuits operating at 60 GHz," *IEEE Journal of Solid-State Circuits*, vol. 40, no.1, pp. 156-167, Jan. 2005.

[4] S. Pinel, et al., "A 90nm CMOS 60GHz radio," *IEEE International Solid-State Circuits Conference*, pp. 130-131, Feb. 2008.

[5] S. K. Reynolds, "A 60-GHz superheterodyne downconversion mixer in silicon-germanium bipolar technology," *IEEE Journal of Solid-State Circuits*, vol. 39, no.11, pp. 2065-2068, Nov. 2004.

[6] U. R. Pfeiffer, et al., "A 77-GHz SiGe power amplifier for potential applications in automotive radar systems," *IEEE Radio Frequency Integrated Circuits Symposium*, pp. 91-94, Jun. 2004.

[7] D.-Y. Kim, et al., "Up-Conversion Mixer for PCS Application using Si BJT," *IEEE 2nd International Conference on Microwave and Millimeter Wave Technology Proceedings*, pp. 424-427, 2000.

[8] S. –H. Lee, et al., "Monolithic SiGe Up-/Down-Conversion Mixers with Active Baluns," *ETRI Journal*, vol. 27, no. 5, pp. 569-578, Oct. 2005.

Millimeter-Wave Low Spurious Quadruple Harmonic Image Rejection Mixer with 90-degree LO Power Divider

Kenji Kawakami, Kazuhiro Nishida, Morishige Hieda, and Moriyasu Miyazaki

Mitsubishi Electric Corporation
Ofuna 5–1–1, Kamakura–shi, Kanagawa, 247–8501 Japan

Kawakami.Kenji@dc.MitsubishiElectric.co.jp

Abstract—**This paper describes a millimeter-wave low spurious quadruple harmonic image rejection mixer (IRM). The even harmonic mixer with the anti-parallel diode pair (APDP) cancels the even harmonics of the input signal and the LO wave. But if we use the higher order harmonic mixer as an up-conversion mixer, there are many odd-order spurious signals that locate nearby a desired RF signal. In order to obtain a low spurious performance of the quadruple harmonic IRM, it employs the 90-degree divider for the LO wave in addition to the RF signal and the IF signal. In the fabricated 60 GHz band IRM MMIC, the 30dB suppression of the 5th order LO harmonics ($5LO$) can be achieved by the proposed IRM configuration compared with the conventional one. Realization of the low spurious harmonic mixer is useful for the low cost millimeter-wave commercial production.**

I. Introduction

In recent years, the wireless systems using the millimeter-wave frequency have been put to practical use in automotive radars, wireless LANs, and so on[1][2]. These millimeter-wave systems require low phase noise performance of the oscillators, in order to correspond to the advanced modulations such as 64 QAM and OFDM. The source with a lower frequency oscillator is suited for the low phase noise[3]. Therefore, many harmonic and sub-harmonic mixers with APDP have been investigated[4][5][6]. These even harmonic mixers can mix the input signal and the even harmonics of the LO without a multiplier. Also, these even harmonic mixers have a technical feature of extreme low virtual LO leakage that locates nearby a desired RF signal.

We have already developed a 40GHz band even harmonic mixer using open and short circuit stubs[6], a 60 GHz band image rejection harmonic mixer[7], a millimeter-wave broadband monolithic even harmonic image rejection mixer[8], a 94 GHz high performance quadruple sub-harmonic mixer[9] and so on.

The most important problem of the harmonic mixer is to reduce the output spurious level caused when up-conversion operates. As mentioned above, the even harmonic mixer with the APDP cancel the even harmonics of the input signal and LO wave. But if we use the higher order harmonic mixer,

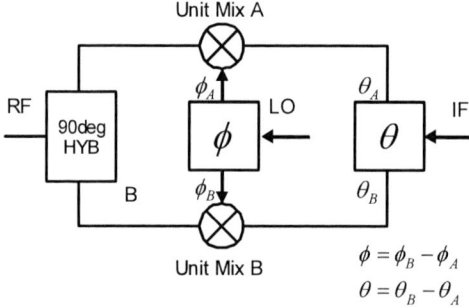

$$\phi = \phi_B - \phi_A$$
$$\theta = \theta_B - \theta_A$$

Conventional harmonic IRM : $\quad \phi = 0, \quad \theta = \pi/2$

Proposed low spurious quadruple harmonic IRM : $\quad \phi = -\pi/2, \quad \theta = \pi/2$

Fig. 1.   The block diagram of the harmonic IRM.

there are many odd-order spurious signals in the output port of the mixer. Especially, the output power of the odd-order harmonics of the LO wave are independent of the input IF level and are very high output level.

In this paper, we focus attention on the phase relations that cancels the undesired signals, and propose the configuration of the low spurious quadruple harmonic image rejection mixer(IRM). In order to obtain the low spurious performance of the quadruple harmonic IRM, it employs the 90-degree divider for LO wave in addition to the RF signal and the IF signal. And also we present the fabricated results of the 60GHz band IRM MMIC.

II. Phase Relations of the Harmonic IRM

Fig.1 shows the block diagram of the harmonic IRM. The conventional IRM consists of two unit mixers, 90-degree hybrid circuit for the RF signal, 90-degree hybrid circuit for the IF signal and in-phase power divider for LO wave.

978-2-8748-7007-1/08 $25.00 © 2008 EuMA

TABLE I
THE SPURIOUS CHART OF THE CONVENTIONAL HARMONIC IRM

	LO	2LO	3LO	4LO	5LO
-5 IF		X		X	
-4 IF	O		O		O
-3 IF		O		O	
-2 IF	O		O		O
-1 IF		X		X	
0 IF	O		O		O
1 IF		O		O	
2 IF	O		O		O
3 IF		X		X	
4 IF	O		O		O
5 IF		O		O	

▨ Canceled in the APDP

X Canceled by the configuration (reverse phase)

O Outputted

In general, the output voltage $v_{out}(t)$ to the input voltage $v_{in}(t)$ in a nonlinear circuit is given by

$$v_{out} = a_1 v_{in}(t) + a_2 v_{in}^2(t) + a_3 v_{in}^3(t) + \cdots . \quad (1)$$

The sine waves of two different frequencies that correspond to the LO frequency and IF frequency are assumed as $v_{in}(t)$.

$$v_{in} = A\cos(\omega_{lo}t + \phi) + B\cos(\omega_{if}t + \theta) \quad (2)$$

The output frequencies and the phases of the each order from the nonlinear circuit can be expressed as follows:

1st order:

$$\omega_{lo}t + \phi$$
$$\omega_{if}t + \theta \quad (3)$$

2nd order:

$$2\omega_{lo}t + 2\phi$$
$$2\omega_{if}t + 2\theta$$
$$(\omega_{lo} + \omega_{if})t + \phi + \theta$$
$$(\omega_{lo} - \omega_{if})t + \phi - \theta \quad (4)$$

$\cdots$

5th order:

$$5\omega_{lo}t + 5\phi$$
$$(4\omega_{lo} + \omega_{if})t + 4\phi + \theta$$
$$(4\omega_{lo} - \omega_{if})t + 4\phi - \theta$$
$$(3\omega_{lo} + 2\omega_{if})t + 3\phi + 2\theta$$
$$(3\omega_{lo} - 2\omega_{if})t + 3\phi - 2\theta$$
$$\cdots . \quad (5)$$

In the case of 2n-th order harmonic mixer, if we assume that the input phases of the unit mixer A are $\phi_A = 0, \theta_A = 0$, the output frequencies and phases of the RF signal and the image signal from the mixer A are as follows:

$$(2n \cdot \omega_{lo} + \omega_{if})t$$
$$(2n \cdot \omega_{lo} - \omega_{if})t \quad (6)$$

where n is integer.

In the same way, if we assume that the input phases of the unit mixer B are $\phi_B = 0, \theta_B = \pi/2$, the output frequencies and phases of the RF signal and the image signal from the mixer B are as follows:

$$(2n \cdot \omega_{lo} + \omega_{if})t + \frac{\pi}{2}$$
$$(2n \cdot \omega_{lo} - \omega_{if})t - \frac{\pi}{2} \quad (7)$$

When the output RF signal from unit mixer A and the output RF signal from unit mixer B are combined via 90-degree hybrid circuit in-phase, the two output image signals from mixer A and B are combined reverse phase and suppressed. This is the principle of the IRM.

Table I shows the spurious chart of the conventional harmonic IRM. The shaded portions in this table are even order, therefore the waves are canceled in the APDP. The waves of the mark X are combined negative phase and suppressed. The waves of the mark O are combined in-phase or in 90-degree phase difference, and outputted. This configuration may have applicability to any harmonic mixer.

Next we discuss the 4th order harmonic mixer. In the case of relatively high IF frequency, there are many spurious signals such as 5LO, 3LO + 2IF and so on, that locate nearby the desired RF signal ($4LO + IF$). Especially, the output power of the odd-order harmonics of the LO wave are very high. Therefore the demand for the filter to suppress these spurious signals becomes severe.

III. PROPOSED CONFIGURATION OF THE LOW SPURIOUS QUADRUPLE HARMONIC IRM

In this section we discuss the phase relations to reduce the spurious signal level such as $5LO$ in the quadruple harmonic IRM. In equation (5), if we assume that the input phases of the unit mixer A are $\phi_A = 0, \theta_A = 0$, and the input phases of the unit mixer B are $\phi_B = \phi, \theta_B = \theta$, the each phase differences of the RF signal, the image, $5LO$ outputted from the mixer A

TABLE II

THE SPURIOUS CHART OF THE PROPOSED QUADRUPLE HARMONIC IRM

	LO	2LO	3LO	4LO	5LO
-5 IF		O		X	
-4 IF	X		O		X
-3 IF		X		O	
-2 IF	O		X		O
-1 IF		O		X	
0 IF	X		O		X
1 IF		X		O	
2 IF	O		X		O
3 IF		O		X	
4 IF	X		O		X
5 IF		X		O	

◩ Canceled in the APDP

X̄ Canceled by the configuration (reverse phase)

Ō Outputted

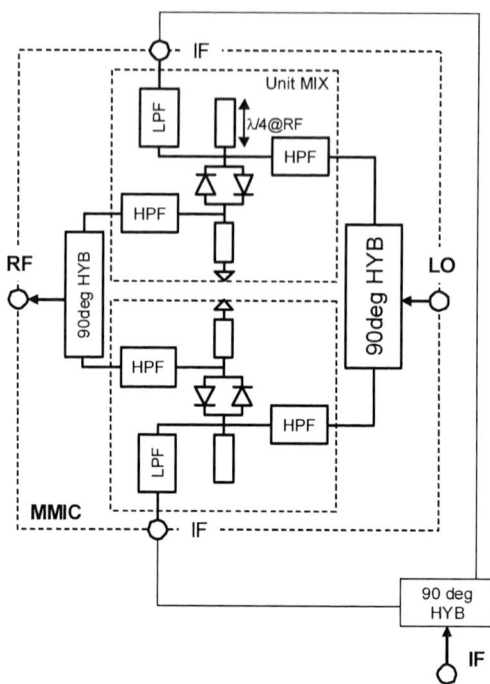

Fig. 2. The configuration of the fabricated 60GHz band low spurious quadruple harmonic IRM MMIC

and B are as follows respectively:

$$RF\ signal : 4\phi + \theta$$
$$Image\ signal : 4\phi - \theta$$
$$Spurious\ signal\ 5LO : 5\phi \tag{8}$$

It is necessary to satisfy the following expression to suppress the $5LO$ with the function of image rejection maintained.

$$4\phi + \theta - \frac{\pi}{2} = 0 + 2\pi n_1,$$
$$4\phi - \theta - \frac{\pi}{2} = \pi + 2\pi n_2,$$
$$5\phi - \frac{\pi}{2} = \pi + 2\pi n_3 \tag{9}$$

where $n_1 \sim n_3$ are integer.

There is $\phi = -\pi/2, \theta = \pi/2$ as one example of the solution that satisfies the equation (9). Thus, the spurious signal $5LO$ can be suppressed by inputting LO wave to two mixers with the phase difference of $-\pi/2$.

Table II shows the spurious chart of the proposed quadruple harmonic IRM. As well as Table I, the shaded portions in this table are even order, therefore the waves are canceled in the APDP. The waves of the mark X are combined reverse phase and suppressed. The waves of the mark O are combined in-phase and outputted. It is understood that not only the $5LO$ but also the $3LO \pm 2IF$, the $5LO \pm 4IF \cdots$ are suppressed. Especially, the $5LO$ and the $3LO + 2IF$ are locate nearby the desired RF signal.

In this configuration, the phase difference of the input LO waves to two unit mixers is $\pi/2$, so the phase difference of the $4LO$ produced in the each APDP is $2\pi (= 0)$. Therefore,

conventional IRM operation becomes possible in the frequency element related to the $4LO$ such as the RF signal and the image signal. And it can be said that this configuration is appropriate for the 4th-order harmonic mixer.

IV. FABRICATED RESULTS OF THE 60GHz BAND QUADRUPLE HARMONIC IRM MMIC

Fig.2 shows the configuration of the fabricated 60 GHz band low spurious quadruple harmonic IRM MMIC. The IRM consists of the two unit mixers with the APDP and the branching filter, the 90-degree hybrid circuit for the RF signal, the LO wave and the IF signal. The Lange couplers are employed for the hybrid circuits of the RF signal and the LO wave. The 90-degree hybrid circuit for the IF signal is connected outside of the monolithic mixer.

Fig.3 shows the photograph of the fabricated IRM MMIC. The chip size of the MMIC is $2mm \times 2.2mm$. We employ the GaAs pHEMT process. Fig.4 shows the conversion gain of the IRM under the condition of the IF frequency is 5.5 GHz, the LO input power is 15 dBm and the IF input power is -10dBm. In this figure, the solid line is calculated data and the dots are measured data. The conversion loss of from 15.6 dB to 16.5dB are obtained within the frequency range from 56 GHz to 62 GHz. The image rejection ratio is 20dB.

Fig.5 shows the output spectrum of the IRM under the condition of the IF frequency is 5.5 GHz, the LO frequency is 13.5 GHz, the RF frequency is 59.5 GHz and the IF input power is -10 dBm. The IRMs of the conventional configuration

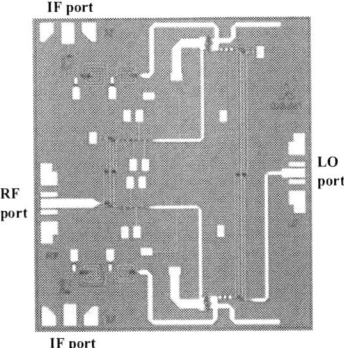

Fig. 3. The photograph of the fabricated IRM MMIC (chip size: $2mm \times 2.2mm$)

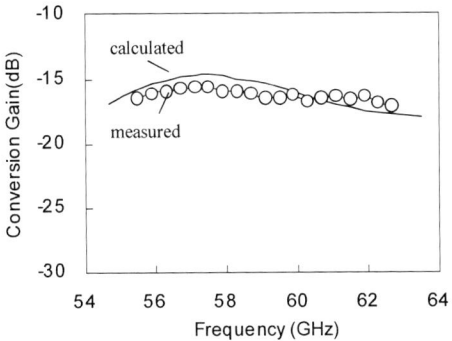

Fig. 4. The conversion gain of the IRM under the conditions of IF frequency is 5.5 GHz, LO input power is 15 dBm, IF input power is -10dBm

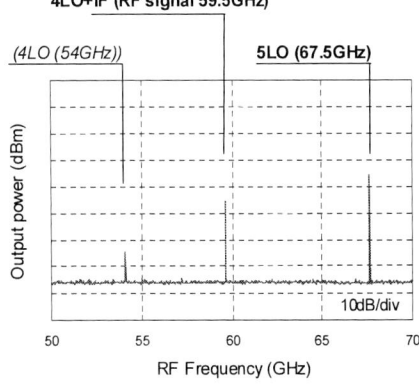

(a) Conventional configuration

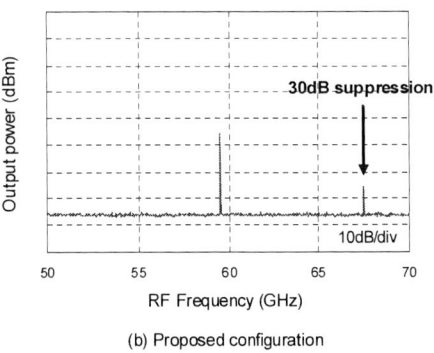

(b) Proposed configuration

Fig. 5. The output spectrum of the IRM under the conditions of IF frequency is 5.5 GHz, LO frequency is 13.5 GHz, RF frequency is 59.5 GHz and IF input power is -10 dBm.

and the proposed configuration are fabricated in the same process. The 30dB suppression of the $5LO$ can be achieved by the proposed IRM configuration (fig.5(b)) compared with the conventional one (fig.5(a)). Since the output level of the $5LO$ is independent of the input IF level, the suppression of the $5LO$ is very effective. Meanwhile the symmetry of the APDPs is a cause in the output level of $4LO$ different in fig.5(a) and (b).

V. CONCLUSION

In this paper, we proposed the configuration of the low spurious quadruple harmonic IRM, and also presented that the proposed configuration was effective by the fabricated 60 GHz band IRM MMIC.

ACKNOWLEDGMENT

This research is one of the results of "Research and development of the antenna system for a broadband communication system on 60GHz millimeter-wave " sponsored by the Japanese Ministry of Internal Affairs and Communications.

REFERENCES

[1] K.Araki, "Millimeter-Wave Activities in Japan, " *IEEE MTT-S Int Microwave Symp. Dig.*, pp. 133-136, June 2007.

[2] H.Zirath, T.Masuda, R.Kozhuharov, and M.Ferndahl, "Development of 60-GHz front-end circuits for a high-data-rate communication system," *IEEE Journal of Solid-State Circuits,*, vol. 39, pp.1640-1649, 2004.

[3] Y.Isota, K.Itoh, A.Iida, and O.Ishida, "Overview Of Millmeter-wave Monolithic Circuits, " *Proceeding of European Microwave Conference* pp.1316-1322, 1997.

[4] M.Cohn, J.E.Degenford, and B.A.Newman, "Harmonic mixing with anti parallel diode pair, " *IEEE Trans. Microwave Theory Tech.*, vol. 23, no.8, pp.667-673, Aug.1975.

[5] D.Blackwell, H.G.Henry, J.E.Degenford, and M.Cohn, "94 GHz Subharmonically pumped MMIC mixer, " *IEEE MTT-S Int Microwave Symp. Dig.*, pp. 1037-1039, June 1991.

[6] K.Itoh, A.Iida, Y.Sasaki, and S.Urasaki, "A 40GHz Band Monolithic Even Harmonic Mixer with an antiparallel diode pair," *IEEE MTT-S Int Microwave Symp. Dig.,*pp. 879-882, June 1991.

[7] M.Hieda, K.Itoh, Y.Horiie, T.Kashiwa, A.Iida, Y.Iyama, O.Ishida, and T.Katagi, "A 60 GHz band image rejection harmonic mixer," *XXVth General Assembly of U.R.S.I,*,p130,1996.

[8] K.Kawakami, M.Shimozawa, H.Ikematsu, K.Itoh, Y.Isota, and O.Ishida, "A millimeter-wave broadband monolithic even harmonic image rejection mixer," *IEEE MTT-S Int Microwave Symp. Dig.*, pp.1443-1446, June 1998.

[9] K.Kanaya, K.Kawakami, T. Hisaka, T. Ishikawa, and S. Sakamoto, "A 94 GHz high performance quadruple subharmonic mixer MMIC," *IEEE MTT-S Int. Microwave Symp. Dig.*, pp.1249-1252 June 2002.

978-2-8748-7007-1/08 $25.00 © 2008 EuMA

A Highly Linear (40.5 – 43.5) GHz MMIC Single Balanced pHEMT Resistive Up-Converter Mixer for LMDS Applications

Antoine Khy [1], Bernard Huyart [1] and Hervé Teillet [2]

[1] *Institut TELECOM –TELECOM ParisTech – LTCI CNRS*

Département de Communications et Electromique (COMELEC)

46 Rue Barrault, 75013 PARIS – FRANCE

Email: antoine.khy@telecom-paristech.fr

bernard.huyart@telecom-paristech.fr

[2] *THALES Communications*

160 Boulevard de Valmy, 92704 COLOMBES - FRANCE

Email: herve.teillet@fr.thalesgroup.com

Abstract — **This paper presents the design and performance of a highly linear (40.5 – 43.5)GHz MMIC single balanced pHEMT resistive up-converter mixer dedicated to LMDS applications. The mixer achieves good performance in terms of conversion loss which is about 6 –7dB and LO/RF isolation that is better than 30dB under LO power equal to 14dBm. The main feature of the mixer is its high linearity since it presents an RF output power @1dB compression point (Pout@1dB) superior to 5dBm in the (39.5 – 43.5)GHz frequency range. The measured OIP3 is equal to 20.2dBm at 40.5GHz and 12.2dBm at 42GHz. The 2mm×1.5mm circuit was fabricated using the D01PH (0.13μm GaAs depletion mode pHEMT) process provided by the OMMIC foundry. To the best of our knowledge, this is the most linear up-converter mixer reported to date in this frequency range associated with as low conversion loss.**

I. INTRODUCTION

The ever increasing demand for higher data rate transmissions in wireless communications leads manufacturers to design broadband systems operating at millimeter-wave frequencies where the spectrum is much less saturated than in the low frequency bands ([1],[2]). As a consequence, RF circuit designers have to adapt to this trend in order to develop transceivers that meet the specifications required by these systems which are quite severe. In this context, LMDS (Local Multipoint Distribution Service) is an example of fixed wireless communications system dedicated to high rate data transmission applications like Internet access, television broadcast and video conference, and also to lower rate applications such as voice. In order to function correctly, this kind of system, which operates in the (40.5 – 43.5)GHz frequency band in Europe, requires high performance and low-cost millimeter-wave circuits. Indeed, linearity of the transmitter and output spectrum mask are of great concern. In particular, the up-converter mixer that is located just before the power amplifier in the transmitter is one of the critical components with regard to the system performance.

So, the up-converter mixer has to present the following features:

– low conversion loss (Lc)
– high linearity (Pout@1dB, OIP3)
– high LO/RF isolation ($I_{LO/RF}$)

An advantage of the use of a highly linear up-converter mixer resides in the fact that it is able to provide a high level RF output signal without distortion, which permits to reduce the constraints on the power amplifier gain. This results in a cost reduction of the whole system. From the specifications mentioned above, we designed a highly linear MMIC up-converter mixer operating in the (40.5 – 43.5)GHz frequency band.

II. MIXER DESIGN

A. Single Resistive Mixer

The first requirement that an up-converter mixer used in a LMDS transmitter has to fulfil concerns linearity. When considering this feature as the most important, the use of a resistive mixer is in theory best suited [3]. This is due to the very linear behaviour of the V–I curves of a FET in the ohmic region when Vds is near 0V. The second requirement imposed to the mixer deals with LO/RF isolation. Especially in up-conversion operations, LO/RF isolation is a very critical parameter since the residual LO power at the RF output port of the mixer is amplified by the following power amplifier. For a single resistive mixer, LO/RF isolation is about 10 – 15dB. This rather poor performance is due to the gate-to-drain Cgd capacitance that is practically equal to the gate-to-source capacitance Cgs when the FET is cold (Vds = 0V) whereas in the saturation region, Cgd is negligible. So, when Vds = 0V, the gate and the drain of the FET are no longer isolated and the LO signal applied to the gate leaks towards the drain through Cgd. However, LO/RF isolation can be improved by connecting an inductance between the gate and the drain of the FET [4]. The association of this inductance in parallel with the Cgd capacitance forms a parallel resonant circuit tuned to present a very high impedance at the LO frequency. In [4], LO/RF isolation of 30dB at 60GHz was achieved with the help of this resonant network. The addition of a capacitance in series with the inductance ensures DC decoupling between gate and drain. Figure 1 shows a single resistive mixer with the parallel resonant circuit we just described.

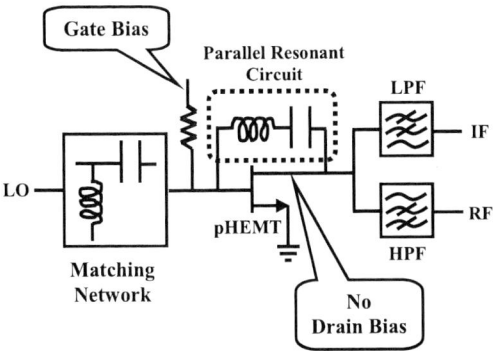

Fig. 1 Schematic of the single resistive mixer

LO signal is applied to the gate of the FET while IF and RF signals are respectively applied and retrieved at the drain, the source being connected to ground. The FET is biased at pinch-off i.e. Vgs = – 0.85V and as the mixer is resistive, no bias is applied to the drain. The reason for this is that minimum conversion loss and minimum inter-modulation distortion are obtained when the FET is biased near its pinch-off region [3]. Since IF and RF signals are present at the same terminal, the mixer requires a diplexer at the drain which is composed of a low-pass filter for the IF path and a high-pass filter for the RF path. The single balanced architecture that we describe in the following is based upon this single mixer topology.

B. Single Balanced Resistive Mixer

In order to improve linearity and LO/RF isolation, we designed a single balanced mixer by associating 2 single mixers as shown in Figure 2. A Marchand balun [5] combining 2 Lange couplers [6] is used to provide the differential LO signals to the mixer while the IF signal is applied through an external off-chip balun. As for the RF signal, it is taken from the output by means of a power combiner. This topology has the advantage to present a strong LO rejection on the RF port since at this point, thanks to the LO balun, the LO signals are out of phase and thus cancel each other. In addition, a theoretical 3dB linearity improvement is expected compared to the single mixer topology due to the balanced configuration at the price of a theoretical 3dB LO power rise, however. The LO Marchand balun topology consists of the combination of 2 Lange couplers connected by a via-hole (that acts as short-circuit) at one end and a transmission line at the other end. This kind of balun ensures amplitude and phase imbalances respectively less than 1dB and 10° between its 2 output ports over a very wide frequency range that can reach an octave ([6], [7]). The main drawback of this structure is its size due to the use of 2

Lange couplers whose lengths are equal to λ/4 at the center frequency. Indeed, the lengths L of the couplers are L = 690μm for operation centred in the middle of the (34.5 – 38.5)GHz LO frequency band i.e. 36.5GHz. The width W of the Lange couplers lines is W=5μm and the spacing S between them is equal to S =14μm.

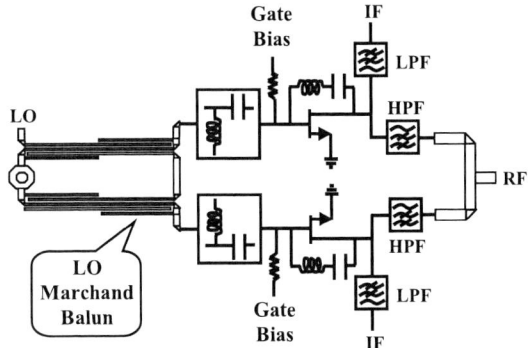

Fig. 2 Schematic of the single balanced resistive mixer

For highly linear performance, large gate area FETs (8×30μm) were used in the design since large devices are better-suited to high power signal handling than small devices, thus they exhibit higher 1dB compression point (Pout@1dB) and third-order intercept point (OIP3). The choice of the FETs size results from a compromise between large signal handling capability, matching network design complexity and LO power level required to drive the FETs in optimal conditions (LO power level increases with FET size). The parallel resonant network used to improve LO/RF isolation is made of a high impedance line that behaves like a small inductance in series with a 1pF capacitance for DC decoupling. The diplexer used to separate IF and RF signals consists of a L – C low-pass filter (series L = 1.1nH and shunt C = 0.8pF) and a C – L high-pass filter (series C = 0.4pF and shunt shorted high impedance line). Figure 3 shows the layout of the single balanced resistive mixer. The size of the circuit is 2mm ×1.5mm.

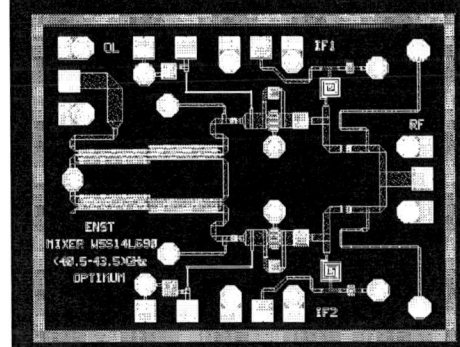

Fig. 3 Layout of the single balanced resistive mixer

The circuit was fabricated using the D01PH (0.13μm GaAs depletion mode pHEMT) process provided by the OMMIC foundry that performed the experimental measurements.

978-2-8748-7007-1/08 $25.00 © 2008 EuMA

III. MEASURED PERFORMANCE

The single balanced resistive up-converter mixer chip was characterized by on-wafer measurements for an RF output signal in the $(39.5 - 45.5)$GHz frequency range. LO frequency was varied from 34GHz to 40GHz and IF frequency was set to a fixed value of 5.5GHz. All the measurements were performed with FETs biased at pinch-off i.e. Vgs= -0.85V and LO power set to P_{LO} =14dBm. The IF signal was applied to the 2 differential IF inputs of the circuit (IF1 and IF2 in Figure 3) with the help of an off-chip balun. Measured and simulated conversion loss Lc of the mixer versus RF frequency is plotted in Figure 4 for IF power level P_{IF}= -30dBm.

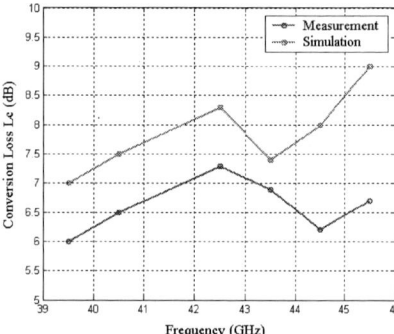

Fig. 4 Measured and simulated conversion loss Lc versus RF frequency @(P_{LO} =14dBm, f_{IF} =5.5GHz and P_{IF} = -30dBm)

Despite some gap (1–2.5dB) between measured and simulated performance the curves show similar frequency responses. The mixer exhibits low and rather constant conversion loss Lc since it varies almost exclusively between 6–7dB in the $(39.5 - 45.5)$GHz RF frequency range. We suppose that conversion loss can be improved by replacing both low-pass filters constituted of lumped elements in the IF1 and IF2 paths (see Figure 3) by some less lossy network. Measured LO/RF isolation versus RF frequency is plotted in Figure 5 for IF power level P_{IF} = -30dBm.

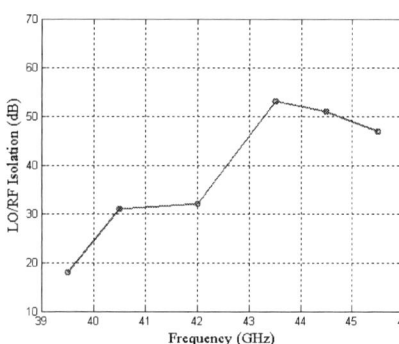

Fig. 5 Measured LO/RF isolation versus RF frequency @(P_{LO} =14dBm, f_{IF} =5.5GHz and P_{IF} = -30dBm)

LO/RF isolation is better than 30dB in the $(40.5 - 43.5)$GHz frequency range and varies between 45dB and 55dB in the $(43.5 - 45.5)$GHz band. This performance is due both to the balanced topology of the mixer and to the resonant parallel

network that connects the gate and the drain of the FET in each of the 2 single mixers.

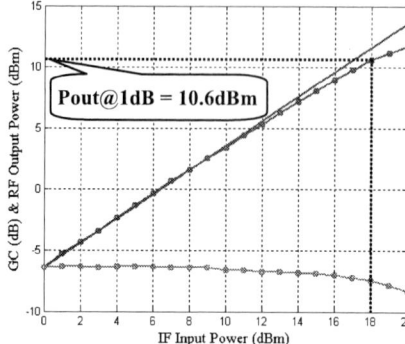

Fig. 6 Measured conversion loss Lc and RF output power versus swept IF input @(P_{LO}=14dBm, f_{RF} = 40.5GHz and f_{IF} = 5.5GHz)

Measured conversion loss Lc and RF output power versus swept IF input are illustrated in Figure 6. To obtain these curves, IF power was varied from 0dBm to +20dBm and RF frequency was set to f_{RF} = 40.5GHz. The RF output power @1dB compression point (Pout@1dB) of the mixer is 10.6dBm at f_{RF} = 40.5GHz. Two-tone OIP3 measurements were performed around f_{RF} = 40.5GHz by applying two equal-amplitude, closely-spaced IF tones (Δf=40MHz) at f_{IF1}=5.48GHz and f_{IF2} = 5.52GHz to the IF input of the mixer. With fixed f_{LO} = 36.5GHz, the spectral components of the corresponding up-converted tones at f_{RF1}= f_{IF1}+ f_{LO} = 41.98GHz and f_{RF2} = f_{IF2} + f_{LO} = 42.02GHz were measured as well as the spectral components of spurious tones at $f_{IMD3, LSB}$ = 2f_{IF1} − f_{IF2} + f_{LO} = 41.94GHz and $f_{IMD3,USB}$ = 2f_{IF2} − f_{IF1} + f_{LO} = 42.06GHz produced by third-order inter-modulation. OIP3 was determined using the following expression:

$$OIP3_{LSB,USB} = P_{RF1,2} + (P_{RF1,2} - P_{IMD3,LSB,USB})/2$$

Table I summarizes measured linearity performance of the mixer, that is to say Pout@1dB and output third-order interception point (OIP3) for different frequencies.

TABLE I

MEASURED LINEARITY PERFORMANCE OF THE MIXER FOR DIFFERENT FREQUENCIES

Frequency	39.5GHz	40.5GHz	42GHz	43.5GHz
Pout@1dB	6.9dBm	10.6dBm	5.4dBm	5.1dBm
OIP3	–	20.2dBm	12.2dBm	–

Measured Pout@1dB is better than 5dBm in the $(39.5 - 43.5)$GHz frequency range while OIP3 is equal to 20.2dBm @40.5GHz and 12.2dBm @42GHz. These high values are due to the use of large FETs ($8 \times 30\mu$m) in the mixer that have better large signal handling capability than small FETs. Table II compares performance of previously reported up-converter mixers with the one presented in this paper. In accordance with theory, the active mixer reported in [8] shows very poor linearity performance. Sub-harmonic diode mixers ([9],[11]) usually present lower linearity than their fundamental counterparts ([2],[10]) while sub-harmonic FET mixers [1] seem to be potentially suitable for high power level operations. The image reject FET mixer in [12] is the only one that presents same linearity than the one depicted in this paper but it shows higher conversion loss.

IV. CONCLUSION

The design and performance of a (40.5 – 43.5)GHz MMIC single balanced pHEMT resistive up-converter mixer dedicated to LMDS applications was presented. As linearity of the transmitter and output spectrum mask are of great concern in LMDS systems, the mixer was optimized for high linearity and LO/RF isolation associated with low conversion loss. In the (40.5 – 43.5)GHz frequency range this mixer achieves good performance in terms of conversion loss which is about 6 – 7dB and LO/RF isolation that is better than 30dB under LO power equal to 14dBm. As for linearity, the mixer exhibits Pout@1dB superior to 5dBm while OIP3 is equal to 20.2dBm at 40.5GHz and 12.2dBm at 42GHz. Considering this performance, this highly linear mixer is to the best of our knowledge the most linear up-converter mixer reported to date in this frequency range exhibiting as low conversion loss.

ACKNOWLEDGEMENT

This work was achieved within the framework of the national research project CONRAHD/OPTIMUM in association with THALES Communications and OMMIC. The authors would like to thank B. Wroblewski at OMMIC for the circuit characterization.

REFERENCES

[1] P. Butterworth, C. Charbonniaud, M. Campovecchio, J. C. Nallatamby, M. Monnier and M. Lajugie, "A Balanced Sub-Harmonic Cold FET Mixer for 40GHz Communication Systems", IEEE Gallium Arsenide Applications Symposium Proceedings, pp. 105 – 108, 2003

[2] B. Lefebvre and A. Bessemoulin, "35 – 45 GHz Image Rejection Star Mixer for Up-and-Down Conversion", IEEE Gallium Arsenide Applications Symposium Proceedings, pp. 381 – 384, 2003

[3] S. A. Maas, "A GaAs MESFET Mixer with Very Low Intermodulation", IEEE Transactions on Microwave Theory and Techniques, Vol. 35, pp. 425 – 429, April 1987

[4] K. S. Ang, M. Chongcheawchamnan and I. D. Robertson, "Monolithic Resistive Mixers for 60GHz Direct Conversion Receivers", IEEE Radio Frequency Integrated Circuits Symposium Proceedings, pp. 35 – 38, 2000

[5] N. Marchand, "Transmission Line Conversion Transformers", Electronics, Vol. 17, N° 12, pp. 142 – 145, December 1944

[6] M. C. Tsai, "A New Compact Wideband Balun", IEEE MTT–S International Microwave Symposium Digest, pp. 141 – 143, 1993

[7] K. S. Ang and I.D. Robertson, "A Monolithic Double-Balanced Upconverter for Millimeter-Wave Point-to-Multipoint Distribution Systems", IEEE Gallium Arsenide Applications Symposium Proceedings, pp. 439 – 442, 2000

[8] P.-C. Huang, R.-C. Liu, J.-H. Tsai, H.Y. Chang, H. Wang, J. Yeh, C.-Y. Lee and J. Chern, "A compact 36 – 65 GHz Up-Conversion Mixer with Integrated Broadband Transformers in 0.18-µm SiGe BiCMOS Technology", IEEE Radio Frequency Integrated Circuits Symposium Proceedings, 2006

[9] K. Hettak, C. J. Verver, G. A. Morin and M. G. Stubbs, "A Novel Uniplanar 44 GHz MMIC Subharmonic Mixer using CPW Series Stubs", IEEE MTT–S International Microwave Symposium Digest, pp. 1157 – 1160, 2004

[10] "HMC560 data sheet", Hittite Microwave Corporation, Chelmsford, Massachusetts, U.S., available at http://www.hittite.com

[11] "XM1003–BD data sheet", MimixBroadband, Inc., Houston, Texas, U.S., available at http://www.mimixbroadband.com, 2008

[12] "XM1005–BD data sheet", MimixBroadband, Inc., Houston, Texas, U.S., available at http://www.mimixbroadband.com, 2007

TABLE II

PERFORMANCE COMPARISON WITH PREVIOUSLY REPORTED UP-CONVERTER MIXERS

Ref.	Technology	Type	Topology	Frequency (GHz)	P_{OL} (dBm)	Lc (dB)	Pout@1dB (dBm)	OIP3 (dBm)	$I_{OL/RF}$ (dB)
[1]	0.15µm GaAs pHEMT	Sub-Harmonic Resistive	Single Balanced	42 – 43.5	12	13.5	0.5@43GHz	–	22*
[2]	GaAs pHEMT	Diode Quads Passive	Image Reject	35 – 45	18	6 –7	2@41.5GHz	–	30 – 35
[7]	0.25µm GaAs pHEMT	Resistive	Double Balanced	40.5 – 43.5	10	16		1	25
[8]	0.18µm SiGe BiCMOS	Active	Gilbert Cell	35 – 65	5	8	–25@40GHz	–16@40GHz	40 – 45
[9]	0.18µm GaAs pHEMT	Sub-Harmonic Passive	Anti-Parallel Diode Pair	43.5 – 45.5	9	8	–11	–	20 – 30
[10]	GaAs MESFET	Passive	Double Balanced	24 – 40	13	8		6@39GHz	25 – 30
[11]	2µm GaAs HBT	Sub-Harmonic Passive	Image Reject	32 – 42	13	10	–	0@40GHz	40 – 45
[12]	0.15µm GaAs pHEMT	Resistive	Balanced Image Reject	37 – 46	15	11 – 12	–	11 – 13	30 – 35
This work	0.13µm GaAs pHEMT	Resistive	Single Balanced	40.5 – 43.5	14	6 – 7	>5	>12	30 – 55

* = 2LO/RF isolation

Proceedings of the 3rd European Microwave Integrated Circuits Conference

A 94-GHz Monolithic Front-End for Imaging Arrays in SiGe:C Technology

Erik Öjefors [1], Ullrich Pfeiffer [2]

Institute of High-Frequency and Communication Technology, University of Wuppertal
Rainer-Gruenter-Str. 21, D-42119 Wuppertal, Germany
[1]erik.ojefors@ieee.org [2]ullrich@ieee.org

Abstract— **A monolithic downconverter for 94 GHz imaging arrays implemented in SiGe:C technology is presented. The downconverter consists of a three stage differential LNA with lumped matching networks and a polyphase subharmonic mixer. It yields 20 dB conversion gain at 94 GHz with an input 1-dB compression point of -31 dB and a current consumption of 45 mA at 3.3 V supply voltage. The total required die area of the complete downconverter (excluding pad frame) is 0.1 mm^2, making it particularly suitable as a front-end in multi-channel receiver systems.**

I. INTRODUCTION

An increased interest in millimeter-wave (mmWave) and THz imaging for security applications has recently created a demand for integrated receiver front-ends, which can be used in phased or focal-plane array configurations. For imaging arrays, the number of channels or pixels present in the system is a critical factor, since it determines the resolution, unless mechanical scanning is used. Hence, the die area and power consumption required by each receiver channel needs to be optimized.

A front-end architecture capable of meeting the requirements of a monolithic integrated receiver array is depicted in Fig. 1. The proposed subharmonic architecture has the advantage that the local oscillator (LO) signal can be generated and distributed at half the operating frequency. By use of differential circuit topology, balanced on- or off-chip antennas, such as folded dipole antennas, can be connected directly to the receiver input without the need for on-chip baluns. An advantage of the combined use of subharmonic mixing with differential architecture is that neither the LO fundamental nor its harmonics fall into the receive band. Provided that quadrature (I/Q) mixers are used, zero-IF or near zero-IF downconversion can thus be implemented without the DC-offset or blocking typically caused by LO self mixing.

A differential implementation of an LNA has been presented at 77 GHz [1]. As transmission line stubs were used in this paper for the interstage tuning a relatively large area was needed. However, it has been shown that spiral inductors can be used as a more compact alternative even at mmWave frequencies, with single-ended amplifier implementations available up to 140 GHz [2].

Subharmonic mixing has previously been demonstrated at 120 GHz in a SiGe:C process [3]. A polyphase concept with two mixer cores was used where the RF signal was

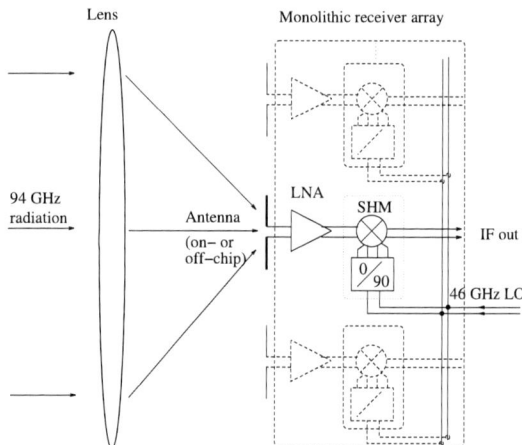

Fig. 1.   Architecture of a subharmonic monolithic array receiver.

successively multiplied by an I and Q-component of the local oscillator signal. The die area, however, was 0.75 mm^2 without pads, primarily due to the area consumption of a 90-degree transmission line hybrid used to generate the necessary local oscillator phase shift for the two mixer cores. More compact solutions as described in this paper are based on conventional lumped RC polyphase networks and have previously been demonstrated primarily at lower frequencies, e.g. up to 60 GHz in [4], [5] and [6].

This paper presents a compact 94-GHz subharmonic front-end based on lumped passive elements and a fully differential circuit topology. Spiral inductors are used in the LNA matching networks and an area efficient RC-polyphase filter is used to provide the subharmonic mixer cores with the required LO signals. Through monolithic integration of the LNA and mixer, the total required die area for the downconverter is 0.1 mm^2, excluding pads.

II. CIRCUIT DESIGN

The downconverter subcircuits were designed and evaluated both as separate components, as well as, in their fully integrated configuration. In the following sections the details of the LNA, mixer, downconverter design, and integration are presented.

978-2-8748-7007-1/08 $25.00 © 2008 EuMA   422

A. Differential 94-GHz amplifier

The amplifier core consists of three differential cascode amplifier stages as illustrated in Fig. 2. The cascode topology uses LC-interstage impedance matching and the transistors in the cascodes are biased at their peak f_T current density. Each

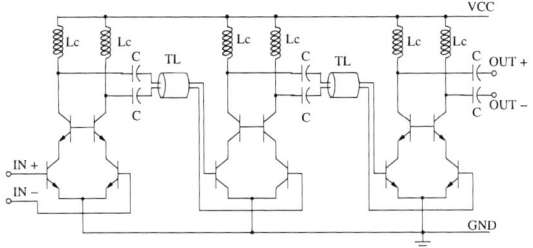

Fig. 2. Simplified schematic of the three-stage cascode LNA

cascode stage uses a lumped element spiral inductor (0.3 nH) and metal-insulator-metal (MIM) capacitor (6 fF) in a LC-tank to match their output impedance to the following stage. The spiral inductors are implemented as a square two-turn coil with the windings in adjacent metal layers. Interstage feedlines (TL) consisting of 55-μm long coupled microstrips with odd mode impedance of 100-Ω are used to cascade the stages of the amplifier.

The simulated gain of the three-stage amplifier is 21 dB, with an input 1-dB compression point of -23 dBm. The three-stage cascode amplifier draws 11 mA from a 3.3 V supply. To simplify design and testing, the biasing network uses separate current mirrors, which increases the total current consumption of the LNA to 28 mA. However, by the use of an optimized and shared biasing network, the total current consumption can be reduced to less than 15 mA.

B. Subharmonic Mixer

The subharmonic mixer, shown in Fig. 3 is based on two cascaded switching quads with the LO fed 90-degree out-of-phase through the help of a RC polyphase network. Multiplication of the RF signal with the in-phase (I) and quadrature component (Q) of the LO signal can mathematically be expressed as

$$
\begin{aligned}
IF_{out} &= RF(t)LO\left(\sin\left(\omega_{LO}t\right)\cos\left(\omega_{LO}t\right)\right) \\
&= RF(t)\frac{LO}{2}\left(\sin\left(2\omega_{LO}t\right)\right) \quad (1)
\end{aligned}
$$

which is identical to mixing of the RF signal with twice the LO frequency. Hence, this type of subharmonic mixing is not based on the use of higher order non-linearities in the devices. A broadband IF output bandwidth of up to 10 GHz was achieved using 200-Ω load resistors and a differential emitter follower buffer. The simulated conversion gain and 1-dB input compression point was 5 dB and -8 dBm, respectively. If desired, a higher conversion gain over a reduced IF bandwidth may be obtained by the use of larger collector load resistors or a tuned output network.

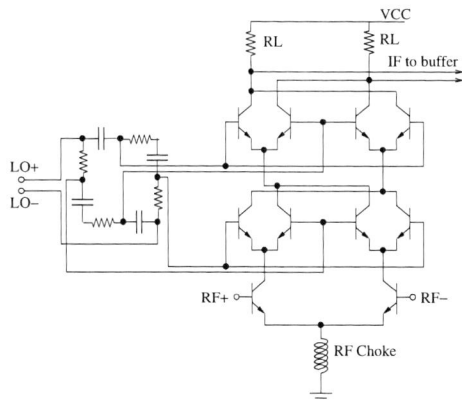

Fig. 3. Schematic of the subharmonic mixer and polyphase network. Biasing and the output buffer is excluded.

In the simulation, 8 dBm LO drive power is required for maximum conversion gain. Most of the LO power is consumed by the 50-Ω resistors in the polyphase network and a higher impedance network can be implemented in order to reduce the LO drive requirement.

The total current consumption of the mixer is 17 mA of which the majority is consumed by the common emitter IF output buffer. Less than 7 mA from a 3.3 V supply is required to power the mixer core.

C. Complete front-end and characterization setup

In the intended application, the differential front-end circuits are interfaced directly to a balanced (dipole) antenna, which can reside either on- or off-chip. Three test-sites with baluns at the RF and LO ports were designed as shown in Fig. 4 a-c) to facilitate characterization in a single ended wafer probing environment. The baluns were implemented as tuned transformers with windings in adjacent metal layers. Associated losses of the RF and LO baluns were extracted from back-to-back test structures.

III. FABRICATION

The circuits were fabricated in the 0.25 μm SiGe:C BiC-MOS technology (SG25H1) from IHP Microelectronics [7] featuring transistors with f_{max} of 220 GHz. The inductors and transmission lines where implemented in a five layer aluminum back-end.

Micrographs of the LNA, mixer, and downconverter test-sites are shown in Fig. 5, Fig. 6, and Fig. 7, respectively.

IV. MEASURED RESULTS

A. LNA characterization

The meausred small-signal gain and the input return-loss of the LNA including the loss of auxiliary input and output test-baluns are shown in Fig. 8. The loss of the 94-GHz auxiliary test-baluns was extracted from back-to-back test structures and measured to be 4 dB each. Hence, the de-embedded gain of the LNA at 94 GHz is 15 dB. The detuning of the LNA gain peak

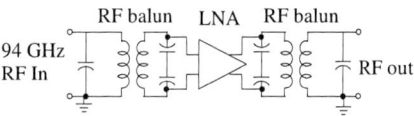

a) LNA test circuit

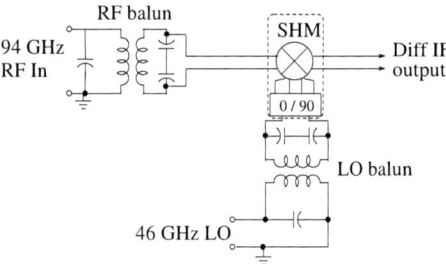

b) Subharmonic mixer test circuit

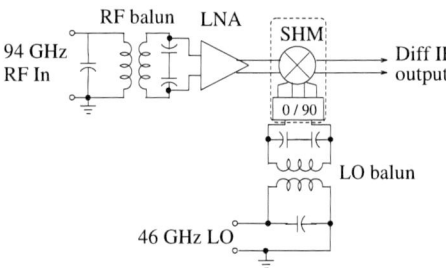

c) Downconverter test chip

Fig. 4.   Test sites designed to allow characterization of the LNA (a), the mixer (b), and the complete front-end (c). Baluns are included to facilitate testing in a single-ended environment.

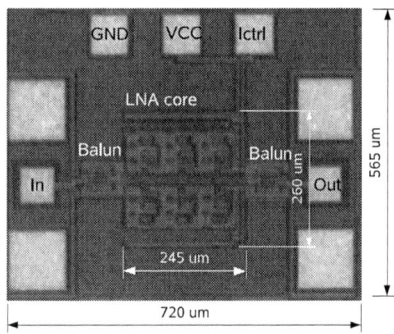

Fig. 5.   LNA micrograph including auxiliary input and output test-balun transformers. The size of the LNA core is 245×260 μm^2.

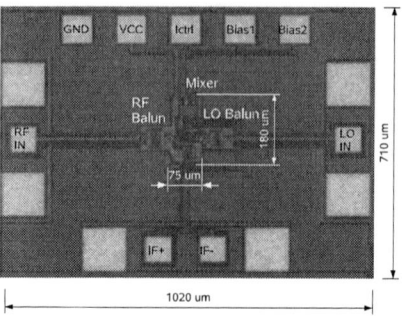

Fig. 6.   Mixer micrograph with auxiliary RF and LO test-baluns at 94 GHz (left) and 46 GHz (right) next to the mixer core. The size of the Mixer core is 245×260 μm^2.

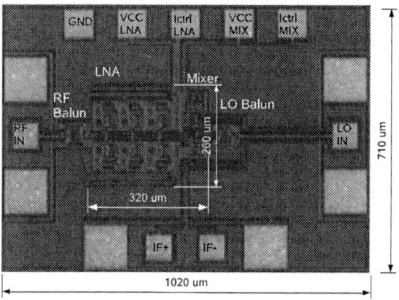

Fig. 7.   Downconverter micrograph with auxiliary 94-GHz RF test-balun (left), LNA, mixer, and 46 GHz auxiliary LO test-balun (right). The size of the downconverter core is 320×260 μm^2.

from 94 GHz to 89 GHz is caused by modeling inaccuracies of the amplifier tank inductor in the initial design. The simulated noise figure is 11.3 dB excluding the losses in the baluns. No noise figure measurements were performed due to limitations in the measurement equipment.

B. Mixer characterization

The simulated and measured mixer conversion gain is shown in Fig. 9 versus the 46-GHz LO drive level at the input of the polyphase network. The measurement used a -25-dBm RF input signal at 94-GHz. The available power at the input to the polyphase network was estimated from the LO generator power, cable losses, and LO test-balun losses. Sufficient power was not available during the characterization to obtain maximum conversion gain, due to an 8-dB insertion loss in the cables and the auxiliary LO test-balun. Hence, the conversion gain was limited to 3 dB, which is in agreement with simulations for this LO drive level.

C. Downconverter characterization

The downconverter has been evaluated with the test circuit shown in Fig. 4 c). Fig. 10 shows the 2-GHz output power versus the 94-GHz input power. A small-signal conversion gain of 20 dB is obtained at 94 GHz, which corresponds to the measured 17-dB gain of the LNA and 3-dB conversion

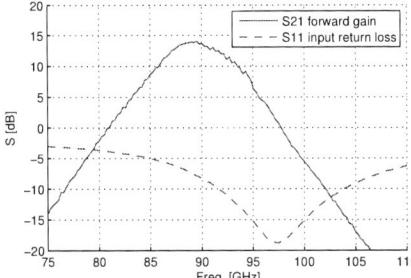

Fig. 8. Measured gain and input-return loss of the three stage differential LNA test chip including baluns (total insertion loss 4+4 = 8 dB at 94 GHz.

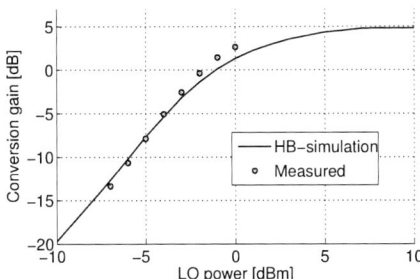

Fig. 9. Measured (circles) and harmonic balance (HB) simulation (line) mixer conversion gain as a function of LO power. RF input level - 25 dBm.

gain of the mixer. The 1-dB compression point is reached at an input level of -31 dBm.

The system noise figure was calculated from the mixer conversion gain G, and the quiescent input (N_0) and output (P_0) noise spectral densities as follows:

$$NF = P_0 - N_0 - G, \qquad (2)$$

where N_0 is -174 dBm/Hz and P_0 = -137 dBm/Hz for a 2-GHz IF respectively. All noise spectral densities were measured with a spectrum analyzer. With a measured conversion gain of $G = 20$ dB, the single-side-band (SSB) system noise figure is 17 dB. The measured output noise level was however close to the noise floor of the available spectrum analyzer, which causes an overestimation of the system noise P_0. Based on the 11.3-dB simulated noise figure of the LNA the expected SSB cascaded system noise figure for a low IF system (where the mirror frequency falls within the LNA bandwidth) should be close to 15 dB, thus corresponding to a 12 dB DSB noise-figure in a superheterodyne or zero-IF (with an additional quadrature mixer core) configuration.

V. CONCLUSION

A subharmonic monolithic downconverter with 20-dB conversion gain at 94 GHz has been implemented in SiGe:C process technology. Despite the use of a fully differential circuit architecture the area requirement of the complete downconverter (excluding pads) is only 0.1 μm, due to the use

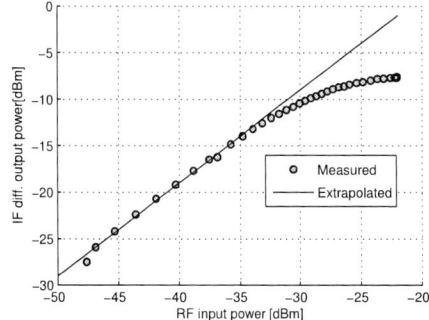

Fig. 10. Measured differential IF output signal as a function of the 94 GHz RF input power to the downconverter with an estimated 46-GHz LO level of 0 dBm at the input of the polyphase network.

of lumped components in the LNA tank and mixer polyphase networks. Future work include array receivers, as well as, extension of the design to a full quadrature (IQ) downconverter to take advantage of the wide IF bandwidth (dc-10 GHz) and the absence of LO-selfmixing problems in a wide-band zero-IF application such as a imaging system with high resolution ranging.

ACKNOWLEDGEMENTS

The authors would like to thank IHP Microelectronics for circuit fabrication as well as Frank Korndoerfer and Johannes Borngräber (IHP) for characterization support. This work was partly supported by the European Commission through the ICT integrated project DotFive and European Young Investigator Award.

REFERENCES

[1] S. Chartier, B. Schleicher, F. Korndorfer, S. Glisic, G. Fischer, and H. Schumacher, "A fully integrated fully differential low-noise amplifier for short range automotive radar using a SiGe:C BiCMOS Technology," in *Microwave Integrated Circuit Conference, 20007. EuMIC 2007. European*, Munich, Germany, Oct. 8–10, 2007, pp. 407–410.
[2] E. Laskin, P. Chevalier, A. Chantre, B. Sautreuil, and S. Voinigescu, "80/160-GHz transceiver and 140-GHz amplifier in SiGe technology," in *IEEE Radio Frequency Integrated Circuits (RFIC) Symposium*, June 2007, pp. 153–156.
[3] A. Müller, M. Thiel, H. Irion, and H. O. Ruoss, "A 122 GHz SiGe active subharmonic mixer," in *13th GAAS Symposium*, Paris, France, October 2005, pp. 57–60.
[4] R. Svitek and S. Raman, "A SiGe active sub-harmonic front-end for 5-6 GHz direct conversion receiver applications," in *IEEE Radio Frequency Integrated Circuits Symposium*, June 2004, pp. 675–678.
[5] P. Lindberg, E. Öjefors, E. Sönmez, and A. Rydberg, "A SiGe HBT 24 GHz sub-harmonic direct-conversion IQ demodulator," in *Topical Meeting on Silicon Monolithic Integrated Circuits in RF Systems*, June 2004, pp. 247–250.
[6] A. Parsa and B. Razavi, "A 60 GHz CMOS receiver using a 30 GHz LO," in *IEEE International Solid-State Circuits Conference*, 2008, pp. 190–191.
[7] B. Heinemann, H. Rucker, R. Barth, J. Bauer, D. Bolze, E. Bugiel, J. Drews, K.-E. Ehwald, T. Grabolla, U. Haak, W. Hoppner, D. Knoll, D. Kruger, B. Kuck, R. Kurps, M. Marschmeyer, H. Richter, P. Schley, D. Schmidt, R. Scholz, B. Tillack, W. Winkler, D. Wolnsky, H.-E. Wulf, Y. Yamamoto, and P. Zaumseil, "Novel collector design for high-speed SiGe:C HBTs," *IEEE Electron Devices Meeting*, pp. 775–778, 2002.

978-2-8748-7007-1/08 $25.00 © 2008 EuMA

Proceedings of the 3rd European Microwave Integrated Circuits Conference

Multiple-Throw Millimeter-Wave FET Switches for Frequencies from 60 up to 120 GHz

I. Kallfass[#1], S. Diebold[#], H. Massler[#], S. Koch[*], M. Seelmann-Eggebert[#], A. Leuther[#]

[#]*Fraunhofer Institute for Applied Solid-State Physics (IAF), Tullastrasse 72, D-79108 Freiburg, Germany*
[1]ingmar.kallfass@iaf.fraunhofer.de

[*]*Sony Deutschland GmbH, Hedelfinger Strasse 61, D-70327 Stuttgart, Germany*

Abstract—**This paper presents the design and performance of various millimeter-wave FET switches realized in a metamorphic HEMT technology. The single-pole multi-throw switch configurations are targeting wireless communication frontends and imaging radiometers at 60, 94 and 120 GHz. In SPDT switches, state-of-the-art insertion loss of 1.4 and 1.8 dB is achieved at 60 and 94 GHz, respectively. Rivalled only by PIN diode switches, an insertion loss of < 2 dB is demonstrated up to 120 GHz. Shorted stubs are used to compensate for parasitic FET capacitance and allow for matching. Linearity data is presented for 60 and 94 GHz SPDT switches. A comprehensive comparison with state-of-the-art planar SPDT switches is included. A 2:6 switch network for multi-antenna transceivers achieves < 4 dB insertion loss at 60 GHz.**

I. INTRODUCTION

Multiple-throw switches fulfill important and often underestimated functions in many millimeter-wave applications. Single-pole-double-throw (SPDT) switches are employed in wireless transceivers at e.g. 60 and 120 GHz, where a single antenna serves both as transmit and receive vehicle (Fig. 1a). The addressing of several transmit/receive antennas is made possible by switches routing signals from two input ports to a number of output ports (2:x switch, Fig. 1b). Using an SPDT switch, millimeter-wave imaging radars and radiometers at 77, 94 and 118 GHz advantageously address two output ports, either for polarisation selectivity in two antennas or for pixel calibration by an integrated termination (Fig. 1c).

Considering size and cost reduction of millimeter-wave systems, it is of capital importance that these switches be integrable with subsequent frontend components, i.e. the switch needs to be implemented in the same semiconductor process as amplifiers and mixers. Being the foremost component in the frontend architecture, the switch limits the receiver noise performance by its non-zero insertion loss in the open branch. Moreover, it needs to be linear enough to pass on the output power of the transmit amplifier. Another important figure of merit of multiple-throw switches is the isolation in the closed branches.

II. FET SWITCH DESIGN

The switches are embedded in a coplanar transmission line environment and use FETs placed in shunt to the signal

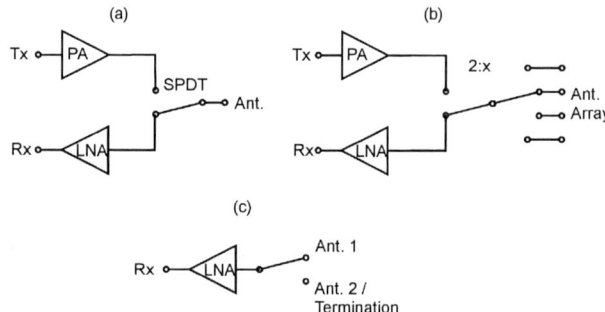

Figure 1. Examples for switch applications in millimeter-wave systems: (a) single antenna transceiver, (b) multiple antenna transceiver, (c) radiometer with polarisation-sensitive antennas or integrated termination.

line. The gate voltage controls the channel conductance, thus realizing a close to short-circuit under above-threshold bias conditions. This is transformed into a close to open-circuit at the branch input by a $\lambda/4$-line (TL1), ensuring isolation of individual branches. Under sub-threshold conditions, the FET channel is pinched-off and merely represents a parasitic capacitance. A shorted stub (TL2), placed in close vicinity to the shunt-FET, achieves a resonant condition at the center frequency which compensates for the parasitic capacitance. In addition to minimizing the switch insertion loss, the shorted stub allows for proper power matching. Moreover, the stub provides the necessary zero drain bias to the FET device. Fig. 2 shows the circuit schematic of a single switch branch, which can be carried out using a single shunt-FET configuration or a double shunt-FET configuration, where the shorted stub is placed between two FET devices. The single shunt-FET configuration results in lowest insertion loss, while the double shunt-FET topology achieves a significantly increased isolation in the closed branch.

The design procedure is straight-forward. Fig. 3 shows a simplified equivalent circuit of the switch branch. In order to minimize the parasitic capacitance of the FET, the gate is terminated in a high resistance, $R_g = 2\,k\Omega$. Under pinch-off channel conditions, the channel resistance r_{ds} represents an

978-2-8748-7007-1/08 $25.00 © 2008 EuMA   426

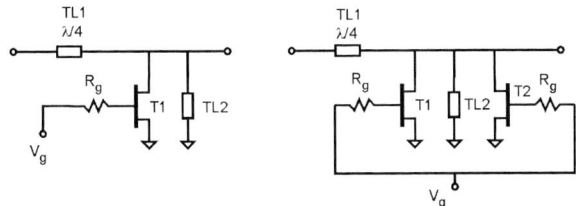

Figure 2. Circuit schematics of individual $\lambda/4$ FET switch branches. Left: single shunt-FET configuration. Right: double shunt-FET configuration.

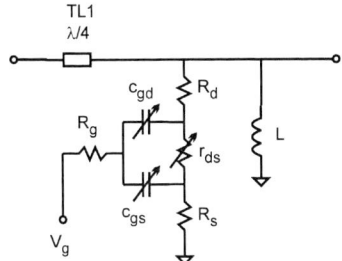

Figure 3. Simplified equivalent circuit of an individual switch branch. The shorted stub is represented by an inductance.

open circuit, and the parasitic capacitance loading the signal line becomes approximately

$$C = c_{gs} \,||\, c_{gd} \qquad (1)$$

where c_{gs} and c_{gd} represent the gate-source and gate-drain capacitance under zero drain bias conditions. Assuming that resonance of the parallel circuit formed by the FET and the shorted stub occurs when its impedance becomes purely resistive, the resonance condition takes on the form of

$$\omega_{res} = \frac{1}{\sqrt{\tau_{LC}^2 + \tau_{RC}^2}} \qquad (2)$$

where $\tau_{LC} = \sqrt{LC}$ and $\tau_{RC} = RC$, with $R = R_d + R_s$ being the sum of the series drain and source contact resistance of the FET. The impedance at resonance becomes $Z_{res} = R\left(\frac{\tau_{LC}^2}{\tau_{RC}^2}\right)$. To realize the particular inductance value L for the desired center frequency, the length l of the inductive stub is chosen according to the well-known relation

$$L = \frac{Z_L \tan\left(\frac{\omega_{res}}{c_0}\sqrt{\varepsilon_{r,eff}}\,l\right)}{\omega_{res}} \qquad (3)$$

where Z_L and $\varepsilon_{r,eff}$ are the characteristic impedance and the effective dielectric constant of the stub line.

III. SINGLE-POLE DOUBLE-THROW SWITCHES

A. SPDT for 60 GHz

The double shunt-FET approach discussed above was adopted in the design of SPDT switches for 60 GHz

operation. Fig. 4(a) shows the chip photograph. The switch MMIC was realized in the IAF metamorphic $In_{0.52}Al_{0.48}As/In_{0.65}Ga_{0.35}As/In_{0.53}Ga_{0.47}As$ HEMT technology with 100 nm gate length [1], featuring maximum cutoff frequencies $f_T = 220\,\mathrm{GHz}$ and $f_{max} = 300\,\mathrm{GHz}$.

The small-signal performance of the switch is evaluated by S-parameter measurements using an Agilent XF network analyzer. Fig. 5 shows the resulting insertion loss and isolation. The switch shows a minimum insertion loss of only 1.4 dB at 60 GHz, while it achieves a respectable isolation of 33 dB. Also evidenced is the good agreement to the simulation results, obtained from an in-house multi-bias small-signal model reflecting the equivalent circuit of Fig. 3. The 1-dB-bandwidth of the insertion loss reaches from 43 to 80 GHz. Within this frequency range, the in- and output matching is better than 9 dB, with 17 dB matching at 60 GHz. The power handling capability was evaluated by a one-tone measurement. The input power could be swept to a power level of 16 dBm. No degradation of the output power was detected (Fig. 6).

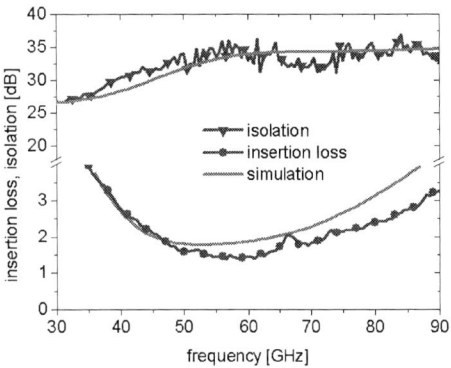

Figure 5. Insertion loss and isolation versus frequency of the 60 GHz SPDT switch.

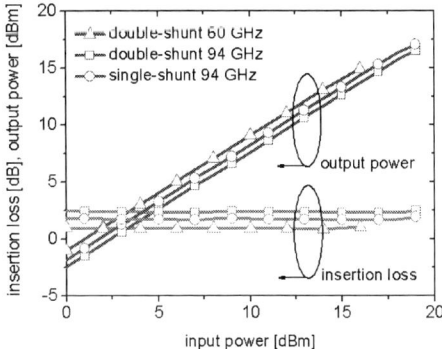

Figure 6. Input power sweeps at 60 and 94 GHz.

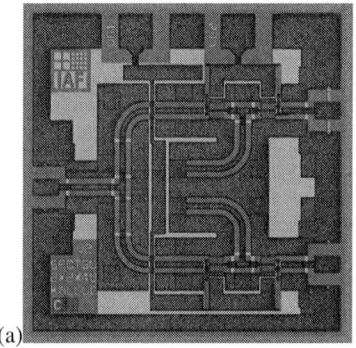 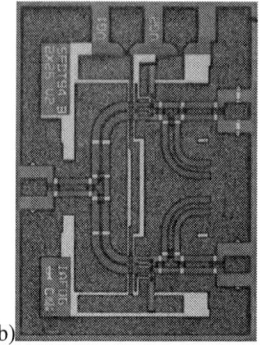 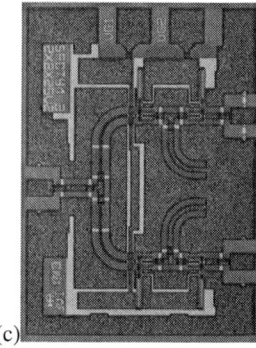

(a)   (b)   (c)

Figure 4. Chip photographs of SPDT switches. (a) 60 GHz SPDT in double shunt-FET configuration with a chip size of 1 x 1 mm². (b) 94 GHz SPDT in single shunt-FET configuration and (c) 94 GHz SPDT in double shunt-FET configuration with a chip size of 1 x 0.75 mm²

B. SPDT for 77 to 120 GHz

SPDT switches for a center frequency of 94 GHz have been designed both in single shunt-FET and double shunt-FET configuration. The related chip photographs are shown in Fig. 4(b) and (c), respectively. They have been realized in the IAF metamorphic HEMT technology with 50 nm gate length [2]. The process uses single-sided doping and a composite channel combining an 80%-In channel with an underlying 53%-In sub-channel for hole confinement and reduced impact ionization. The transistors achieve typical cutoff frequencies of $f_T = 400$ GHz and $f_{max} = 420$ GHz. For the realization of FET switches, their low source- and drain access resistance is of prime importance.

Fig. 7 shows the measured insertion loss and isolation of both SPDT switches. They perform over a broad frequency range encompassing 77 GHz and 94 GHz up to 120 GHz. The insertion loss has increased by 1 dB below 66 GHz. The SPDT in single shunt-FET configuration achieves an insertion loss of 1.8 dB at 77 and 94 GHz, and 1.9 dB at 120 GHz. Isolation values are better than 20 dB over the full frequency range. The double shunt-FET switch shows a > 9 dB better isolation, but has 0.4 dB higher insertion loss on average. In both switches, the in- and output matching of the opened branch is better than 8 dB between 77 and 120 GHz. Linearity has been evaluated in a one-tone power measurement at 94 GHz up to a maximum power level of 19 dBm. Again, no significant degradation of the output power was detected, both, in the single and double shunt-FET configurations (Fig. 6).

C. Comparison to Other Published SPDT Switches

In Fig. 8 the achieved results are compared to published data of SPDT switches realized in CMOS, PIN diode, GaAs FET and MEMS technologies. We include the application frequencies at 60, 77, 94 and 120 GHz. To our knowledge, the only MEMS-based SPDT in this frequency range was reported at 60 GHz in [3]. The best reported millimeter-wave performance of a CMOS-based SPDT switch is from [4]. The best data on PIN-diode-based SPDT switches is found in [5] for 60 GHz and 120 GHz, [6] for 77 GHz, [7] for 94 GHz. Ref.

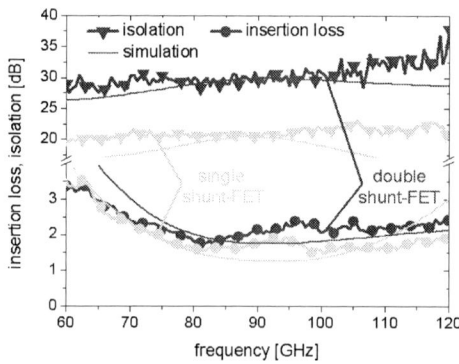

Figure 7. Insertion loss and isolation versus frequency of single and double shunt-FET SPDT switches operating from 77 up to 120 GHz.

[8] reports an InGaAs PIN diode SPDT with an estimated 3 dB insertion loss but < 10 dB isolation at 145 GHz. Record GaAs FET-based SPDT switches have been reported in [9] for 60 GHz, [10] for 77 GHz, [11] for 94 GHz and 120 GHz.

IV. MULTIPLE-THROW TRANSCEIVER SWITCHES (2:X SWITCH)

Previously, the 100 nm mHEMT process was used to realize several 2:2 switches with 3.0 dB insertion loss, which have been incorporated in a 60 GHz transmit/receive amplifier MMIC, capable of routing and amplifying signals from two input/output ports to four transmit/receive antennas [12].

Here, we demonstrate an MMIC realizing a 2:6 switch network (cp. Fig. 1b), thus allowing to address six antennas in 60 GHz transceivers for broadband wireless communication. Fig. 9 shows the chip photograph of the MMIC, realized in our 100 nm mHEMT technology. The signal passes through three consecutive switch branches before reaching one of the six antenna ports.

Fig. 10 shows the measured insertion loss and isolation for all six antenna ports. At 60 GHz, the insertion loss lies

978-2-8748-7007-1/08 $25.00 © 2008 EuMA   428

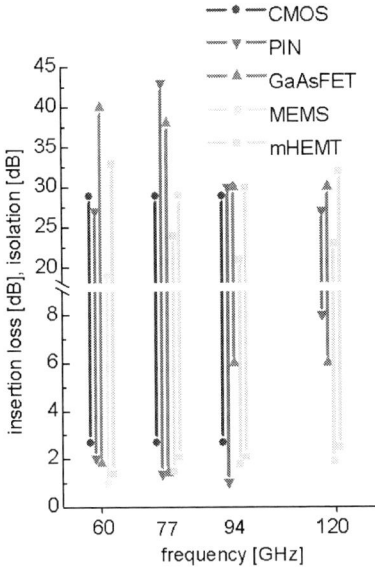

Figure 8. Compilation of published SPDT switches in the frequency range from 60 to 120 GHz.

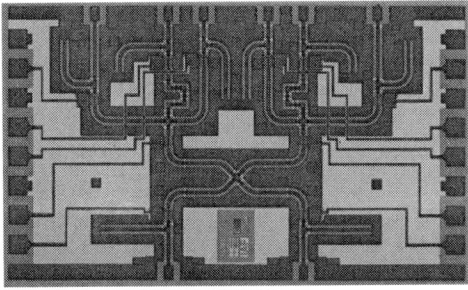

Figure 9. Chip photograph of a 60 GHz 2:6 switch network. The chip size is 1.5 x 2.5 mm².

between 3.7 and 4.0 dB, while the isolation of this single shunt-FET topology varies between 18.3 and 19.9 dB. Both figures of merit don't show any significant degradation in the frequency range from 55 to 65 GHz, which will be exploited in future broadband wireless communication systems.

V. CONCLUSION

Being implemented in a conventional transistor technology, the presented multiple-throw switches can easily be combined with amplifier and mixer stages to form fully monolithic integrated, multifunctional frontends for millimeter-wave communication and radar systems operating up to and beyond 100 GHz.

VI. ACKNOWLEDGEMENT

We express our gratitude to our colleagues in the IAF technology department for their excellent contributions during epitaxial growth and wafer processing.

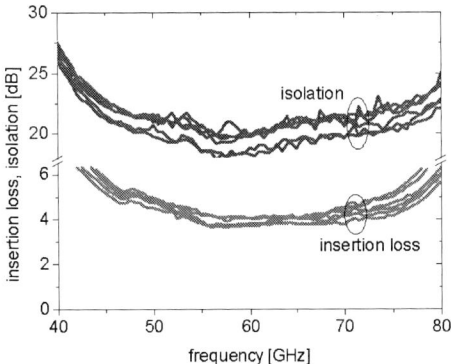

Figure 10. Insertion loss and isolation versus frequency for all transmission paths of the 2:6 switch.

REFERENCES

[1] A. Tessmann, "220-GHz metamorphic HEMT amplifier MMICs for high-resolution imaging applications," *IEEE Solid-State Circuits*, vol. 40, pp. 2070–2076, Oct. 2005.

[2] A. Leuther, A. Tessmann, M. Dammann, C. Schwörer, M. Schlechtweg, M. Mikulla, R. Lösch, G. Weimann, "50 nm MHEMT Technology for G-and H-Band MMICs," *Int. Conf. on Indium Phosphide and Related Materials (IPRM), Matsue, Japan*, pp. 24–27, May 2007.

[3] J.-H. Park, S. Lee, J.-M. Kim, Y. Kwon, Y.-K. Kim, "A 35-60 GHz single-pole double-throw(SPDT) switching circuit using direct contact MEMS switches and double resonance technique," in *Proc. 12th International Conference on Transducers, Solid-State Sensors, Actuators and Microsystems, Boston*, vol. 2, pp. 1796–1799, June 2003.

[4] S.-F. Chao, H. Wang, C.-Y. Su, J. G. J. Chern, "A 50 to 94-GHz CMOS SPDT Switch Using Traveling-Wave Concept," *IEEE Microwave and Wireless Components Letters*, vol. 17, pp. 130–132, Feb. 2007.

[5] V. Ziegler, C. Gaessler, C. Wolk, F.-J. Berlec, R. Deufel, M. Berg, J. Dickmann, H. Schumacher, E. Alekseev, D. Pavlidis, "InP-based and metamorphic devices for multifunctional MMICs in mm-wave communication systems," in *Proc. Int. Conf. on Indium Phosphide and Related Materials, Williamsburg*, pp. 341–344, May 2000.

[6] E. Alekseev, D. Pavlidis, V. Ziegler, M. Berg, J. Dickmann, "77 GHz high-isolation coplanar transmit-receive switch using InGaAs/InP PIN diodes," in *Proc. 20th Gallium Arsenide Integrated Circuits Symposium, Atlanta*, pp. 177–180, Nov. 1998.

[7] J. Putnam, M. Fukuda, P. Staecker, Y.-H. Yun, "A 94 GHz monolithic switch with a vertical PIN diode structure," in *Proc. 16th Gallium Arsenide Integrated Circuits Symposium, Philadelphia*, pp. 333–336, Oct. 1994.

[8] L. Samoska, P. Kangaslahti, D. Pukala, G. Sadowy, B. Pollard, R. Hodges, "A G-Band 160 GHz T/R Module Concept for Planetary Landing Radar," in *Proc. 36th European Microwave Conference, Manchester*, pp. 757–760, Sept. 2006.

[9] Y. Tsukahara, T. Katoh, Y. Notani, T. Ishida, T. Ishikawa, M. Komaru, Y. Matsuda, "Millimeter-wave MMIC switches with pHEMT cells reduced parasitic inductance," in *Proc. IEEE MTT-S Int. Microwave Symp., Philadelphia*, vol. 2, pp. 1295–1298, June 2003.

[10] J. Kim, W. Ko, S.-H. Kim, J. Jeong, Y. Kwon, "A high-performance 40-85 GHz MMIC SPDT switch using FET-integrated transmission line structure," *IEEE Microwave and Wireless Components Letters*, vol. 13, pp. 505–507, Dec. 2003.

[11] Z.-M. Tsai, M.-C. Yeh, M.-F. Lei, H.-Y. Chang, C.-S. Lin, H. Wang, "DC-to-135 GHz SPST and 15-to-135 GHz SPDT Traveling Wave Switches Using FET-Integrated CPW Line Structure," in *Proc. IEEE MTT-S Int. Microwave Symp., Long Beach*, pp. 1393–1396, June 2005.

[12] S. Koch, I. Kallfass, R. Weber, A. Leuther, M. Schlechtweg and M. Uno, "An Analogue, 4:2 MUX/DEMUX Front-End MIMIC for Wireless 60 GHz Multiple Antenna Transceivers," in *Proc. IEEE MTT-S Int. Microwave Symp., Honolulu, Hawai*, pp. 1121–1124, June 2007.

978-2-8748-7007-1/08 $25.00 © 2008 EuMA

120-GHz-band Low-noise Amplifier with 14-ps Group-delay Variation for 10-Gbit/s Data Transmission

Hiroyuki Takahashi[#], Toshihiko Kosugi[*], Akihiko Hirata[#], Koichi Murata[*],

and Naoya Kukutsu[#]

[#]*NTT Microsystem Integration Laboratories, NTT Corporation*

3-1, Morinosato Wakamiya, Atsugi-shi, Kanagawa Pref., 243-0198 Japan

t-hiro@aecl.ntt.co.jp

[*]*NTT Photonics Laboratories, NTT Corporation*

3-1, Morinosato Wakamiya, Atsugi-shi, Kanagawa Pref., 243-0198 Japan

Abstract— This paper presents a 120-GHz-band low-noise amplifier (LNA) for a receiver microwave monolithic integrated circuit (MMIC), which is used for a 10-Gbit/s wireless link. The LNA was designed for low-noise performance, a high gain, and low group-delay variation. To achieve enough stability with low-noise performance and low group-delay variation, we introduce a new stabilizing circuit consisting of two coplanar-waveguide stubs. The LNA MMIC was fabricated using 0.1-μm-gate InP high-electron-mobility transistors (HEMTs). We integrated the LNA into a WR-8 (90-140 GHz) waveguide module and evaluated it. The LNA module achieved an averaged noise figure of 5.6 dB, a gain of > 19.5 dB, and group-delay variation of < 14 ps from 117.5 to 132.5 GHz.

I. INTRODUCTION

With the increasing use of high-speed local area networks, a strong demand has arisen for gigabit-class wireless links. In particular, there is a big demand for the transmission of high-definition-television (HDTV) data in broadcasts. To transmit HDTV over wireless links, data-compression techniques are generally used. However, to avoid quality degradation and long delay times, it is necessary to transmit uncompressed HDTV signal. Furthermore, a wireless link that can transmit multiplexed uncompressed HDTV signals is desired in large-scale live relay broadcasting.

To cope with these demands, we are developing a 10-Gbit/s wireless link [1] and MMICs [2] using the 120-GHz band. The 120-GHz band is promising for wideband wireless links because of it provides sufficient bandwidth and small atmospheric absorption. The required communication distance to relay a broadcast of HDTV is about 2.0 km under fair condition. To extend the communication distance, we need high-performance low-noise amplifier (LNA) in order to improve sensitivity. In wideband data transmission, an LNA needs both of low-noise performance and small group-delay variation in the occupied frequency band (OFB). Some LNAs with low-noise performance at over 100 GHz have been reported [3-5]. However, there has been no LNA designed in consideration of the need for small group-delay variation.

In this paper, we describe a 120-GHz-band five-stage LNA with small group-delay variation for 10-Gbit/s wireless links. The OFB of the wireless link ranges from 117.5 to 132.5 GHz. First, we designed the LNA to achieve a noise figure of < 5.5 dB, high gain of 20-dB, and a group-delay variation of < 20 ps, which is 20% of the symbol length in the OFB. We introduced a new stabilizing circuit with a dual-open-stub resonator for stable operation. Next, we integrated the LNA MMIC into a WR-8 waveguide module and evaluated the S-parameters, group-delay variation, noise figure, and input-output characteristics. We then investigated the implementation of the LNA for 10-Gbit/s wireless link.

II. CIRCUIT DESIGN

A. Design for low-noise performance and small group-delay variation

Figure 1 shows a block diagram of the 120-GHz-band five-stage LNA MMIC. To achieve low-noise performance and small group-delay variation for wideband data transmission, we designed stages 1, 2 and 3 and stages 4 and 5 using different concepts. The design of stages 1-3 was optimized for low-noise performance in the OFB. Stages 4 and 5 compensate the variation of group-delay and gain. The LNA has a level-shift circuit for applying different voltages to the stages 1-3 and stages 4 and 5.

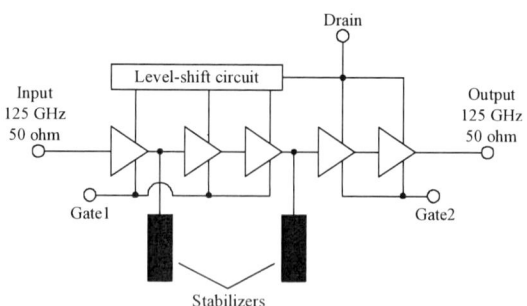

Fig. 1. Block diagram of the LNA MMIC.

978-2-8748-7007-1/08 $25.00 © 2008 EuMA

We determined the drain voltage of the stages 1-3 experimentally to achieve low-noise performance of the HEMT. The minimum noise figure of the HEMT was 3.5 dB at 125 GHz when the applied voltage was 0.8 V. To decrease the transmission loss in the circuit, simple matching networks were used. We optimized the design of the stages 1-3 to attain an average noise figure of less than 5 dB in the OFB. The group-delay flatness and gain flatness of these stages were poor because they were not included in the optimization.

In the stages 4 and 5, matching networks were designed to compensate for variations of the group delay and gain, which come from the stages 1-3. The drain voltage of the HEMTs was 1.5 V for high-gain performance. The total gate widths of 4th and 5th stages were determined to achieve saturated output power of > 0 dBm, which is required for our wireless link. The noise figure of the stages 4 and 5 was over 6.0 dB, which was worse than that of stages 1-3. However, the influence of this noise figure is small because the gain of the stages 1-3 is over 10 dB in this design.

B. Stabilizing circuit

An LNA is normally designed to have enough stability at all frequencies, and stability is often given excessively in the operation band. In our LNA, excess stability directly makes the noise figure worse because the associated gain of a FET is small at over 100 GHz. In that case, a frequency-dependent circuit, such as a $\lambda/4$ short stub, is often used as a stabilizing circuit. The circuit has a low insertion loss in the operation band. And the loss increases outside of the operation band. However, the steepness of the change is not enough at over 100 GHz, because the Q-value of a transmission line is small. To overcome this problem, we introduce a dual-stub stabilizer for the stabilizing circuit. The circuit has a high-Q resonator and achieves a steep change of the insertion loss. Figure 2 shows an equivalent circuit of the dual-stub stabilizer. The circuit consists of two coplanar lines, a resistor, and a MIM capacitor. An open stub and a stub with a MIM capacitor act as a high-Q resonator. The circuit resonates at a frequency that satisfies the equation,

$$-Z_1 \cot(\beta_1 L_1) = Z_2 \cot\{\beta_2(L_2 + \delta)\} \quad (1)$$

where Z, β, and L are the intrinsic impedance, and the phase-constant ($2\pi/\lambda$) and length of coplanar lines, respectively. The δ is given by using the impedance of a MIM capacitor jX:

$$\delta = -\frac{1}{\beta_2}\tan^{-1}\left(-\frac{Z_2}{X}\right) \quad (2)$$

The size of the dual-stub stabilizer was decreased by using a shunt capacitor at the end of the stub. We compared the insertion loss of a $\lambda/4$ short stub with that of the dual-stub stabilizer by simulation. Figure 3 shows the calculated S_{21} characteristics of a transmission line with each stabilizing circuit. In this simulation, Z and β of each stabilizing circuit were the same value. Insertion losses of both circuits were < 1 dB in OFB. However, outside of the band, the insertion loss of the dual-stub stabilizer increased rapidly. At 90 GHz, insertion losses of $\lambda/4$ short stub and the dual-stub stabilizer were

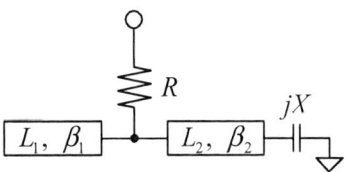

Fig. 2. Equivalent circuit of the dual-stub stabilizer.

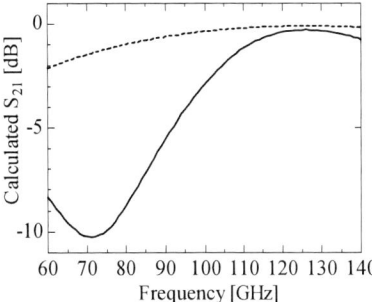

Fig. 3. Calculated S_{21} with the dual-stub stabilizer （—） and with $\lambda/4$ short stub （...） .

-0.6 dB and -5.5 dB, respectively. This indicates that the dual-stub stabilizer is more effective for stability in our LNA. The stability of the LNA was poor at below 105 GHz in stability calculation. We set the parameters of dual-stub stabilizer to introduce large insertion loss below 105 GHz.

III. CIRCUIT PERFORMANCE

A. MMIC fabrication and integration of LNA module

We used 0.1-μm HEMT technology on InP substrate. The devices have a current-gain cut-off frequency (f_t) of 170 GHz and a maximum oscillation frequency (f_{max}) of 350 GHz. Figure 4 shows a photograph of the LNA MMIC. The LNA chip is 1 mm × 2 mm. The power consumption is 120 mW. Before measurements, we integrated the LNA MMIC into a WR-8 waveguide module (Fig. 5). The module is 11 cc in size. In the module, the dominant mode of an input signal is changed from the waveguide mode to coplanar mode by a coupler fabricated on a quartz substrate. We estimated the transmission loss from the input port to the MMIC by electromagnetic field simulation. The loss was 0.7 dB and the return loss was smaller than -15 dB from 115 to 135 GHz.

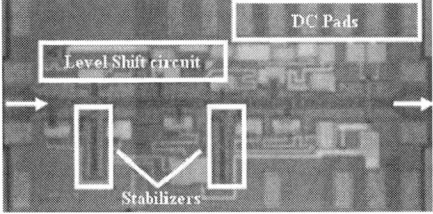

Fig. 4. Photograph of the LNA MMIC.

Fig. 5. Photograph of the LNA module.

B. Measurement results of LNA module

Measured frequency characteristics of the LNA module are shown in Fig. 6. For the measurement, we used an F-band (90-140 GHz) vector network analyser. In the OFB, from 117.5 to 132.5 GHz, the small-signal gain was > 19.8 dB and the variation was 3.5 dB. The gain was more than 10 dB from 100 to 140 GHz. S_{11} and S_{22} were smaller than -5 dB. We then measured the frequency characteristics of the group-delay and noise figure as shown in Fig. 7. The group-delay was calculated from S-parameters and was averaged (the aperture range was 2-GHz) to suppress variations caused by calibration error of vector network analyser. The group-delay variation was 14 ps and the averaged noise figure was 5.6 dB within OFB. The frequency characteristics of the group delay and noise figure were flat in the OFB. As mentioned, the estimated transmission loss of the module from the input port to MMIC was 0.7 dB in simulation. So, the noise figure of the LNA MMIC can be estimated to be 4.9 dB. Figure 8 shows input-output characteristics at a frequency of 125 GHz. The saturated output power was 5.5 dBm. The input 1-dB gain compression point was -16 dBm, and the output power at that point was 2.8 dBm. We also evaluated the stability of the LNA module. The module worked without oscillation under the setup conditions.

IV. IMPLEMENTATION FOR SYSTEM

The measurement results indicate that the characteristics of the LNA, such as small group-delay variation of 14 ps and low-noise figure of 5.6 dB, are sufficient for 10-Gbit/s data transmission using the 120-GHz band. The LNA has a gain of >10 dB outside of the OFB that are from 100 to 117.5 GHz and from 134.5 to 140 GHz. So, when the LNA was set in a receiver, it is necessary to put a band-pass filter in the back of the LNA to limit noise bandwidth. Equivalent input noise (P_{noise}) of a receiver is given by

$$P_{noise} = N \times k_B TB \qquad (3)$$

where N is the noise figure of a receiver, k_B is Boltzmann's constant, T is temperature, and B is noise bandwidth. We have reported the tunable band-pass filter that can be integrated into a MMIC [6]. The filter can control the bandwidth from 8 to 16.5 GHz. When the filter that sets the bandwidth to be 15 dB

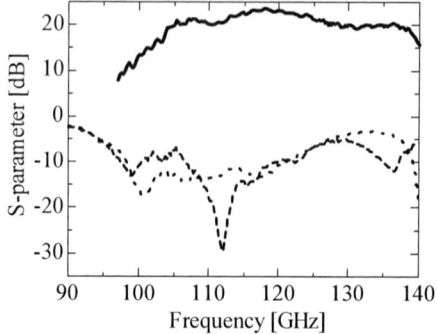

Fig. 6. S_{21} (—) , S_{11} (...) , and S_{22} (---) of the LNA module..

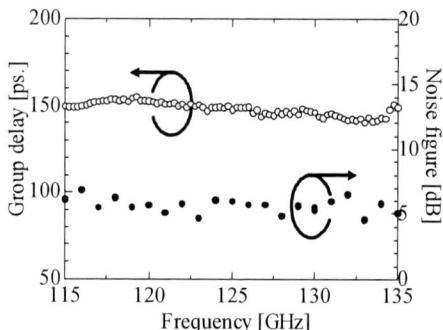

Fig. 7. Noise figure and group delay of the LNA module.

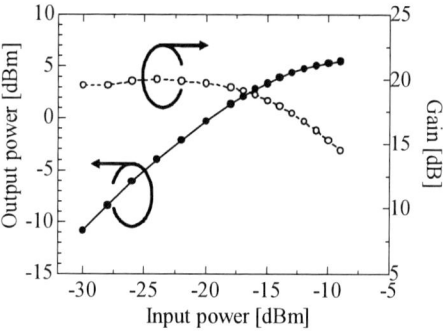

Fig. 8. Input-output characteristic of the LNA module

is used for a receiver, we can estimate P_{noise} to be -66.4 dB. If the requested bit error rate is smaller than 10^{-12} for 10-Gbit/s, the carrier-to-noise (C/N) ratio must be greater than 20.2 dB, when ASK modulation and envelope detection are employed. Therefore, the input signal power of the LNA must be greater than -46.2 dBm in 10-Gbit/s data transmission. The input 1-dB gain compression point of the LNA is -16 dBm. This means that the LNA can amplify the received signal with enough back-off.

V. CONCLUSIONS

We designed and fabricated a five-stage LNA MMIC for a 10-Gbit/s wireless receiver using the 120-GHz band and integrated the LNA into a WR-8 waveguide module. The group-delay variation of the LNA module is 14 ps and the averaged noise figure is 5.6 dB from 117.5 to 132.5 GHz. The gain is >19.8-dB and the return loss is < -5 dB. The input 1-dB gain compression point is -16 dBm, and the saturated output power is 2.6 dBm. Judging from the input 1-dB gain compression point, the LNA accomplished required input power of -46.2 dBm with enough margins. These results indicate that the characteristics of the LNA are good enough for 10-Gbit/s data transmission. In the future, we will apply this LNA to a 10-Gbit/s wireless link and evaluate the data transmission characteristics.

ACKNOWLEDGMENT

The authors thank Drs. T. Enoki, Y. Kado, and Prof. T. Nagatsuma for their continuous encouragement throughout this research.

Part of this work was supported by "The Research and Development Project for the Expansion of Radio Spectrum Resources" made available by the Ministry of Internal Affairs and Communications, Japan.

REFERENCES

[1] A. Hirata, T. Kosugi, T. Furuta, H. Ito, M. Tokumitsu, and T. Nagatsuma, "Photonic Devices for Ultra-Broadband Wireless Link, Sensing and Measurement System," in Tech. Dig. International Topical Meeting on Microwave Photonics 2005, pp.67-70.

[2] T. Kosugi, M. Tokumitsu, T. Enoki, and M.Muraguchi, "120-GHz Tx/Rx Chipset for 10-Gbit/s Wireless Applications Using 0.1-μm-gate InP HEMTs," IEEE CSIC 2004 Digest, pp.171-174.

[3] H. Wang, R. Lai, D.C.W. Lo, D.C.Streit, P.H. Liu, R.M. Dia, M.W. Pospieszalski, and J. Berenz, "A 140-GHz monolithic low noise amplifilier," Microwave and Guided Wave Letters, IEEE, vol.5, Issue 5, pp.150-152, May 1995.

[4] H. Wang, R. Lai, Y.-L. Kok, T.-W. Huang, M.V. Aust, Y.C. Chen, P.H. Siegel, T. Gaier, R.J. Dengler, and B.R. Allen, "A 155-GHz monolithic low-noise amplifier," Microwave Theory and Techniques, IEEE Transaction, vol. 46, Issue 11, Part1, pp.1660-1666, Nov. 1998.

[5] R.Raja, M. Nishimoto, B. Osgood, M. Barsky, M, Sholley, R. Quon, G. Barber, P. Liu, R. Lai, F. Hinte, G. Haviland, and B. Vacek, "A183 GHz low noise amplifier module with 5.5 dB noise figure for the conical-scanning microwave imager sounder (CMIS) program," Microwave Symposium Digest, 2001 IEEE MTT-S International, vol. 3, 20-25 May 2001, pp.1955-1958

[6] H. Takahashi, T. Kosugi, A. Hirata, K. Murata, and T. Nagatsuma, "Tunable Coplanar Filter for F-band Wireless Receivers," Asia-Pacific Microwave Conference 2006, 12-15 Dec. 2006, pp.15-18.

Proceedings of the 3rd European Microwave Integrated Circuits Conference

A rigorous assessment of electro-thermal device instabilities via Harmonic Balance modeling

Federica Cappelluti[1], Fabio L. Traversa[2], Fabrizio Bonani[3]

Dipartimento di Elettronica, Politecnico di Torino
Corso Duca degli Abruzzi 24, I-10129 Torino, Italy
[1]federica.cappelluti@polito.it
[2]fabio.traversa@polito.it
[3]fabrizio.bonani@polito.it

Abstract—**This paper presents a rigorous numerical approach to the assessment of electro-thermal instabilities arising in high-power semiconductor devices operating under time-periodic conditions. The methodology is entirely developed in the frequency domain with reference to the Harmonic Balance technique, i.e. no time-domain calculations are required for the determination of the Floquet multipliers exploited for the stability analysis. As an example of application, the current gain collapse occurring in multifinger AlGaAs/GaAs HBTs is studied and compared to the customary stability criterion based on a purely static analysis.**

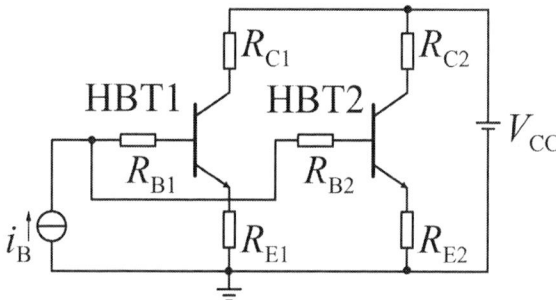

Fig. 1.   Electrical circuit for a two-finger HBT.

I. INTRODUCTION

Electro-thermal simulation is a very important issue in device modeling, mainly because of the strong impact of self-heating on the device and, therefore, system performance in high-power circuits. In particular, this poses severe limitations to the applicabilty of power bipolar devices, wherein the usual approach of exploiting a multiple emitter finger device layout (called for by technological requirements) coupled to the current gain variation as a function of temperature may lead to an uneven temperature distribution over the device fingers and, ultimately, to thermal instability phenomena which, in the worst case, may ultimately drive the device to destruction (see e.g. [1], [2], [3] and [4]). The macroscopic effect of such instability on the current-voltage characteristics depends on the device considered, mainly beacuse standard bipolar transistors (BJT) exhibit a current gain which increases as a function of temperature, while heterostructure devices (HBT) have a negative current gain temperature coefficient: this leads to the current collapse phenomenon discussed, e.g., in [4].

The thermal instability issue is often fought with various stabilization techniques, either electrical (such as emitter [5] or base [6] ballasting) or thermal (with the use of thermal shunts, i.e. of increased thermal coupling among the fingers in order to minimize temperature differences [7]).

In the vast majority of cases, the stability analysis performed either to verify the safety of the design, or to define the ballasting resistances (or, less often, to optimize the device layout to increase thermal coupling between the fingers) is performed in static conditions only, mainly because of the lack of a well established technique for the stability analysis of dynamical systems. The aim of this contribution is to provide such a technique, based on the frequency domain

determination of the relevant parameters assessing system stability presented in [8].

II. FREQUENCY-DOMAIN STABILITY ANALYSIS

To fix the ideas, let us consider the electro-thermal dynamic model of a multifinger bipolar device. Of course, the same approach can be easily extended to any kind of semiconductor device, provided that the final equations can be stated in the form of a Differential Algebraic Equation (DAE) system [8], i.e. no distributed elements are allowed for:

$$\frac{\mathrm{d}}{\mathrm{d}t}q(x) + f(x,t) = 0, \tag{1}$$

where x is the vector collecting the model unknowns (electrical and thermal), and $q(x)$ and $f(x,t)$ are nonlinear functions.

The electrical part of the model is sketched in Fig. 1 for the case of two fingers, where bias resistors and generators are also included. Each device finger is represented by a proper description of the relationship between currents and voltages, for instance in this case the standard Gummell-Poon model [9] for forward DC operation, possibly complemented by (generally) nonlinear capacitances between emitter and collector if dynamic effects are to be included in the model. The electrical model must account for thermal effects, typically by introducing the (finger) temperature as a parameter of the model itself: this means that a further variable T_k is added for each of the $k = 1, \ldots, N_\mathrm{f}$ fingers of the device.

The finger temperatures are in turn estimated exploiting an equivalent thermal impedance approach, whereby the temperature raises $\Delta T_k = T_k - T_\mathrm{ref}$ (T_ref is the reference

978-2-8748-7007-1/08 $25.00 © 2008 EuMA   434

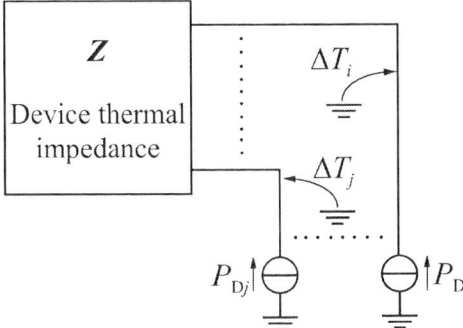

Fig. 2. Equivalent thermal circuit.

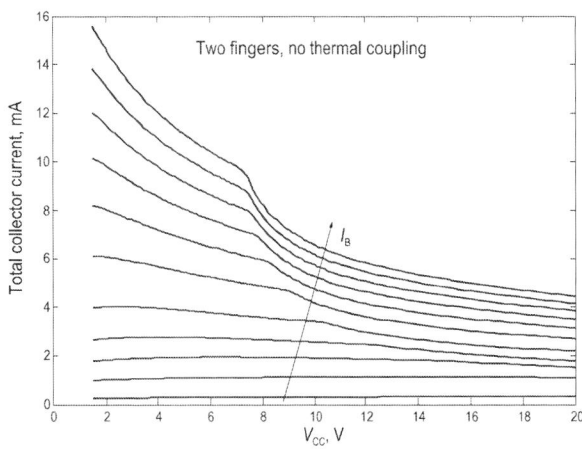

Fig. 3. Total DC collector current as a function of V_{cc} for a 2 finger HBT. The total base current I_B ranges from 10 μA to 400 μA. No thermal coupling between the fingers.

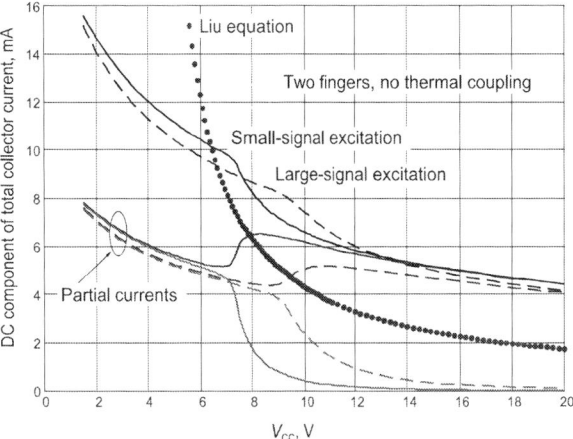

Fig. 4. DC component of the total collector current as a function of V_{cc} for a 2 finger HBT. No thermal coupling between the fingers. The total base current is made of a 400 μA DC component and a very small (small-signal) or 400 μA (large-signal) input tone at 1 Mhz.

temperature of the model, most often assumed equal to 300 K) are related to the (finger) dissipated power P_{Dj} through a thermal impedance matrix as shown in Fig. 2. According to the request of a nonlinear dynamical system in DAE form, the thermal impedance must be represented through a lumped element approximation, for instance extracted from numerical thermal FEM simulations or by less numerically intensive analytical or semi-analytical approaches, such as that exploited in [9]. Notice that, in principle, nonlinearity due to temperature dependent thermal parameters may be taken into account, as far as leading to a thermal model of the form of equation (1).

Once the electro-thermal model has been established, its solution in the case of time-periodic excitation can be obtained by exploiting the standard harmonic balance approach [10], thus leading to the determination of the corresponding limit cycle. The stability of the limit cycle is finally assessed by computing the corresponding Floquet multipliers (FMs) [11], [12], which must all be confined into the interior of the unit circle of the complex plane in order to guarantee the asymptotic stabilty of the limit cycle itself. The FMs are effectively determined in the frequency domain exploiting the algorithm proposed in [8], which is based only on the knowledge of the jacobian matrix of the harmonic balance system, typically already available at the final step of the limit cycle determination since the latter is almost always performed by means of Newton iterations.

III. CASE STUDY

The previous stability analysis technique has been applied to the simulation of current gain collapse of AlGaAs/GaAs HBTs, whose dynamic electro-thermal model was previously discussed in [13]. With respect to the model in [13], the thermal model had to be simplified, since the algorithm for the computation of FMs requires to approximate the thermal impedance matrix of the device with a lumped element circuit: for the sake of simplicity, the results here presented are derived assuming a single pole approximation (i.e., a parallel RC group) of the frequency dependence of the thermal impedance. Furthermore, when thermal coupling is included it is assumed to be present between first neighbouring fingers only, again under a single pole approximation.

The HBT is current-driven by the total base current generator shown in Fig. 1, while the collector bias V_{cc} is used as a parameter in the device simulation. For a purely DC base bias, the output characteristics shown in Fig. 3 have been derived. In this case, a 2-finger device has been considered, without any thermal coupling between the fingers. The current collapse is clearly observed for a V_{cc} value slightly lower than 8 V.

In order to rigorously analyze the collapse phenomenon, a fully dynamic simulation has been performed via the application of the harmonic balance approach. The DC component of the total base current is 400 μA, while two input tones at 1 MHz have been considered: a very small amplitude one, thus effectively operating the device in small-signal conditions, and a large-signal excitation leading to the turn-off of the device, since the tone amplitude is 400 μA. Simulation exploits 10

978-2-8748-7007-1/08 $25.00 © 2008 EuMA

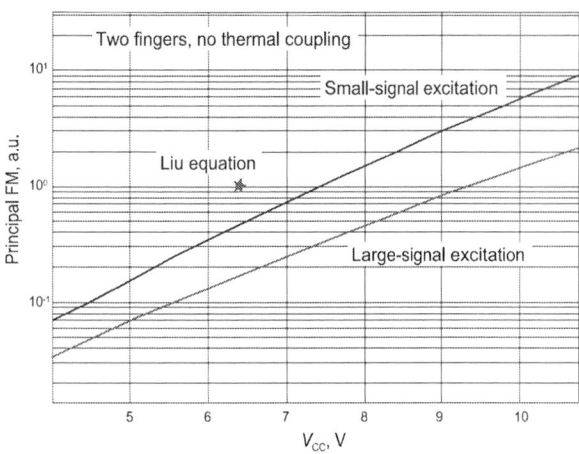

Fig. 5. Principal FM for the 2 finger HBT without thermal coupling between the fingers.

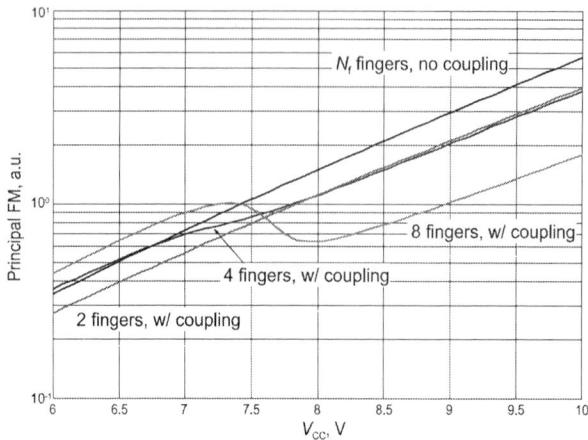

Fig. 6. Principal FM for a decoupled N_f finger HBT, and for 2, 4 and 8 finger HBTs with thermal coupling. The total base current is made of a $200 \times N_f$ μA DC component and a very small input tone at 1 Mhz.

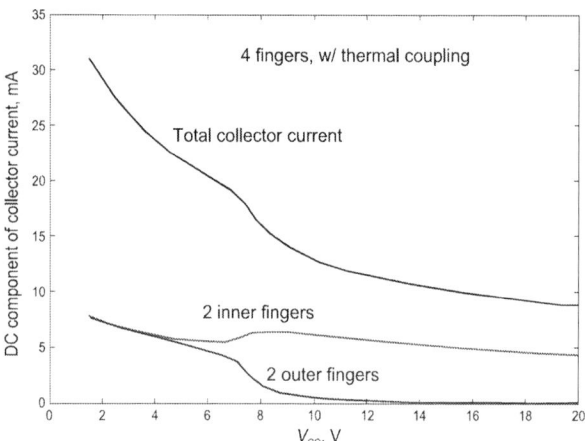

Fig. 7. DC component of the collector current as a function of V_{cc} for a 4 finger HBT with thermal coupling between the fingers. The total base current is made of a 800 μA DC component and a very small input tone at 1 Mhz.

and 30 harmonics plus DC in the small-signal and large-signal case, respectively. Notice that the very low frequency exploited makes the electrical part of the device to be excited practically in static conditions (the device cutoff frequency is over 30 GHz): this choice has been motivated by the requirement to fully highlight the importance of dynamic effects on the onset of the collapse.

Fig. 4 shows the DC component of the total collector current (and its two partial components, i.e. the finger currents) as a function of collector bias, for both the small- and large-signal excitations. As expected, the small-signal curve closely resembles the DC case, while even for such a small frequency the large-signal case exhibits a significant shift of the collapse onset towards larger V_{cc} values. In the same figure, symbols represent the critical collector current as evaluated through the analytical stability criterion in [4], derived in purely static conditions. Even in the small-signal case, the latter underestimates the V_{cc} limit value, thus providing a worst-case bound to ensure stable operation.

The current approach allows for a more rigorous treatment of instability, by means of the explicit computation of the FMs. As devised in [14], [3], the gain collapse instability can be traced back to a bifurcation in the solution of the electro-thermal model. In fact, examining the FMs of the 2-finger device the result is as follows: all the FMs are always stable (i.e., they remain within the unit circle in the complex plane for all V_{cc} values), except one (that we shall call the "principal FM" according to [8]). The principal FM (see Fig. 5) is always real, and grows as V_{cc} increase, ultimately crossing the unit circle with a value equal to +1, thus corresponding to a fold bifurcation [11], [8]. This approach thus allows for a precise definition of the onset of instability, showing that for the large-signal excitation the onset of collapse moves to $V_{cc} \approx 9.2$ V from the small-signal $V_{cc} \approx 7.4$ V, while Liu's criterion yields $V_{cc} \approx 6.4$ V.

The model has been also applied to the simulation of multi-finger devices with a spatially symmetric layout, including the effect of a small thermal coupling (the off-diagonal, coupling term is less than 4% of the diagonal value) between first neighbouring fingers only. The applied base current is scaled with the number of fingers, so as to maintain a constant dissipated power density on the device. The principal FM calculated for a 2-, 4- and 8-finger device is shown in Fig. 6. In the same figure, we also include the curve corresponding to N_f non-interacting fingers, for which of course the curve is independent of the number of fingers. The 2- and 4-finger layouts yield the same critical value of the V_{cc} parameter, around 7.8 V (a 5.4% increase with respect to the 7.4 V of the non interacting fingers). The collector current for the 4-finger device is shown in Fig. 7, along with the partial components that show how the collapse corresponds to the fact that the two inner fingers of the layout draw most of the current.

978-2-8748-7007-1/08 $25.00 © 2008 EuMA

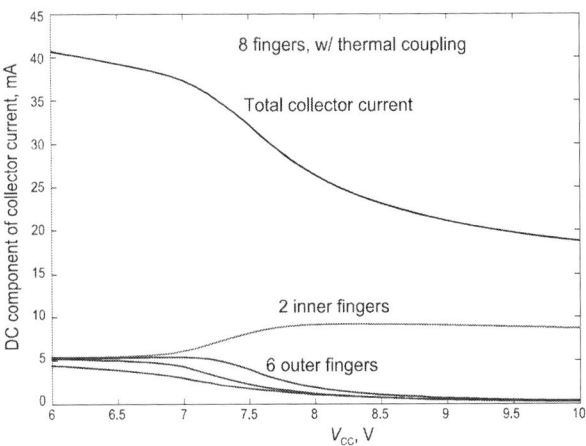

Fig. 8. DC component of the collector current as a function of V_{cc} for a 8 finger HBT with thermal coupling between the fingers. The total base current is made of a 1600 μA DC component and a very small input tone at 1 Mhz.

The 8-finger layout exhibits a peculiar behaviour, wherein a first bifurcation takes place for $V_{cc} \approx 7.3$ V, followed by a second one around 9 V. According to the collector current graph in Fig. 8, after the first bifurcation most of the current is drawn by the two inner fingers of the device. The second bifurcation probably corresponds to the fact that one of the inner fingers ultimately sustains all the device current.

IV. CONCLUSIONS

In this paper we have presented the first full implementation of the harmonic balance-based calculation of the Floquet multipliers for the assessment of the onset of thermal instabilities of multi-finger power semiconductor devices. The methodology is general, and can be applied to any electro-thermal device model, provided that the electrical and thermal parts of the system are represented by lumped element circuits, i.e. no distributed elements are included.

As an example of application, we have discussed the current gain collapse phenomenon taking place in GaAs-based HBTs, typically studied by means of purely static considerations only. According to our results, the latter provide a worst-case bound for the onset of thermal instability, since a significant dispersion effect is shown even for very small frequency. Furthermore, the presence of thermal coupling among the device finger further improves the stability of the whole system.

ACKNOWLEDGMENT

This work was partially supported by the research project CNT4SIC funded by Regione Piemonte under the "Bando regionale sulla ricerca industriale e lo sviluppo precompetitivo per l'anno 2006".

REFERENCES

[1] R.H. Winkler, "Thermal properties of high-power transistors", *IEEE Trans. Electron Devices,* vol. ED-14, pp. 260–263, May 1967.
[2] N. Rinaldi, V. D'Alessandro, "Theory of electrotrhermal behavior of bipolar transistors: Part I–Single-finger devices", *IEEE Trans. Electron Devices,* vol. ED-52, pp. 2009–2021, Sept. 2005.
[3] N. Rinaldi, V. D'Alessandro, "Theory of electrotrhermal behavior of bipolar transistors: Part II–Two-finger devices", *IEEE Trans. Electron Devices,* vol. ED-52, pp. 2022–2033, Sept. 2005.
[4] W. Liu, S. Nelson, D.G. Hill, A. Khatibzadeh, "Current gain collapse in microwave multifinger heterojunctionbipolar transistors operated at very high power densities", *IEEE Trans. Electron Devices,* vol. ED-40, pp. 1917–1927, Nov. 1993.
[5] C.-H. Liao, C.-P. Lee, N.L. Wang, B. Lin, "Optimum Design for a Thermally Stable Multifinger Power Transistor", *IEEE Trans. Electron Devices,* vol. ED-49, pp. 902–908, May 2002.
[6] W. Liu, A. Khatibzadeh, J. Sweder, H.F. Chau, "The use of base ballasting to prevent the collapse of current gainin AlGaAs/GaAs heterojunction bipolar transistors", *IEEE Trans. Electron Devices,* vol. ED-43, pp. 245–251, Feb. 1996.
[7] D.J. Walkey, D. Celo, T.J. Smy, R.K. Surridge, "A Thermal Design Methodology for Multifinger Bipolar Transistor Structures", *IEEE Trans. Electron Devices,* vol. ED-49, pp. 1375–1383, Aug. 2002.
[8] F.L. Traversa, F. Bonani, S. Donati Guerrieri, "A frequency-domain approach to the analysis of stability and bifurcations in nonlinear systems described by differential-algebraic equations", *Int. J. Circuit Theory & App.,* Vol. 36, No. 4, pp. 421–439, June 2008.
[9] F. Cappelluti, F. Bonani, S. Donati Guerrieri, G. Ghione, C.U. Naldi, M. Pirola, "Dynamic, self consistent electro-thermal simulation of power microwave devices including the effect of surface metallizations", *Proc. GAAS 02,* pp. 209–212, Milan, Italy, 23–24 September 2002.
[10] K. S. Kundert, A. Sangiovanni-Vincentelli, J. K. White, *Steady-state methods for simulating analog and microwave circuits,* Boston: Kluwer Academic Publisher, 1990.
[11] A. Suárez, R. Quéré, *Stability analysis of nonlinear microwave circuits,* Norwood: Artech House, 2003.
[12] M. Farkas, *Periodic motions,* New York: Springer-Verlag, 1994.
[13] F. Cappelluti, F. Bonani, S. Donati Guerrieri, G. Ghione, M. Peroni, A. Cetronio, R. Graffitti, "Self-consistent fully dynamic electro-thermal simulation of power HBTs", *Proc. GAAS 01,* pp. 199–202, London, United Kingdom, 24–25 September 2001.
[14] Ke Lu, C.M. Snowden, *Analysis of thermal instability in multi-finger power AlGaAs/GaAs HBTs,* IEEE Trans. Electron Devices, Vol. 43, p. 1799, 1996.

Proceedings of the 3rd European Microwave Integrated Circuits Conference

Efficient Circuit-Level Nonlinear Analysis of Interference in UWB Receivers

Vittorio Rizzoli[#1], Franco Mastri[#2], Alessandra Costanzo[#1], Diego Masotti[#1] and Francesco Donzelli[#1]

[#1] *DEIS, University of Bologna, Viale Risorgimento 2, 40136 Bologna, Italy*

`vittorio.rizzoli (alessandra.costanzo, diego.masotti, francesco.donzelli)@unibo.it`

[#2] *DIE, University of Bologna, Viale Risorgimento 2, 40136 Bologna, Italy*

`franco.mastri@unibo.it`

Abstract— **The paper demonstrates for the first time a circuit-level approach to the analysis of pulse-UWB receiver front ends in the presence of interfering communication signals. The procedure is based on a model-order reduction harmonic balance technique (MORHB) that has been especially devised to efficiently handle signal spectra including very large numbers of arbitrarily spaced lines. At each step of the nonlinear solution loop the GMRES iteration is used to find an approximate Newton update belonging to a suitable Krylov subspace, and a novel efficient algorithm for performing matrix-vector multiplications is exploited. The resulting simulation tool allows rigorous computation of interference effects on the nonlinear regime of UWB receivers and accurate circuit-level prediction of receiver sensitivity and channel capacitance. Simplifying assumptions typical of system-level approaches are overcome in this way, while keeping computational time at acceptable levels.**

I. INTRODUCTION

Pulse-based ultra-wideband links are vastly used for an increasing number of short-range applications, mostly indoor, both in personal and in industrial communications environments. Such systems make use of signals with very low power spectral densities and very broad bandwidths. Due to the huge frequency band involved, from 3.1 to 10.6 GHz, they suffer from active and passive interference with other coexisting wireless systems. A systematic approach to the prediction of UWB front end performance in the presence of interference is thus a key issue. A number of studies aimed at the definition of pulse shapes and their associated spectra suitable for mitigating interference effects are available in the literature (e.g., [1]). They usually consist of theoretical signal analyses coupled with system simulations based on some sort of black-box models of the individually designed subsystems to be interconnected. Such methods cannot adequately capture complex subsystem interactions that take place in actual front ends due to mismatches and nonlinear couplings. In order to develop an exact approach to nonlinear UWB front-end simulation, one has to face two contradictory requirements. On the one hand, ultra-short waveforms such as UWB pulses are best handled in the time domain, but time-domain techniques cannot adequately handle the strong frequency dispersion that unavoidably takes place across ultra-wide bands in all linear components, from antennas [2] to dielectric substrates. On the other

hand, ordinary frequency-domain methods become rapidly unaffordable when the number of spectral lines reaches up to the order of several hundreds, or even thousands, as is the case for the intermodulation (IM) of UWB pulses with interfering signals.

In this paper, we show that all the above difficulties may be effectively overcome by an evolution of the harmonic balance analysis technique based on model-order reduction (MORHB). Indeed, we show that the nonlinear analysis of a typical UWB receiver driven by a sequence of UWB pulses of arbitrary shapes in the presence of one or several interferers, can be carried out on ordinary PCs with CPU times compatible with the needs of an industrial R&D lab.

II. MORHB ANALYSIS OF A UWB RECEIVER

The key problem of HB analysis with large spectra is that the memory and CPU time required to store and factorize the Jacobian matrix become quickly unaffordable as the number N of unknowns is increased. The way out is to resort to MOR techniques based on the search for an approximate Newton update within a suitable low-dimensional subspace of R^N [3]. This technique will be briefly summarised in the present section. Let the nonlinear solving system be denoted by

$$\mathbf{E}(\mathbf{X}) = \mathbf{0} \tag{1}$$

where $\mathbf{E} : R^N \to R^N$ is continuously differentiable. In order to efficiently solve (1) we generate a sequence of iterates by finding at each step some update $\mathbf{s}_i$ and some *forcing term* f_i ($0 \le f_i < 1$) such that

$$
\begin{aligned}
\left\| \mathbf{E}(\mathbf{X}_i) + \mathbf{J}(\mathbf{X}_i)\,\mathbf{s}_i \right\| &\le f_i \left\| \mathbf{E}(\mathbf{X}_i) \right\| \\
\mathbf{X}_{i+1} &= \mathbf{X}_i + \mathbf{s}_i
\end{aligned}
\tag{2}
$$

where $\|\cdot\|$ denotes the Euclidean norm. For $f_i = 0$ (2) reduces to the ordinary Newton iteration; otherwise, the forcing term is in some way a measure of the allowed deviation of the actual update $\mathbf{s}_i$ from the exact Newton update $\mathbf{n}_i$. The forcing term is updated at each step making use of the formula suggested in [4],

978-2-8748-7007-1/08 $25.00 © 2008 EuMA

$$f_i = \frac{\left\| \mathbf{E}(\mathbf{X}_i) - \mathbf{E}(\mathbf{X}_{i-1}) - \mathbf{J}(\mathbf{X}_{i-1})\, \mathbf{s}_{i-1} \right\|}{\left\| \mathbf{E}(\mathbf{X}_{i-1}) \right\|} \qquad (3)$$

where $\mathbf{J}(\mathbf{X}) = \partial \mathbf{E}/\partial \mathbf{X}$ is the Jacobian matrix, and $f_0 = 0.5$. Note that the vector $\mathbf{E}(\mathbf{X}_{i-1}) + \mathbf{J}(\mathbf{X}_{i-1})\, \mathbf{s}_{i-1}$ may be considered as a linearised model of the nonlinear map $\mathbf{E}(\mathbf{X})$ in the neighborhood of the $(i-1)$-th iterate. (6) is thus a normalized measure of the deviation between $\mathbf{E}(\mathbf{X})$ and its local linear model at the $(i-1)$-th step. After choosing a forcing term, an approximation $\mathbf{s}_i$ to the exact solution of the Newton equation is generated in such a way that $\mathbf{s}_i$ satisfies the first of (2) for the given f_i. An efficient way to do so is to carry out the search for $\mathbf{s}_i$ within a Krylov subspace of R^N [3]. As a first step, we select a suitable preconditioner $\mathbf{P}_i$. $\mathbf{P}_i$ should provide the best possible approximation to $\mathbf{J}(\mathbf{X}_i)^{-1}$, compatibly with the conditions that the computation and storage of $\mathbf{P}_i$ and its multiplication by a real vector be relatively inexpensive. The Krylov subspace of dimension m associated with the i-th iteration is then defined by

$$\mathbf{K}_i = span\left[\mathbf{K}_i^{(1)}, \mathbf{K}_i^{(2)}, \dots, \mathbf{K}_i^{(m)} \right] \qquad (4)$$

where the basis vectors are computed by the recursive relation [3]

$$\begin{aligned} \mathbf{K}_i^{(1)} &= \left[\mathbf{J}(\mathbf{X}_i)\, \mathbf{P}_i - \mathbf{1}_N \right] \mathbf{E}(\mathbf{X}_i) \\ \mathbf{K}_i^{(q+1)} &= \mathbf{J}(\mathbf{X}_i)\, \mathbf{P}_i\, \mathbf{K}_i^{(q)} \\ &(1 \le q \le m-1) \end{aligned} \qquad (5)$$

and $\mathbf{1}_N$ is an identity matrix of order N. The m-th order approximation to $\mathbf{n}_i$ is now defined as

$$\sigma_i^{(m)} = -\mathbf{P}_i\, \mathbf{E}(\mathbf{X}_i) - \mathbf{P}_i \sum_{q=1}^{m} \alpha_i^{(q)}\, \mathbf{K}_i^{(q)} \qquad (6)$$

The combination coefficients α are found by a least-squares technique in such a way as to minimise the Euclidean norm of the m-th order residual

$$\mathbf{r}_i^{(m)} = -\mathbf{E}(\mathbf{X}_i) - \mathbf{J}(\mathbf{X}_i)\, \sigma_i^{(m)} = \mathbf{K}_i^{(1)} + \sum_{q=1}^{m} \alpha_i^{(q)}\, \mathbf{K}_i^{(q+1)} \qquad (7)$$

In order to solve the problem we now carry out a GMRES iteration [5], i.e., starting from m = 1 we repeat the entire procedure with increasing values of m until the minimised norm of the residual (7) satisfies the first of (2). This always happens for some m, because it can be shown [5] that

$$\lim_{m \to \infty} \sigma_i^{(m)} = \mathbf{n}_i \qquad (8)$$

Note that in this way only the preconditioner and the Krylov subspace basis vectors need be stored, so that this procedure is advantageous only if m $\lll$ N. The Krylov subspace dimension is thus always limited to some upper value M. In most practical cases it is found that M $\approx$ 50 is sufficient to achieve the desired result even if N is of the order of millions. If for some reason the first of (2) is not satisfied when M is reached, the whole procedure is restarted according to the technique discussed in [6].

With the above method, the bulk of the CPU time is spent in performing matrix-vector multiplications of the form $\mathbf{J}(\mathbf{X})\mathbf{v}$, where $\mathbf{v}$ is a real vector of dimension N. By an accurate exploitation of the peculiar architecture of the harmonic Jacobian matrix [7], each matrix-vector product may be expressed under the form of a (complex) discrete convolution, and may thus be computed by the fast Fourier transform (FFT) [8]. As a consequence, if the spectrum is dense, that is, consists of a fundamental frequency and its harmonics up to a prescribed order, the analysis time increases slowly (asymptotically as $n \cdot \log(n)$) with the number n of spectral components. However, UWB signal spectra consist of groups of lines separated by relatively large gaps (even more so when the basic signals intermodulate with external interferers). The procedure then becomes inefficient, because the FFT algorithm is forced to carry out a large number of useless operations (multiplications by zero). In our algorithm we circumvent this problem by resorting to a mapping technique whereby the original sparse spectrum is mapped onto a fictitious dense spectrum based on an artificial fundamental. Note that the use of a mapping technique in the solution of the HB system was previously proposed by other authors [9]. However, the results of this approach are not satisfactory for strongly nonlinear situations, because existing frequency components that are not considered in the HB analysis map onto the lines of the dense spectrum, thus strongly increasing the aliasing error. With our method the mapping is only used for matrix-vector multiplications (i.e., linear operations that do not produce aliasing), while all the nonlinear analysis is carried out by the multidimensional Fourier transform. This provides exact results at all drive levels except for the normal HB truncation error.

III. ANALYSIS RESULTS OF INTERFERENCE FROM CW SIGNALS

This section discusses a first set of analysis results carried out for an UWB receiver based on the transmitted reference principle [10]. Fig. 1 shows its block schematic: a nine-FET distributed amplifier follows the antenna, a ring-diode mixer is used for multiplying the received signal with a delayed copy of itself, and the result of such multiplication is input to an integrator and sampled to produce the output data. The total number of device ports is $n_D = 24$. The transmitted signal is structured as a sequence of frames of duration $\tau_f = 4$ ns which consist of two pulses separated by a time delay $\tau_D = 1$ ns. A conventional DPSK modulation is considered and its UWB equivalent is obtained by representing two pulses of the same polarity as a logical "one" and two pulses of opposite polarity as a logical "zero". The corresponding waveforms are plotted in Fig. 2 with each pulse defined as the second derivative of a Gaussian pulse with a duration time of 200 ps and amplitude dynamics designed to fit the FCC requirements.

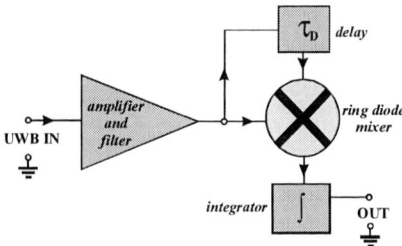

Fig. 1 Block representation of a transmitted-reference UWB receiver.

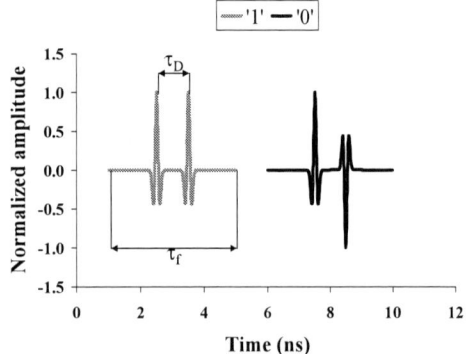

Fig. 2 Pulse pairs waveforms.

The associated spectra are plotted in Fig. 3 and are derived from a periodic sequence of pulse pairs having a fundamental frequency of 0.25 GHz, which also defines the signal spectral resolution (i.e. the gap between two consecutive spectral lines). The two plots show complementary frequency nulls that may result in different receiver performance with respect to different interferer carrier frequencies and modulation laws.

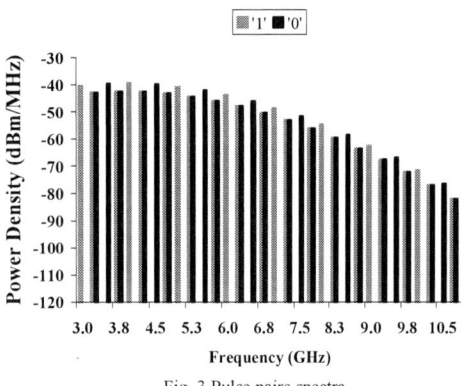

Fig. 3 Pulse pairs spectra.

The receiver output is first evaluated in the absence of interferers with an UWB signal representing a periodic sequence of "ones" described by 128 harmonics of a fundamental angular frequency $\omega_{UWB} = 2\pi \cdot 0.25$ GHz. In Fig. 4 the voltage waveforms at the interconnection ports of the receiver subcircuits are compared. The mixer output shows a slightly distorted waveform with respect to the amplifier output, thus the integrator output may be correctly detected within the designed integration time. The CPU time is only 0.55 s on a 3.8 GHz PC. Similar results are obtained with the UWB signal representing a sequence of "zeroes".

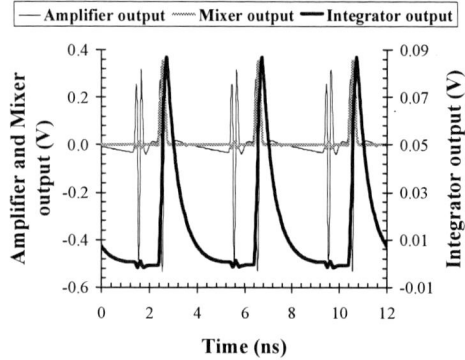

Fig. 4 Amplifier, mixer and integrator output waveforms of the UWB receiver in the absence of interference.

As a first interference test, we assume that the UWB receiver is simultaneously excited by the previous UWB signal and by a sinusoidal interferer (e.g., produced by a nearby WLAN device) of angular frequency $\omega_{INT} = 2\pi \cdot 5.18$ GHz. The resulting large-signal regime is quasi-periodic with spectral lines at all the intermodulation products

$$\Omega_{T,\mathbf{k}} = k_1 \omega_{UWB} + k_2 \omega_{INT}$$
$$\mathbf{k} = \begin{bmatrix} k_1 & k_2 \end{bmatrix}^{tr}$$
$$|k_1| \leq N_{H1} \qquad (9)$$
$$|k_2| \leq N_{H2}$$

where k_1 and k_2 are integer harmonic numbers and $N_{H1} = 128$, $N_{H2} = 5$. The total number of spectral lines for the MORHB analysis is 1413 and the number of equations is 67878. The analysis is first performed with an interferer power level (P_{INT}) 10 dB below the total UWB power. In Fig. 5 the mixer and integrator output waveforms are plotted against time. The nonlinear distortion effect at the mixer output is now relevant but does not significantly affect the integrator output, so that the peaks may still be correctly detected. The CPU time required is about 102 s. Further simulations have been carried out for P_{INT} 10 dB above the total UWB power. Severe intermodulation distortion is now observed at the mixer output (Figs. 6-7). This is mainly caused by the strong nonlinear behaviour of the mixer diodes when driven in forward conduction. Moreover, due to the interferer, a relevant DC offset is now generated at the mixer output. In such conditions the integrator block is no more able to correctly detect the real maxima and minima. The simulation time is only increased by a 10% with respect to previous analyses.

As a final test, in view of a realistic computation of the receiver sensitivity, a pseudorandom periodic sequence of 64 pulse pairs is used as the UWB excitation. In this case the total number of spectral lines rises to 45061 with a CPU time of only 770 s. The interference is still 10 dB above the UWB power. The integrator output waveforms in Fig. 8 clearly show that a simple detector is not sufficient for understanding the received UWB sequence.

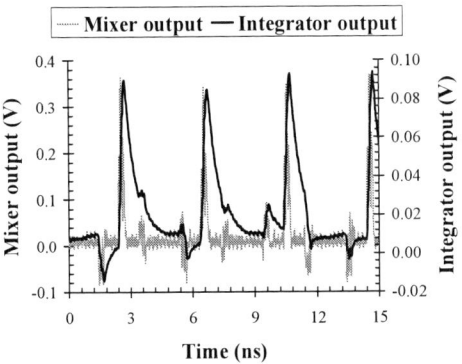

Fig. 5 Mixer and integrator output waveforms in the presence of a sinusoidal interferer 10 dB below the UWB signal power.

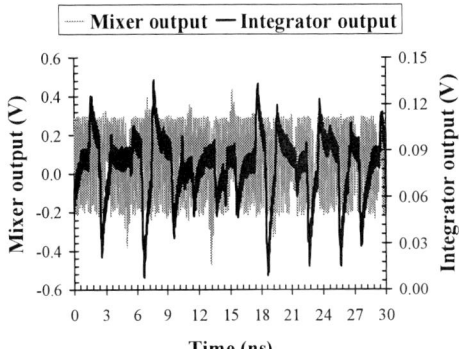

Fig. 8 Mixer and integrator output waveforms with an UWB signal representing a random sequence of bits in the presence of interference.

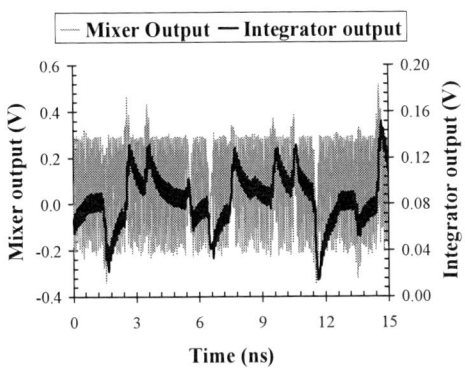

Fig. 6 Mixer and integrator output waveforms in the presence of an interferer at 10 dB above the UWB signal, representing a periodic sequence of "ones".

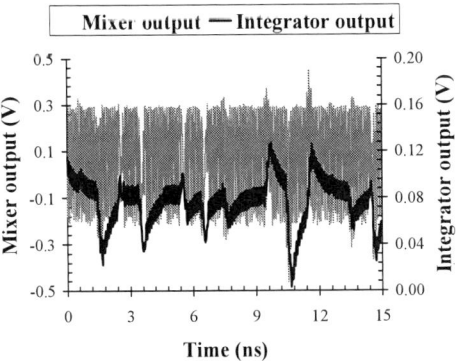

Fig. 7 Mixer and integrator output waveforms in the presence of an interferer at 10 dB above the UWB signal, representing a periodic sequence of "zeroes".

REFERENCES

[1] K. Ohno and T. Ikegami, "Interference mitigation study for UWB radio using template waveform processing," *IEEE Trans. Microwave Theory and Techniques*, Vol. 54, no. 4, pp. 1782–1792, Apr. 2006.

[2] A. Boryssenko and D. Shaubert, "On optimal port loading conditions for maximizing product of energy gain and bandwidth in broadband antenna links", *IEEE Trans. Antennas and Propagation*, Vol. 55, no. 12, pp. 3668–3676, Dec. 2007.

[3] V. Rizzoli *et al.*, "Harmonic-balance simulation of strongly nonlinear very large-size microwave circuits by inexact Newton methods", *1996 IEEE MTT-S Int. Microwave Symp. Digest*, San Francisco, CA, pp. 1357-1360, June 1996.

[4] S. C. Eisenstat and H. F. Walker, "Choosing the forcing terms in an inexact Newton method", *Research Report 6/94/75*, Department of Mathematics and Statistics, Utah State University, June 1994.

[5] Y. Saad and M. H. Schultz, "GMRES: a generalized minimal residual method for solving nonsymmetric linear systems", *SIAM J. Sci. Stat. Comput*, Vol. 7, no. 7, pp. 856-869, July 1986.

[6] V. Rizzoli *et al.*, "Multitone intermodulation and RF stability analysis of MEMS switching circuits by a globally convergent harmonic-balance technique", *Proc. European Microwave Association*, Vol. 1, no. 1, pp. 45-54, Mar. 2005.

[7] V. Rizzoli *et al.*, "State-of-the-art harmonic-balance simulation of forced nonlinear microwave circuits by the piecewise technique", *IEEE Trans. Microwave Theory Tech.*, Vol. 40, no. 1, pp. 12-28, Jan. 1992.

[8] H. J. Nussbaumer, *Fast Fourier Transform and Convolution Algorithms*. Springer-Verlag: Berlin, 1981.

[9] D. Hente and R. H. Jansen, "Frequency-domain continuation method for the analysis and stability investigation of nonlinear microwave circuits", *Proc. Inst. Elec. Eng.*, pt. H, Vol. 133, no. 10, pp. 351-362, Oct. 1986.

[10] R. Hoctor and H. Tomlinson, "Delay-hopped transmitted reference RF communication", *2002 IEEE Ultra Wideband Systems and technologies Digest*, Boston, MA, June 2005, pp. 265-269.

Large-Signal Performance of Resonant Tunnelling Diodes in K-Band Oscillators

B. Münstermann, A. Matiss*, W. Brockerhoff and F.-J. Tegude

Solid-State Electronics Department, University of Duisburg-Essen
Lotharstrasse 55, 47057 Duisburg, Germany
*now at u²t photonics, Reuchlinstrasse 10/11, 10553, Berlin
benjamin.muenstermann@uni-due.de

Abstract—**The large signal behaviour of resonant tunnelling diodes (RTD) in K-Band oscillators is investigated in order to optimize the RF-output power of RTD-based voltage controlled oscillators. Circuit simulations based on a scaleable large-signal RTD model are presented and different approaches to increase the RF-power are proposed. A new differential RTD-VCO-Circuit in InP RTD/HBT technology with a wide tuning range is introduced, employing balanced RTD-pairs.**

I. INTRODUCTION

Because of their application in satellite communication low power consumption of K-Band voltage controlled oscillators (VCO) is crucial. Resonant tunnelling diodes exhibit non-linear current-voltage characteristics with a negative differential resistance regime at low voltages and are therefore applicable as low-power active devices in oscillators. Combined with InP-HBT technology VCO's with oscillation-frequencies up to 23 GHz have been reported [1-4]. Apart from low DC-power consumption these oscillators showed low phase-noise but achieved only low signal power levels. Improving the oscillation power does also yield better phase-noise values. In this work the time-domain behaviour of RTD's in K-Band oscillators is investigated and methods to increase the available RF-Power by using alternative circuit topology and optimizing the RTD-characteristics are proposed.

II. RTD-DESIGN AND MODELLING

The model used in this work is based on measurements on resonant tunnelling diodes with a double-barrier structure, consisting of a thin InGaAs layer (3 nm) sandwiched between two InAlAs layers (5 nm), grown by Metal Organic Vapour Phase Epitaxy (MOVPE).

The measured and modelled I/V characteristics of RTD's with different sizes are shown in Fig.1. The RTD's have a peak current density of about 70 kA/cm² at a peak voltage $V_p = 0.35$ V. Above this voltage the current through the device decreases as a result of the decreasing tunnelling probability until the valley voltage of $V_v = 0.51$ V is reached. Between these two voltages the RTD characteristic exhibits a negative differential resistance, which can be utilized to enforce oscillation. At voltages above V_v the current increases exponentially and shows positive differential resistance behaviour.

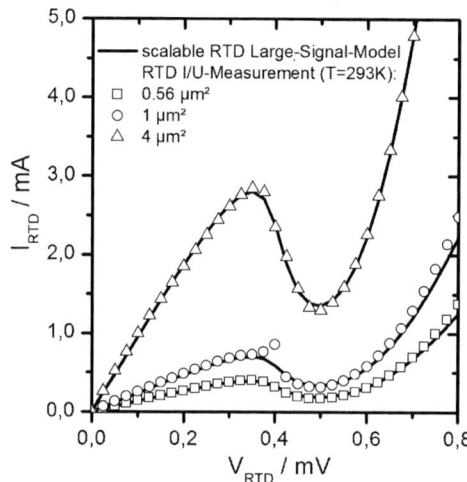

Fig. 1 I/V-characteristics of realized and modelled RTDs

Basically, the large-signal behaviour of RTD's can be described by a voltage-controlled current source in parallel to a capacitor C_{Rtd}, depending on the voltage V_c across the barrier structure. The resistor R_s is added with respect to the resistive losses of the contact layers. The parameters to fit the DC- and AC- curves have been described in [5], using the semi-physical current equations of Tsu and Esaki[6].

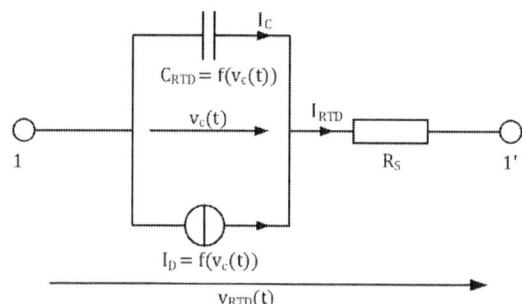

Fig. 2 Large-signal equivalent circuit model

III. RTDs IN MMIC-OSCILLATORS

To describe the behaviour of RTD-oscillators the simplified model shown in Fig.3 can be used. The ohmic losses of the resonator are represented by the conductance g_L. To enable oscillation these losses have to be compensated by the negative conductance g_a of the active device. If condition (1) is met the RF-power at the oscillation frequency originated from noise will be amplified until the quasi-stable state (2) of the oscillator is reached.

$$g_a(\omega_{osc}) + g_L(\omega_{osc}) < 0 \qquad (1)$$

$$g_a(\omega_{osc}) + g_L(\omega_{osc}) = 0 \qquad (2)$$

When using RTD's as active devices, the limitation of amplitude depends on the width of the NDR-region. As seen in Fig.4, the differential conductance g_{RTD} fulfils condition (1) for voltages in a small voltage-range around the optimal bias point $V_{RTD,opt}$.

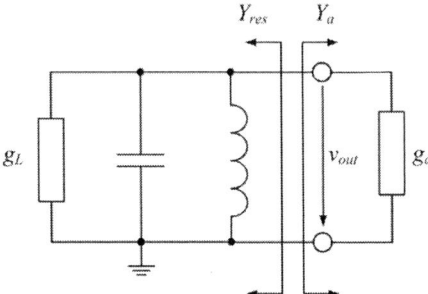

Fig. 3 Simplified model of oscillators using negative differential devices

For K-band-frequencies a resonator, consisting of lumped inductor and capacitor in parallel, with a quality factor of 22 at 20 GHz was used. The dynamic I/V characteristic of the RTD, plotted in Fig.4, has been simulated at a bias-voltage $V_{RTD,opt} = 0.41$ V and showed a peak to peak voltage $V_{pp} = 240$ mV.

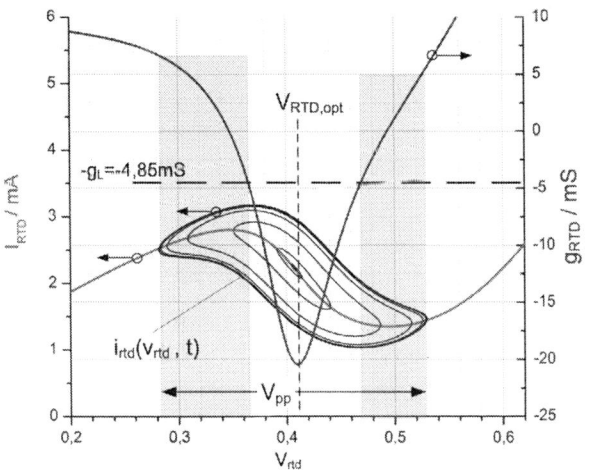

Fig. 4 Time domain behaviour of the RTD oscillating at 20 GHz

The influence of the bias-voltage on the build-up time of oscillations can be seen in Fig.5. The VCO reaches the quasi-stable state after 600 ps, when the RTD is biased at 0.41 V where $|g_{RTD}|$ is maximal.

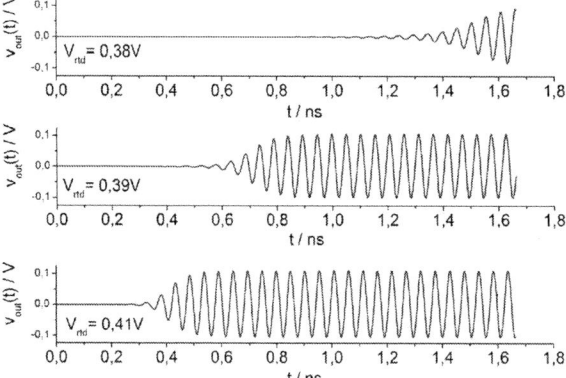

Fig. 5 Transient response of the single-RTD-VCO

IV. OPTIMIZATION OF AVAILABLE RF-POWER

The signal power available from RTD-oscillators depends on the available g_{RTD} at the bias point and on the width of the NDR-region. One way to improve these properties is to use an RTD-pair as proposed in [7]. Here, two RTD's of identical areas, connected in series and biased with opposite voltages, are used to create a negative conductance at the center node P (Fig.6).

In Fig.7 the current I_p flowing into node P is plotted as a function of the center node voltage V_p. This I/V-characteristic of the RTD-pair can be influenced by varying the voltage $|V_{RTD}|$ to improve the differential conductance as needed for optimizing output power. Transient simulations of the RTD-pair oscillator provide an improvement of 3.4 dB signal power compared to the single RTD configuration at the optimized bias voltage $|V_{RTD}| = 0.41$V.

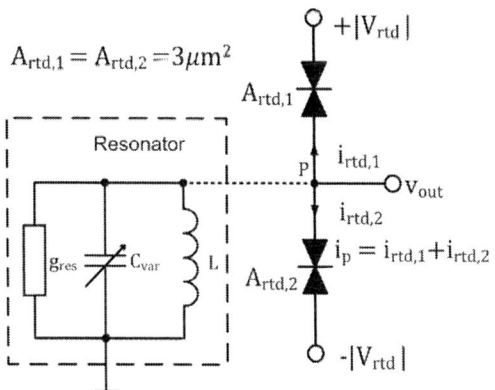

Fig. 6 RTD-VCO using a RTD-pair

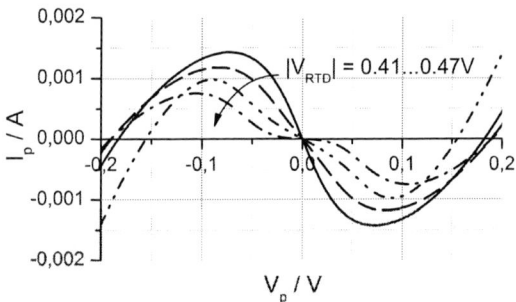

Fig. 7 DC-characteristics of the RTD-pair

The bias instability in conventional RTD-oscillators, as discussed in [7,8], is caused by spurious oscillation in the bias line. This effect is more critical for high values of $|g_{RTD}|$, and therefore restricts the usable RTD-areas and available RF-power. An additional advantage of the RTD-pair is the separation of the oscillation node from the DC-supply lines, thus large shunt capacitors can be added to reject unwanted spurious oscillation without influencing the oscillation frequency.

For the use in K-Band applications the balanced VCO-circuit [1] is extended by the RTD-pair as shown in Fig.8. Two identical RTD-VCO circuits are combined to form a virtual ground [9] at the node between the varactor-diodes. The differential outputs are isolated from the measurement environment by an output buffer stage, consisting of an HBT in collector configuration.

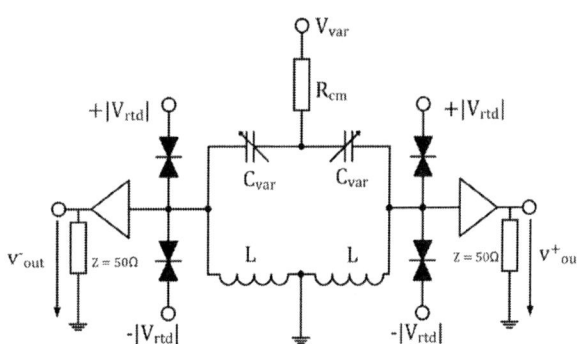

Fig. 8 Proposed differential VCO using RTD-pairs

The resonator was designed to allow a wide tuning range of 2.5 GHz around the center frequency of 20 GHz and is based on a fitted model of a varactor diode using the base-collector junction of the InP-HBT technology. The inductance-model was designed, based on EM-simulations of a spiral inductor, with 0.3 nH and a quality factor of 26 at 20 GHz. The results of the circuit simulation performed in Advanced Design System 2006A (ADS2006A) with the harmonic balance method are presented in Fig.9. The simulated amplitude of oscillation of a single output is increased by 20% in comparison to the single RTD-oscillator, leading to a differential signal power of -0.2 dBm.

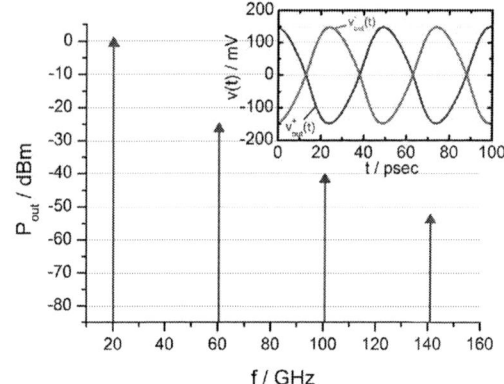

Fig. 9 Circuit-simulation results for the proposed VCO-Circuit

To investigate the influence of the I/V-characteristics of the RTD on the signal power, the RTD-model was varied to reveal possible improvements available by modification of the device design. In order to increase the negative conductance g_{RTD}, a high peak current I_{peak} in combination with low valley currents I_{valley} has to be realised. Enlarging the peak current by increasing device area comes along with additional RTD-capacitance and thus limits the oscillation frequency range. Therefore, only modification of the layer composition can achieve further improvements.

Apart from optimizing the peak-to-valley current ratio (PVCR), the shift of the valley-voltage to higher values allows the creation of wider NDR-regions with the RTD-pair. The results presented above are achieved by employing the standard RTD represented in Fig. 10. Optimizing the layer structure yields similar RTD characteristics as used in [1] (dotted line in Fig. 10).

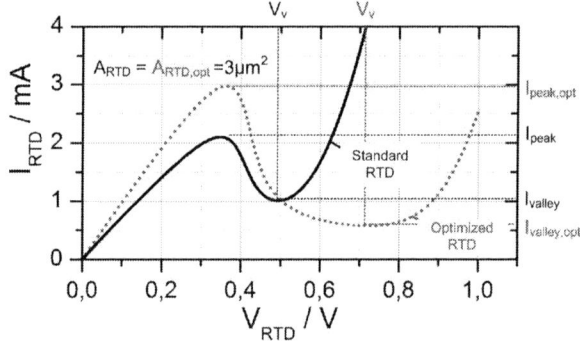

Fig. 10 I/V characteristics of the investigated RTD-models

In Fig.11, the dynamic behaviour of the optimized RTD-pair is shown to emphasize the effect of enlarging NDR-region. At a supply voltage of $|V_{RTD}| = 0.45$ V, a wide voltage range can be achieved in which the RTD amplifies the oscillation, resulting in a peak to peak voltage of $v_{pp} = 710$ mV. In comparison to the single-RTD behaviour the simulated build-up time increased because of the low g_{RTD} at $v_{rtd}(t = 0)$.

978-2-8748-7007-1/08 $25.00 © 2008 EuMA

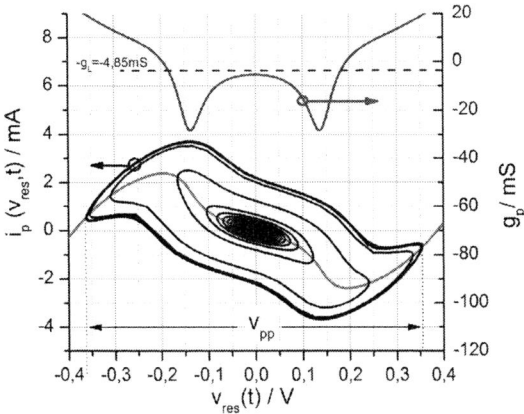

Fig. 11 Dynamic behaviour of the RTD-pair with modified RTDs

In table I the results of the circuit-simulations of RTD-VCO's are summarized. The observed signal output power of the simple single-ended RTD-VCO is limited to -10 dBm. The combination of the RTD-pair topology and the balanced VCO-design enables to increase the output power to -0.2 dBm, which can be further increased by modifying the RTD-layer structure to at least 6.3 dBm.

TABLE I
CIRCUIT-SIMULATION RESULTS

	Signal Power	**Tuning-Range**
Single-RTD-VCO	-10 dBm	18,5 GHz-21 GHz
RTD-Pair	-6,6 dBm	18,5 GHz-21 GHz
Proposed Balanced VCO	-0.2 dBm	18,5 GHz-21 GHz
Proposed Balanced VCO w. optimized RTD	6.3 dBm	18 GHz-21 GHz

V. CONCLUSIONS

The transient response behaviour of resonant tunnelling diodes in different oscillator circuit topologies was simulated to investigate possible improvement of output power. Based on the combination of the RTD-pair with the balanced-VCO concept in InP technology a monolithic integrated VCO-circuit is proposed that significantly improves output power level, ready for use in K-Band communication systems. Furthermore potential to increase the signal power by modifying the RTD-characteristics is presented.

REFERENCES

[1] S. Choi, Y. Jeong, and K. Yang, *"Low DC-Power Ku-Band Differential VCO Based on an RTD/HBT MMIC Technology"*, IEEE Microw. And Wirel. Comp., vol. 15, no. 11, pp.742-744, 2005

[2] S. Choi, K. Yang, *"Low-voltage low-power K-band balanced RTD-based MMIC VCO"*, Microw. Symp. Digest, p.743-746, 2006

[3] Y. Jeong, S. Choi, K. Yang, "Performance Improvement of InP-based Differential HBT VCO using the Resonant Tunneling Diode", Int. Conf. on IPRM, pp. 42-45, 2006

[4] H. J. Santos, K. K. Chui, D. H. Chow, and H. L. Dunlap "An efficient HBT/RTD oscillator for wireless applications", IEEE Microwave and wireless components letters, vol. 11, no. 5, pp.193-195, 2001

[5] A. Matiss, A. Poloczek, W. Brockerhoff, W. Prost and F.- J. Tegude *"Large-Signal Analysis and AC Moedlling of Sub Micron Resonant Tunneling Diodes"*, Europ. Microw. Conf., pp.207-210, 2007

[6] R. Tsu & L. Esaki *"Tunneling in a finite superlattice"*, Applied Physics Letters, vol.22, no. 11, pp. 562-564, June 1973

[7] Y. Ookawa, S. Kishimoto, K.Maezawa, T. Mizutani, *"Novel Resonant Tunneling Diode Oscillator Capable of Large Output Power Operation"* IEICE Trans. Electron., vol. E89-C, No.7, pp.999-1004, 2006

[8] C. Kidner, I. Mehdi, J. R. East, and G. I. Haddad, *"Power and stability limitations of resonant tunneling diodes,"* IEEE Trans. Microw. Theory Tech., vol.38, pp.864-872, 1990

[9] B. Jung, R. Harjani *"High frequency LC VCO design using capacitve degenereation"* IEEE J. Solid-State Circ., vol. 39, no. 12, pp.2359-2370, 2004

Proceedings of the 3rd European Microwave Integrated Circuits Conference

Analysis and Synthesis of the Microstrip Lines Based on Support Vector Regression

Nurhan TÜRKER TOKAN[#1], Filiz GÜNEŞ[#2]

[#]Electronic and Communication Engineering Department, Yıldız Technical University
Beşiktaş, Istanbul/TURKEY
[1]nturker@yildiz.edu.tr, [2]gunes@yildiz.edu.tr

Abstract— In this work, the Support Vector Regression is adopted to the analysis and synthesis of microstrip lines on all isotropic/anisotropic dielectric materials, which is a novel technique based on the rigorous mathematical fundamentals and the most competitive technique to the popular artificial neural networks. In this design process, accuracy, computational efficiency and number of support vectors are investigated in detail and the Support Vector Regression performance is compared to an artificial neural network performance. It can be concluded that the artificial neural network may be replaced by the Support Vector Machines in the regression applications due to its high approximation capability and much faster convergence rate with the sparse solution technique. Synthesis is achieved by utilizing the analysis black-box bidirectionally by reverse training. Furthermore, by using the adaptive step size, a much faster convergence rate is obtained in the reverse training. Besides, design of microstrip lines on the most commonly used isotropic/ anisotropic dielectric materials are given as the worked examples.

I. INTRODUCTION

Support vector machines and kernel methods, which enable to generalize 'discrete' data into the 'continuous' domain have become one of the most popular learning machines in the last few years. In particular, support vector machines are based on a judicious and rigorous mathematics combining the generalization and optimization theories together and verified to be computationally very efficient (the so-called Vapnik-Chervonenkis theory [1]–[2]). This learning machine has found many fruitful applications in science and engineering, recently with the typical application in modeling of microwave transistor [3].

In this work, Support Vector formulation of the quasi-TEM mode of propagation at low frequencies is completed and compared with its artificial neural network competent. The term low frequency is a relative one. It is the ratio of the line dimensions to wavelength that determines whether a microstrip line can be adequately described in terms of the quasi -TEM mode of propagation. In MIC circuits with line widths as small as 100μm, the low- frequency region can extend as high as 20 to 30 GHz with %1 worst –case accuracy, replaced the complicated and computationally inefficient EM full-wave solution.

Originalities of this work may briefly be summarized as follows: (i) The Support Vector formulation for the analysis of the microstrip lines is completed, where the characteristic impedance $\{Z_o\}$ and effective dielectric constant $\{\varepsilon_{eff}\}$ are

outputted with the given microstrip transmission line $\{W, H, T, \varepsilon_r, \varepsilon_y\}$ system; (ii) Synthesis procedure which is to determine the strip width $\{W\}$ for the required $\{Z_o\}$ on the isotropic and/or anisotropic dielectric substrate $\{H, T, \varepsilon_r, \varepsilon_y\}$ is achieved by using the analysis support vector regressor/ artificial neural network unit bidirectionally, where very fast updating process is obtained by applying adaptive step sizes in the algorithm; (iii) Accuracy, computation time and number of support vectors of the support vector regression performance are investigated with ε -parameter of the Vapnik's ε -insensitive loss function and is compared with the artificial neural networks' performance in the simple structures; (iv) Changes in the strip width ΔW for a certain Z_o are determined due to the changes in the substrate thickness ΔH, dielectric permeabilities $\Delta\varepsilon_r$, $\Delta\varepsilon_y$ by using the support vector regressor and/or artificial neural network design models.

The paper is organized as follows: Definition of the problem takes place in section II, which includes the black-box characterization of the design, the data generation, fundamentals of synthesis. In section III, the theory of the SVR is reported. In section IV, worked examples are given. Section V is the final section that includes conclusions.

II. DEFINITION OF THE PROBLEM

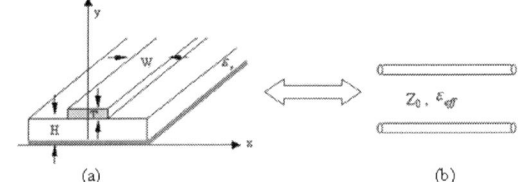

(a)　　　　　　　　　　(b)
Fig. 1 (a) Microstrip line (M); (b) Its transmission line equivalence (T)

A microstrip transmission line given in Fig. 1.a can be characterized in **M** – space defined by **M**= $\{$**G**, $\varepsilon\} \in \mathbb{R}^5$ where **G** and ε are the geometry and permeability vectors, respectively defined as:

$$\mathbf{G}'= [\ W, H, T\ ], \quad \varepsilon' = [\ \varepsilon_r,\ \varepsilon_y\ ]$$

which are the width of the strip, height of the dielectric substrate, and strip thickness respectively, and $\varepsilon_r, \varepsilon_y$ are the dielectric constants. Similarly, Z_o and ε_{eff} are the

978-2-8748-7007-1/08 $25.00 © 2008 EuMA

characteristic impedance and effective dielectric parameters of the equivalent transmission line (Fig. 1.b) and can be represented in a $\mathbf{T}$ – space: $\mathbf{T} = \{ Z_o, \varepsilon_{eff} \} \in \mathrm{R}^2$

So analysis of the microstrip lines can be achieved by mapping: $A : \mathbf{M}\,(\,\mathrm{R}^5\,) \rightarrow \mathbf{T}\,(\,\mathrm{R}^2\,)$. Similarly, synthesis of the microstrip line can be defined by mapping $S : \mathbf{T}\,(\,\mathrm{R}^2\,) \rightarrow \mathbf{M}\,(\,\mathrm{R}^5\,)$. Analysis and synthesis black-box models are given in Fig. 2.a and b, respectively.

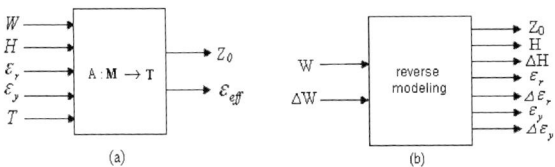

Fig. 2 Black- Box model of (a) Analysis $(A\,:\,\mathbf{M}\,(\,\mathrm{R}^5\,) \rightarrow \mathbf{T}\,(\,\mathrm{R}^2\,))$; (b) Synthesis $(S\,:\,\mathbf{T}\,(\,\mathrm{R}^2\,) \rightarrow \mathbf{M}\,(\,\mathrm{R}^5\,))$ of the microstrip lines

Here for the mapping $A:\,\mathbf{M} \rightarrow \mathbf{T}$, either data based on measurements or formulas can be used. We prepared a "data generation" programme exploiting the formulas based on the approximate analytical solutions along with the empirical adjustment of various numerical constants so as to achieve the desired accuracy [4]-[5].

In the mapping $S:\,\mathbf{T} \rightarrow \mathbf{M}$, the reverse training is used, where the weight vector obtained in the analysis process is used. While this weight vector remains constant, the inputs W and $\varDelta W$ change themselves to adapt to the desired output. Thus update equation for the input W can be given as:

$$W_{new} = W_{old} + \varDelta W.\lambda \qquad (1)$$

where adaptive step size is used:

$$\lambda = e^{(-\frac{\tau}{E})} \quad (2), \qquad E = \frac{1}{2}(Z_o - Z_{oref})^2 \qquad (3)$$

Here τ is the adaptation parameter, E is the squared error.

III. SUPPORT VECTOR REGRESSION

The regression problem related to the estimation of the $\mathbf{T}\{Z_o, \varepsilon_{eff}\}$ can be stated as follows: Firstly, let us consider Z_o. In the training phase a set of L training pairs $\{(\,\mathbf{m}^0, Z_o^0), (\,\mathbf{m}^1, Z_o^1),...,(\mathbf{m}^{L-1}, Z_o^{L-1})\}$ is constructed by considering various strip widths in the range of $0.1\mathrm{mm} \leq W \leq 4\mathrm{mm}$ on various isotropic and/or anisotropic dielectric substrates in the range of $0.1\mathrm{mm} \leq H \leq 4\mathrm{mm}$; $1 \leq \varepsilon_r, \varepsilon_y \leq 12.9$. Starting from these samples of the input/output values of Z_o, the goal is to find a function $\tilde{Z}_o$ which approximates as well as possible unknown function $Z_o\,(\mathbf{m})$. By using the SVR, $\tilde{Z}_o$ is defined as:

$$\tilde{Z}_o(\mathbf{m}) = \langle w, \varphi(\mathbf{m}) \rangle + \mathrm{b}, \qquad (4)$$

where $\langle .,. \rangle$ denotes the inner product, φ is a nonlinear mapping vector that performs a transformation of the input

vector from the space $\mathbf{M}$ to a high-dimensional space. w and b are the weighting vector and bias, respectively which are obtained by minimizing the primal convex objective function (Regression Risk), defined as [2]:

$$R_{reg} = \frac{1}{2}\| w \|^2 + C\sum_{i=0}^{L-1} L^\varepsilon (\mathbf{m}, Z_o, \tilde{Z}_o) \qquad (5)$$

where C is the regularization constant and $L^\varepsilon(\mathbf{m}, Z_o)$ is a general loss function. Since the given objective function given in (5) has no local minima and it guarantees the global minimum which is one of the superiorities of SVM on the other pattern recognition methods, particularly neural networks. In our work, so-called ε – insensitive loss function developed by Vapnik [1] is used:

$$L^\varepsilon (\mathbf{m}, Z_o, \tilde{Z}_o) = \begin{cases} 0, & \text{if } \left| Z_o{}^i - \tilde{Z}_o(\mathbf{m}^i) \right| \leq \varepsilon \\ \left| Z_o{}^i - \tilde{Z}_o(\mathbf{m}^i) \right| - \varepsilon, & \text{otherwise} \end{cases} \qquad (6)$$

this defines an ε tube so that if the predicted value is within the tube the loss is zero, while if the predicted point is outside the tube, the loss is the magnitude of the difference between the predicted value and the radius ε of the tube.

It is possible to recast the minimization of the regression risk as a dual optimization problem, in which the vector can be written in terms of the input data $\mathbf{m}$ as:

$$w = \sum_{i=0}^{L-1} (\alpha_i - \alpha_i^{'})\, \varphi(\mathbf{m}^i) \qquad (7)$$

where α_i and $\alpha_i^{'}$ are the unknown Lagrange multipliers. By substituting (7) in (4), $\tilde{Z}_o$ is rewritten as

$$\tilde{Z}_o(\mathbf{m}) = \sum_{i=0}^{L-1} (\alpha_i - \alpha_i^{'})\, \langle \phi(\mathbf{m}^i), \phi(\mathbf{m}) \rangle + \mathrm{b} \qquad (8)$$

$$= \sum_{i=0}^{L-1} (\alpha_i - \alpha_i^{'})\, K(\mathbf{m}^i, \mathbf{m}) + \mathrm{b}$$

and the coefficients $\alpha_i, \alpha_i^{'}$ and b must be chosen in order to minimize the regression risk in the dual problem. In (8), the kernel function $K(\mathbf{m}^i, \mathbf{m}) = \langle \phi(\mathbf{m}^i), \phi(\mathbf{m}) \rangle$ works on the original space $\mathbf{M}$. Commonly used kernels are polynomial and radial kernels [1].

Applying the standard Lagrange multiplier technique results in the equivalent maximization of the dual space objective function:

$$W(\alpha, \alpha^{'}) = -\varepsilon \sum_{i=0}^{L-1} (\alpha_i^{'} + \alpha_i) + \sum_{i=0}^{L-1} Z_o^i (\alpha_i^{'} - \alpha_i) \qquad (9)$$

$$- \frac{1}{2}\sum_{i,j=0}^{L-1} (\alpha_i^{'} - \alpha_i)(\alpha_i^{'} + \alpha_i) K(\mathbf{m}^i, \mathbf{m})$$

with the constraints:

$$0 \leq \alpha_i^{'}, \alpha_i \leq C \quad (10.a), \qquad 0 \leq \alpha_i^{'}, \alpha_i \leq C \qquad (10.b)$$

The dual variables $\alpha_i, \alpha_i^{'}$ and b are computed using the Karush-Kuhn-Tucker conditions to maximize (9) subject to the constraints given by (10.a), (10.b). From Karush-Kuhn-Tucker conditions, it follows that only for $\left| \tilde{Z}_o(\mathbf{m}^i) - Z_o{}^i \right| \geq \varepsilon$, the Lagrange multipliers may be nonzero, or in the other words for all samples inside the ε - tube $\Leftrightarrow \left| \tilde{Z}_o(\mathbf{m}^i) - Z_o{}^i \right| < \varepsilon$ the

$\alpha_i, \alpha_i^{'}$ vanish. Therefore we have a sparse expansion of w in terms of m^i (i.e. we do not need all m^i to describe w). The samples that come with nonvanishing coefficients are called Support Vectors. The idea of representing the solution by means of a small subset of training points has also enormous computational advantages. This reduced number of non-zero parameters together with the guaranteed global minimum gains superiority to SVM over the alternative methods. A detailed mathematical background together with the literature can be found in [2].

The regularization parameter C has been found to represent a measure of the tradeoff between the capabilities of the approach in estimating characteristic impedance of the microstrip line using training and test sets. In order to correctly estimate the Z_o, L data pairs are generated in the form: $\{(m^o, Z_0^{o}), (m^1, Z_0^{1}),...,(m^{L-1}, Z_0^{L-1})\}$ for isotropic and anisotropic substrates in the training phase. At the end of the training phase, in the so-called test phase Z_o is estimated for the dielectric substrates not included in the training set. Similarly, SVR can be applied to the regression of the ε_{eff} property of the equivalent transmission line. Numerical details of the SVR analysis of the microstrip lines and synthesis by the reverse modeling will be given in the next section.

IV. NUMERICAL RESULTS

Radial basis function kernel is chosen as a suitable kernel function in our application:

$$K(m^i, m) = \exp(-\gamma \|m - m^i\|^2) \qquad (11)$$

where, the width parameter γ is set to 0.1 for the optimum performance. Similarly, the regularization parameter, C is set to 1. 886 data pairs for the dielectric substrates within the ranges of $0.1mm \leq H \leq 4mm$; $1 \leq \varepsilon_r, \varepsilon_y \leq 12.9$, and the strip widths in the ranges of $0.1mm \leq W \leq 4mm$ are generated for the aim of training of SVR and 120 data pairs are also generated for the testing. However, 466 and 479 support vectors ($\Leftrightarrow \varepsilon = 0.1$) of 886 data pairs are used in the training of SVR model for the determination of characteristic impedance and effective dielectric constant with the accuracy of 99.26 % and 99.46 %, respectively, as given in Table I. From Table I and Fig. 3, it can be seen that reasonable accuracies can be achieved with small number of SVs within the short elapsed times.

TABLE I

PERFORMANCE PARAMETERS FOR THE SVR OF THE Z_o AND ε_{eff}

$\varepsilon = 0.1$	Z_o	ε_{eff}
Number of SVs (%)	52.59	54.06
Elapsed time (average in sec)	0.5	0.5028
Accuracy (%)	99.26	99.46

Furthermore, the number of the SVs with the accuracy and the elapsed time of the regression with respect to the ε

parameter of the insensitive loss function within the range of $0.1 \leq \varepsilon \leq 50$ are investigated for the functions of Z_o and ε_{eff} and here the graphs for Z_o are given in the Figs 3.a and b.

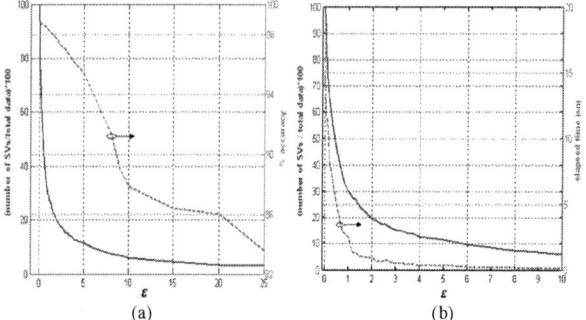

(a)   (b)

Fig. 3 Variation of (a) accuracy; (b) elapsed time and number of SVs with respect to ε parameter for characteristic impedance of microstrip lines

Performance of the SVR is compared with the performance of the other popular learning machine, ANN. For this purpose, two Multilayer Perceptron networks are used within the black-box in the Fig. 2.a and b: The first one has a single hidden layer with 7 neurons, the second has two – hidden layers with 7 and 4 neurons, respectively. Both ANNs are activated by hyperbolic tangent sigmoid and linear transfer functions and are trained by Levenberg-Marquardt backpropagation algorithm using the same data set with the learning rate 0.1. Results of the comparison for the accuracy and computation efficiency are given in Table II and III, respectively.

TABLE II

ACCURACIES OF ANN AND SVR MODELS FOR MICROSTRIP LINES

% Accuracy	ANN 1*	ANN 2**	SVR
Z_o	97.64	99.13	99.26
ε_{eff}	98.12	99.55	99.46

* with one hidden layer   ** with two hidden layers

TABLE III

TIME ANALYSIS OF ANN AND SVR MODELS FOR MICROSTRIP LINES

	ANN 1	ANN 2	SVR
Training time(sec)	27.953*	44.156*	1.015
Test time (sec)	0.015	0.017	0.013
Total time (sec)	27.968	44.173	1.028

*trained for 300 epochs

Worked example for the analysis process, Z_o variations with the width W are given in Fig. 4 for the commonly used anisotropic and isotropic dielectric substrates which are PTFE/microfiber glass ($\varepsilon_r = 2.26$, $\varepsilon_y = 2.2$), RT/Duroid 6006 ($\varepsilon_r = 6.36$, $\varepsilon_y = 6$), Alumina ($\varepsilon_r = 9.7$), Sapphire ($\varepsilon_r = 9.4$, $\varepsilon_y = 11.6$) and Gallium arsenide ($\varepsilon_r = 12.9$).

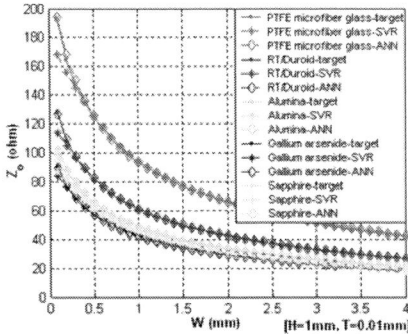

Fig. 4 Variation of characteristic impedance of a microstrip line with respect to the strip width

Investigating attitude of the system against the adaptation parameter from the Fig. 5, τ is chosen 0.1 for the support vector regression modeling and by the similar consideration, 0.3 for the artificial neural network synthesis black-box. Moreover, the performances of the constant and adaptive step sizes are compared for the support vector regressor and artificial neural network in the Table IV where it can be seen that 20-70 time faster performances can be obtained by the adaptive step sizes.

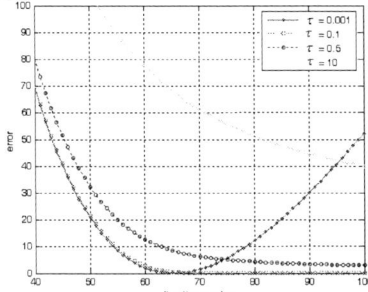

Fig. 5. Error with iteration number for various adaptation parameters for the SVR modeling.

TABLE IV
COMPARISON OF THE PERFORMANCES WITH CONSTANT AND ADAPTIVE STEP SIZES

	constant step size			adaptive step size		
	W (mm)	Z_o (Ω)	elapsed time(sec)	W (mm)	Z_o (Ω)	elapsed time(sec)
SVR	3.658	42.93	**11.234**	3.646	43	**0.45**
ANN	3.625	43.13	**36.188**	3.612	43.19	**0.64**

In the synthesis process, the strip width, Ws are determined for the characteristic impedance Z_0s in the range of 15- 170 Ω for the commonly used isotropic and anisotropic dielectric materials as given in Figs. 6 and 7. In Fig. 6, the substrate and strip thicknesses are kept fixed, which are H=1mm, T=0.01mm, while effect of the substrate thickness is examined for the isotropic alumina and anisotropic RT/Duroid 6006 materials.

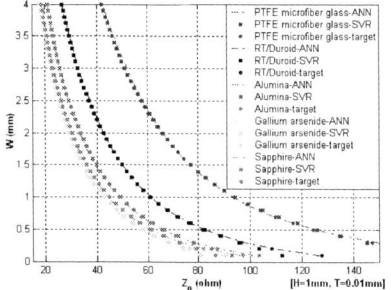

Fig. 6 Strip width with respect to the characteristic impedance

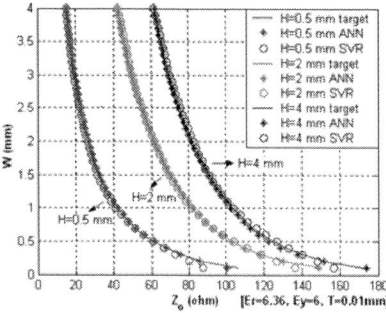

Fig. 7 Strip width with respect to the characteristic impedance for microstrip lines with anisotropic substrate having different thicknesses

V. CONCLUSIONS

In this work, Support Vector Regression is applied to the analysis and synthesis of microstrip lines for widespread isotropic/ anisotropic dielectric materials. The synthesis process is performed by utilizing the analysis black-box bidirectionally by reverse training. This immediately calls that analysis and synthesis can be achieved by using only the analysis black-box. Furthermore, by using the adaptive step size, a much faster convergence rate is obtained. The SVR performance is compared with its ANN competent performance and so it may be concluded that the ANN can be replaced by the SVR in the regression applications due to its high approximation capability and much faster convergence rate with the sparse solution technique. Thus, the resulted formulation can be conveniently exploited for the analysis and synthesis of the microstrip lines in MIC up to 20-30 GHz.

ACKNOWLEDGMENT

This work was supported by The Scientific and Technological Research Council of Turkey. (Project Number: 106E171)

REFERENCES

[1] V. N. Vapnik, *Statistical Learning Theory*, New York: Wiley, 1998.
[2] N. Cristianini, and J. Shawe-Taylor, *An introduction to support vector machines (and other kernel-based learning methods)*, Cambridge University Press, 2000.
[3] F. Güneş, N. Türker, and F. Gürgen, "Signal-Noise Support Vector Model of a Microwave Transistor", *Int J RF and Microwave CAE*, Vol. 17, pp. 404–415, 2007.
[4] R.E. Collins, *Foundations for Microwave Engineering*, McGraw-Hill, New York, 1992.
[5] T.C. Edwards, *Foundations for Microstrip Circuit Design*, Wiley–Interscience, New York, 1981.

Proceedings of the 3rd European Microwave Integrated Circuits Conference

Global Digital-Analog Co-Simulation Methodology for Power and Signal Integrity aware Design and Analysis

Sidina Wane and Guillaume Boguszewski

NXP-Semiconductors, 2, Esplanade Anton Philips, Campus EffiScience, Colombelles
BP 2000, 14906 CAEN Cedex 9 France
Email Contact: Sidina.Wane@ieee.org

Abstract — **In this paper a global co-simulation methodology for concurrent/simultaneous analysis of passive and active analog/digital parts is proposed. An original** *power-signature* **concept is introduced to model high-speed digital modules temporal and spatial distribution of their power switching activity through specified chip partitions. Dedicated real-world NXP-Philips-Semiconductors active modules mounted on test-board have been designed and measured for validation of the proposed co-simulation methodology. Full-wave electromagnetic modeling, broadband SPICE compact model extractions and measurement results are successfully compared.**

I. INTRODUCTION

Co-simulation of mixed signal integrated circuits (IC) requires very challenging techniques and methodologies to accurately combine passive and active parts for system level analysis. Distributed «co-design» based on simultaneous/concurrent real-time analysis at IC-level, Package-level and even Board-level requires «open-architecture» driven tooling flows controlled/emulated by appropriate and flexible methodologies for hybrid coupling of global microwave and high-speed digital circuit modeling and simulation. To meet the requirement of active analog and digital co-simulations, scalable hierarchical approaches that allow for coupled analysis between different abstract views (schematic/symbol/physical) are necessary. In common practices, with classical co-simulation approaches, passive and active parts are generally treated separately following "chaining" procedures where sequential/cascade/devide-and-conquer reductions are assumed. In such devide-and-conquer reductions passive circuitry parts including interconnect lines, functional components (self inductors, baluns, filters) are analyzed using quasi-static/full-wave electromagnetic solutions, active parts being represented by Spice-like lumped element representations or behavioral models when associated complexities can be afforded using existing/available simulation technologies and use-models. Transistor level description or behavioral-modeling for particular noisy block could be sufficient in capturing analog active parts intrinsic responses. However for digital dies - generally considered as aggressors (noise injectors)- additional details on their power consumption and dynamic switching activities are important to properly deal with global power and signal integrity analysis and time-budgeting considerations. In the published

research work various approaches have been proposed for the estimation of time-domain switching activity profiles for digital active modules, with restriction to microprocessors and micro-controllers. Among such approaches are analytical waveform profile calculation, numerical macro-modeling and/or statistical techniques, and measurement methods [1-2]. Analytical calculations, based on peak-value assumption referring to simple canonical waveforms (triangular/trapezoidal shapes) for rough model representation of digital dies internal current profile, are unable to derive temporal and spatial distribution of power activity through chip partitions accordingly to multi-clock frequency domains. Main limitation of macro-modeling techniques concerns difficulties to extract time-domain current activity phase information of derived waveform profiles. Extraction of current activity phase information in complement to magnitude responses requires the use of advanced numerical techniques such as wavelet transforms [3]. Measurement based methods, in order to be exploitable and easily correlated with simulation results, requires efficient on-board setting protocols and efficient numerical de-embedding algorithms backed up by signal processing analysis.

In this paper a global distributed iterative co-simulation methodology for concurrent/simultaneous analysis of passive and active parts is proposed. The originality of the proposed contribution lies in the two following major attributes drawing improvements over state of the art co-simulation methodologies:

1) A *power-signature* concept is introduced to model high-speed digital modules temporal and spatial distribution of their power switching activity through specified chip partitions accordingly to multi-clock frequency domains with asynchronous multi-port stimuli.

2) Dynamic frequency-domain multi-port analog dies impedance/admittance extraction is proposed for iterative coupling analysis between active parts and their embedding PDN (Passive Distribution Network).

The first point is investigated based on comparative analysis of different approaches: semi-analytical, analytical, macro-modeling/statistical techniques, including use of advanced digital activity simulation tools supported by state of

978-2-8748-7007-1/08 $25.00 © 2008 EuMA

the art Electronic Design Automation (EDA) tooling (e.g. Cadence, Optimal, Apache). The second point is carefully analysed through combined comparisons of full-wave electromagnetic modeling on one hand, broadband SPICE compact model extractions and measurement results on the other, for dedicated real-world NXP-Philips-Semiconductors active module carriers reported on dedicated test-board structures.

II. Co-Simulation Methodology, Test Carriers Description, Analysis Results, Comparisons and Discussions

A. Concept of Power-Signature for Co-Simulation of Digital and Analog Modules through Multiport PDN

The proposed Co-Simulation methodology is articulated around three principal approaches: digital high-speed dynamic switching power activity modeling, analog active Behavioral Model (BM) representation and Electromagnetic extraction of generalized multi-port PDN. The analog and digital active models are coupled through passive embedding environments. The passive embedding environments are referenced in Fig.1 as D-PDN for the digital module, A-PDN for the Analog blocks and PB-PDN for the package/board following generalized multiport topology/functionality-based partitioning. The coupled quasi-static and full-wave electromagnetic-based partitioning approach proposed in [4]-[5] is applied to accurately derive the multi-ports D-PDN, A-PDN and PB-PDN. The resulting multi-ports are iteratively coupled to macro-model representations of the digital and analog modules, in circuit-simulator framework.

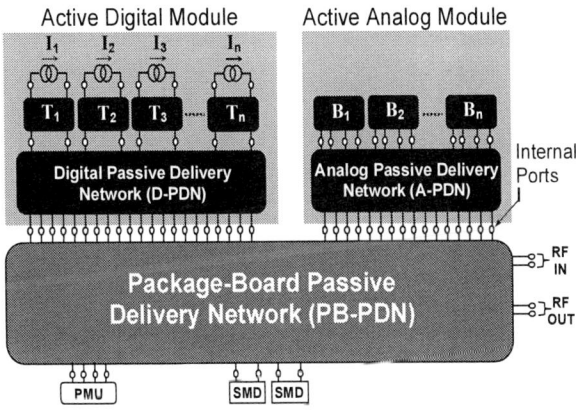

Fig.1 Schematic representation of coupled active digital-analog modules and passive multi-ports (IC-level, Package/board levels).

A concept of «power-signature» is introduced to characterize high-speed digital modules switching activities. In Fig.1, two-port matching elements (T_1, T_2, ... T_n) are introduced to couple current switching sources for the digital module to their embedding PDN multi-ports. The analog blocks (B_1, B_2, ... B_n) are represented by behavioral models (when available) or transistor-level descriptions. The proposed «power-signature» concept captures in a macro-model (reference to Apache CPM approach) a multi-clock frequency domains the switching

activity of standard cells (gates), input-output buffers, decoupling capacitors and IP memories. The switching activity is extracted at the level of the input/output pads of the digital active module. Different stimuli scenarios are considered to emulate both worst-case (WC) activity (assuming all gates switching at the same time) and realistic-case (RC) activity. A realistic case could be emulated based on timing analysis conducted during initial design steps (including VCD inputs).

B. Test Carriers Description, Discussion of Measurement/Simulation Results and Validation

To investigate the proposed global active-passive co-simulation methodology, a satellite-TV-system and TV-on-Mobile [4] real-world NXP-Philips-Semiconductors test case carriers including two active analog (in BiCMOS technology) and digital (in CMOS technology) dies packaged and reported on test-board is considered. Different integration configurations (satellite carrier) are studied: a first configuration where the two active dies are separately reported on the test-board (case of Fig.2(a)) and a second configuration where the two dies are stacked in a single package (SiP: System-in-Package) which is reported on test-board (case of Fig.3(a),(b)). In the single package option two variants (with and without active SiP module) are introduced to analyze the coupling between the active SiP and the package-board at the interfacing junctions. Both integration alternatives, addressing same application, can be compared in terms of size, electrical performances and cost reduction.

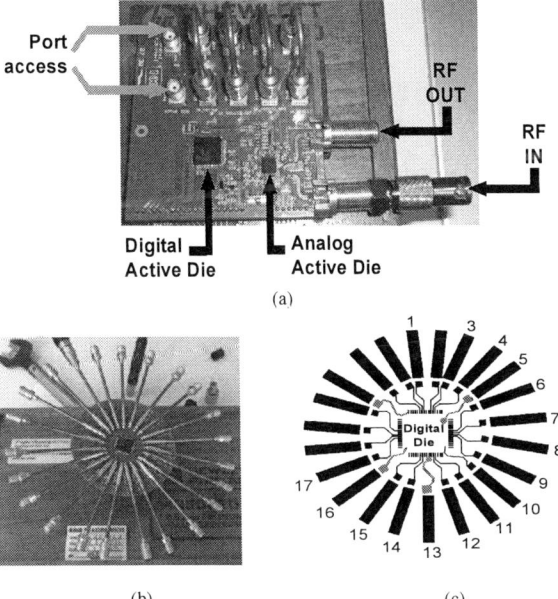

Fig.2 Photograph of a satellite-TV-system developed by NXP-Semiconductors including active analog (BiCMOS) and digital (CMOS) dies (a) and dedicated multi-port carrier for measurement of internal switching activity (b). Input/Output pins numbering only illustrative not in correlation with digital activity pinning.

978-2-8748-7007-1/08 $25.00 © 2008 EuMA

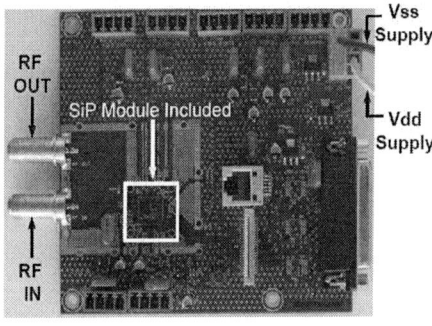 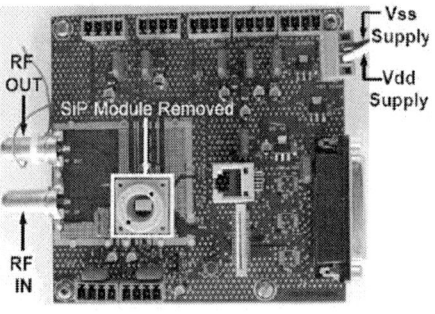

(a)   (b)

Fig.3 Measured test carriers with active SiP module (a) and without active SiP module (b) for global Co-simulation analysis.

Fig.4 shows the power-activity normalized spectrum against frequency for 17 IOs (Input/Output pins), in reference to the TV-on-Mobile carrier (photograph not shown) analyzed using Cadence-SiP and Apache solutions [4]. The measured average power of the active digital die is 140mW. The considered digital die comprises 4 principal clock domains at 20MHz, 60MHz, 120MHz and 480 MHz. Assuming a constant voltage supply for each frequency domain, within a certain margin, at the delivering PMU (Power Management Unit) sources, current switching activity waveforms are extracted (using Apache CPM approach methodology) in the time-domain as depicted in Fig.5(a), up to 50nS.

necessarily cancelling due to distribution of ground paths within the chip between different partitions and IOs.

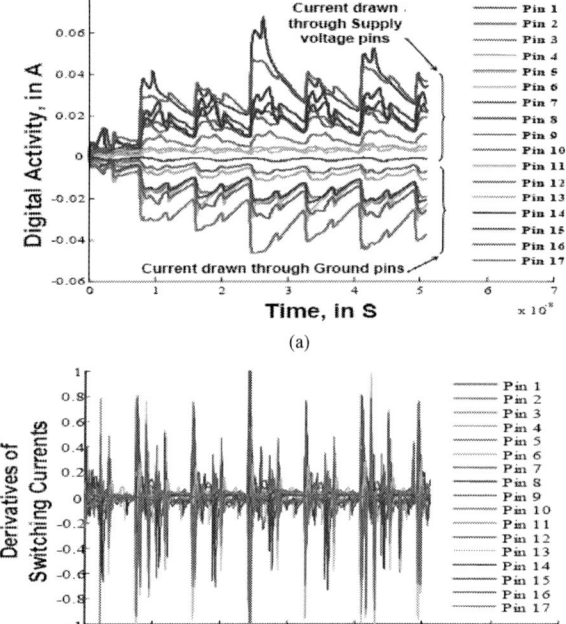

(a)

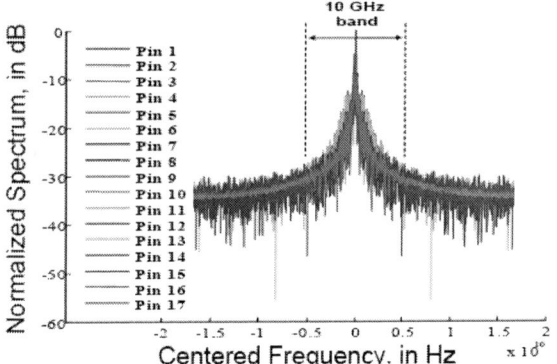

Fig.4 Active digital die normalized power spectrum in frequency domain 17 IOs (Input/Output pins). Pins numbering are not referring to the photograph of Fig.2, which uses different chip design/circuits.

To evaluate important requirement for current waveform models concerning IR-drop analysis capability, current derivatives are extracted based on numerical derivation from Fourier transform expansion as shown in Fig.5(b). In classical approaches based on triangular waveform assumptions (shown in Fig.6) the derivative of current models is not properly calculated since it leads to a pulse-like representation (Dirac type of discontinuity). It is essential to notice the polarity attribute of the extracted current activity waveforms in the time-domain (Fig.5(a)): negative polarity refers to ground pins while positive polarity is associated to supply pins (drawing currents convention). It is observed for a pair of ground/supply pins the sum of the drawn current is not

(b)

Fig.5 Extracted current signatures (using Apache solutions) behavior against time (a) and associated derivatives (b) for 17 IOs (Input/Output pins).

The obtained simulation results are compared to analytical, semi-analytical calculations estimating average power consumption of a digital high speed active die based on estimated number of flip-flops and current profiles at gate level when average power measurement is not available.

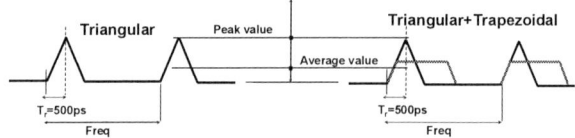

Fig.6 Classical triangular or combined triangular trapezoidal waveform profile for current activity profile.

978-2-8748-7007-1/08 $25.00 © 2008 EuMA   452

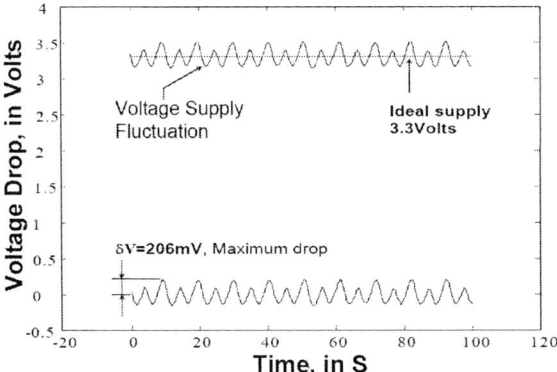

Fig.7 Measured time-domain Voltage drop fluctuations for the 3.3 V supply domain.

To evaluate the accuracy of constant voltage supply assumption at PMU level the fluctuations of the voltage waveform are measured. Fig.7 represents the variation against time up to 100 nS of supply voltage and relative fluctuations with 206mV margin value. To investigate coupling effects at system level including interferences between analog and digital dies S-parameters in two-port configuration are measured in reference to RF-IN and RF-OUT access ports in Fig.3(a),(b). Both configurations in Fig.3(a) and Fig.3(b) have been measured and modeled based PDN extractions. Fig.8 shows comparisons between electromagnetic simulation model, Broadband SPICE extraction model and measurement data for both configurations with and without active SiP module demonstrating excellent agreement up to 6 GHz. Some spurs are noticed in the configuration with active SiP module around 1GHz and 3.4GHz.

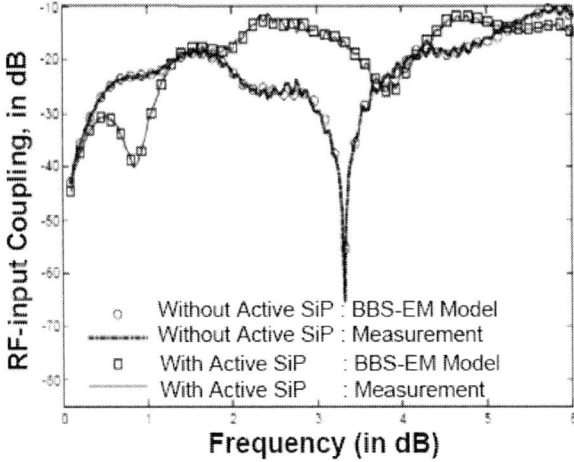

Fig.7 Coupling between RF-IN and RF-OUT access ports with and without active SiP module against frequency: comparison between full-wave EM model, BBS (BroadBand Spice) model and measurement.

In Fig.8 comparison between electromagnetic model and measurement data for input admittance (RF-IN) is shown. Satisfactory agreement is obtained both for the imaginary and real parts and resonant frequency locations up to around 6GHz.

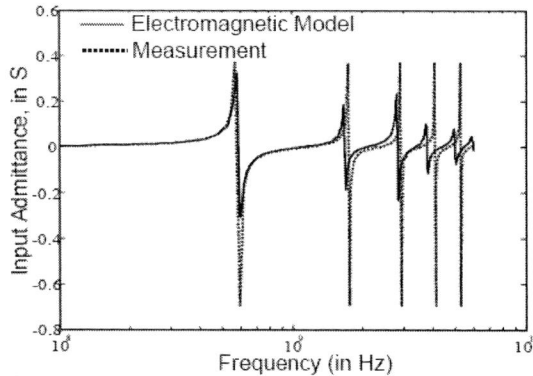

Fig.8 RF Input admittance imaginary part against frequency: comparison between Electromagnetic model and on board system-level measurement.

IV. CONCLUSION

A global distributed co-simulation methodology for concurrent/simultaneous analysis of passive and active parts have been proposed and successfully applied to different real-world NXP-Philips-Semiconductors test carrier modules. An original «power-signature» concept is used to model high-speed digital modules temporal and spatial distribution of their power switching activity for analog-digital co-simulations. The proposed concept has been validated by comparison with average power measurement showing satisfactory agreement. Analysis of coupling effects at system level through RF input/output signal access ports have been presented with and without active SiP modules. The measured coupling is in agreement with predicted simulation results based on electromagnetic simulations and broadband SPICE extractions. Work under investigation concerns correlations between observed spurs in presence of SiP active modules and the behavioral response (transfer function) of the active die multi-port, and multi-port de-embedding analysis.

ACKNOWLEDGMENT

The authors are grateful to fruitful collaboration with Apache solutions, with particular thanks to An-Yu Kuo for co-work on Power and Signal integrity methodologies and flows. The authors thank Dominique Bosquet for the measurement efforts and Jerôme Toublanc, Andries Van Der Veen, Jean-Marc Yannou, Alain Meresse with Genevieve Duchamp for exchanges.

III. REFERENCES

[1] J. Mao, B. Archambeault, J.L. Drewniak, T.P. Van Doren, "Estimating DC Power Bus Noise", IEEE Int. Symp. on EMC, Minneapolis, Minnesota, USA, August 19-23, 2002.

[2] M. Leone, V. Ricchiuti, G. Antonini and A. Orlandi ,"Measurement and Modeling of noise current spectrum for large ASICs", IEEE 7th Workshop on Signal Propagation on Interconnects, 2003.

[3] S. Mallat, "A theory for multiresolution signal decomposition: The wavelet representation," *IEEE Trans. Pattern Anal. Mach. Intell.*, vol. 11, no. 7, pp. 674–693, Jul. 1989.

[4] J. Mao, G. Fitzgerald, A. Kuo and S. Wane, "Coupled Analysis of Quasi-static and Full-Wave Solution towards IC, Package and Board Co-design", published in Electrical Performance of Electronic Packaging (EPEP), Atlanta October 2007.

[5] S. Wane " Partition and Global Methodologies for IC, Package and Board Co-Simulation in SiP Applications," presented at European Microwave Week, Munchen October 2007.

978-2-8748-7007-1/08 $25.00 © 2008 EuMA

Proceedings of the 3rd European Microwave Integrated Circuits Conference

24 GHz LTCC I/Q Mixer Using Packaged HEMTs

Veljko Napijalo [#1], Vicentiu Cojocaru [#2]

[#] TDK Electronics Ireland

3022 Lake Drive, City West Business Campus, Dublin 24, Ireland

[1]Napijalo@tdk.de

Abstract—**An I/Q active mixer in LTCC technology using packaged HEMTs as mixing devices is described. A mixer is designed for use in the 24 GHz automotive radar application. An on-tile buffer amplifier was added to compensate for the limited power available from the system oscillator. Careful choice of the type or topology for each of the passive circuits implemented resulted in an optimal mixer layout, so a very small size for a ceramic tile of 15x15x0.8 mm³ was achieved. The measured conversion gain of the mixer for a 0 dBm LO level was -6.7 dB for I and -5.2 dB for Q. The amplitude imbalance between I and Q signals resulting from the aggressive miniaturization of the quadrature coupler could be compensated in the DSP stages of the system at no additional cost. The measured I-Q phase imbalance was around 3 degrees. The measured return losses at mixer ports and LO-RF isolations are also very good.**

I. INTRODUCTION

In recent years a number of applications requiring high volume and low-cost manufacturing emerged at frequencies between 20-26 GHz. The important aspects in the development of such an application are the choice of system architecture and an optimum realization strategy for each block in the system so that required functionality is achieved at low cost.

The homodyne receiver type could be a convenient choice for 24 GHz ISM band automotive radar applications [1]-[2]. It uses an I/Q mixer to directly convert RF signal to the base band, therefore the receiver chain is very simple. An I/Q mixer circuit can be built using a variety of mixing devices, namely diodes, discrete transistors (packaged or bare dies) or MMICs. Discrete transistors can be used to realize either active or resistive mixers. An active mixer provides higher conversion gain then a resistive or a diode mixer, so it can contribute to the reduction in the number of gain blocks in a system [3]. Flip chip mixer diodes covering 24 GHz ISM band are currently more expensive than packaged HEMT alternatives. At the same time, as opposed to a bare die, a packaged device can be used with standard pick-and-place assembly followed with a solder reflow process. Thus, using an active mixer with packaged devices in a system can potentially lead to a significant cost reduction.

To reduce the cost even further, the LTCC technology can be utilized for mixer manufacturing. LTCC technology combines low-cost and well established screen printing techniques to form a highly integrated multilayer circuit. Two LTCC mixer applications have been reported recently at K band, an MMIC up converter [4] that utilizes a multilayer LTCC environment to embed the necessary filters, and a diode star mixer [5] with very good performance achieved through implementation of a multilayer Marchand balun. This paper describes low-cost, small size LTCC I/Q mixers with high performance packaged HEMTs for the 24 GHz automotive radar applications.

II. MIXER DESIGN

The general block schematic of the active I/Q mixer is presented in Fig. 1. It can be viewed as comprising two single device gate transconductance mixers connected to a phasing network. The phasing network provides the required phase balance of LO and RF signals at the device inputs. Basic principles of operation of an active mixer are explained in [3].

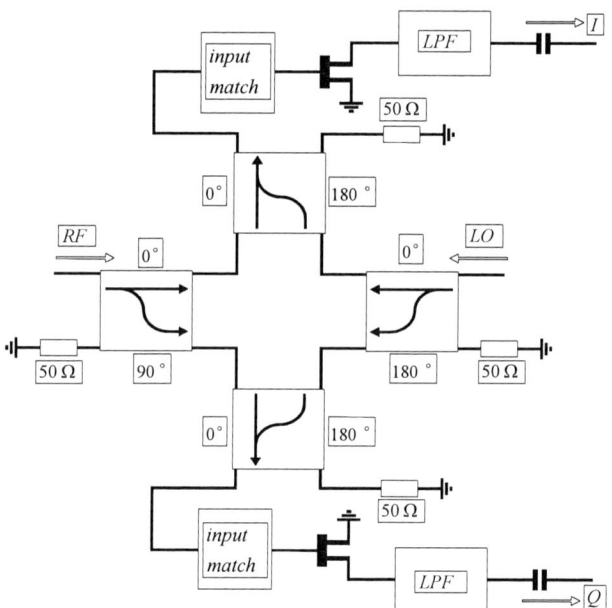

Fig. 1 Block schematic of I/Q mixer

An active mixer represents a strongly nonlinear circuit, so it is important to accurately characterize and model the nonlinearities in order to obtain reliable results from nonlinear circuit simulations. Therefore, special attention was paid to the development of an accurate device model. The parameters of the nonlinear model have been extracted in-house for the particular HEMT device selected for use in the mixer.

As a first design step, the elements of an I/Q phasing network were designed as described in the next section. The second step involved designs of two other passive circuits i.e. input matching and output low pass filter (LPF). Each of the circuits was simulated by connecting together the required

978-2-8748-7007-1/08 $25.00 © 2008 EuMA

microstrip transmission line elements and using linear circuit simulations from [6]. Integral mixer schematic according to the block diagram shown in Fig. 1 was set-up next using blocks designed in previous steps. Parameters of the microstrip lines were optimized through nonlinear harmonic balance simulations from [6] to achieve the optimal characteristics for the mixer. The final layout of both passive circuits was analysed using electromagnetic simulations (EM).

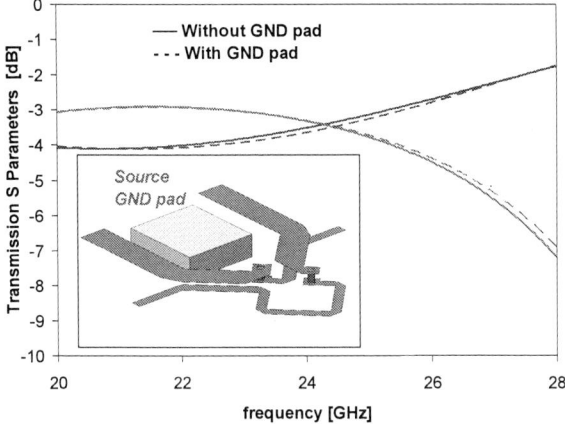

Fig. 2 EM simulated transmission S parameters of the 180° hybrid with and without source grounding pad; actual pad possition is given in the inset

III. MIXER PASSIVE CIRCUITS

A. I/Q Phasing Network

As shown in Fig. 1, the chosen topology for the phasing network comprises three 180° hybrids and a 90° hybrid.

A convenient choice for the 180° hybrid is the multilayer coupled line hybrid described in [7]. It uses the benefits of multilayer LTCC environment to achieve non-interspersed inputs and outputs and has reduced size compared to the classical rat-race hybrid. The use of the multilayer hybrid allows further reduction of a mixer size as the transistor source pad can be conveniently positioned between the hybrid ports as shown in Fig. 2. The transmission S parameters of the hybrid with and without source grounding pad obtained by EM simulations are presented in Fig. 2. It is obvious that the presence of the pad does not significantly alter the transmission S parameters of the hybrid.

A classical branch line coupler was considered for the quadrature hybrid. For the microstrip substrate used the length/width ratios of the lines required to construct that type of hybrid were 1.67 for the case of 35 Ω line and 3.37 for the 50 Ω line. Branch line hybrid constructed with lines having such a small aspect ratio includes strongly coupled microstrip T junction discontinuities. The attempted optimization of such a circuit was found very difficult as the coupler characteristics were highly sensitive to very fine variations of the line dimensions. It was concluded that such a hybrid would have very poor yield in manufacturing.

The rectangular microstrip patch coupler [8] was used instead. The hybrid was designed through electromagnetic optimization. The optimized and the measured transmission S parameters of the patch hybrid are shown in Fig. 3 while the layout is shown in Fig. 4 (a). The measured amplitude imbalance between the hybrid transmission paths was ~1.9 dB, much larger that simulated, while the phase between the outputs was 90±0.5° which was acceptable. As shown in Fig. 4 (b), a pair of the output microstrip connecting lines was bent very close to the hybrid rectangle and this modification resulted in the increased value of the amplitude imbalance. As the compactness of the mixer circuit was very important and, on the other hand, the negative impact of the measured imbalance could be minimised in the BB/DSP stages without any increase of the complexity of the system, it was decided to keep the patch hybrid circuit as designed.

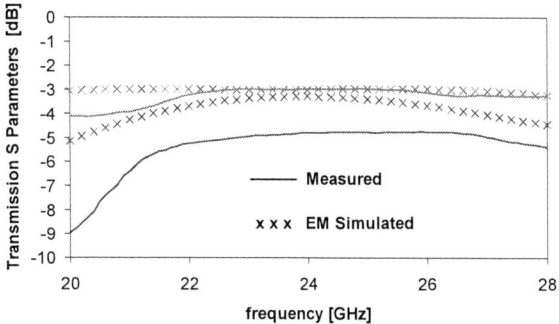

Fig. 3 EM simulated and measured transmission S parameters of the 90° hybrid

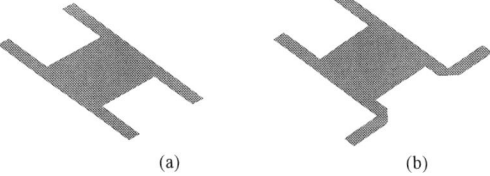

(a)   (b)

Fig. 4 Microstrip layout of the patch hybrid: (a) as optimized through EM optimization; (b) as implemented in the mixer circuit

Test circuits of I/Q phasing network have been built and measured. The measured insertion losses of the RF and LO paths were 7.8±0.7 dB and 6.7±0.2 dB respectively. The measured phase imbalances of the RF and LO paths were within 3° margin with respect to the required values.

B. Input Matching Network

The gate matching circuit for the mixing devices was designed to match a single device mixer for LO/RF frequency in large-signal operating conditions. It also provides short circuit to any IF signal leaking from the device drain, ensuring the stable operation of the device [3]. The gate biasing distribution elements for the voltage and for blocking the flow of biasing current through the phasing network are also included in the matching network.

978-2-8748-7007-1/08 $25.00 © 2008 EuMA

C. Low-Pass Filter / Impedance Transformer

LPF should represent a short circuit termination for the LO signal and provide an adequate IF terminating impedance seen by the device [3]. The filter topology used was of high-low impedance type [9] where low impedance sections were replaced with radial stubs. This modification allows convenient meandering of the high impedance lines so that the area occupied with the output circuit is minimized. The recommended short circuiting of the LO signal and its higher harmonics is retained with the implemented modification.

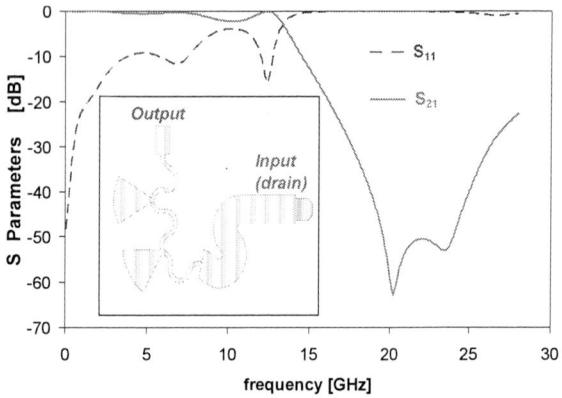

Fig. 5 S parameters of the LPF/impedance transformer as obtained from EM simulations; actual microstrip layout of the circuit is given as the inset

The LPF/transformer microstrip circuit is given as inset in Fig. 3. The input of the circuit is connected to the transistor drain. The reason for the presence of the hexagon metal shape at transistor terminal was explained in [10]. The output of the circuit is connected to the drain bias supply. The same node is connected to the IF decoupling capacitor as shown in Fig. 1.

The optimum wideband response of the LPF/transformer obtained from EM simulations is presented in Fig. 5. The simulated rejection of the LO signal is better than 50 dB.

D. Buffer Amplifier

In the particular radar system considered, the power available from the system oscillator was limited to the range of 0 ± 1 dBm. According to the nonlinear simulations, the device requires higher LO power level, so the oscillator signal has to be amplified.

A single stage buffer amplifier has been designed for the implementation on the same LTCC tile with the mixer. As for the other passive circuits, special care has been taken to find an optimum topology for the matching networks so that the amplifier occupies the minimum area of an LTCC tile. It was found that the matching networks can take the form of the same two-element structure for both, gate and drain circuit. First element is a high characteristic impedance line used as an impedance transformer. Second element is a DC block in a form of the coupled line section, which can also perform an impedance transformation. The high impedance lines are narrow, so they could be easily bent, and as a result the amplifier layout has high flexibility to fit into very small area, as shown in Fig. 6.

The initial amplifier circuit was designed using microstrip elements and a linear simulator [6]. A standard linear design procedure was carried out using transistor S parameters for sufficiently high drain current expected under the real, large signal working conditions of the buffer. The final layout was obtained through EM simulations. The optimized response is presented in Fig. 7.

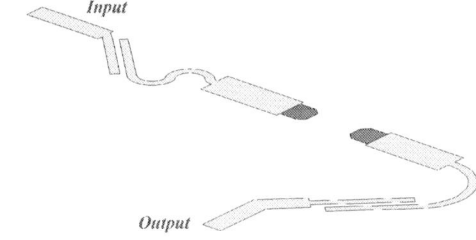

Fig. 6 Optimized microstrip circuit of the buffer amplifier (bias circuit not shown)

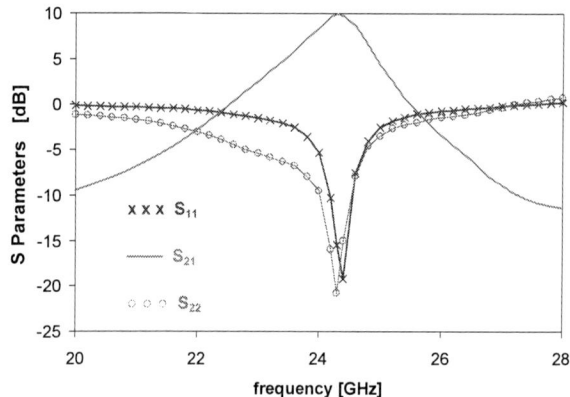

Fig. 7 Simulated S parameters of the buffer amplifier

IV. MEASURED MIXER CHARACTERISTICS

The multilayer mixer circuit was laid out on LTCC. All the necessary surface mount (SMD) discrete components required have been added as on-tile elements. The total size of the LTCC tile was only $15\times15\times0.8$ mm^3 including the buffer amplifier. Test mixer circuits were manufactured using TDK LTCC technology and measured using CPW G-S-G probes.

The measured frequency response of the conversion gain is presented in Fig. 8, while the measured dependency of mixer conversion gain on the power level of the LO signal applied to the input of the buffer amplifier is presented in Fig. 9. With the LO signal level of 0 dBm at the buffer amplifier input, the measured conversion gains were -6.7 dB for I and -5.2 dB for Q. It is obvious that the imbalance of the quadrature coupler translates directly to the imbalance between I and Q outputs. Fig. 10 shows the measured I-Q phase difference, which is better than 3 degrees.

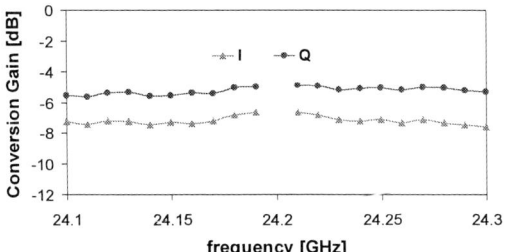

Fig. 8 Measured frequency dependency of the mixer conversion gain; f_{LO}=24.2 GHz, P_{LO}=0 dBm

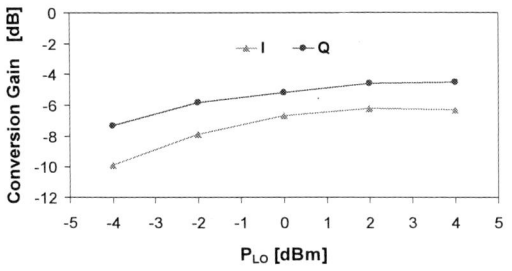

Fig. 9 Measured dependency of the conversion gain vs. oscillator power at buffer amplifier input; f_{LO}=24.2 GHz, f_{RF}=24.175 GHz

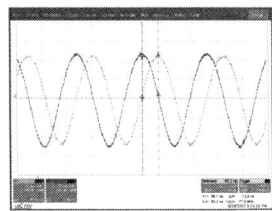

Fig. 10 Measured phase difference between the I and Q signals (f_{IF} ~ 20MHz)

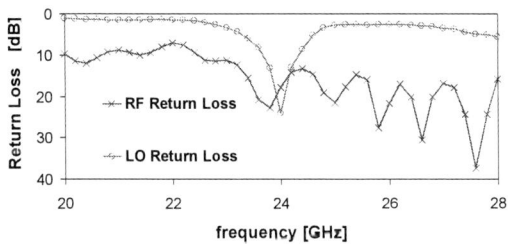

Fig. 11 Measured return losses at mixer ports; f_{LO}=24.2 GHz

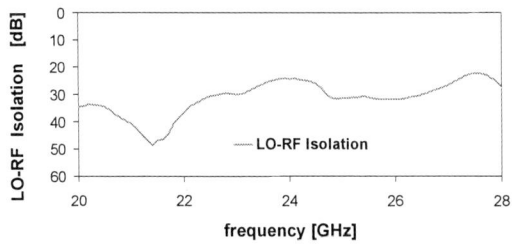

Fig. 12 Measured LO-RF isolation between mixer ports; f_{LO}=24.2 GHz

The measured return losses and isolations between the mixer ports are presented in Fig. 11 and Fig. 12. Measured RF return loss is 14.2 dB while the return loss measured at the input of the buffer amplifier is 13 dB. There is a slight downward shift of the LO port return loss with respect to the simulated response of the buffer amplifier shown in Fig. 7 which can be attributed to difference in performance for large signal operating conditions of the amplifier with respect to linear, high current regime assumed in the simulations. Also, the actual loading of the amplifier is a nonlinear input impedance of the mixer which is different from 50 Ω impedance. Thus, it can be expected that extending the nonlinear design to the buffer amplifier will result in much better agreement between measured and simulated values. The measured LO-RF isolation is ~ 25 dB.

V. CONCLUSIONS

The design of an LTCC compact size, low-cost, active I/Q mixer for 24 GHz automotive radar applications has been described. The size of the LTCC tile of only 15x15x0.8 mm³ was achieved including a LO buffer amplifier stage. The key factor to enable a successful reduction of the mixer size was the careful choice of the types and topologies of all of the passive circuits used, namely coupled line 180° hybrid, 90° patch hybrid, radial stub low pass filter for the mixer output circuit and impedance transformers with high characteristic impedance lines for the buffer amplifier matching networks. The measured characteristics of the mixers are very good. The value of the I/Q phase imbalance is very good, while the gain imbalance resulting from the aggressive miniaturization of the patch quadrature coupler can be compensated for in the DSP stages of the system.

REFERENCES

[1] V. Cojocaru, V. Napijalo, D. Humphrey, and B. Clarke, "24GHz Low-cost UWB front-end design for short range pulse radar applications," in *Proc. EuMC 2004*, paper 40.2, p. 1273.

[2] V. Cojocaru, et. al., "A 24GHz low-cost, long-range, narrow-band, monopulse radar front end system for automotive ACC applications", in *Proc. MTT-S 2007*, 2007, paper TH1C.6, p. 1327.

[3] S. Maas, *Microwave Mixers*, 2nd ed., Norwood MA, USA: Artech House, 1993.

[4] J.-C. Jeong, I.-B. Yom, H.-J. Lee and K. W. Yeom, "Filter embedded K-band LTCC Upconverter," in *Proc. EuMC 2007*, 2007, p. 1314.

[5] T. Baras, J. Muller, and A. F. Jacob, "K-Band LTCC star mixer with broadband IF output network," in *Proc. MTT-S 2007*, 2007, paper TH1F.5, p. 1405.

[6] *Microwave Office*, ver. 7, Applied Wave Research, El Segundo, CA, 2006.

[7] V. Napijalo, "Multilayer 180° Hybrid Coupler in LTCC Technology for 24GHz Applications", in *Proc EuMC 2007*, 2007, paper 30-2, p. 552.

[8] Burns, I. T. W., "Planar, quadrature microwave coupler," U.S. Patent 4 492 939, Jan. 8, 1985.

[9] R. Gilmore and L. Besser, *Practical RF Circuit Design for Modern Wireless Systems Volume II - Active Circuits and Systems*, Norwood MA, USA: Artech House, 2003.

[10] V. Napijalo, V. Cojocaru, T. Yokoyama, and T. Young, "Accurate substrate-related packaged transistor modeling for 24GHz circuit design", in *Proc EuMC 2006*, 2006, paper 28.3, p. 522.

Proceedings of the 3rd European Microwave Integrated Circuits Conference

A Triple Tuned Ultra-Wideband VCO in X-K Band

Masaomi Tsuru, Kenji Kawakami, Ken'ichi Tajima, Kazuhiro Miyamoto, Masafumi Nakane, Morishige Hieda, Moriyasu Miyazaki

Mitsubishi Electric Corporation, 5-1-1 Ofuna, Kamakura, Kanagawa, 247-8501, JAPAN
Tsuru.Masaomi@eb.MitsubishiElectric.co.jp

Abstract— We have already proposed a triple tuned Voltage Controlled Oscillator (VCO) for achieving an ultra-wideband characteristic. The fabricated VCO in C-Ku band has achieved the oscillation bandwidth of 5.6GHz-16.8GHz.

In this paper, for achieving the wide oscillation bandwidth in the higher frequency band, we proposes a VCO that consists of a MMIC working as an active circuit and external hyper-abrupt junction varacter diodes connect to the MMIC with Au wires. The fabricated VCO in X-K band achieves the oscillation bandwidth of 9.9GHz-21.9GHz and the phase noise of − 109.4dBc/Hz +/-2.4dB at 1MHz offset from the carrier. In addition, it is shown that the fabricated VCO achieves the widest oscillation bandwidth and the lowest phase noise in X-K band compared with other reported wideband VCOs.

I. INTRODUCTION

Wideband tunable oscillators are useful for reducing the size of microwave measurement systems and communication systems. Thus the wideband tunable oscillators using YIG (Yttrium-Iron-Garnet) resonators [1]-[3] have been applied for many microwave systems. However, the oscillators using YIG resonators have some disadvantages of slow tuning speed, the large magnetic circuits, and so on. Accordingly, wideband VCOs using planar circuits have been required. In the past, some configurations such as double tuned VCOs [4]-[9], a doubling oscillator [10], and VCOs with variable inductors [11]-[14] have been proposed for achieving the wide oscillation bandwidth, which has been up to 74% at C-Ku band [5]. We have developed a double tuned VCO and achieved the oscillation bandwidth of 69.8% at C-X band [9].

In order to cover the wider bandwidth than that of the double tuned VCO, we have already proposed the triple tuned VCO and achieved the oscillation bandwidth of 100% in C-Ku band [15]. Although the tuned circuit at the collector is different, U. L. Rohde reported on the other triple tuned VCO that achieved the oscillation bandwidth of 100% in S-C band [16]. Thus these configurations are effective for achieving the ultra-wideband characteristic.

To cover the higher frequency band, the VCO will be made with MMIC because of a reduction of parametric capacitances or inductances. However, as capacitance ratio of a varactor diode on MMIC is small, it is difficult that the VCO achieves the wide oscillation bandwidth.

In this paper, for achieving the wide oscillation bandwidth in the higher frequency band, we proposes a triple tuned VCO that consists of a MMIC working as an active circuit and

external hyper-abrupt junction varacter diodes connect to the MMIC with Au wires and shows measurement results of the fabricated VCO in X-K band.

In addition, the performances of the proposed VCO are compared with other reported wideband VCOs.

II. CONFIGURATION

Fig.1 shows a configuration of the triple tuned VCO. This VCO consists of an HBT and three LC tuned circuits as shown in Fig.1. Z_a is the input impedance of an active circuit at the base terminal of the HBT. The active circuit consists of the HBT, an emitter tuned circuit and a collector tuned circuit.

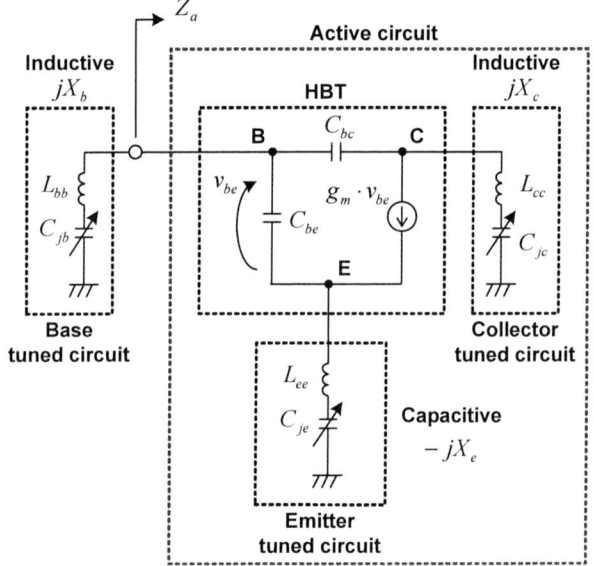

Fig. 1. Configuration of the triple tuned VCO.

III. CONDITIONS FOR THE OSCILLATION

Y-parameter Y_{HBT} of the HBT in Fig.1 is as follows:

$$Y_{HBT} \equiv \begin{bmatrix} Y_{11} & Y_{12} \\ Y_{21} & Y_{22} \end{bmatrix} \tag{1}$$

$$Y_{11} = j\omega(C_{bc} + C_{be}) \tag{2}$$

$$Y_{12} = -j\omega C_{bc} \tag{3}$$

$$Y_{21} = g_m - j\omega C_{bc} \tag{4}$$

$$Y_{22} = j\omega C_{bc} . \tag{5}$$

978-2-8748-7007-1/08 $25.00 © 2008 EuMA   458

We assume that tuning circuits are lossless components. Conditions for oscillation are as follows:

$$\mathrm{Re}(Z_a) < 0 \tag{6}$$

$$\mathrm{Im}(Z_a) + X_b = 0 . \tag{7}$$

The condition for generating negative resistance is as follows from (6):

$$X_e > 0 \tag{8}$$

and

$$0 \le X_c < \frac{1}{\omega C_{bc}} . \tag{9}$$

Here, X_e, X_c and X_b are as follows:

$$-X_e = \omega L_{ee} - \frac{1}{\omega C_{je}} \tag{10}$$

$$X_c = \omega L_{cc} - \frac{1}{\omega C_{jc}} \tag{11}$$

$$X_b = \omega L_{bb} - \frac{1}{\omega C_{jb}} . \tag{12}$$

The equation (9) leads to the follow conditions for L_{ee} and L_{cc}:

$$L_{ee} < \frac{1}{\omega_o^2 C_{je}} \tag{13}$$

and

$$\frac{1}{\omega_o^2 C_{jc}} < L_{cc} < \frac{1}{\omega_o^2}\left(\frac{1}{C_{jc}} + \frac{1}{C_{bc}}\right) \tag{14}$$

where ω_o is an oscillation angular frequency.

In the high oscillation frequency, L_{ee} and L_{cc} have to be small. Therefore, to reduce a parasitic element, components except varactor diodes are made on MMIC. The varactor diodes are discrete devices that have the large capacitance ratio to achieve the wide oscillation bandwidth.

Fig. 2 shows calculation results of L_{ee} and L_{cc} versus C_{je} and C_{jc} using (13) and (14). Considering the feasibility of L_{ee} and L_{cc}, it is better that C_{je} is small as much as possible and C_{jc} is larger than C_{bc}.

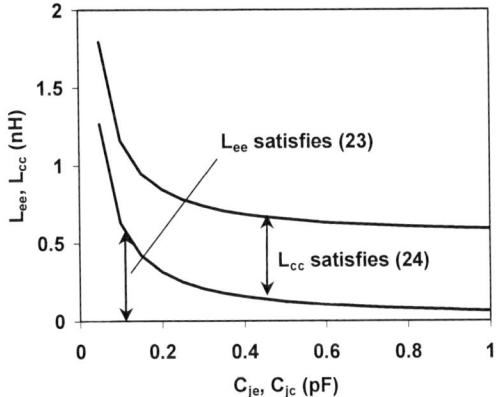

Fig. 2. Calculation results of L_{ee} and L_{cc} versus C_{je} and C_{jc}: C_{bc}=0.12pF, f_o=20 GHz.

IV. EXPERIMENTAL RESULTS

Fig. 3 shows a configuration of the fabricated triple tuned VCO in X-K band. The active device is an InGaP/GaAs HBT. The emitter size is 120um². The collector current density is 25kA/cm². The cut-off frequency is 31.6GHz. The flicker corner frequency is 20kHz. Varactor diodes are GaAs hyper-abrupt junction diode. A capacitance ratio of the varactor diodes is about 13.6 in the voltage range from 0 to -16V. The capacitance of the collector tuned circuit is the twice larger than that of the emitter tuned circuit.

Moreover, all varactor diodes are controlled by a single voltage source for the purpose of incorporating into a synthesizer without difficulty.

Fig. 4 shows a photograph of the fabricated VCO. The varactor diodes connect to the MMIC with Au wires. L_{ee}, L_{cc} and L_{bb} consist of the Au wires. The size of the VCO is 1.4mm × 1.9mm. Fig. 5 shows measurement and simulation results of the fabricated VCO. The bias voltages of the HBT were V_c=2.0V and V_b=1.9V. The consumption current was less than 44.7mA. At the tuning voltage from 0V to -15V, the VCO covered the oscillation bandwidth of 9.9GHz-21.9GHz, or 75.5%. The phase noise was -109.4 dBc/Hz +/-2.4 dB at 1 MHz offset from the carrier. The output power was -1.8 dBm +/-3.3 dB. The measurement results almost agreed with simulation results. In addition, although the fluctuation of the Au wires affects a manufacturing error, the fluctuation of the oscillation frequency was less than +/- 2.8%.

Fig. 6 shows the oscillation bandwidth and the phase noise of the reported wideband VCOs. Fig. 6(a) shows that the fabricated VCO has covered the widest oscillation bandwidth in X-K band. Fig. 6(b) shows the fabricated VCO has achieved the lowest phase noise in X-K band.

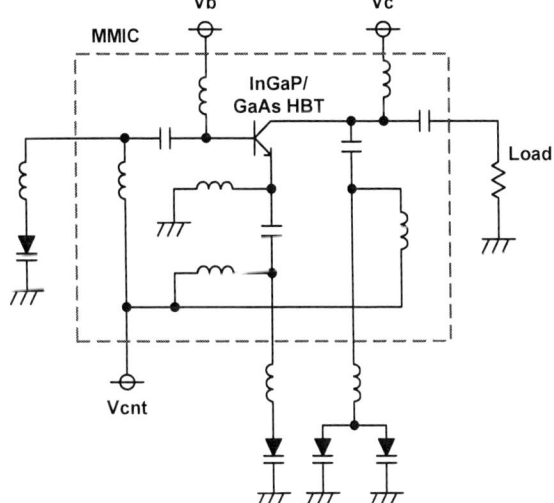

Fig. 3. Configuration of the fabricated triple tuned VCO in X-K band.

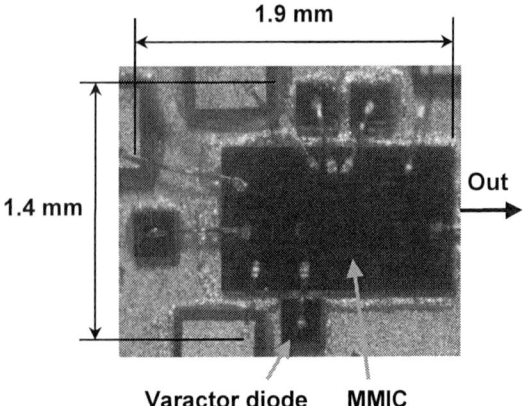

Fig. 4. Photograph of the fabricated triple tuned VCO.

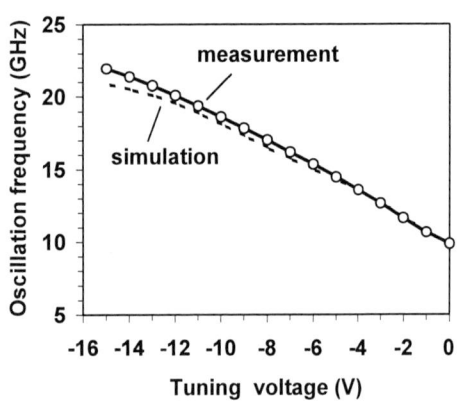

(a)

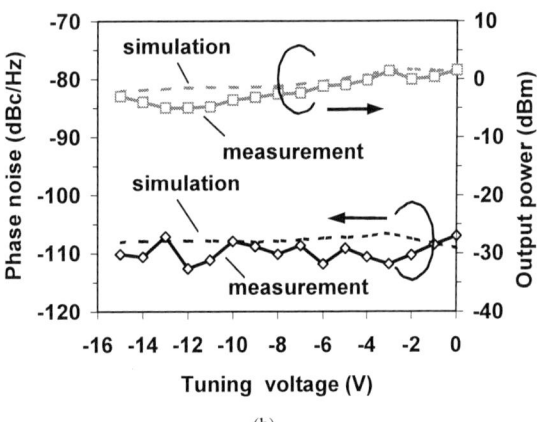

(b)

Fig. 5. Measurement and simulation results of the triple tuned VCO in X-K band: (a) Oscillation frequency. (b) Phase noise at 1MHz offset from the carrier and output power.

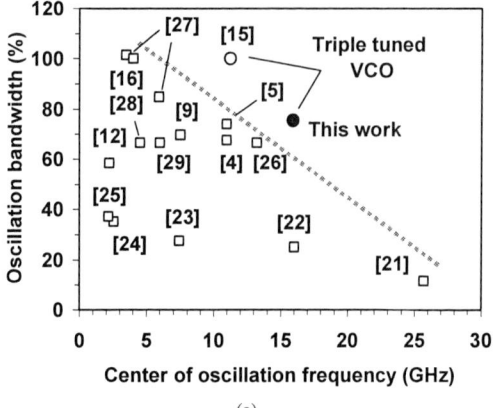

(a)

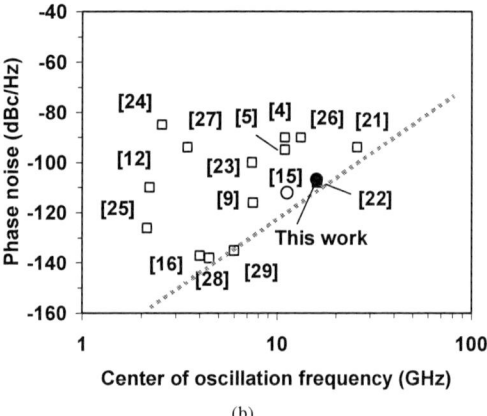

(b)

Fig. 6. Oscillation bandwidth and phase noise of the reported wideband VCOs: (a) Oscillation bandwidth versus center of the oscillation frequency. (b) Phase noise versus center of the oscillation frequency.

V. CONCLUSIONS

For achieving the wide oscillation bandwidth in the higher frequency band, we proposed the triple tuned VCO that consists of a MMIC working as an active circuit and external hyper-abrupt junction varacter diodes connect to the MMIC with Au wires. The fabricated VCO in X-K band achieved the oscillation bandwidth of 9.9GHz-21.9GHz and the phase noise of −109.4dBc/Hz +/-2.4dB at 1MHz offset from the carrier.

In addition, it was shown that the fabricated VCO achieved the widest oscillation bandwidth and the lowest phase noise in X-K band compared with other reported wideband VCOs.

REFERENCES

[1] R.Souares, "GaAs MESFET Circuit Design," *Artech House*, pp.382-399, 1988.

[2] G. R. Basawapatna, and R. B. Stancliff, "A Unified Approach to the Design of Wide-Band Microwave Solid-State Oscillators," *IEEE, Trans. MTT*, vol. 27, no.5, pp.379-385, May. 1979.

[3] C. F. Schiebold, "An Approach to Realizing Multi-Octave Performance in GaAs-FET YIG-Tuned Oscillators," *IEEE, MTT-S International Microwave Symposium Digest*, vol.85, no.1, pp.261-263, Jun. 1985.

[4] J. Kitchen, "Octave Bandwidth Varactor-Tuned Oscillators," *Microwave Journal*, pp.347-353, May. 1987.

[5] A. Adar, and R. Ramachandran, "An HBT MMIC Wideband VCO," *IEEE, MTT-S International Microwave Symposium Digest*, vol.1, pp.247-250, Jun. 1991.

[6] B. N. Scott, M. Wurtele, and B. B. Cregger, "A Family of Four Monolithic VCO MIC's Covering 2-18 GHz," *IEEE, Microwave and Millimeter-Wave Monolithic Circuits*, vol.84, no.1, pp.58-61, May. 1984.

[7] W. El-Kamali, J.-P. Grimm, R. Meierer, and C. Tsironis, "New Design Approach for Wide-Band FET Voltage-Controlled Oscillators," *IEEE, Trans. MTT*, vol. 34, no.10, pp.1059-1063, Oct. 1986.

[8] A. Grebennikov, "Wideband VCO Designs Are Independent of Circuits Parameters," *Microwave & RF*, pp.147-155, Aug. 2001.

[9] K. Tajima, Y. Imai, Y. Kanagawa, and K. Itoh, "A 5 to 10GHz Low Spurious Triple Tuned Type PLL Synthesizer Driven by Frequency Converted DDS Unit," *IEEE, MTT-S International Microwave Symposium Digest*, vol.3, pp.1217-1220, Jun. 1997.

[10] R. G. Winch, "Wide-Band Varactor-Tuned Oscillators," *IEEE, Journal of Solid-State Circuits*, vol. 17, no.6, pp.1214-1219, Dec. 1982.

[11] A. J. Brodersen, E. R. Chenette, and W. L. Engl, "Wide-Range Tuning of Solid-State Microwave Oscillators," *IEEE, Journal of Solid-State Circuits*, vol. 5, no.2, pp.82-84, Apr. 1970.

[12] H. Hayashi, and M. Muraguchi, "A novel broad-band MMIC VCO using an active inductor," *IEICE transaction fundamentals*, vol. E80-A, no.6, pp.1-6, Jun. 1991.

[13] W. Michielsen, L. R. Zheng, and H. Tenhunen, "Analysis and Design of a Double Tuned CLAPP Oscillator for Multi-Band Multi-Standard Radio," *IEEE, International Symposium Circuits and Systems*, vol.1, pp. I-681-I-684, May. 2003.

[14] C.-C. Wei, H.-C. Chiu, and W.-S. Feng, "An Ultra-Wideband CMOS VCO with 3-5 GHz Tuning Range," *IEEE, International Workshop on Radio-Frequency Integration Technology*, pp. 87-90, Nov. 2005.

[15] M. Tsuru, K. Kawakami, K. Tajima, K. Miyamoto, M. Nakane, K. Itoh, M. Miyazaki, and Y. Isota, "A Triple-Tuned Ultra-Wideband VCO," *IEEE, Trans. MTT*, vol. 56, no.2, pp.346-354, Feb. 2008.

[16] U. L. Rohde, and A. K. Poddar, "Impact of Device Scaling on Phase Noise in SiGe HBTs UWB VCOs," *IEEE, MTT-S International Microwave Symposium Digest*, pp. 1793-1796, THP, Jun. 2006.

[17] J.-C. Nallatamby, M. Prigent, M. Camiade, and J. J. Obregon, " Extension of the Leeson Formula to Phase Noise Calculation in Transistor Oscillators With Complex Tanks," *IEEE, Trans. MTT*, vol. 51, no.3, pp.690-696, Mar. 2003.

[18] T. Ohira, " Comments on Extension of the Leeson Formula to Phase Noise Calculation in Transistor Oscillators With Complex Tanks," *IEEE, Trans. MTT*, vol. 55, no.1, pp.185-185, Jan. 2007.

[19] D. B. Leeson, " A Simple Model of Feedback Oscillator Noise Spectrum," *Proc. IEEE*, vol. 54, pp.329-330, Feb. 1966.

[20] U. L. Rohde, and A. K. Poddar, "Novel Multi-Coupled Line Resonators Replace Traditional Ceramic Resonators in Oscillators/VCOs," *IEEE, International Frequency Control Symposium and Exposition*, pp. 432-442, Jun. 2006.

[21] T. Takenaka, A. Miyazaki, and H. Matsuura, "Wideband MMIC voltage control oscillator using active impedance load matching," *IEEE, MTT-S International Microwave Symposium Digest*, vol.3, pp.1503-1506, Jun. 1996.

[22] R. Martin, and F Ali, "A Ku-band Oscillator subsystem using a broadband GaAs MMIC push-pull amplifier/doubler," *IEEE, Microwave and Guided Wave Letters*, vol.1, no.11, pp.348-350, 1991.

[23] Y.Sun, T.Tieman, H.Pflug, and W.Velthuis, "A Fully Integrated dual-frequency push-push VCO for wideband wireless applications," *GaAs2000, Conference Proceedings*, pp.460-463, 2000.

[24] J.-S. Ko, and K. Lee, "Low Power, tunable active inductor and its applications to monolithic VCO and BPF," *IEEE, MTT-S International Microwave Symposium Digest*, vol.2, pp.929-932, Jun. 1997.

[25] F. Herzel, H. Erzgräber, and P.Weger, "Integrated CMOS wideband oscillator for RF applications," *Electronics Letters*, vol.37, no.6, pp.330-331.

[26] M. Kimishima, S. Ohmura, and T. Ashizuka, "A Semi-Monolithic Wideband VCO with Output Power Control Capability Using an Active Power Splitter," *IEEE, MTT-S International Microwave Symposium Digest*, vol.3, pp.1317-1320, Jun. 1992.

[27] R.Mukhopadlhyay, Y.Park, S.W.Yoon, C.-H.Lee, S.Nuttinck, J.D.Cressler, and J.Laskar, "Active-Inductor-based Low-Power Broadband Harmonic VCO in SiGe Technology for Wideband and Multi-Standard Applications," *IEEE, MTT-S International Microwave Symposium Digest*, TH1A-2, Jun. 2005.

[28] U. L. Rohde, A. K. Poddar, Juergen Schoepf, Reimund Rebel, and Parimal Patel, "Low Noise Low Cost Ultra Wideband N-Push VCO," *IEEE, MTT-S International Microwave Symposium Digest*, pp. 1171-1174, WEPG-1, 2005.

[29] U. L. Rohde, and A. K. Poddar, "Configurable Adaptive Ultra Low Noise Wideband VCOs," *IEEE, International Conference on Ultra-Wideband*, pp. 452-457, Sep. 2005.

24 GHz LTCC Amplifier Using Packaged HEMTs

Veljko Napijalo [#1], Vicentiu Cojocaru [#2], Takeshi Yokoyama [*3]

[#] *TDK Electronics Ireland*

3022 Lake Drive, City West Business Campus, Dublin 24, Ireland

[1]Napijalo@tdk.de

[2]Cojocaru@tdk.de

[*] *TDK Corporation*

2-15-7, Higashi-Ohwada, Ichikawa-shi, Chiba, 272-8558, Japan

[3]tayoko@mb1.tdk.co.jp

Abstract— The design of an LTCC self-packaged amplifier for the 24 GHz ISM band using packaged transistors is described. The amplifier utilizes to a full extent the benefits of the multilayer environment as all matching networks are realized by employing multilayer impedance transformers. Advantages of using such transformers are the elimination of dedicated DC blocks, flexible layout routing and a milder discontinuity in the vertical transitions towards the PCB connection. A combination of on-tile and off-tile distributed matching has been implemented allowing for the amplifier to be used on various PCB substrate materials. The amplifier can be used in both the receiver and transmitter chains, in various application scenarios. Examples of the measured amplifier characteristics designed on two different substrate materials are shown, illustrating the trade-off between the cost and the performance of the materials used.

I. INTRODUCTION

Multilayer technologies offer the advantage of reducing the size of microwave circuits in a low cost, high volume production environment. In particular, the LTCC technology combines low-cost and well established screen printing techniques to form a multilayer circuit that can be highly integrated and have complex functionality. The elements of a circuit can be fully or partially embedded in a multilayer ceramic substrate, while a large variety of passive and active components could be mounted on the top of the substrate.

In recent years a number of applications requiring high volume and low-cost manufacturing emerged at frequencies between 20-26 GHz. One such application is the automotive radar in the 24 GHz ISM band [1]-[2]. Key sub-circuits for such a system are 24GHz amplifier blocks, needed for both the receiver and the transmitter chain. A convenient low-cost solution for the active device used in the amplifier design is a packaged transistor. This choice can reduce also the cost in mass production as the standard pick-and-place assembly followed by solder reflow technology can be used in manufacturing. At the same time the performance of packaged devices is constantly improving, and the designs realized with such components can be comparable with that of the higher-cost MMIC technologies.

The range of automotive radar applications has been widening, from comfort to safety, and this usually translates into different requirements for the front end system. A good strategy for the circuit design solutions should aim towards the development of components with a high level of flexibility.

Such an approach allows for simple modifications and variations of a basic front-end system without the need for expensive redesigns of all the circuits.

This paper describes the realization of a low-cost, general purpose, linear LTCC amplifier for the 24 GHz automotive radar applications utilizing high performance packaged HEMT devices.

II. AMPLIFIER DESIGN

As the maximum EIRP for the 24 GHz applications is limited by regulations to relatively low values, the same low to medium power active device can be used for all of the amplifiers throughout the system. Good noise performance for the receiver and a sufficiently high 1-dB compression point for the transmitter can be achieved through a single two stage design offering a good level of flexibility. The block schematic of the two stage amplifier is presented in Fig. 1. The general procedures for linear amplifier design are well known ([3]-[4]) and will readily lead to a successful design. The main specific challenge to the design described here has been the optimal use of the multilayer LTCC environment.

The LTCC technology can be applied to realize a microstrip-like circuit for the amplifier matching networks. An example of such a design has been illustrated in previous work [5]. The microstrip interstage matching network from [5] is shown in Fig. 2 (a). It is obvious that the LTCC substrate has a very low level of utilization of its multilayer structure, with only the biasing being distributed to transistors through lines located inside the tile, underneath the microstrip ground plane.

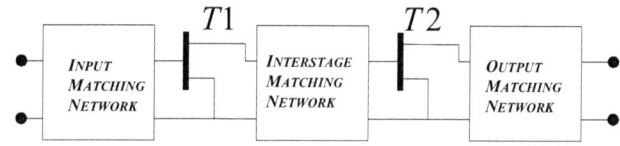

Fig. 1 General schematic of a two-stage amplifier

Although the use of conventional microstrip technology can certainly lead to a successful design as confirmed in [5], that is not the optimal design possible in an LTCC environment. An example of advanced interstage matching network design utilizing broadside coupled lines used as multilayer impedance transformers is presented in Fig. 2 (b). The reduction of the area occupied by the network is obvious. Only the transistor lead footprints and the footprints of other passive SMD components required are located at the top of the substrate. This approach leads to a self-packaged solution similar to the one described in [6], i.e. the amplifier itself does not need any environmental protection before or after the final system assembly as the conductors are all embedded, thus the final package could be in fact the housing of the overall system.

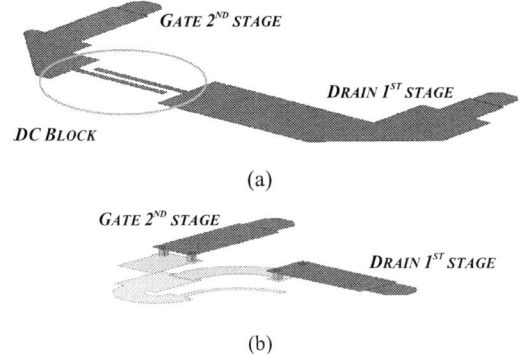

(a)

(b)

Fig. 2 Details of the interstage matching network as simulated in the EM simulator (a) microstrip from [5] and (b) present multilayer solution. The ground planes and bias circuits are not shown

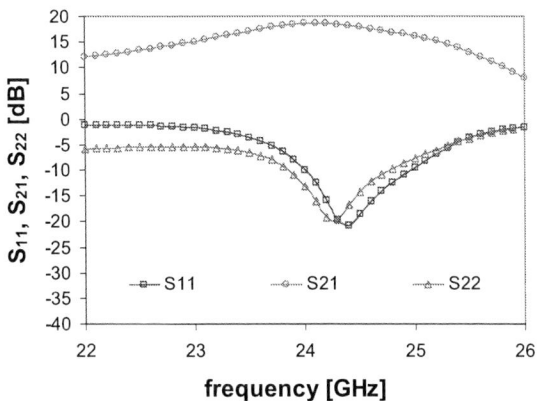

Fig 3 Simulated results for the "*intrinsic*" amplifier with input and output terminations taken on 50 Ω embedded lines. Total drain current Id $_{TOT}$=40mA

The interstage matching network shown in Fig. 2 (b) includes two broadside coupled multilayer impedance transformers connected through an embedded transmission line acting itself as an additional impedance transforming element. This eliminates the necessity for dedicated DC blocks as the transformers themselves mutually isolate the biasing circuits of the two stages. The network geometry is very flexible for layout routing as the bending of a broadside coupled line is straightforward, unlike the edge coupled line in Fig. 2 (a). This contributes to a significant size reduction of

the design. The hexagons shown on the top layer as extensions of the transistor leads are EM simulation generated, specific additions to the actual circuit layout and were inserted in order to include the effects of the additional length of the transistor leads not accounted for during S parameter measurements in a test fixture, as described in [5].

For the input and output matching, a similar type of network as the one shown in Fig. 2 (b) was used. For the interstage network, the impedance was transformed through the first transformer from the top to an embedded layer and then through the second one back to the top layer. This arrangement is natural as both ends of the interstage matching network have to be on the top layer. In the input and output matching networks, transformers were used in succession to embed the signal deeper into the substrate. As a result, the final embedded 50 Ω lines at both amplifier ends are narrow and their height with respect to the ground plane is small. The amplifier uses a BGA ball on the bottom side of the tile to connect to a PCB. The motivation for embedding the final 50 Ω lines deeper into the substrate was to make the discontinuity of the vertical transition trough the ground plane and the remaining layers of the LTCC substrate towards the BGA ball connection, as small as possible. The simulated S parameters of the "*intrinsic*" amplifier with the input and output taken on the 50 Ω lines are shown on Fig. 3. In order to prevent the coupling between the terminations trough the substrate, the transition chosen was of a modified coaxial type, i.e. a combination of those presented in [7]-[8].

In line with the flexible design goal, we chose to match the amplifier to 50 Ω inside the LTCC tile, while the matching circuit for the *connecting arrangement*, including the vertical transition, the BGA ball and the associated PCB pad, was implemented on the PCB itself. This distributed matching approach makes the amplifier design flexible with respect to the PCB material parameters, which are predominantly determining the electrical characteristics of the *connection arrangement*, in particular the ball landing pad capacitance.

The bias of the transistor gates is flexible as both can be biased from the same source or independently. This allows for the use of the amplifier as a pulse modulator, as a modulating signal can be easily applied to one of the stages. Bias configuration is set on-tile by choosing between two positions of a bias resistor during assembly. The quiescent point of each transistor can be set independently by on-tile resistors.

The size of the LTCC amplifier tile is 11x10x0.8 mm³. The actual BGA balls used to connect the amplifier input and output to the PCB are located at inner rows/columns of the array. The addition to the total amplifier footprint introduced by the PCB matching circuits is only in the range of 7%, as most of the matching circuit is located underneath the tile.

III. AMPLIFIER EVALUATION

To evaluate the concept of distributed matching, off-tile matching circuits of the amplifier were design for the case of two different substrate materials. The main properties of the substrates are listed in Table I. The most important differences between these materials are the loss tangent and price, with the cheaper material B having much higher dielectric losses.

The design of the PCB matching networks has been done with a 3D EM simulator [9]. Considering the computational effort required, the distributed matching concept adopted has an additional benefit as it allows the partitioning of the EM structure. The *connection arrangement* together with the PCB matching can be analysed independently from the matching networks of the on-tile amplifier.

TABLE I
PROPERTIES OF PCB SUBSTRATES

material	h [mm]	ε_r	tan δ	price
A	0.254	2.2	0.0009	high
B	0.260	3.47	0.0077	low

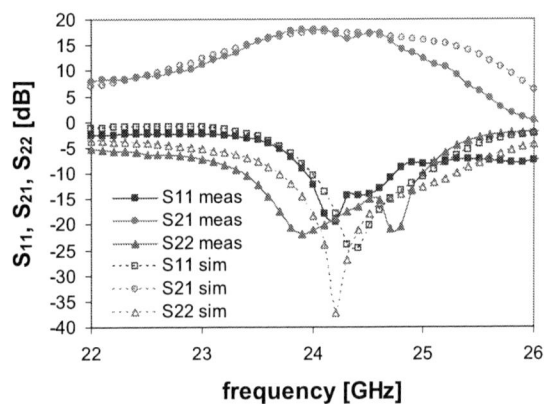

Fig. 4 Measured and simulated results for amplifier mounted on the PCB substrate material A. Total drain current Id $_{TOT}$ = 40mA

The substrate material A can be used for performance demanding applications where price is of secondary importance. The measured S-parameters of the amplifier mounted on this substrate with respect to frequency are presented in Fig. 4, together with simulation results. The results for the frequency of 24.2 GHz are listed in Table II. The measured values are in good agreement with simulations and the overall performance of the amplifier is very good.

For the applications where the lower performance level is acceptable and the low-cost is essential, the substrate material B can be used. According to the distributed matching concept, the design of the PCB matching should be repeated as it has to be specific to the particular material used. In addition to the change of material, it was decided to use the underfill to mechanically strengthen the connection between the LTCC tile and PCB. This introduced an additional design challenge since the matching elements are predominantly located under the tile, and so the dielectric properties of the underfill material will influence the amplifier characteristics and must be accounted for. The underfill resin is typically applied on two adjacent sides of the tile, i.e. in an L shaped pattern. The

capillarity effect in a gap formed by the LTCC tile and PCB pulls the resin underneath the tile. During a short curing process resin melts and flows further along the tile so that at the end of the process it appears on remaining two sides of the tile with the actual profile of the overflow being to some extent random. For that reason the PCB matching network should be designed to be robust and retain the acceptable level of performance as the overflow profiles varies. In our case an additional challenge has been the fact that dielectric properties of the underfill were not known at 24 GHz. The dielectric constant was estimated from provisional measurements while the loss tangent remained unknown, so in simulations the material was considered as lossless.

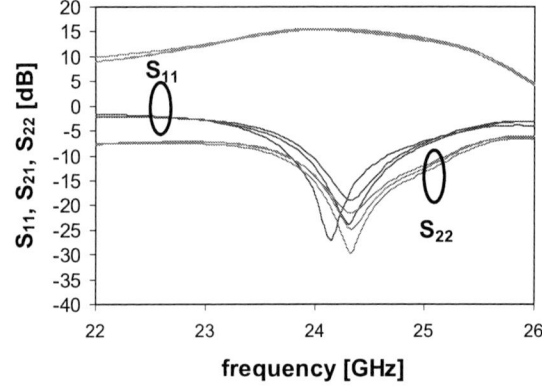

Fig. 5 Simulated variations of the amplifier S-parameters with different underfill overflow profiles for the substrate material B. Total drain current Id $_{TOT}$ =30mA

TABLE III
MEASURED VS. SIMULATED S-PARAMETERS OF THE AMPLIFIER AT 24.2 GHz

	Material A		Material B with underfill	
$I_{d TOT}$ [mA]	40		30	
	measured	simulated	measured	simulated
S_{11} [dB]	-19.4	-18	-10.3	-19.9
S_{21} [dB]	17	17.5	13.9	15.6
S_{22} [dB]	-18.8	-37	-14.6	-20.5

The simulated performance of the amplifier with three underfill overflow profiles for both the input and the output PCB matching networks are presented in Fig. 5. As the profile changes, the characteristics of the amplifier change too, but remain within satisfactory tolerances in all cases.

The measured S-parameters of the amplifier on the substrate material with the underfill applied are presented in Fig. 6 while the values at 24.2 GHz are listed in Table II. From Fig. 6 it is seen that the measured gain is different from the simulated one, which is the consequence of neglecting the underfill loss. The small shift in frequency seen is attributed to the particular shape properties of the overflow profile of a measured sample. Nevertheless, the performance of the amplifier is good considering the choice of the PCB material.

For both materials considered the simulated and measured performance is not significantly different from the *intrinsic* amplifier performance shown in Fig. 3, except for the reduction in the gain due to the substrate and underfill loss, showing the good quality of the designed PCB matching networks and validating the distributed matching concept.

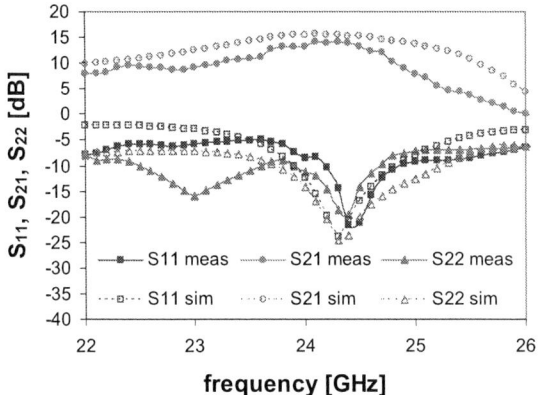

Fig. 6 Measured and simulated results for the amplifier mounted on the PCB substrate using material B and underfill. Total drain current Id $_{TOT}$=30mA

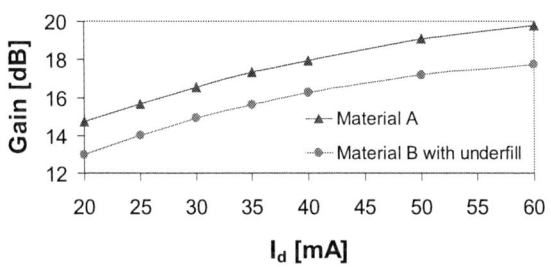

Fig. 7 Measured amplifier gain vs. total drain current for the two substrate materials used

The results presented in Fig. 4 and Fig. 6 are measured with the test PCBs mounted in a jig. Since the lines used for connections in the jig were considerably longer than those required in the actual front end system, the final amplifier gain performances were adjusted accordingly. The results of the amplifier gain vs. total drain current are shown on Fig. 7. For the purpose of this particular test, the amplifier was biased so that the total drain current was equally divided between the

stages. For all the drain currents, the tested amplifier stayed well matched with only minor change in values of S_{11} and S_{22}.

The amplifier has been used in both, the receiver and the transmitter blocks, so the design was not optimized for noise. As the extrapolation of the transistor noise parameters to the frequencies of interest was found insufficiently accurate, the simulated values are predicting worse noise performance than the one measured. Based on subsequent system measurement results, the estimated noise figure of the amplifier when biased with 15mA drain current in both stages was around 3.5 dB when using material A.

IV. CONCLUSIONS

Multilayer LTCC environment provides an attractive solution for the realization of a 24 GHz amplifier with packaged devices. Reduced size and self-packaged properties have been achieved through the utilization of multilayer broadside coupled impedance transformers to realize all impedance matching networks. The distributed matching concept adopted with distinct on-tile and off-tile parts enables easy implementation of the amplifier on various substrates with very small additional area of the PCB substrate occupied by the matching circuit.

The amplifier concept was designed and evaluated on two different substrate materials chosen as opposite cases of the cost vs. performance trade-off. The measured characteristics of the amplifier are good and in agreement with simulations..

REFERENCES

[1] V. Cojocaru, V. Napijalo, D. Humphrey, and B. Clarke, "24GHz Low-cost UWB front-end design for short range pulse radar applications," *Proc. EuMC 2004*, paper 40.2, p. 1273.

[2] V. Cojocaru, et. al., "A 24GHz low-cost, long-range, narrow-band, monopulse radar front end system for automotive ACC applications", *Proc. MTT-S 2007*, 2007, paper TH1C.6, p. 1327.

[3] G. Gonzalez, *Microwave Transistor Amplifiers - Analysis and Design*, 2nd ed., Prentice Hall: New Jersey, 1997.

[4] G. D. Vendelin, A. M. Pavio and. U. L. Rohde, *Microwave Circuit Design Using Linear and Nonlinear Techniques*, J. Wiley & Sons: New York, 1990.

[5] V. Napijalo, V. Cojocaru, T. Yokoyama, and T. Young, "Accurate substrate-related packaged transistor modeling for 24GHz circuit design", *Proc EuMC 2006*, 2006, paper 28.3, p. 522.

[6] A. Darwish, A. Ezzeddine, M. Mah, and J. Cook, "Inexpensive X-band ½ watt PA using 3D LTCC technology", *Proc. MTT-S 2004*, 2004, paper WEIF.45, p. 1205.

[7] S. A. Wartenberg, and Q. H. Liu,, "A coaxial-to-microstrip transition for multilayer substrates," *IEEE Trans. Microwave Theory & Tech.*, vol. 52, no. 2, pp. 584-588, Feb. 2004.

[8] J. Heyen, A. Gordiyenko, P. Heide, and A. F. Jacob, "Vertical feedthroughs for millimeter-wave LTCC modules", *Proc. EuMC2003*, 2003, paper 7.1, p. 411.

[9] *CST Microwave Studio 2006*, Darmstadt, Germany: CST GmbH, 2006.

Proceedings of the 3rd European Microwave Integrated Circuits Conference

K-Band Frequency Synthesizer with Subharmonic Signal Generation and LTCC Frequency Tripler

Torben Baras and Arne F. Jacob

Institut für Hochfrequenztechnik - Tech. Univ. Hamburg-Harburg
21073 Hamburg, Germany, Tel: +49 40 42878 3370, Fax: +49 40 42878 2755
t.baras@tuhh.de

Abstract— A subsystem implementation study of a fractional-N frequency synthesizer for K-band satellite communications is presented. The architecture is based on a subharmonic frequency generation implemented with commercially available voltage-controlled oscillator, phase-locked loop chip and a custom designed frequency tripler implemented in low temperature co-fired ceramics with vertical integration techniques. As a key component, the frequency multiplier is analyzed in terms of conversion efficiency and suppression of undesired harmonic signals. With the entire subsystem, a reference-stabilized signal in the range of 19...21 GHz with phase noise of -80 dBc/Hz at 10 kHz offset and a 25 kHz channel spacing is achieved. The output power is -1 dBm and the spectral purity better than 25 dBc over the entire range. With the presented approach it is demonstrated that the employed technology can be used to successfully design customized, packaged modules and subsystems in the upper microwave frequency range.

I. INTRODUCTION

Tomorrow's communication systems do not only require innovative circuit and system designs, but also depend on a developing technology to implement them. One possible approach is the vertical integration utilizing multilayer substrates. This way, high integration levels are achievable helping to reduce volume and weight of components.
LTCC (low temperature co-fired ceramics) is one promising technology commonly known from applications employing frequencies of a few GHz [1], [2]. Within the scope of a government funded research project, we pursue the implementation and verification of highly integrated components for satellite applications at K-Band frequencies for upcoming multimedia services using this ceramic multilayer technology. In the past we could already demonstrate the successful realization of broadband, double-balanced mixers [3]. Towards the target of realizing LTCC-based communication systems, a compact signal generator with low phase noise is required. The design and implementation of such a subsystem is discussed in the following.

II. SYNTHESIZER ARCHITECTURE

The general architecture of a synthesizer is depicted in Fig. 1(a). The voltage controlled oscillator (VCO) generates the signal at the frequency f_{out}/N. One part of the signal is fed into a multiplier circuit that performs the required multiplication by a factor of N to achieve f_{out} at the output of the subsystem. A second part is applied to a prescaler to divide

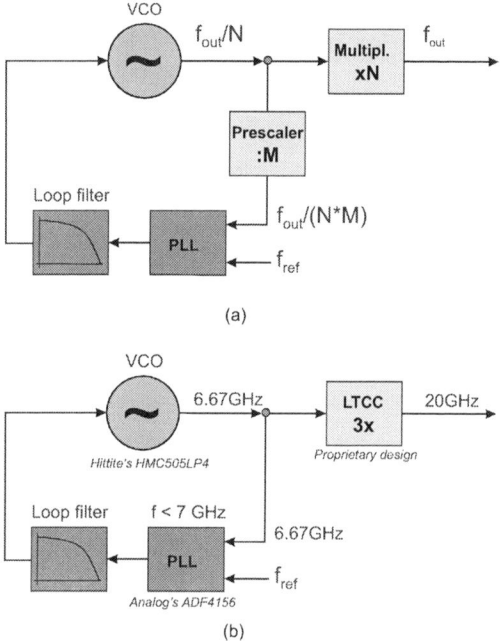

Fig. 1. (a) General concept of stabilized frequency generation. (b) Simplified concept omitting the prescaler for generating a 20 GHz signal.

the signal to a frequency that can be processed by the phase-locked loop (PLL) circuit. The loop is closed by connecting the charge pump output of the PLL via a low pass filter with the tuning input of the VCO.
For our target application a stable signal at $f_{out} = 20$ GHz is required. Recently, integrated PLL circuits became available that can be operated up to 7 GHz at the RF input. Thus, the prescaler is omitted (M=1), the signal is generated at 6.67 GHz and solely a frequency tripler (N=3) is required to achieve the desired output frequency. At the same time power consumption and circuit complexity can be reduced, yielding the simplified architecture shown in Fig. 1(b). For the PLL chip we chose *Analog's ADF 4156*, a fractional-N type integrated circuit. Even though this particular device is specified up to 6 GHz, it has sufficient RF sensitivity to be operated up to 7 GHz as per manufacturer's datasheet. Due to numerous commercial applications, a large variety of VCOs up to 8 GHz is available, as e.g. *Hittite's HMC505LP4*. This device is specified with an output frequency between 6.8 ... 7.4 GHz and an output

978-2-8748-7007-1/08 $25.00 © 2008 EuMA

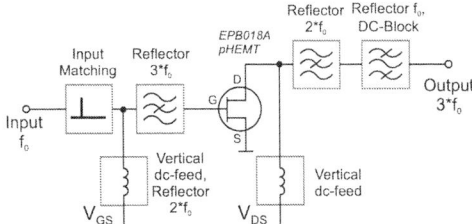

Fig. 2.   Topology of the custom designed LTCC tripler module.

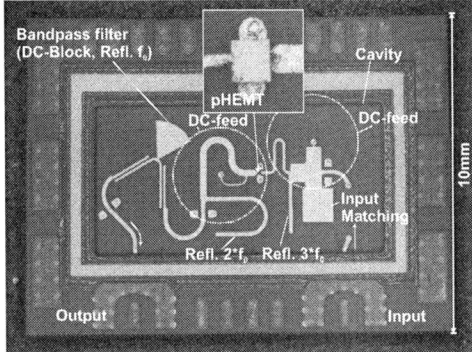

Fig. 3.   Bottom view of the frequency tripler.

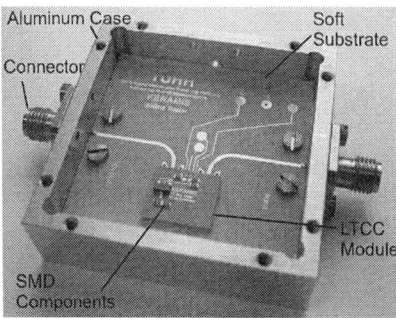

Fig. 4.   Mounted tripler module in an Aluminum case for test purposes.

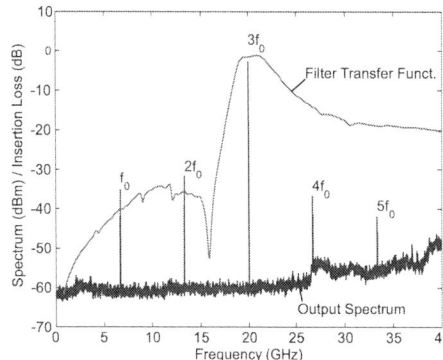

Fig. 5.   Measured output spectrum of the tripler module at $f_0 = 6.67\,\mathrm{GHz}$ and overlayed measured transfer function of the output filter.

power of $11\,\mathrm{dBm}$. The last component required to complete the subsystem is the frequency tripler. This device is realized as a custom designed LTCC module with vertical integration techniques, the circuit design and philosophy being described next.

III. LTCC Frequency Tripler

The topology of the designed tripler module is shown in Fig. 2. It is based on the method of reflector networks to realize idler circuits [4] and contains a single discrete transistor operated in a strongly nonlinear mode. The input comprises a matching network at the driving frequency f_0 followed by a stopband filter for the third tone. A filter for the second tone, which is generated due to the nonlinearities in the active device is omitted at the input port, since the dc-feed provides this function simultaneously. In our implementation, the bias networks were not realized as planar structures but vertical feeds, a detailed study being presented in [5]. They are buried in the inner layers of the module. The passive input network is followed by a discrete pHEMT (*Excelics' EPB018A*) in Source configuration. The output network at the Drain output terminal is composed of a cascade of the reflector for $2f_0$ and a passband filter serving for multiple purposes. The signal at f_0 is reflected with proper phase, the leaking harmonic at $2f_0$ is further suppressed, and the bandpass blocks the dc from the drain terminal.

The bottom view of the realized module with a size of $10\,\mathrm{mm} \times 14\,\mathrm{mm}$ is depicted in Fig. 3. The manufacturing of the LTCC circuit was carried out by our partner *MicroSystems Engineering GmbH*, Berg, Germany. The foundry process

offers feature sizes as low as $50\,\mu\mathrm{m}$, multi-step cavities to host the bare dies and advanced features such as embedded high-k dielectrics. The multilayer substrate consists of eight dielectric layers of *DuPont's 951A2* tape with a fired layer thickness of $130\,\mu\mathrm{m}$. Previously, we could demonstrate that this particular process delivers very precise and reliable circuits with low manufacturing variations [6].

The pHEMT with a size of $320\,\mu\mathrm{m} \times 400\,\mu\mathrm{m}$ is mounted in the cavity using thermo-compression flip-chip attachment (see inset in Fig. 3). The planar part of the circuit is realized in a microstrip environment occupying one dielectric layer, all lines being located in a cavity two layers deep. Dashed circles indicate the positions and dimensions of the vertical dc feeds. The microstrip circuit above thus can be folded to achieve a compact size. The input matching was realized by an area capacitance and a low impedance line. Wherever possible the open circuit stubs and delay lines are implemented as meanders and curves to efficiently occupy the area. To the output at the left side the bandpass filter is clearly visible. This element provides a passband at $20\,\mathrm{GHz}$ with $1.6\,\mathrm{dB}$ insertion loss and $35\,\mathrm{dB}$ suppression below $16\,\mathrm{GHz}$. Since the dc-feeds are buried below the planar circuit, the only visible part of them is the emerging vertical transition of the feed point. At the edge of the cavity on the bottom layer a brazing ring of gold in combination with a solder stop is printed. This way, the module can be hermetically sealed using AuSn solder and

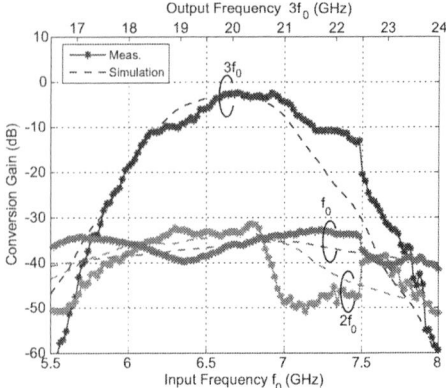

Fig. 6. Conversion gain vs. input frequency for the first three tones. The input drive level is 3 dBm.

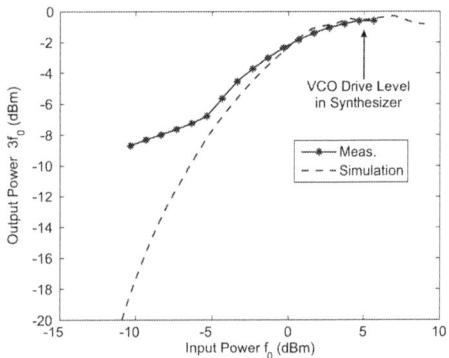

Fig. 7. Measured output power at $3f_0$ vs. input power drive at f_0 at 6.67 GHz input frequency.

a *Kovar* lid. The module is further equipped with pads to supply dc and RF signals. The RF transitions are a proprietary design capable of guiding signals in and out of the module up to 45 GHz [3]. This ensures the module's compatibility with surface mount technology.

For verifying the approach we have assembled and mounted several circuits in a housing as displayed in Fig. 4. The tripler is soldered on a soft-substrate (*Roger's RT/duriod 6002*). Due to the face-down module concept, the active bias circuitry could be conveniently placed on top of the module helping the miniaturization. The typical measured output spectrum in this environment is revealed in Fig. 5 together with the average measured output filter performance. The first two tones at 6.67 GHz and 13.33 GHz are greatly suppressed due to the high insertion loss of the filter. The harmonics at 26.67 GHz and 33.33 GHz exhibit even a lower power level. The measured conversion gain versus the input frequency with a fixed input power is depicted in Fig. 6 together with the simulation data. Here, we have omitted the fourth and fifth tone for reasons of clarity. In general a good agreement is seen between the curves apart from an instability occurring around 7.5 GHz. The differences are attributed to uncertainties in the pHEMT model rather than to manufacturing issues of the LTCC, the latter

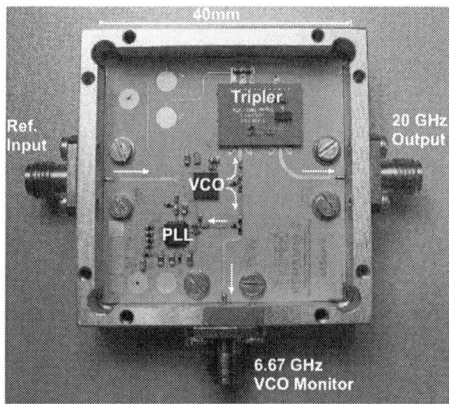

Fig. 8. Complete *K*-band fractional-N frequency synthesizer.

having proven to be reliable in the past.

To obtain the saturation characteristics, a sweep over the input power at a fixed input frequency of 6.67 GHz was performed and is presented in Fig.7. Also here, an agreement between simulation and measurement is seen. However, the circuit tends to oscillate at approximately 7 GHz with a low output power in the absence of an input signal. This oscillation can be injection locked with a driving generator already with low levels. For this reason the output power does not decline below -5 dBm input power levels as in the simulation. Fortunately, this unwanted behavior does not have a negative influence on the system operation, because the VCO provides sufficient output power to operate the tripler in the saturated region.

IV. FRACTIONAL-N SYNTHESIZER

In a final step we have integrated the tripler module together with the VCO and PLL chip on a circuit board as shown in Fig. 8. The VCO with a nominal output power of 11 dBm is connected to a resistive 6 dB-power splitter, one branch leading directly to the tripler yielding 5 dBm input drive power. The second branch is connected to another 6 dB-power splitter, delivering -1 dBm to the RF input of the PLL-chip and another -1 dBm to a monitor output. To the left side in this image, the input for the reference source as well as the digital inputs for configuring the PLL-chip are located. The tripler is placed in the upper half of the test fixture. It delivers the 20 GHz signal to the connector on the right side. In order to reduce the size of this synthesizer, the VCO and PLL could be integrated on the top surface of the tripler to realize a single subsystem-in-package.

The spectrum of the locked synthesizer is displayed in Fig. 9. The stable reference was a 56 MHz temperature compensated oscillator and the PLL was programmed to an integer value close to 6.67 GHz. The graph shows a very clean and low noise spectrum at both VCO monitor and tripler output. In a second step the synthesizer was programmed to a fractional value of 20/3360, showing typically occurring integer boundary spurs. The measured spectrum of this mode is provided in Fig. 10. The occurring spurs are in agreement with the values supplied in the manufacturer's datasheet of the PLL and decrease with

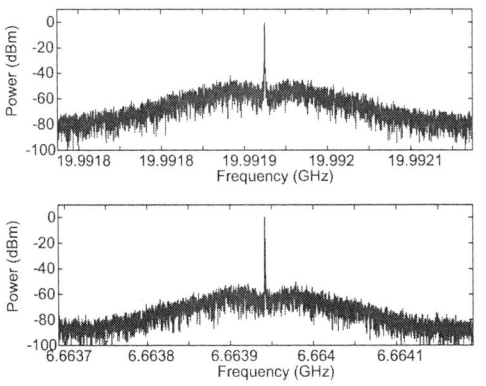

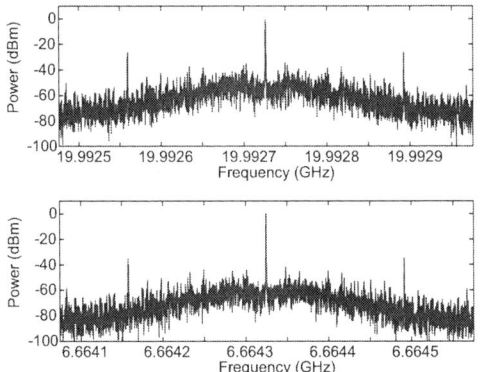

Fig. 9. Measured spectrum at the VCO output at 6.7 GHz and after the LTCC tripler at 20 GHz. The PLL was set to a value int=238, frac=0/3360.

Fig. 10. Measured spectrum at the VCO output at 6.7 GHz and after the LTCC tripler at 20 GHz. The PLL was set to a value int=238, frac=20/3360.

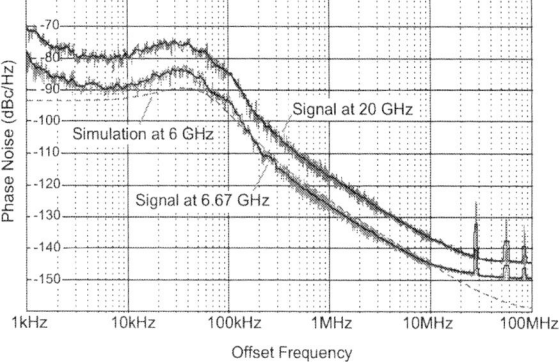

PHASE NOISE				
Settings		Residual Noise		Spot Noise [T1]
Signal Freq:	19.991976 GHz	Evaluation from 1 kHz to 100 MHz	1 kHz	-71.27 dBc/Hz
Signal Level:	-0.81 dBm	Residual PM	3.207 °	10 kHz	-79.99 dBc/Hz
		Residual FM	63.593 kHz	100 kHz	-84.72 dBc/Hz
		RMS Jitter	0.4455 ps	1 MHz	-117.33 dBc/Hz
Signal Freq:	6.663992 GHz	Evaluation from 1 kHz to 100 MHz	1 kHz	-77.93 dBc/Hz
Signal Level:	-0.03 dBm	Residual PM	1.167 °	10 kHz	-88.70 dBc/Hz
		Residual FM	32.421 kHz	100 kHz	-93.30 dBc/Hz
		RMS Jitter	0.4864 ps	1 MHz	-126.45 dBc/Hz

Fig. 11. Measured and simulated synthesizer phase noise at monitor output at 6.7 GHz and after the LTCC tripler at 20 GHz.

source was realized at 20 GHz entirely using surface mount technology and thus maintaining compatibility to high volume production needs. With this study it is demonstrated that LTCC technology can be pushed to design ready-packaged, custom designed components for applications in the upper microwave or even millimeter wave range. Future investigations will be concerned with the evaluation of other possible concepts such as fundamental frequency generation.

higher fractional values, which is common for this type of PLL. More important, the spectrum at 20 GHz shows a very similar performance with regard to this aspect.

Finally the phase noise is presented in Fig. 11, showing the performance at both frequencies. This parameter was measured using a *Rohde & Schwarz' FSQ40* signal analyzer with a phase noise utility. In addition, the simulation result of *ADIsimPLL* is shown in the graph. Unfortunately, the PLL could not be simulated at 6.7 GHz, since the software allows the highest frequency to be 6 GHz. Nevertheless, a good agreement is seen between the curves. The phase noise at 20 GHz is located approximately 10 dB higher than the phase noise at 6.67 GHz, being close to the theoretical value of 9.54 dB and thus showing that the LTCC frequency tripler provides very low intrinsic phase noise. As indicated in the tables above, the residual phase modulation amounts to 3.2° at 20 GHz, a negligible implementation loss for a communication system as reported in [7].

V. CONCLUSIONS

A subsystem study of a K-band fractional-N frequency synthesizer has been presented. The system is based on a stabilized signal generation at one third of the desired output frequency and subsequent tripling. A hybrid, low phase noise

ACKNOWLEDGEMENT

The authors wish to acknowledge funding and support of this work by the German Aerospace Center (DLR) on behalf of the German Federal Ministry of Economics and Technology (BMWi) under research contracts 50YB0314 and 50YB0623.

REFERENCES

[1] R. Lucero, W. Qutteneh, A. Pavio, D. Meyers, and J. Estes, "Design of an LTCC switch diplexer front-end module for GSM/DCS/PCS applications," in *Radio Frequency Integrated Circuits (RFIC) Symposium, proceedings,* Phoenix, AZ, USA, 2001, pp. 213 – 216.

[2] G.-A. Lee, M. Megahed, D. Agahi, and F. de Flaviis, "Embedded passives design and optimization for quadband EGSM single package RF module," in *Europ. Microw. Confer., Proceedings,* Amsterdam, The Netherlands, 2004, pp. 77 – 80.

[3] T. Baras, J. Mueller, and A. F. Jacob, "K-Band LTCC star mixer with broadband IF output network," *IEEE Trans. on Microw. Theor. and Techn.,* vol. 52, no. 12, pp. 2766–2771, Dec. 2007.

[4] Y. Iyama, A. Iida, T. Takagi, and S. Urasaki, "Second-harmonic reflector type high-gain FET frequency doubler operating in K-band," *Microwave Symposium Digest, 1989., IEEE MTT-S International,* vol. 3, pp. 1291–1294, 1989.

[5] T. Baras and A. F. Jacob, "Compact vertical bias feed networks for LTCC millimeter wave circuits," in *Europ. Microw. Conf., proceedings,* Manchester, Great Britain, September 2006, pp. 60–64.

[6] ——, "Design and manufacturing reliability of passive components for LTCC millimeterwave hybrid circuits," in *Europ. Microw. Conf., proceedings,* Munich, Germany, October 2007, pp. 660–663.

[7] S. Michelson, S. Fouche, N. Bornman, and W. V. Niekerk, "The design of a modular microwave frequency synthesiser: synthesisers for an 8 GHz and 23 GHz digital radio," in *4th Africon, Proceedings,* Sept. 1996, pp. 734–738.

978-2-8748-7007-1/08 $25.00 © 2008 EuMA

Proceedings of the 3rd European Microwave Integrated Circuits Conference

Demonstration of Heterogeneous Integration of Technologies for a Ku-Band SiP Doppler Radar

X. Sun[#1], S. Brebels[#1], S. Stoukatch[#1], R. Jansen[#1], L. Dussopt[#2],

M.-A. Dubois[#3], C. O'Mahony[#4], S. Berberich[#5], R. Houlihan[#4] and W. De Raedt[#1]

[#1] IMEC, HRFP group, Kapeldreef 75, 3001 Leuven, Belgium
[#2] CEA Grenoble, LETI/DCIS/SMOC, 17, rue des Martyrs, 38054 Grenoble, France
[#3] CSEM, RF & Piezo components, Microelectronics, SA Jaquet Droz 7, CH 2002 Neuchâtel, Switzerland
[#4] Tyndall National Institute, Lee Maltings, Cork, Ireland
[#5] Fraunhofer Institut Integrierte Systeme und Bauelementetechnologie, Schottkystr. 10, 91058 Erlangen, Germany

Abstract— **Heterogeneous integration of different technologies is increasingly required in microsystems integration. This entails the need of both technology development and new design methodologies. This work realized a high-performance Ku-band SiP Doppler radar demonstrator based on the combination of technologies and design methodologies.**

I. INTRODUCTION

Microsystems integration typically involves the use of several different technologies and this is particularly true for RF systems. Although there is a continuous evolution towards SoC (system on Chip) integration, obtaining different functionalities of the system requires different technologies. Therefore, the heterogeneous integration of different technologies becomes important and research on both technology development and on design methodologies is needed. The main aim of this work is to show the technological capability of integrating different technologies on one carrier substrate. This is demonstrated by a Doppler radar module based on integrating technologies and devices from the different partners (IMEC, LETI, CSEM, FHG, and Tyndall) in a single package.

This paper begins with the principle of the Doppler radar module in part II. The different technologies for realizing and assembling the radar module will be introduced in part III. In part IV, the individual heterogeneous functional blocks of the module will be explained; finally the actual functionality is confirmed by a test of the entire system in part V.

II. KU-BAND DOPPLER RADAR MODULE ARCHITECTURE

A basic diagram for a novel Doppler radar module is sketched in Fig.1, which is composed of a 13 GHz oscillator, a buffer amplifier, a 3 dB coupler, an antenna, a LNA, a mixer, a low-pass filter and a RF MEMS phase shifter. A forward RF signal is generated by the oscillator. After amplification, coupling, and phase shifting, it is transmitted through the antenna. By a moving target, the signal is reflected with a frequency shift due to the Doppler effect [1]. With the 3-dB directional coupler, the forward RF signal and Doppler-shifted reflected signal are separated and then down-converted to intermediate frequency (IF) on the mixer. Thus, the Doppler information is extracted through coherent demodulation.

Unlike traditional designs for a continuous wave (CW) Doppler radar, pulse modulation is introduced through a MEMS 45-dgree phase shifter to provide in-phase and quadrature modulation [2]. The benefit lies in two aspects: first, full information about direction, velocity, and displacement can be obtained. Second, pulse modulation greatly enhances Doppler detection sensitivity, especially for the case of a slow-moving target where flicker noise from the oscillator deteriorates the signal-to-noise ratio for Doppler signal detection.

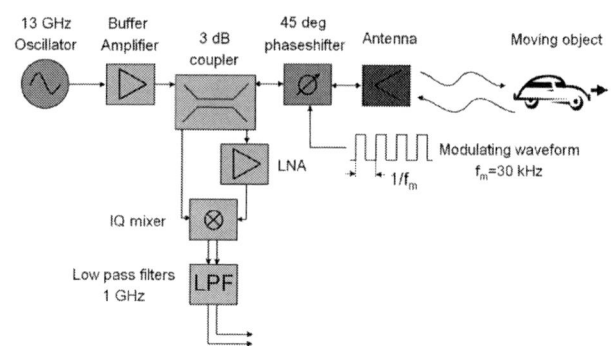

Fig. 1   Block diagram of the Doppler radar module

III. SiP BASED SYSTEM AND INTEGRATION TECHNOLOGIES

To integrate the entire Doppler radar functionality on a single carrier substrate, MCMD thin film substrates are chosen for their good performance in RF applications. As shown in Fig. 2, the thin film substrate consists of 725 μm high-resistivity silicon and two layers of BCB.

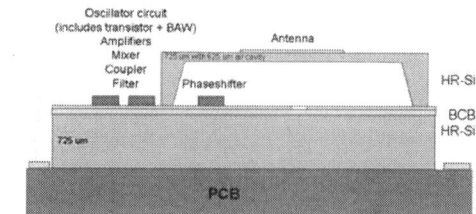

Fig. 2   Cross-sectional view of the thin film substrate and the different function blocks for Ku-band Doppler radar module

978-2-8748-7007-1/08 $25.00 © 2008 EuMA   470

Our Doppler radar demonstrator consists of five heterogeneous technologies integrated on the carrier substrate, which are represented by different colors in Fig. 1: the oscillator (orange) includes a flip-chipped PHEMT transistor and a wire-bonded BAW resonator; the LPF and coupler (green color) were integrated directly on the carrier substrate; the amplifier, LNA, and mixer (blue color) are commercial MMICs and were wire-bonded on the carrier substrate; the MEMS phase shifter (pink color) will be flip-chipped and the antenna (brown color) will be BCB-bonded to thin film substrate at low temperature; Fig. 3 shows a photograph of the first Doppler radar system realized on the thin-film carrier substrate. The HEMT transistor, BAW, MMICS (mixer, LNA, amplifier) are mounted in the indicated places. The details of each individual function block will be discussed separately in the following sections.

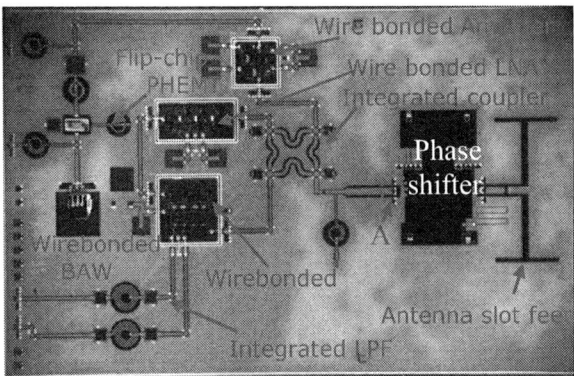

Fig. 3 Photograph of the entire Doppler radar module with different building blocks assembled on a thin-film substrate

IV. DOPPLER RADAR HETEROGENEOUS TECHNOLOGY BUILDING BLOCKS MOUNTED ON THE CARRIER SUBTRATE

A. Thin-film integrated passive components: LPF and coupler

The LPF and the coupler were directly integrated on the carrier substrate using low-loss passive components.

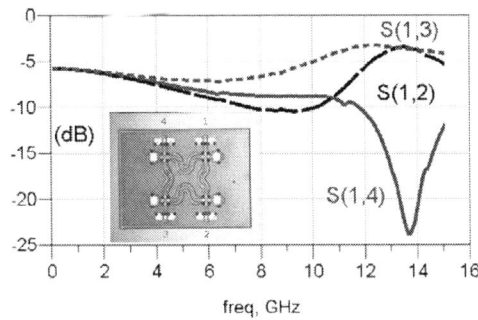

Fig. 4   Measured 3-dB coupler performance

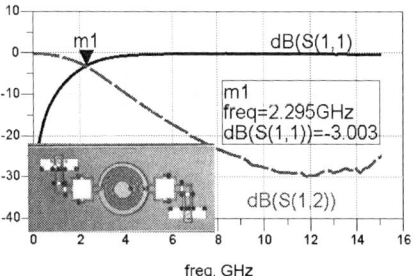

Fig.5 Measured low pass filter performance

A 13.4 GHz centered hybrid coupler (Fig. 4) and a 2.3 GHz low pass filter (Fig. 5) were realized: the isolation of the coupler is higher than 15 dB and the 3 dB power split was reached at a center frequency of 13.4 GHz.

B. Commercial MMIC components

1) Amplifier

An Avagotech AMMC-5618 (6-20 GHz Amplifier) was selected. The bare die is attached directly to a dedicated place on the metal plane of the carrier substrate using electrically conductive epoxy (Fig. 6). The backside of this MMIC chip is RF ground. Since the dedicated thin film metal plane is connected to ground, the backside of the chip is also connected to ground. The chip RF and DC signal pads are wire-bonded to the appropriate RF and DC signal pads on the substrate. The wire-bonds are kept as short as possible to reduce RF signal losses. The measurement (Fig. 7) demonstrates that the amplifier still shows good performance after mounting with the gain being close to that of the bare die amplifier.

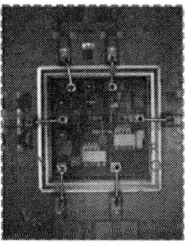

Fig. 6 Photograph of the amplifier after mounting and wire-bonding

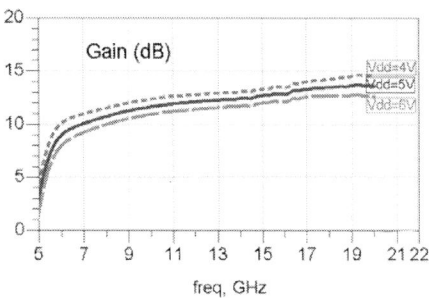

Fig. 7 Measured amplifier performance after mounting on the substrate

978-2-8748-7007-1/08 $25.00 © 2008 EuMA

2) LNA

The same assembly processing was applied to an Avagotech AMMC-6220 (6-20 GHz Low-Noise Amplifier) bare die chip. Fig. 8 shows the mounted and wirebonded LNA on the thin-film substrate.

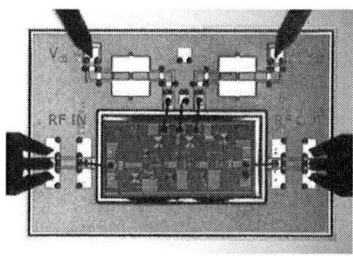

Fig.8 Photograph of the LNA after mounting and wire-bonding

3) Mixer

An Avagotech AMMC-65300 (5-30 GHz image rejected mixer) was mounted and wire-bonded on the substrate (Fig. 9). A measured IF spectrum of 10 MHz with an output power of -21.93 dBm is shown in Fig. 10. The output power of the IF spectrum as a function of LO power is given in Fig. 11.

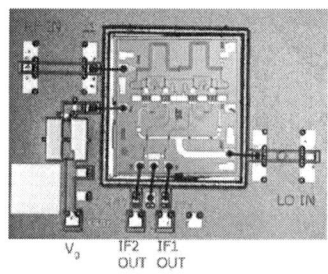

Fig. 9 Photograph of the mixer on the thin-film substrate after mounting and wire-bonding

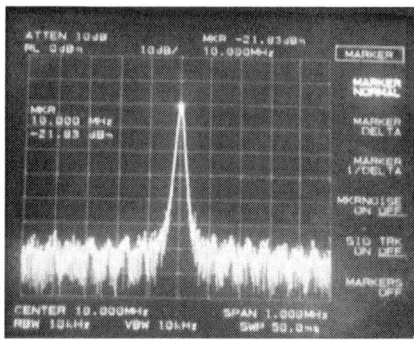

Fig.10 Typical measured IF spectrum (RF power -10 dBm, LO power 10 dBm, f_{RF}=13.4 GHz, f_{LO}=13.41 GHz) of the mixer.

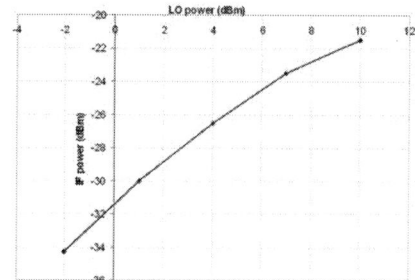

Fig.11 Measured IF power as a function of LO power.

C. Trench capacitor

Since the mixer in a Doppler radar has to operate at very low IF frequencies, a very stable bias is required. A 10 nF trench capacitor was wire-bonded onto the substrate for gate bias stabilization. Fig. 12 shows a top view and a SEM cross-sectional image of the capacitor. The trench capacitor is a 3D-type high-density capacitor, which leads to a capacitance increase by a factor of 15 as compared to planar capacitors of the same dielectric thickness.

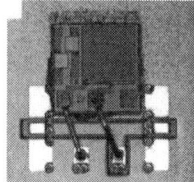 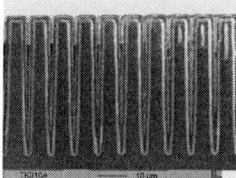

Fig.12 Top view (left) and cross-section SEM image (right) of a 3D trench capacitor

D. Microwave Oscillator

The design of the microwave oscillator (Fig.13) uses a commercial UMS2612 GaAs HEMT, which was flip-chipped on the carrier substrate. A BAW resonator was mounted and wire-bonded onto the substrate. The measurement in Fig.14 shows that the circuit oscillates at 13.44 GHz. The LO power is 1.17 dBm at 13.44 GHz.

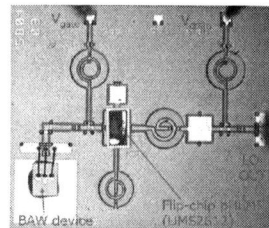

Fig.13 Photograph of the 13.4GHz oscillator

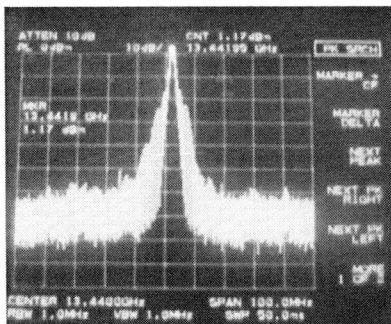

Fig. 14 Measured performance of the oscillator

V. Entire Doppler Radar Module Measurement

The Doppler radar module was next mounted and wire bonded to a test PCB. A flat cable connector was added to transmit all biases to the substrate (Fig. 15). Since the integrated patch antenna is not yet available, an external standard gain horn antenna was used in the test setup. Fig. 16 shows the detection of the Doppler signal when a metal plate was moved away from the horn antenna. From the analysis of the data, we obtain a speed of the metal plate of 0.37 m/s.

Fig.15 Photograph of the assembled Doppler radar module on PCB with the flat cable connector

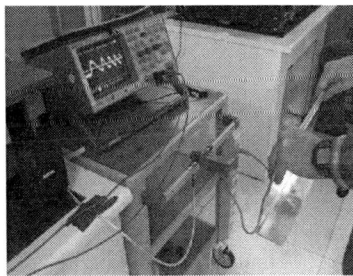

Fig.16 Detection of the Doppler signal from a moving metal plate

VI. Future work

Although the system is already fully functional, the phase shifter (Fig.17) and patch antenna (Fig.18) will be integrated into the system in the near future. The 0/45° two-state phase shifter will be based on MEMS switches. Two different de-

signs (loaded-line design and switched-line design) were fabricated. The loaded-line phase shifter uses tethered membrane switches (lower on-off capacitance ratio, lower actuation voltage, and faster switching) and is depicted in Fig. 17.

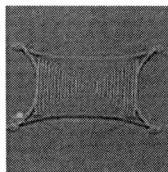 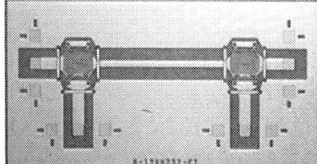

Fig. 17 Loaded-line phase shifter

Fig. 18 shows a schematic of the patch antenna, which was processed on a silicon cap (8x8x0.725mm, 100 µm Si, 625 µm air cavity patch). The patch (6 x 6 mm^2) is coupled to the circuit by a slot etched out of the thin film substrate. It will be attached to the carrier substrate by low-temperature BCB bonding.

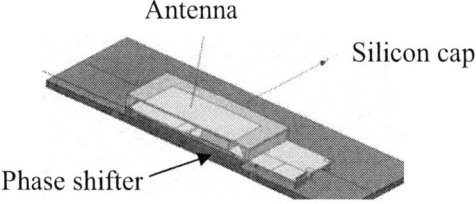

Fig. 18 Half view of the antenna on the thin-film substrate

VII. Conclusions

Heterogeneous integration of different technologies was demonstrated in a SiP Doppler radar system. The actual functionality is confirmed by a test of the assembled system. This established the technological capability of integrating wafers and chips from several partners into a working RF SiP. As a general technology, it can be applied to more extensive RF integrated applications.

Acknowledgment

This project MNT (RII3-CT-2004-506231) was funded by the European Community under the "Structuring the European Research Area" specific programme of the research infrastructure action. The authors wish to thank Nele Van Hoovels and Kristof Vaesen for their help with measurements.

References

[1] A. Droitcour, et al., " A microwaves Radio for Doppler Radar Sensing of. Vital Signs," *2001 IEEE MTT-S Digest,* vol. 1, pp. 175-178

[2] P. Heide, V. Magori and R. Schwarte, "Coded 24 GHz Doppler radar sensors: a new approach to high-precision vehicle position and ground-speed sensing in railway and automobile applications," Microwave *Symposium Digest, 1995.,* IEEE MTT-S international, PP. 965-968

A Wide Tuning Range MEMS Varactor Based on a Toggle Push-Pull Mechanism

Paola Farinelli[#1], Francesco Solazzi[*2], Carlos Calaza[*3], Benno Margesin[*4], Roberto Sorrentino[**5]

[#]*RF Microtech, Via G. Duranti, 93, 06125, Perugia, Italy*
[1] farinelli@rfmicrotech.com

[*]*MEMSRaD Research Unit, FBK-irst, Via Somarive 14, 38050 Trento, Italy*
[2] solazzi@fbk.eu
[3] carlos@fbk.eu
[4] margesin@fbk.eu

[**]*University of Perugia, Dept. of Electronic and Information Engineering, Via G. Duranti, 93, 06125, Italy*
[5] sorrentino@diei.unipg.it

Abstract — **This paper presents a novel wide tuning range MEMS varactor based on a toggle push – pull mechanism for high RF power applications and improved reliability. The device anchoring utilizes a torsion spring mechanism which virtually allows for a full capacitance tuning range. Improved mechanical stability is also provided by the actively controlled pull-out implementation that is realized without increasing the MEMS manufacturing complexity. As a proof of concept, a toggle MEMS varactor has been modeled, designed and manufactured in shunt configuration on a 50 Ω coplanar transmission line. Analytical and full wave electromechanical models are provided as well as electromagnetic characterization. The device has been manufactured on HR Silicon substrate by using the standard FBK-irst RF MEMS process. Optical profile, DC and RF measurements are presented in the 0-40 GHz frequency band. Excellent RF performance as well as a capacitance tuning ratio of 2.5 has been obtained.**

I. INTRODUCTION

The importance of RF MEMS technology in high value space and commercial telecommunication applications is demonstrated by the vast amount of literature published over the last years [1]. Superior electrical performance, low power consumption, high linearity and high level of integration are just a few of the numerous advantages that RF MEMS switches offer over solid-state devices. However high actuation voltage, low RF power handling capability, switching speed and reliability still limit their potential applications.

A key component for an extremely broad range of applications is the MEMS varactor. It is the basic element of a large number of tunable components such as voltage controlled oscillators (VCO), filters, matching networks, phase shifters, couplers and reconfigurable patch antennas [1]. Most MEMS varactors are based on a parallel plate configuration and use electrostatic actuation. Compared to digitally controlled

MEMS capacitive switches, MEMS varactors have the advantage of not suffering from sticking since the intimate contact between the actuation electrodes and the movable plate is avoided. In addition they enable a fine and continuous capacitance tuning with no need of large capacitance banks. The capacitance tuning range of such components, however, is often limited by the pull-in effect. Several solutions have been presented in the literature to enhance the tuning capability by properly modifying the mechanical design of the varactor, as shown in [2]. Still the control of the MEMS capacitance is often not fully predictable due to its high susceptibility to variations of the material stresses, temperature, external shocks and vibrations as well as self-biasing problems. To make the design more robust with respect to self-actuation and vibrations a third electrode (pull-out electrode) on top of the movable bridge has been realized in [3]. By using the pull-out electrode all switch states can be actively controlled and the mechanical stability of the device is increased. This often implies however a more complex fabrication process (double sacrificial layer process) and possible performance degradation due to additional RF parasitics. Alternatively the pull-out electrode can be realized by implementing a cantilever beam based on push-pull or toggle switch, as presented in [4]. However the stress management in cantilever fabrication design is very critical. Bending and torsion components can significantly affect the beam flatness [5].

This work presents a novel MEMS varactor based on a torsion spring structure applied at both beam anchors. This structure provides active actuation for all switch state commutations by using a simple single sacrificial layer process. In addition it allows in principle a full tuning range since the central part of the bridge can be completely lowered down to the RF line without pull-in. This is better explained in

Section III where the varactor electromechanical model is described. The electromagnetic model is briefly presented in Section IV whereas fabrication and measurement data are presented in Section V.

II. TOGGLE MEMS VARACTOR OPERATION

The device consists of a movable gold bridge (640 µm long and 90 µm wide) mechanically anchored by four identical torsion springs and electrically shorted to CPW ground (Fig.1). The bridge central area realizes the varactor top electrode (90 x 90 µm²) and is suspended 2.07 µm above the CPW central conductor. As depicted in the side view of Fig.1.b the pull-out electrodes (outer pair), electrically connected through a poly-silicon line, are biased to keep the beam clamped in the up state. When biasing the pull-out electrodes the bridge central part raises up and the up-state capacitance is reduced. In addition the pull out mechanism actively keeps the bridge in the up state allowing a higher robustness to external vibrations and self-actuation.

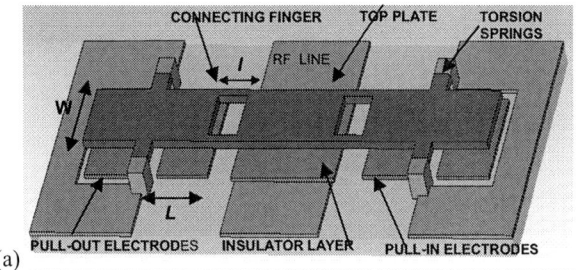

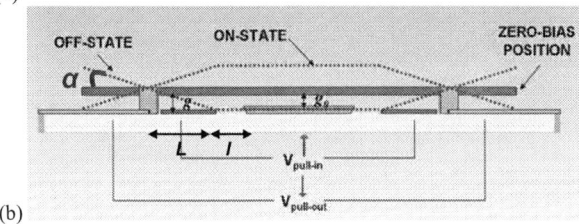

Fig. 1 Schematic of the toggle MEMS varactor operation. Prospective view (a). Lateral view (b).

Conversely the pull-in electrodes (inner pair) are biased to a voltage lower than the pull-in threshold in order to pull down the beam central part thus increasing the bridge capacitance. As shown in the next section, a proper design can make the bridge central part touch the RF line (covered with 100 nm thick Low Temperature Oxide (LTO) for safety) at applied voltages smaller than the pull-in threshold voltage. Consequently the tuning range of the varactor is virtually arbitrarily high. In the following section an analytical expression for the actuation voltage is derived as well as the ANSYS® electromechanical model.

III. ELECTRO-MECHANICAL MODEL

In a first step the toggle varactor has been modelled and designed by using the analytical equations described in detail in [6]. In this Section only the most relevant formulas are reported. The bridge snaps down when the angle α between the pull-in electrode and the membrane is higher than the limit value of $\alpha_{pull-in}$. For the present structure $\alpha_{pull-in}$ was found to be equal to:

$$\alpha_{pull-in} = 0.44949 \cdot \frac{g}{L} \tag{1}$$

where g is the initial air gap between the membrane and the electrodes and L is the length of the pull-in electrode.

The pull-in voltage is given by the following expression:

$$V_{pi} = \sqrt{\frac{E}{2.3914\varepsilon_0 W}\left(\frac{g}{L}\right)^3\left(\frac{0.33}{1-v}\frac{b_t h_t^3}{l_t} + \frac{L}{l^2}\frac{bh^3}{6}\right)} \tag{2}$$

where h_t, b_t and l_t are the thickness, width and length of the torsion spring and h, b and l are the thickness, width and length of the connection fingers between the rigid plates. E and v are the Young modulus and the Poisson ratio of the gold membrane, W and L are the width and length of the pull-in electrode. The maximum displacement of the membrane before the pull-in is given by:

$$H_{max} = \left(L + \frac{1}{3}l\right)\alpha_{pull-in} = \left(L + \frac{1}{3}l\right)0.44949\frac{g}{L} \tag{3}$$

Note that this simplified model does not take into account the non-linear behaviour of the torsion spring and the presence of a residual stress in the structural material. According to these expressions the present varactor exhibits a pull-in voltage of 68 V and a maximum beam displacement of 1.32 µm with respect to the zero bias position.

These results have later been verified by using the FEM simulator ANSYS®. The simulator has been utilized to compute the static deflection of the toggle structure caused by different actuation voltages, taking into account the initial deformation and rigidity of the structure due to residual stress of the materials. For the Young modulus and the residual stress of the gold membrane the measured values of 98.5 GPa and 180 MPa were adopted. The gold density was equal to 19.3 x 10^{-15} Kg/µm³. Non-linear stress-strain behaviour has been considered in order to take into account the gold plasticity [7]. Fig. 2a shows the toggle displacement just after the release of the structure, with the membrane about 300 nm warped due to the residual stress in the gold layer. The displacement just before pull-in is shown in Fig. 2b.

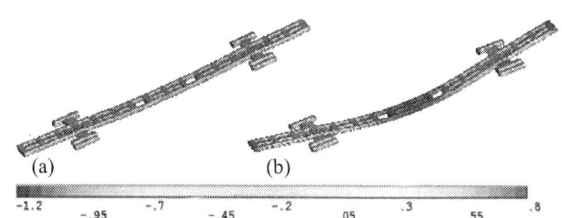

Fig. 2 Beam displacement (µm) after release (a) and in the down state just before the pull-in (b) (deformation scale factor: 25).

978-2-8748-7007-1/08 $25.00 © 2008 EuMA

According to this finite element analysis the toggle actuation voltage is 56 V and the maximum displacement before pull-in is 1.1 μm. These values do not perfectly agree with those provided by the analytical model since the latter did not account for the initial beam deformation due to residual stresses.

IV. ELECTROMAGNETIC MODELING

The analytical model has shown that the beam should be able to move symmetrically 1.32 μm upwards or downwards with respect to its zero bias height when biasing the pull-out or the pull-in electrodes respectively. The corresponding capacitance variation C_b can be calculated by the following expression [1]:

$$C_b = \frac{\varepsilon_0 A}{g_v}\left(1.25 - 0.20\frac{g_0 - g_v}{g_0}\right) \qquad (4)$$

where A is the area of the bridge top electrode, g_v is the variable bridge height, g_0 is the height at zero bias and ε_0 is the vacuum permittivity. Note that (4) takes into account fringing effects. The varactor is electromagnetically modelled as a variable capacitance in parallel to a 50 Ω coplanar line as shown in Fig. 3. The resulting theoretical capacitance tuning ratio is thus equal to 3.7. Such a simple model fits very well with the full-wave ADS Momentum® simulation of the device. Comparison between equivalent circuit and full wave simulations is shown in Fig. 4 as an example.

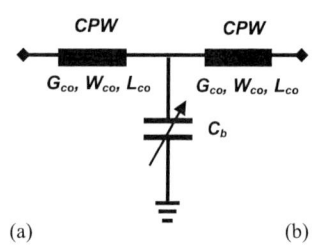

H_{sub}	525 μm
ε_r	11
W_{co}	160 μm
G_{co}	96 μm
L_{co}	605 μm
C_{b0}	42 fF
C_{bmin}	29 fF
C_{bmax}	108 fF

(a)   (b)

Fig. 3 Equivalent circuit of the toggle MEMS varactor (a). Equivalent circuit parameters (b).

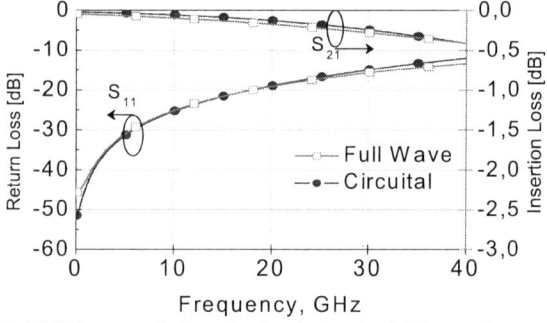

Fig. 4. ADS Momentum (red curve) and equivalent circuit (blue curve) simulations of the MEMS varactor at g_0 height.

V. EXPERIMENTAL RESULTS

Fig. 5 shows the MEMS toggle varactor manufactured on 525 μm HR Silicon substrate by using the FBK-irst MEMS process described in [8]. The bridge consists of a 1.8 μm thick electrodeposited gold layer locally reinforced with 3 μm thick gold to ensure high beam flatness [9]. Note that no additional process steps with respect to the FBK-irst standard MEMS process are needed.

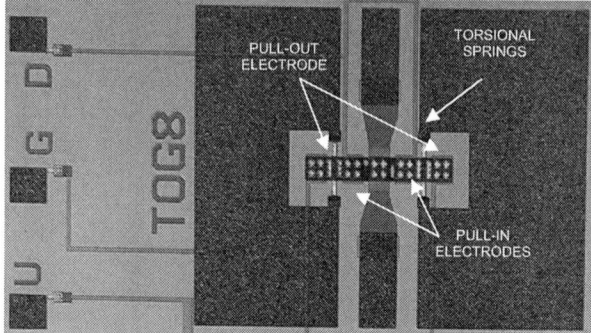

Fig. 5. Photo of the MEMS toggle varactor

An Optical profiler has been used to measure the actual beam height over the RF line as well as to verify possible out of plane bending or torsions of the beam due to residual stresses and stress gradients. Fig. 6 shows the profiler image of the manufactured device when no biasing is applied.

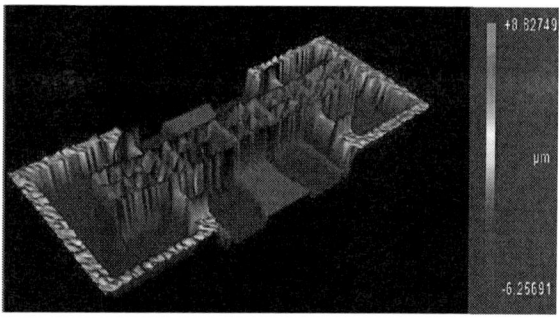

Fig. 6 Optical profiler image of the toggle MEMS varactor

As shown in the profiler image, the membrane ends are about 400 nm bended upwards at the position of the pull-out electrodes due to residual stress gradients. The central part of the beam however has a very flat shape and an actual bridge height of 2.07 μm was measured on top of the CPW central line as designed.

Both DC and RF characterizations have been performed for different voltages applied to the pull-out or to the pull-in electrodes in order to measure the MEMS capacitance tuning capability. The RF measurements in the 0-40 GHz frequency band are shown in Fig. 7. When biasing the pull-out electrodes the bridge central part raises up and the up-state capacitance is

978-2-8748-7007-1/08 $25.00 © 2008 EuMA

reduced. Excellent transmission performance is thus obtained, featuring a return loss better than 25 dB and an insertion loss better than 0.2 dB up to 20 GHz, and better than 0.5 dB up to 40 GHz, line loss included (1.21 mm long CPW line). Conversely, when biasing the pull-in electrodes central part of the beam is lowered down increasing the shunt capacitance. The beam collapses on top of the dielectric layer at a pull-in voltage higher than 35 V.

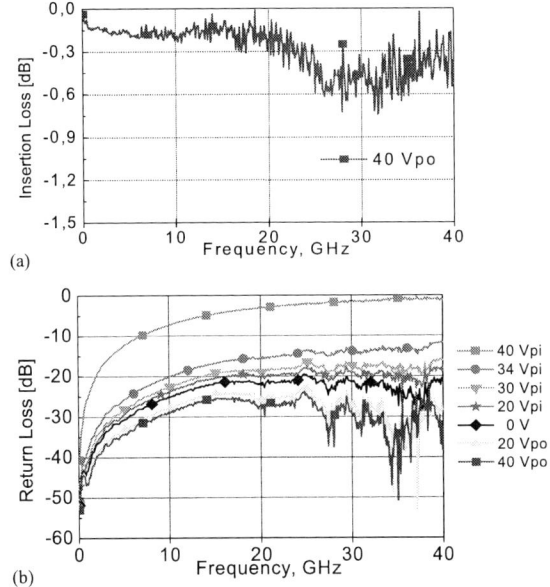

(a)

(b)

Fig. 7 Measured S-parameters of the toggle MEMS varactor. Insertion loss when 40V are applied on the pull-out electrodes (a) and return loss for different applied voltages (b).

Fig. 8 shows the comparison of the C-V characteristic obtained by DC and RF measurements for a 0-40 V voltage sweep on the pull-out electrode (left x-axis) and on the pull-in (right x-axis). The equivalent circuit of Fig. 3 was used to fit the RF capacitances. After subtracting the DC measurements capacitance offset, the DC and RF capacitance measurements are in a very good agreement and clearly show a pull-in voltage at V_{pi} = 35 V and a pull-out actuation voltage at V_{pu} = 38 V.

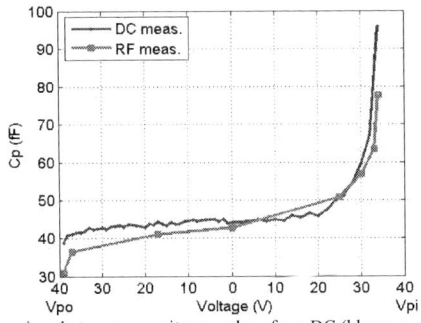

Fig. 8 Comparison between capacitance values from DC (blue curve) and RF (red curve) measurements

The toggle MEMS tunable capacitance can be varied from a minimum of 31.4 fF to a maximum of 77.6 fF resulting in a tuning ratio of 2.5. This value is lower than the theoretical one (3.7) because the connection the mechanical connection between the centre part and the actuation electrodes resulted to be too stiff. The manufactured membrane is able to move symmetrically 1 µm upwards or downwards with respect to its zero bias height instead of 1.32 µm as theoretically calculated.

VI. CONCLUSIONS

A novel wide tuning range MEMS varactor based on a toggle push – pull mechanism has been presented. The electromechanical and electromagnetic models showed the great potentialities of such a very low loss structure, which provides a high mechanical stability without increasing the manufacturing complexity. As a proof of concept a toggle MEMS varactor was manufactured showing insertion loss better than 0.5 dB in the 0–40 GHz frequency band and a capacitance tuning ratio of 2.5.

ACKNOWLEDGEMENT

This work has been developed and supported in the framework of the European Network of Excellence AMICOM.

REFERENCES

[1] W. G. M. Rebeiz " RF MEMS, Theory, Design and Technology", Ed. Wiley Interscience, 2003.

[2] Y.Lu, L..P.B.Katehi, D.Peroulis "High-Power MEMS Varactors and Impedance Tuners for Millimeter-Wave Applications" *IEEE Trans. On Microwave Theory and Techniques*, Vol. 52, No. 11, Nov. 2005

[3] D. Peroulis, S.P. Pacheco, L.P.B. Katehi, "RF MEMS Switches With Enhanced Power-Handling Capabilities," *IEEE Trans on Microwave Theory and Techniques*, Vol. 52, No.1, pp.59-68 Jan 2004

[4] W. Simon, et Al., "Toggle switch: investigations of an RF MEMS switch for power applications," *IEEE Proc.-Microw. Antennas Propag.*, Vol. 152, No. 5, October 2005, pp. 378-384

[5] C. Venkatesh, P. Shashidhar, N. Bhat, R. Pratap, "A torsional MEMS varactor with wide dynamic range and low actuation voltage". *Sensors and Actuators A: Physical* 121(2):pp. 480-487, Aug. 2005.

[6] K. J. Rangra, B. Margesin, L. Lorenzelli, F. Giacomozzi C. Collini* M Zen, G. Soncini, L del Tin, R. Gaddi, "Symmetric Toggle Switch - A New type of RF MEMS Switch for Telecommunication Applications: Design and fabrication", *Sensors and Actuators* Vol. 123-124 pp. 505-514, Sept. 2005

[7] Margesin, B.; Bagolini, A.; Guamieri, I.; Giacomozzi, F.; Faes, A. "Stress characterization of electroplated gold layers for low temperature surface micromachining" Design, Test, Integration & Packaging of MEMS-MOEMS, Cannes (France), May 2003

[8] Ocera, A.; Farinelli, P.; Cherubini, F.; Mezzanotte, P.; Sorrentino, R.; Margesin, B.; Giacomozzi, F.; , "A MEMS-Reconfigurable Power Divider on High Resistivity Silicon Substrate", International Microwave Symposium IEEE/MTT-S 2007, 3-8 June 2007 pp.501 – 504

[9] P.Farinelli, B. Margesin, F. Giacomozzi, P. Rantakari, T. Vähä-Heikkilä, "Continuously Tunable Millimeter Wave MEMS Phase Shifter", MINT-MIS2007, the 8th MINT Millimeter-Wave International Symposium, Seoul (Korea) 26-27 Feb. 2007

Proceedings of the 3rd European Microwave Integrated Circuits Conference

Phase Shifter Design Based on Fast RF MEMS Switched Capacitors

Benjamin Lacroix [1], Arnaud Pothier [2], Aurelian Crunteanu [3], Pierre Blondy [4]

XLIM UMR 6172 - Université de Limoges / CNRS
123 avenue Albert Thomas 87060 Limoges cedex FRANCE
[1]lacroix@xlim.fr
[2]pothier@xlim.fr
[3]crunteanu@xlim.fr
[4]pblondy@xlim.fr

Abstract— This paper presents the design and fabrication of fast DMTL RF MEMS phase shifters. Distributed MEMS Transmission Lines are being used with miniature RF MEMS switched capacitors (40x40 μm^2), actuating at 25 V with a switching time around 1 μs. Both 90 and 180 degree phase shifters presented here operate at 20 GHz, are respectively less than 4.5 mm and 8.5 mm long. They are designed with 6 and 12 unit cells to achieve the desired phase shift. Measured return loss is respectively better than -13 dB and -11 dB for the 90 and the 180 degree phase shifters, and insertion loss is respectively less than 0.8 dB and 1.8 dB at 20 GHz.

I. INTRODUCTION

The design and optimization of Distributed MEMS Transmission Lines have been widely explained in the past few years [1]–[5]. DMTL may be intensively used for microwave systems such as phased array antennas [6], broadcasting systems [7], matching networks [8] [9] and impedance tuners [10], or tunable filters applications [11]. The principle is to periodically load a coplanar transmission line (CPW), using MEMS air bridges between ground planes. When MEMS capacitors are actuated, a phase shift is obtained since the wave velocity of the CPW line is decreased.

However, when fast phase shifting is required, MEMS cannot achieve fast switching times as semiconductors components like FET or PIN diodes, even if they don't offer as good performance such as RF MEMS.

Sub-microsecond fast switching times have been demonstrated by miniature RF MEMS switched capacitors [12]–[14]. This work presents a topology of two 90 and 180 degree DMTL phase shifters loaded with miniature MEMS capacitors actuating at 25 V with a measured switching time of 1.2 μs.

II. MINIATURE RF MEMS SWITCHED CAPACITORS

A. Miniature RF MEMS and switching speed

A miniature RF MEMS switched capacitor has been designed and fabricated. Fig. 1 presents a photograph of two switched capacitors implemented on a coplanar waveguide line and a cross-view section of the capacitor. These beams are 40x40 μm^2 and 350 nm thick, and are suspended 1 μm above a contact electrode covered by a 400 nm thick alumina (Al_2O_3) dielectric layer.

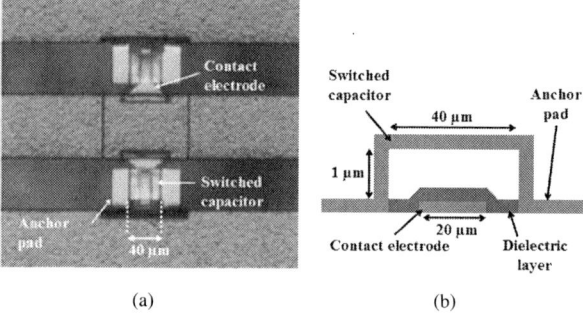

Fig. 1. Miniature RF MEMS switched capacitors (40x40 μm^2): (a) photograph and (a) cross-view section

As shown in [13], one way to achieve submicrosecond switching times for RF MEMS switched capacitors is to miniaturize their mechanical structure. Miniature beams are less senstive to temperature variation and charging in the dielectric layer since their spring constants around 10 times higher than standard beams ensure high pull-up pressure. They are also less sensitive to residual stress. Indeed, for a fixed-fixed beam, the spring constant k is the sum of a component k' mainly depending on the bridge geometry and a component k'' relying on the residual stress induced by the fabrication process. Fig. 2 presents the weight of k' and k'' for 10 μm wide and 0.35 μm thick bridges, with a residual stress of 10 MPa. The residual stress component is only 25% of the total spring constant for 40 μm long beams whereas it is 90% for 300 μm long beams. Miniature structures can also achieve fast switching times as shown by Fig. 3, where a 40 μm long bridge is in theory able to switch around 700 ns.

B. Switching time measurement

The switching time of these miniature MEMS RF switched capacitors has been measured with a specific bench, presented in [13] and [15]. The principle consists in detecting an amplitude modulation of the continuous wave signal applied to the component since the impedance of the line is decreased when the bridges actuate. First, a 10 Hz monopolar square signal is applied to the bridge and is increased until the pull-

978-2-8748-7007-1/08 $25.00 © 2008 EuMA   478

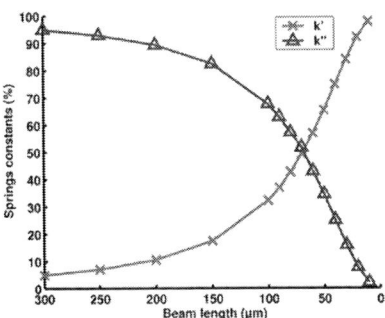

Fig. 2. Variation of the spring constant components

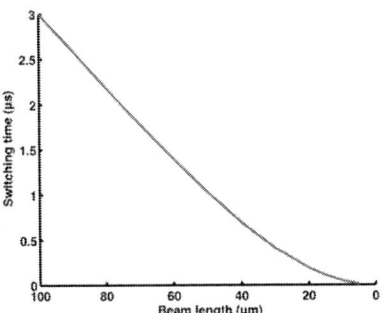

Fig. 3. Switching time versus bridge length

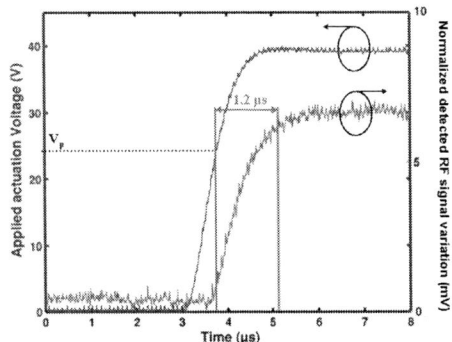

Fig. 4. Measured switching time of a unit cell

TABLE I
COPLANAR WAVEGUIDE PARAMETERS WITH $Z_0 = 85\Omega$

W/G	80/60 μm
$\varepsilon_r/\varepsilon_{eff}$	3.78/2.39
RL_{max}	$-15\ dB$
f_b	46 GHz
s	660 μm
Z_u/Z_d	59.8/41.8
C_{up}/C_{down}	126/40.7 fF
$\Delta\phi$	15 deg.
n_{90}/n_{180}	6/12

The length s of a unit cell depends on the Bragg frequency f_b chosen by the designer and is given by Eq. 2:

$$s = \frac{Z_d c}{\pi f_b Z_0 \sqrt{\varepsilon_{eff}}} \qquad (2)$$

where :

- Z_d is the loaded-line impedance at the down-state position,
- c is the speed of light in vacuum,
- f_b is the Bragg frequency,
- ε_{eff} is the effective dielectric constant.

Table I presents the different parameters of the designed phase shifter operating at 20 GHz. The coplanar waveguide line is built on a quartz substrate ($\varepsilon_r = 3.78$, $tan\ \delta = 0.0009$) with a characteristic impedance Z_0 of 85 Ω ($W/G = 80\mu m/60\mu m$). The maximum return loss desired RL_{max} and the Bragg frequency f_b are chosen to be respectively $-15\ dB$ and 46 GHz, resulting in a spacing s of 660 μm, found with Eq. 2. The up-state and down-state loaded-line impedances are 59.8 Ω and 41.8 Ω with up- and down- state capacitances C_{up} and C_{down} respectively of 40.7 fF and 126 fF for each unit cell, resulting in a phase shift $\Delta\phi$ of 15 degrees, calculated with Eq. 1. 6 and 12 cascaded unit cells are respectively required to achieve the total desired 90 and 180 degree phase shift.

A. Unit Cell

40 μm miniature bridges cannot link ground planes over the transmission line as in standard designs, due to their small

in voltage V_p is measured ($V_p = 24V$). Next, the signal amplitude is amplified to 40 V and the modulation induced by the actuation of the bridge is detected. The accuracy of the measured switching time depends on the microsecond range rise time on the applied actuation voltage induced by the low frequency amplifier. Fig. 4 shows a typical switching time measurement for our switched capacitor. We estimate the switching time is around 1.2 μs.

In the next part, we show how to implement these miniature RF MEMS switched capacitors to fabricate two 1-bit 90 and 180 degree Distributed MEMS Transmission Lines phase shifters.

III. DESIGN OF A FAST DMTL PHASE SHIFTER

The principle of the DMTL phase shifter presented here is to cascade n (6 or 12) unit cells each loaded by two fast miniature RF MEMS switched capacitors to provide an impedance transformation, allowing a phase shift along the total length of the line. The desired phase shift ϕ (90 or 180 degrees) is obtained by cascading $n.\Delta\phi$ phase shift unit cells, where $\Delta\phi$ is given by Eq. 1:

$$\Delta\phi = \frac{360 f s Z_0 \sqrt{\varepsilon_{eff}}}{c} \left(\frac{1}{Z_u} - \frac{1}{Z_d} \right) degrees/section \quad (1)$$

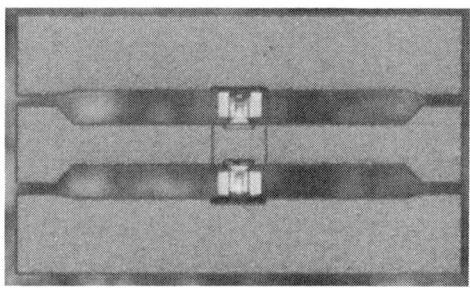

Fig. 5. The designed unit cell

Fig. 6. Picture of the designed 90 degree phase shifter (6 cells)

length. Fig. 5 shows a picture and a SEM photograph of the designed 660 μm long unit cell. Two miniature RF MEMS switches capacitors allow to achieve the desired up- and down-state capacitances. In down state, each capacitor contacts a 20 μm wide finger of the signal line covered by a dielectric layer. The CPW is terminated with 50 Ω tappers for probe pads. This unit cell is designed to achieve a 15 degree phase shift at 20 GHz.

B. 90 and 180 degrees sections

By cascading 6 unit cells, a 90 degree phase shift section is obtained, as shown Fig. 6. A 180 degree phase shift section is also obtained by cascading 12 unit cells. The dimensions of both 90 and 180 degree phase shift sections are respectively 4.43x0.49 mm^2 and 8.39x0.49 mm^2. All bridges are simultaneously actuated through the CPW transmission line, but 90 and 180 degree sections can be cascaded to create a multibit phase shifter.

IV. RF MEASUREMENTS

Insertion loss, return loss and phase shift measurements for both 90 and 180 degree phase shifters have been done with a HP 8722ES network analyzer using a Single-Open-Load-Thru calibration technique. The results are presented Fig. 7 and Fig. 8. For the 90 degree phase shifter, return loss less than -13 dB and a 95.3 degree phase shift have been measured at 20 GHz. Insertion loss is better than 0.8 dB at up- and down-state positions.

For the 180 degree phase shifter, return loss is less than -11 dB for a 173.5 degree measured phase shift at the operation frequency. Insertion loss is better than 1.8 dB at both states. For each phase shifter, miniature RF MEMS switched capacitors are actuated with a 25 V unipolar bias voltage.

V. CONCLUSIONS

We have demonstrated a novel topology of Distributed MEMS Transmission Lines phase shifters with miniature

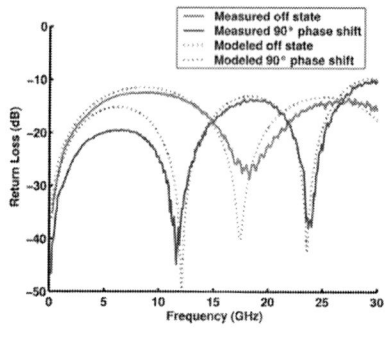

(a)

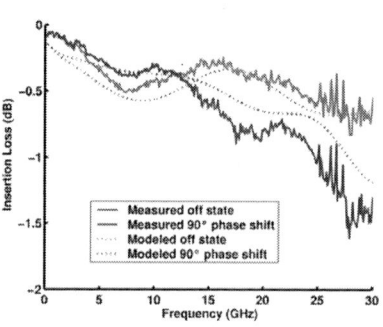

(b)

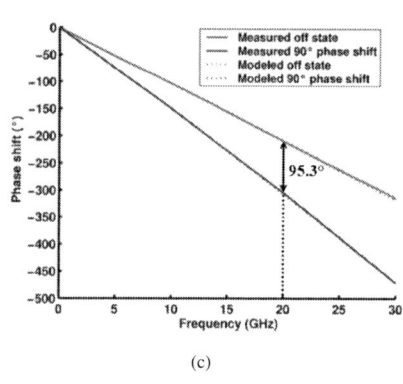

(c)

Fig. 7. Measured and modeled (a) return loss, (b) insertion loss, and (c) phase shift of the 90 degree phase shifter (6 cells)

switched RF MEMS capacitors actuating at 25 V with a measured switching time of 1.2 μs. We have fabricated two 90 and 180 degree 1-bit phase shifters and measured good return loss (respectively -13 dB and -11 dB) and insertion loss (respectively better than 0.8 dB and 1.8 dB). These 1-bit sections can be cascaded to obtain a multibit phase shifter. The principle of fast DMTL shown in this paper can also be applied to matching networks or impedance tuners.

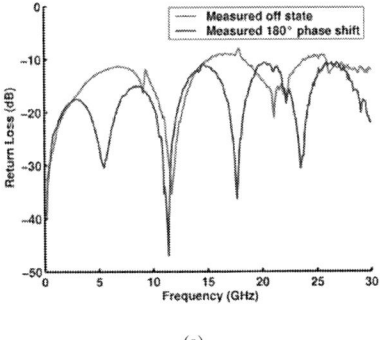

(a)

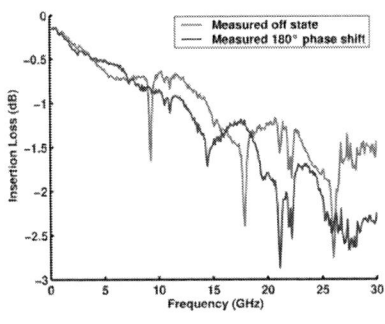

(b)

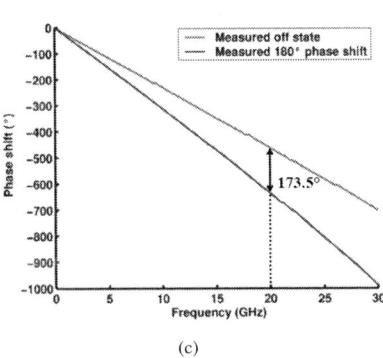

(c)

Fig. 8. Measured (a) return loss, (b) insertion loss, and (c) phase shift of the 180 degree phase shifter (12 cells)

REFERENCES

[1] N. S. Barker and G. M. Rebeiz, "Distributed MEMS True-Time Delay Phase Shifters and Wide-Band Switches," *IEEE Trans. on Microwave Theory and Techniques*, vol. 46, pp. 1881–1890, Nov 1998.

[2] N. Barker and G. Rebeiz, "Optimization of Distributed MEMS Phase Shifters," *IEEE MTT-S International Microwave Symposium Digest*, vol. 1, pp. 299–302, Jun 1999.

[3] G. M. Rebeiz, G. L. Tan, and J. S. Hayden, "RF MEMS Phase Shifters: Design and Applications," *IEEE Microwave Magazine*, vol. 3, pp. 72–81, Jun 2002.

[4] J. S. Hayden, A. Malczewski, J. Kleber, C. L. Goldsmith, and G. M. Rebeiz, "2 and 4-Bit DC-18 GHz Microstrip MEMS Distributed Phase Shifters," *IEEE MTT-S International Microwave Symposium Digest*, vol. 1, pp. 219–222, May 2001.

[5] G. L. Tan, R. E. Mihailovich, J. B. Hacker, J. B. DeNatale, and G. M. Rebeiz, "A Very-Low-Loss 2-Bit X-Band RF MEMS Phase Shifter," *IEEE MTT-S International Microwave Symposium Digest*, vol. 1, pp. 333–335, Jun 2002.

[6] Y. J. Ko, J. Y. Park, and J. U. Bu, "Integrated 3-Bit RF MEMS Phase Shifter With Constant Phase Shift for Active Phased Array Antennas in Satellite Broadcasting Systems," *12th International Conference on Transducers, Solid-State Sensors, Actuators ans Microsystems*, vol. 2, pp. 1788–1791, Jun 2003.

[7] Y. J. Ko, J. Y. Park, H. T. Kim, and J. U. Bu, "Integrated Five-Bit RF MEMS Phase Shifter for Satellite Broadcasting/Communication Systems," *IEEE The Sixteenth Annual International Conference on Micro Electro Mechanical Systems*, pp. 144–148, Jan 2003.

[8] T. Vähä-Heikkilä and G. M. Rebeiz, "A 20-50 GHz Reconfigurable Matching Network for Power Amplifier Applications," *IEEE MTT-S International Microwave Symposium Digest*, vol. 2, pp. 717–720, Jun 2006.

[9] S. Qin and N. S. Barker, "Distributed MEMS Tunable Matching Network Using Minimal-Contact RF-MEMS Varactors," *IEEE Trans. on Microwave Theory and Techniques*, vol. 54, pp. 2646–2658, Jun 2006.

[10] T. Vähä-Heikkilä, J. Varis, J. Tuovinen, and G. M. Rebeiz, "A 20-50 GHz RF MEMS single-stub impedance tuner," *IEEE Microwave and Wireless Components Letters*, vol. 15, pp. 205–207, Apr 2005.

[11] Y. Liu, A. Borgioli, A. Nagra, and R. A. York, "Distributed MEMS Transmission Lines for Tunable Filter Applications," *Int. J. RF Microwave CAE*, vol. 11, pp. 254–260, Aug 2001.

[12] D. Mercier, K. V. Caekenberghe, and G. M. Rebeiz, "Miniature RF MEMS switched capacitors," *IEEE MTT-S International Microwave Symposium Digest*, vol. 3, pp. 1931–1934, Jun 2005.

[13] B. Lacroix, A. Pothier, A. Crunteanu, and P. Blondy, "Sub-Microsecond RF MEMS Switched Capacitors," *IEEE Trans. on Microwave Theory and Techniques*, vol. 55, pp. 1314–1321, Jun 2007.

[14] B. Lakshminarayanan and G. M. Rebeiz, "High-Power High-Reliability Sub-Microsecond RF MEMS Switched Capacitors," *IEEE MTT-S International Microwave Symposium Digest*, pp. 1801–1804, Jun 2007.

[15] D. Mercier, A. Pothier, and P. Blondy, "Monitoring mechanical characteristics of MEMS switches with a microwave test bench," *4th round table on micro and nano technologies for space*, Jun 2003.

Proceedings of the 3rd European Microwave Integrated Circuits Conference

Microwave Tunable Bandpass Filter with MEMS Thermal Actuators

Siamak Fouladi, Winter Dong Yan, and Raafat R. Mansour

Centre for Integrated RF Engineering (CIRFE)

University of Waterloo

Waterloo, Ontario, N2L 3G1, Canada

siamakf@mems.uwaterloo.ca

Abstract— **This paper presents an approach to linearly tune the center frequency of a microwave bandpass filter using MEMS thermal actuators. A three-pole microwave tunable bandpass filter operating at Ka-band is designed, fabricated and tested. The fabricated filter has a center frequency at 33 GHz and a 14% relative bandwidth. The mid-band insertion loss is 2.7 dB and the return loss is better than 10 dB. A 15% continuous linear tuning range from 28 GHz to 33 GHz is achieved with a mid-band insertion loss better than 3.3 dB over the tuning range.**

I. INTRODUCTION

Low-loss, high tuning range microwave tunable filters are the key elements for multiband millimeter-wave systems. Digital and analog type microwave tunable filters have been previously reported using MEMS technology [1-3]. Ohmic contact MEMS switches and MEMS switched capacitor banks are well suited for the fabrication of digital type tunable filters. These filters can achieve low loss and discrete center frequencies with wide tuning range [4-5]. However, the main drawback of using contact switches for microwave tunable filters is their limited reliability for high power applications. High-performance MEMS varactors are used as the tuning elements of analog type tunable filters in the microwave band [6-7]. Continuous filter tuning with low insertion loss can be achieved by employing MEMS parallel-plate capacitors, but the maximum frequency shift is limited to 13.4% due to the restricted deflection range before pull-in. An analog type MEMS tunable bandpass filter is reported at V-band using MEMS variable capacitors [6]. The filter achieves 10% tuning at 65.5 GHz with a low insertion loss of 3.3 dB. The filter reported by Tamijani et al. [7] consists of half-wavelength coplanar waveguide resonators loaded with MEMS bridge capacitors and tunes over a 14% bandwidth from 18.6 to 21.4 GHz with a mid-band insertion loss better than 4.15 dB. In addition to the limited tuning range, in all the previously reported analog MEMS tunable filters, the maximum RF power handling is limited by the self-actuation of the varactors that utilize electrostatic actuators [8].

This paper reports experimental and simulation results on a Ka-band microwave tunable bandpass filter implemented using varactors with MEMS thermal actuators. The use of MEMS thermal actuators allows a continuous tuning with more linear and wider tuning range compared to the MEMS varactors with electrostatic actuation. The tuning elements have been designed so that the filter can handle relatively high RF power levels. The filter tunes over a 15% bandwidth from 28 to 33 GHz with an insertion loss better than 3.3 dB.

II. THREE-POLE FILTER DESIGN

Fig. 1 presents a schematic view and photograph of the three-pole filter. The filter consists of three half-wavelength coplanar waveguide (CPW) resonators capacitively coupled to each other. Capacitive coupling is used instead of magnetic coupling in order to reduce the insertion loss at higher microwave frequencies. As shown in Fig. 1, each resonator section is loaded with two interdigital MEMS varactors between the signal line and ground. This loading effect forms a slow-wave structure with higher effective propagation constant and hence reduces the overall filter size. Also by tuning these MEMS varactors, the effective electrical length of each resonator is adjusted and the filter center frequency can be tuned. The filter is fabricated using the MetalMUMPs® process. This process features a 20 μm thick electroplated nickel layer with 0.5 μm of gold on it as the primary structural layer of the filter.

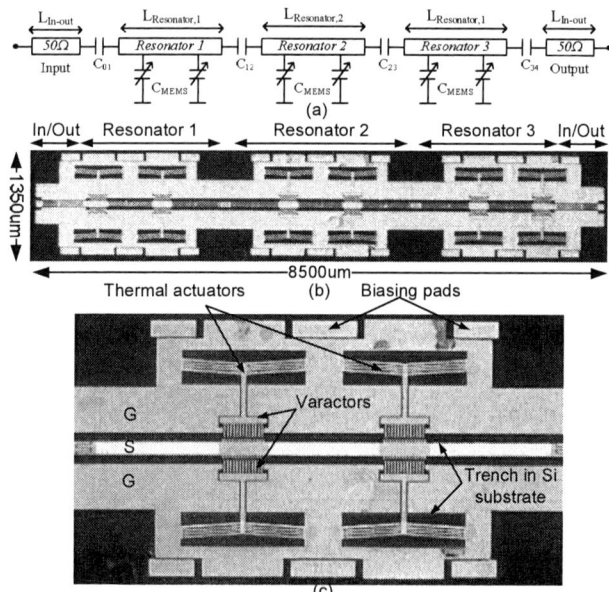

Fig. 1 (a) Filter schematic view, (b) photograph of the fabricated three-pole filter, (c) photograph of the resonator section.

978-2-8748-7007-1/08 $25.00 © 2008 EuMA 482

The main design parameters are the characteristic impedance of the CPW transmission line, physical length of each resonator section, inter-resonator and input/output coupling capacitors, and the initial value of MEMS varactors listed in Table I. The initial dimensions of the elements were calculated analytically and then verified using EM simulations. The dimensions are tuned to achieve the desired filter characteristics. The conductor loss and hence the mid-band insertion loss of the filter depends on the dimensions of the CPW lines. Based on EM simulation results for different dimensions of CPW lines and different characteristic impedances the minimum loss is obtained to be 0.49 dB/cm for a CPW line with dimensions of 60/80/60 μm (G/S/G) and a characteristic impedance of 77 Ω. The filter is fabricated on a 675 μm thick high-resistivity silicon substrate (ε_r = 11.4, ρ > 5 kΩ.cm) and there is a 25 μm deep trench under the CPW signal lines. This trench helps to improve the insertion loss and the quality factor of the transmission line. The corresponding quality factor and effective permittivity of the CPW line before loading with MEMS varactors are Q = 62.12 and $\varepsilon_{r\text{-eff}}$ = 1.81 at 30 GHz.

TABLE I
FILTER DIMENSIONS AND PARAMETERS

In-out section length (μm)	$L_{\text{In-out}}$	250
In-out coupling capacitors (fF)	C_{01}, C_{34}	104
In-out characteristic impedance (Ω)	$Z_{\text{In-out}}$	50
Resonator section length (μm)	$L_{\text{Resonator,1}}$	2215
Resonator section length (μm)	$L_{\text{Resonator,2}}$	2980
CPW characteristic impedance (Ω)	$Z_{\text{Resonator}}$	77
Inter-resonator coupling capacitors (fF)	C_{12}, C_{23}	20
MEMS varactor at 0 V bias voltage (fF)	C_{MEMS}	35

III. MEMS THERMAL ACTUATOR AND VARACTOR DESIGN

MEMS electro-thermal actuators which are based on joule heating and mechanical expansion of a conductor, can produce sufficient amount of displacement and are very compact compared to the MEMS electrostatic actuators. The MEMS thermal actuators, presented in this work, consist of two sets of four 375×8μm² in-plane v-shaped nickel beams with a bend angle of ±4° packed close together between two anchor points and a movable arm as shown in the SEM image of Fig. 2. There is a 0.7 μm thick poly-silicon layer available in the MetalMUMPs® process which can be used as a resistive layer to fabricate the heater element under the array of beams as shown in Fig. 2. The applied DC bias voltage on the poly-silicon heater causes ohmic heating and expansion of the v-shaped beams, pushing the apex outward. A 25 μm cavity in the silicon substrate was also used to improve thermal isolation and reduce the heat loss through the substrate.

Coupled-filed finite element simulation of the actuator is performed using ANSYS. Thermal conduction through the air-gap between the poly-silicon heater and the beams is assumed to be the only means of heat transfer. Fig. 3 shows the finite element simulation results for a 50 V DC bias

voltage and 1.8 mA current passing through the poly-silicon heater. The maximum nodal temperature and the lateral displacement of the actuator are 501 °K and 5.2 μm, respectively.

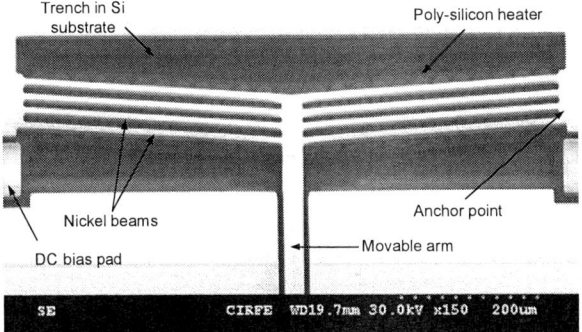

Fig. 2 SEM image of the thermal actuator with poly-silicon heater.

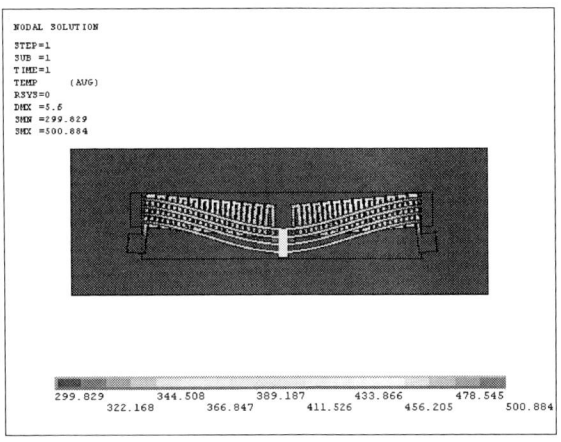

Fig. 3 FEM simulation results for temperature distribution and actuator displacement at V_{DC}=50 volt.

The tuning speed of the proposed microwave tunable bandpass filter depends on the dynamic thermal response of the MEMS actuator and how fast it heats up and cools down. Actuator response time was analyzed using finite element transient simulations. Fig. 4 illustrates the maximum displacement when a 50 V step bias voltage is applied to the actuator. The thermal time constant for temperature rise and fall are 20 ms and 30 ms, respectively. Therefore the maximum tuning speed of the proposed varactor is less than 30 ms.

Fig. 5 shows a close-up SEM image of the varactor. It has an interdigital structure with two sets of movable fingers. The effective capacitive area between the signal and ground is adjusted by moving the tuning fingers using thermal actuators. The 20μm thick metal layer allows the fabrication of the desired capacitance value with a less number of fingers.

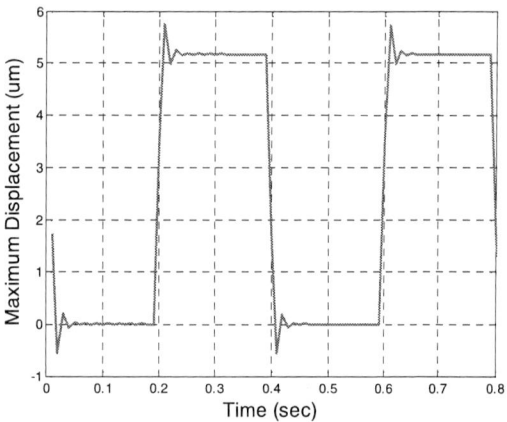

Fig. 4 Transient response of the MEMS thermal actuator.

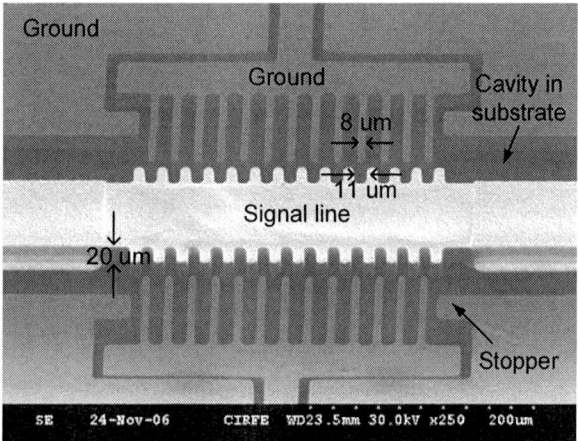

Fig. 5 SEM image of the varactor used for filter tuning.

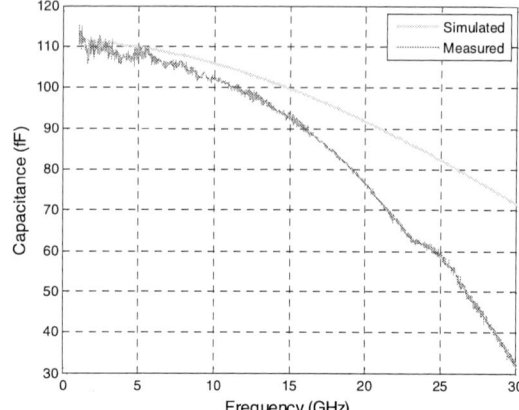

Fig. 6 Simulated and measured capacitance at V_{DC}=0 volt.

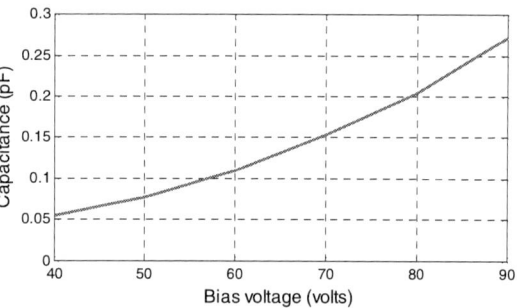

Fig. 7 Measured tuning characteristic of the varactor at 33 GHz.

The cavity in the substrate and underneath the signal line reduces the parasitic effects of the substrate and hence increases the quality factor and self-resonance frequency of the varactor. Fig. 6 shows the EM simulation result with HFSS and the measured capacitance over a frequency range up to 30 GHz. The difference between the simulation and measurement is due to the slight vertical bending of the tuning fingers and reduced effective area of the varactor. The variable loading effect of the varactor is measured using a single resonator section. A DC voltage sweep from 40 volt up to 90 volt is applied to the actuators. Fig. 7 shows the measured tuning characteristic at 33 GHz. The linear tuning range of the proposed varactor is found to be 396%.

The maximum power handling of the proposed filter is limited by the self-actuation of the varactors. Although the actuation mechanism is thermal, the structure will respond to any electrostatic force with frequencies lower than the mechanical self-resonance.

When an RF voltage $V(t)=V_o cos(\omega t)$ is applied between the signal line and ground, it will induce an electrostatic force

$$F_{RF} = \frac{1}{2} k \frac{dC(z)}{dz} \left(\frac{V_o}{\sqrt{2}} \right)^2 \left(1 + \cos(2\omega t)\right) \qquad (1)$$

where C is the total capacitance of the varactor, z indicates the direction of motion and k is a frequency dependent factor. The DC component of (1) is equivalent to an electrostatic force caused by a dc voltage [9]

$$V_{eq} = \sqrt{k} \frac{V_o}{\sqrt{2}} = \sqrt{k} V_{RF}^{rms} \qquad (2)$$

applied between the tuning fingers and the signal line. If the equivalent voltage V_{eq} exceeds the pull-in voltage then RF power will cause the actuation of the varactor. EM simulation with HFSS is used to find the frequency dependent factor k=0.91 at 33 GHz. The pull-in voltage (V_p) for the deformed structure, when V_{DC}=90 volt was applied on the poly-silicon heater, was found to be 34 volt. Then using (2) the maximum RF power handling of the proposed filter is obtained to be 25 Watt at 33 GHz.

978-2-8748-7007-1/08 $25.00 © 2008 EuMA   484

IV. FILTER SIMULATION AND MEASUREMENT RESULTS

Fig. 8 illustrates a comparison between the simulated and measured S-parameters of the proposed tunable bandpass filter at V_{DC}=0 volt. From the simulation results, the filter is predicted to have a center frequency at 33 GHz with a relative 3 dB bandwidth of 15% and the simulated mid-band insertion loss is 2.1 dB. The measured center frequency is slightly higher. This is because of the vertical bending of the tuning fingers. The measured mid-band insertion loss and bandwidth are approximately 2.7 dB and 14% respectively.

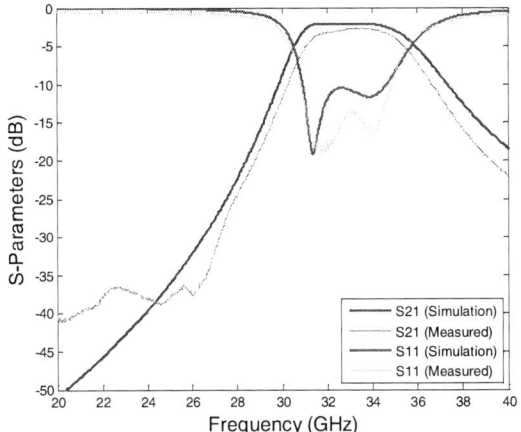

Fig. 8 Simulated and measured S-parameters of the filter at V_{DC}=0 volt.

The measured filter tuning response for various bias voltages from 0 up to 90 volt is presented in Fig. 9. The center frequency varies as much as 5 GHz (15% tuning) from 33.22 to 28.17 GHz. The 3 dB relative bandwidth also changes from 14% at V_{DC}=0 volt to 11.6% at V_{DC}=90 volt. This is due to the fixed input/output coupling of the filter while the center frequency is tuned. The insertion loss at the center frequency is increased from 2.7 dB to 3.3 dB and also the return loss is better than 10 dB throughout the entire tuning range.

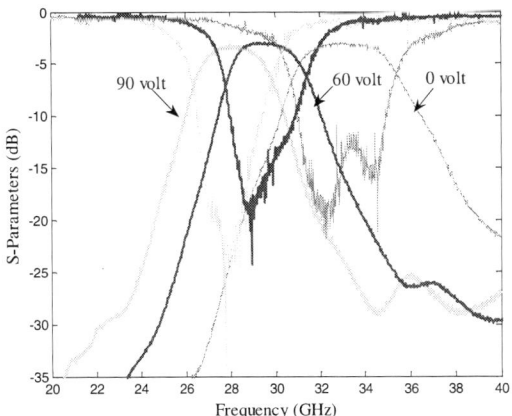

Fig. 9 Measured tuning characteristic of the proposed filter.

V. CONCLUSIONS

In this paper a novel approach to construct microwave tunable bandpass filters employing MEMS varactors with thermal actuators has been presented. The application of the proposed thermal actuators results in improved tuning range and also increased power handling capability of the filter. Electro-thermo-mechanical study of the MEMS thermal actuators has been performed to find the maximum tuning speed and power handling capability of the filter. A three-pole microwave tunable bandpass filter at Ka-band has been designed and fabricated using the MetalMUMPs® technology. The filter achieves a continuous tuning range of 15% from 33 to 28 GHz with a mid-band insertion loss less than 3.3 dB and a return loss better than 10 dB.

REFERENCES

[1] K. Entesari, and G.M. Rebeiz, "A 12-18-GHz Three-Pole RF MEMS Tunable Filter", IEEE MTT-Trans., vol. 53, no. 8, August 2005, pp. 2566-2571.

[2] R.D. Streeter, C.A. Hall, R. Wood, and R. Mahadevan, "VHF High-Power Tunable RF Bandpass Filter Using Microelectromechanical (MEMS) Microrelays", Int. J. of RF and Microwave Computer-Aided Engineering, vol. 11, issue 5, August 2001, pp. 261-275.

[3] D. Mercier, J.-C. Orlianges, T. Delage, C. Champeaux, A. Catherinot, D. Cros, and P. Blondy, "Millimeter-Wave Tune-All Bandpass Filters", IEEE MTT-Trans., vol. 52, no. 4, April 2004, pp. 1175-1181.

[4] A. Potheir, J.C. Orlianges, G. Zheng, C. Champeaux, A. Catherinot, D. Cros, P. Blondy, and J. Papapolymerou, "Low-Loss 2-Bit Tunable Bandpass Filters Using MEMS DC Contact Switches", IEEE MTT-Trans., vol. 53, no. 1, January 2005, pp. 354-360.

[5] S. Lee, J.-H. Park, J.-M. Kim, H.-T. Kim, Y.-K. Kim, and Y. Kwon, "A Compact Low-Loss Reconfigurable Monolithic Low-Pass Filter Using Multiple-Contact MEMS Switches", IEEE Microwave and Wireless Comp. Letters, vol. 14, no. 1, January 2004, pp. 37-39.

[6] H.-T. Kim, J.-H. Park, Y.K. Kim, and Y. Kwon, "Low-Loss and Compact V-Band MEMS-Based Analog Tunable Bandpass Filters" IEEE Microwave and Wireless Comp. Letters, vol. 12, no. 11, November 2002, pp. 432-434.

[7] A. Abbaspour-Tamijani, L. Dussopt, and G.M. Rebeiz, "Miniature and Tunable Filters Using MEMS Capacitors", IEEE MTT-Trans., vol. 51, no. 7, July 2003, pp. 1878-1885.

[8] L. Dussopt, and G.M. Rebeiz, "Intermodulation Distortion and Power Handling in RF MEMS Switches, Varactors, and Tunable Filters", IEEE MTT-Trans., vol. 51, no. 4, April 2003, pp. 1247-1256.

[9] D. Peroulis, S.P. Pacheco, and L.P.B. Katehi, "RF MEMS Switches With Enhanced Power-Handling Capabilities", IEEE MTT-Trans., vol. 52, no. 1, January 2004, pp. 59-68.

Proceedings of the 3rd European Microwave Integrated Circuits Conference

Monolithic MEMS T-type Switch for Redundancy Switch Matrix Applications

King Yuk (Eric) Chan[#*], Mojgan Daneshmand[*], Arash A. Fomani[*], Raafat R. Mansour[*], Rodica Ramer[#]

[#]*School of Electrical and Telecommunications Engineering, University of New South Wales*
UNSW Sydney NSW 2052 Australia

z3025962@student.unsw.edu.au

ror@unsw.edu.au

[*]*Centre for Integrated RF Engineering (CIRFE), University of Waterloo*
200 University Ave. W., Waterloo, Ontario, Canada, N2L 3G1

rrmansou@ecemail.uwaterloo.ca

Abstract— This paper presents a novel approach to monolithically implementing RF MEMS T-type switches for redundancy switch matrix applications. The T-type switch performs three operational states: two turning states and one crossover state. A six-mask fabrication process is adapted to fabricate the proposed design. Novel RF circuits were used to implement the entire system, including series contact cantilever beams, RF crossover, 90 degree turns and four-port cross junctions. The measured results for the entire T-type switch demonstrate an insertion loss of 1.5dB, a return loss of better than -20dB and an isolation higher than 28dB for all states for frequencies up to 30GHz. To our knowledge, this is the first time an RF MEMS T-type switch has ever been reported.

I. INTRODUCTION

Microwave switches are very often used as basic building blocks for redundant networks in satellite communication where high reliability is desired. A redundant network provides the required connection to spare components such that a full functionality of the system is maintained in case of any failure. A common example of that is the malfunctioning of an RF amplifier in satellite payloads where a redundancy network is used to route the signal to a spare component. Satellite payload switching networks currently use semiconductor and electromechanical technologies. In general, RF electro-mechanical switches are used as they can provide very good RF performance compared to their semiconductor counterparts. These switches are usually very bulky, heavy and capable of switching only at a low speed. Although switching time is not a concern in redundant networks, but the weight and size are major drawbacks especially in satellites where mass reduction is critical.

The majority of published work on RF MEMS switches has been on simple SPST switches [1]. More recently, several papers have been published on RF MEMS C-type, R-Type switches [2] – [4]. To our knowledge, this is the first time an RF MEMS T-type switch has ever been reported.

The T-type switch provides a high degree of flexibility for Redundancy network design. It consists of four ports, six signal paths with 3 operational states as shown in Fig 1. Each state has two conducting paths connecting two pairs of ports.

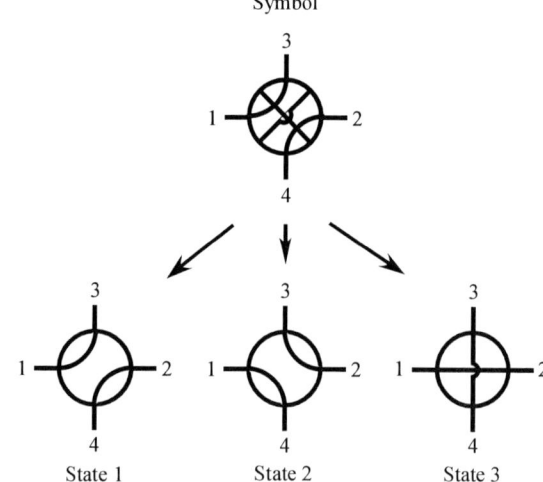

Fig 1 The symbolic representation of a T-type switch and its operational states. State 1, connection between ports 1, 3 and ports 2, 4. State 2, connection between ports 1, 4 and ports 2, 3. State 3, connection between ports 1, 2 and ports 3, 4.

In state 1, ports 1, 3 and ports 2, 4 are connected. In state 2, ports 1, 4 and ports 2, 3 are connected and in state 3, the two RF signals crossover and provides connectivity between ports 1 and 2 and ports 3 and 4.

Application of the T-type switch in redundant networks is very attractive as it provides more and unique routing options in case of path failure in comparison with C-type and R-type switches [3]–[4]. Fig. 2 shows an example of redundant ring that consists of T-switches and amplifiers. This redundant system allows rerouting any input to a spare amplifier (S1), and then to the corresponding output, in case of any amplifier failure. T-type switch provides the chance for redundant network to form multiple topologies of rerouting faulty paths to maintain an overall functional system [5]–[6]. It gives the options of either routing only the faulty signal path without disturbing the operation of other amplifiers or routing all the signal paths in order to maintain a uniform electrical length with similar RF performance for all the connections [7].

978-2-8748-7007-1/08 $25.00 © 2008 EuMA

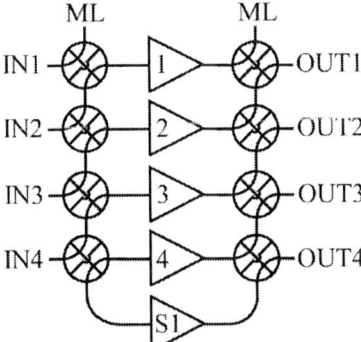

Fig 2 An example of T-type switches used for redundancy application

II. T-TYPE SWITCH CIRCUIT DESIGN

Fig 3 shows a schematic of a switch configuration that is capable of realizing the operation of the three states of the T-type switch. It consists of four single ended SPSTs, four double ended SPSTs, four four-port cross junctions, one RF crossover and interconnects. The four inner SPSTs (marked as S3) are operating as a single unit whereas the double ended SPST pairs at opposite corners are operating together as marked in red (S1) and green (S2) as shown in Fig. 3. Note that the red, green and blue circles indicate the switches required for state 1, 2 and 3 respectively. To achieve broad band design, for all the turn paths, double ended SPSTs are used and all the switch openings are placed around the junctions in order to minimize open stubs length when switches are at their off state.

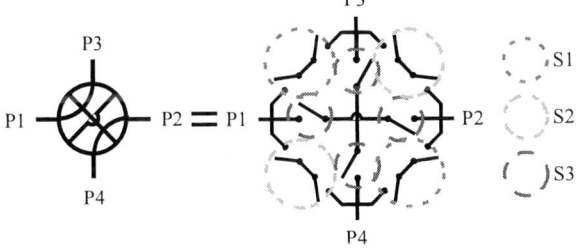

Fig 3 T-type switch symbol and proposed circuit model

III. T-TYPE SWITCH BUILDING BLOCKS DESIGN.

To solve this complex microwave system, the T-type switch is broken up into more manageable building blocks. Each block is solved individually with its corresponding boundary conditions. The T-type switch is broken up to the following subsystems: MEMS switch, four-port cross junction, RF crossover and interconnection networks that include 90° turns as shown in Fig. 4. The MEMS switch used in the T-type switch is a cantilever beam series contact switch published in [8].

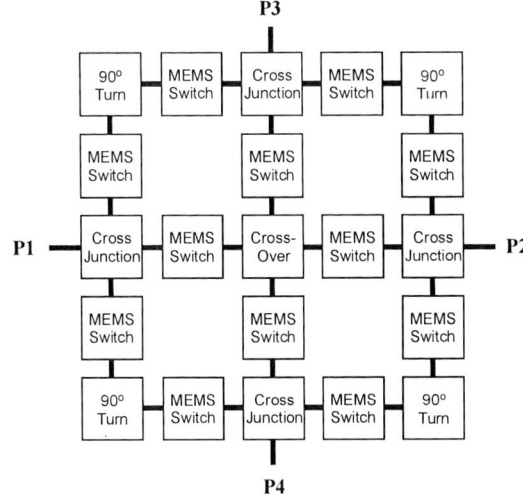

Fig 4 T-type switch block diagram

In the T-type switch design, capacitive switches are avoided in order to increase the overall operation bandwidth by avoiding quarter wavelength stubs. The double ended SPSTs are realized by two back to back cantilever switches with a 90° turn in between as shown in Fig. 5. To optimize the matching, it is essential to ensure that the switch and the 90° turn are both matched to 50Ω for a wide frequency band.

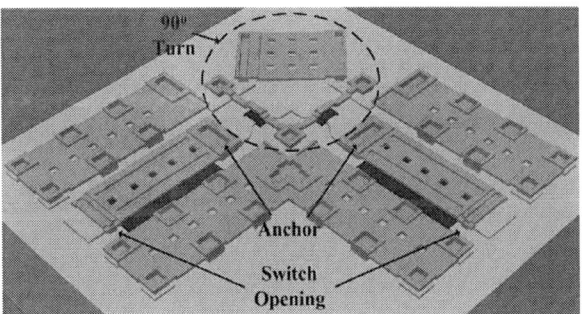

Fig 5 3D schematic of the proposed double ended SPST.

Another critical section in the design of T-type switch is the four ports cross junction at the input of each port. The cross junction is designed such that it exhibits good matching from all ports in all conditions. By analysing the operations of different states of the T-type switch, one can conclude that the cross junction with the switches around it faces only two loading modes as shown in Fig. 6. In each state, we can treat the junction with the switches around as a two port network where the other two ports have capacitive loadings only. Therefore the cross junction ports in any state needs to be matched considering the effect of the capacitive loads. The cross junction is designed with a very narrow center conductor with some out of plane discontinuities to increase the overall inductance. Conventionally, air bridges are required around

978-2-8748-7007-1/08 $25.00 © 2008 EuMA

the cross junction to remove parasitic modes. In this design, in order to minimize the loading capacitance from the ground bridges, they are connected at one point at the center rather than using four single bridges, as shown in Fig. 7.

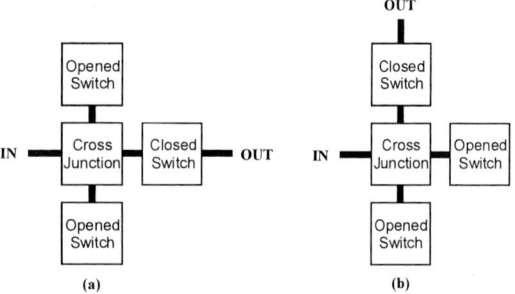

Fig 6 Proposed four-port cross junction loading modes, (a) thru loading (in state 3), (b) turn loading (in state 1 and 2)

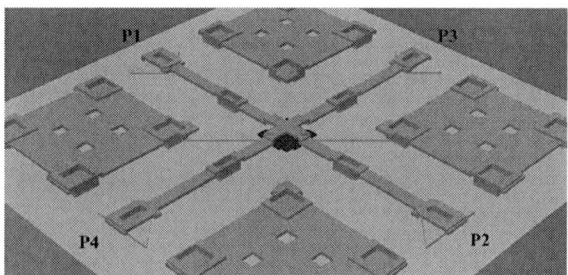

Fig 7 3D schematic of the proposed. four-port cross junction

As mentioned before, an RF crossover is required in the centre of the T-type switch. Here, the crossover is designed such that the signal travelling in the signal path connecting ports 1 and 2 is not strongly coupled to the other line connecting ports 3 and 4, as shown in Fig. 8.

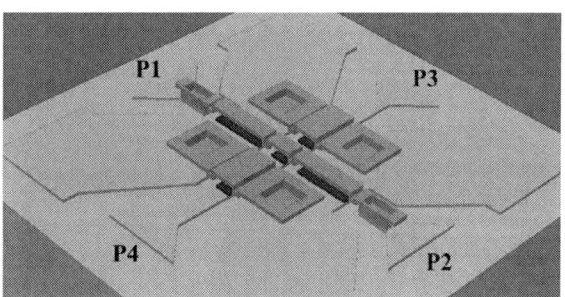

Fig 8 3D schematic of the proposed. RF crossover

The overall switch is configured with the sub-circuits described and the layout is shown in Fig. 9. Each sub-circuit is modelled, simulated and optimized with microwave software Ansoft HFSS, and the overall T-type switch simulation results are obtained by combined the HFSS results with Agilent ADS and they are shown in Figs. 10 and 11.

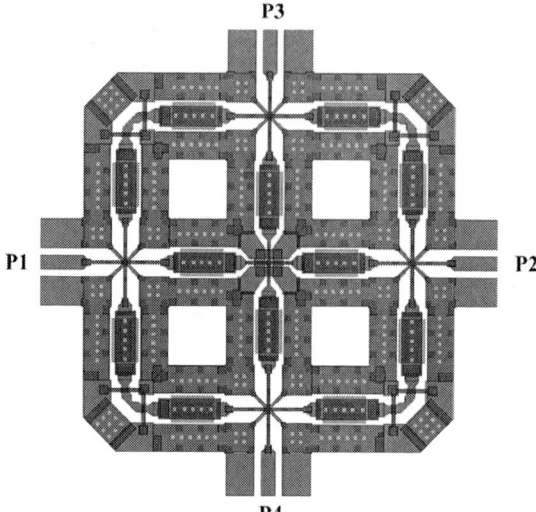

Fig 9 Proposed T-type switch layout

The simulation results show that, in all operation states, the T-type switch design can achieve excellent RF performance. Fig. 10 shows the RF performance in state 3. Since it only has two signal paths and the switch is symmetrical, only ports 1 and 3 return and insertion losses are shown. Moreover, as the proposed T-type switch is symmetrical, states 1 and 2 have the same RF performance, as it is shown in Fig. 11.

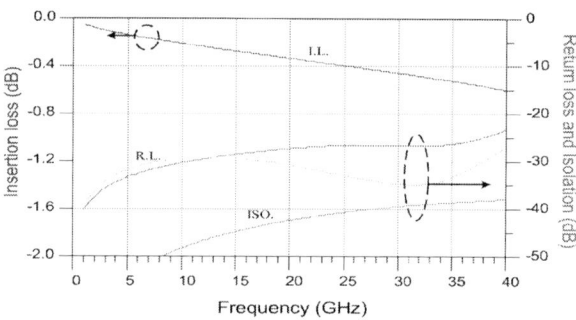

Fig 10 State 3 S-parameters in ports 1 and 3.

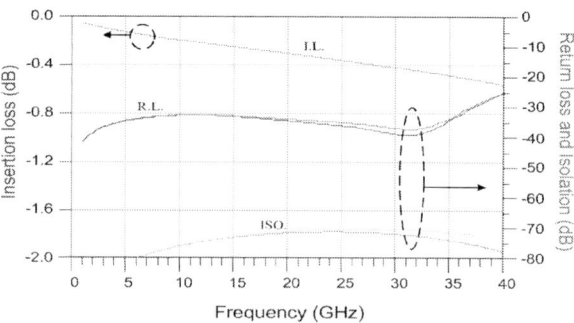

Fig 11 States 1 and 2 S-parameters in ports 1 and 4.

978-2-8748-7007-1/08 $25.00 © 2008 EuMA

IV. THE FABRICATION PROCESS

The proposed structure is fabricated using the UWMEMS process [9]. The UWMEMS process is a multi-gold layer process developed at the University of Waterloo. To achieve good RF performance devices, a low loss ceramic, alumina wafer with dielectric constant of 9.9 is selected as the base substrate. The fabrication process starts with an E-beam deposition and a lift-off patterning of a 1um thick gold (Au) layer to define the CPW transmission lines. A 0.7μm oxide layer is deposited and patterned to form an insulation layer. To improve the adhesion of the oxide to the gold, a layer of Titanium-Tungsten (TiW) alloy (0.04μm) is introduced between the two layers. A 2.5um of PI2562 polyimide is spin-coated and cured as the sacrificial layer. Subsequently, anchors and dimples are etched in the polyimide layer using an oxygen plasma etching process. The anchor provides the connection of the top to the bottom metal layers while the dimples are mainly used to enhance the contact force which reduces the contact resistance. Next, the top gold layer is formed by a 1.2um gold electroplating inside a mold on top of a 50nm E-beam gold seed layer. The last step of the fabrication is an oxygen plasma dry etching of the polyimide sacrificial layer to release the MEMS devices.

V. THE FABRICATED T-TYPE SWITCH

The T-type switch is fabricated, and the SEM picture is shown in Fig. 15. The measurement results of the fabricated device demonstrate excellent RF performance.

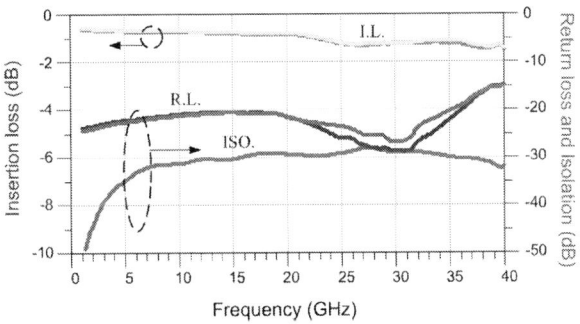

Fig 13 Measured State 3 S-parameters in ports 1 and 3.

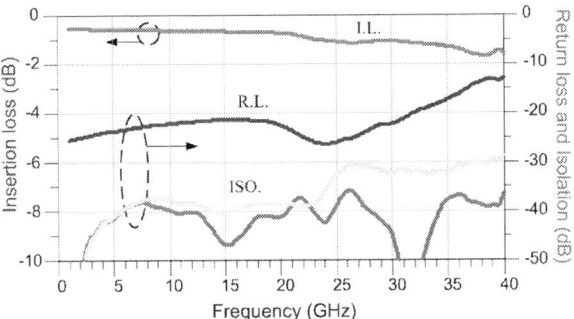

Fig 14 Measured States 1 and 2 S-parameters in ports 1 and 4.

In all the states, the measured insertion loss and return loss are better than 1.5dB and -20dB respectively. The isolation is also better than 28dB below 30GHz. The fabricated device validates the concept of designing T-switch with optimized sub-circuits and the simulation models.

VI. CONCLUSIONS

A novel monolithically integrated MEMS T-type switch is presented for the first time. The device was designed to operate for a very wideband frequency range (0 to 30GHz) with excellent RF performance. The device was fabricated using the six mask process. This is the first monolithically integrated MEMS T-type switch ever reported. The proposed T-type switch promises to be useful in the realization of highly compact redundancy switch matrices for satellite applications.

Fig 15 SEM image of the fabricated T-type switch

REFERENCES

[1] G. M. Rebeiz, *RF MEMS Theory, Design, and Technology*: John Wiley & Sons, 2003.

[2] Daneshmand, M.; Mansour, R.R., "Redundancy RF MEMS Multiport Switches and Switch Matrices," *Microelectromechanical Systems, Journal of*, vol.16, no.2, pp.296-303, April 2007

[3] Daneshmand, M.; Mansour, R.R., "Hybrid integration of RF MEMS multiport switches," *Electrical and Computer Engineering, Canadian Journal of*, vol.31, no.2, pp.65-70, Spring 2006

[4] Daneshmand, M.; Mansour, R.R., "C-type and R-type RF MEMS Switches for Redundancy Switch Matrix Applications," *Microwave Symposium Digest, 2006. IEEE MTT-S International*, vol., no., pp.144-147, June 2006

[5] R. Kwiatkowski, "Bi-Planar Microwave Switches and Switch Matrices," U.S. Patent 6 951 941, Oct. 4, 2005.

[6] C. J. Pentlicki, "Electromechanical Switch" U.S. Patent 4 330 766, May. 18, 1982.

[7] P. G. Petrelis, "Redundancy Switching System", U.S. Patent 4 061 989, Dec. 6, 1977.

[8] K. Y. Chan, M.Daneshmand, R.R. Mansour, R. Ramer, "Novel Beam Design for Compact RF MEMS Series Switches", Asian Pacific Microwave Conference, Dec 2007.

[9] Center for Integrated RF Engineering, University of Waterloo, Waterloo, Ontario, Canada, http://www.cirfe.uwaterloo.ca/

Reliability of Dielectric Less Electrostatic Actuators in RF-MEMS Ohmic Switches

D. Mardivirin[#1], A. Pothier[#2], M. El Khatib[#3], A. Crunteanu[#4], O. Vendier[*5] and P. Blondy[#6]

[#] *XLIM UMR 6172 – Université de Limoges/CNRS - 123 Avenue Albert Thomas - 87060 LIMOGES Cedex FRANCE*

[1] david.mardivirin@xlim.fr
[2] arnaud.pothier@xlim.fr
[3] mohammed.el-khatib@xlim.fr
[4] aurelian.crunteanu@xlim.fr
[6] pierre.blondy@xlim.fr

[*] *THALES ALENIA SPACE, 26 Avenue J.F Champollion Toulouse F-31037, France*

[5] olivier.vendier@thalesaleniaspace.com

Abstract— **This paper presents the effects of residual charging in dielectric less actuators of RF-MEMS ohmic switches. Indeed, in order to strongly reduce component sensitivity to charging, a dielectric less electrostatic actuator has been introduced in a conventional DC contact series MEMS relay design, resulting both in strong improvement in reliability and preservation of its intrinsic RF performance. Under various stress applied, the pull-in and pull-out voltages drift over time of these components have been observed and analyzed. Hence, based on component pull-in and pull-out voltage measurements during only few minutes of a given stress, an efficient model able to accurately predict the actuator reliability up to 60 days with good agreement will be presented.**

I. INTRODUCTION

RF-MEMS switches are one of the most promising applications in micro-technologies development for telecommunication. Since few years, their low loss, low power consumption and small size, have made them a serious potential alternative to existing semiconductor switches as field-effect transistors and pin-diodes. But, their recurrent not fully solved reliability problems made their commercialization still challenging and require again strong research efforts.

Especially, the reliability of the electrostatic actuator remains a major issue, since all mechanical and electrical performance of the device relies on it. The low power requirement and the easiness of integration of design and fabrication process of parallel plate electrostatic actuator made with this actuation mechanism the most commonly used in RF-MEMS devices up to now [1]. However it generally suffers from charging phenomenon attributed to parasitic charge injection in the dielectric layer that is used to prevent a direct contact between the metallic electrodes of the actuator [2-3]. Especially in DC contact MEMS switch case, charging of the actuator is problematic. It may induce consequent drifts in component RF performances since a highly uniform and repeatable force is required on the metal to metal contact areas, to maintain a low contact resistance over cycles.

One approach to solve this problem is to suppress the dielectric layer where most of parasitic charges are trapped. As one can see, this solution is already used in some commercially available MEMS DC relay design [4-5] which have demonstrated impressive reliability.

In this paper, we will see that charging still occurs even if the dielectric layer have been removed, but following others mechanisms, a lot less harmful for the component performance. Moreover it will be shown that the lifetime of the switch is strongly increased, especially by using optimized bias waveforms. A model of the long-term actuation voltage drift prediction will be also introduced with a good agreement with experimental measurement, as an efficient tool to have an estimation of the component lifetime.

II. SWITCH DESIGN AND PERFORMANCES

The micro-relay studied in this paper, presented on Fig. 1, is based on MEMS series DC contact switch. It is fabricated with conventional gold cantilever beam, similar to that used in [6], and which is actuated to close a metal to metal contact between the cantilever beam free end and two narrow contact electrodes.

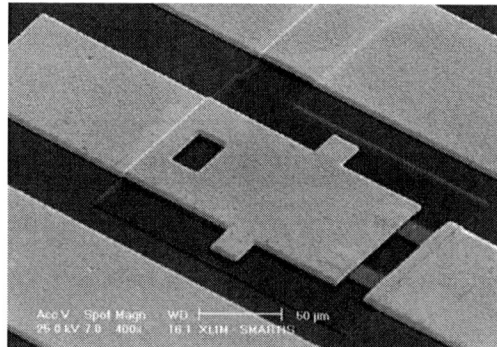

Fig. 1 SEM picture of fabricated DC contact switch with dielectric less electrostatic actuator (designed using a 130µm length, 90µm wide and 4µm thick beam).

In the present case, the contact material is pure gold and the micro relay is actuated thanks to a vertical electrostatic actuator fabricated without any insulating layer. Two metallic mechanical stoppers, made under the cantilever, have been

978-2-8748-7007-1/08 $25.00 © 2008 EuMA

introduced on the sides of the switch beam (Fig. 2) in order to stop it before any contact with the actuation electrode could occur. Their location and thickness have been optimized to keep a sufficient contact force on the metal to metal contact areas. The actuation electrode is fabricated using a thin resistive layer (typically 2kΩ square) and designed to maximize force on the contact areas with moderate actuation voltage.

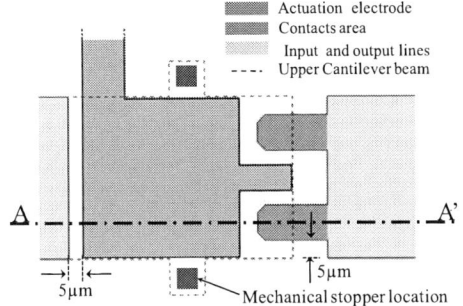

Fig. 2 Top view of the electrostatic actuator layout.

Typical RF performance measured on this switch is shown on Fig. 3. The switch open state capacitance is 16fF, allowing an isolation level better than 15dB up to 20GHz. To pull down the cantilever beam, 65 to 70V is typically required, but 80V to 100V are applied to obtain a reproducible and low value contact resistance resulting in low insertion loss. The beam geometry allows obtaining a stiff mechanical structure resulting in a switch pull-out voltage only a few volts lower than its pull-in voltage (Fig. 4). As can be seen on Fig. 3, the dielectric less actuator implementation does not alter the device microwave characteristics compared to conventional electrostatically actuated switch design [6].

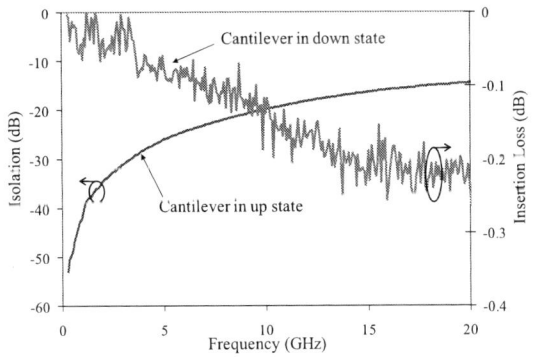

Fig. 3 Measured RF performances of MEMS switch fabricated on sapphire substrate with dielectric less electrostatic actuator.

The switch impedance change versus applied actuation voltage characteristic (C(V) for a capacitive switch) has been extracted from Vector Network Analyser (VNA) calibrated measurement at 10 GHz. This device characteristic has been periodically measured while the switch was held in down

position during a 1 hour with 80V DC bias voltage. As shown on Fig. 4, a significant drift has been observed as an evidence of residual charging phenomenon. Indeed, the applied stress induces a decrease of the switch pull-in and pull-out voltage value symmetrically in the positive and negative voltage domain, narrowing the C(V) plot. The resulting voltage drift is less than 5 volts after one hour of stress. The actuator charging mechanism will be more intensively studied in the next section, but we can underline that its behavior is more favorable to component reliability. Indeed in this case, we can expect that if a sufficient biasing voltage (at least higher than the initial pull-in voltage) is used, it will be still efficient to reach the required contact force on the switch contact area after a long actuation period. Accordingly, such actuators will decouple the reliability problems due to electrical contact degradation from those related to the component actuation.

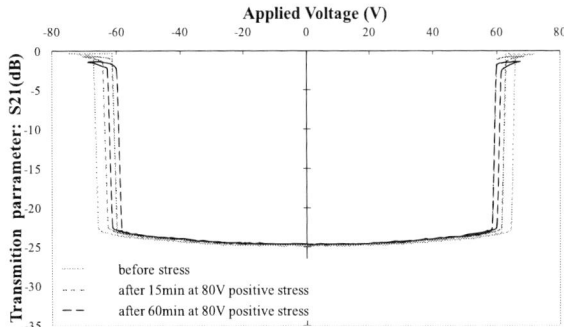

Fig. 4 Measured drift of the impedance versus applied voltage characteristic of a switch held in the down state up to one hour.

III. RELIABILITY STUDY

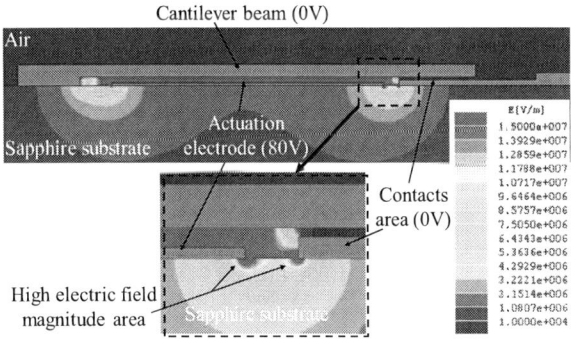

Fig. 5 Electrical field distribution in the dielectric less electrostatic actuator under 80V considering the switch in open state position (following the Fig. 2 A-A' cut line).

All components measured in this paper were fabricated on low loss single crystal sapphire substrates. To understand where residual trapped charges are located, 2D electrostatic simulations have been performed using ANSOFT Maxwell SV simulator. As presented on Fig 5, most of applied electric field is located in the air gap between the cantilever and the

978-2-8748-7007-1/08 $25.00 © 2008 EuMA
491

actuation electrode. But there are also some areas in the sapphire substrate where the computed electric field magnitude is higher than 50MV/m, which can be sufficient to induce charge injection in these localized zones.

A. Charging over time

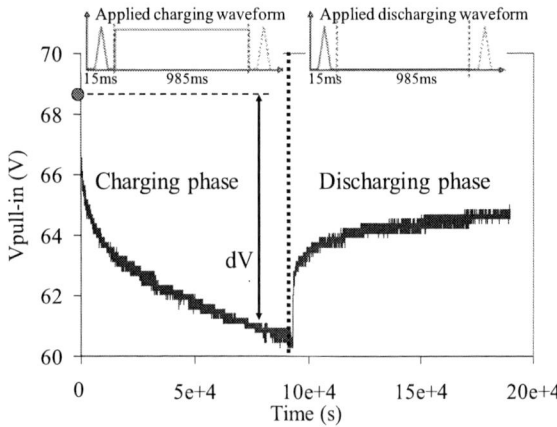

Fig. 6 Typical recording sequence for lifetime testing on DC contact switches for 70V applied DC stress.

To study specifically the actuator reliability, a dedicated test bench has been used allowing monitoring the evolution of the switch pull-in and pull-out voltage over time, for different stresses [7]. Since, tested switches are not packaged; all tests were performed under dry nitrogen in a controlled atmosphere chamber. A typical measurement sequence is presented on Fig. 6.

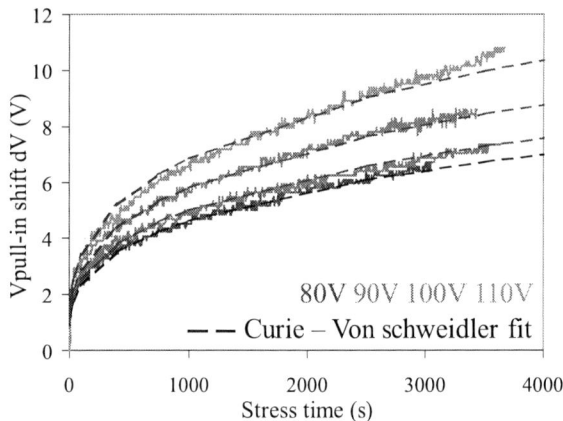

Fig. 7 Measured DC contact switch pull-in voltage shift over time for various magnitude DC stress.

As shown, the actuator is biased with a specific periodic actuation waveform. Thanks to a quick (15ms) positive triangular pulse, the switch impedance versus applied voltage characteristic is recorded. Then, during the remaining 985 ms,

a chosen stress is applied. In the case of a continuous positive DC stress, the switch stays 98% of the time actuated which can be considered as a "worst case testing". Another possible stress is to apply a 10 Hz unipolar square waveform with a user specified duty cycle, as conventionally used cycling waveforms.

The observed actuation voltage drift "dV" both in charging and discharging test sequences is much quicker in the first minutes and become more progressive with increasing time (Fig. 6). Furthermore, the observed charging phenomenon is reversible but taking longer time to fully discharge the actuator.

The pull-in voltage drift has been also recorded for 1 hour stress periods where the switch was held in down state position considering different magnitudes of applied DC stress on the actuator. As presented on Fig. 7, stress magnitude dependence has been observed that will be important to consider in the switch lifetime evaluation as we will see later.

B. Modelling and lifetime prediction

The charge buildup mechanism observed in the tested components is not fully understood yet [8-9]. In our case, all recorded voltage drifts can be fitted using a Curie-Von schweidler equation (1) as shown on Fig. 7. This equation is intensively used to describe voltage retention in dielectric materials [10]. It is based on a power-law time dependence of the stored charge following approximately t^n (where n < 1). Considering that the up state capacitance of the switch stays constant during the measurement, the stored charge in the actuator is then proportional to the observed voltage drift dV. As the result, the equation to be considered becomes:

$$dV = A.t^n \qquad (1)$$

Where dV is the measured voltage drift, t is the stress time and A and n are model-building coefficients.

TABLE I
EXTRACTED MODEL PARAMETERS AS FUNCTION OF APPLIED DC STRESS MAGNITUDE

Stress magnitude	A	n
80V	0.58	0.3
90V	0.63	0.3
100V	0.73	0.3
110V	0.86	0.3

The A and n parameters are then adjusted to find the more appropriated and efficient fit with measured data. The used modeling coefficients in the Fig. 7 are reported in Table I. The stress magnitude dependence of A is not completely understood, but we can observe that n is clearly independent of it.

The influence on the pull-in voltage drift characteristic of a periodic square signal as biasing waveform stress has been also evaluated. Experimental results, presented on Fig. 8 with

978-2-8748-7007-1/08 $25.00 © 2008 EuMA

a double logarithm axes plot scale, also follow a Curie- Von Schweidler behavior with a noticeable property. Indeed, measured voltage drifts strongly depend on the stress waveform duty cycle.

As shown, the same n coefficient can be used to accurately model the measured voltage shift whereas the A parameters follows a simple proportional law with the duty cycle value (i.e. for this switch A=0.9 × Duty cycle %). This property has already been observed in [7] on another switch design using dielectric less actuator. In the present case, with n=0.28, the time for dV to reach a given value is (1/0.4)1/0.28=26.4 times more when going from 100% duty cycle to 40% duty cycle.

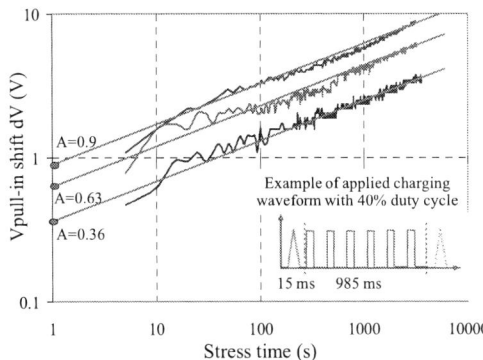

Fig. 8 Double logarithm axes plot of the measured voltage shift over time for 97.7%~100% duty cycle (brown), 70% (green) and 40% (blue) and the corresponding modeling fit (red) with n=0.28.

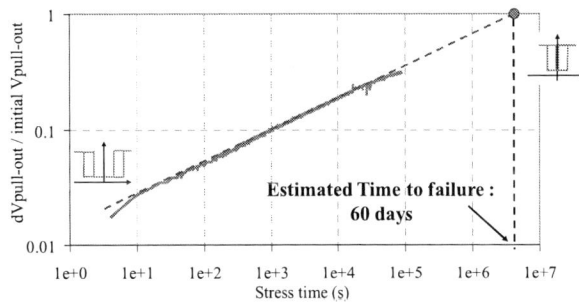

Fig. 9 Normalized measured pull-out voltage shift versus time (red) and corresponding model (blue), for a switch left in the down state with 80 volts DC applied.

Thanks to its good agreement with experimental data, the presented model appears as an efficient tool to obtain a fine estimation of the actuator lifetime. Actually, only few minutes (at least two to three decades) of switch pull-out voltage monitoring is sufficient to predict when the actuator will fail. As seen on Fig 9, after about 24 hours hold in down state, the tested switch has shown a pull-out voltage reduction of 25V whereas the first 10 volts are lost in only 1 hour. By extrapolating this behavior, the model predicts that the pull-out voltage shift reaches 30V after 50 hours, and the failure of the switch actuator will occur after 1400 hours (~59 days),

with the voltage shift reaching 62V: the initial pull-out voltage. It is also interesting to note that this model predicts that cycling the switch with a square positive voltage with duty cycle of 50%, will result in failure after 59 days x 11.9, about 700 days (with n=0.28). It is likely that under this type of stress, contact ageing may be the prominent failure mechanism of the component.

IV. CONCLUSIONS

Residual charging in dielectric less electrostatic actuators for RF-MEMS switches has been studied and the benefits on lifetime have been demonstrated. This type of actuator allows preserving the intrinsic device RF performance and strongly extending its lifetime. We have seen that the observed drift in pull-in or pull-out voltages under various stress can easily be modelled using a simple Curie-Von Schweidler equation with good accuracy. It allows predicting the long term behavior of such actuator, based on an only short period of test (less than one hour). It has been shown that hold down periods longer than one month without failure of the switch actuator are today feasible, without applying any special bias waveform. It is also shown that reducing the duty cycle of the bias signal from 100% to 50% results in more than 10 times lifetime improvement of the electrostatic actuator, shifting the component lifetime study on other reliability issues such as metal to metal contact degradation.

ACKNOWLEDGMENT

This work has been led in the frame of the European SMARTIS project and the authors wish to acknowledge the French MoD for its support and founding for this study.

REFERENCES

[1] G. M. Rebeiz, RF MEMS Theory, Design, and Technology, Willey and sons 2003.
[2] M.van Spengen, W.M.; Puers, R.; De Wolf, I.; "The prediction of stiction failures in MEMS", Device and Materials Reliability, IEEE Transactions on Volume 3, Issue 4, Dec. 2003 Page(s):167 - 172
[3] X Yuan, Z. Peng, and all, "Acceleration of Dielectric Charging in RF MEMS Capacitive Switches", IEEE transactions on device and materials reliability, vol. 6, no. 4, Dec 2006, pp 556 -563.
[4] I McKillop, T. Fowler, D. Goins and R. Nelson, "Design, Performance and Qualification of a Commercially Available MEMS Switch", 36th European Microwave Conf, Sept 2006 pp. 1399–1401.
[5] S. Majumder, J. Lampen, R. Morrison and J. Maciel; "A Packaged, High-Lifetime Ohmic MEMS RF Switch", 2004 IEEE MTT-S Int. Microwave Symp. Dig, June 2003 pp:1935 – 1938.
[6] A Pothier ,"MEMS DC contact micro relays on ceramic substrate for space communication switching network", 35th Microwave Conf, Sept. 2005.
[7] D. Mardivirin, D. Bouyge, A. Crunteanu, A. Pothier, P. Blondy, "Study of Residual Charging in Dielectric Less Capacitive MEMS Switches", 2008 IEEE MTT-S Int. Microwave Symp, to be published.
[8] G.J Papaioannou, G. Wang, D. Bessas, J. Papapolymerou, "Contactless Dielectric Charging Mechanisms in RF-MEMS Capacitive Switches", 36th European Microwave Conf, Sept 2006 pp:1739 – 1742.
[9] R.W. Herfst, P.G. Steeneken, and J. Schmitz, "Time and voltage dependence of dielectric charging in RF MEMS capacitive switches", IEEE 45th International Reliability Physics Symposium, 2007, pp 417-421.
[10] A.K. Jonscher, "Dielectric Relaxation in Solids", J. Phys. D: Appl. Phys.32, 1999.

A 400 MHz – 1600 MHz SiGe MMIC beam-former for the Square Kilometre Array

Klaas Visser[1], Erik van der Wal, Mark Ruiter and Dion Kant[2]

ASTRON, Netherlands Institute for Radio Astronomy
Dwingeloo, The Netherlands
[1] `visser@astron.nl`
[2] `kant@astron.nl`

Abstract—**This paper describes an implementation of a beam former Monolithic Microwave Integrated Circuit (MMIC) for the SKA demonstrator project known as EMBRACE using the SG25H1 Silicon Germanium process of IHP microelectronics. This process is qualified as a high-performance technology with npn-HBTs up to ft/fmax = 180/220 GHz.**

I. INTRODUCTION

Astronomers and engineers are aiming to build the world's biggest telescope, called Square Kilometre Array (SKA), capable of peering deep into space at such things as gamma-ray bursts, extrasolar planets, evolving galaxies, dark matter and possibly even back to the Big Bang, where it all began. The SKA project will create a huge radio telescope. The Netherlands Institute for Radio Astronomy (ASTRON) is working on design of the European demonstrators for the SKA project.

Several demonstrator design studies are carried out all over the world. ASTRON focuses on the development of Aperture Array concept called EMBRACE. In contrast to conventional radio telescopes, which are mechanically steered, the EMBRACE array is controlled by pure electronics. No mechanical movement is required. Aperture Array technology offers the unique capability to adaptively suppress unwanted RFI signals effectively. In addition, it will be able to detect signals from different astronomical sources simultaneously (multi-beaming). Currently, EMBRACE implements two independent antenna beams.

Embrace is a concept in which 72 elementary antennas are densely packed into a tile and connected with beam-forming circuitry. A tile provides a physical collecting area of around one square metre. Two SKA demonstrators will be built with a total collecting area of around 400 m^2.

For cost reasons, analogue beam forming and combiners in the signal chain of the total system are used to limit the number of digital receivers. In order to achieve the EMBRACE cost targets, a higher integration density and a lower cost technology are required compared to earlier designed discrete prototypes [1]. Apart from gain-stages the beam-former chip (BFC) will replace roughly 90 percent of the electronics of the current prototype boards. The beam-former chip replaces eight vector modulators.

The chip is designed for the beam control by means of phase-shift and amplitude control. The design includes dual loop feedback amplifiers, phasing and combining networks as well as amplitude correction capability. The large ft/fmax is required to accomplish the specified bandwidth of the negative feedback amplifiers implemented in the design. All settings of the chip are under full digital control by a SPI interface. A total of 40 buffered registers are interfaced internally to the analogue part, of which some registers are combined as input for D/A conversion.

The BFC is a tiny building block of a receiver and signal processing system which should handle signal frequencies from 400 MHz till 1600 MHz. The number of phase steps is eight (3 bits) and are equally distributed, consequently a phase error of the total system should be less than 22.5 degrees. A phase error of 10 degrees is appointed to the BFC and 15 degrees to the rest of the system. Amplitude control of 3 dB is implemented to accomplish a phase dependent amplitude variation of less then 1 dB.

The main reason for an analogue BFC is its processing power. The analogue processing power of the chip can be expressed in equivalent number of Flops, when a digital solution would need to perform a similar task. A conservative number leads to 64 GFlops of processing power for signal processing of four antennas. A tile of approximately one square metre consisting of 72 antennas leads to an equivalent digital processing power of 1.2 TFlops. A large number of these tiles will form the collecting area. Imagine the digital processing power required for a square kilometre array.

First in this paper the architecture (II) of the BFC will be explained, after which the gain stage topology (III), the phasing network (IV), vector summation (V) and the combining network (VI) of the BFC will be described. Finally, some measurement results (VII) will be presented and some conclusions (VIII) will be drawn.

II. THE BEAM-FORMER ARCHITECTURE

Four antennas including LNAs (RF inputs) are connected to the chip, in which two independent beams are generated by a beam-forming technique. The BFC architecture can be divided in four channels which generate two beams. Each channel contains a combining network to add the desired signal stream to an analogue signal bus. The power supply of each channel is carried out separately to the outside world due to a stringent demand on channel isolation (ISO$_{ch}$). The channel isolation specification is also the reason to implement an inductor-less BFC architecture. Each channel contains its own digital control circuit to set the phase states of the beams. Figure 1 shows the block diagram of a channel, which consists of gain and buffer stages, phasing and combining networks.

The implementation of the BFC is differential, despite the fact the antenna and LNA will be carried out single ended in first instance. Apart from isolation the motivation to implement the differential concept is the cancellation of the even-order intermodulation products. Also for suppression of interference sources at the horizon, large common mode rejection is desired which is achieved by a proper differential design.

The first gain stage is a dual loop differential feedback amplifier to accomplish with the first loop the gain and with the second loop to ensure input matching with the output impedance of the discrete gain-stage in front. Furthermore the amplifier should have low noise and should take care of most of the gain. The main reason is to minimise second stage noise contribution (Friis's equation), despite the fact noisy components such as the polyphase, switching and combining network will be included

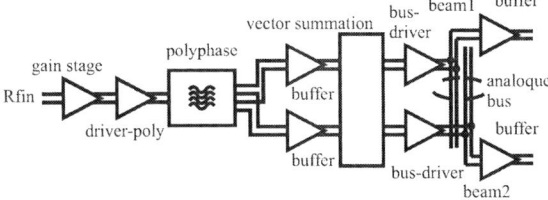

Fig. 1 Block diagram of a channel.

The output of the gain amplifier is fed to a driver circuit which drives a polyphase network. The polyphase circuit generates four cardinal vectors 0°, 90°, 180° and 270° from the input signal. The four polyphase signals are four vectors having the same frequency and amplitude but differs in phase. The motivation to implement a polyphase filter instead of several RC networks is the inherent insensitivity to process variation. The implemented polyphase consists of identical resistors and capacitors, while two RC networks possess different components in value and area.

The four vectors generated by the polyphase will be used to form the desired phase of the output signal by vector summation. The combining circuit combines the channel output with the outputs of the three other channels on an analogue bus shown in figure 1. The bus is buffered and an output driver interfaces to the outside world.

III. GAIN STAGE TOPOLOGY

A low noise differential amplifier, in combination with negative feedback is implemented. The closed loop gain depends on a resistance ratio resulting to a process spread insensitive design under assumption of sufficient loop-gain ($A\beta$). The overall gain of a stable feedback system ($A_{fb}=A/(1+A\beta)$) will be sufficiently smaller than the forward gain (A). Infinite forward gain is desired to suppress completely the error in the total system.

Early in the design process of the BFC, a three stage differential gain stage with a single feedback loop was designed, processed and measured. Figure 2 shows the

measurement results of the gain versus frequency, which are in line with simulations.

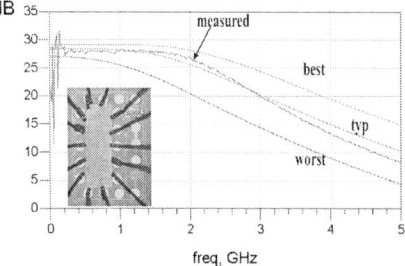

Fig. 2 Gain versus frequency of a differential three stage gain stage with a single feedback loop.

Implementing a second loop to ensure accurate input impedance incorporates an additional phase shift which resulted in instability in the earlier three stage differential amplifier. To ensure unconditional stability one stage needs to be removed, despite the fact of the negative consequences of reducing the forward gain. In figure 3 the topology of the differential two stage dual feedback amplifier is shown. [1]

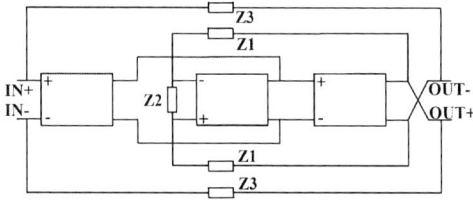

Fig. 3 The differential two stage dual feedback amplifier.

Several buffers with equal topology are used. The feedback impedances are changed to reduce the gain of the buffer. Furthermore the second feedback loop, which determines the input impedance has been omitted.

IV. PHASING NETWORK

Sequence asymmetric polyphase networks are common in designs of radio-frequency integrated transceiver and receivers to provide an efficient way for wide band quadrature signal generation. In the recent years these networks are thoroughly investigated and published. Equations are presented to calculated side-band suppression, the number of segments, selecting the RC values for the segments etc. [3] and [4]

In the beam-former design the polyphase will be used to generate the four cardinal vectors (0°, 90°, 180° and 270°). Vector summation of these will generate four additional vectors (45°, 135°, 225° and 315°). Figure 4 shows idealized the construction of the vector. It also highlights a problem with this topology. Even with an idealized polyphase network, some vectors have larger magnitude than others under assumption that both input vectors are equal in magnitude. This error appears structurally at the summed vectors independent to the number of segments placed in series. These states are a factor of √2 (see figure 4C) too large in magnitude. Consequently, the summed vectors determine the phase and amplitude mismatch of the polyphase filter. Additional design considerations should be taken into account due to a stringent demand on amplitude variation (± 0.5 dB for the total chip) between the several phase states.

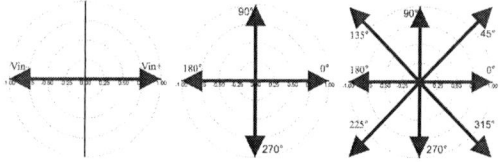

Fig. 4 Vector diagrams showing (a) the input of the polyphase network. (b) the idealized output of the polyphase network. (c) the phase states after vector summation.

Actual implementation of the polyphase network is an additional source of errors. The bandwidth of the polyphase network with vector summation depends on the number of polyphase segments used in series. Figure 5 shows the effect of the number of segments on the error.

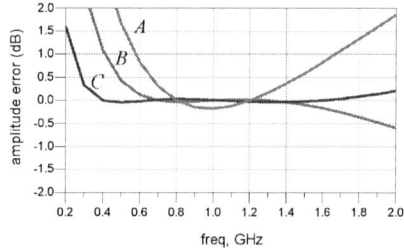

Fig. 5 Additional amplitude error versus frequency versus the number of segments. (A), 2 segments, (B): 3 segments and (C): 4 segments.

A fourth order polyphase filter fulfils the requirements. The poles of each segment are evenly distributed over the desired bandwidth. During the synthesis of the network, the order in which the poles are extracted can be chosen arbitrarily. The filter response will not be influenced, however the impedances will be. And through noise mismatch, this will influence the noise performance of the network. Extracting the low frequency poles first, gave a good compromise between the drive impedance of the network and the noise performance.

V.   VECTOR SUMMATION

As explained in the previous section, the four cardinal vectors are supplied to a summation circuit. This circuit is responsible for selecting one of four inputs or to construct a vector by vector summation of two selected inputs and pass it to the output. Consequently, the vectors 0°, 45°, 90°, 135°, 180°, 225°, 270° and 315° can be achieved.

The differential vector summation network consists of emitter followers, combined with a resistive network. The topology integrates both the summation and the amplitude correction functionality. For clarity, only half of the network is shown in figure 6A. The 180° and 270° branches are connected in parallel.

The network can be biased in three different ways. First is the state where all current sources are off. This will not pass any signal. Although not used in normal operation, it will be used during verification measurements and system calibration.

To pass the cardinal states, only the current source of that particular vector will be biased. The input voltage will appear at the output, attenuated by the resistive network

$$V_{out} = V_i \cdot \frac{R_L}{R_S + R_L} \quad (1)$$

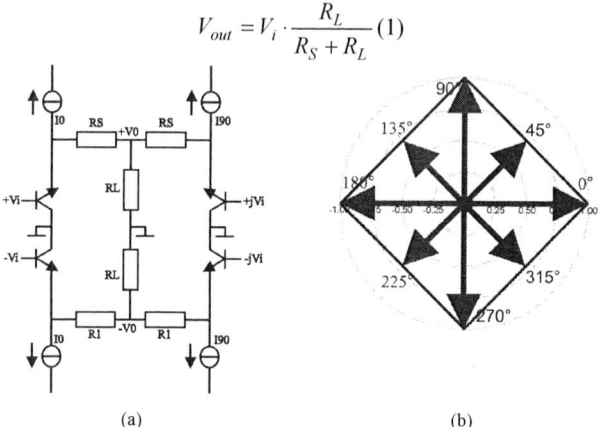

Fig. 6 (a) Vector summation network (b) Vector diagram

To form the summed vector states, the current sources will be biased pair-wise. In case a 45° shifted vector is desired, a construction of two inputs should be carried out by turning on two relevant current sources, I_0 and I_{90}. The output voltage is then formed by the sum of the 0° (v_i) and the 90° (j v_i) vector.

$$V_{out} = \frac{V_i + jV_i}{2} \cdot \frac{R_L}{R_S/2 + R_L} = \frac{V_i}{\sqrt{2}} \cdot \frac{R_L}{R_S/2 + R_L} \angle 45^o \quad (2)$$

Assuming a high load impedance, the network will generate the vector diagram as shown in figure 6B. When examining the equations of the voltage magnitudes, it can be seen that they are different. It is this difference, which will be used to our benefit. As mentioned in the previous section, a ratio of √2 between the cardinal states and the summed states is structural. Setting the voltage ratio equal to the inverse of this factor, gives:

$$\frac{\left(V_{O_1}/V_{I_1}\right)}{\left(V_{O_2}/V_{I_2}\right)} = \frac{\frac{1}{2}R_S + R_L}{R_S + R_L} = \frac{1}{\sqrt{2}} \Rightarrow R_L = \frac{R_S}{\sqrt{2}} \quad (3)$$

Consequently, by selecting this load impedance intrinsic compensation of the amplitude variation is accomplished.

VI.   COMBINING NETWORK

After the signals have been properly aligned by the phasing network, they need to be added. This is performed by means of four emitter followers driving the output bus. Figure 7 shows the bus driving topology. The real implementation of this network is just like all other sub-blocks done in a differential manner. First, ignoring the bottom transistor Q_C, the network performs the summation of the input voltages, with equal weights, assuming they are all biased equally. The overall gain of the BFC might change a little due to process tolerances. By placing an additional transistor (Q_C) at the bus network, allows us to accomplish amplitude compensation using a minimal number of additional components. The gain compensation is achieved by allowing a varying bias current through the compensation transistor, which controls the device transconductance. Equation 4 shows the generic equation for the summation.

$$Vout = \frac{\sum\limits_{i=1}^{4} v_i \cdot g_{m_i}}{\sum\limits_{i=1}^{4} g_{m_i} + g_{m_C}} = \frac{gm_i}{gm_i + gm_c} \sum\limits_{i=1}^{4} v_i \quad (4)$$

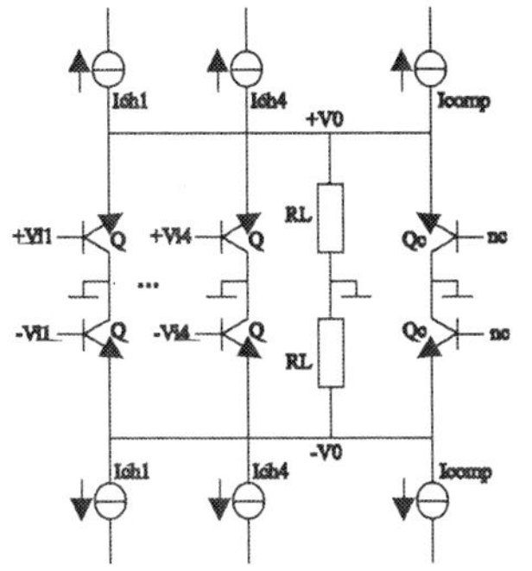

Fig. 7 Combining network without bias (de)coupling networks.

The last expression in this equation shows that the actual summation will not be influenced by the gm. All channels still have equal weights. The magnitude however is controlled by a scale factor. The control of this network is done digitally, using a 2 bit D/A convertor, allowing for 0.5 dB steps in gain.

VII. MISCELLANEOUS

The SG25H1 process with 5th thick metal layer of IHP microelectronics was used. This process is a more or less standard Silicon process with Germanium added for high frequency performance while maintaining all the benefits of mainstream Silicon technology.

Fig. 8 The Beam-Former Chip

The metal 5 layer was used due to a dense layout and to supply the die with 1500 mA @ 2V. The total current consumption of the BFC can be reduced by at least 500 mA when SKA is finally realized. SKA will be located in an area where interference will be absolutely minimum. Embrace has to deal with a broadcast station within 25 km, which gives tighter specifications for radio frequency interference.

MOS devices were used for the digital part of the chip to control the phase states of each channel separately, the two

independent beams and amplitude control. The digital input and output pins fulfil 4-kV (human-body model) ESD protection. Figure 8 shows a photograph of the Beam-Former Chip. The total BFC chip occupies an area of 10.2 mm² of silicon including the pads. The chip is mounted in a QFN56 package with ground paddle. The paddle is needed to conduct the heat from the chip to the PCB. The package is used for debugging and functional tests. Also several functional blocks of the BFC were processed independently for debugging. The dies were mounted on a PCB board, bonded and tested.

The overall performance of the phase settings are represented in figure 9. Eight phase states of one channel are noticeable at the output of the BFC.

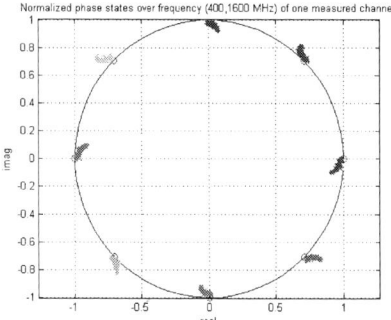

Fig. 9 Eight measured phase states of one channel

VIII. CONCLUSION

A SiGe beam-former for the SKA demonstrator project EMBRACE is reported. The functionality of this prototype is proven by measurement. Common mode suppression is accomplished to reject interference sources like broadcast stations at the horizon. In case interference signals are negligible the current consumption of the presented BFC can be reduced.

IX. ACKNOWLEDGEMENT

The authors acknowledge Dr. René Scholtz and his team from IHP-microelectronics for the support of the SG25H1 Design Kit, respectively, Albert van Duin and Eim Mulder of the support team within Astron for their contribution in the design, mounting and bonding of the test PCBs. This work is supported under the funds provided by the European Commission's FP6 research programme through a project known as SKADS.

X. REFERENCES

[1] J.G. bij de Vaate, G.W. Kant, "The Phased Array Approach to SKA, Results of a Demonstrater Project", European Microwave Conference, Milaan, Sept. 2002.

[2] P.T.M. van Zeijl, "A New High-Dynamic Range Dual-loop Power-to-Current Amplifier", IEEE Journal of Solid-State Circuits, vol. 24, no. 3, June 1989.

[3] Farbod Behdahani, Yoij Kishigami, John Leete and Asad A. Abidi, "CMOS Mixers and Polyphase Filters for Large Image Rejection", IEEE Journal of Solid-State Circuits, vol. 36, no. 6, June 2001.

[4] M.J. Gingell "Single Sideband Modelation using Sequence Asymmetrical Polyphase Networks", Electrical Cummunicatons, Vol. 48 (1973) – No.1 and 2

Proceedings of the 3rd European Microwave Integrated Circuits Conference

X- and K-band Tunable Phase Generation Circuits for Monolithic mm-Wave Phased Arrays

Corrado Carta [1], Munkyo Seo, Upamanyu Madhow and Mark Rodwell

Department of Electrical and Computer Engineering
University of California at Santa Barbara, CA 93106-9560, USA
[1]`corrado.carta@ieee.org`

Abstract— This paper presents design and characterization of two polyphase generation circuits, operating at X and K bands. These are crucial building blocks of an intended mixed-signal scalable architecture, suitable for integration of large beamformers on silicon technologies, operating at mm-wave frequencies. Circuits are designed with a mixed-signal approach: signals and topologies are digital, but performance and accuracy are achieved with analog-IC techniques. The novel topology proposed for the phase generator bases its operation on the time delay of matched ECL inverters loaded with quasi-matched varactors. One core phase shifter generates the phase gradient step, and it is reused several times to generate eight output differential signals of different phases. Without using any inductive component, the two phase generators provide 4 bit phase resolution in the 5-17 GHz and 11-26 GHz bands requiring 534 mW and 615 mW respectively, and are excellent candidates for the silicon integration of high-density beamformers.

I. INTRODUCTION

The large bandwidth available for mm-wave radio communications and the need of directivity for extending the usable link range motivate research interest in the integration of large phased array systems. Medium- to small-size phased arrays offer beam steering capabilities, but large-size arrays are necessary to enable WAN distances: e.g., radiating 15 dBm of signal power at 44 GHz, a pair or 10×10 arrays can transmit 1 Gbit/s over a 500 m link. While most of electronics complexity and cost of such systems is in the beamforming circuitry, modern and non-expensive silicon technologies offer performances sufficient for the 30-100 GHz range and are, thus, suitable for cost reduction by means of monolithic integration of mm-wave beamformers. This is, however, very difficult with present beamsteering IC architectures [1], [2], mostly because of their intrinsic unidimensional nature. The problem can be approached efficiently with a bi-dimensional row-column architecture [3], as its complexity and size grow with $\sqrt{N}$ for N-element arrays. The integration of very large monolithic beamformers can also benefit from the use of mixed-signal and digital blocks, instead of reactively-tuned ones: small die-area per pixel requires inductor-free designs; given the large number of array elements, the number of cascaded analog-RF stages must be minimized, as the gain variation accumulates and is very expensive to be controlled at high frequency. Larger array sizes are possible distributing LO signals as digital signals on long lines, and keeping the operation frequency of long signal/analog lines (IF) as low as possible, since attenuation is smaller and precise gain control

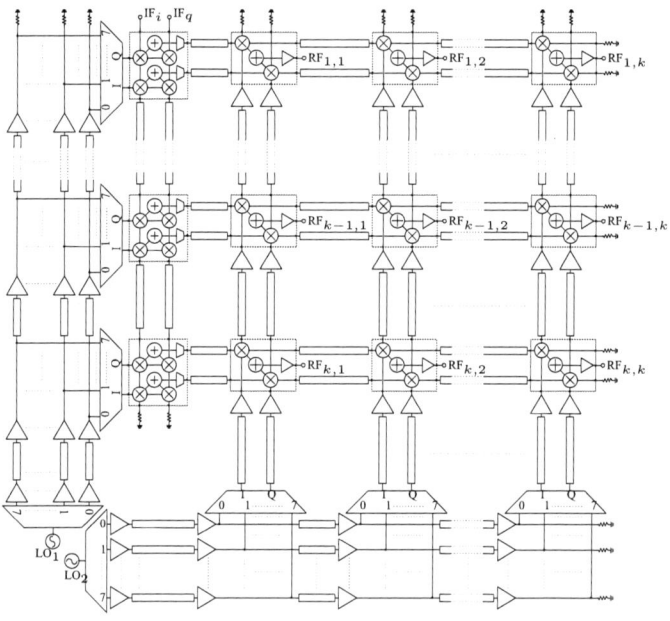

Fig. 1. The proposed architecture for large monolithic phased arrays. Phase generation blocks are shown in the bottom-left corner and drive the LO distribution buses on the bottom and left-hand edges.

is of simpler implementation.

The beamformer system architecture shown in Fig. 1 and described in detail in [3] bases its operation on steering the beam in altitude and azimuth by separately imposing vertical and horizontal phase gradients. This simple assumption enables a neat separation of IF and LO distribution lines, and allows improvements in both array size and maximal operation frequency by means of two techniques: shifting most of wiring complexity to the edges of the system; distributing fast LO signals on digital buses and low-frequency IF signals on analog buses. This mixed-signal technique enables RF output frequencies higher than those possible on analog transmission lines of given length and technology [3].

Making use of two LO signals and distributed double-upconversion double-quadrature transmitters, the input I-Q IF signals are first multiplied by a vertical phase gradient, then an independent horizontal gradient is applied and summed by means of the second upconversion mixing. At each row and

978-2-8748-7007-1/08 $25.00 © 2008 EuMA   498

column front-end, the best LO phase is selected from an n-phase LO signal, distributed on digital LO buses on the south and west edges of the array.

The limit to scalability is set by the maximal length of analog and digital buses, and depends on the system frequency plan and available technology. However, other fundamental system limits are related to the performance of the crucial blocks generating, distributing and selecting the multiphase LO signals: this paper describes in detail design and characterization of circuits suitable for polyphase LO-signal generation, which are crucial for the integration of large beamformers based on the intended architecture and requiring LOs in X and K bands.

II. CIRCUIT DESIGN

The phase-generator function consists in creating an n-phase signal from an input single-phase LO; this polyphase signal is then fed to the first LO digital buses. The generation of signals of equally spaced phases must rely on the use of a phase shifter suitable for use in a periodic structure. Many phase shifters base their operation on periodic structures, such as synthetic transmission lines using lumped elements [4] or actual transmission line sections [5], with the purpose of extending the tuning range of a core phase shift block. These topologies would be able to generate poly-phase signals, but the amplitude-loss imbalance at output ports limits their use to a small number of generated phases. Moreover, the use of inductors or transmission-line sections always increases the area required for the implementation of the phase generation blocks, specially when a large number of phases, e.g. 16, has to be available simultaneously.

For generating poly-phase LO signals, which in most cases do not carry amplitude information, digital electronics building blocks can be exploited: as shown in Fig. 2(a) for a single section, the cascade of N inverters driven by an input LO signal can provide N different phases at each interface. The input-output phase difference depends directly on the output capacitive load and can be adjusted with a small varactor or, as shown, with a pair driven by one control voltage V_{ctrl}. A similar structure, in closed-loop configuration, is known as *ring VCO* and may use tuned LC loads to boost inverter performance at a certain resonance frequency [1]. In open loop configuration, an input LO signal is required, but the operation frequency depends only on the phase gradient required and on the cut-off frequency of the inverters. Load-capacitance control allows to compensate process tolerances and select the desired phase shift on a frequency range which depends on the varactor tuning ratio. At very high frequencies, this approach is severely limited by the minimum delay achievable with the available inverter: fixed the frequency, there will be a minimum phase step; for a given phase step, there will be a maximal operation frequency.

For high-frequency *and* high-resolution applications, e.g. 16 phases at 30 GHz, we propose to generate different phases exploiting the phase difference at the output of two matched inverters loaded with quasi-matched varactors. As shown in

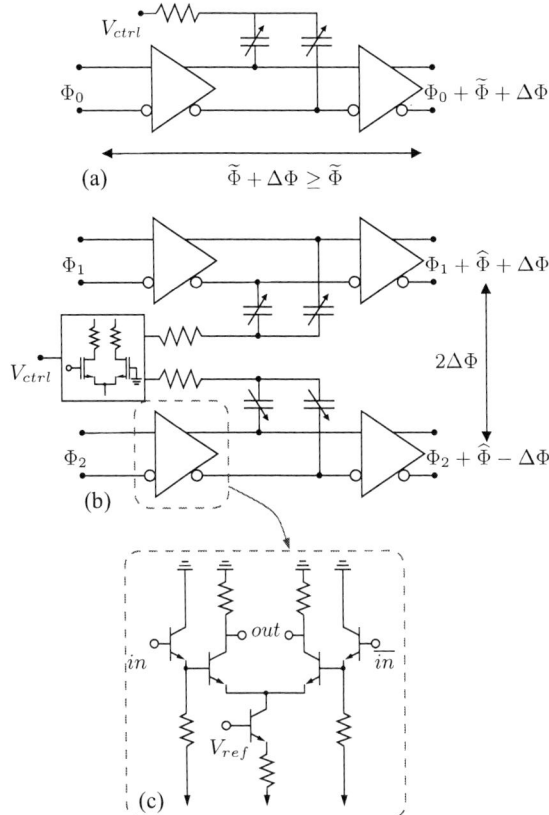

Fig. 2. Polyphase LO signal generation realized by means of tuned inverter delays: the cascade of N inverters can provide N different phases of the same signal (a); two matched inverters loaded with quasi-matched capacitors can provide small phase differences (b); implementations presented in this paper exploit ECL-logic inverters (c).

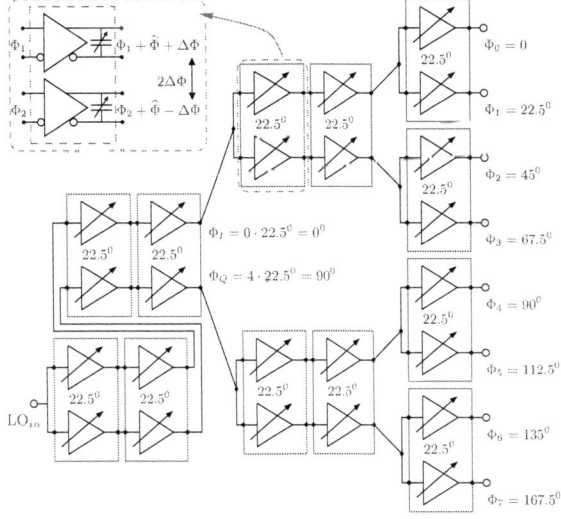

Fig. 3. System-level schematic of a 16-phase generator. Dashed box shows the details of the differential core shift element.

Fig 2(b), it is possible to increase the phase difference of two signals by an arbitrarily-small phase delay: two signals of different phases are fed into two matched inverters, loaded with two pairs of matched varactors of slightly different value. By means of a differential pair, the time delay difference can be increased and reduced around the nominal value with one control voltage. At a given frequency of operation, the phase difference between the two signals depends on the varactor control voltages and size ratio. Exploiting this basic idea, it is possible to generate 2^n phases arranging in a binary tree several blocks, each producing a $360°/2^n$ phase shift. Fig. 3 shows the detail of a 16-phase design, based on a $22.5°$ building block. The circuit bases its operation on the use of one basic phase-shift block: phase matching among different instances relies on same-die matching; phase accuracy is achieved by analog continuous adjustment of the varactor, controlled by the same voltage in each instance.

Implementing the topologies shown in Fig. 2(b) and Fig. 3, two phase generators were design and fabricated, optimized for X and K bands, in order to be capable of generating 16 phases of a differential LO signal provided externally. Digital inverters are in ECL logic (Fig. 2(c)), optimized for operating at peak f_t collector current density, except the last buffer stages, designed to feed the $\pm300\,$mV ECL digital signal to the output $50\,\Omega$ transmission-lines. The X-band version consists of 252 HBTs, 192 $r$esistors and 48 $v$aractors, while the K-band version employs 372 HBTs, as in each core shift block an extra inverter is used before driving the varactor: this extend the phase-shifter cutoff frequency.

III. EXPERIMENTAL RESULTS

A. Fabrication

The phase generation ICs were fabricated on a commercially-available $0.18\,\mu$m BiCMOS technology, featuring f_t=155 GHz, f_{max}=200 GHz HBTs and high-quality passive elements.

Available die area limits the number of outputs to two differential signals; the ICs were fabricated with contact pads on two quadrature signals, for characterization ease. Phase accuracy of the remaining outputs relies on same-die process tolerances.

A photograph of the X-band version is shown in Fig. 4(a): IC measures $1.34\,$mm$\times0.68\,$mm, and circuit active area occupies $0.13\,$mm^2. A photograph of the the K-band phase generator is in Fig. 4(b): in the same pad-limited IC area, the K-band version active area is $0.15\,$mm^2.

B. Circuit Characterization

The two phase generation ICs were tested on wafer. The X-band and K-band versions require $181\,$mA and $205\,$mA of bias current, respectively, from -3 V. Output signals are two differential ECL waveforms, whose phase difference depends on a control voltage provided externally. The two output signals were observed by means of a two-channel 50 GHz sampling scope and phase differences derived as time

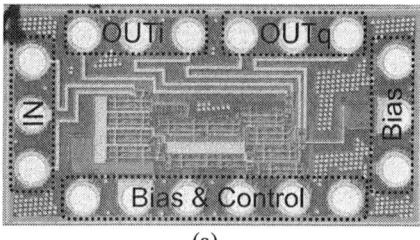

(a)

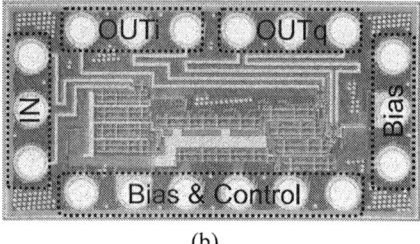

(b)

Fig. 4. IC photographs: phase generation ICs operating at X-band (a) and K-band (b). Die area shown measures $1.34\,$mm$\times0.68\,$mm in both cases, while active are used is $0.13\,$mm^2 for the X-band version and $0.15\,$mm^2 for the K-band version. At the highest frequency of operation, the core phase shifter needs one extra inverter to restore digital levels between stages.

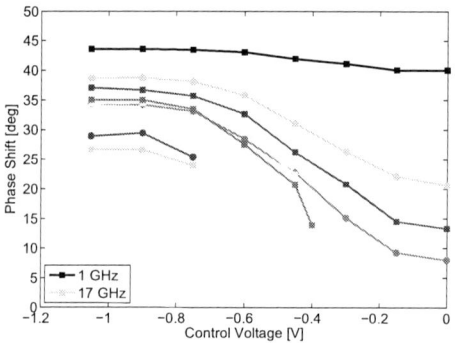

Fig. 5. Unit phase shift as function of control voltage for the X-band version of the phase generator. Frequencies shown are spaced by 2 GHz.

delay measurements or Fourier analysis; the two approaches returned phase measurements consistent within $\pm1°$ in the band of operation, as higher order harmonics are filtered by the inverter frequency response. Results presented in this paragraph are elaboration of time delay measurements.

Fig. 5 shows the phase shift of the core shift element reused hierarchically in the X-band phase generator. The measured phase shift as function of control voltage is shown for a set of frequencies spaced by 2 GHz. In the usable band, the planned 1.1 V control range results in a 20-25° tuning range, centered around the 22.5° target. Maximal operation frequency for this circuit is 17 GHz: since the additional capacitive load of the varactors reduces inverter cutoff frequency, at frequencies higher than 17 GHz digital voltage

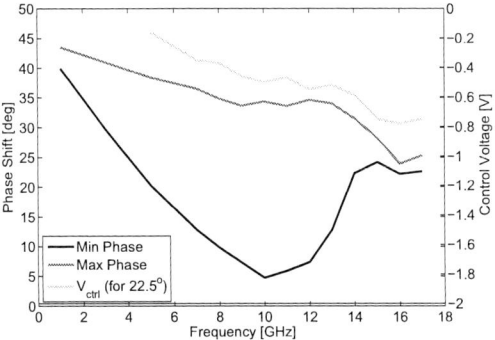

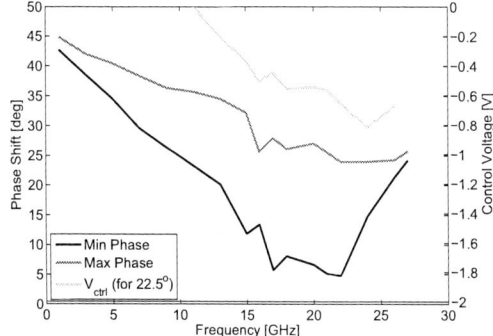

Fig. 6. Measured phase difference produced by the core phase shifter of the K-band phase generator. The target phase of 22.5° is selectable from 5 GHz to 17 GHz.

Fig. 8. Measured phase difference for the core phase shifter of the X-band phase generator. Useable band is 11 GHz to 26 GHz.

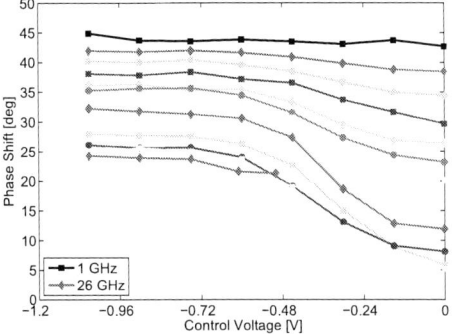

Fig. 7. Unit phase shift as function of control voltage for the K-band version of the phase generator. Frequencies are spaced by about 2 GHz.

levels are not reached at the output for any value of the control voltage; at the highest useable frequencies, the tuning range is progressively reduced, because of the same effect. Fig. 5 shows the frequency behavior of maximal and minimal selectable phases, together with the control voltage required for the target phase of 22.5°. The phase generator is capable of providing the required phase gradient in the 5-17 GHz band.

Fig. 7 shows measured phase shifts for the K-band version. A tuning range similar to that of the X-band version is available at an higher frequency: maximal frequency of operation for this circuit is 27 GHz. Fig. 8 shows the measured phase difference produced in the K-band IC by the core phase shifter as function of frequency; for each frequency, minimal and maximal phase shift correspond to the ends of the range achievable by tuning the varactor. The K-band phase-shift block can be set to the required 22.5° in the 11-26 GHz band.

IV. CONCLUSION

The proposed system architecture for a row/column beam-former IC bases its operation on the availability of on-chip generated polyphase LO signals. The design and implementation of building blocks providing the phase generation function has been presented. As the row/column architecture exploits analog and digital design techniques in order to extend its scalability and operation frequency limits, the design of the phase generation components similarly benefits of both approaches. Polyphase LO signals are generated combining digital ECL inverters, but phase accuracy and control is achieved by means of on-chip device matching and continuous capacitive load tuning, i.e. analog techniques. This innovative design approach results in satisfactory performance at X and K band: without employing any inductive component, presented phase generation circuits are excellent candidates for the integration of very large mixed-signal phased arrays operating in the 30-40 GHz bands.

ACKNOWLEDGEMENT

This work was funded by the DARPA SMART program and National Science Foundation, under grants *ECS-0636621* and *CNS-0520335*. The authors wish to thank N. Desai and A. Ballinger for their assistance in characterizing the circuits.

REFERENCES

[1] G. Xiang, G. Xiang, H. Hashemi, A. Komijani, and A. A. H. A. Hajimiri, "Multiple phase generation and distribution for a fully-integrated 24-GHz phased-array receiver in silicon," in *Radio Frequency Integrated Circuits (RFIC) Symposium, 2004. Digest of Papers. 2004 IEEE*, H. Hashemi, Ed., 2004, pp. 229–232.

[2] K. Kwang-Jin, K. Kwang-Jin, and G. M. Rebeiz, "0.13-um CMOS Phase Shifters for X-, Ku-, and K-Band Phased Arrays," *Solid-State Circuits, IEEE Journal of*, vol. 42, no. 11, pp. 2535–2546, 2007, 0018-9200.

[3] C. Carta, M. Seo, and M. Rodwell, "A Mixed-Signal Row/Column Architecture For Very Large Monolithic mm-Wave Phased Arrays," in *Lester Eastman Conference on High Performance Devices*, 2006, pp. 111–114.

[4] F. Ellinger, H. Jackel, and W. Bachtold, "Varactor-loaded transmission-line phase shifter at C-band using lumped elements," *Microwave Theory and Techniques, IEEE Transactions on*, vol. 51, no. 4, pp. 1135–1140, 2003, 0018-9480.

[5] K. Dong-Woo, K. Dong-Woo, and H. Songcheol, "A 4-bit CMOS Phase Shifter Using Distributed Active Switches," *Microwave Theory and Techniques, IEEE Transactions on*, vol. 55, no. 7, pp. 1476–1483, 2007, 0018-9480.

Proceedings of the 3rd European Microwave Integrated Circuits Conference

A Four-Antenna Transceiver MIMIC for 60 GHz Wireless Multimedia Applications

S. Koch [#1], I. Kallfass [*2], A. Leuther [*2], M. Schlechtweg [*2], S. Saito [#1] and M. Uno [#3]

[#1] *Sony Deutschland GmbH, Hedelfinger Strasse 61, D-70327 Stuttgart, Germany*
Email: koch@sony.de

[*2] *Fraunhofer Institute for Applied Solid-State Physics (IAF), D-79108 Freiburg, Germany*

[#3] *Sony Corporation, 5-1-12 Kitashinagawa Shinagawa-ku, Tokyo, 141-0001, Japan*

Abstract— **A transceiver MIMIC (millimeter wave monolithic integrated circuit) with four antenna ports functionality for 55 to 65 GHz wireless multimedia applications has been developed. The chip has been realized using 100 nm gatelength metamorphic InAlAs / InGaAs HEMT (high electron mobility transistor) technology on GaAs substrate together with GCPW (grounded coplanar waveguide) technology. The novel transceiver topology for the MIMIC is described. The different building blocks, namely the 2:4-switch design, sub-harmonic IQ-resistive mixer and variable gain amplifiers, are explained in details. Simulated and measured results are compared. Finally, the overall performance for the integrated transceiver chip is presented.**

I. INTRODUCTION

During the last years, the usage of wireless Gbit/s data rate system is heavily investigated for future HD (high definition) multimedia applications (such as TV- or video-streaming). Within the IEEE 802.15.3c standardization body, applications up to 5 Gbps are suggested using the 60 GHz ISM (industrial, scientific and medical) band.

To cope with propagation channel distortion and system gain margins a multi-beam antenna system is the target for the MIMIC investigated in this paper. Such a system supports adaptively selectable antenna beams of a beam switched antenna with certain gain and half-power beam width to realize a sectorization of the antenna view area. For the support of such systems, an antenna port switching capability must be included in the mm-wave system design.

There have been many publications within the last years dealing with 60 GHz transceivers, transmitters, receivers or building blocks, using compound semiconductors [1] - [5] or silicon based processes [6] - [8]. None of these publications deal with multiple antenna switching capabilities. The MIMIC presented in this paper offers such antenna beam switching functionality.

The design is a rigorous enhancement of the MIMIC presented in [9] towards a novel, fully integrated transceiver offering four antenna ports switching. By using a new type of 2:4 switch, the chip size could be reduced by ~ 30 %. Additionally, frequency conversion capability realized by an IQ-sub-harmonic resistive mixer has been added.

A detailed explanation of the MIMIC topology will be given. The results for the individual building blocks as

variable gain amplifiers, a new type of 2:4 switch, the IQ-sub-harmonic resistive mixer, and finally the overall conversion performance of the chip will be explained.

II. TRANSCEIVER TOPOLOGY

The novel transceiver topology of the MIMIC is given in Fig. 1. At the output, the MIMIC comprises a 2:4 antenna switch. With this switch, the signals at the antenna ports (ANTi, i = 1...4) are routed to either a low-noise amplifier (LNA) or a medium power amplifier (MPA) according Rx- or Tx-operation, respectively.

The I- and Q-baseband signals are down- or up-converted using a sub-harmonic pumped resistive mixer. Both, the receiver as the transmitter amplifier are carried out as variable gain amplifiers to be capable of gain adjustment during system optimization and operation.

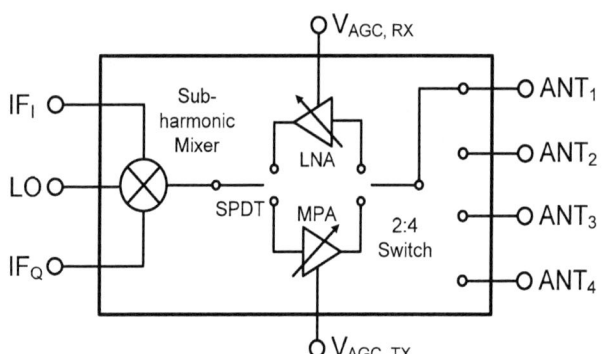

Fig. 1: Novel topology of the four-antenna transceiver MIMIC.

The mixer is driven by an off-chip 30 GHz voltage controlled oscillator. A chip photograph is given in Fig. 2. The chip has a size of 2.5 x 5.5 mm².

The transceiver is realized using a 100 nm gate length metamorphic InAlAs / InGaAs HEMT process [10] offered by Fraunhofer Institute for Applied Solid-State Physics (IAF). The process technology offers a transit frequency $f_T > 200$ GHz and $f_{MAX} > 300$ GHz. Therefore the technology is well suited for 60 GHz applications.

978-2-8748-7007-1/08 $25.00 © 2008 EuMA

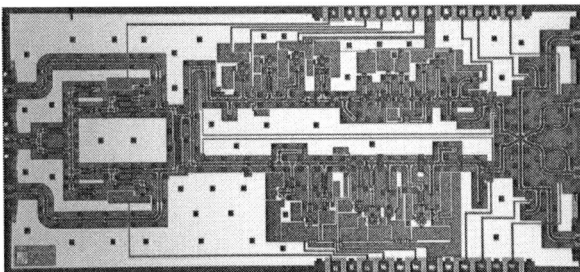

Fig. 2: Chip photograph of the integrated four-antenna MIMIC-transceiver; Chipsize 2.5 x 5.5 mm2

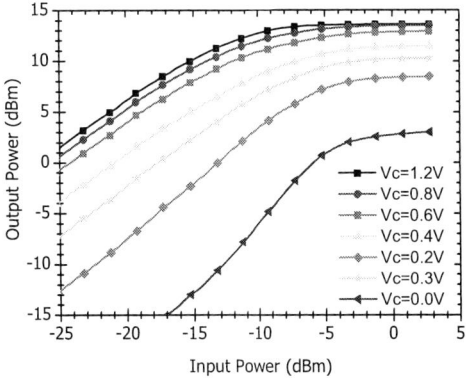

Fig. 3: MPA measured output power versus input power levels for different cascode control voltage settings (Vc, second stage) at frequency of 60 GHz. The drain voltages are set to 1.2 V and 2.2 V for the common-source and cascode stages, respectively. The gate voltage is set for maximum transconductance gM,MAX.

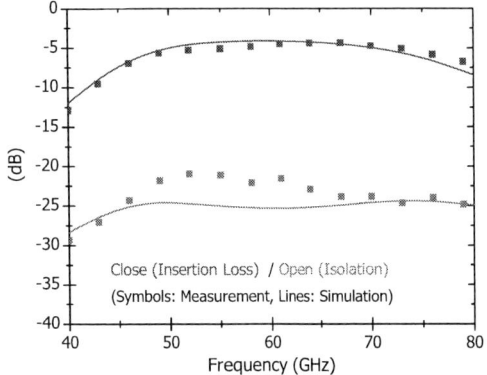

Fig. 4: Measurement results versus simulations for insertion loss and isolation of the 2:4-switch.

III. TRANSCEIVER BUILDING BLOCKS

A. Amplifiers

The design for the receiver LNA was modified from the design presented in [9]. In addition to the three-stage LNA with fixed gain, a two-stage variable gain amplifier was added at the output of the LNA. This adds additional 17 dB gain with a gain control in the order of > 10 dB. The gain control capability was added for system optimization. The excellent noise figure (without switching losses), determined by the LNA, is in the order of NF < 2.6 dB for a biasing voltage of 1.0 V at maximum transconductance ($g_{M,MAX}$).

The MPA design was changed from a three-stage constant gain amplifier to a four-stage variable gain design to adopt different TX input power levels. The gain control was realized by incorporating a cascode transistor as second stage.

The achieved gain control is again more than 10 dB and only limited due to bias constraints of the FET elements. The measured output power behaviour versus input power level for different settings of the control voltage Vc of the modified MPA is given in Fig. 3.

B. Switch Design

Two different switches are incorporated into the transceiver itself. A single pole double throw (SPDT) switch is used for switching receive- or transmit-path from the mixer towards the respective amplifier stages (LNA or MPA). The second switch implementation is a 2:4-switch employed as an antenna selection switch. The design of the switches is based on FETs placed in shunt with the signal lines. With the gate control the FET channel conductance is controlled for on-off states switching.

For quasi "open circuit" of a switch branch, a λ/4 input line ensures high isolation. On the other hand, during quasi "short circuit" of the FET, a parallel shorted stub achieves a resonant circuitry to minimize switch insertion loss and power matching. More detailed information about the switch design methodology can be found in [13]. The 2:4-switch achieves an insertion loss < 5 dB and isolation > 20 dB from 55 to 65 GHz. The measured versus the simulated results are given in Fig. 4.

C. Sub-Harmonic IQ-Resistive Mixer

For the mixer design it was decided to adopt a sub-harmonic IQ-resistive mixer topology to achieve high LO-RF-isolation, thus avoiding LO-filter networks. Filter networks are difficult to realize on MIMIC level and would yield a reduction of flexibility and increase of chip size.

Furthermore the resistive mixer design was chosen to offer the advantage of good intermodulation behaviour and low power consumption compared to e.g. active mixer types.

The principal design of the mixer core-cell is explained in literature [11]. By combining two core mixer cells together with a Wilkinson power divider and a 90-degree Lange-coupler on the LO- and RF-side respectively, an IQ-mixer was realized. To minimize chip area, transmission lines (TRLs) in the LO-matching circuitry as well as in the Wilkinson power divider have been replaced by "spiral folded" type of transmission lines (TRLs). Due to this approach, the overall mixer core occupies a chip area of only 1.5 x1.5 mm². The schematic of the mixer, incorporating the spiral folded TRLs, is shown in Fig. 5. A photograph of the detailed mixer section is given in Fig. 6.

The spiral folded TRLs are realized in a way that ground-to-ground spacing of the GCPW transmission line did not change. For implementation, the spiral folded TRLs have been simulated using the IE3D field simulator. In Fig. 7 the measured versus simulated results are presented for a spiral

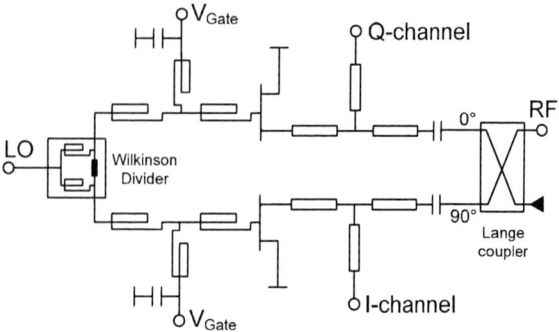

Fig. 5: Topology of sub-harmonic IQ-resistive mixer. Spiral folded transmission lines (TRLs) are indicated on LO-side (used in Wilkinson power divider and matching circuitry).

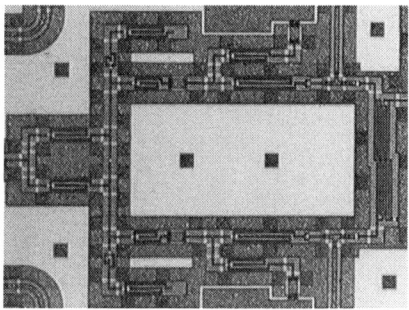

Fig. 6: Detailed photograph of the mixer section.

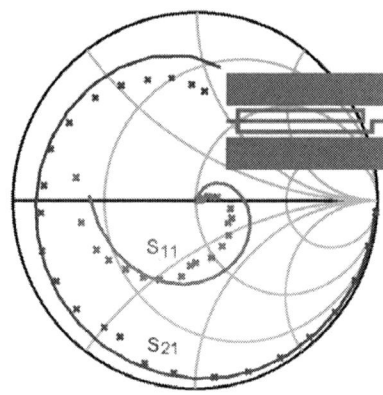

Fig. 7: Measured versus simulated (IE3D) results for a spiral folded transmission line (TRL) (equivalent to a transmission line with a geometrical length of 430 μm and 70 Ω line impedance). Frequency range is 1 to 120 GHz. Lines are simulations and symbols are measurements. Blue colour represents S_{21} and red colour S_{11}.

folded TRL replacing a TRL with a geometrical length of 430 μm and 70 Ω line impedance. There is a good agreement between measured and simulated results in the frequency range 1 to 120 GHz. The spiral folded TRL reduces the geometrical length from 430 μm to 180 μm in this specific case. Different spiral folded TRLs have been employed in the mixer design.

The measured conversion loss for the mixer in Rx-operation is between 13 -15 dB (per I- or Q-channel) within the LO-frequency range from 27.5 to 32.5 GHz. The LO-power level was set to + 10 dBm and IF-frequency was fixed to 100 MHz. The observed IQ-imbalance is smaller than 1 dB. For up-conversion negligible differences in conversion loss were observed.

The LO-suppression measured during up-conversion was in the order of 25 dB for the same LO-frequency range. For a LO-power sweep the measured versus the simulated conversion gain results (down-conversion) of the mixer are plotted in Fig. 8. The LO-frequency is set to 30 GHz, the RF-frequency to 60.1 GHz. The LO-power is swept from 0 to 12 dBm power level. There is a good agreement between measured and simulated data.

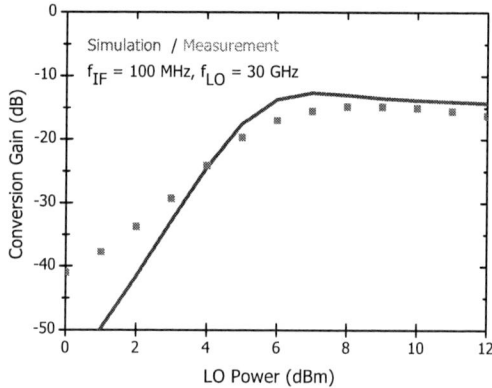

Fig. 8: Measurement results versus simulations for down conversion gain of the sub-harmonic resistive mixer as a function of LO-power. LO-frequency is fixed to 30 GHz. IF-frequency is 100 MHz. The single carrier RF input power level is set to - 15 dBm.

IV. TRANSCEIVER PERFORMANCE

The overall performance of the transceiver was measured on-wafer using GSG- (Ground-Signal-Ground) and DC-probe heads to support biasing voltages. The drain voltages were set to 1.0...1.2 V for common source FETs. For cascode FET configurations the drain voltages were set to 2.0...2.2 V. The gate voltages were adjusted for maximum transconductance which is achieved for voltages in the order of 0.1...0.15 V. Based on these settings, the overall power dissipation of the transceiver is 150 mW / 200 mW under receive / transmit operation condition.

In Fig. 9 the conversion gain for transmit- and receive-operation is given in comparison to the simulated data. The conversion gain was measured on I- or Q-channel of the transceiver with the other channel terminated by a 50 Ω load.

The input power levels were chosen small enough to avoid saturation effects in the building blocks and therefore were set to 0 dBm for up- and -20 dBm for down-conversion.

The comparison of the measured data with simulation gives good agreement which validates modelling, simulation and design methodology.

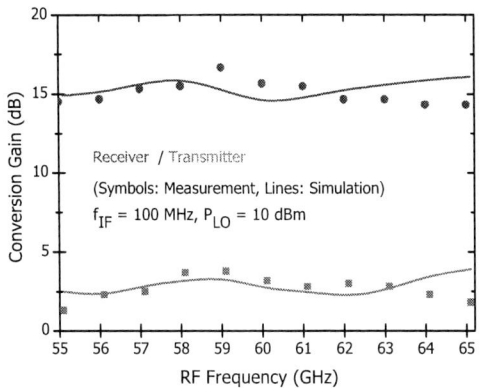

Fig. 9: Measurement versus simulation results for conversion gain of transceiver. LO-power level is + 10 dBm and IF-frequency is 100 MHz. LO-frequency is swept and RF is set / measured accordingly.

V. CONCLUSIONS

The conversion performance of a novel four-antenna transceiver MIMIC for 55 - 65 GHz frequency range has been presented. Different building blocks of the transceiver, namely the 2:4-switch design, sub-harmonic IQ-resistive mixer, variable gain amplifiers and spiral folded transmission lines, have been described and results given. Measured results agree well with simulations, especially when taking the circuit complexity into account. The chip size has been reduced by 30% compared to earlier results in combination with enhanced functionality of the MIMIC.

VI. ACKNOWLEDGEMENT

The authors wish to acknowledge the contributions of H. Massler and A. Tessmann at IAF for continuous support during design and measurements.

REFERENCES

[1] S. Gunnarsson, C. Kärnfelt, H. Zirath, R. Kozhuharov, D. Kuylenstierna, A. Alping, and C. Fager, "Highly Integrated 60 GHz Transmitter and Receiver MMICs in a GaAs pHEMT Technology," *IEEE Journal of Solid-State Circuits.*, vol. 40, no. 11, pp. 2174-2186, November 2005.

[2] K. Fujii, M. Adamski, P. Bianco, D. Gunyan, J. Hall, R. Kishimura, D. Lesko, M. Schefer, S. Hessel, H. Morkner, A. Niedzwiecki, "A 60GHz, MMIC Chipset for 1-Gbit/s Wireless Links", *2002 IEEE MTT-S Int. Microwave Symp. Dig.*, pp. 1725-1728.

[3] O. Vaudescal, B. Lefebvre, V. Lehoué, and P. Quentin, "A Highly Integrated MMIC Chipset for 60 GHz Broadband Wireless Applications," *2002 IEEE MTT-S Int. Microwave Symp. Dig.*, pp. 1729-1732.

[4] Y. Mimino, K. Nakamura, Y. Hasegawa, Y. Aoki, S. Kuroda, and T. Tokumitsu, "A 60 GHz Millimeter-wave MMIC Chipset for Broadband Wireless Access System Front-end", *2002 IEEE MTT-S Int. Microwave Symp. Dig.*, pp. 1721-1724.

[5] K. Ohata, K. Maruhashi, M. Ito, S. Kishimoto, K. Ikuina, T. Hashiguchi, N. Takahashi, S. Iwanagam, "Wireless 1.25Gh/s Transceiver Module at 60GHz-Band", *2002 IEEE International Solid-State Circuits Conference Dig.*, February 2002.

[6] B. Floyd, S. Reynolds, U. Pfeiffer, T. Beukema, J. Grzyb, and C. Haymes, "A Silicon 60 GHz Receiver and Transmitter Chipset for Broadband Communications", *2006 IEEE International Solid-State Circuits Conference Dig.*, vol. 49, pp. 130-131, February 2006.

[7] C-H. Wang, Y-H. Cho, C-S. Lin, H. Wanf, C-H. Chen, D-C. Niu, J. Yeh, C-Y. Lee, and J. Chern, "A 60 GHz Transmitter with Integrated Antenna in 0.18μm SiGe BiCMOS Technology", *2006 IEEE International Solid-State Circuits Conference Dig.*, vol. 49, pp. 132-133, February 2006

[8] B. Gaucher, T. Beukema, S. Reynolds, B. Gloyd, T. Zwick, U. Pfeiffer, D. Lin, J. Cressler, "MM-Wave Transceivers Using SiGe HBT Technology", *2004 Topical Meeting on Silicon Integrted Circuits in RF Systems*, pp. 81 – 84.

[9] S. Koch, I. Kallfass, R. Weber, A. Leuther, M. Schlechtweg, M. Uno, "An Analogue, 4:2 MUX/DEMUX Front-End MIMIC for Wireless 60 GHz Multiple Antenna Transceivers", 2007 IEEE MTT-S Int. Microwave Symp. Dig., pp. 1121-1124.

[10] A. Leuther, A. Tessmann, M. Dammann, W. Reinert, M. Schlechtweg, M. Mikulla, M. Walther, and G. Weimann, "70 nm low noise metamorphic HEMT technology on 4 inch GaAs wafers," in *Proc. Indium Phosphide Related Materials Conf. (IPRM)*, May 2003, pp. 215–21

[11] N. A. Rahman, B. Y. Majlis, A. Ariffin, " A 28 GHz PHEMT GaAs MMIC Single-ended Resistive Mixer", ICSE7.002 Proc. 2002, Penang, Malaysia, pp. 511 – 513.

[12] M. Lang, A. Leuther, W. Benz, U. Nowotny, O. Kappeler, and M. Schlechtweg, "66 GHz 2:1 static frequency divider using 100 nm metamorphic enhancement HEMT technology", *Electron. Lett.*, vol. 38, no. 14, pp. 716-717, July 2002.

[13] I. Kallfass, S. Diebold, H. Massler, S. Koch, M. Seelmann-Eggebert, A. Leuther, "Multiple-Throw Millimeter-Wave FET Switches for Frequencies from 60 up to 120 GHz", *submitted for EuMW 2008 acceptance.*

A multi-channel S-band FMCW radar front-end

A.P.M. Maas[1], F.E. van Vliet[2]

TNO Defence, Security and Safety, Oude Waalsdorperweg 63, 2597 AK Den Haag, Netherlands
[1]noud.maas@tno.nl
[2]frank.vanvliet@tno.nl

Abstract— **This paper describes the design and performance of a low-cost synthesized FMCW radar module, operating in S band. The bi-layer PCB contains a frequency-agile low phase-noise synthesizer and three identical coherent receive-channels. The transmit channel has an automatic power control system that reduces the output power when a large reflection causes the receiver input level to exceed the linear input range. Standard surface-mount components from commercial WLAN applications have been used to create a versatile programmable radar module.**

The DDS-based PLL synthesizer achieves a SSB phase-noise level of -101 dBc/Hz @ 10 kHz offset from a 2.4 GHz carrier. A basic serial PC interface enables control of the FMCW radar parameters that can be stored in the on-board non-volatile memory. The complete front-end operates from a single 3.6 - 5.5 Volts supply, drawing 220 mA and measures only 55 x 100 mm.

I. INTRODUCTION

FMCW radar systems are often the preferred choice for low-cost, simple radar systems, for applications as diverse as height sensing and presence detection.

This paper describes the implementation of one, typical, front-end which was originally designed for short-range 3-D "through-the-wall" radar imaging applications, although it has found much wider applications.

The original application required a USB-powered (mobile) multi-channel front-end, which could later be interfaced with a portable signal processing device. Different antennas and measurement set-ups were envisaged.

The design needed a careful balancing between performance and production cost. The wide application area forced relatively stringent performance requirements and at the same time forced absolute cost limits.

II. DESIGN FEATURES

To accommodate for the wide application range, several features were added to make the front-end fail-safe and flexible:

- One example of this flexibility is the fact that the maximum output power can be selected by changing the value of the on-board attenuators, to adapt the transmit power for the antenna used, keeping the EIRP below the maximum value according to the IEEE802.11b standard.
- A second example of this flexibility is the implemented ability to program the FMCW radar parameters through a simple RS232 connection and store them in the internal non-volatile memory.

Direct switching between two different pre-defined sweep-profiles is provided by an on-board DIP-switch.

III. DETAILED DESIGN

A block schematic of the complete module is shown in Figure 1.

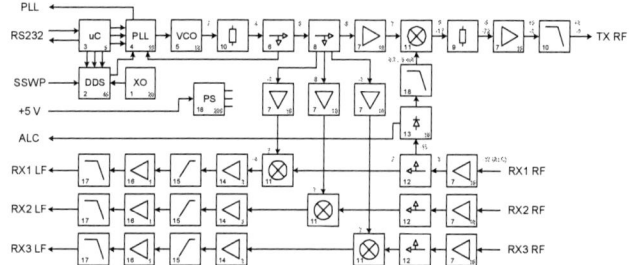

Fig. 1 Block diagram of the implemented front-end

As can be seen in the block diagram, a single DDS-PLL synthesizer combination was chosen to generate the microwave signal. Signal levels are indicated in the block diagram in red. Due to the integration of receiver and transmitter on a single board, re-use of the transmitted signal for down-conversion in the receiver could be implemented in a straightforward manner.

Multiple identical receivers were implemented in such a way that additional receivers could be added without major redesigns.

A. Frequency Synthesizer design

A 400 MHz commercial Direct Digital Synthesizer (DDS) IC is used, running at an on-board generated 125 MHz clock-frequency for a reduced the supply current. The DDS can now be programmed to generate sinusoidal output signals up to 50 MHz with a very low spurious content of typically less than -70 dBc.

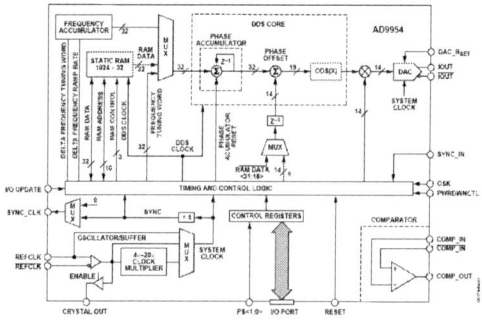

Fig. 2 Block-diagram of the DDS used (from AD9954 datasheet)

An external low-pass filter is required to remove the image frequency from the 14-bit resolution DAC output.

The particular DDS used also has the ability to generate a programmable automatic linear sweep. The sweep sequence is started immediately after the detection of a trigger signal on the trigger input of the device.

The DDS output signal is fed to the reference frequency input of a standard integrated Phase Locked Loop (PLL) IC. A low-voltage commercial 2400 MHz VCO is used which is then phase-locked to the DDS output signal, using a fixed divider ratio. In this way, the DDS output is effectively multiplied in frequency from 50 MHz to 2400 MHz.

PSPICE was used to simulate the critical parts of the circuit, in particular for the synthesizer system. A behavioural simulation model was made (shown in figure 3) for the PLL system to evaluate the transient behaviour of the loop as well as the phase-noise level of the output signal.

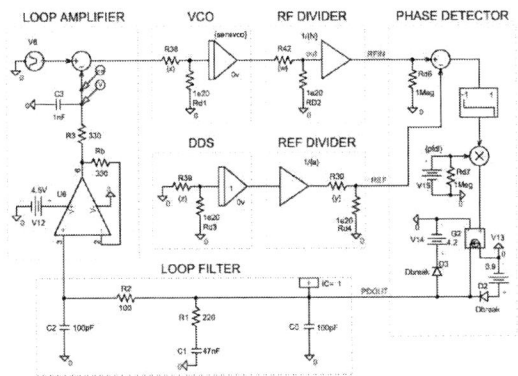

Fig. 3 PSPICE behavioural model for the PLL system

The circuit was optimised for sufficient transient response and lowest possible phase-noise level. The final loop bandwidth was set a 250 kHz, enabling sufficient tracking capability of the PLL at DDS sweep durations down to 50 µs. The simulated phase-noise level is shown in figure 4 and includes the noise-contributions from the 125 MHz Crystal Oscillator, the VCO, the DDS, the digital phase-detector of the PLL, and the loop-amplifier and loop-filter electronics.

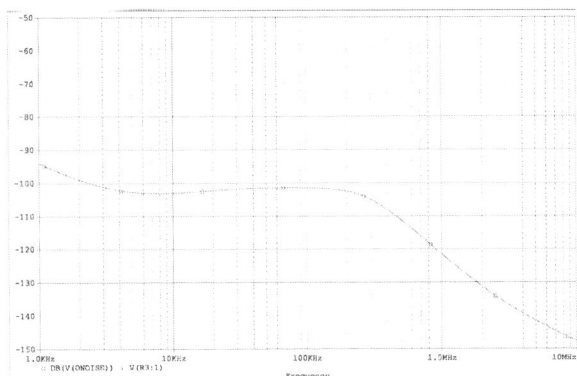

Fig. 4 Simulated SSB phase-noise of the 2400 MHz DDS-PLL

Since both the DDS and the PLL circuits need to be programmed and have a serial interface, a simple microcontroller is selected to perform this essential task at the start-up of the power-supply. Multiple other functions have been added, such as:

- An RS232 interface to remotely change the PLL and DDS parameters and store them in the non-volatile memory
- A buffered PLL Lock Detect Error output
- Sensing of the on-board DIP-switch to directly select between 2 different sweep-profiles without using the RS232 interface
- On-board generation of a programmable DDS sweep trigger signal (alternative to connecting an external trigger input signal)

B. RF & DC distribution

To be able to distribute both RF and DC from the synthesizer and the power supply circuit to the different channels of the module using a single metal layer, the DC signal from the power supply is superimposed on the RF output signal of the synthesizer. The RF signal is then split into 4 RF output signals using small surface-mount Wilkinson power splitters, capable of carrying the required DC current.

C. Receiver channel design

Each of the three receiver channels is basically identical and has a Local Oscillator (LO) & DC input, an RF Receive (RX) input and a Low Frequency (LF) output containing the down-converted beat-note signal of the radar. A coupled RF output port is designed, making it possible to measure the received signal power at the input of the mixer, which will be used to prevent the receiver channel from being overloaded by the signal from a very strong reflection. The schematic is shown is figure 5.

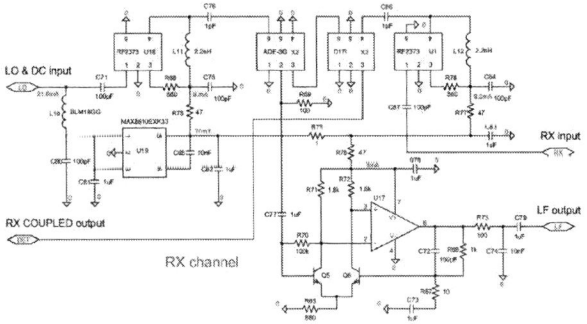

Fig. 5 Single receive channel schematic

The amplification of RX and LO signals is performed by low-power low-cost RF ICs from RF Micro Devices. The impedance matching is done using lumped inductors and capacitors (0603-size).

The current consumption of the complete receive channel is approximately 21 mA, at a supply voltage of 3.6 Volts. The overall conversion gain (RF in to LF out) is approximately

50 dB and the double side-band (DSB) noise-figure is around 7 dB.

D. Transmit channel design

Part of the transmit channel is identical to the receive channels; both the LO buffer-amplifier, the mixer and the power supply circuits are the same. In the transmit channel, the mixer is going to be used as an amplitude modulator for the transmit signal, forming part of the Automatic Level Control (ALC) system which will be discussed in the next paragraph. For this system, the necessary RF detector circuit (AD8314) is included in the transmit channel. An additional "ALC active" on-board indicator LED and a digital output signal are provided to warn that the input signal of the RX channel has exceeded its linear input range. The exact power level for the loop to be triggered is settable by a DC voltage (selected by choosing a resistor-value).

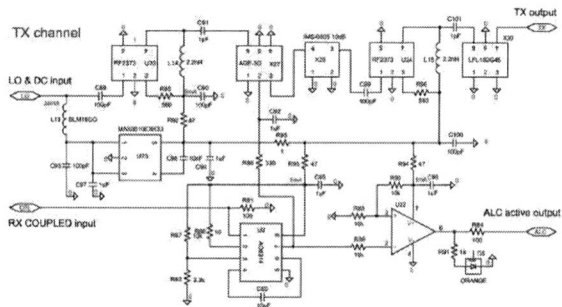

Fig. 6 Transmitter channel schematic

The basic transmit chain consists of a buffer amplifier, an amplitude modulator, a fixed attenuator to set the nominal output power level, an output amplifier and a low-pass filter to be able to both comply with CE regulations concerning maximum EIRP levels and spurious output. The settable output power is used to adapt the generated power to the gain of the specific antenna used, keeping the EIRP close to the maximum allowed value for maximum system dynamic range.

The total current consumption of the transmit channel is 24 mA from +3.6 Volts, and the output power can be set from -2 to +8 dBm. The amplitude modulator is able to further reduce the output power by 15 to 20 dB.

E. Automatic Level Control system

The Automatic Level Control (ALC) system has been implemented to avoid RF overload of the receiver input caused by reflections from large objects close to the antenna. When the input signal of the receiver exceeds its linear range, the gain will be decreased causing an increase in the effective noise figure and the radar will become less sensitive. In order to avoid this situation, part the RF input signal of the receiver is fed to a rms detector IC, and compared with the maximum linear input level for the receiver. When the input level exceeds this value, the transmitter power is automatically lowered (which offcourse directly influences the received

signal level) until the receiver signal level is back to the linear input range.

In order to model the behaviour of the ALC and prevent unstable operation, a behavioural model has been designed and implemented in PSPICE, as shown in figure 7.

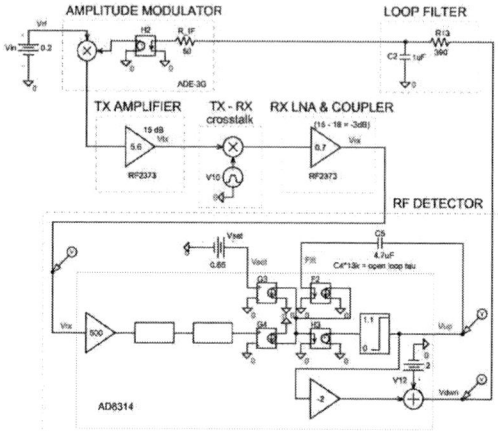

Fig. 7 Behavioural model of the ALC system

In the model, DC voltages are used to describe the envelope of the RF signal, making the simulation both fast and simple to implement. The loop gain is dependent on the signal level itself since the amplitude modulator (consisting of a balanced mixer) works completely linear while the RF detector IC (AD8314) only has a logarithmic output (V/dB). However, stable operation has been demonstrated with the set-up as shown, yielding a very simple system with a low number of components.

IV. REALISATION

The complete circuit was implemented on a bi-layer printed circuit board (PCB) using 8 mil thickness RO4003 substrate material, to make the module both low-cost and easily accessible. The synthesizer part is fully separate from the transmit and multi-channel receiver part. The only interconnection is a single micro-strip line, carrying both the RF signal from the synthesizer as well as the DC bias voltage for the TX and RX channels. This set-up makes modification of the front-end to a larger number of channels quite easy.

A photograph of the module is shown in figure 8; the individual blocks can easily be recognised; frequency synthesizer, receive and transmit channels and the LO & DC distribution.

Although the circuit uses relatively large components (all passives are 0603-size or larger), each complete receiver channel and the transmit channel only measure 15 x 30 mm (excluding connectors). The placement of metal shielding parts is facilitated by several grounded metal strips surrounding the circuits. The actual required shielding will be dependent on the mechanics and the specifications of the system in which the module is used.

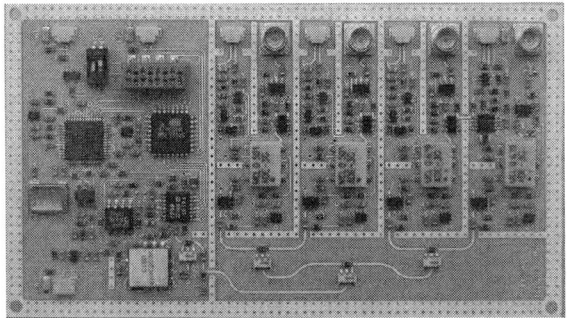

Fig. 8 Photograph of the assembled front-end

SMP surface-mount RF connectors are used because of their small size and high reliability. Furthermore, a very good and reproducible impedance-matched transition from micro-strip to coax is guaranteed by the design.

V. MEASUREMENTS

The full module was characterised in-house; the module was functioning in one pass. Its main specifications are summarised in the table below.

TABLE I
S-BAND MODULE PERFORMANCE SUMMARY

Parameter	Specification
Frequency range (programmable)	2290 – 2500 MHz
Sweep time (programmable)	50 to 100000 µs
Transmit power (setting range)	-2 .. +8 dBm
Harmonics (DC - 26.5 GHz)	< -47 dBc
Spurious output signals	< -70 dBc
Transmit power reduction range by ALC	> 17 dB
Receiver typical DSB noise figure	7 dB
Maximum receiver RF input level (P1dB)	-12 dBm
Maximum receiver RF output level (P1dB)	+10 dBm
Receiver output LF bandwidth (-3 dB)	200 kHz
Power Supply voltage	3.6 - 5.5 Volts
Power Supply current	220 mA

The SSB phase-noise of the transmit signal was measured and found to be close to the simulated value. The result is shown in figure 9.

The measured conversion (voltage) gain from RF input to LF output, and the Double Side-Band (DSB) Noise Figure is shown in figure 10. Note that the plot uses a logarithmic scale for the x-axis. The roll-off in conversion gain for lower frequencies is deliberately done, to avoid clipping of the output signal for close-by reflections (low frequency signals).

The RX channel-to-channel amplitude and phase matching makes accurate direction finding applications possible, as well as small phased array applications using digital beam forming techniques.

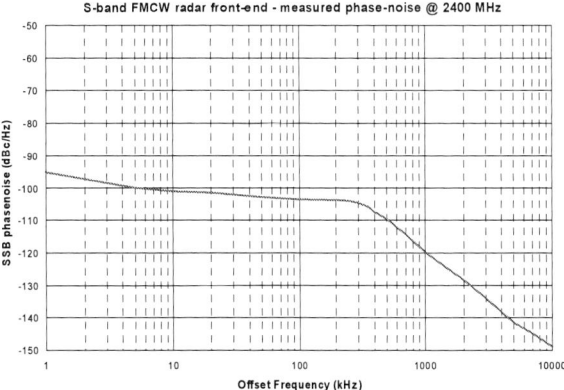

Fig. 9 Measured SSB phase-noise of the transmit signal

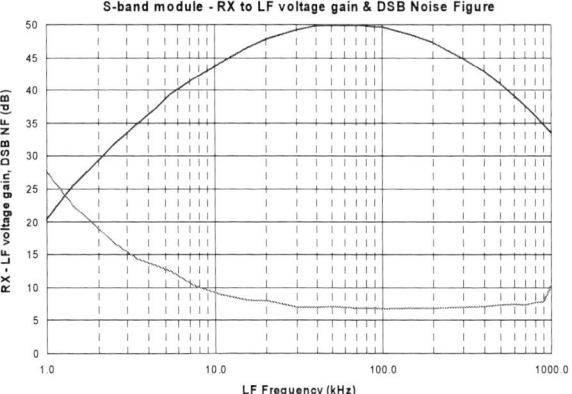

Fig. 10 Measured conversion gain and DSB noise-figure

Typical sweep-settings for the module are a linear up-sweep from 2400 to 2483.5 MHz in 900 µs, yielding a maximum range (@ 200 kHz LF) of 320 meters. Several other settings are possible, creating a large number of possible applications, ranging from short to medium range systems.

VI. CONCLUSIONS

A versatile S-band FMCW radar module has been designed and realized on a single board. Low-cost technology and components deliver high performance, due to careful design, dimensioning and layout. With the ability to be "USB-powered" the front-end is especially useable in portable multi-channel radar-systems.

ACKNOWLEDGMENT

The authors wish to thank the mechanical prototype department for the fast and accurate assembly of the front-end.

REFERENCES

[1]   *AD9954 Direct Digital Synthesizer*, datasheet, Analog Devices, 2003.

A 0.13-μm SiGe BiCMOS LNA
for 24-GHz Automotive Short-Range Radar

E. Ragonese[#1], A. Scuderi[*2], and G. Palmisano[#3]

[#]*Dipartimento di Ingegneria Elettrica Elettronica e dei Sistemi, Facoltà di Ingegneria, Università di Catania*
V.le. A. Doria 6, 95125, Catania, Italia
[1]eragonese@diees.unict.it
[3]gpalmisano@diees.unict.it

[*]*STMicroelectronics*
Str.le Primosole 50, 95121, Catania, Italia
[3]angelo-apg.scuderi@st.com

Abstract — **In this paper a 24-GHz low-noise amplifier for automotive short-range radar applications is presented. The circuit was fabricated in a 0.13-μm SiGe BiCMOS process and includes three fully differential transformer-loaded cascode stages with variable gain functionality. The amplifier provides an outstanding power gain of 35 dB and a noise figure as low as 3.4 dB, exhibiting a reverse isolation better than –60 dB. The circuit guarantees an input 1-dB compression point of –12 dBm, while drawing 56 mA from a 2.4-V supply.**

I. INTRODUCTION

Radar-based advanced safety systems are considered essential to reduce road accidents caused by driver inattention. However, due to the high cost of *mm*-wave electronics traditionally addressed using III-V technologies, today only high-end cars are equipped with radar sensors. Implementation of such systems using silicon-based processes can reduce the cost by an order of magnitude, enabling ubiquitous and pervasive adoption of this technology and thus increasing the safety and security of transportation. In this scenario a global standardization is not yet achieved. Indeed, in 2002 the Federal Communication Commission allocated a 7-GHz-wide band between 22 to 29 GHz for the ultra-wide band (UWB) short-range radar (SRR). On the other hand, the European Telecommunications Standards Institute temporarily assigned the band between 22 to 26.625 GHz to support the development of UWB SRR systems in commercial silicon-based technology.

The bottleneck for the development of a low-cost silicon-integrated radar system is still the *mm*-wave front-end. In particular, the low-noise amplifier (LNA) is the core block in the receiver (RX) section, since it must guarantee high performance in terms of both noise figure and power gain to fulfil typical radar sensor specifications. Indeed, the signal-to-noise ratio at the output of the receiver chain, whose value is mainly determined by the noise figure and gain of the LNA, is a key-parameter since it is related to both detection probability (P_d) and probability of false alarm (P_n). To comply with typical values of P_d and P_n, which are 90% and 1% respectively, a receiver noise figure lower than 8 dB is required. Consequently, an LNA with a power gain higher than 30 dB, which allows noise contributions of the cascaded

circuitry to be extremely reduced, should guarantee a noise figure lower than 6 dB. Moreover, since echo signals exhibit a very wide dynamic range, the LNA should also provide high linear range and variable-gain functionality.

This paper presents the design and measurements of a 24-GHz silicon-integrated LNA for vehicular SRR, which was developed for the RX radar front-end, adopting a zero-IF architecture, shown in Fig. 1. The amplifier exhibits excellent power gain and noise figure, while providing high linearity and variable-gain functionality.

II. CIRCUIT DESIGN

The simplified schematic of the LNA is shown in Fig. 2. The amplifier exploits a fully differential topology in order to avoid detrimental effects of parasitic ground inductance, while providing high rejection of common-mode spurious signals and substrate noise. It consists of three gain stages adopting a transformer-loaded cascode topology, which allows excellent power gain, first-rate reverse isolation, and high linearity to be achieved. The first stage, LNA1, was designed to achieve simultaneous noise and 50-Ω input impedance matching by means of proper transistor sizing and emitter degeneration inductors (L_{E1}). In both LNA2 and LNA3 inductive emitter degeneration (L_{E2}, and L_{E3}) was also employed to trade-off both gain and linearity performance. Transformer resonant loads (T_{L1}, T_{L2}, and T_{L3}) were properly designed to optimize the overall power gain performance. It is worth noting that the inter-matching networks (T_L, C_P, and C_S) were properly tuned to guarantee the required gain bandwidth flatness. The use of transformer load T_{L3} allows a differential direct connection to

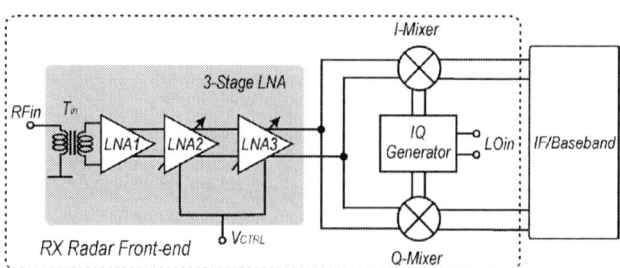

Fig. 1 Block diagram of the RX radar front-end.

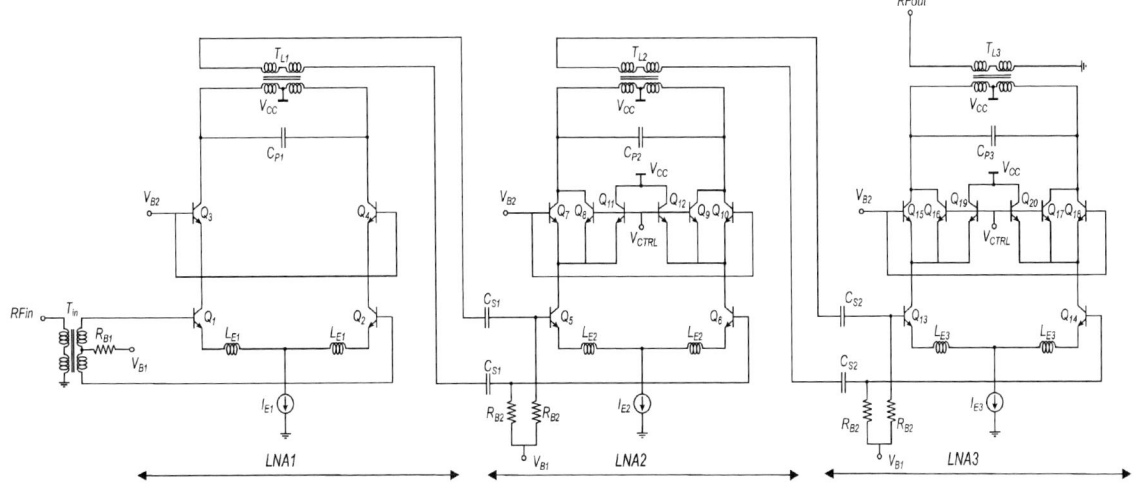

Fig. 2 Simplified schematic of the 24-GHz LNA.

the I/Q Gilbert quads to be easily obtained, which avoids the use of voltage-to-current converters, thus maximizing the linearity of the overall RX front-end.

To further relax linearity requirements of the I/Q mixers and subsequent blocks in the RX chain, a variable-gain approach is also adopted. As shown in Fig. 2, a single-bit gain control is implemented by means of the control voltage, V_{CTRL}. In particular, at low-gain setting (V_{CTRL} = 2.4 V) the current delivered to the resonant load depends on the emitter area ratio of transistors $Q_{8,9}$ - $Q_{11,12}$ ($Q_{16,17}$ - $Q_{19,20}$), while at high-gain setting (V_{CTRL} = 0 V) the stages work as traditional cascode amplifiers. This technique significantly increases the receiver linearity, still maintaining excellent noise figure performance. At the LNA output, differential-to-single-ended conversion is achieved by means of the transformer T_{L3}, while at the input a balun (T_{in}) was included for testing purpose.

III. INDUCTIVE COMPONENT DESIGN AND LAYOUT

The amplifier was integrated in a 0.13-μm SiGe BiCMOS technology featuring high-speed npn transistors with f_T/f_{max} of 166/175 GHz, and 6-level metal copper back-end [1]. The layout drawing of *mm*-wave Si-based blocks is a crucial phase in the overall design flow. Indeed, geometrical asymmetries, interconnection-path parasites, and EM coupling produce a considerable degradation of the expected results. Moreover, a poor on-chip ground reference, due to the stringent metal density rules of modern processes further deteriorates both circuit reliability and performance. For these reasons, much attention was paid to the design of differential-signal paths in terms of symmetry and length. This was also accomplished by means of a proper floor plan, which allowed horizontal axial symmetry to be obtained, while reducing the use of lossy inter-connections. To this aim, a symmetric interleaved configuration was adopted to implement the transformer loads T_L, which provided straightforward differential input / output terminals placed on opposite sides, as shown in Fig. 3. The design of an integrated transformer is quite critical at

mm-wave frequencies since the adopted inductance values fall in the sub-nH range [2] and modeling, measurement, and simulation matters arise. Moreover, the transformer optimization procedure requires iterative steps and time-consuming electromagnetic (EM) simulations. For these reasons, scalable lumped model was first exploited to define the spiral geometrical parameters, and then EM simulators were used to validate the transformer layout structure and refine its EM behaviour, taking into account the connection paths. Primary and secondary coils have widths of 8 and 5 μm, respectively, while the inner diameter is 73 μm. To minimize the resistive losses both windings exploited a multi-layer structure consisting of two Cu metals (metal 6 and metal 5) plus top Al layer, while substrate losses were reduced by means of a conductive patterned ground shield. Inductance (L) and quality factor (Q) for both the primary and secondary coils of T_L are reported in Fig. 4. Transformers T_L were properly optimized to maximize the transformer characteristic resistance (TCR) [3], which is strictly related to the available output power of each LNA stage, according to equation (1):

$$TCR = \omega Q_{EQ} L_{EQ} =$$
$$= \omega \left(Q_P \frac{k^2 Q_P Q_S}{1 + k^2 Q_P Q_S} \right) \cdot \left(L_P + \frac{L_P}{Q_P^2} + k^2 \frac{L_P \cdot Q_S}{Q_P} \right) \quad (1)$$

where k is the magnetic coupling factor. Fig. 5 depicts the TCR of T_L as a function of frequency.

Degeneration inductors L_E were implemented by using differential folded microstrips. This choice allowed further optimizing the circuit floor plan since folded microstrips are highly area-saving compared with spiral inductors, still providing excellent performance at low-value inductance.

Finally, a low-resistance / low-inductance ground plane was adopted, which makes use of an appropriate metal 2 / metal 3 pattern to meet density requirements. Extensive EM post-layout simulations were carried out to take into account RLC parasitics and coupling effects.

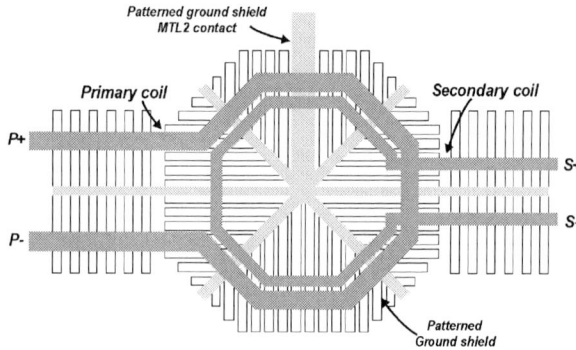

Fig. 3 Symmetric interleaved transformer T_L.

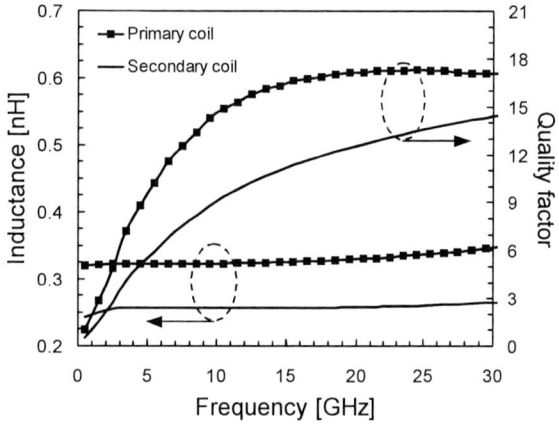

Fig. 4 Inductance and quality factor of primary and secondary coils of T_L.

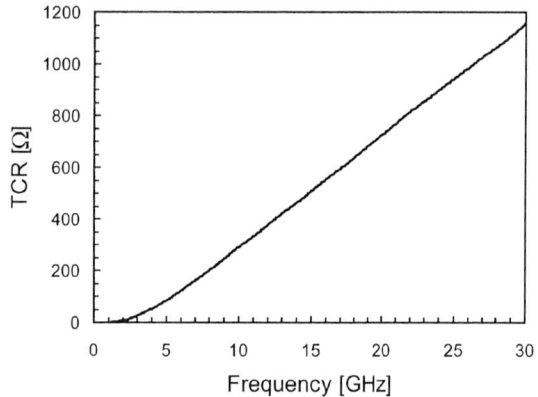

Fig. 5 Transformer characteristic resistance (TCR) of T_L.

IV. EXPERIMENTAL RESULTS

The LNA was mounted on a 400-μm thick FR4 substrate adopting chip-on-board assembly technique. The die photograph of the LNA is shown in Fig. 4. The measurement setup consists of a two-port Agilent E8364B vector network analyzer, a Summit 12000 Cascade Microtech prober and an Agilent N8975A noise figure analyzer. Testing was carried out at 2.4-V supply. Raw data were de-embedded only for the input balun loss, which was used only for testing purpose.

The power gain (S_{21}) and noise figure at high-gain setting are reported in Fig. 4. The LNA achieves a maximum gain of 35 dB and a 3.4-dB noise figure. At low-gain setting it provides a 14.5-dB power gain and a noise figure of 4.5 dB. By means of the gain-control functionality, the LNA achieves an input 1-dB compression point (IP_{1dB}) of –12 dBm, as shown in Fig. 6. Both input and output return losses are shown in Fig. 7.

Fig. 6 Die photograph of the 24-GHz LNA.

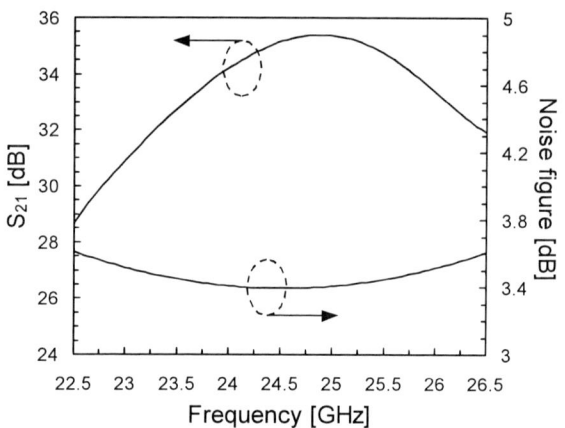

Fig. 7 Power gain (S_{21}) and noise figure at high-gain setting.

In Table I measurements are summarized and compared with published K-band silicon-based LNAs. The proposed amplifier demonstrates excellent achievements in terms of gain, noise figure and dynamic range. It also provides first-rate reverse isolation performance, thanks to the use of the cascode topology and the adoption of proper layout approaches. As far as the power consumption is concerned, the LNA draws 56 mA from a 2.4-V supply. This performance is mainly determined by the use of a fully differential topology, as well as by the linearity specifications for both the second and third stages.

TABLE I
COMPARISON WITH PUBLISHED *K*-BAND SILICON-BASED LNAs.

Process	f_T / f_{MAX} [GHz]	Freq [GHz]	NF [dB]	S_{21} [dB]	S_{12} [dB]	IP_{1dB} [dBm]	V_{CC} [V]	I_{CC} [mA]	Gain control [dB]	Circuit Topology	Ref.
90-nm SOI CMOS	149 / n.a.	35	3.6	11.9	–25	4	2.4	17	—	Cascode	[4]
0.18-*μ*m CMOS	70 / 58	24	3.9	13.1	n.a.	–12.2	1	14	—	2-stage CS	[5]
0.5-*μ*m SiGeHBT	80 / n.a..	24	5	25	n.a.	–18	5	50	—	3-stage Differential Cascode	[6]
90-nm CMOS + WLP inductors	170 / 240	24	3.2	7.5	–15	n.a.	1	10.6	—	CS	[7]
0.25-*μ*m SiGe:C BiCMOS	130 / 140	24	4.2	11	–45	n.a.	3.3	6	—	Cascode	[8]
0.8-*μ*m SiGe HBT	80 / 80	24	6.6	20	–60	n.a.	4	85	—	3-stage Differential Cascode	[9]
90-nm CMOS	100 / 180	28.5	4	20	n.a.	–17	1	16.3	—	CS + Cascode	[10]
0.13-*μ*m SiGe:C BiCMOS	**166 / 175**	**24.5**	**3.4**	**35**	**–62**	**–12**	**2.4**	**56**	**20**	**3-stage Differential Cascode**	**This work**

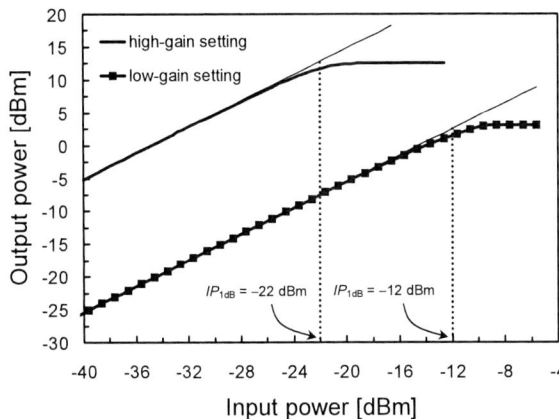

Fig. 8 Output power versus input power at 24.5 GHz.

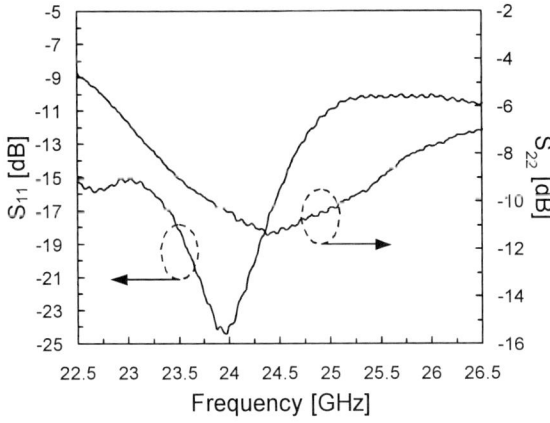

Fig. 9 Input (S_{11}) and output (S_{22}) return losses.

V. CONCLUSION

A silicon-integrated LNA for a 24-GHz radar front-end has been presented. The circuit adopts a three-stage fully differential cascode topology with single-bit gain control. Transformer loads and differential folded microstrips have been largely used, which allowed a fully symmetric layout to be implemented, thus maximizing common-mode rejection and isolation. Extensive measurements displayed 35-dB power gain, a noise figure of 3.4 dB, a reverse isolation better than –60 dB, and IP_{1dB} of –12 dBm at low-gain setting.

REFERENCES

[1] M. Laurens *et al.*, "A 150 GHz f_T/f_{max} 0.13 *μ*m SiGe:C BiCMOS technology," in *Proc. IEEE Bipolar/BiCMOS Circuits Technol. Meeting*, Oct. 2003, pp. 199-202.

[2] T. Biondi, A. Scuderi, E. Ragonese, and G. Palmisano, "Sub-nH inductor modeling for RF IC design," *IEEE Microwave Wireless Comp. Lett*, vol. 15, pp. 922-924, Dec. 2005.

[3] F. Carrara, A. Italia, E. Ragonese, and G. Palmisano, "Design methodology for the optimization of transformer-loaded RF circuits, *IEEE Trans. Circuits Syst. I*, vol. 53, pp. 761-768, April 2006.

[4] F. Ellinger, "26-42 GHz SOI CMOS low noise amplifier," *IEEE J. Solid-State Circuits*, vol. 39, pp. 522-528, Mar. 2004.

[5] S. Shin, M. Tsai, R. Liu, K. Lin, and H. Wang," A 24-GHz 3.9-dB NF low-noise amplifier using 0.18-*μ*m CMOS technology," *IEEE Microwave Wireless Comp. Lett.*, vol. 15, pp. 448-450, Jul. 2005.

[6] I. Gresham *et al.*, "A fully integrated 24 GHz SiGe receiver chip in a low-cost QFN plastic package," *in IEEE Radio Frequency Integrated Circuits Symp. Dig.*, Jun. 2006, pp. 371-374.

[7] X. Sun *et al*, "High-Q above-IC inductors using thin-film wafer-level packaging technology demonstrated on 90-nm RF-CMOS 5-GHz VCO and 24-GHz LNA," *IEEE Trans. Adv. Packag.*, vol. 29, pp. 810-817, Nov. 2006.

[8] E. van der Heijden, H. Veenstra, R. Havens, "16-26 GHz low noise amplifier for short-range automotive radar in a production SiGe:C technology," in *Proc. IEEE Topical Meeting on Silicon Monolithic Integrated Circuits in RF Systems*, Jan. 2007, pp. 241-244.

[9] E. Öjefors *et al.*, "Monolithic integration of a folded dipole antenna with a 24-GHz receiver in SiGe HBT technology," *IEEE Trans. Microwave Theory Tech.*, vol. 55, pp. 1467-1475, July 2007.

[10] E. Adabi, B. Heydari, M. Bohsali and A. M. Niknejad, "30 GHz CMOS low noise amplifier," in *IEEE Radio Radio Frequency Integrated Circuits Symp. Dig.*, Jun. 2007, pp. 625-628.

Optimization of Class E Power Amplifier Design above Theoretical Maximum Frequency

Elisa Cipriani, Paolo Colantonio, Franco Giannini, Rocco Giofré

Electronic Engineering Department, University of Roma Tor Vergata - Via del Politecnico 1, 00133 Roma, Italy
giofr@ing.uniroma2.it

Abstract— **In this contribution, the analysis on high frequency Class E design approach is presented. Starting from the classical theory, a numerical analysis is performed to extend class E feasibility at higher frequencies. The design of hybrid Class-E amplifier in LDMOS technology for UMTS base-station applications will be presented, in order to validate the theoretical results. The simulated PA reaches an output power of 40.7dBm in correspondence of 56% drain efficiency.**

I. INTRODUCTION

In the last decades, class E power amplifier (PA), firstly proposed by Sokal in '70s [1], has become very popular in wireless communication systems, thanks to both its simple closed form design relationships and its high obtainable performances. As diffusely explained in [1], class E operation is based on the assumption that the active device acts as a switch, so minimizing DC power consumption. Additionally, its output matching network (OMN) is designed to properly shape the drain waveform, ideally nulling the power delivered at the higher harmonics of the fundamental frequency. Moreover, the OMN topology is fixed and the values of its components are determined to satisfy the well-known operating conditions formulated by Sokal in [1].

To date, many publications have already demonstrated the possibility to integrate the Class-E PA either in the new wireless systems such as GSM/GPRS, UMTS and WLAN terminals [2] or in the PA architectures, like Envelope Elimination and Restoration (EER) [3]. At the same time, however, the intrinsic limitation of Class E approach, due to the non ideal switching operation of the active device adopted, is still an issue. In fact, the device output capacitance decreases the maximum operation frequency (f_{Max}) of Class-E up to few GHz, as demonstrated in [4]. Moreover, it can be demonstrated that f_{Max}, is directly proportional to the device maximum output current I_{max}, and inversely proportional to the selected drain bias voltage (V_{DC}) and the device output capacitance C_{ds}. Nonetheless, the development of high frequency Class E design had recently registered an improvement related to the use of emerging technologies, like SiGe-HBT and Si-LDMOS [5]. The former, thanks to the high cut-off frequency (greater than 40 GHz), represents a very attractive solution to minimize the problems related to f_{Max}. In addition, SiGe BiCMOS process allows also the integration on the same chip of the RF power and digital-control sections of entire wireless system [7]. However, for base station applications a very high output power level is required, thus making the SiGe HBT devices unfeasible if compared to LDMOS ones. The latter, in fact, thanks to their higher

breakdown voltage and maximum allowable current, appear as a proper solution for base station amplifiers in Class E configuration [5]. However, in this case f_{Max} is limited to hundreds of MHz, thus representing a critical issue.

The aim of this contribution is to point out how it is possible to extend the class E maximum operating frequency, through the optimization of the drain voltage waveform and the proper choice of the fundamental impedance. The approach is based on a numerical optimisation and to validate the proposed method, the design of a 2.14 GHz class E PA in LDMOS technology is finally presented.

II. FREQUENCY DOMAIN ANALYSIS

A. Low Frequency-domain Analysis

The basic topology of a class E PA employs a single active device, is depicted in Fig. 1.

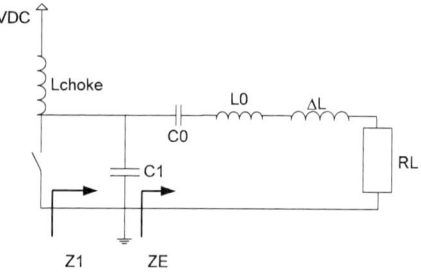

Fig. 1 Classical class E topology

The device output capacitance C_{ds} is included in C_1. As mentioned in the introduction, the OMN components are determined to satisfy the following operating conditions formulated by Sokal in [1]:

a) purely sinusoidal current across the load R_L realised through the ideal filter C_0-L_0;

b) zero voltage switching conditions (ZVS) to prevent simultaneous non-zero voltage and current across the device;

c) zero voltage derivative switching conditions (ZVDS) to assure that the current starts to flow after the voltage has reached its null value.

Under these assumptions, the resulting ideal output voltage and current waveforms are the ones reported in Fig. 2.

978-2-8748-7007-1/08 $25.00 © 2008 EuMA

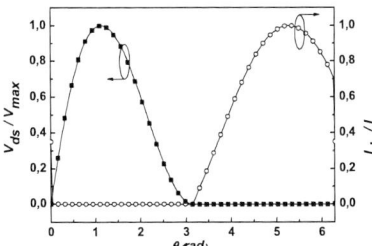

Fig. 2 Ideal class E waveforms

Performing a Fourier analysis of the resulting drain current waveform, it is possible to express its harmonic coefficients in a closed form [4]:

$$I_n = \begin{cases} I_{DC} & n=0 \\ \dfrac{I_{DC}}{8\pi}\left(2\pi + 8i + i\pi^2\right) & n=1 \\ \dfrac{I_{DC}}{2} i \dfrac{2n+in}{\pi\left(n^2-1\right)} & n>1,\,even \\ \dfrac{I_{DC}}{n\pi} i & n>1,\,odd \end{cases} \tag{1}$$

where I_{DC} is the drain current DC component. Referring to Fig. 1, the drain voltage components across the device result in:

$$V_n = \begin{cases} V_{DC} & n=0 \\ \left|\Psi_n\right| \cdot e^{j\arg(\Psi_n)} & otherwise \end{cases} \tag{2}$$

where Ψ_n components can be written as following:

$$\Psi_n = \begin{cases} \dfrac{I_{DC}}{2\pi^2 f \cdot C_1}\left(\dfrac{\pi^2}{8}-1-i\dfrac{\pi}{4}\right) & n=1 \\ \dfrac{I_{DC}}{2\pi^2 f \cdot C_1}\left(\dfrac{2n+in}{2n(1-n^2)}\right) & n>1,\,even \\ \dfrac{I_{DC}}{2\pi^2 f \cdot C_1}\left(-\dfrac{1}{n^2}\right) & n>1,\,odd \end{cases} \tag{3}$$

As can be noted, the dependence on frequency appears only in the expression of the voltage coefficients, while it has not influence on the current ones.

Therefore, the optimum fundamental impedance Z_E (see Fig. 1) to be synthesised at the device terminals, to fulfil the Class E requirements is:

$$Z_E = \frac{0.28}{2\pi f \cdot C_1} e^{j49^\circ} \tag{4}$$

and the corresponding Z_1 is given by:

$$Z_1 = \frac{V_1}{I_1} = \frac{0.35}{2\pi f \cdot C_1} e^{j36^\circ} = Z_E \,//\, \frac{1}{j2\pi f \cdot C_1} \tag{5}$$

B. High Frequency-domain Analysis

The availability of such closed form expressions, useful for the design of a very efficient and mainly non-linear amplifier, justify the attractiveness of Class E approach. However, from the above relationships, a limit in Class E maximum operating frequency rises [4,6], also considering the device physical constraints:

$$f_{Max} = \frac{I_{DS}}{2\pi^2 C_1 V_{DC}} \simeq \frac{I_{Max}}{56.5 C_1 V_{DC}} \tag{6}$$

where the lower limit of C_1 is represented by C_{ds}.

As a consequence, the Class E operating frequency is intrinsically limited by the active device switching behaviour. Therefore, exceeding this maximum value, the ideal class E conditions and wave shaping (ZVS and ZVDS) cannot be further satisfied [7]. Moreover, for micro and millimetre-wave range applications, the available active devices usually do not act as ideal switches. In fact, their not negligible parasitic effects increase switching time transitions resulting in low pass filters behaviours, practically shorting the higher frequency drain voltage components and consequently not allowing the desired wave-shaping. Thus, to modify ideal Class E conditions and to infer other optimum design criteria for high frequency applications, the above mentioned limitations have to be taken into account.

Therefore, the higher voltage harmonics are considered as effectively shorted by C_1 and only the harmonic loads up to the third order can be effectively controlled. Under these hypothesis, and considering the current unaffected the resulting drain voltage and current waveforms are depicted in Fig. 3.

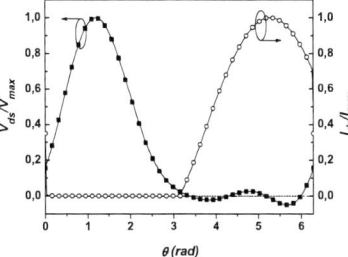

Fig. 3 Drain waveforms considered in the analysis: the current is unaffected, while the voltage is truncated at the 3rd order.

As can be noted, the resulting drain voltage waveform does not respect the ZVS and ZVDS conditions [7] and the device physical constrains, since negative values are observed.

As pointed out in [6], two solutions can be adopted to prevent these negative values in the drain voltage. Obviously, it is possible to increase drain bias voltage, but it would mean a non negligible increase in the DC dissipated power with a consequent drain efficiency decrease. In addition, an increasing on peak voltage value could exceed breakdown limitations of the transistor.

A second strategy consist in the optimisation of the Z_1 value, assuming unchanged the higher voltage components expressed by (1), in order to meet the device limits and without loosing the Class E behaviour. This becomes a nonlinear problem, which can be solved by numerical approach. In the following section, the developed algorithm to extend the Class E approach above its maximum operating frequency will be described.

III. DEVELOPED ALGORITHM

Starting from the frequency domain approach, the following assumptions are possible, without loss of generalization:

- the device has zero on-resistance and infinite off-resistance: this leads to an unaffected current waveform, which is fixed and equal to the ideal case;
- a purely sinusoidal current flows into the load R_L (i.e. the C_0-L_0 filter in the OMN is ideal).

As a consequence, the harmonic loads above the fundamental at the drain section are only due to the capacitance C_1. Therefore, the drain voltage can be expressed by a Fourier series, truncated at the 3rd order. In order to extend the analysis above f_{Max}, it is useful to introduce the normalized frequency, k, as a parameter. Hence the drain voltage can be written as:

$$v_{DS}(\vartheta) = V_{DD} - 2 \cdot \text{Re}\left(Z_1 I_1 \cdot e^{jk\vartheta}\right)$$
$$- 2 \cdot \text{Re}\left(\sum_{n=2}^{3} \frac{1}{jn \cdot k \cdot 2\pi f_{Max} \cdot C_1} I_n \cdot e^{jnk\vartheta}\right) \quad (7)$$

being $k = f/f_{Max}$. Now, assuming the capacitance C_1 and the bias point (V_{DC}, I_{DC}) to be fixed, the proposed solution implies to find the optimum fundamental load impedance Z_1 in order to satisfy the following two conditions:

- prevent negative values of drain voltage;
- maximise the efficiency (hence output power at fundamental frequency).

As a first step, the voltage function was posed in a discrete form. Then, the unknown fundamental impedance, Z_1, was written in magnitude and phase form. The former was swept around the Sokal's value given by (5), in a range of $\pm 90\%$ while the latter was swept from $-\pi/2$ to $+\pi/2$ so assuming whatever realisable value using passive components. Therefore:

$$\begin{cases} Z_1 = |Z_{1,ideal}| \cdot \Delta Z \cdot e^{j \cdot \Delta \Phi} \simeq \dfrac{0.35}{\omega C_1} \cdot \Delta Z \cdot e^{j \cdot \Delta \Phi} \\ \Delta Z \in \left[\dfrac{1}{10}, \dfrac{19}{10}\right] \\ \Delta \Phi \in \left[-\dfrac{\pi}{2}, +\dfrac{\pi}{2}\right] \end{cases} \quad (8)$$

Consequently, the function (7) was calculated for every value of the fundamental impedance taking as a result only the couples of ΔZ and $\Delta \Phi$ which respect the conditions previously remarked. The process was iterated for k going from 1 to 5 in steps of 0.1. Fig. 4 reports the ΔZ and $\Delta \Phi$ behaviours as a function of k. As can be seen, a quasi monotonically decrease in the variation of the magnitude of fundamental impedance is observed, leading to a consequent decrease in the fundamental voltage amplitude, as well as in maximum drain voltage. On the other hand, the phase behaviour is monotonically decreasing leading almost purely resistive impedance at fundamental for the higher values of k.

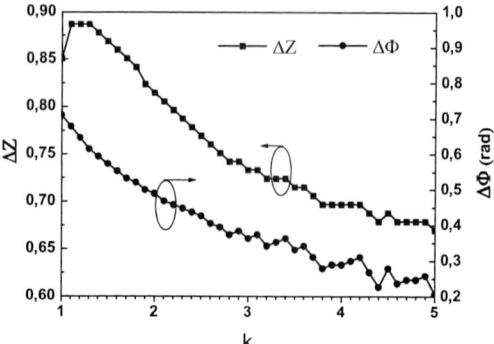

Fig. 4 ΔZ and $\Delta \Phi$ behaviours as a function of k.

The resulting drain efficiency, reported in Fig. 5, is significantly different from the unity, due to the non ideal operating conditions. The main reasons of efficiency decrease are due to the lower value of fundamental voltage amplitude with respect to the ideal one and the different phase shifting between voltage and current.

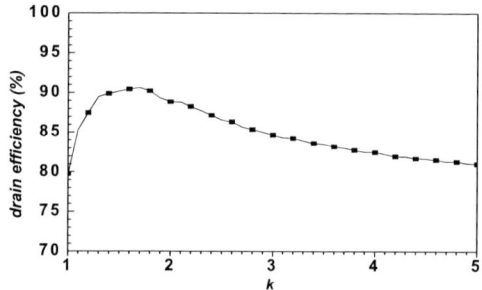

Fig. 5 Drain efficiency as a function of normalized frequency.

The peak in drain efficiency is observed at 1.6 times the maximum frequency, resulting in a combination of voltage and current which effectively minimizes the power dissipated in the transistor. This analysis suggests a useful and immediate approach to design high frequency Class E PA, improving the closed form relationships of the classical approach.

IV. EXPERIMENTAL RESULTS

In order to validate the results obtained, a Class E PA @2.14 GHz was designed, using a medium power N-Channel LDMOS device. The maximum current and breakdown voltage are respectively 2.5A and 70V, evaluated by DC simulation on the nonlinear model provided by the foundry. The PA bias voltage was selected to maintain the operating point within the physical limits of the device. In particular, the dc current and drain voltage were selected according to the theoretical Class E boundary conditions [6]:

$$\begin{aligned} V_{DC} &\leq V_{Break}/3.562 \\ I_{DC} &\leq I_{Max}/2.862 \end{aligned} \quad (9)$$

Thus a V_{DC} of 20V and V_{GS} of 3.3V (I_{DC}=800 mA) were adopted. The device has an output capacitance estimated by simulation in 4.2pF, resulting in a f_{Max} in pure Class E conditions of about 520MHz, which leads to an ideal Class E impedance $Z_1 = 25.1e^{j36°}$. Therefore, the design frequency is almost 4.1 times greater than f_{Max}: hence as can be deduced from Fig. 4, a decrease of 30% in magnitude and an absolute phase of 17° are expected for high frequency Class E impedance. The preliminary values of impedance were found with a harmonic tuner and effectively demonstrate the pertinence of the performed analysis. An optimum impedance of $Z_{1,opt} = 17.5e^{j17°}$ was obtained. The expected drain efficiency was nearly 76%, slightly less than the expected from Fig. 5. In fact, it should be remarked that knee voltage has been omitted in the numerical algorithm for sake of simplicity although it has some influence on the overall performances.

Consequently, the output matching network (OMN) was synthesized using a mixed distributed and lumped components approach. Additional work was performed on the input matching network (IMN) to assure unconditional stability and conjugate input matching of the PA. Then, both networks were implemented on ceramic substrate ($\varepsilon_r = 10$, $h = 640\mu m$) and using SMD components.

Simulated drain voltage and current waveforms together with the load line are reported in Fig. 6, demonstrating a good approximation of ideal class E behaviour even above f_{Max}.

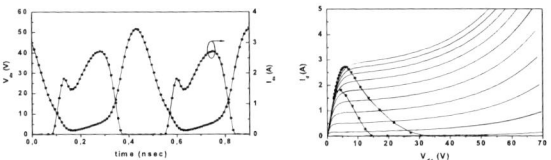

Fig. 6 Simulated drain waveforms and load line.

At the nominal bias voltage and operating frequency, the PA reaches an output power of 40.7dBm with a drain efficiency of 56% as reported in Fig. 7.

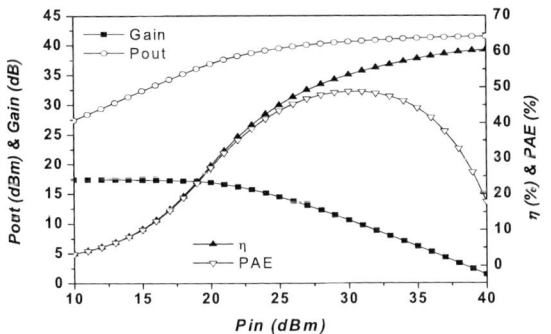

Fig. 7 Output performance of the designed PA.

The efficiency is lower with respect to the one estimated using ideal tuners due to the losses introduced by the real components in the networks. In order to investigate the possible integration of the PA in EER systems, additional nonlinear simulation were performed varying the drain bias voltage from 15V to 25V. Results are given in Fig. 8, showing a quasi constant efficiency with V_{DC} variations.

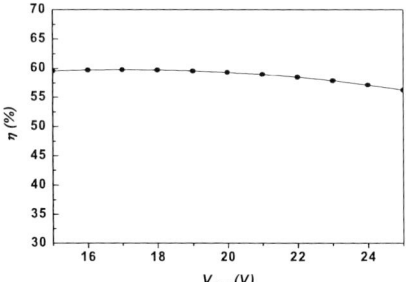

Fig. 8 Maximum drain efficiency as a function of V_{DC}

V. CONCLUSIONS

In this contribution the analysis on high frequency Class E design approach was presented. Starting from the classical approach, a numerical analysis, based on the optimisation of the fundamental impedance was performed to extend Class E feasibility at higher frequencies. The method developed was validated designing a PA based on LDMOS device, at a frequency 4.1 times higher than the maximum allowable for a traditional class E design. The PA reaches an output power of 40.7dBm in correspondence of 56% drain efficiency.

ACKNOWLEDGMENT

Research reported here was performed in the context of the network TARGET– "Top Amplifier Research Groups in a European Team" and supported by the Information Society Technologies Programme of the EU under contract IST-1-507893-NOE, www.target-net.org.

REFERENCES

[1] N. O. Sokal, A. D. Sokal, "Class E – A new class of high efficiency tuned single ended switching power amplifiers", *IEEE Journal of Solid State Circuits*, Vol. 10, No. 6, pp. 168-176, June 1975.

[2] N. Ui and S. Sano, "A 45% Drain Efficiency, -50 dBc ACLR GaN HEMT Class-E Amplifier with DPD for W-CDMA Base Station", 2006 IEEE International Microwave Symposium Digest, June 2006, pp. 718 – 721.

[3] Narisi Wang, Xinli Peng, V. Yousetzadeh, D. Maksimovic, S. Pajic, Z. Popovic, "Linearity of X-band class-E power amplifiers in EER operation", *IEEE Transactions on Microwave Theory and Techniques*, Vol. 53, No. 3, Part 2, March 2005, pp. 1096 - 1102

[4] T. B. Mader, E. W. Bryerton, M. Markovic, M. Forman, Z. Popovic, "Switched-mode high-efficiency microwave power amplifiers in a free-space power-combiner array," *IEEE Transactions on Microwave Theory and Techniques*, MTT-46, No.10, Part 1, Oct.1998, pp.1391-1398.

[5] A. Litwin, Q. Chen, J. Johansson, G. Ma, L.A. Olofsson, P.Perugupalli, "High Power LDMOS Technology for Wireless Infrastructure", *Gallium Arsenide applications symposium. Gaas 2001*, 24-28 September 2001, London

[6] P. Colantonio, F. Giannini, R. Giofrè, M. A. Yarleque Medina, D. Schreurs, and B. Nauwelaers, "High frequency Class E design methodologies," *Proceedings of the Gallium Arsenide Applications Symposium*, GAAS 2005, Paris, France, October 2005, pp. 329-331

[7] F.H. Raab, "Class-E, Class-C and Class-F power amplifiers based upon a finite number of harmonics", *IEEE Transactions on Microwave Theory and Techniques*, MTT-49, No.8, August 2001, pp.1462-1468.

Proceedings of the 3rd European Microwave Integrated Circuits Conference

Compact Concurrent Dual-Band Power Amplifier for 1.9GHz WCDMA and 3.5GHz OFDM Wireless Systems

Alessandro Cidronali, Niccolò Giovannelli, Iacopo Magrini, Gianfranco Manes

Department of Electronics and Telecommunications, University of Florence, V. S. Marta, 3, Florence, I-50139, Italy

{name.surmane}@unifi.it

Abstract—**The aim of this paper is to focus on the dual band power amplifier design and characterization as an enabling technology for beyond 3G wireless systems. The design approach within a full characterization of the dual band power amplifier is given in the paper. For 1.98 GHz 3GPP UL WCDMA and 3.42 GHz 5 MHz 16QAM WiMAX digital systems, the dual-band concurrent exhibited simultaneous peak power levels of 24 dBm and 17dBm respectively, to maintain ACPR and EVM within the regulatory requirements. A performance discussion is then outlined with respect to multi band multi module PA architectures for software defined radio wireless transmitters.**

I. INTRODUCTION

In beyond 3G voice/data systems, users may be moving, while simultaneously performing broadband data access or a multimedia streaming session [1]. A radio technology that is expected to interact with a multi-services network should be able to change between different operating bands and adapt its features according with the different available standard and requirements. Most of the research efforts performed during the last years dealt with issues related to the physical layer of the communication stack; however, despite the growing interest in multi-standard operation, less attention has been devoted to the radio-frequency front-end, which therefore remains one of the most challenging parts of a multi-band radio, [2]-[3]. One main reason for the delay in effectively implementing multi-standard transceivers can be attributed to the implementation of the RF transmit power amplifier (PA). Today, dedicated, single standard PAs achieve very good power added efficiency (PAE) and, in this way, long battery lifetime. Any reconfigurable PA, needed for the support of different, not always predefined, communication systems, should compete with such dedicated solutions. In this paper, we investigate a new PA design methodology for effective dual-band operation that would enable the implementation of the above described scenario, providing a full system level characterization considering as reference test-bed a pair of dedicated single-band PAs.

II. CONCURRENT DUAL-BAND PA DESIGN

The investigation carried out in this paper relies on prototypes designed and fabricated using low cost off the shelf active devices along with discrete SMD passive components assembled on FR4 0.8 mm thick evaluation printed circuit board designed with microstrip technology.

The selected active device was the ATF50189 from AVAGO Technologies, a medium power Enhanced mode p-HEMT with a cut off frequency of 6 GHz and a 1-dB compression point of 29 dBm at 2 GHz. Optimum bias point for efficiency, linearity and gain was found to be with a 4.5V drain supply voltage with a corresponding quiescent current of 200mA. The chosen bias point drives the ATF50189 transistor in the AB class operation.

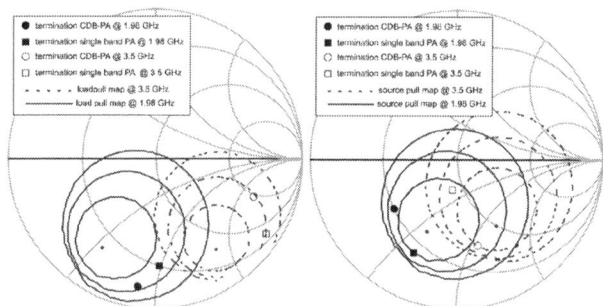

Fig. 1. Simulated load-pull (left) and source-pull (right) contours at 1.98 GHz and 3.42GHz, 1 dB steps, and terminations at fundamentals for the single band and dual-band prototypes

The design was based on load and source pull simulations carried out at the two fundamental frequencies of 1.98GHz and 3.42GHz, adopting a nonlinear device model which included the package parasitics; Simulations provided saturated output power of 28 dBm and 26 dBm respectively in the lower and higher frequency bands with a power added efficiency of approximately 40% and 35% at the 1-dB gain compression point. The resulting load and source constant power contours are shown respectively in Fig. 1.

The implementation of the source and load terminations defined by the source- and load-pull analysis were obtained by using lumped elements matching networks. This technique, by employing a different approach with respect to standard microstrip technology [4]-[5], enabled the achievement of highly compact prototypes. All the designed PAs adopt the same general topology for the input and output matching networks, whose schematics are represented in Fig. 2. The presence of shunt capacitors at both the gate and drain terminals, C2 and C4 in the figure, provide a short circuit to the second harmonic and an open circuit at the fundamental frequency.

978-2-8748-7007-1/08 $25.00 © 2008 EuMA

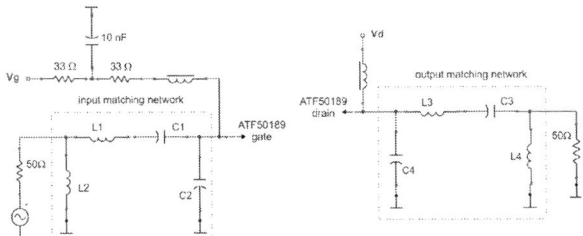

Fig. 2. Prototypes input (left) and output (right) matching network circuit schematic

The selected pi-topology exhibits high out of band frequency roll off and a null in the transfer characteristic between the two fundamental frequency bands at 1.98 GHz and 3.42 GHz. In order to properly define the networks the additional conditions for maximum efficiency and 1-dB compression output power under large signal excitations were taken into account. The resulting optimum values for the SMD capacitances and inductances are those indicated in Table 1 and Table 2 respectively for the input and output matching networks. The achieved impedances are reported in the Fig. 1, together with the mismatch between the actual achieved values and the simulated values. This impedance mismatch takes into account for the additional condition of small signal gain at the two centre band frequencies, the commercial availability of the nominal values and the component tolerances.

TABLE I
INPUT MATCHING NETWORK L-C VALUES

	L1	C1	L2	C2
Concurrent dual-band	1.7 nH	0.33 pF	7.6 nH	0.3 pF
1.98 GHz single band	0.6 nH	1.47 pF	n.a.	n.a.
3.42 GHz single band	1.6 nH	0.26 pF	n.a.	n.a.

TABLE II
OUTPUT MATCHING NETWORK L-C VALUES

	C4	L4	C3	L3
Concurrent dual-band	0.6pF	3.2nH	0.77pF	2.88nH
1.98 GHz single band	n.a.	n.a.	1.7nH	1.9pF
3.42 GHz single band	n.a.	n.a.	1.1nH	0.5pF

III. EVALUATION RESULTS

Three PA modules were fabricated using in-house facilities according with the values reported in Table I and in Table II, and tested against small signal, large signal and digitally modulated signals.

A. Small signal evaluation

The measured small signal gains associated with the concurrent dual-band PA and with the two single-band PAs prototypes are compared in Fig. 3. The figure indicates that at

1.98 GHz the maximum linear gain is approximately 11dB for both single band and dual band circuits. In the 3.4 GHz band the PA prototypes exhibit a maximum linear gain of 6 dB at 3.42 GHz, and present a 0.5dB gain bandwidth of approximately 60 MHz. Input and output return losses are not reported but are below −15 dB in the respective frequency bands for all the PAs. Small signal characterization has indicated a very close correspondence between the single band circuits and the concurrent dual band PA in terms of both input/output return loss and gain. This results show that the design of a concurrent dual-band PA using compact lumped elements is feasible without loss of performance at small-signal and makes coherent the characterization and comparison with large and modulated signals.

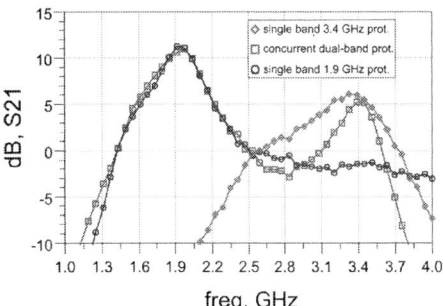

Fig. 3. Measured small-signal gain for the concurrent dual-band and the two single band prototypes

B. Large-signal and modulated signal characterization.

The base-band signals were down-loaded in the arbitrary signal generators (Agilent ESG 4438C) by using the tools available in the Agilent ADS2006A systems. Two different digitally modulated signals were employed to evaluate PA performance: a 3GPP up-link W-CDMA 3.84MHz chip rate signal at 1.98GHz and a 5MHz OFDM 16-QAM signal at 3.42GHz corresponding to one of the WiMAX modes. The output of the PA under test was connected to the VSA (Agilent N9020, 26MHz bandwidth) which was synchronized with the two arbitrary signal generators.

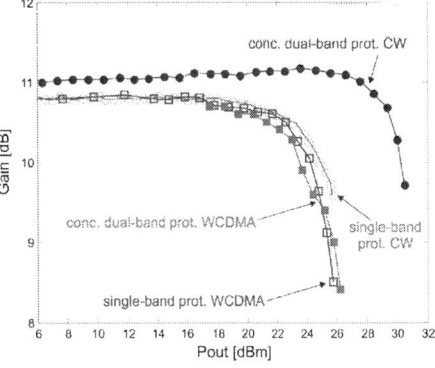

Fig. 4. Gain curve versus output power for CW and WCDMA modulated excitations, both with carrier at 1.98 GHz.

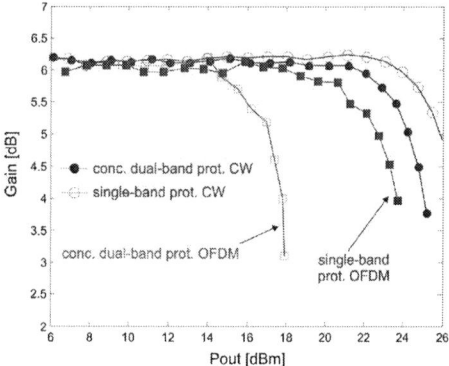

Fig. 5. Gain curve versus output power for CW and OFDM modulated excitations, both with carrier at 3.42 GHz.

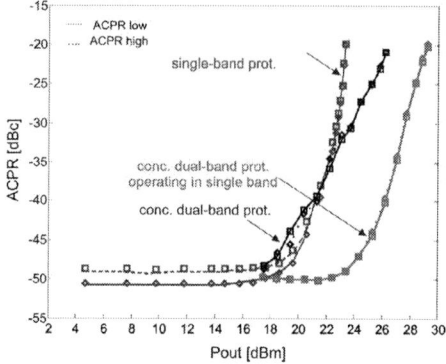

Fig. 6. Adjacent channel power ratio measured at 5MHz offset and integrated over the bandwidth, for the single band and the dual-band prototypes with the WCDMA signal at 1.98 GHz.

The first set of data refers to the large signal gain plotted against the output power for the three PA modules; the comparisons between several operating conditions are shown in Fig. 4 and in Fig. 5 for the lower and higher frequency bands respectively. From Fig. 5 it is possible to observe that the single-band PA exhibits 25.5dBm output power at 1dB gain compression point with a CW signal. The concurrent dual-band module reaches instead a 1dB gain compression output power of 28.2dBm; the different values for peak output power have to be ascribed mainly to a beneficial effect due to the matching conditions obtained with the dual-band PA. In the higher frequency band both single band and dual band amplifiers show a reduced gain of approximately 6 dB. The 1-dB output powers in this case are 25 dBm and 24 dBm respectively for the single-band and dual-band PAs, both measured against a single CW tone at 3.42 GHz. It is observed that when the amplifiers are driven by a single modulated signal peak power at 1dB gain compression point decreases: this effect is explained by the fact that gain compression in PAs driven by digitally modulated signals occurs at lower power levels than for 1-tone CW signals, as described in [6]. In addition, load pull CAD analysis and successive design were performed based on a CW test signal, while experimental results show that the optimum load impedance for maximum linear output power as well as peak efficiency varies depending on the characteristics of the input signal, i.e. pulsed, modulated or CW [7]. Concurrent mode was then operated by simultaneously feeding the dual-band PA with OFDM and WCDMA signals at the two center band frequencies; a resulting 4dB and a 4.5dB peak power reduction at 1.98 GHz and at 3.42 GHz respectively was measured with respect to the single channel cases.

Fig. 6 shows how at 1.98 GHz the maximum achievable output power, due to ACPR constrains, is 27.5 dBm when the dual-band PA is working in single-channel mode, while for the concurrent dual-band case this limit decreases to 23 dBm. Data in Fig. 7 show EVM vs output power results for single channel operation and dual band mode at 3.42 GHz: a maximum output power of 20 dBm is achieved in the first case while when the dual-band PA is working in concurrent mode, maximum output power settles to 17 dBm.

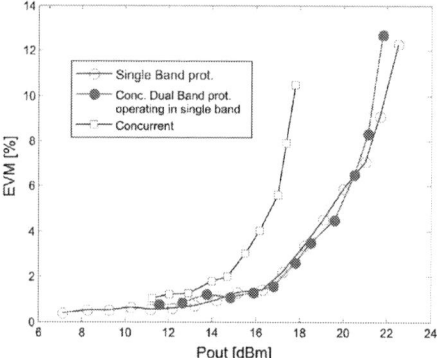

Fig. 7. Error vector magnitude measured for the single band and the dual-band prototypes with the OFDM signal at 3.42 GHz.

The above data indicate that a significant change in performance arises when the PA is driven in the concurrent dual-band mode, specifically resulting in a peak power back off of about 4.5 dB and 3dB respectively for the lower and higher frequency bands due to meet the EVM and ACPR restrictions. The system level performance of the concurrent dual-band PA is summarized by Fig. 10 and Fig. 11, where the output envelope spectrum for the 1.98 GHz 3GPP UL WCDMA signal and the received constellation for the 3.42 GHz WiMAX 5MHz 16-QAM signal have been reported. In particular Fig. 10 shows the comparison between the RF spectrums in single channel and concurrent operation modes, considering a constant output power settled to 24.6 dBm. We recognize the strong impact of the concurrent signal operating at the higher frequency which determines a third-order spectral regrowth which sensibly degrades the adjacent channel power ratio, where this value moves from 51 dBc to 37 dBc as reported in the figure. This data are consistent with those reported by Fig. 8, taking into account that the latter were calculated integrating the power density over the channel bandwidth. The comparison between the received WiMax signal constellations is shown in Fig. 9, where the received symbols have been related to the corresponding output power levels required to reach similar EVM values.

978-2-8748-7007-1/08 $25.00 © 2008 EuMA

Experimental data showed that a 2 dB back-off is necessary with concurrent operation to maintain the EVM at 4.1%.

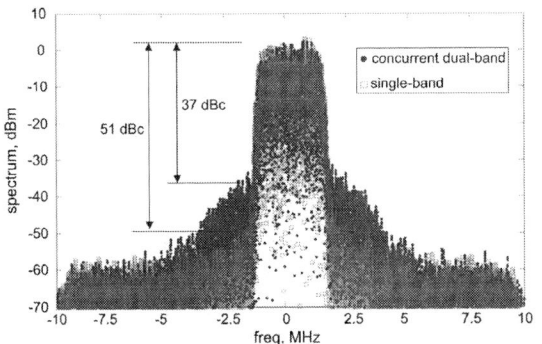

Fig. 8. WCDMA envelopes spectrum measured at the concurrent dual-band PA prototype output when operated in single-band and dual-band, for a constant output power of 24 dBm.

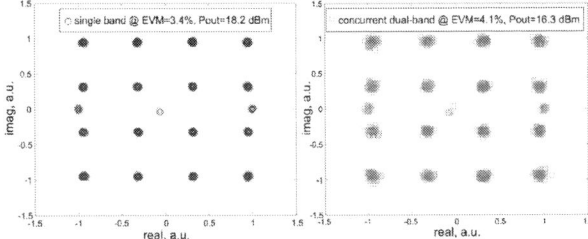

Fig. 9. WiMAX constellation measured at the concurrent dual-band PA prototype output when operated in single-band (left) and dual-band (right).

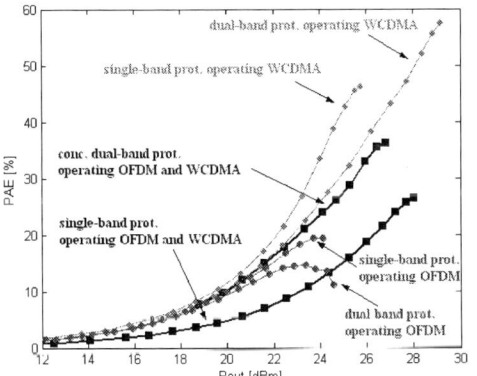

Fig. 10. Power added efficiency for the prototypes in several operative conditions.

A final efficiency comparison is reported in Fig. 10. As expected due to the higher gain and peak output power, the efficiency for the WCDMA signal is higher than for the WiMax signal. When the dual band PA is operated concurrently the power added efficiency is calculated by means of eq. (1), which describes the power supply current re-use for the both the frequency bands.

$$PAE = \frac{\left(P_{load}^{OFDM} - P_{av}^{OFDM}\right) + \left(P_{load}^{WCDMA} - P_{av}^{WCDMA}\right)}{P_{dc}} \quad (1)$$

Figure 10 shows device efficiency curves and system efficiency curves (black), where performance of the two single band PAs has been combined together considering a no loss diplexer. Dual-band PA yields an overall 20% system efficiency with 22.5dB peak output power against a 14% system efficiency with 24dB output power delivered by the combined single band PAs. Efficiencies have been calculated for output power levels consistent with standard limits for ACPR and EVM.

IV. CONCLUSION

This paper has dealt with a feasibility study for a concurrent dual-band PA design and characterization for 1.98 GHz 3GPP UL WCDMA and 3.42 GHz 5 MHz 16QAM WiMAX digital systems. The amplifier exhibited simultaneous peak power levels of 24 dBm and 17dBm in the lower and in the higher bands to maintain ACPR and EVM requirements below 33 dBc and 5% respectively. With respect to a combined PA architecture, the concurrent dual-band PA delivers a 6% higher system efficiency with a 1.5dB back off in peak output power, also thanks to the bias current re-use across frequency bands.

ACKNOWLEDGEMENT

The authors wish to thank Agilent Technology Italy which provided part of the equipment for the system level characterization. Research reported was performed in the context of the network TARGET– "Top Amplifier Research Groups in a European Team" and supported by the Information Society Technologies Programme of the EU under contract IST-1-507893-NOE, www.target-net.org.

REFERENCES

[1] M. Steer, "Beyond 3G", *IEEE Microwave Magazine*, Feb. 2007, pp. 76-82
[2] Earl McCune, "High-Efficiency, Multi-Mode, Multi-Band Terminal Power amplifiers" *IEEE Microwave magazine*, 44 March 2005, pp. 44-55
[3] A. Abidi, "The path to the Software Defined Radio Receiver", *IEEE Journal of Solid-State Circuits*, Vol. 42, May 2007, pp. 954-966
[4] D. Bespalko, N. Messaoudi, S. Boumaiza, 'Concurrent dual-band GaN Power Amplifier with compact micro-strip matching network' 2008 IEEE Topical Symposium on Power Amplifier for Wireless Communications, Jan. 21-22, 2008, Orlando, Florida
[5] P. Colantonio, F. Giannini, R. Giofre, L. Piazzon, "Simultaneous dual-band high efficiency harmonic tuned power amplifier in GaN technology", European Microwave Integrated Circuit Conference, 8-10 Oct. 2007 pp.127 - 130
[6] H. M. Gutierrez, K. G. Gard, M. B. Steer "Nonlinear gain compression in microwave amplifiers using generalized power series analysis and transformation of input statistics" IEEE Trans. on Microwave Theory and Tech., vol. 48, Oct. 2000, pp. 1774–1777.
[7] P. Ghanipour, S. Stapleton, J. H. Kim, 'Load-pull characterization sing different digitally modulated stimuli' IEEE Microwave and Wireless Components Letters, Vol. 17, May 2007 pp. 400-402

978-2-8748-7007-1/08 $25.00 © 2008 EuMA

Proceedings of the 3rd European Microwave Integrated Circuits Conference

Doherty amplifier design for 3.5 GHz WiMAX considering load line and loop stability

M. A. Yarleque Medina[#1,*2], D. Schreurs[#3], I. Angelov[&], B. Nauwelaers[#5]

[#]*Katholieke Universiteit Leuven, Div. ESAT-Telemic, Kasteelpark Arenberg 10, B-3001 Leuven-Heverlee, Belgium*
[1]myarlequ@esat.kuleuven.be
[3]Dominique.Schreurs@esat.kuleuven.be
[5]Bart.Nauwelaers@esat.kuleuven.be

[*]*Pontificia Universidad Católica del Perú, Sección Electrónica, Apdo. 1761, Lima-100, Perú*
[2]myarleq@pucp.edu.pe
[&]*Chalmers University of Technology, Goteborg, Sweden*

Abstract— **Legacy Doherty amplifier is characterized by using a larger transistor for the peak amplifier such that this reaches saturation with a smaller excitation signal. However due to device availability and modelling considerations, this is not often feasible. In this paper, the design and measurement of a Doherty amplifier utilizing only single sized device is realized. Unlike previous research works, intrinsic load line is utilized to tune the offset lines, as well as to verify the actual dynamic load principle. Stability aspects are covered for this type of amplifier, which are not normally included in earlier works. Finally an assessment of its applicability and benefits for WiMAX at 3.5 GHz is realized using a class AB amplifier as a comparison basis.**

I. INTRODUCTION

WiMAX has arisen as a very attractive technological solution for the last-mile subscriber access. It will be applied on sceneries where DSL and cable networks are not cost-effective or difficult to be deployed. Currently two WiMAX standards exist, one for fixed wireless access (802.16) and one for mobile subscribers (802.16e). In any of these cases the power efficiency is a vital factor since low electrical consumption in rural areas deployment (fixed access) or longer battery life (mobile context) is essential nowadays. On the other hand, this technology is very robust in multipath channels due to the utilization of OFDM technique. However, this OFDM signal exhibits large peak to average power ratio (PAPR), and in the particular case of WiMAX, this varies between 8 to 10 dB. As a result, classical linear amplifiers have to be operated at 8 to 10 dB backoff from the 1 dB compression point in order to amplify linearly over the full signal dynamic range. Nonetheless, this comes at the cost of a very poor efficiency. Hence, an amplifier that works linearly over this 8-10 dB PAPR and still keeps an acceptable efficiency over this range has become a challenge.

Several techniques are under research to tackle this problem. Among these, the Doherty power amplifier (DPA) has received lots of attention lately due to its simple architecture, efficiency range and inherently improved linearity [1-3]. In its basic configuration this amplifier comprises a carrier amplifier (CA) and a peak amplifier (PA). The transistor in the peak amplifier is larger than the one in

the carrier amplifier. This may not be an issue when different sizes of devices are available; however a scalable non-linear model or two different non-linear models are not always available. Moreover in order to verify proper Doherty operation it is important to screen the intrinsic device load line. Normally, this is not presented in research works, and it may be the reason why typical efficiency enhancements are not observed in the results. In this paper, the design of DPA using only single-sized devices is developed. Despite this amplifier provides only 6 dB of efficiency range, it can be applied to WiMAX requiring 8-10 dB efficiency range. Its actual efficiency improvement in comparison with legacy class-AB is assessed at the end of this paper. The device is a Filtronic GaAs HEMT FPD750 with an in-house Angelov non-linear model. This model is developed as SDD component, which permits the reading of its intrinsic variables.

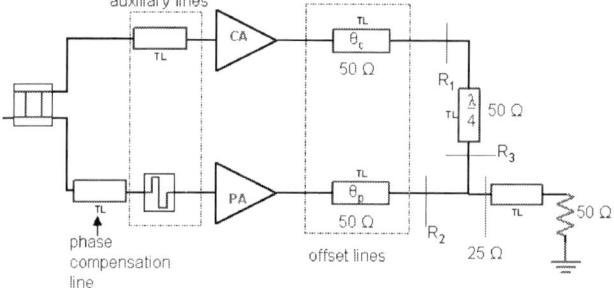

Fig. 1 Doherty amplifier configuration

II. DESIGN OF DOHERTY AMPLIFIER

A class-AB and class-C amplifier, utilizing the same GaAs device, was utilized as peak and carrier amplifier. These amplifiers were designed using the concept of equivalent capacitance [4] and were both matched for maximum output power and 50 Ω output load. For the following analysis, the load seen by the device for maximum output power will be nominated as $R_{opt\text{-}device}$, and the reference load used in the DPA will be named as $R_{opt\text{-}Doherty}$, which in this case is 50 Ω.

The Doherty amplifier configuration is shown in Fig. 1. During impedance inversion condition, only the carrier

978-2-8748-7007-1/08 $25.00 © 2008 EuMA

amplifier works and sees a resistance R_1 equal to $50^2/25 = 100$ Ω ($2R_{opt\_Doherty}$). During load pull condition, the peaking amplifier injects current, pulling up the resistance R_3 from 25 to 50 Ω, by which the last value is reached when the currents are equal and maximum. This resistance R_3 keeps its value after the impedance inversion, i.e., R_1 equals to 50 Ω. Likewise, the peaking amplifier will see an impedance R_2 of $25/(1-0.5) = 50$ Ω at the maximum current.

Some points have been missed in the analysis described above. The matching of the main amplifier has been sized for 50 Ω. This is the value which must be reflected internally, at the drain, as a resistance for device maximum power ($R_{opt-device}$). Nonetheless, when 100 Ω is presented to the carrier amplifier, this will not be reflected as twice the resistance for device maximum power ($2R_{opt-device}$), due to the effect of all the lumped components in between. This has been simulated and can be observed in the load line shown in Fig. 2. In order to correct this response an offset transmission line [1-3] can be added at the output. The length of this transmission line can be tuned until a proper load line is observed as shown in Fig. 2. The characteristic impedance of this line is set to 50 Ω. This line is transparent when R_1 becomes 50 Ω due to the load pull effect. The same correction can be performed observing the intrinsic load value on the Smith chart [1].

An offset line is also required at the peaking amplifier. During impedance inversion condition, the load seen into the peaking amplifier must be infinite such that no power is leaking from the main to the peaking amplifier. To this purpose an offset-line is tuned to assure that the impedance is large enough to resemble an open circuit. This is realized by observing the output return loss (S_{22}) in the Smith chart [1]. Like the carrier amplifier, the impedance of this offset-line is set to 50 Ω in order not to affect the peaking amplifier during load pull condition. The offset length was found to be -20° (or 340°) for the carrier amplifier. Since this line has the same impedance as the impedance inverter, these lines can be joined in one transmission line of 70°. The offset line for the peaking amplifier was around -3° and it was neglected.

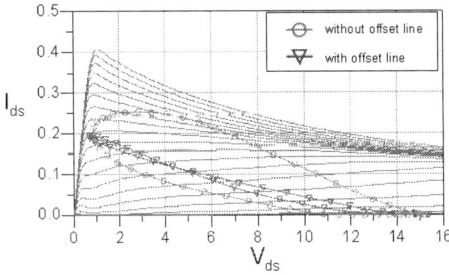

Fig. 2 Tuning of offset line for carrier amplifier

Based on the formulation developed in [5] the power divider should have a 1/3 ratio for DPA with devices of the same size. However, this ratio is too high as it would mean a drastic drop in the gain and PAE of the complete amplifier. The class AB has a gain of about 9 dB; the power divider would introduce 6 dB power drop, resulting in a DPA of only 3 dB of gain. Therefore this power ratio was reduced to the minimum 1/1, although this still introduces a detrimental 3 dB

power drop. To implement this power divider, a branch line divider configuration was utilized.

Theoretically, the phase compensation line at the peak amplifier has to be as long as the impedance inverter, i.e., 90°. This would be a good approximation if the matching circuits of the amplifiers were implemented with lumped components; nevertheless this was not the case. The matching circuits introduce additionally delays that should be accounted as well. The best way to estimate the phase difference of the branches is by simulating the phase output of both amplifiers. This difference is compensated with a line at the peaking amplifier. This turns out to be a 112° line. Additionally the branch line introduces a 90° delay difference between outputs, and then the total compensation line would be 202 °.

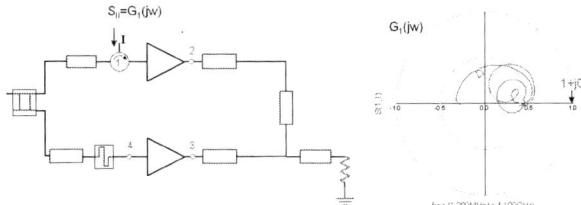

Fig. 3 Verification of Doherty amplifier stability by Nyquist criterion

Stability analysis

The necessary and sufficient condition for unconditional stability in power amplifiers were restated in [6-7] as follows:

1. No Right Hand Plane (RHP) poles in a network terminated with Z_{01}, Z_{02}.

2. $k > 1$ and $\Delta < 1$, or alternatively $\mu > 1$, for all frequencies.

The first condition assures the stability of the system when it is terminated with the impedance that defines the scattering matrix. The second condition assures stability for all other passive terminations provided the first condition is true.

The first condition can be simply fulfilled if the S parameters can be measured, i.e., the device is stable in the measurement system. This is perhaps the reason why this condition has been normally omitted. Nonetheless, this omission proceeds when the design involves only one device (the measured device). When more than one active device is involved in the network (or amplifier) the first condition can not be assured [7, 8], and alternatives methods have to be applied [6, 9, 10]. On the other hand, specific method to detect instability condition (first condition) has been developed for the specific case of an amplifier with parallel devices exhibiting symmetric layout. In this case, the first condition is analysed by checking the presence of potential odd-modes instabilities [11-14]. However this method can not be applied to a DPA, which presents an asymmetric structure per se.

A method developed in [9] can be applied for Doherty Amplifier. This method is based on Nyquist criterion. This is a graphical method that verifies the feedback loop stability from how the complex locus (or Nyquist plot) of the open loop G(jw) encloses the point 1+j0. If the number of poles in the RHP of G(jw) is zero, then the system is stable if the number of clockwise revolutions of G(jw) around the 1+j0 is zero, i.e., it does not enclose the point 1+j0 clockwise. The condition of

RHP poles equal to zero is superseded in the method by considering that the active devices are stable and by performing Nyquist plots several times under different conditions. For instance, this test has to be performed four times, as it is indicated in Fig 3. An ideal circulator is introduced at the interface port 1 and the parameter S_{11} looking from the port I of this circulator corresponds to $G_1(jw)$. Then, this ideal circulator is moved to the interface port 2 and an isolator is introduced at port 1. Again S_{11} looking from port I of the circulator is $G_2(jw)$, and so forth. At the end, none of these $G_i(jw)$ has to enclose $1+j0$ in order to guarantee amplifier stability. The ideal isolator can be implemented with a circulator, in which port I is terminated with the reference impedance (50 Ω). This method was applied to the designed DPA to verify the stability. There was no indication of instability, as indicated in one of the outputs reported in Fig. 3.

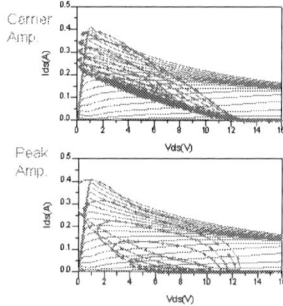

Fig. 4 Intrinsic load line of carrier and peak amplifier

A Harmonic Balance simulation of the complete amplifier was realized to verify Doherty operation. The intrinsic device loadlines of the carrier and peaking amplifier are reported in Fig. 4. As it can be observed, the carrier amplifier reaches maximum voltage excursion at half of the maximum current, then it changes its load line due to the load pull effect of the peaking amplifier. A notorious looping on the intrinsic device loadline of the peaking amplifier can be observed. This could be justified considering that the load presented to the peaking amplifier is only 50 Ω at maximum current.

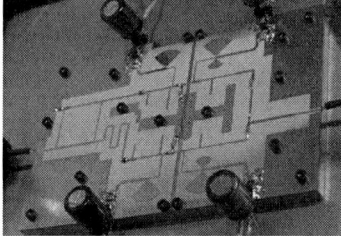

Fig. 5 Doherty amplifier during LSNA testing

III. MEASUREMENTS AND PERFORMANCE COMPARISON

The final implementation of the DPA can be observed in Fig. 5. The substrate material utilized for the fabrication of this circuit is Rogers RO4003C (ε = 3.55, thickness=200 um). As the FET devices on the board would be placed on a metallic strip, an additional straight transmission line at the carrier amplifier branch and a meander line at the peaking

amplifier branch were needed to have the FETs in line. These straight and meander lines have the same electrical length.

Simulated and measured power/efficiency figures of merit are reported in Fig. 6. The correspondence between simulation and measurements is acceptable over most of the range, which validates the load line approach and stability analysis. An efficiency range of about 6 dB is obtained. The PAE at peak power is 31.3 % and it is 32% at 6 dB input backoff (IBO). The maximum output power is 30 dBm.

EVM and ACPR can be used as linearity FoM. However EVM is more critical and it will be used as comparison parameter. For this evaluation, the WiMAX 802.16e simulation toolkit of Agilent is employed, since actual WiMAX instruments were not available at the moment of the testing. The amplifier was cosimulated with Ptolemy and Envelope simulators of ADS. This simulation was performed for the DPA and compared with the class AB amplifier. The signal used for this testing is a 64QAM with 1/2 code rate, 10 MHz bandwidth and 8.2 dB PAPR at 0.1 % CCDF. The input powers for the class AB and 6 dB DPA are set to 14 dBm and 17 dBm, such that both amplifiers have approximately the same average peak power (27.5 dBm) and average output power (20 dBm).

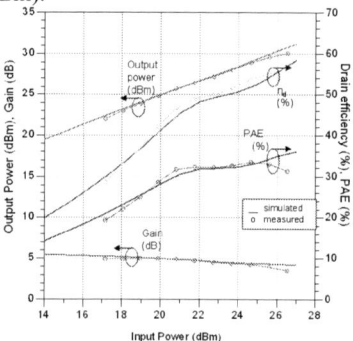

Fig. 6 DPA performance, VDS = 8.1 V, VGS$_c$ = −0.85 V and VGS$_p$ = −3.6 V

The output constellation with the corresponding relative constellation RMS error (RCE or EVM), averaged over subcarriers, OFDMA frames, and packets are reported in Fig. 7 for both amplifiers. This relative constellation error (RCE) in dB should not exceed −26 dB for the testing conditions. Both amplifiers comply this specification, however the DPA has a better RCE value (-34 dB) than the class AB amplifier (-26 dB). This improved linearity performance of the DPA was also reported in a recent paper [3], in which the results were based on measurements. The reason of this linearity improvement yields on the cancellation of the third-order harmonic generation (coefficient gm3) between the carrier and peaking amplifier [1-3].

Finally the average PAE of the class-AB and DPA will be calculated considering the WiMAX signal statistics and the amplifier efficiency. The average PAE can be calculated as

$$PAE_{ave} = \frac{\int_0^\infty (P_{out} - P_{in}) \cdot p(P_{out}) \cdot dP_{out}}{\int_0^\infty P_{DC}(P_{out}) \cdot p(P_{out}) \cdot dP_{out}} \qquad (1)$$

where *p(Pout)* corresponds to the probability distribution function (pdf) of the output power.

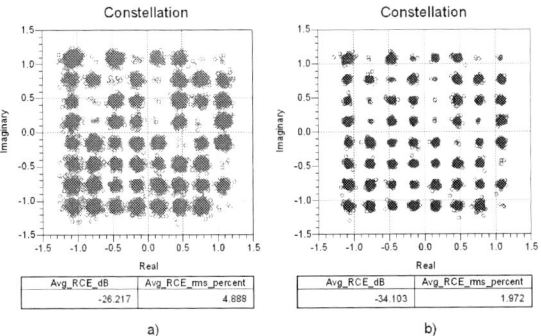

Fig. 7 EVM performance of the class AB amplifier a) and DPA b)

The probability distribution is obtained from simulations (see Fig. 8) and using the same excitation signal as the one for the linearity testing. The values of DC power and output power are measured values. The average efficiency values are reported in Table 1 (rows AB and DPA1) along with the linearity figures.

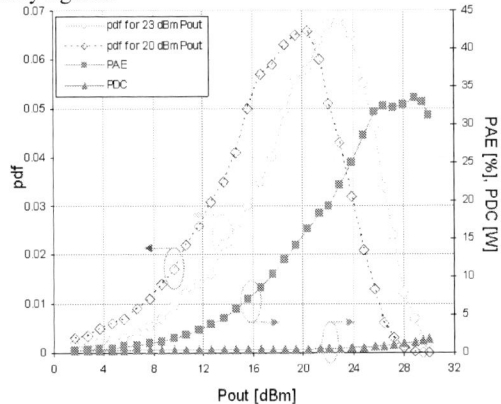

Fig. 8 Probability distribution function of the DPA output power

TABLE I
COMPARISON BETWEEN CLASS AB AND DPA FIGURES OF MERIT FOR TWO DIFFERENT TESTING CONDITIONS (DPA1 AND DPA2)

Amp.	Pin (dBm)	Pout_ave (dBm)	Pout_peak (dBm)	η_ave (%)	PAE_ave (%)	RCE (dB)
AB	14	20	27.7	16.7	14.7	-26
DPA1	17	20	27	20.7	14.3	-34
DPA2	20	23	30	31	21	-26

As can be observed, the average PAE is similar for both amplifiers, though the linearity is better in case of DPA. This low PAE in DPA is because of the lower gain and higher power drop of the DPA in comparison with the class AB. Another important reason for the low efficiency is that the peak output power only reaches 27 dBm while its maximum is about 30 dBm, losing 3 dB of optimal efficiency range (see Fig. 8, pdf for 20 dBm output power). When the DPA is driven such that the output signal covers most of its optimal

efficiency range (see Fig. 8, pdf for 23 dBm output power), better efficiency figures are obtained, as indicated in Table 1 (row DPA2). In this case, DPA exhibits a more pronounced improvement with an average PAE of 21%. However, its linearity performance in terms of RCE drops to -26 dB, although this still complies with the WiMAX standard and is comparable with the class-AB amplifier.

IV. CONCLUSIONS

In this paper, the design of a DPA with single-sized devices was realized successfully considering intrinsic load line. This permits to tune the offset line and to verify the impedance inversion and load pull principles of DPA. The stability of this amplifier was analyzed using Nyquist criterion. Efficiency and linearity assessment were realized and compared with a class AB amplifier, showing the superiority of DPA for WiMAX at 3.5 GHz.

ACKNOWLEDGMENT

This work is a result of the former NoE Target.

REFERENCES

[1] B. Kim, J. Kim, I. Kim, and J. Cha, "The Doherty Power Amplifier," *IEEE Microw. Mag.*, vol. 7, no. 5, pp. 42–50, Oct. 2006.

[2] B. Kim, J. Kim, I. Kim, J. Cha, and S. Hong, "Microwave Doherty Power Amplifier for High Efficiency and Linearity," in *Proc. Int. Workshop on Int. Nonlinear Microw. and Millim.-Wave Circuits*, 30-31 Jan. 2006, pp. 22–25.

[3] J. Moon, J. Kim, I. Kim, Y. Y. Woo, S. Hong, H. S. Kim, J. S. Lee, and B. Kim, "GaN HEMT Based Doherty Amplifier for 3.5-GHz WiMAX Applications," in Proc. 37th Eur. Microw. Conf, 8-10 Oct. 2007, pp. 395–398.

[4] M. Yarleque Medina, D. Schreurs, and B. Nauwelaers, "RF Class-E Power Amplifier Design based on a Load Line-Equivalent Capacitance method," to be published in *IEEE Microw. Guided Wave Lett*, Mar. 2008.

[5] J.-Y. Lee, J.-Y. Kim, J.-H. Kim, K.-J. Cho, and S. Stapleton, "A High Power Asymmetric Doherty Amplifier with Improved Linear Dynamic Range," in *IEEE MTT-S Int. Dig.*, 11-16 June 2006, pp. 1348–1351.

[6] R. Jackson, "Rollett Proviso in the Stability of Linear Microwave Circuits A Tutorial," *IEEE Trans. Microw. Theory Tech.*, vol. 54, no. 3, pp. 993 – 1000, Mar. 2006.

[7] M. Ohtomo, "Proviso on the Unconditional Stability Criteria for Linear Two Port," *IEEE Trans. Microw. Theory Tech.*, vol. 43, no. 5, pp. 1197 – 1200, May 1995.

[8] A. Platzker, W. Struble, and K. Hetzler, "Instabilities Diagnosis and the Role of K in Microwave Circuits," in *IEEE MTT-S Int. Dig.*, 1993, vol.3, pp. 1185–1188,

[9] M. Ohtomo, "Stability Analysis and Numerical Simulation of Multidevice Amplifiers," *IEEE Trans. Microw. Theory Tech.*, vol. 41, no. 6, pp. 983 – 991, Jun.-Jul. 1993.

[10] W. Struble and A. Platzker, "A Rigorous Yet Simple Method for Determining Stability of Linear N-port Networks," in *GaAs IC Symp. Tech. Dig.*, 10-13 Oct. 1993, pp. 251–254.

[11] A. Costantini, G. Vannini, F. Filicori, and A. Santarelli, "Stability Analysis of Multi-transistor Microwave Power Amplifiers," in *proc. GAAS 2000 Symp.*, Paris, Oct. 2000.

[12] R. Freitag, S. Lee, D. Krafcsik, D. Dawson, and J. Degenford, "Stability and Improved Circuit Modeling Considerations for High Power MMIC Amplifiers," in *IEEE Microw. and Millim.-Wave Monolit. Circuits Symp. Dig.*, 24-25 May 1988, pp. 125–128.

[13] R. Freitag, "A Unified Analysis of MMIC Power Amplifier Stability," in *IEEE MTT-S Int. Dig*, 1-5 June 1992, vol.1, pp. 297–300.

[14] R. Weber, "Even Mode versus Odd Mode Stability [Microwave Networks]," in *Proc. of the 40th Midwest Symp. on Circuits Syst.*, 3-6 Aug. 1997, vol. 1, pp. 607–610.

GaN Doherty Amplifier With Compact Harmonic Traps

Paolo Colantonio, Franco Giannini, Rocco Giofrè, Luca Piazzon

Electronic Engineering Department, University of Roma Tor Vergata - Via del Politecnico 1, 00133 Roma, Italy
giofr@ing.uniroma2.it

Abstract—**In this contribution, the design of an uneven AB-C Doherty power amplifier (DPA) in GaN technology, implementing a new approach to control the higher device harmonics, is presented. The DPA was designed to operate at 2.14GHz and with the aim to reduce as much as possible the chip size, without losing the Doherty operating principle. The measurement results in CW conditions at 2.14GHz had shown average drain efficiency higher than 55% at 6dB of back-off, with a saturated output power of 37dBm.**

I. INTRODUCTION

The increasing complexity of modulation schemes adopted in the new wireless systems to achieve highest data rate transfer, are requiring PAs able to manage signals with a large time-varying envelope. The resulting peak-to-average power ratio (PAPR) of the involved signals critically affects the average efficiency achievable with traditional PAs. For instance, in the European UMTS standard with W-CDMA modulation, a PAPR of 7-10 dB is typical registered. Such high values of PAPR imply usually a great back-off operating condition, thus dramatically reducing the average efficiency levels attained by using traditional solutions (Class AB, B, F). Consequently, in order to improve the average efficiency levels, different PA architectures, like the Envelope Elimination and Restoration (EER), the Envelope Tracking (ET), the Output Phasing System and the Doherty were revisited and extensively investigated [1,2]. In particular, the Doherty amplifier (DPA), due to its relative simple implementation as compared as to the efficiency benefits attainable, represents an attractive solution for systems with high PAPR [3].

The typical scheme of a DPA is constituted by two PAs, namely *Main* and *Auxiliary*, an Impedance Inverter Network, a Phase Control Network and a Power Divider, as graphically depicted in Fig. 1 [3]. Usually the design of a DPA is performed designing the *Main* and *Auxiliary* amplifiers separately, and then optimising their combination using the remaining components, i.e. the input and output quarter-wave transmission line (λ/4-TL) and the input power splitter. The resulting DPA is characterised by a large chip area, due to the design of two almost identical output matching network required by both the amplifiers.

In this contribution, a new solution to design a more compact uneven AB-C DPA is presented. The proposed solution allows the possibility to dramatically reduce the size of the resulting DPA, by using a single output matching

network for both *Main* and *Auxiliary* devices, without drop the operating criteria. The proposed approach, based on a new harmonic trapping network, has been validate through the design of a DPA based on GaN HEMT devices, biasing the *Main* and *Auxiliary* devices in Class AB and C respectively. The measurement performed on the realised DPA shown average drain efficiency higher than 55% in a 6dB of output back-off (OBO), with a saturated output power of 37dBm. The DPA was also optimised for linearity behaviour, resulting in gain flatness contained in 1 dB in the 6dB OBO (i.e. in the Doherty region).

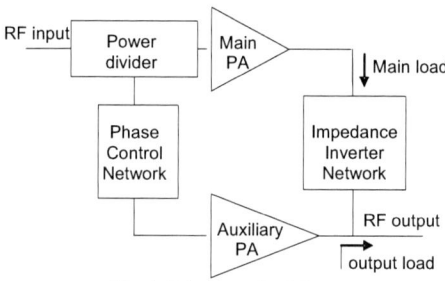

Fig. 1 Scheme of the DPA.

II. THE UNEVEN AB-C DOHERTY PA

To describe the DPA behaviour, is useful to consider two operating regions. For low input power levels, only the *Main* PA is active, while the *Auxiliary* is kept off, up to the achievement of *Main* PA maximum efficiency. As a consequence, the DPA works as a common PA biased in Class AB. Then, increasing the input power levels, the *Auxiliary* PA is turned on, modulating the load seen by the *Main* PA from $2R_{opt}$ to R_{opt} until the overall DPA saturation is achieved [[1,3]. The switching on condition of the *Auxiliary* device is typically referred as a breaking point, while the power range where both devices are active is named Doherty region. It is to note that in this region the *Main* device is forced by the *Auxiliary* to modify its behaviour from current to voltage source (limited by the knee voltage), and it is pushed to achieve its maximum output current value [4].

A different bias condition for both PAs is needed to avoid any additional control circuitry to properly turn on the *Auxiliary* amplifier, i.e. to control the breaking point [4]. At the same time, an uneven power splitter is required to deliver more power to the *Auxiliary* PA in order to completely and properly modulate the load seen by the *Main* PA [4,5]. Finally

978-2-8748-7007-1/08 $25.00 © 2008 EuMA

the input phase control network, implemented using a second quarter-wave transmission line (λ/4 TL), is needed to properly reconstruct at the output the signals rising from the *Main* and *Auxiliary* devices.

Typically a Tuned Load configuration (short circuit at higher harmonics at the intrinsic current source of the active device equivalent circuit model) for both devices is adopted [6], even if a different harmonic termination could be adopted [7]. As well know in literature, the active device is typically modelled as reported in Fig. 2, through a non linear current source (I_{ds}) a shunt resistor (R_{ds}) and capacitor (C_{ds}).

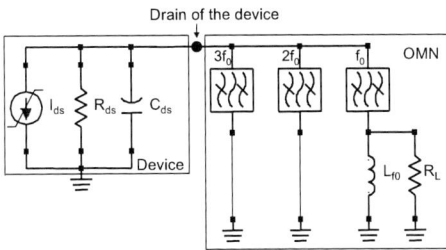

Fig. 2 Basic scheme of the actual TL configuration.

In order to fulfil the Tuned Load configuration, the output matching network (OMN) has to guarantee at the nonlinear current source (NCS) section a purely resistive load at fundamental frequency and a short circuit condition at the higher harmonics. Therefore, as depicted in Fig. 2, where the harmonics were limited up to third order, an external inductive complex load at the fundamental frequency is needed to compensate the effect of C_{ds}. The value of the inductance L_{f0} can be easily evaluated imposing the resonance with C_{ds} at the fundamental frequency, while the value of R_L in Fig. 2 is selected to maximise the voltage and current swing and thus the output power [6].

Referring to the DPA design, where two active devices are involved, the typical approach is based on the realisation of two single OMNs, one for the *Main* and one for the *Auxiliary* device, as schematically reported in Fig. 3.

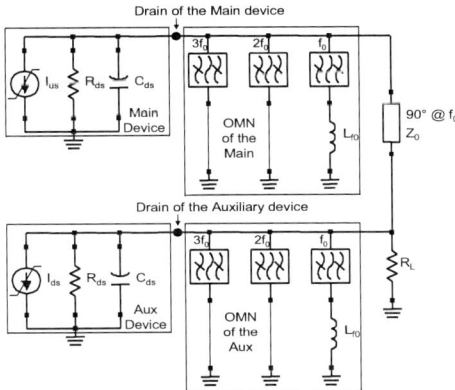

Fig. 3 Basic scheme of a typical DPA.

These OMNs, have a large impact on the size of the overall DPA, while a reduction becomes important and useful for

instance if a DPA has to be integrate in a monolithic solution. Moreover, since both Main and Auxiliary devices present similar parasitic elements usually these two networks are almost the same.

III. The Proposed Harmonic Traps Solution

The approach here presented was developed to reduce the AB-C DPA overall size, avoiding the duplication of the OMNs, realising a unique network to synthesise the optimum harmonic loads for both devices, as schematically reported in Fig. 4. Referring to this figure, at the common node (C.N.) of the two PAs an ideal short circuit and open circuit conditions can be realised at second and third harmonic respectively.

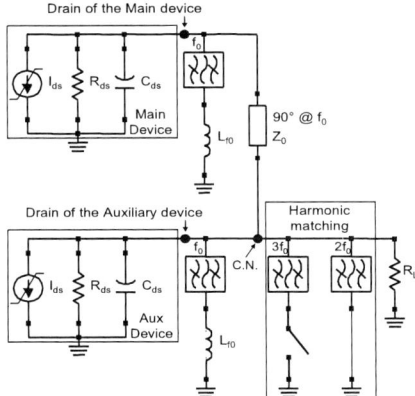

Fig. 4 Scheme of the proposed AB-C DPA.

Consequently, a perfect short circuit condition can be achieved at both Main and Auxiliary NCSs, since the quarter-wave TL on the Main path becomes transparent at second harmonic. For the third harmonic, the open circuit condition at the C.N. is still transformed in a perfect short circuit at the NCS of the Main device, still due to the λ/4-TL, achieving a perfect Tuned Load configuration. Conversely, the NCS of the Auxiliary device sees the shunt combination of the external open circuit and the internal parasitic capacitor C_{ds}. Thus, the latter is the actual element sees by the NCS of the Auxiliary at third harmonic, which realised a load not far from the short circuit condition (required for a Tuned Load condition). In fact, the value of C_{ds} for the device used is 4.3pF resulting at 6.42GHz in an impedance value around -j5Ω.

The reduction in the unique resulting OMN is thus paid with a little lack in the control of only one harmonic termination. In particular, the constraints required by the Tuned Load configuration are totally respected for the Main device, while for the third harmonic of the Auxiliary, the corresponding load is slightly capacitive instead of a perfect short circuit. Of course, the perfect Tuned Load condition could be achieved for the Auxiliary device instead of the Main one changing the third harmonic load seen at the C.N. from open circuit to short circuit condition. However, since the Main device operates for a longer period respect to the

Auxiliary, the adopted approach seems to be better to achieve the best overall DPA performances.

In the proposed solution, the DPA was designed to operate at f_0=2.14 GHz for UMTS applications, by using two GaN HEMT with 1 mm of gate periphery, provided by Selex-SI. The device was in-house extensively characterised in linear and non-linear regime, developing a large signal equivalent circuit model [8].

The *Main* PA was biased in Class AB condition, resulting in V_{DD}=18V and V_{GG}=-4.2V (I_{DC}=47mA$\approx$7% of the *Main* maximum drain current). The resulting optimum resistance value (R_{opt}) under a Tuned Load condition was 35Ω. In order to have a 6dB of OBO, the resistor at the break point should be 70Ω ($2R_{opt}$). Consequently the values of the output resistor (R_L) and the characteristic impedance of the output λ/4 TL (Z_0) are respectively 17.6Ω (R_{opt}/2) and 35Ω (R_{opt}) [4].

In order to design the OMN for the harmonic termination control, the first step was to identify the optimum harmonic loading conditions for both devices at the intrinsic current source, i.e. taken into account the extrinsic parasitic elements. For this purpose, an ideal tuner was used as DPA output termination, i.e. connected to the common node (C.N.) reported in Fig. 4, resulting in the optimised load values of 17.6Ω@f_0, -2.8jΩ@$2f_0$ and 130.7jΩ@$3f_0$.

Then, the ideal values were implemented by using a combination of lumped and distributed elements. In particular, for the fundamental frequency it is required to resonate the devices output capacitance. For this purpose, the required inductance value was synthesised through a shorted circuit stub, also used to bring the drain bias voltages. The harmonics loads were synthesised with the network shown in Fig. 5, adopting the impedance buffer methodology [9]. Starting from left to right of this network, the first cell, composed by the series TL and the shunt open stub, actualises the desired load conditions at $3f_0$, while the second cell realises the load at $2f_0$. Finally, the last LC resonator is used to match the R_L=17.6Ω in the point C.N. in the Fig. 4 to the external standard 50Ω termination.

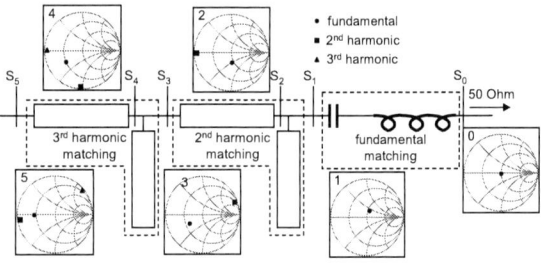

Fig. 5 DPA Output matching networks.

For the input matching network (IMN) design, a traditional approach was adopted, i.e. two separate IMNs were designed. In fact, even if also in this case a possible integration in an unique network could be feasible, the presence of the input power splitting could complicate the design. Therefore in this work it was preferred to investigate only the effects of the

unique OMN, remanding to further work the analysis of the IMN integration also. For the input power splitter, a Wilkinson solution was adopted [10], implementing a splitting power factor of 1/3, i.e. 24% of the input power delivered to the *Main* device, while the remaining 76% to the *Auxiliary* one.

The simulated load curves for both *Main* and *Auxiliary* devices at break point (case A) and saturation conditions (case B) are depicted in Fig. 6, confirming the DPA operating principle.

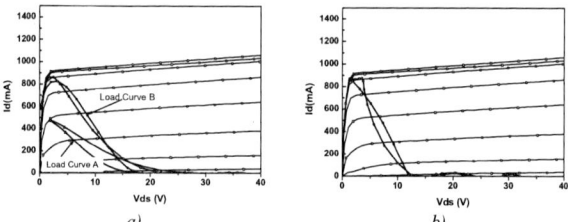

Fig. 6 Output I-V load curves for the *Main* (a) and *Auxiliary* device (b).

IV. REALIZATION AND EXPERIMENTAL RESULTS

The AB-C DPA was realised on 3010 Taconic substrate characterised by a dielectric constant of ε_r=10 and a height of h=0.635 mm. The amplifier was mounted on an aluminium carrier, whose photo is reported in Fig. 7. Two SMA connectors were used to interface the DPA, while two capacitors were inserted in the DC pads to filter out any further low frequency DC signals.

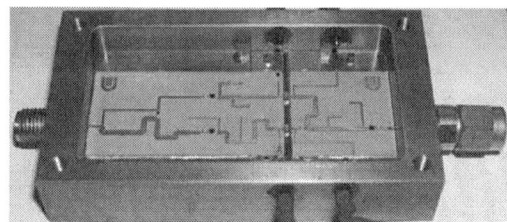

Fig. 7 Photo of the realised DPA.

The DPA was preliminary characterized in small signal condition, at the designed nominal bias conditions, resulting in V_{DS_Main}=18V, V_{GS_Main}=-4.2V, V_{DS_Aux}=18V, V_{GS_Aux}=-8.7V. The comparisons between the measured and simulated reflection and transmission coefficients of the amplifier are shown in Fig. 8.

Then the DPA was characterised in large signal conditions, i.e. performing several continuous wave (CW) power sweeping measurements.

Moreover, an investigation of the effects of the *Auxiliary* biasing condition was performed. In particular, fixing the *Main* device bias point at the same level of simulations, resulting in the bias current I_{d_Main}=47mA, several CW power sweep measurements were performed at 2.14 GHz. The resulting measurements are reported in Fig. 9.

From this figure it can be noted that while the gain behaviour seems to be slightly affected by the *Auxiliary* bias voltage variations, a different result arises for the efficiency curves. In particular, being the overall DPA gain basically related to the *Main* PA, its behaviour is scarcely dependent on the *Auxiliary* biasing point. On the contrary, the detrimental effects in the efficiency behaviour observed in the Doherty region can be ascribed to the earlier *Auxiliary* turning on condition, which does not allow to the *Main* device to reach its maximum efficiency.

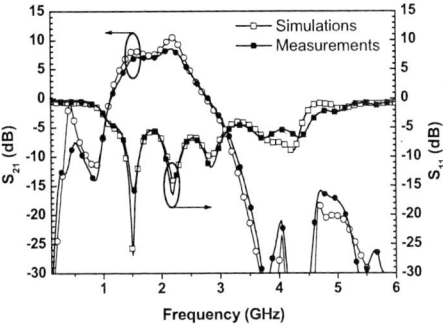

Fig. 8 Comparisons between measured and simulated DPA S-Parameters.

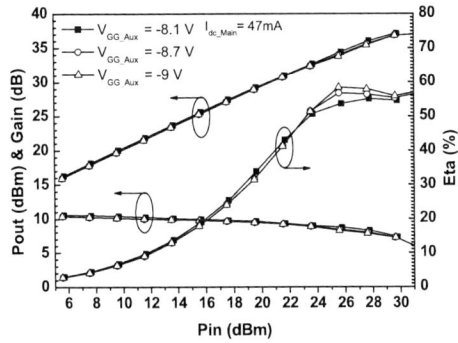

Fig. 9 DPA performances for different *Auxiliary* biasing condition.

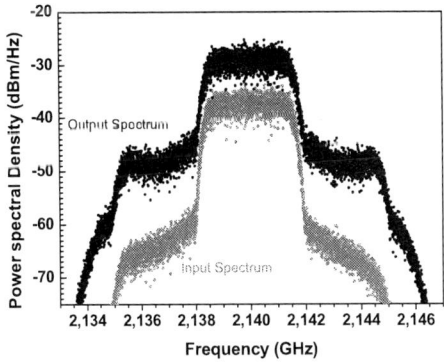

Fig. 10 Measured output spectra for 3GPP WCDMA signal.

The optimum *Auxiliary* bias condition $V_{gs\_Aux}=-8.7V$ assure an almost constant efficiency for 6dB of input power swing (from 24dBm to 30dBm). The maximum output power

measured was 37dBm, with a drain efficiency of 58%, while an average value as higher as 55% in the 6dB range was also achieved.

Finally, the DPA was also tested with complex modulation, and in particular using a 3GPP WCDMA signal @2.14GHz with 4MHz of relative bandwidth and 7.4dB of PAPR. Fig. 10 shows the input and output power spectral density at 27.7 dBm of average input power. As can be seen from the picture, the power in the adjacent channel is roughly 20dB lower than the power in the operating band.

V. CONCLUSIONS

In this contribution, the design of an uneven AB-C DPA in GaN technology implementing a new method to control the higher device harmonic terminations was presented. The DPA was designed with the aim to reduce as much as possible the required area, without losing the Doherty principle of operating. The CW measurements @2.14GHz had shown average drain efficiency higher than 55% at 6dB of back-off, with a saturated output power of 37dBm.

ACKNOWLEDGMENT

The authors wish to acknowledge Dr. V. Camarchia for the measurements with 3GPP WCDMA signal.

Research reported here was performed in the context of the network TARGET– "Top Amplifier Research Groups in a European Team" and supported by the Information Society Technologies Programme of the EU under contract IST-1-507893-NOE, www.target-net.org.

REFERENCES

[1] S. C. Cripps, "Advanced Techniques in RF Power Amplifiers Design," Norwood, MA, Artech House, 2002.

[2] F. H. Raab, P. Asbeck, S. C. Cripps, P. B. Kenington, Z. B. Popovic, N. Pothecary, J. F. Sevic, amd N. O. Sokal, "Power Amplifiers and Transmitters for RF and Microwave," IEEE Trans. on MTT, Vol. 50, pp. 814 – 826, Mar. 2002.

[3] W. H. Doherty, "A New High Efficiency Power Amplifier for Modulated Waves," Proc. of the IRE, Vol. 24, pp. 1163 – 1182, 1936.

[4] A. Z. Markos , P. Colantonio, F. Giannini, R. Giofrè, M. Imbimbo, G. Kompa "A 6W Uneven Doherty Power Amplifier in GaN Technology" 2nd EuMIC 2007. Munich, Germany, October 2007, pp. 299-302.

[5] Ui, Norihiko; Sano, Hiroaki; Sano, Seigo;"A 80W 2-stage GaN HEMT Doherty Amplifier with 50dBc ACLR, 42% Efficiency 32dB Gain with DPD for W-CDMA Base station" IEEE/MTT-S International Microwave Symposium, 2007. 3-8 June 2007 Page(s):1259 - 1262

[6] L.J. Kushner, "Output Performances of Idealised Microwave Power Amplifiers", Microwave Journal, October 1989, pp.103-116.

[7] J. Kim, J. Moon, Y.Y. Woo, S. Hong, I. Kim, J. Kim, B. Kim "Analysis of a fully matched saturated Doherty amplifier with excellent efficiency", IEEE MTT., Vol. 56, No.2, February 2008, pp. 328-338.

[8] I. Angelov, L. Bengtsson, M. Garcia, "Extension of the Chalmers Nonlinear HEMT and MESFET Model", IEEE MTT., Vol. 44, No.10, October 1996, pp. 1664-1674.

[9] R. Giofrè, P. Colantonio, F. Giannini, L. Piazzon "A new design strategy for multi frequencies passive matching networks" 37th EuMC 2007, Munich, Germany, October 2007, pp. 838-841.

[10] H. Oraizi and A-R. Sharifi, "Design and Optimisation of Broadband Asymmetrical Multi-Section Wilkinson Power Divider," IEEE Trans. on MTT, Vol. 54, pp. 2220- 2231, May 2006.

Proceedings of the 3rd European Microwave Integrated Circuits Conference

An Innovative Time-Domain Simulation Technique for Strongly Nonlinear Heterogeneous RF Circuits Operating in Diverse Time-Scales

Jorge F. Oliveira[#1], José C. Pedro[*2]

[#]*Department of Electrical Engineering, Technology and Management School - Polytechnic Institute of Leiria*
Morro do Lena, Alto Vieiro, Apartado 4163, 2411-901 Leiria, Portugal, Phone: +351 244 820300
[1]oliveira@estg.ipleiria.pt

[*]*Institute of Telecommunications – University of Aveiro*
3810-193 Aveiro, Portugal, Phone: +351 234 377900
[2]jcpedro@det.ua.pt

Abstract— **With the advent of wireless transceiver reconfigurability, a need has been felt to take profit of digital signal processing tools, this way increasing RF circuits' complexity and heterogeneity. Having this objective in mind, this paper presents an analytical formulation and a novel numerical method for simulating, in a very efficient way, strongly nonlinear heterogeneous RF circuits running in three different time-scales. In order to reduce the computational workload, a new multi-line double multi-rate shooting technique is proposed to operate within a multi-dimensional warped time framework. Obtained results of an illustrative circuit, reveal significant advantages in speed over previous methods recently proposed for the simulation of the same category of circuits.**

I. INTRODUCTION

The integration of digital signal processors with traditional analog RF circuits [1] has created a new range of challenges to circuit simulation, namely in terms of the resulting circuits' heterogeneity. The way that has been followed to circumvent this problem is to combine system level and circuit level simulation, whenever the sub-circuits and the sub-system blocks of the whole network do not interact. Unfortunately, this methodology can not be applicable whenever the conceptual separability of the various functional blocks is lost. A clear example of this is the all-digital PLL and transmitter SoC (system on chip) recently demonstrated for GSM/EDGE hand-sets [2]. In this example, most of all traditional analog functions were substituted by digital algorithms, as was the case of the power amplifier driver, there replaced by a high-speed DAC named a digital-to-RF-amplitude converter.

Moreover, it is curious to note that, from a simulation point of view, this situation can not even be said to be new. The combination of base-band and RF signals, and corresponding components in the same circuit, is as old as the existence of modulators and demodulators or even mixers, as is illustrated by the drain mixer reported in [3], or by the simplified power amplifier, PA, schematic of a polar transmitter shown in Fig. 1 [4].

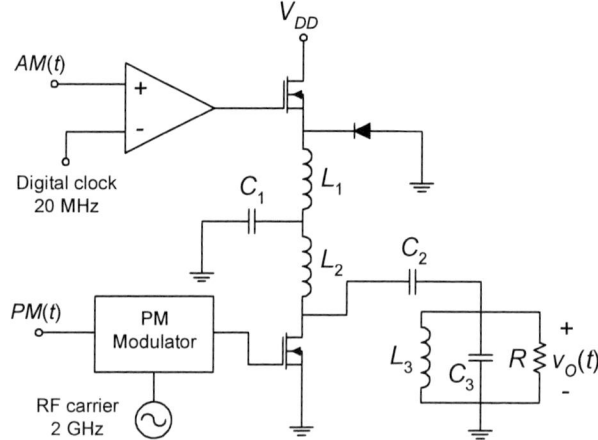

Fig. 1   Simplified power amplifier schematic used in wireless polar transmitters [4]

In this heterogeneous circuit we have a mixture of periodic (RF carrier and digital clock) and aperiodic [$AM(t)$ and $PM(t)$] forcing functions, running in three distinct time-scales. The $AM(t)$ and the $PM(t)$ signals are base-band excitations, whose bandwidth is about $200\,\mathrm{kHz}$ (the slowest time-scale). The $AM(t)$ signal is over-sampled with a digital clock of $f_0 = 20\,\mathrm{MHz}$ (an intermediate time-scale), to get a pulse-width modulation format, while the $PM(t)$ signal modulates the phase of a CW RF carrier of $f_C = 2\,\mathrm{GHz}$ (the fastest time-scale). Beyond this stimulus disparity, we have also node voltages and branch currents with widely disparate rates of variation, reason why this circuit will be described by a system of warped multi-rate partial differential algebraic equations, WaMPDAE [4], [6]. Some of its components evidence no fluctuations in the digital clock intermediate time-scale, while others have no fluctuations in the fast carrier warped time-scale. As shown in the next sections, this paper proposes a new simulation technique that takes profit of these characteristics to efficiently simulate RF circuits of this category.

978-2-8748-7007-1/08 $25.00 © 2008 EuMA   530

II. THEORETICAL BACKGROUND

In general, an RF circuit can be described by the following nonlinear system of differential algebraic equations, DAEs, in time,

$$p[y(t)] + \frac{dq[y(t)]}{dt} = x(t), \qquad (1)$$

where $x(t) \in \mathbb{R}^n$ and $y(t) \in \mathbb{R}^n$ stand for the excitation and state-variable vectors, respectively. $p[y(t)]$ represents memoryless linear or nonlinear elements, while $q[y(t)]$ models dynamic linear or nonlinear elements (capacitors or inductors). Since we are interested in a case where $x(t)$ and the state-variables $y(t)$ combine diverse types of periodic and aperiodic stimulus running in three distinct time-scales, the system of (1) will be converted in the following MPDAE, system [5]:

$$p[\hat{y}(t_1, t_2, t_3)] + \frac{\partial q[\hat{y}(t_1, t_2, t_3)]}{\partial t_1}$$

$$+ \frac{\partial q[\hat{y}(t_1, t_2, t_3)]}{\partial t_2} + \frac{\partial q[\hat{y}(t_1, t_2, t_3)]}{\partial t_3} = \hat{x}(t_1, t_2, t_3). \qquad (2)$$

Because PM signals are not compact in the straightforward multivariate representation mentioned above, we now make use of the concept of warped time [6]. By using this technique, we dynamically rescale the fast time axis to significantly reduce the number of fluctuations of the PM signals. In this work we will adopt the following procedure: for the slowly-varying parts of the expressions of $x(t)$ and $y(t)$ t is replaced by $\tau_1 = t$; for the intermediate-varying parts t is replaced by $\tau_2 = t$; finally, for the fast-varying parts, t is replaced by $\tau_3 = \phi(t)$, where $\phi(t)$ is any appropriate rescaling function (the so-called warped time [6]). In this context we will have the new warped multivariate forms $\hat{x}(\tau_1, \tau_2, \tau_3)$ and $\hat{y}(\tau_1, \tau_2, \tau_3)$, and the DAE (1) will be now converted into the WaMPDAE, system [4], [6]:

$$p[\hat{y}(\tau_1, \tau_2, \tau_3)] + \frac{\partial q[\hat{y}(\tau_1, \tau_2, \tau_3)]}{\partial \tau_1} +$$

$$\frac{\partial q[\hat{y}(\tau_1, \tau_2, \tau_3)]}{\partial \tau_2} + \omega(\tau_2) \frac{\partial q[\hat{y}(\tau_1, \tau_2, \tau_3)]}{\partial \tau_3} \qquad (3)$$

$$= \hat{x}(\tau_1, \tau_2, \tau_3),$$

where $\omega(t) = d\phi(t)/dt$.

Now, the mathematical relation between (1) and (3) establishes that, if $\hat{x}(\tau_1, \tau_2, \tau_3)$ and $\hat{y}(\tau_1, \tau_2, \tau_3)$ satisfy (3), then the corresponding univariate forms $x(t) = \hat{x}(t, t, \phi(t))$ and $y(t) = \hat{y}(t, t, \phi(t))$ satisfy (1) [6]. Consequently, the univariate solutions of (1) are available on paths of parametric equations $\tau_1 = t$, $\tau_2 = t$, $\tau_3 = \phi(t)$, along the multivariate solutions $\hat{y}(\tau_1, \tau_2, \tau_3)$ in the τ_1, τ_2, τ_3 space. Since in our case study we will have periodicity in τ_2 and τ_3 dimensions,

and so we will have an enclosed domain in those time axes, the original univariate solution will be recovered from its multivariate form by setting

$$y(t) = \hat{y}(t, t \bmod T_2, \phi(t) \bmod T_3), \qquad (4)$$

where T_2 and T_3 are the period of the excitation and the solution in τ_2 and τ_3 dimensions, respectively.

III. INNOVATIVE SIMULATION METHOD

Contrary to what we previously did in [4], in which a bi-dimensional problem had to be assumed, now we do not need to treat the digital clock and the base-band signals in the same time scale, regardless of their periodic or aperiodic nature, and disparate time evolution rates. Indeed, now we will not only select a more appropriate time-step size for the τ_1 time, as we will solve (3) taking profit of the periodicity of the problem in τ_2 and τ_3 dimensions (the digital clock intermediate time-scale and the fast carrier warped time-scale, respectively). By using a 3-D envelope transient oriented technique, we replace each derivative of (3) in τ_1 dimension (the aperiodic base-band slow time-scale of our example) with a finite-differences approximation, to obtain, for each level $\tau_{1,i}$, the partial differential algebraic system

$$p[\hat{y}(\tau_{1,i}, \tau_2, \tau_3)] +$$

$$\frac{q[\hat{y}(\tau_{1,i}, \tau_2, \tau_3)] - q[\hat{y}(\tau_{1,i-1}, \tau_2, \tau_3)]}{h_{1,i}} +$$

$$\frac{\partial q[\hat{y}(\tau_{1,i}, \tau_2, \tau_3)]}{\partial \tau_2} + \omega(\tau_2) \frac{\partial q[\hat{y}(\tau_{1,i}, \tau_2, \tau_3)]}{\partial \tau_3} \qquad (5)$$

$$= \hat{x}(\tau_{1,i}, \tau_2, \tau_3),$$

where $h_{1,i} = \tau_{1,i} - \tau_{1,i-1}$, with the bi-periodic boundary conditions

$$\hat{y}(\tau_{1,i}, 0, \tau_3) = \hat{y}(\tau_{1,i}, T_2, \tau_3),$$

$$\hat{y}(\tau_{1,i}, \tau_2, 0) = \hat{y}(\tau_{1,i}, \tau_2, T_3), \qquad (6)$$

on the rectangle $\tau_{1,i} \times [0, T_2] \times [0, T_3]$. In other words, we start with some initial condition for $\tau_1 = 0$ and then solve a set of 2-D boundary value problems with periodic boundary conditions in both τ_2 and τ_3 dimensions. In the following we will present an innovative technique to solve these 2-D bi-periodic boundary value problems in a very efficient way.

A. New Multi-Line Shooting Technique

Let us consider the system of (5) with n state-variables. Let us also consider the semi-discretization of the rectangular domain $[0, T_2] \times [0, T_3]$ in τ_2 dimension defined by

$$0 = \tau_{2,0} < \tau_{2,1} < \cdots < \tau_{2,j} < \cdots < \tau_{2,K_2} = T_2,$$

$$h_{2,j} = \tau_{2,j} - \tau_{2,j-1}. \qquad (7)$$

By using a finite-differences scheme to approximate the derivatives of (5) in τ_2 dimension, we can obtain a set of $n \times K_2$ ordinary differential algebraic equations in τ_3

$$p\left[\hat{y}_{i,j}\left(\tau_3\right)\right] + \frac{q\left[\hat{y}_{i,j}\left(\tau_3\right)\right] - q\left[\hat{y}_{i-1,j}\left(\tau_3\right)\right]}{h_{1,i}} +$$

$$\frac{q\left[\hat{y}_{i,j}\left(\tau_3\right)\right] - q\left[\hat{y}_{i,j-1}\left(\tau_3\right)\right]}{h_{2,j}} + \tag{8}$$

$$\omega\left(\tau_{2,j}\right) \frac{dq\left[\hat{y}_{i,j}\left(\tau_3\right)\right]}{d\tau_3} = \hat{x}_{i,j}\left(\tau_3\right),$$

$$j = 1, 2, \ldots, K_2, \quad \hat{y}_{i,0}\left(\tau_3\right) = \hat{y}_{i,K_2}\left(\tau_3\right),$$

where $\hat{x}_{i,j}\left(\tau_3\right) \equiv \hat{x}\left(\tau_{1,i}, \tau_{2,j}, \tau_3\right)$, $\hat{y}_{i,j}\left(\tau_3\right) \equiv \hat{y}\left(\tau_{1,i}, \tau_{2,j}, \tau_3\right)$. This whole system can be time-step integrated with any appropriate initial value solver (e.g. Runge-Kutta). In addition, we must note that, by taking into account $\hat{y}_{i,0} = \hat{y}_{i,K_2}$, the first periodic boundary condition of (6) will be automatically satisfied throughout the integration process.

This way, in opposition to the "single-line" shooting technique presented in [3], [4], (and we must note that due to the bi-periodicity of the problem here we would have to do 2-D shooting), in this work we will integrate simultaneously all the equations in (8). After that, we have to look only at the solution $\hat{y}_{i,j}\left(T_3\right)$ on the top of the rectangle $\left[0, T_2\right] \times \left[0, T_3\right]$, to then wisely update the initial solution on the bottom, $\hat{y}_{i,j}\left(0\right)$, until the second condition of (6) is satisfied. This is a "multi-line" shooting strategy, in which the final bi-periodic solution is quickly achieved.

B. Double Multi-Rate Strategy

The polar transmitter PA presented in the Introduction is a highly heterogeneous nonlinear RF multi-rate circuit, whose state-variables have different rates of variation. For this reason, when this circuit is described by the WaMPDAE system of (3) some of its components practically evidence no fluctuations in τ_2 dimension (the digital clock time-scale), while others have no fluctuations in τ_3 dimension (the carrier time-scale). If we take into account these features when solving each one of the (5)-(6) 2-D bi-periodic boundary value problems we can benefit from them.

First, since we have to time-step integrate (8) in τ_3 dimension, instead of using standard Runge-Kutta, RK, solvers we can use modern multi-rate Runge-Kutta schemes, MRK, [7], [8]. With these methods we can apply a large time-step size (macrostep) to the components that have no fluctuations in τ_3 dimension and a small step size (microstep) to the remaining components. As seen from [3], [4], this multi-rate technique significantly reduces the simulation time.

Next, because we have components that evidence no fluctuations in τ_2 dimension, instead of considering a fine grid in (7), we can use a coarse grid with a large grid spacing. This will considerably reduce the number of equations in (8)

and will prevent a lot of unnecessary computation work. So, it will further reduce the required simulation time.

IV. ILLUSTRATIVE APPLICATION EXAMPLE

In order to show the capabilities of the method proposed in Section III, we will now apply it to the illustrative polar transmitter PA example presented in Fig. 1 of the Introduction.

This circuit was simulated in MATLAB® with the novel numerical method (3-D WaMPDAE envelope transient over MRK multi-line shooting), and compared with the corresponding 3-D version of the method proposed in [4] (3-D WaMPDAE envelope transient over MRK single-line shooting). Numerical computation times (in seconds) for simulations in the [0, 125 ns] and [0, 1.25 µs] intervals are presented in Table I. No comparison was made with any classical time-marching engine, as SPICE, because the 3-D nature of the circuit operation would determine an unbearably large simulation time. No comparison was also made with any frequency-domain, e.g. harmonic balance [9], or hybrid solvers, because the highly nonlinear regimes of the circuit would lead to an intolerably large number of harmonics.

TABLE I
COMPUTATION TIMES (AMD 1.8 GHz, 700 Mb RAM)

Simulation time intervals	3-D WaMPDAE envelope transient over MRK shooting		Speedup gains (approx.)
	Multi-line shooting (new method)	Single-line shooting	
[0, 125 ns]	0.8 s	6.3 s	8
[0, 1.25 µs]	6.7 s	52.3 s	8

Figs. 2 and 3 depict the solutions for the AM branch transistor source voltage and the L_1 inductor current, on $\left[0, T_2\right] \times \left[0, T_3\right]$, with $\tau_1 = 125\,\text{ns}$. T_2 is the 50 ns digital clock period and we chose a warped time function $\phi\left(t\right)$ so that $T_3 = 1$. As we can observe, there are no undulations in the

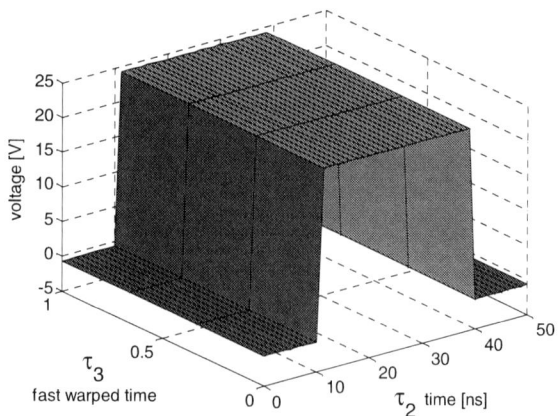

Fig. 2 Source voltage of the AM envelope modulator transistor

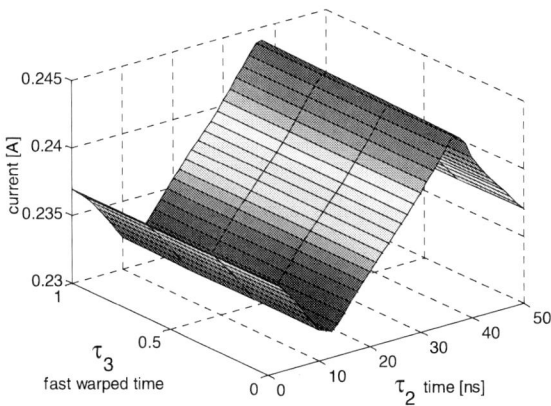

Fig. 3 Envelope reconstruction filter L_1 inductor current

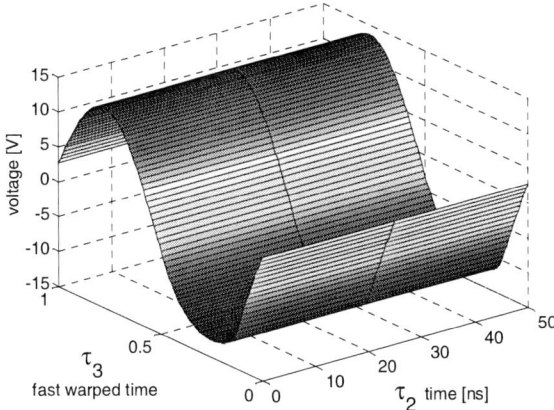

Fig. 4 Output voltage of the polar transmitter PA circuit

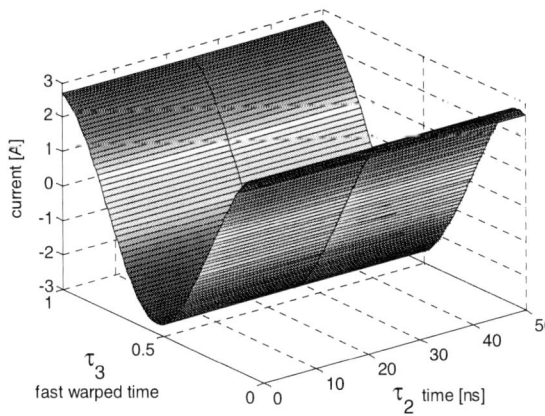

Fig. 5 Output band-pass filter L_3 inductor current

τ_3 dimension, which allowed us to use in both cases a large time-step integration size in that dimension.

The solutions for the output voltage of the circuit and the L_3 inductor current, on the same rectangular domain, are shown in Figs. 4 and 5, respectively. In these cases no undulations can be seen in the τ_2 dimension, which allowed us to use a coarse grid with a large grid spacing in τ_2, i.e., only a few shooting lines.

Finally, it must be added that we have conducted several other simulations of the circuit, under different conditions and different time intervals, and the results obtained were always analogous to the ones shown in Table I.

V. CONCLUSIONS

In this paper an innovative and very efficient time-domain simulation method was proposed. This method is particularly appropriate for highly nonlinear heterogeneous RF circuits running in three distinct time-scales, and introduces a new multi-line double multi-rate shooting technique within a 3-D warped time framework. Significant computational work reductions were obtained, without compromising accuracy, and speedup gains of about 8 times were achieved for the tested circuit example.

ACKNOWLEDGMENT

The authors wish to acknowledge the financial support provided by the Portuguese Foundation for Science and Technology, FCT, through the PhD grant provided to the first author, and by the Network of Excellence TARGET – "Top Amplifier Research Groups in a European Team" supported by the information Society Technologies Program of the EU under contract IST-1-507893-NOE, www.target-org.net.

REFERENCES

[1] P. Asbeck, L. Larson, and I. Galton. "Synergistic Design of DSP and Power Amplifiers for Wireless Communications," *IEEE Trans. on Microwave Theory and Tech.*, vol. MTT-49, no. 11, pp. 2163-2169, Nov. 2001.

[2] R. Bogdan et. al., "All-digital PLL and transmitter for mobile phones", *IEEE Jour. of Solid-State Circuits*, vol. JSSC-40, no. 12, pp. 2469-2482, Dec. 2005.

[3] J. F. Oliveira and J. C. Pedro, "A new time-domain simulation method for highly heterogeneous RF circuits", in *Proc. 37'th European Microwave Conf.*, München, Oct. 2007, p. 1161-1164.

[4] J. F. Oliveira and J. C. Pedro, "An efficient time-domain simulation method for multi-rate RF nonlinear circuits", *IEEE Trans. Microw. Theory Tech.*, vol. 55, no. 11, pp. 2384-2392, Nov. 2007.

[5] J. Roychowdhury, "Analyzing circuits with widely separated time scales using numerical PDE methods," *IEEE Trans. on Circuits and Systems*, vol. 5, no. 48, pp. 578-594, May 2001.

[6] O. Narayan and J. Roychowdhury, "Analyzing oscillators using multitime PDEs," *IEEE Trans. on Circuits and Systems*, vol. 50, no. 7, pp. 894-903, Jul. 2003.

[7] A. Kværnø and P. Rentrop, "Low order multirate Runge-Kutta methods in electric circuit simulation," *IWRMM Universität Karlsruhe*, Preprint No.99/1, 1999.

[8] M. Günther, A. Kværnø and P. Rentrop, "Multirate partitioned Runge-Kutta methods," *BIT*, vol. 41, no. 3, pp. 504-514, Jun. 2001.

[9] K. Kundert, J. White and A. Sangiovanni-Vincentelli, *Steady-State Methods for Simulating Analog and Microwave Circuits*. Norwell: Kluwer Academic Publishers, 1990.

Vertical RF Transition with Mechanical Fit for Three-Dimensional Heterogeneous Integration

Lihan Chen[*1], Joe Wood[†2], Sanjay Raman[†3], N. Scott Barker[*4]

[*]*Charles L. Brown Dept. of ECE, University of Virginia, Charlottesville, VA 22904, USA*
[1]lc4kh@virginia.edu, [4]barker@virginia.edu

[†]*Bradley Department of ECE, Virginia Tech, Blacksburg, VA 24061, USA*
[2]jowood4@vt.edu, [3]sraman@vt.edu

Abstract—This paper presents the design, simulation and measurement of a vertical interconnect with mechanical fit for three-dimensional heterogeneous integration. The mechanical fit is a strategy employing interlocking SU-8 structures to transition between flip-chip style stacked chips through vertical CPW transmission lines. The mechanical fit is introduced in this paper to reduce flip-chip alignment difficulty and increase the reliability of the interconnects. This paper also describes a process for using pre-fabricated active ICs in a mechanical fit vertical configuration. Experimental results show excellent RF performance up to 50 GHz, with extremely low insertion loss (better than 0.25 dB at 40 GHz per transition). The transitions have been fabricated and tested for 380 μm-thick silicon substrates with passive components and experiments are being conducted on active components.

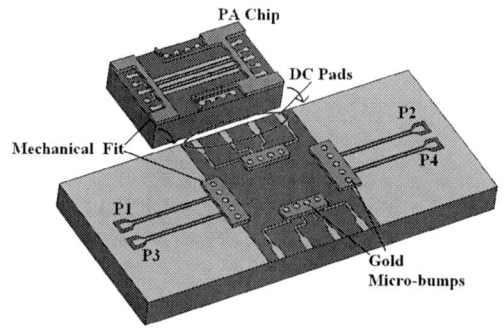

Fig. 1. The structure of proposed vertical transition with mechanical fit.

I. INTRODUCTION

High-density packaging requires vertical transitions that can transfer signals between stacked wafers. Low-loss vertical transitions can significantly reduce the volume and cost of multiwafer circuits. Most vertical transitions developed previously fall into two categories. One is direct contact and the other is indirect contact. The indirect contact transitions are realized by coupling of electromagnetic energy. The most common structures are slot line coupling, aperture coupling and cavity coupling [1]-[3]. Although the mechanisms of coupling are different, the sizes of indirect contact interconnects are related to the wavelength in which they work and are therefore band-limited. Direct contact transitions, on the other hand, require much less area to realize. Furthermore, direct contact transitions can operate down to DC and are therefore preferred over indirect transitions. Unfortunately, direct contact vertical transitions are not easy to realize, especially at the chip level. Herrick et al. describe a vertical interconnect using via-holes for multiwafer microstrip-based circuits [4]. This structure is demonstrated from DC to 20 GHz, but is not suitable for mixed substrates, which will suffer from Coefficient of Thermal Expansion (CTE) mismatch. CTE mismatch causes adjacent stacked chips to expand or contract at different rates, creating shear stress on the interface that could ultimately lead to interconnect failure.

An increasingly popular direct contact transition configuration is flip-chip interconnect, which has been demonstrated up to *W*-band [5]. Due to its small dimensions, the flip-chip

interconnect is attractive for high-performance, high-frequency and broadband applications [6]. This paper describes the concept of including a mechanical fit as part of the flip-chip vertical interconnect, which would enable the heterogeneous integration of different microsystem layers fabricated in different technologies. This process is best suited for direct contact CPW traces through vertical micro-bumps as illustrated in Fig. 1. The benefits of mechanical fit include:

- *Reducing the difficulty in flip-chip alignment.*
  The assembly can be accomplished by hand with the resulting interconnects demonstrated to work up to 50 GHz.
- *Improving reliability by supporting the micro-bumps with SU-8.*
  The SU-8 structures add strength to the tall gold micro-bumps against the shear stress caused by CTE mismatch between the stacked chips.
- *Separating adjacent chips for improved RF and thermal isolation.*
  The shape of the mechanical fit structures is specifically designed to avoid sensitive areas that could detune active RF chips while serving as a thermal underfill in less sensitive areas. The additional space in between chips ($\sim$55 μm) due to the height of the mechanical fit structures helps to reduce cross-coupling and also enable the integration/self-packaging of high aspect ratio 3D and mems structures between assembled layers.

To demonstrate the viability of this technique, this paper

978-2-8748-7007-1/08 $25.00 © 2008 EuMA   534

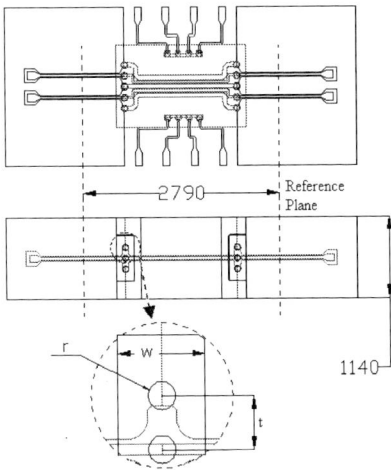

Fig. 2. Dimension of the designed vertical transition(in microns) with r=35 μm, t=150 μm, w=245 μm, CPW dimensions are 15/30/15 μm

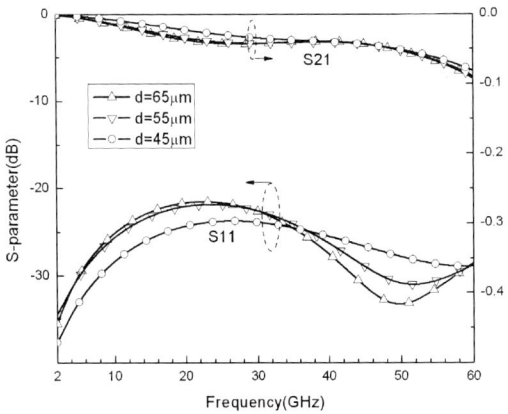

Fig. 3. HFSS simulation of a single transition for various transition heights with other dimensions given in Fig. 2.

also describes the integration of the mechanical fit with a previously designed power amplifier chip. This integration is conducted by post-processing the mechanical fitting structure onto a BiCMOS power amplifier chip. The chip with the fitting structure is then mounted onto the micro-bumps through the interlocking sections. With the chip in place, the top and bottom substrates are bonded through thermocompression to electrically connect the power amplifier chip pads with the micro-bumps.

II. DESIGN

The goal of the vertical transition with mechanical fit is to allow RF circuits on separate substrates to be vertically integrated with ease. To validate the design, the top chip was chosen to be a previously-designed CPW-based 30 GHz BiCMOS power amplifier chip [7]. The power amplifier chip is processed in the IBM 8HP technology. A passive chip that emulates the pad layout and impedance of the PA chips was also created to develop the micro-bump and mechanical fit process and validate the transition design. The pattern of the mechanical fit structure (shown in Fig. 1) is chosen to avoid interference with the matching network on the PA chips.

The characteristic impedance of the CPW is 50 Ω with 30 μm wide signal lines and 15 μm gaps. The transition is designed to be compatible with the CPW testing pads located on the PA chip. The space between the signal line and ground line (t in Fig. 2) is 150 μm to match the PA chip layout. To obtain a better characterization of the interconnects, a simplified testing structure is also designed and tested (as shown in Fig. 4). The dimensions of the transition in both designs are the same. The other dimensions are shown on Fig. 2.

The challenge in designing RF vertical transitions is to propagate the wave from one wafer to another with minimal degradation in overall circuit performance. This requires impedance matching and minimization of radiation. The 30 μm wide signal lines and 15 μm gaps are increased gradually into the size of the transition to minimize radiation. To simplify the mask design and fabrication, the radius of all of the microbumps, signal and ground, are the same. The diameter of the micro-bump is set to 70 μm to minimize the impedance mismatch. Another critical dimension is the distance between the two chips. The transition is simulated using Ansoft HFSS V.10.1. The simulation results (Fig. 3) show that the vertical interconnect works from DC to 50 GHz and that the return loss is reduced when the distance is decreased. However, placing the chips too close would result in unwanted coupling between them. Therefore, the distance is chosen to be 55 μm. The back-to-back configuration of the interconnections (simplified design as shown in Fig. 4) is simulated using HFSS and the results are plotted together with measured results in Fig. 7.

III. FABRICATION AND ASSEMBLY

There are two distinct process flows for the fabrication of the mechanical fit process: the bottom substrate with the micro-bumps and the top substrate with mating structure on the PA chip. For the test structures presented in this work, the substrates are 380 μm thick high-resistivity silicon ($\rho >$ 2000 $\Omega \cdot$cm). The bottom substrate fabrication mainly consists of the following three steps (as detailed in Fig. 4):

1) Deposit and pattern metal layers (Ti-Au-Cr) on silicon substrate by lift-off.
2) Spin and define the SU-8 layer.
3) Form gold microbumps by electroplating into SU-8 mold.

A PECVD Si_xN_y layer is deposited upon the metal layers and serves as a plating mask to prevent Au plating outside the SU-8 mold and is etched using a CF_4/O_2 RIE after plating. The SU-8 layer is deposited by a multi-layer SU-8 process [8]. The first spin of SU-8 is 1000 RPM for 60 sec and the second is 2000 RPM for 60 sec. The combination of the two spins creates a smooth 55 μm thick layer. The micro-bumps are

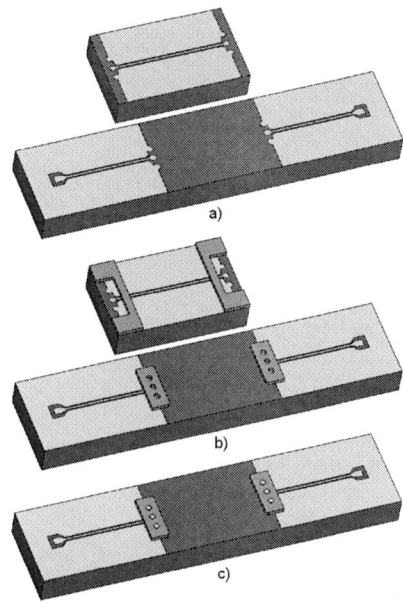

Fig. 4.   Process flow for the vertical transition.

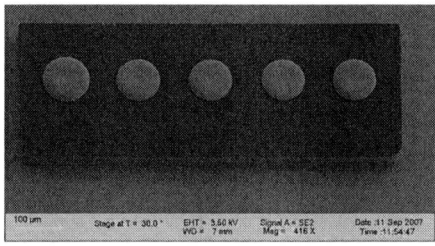

Fig. 5.   SEM picture of a test sample with 57 μm tall gold micro-bump in a 55 μm SU-8 mold.

plated to just over the thickness of the SU-8 mold (57 μm) to aid with bonding. The plating procedure is carefully controlled to ensure each micro-bump is the same height and the top shape of the micro-bumps is suitable for bonding. Fig. 5 shows a test sample with the micro-bumps formed. The centers of the micro-bumps are slightly higher than the edges, making assembly easier. The top and bottom chips can be assembled together by hand since the mechanical fit provides the required placement accuracy. A gold-to-gold thermosonic bonding [9] is used to electrically connect the two chips together.

The top substrate fabrication mainly consists of the following three steps:

1) Etch a cavity into a carrier wafer in order to post-process the diced PA chips.
2) Spin and define SU-8 layer on PA chip held by carrier wafer.
3) Remove PA chip with patterned SU-8 from carrier wafer.

A carrier wafer is necessary to planarize the surface of the PA chips in order to create the SU-8 pattern for mechanical fit.

However, this step could be eliminated by adding the SU-8 structures as the final processing step in the foundry process flow before dicing the wafer. The carrier wafer is fabricated by etching a cavity in the silicon surface using TMAH etching techniques. The slanted sidewalls of a TMAH-etched cavity aid in the removal of the chip after the SU-8 processing is completed. Matching the cavity depth is critical in establishing the SU-8 height on the chip. With the PA chip inside the carrier wafer, SU-8 is spun across the surface in the same manner as previously described. The final SU-8 spin speed is increased to 2500 RPM to ensure that the height of the top structure is less than the height of the bottom structure. To release the PA chip after processing, the carrier wafer is shaken ultrasonically during the final minute of the SU-8 development. The processed PA chip is show in the Fig. 6

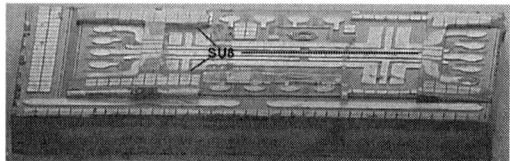

Fig. 6.   PA chip with SU-8 mechanical fit structure.

IV. MEASUREMENT RESULTS

The fabricated structures have been measured using an HP 8510 network analyzer and two GSG probes (Picoprobe Model 50A, DC-50GHz) with a on wafer thru-reflect-line (TRL) calibration for 2-50 GHz [10]. The CPW line attenuation is determined from the calibration lines to be 6.25 dB/cm at 40 GHz. Fig. 7 shows the measured and simulated results of the design shown in Fig. 4. As seen from this data, the measured and simulated results both show return loss below -15 dB. To get a more accurate characterization of the transition, we de-embed the loss of the CPW line sections included in the measurement. We calculate the loss by the following formula:

$$Loss(dB) = -10log(\frac{|S_{21}|^2}{1 - |S_{11}|^2}) - \alpha \cdot L$$

where α (dB/cm) is the attenuation of the CPW line and L (0.2504 cm) is the length of the CPW line de-embedded from the measurement. From Fig. 8, the loss per transition varies form 0.1 dB at 2 GHz to 0.23 dB at 40 GHz. At frequencies higher than 42 GHz, the calibration is less accurate causing the loss to drop.

The design shown in Fig. 1 was also tested and the measurement of both sets of ports is shown in Fig. 9. The testing results are similar to that of the simplified design. To measure the isolation, we test the insertion loss between P1 and P4 (as shown in the Fig. 1), leaving the other ports open. Fig. 10 shows the isolation measurement results. The isolation between the two channels is better than 25 dB from 2-50 GHz.

978-2-8748-7007-1/08 $25.00 © 2008 EuMA   536

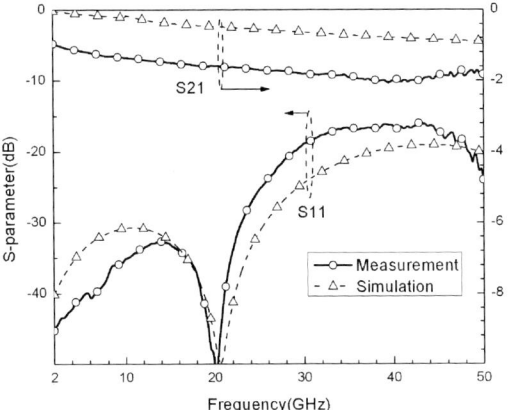

Fig. 7. The simulation and measurement results with reference plane shown in Fig. 2.

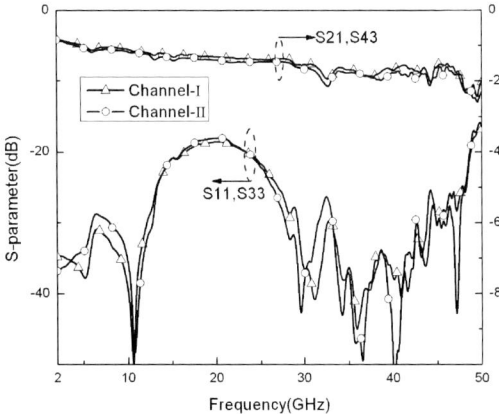

Fig. 9. Measurement results of the double channel back-to-back transitions shown in Fig. 1.

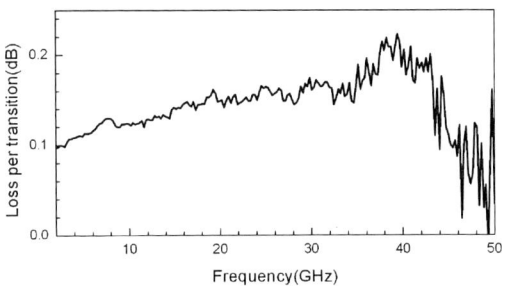

Fig. 8. The loss per transition after deembedding the CPW line loss from the measured results.

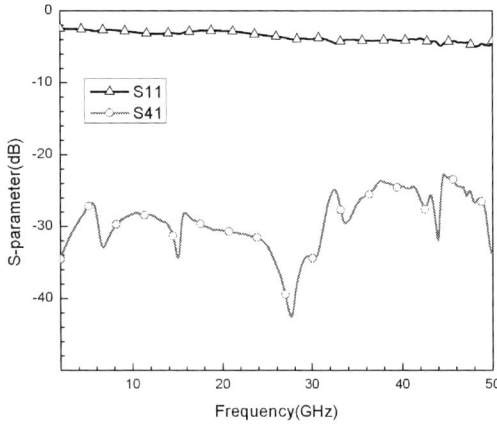

Fig. 10. The isolation between channels P1 and P4 as shown in Fig. 1. Ports P2 and P3 are left open during the measurement.

V. CONCLUSIONS

A structure for vertical interconnects with integrated mechanical fit is demonstrated in this paper. The measurement results shows excellent performance up to 50 GHz with insertion loss less than 0.25 dB per interconnect at 40 GHz. This paper also outlines techniques for assembling planar chips to be stacked in a vertical configuration.

ACKNOWLEDGEMENT

The work is funded by the Army Research Office under contract number W911NF-06-1-0368 and W944NF-06-1-0422. The authors would like to acknowledge the support of Dr. Dev Palmer, U.S. Army Research Office. The authors would like to acknowledge the help of C. Smith with SU-8 processing and also want to thank K. Vummidi and T. Haque.

REFERENCES

[1] Ellis, T.J.; Raskin, J.P.; Katehi, L.P.B.; Rebeiz, G.M., "A wideband CPW-to-microstrip transition for millimeter-wave packaging," *1999 IEEE MTT-S International Microwave Symposium Digest*, vol.2, pp.629-632, 1999.

[2] Zhu, L.; Wu, K., "Ultrabroad-band vertical transition for multilayer integrated circuits," *Microwave and Guided Wave Letters*, IEEE, vol.9, pp.453-455, Nov 1999

[3] Li, E.S.; Chih Che Lai, "Designs for broadband cavity-coupled microstrip vertical transitions," *2005 IEEE Antennas and Propagation Society International Symposium*, vol.3B, pp. 385-388, July 2005

[4] Herrick, K J; Lee, Y; Margomenos, A; Mohammadi, S; Katehi, L P B, "Multiwafer Vertical Interconnects for Three-Dimensional Integrated Circuits," *IEEE Transactions on Microwave Theory and Techniques*, Vol. 54, no. 6, pp. 2699-2706, June 2006.

[5] Jentzsch, A.; Heinrich, W., "Theory and measurements of flip-chip interconnects for frequencies up to 100 GHz," *Microwave Theory and Techniques, IEEE Transactions on*, vol.49, no.5, pp.871-878, May 2001

[6] Heinrich, W., "The flip-chip approach for millimeter wave packaging," *Microwave Magazine, IEEE*, vol.6, no.3, pp. 36-45, Sept. 2005

[7] Haque, T.; Studtmann, G.; and Raman, S., "A High Linearity 30-GHz Range SiGe Differential Power Amplifier IC," *2007 IEEE Topical Meeting on Silicon Monolithic Circuits in RF Systems Digest, Long Beach, CA*, pp. 115–118. Jan, 18-20, 2007.

[8] Smith, C.H., III; Haiyong Xu; Barker, N.S., "Development of a multi-layer SU-8 process for terahertz frequency waveguide blocks," *2005 IEEE MTT-S International Microwave Symposium Digest*, vol.3, pp. 439-442, June 2005

[9] Cheah, L. K.; Tan, Y. M.; Wei J. and Wong, C. K., "Gold to Gold Thermosonic Flip-chip Bonding," *Proceedings HDI 2001*, pp.165-175, April 2001.

[10] Marks, R.B., "A multiline method of network analyzer calibration," *IEEE Transactions on Microwave Theory and Techniques*, vol.39, no.7, pp. 1205-1215, Jul 1991

Proceedings of the 3rd European Microwave Integrated Circuits Conference

The Effect of Dielectric Height and Ground Plane Width on Multilayer MCM FGCPW Lumped Elements

K. K. Samanta & I. D. Robertson

Institute of Microwave and Photonics, School of Electronics and Electrical Engineering,
University of Leeds, Leeds, LS2 9JT, UK

Abstract— **This paper investigates the effect of dielectric height and ground plane width on the performance of multilayer CPW transmission lines, lumped elements and circuits fabricated using photoimageable thick film technology. The effective dielectric constant and characteristic impedance are extracted to investigate the effect on transmission line dispersion. The equivalent circuit parameters for lumped elements are investigated for various dielectric thicknesses and ground plane widths. Finally, the practical effect of dielectric height is investigated for a 3GHz lumped element low pass filter.**

I. INTRODUCTION

The need for reduced size, weight and cost, have been driving the trend towards multilayer multichip modules (MCM) for microwave applications. Coplanar Waveguide (CPW) is often preferred at high frequency as it has several advantages over microstrip such as not requiring through-substrate vias or backside processing. Multilayer technology enables the designer to use a number of different transmission line media with a wide range of achievable characteristic impedance. So, an optimum 3D system-in-package (SIP) architecture will incorporate thin film microstrip (TFMS), coplanar waveguide (CPW), and substrate integrated waveguide (SIW) side-by-side in a multilayer substrate. The use of finite-ground coplanar waveguide (FGCPW) [1-3] is essential to make the MCM compact and suitable for high frequency applications. Compact and high quality lumped elements are one of the most critical requirements for a MCM. Fig. 1 shows the cross sectional view of the multilayer MCM process used in this work, with several types of transmission lines, including SIW, CPW and TFMS, and integrated passive components formed on the multilayer substrate to optimize the performance of the SIP module. The test structures were fabricated using Hibridas photoimageable technology [4] on an Alumina base substrate (CoorsTek ADS-96R) using a conventional off-contact screen printer. This technology has been shown to be useable to 110 GHz [5] and suitable for mm-wave MCMs [6-7]. The compactness and the performance of a multilayer CPW based MCM is highly dependant on the dielectric height and ground plane width of the CPW lines and lumped elements.

To the best of the authors' knowledge, in this paper for the first time the effect of ground width and dielectric height is studied simultaneously on multilayer finite ground CPW line dispersion and finite ground CPW lumped elements on a ceramic substrate. Finally, the effect of these parameters are demonstrated experimentally for a CPW lumped element low pass filter.

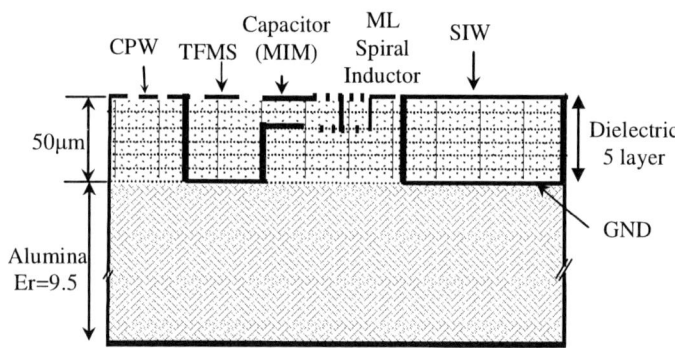

Fig. 1 Cross-sectional view of the multilayer with integrated passives using multilayer photoimageable thick film technology.

II. THE EFFECT OF DIELECTRIC HEIGHT ON MULTILAYER FGCPW LINES

Here, multilayer CPW is fabricated on a dielectric layer coated onto Alumina base substrate using a conventional off-contact screen printer. Layers of photoimgeable Hibridas dielectric paste (HD1000) and low loss silver conducting paste (HC4700) are printed alternately using screen printing technique to form the required height of dielectric. Profilometer readings showed that printing of each dielectric layers gives a height of about 10 μm where as that for conducting layer is 5 μm. The two, three, five and six layers of dielectric are coated to form a total Hibridas dielectric height (h_l) of 20μm, 30μm, 50μm and 60μm respectively above the Alumina substrate.

With the help of high frequency structure simulator (HFSS) and Momentum^{TM}, the effect of dielectric height and ground plane width on the performance of FGCPW lines for ground plane width 280 μm and 750 μm and dielectric height of 60 μm and 20 μm were studied. To obtain the required parameters, the *ABCD* parameters were used, and the transmission line parameters extracted using (1) [8].

978-2-8748-7007-1/08 $25.00 © 2008 EuMA   538

$$\begin{pmatrix} A & B \\ C & D \end{pmatrix} = \begin{pmatrix} \cosh \gamma l & jZ_{ch} \sinh \gamma l \\ jY_{ch} \sinh \gamma l & \cosh \gamma l \end{pmatrix} \quad (1)$$

The characteristic impedance of the CPW line Z_{ch} can be expressed as:-

$$Z_{ch} = \sqrt{\frac{B}{C}} \quad (2)$$

$$= Z_o \sqrt{\frac{(1+S_{11})(1+S_{22}) - S_{12}S_{21}}{(1-S_{11})(1-S_{22}) - S_{12}S_{21}}} \quad (3)$$

Where, Z_0 is the normalisation resistance of S-parameters. Then, the effective dielectric constant is extracted from:-

$$\varepsilon_{eff} = \left(\frac{c_0}{2\pi f} \, imag(\gamma) \right)^2 \quad (4)$$

Where,

$$\gamma = \frac{\sinh^{-1} \sqrt{B.C}}{l \times e^{-3}} \quad (5)$$

The length of transmission line, l, and,

$$B = Z_o \frac{(1+S_{11})(1+S_{22}) - S_{12}S_{21}}{2S_{21}} \quad (6)$$

$$C = Y_o \frac{(1-S_{11})(1-S_{22}) - S_{12}S_{21}}{2S_{21}} \quad (7)$$

The extracted characteristic impedance (Z_{ch}), using (3), for variation of ground width (g) and dielectric height as parameters is shown in Fig. 2. It shows that the characteristic impedance decreases with decrease of dielectric height for the same ground plane width. For a change of dielectric thickness the variation in characteristic impedance is more significant than that for a change in ground plane width from 280μm to 750 μm.

The extracted effective dielectric constant (ε_{eff}), using (4), with variation of width (g) from 280 μm to 750 μm and dielectric height from 20 μm to 60 μm is depicted in Fig. 3. For a fixed ground plane width, the effective dielectric constant decreases with increasing dielectric height and is

much more pronounced than the variation due to ground plane width.

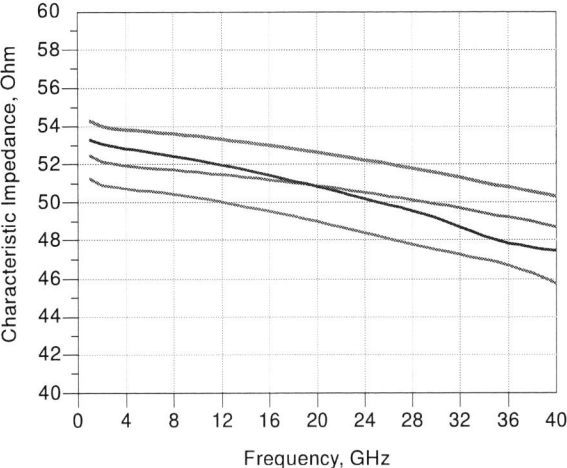

g = 280 μm, h=60 μm   g = 280 μm, h=20 μm
g = 750 μm, h=60 μm   g = 750 μm, h=20 μm

Fig. 2 The variation of characteristic impedance with frequency for ground plane widths of 280 μm and 750 μm, and with dielectric height of 20 μm and 60 μm

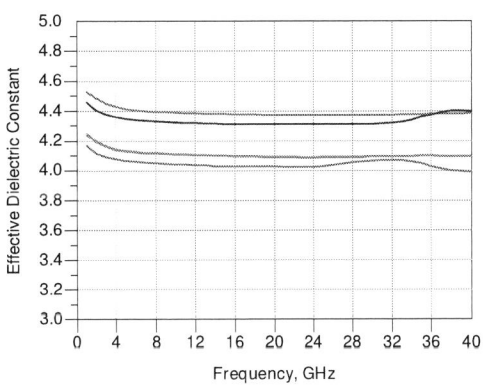

g = 280 μm, h=20 μm   g = 280 μm, h=60 μm
g = 750 μm, h=20 μm   g = 750 μm, h=60 μm

Fig. 3 Variation of effective dielectric constant with frequency for ground plane widths of 280 μm and 750 μm, and for dielectric height of 20 μm and 60 μm

III. THE EFFECT OF DIELECTRIC HEIGHT AND GROUND WIDTH ON MULTILAYER CPW LUMPED ELEMENTS

The performance of a circular multi-turn spiral inductor and a MIM capacitor designed on FG coplanar waveguide with different ground plane widths (g) and dielectric heights was

978-2-8748-7007-1/08 $25.00 © 2008 EuMA

studied. For the inductor, the coil centre was connected to the input using the bottom metallization layer and two dielectric vias, while the other end is on the top layer and connected to the output. To keep the size minimum, the track width (*W*) was set to 25 μm keeping the spacing (*S*) at 100 μm. The design details of the inductor and MIM capacitor are described in [9].

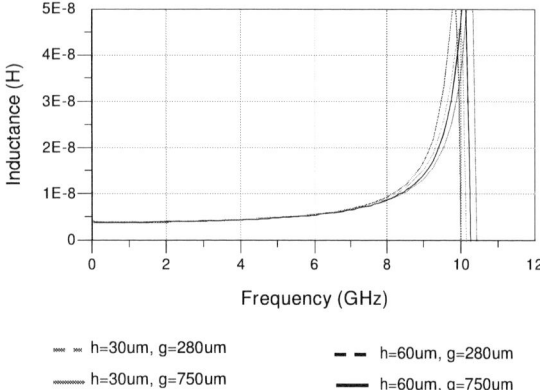

Fig. 4 Inductance of a spiral inductor of 2.5 turns with ground plane widths of 280 μm, and 750 μm for a dielectric height of 30 μm and 60 μm

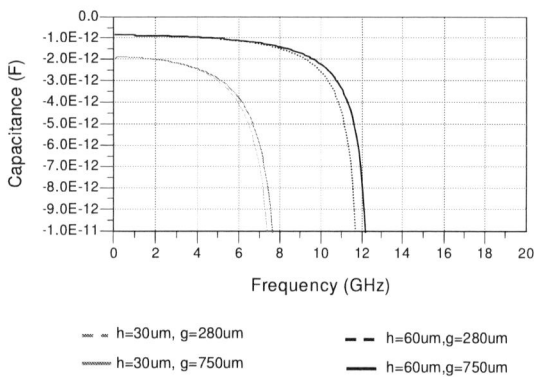

Fig. 5 Capacitance of a MIM capacitor with ground plane widths of 280 μm, and 750 μm for a dielectric height of 30 μm and 60 μm

To predict the behavior, as several high frequency phenomena like surface roughness, finite conductor thickness, etc., start to play a significant role, the equivalent circuit parameters were extracted for the equivalent circuit model described in [9]. Fig. 4 shows the variation of equivalent inductance for two different ground plane widths (*g*), of 280 μm and 750 μm and for dielectric height of 30 μm and 60 μm. The equivalent capacitance variation is shown in Fig. 5. In case of an inductor, as the height reduces the parallel ground parasitic capacitance increases, which reduces the SRF. In a MIM capacitor, variation of parasitic is greater for a

variation of ground plane width from 280 μm to 750 μm at a dielectric height of 30 μm than at 60μm.

IV. THE EFFECT OF DIELECTRIC HEIGHT AND GROUND WIDTH ON CPW LUMPED ELEMENT LOW PASS FILTER

A miniaturized, multilayer FGCPW low pass filter is used to demonstrate the effect of the two aforementioned parameters on the behavior of a lumped-element circuit in a multilayer environment. Using the model developed for capacitors and inductors, the 5-element low-pass filters were fabricated on dielectric of heights of 50 μm and 30 μm, each with two different ground plane widths. The layouts of the two CPW-based low-pass filters are shown in Fig. 6 (a) and

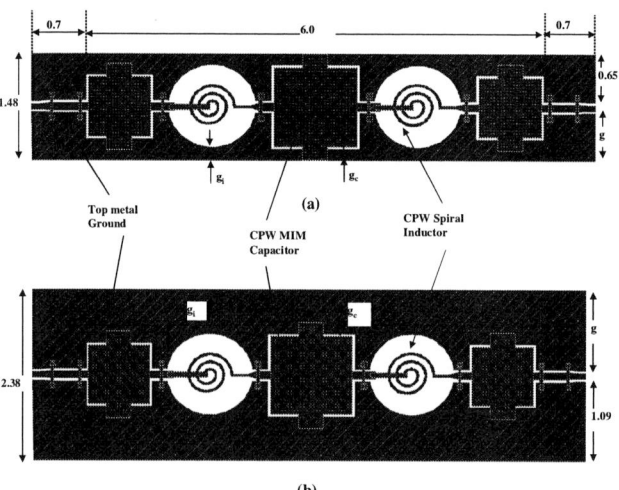

Fig. 6 Layout of the fabricated CPW lumped-element filters with dielectric height of 50μm/30 μm for (a) ground plane width (gc/gi) 160 μm and (b) ground plane width (gc/gi) 600 μm

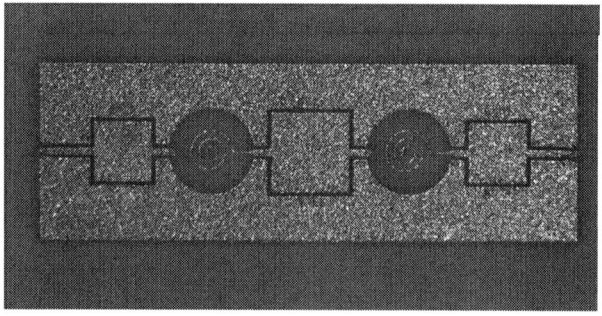

Fig. 7 Microphotograph of a fabricated multilayer CPW lumped-element filter for ground plane width 600 μm with dielectric height of 50 μm/30 μm

6(b) for ground plane width (gc/gi) 160 μm and 600 μm respectively. The circuit consists of two series inductors and three shunt capacitors. Layer 0 and 5/3 were used for the bottom and top metallization of the parallel capacitor electrodes of side 800μm and 1075 μm. The inductors were

978-2-8748-7007-1/08 $25.00 © 2008 EuMA   540

realized by circular spirals of 2.5 turns, with 25 μm track width and 100 μm spacing, fabricated on the 5/3 th layer.

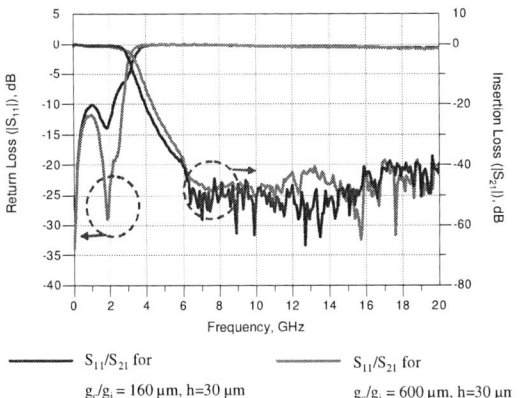

Fig. 8 The measured response of the CPW multilayer lumped-element lowpass filter with ground plane widths (gc/gi) 160 μm and 600 μm and dielectric height of 30 μm

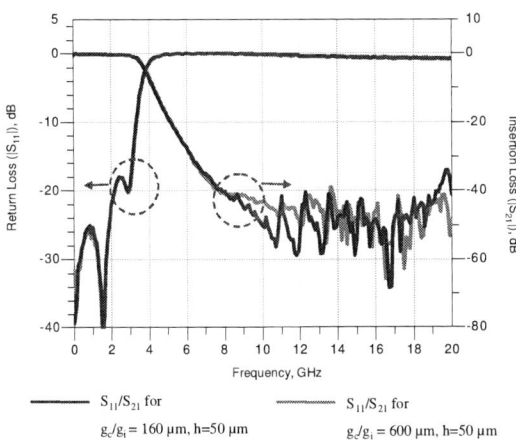

Fig. 9 The measured response of the CPW multilayer lumped-element lowpass filter with ground plane width (gc/gi) 155 μm and 600 μm and dielectric height of 50 μm.

Fig. 6(b) is with capacitor ground plane width (g_c) and inductor ground plane widths (g_i) of 600 μm each, whereas for the other filter shown in Fig. 6 (a), the g_c and g_i values are 150 μm and 160 μm respectively. For both the filters, the centre conductor width of the intermediate CPW line is 90 μm, whereas its associated ground plane widths, as well as the ground plane width of the input and output CPW lines were 650 μm and 1090 μm, respectively. The microphotographs of a fabricated multilayer LPF is shown in Fig. 7. The overall physical dimensions of the filters are 6.0 mm x 1.5 mm and 6.0 mm x 2.4 mm, respectively.

A comparison of the measured insertion loss (S_{21}) and return loss (S_{11}) for dielectric height of 30 μm and with ground plane widths (g_c/g_i) 160 μm and 600 μm is shown in Fig. 8, whereas that for a dielectric height of 50μm with same variation of ground plane width is depicted in Fig. 9. The insertion loss of the filter is around 0.3 dB, with input and output return loss better than 20 dB and attenuation > 40 dB. For a dielectric height of 50μm the variations in performance are negligible. Whereas, there is significant variation in measured results of both the parameters for a lower dielectric height of 30 μm.

V. CONCLUSION

For the first time, the effects of dielectric height, ground plane width and their interrelationship on conductor backed multilayer FGCPW lumped elements supported on a multilayer MCM substrate are studied. For CPW lumped elements, a low pass filter has been characterized to investigate the effect of ground plane width. The effect of the dielectric height has also been investigated and is shown to have an important influence on parasitics.

REFERENCES

[1] M. Cai, P. S. Kooi, M. S. Leong, and T. S. Yeo, ''Symmetrical Coplanar Waveguide with Finite Ground Plane,'' *Microwave Optical Tech. Lett.,* Vol. 6, No.3, pp. 218—220, March 1993.

[2] R. Jackson, "Mode conversion at discontinuities in Finite-width conductor-backed coplanar waveguide", *IEEE Trans. on Microwave Theory and Techn.,* vol. 37, no. 10, pp. 1582-1589,Oct. 1989.

[3] W.-T. Lo, C.-K. C. Tzuang, S. T. Peng, C.-C. Tien, C.-C. Chang, and J.-W. Huang, ''Resonant Phenomena in Conductor-Backed Coplanar Waveguides (CBCPW's),'' *IEEE Trans. Microwave Theory Tech.,* Vol. 41, No. 12, pp. 2099—2107, Dec. 1993.

[4] Stephens D. ; Young P. R. ; Robertson I. D., "Millimeter-Wave Substrate Integrated Waveguides and Filters in Photoimageable Thick-Film Technology", IEEE Transactions on Microwave Theory and Techniques, Issue 99, 2005, pp:1– 7.

[5] K. K. Samanta, D. Stephens and I. D. Robertson, "Ultrawideband Characterisation of Photoimageable Thick Film Materials for Microwave and Millimetre-wave Design", IEEE MTTS, International Microwave Symposium 2005, Long Beach, USA, 11-17 June, [CD ROM].

[6] K. K. Samanta, D. Stephens and I. D. Robertson, "60 GHz Multi-Chip-Module Receiver with Substrate-Integrated-Waveguide Antenna and Filter", Electronics Letters, 8th June 2006 Vol. 42, Issue 12, p. 701-702.

[7] K. K. Samanta, D. Stephens and I. D. Robertson, "Design and Development of a 60GHz Multi-Chip Module Receiver Employing Substrate Integrated Waveguides" IEE Proc. on Microwaves, Antennas & Propagation, October 2007, Vol.1, issue 5, p. 961-967.

[8] D.M. Pozar, "Microwave Engineering", *John Wiley & Sons,* 2005, ISBN 0-471-44878-8.

[9] K. K. Samanta and I.D.Robertson, "High Performance CPW Embedded Passive Components Using Photoimageable Thick Film Technology", *European Microwave Conference,* Paris, France, 3-7 October, 2005

Proceedings of the 3rd European Microwave Integrated Circuits Conference

3D Packaging Technology for Integrated Antenna Front-Ends

Barbara Bonnet[1], Philippe Monfraix[1], Renaud Chiniard[2], Jérôme Chaplain[3],
Claude Drevon[1], Hervé Legay[1], Pascal Couderc[3], Jean-Louis Cazaux[1]

[1]Thales Alenia Space
26 av. JF. Champollion, BP1187, F-31037 Toulouse Cedex 1, France
barbara.bonnet@thalesaleniaspace.com

[2]Thales Services
Parc Technologique du canal, 3 av Europe 31400 Toulouse, France

[3]3D Plus
641, rue Hélène Boucher Z.I. 78532 BUC Cedex, France

Abstract— **Thanks to Vertical Multi-Chip Module packaging technology (MCM-V), a novel concept of integrated antenna feed in Ka band has been developed. This technology enables the integration of active elements very close to the radiating surface, which reduces dramatically the weight and volume of the antenna. In this paper the different technological building blocks are described, and the measurements obtained on the first breadboard are discussed. The promising results obtained should lead to a major breakthrough for active receive antennas, driving down cost and complexity.**

I. INTRODUCTION

The improvement of organic materials now allows the introduction of quasi hermetic encapsulation for space microwave applications. Thales Alenia Space has been working on 3D Vertical Multi-Chip Module (MCM-V) technology for more than 10 years [1-5], developing for instance vertical interconnection up to 60GHz [6] and also 3D active building blocks (including LNA and HPA functions) for T/R module applications through the MARCOS program funded by the French Department of Defense (DGA).

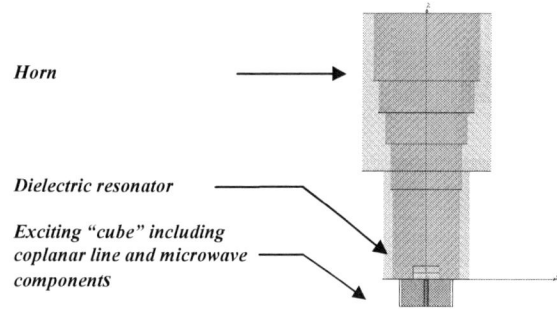

Fig. 1 The integrated feed concept

In this paper a novel feeding concept in Ka band (30GHz) based on an exciting "cube" realized in 3D packaging technology is described. It allows the integration of active components very close to the radiating plane, and reduces considerably the mass and volume compared to classical waveguide solutions.

After a first paragraph on MCM-V packaging technology, the concept of integrated feed will be described. Some information on the technological building blocks developed for manufacturing will then be given, and finally the measurements on the first breadboard will be commented.

II. 3D PACKAGING TECHNOLOGY

The 3D technology – also called MCM-V - is based on the full encapsulation in a resin, with an interconnection system etched on the surface of the "cube". The main steps are:

- manufacturing of the individual levels made on printed circuit board populated with passive and bare components (1),
- stacking in a mold with spacers (2),
- molding in an epoxy resin and polymerization (3),
- machining to the final size (4),
- plating and etching (laser routing) to define the interconnection between all the levels of the stack (5).

The different steps are resumed on Fig. 3.

The technology is available at 3D Plus company [7] and is already used in satellite payloads for digital equipments.

III. THE CONCEPT OF INTEGRATED FEED

The radiating element is a compact horn excited with a dielectric resonator. The innovation resides in the feeding by a line that is orthogonal to the resonator [8,9], etched on a printed circuit board (PCB) on which active devices can be mounted with standard chip on board technology.

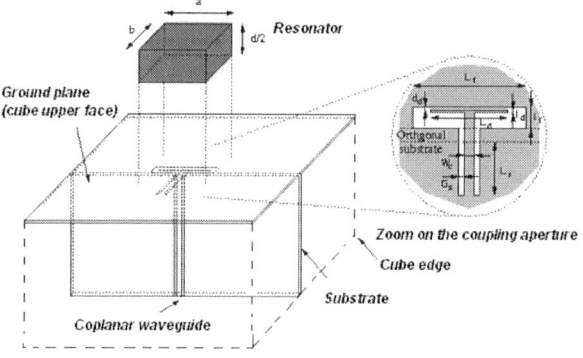

Fig. 2 Resonator fed by an orthogonal CPW line

978-2-8748-7007-1/08 $25.00 © 2008 EuMA

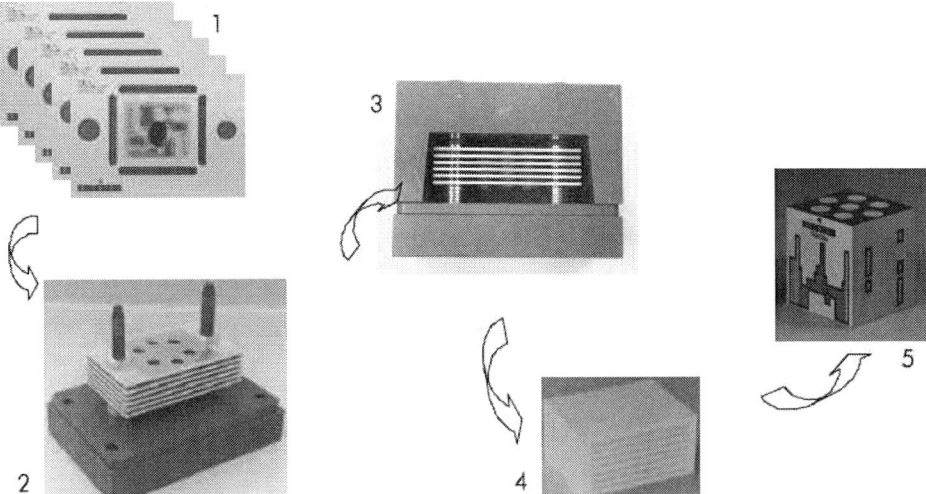

Fig. 3   Principle of 3D technology

The PCB is moulded in a "cube" of resin using MCM-V packaging technology. The transition from the PCB to the dielectric resonator is realized through a coupling aperture (see Fig. 2). The MCM-V module is called the exciting "cube" of the horn, and these two elements, plus the resonator, constitute the integrated feed.

IV. TECHNOLOGICAL BUILDING BLOCKS

Prior to the realization of the breadboard, the manufacturing process needed some improvements in order to comply with the constraints of an operating frequency in Ka band:

- Upgrading of the electrical interconnection through the coupling aperture
- Development of a connector access at 30 GHz

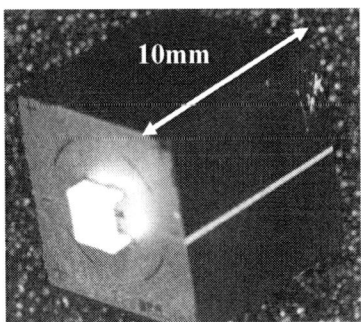

Fig. 4   Breadboard of the exciting "cube"

A. Etching process on MCM-V modules for Ka band application

One faces some limitations with standard laser etching process on MCM-V modules for microwave application. The patterns are etched too deeply in the resin, and due to the point of impact of the laser, the edges of the pattern are indented.

These two defaults lead to electrical losses enhanced with the increase in frequency. In the case of the integrated feed concept, the accuracy of the dimensions of the coupling aperture directly impacts the transmitted power from the PCB to the resonator.

For the manufacturing of the integrated feed breadboard, chemical etching on the face of the "cube" was investigated in collaboration with 3D Plus company. Electrical performances of coplanar lines etched by chemical and laser etching have been compared. On the same metallization, one gets a 50% decrease of the RF losses with chemical etching (0.023dB/mm versus 0.046dB/mm at 15GHz).

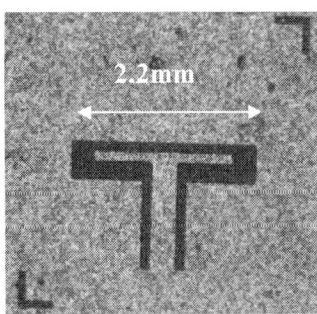

Fig. 5   Coupling aperture pattern realized by chemical etching on the face of the "cube"

B. The connector

The connector selected is a Rosenberger push-on, well match for Ka band (mini SMP).

The connector is mounted on the face of the "cube" directly in contact with the PCB line that is accessible on the edge of the "cube". The face is entirely metallized and the signal line is isolated from the ground by etching, as shown in Fig. 6.

For mechanical reasons, instead of a "Surface Mount Technology" as in [5], it was chosen to have a machining of the module. A hole is drilled at the location of the pad, and the pin of the connector is pasted with conductive glue inside the hole, i.e. at the location of the signal line.

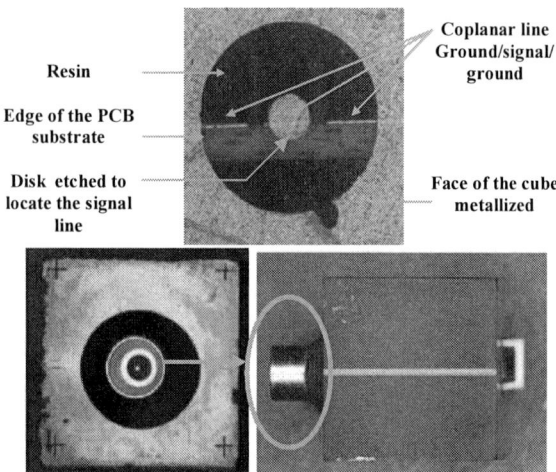

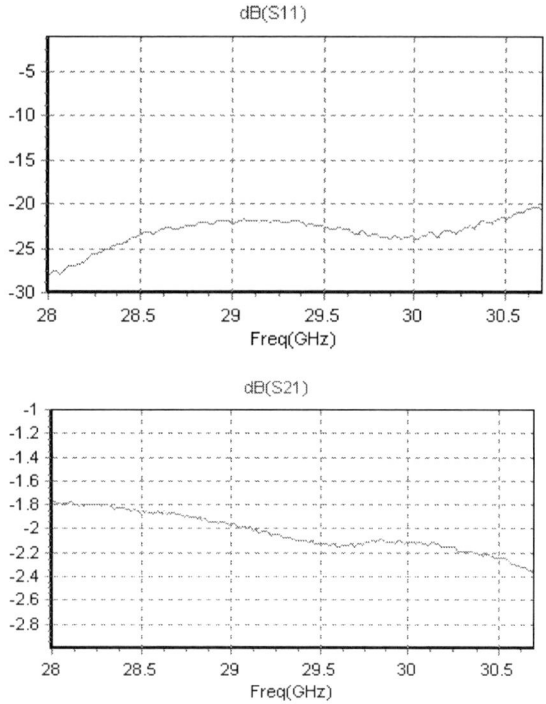

Fig. 6 a) Mark etched on the face of the cube to repair the location of the signal line – b) Connector mounted on the "cube"

To characterize the transition from the connector to the PCB, the [S] parameters of a module with a connector at each end have been measured, and are shown on Fig. 7.

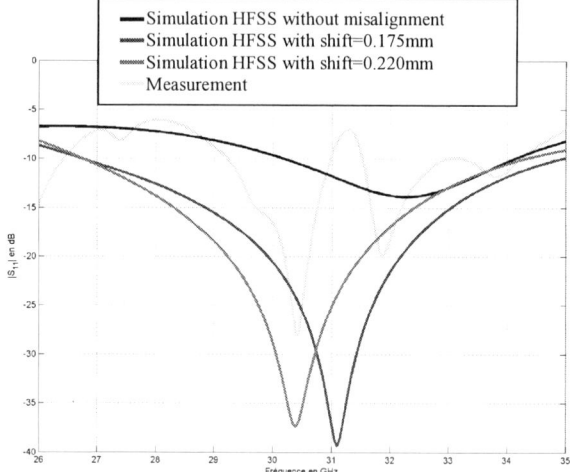

Fig. 7 Measured S parameters of a module with a connector at both ends

These parameters have been measured with a network analyzer calibrated in V band using V/mini SMP adapters. The elements measured are the adapters, the connectors and a 10mm long coplanar line altogether. In the range 28-31GHz the reflection coefficient |S11| is inferior to –20dB. Losses of about 2dB have been measured in transmission but this measurement includes the impact of the adapters that are not taken into account in the calibration step.

V. TEST OF THE BREADBOARD

The performances of the integrated feed mounted in a compact horn have been measured in an anechoic chamber. The configuration is shown on Fig. 8.

Fig. 8 Measurement configuration

The reflection parameter |S11| has been characterized. The difference observed between simulation and measurement is explained by a slight misalignment of the coupling aperture etched on the module compared to the edge of the PCB, which is confirmed by back simulations (see Fig. 9). This problem of misalignment of the etching mask has been studied and solved after the manufacturing of these first breadboards.

Fig. 9 Reflection parameter |S11| – Comparison of measurement and simulation

978-2-8748-7007-1/08 $25.00 © 2008 EuMA   544

When comparing the measured radiation patterns at 30GHz to the radiation patterns obtained by simulation with an ideal excitation of the horn, one can see that the expected performances have been reached. The level of cross polarization is higher in the experiment compared to the ideal case, but it stays under –20dB. This phenomenon is due to the slight misalignment of the coupling aperture.

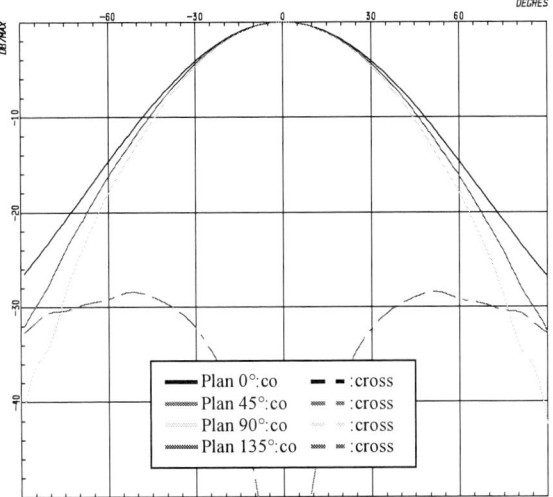

Fig. 10 Simulated radiation patterns of the horn excited by an ideal feed at 30 GHz

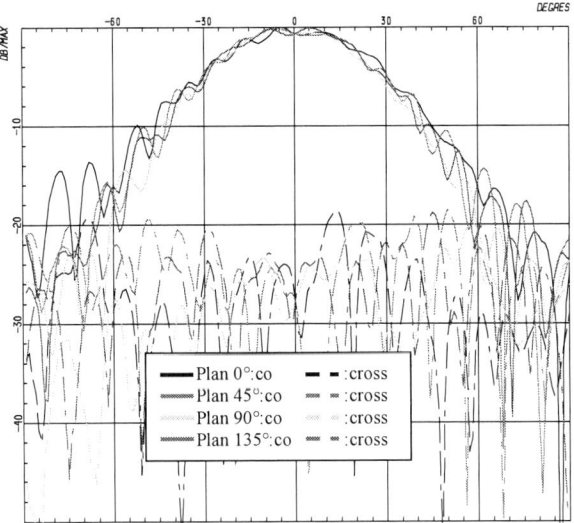

Fig. 11 Measured radiation patterns of the horn excited by the integrated feed at 30GHz

VI. CONCLUSION

In this paper the concept of an integrated feed in Ka band using 3D packaging technology has been successfully validated. The performances of the tested modules meet the expectations.

After a presentation of the concept, the technological improvements required for the manufacturing of the breadboard have been detailed. The design of microwave patterns on the face of 3D modules has been improved using chemical etching, compared to the standard laser routing. The implementation of a microwave mini SMP connector on the "cube" for an application at 30GHz has been validated.

Future work will consist in designing and testing an active integrated feed module containing microwave active and passive components. The design of a passive integrated feed in dual polarization is also an important milestone, since this property will bring promising capabilities to Focal Array Fed Reflector active antennas that are not achievable with standard feeding technology.

ACKNOWLEDGMENT

The authors wish to acknowledge the European Union for funding parts of the work mentioned above through the European project e-CUBES (contract n° 026461), as well as the French Space Agency CNES for its support under contract n°R-506/TC-0003-028.

REFERENCES

[1] P. Ulian et al., "3D active modules for high integration active antennas", in *Proc. 28th EuMC 98*, 1998, vol.1, pp. 271-276

[2] P. Monfraix et al., "3D Microwave Modules for Space", in *1998 IEEE MTT-S International Microwave Symposium Digest*, 1998, p. 1289-1292

[3] C. Drevon., "Multilayer Printed Circuit Board at 12- 14 GHz with MCMs and MMICs " in *Wireless Workshop Proceedings*, Gold Canyon - Arizona, 1999, p. 6-9

[4] O. Vendier and al., "3D Beam Forming Network for High Integration Active Antennas", in *Proc. 29th EuMC 99*, 1999, vol.3., pp. 379-381

[5] O. Vendier, C. Drevon, P. Monfraix, "Microwave 3D concept for Beam Forming Networks", in *Proc. ESCCON 2002*, 2002, p. 221

[6] P. Monfraix and al., "New 3d low loss, wide band microwave interconnection", in *Proceeding of the IMS conference*, Denver, June 1997

[7] C.Val , T.Lemoine, "3D Interconnexion for Ultra Dense Multichip Modules" in *IEEE Transition on Components, Hybrids and Manufacturing thechnology*, Vol 12 N°4, December 1990

[8] G. Lathiere, R. Gillard, H. Legay, "Dielectric resonator antenna with orthogonal slotline feed", in *Electronics Letters*, vol. 40, p. 715-716, June 2004

[9] H. Legay, G. Lathiere, R. Gillard, "Radiating device with orthogonal feeding", European Patent EP1605546, Dec. 14, 2005

978-2-8748-7007-1/08 $25.00 © 2008 EuMA

Proceedings of the 3rd European Microwave Integrated Circuits Conference

Design of 77 GHz Interconnects for Buried SiGe MMICs Using Novel System-in-Package Technology

Marius D. Richter #1, Karl-F. Becker *2, Lars Böttcher *3, Martin Schneider #4

#*Department of RF & Microwave Engineering, University of Bremen*
Otto-Hahn-Allee, 28359 Bremen, Germany
[1]marius.richter@hf.uni-bremen.de
[4]martin.schneider@hf.uni-bremen.de

*Fraunhofer Institute for Reliability and Microintegration
Gustav-Meyer-Allee 25, 13355 Berlin, Germany
[2]karl-friedrich.becker@izm.fraunhofer.de
[3]lars.boettcher@izm.fraunhofer.de

Abstract— **This paper describes the design, simulation and measurement of interconnects of buried active 77 GHz chips to a high frequency substrate using microvia technology. The embedding technology proposed offers great opportunities for a very broad range of frequencies and applications as well as a large potential for cost reduction.**

I. INTRODUCTION

This paper presents a part of a project which aims at integrating 77 GHz components (SiGe monolithic microwave integrated circuits (MMICs)) into printed circuit boards (PCBs), thereby combining driver and radio frequency (RF) circuitry and integrating antenna elements. Within the scope of this project, we have developed a new manufacturing process which is based on a combination of the so-called chip in polymer (CiP) and chip in duromer (CiD) technology ([1]–[5]). The active integrated chips (ICs) are first embedded into a duromer, enabling the integration of several ICs into one module (CiD), then embedded into a PCB (CiP) and connected to the RF substrate with microvias fabricated by laser beam drilling and copper plating. In this paper, we specifically consider a voltage-controlled oscillator (VCO) SiGe chip for 77 GHz which is embedded and whose differential output is connected using this technology. To evaluate the performance of the interconnect itself, daisy chain chips with coplanar and differential coplanar lines are connected using the same packaging technology.

II. TECHNOLOGY

The main advantage of this novel technology is that standard processes like laser drilling and metal plating can be used to replace wire bonding for connecting active chips, enabling potentially cheaper and more reliable interconnects. Furthermore, three-dimensional integration by stacking several layers of active and passive chips and wiring layers can drastically reduce area consumption.

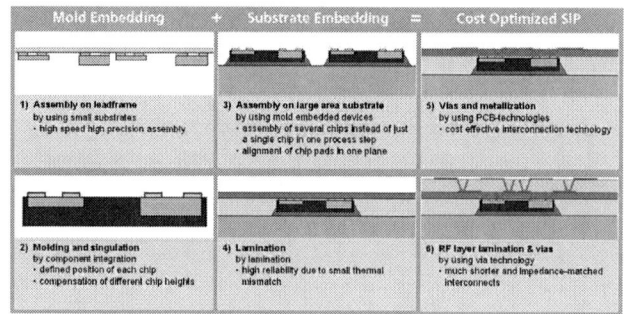

Fig. 1. Process flow of the used technology based on combination of CiP and CiD technology

The embedding process (Fig. 1) is based on a 125 μm thick copper leadframe with a 30 μm thick adhesive carrier tape that contains openings for the active chips. In the first step of the process, the chips are placed active side down on the carrier film inside the openings of the leadframe (Fig. 1(1)). Second, the leadframe is overmolded using highly filled transfer molding compound in a map molding process (Fig. 1(2)). Third, the overmolded packages are singularized using wafer sawing and are then bonded onto a core substrate using a thin die attach layer (Fig. 1(3)). Fourth, these modules are embedded with a combination of epoxy prepreg and a resin-coated copper layer (RCC), using PCB manufacturing processes (Fig. 1(4)). Fifth, vias connecting the pads of the chips are created by laser drilling and a subsequent metallization process [6]. The RCC metal layer is structured using a laser direct imaging (LDI) process that uses Sn-masking and subsequent etching (Fig. 1(5) and Fig. 2). In the last step, a high frequency RCC is laminated onto the embedded module and again vias are formed by laser drilling and subsequent metallization. The high frequency metal layer is then structured using standard photolithography and etching processes (Fig. 1(6)).

978-2-8748-7007-1/08 $25.00 © 2008 EuMA   546

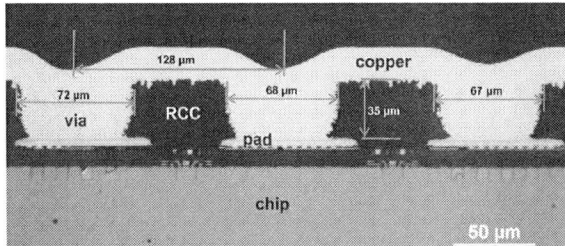

Fig. 2. Cross cut through buried chip and RCC layer

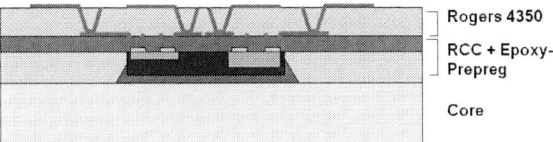

Fig. 3. Cross section of embedded and connected chips

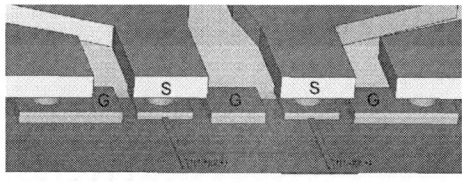

Fig. 4. Differential microstrip lines on the SiGe VCO chip

The final metal layers have a thickness of approximately 30 μm. As a compromise between processability and RF performance, we selected RO4350 as high frequency laminate. It has a permittivity of $\varepsilon_r = 3.4$ and a thickness of 100 μm. A sketch of this setup is depicted in Fig. 3.

The active chips embedded into the duromer encapsulate produce significant thermal power, which is dissipated by thermal vias connecting the backside of the chips to heat sinks.

III. INTERCONNECTS

Today, the integration of several wiring layers inside a single module, like multi-layer PCBs or low-temperature co-fired ceramics (LTCC), is standard. To obtain a multi-layer PCB stack, single layers are structured and then laminated together. Interconnects between different wiring layers are built by fabricating microvias, using conventional drilling or laser drilling and subsequent metal plating. Additionally, passive devices like capacitors and resistors can be embedded into PCBs.

Semiconductor chips are usually connected to a circuit board using either wire bonding, flip chip bonding with solder bumps [7], or tape-automated bonding. In contrast, embedding chips into a substrate and connecting the chips using etching or drilling processes and subsequent metal plating processes, being a comparatively new technology, is still rarely used. One example of this technology is General Electric's "chip-first" technology [8] for fabricating multi-chip modules or the redistributed chip package (RCP) technology [9]. The technology to embed chips into multilayer PCBs offers great opportunities for stacking active, passive and RF components and for reducing fabrication costs. And while wire bonds exhibit considerable inductance and resistance values because of their small diameters and non-negligible lengths, this is less the case for very short microvias used to connect buried chips. Moreover, microvias can be more easily trimmed to obtain impedance optimized interconnects.

The objective of the work described here is to design interconnects to connect buried active chips to a high frequency laminate. In the following section, three types of interconnects are considered: direct current (DC), high frequency differential, and high frequency single-ended interconnects.

A. Direct Current Interconnects

Direct current interconnects have already been fabricated using the above-described technology showing very promising results regarding low resistance and high reliability.

B. High Frequency Differential Interconnects

For the RF front end of the radar system developed within the project, different active SiGe chips are integrated into a PCB using the above-described technology. One of the chips is a VCO for 77 GHz. This SiGe chip has an area of roughly $2 \times 2 \text{ mm}^2$, the smallest pad size is $80 \times 65 \text{ μm}^2$ and the smallest pad pitch is 100 μm. To operate the chip, several direct current pads as well as the RF output must be connected to DC and transmission lines on the RF substrate, respectively. The RF output of the VCO has a GSGSG configuration (Fig. 4), where the two signal outputs have a phase difference of 180°.

For standard wire bonding, just the two signal lines are bonded to the RF substrate, forming a microstrip-microstrip transition (Fig. 5(a)). Even if no metal connection between the chip ground and the RF ground exists, a good transition of the signal from the chip to the RF substrate is enabled. However, simply connecting the signal pads is not sufficient for the case considered here, as the ground of the RF substrate lies above the chip (Fig. 5(b)). The setup is comparable to a measurement setup where the chip is directly contacted using GSGSG probes. Therefore, the ground pads have to be connected to the ground of the RF substrate to enable a transition of the microstrip mode on the chip to a differential coplanar mode and back to a microstrip mode on the RF substrate. As the minimum realizable pitch with the technology used is currently 125 μm, it is not feasible to contact all five pads as they have a pitch of just 100 μm. Furthermore, the ground pads are RF but not DC grounds. Therefore, these pads cannot be directly connected to the ground of the RF substrate, but have to be capacitively coupled.

The optimized interconnect is shown in Fig. 6; the solid curves in Fig. 7 show the scattering parameters of the optimized interconnect. The two chip lines are excited with a differential microstrip mode, and the signals on the microstrip lines on the RF substrate are measured. As the energy is

978-2-8748-7007-1/08 $25.00 © 2008 EuMA

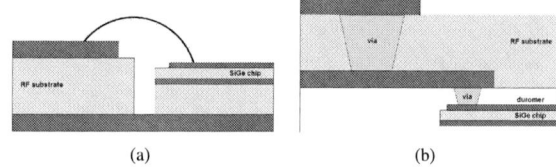

(a)   (b)

Fig. 5.   Cross section of wire-bonded chip (a) and cross section of buried chip (b)

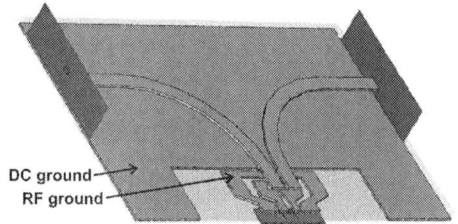

Fig. 6.   Layout of the optimized interconnect for the SiGe VCO chip

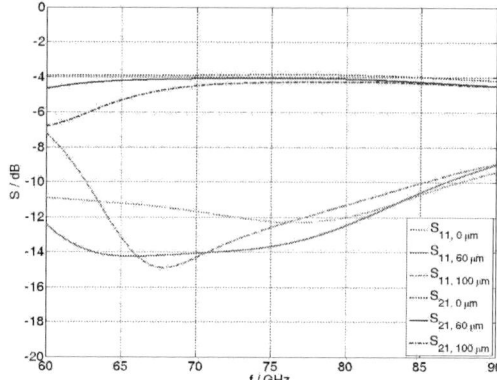

Fig. 7.   Simulated scattering parameters of the optimized interconnect for the SiGe VCO chip; port 1 is attached to the differential microstrip line on the chip, port 2 to one of the microstrip lines on the RF substrate. The specified lengths refer to the gap between RF and DC ground.

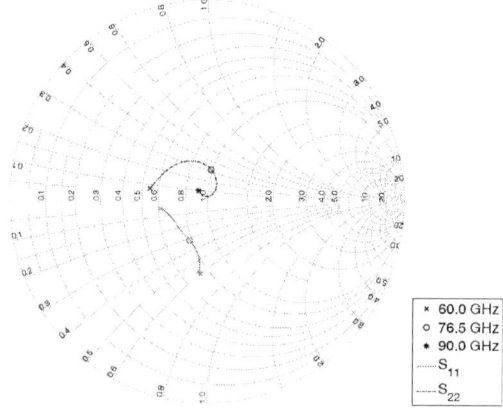

Fig. 8.   Smith Chart showing return loss of the VCO differential output (S_{11}) and the microstrip line on the RF substrate (S_{22})

split between the two output ports, S_{21} is -3 dB for perfect transmission. The simulations show additional insertion loss of approximately 1 dB, which stems from the following sources:

- The lossy RF substrate and metal account for approximately 0.2 dB.
- The microstrip lines on the chip are designed for 50 Ω. But as the active side of the chip is covered with a dielectric film, the impedance of the line is reduced, resulting in an impedance mismatch between the chip line and the 50 Ω-microstrip line on the RF substrate (Fig. 8). This accounts for approximately 0.15 dB of the insertion loss.
- Additional reflections are caused by the gap below the microstrip line. The minimum gap size that can be fabricated is 60 μm resulting in an additional insertion loss of 0.2 dB, as can be seen from the dotted lines in Fig. 7.
- The mode mismatch because of the missing interconnect to the center ground pad.

Different matching structures such as coplanar waveguide shorts and opens and slots in the transmission lines have been designed to compensate for the impedance mismatch. But as these matching structures add additional loss, the matching but not the insertion loss is improved. Therefore, matching structures are not incorporated in the design presented here.

Overall, the interconnect is very broadband and the performance comparable to wire bonds, but can be further improved e. g. by using low-loss materials, or customizing chips for the needs of the used technology.

To evaluate further the above-described interconnect between VCO output and the RF substrate, Si daisy chain chips were fabricated with the same pad geometry, but with 50 Ω-coplanar lines instead of microstrip lines because ground metallization was not available.

C. High Frequency Single Ended Interconnects

In addition to the VCO chip, mixer chips are needed to build the radar front-end. These chips feature single-ended coplanar outputs. To evaluate the interconnects between the chip and the RF substrate, daisy chain chips were fabricated with the same pad geometry but 50 Ω-coplanar lines instead of microstrip lines (Fig. 9) as ground metallization was not available. First simulation results show similar behavior as for the differential coplanar lines.

IV. MEASUREMENTS

Several modules containing either a buried daisy chain chip or a VCO were fabricated. The first measured module containing a buried differential coplanar daisy chain chip and different test structures is shown in Fig. 10. The buried chip is connected to the RF substrate using the above-described technology. Rat race power combiners on the RF substrate are used in order to measure the line with GSG probes. A comparison between the simulation and the measurement results

Fig. 9. Single ended daisy chain chip interconnect

is shown in Fig. 11 and Fig. 12. In contrast to Fig. 7, here two interconnects in series with a differential coplanar line on the Si chip and two power combiners are considered. Therefore, the effect of multiple reflections on the differential coplanar line can be seen as ripples in the scattering parameters. Nonetheless, the interconnects show again fairly broadband behavior, and insertion loss is consistent with the results from Fig. 7. Accounting for transmission loss of the coplanar line on the chip (≈ 1 dB), loss of the power combiners (each ≈ 1 dB), each transition shows an insertion loss of $2-3.5$ dB, which is consistent with the first measurement results obtained from measuring the output power of the buried VCO chips. We are working on further reducing the loss of the interconnects by improving the packaging technology.

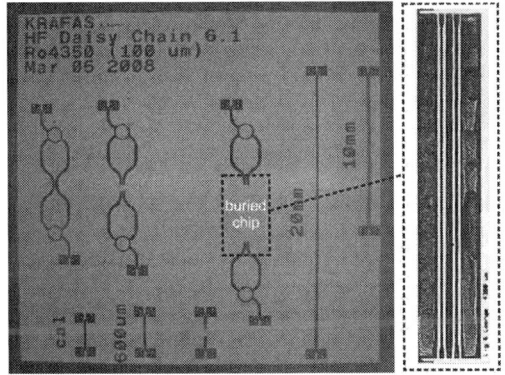

Fig. 10. CiP module with one buried Si chip and different test structures

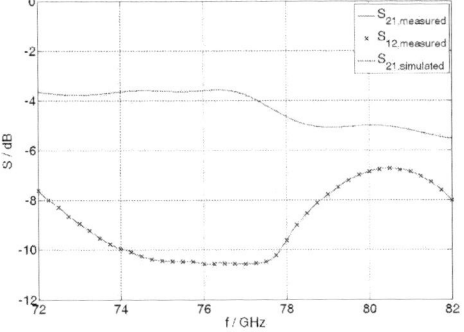

Fig. 11. Insertion loss of differential coplanar line on buried Si chip connected to microstrip lines on RF substrate via power combiners (Fig. 10)

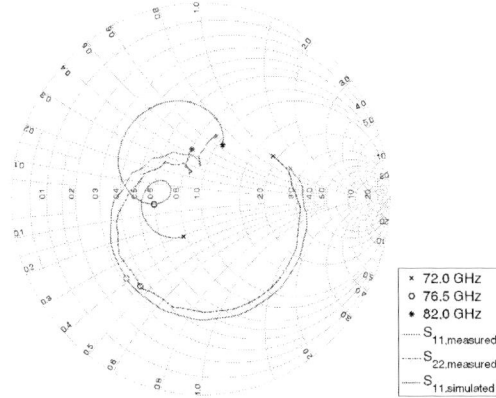

Fig. 12. Return loss of differential coplanar line on buried Si chip connected to microstrip lines on RF substrate via power combiners (Fig. 10)

V. CONCLUSIONS

A novel integration technology is used to embed 77 GHz SiGe chips. RF interconnects are designed to connect a VCO to an RF substrate showing very promising results. The characteristic of the interconnects can be further improved by customizing chips for the needs of this technology.

ACKNOWLEDGMENT

The authors wish to thank the Bundesministerium für Bildung und Forschung (BMBF) for financial support as the work presented here was partly funded by the KRAFAS project (Förderkennzeichen: 16SV2175) and Robert Bosch GmbH for fabrication of the daisy chain chips and measurement support. Furthermore, we wish to thank Mathias Koch for mold tool design and molding experiments and Ruben Kahle for lamination process development and demonstrator realization.

REFERENCES

[1] K.-F. Becker et al., "Duromer MID technology for system-in-package generation," IEEE Trans. Electron. Packag. Manufact., vol. 28, pp. 291–296, Sept. 2005.
[2] R. Aschenbrenner et al., "Process flow and concept for embedding active devices," in Electron. Packag. Technol. Conf., Singapore, Dec. 2004, pp. 605–609.
[3] A. Ostmann et al., "Realization of a stackable package using chip in polymer technology," in Polytronic Conf., Zalaegerszeg, Hungary, June 2002, pp. 160–164.
[4] K.-F. Becker et al., "Stackable system-on-packages with integrated components," IEEE Trans. Adv. Packag., vol. 27, pp. 268–277, May 2004.
[5] T. Braun et al., "Reliability potential of epoxy based encapsulants for automotive applications," in Europ. Symp. Rel. of Electron Dev., Fail. Phys. and Anal., Arcachon, France, Oct. 2005.
[6] L. Boettcher et al., "Electroless wafer-level redistribution-Further development and new approaches for system integration," in Surface Mount Technol. Assoc. Int., Chicago, Il, USA, Sept. 2003.
[7] M. Megahed et al., "Flip chip design of dual band/tri mode SiGe BiCMOS transmitter IC for CDMA wireless applications," in Int. Microwave Symp., Philadelphia, PA, USA, June 2003, pp. 1547 – 1550.
[8] R. Fisher et al., "High-frequency, low cost, power packaging using thin film power overlay technology," in Appl. Power Electron. Conf. and Exp., Dallas, TX, USA, Mar. 1995, pp. 12–17.
[9] B. Keser et al., "Advanced packaging: The redistributed chip package," IEEE Trans. Adv. Packag., vol. 31, pp. 39–43, Feb. 2008.

978-2-8748-7007-1/08 $25.00 © 2008 EuMA

Proceedings of the 3rd European Microwave Integrated Circuits Conference

Packaging Aspects of Photodetector Modules for 100 Gbit/s Ethernet Applications

C. Jiang[#1], G. G. Mekonnen[*2], V. Krozer[#], T. K. Johansen[#], H-G. Bach[*]

 # Department of Electrical Engineering, Technical University of Denmark

Oersteds Plads building 348, Lyngby, 2800, Denmark

[1]cj@elektro.dtu.dk

*Fraunhofer Institute for Telecommunications, Heinrich-Hertz-Institut (HHI)

Einsteinufer 37, D-10587 Berlin, Germany

[2]mekonnen@hhi.fhg.de

Abstract — **Packaging is a major problem at millimetre-wave frequencies approaching 100 GHz. In this paper we present that insertion losses in a multi-chip module (MCM) can be less IL < 0.6 dB at 100 GHz. The paper also analyzes in detail resonance modes in the packages. The characteristic of conductor-backed coplanar waveguides (CBCPWs) with vias is accurately analyzed using 3D electromagnetic (EM) simulation over a wide frequency range. Patch antenna mode resonances are identified as a major origin of resonances in simulated and measured transmission characteristics of the CBCPW with vias. Based on EM simulations, we propose several optimized arrangements for vias and bonding wires placement, to efficiently suppress the resonances and achieve excellent transmission performance of the PD module packaging. Based on our simulated results we postulate that it is possible to obtain resonance-free electrical transmission in the PD package with IL < 0.6 dB over a frequency from DC to 110 GHz.**

I. INTRODUCTION

High-speed multi-gigabit networks are essential for future large volume data transmission. 100 Gbit/s single channel Ethernet systems is considered to be the next generation of Ethernet application after the 10 Gbit/s Ethernet standard [1]. The optoelectronic transceiver working at the rate of 100Gbit/s is crucial for realizing 100 Gbit/s Ethernet systems. The EU FP6 project, GIBON, aims at developing the components for the 100 Gbit/s transmitter and receiver Ethernet system, partial results of which are shown here. The packaging of these high-speed components is very challenging when aiming at the rate of 100 Gbit/s, especially due to the multi-chip module (MCM) structure involving several chip-to-chip and/or chip-to-substrate transitions. In this paper we concentrate on the packaging aspects of the high-speed receiver front-end comprised of a high-speed PD with monolithically integrated fibre-connection and a MCM electrical connection.

There exist several methods to realize high-speed interconnects inside an MCM package. The most frequent solutions are wire-bonding and flip-chip interconnects. Generally, wire-bonding is regarded to be technologically easier to process, however, is not considered to be efficient at

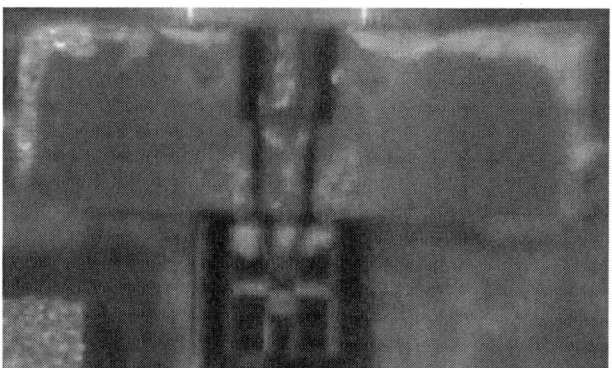

Fig. 1 Micrograph of the packaging of the photodetector module as well as the photodetector chip.

millimetre-wave frequencies. On the other hand, flip-chip interconnect technology has proven to exhibit low insertion losses, but at the expense of more complex technology. In this paper we show that wire bonding can be efficiently used up to 110 GHz with very low losses, comparable or even lower than the flip-chip insertion losses.

Photodetectors (PD) modules are typically packaged using conductor-backed coplanar waveguides (CBCPWs) to connect the PD chip to a coaxial connector. An illustration of the whole packaging structure is shown in Fig. 1.The pin of the 1 mm coax connector [2] is directly soldered onto the centre conductor of the CBCPW with vias to suppress losses. The upper ground planes are soldered to the outer conductor of the connector. This contact is important to obtain good transmission characteristic in the low frequency range and also shorts the gap between the CBCPW and the connector to reduce unexpected coupling effects. Bonding wires are utilized to connect the PD chip and the CBCPW.

The packaging structure including the bonding wires, the CBCPW with vias and the transition between the coaxial connector and the CBCPW, is currently limiting the frequency performance of packaged PD modules. Down-scaling of the geometrical dimension of the CBCPW for improved millimetre-wave performance is limited by the coaxial connector dimensions and by fabrication restrictions.

978-2-8748-7007-1/08 $25.00 © 2008 EuMA   550

Developing vias in the CBCPWs is considered to be an alternative way to effectively improve the bandwidth of the packaging structure. A branch of bonding wires is another main transition in the packaging structure influencing the bandwidth. Therefore, the CBCPW with vias and bonding wires should be optimized for the wideband requirement.

II. EM ANALYSIS OF CBCPWs WITH VIAS

The backside metallization of CBCPWs increases the ability of heat sinking, shielding and mechanical strength of coplanar circuits [3]. However, the backside metallization can couple to the upper ground planes of the CBCPW through undesired parallel plate mode, which propagates along the waveguide [4] or generates patch antenna mode resonances [5]. These higher order propagating modes and resonances seriously degrade the transmission performance for wideband applications.

A popular way to suppress the coupling effect is to utilize metallic vias to short-circuit the upper ground planes with the backside metallization. However, higher order resonances can not be suppressed efficiently if vias are placed randomly [6], which requires systematic design of the via arrangement.

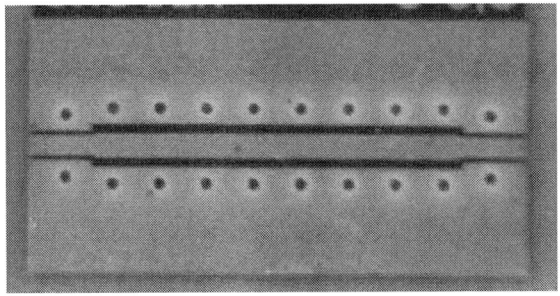

Fig. 2 The CBCPW with vias under investigation

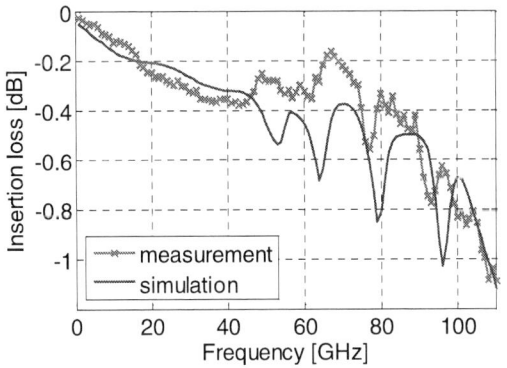

Fig. 3 Measured and simulated insertion losses of the CBCPW with vias

A CBCPW with vias, part of which will be used later in the packaging of the PD module, is shown in Fig. 2. The material of the substrate is quartz, and the conductor is gold. It is seen that the vias are placed close to the gap in the CBCPW structure and the distance between vias from center to center is kept minimum. Such an arrangement is designed to confine the transmitted electromagnetic (EM) energy to within the

CBCPW structure and achieve optimal transmission characteristic [6]. However, this approach changes the characteristic impedance of the CBCPW, which has to be readjusted to achieve 50 Ω transmission line properties.

A 3D EM simulation tool, Ansoft HFSS, is utilized to perform the optimization of the via placement in the CBCPW structure. The accuracy of the EM simulation results is evaluated by comparing measured and simulated insertion loss of the CBCPW with vias is shown in Fig. 3. It is crucial to employ lumped ports in HFSS simulations in order to accurately excite the CBCPW resembling the measurement setup with GSG probes. The simulation results have been calibrated using the method described in [7].

Although the simulated insertion loss does not exactly overlap the measured one, the main features of the measured insertion loss are still accurately captured by the simulations. Some notch frequencies are observed in the both measured and simulated characteristics. Fig. 4 shows the electric fields (E-fields) in the substrate at two notch frequencies in the simulated insertion loss, which are at 79 GHz and 96 GHz, respectively. The E-field patterns exhibit typical patch antenna mode resonances.

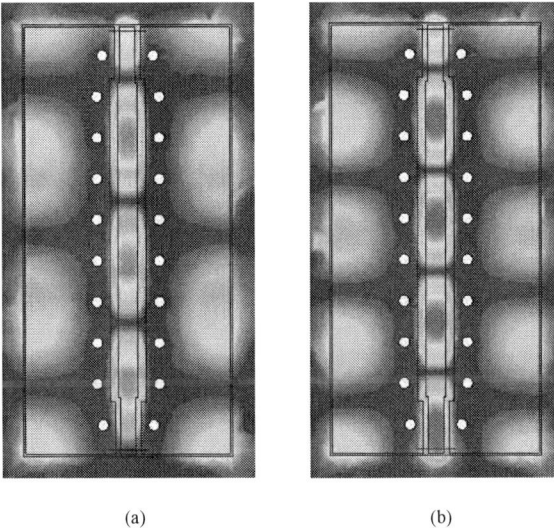

(a)   (b)

Fig. 4 The electric-field pattern in the substrate of the CBCPW with vias: (a) at 79 GHz; (b) at 96 GHz;

The resonance frequencies can be predicted analytically by

$$f_{mn} = \frac{c}{2\sqrt{\varepsilon_r}} \left[\left(\frac{m}{w} \right)^2 + \left(\frac{n}{l} \right)^2 \right]^{0.5} \quad (1)$$

where ε_r is the relative permittivity of the substrate, c is the velocity of the light in vacuum, w and l are the width and length of effective patch antennas, and m and n are the order numbers of resonances. For the cases shown in Fig. 4, m is 0.5 and n is 3 for 79 GHz and 4 for 96 GHz, respectively. The measured, simulated and analytical resonance frequencies are listed in the Table I, using $w = 700$ μm representing the

978-2-8748-7007-1/08 $25.00 © 2008 EuMA

distance from the center of the vias to the edge of the upper ground planes, and $l = 3950$ µm. The resonance frequencies derived from the three different methods agree well with each other. It is evident that the vias can not eliminate the resonances completely, although they are placed very close to the slot of the CPW structure.

TABLE I
RESONANCE FREQUENCIES COMPARISON

Freq.	Measurement	HFSS	Analytical
$f_{0.5,3}$ (GHz)	77	79	80.4
$f_{0.5,4}$ (GHz)	93	96	95.6

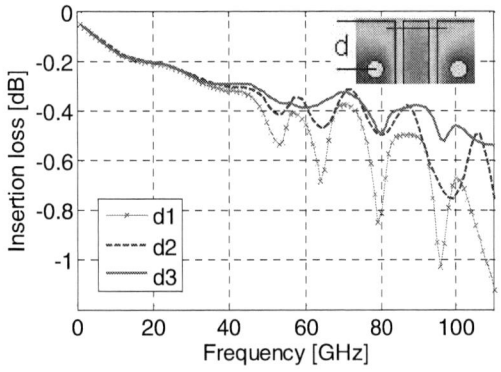

Fig. 5 Simulated insertion losses of the CBCPW with vias for decreasing distance (d1 > d2 > d3) between the end vias and the end of CBCPW is parametrically analysed.

Fig. 4 reveals that the CBCPW mode energy leaks at the end of the CBCPW with vias to the patch antenna resonances. The impact of the distance d, which is indicated in the inset of Fig.5, between the end-via to the edge of the CBCPW, has been carried out, as depicted in Fig.5. It demonstrates that the closer the end-vias are placed to the edge of the CBCPW structure, the lower insertion loss and less significant resonances are visible. It is concluded that the energy leakage can be inhibited by placing vias close to the edge of the CBCPW structure. Moreover, the energy leakage can also take place between vias if the distance between these is sufficiently large and should therefore be also chosen appropriately for the highest frequencies of operation.

III. EM ANALYSIS OF THE FULL PACKAGED MODULES

Our previous work, [8], investigated potential resonances due to the CBCPW alone without taking into account of the PD chip itself. Fig. 6 shows the HFSS simulation model for the new PD module packaging including the PD chip, bonding wires, the CBCPW with vias and the 1 mm coaxial connector. It resembles the real packaging structure as shown in Fig. 1. The photodiode in the PD chip is modelled by a lumped port as illustrated in the inset of Fig. 6. The bonding wire interconnect is modelled using rectangular Au structures of appropriate dimensions. Vias arrangement in the CBCPW can be recognized in Fig. 6. The 1mm coaxial connector is modelled by ideal coaxial cable line. Therefore, the loss due to the connector itself is ignored in the packaging model.

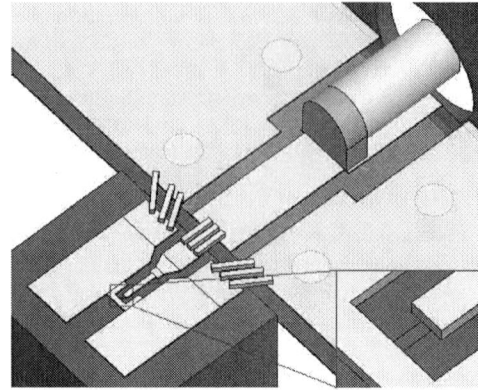

Fig. 6 HFSS model to investigate the insertion loss of the photodetector module packaging. The inset shows how a lumped port sits between anode and cathode acting as a photodiode.

Several cases have been analyzed using EM simulations and the results are presented in Fig.7. As outlined above, vias can not completely eliminate the resonances in the CBCPW transmission characteristics due to the energy leakage at the end of the CBCPW. The simulated insertion loss for the packaging structure illustrated in Fig.6 is shown in Fig.7 as the case1. Although no obvious resonances exist in the insertion loss characteristic a strong E-field at the edge of CBCPW is visible at 77 GHz, as shown in Fig. 8 (a). The pattern can be recognized as the patch antenna mode with both order numbers being 0.5. In spite of the leakage the insertion loss is less than IL<1dB up to 110 GHz.

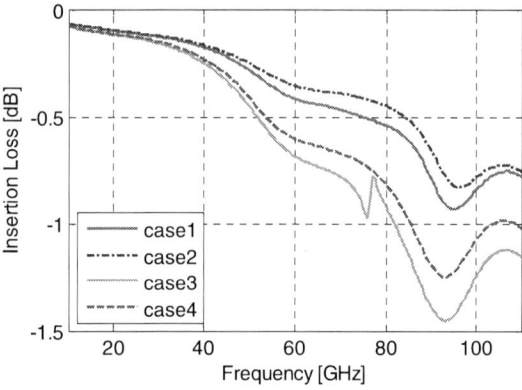

Fig. 7 The simulated insertion losses for the different packaging structures; case1: originally from Fig.6; case2: the end-vias are moved to the end of the CBCPW; case3: only one bonding wire connects each output pad of the chip to the CBCPW; case4: only one bonding wire connects each output pad of the chip to the CBCPW while the end-vias are at the end of CBCPW.

In order to prevent the energy leakage from the CBCPW mode to the substrate resonances, the end-vias are moved to the edge of CBCPW while keeping the distance between vias at the minimum value limited by fabrication technology. The insertion loss of the packaging structure with such optimization is shown in Fig. 7 as case2 and improves the insertion loss of the package by around 0.1 dB from 40 GHz

978-2-8748-7007-1/08 $25.00 © 2008 EuMA

to 110 GHz. However, the resonance is not eliminated completely by the optimization as shown in Fig. 8 (b). The EM energy still leaks between vias as well as the gap between the PD chip and the CBCPW. Further EM simulations demonstrate that EM energy leakage between vias can be completely eliminated by placing more vias in the substrate.

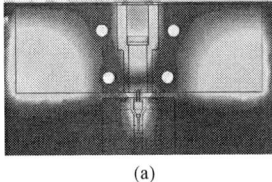 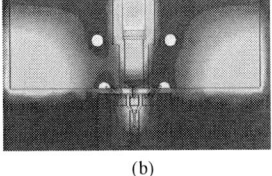

(a)   (b)

Fig. 8 (a) E-field of the original packaging structure at 77 GHz; (b) E-field of the packaging structure with the optimized vias arrangement at 77 GHz.

In the above simulations we have employed three bonding wires on each output pad of the PD chip for the chip-to-chip transition. The number of bonding wires is critical to the insertion loss of the package. Instead of three bonding wires on each pad, an interconnect with one bonding wire has also been studied in Fig. 7 as case3. Compared to the cases with three bonding wires, the insertion loss in this case is at least 0.5 dB higher independent of the optimization of the vias in the substrate, as indicated by the case 4 in Fig. 7.

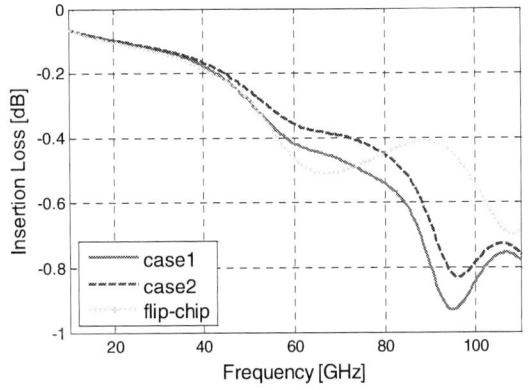

Fig. 9 Simulated insertion loss of the package with flip-chip transition instead of bonding wires. It is compared directly with the insertion losses in both case1 and case2 described in Fig.7.

The packaging structure with flip-chip transition from the PD chip to the CBCPW is investigated. An EM simulation model of the structure is built in HFSS and the simulated insertion loss is illustrated in Fig. 9. In general, the insertion loss of the package with flip-chip transition is comparable to the ones with bonding wires. The optimized package with bonding wires even exhibits lower insertion loss below 80 GHz.

IV. CONCLUSION

The paper presents very low-loss interconnect techniques for transition from chip to coaxial connectors up to 110 GHz in a MCM package. A notch free insertion loss characteristic of the package structure is achieved with a maximum insertion loss IL < 0.8 dB up to 110 GHz. The notch free insertion loss characteristic is achieved by optimizing the bonding wires and CBCPW structures. The paper reveals that the patch antenna mode resonance is the dominant loss mechanism for the CBCPW at higher frequencies. The resonance can be suppressed by reducing the gap between the end-vias and the end of CBCPW structure. With additional vias and bond wire structure optimization it is believed that an insertion loss IL < 0.6 dB is achievable for such a package structure.

ACKNOWLEDGMENT

The authors would like to thank the European Commission for support under the 6th framework programme to the project "Opto-electronic integration for 100 Gigabit Ethernet Optical Networks (GIBON)".

REFERENCES

[1] A.Zapata, M.Duser, J.Spencer, P. Bayvel,I. Miguel, D. Breuer, N. Hanik and A. Gladisch, "Next-generation 100-Gigabit Metro Ethernet (100 GbME) using multiwavelength optical ring," *J. Lightwave Technology*, vol. 22, Issue. 11, pp 2420-2434, November 2004.

[2] Agilent Technology, "11923A Launch Assembly Operating and Service Manual", 2005

[3] Y. Liu, K. Cha. and T. Itoh, "Non-leaky coplanar (NLC) waveguides with conductor backing," *IEEE Trans. Microwave Theory Tech.*, vol. 43, pp 1067-1072, May 1995

[4] H. Shigesawa, M. Tsuji, and A. A. Oliner, "Conductor-backed slot line and coplanar waveguide: Dangers and full-wave analyses," in *IEEE MTT-S Int. Mirowave Symp. Dig.*, June 1988, pp. 199-202

[5] J. A. Navarro and K. Chang, "Active microstrip antennas," in *Advances in Microstripand Printed Antenna*, K. F. Lee and W. Chen, Eds. New York: Wiley, 1997, ch. 8

[6] W. H. Haydl, "On the Use of Vias in Conductor-Backed Coplanar Circuits", *IEEE Trans. Microwave Theory Tech.*, vol. 50, pp 1571-1577, June 2002

[7] T. K. Johansen, C. Jiang, D. Hadziabdic, V. Krozer, "EM Simulation Accuracy Enhancement for Broadband Modeling of On-Wafer Passive Components", in *Proc. EuMIC2007 Munich*, Oct. 2007, pp.447-450

[8] C. Jiang, T.K. Johansen, V. Krozer, G. G. Mekonnen and H-G. Bach, "Optimization of Packaging for PIN Photodiode Modules for 100Gbit/s Ethernet Applications", in *Proc. APMC2007 Bangkok*, Dec. 2007,

Proceedings of the 3rd European Microwave Integrated Circuits Conference

A 75 – 95 GHz Wideband CMOS Power Amplifier

Byron Wicks[#1], Efstratios Skafidas[#2], Rob Evans[#3]

[#]National ICT Australia, Department of Electrical and Electronic Engineering, University of Melbourne, Parkville, Victoria
3010, Australia
[1]bnw@ee.unimelb.edu.au
[2]s.skafidas@ee.unimelb.edu.au
[3]r.evans@ee.unimelb.edu.au

Abstract— **In this paper, a 75 – 95 GHz wideband power amplifier (PA) in 0.13-μm CMOS is implemented to explore the feasibility of low-cost CMOS technology for use in 71 – 76 GHz, 81 – 86 GHz and 92 – 95 GHz fixed point-to-point link bands and the 77-GHz vehicular radar band. The fully-integrated design incorporates the power amplifier, matching networks, and input and output transmission line networks on-chip. The designed amplifier achieves small signal gain of 6.0 dB at 77 GHz, a 3-dB bandwidth of 75 – 95 GHz and delivers a saturated output power of 8.1 dBm.**

I. INTRODUCTION

Currently, the monolithic millimeter-wave integrated circuits (MMIC) used in the 71 – 76 GHz, 81 – 86 GHz and 92 – 95 GHz fixed point-to-point link bands and the 77-GHz vehicular radar band, shown in Fig. 1, are the domain of type III-V semiconductor technologies such as the Gallium Arsenide (GaAs), Indium Phosphide (InP) PHEMT or Silicon Germanium (SiGe) processes [1], [2]. These technologies possess superior performance, compared to CMOS technology, due to characteristics such as higher carrier mobility, higher transistor breakdown voltage and lower process parasitics.

Realizing these 75 – 95 GHz band MMIC devices on the CMOS process promises the integration of baseband signal processing, mixed signal and radio frequency (RF) components on a single die, with this integration resulting in significant cost benefits and the facilitation of a range of new lost-cost applications.

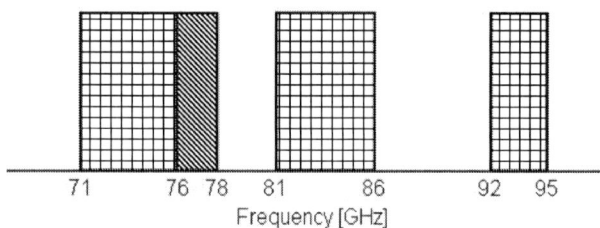

Fig. 1. The United States fixed point-to-point link (⊞) and radar (▧) band allocation in the 65 – 100 GHz spectrum.

Recent advancements in CMOS technology have allowed the implementation of MMIC components on the CMOS process for frequencies up to 77 GHz [3]-[7] and CMOS oscillators at higher frequencies [8].

The power amplifier is an important but difficult to integrate component in building an integrated transceiver. Some of the difficulties encountered are due to the low

breakdown voltage, low power density and operation at a significant portion of the transistors maximum oscillating frequency.

In this paper, we present a two stage design and optimisation technique for high frequency PA design on CMOS. The methods outlined in this paper were used to build the power amplifier using a 0.13-μm RF-CMOS process, provided by IBM.

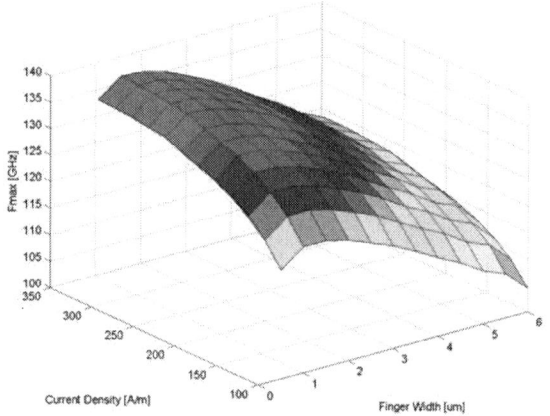

Fig. 2. Simulated f_{max} for varying device finger size and current density. Total device size was kept constant at 60 μm.

II. TRANSISTOR DESIGN

An important parameter when designing high frequency amplification components is the transistors maximum oscillating frequency (f_{max}), defined as the frequency where the Mason's unilateral gain of the transistor becomes unity (U = 1). A transistors f_{max} is most significantly dependent on the channel length, width, number of fingers (which influences the gate resistance), biasing conditions and transistor parasitic resistances and capacitances. The dominant parasitic losses are due to the gate resistance, source resistance, drain resistance, channel resistance and the substrate resistance networks. It is common practice for CMOS circuits to be designed for operation at frequencies of less then one-fifth of the transistors peak f_{max}. At higher frequencies, especially those approaching the performance limits of a process, these loss mechanisms must be carefully considered and optimised to increase f_{max}, thus increasing the frequency range operation of the designed components.

978-2-8748-7007-1/08 $25.00 © 2008 EuMA   554

A critical step of the power amplifier design process is the design and optimisation of transistors for optimal performance. In this paper, NMOS transistors are used as the active component building block. NMOS transistors have higher gain then a PMOS counterpart because of the higher mobility of electrons in silicon. As a consequence lower current densities are required, less power is dissipated and higher output power ca be achieved.

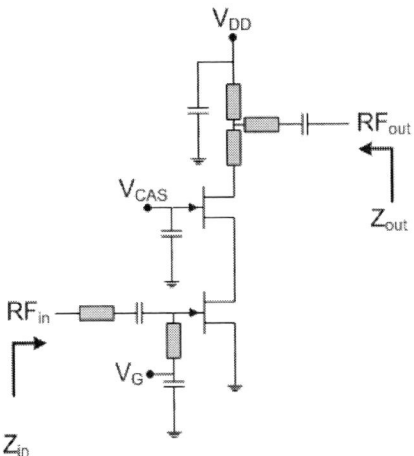

Fig. 3.   Schematic of a single stage of the power amplifier.

The dimensions of the implemented NMOS devices in this paper were chosen with the aid of CAD simulations and verified against multiple fabricated devices with varying lengths, widths and fingers. The f_{max} of the NMOS transistors increases as the finger width reduces from 8 μm to 2 μm due to a significantly reduced gate resistance. Below 1 μm the substrate resistance network becomes a significant factor and f_{max} begins to decrease, Fig. 2. From our device simulations, a transistor with gate length of 0.12 μm, finger width of 2.5 μm (to increase source-drain current and allow ease of matching), and 32 fingers per device for a total device size of 80 μm was chosen. With these parameters the device possessed an f_{max} of greater than 135GHz at a biasing of 350 μA/μm of gate width.

At millimetre wave frequencies, the layout of a transistor can have a significant impact on the characteristics of the device due to the major effect the layout has on the parasitic elements especially gate resistances, metal layer inductances, losses in the substrate and capacitive coupling between transistor components. This requires each transistor to be implemented (laid out) carefully to minimise the parasitics in the gate (including connection of both gate contacts), source, drain, and substrate resistances (including minimising the resistance to ground of the substrate contact ring).

III. CIRCUIT DESIGN

A. Architecture

The PA presented in this paper, is a five-stage amplifier with each stage operating as a Class-A amplifier as shown in Fig. 3, with the input, output and interstage matching networks all integrated onto one chip. A cascode topology was

employed at each stage to increase the gain, reduce the Miller Capacitance, and to increase the stability of each stage. To allow the device to operate at 77 – 95 GHz strong biasing into the inversion region is required.

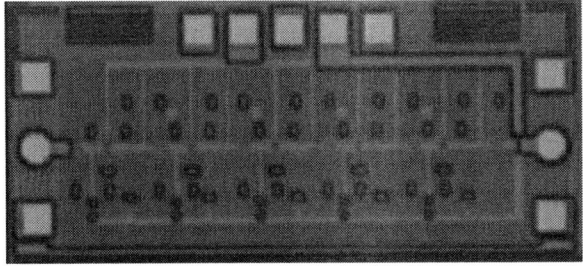

Fig. 4. Die micrograph of the 0.13μm CMOS 77-GHz PA with, dimensions 1230 μm by 520 μm.

When biased with a current density of 350 μA/μm of gate width, each cascode pair possessed a simultaneously conjugate matched maximum gain (G_{MAX}) of 4 dB at 75 GHz and 3 dB at 95 GHz.

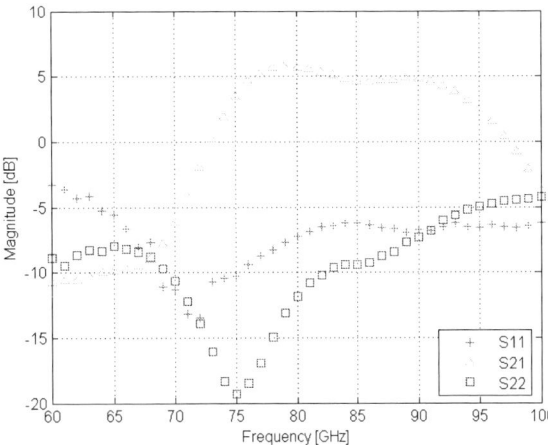

Fig. 5.   Measured S-parameters of the PA.

B. Matching Networks

Such a limited G_{MAX} and the utilization of large devices with limited output impedances requires precise matching networks with losses in the passive components needed to be kept to a minimum to maintain adequate gain. Careful selection of metal layers and ground planes was required to build microstrip transmission lines which were used for the design of the impedance matching networks, interconnects and for biasing the common gate and common source devices. These transmission lines were optimized to increase bandwidth and minimize losses (both resistive and radiation losses due to discontinuities), and are significantly shorter than λ/4 (< 80 μm) to reduce the occurrence of transmission line effects. AC coupling has been used between each amplifier stage, as well as the input and output connections between the pads and the PA. This was accomplished by using a metal-insulator-metal (MIM) capacitor. The input and output of the PA were both designed to be matched to 50Ω.

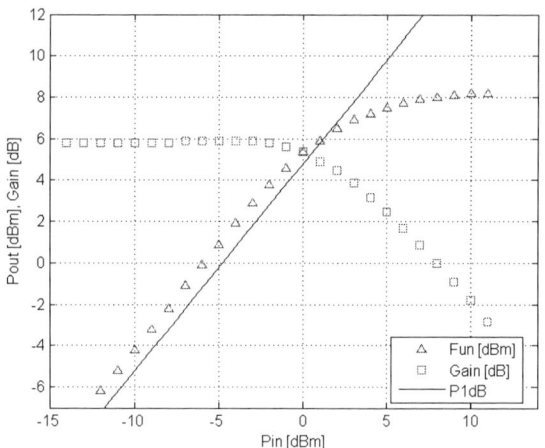

Fig. 6.   Measured large-signal characteristics at 77 GHz.

IV. EXPERIMENTAL RESULTS

The S-parameter results of the power amplifier measured using a probe station with 110 GHz Ground-Signal-Ground (G-S-G) coplanar probes and a 110 GHz Vector Network Analyser. Power measurements were performed using an power meter and a signal generator. The micrograph of the fabricated power amplifier is shown in Fig. 4, and has dimensions 1230 μm by 520 μm.

At 77 GHz the power amplifier achieves a measured small signal gain of 6.0 dB, input and output return losses better then 9 and 15 dB respectively and a 3-dB bandwidth of 75 – 95 GHz as shown in Fig. 5, whilst drawing 135 mA from a 2.5 V power supply. It demonstrated a saturated output power of 8.1 dBm, an output referred power at the 1-dB compression point of 6.3 dBm, shown in Fig. 6, and maximum power added efficiency (PAE) of 0.5% at P_{1dB}.

The ITRS has defined a PA figure of merit (FoM$_{PA}$), which links the output power (P_{OUT}) with the Gain (G), PAE and frequency (f) as a standard to compare power amplifiers:

$$\text{FoM}_{PA} = P_{OUT} * G * PAE * f^2 \text{ [6]} \qquad (1)$$

Table I provides a comparison of this PA with other published millimeter-wave PAs in terms of the ITRS FoM$_{PA}$. To the best of the authors' knowledge, this power amplifier is the highest frequency CMOS PA reported and also the only CMOS PA reported for the 81 – 86 GHz and 91 – 95 GHz ISM bands.

V. CONCLUSION

In this paper, a wideband 75 – 95 GHz power amplifier was designed and implemented on 0.13 μm RF-CMOS to explore the feasibility of low-cost CMOS technology for use for 75 – 95 GHz ISM bands, and automotive radar band. The fully-integrated design implements the main and auxiliary

amplifiers, matching networks, and input and output transmission line networks on-chip. The designed amplifier achieves small signal gain of 6.0 dB at 77 GHz, a 3-dB bandwidth of 75 – 95 GHz and delivers a saturated output power of 8.1 dBm at 77-GHz.

ACKNOWLEDGEMENT

The authors would like to acknowledge support from IBM, MOSIS and National ICT Australia.

National ICT Australia is funded by the Australian Government's Department of Communications, Information Technology, and the Arts and the Australian Research Council through Backing Australia's Ability and the ICT Research Centre of Excellence programs.

TABLE I

COMPARISON WITH PUBLISHED CMOS MILLIMETER-WAVE PA RESULTS

PA Technology	f [GHz]	G [dB]	P_{sat} [dBm]	P_{1dB} [dBm]	PAE [%]	FoM [6]
84 GHz f_{max} 0.18-um CMOS [9]	40	7	10.4	-	2.9	2.6
130 GHz f_{max} 0.130-um [3]	60	12	-	2.0	-	-
200 GHz f_{max} 90nm CMOS [4]	60	5.2	9.3	6.4	7.4	7.5
200 GHz f_{max} 90-nm CMOS [7]	60	8	10.6	8.2	-	-
200 GHz f_{max} 90-nm CMOS [7]	77	9	6.3	4.7	-	-
130 GHz f_{max} 0.130-um [10]	60	13.5	7.8	7.0	3.0	15.2
130 GHz f_{max} 0.130-um (this work)	77	6	8.1	6.3	0.5	2.1

REFERENCES

[1]   H.-Y. Chang, H. Wang, M. Yu, and Y. Shu, "A 77-GHz MMIC Power Amplifier for Automotive Radar applications," *IEEE Microwave Wireless Compon. Lett.*, vol. 13, no. 4, pp. 143-145, Apr. 2003.

[2]   P. P. Huang, T.-W. Huang, H. Wang, E. W. Lin, Y. Shu, S. D. Gee, R. Lai, H. M. Biedenbender, and J. H. Elliott, *"A 94-GHz 0.35-W power amplifier module,"* IEEE Trans. Microwave Theory Tech., vol. 45, pp. 2418–2423, Dec. 1997.

[3]   C. H. Doan, S. Emami, A. M. Niknejad, and R. W. Brodersen, "Millimeter-wave CMOS design," *IEEE J. Solid-State Circuits*, vol. 40, no.1, pp. 144–155, Jan. 2005.

[4] C. M. Ta, B. Wicks, F. Zhang, B. Yang, Y. Mo, K. Wang, Z. Liu, G. Felic, P. Nadagouda, T. Walsh, R. J. Evans, I. Mareels and E. Skafidas, "Issues in the Implementation of a 60GHz Transceiver on CMOS," *The 2nd IEEE International Workshop on Radio-Frequency Integration Technology*, Singapore, 09-11 Dec., 2007.

[5] T. Yao, M. Q. Gordon, K. K. W. Tang, K. H. K. Yau, M.-T. Yang, P. Schvan, S. P. Voinigescu, "Algorithmic Design of CMOS LNAs and PAs for 60-GHz Radio", *IEEE J. OF Solid-State Circuits*, Vol. 42, No. 5, May 2007.

[6] B. Wicks, E Skafidas, I Mareels, R Evans, "A 46.7-46.9-GHz CMOS MMIC Power Amplifier for Automotive Radar Applications", *IASTED ARP 2007*

[7] T. Suzuki, Y. Kawano, M. Sato, T Hirose, K Joshin, "60 and 77GHz Power Amplifiers in Standard CMOS", *IEEE Int. Solid-State Circuits Conf. Dig. Tech. Papers*, Feb. 2008, pp. 562-563.

[8] L. M. Franca-Neto, R. E. Bishop, and B. A. Bloechel, "64 GHz and 100 GHz VCO's in 90 nm CMOS using optimum pumping method," in *IEEE Int. Solid-State Circuits Conf. Dig. Tech. Papers*, Feb. 2004, pp. 444–445.

[9] H. Shigematsu, T. Hirose, F. Brewer, and M. Rodwell, "Millimeterwave CMOS circuit design," *IEEE Trans. Microw. Theory Tech.*, vol. 53, no. 2, pp. 472–477, Feb. 2005.

[10] B. Wicks, E. Skafidas, R. Evans, , "A 60-GHz Fully-Integrated Doherty Power Amplifier Based on 0.13-μm CMOS," *IEEE Radio Frequency Integrated Circuits (RFIC) Symposium, June*, 2008, to be published.

Proceedings of the 3rd European Microwave Integrated Circuits Conference

A 19.1-dBm Fully-Integrated 24 GHz Power Amplifier Using 0.18-μm CMOS Technology

Jing-Lin Kuo[1], Zuo-Min Tsai[2], Huei Wang[3]

Dept. of Electrical Engineering and Graduate Institute of Communication Engineering, National Taiwan University

No. 1, Sec. 4, Roosevelt Road, 10617 Taipei, Taiwan.

[1]r95942083@ntu.edu.tw
[2]f90006@ew.ee.ntu.edu.tw
[3]hueiwang@ew.ee.ntu.edu.tw

Abstract— **A 24 GHz, 19.1 dBm fully-integrated power amplifiers (PA) was designed and fabricated in the 0.18-μm deep n-well (DNW) CMOS technology. This power amplifier is a 2-stage design using cascode RF NMOS configuration and has a maximum measured output power of 19.1 dBm, an OP_{1dB} of 13.3 dBm, a power added efficiency (PAE) of 15.6%, and a linear gain of 18.8 dB when V_{DD} and DNW are both biased at 3.6 V. The chip size is only 0.56 x 0.58 mm². To the author's knowledge, this PA demonstrates the highest output power of +19.1 dBm among the reported PAs above 15 GHz in CMOS processes.**

I. INTRODUCTION

There have been several applications in the industrial, scientific, and medical (ISM) band at 24 GHz such as wireless communication transceivers [1], and automotive vehicular radar [2] recently. In these applications, power amplifiers play an important role in high-frequencies transmitter front-end circuits. At frequencies around 20 GHz of power amplifier, GaAs-based pHEMT [3] and SiGe BiCMOS [4]-[5] MMICs occupied most of the applications. Recent works have illustrated the rapid development of CMOS devices are still attractive because of high level of integration, small size, and hence cost consideration [6]-[7].

In sub-micron CMOS technology, it is still challenging to design a fully-integrated power amplifier of high output power. The major limitations in sub-micron CMOS technology are the low breakdown voltage, high substrate loss, and low power density. Especially, the voltage swing is limited by the low breakdown voltage. In order to achieve high gain at high frequency, the cascode device configuration is commonly used, and the boost in power also benefits the CMOS power amplifier designs [8]-[11]. There were some reports of CMOS cascode amplifiers designed for frequencies above 15 GHz. A 24 GHz amplifier consists of two-stage cascode cells demonstrated 7 dB gain [12]. It is observed that the CMOS circuits demonstrated good performance in high frequency range.

It is reported that NMOS on a DNW improves linearity [13], f_T, and f_{max} [14], by effectively reducing intrinsic gate-source capacitance (C_{gs}). In this paper, we present a 24 GHz power amplifier fabricated by 0.18-μm CMOS technology under different deep n-well (DNW) biasing. The 2-stage cascode RF NMOS configuration using of spiral inductors for on-chip matching was implemented in a very compact chip size.

Table I summarizes the performances of the previously reported power amplifiers designed for frequencies around 20 GHz. This power amplifier demonstrates the highest output power in the reported CMOS PAs above 15 GHz, and represents state-of-the-art results.

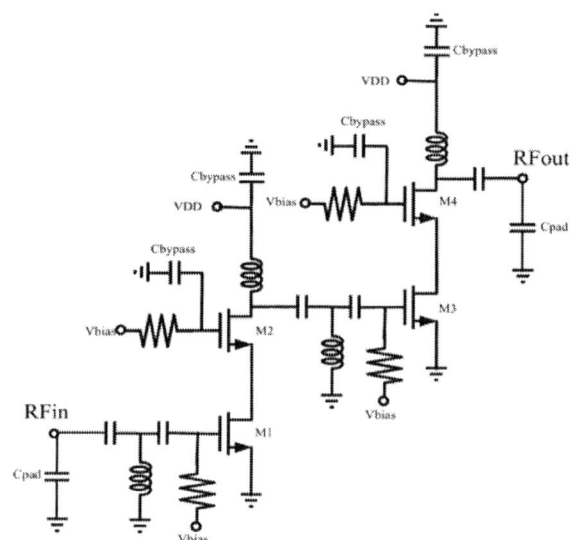

Fig. 1. Schematic of the PA.

II. CIRCUIT DESIGN

The MMIC power amplifier was designed in 0.18-μm deep n-well (DNW) CMOS technology. This technology provides single poly layer for the gates of the MOS and six metal layers for inter-connection [15]. The transistor of 64 fingers with total gate width of 386-μm has a typical unit current gain cutoff frequency (f_T) of higher than 49 GHz and maximum oscillation frequency (f_{max}) of greater than 47 GHz, respectively.

The power amplifier is targeted to deliver the most power from the device. This means that the voltage and current swings of the device should maximum. Since there is a physical limit to the maximum voltage swing of a device, in order to obtain higher output power, the supply voltage must be increased to gain extra voltage swing. The cascode

978-2-8748-7007-1/08 $25.00 © 2008 EuMA

configuration is usually used to get higher output voltage swing. Figure 1 shows the schematic diagram of the power amplifier. The amplifier consists of two stages of cascode devices with input, output, and inter-stage matching networks. In order to achieve maximum output power, the transistors (M3, M4) in the second stage are selected to be 64 fingers with each finger length of 6-μm. In order to achieve higher power added efficiency (PAE), the transistors (M1, M2) in the first stage are selected to be 16 fingers with each finger length of 6-μm. As the CMOS power amplifier operates in a high voltage swing, the deep n-well of the transistors should be forward biased, therefore, which can keep the parasitic diodes from being positive biased. In this design, all the transistors are biased via 5-KΩ resistors in order to investigate the effect of bias on the DNW. Output matching network is designed for maximum output power, and the inter-stage and input matching networks are designed for gain and input return loss, respectively. In order to reduce the unwanted coupling effects, the inductors are separated apart with at least 50-μm.

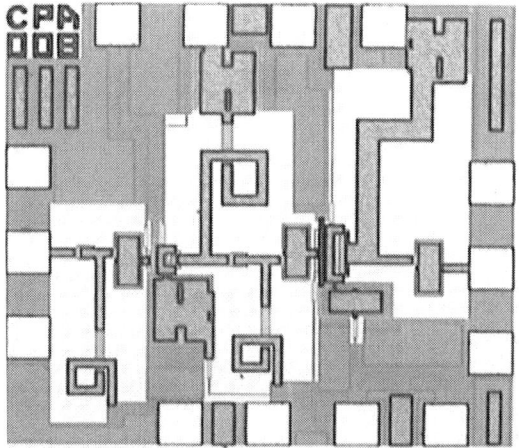

Fig. 2. The chip photo of the PA with chip size of 0.56 x 0.58 mm².

III. SIMULATION AND MEASUREMENT RESULTS

The amplifier was measured through on-wafer probing calibrated to the probe tips. All of the components in this design, including spiral inductors and MIM capacitors are on-chip implemented. Figs. 3-5 show the simulated and measured S-parameters of this amplifier when the V_{DD} and DNW are both biased at 3.6 V.

When the drain and DNW voltage is biased at 3.6 V, the amplifier has peak gain of 18.8 dB at 24 GHz, and the minimum input and output return losses are 21 and 9 dB, respectively. Power performances between different DNW biased of this power amplifier are shown in Fig. 6. When DNW is biased at 3.6 V, the power amplifier delivers a saturated power (P_{sat}) of 19.1 dBm and output 1-dB compression point (P_{1dB}) 13.3 dBm. The measured maximum PAE of the cascode power amplifier is 15.6%. When DNW is biased at 0 V, the power amplifier delivers a saturated power (P_{sat}) of 17.8 dBm and output 1-dB compression point (P_{1dB}) 11.9 dBm. The measured maximum PAE of the cascode power amplifier is 10.5%. It is observed that DNW with biasing at V_{DD} can improve not only linear gain but also power performance.

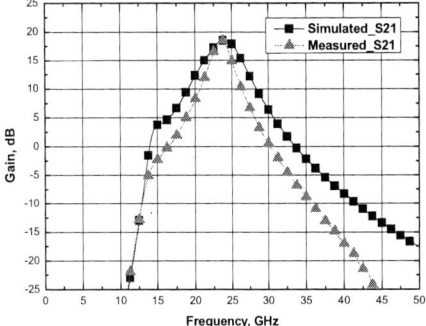

Fig. 3. Simulated and measured small-signal gain of the PA.

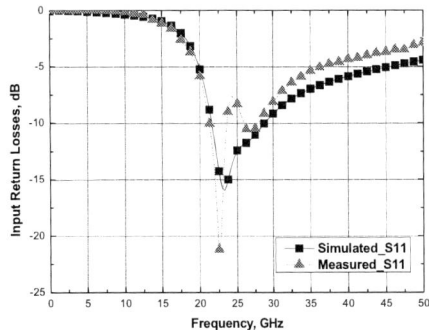

Fig. 4. Simulated and measured input return losses of the PA.

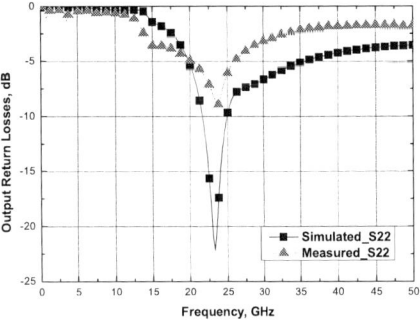

Fig. 5. Simulated and measured output return losses of the PA.

978-2-8748-7007-1/08 $25.00 © 2008 EuMA

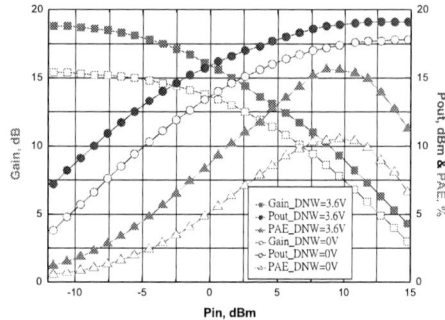

Fig. 6. Measured power performance with different DNW biased of the propose PA.

like to thank Kun-You Lin, Ming-Fong Lei, Che-Chung Kuo, and Yu-Sian Jiang for their helpful suggestions.

IV. CONCLUSIONS

A 24 GHz fully-integrated power amplifier using 0.18-μm deep n-well CMOS process has been designed, fabricated and measured. Due to two-stage cascode circuit topology, the MMIC achieves a high gain of 18.8 dB, and delivers a maximum output power of 19.1 dBm with 15.6% PAE at 24 GHz and OP_{1dB} is 13.3 dBm under proper deep n-well biasing. In addition, using less matching components, the power amplifier has a miniature chip size of 0.56 x 0.58 mm^2. The MMIC demonstrates miniature chip size, high gain, high power added efficiency (PAE), and the highest output power among the CMOS PAs above 15 GHz.

ACKNOWLEDGMENT

The chip is fabricated by TSMC through Chip Implementation Center (CIC), Taiwan, R.O.C. This work was supported by the National Science Council of Taiwan, R.O.C. (contact no. NSC 96-2752-E-002-003-PAE, NSC 96-2219-E-002-015 and NSC 96-2219-E-002-020). The authors would

REFERENCES

[1] A. Ghazinour, P. Wennekers, J. Schmidt, Y. Yi, R. Reuter, and J. Teplik, "A fully-monolithic SiGe-BiCMOS transceiver chip for 24 GHz applications," *IEEE BCTM*, pp. 181-184, 2003.

[2] A. Natarajan, A. Komijani, and A. Hajimiri, "A fully integrated 24-GHz phased-array transmitter in CMOS," *IEEE Journal of Solid-State Circuits*, vol. 40, no. 12, Dec. 2005.

[3] D. Lu, M. Kovacevic, J. Hacker, and D. Rutledge, "A 24 GHz active phased-array with a power amplifier/low-noise amplifier in MMIC," *Int. J. Infrared Millimeter Waves* vol. 23, pp. 693-704, May. 2002.

[4] I. Gresham, A. Jenkins, R. Egri, C. Eswarappa, F. Kolak, R. Wohlert, J. Bennett, and J. P. Lanteri, "Ultra wide band 24GHz automotive radar front-end," *in IEEE MTT-S Int. Microwave Symp. Dig.*, Jun. 2003, pp. 369-372.

[5] E. Sonmez, A. Trasser, K.-B. Schadd, P.Abele, and H. Schumacher, "A single-chip 24GHz receiver front-end using a commercially available SiGe HBT foundry process," *in Radio Frequency Integrated Circuits (RFIC) Symp. Dig.*, Jun. 2002, pp. 159-162.

[6] Y. S. Jiang, Z. M. Tsai, J. H. Tsai, H. T. Chen, and H. Wang, *et al.*, "A 86 to 108 GHz amplifier in 90-nm CMOS," *IEEE Microwave and Wireless Components Lett.*, to appear in Feb. 2008.

[7] T. Yao, M. Q. Gordon, K. K. W. Tang, K. H. K. Yau, M. T. Yang, P. Schvan, S. P. Voinigescu, "Algorithmic design of CMOS LNAs and PAs for 60-GHz radio," *IEEE Journal of Solid-State Circuits*, vol. 42, no. 5, May. 2007.

[8] T. Sowlati, and D. M.W. Leenaerts, "A 2.4 GHz 0.18-μm CMOS self-biased cascode power amplifier," *IEEE Journal of Solid-State Circuits*, vol. 38, no. 8, pp.1318-1324, Aug. 2003.

[9] C. Park, Y. Kim, H. Kim, and S. Hong, "A 1.9 GHz CMOS power amplifier using three-port asymmetric transmission line transformer for a polar transmitter," *IEEE Trans. Microwave Theory Tech.*, vol.55, no. 2, pp. 230-238, Feb. 2007.

[10] A. Mazzanti, L. Larcher, R. Brama, and F. Svelto, "Analysis of reliability and power efficiency in cascode class-E PAs," *IEEE Journal of Solid-State Circuits*, vol. 41, no. 5, pp.1222-1229, May. 2006.

[11] C. M. Lo, C. S. Lin, and H. Wang, "A miniature V-band 3-stage cascode LNA in 0.13-μm CMOS," *in IEEE Int. Solid-State Circuits Conf. Tech. Dig.*, Feb. 2006, pp. 1254-1263.

[12] A. Komijani and A. Hajimiri, "A 24 GHz, +14.5 dBm fully-integrated power amplifier in 0.18um CMOS," *in Proc. IEEE 2004 Custom Integrated Circuits Conf.*, Oct. 2004, PP. 561-564

[13] J. Kang, D. Yu, Y. Yang, and B. Kim, "Highly linear 0.18-μm CMOS power amplifier with deep-n-well structure," *IEEE Journal of Solid-*

TABLE I

COMPARISON WITH PREVIOUSLY REPORTED PAs FOR FREQUENCIES AROUND 20 GHz.

Reference	Process	Note	Operation Frequency [GHz]	Gain [dB]	Psat [dBm]	PAE [%]	Chip Size [mm^2]
This work (DNW = 3.6 V)	0.18-μm CMOS	2-stage cascode	24	18.8	19.1	15.6	0.325
This work (DNW = 0 V)	0.18-μm CMOS	2-stage cascode	24	15.4	17.8	10.5	0.325
[12]	0.18-μm CMOS	2-stage cascode	24	7	14.5	5 ~ 6	1.26
[16]	0.18-μm CMOS	3-stage cascode	27	17	14	NR	2.04
	0.18-μm CMOS	3-stage cascode	40	7	10.4	NR	2.04
[17]	0.13-μm CMOS	4-stage Class-E	18	30	10.9	23.5	0.782
	0.13-μm CMOS	4-stage Class-E	20	26	10.2	20.5	0.782
[18]	0.13-μm CMOS	Off-chip matching	17	11.5	17.8	15.6	0.8
	0.13-μm CMOS	On-chip matching	17	14.5	17.1	9.3	0.9
[19]	150 GHz SiGe HBT	1-stage cascode	24	12	20@1dB	14@1dB	1.02
[20]	80 GHz SiGe HBT	3-stage balanced	24	18	12	4.5	NR

State Circuits, vol. 41, no. 5, May. 2006.

[14] J. G. Su, H. M. Hsu, S. C. Wong, C. Y. Chang, T. Y. Huang, and J. Y. C. Sun, "Improving the RF performance of 0.18-μm CMOS with deep n-well implantation," *IEEE Electron Device Lett.,* vol. 22, no. 10, pp.481-483, Oct. 2001.

[15] S. C. Shin, M. D. Tsai, R. C. Liu, K. Y. Lin, and H. Wang, *et al.*, "A 24-GHz 3.9-dB NF low-noise amplifier using 0.18-μm CMOS technology." *IEEE Microwave and Wireless Components Lett.,* vol. 15, no.7, July. 2005.

[16] H. Shigematsu, T. Hirose, F. Brewer, and M. Rodwell, "Millimeter-Wave CMOS circuit design," *IEEE Trans. Microwave Theory Tech.,* vol.53, no. 2, Feb. 2005.

[17] C. Cao, H. Xu, Y. Su , and K. K. O, "An 18 GHz, 10.9 dBm fully-integrated power amplifier with 23.5% PAE in 130-nm CMOS." *in Proceeding of ESSCIRC, Grenoble, France, 2005.*

[18] A. V. Vasylyev, P. Weger, W. Bakalski, and W. Simbuerger, "17 GHz 50-60 mW power amplifier in 0.13-μm standard CMOS." *IEEE Microwave and Wireless Components Lett.,* vol. 16, no.1, Jan. 2006.

[19] J. P. Comeau, J. M. Andrews, and J. D. Cressler, "A monolithic 24 GHz, 20 dBm, 14% PAE SiGe HBT power amplifier," *in Proceeding of the 36th European Microwave Conference.* Sept. 2006.

[20] N. Kinayman, A. Jenkins, D. Helms, and I. Gresham, "Design of 24 GHz SiGe HBT balanced power amplifier for system-on-a-chip ultra-wideband applications," *in Radio Frequency Integrated Circuits (RFIC) Symp. Dig.,* June. 2005, pp. 91-93.

[21] *"Sonnet User's Manual, Release 9.0,"* Sonnet Software, Inc., May 2003, Syracuse, NY.. M. Metev and V. P. Veiko, *Laser Assisted Microtechnology,* 2nd ed., R. M. Osgood, Jr., Ed. Berlin, Germany: Springer-Verlag, 1998.

10 Watt High Efficiency GaAs MMIC Power Amplifier for Space Applications

Francesco Scappaviva[1,2], Rafael Cignani[2], Corrado Florian[2],
Giorgio Vannini[3], Fabio Filicori[2], Marziale Feudale[4]

[1] MEC S.r.l –Microwave Electronics for Communications, Bologna, Viale Pepoli 3/2, 40123, Italy

[2] University of Bologna, DEIS, Bologna, Viale Risorgimento 2, 40136, Italy

[3] University of Ferrara, Department of Engineering, Via Saragat 1, 44100, Italy

[4] TAS-I Thales Alenia Space Italia, Roma, Via Saccomuro 24, 00131, Italy

Abstract— **This paper describes the design of a GaAs monolithic high power amplifier at Ku band. The chip delivers about 40 dBm of saturated output power, in CW operating conditions, at 11.7 GHz central frequency, with 17% of bandwidth. The saturated power gain is 12.4 dB with 2 dB gain flatness across the application bandwidth while the chip power added efficiency is estimated between 33% to 47%. The amplifier is designed to be used as final stage of a downlink satellite transmitter for Tracking Telemetry & Command system. A commercial power p-HEMT process capable of handling a power density higher than 1 W/mm has been selected for the MMIC design. Due to the space application, special attention must be put on the process and MMIC reliability: to this aim performances must be guaranteed in de-rated conditions respect to the process maximum ratings and, in addition, the channel temperature of the active devices must be kept within the value established by Space Requirements and carefully controlled. This makes the design objective very tight. The MMIC power amplifier design and some measurement results are presented in the paper.**

I. INTRODUCTION

The evolution of satellite services in the recent years pushed the development of new equipment with improved performances in order to fulfil the requirements coming from market and services providers. This reflected in the improvement of the integration level and relevant technologies used in the production. The equipment in which the mentioned improvements result effective is the transmitter used for telemetry and telecommand, the downlink path which provides the communication channel for geostationary satellites. For these equipment, indeed, until the end of the past decade, the final power stage has usually been provided by hybrid amplifiers [1] or pre-matched FET mini-modules [2], delivered in hermetic sealed ceramic packages, while availability of GaAs p-HEMT and HBT processes with very high power densities gives nowadays the opportunity to synthesize these functions in a monolithic die, with obvious advantages in terms of cost, space, reliability and performance reproducibility.

In this context a 10 Watt monolithic power amplifier at Ku band has been designed and developed exploiting a commercial p-HEMT power process to be used as the final stage of an on-board transmitter.

The state of the art maximum power level achieved with the mentioned technologies for a monolithic amplifier at X and Ku band is around 40 dBm [3]-[4] with fabrication process used near the maximum ratings (i.e. without the space de-rating compliance). In this work comparable performances are achieved for a space product, by exploiting a high performance p-HEMT process and applying an accurate design strategy for the optimization of output power and efficiency.

To reach high output power a large number of active devices have to be adopted in the amplifier output stage, thus a special attention must be focused on the stability analysis (especially the odd-mode instability must be prevented). Another fundamental point in the design for space applications is to guarantee high reliability and to this aim the performances have to be reached by keeping the device operating conditions at a proper de-rating with respect to the process maximum ratings provided by the foundry and imposed by the particular application. Since one of the fundamental goal is the highest output power, the V_{DS} breakdown voltage and the maximum channel temperature of the devices are the constrains which represent the biggest challenge. In fact, on one side the reliability is strongly dependent on the channel temperature and the peak voltages, while the maximum output power is limited by the V_{DS} breakdown. The design focus is on the choice of a proper bias and dynamic load line which are able to optimize circuit performances like power and efficiency within the de-rated operating conditions.

The paper describes the commercial p-HEMT technology adopted, the single cell performances in terms of power added efficiency (PAE), output power and gain and the amplifier design flow. Small-signal and large-signal measurements are finally presented.

II. AMPLIFIER DESIGN

A. Process Characteristics

A space qualified 0.35-um gate length GaAs p-HEMT process from Triquint Semiconductors Texas was selected for this application. This process is characterized by a V_{DS} breakdown exceeding 24 Volts, a maximum current density of

650 mA/mm and can reach 1.2 W/mm power density at 10 GHz at maximum ratings; due to the space application constraints, MMIC performances must be obtained using devices at de-rated conditions. For this application these were identified as 20% de-rating on breakdown voltages and current densities. A power cell with 1440 um of active area (12 gate fingers, 120 um width) has been selected for both the final and driver stages. In order to achieve the best power performances and to assure a junction temperature to guarantee reliability (MTBF greater than $5x10^7$), a strongly symmetrical two stage amplifier solution was adopted, with 8 devices in the final stage and 4 devices as drivers, maintaining a 2:1 active periphery ratio between the two stages.

B. Single Cell Performances

A sample of the power cells chosen for the HPA has been fully characterized in small and large signal conditions in order to verify the electrical models accuracy. Small signal measure has been performed in a de-rated bias condition in order to maintain the temperature of the device under the limits suggested by the foundry and also to adopt the same bias point available for the linear electrical model.

Figure 1 shows the Source/Load Pull system used to define the best load and source impedances for both maximum output power and maximum PAE of the single power cell; even though the system allows to carefully control the impedances at one frequency up to 26.5 GHz, it gives the opportunity to track also the second and the third harmonics. The results of large signal measurements fully match the optimum impedance analysis performed with the Harmonic Balance simulations using Agilent's ADS (Advanced Design System) and the TOM3 model provided by the Foundry.

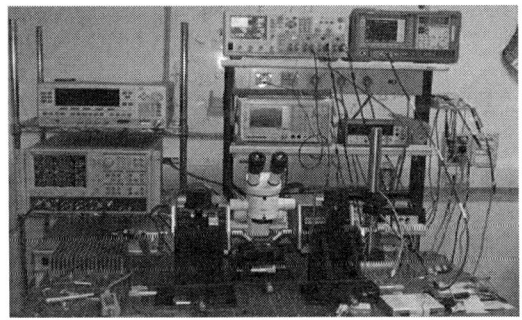

Fig. 1. Source/Load Pull measurement bench. The bench includes also instrumentations needful for the calibration procedures.

The stability of each active device inside the MMIC has been assured with a lumped RC network in series with the gate: in this way the device is made unconditionally stable from 200 MHz upward, ensuring a good trade-off with the achievable gain. This network can avoid common mode instabilities, but it is not adequate to prevent also differential mode instabilities. After a deep analysis of such a problem, carefully dimensioned resistive branches have been introduced

between the symmetrical points of the circuit identified as more sensitive to potential differential instability start-ups. The quiescent bias point of the final stage devices is: $V_{DS} = 9.6$ V, $I_D = 115$ mA (about Idss/3) obtained with a $V_{GS} = -0.65$ V. The final stage device works in AB Class and the optimum output impedance (Fig. 2), starting from the Load Pull measures, has been computed for the fundamental frequency; moreover a tuning on the second and third harmonics has been also performed in order to reach a good trade off between power, efficiency and channel temperature and to maintain the de-rated electrical constrains.

Fig. 2 shows also the resulting dynamic load line over the I-V pulsed characteristic and the DC working point. It is worth noticing that by means of a suitable manipulation of the nonlinear device model it has been possible to compute the intrinsic drain current and control the voltages at the intrinsic device terminals. Hence the dynamic load line is indeed computed at the intrinsic device terminals to correctly optimize the output power and PAE and control the maximum peak voltage with respect to the breakdown value.

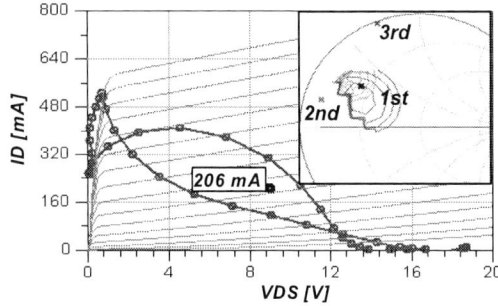

Fig. 2. In the inset it is shown the optimum device output impedances computed for the final stage device. IV Pulsed Curves, the intrinsic dynamic load line and the average drain current value of the final stage device are plotted in the graph.

The situation depicted in Figure 2 refers to an operating condition of 2 dB gain compression: by proper selection of drain and gate bias and an adequate tailoring of the load line (fundamental frequency impedance and harmonic tuning) the simulated maximum V_{DS} voltage is below 19 V, which is more than 20% lower than breakdown (24V).

Moreover, under large signal conditions, due to non-linear DC conversion, the drain average current I_D rises form 115 mA to 206 mA. This condition allows to obtain from a single cell an output saturated power of about 1.33 Watts with 56% PAE at 2 dB compression point, while the junction temperature, for both final and driver active devices, is kept under 130°C with a base plate temperature of 70°C.

To accurately simulate the thermal behavior of the active devices, a thermal simulator based on a finite difference solver has been adopted. By analysis, a thermal resistance of about 65.5°C/W for the 12x120 um device has been extracted.

Both driver and final stages are biased in the same quiescent point, so that the MMIC needs only two bias voltages: $V_{DS} = 9.6$ V and $V_{GS} = -0.65$ V.

C. HPA Layout

The MMIC layout, which dimensions are 4.7 X 4 mm, is shown in Fig. 3. It is possible to notice that the output matching network, used to combine the power from the 8 final cells, is composed by a bus bar structure, cascaded with a 4:1 tree combining network. The outputs of the last stage devices are connected together through the wide bus bar structure which feeds the DC current to all the drains. Shunt MIM capacitors are placed at proper symmetric points on the bus bar and the RF power is tapped off at other symmetric points along the bar. In this way the drain bias path does not need a conventional RF choke, and a decoupling shunt capacitor is sufficient to decouple the path to the external DC circuitry [5].

The shunt capacitors and the cascaded combiner network must be dimensioned and designed in order to impose an identical optimum output impedance (Fig. 2) to all the devices of the final stage. The tree combining network completes the impedance transformation by means of distributed (microstrip lines) and lumped (shunt capacitors) elements. The thick metal layer provided by the process (three metal layers are available for a total thickness of 6.7 um) and the good Q factor of the lumped elements adopted are able to guarantee a very low loss output network (about 0.5 dB) which improves the total circuit efficiency and the amplifier output power.

The inter-stage network design has been carried out considering a trade-off between different factors: layout dimension and topology, load impedance of the driver cells and gain flatness: a flat gain has to be obtained in a 20% bandwidth. The 2:1 combining network and the 1:4 splitting network have been designed to preserve a symmetrical structure, while the choice to carry the RF signal to one central point and then to split it to the final cells has the aim to avoid some possible spurious paths from the driver to the final stage, which can trigger some differential mode oscillations. In the inter-stage network some losses have been necessarily added in order to obtain a good flatness of the gain. A 2 dB resistor attenuator has been added to the RF path in the input matching network, composed by a 1:4 power splitter, with the scope to improve the input return loss.

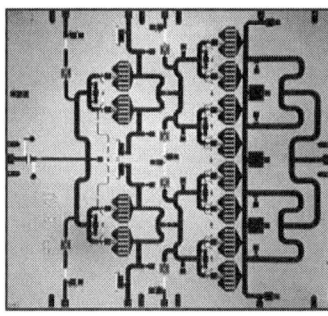

Fig. 3.   MMIC High Power Amplifier Layout.

III. POWER AMPLIFIER CHARACTERIZATION

A preliminary small-signal characterization of the MMIC was carried out directly on wafer in order to test the chip performance without the influence of the external circuitry. Due to the poor thermal dissipation of the on-wafer measurement system, the scattering parameters have been first measured at a de-rated bias condition to keep the device temperature within the maximum limits.

For proper characterization (small signal and power) at the nominal operating point, the test jig shown in Fig. 4 has been used. The die has been attached on a gilded brass carrier using an epoxy glue, with high thermal and electrical conductivity, and then bonded to the test circuit that includes off-chip decoupling capacitors implemented on a high frequency laminate (Rogers TMM10i). The test board has been then mounted on a cooling system in order to have good control on the base plate temperature. Due to the high operating frequencies, the effect of the access structures (including bonding wires) was considered during the design of input and output matching networks. A TRL calibration kit, carried out on the same substrate of the test board, allowed the proper de-embedding of the whole test jig.

Fig. 4.   Test jig developed for the Amplifier characterization.

As shown in Figure 5, measured small-signal gain is not less than 11.5 dB with a flatness of about ±1.5 dB over the frequency range. The input VSWR does not exceed 2:1 between 11.3 GHz and 12.7 GHz, while the output VSWR does not exceed 1.6:1 in the bandwidth of interest.

A measurements set up has been implemented at the University of Bologna Labs (Laboratory of Electronics for Communications) for large signal characterization. It consists of a signal synthesized sweeper followed by a linear solid state amplifier to drive the DUT with sufficient microwave power.

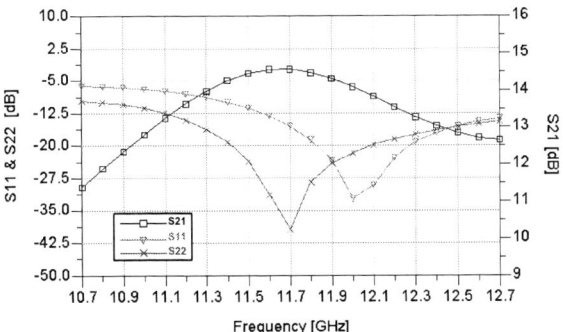

Fig. 5 Small Signal amplifier measurements: S(1,1) triangles, S(2,2) cross and S(2,1) squares. These measurements have been done in the nominal quiescent bias condition.

Fig. 6 shows power gain, output power and PAE plotted as a function of the available input power at 11.7 GHz. As can be seen an output power of 10 Watt is obtained at 2 dB gain compression with a PAE higher than 47%. An important result is the high power density obtained despite the tight constrains and dimensions: indeed, taking into account 11.52 mm of active periphery, the computed power density is 0.87 W/mm.

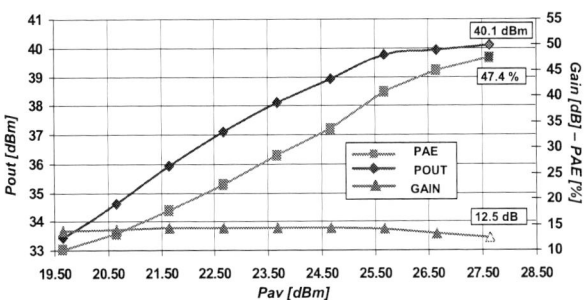

Fig. 6. Power Gain, Output Power and PAE @11.7 GHz as a function of available power.

Figure 7 shows the output power at 2 dB gain compression for the entire bandwidth. Measurements show 8 Watt of output power over the bandwidth with a peak of 10 Watts at 11.7 GHz. The power gain ripple is within 2 dB.

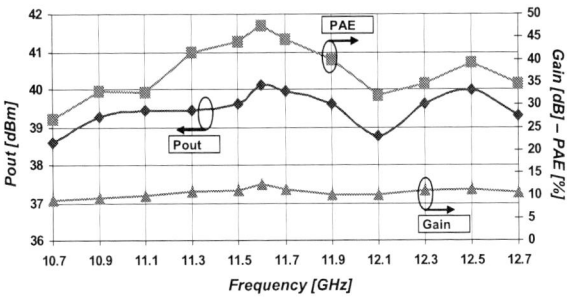

Fig. 7. Power Gain, Output Power and PAE slope within the frequency band at 2 dB gain compression point.

IV. APPLICATION OF DEVELOPED MMIC

As mentioned, the described MMIC has been developed mainly for transmitter section of downlink / telemetry equipment. Once the device will be space validated, it will allow the implementation of power sections based on single hybrid solutions simplifying the whole equipment with consequent cost reduction and reliability improvement. To this purpose, a study of combined solutions has been also carried out to fulfil different requirements in terms of power and bandwidth. A very compact solutions based on MMIC's combined through Multilayer Ceramic (LTCC) structures mounted on a single carrier is envisaged to replace big and expensive modules based on discrete devices. In particular LTCC allows to route both RF and DC lines signals avoiding dangerous crossing. In addition capacitors and inductances can be easily implemented on the top layer as did for the prototypes on Roger TMM10. MMIC's can instead be brazed directly on a high dissipative carrier.

V. CONCLUSION

In this paper the design of a fully monolithic high power amplifier for space applications has been described. The use of a specific p-HEMT process, tailored for high power density in Ku band, and the proper design technique, allowed the development of a single MMIC that can potentially replace the functionality of big and expensive modules based on discrete devices. The amplifier is capable to deliver up to 10 Watts of output power with a 47 % of power added efficiency at 11.7 GHz. Along the 2 GHz bandwidth (10.7 to 12.7 GHz) an output power higher than 8 Watts was obtained in conjunction with a minimum of 33 % PAE.

ACKNOWLEDGEMENT

The authors wish to acknowledge Thales Alenia Space Italia for supporting this activity and Professor Vito A. Monaco for useful discussions and advices.

REFERENCES

[1]   [1] J. Czech, A.-M. Khilla, M. Schunzel, "A 10 Watt C-Band GAAS FET Power Amplifier for Satellite Down-Link" European Microwave Conference, 1984. 14th Oct 1984.

[2]   [2] J. Fukaya, M. Ishii, Matsumoto, M.; Hirano, Y.; "A C-Band 10 Watt GaAs Power FET," Microwave Symposium Digest, MTT-S International Volume 84, Issue 1, May 1984.

[3]   [3] Qi Zhang and Steven A. Brown, "Fully Monolithic 8 Watt Ku-Band High Power Amplifier", IEEE MTT-S Digest, 2004.

[4]   [4] UMS, "X-band GaInP HBT 10 W High Power Amplifier including on-chip bias Control Circuit"

[5]   [5] S.P. Marsh, "MMIC Power Splitting and Combining Techniques", Design of RFIC's and MMIC's, IEE Tutorial Colloquium on 26 Nov. 1997.

A 20 Watt Micro-strip X-Band AlGaN/GaN HPA MMIC for Advanced Radar Applications

C. Costrini[1], M. Calori[1], A. Cetronio[1], C. Lanzieri[1], S. Lavanga[1], M. Peroni[1], E. Limiti[2], A. Serino[2],
G. Ghione[3], G. Melone[4]

[1]SELEX Sistemi Integrati S.p.A., Via Tiburtina Km. 12,400, 00131 Rome, Italy

[2]Dipartimento di Ingegneria Elettronica, Università di Roma 'Tor Vergata', via del Politecnico, 1, 00133 Rome, Italy

[3]Dipartimento di Elettronica, Politecnico di Torino, Corso Duca degli Abruzzi, 24, 10129 Torino, Italy

[4]Consorzio OPTEL, SS n.7 per Mesagne, Km 7,300, 72100 Brindisi, Italy

[1]ccostrini@selex-si.com

Abstract— In this paper a first iteration design, fabrication and test of a two-stage X-Band MMIC HPA in micro-strip AlGaN/GaN technology is reported. With 20 V drain voltage operating bias point, at 3 dB compression point, the HPA delivers a pulsed output power ranging from 21 to 28.5 W, an associated gain from 12.9 to 16.5 dB and an associated PAE from circa 30% to 40%, over the 8-10.5 GHz frequency bandwidth. In the best performance frequency points (8.5 and 9 GHz) the HPA exhibits a saturated output power of 30 W with an associated PAE of 40%.

I. Introduction

Gallium Nitride (GaN) High Electron Mobility Transistor (HEMT) is a new pacing technology mainly targeted for high power applications at microwave and millimetre-wave frequencies, due to its high critical breakdown field, its one-order of magnitude higher power density, its saturation drift velocity and availability of a high thermal conductivity semi-insulating SiC substrate [1].

Furthermore, the recent improvements in GaN technology, and in particular the introduction of through substrate via-hole ground interconnection, allow to employ the well-known micro-strip design approach that has been successfully adopted with standard GaAs technology for power applications.

For radar applications, micro-strip MMIC fabrication of the high power amplifiers is preferred due to the compatibility with current Transmit/receive Modules (TRMs) [2], [3]. For this reason, the micro-strip technology appears to be the most suitable to develop a GaN based Front-End chip-set, which could be composed by a robust low-noise amplifier and an high power switch.

This paper presents a micro-strip two-stage high power MMIC amplifier operating between 8 and 10.5 GHz suitable for phased array radar applications in X-Band with a chip size of approximately 18 mm². In describing the design, fabrication and test of said MMIC such aspects as AlGaN/GaN HEMT fabrication process, active device "unit cell" RF characterisation and modelling, HPA circuit design, and MMIC measurements will be presented.

II. AlGaN/GaN HEMT Technology

The HPA MMIC reported below has been fabricated with the current SELEX Sistemi Integrati GaN-HEMT micro-strip (MS) MMIC technology. Said process is based on an epi-layer structure of GaN/AlGaN/GaN deposited on semi insulating SiC substrates by either MOCVD or MBE techniques. The mask levels for MMIC fabrication are based on a mix and match procedure utilising both I-Line Stepper and Electron Beam Lithography (EBL) processes. The latter is used only for the fabrication of quarter micron gate-length HEMT devices when necessary. For this particular X-Band HPA 0.5 μm gate-length devices have been used and as such all mask levels have been defined by stepper lithography. Drain and Source electrodes of the devices are made by ohmic contact formation of a Ti/Al/Ni/Au metallisation to the GaN/AlGaN epi-layer via a high temperature alloying cycle. Wafer passivation for surface protection is carried out by SiN plasma-enhanced chemical vapour deposition (PE-CVD), while the active device isolation is achieved via Fluorine ion implantation. The SiN passivation film deposition has been optimised to minimize the carrier trap concentration at the interface with the semiconductor in order to keep as low as possible the detrimental drain dispersion phenomenon of the transistors.

After active device formation the MMIC fabrication process comprises: the deposition of NiCr thin-film resistors, Metal-Insulator-Metal (MIM) capacitors and electro-plated inductors and interconnect transmission lines, with air-bridges where necessary.

The fabrication process is concluded with back-side wafer processing for the fabrication of through substrate via-hole interconnects. Said process comprises: wafer thinning down to circa 70 μm, via-hole etching by means of an Inductively Coupled Plasma (ICP) dry etch process and finally back-side (substrate and via-hole) metallisation with a 10 μm thick electro-plated Au film deposited on an appropriate barrier metal layer.

III. Device Characterisation and Modeling

A 1 mm gate periphery active device "unit cell" has been extensively characterized in DC, small-signal, large-signal, and pulsed operating conditions. The resulting measured data have been used to evaluate the technology performance and to extract a nonlinear model of the device.

On wafer S-parameters have been measured up to 30 GHz using a set-up comprising a probe station and a vector network analyzer. S-parameter at "Cold-FET" (V_{DS}=0 V) and "Hot-FET" (V_{DS}>0 V) bias conditions have been measured, taking into account the device power ratings. In particular, bias points with V_{DS} from 0 V up to 30 V and V_{GS} from -9 V up to 1.2 V has been considered. Moreover, DC I-V measurements have been simultaneously performed together with S-parameters.

On wafer load pull measurements of the device have been performed at 10 GHz utilising a GaN-oriented active load-pull test set operating up to 18 GHz [4]. The system allows harmonic control, and DUT temperature control and it is essentially composed by a probe station, two home made wide band couplers, a VNA, a MTA, a power meter and a spectrum analyser.

Pulsed I-V measurements were performed using an on-wafer measurement set-up comprising a probe station and a GaAsCode [5] pulsed measurement system. Many quiescent bias points have been investigated applying pulses having a duration of 500 ns with a separation of 0.5 and 5 ms at the device terminals.

The device has been modelled by using a nonlinear equivalent circuit model in which the nonlinear elements are described by the expressions proposed by Angelov et al. [6], [7]. The parasitic elements were extracted from Cold-FET S-parameters measured changing V_{GS} from -9 V up to 1.2 V [8]. In particular, resistive and inductive parasitic elements have been estimated from measurements with the gate junction biased in forward condition, while measurements with the gate in pinch-off bias condition were used to determine the values of the parasitic capacitances.

The adopted nonlinear equivalent circuit includes three nonlinear elements: a voltage-controlled drain current source, the gate–source and the drain–source capacitances. The nonlinear current source represents the AM–AM distortion of the device; the parameters of the function describing this element have been determined by a fitting of the DC and pulsed I-V measurements. Voltage-controlled gate–source and gate–drain capacitances contribute to the AM–PM distortion phenomena in the device. Analytical functions parameters describing the latter elements have been determined by a fitting of the values of the bias dependent gate–source and gate–drain capacitances, as determined from the bias-dependent Hot-FET S-parameter measurements.

The model has been implemented as a component of a library on commercial CAD design environment and, as final step of the extraction procedure, it has been tested and optimised utilising the performed nonlinear measurements. Fig. 1 shows a comparison between simulated and measured nonlinear performance at 10 GHz when the device is biased at V_{DS}=25 V and V_{GS}=-5.5 V, and a load maximising the output power, obtained from load pull maps, is connected at the drain of the device. A good agreement is obtained between

measurements and simulations demonstrating a good accuracy of the extracted nonlinear model.

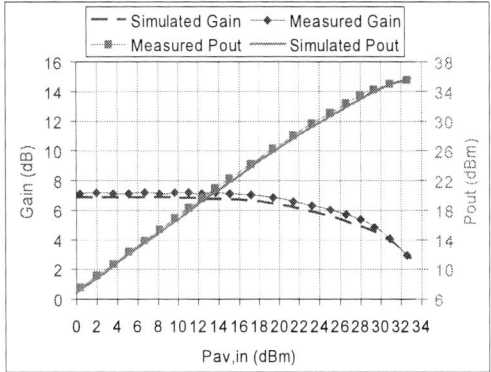

Fig. 1   Simulated and measured output power and gain versus input power @10 GHz.

IV. CIRCUIT DESIGN

The HPA consists of two stages, with a first stage of 4 mm gate-width periphery, and a final stage of 8 mm gate-width periphery. The first stage is composed of 4 unit cells with 1 mm (10 × 100 µm) gate-width, whereas the final stage is composed of 8 unit cells with 1 mm (10 × 100 µm) gate-width. The binary number of cells for each stage was chosen to facilitate power combining and splitting.

The design bandwidth ranges from 8.5 to 10 GHz, with a design goal up to 10.5 GHz.

Concerning the amplifier stabilization, this function is performed by a parallel RC cell [9] in series with the gate of two transistors, that can also prevent parametric oscillations ([10]-[11]) typical of High Power MMIC Amplifiers.

The HPA stability was analysed during design by using the Ohtomo method [12].

With regard to the output network design, the aim was to synthesize the optimum power impedance extracted from active device load pull measurements and non linear model simulations.

The subsequent design step has concerned the inter-stage network that not only has to provide the first stage devices with the proper power load to drive the final stage, but also has to ensure an appropriate gain: as a consequence, the design target was to achieve a suitable trade-off.

Finally, the input network has been designed to match the amplifier input to 50 ohm.

V. MEASURED RESULTS

A photograph of the chip is shown in Fig. 2. The chip size of the MMIC is 5.33 mm×3.5 mm, and the SiC substrate thickness is 70 µm. In this first iteration design, the layout has not been fully optimized for chip-size.

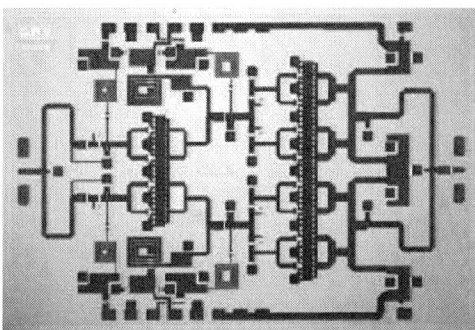

Fig. 2 A photograph of the chip. Circuit dimensions are: 5.33 mm×3.5 mm, substrate thickness is 70 μm.

The HPA power performance has been measured on-wafer in pulsed condition with 1% duty cycle and 10 μs pulse width.

The drain voltage is 20 V and the quiescent current is 1.8 A, that corresponds to class AB operating condition.

Fig. 3, Fig. 4 and Fig. 5 illustrate output power and associated gain versus input power at respectively 8.5, 9.5 and 10.5 GHz.

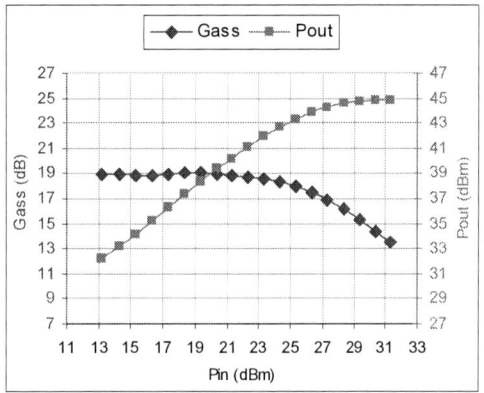

Fig. 3 Measured Pout and Gass vs Pin @ Freq=8.5GHz, Vds=20V, Idsq=1.8A, pulse width=10μs, duty cycle= 1%.

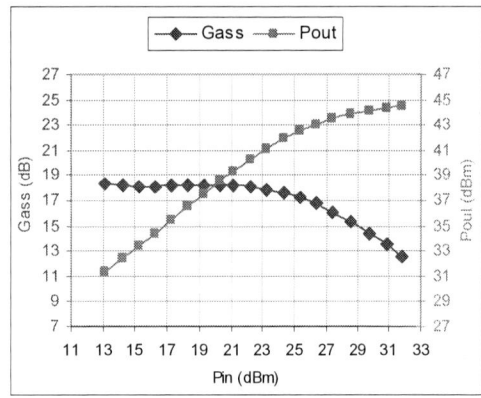

Fig. 4 Measured Pout and Gass vs Pin @ Freq=9.5GHz, Vds=20V, Idsq=1.8A, pulse width= 10μs, duty cycle= 1%.

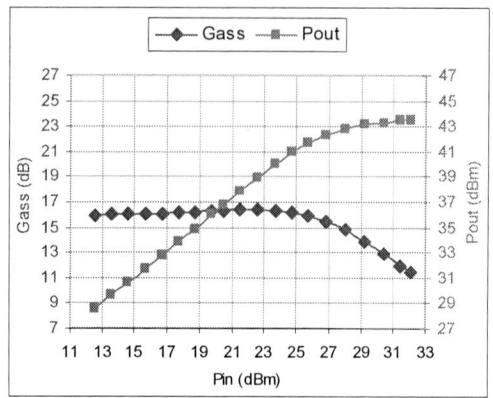

Fig. 5 Measured Pout and Gass vs Pin @ Freq=10.5GHz, Vds=20V, Idsq=1.8A, pulse width = 10 μs, duty cycle = 1%.

As shown in the graphs, best output power and associated gain performances has been achieved in the lower frequency bandwidth range: nevertheless, even at the frequency of 10.5 GHz, which is the bandwidth worst case, the Pout at 3 dB compression point is more than 20 W.

The HPA power performances are summarized in Fig. 6, that reports output power, associated gain and power added efficiency at 3 dB compression point in the frequency bandwidth 8.0 ÷ 10.5 GHz.

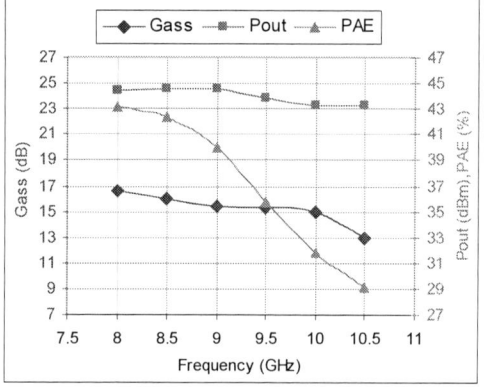

Fig. 6 Measured Pout, Gass, and PAE vs Freq @ 3dBc, Vds=20V, Idsq=1.8A, pulse width=10μs, duty cycle= 1%.

In said frequency bandwidth the HPA delivers an output power from 21 to 28.5 W, an associated gain from 12.9 to 16.5 dB and a power added efficiency from 29 to 43%.

The graph of Fig. 7 shows Pout, Gass and PAE vs frequency at input power level of 30.5 dBm: in this condition, for the 8.5 and 9 GHz frequency points, the HPA exhibits a saturated output power of 30 W with an associated PAE of 40%.

978-2-8748-7007-1/08 $25.00 © 2008 EuMA

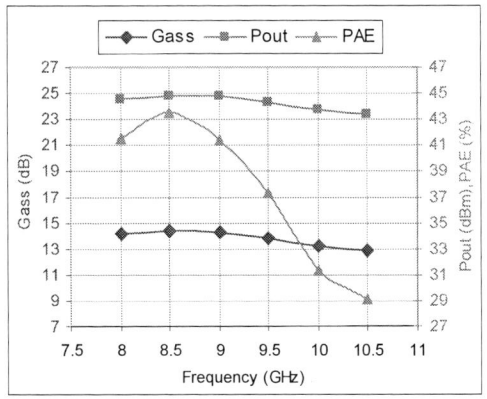

Fig. 7 Measured Pout, Gass, and PAE vs Freq @ Pin=30.5dBm, Vds=20V, Idsq= 1.8 A, pulse width= 10μs, duty cycle= 1%.

VI. CONCLUSIONS

RF performances of a MMIC micro-strip GaN HPA designed for X-band radar applications have been reported.

The chip fabrication has been based on an AlGaN/GaN HEMT process with a 0.5 μm gate-length defined by stepper lithography.

With 20 V operating drain voltage, said amplifier is capable of delivering more than 20 W of pulsed output over a 25% bandwidth (8÷10.5 GHz) at 3 dB compression point, with an associated PAE ranging from 29 to 43%.

In the best performance frequency points (8.5 and 9 GHz) the HPA exhibits a saturated pulsed output power of 30 W with an associated PAE of 40%.

ACKNOWLEDGMENT

This research activity has been carried out in the KORRIGAN RPT N° 102.052 Project funded within the EUROPA framework in the CEPA 2 priority area. The authors wish to acknowledge the consortium members for their contribution and the European Defence Agency for its financial support.

REFERENCES

[1] M. van Heijningen, F.E. van Vliet, R. Quay, F. van Raay, R. Kiefer, S. Muller, D. Krausse, M. Seelmann-Eggebert, M. Mikulla, and M. Schlechtweg, "Ka-Band AlGaN/GaN HEMT High Power and Driver amplifier MMICs", in European Microwave Conference Proceedings, 2005, pp. 237 – 240.

[2] F. van Raay, R. Quay, R. Kiefer, H. Walcher, 0. Kappeler, M. Seelmann-Eggebert, S. Muller, M. Schlechtweg, and G. Weimann, "High Power/High Bandwidth GaN MMICs and Hybrid Amplifiers: Design and haracterization", in European Microwave Conference Proceedings, 2005, pp. 373-376.

[3] J. W. Palmour, S. T. Sheppard, R. P. Smith, S. T. Allen, W. L. Pribble, T. J. Smith, Z. Ring, J. J. Sumakaris, A. W. Saxler, and J. W. Milligan, "Wide Bandgap Semiconductor Devices and MMICs for RF Power Applications", IEDM 2001, Washington DC, USA, Dec. 2-5, 2001, pp. 385-388.

[4] A. Ferrero and U. Pisani, "An improved calibration technique for on-wafer large-signal transistor characterization", IEEE Trans. Instrum. Meas., vol. IM-47, Apr. 1993, pp. 360-364.

[5] Manual for Pulsed-Measurement Instrument. Cambridge, U.K.: GaAs Code Ltd., 2000.

[6] I. Angelov, H. Zirath, and N. Rorsman, "A new empirical nonlinear model for HEMT and MESFET devices," IEEE Trans. Microw. Theory Tech., vol. 40, no. 12, pp. 2258–2266, Dec. 1992.

[7] I. Angelov, L. Bengtsson, and M. Garcia, "Extension of the Chalmers nonlinear HEMT and MESFET model," IEEE Trans. Microw. Theory Tech., vol. 44, no. 10, pp. 1664–1674, Oct. 1996.

[8] G. Dambrine, A. Cappy, F. Heliodore, and E. Playez, "A new method for determining the FET small-signal equivalent circuit," IEEE Trans. Microw. Theory Tech., vol. 36, pp. 1151–1159, Jul. 1988.J. Padhye, V. Firoiu, and D. Towsley, "A stochastic model of TCP Reno congestion avoidance and control," Univ. of Massachusetts, Amherst, MA, CMPSCI Tech. Rep. 99-02, 1999.

[9] A.P. de Hek, A. de Boer, T. Svensson, "C-band 10-Watt HBT High-power Amplifier with 50% PAE", GAAS'01 Symposium Digest, 2001.

[10] D. Teeter, A. Platzker and R. Bourque, "A Compact Network for Eliminating Parametric Oscillations in High Power MMIC Amplifiers", IEEE MTT-S Symposium Digest, pp. 967-970, June 1999.

[11] Steve Nelson's section, "Power Amplifiers: From Millliwatts to Kilowatts . . . Cool Devices with Hot Performance", Short Course Notes, 1998 GaAs IC Symposium.

[12] M. Ohtomo, "Stability Analysis and Numerical Simulation of Multidevice Amplifiers", IEEE Trans. Microwave Theory Tech., vol. MTT-41, pp. 983-991, June/July 1993.

A Novel Silicon High Voltage Vertical MOSFET Technology for a 100W L-Band Radar Application

Brian Battaglia, Dave Rice, Phuong Le, Bishnu Gogoi, Gary Hoshizaki, Mike Purchine, Robert Davies, Walt Wright, Dave Lutz, Mike Gao, Dan Moline, Alex Elliot, Son Tran, Robert Neeley

HVVi Semiconductors, Inc.
Phoenix, Arizona, 85044, USA
Brian.Battaglia@hvvi.com
Dave.Rice@hvvi.com
Phuong.Le@hvvi.com

Abstract — **The silicon vertical MOSFET RF power amplifier described in this paper is the industry's first to utilize high voltage vertical technology. Operating under pulse conditions of 200µsec pulse width and 10% duty cycle it delivers more than 100W of peak power. Operating in Class AB with only 50mA of bias current the device achieves more than 20dB of gain and 47% power added efficiency at P1dB compression across 200MHz of bandwidth at L-Band from 1.2GHz to 1.4GHz. The DC characteristics include a BVdss of 115 volts enabling high voltage operation with a 48V power supply.**

I. INTRODUCTION

The HVVFET™ (High Voltage Vertical Field Effect Transistor) is an advanced vertical MOSFET structure that expands the operating frequencies of vertical silicon RF power MOSFETs well into the microwave spectrum. The device architecture also yields an extremely rugged device by suppressing the activation of the parasitic device that destroys other pulsed transistors. The HVVFET is rated at 20:1 VSWR with an operating voltage of 48V. Other device attributes such as gain and efficiency are described in this paper. One example of an application that is addressed by the new transistor is in L-Band pulsed radar systems.

Two common methods of achieving high output power are increasing gate periphery and increasing operating supply voltage. Increasing the gate periphery for higher output power reaches a point of diminishing return because of the reduction in output impedance. High voltage operation has several benefits:

1) transistors operating at higher voltages dissipate less heat due to lower current consumption at a given power level;

2) transistor impedances increase with higher voltage operation, making it easier to design an effective impedance matching network; and

3) higher output impedances aid wider bandwidth designs.

One key aspect of a vertical MOSFET is that the current flows vertically in the drain region while, conversely, in a LDMOS structure the current flow is lateral. The vertical current flow allows the transistor cells of the device to be tightly spaced. The architecture of the HVVFET is optimized

for this attribute yielding a power density approximately 3 times that of an equivalent lateral transistor device (e.g. LDMOS). The increased power density allows high power components to be fitted into the smallest package footprint.

Avionics and military pulsed applications have traditionally used bipolar transistors as the primary device technology because of their high power capability. Modern system requirements are demanding additional transistor performance. Electrical performance improvements in the HVVFET technology such as higher power density, increased gain and better gain flatness offer significant benefits to power amplifier designers, and will produce higher performance amplifiers having reduced size and weight.

This technology offers many system level advantages. The gain of the device is at least 3dB greater than competitor devices which reduces the driver output power requirement by 50%. The combination of high power packing density and smaller driver output power reduces both the size of the PCB and the heatsink and cooling requirements. The extreme ruggedness of the transistor and the inherent reliability of a silicon based technology leads to lower field failure rates and a reduction in maintenance cost. All of the above factors improve system performance and lower total system cost.

II. VERTICAL MOSFET TECHNOLOGY

The HVVFET device is fabricated using standard silicon wafer process technologies. The performance of the device is enhanced by using a number of innovative features which are briefly described. The high operation voltage of the device is enabled by using a vertical configuration that uses the epi thickness to determine the breakdown voltage while maintaining small cell pitch on the top device structure, thereby achieving high power density without sacrificing performance. By utilizing an innovative termination scheme, near ideal planar breakdown voltages are achieved while using the optimal epi doping and thickness to reduce $R_{DS(on)}$. The feedback capacitance of the device is reduced significantly by using an integrated device shield to minimize the coupling of the gate to the drain. In addition, by utilizing the same shield structure, the intrinsic and extrinsic feedback capacitance is reduced by a factor of 20 compared to similar vertical devices.

III. DEVICE CONFIGURATION

This single-ended high power transistor is a first-generation HVVFET device. The discrete silicon N-Channel enhancement mode transistor is implemented in common source configuration for high power operation.

The single die is attached directly to the flange material allowing the generated heat to be quickly extracted to the heatsink for maximum thermal performance. An optimized die attach process is achieved without solder preforms. Based upon simulations and measured results the thermal resistance of the package part is 0.85°C/W.

The device is housed in a RF high power bolt-down package with an industry standard HV400 footprint as shown in Figure 1. The package can be mounted with screws or soldered to the heatsink for optimum attach to the thermal interface.

Fig. 1 HV400 Power Package

IV. MATCHING NETWORKS

The internal matching networks within the device package transform the low impedance of the die to higher impedance at the terminal leads of the package. The matching networks are designed to achieve a minimum 15% fractional bandwidth performance at 1300MHz. At any given power level the high drain voltage bias scheme creates higher devices impedances than similarly power rated devices. The nature of the vertical device structure produces low intrinsic capacitances. The inherently low input and output die capacitance makes the device easy to match. In fact, only a single stage of input and output matching is required internal to the package as seen in Figure 2. All matching is done using reliable gold wires. A single matching section of a low pass network is formed at the input with wire LG1 shunted to ground through MOS capacitor C1. The series wires LG2 connect the low pass filter to the gate terminal of the die. The output impedance match is realized by bonding wires from the drain of the transistor to a large MOS capacitor which acts as a DC block. Intrinsic die output capacitance is resonated with inductance LD2 which are shunted to ground. The MOS capacitor C2 in series with the shunt LD2 wires allows only RF current to flow to ground effectively presenting high impedance to low frequency and DC components of the current. The simple, single-plate MOS capacitors are easy to manufacture, low cost and are the only additional elements internal to the package. The internal matching effectively allows the device to achieve flat gain and efficiency and the IRL response across the 200MHz bandwidth at the high end of the L-Band.

The wirebond profiles are not complicated, allowing the devices to be manufactured accurately and repeatably, using automatic wirebonders. The wires are evenly spaced for uniform current distribution preventing thermal issues such as hotspots on the die resulting from phase mismatching.

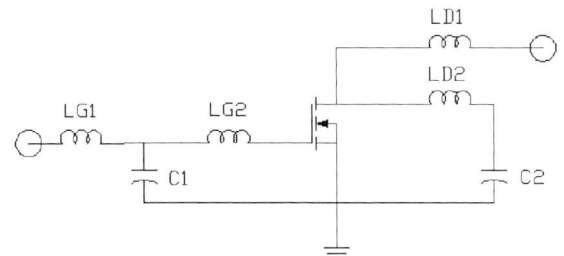

Fig. 2 Schematic Diagram of the Input and Output Matching Networks

The single-ended input and output impedances achieved with the low-loss internal matching elements are listed in Table 1. External matching was accomplished through lumped elements on a PCB raising the impedance to 50 ohms.

TABLE I
SUMMARY OF PACKAGED DIE IMPEDANCES

Frequency	Zin	Zout
1200 MHz	2.9-j4.1	6.3-j4.2
1300 MHz	2.5-j2.8	5.8-j2.5
1400 MHz	2.2-j0.8	5.6-j0.9

V. RF PERFORMANCE CHARACTERISTICS

The test fixture was optimized for pulsed power performance from 1200MHz to 1400MHz. At least 15% of fractional bandwidth was attained, centered at 1300MHz. the transistor was matched not only for high power performance metrics but also for frequency response. The test circuit was tuned with microstrip lines and surface mount chip components for flat gain and efficiency across the frequency band of operation. Figures 3 and 4 show the flat response of the device over frequency.

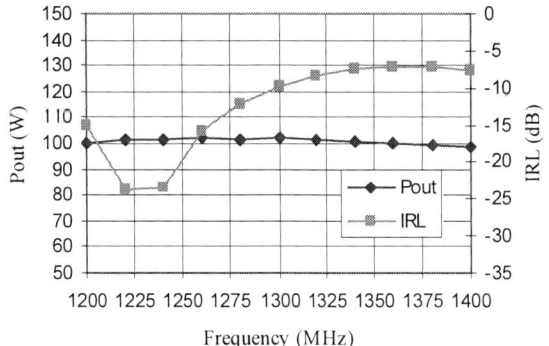

Fig. 3 Measured RF performance of peak output power and input return loss over frequency taken with pulsed signal conditions of 200μsec pulse width, 10% duty cycle, VDD = 48V, Class AB bias circuit. Peak output power exceeds 100W across the band with minimum 47% power added efficiency.

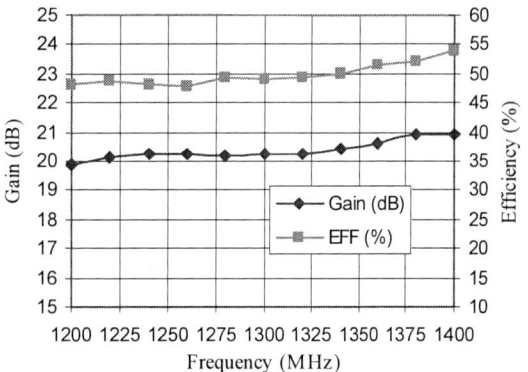

Fig. 4 Measured RF performance of gain and efficiency versus frequency taken with pulsed signal conditions of 200μsec pulse width, 10% duty cycle, VDD = 48V, Class AB bias circuit. Gain exceeds 20dB and is less than 1dB flat across 200MHz of bandwidth.

The device achieves greater than 20dB of gain at the P1dB compression point. At this power level the power transistor delivers greater than 50% drain efficiency. With gain as high as 20dB the power added efficiency is essentially the same as the drain efficiency since the required input power level is so small compared to the output power. [1]

High power design is rigorous since not only are high power performance metrics such as peak power, gain and efficiency important but this performance must be achieved over an entire frequency band of operation. The gain flatness is less than 1dB and will meet the minimum system specification without having additional circuitry like AGC (automatic gain control) modules to compensate the system gain. The flat efficiency response results in constant power supply consumption. The balanced power consumption prevents any fluctuations in the power supply lines, resulting in stable performance.

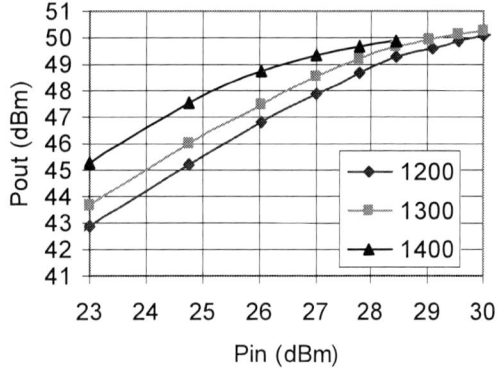

Fig. 5 Output power versus input power at VDD = 48V under pulsed conditions: 200μsec pulse width, 10% duty cycle.

The device achieves over 100W of peak power at 200μsec pulse width and 10% duty cycle. Figure 5 shows the

performance over power drive displaying the linear range of operation and driving into the saturation region.

One of the most challenging test conditions for any RF transistor is being able to handle load mismatches at all phases as in [2]. Under nominal operating voltage of 48V at rated power, the device is able to withstand a 20:1 VSWR without exhibiting any lasting performance degradation.

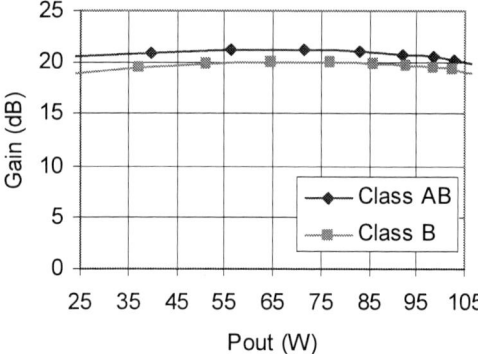

Fig. 6 Measured gain flatness across output power at 1300MHz with VDD = 48V and pulse signal conditions of 200μsec pulse width, 10% duty cycle, showing Class B versus Class AB bias levels.

Figure 6 displays the gain which is greater than 20dB with Class AB current bias. When biased in class B mode for high efficiency applications the device maintains the high gain characteristic of the classical Class AB mode of operation. The advantage of the Class B design is that the high efficiency approach draws zero current and consumes zero power when the transmitter is off. The Class AB bias scheme has higher gain but consumes DC power throughout the pulse cycle without regard to whether the RF signal is applied or not. [3] The HVVFET technology maintains gain above 20dB across the band at Class B which maximizes efficiency.

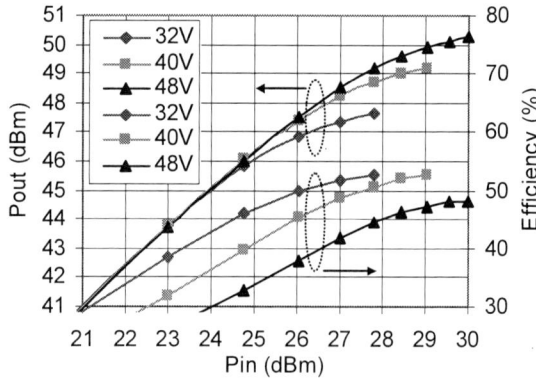

Fig. 7 Measured output power at 1300MHz, VDD = 48V with pulse signal 200μsec pulse width and 10% duty cycle displaying performance across varying voltage supply bias.

Figure 7 shows the tradeoffs of voltage bias supply and RF performance. It is clear that a lower operating voltage achieves greater efficiency at the cost of maximum output power. Some applications may require higher efficiency and are willing to sacrifice some power in order to attain the proper system specification. The device gain is not affected by variations in the drain power supply.

Fig. 8 Picture of the test fixture matching at L-Bamd

The external matching circuitry covers over 200MHz of bandwidth in a single printed circuit board. Figure 8 shows the external input and output matching networks which are comprised of a combination of both microstrip transmission lines and lumped elements. The power gain flatness is less than 1.0dB across the frequency band. On the DC bias lines of both the gate and drain are large capacitors in the µF range that suppress the low frequency components on the DC supply lines preventing stability issues. RF shunt capacitors are placed at one quarter of a wavelength of the fundamental frequency away from the leads of the device to isolate the DC bias feed line from the RF impedance matching network. These capacitors presents a null or low impedance at the fundamental frequency shunting the residual RF signal to ground with the capacitor effectively isolating the DC current. DC blocking capacitors on the RF lines isolate the bias supply from the connectors.

The high impedance at the package leads, because of the internal matching, makes the test fixture circuit easy to match to 50 ohms. The match does not require expensive high dielectric material but uses low cost standard dielectric material from Rogers Corporation, resulting in a significant cost savings. The use of only standard values of capacitance for the piece parts (chip capacitors) for both RF match and DC bias networks results in a cost effective solution.

VI. FUTURE ROADMAP

Simulations indicate that higher performance at frequencies as high as 10GHz and breakdown voltages exceeding 200V are feasible with the HVVFET structure once optimized layout for CW and linear applications is pursued. The performance advantages achievable with future generations of HVVFETs make it a likely candidate for many high power RF and microwave applications.

VII. CONCLUSIONS

The HVVFET is the first new silicon high frequency RF power transistor introduced in more than a decade. This paper described a unique approach to high power amplifier design utilizing technological breakthroughs that result in increased RF performance and many system benefits. The first generation of this technology demonstrates state-of-the-art performance with a clear path to higher breakdown voltages to enhance performance advantages in future generations.

ACKNOWLEDGMENT

The authors wish to acknowledge the assistance and support of Kimberly Romine of HVVi Semiconductors in the assembly and test area.

REFERENCES

[1] S. C. Cripps, *RF Power Amplifiers for Wireless Communications*, Norwood, MA: Artech House, 1999.

[2] N. Dye and H. Granberg, *Radio Frequency Transistors: Principles and Practical Applications*, Stoneham, MA: Reed, 1993.

[3] B. Battaglia, "The ABCs of Device Biasing," *Microwave Journal*, pp. 136-146, Nov. 2000.

AUTHOR INDEX

Abbasi, M. 9
Abele, J. 151
Abele, P. 56
Alderman, B. 202
Alleaume, P. F. 52
Alleva, V. 194
Amasuga, H. 282
Ambacher, O. 87, 95, 210
Angelov, I. 522
Astre, G. 187
Auvinet, C. 52
Bach, H. G. 550
Bandler, J. W. 310
Bao, M. 75
Barale, F. 5
Baras, T. 466
Barker, N. S. 382, 534
Bartolucci, G. 362
Bastioli, S. 338
Battaglia, B. 570
Becker, K. F. 546
Behammer, D. 56
Benech, P. 40
Bengtsson, O. 222
Berberich, S. 470
Berg, H. 266, 322
Bettidi, A. 194
Betts, L. 135
Bhatta, D. 143
Biglarbegian, B. 394
Bilbro, G. L. 286
Blaakmeer, S. C. 163
Blondy, P. 478, 490
Boeck, G. 13
Boguszewski, G. 450
Boles, T. 258

Bolognesi, C. R. 107
Bonani, F. 434
Bonnet, B. 542
Bordas, C. 398
Borremans, J. 234
Bottcher, L. 546
Boueri, F. 28
Bozzi, M. 230
Brazil, T. J. 290
Brebels, S. 470
Brockerhoff, W. 442
Bronner, W. 87, 95
Bryllert, T. 206
Bu, G. 71
Burger, J. 28
Burke, D. R. 290
Buttiglione, R. 366
Calaza, C. 474
Calori, M. 566
Camiade, M. 52
Cappelluti, F. 434
Carchon, G. 250
Carnez, B. 119
Carroll, M. S. 36
Carta, C. 498
Catoni, S. 362, 366
Cazaux, J. 542
Cetronio, A. 83, 194, 218, 566
Chan, K. Y. 486
Chang, C. 119, 198
Chang, D. 151
Chang, E. Y. 198
Chang, H. 358
Chang, S. 350
Chaplain, J. 542
Chen, C. 68, 91

AUTHOR INDEX

Chen, Chung-Chun 17
Chen, L. 534
Chen, M. 358
Chen, T. 91
Cheng, W. 28
Chevallier, J. 187
Chini, A. 218
Chiniard, R. 542
Chiong, C. 358
Chow, Y. H. 306
Ciccognani, W. 194, 314, 334
Cidronali, A. 518
Cignani, R. 562
Cipriani, E. 514
Cojocaru, V. 454, 462
Colantonio, P. 514, 526
Comeau, J. P. 374
Cordier, Y. 330
Costanzo, A. 438
Costrini, C. 566
Couderc, P. 542
Cressler, J. D. 374
Crunteanu, A. 478, 490
Cubilla, P. 83
Cumana, J. 179
Dahmani, S. 226
Dambrine, G. 214
Dammann, M. 87
Daneshmand, M. 486
Davies, R. 570
Davies, R. A. 44
Davies, S. 202
Dawe, G. 151
Dawn, D. 5
De Angelis, G. 366
De Boet, J. A. M. 24

De Dominicis, M. 194
De Groote, F. 119
De Hek, A. P. 262
De Jaeger, J. C. 330
De Raedt, W. 250, 470
De Souza, A. A. Lisboa 123, 318
Decoutere, S. 103, 234
Defrance, N. 330
Dehan, M. 103, 234, 250
Delage, S. 187
Di Giacomo, V. 294
Di Maggio, F. 338
Di Paolo, F. 334
Diebold, S. 426
Ding, H. 342
Dispenza, M. 366
Do, V. 13
Domnesque, D. 52
Donzelli, F. 438
Drevon, C. 542
Dubois, M. A. 470
Dubuc, D. 398
Dupuis, O. 250
Dussopt, L. 470
Eickelkamp, M. 179
El Khatib, M. 490
Elfgren, S. 322
Elliot, A. 570
Eng, Y. W. 306
Entesari, K. 71
Eriksson, J. 135
Esposto, M. 218
Evans, R. 64, 302, 554
Fager, C. 9
Farinelli, P. 338, 362, 474
Fattorini, A. 386

AUTHOR INDEX

Fernandez-Barciela, M. 131
Ferndahl, M. 254
Ferrari, M. 194, 334
Feudale, M. 562
Filicori, F. 294, 562
Fiorello, A. 366
Fleckenstein, A. 274
Florian, C. 562
Fomani, A. A. 486
Formicone, G. 28
Fouladi, S. 482
Fujimori, K. 370
Fujishima, M. 32
Fusco, V. F. 242
Gao, M. 570
Gaquiere, C. 330
Gasseling, T. 318
Gebara, E. 143
Gerbedoen, J. C. 330
Gering, J. M 36
Ghione, G. 566
Giacomozzi, F. 338, 362
Gianesello, F. 40
Giannini, F. 314, 334, 514, 526
Giofre, R. 514, 526
Giovannelli, N. 518
Giovine, E. 194
Gloria, D. 40
Gogoi, B. 570
Goliasch, J. 179
Gong, S. 382
Gonzalez-Garrido, M. A. 83
Goto, S. 282
Gourary, M. M. 326
Grajal, J. 83
Grenier, K. 398

Groeseneken, G. 103
Gullapalli, K. K. 326
Gunes, F. 446
Gunnarsson, S. E. 9
Gunyan, D. 135
Guo, J. 171, 390
Gustat, H. 147
Gutta, V. 386
Hadiashar, A. 151
Hajimiri, A. 159
Halder, S. 147
Halonen, K. 167
Halonen, K. A. I. 115
Harm, L. 87
Harvey, J. T. 386
Hasan-Abrar, Z. 306
Hedman, J. 266
Hellen, J. 322
Hemmendorff, E. 266
Hieda, M. 414, 458
Hilton, K. P. 190
Hirata, A. 430
Holmberg, J. 60
Horio, K. 183
Horn, J. M. 135
Hoshizaki, G. 570
Houlihan, R. 470
Hsu, H. 198
Huang, Bo-Jiun 17
Huang, Bo-Jr 17
Huyart, B. 418
Inoue, A. 282
Inui, C. 32
Itagaki, K. 183
Italia, A. 246
Jacob, A. F. 466

AUTHOR INDEX

Jansen, R. 470
Jansen, Rolf H. 179
Janssen, J. 190
Jardel, O. 318
Javvadi, S. 151
Jiang, C. 550
Jiao, G. 91
Joblot, S. 330
Johansen, T. K. 550
Johansson, T. 254
Jussila, J. 167
Kallfass, I. 210, 426, 502
Kant, D. 494
Kantanen, M. 60
Karkkainen, M. 115
Karnfelt, C. 9
Karttaavi, T. 60
Kasper, E. 20
Kautio, K. 366
Kawakami, K. 414, 458
Keehr, E. A. 159
Kerr, D. C. 36
Kessler, J. 48, 238
Keusgen, W. 1, 13
Khan, H. A. 346
Khy, A. 418
Kiefer, R. 87, 95, 274
Kim, H. 155, 410
Kim, H. S. 143
Kim, Y. 28
Klappe, J. G. E. 24
Klimashov, O. 298
Klumperink, E. A. M. 163
Koch, S. 426, 502
Kompa, G. 226
Kosugi, T. 430

Kozhuharov, R. 9
Koziel, S. 310
Krozer, V. 550
Kuang, W. 286
Kuhn, J. 87, 95
Kukutsu, N. 430
Kumar, S. 238
Kuo, C. 198
Kuo, J. 558
Kuo, W. L. 374
Kuylenstierna, D. 9
Lacroix, B. 478
Ladhes, M. 366
Lanzieri, C. 83, 194, 218, 566
Laskar, J. 5, 139, 143
Lautensack, C. 179
Lavanga, S. 218, 566
Le, P. 570
Leberer, R. 274
Leclerc, E. 119
Lederer, D. 36
Lee, J. 155, 410
Lee, K. H. 143
Lee, S. 410
Leenaerts, D. M. W. 163
Legay, H. 542
Leuther, A. 210, 426, 502
Leuzzi, G. 354
Li, B. 298
Li, Y. 75
Limiti, E. 194, 314, 334, 566
Lin, Kun-You 17
Lin, Y. 390
Lindfors, S. 115
Liu, H. 107
Liu, Y. 286, 350

AUTHOR INDEX

Liu, Z. ...302

Lonac, J. ...294

Longhi, P. E. ...314

Lu, J. ...378

Lu, S. ...171

Lucibello, A. ...362

Lugin, M. ...151

Luo, J. ...1

Lutz, D. ...570

Maas, A. P. M. ...506

Maclean, J. O. ...190

Madhow, U. ...498

Maestrini, A. ...202

Magrini, I. ...518

Mancuso, Y. ...270

Manes, G. ...518

Mansour, R. R. ...482, 486

Manstretta, D. ...175

Manzawa, Y. ...32

Marcelli, R. ...362, 366

Mardivirin, D. ...490

Mareels, I. ...64

Margesin, B. ...338, 362, 474

Martin, T. ...190

Masotti, D. ...438

Massler, H. ...210, 426

Mastri, F. ...438

Matheson, D. ...202

Matiss, A. ...442

McKay, T. G. ...36

Medina, M. A. Y. ...522

Megna, A. ...194

Mekonnen, G. G. ...550

Melone, G. ...566

Mengistu, E. S. ...226

Mercha, A. ...103

Mikulla, M. ...87, 95, 274

Mirzaei, H. ...394

Miyamoto, K. ...458

Miyazaki, M. ...282, 414, 458

Mo, Y. ...64

Mojon, O. ...131

Moline, D. ...570

Monfraix, P. ...542

Morkner, H. ...48, 238

Morschbach, M. ...20

Mulloni, V. ...362

Mulvaney, B. J. ...326

Munoz, C. ...151

Munstermann, B. ...442

Murata, K. ...430

Musser, M. ...87

Nakajima, A. ...183

Nakane, M. ...458

Nallatamby, J. C. ...123

Napijalo, V. ...454, 462

Narhi, T. ...202

Nauta, B. ...163

Nauwelaers, B. ...522

Neeley, R. ...570

Neve, C. R. ...36

Nezhad-Ahmadi, M. ...394

Nguyen, L. ...48

Nishida, K. ...414

Noculak, A. ...179

Nogi, S. ...370

Notten, M. ...99

Ocera, A. ...338

Ojefors, E. ...422

Oliveira, J. F. ...530

Olsen, A. O. ...206

Olsson, J. ...222

AUTHOR INDEX

O'Mahony, C. 470

Oppermann, M. 274

Orlhac, J. 214

Ostinelli, O. 107

Pagani, M. 294

Pailloncy, G. 36

Palmisano, G. 246, 406, 510

Parker, A. E. 386

Parvais, B. 103, 250

Pastore, C. 40

Pavageau, C. 250

Pedro, J. C. 530

Pekarik, J. J. 111

Peroni, M. 194, 566

Perregrini, L. 230

Perunama, B. 5

Peter, M. 1

Pfeiffer, U. 422

Phan, K. 48

Piazzon, L. 526

Pinel, S. 5, 139

Pochiraju, T. 242

Pomona, I. 338

Pothier, A. 478, 490

Powell, J. 190

Prigent, M. 123

Purchine, M. 570

Quay, R. 87, 95, 274

Quentin, P. 52

Quere, R. 119, 318

Quintal, R. 151

Raffo, A. 294

Ragonese, E. 510

Raman, S. 534

Ramer, R. 486

Raskin, J. P. 36, 127

Reber, R. 274

Ren, C. 91

Resca, D. 294

Rezazadeh, A. A. 44, 346, 378

Rice, D. 570

Richter, M. D. 546

Riepe, K. 87

Rizzoli, V. 438

Robertson, I. D. 402, 538

Rochette, S. 318

Rodenburg, M. 262

Rodle, T. 87

Rodriguez-Testera, A. 131

Rodwell, M. 498

Roland, J. 48

Romanini, P. 194

Ronka, K. 366

Root, D. E. 135

Rousseau, M. 330

Rudolph, M. 278

Ruiter, M. 494

Rusakov, S. G. 326

Russo, M. 338

Ryynanen, J. 167

Saad, M. 87

Saari, V. 167

Safavi-Naieini, S. 394

Saito, S. 502

Samanta, K. K. 402, 538

Sanagi, M. 370

Sanchez, E. 131

Sandstrom, D. 115

Sanghera, H. 202

Sansen, W. 103

Santarelli, A. 294

Sapone, G. 406

AUTHOR INDEX

Sarkar, S. 5, 139

Scappaviva, F. 562

Schafer, M. 56

Scheytt, C. 147

Schlechtweg, M. 87, 95, 210, 502

Schneider, M. 546

Scholz, C. 143

Schreurs, D. 294, 522

Schuh, P. 274

Scuderi, A. 510

Seelmann-Eggebert, M. 87, 95, 274, 426

Sen, P. 5, 139

Seo, M. 498

Serino, A. 566

Serret, E. 40

Shao, K. 91

Shen, H. 382

Sidiropoulos, G. 266

Siles, J. V. 202

Skafidas, E. 64, 302, 554

Sledzik, H. 274

Smith, D. 214

Sneijers, W. J. A. M. 24

Solazzi, F. 474

Soltani, A. 330

Sommet, R. 119, 318

Sorrentino, R. 338, 474

Splettstosser, J. 56

Stagni, C. 246

Stake, J. 206

Stieglauer, H. 56

Stoij, C. 9

Stornelli, V. 354

Stoukatch, S. 470

Subramaniam, S. C. 346

Subramanian, V. 13, 103

Sun, Q. 44, 378

Sun, X. 470

Tajima, K. 458

Takahashi, H. 430

Tan, J. 44

Tartarin, J. G. 187

Tavakoli, A. R. 71

Tegude, F. J. 442

Teillet, H. 418

Tessmann, A. 210

Teyssandier, C. 119

Teyssier, J. 119

Theeuwen, S. J. C. H. 24

Thiede, A. 147

Thiesies, H. 266

Thinnes, M. 56

Thomas, B. 202

Thorpe, J. 87

Thrivikraman, T. K. 374

Thumm, M. 95

Tikka, T. 115, 167

Tinoco, J. C. 127

Titizian, J. 28

Tkachenko, Y. 298

Tokan, N. T. 446

Toussain, C. 52

Tran, S. 570

Traversa, F. L. 434

Traverso, P. A. 294

Treuttel, J. 202

Trew, R. J. 286

Tsai, Y. 390

Tsai, Z. 558

Tsuru, M. 458

Tzuang, C. C. 68

Ulyanov, S. L. 326

AUTHOR INDEX

Uno, M. 502

Uren, M. 83, 190

Van Der Wal, E. 494

Van Heijningen, M. 79, 190

Van Raay, F. 87, 95, 274

Van Rijs, F. 87

Van Vliet, F. E. 79, 190, 262, 506

Vannini, G. 294, 562

Varonen, M. 115

Veenstra, H. 99

Vendier, O. 490

Verspecht, J. 135

Verzellesi, G. 218

Vescan, A. 179

Vestling, L. 222

Vice, M. 48

Visser, G. C. 79

Visser, K. 494

Vo, V. T. 44, 378

Volotinen, J. 60

Vukusic, J. 206

Wallin, T. 322

Wallis, D. J. 190

Waltereit, P. 87

Wambacq, P. 234

Wane, S. 450

Wang, C. 242

Wang, G. 342

Wang, H. 202, 558

Wang, Huei 17

Wei, C. 298

Wei, M. 350

Werner, J. 20

Wicks, B. 554

Wood, J. 534

Woods, W. 342

Wright, W. 570

Wu, K. 230

Wurfl, J. 79

Xiong, A. 318

Xu, H. 20

Yamamoto, T. 370

Yan, W. D. 482

Yang, N. 91

Yeh, D. A. 5, 139

Yokoyama, T. 462

Yu, H. 155

Yuan, J. 378

Zeng, Y. 107

Zhang, C. 298

Zharov, M. M. 326

Zhong, S. 91

Zhu, Y. 298

Zirath, H. 9, 254

Zorcic, M. 87